Index of Study Strategies

About the Author

George Woodbury,
College of the Sequoias

George Woodbury earned a bachelor's degree in mathematics from the University of California, Santa Barbara, in 1990 and a master's degree in mathematics from California State University, Northridge, in 1994. He currently teaches at College of the Sequoias in Visalia, California—just outside of Fresno. George has been honored as an instructor by his students as well as his colleagues. Aside from teaching and writing, he served as the department chair for College of the Sequoias' math/engineering division from 1999–2004. Additionally, he is the author of *Elementary Algebra* and *Elementary and Intermediate Algebra*, Second Edition.

Intermediate Algebra

George Woodbury

College of the Sequoias

Boston San Francisco New York
London Toronto Sydney Tokyo Singapore Madrid
Mexico City Munich Paris Cape Town Hong Kong Montreal

Editorial Director: Christine Hoag
Editor in Chief: Maureen O'Connor
Executive Editor: Jennifer Crum
Executive Project Manager: Kari Heen
Senior Project Editor: Lauren Morse
Assistant Editor: Antonio Arvelo
Production Manager: Ron Hampton
Cover Designer: Joyce Cosentino Wells
Media Producers: Sharon Tomasulo, Carl Cottrell, Ceci Fleming
Software Development: Monica Salud, MathXL; Ted Hartman, TestGen
Senior Marketing Manager: Michelle Renda
Marketing Coordinator: Nathaniel Koven
Market Development Manager: Dona Kenly
Senior Author Support/Technology Specialist: Joseph K. Vetere
Senior Prepress Supervisor: Caroline Fell
Senior Manufacturing Buyer: Carol Melville
Senior Media Buyer: Ginny Michaud
Composition/Production Coordination: Pre-Press PMG
Cover photo: © Keiji Watanabe/Getty Images

Library of Congress Cataloging-in-Publication Data

Woodbury, George 1967–
Intermediate algebra / George Woodbury.
p. cm.
ISBN-10: 0-321-16641-8 ISBN-13: 978-0-321-16641-8 (student)
1. Algebra—Textbooks. I. Title.
QA154.3.W66 2008
512.9—dc22 2007060832

4 5 6 7 8 9 10—VHP—11 10

To Tina, Dylan, and Alycia
You make everything meaningful and worthwhile—
yesterday, today, and tomorrow.

Contents

Preface

George Woodbury's primary goal as a teacher is to empower his students to succeed in algebra and beyond. To successfully teach developmental algebra courses, instructors must find a way to reach two equally important goals: 1) fully prepare students for collegiate-level mathematics, and 2) build students' confidence in their mathematical abilities.

One of the ways George prepares his students for future mathematics courses is by introducing the fundamental algebraic concepts of graphing and functions early in *Intermediate Algebra.* He then consistently and frequently incorporates these concepts throughout the text. George also understands that strong exercise sets that provide both volume and variety are critical to student success, so he has written extensive sets that incorporate not only a substantial amount of skill and drill, but also plenty of writing exercises to encourage critical thinking and creativity. The key to success for developmental students is often not just about the math—it's about making sure that students understand how to study and prepare for a math class. To help give students the confidence they require to succeed, George fully integrates targeted study strategies throughout the text and also provides further guidance via a *Math Study Skills Workbook,* written specifically for his texts by Alan Bass.

Success breeds further success. With the content, features, and ancillaries of *Intermediate Algebra,* George Woodbury strives to help his students not simply to "make it through" to college algebra or precalculus, but to truly gain the skills and conceptual understanding necessary for further success in those courses.

How Does Woodbury Empower Students for Success?

Early-and-Often Approach to Graphing and Functions

Woodbury introduces the primary algebraic concepts of graphing and functions early in the text (Chapter 2) and then consistently incorporates them throughout the text, providing optimal opportunity for their use and review. Introducing functions and graphing early helps students become comfortable with reading and interpreting graphs and function notation. Working with these topics throughout the text establishes a basis for understanding that better prepares students for future math courses.

Practice Makes Perfect!

Examples Based on his experiences in the classroom, George Woodbury has included an abundance of clearly and completely worked-out examples.

Quick Checks The opportunity for practice shouldn't be designated only for the exercise sets. Nearly every example in this text is immediately followed by a *Quick Check* exercise, allowing students to practice what they have learned and assess their understanding of newly learned concepts. Answers to the *Quick Check* exercises are provided in the back of the book.

Exercises Woodbury's text provides more exercises than most other texts, allowing students ample opportunity to develop their skills and increase their understanding. The exercise sets are filled with both traditional skill- and drill-type exercises as well as unique exercise types that require thoughtful and creative responses.

EXAMPLE 1 Find the equation of a line whose slope is 2 and that passes through the point $(4, -3)$. Write the equation in slope–intercept form.

Solution

We will substitute 2 for m, 4 for x_1, and -3 for y_1 in the point–slope form. We will then solve the equation for y in order to write the equation in slope–intercept form.

$y - (-3) = 2(x - 4)$ Substitute into point–slope form.
$y + 3 = 2(x - 4)$ Eliminate double signs.
$y + 3 = 2x - 8$ Distribute 2.
$y = 2x - 11$ Subtract 3 to isolate y.

The equation of a line with slope 2 that passes through $(4, -3)$ is $y = 2x - 11$. Here is the graph of the line, showing that it passes through the point $(4, -3)$.

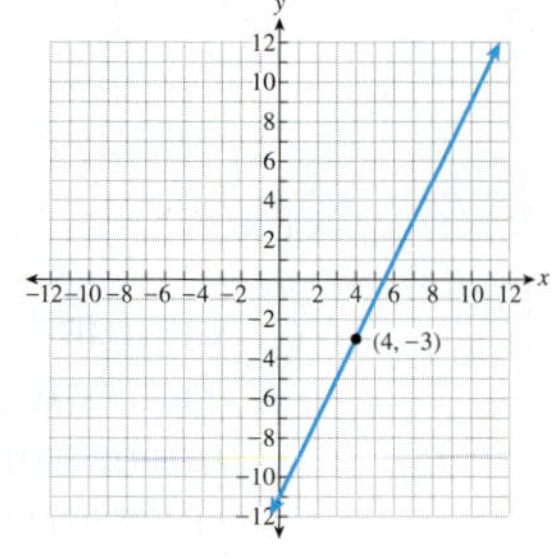

Quick Check 1 Find the equation of a line whose slope is 3 and that passes through the point $(-2, 1)$. Write the equation in slope–intercept form.

EXAMPLE 2 Find the equation of a line whose slope is -1 and that passes through the point $(-7, 4)$. Write the equation in slope–intercept form.

Solution

We begin by substituting -1 for m, -7 for x_1, and 4 for y_1 in the point–slope form. After substituting, we solve the equation for y.

$y - 4 = -1(x - (-7))$ Substitute into point–slope form.
$y - 4 = -1(x + 7)$ Eliminate double signs.
$y - 4 = -x - 7$ Distribute -1.
$y = -x - 3$ Add 4 to isolate y.

Quick Check 2 Find the equation of a line whose slope is -2 and that passes through the point $(-1, -5)$. Write the equation in slope–intercept form.

The equation of a line with slope -1 that passes through $(-7, 4)$ is $y = -x - 3$.

In the previous examples, we converted the equation from point–slope form to slope–intercept form. This is a good idea in general. We use the point–slope form because it is a convenient form for finding the equation of a line if we know its slope and the coordinates of a point on the line. We convert the equation to slope–intercept form because it is easier to graph a line when its equation is in this form.

Types of Exercises

- **Vocabulary Exercises:** Each exercise set starts out with a series of exercises that check student understanding of the basic vocabulary covered in the preceding section. (See pg. 205)

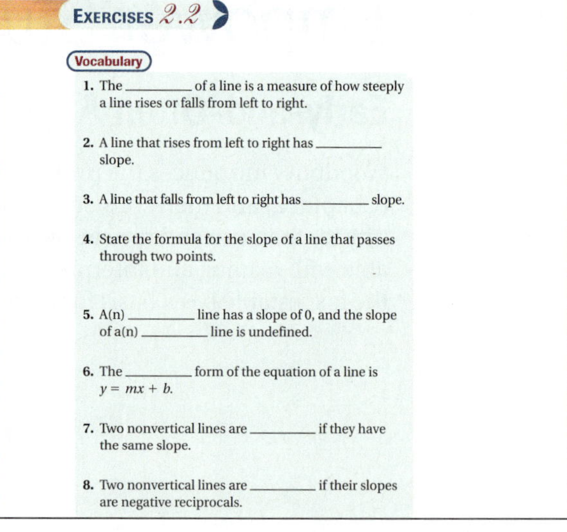

EXERCISES 2.2

Vocabulary

1. The ________ of a line is a measure of how steeply a line rises or falls from left to right.
2. A line that rises from left to right has ________ slope.
3. A line that falls from left to right has ________ slope.
4. State the formula for the slope of a line that passes through two points.
5. A(n) ________ line has a slope of 0, and the slope of a(n) ________ line is undefined.
6. The ________ form of the equation of a line is $y = mx + b$.
7. Two nonvertical lines are ________ if they have the same slope.
8. Two nonvertical lines are ________ if their slopes are negative reciprocals.

- **Mixed Practice Exercises:** Mixed Practice exercises are provided as appropriate throughout the book to give students an opportunity to practice multiple types of problems in one setting. These exercises require students to focus on determining the correct method to use to solve the problem and reduce their tendency to simply memorize steps to solving the problems for each objective. (See pg. 206)

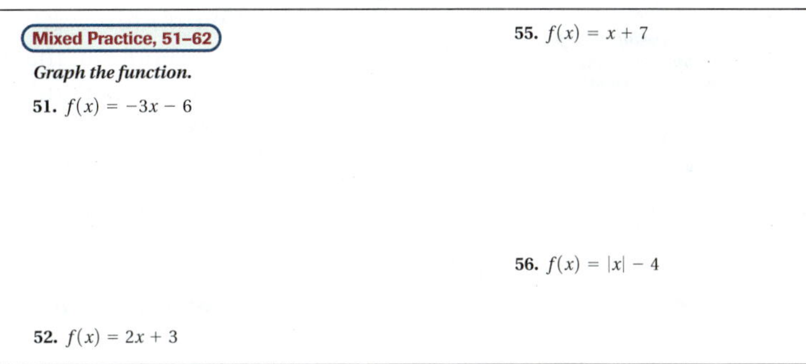
Mixed Practice, 51–62

Graph the function.

51. $f(x) = -3x - 6$

52. $f(x) = 2x + 3$

55. $f(x) = x + 7$

56. $f(x) = |x| - 4$

- **Writing in Mathematics Exercises:** Asking students to explain their answer in written form is an important skill that often leads to a higher level of understanding as acknowledged by the *AMATYC Standards.* (See pg. 208)
  - **Solutions Manual Exercises:** Solutions Manual exercises require students to solve a problem completely with step-by-step explanations as if they were writing their own solutions manual. (See pg. 223)
  - **Newsletter Exercises:** As a way to encourage students to be creative in their mathematical writing, students are asked to explain a mathematical topic. The explanation should be in the form of a short, visually appealing article that might be published in a newsletter read by people who are interested in learning mathematics. (See pp. 247)

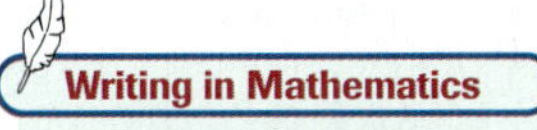

Answer in complete sentences.

73. True or false: every ordered pair is a solution to either $3x - 8y < 24$ or $3x - 8y > 24$. Explain your answer.

74. Explain why we use a dashed line when graphing inequalities involving the symbols $<$ and $>$, and why we use a solid line when graphing inequalities involving the symbols $\leq$ and $\geq$.

75. ***Solutions Manual**** Write a solutions manual page for the following problem:

Graph the linear inequality $2x + 3y < 12$.

76. ***Newsletter**** Write a newsletter explaining how to graph linear inequalities in two variables.

*See Appendix B for details and sample answers.

Applying Skills and Solving Problems

Problem solving is a skill that is required daily in the real world and in mathematics. Based on George Pólya's text, *How to Solve It,* George Woodbury presents a **six-step problem-solving strategy** in Chapter 1 that lays the foundation for solving applied problems. He then expands on this problem-solving strategy throughout the text by incorporating **hundreds of applied problems** on topics such as motion, geometry, and mixture problems. Interesting themes in the applied problems include investing and saving money, understanding sports statistics, landscaping, home ownership, and cell phone usage.

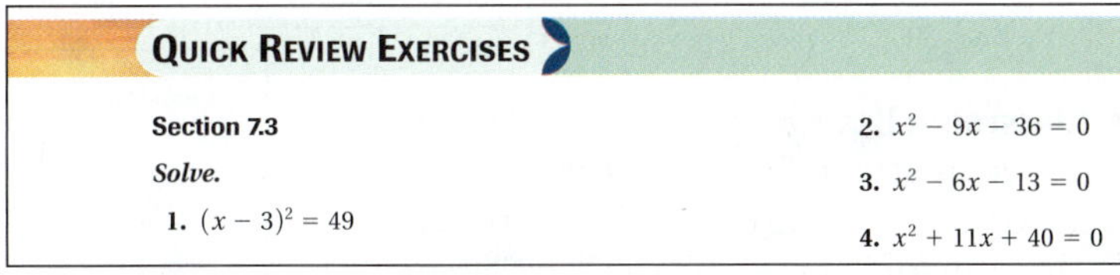
QUICK REVIEW EXERCISES

Section 7.3

Solve.

1. $(x - 3)^2 = 49$

2. $x^2 - 9x - 36 = 0$

3. $x^2 - 6x - 13 = 0$

4. $x^2 + 11x + 40 = 0$

- **Quick Review Exercises:** Appearing once per chapter, Quick Review Exercises are a short selection of review exercises geared toward helping students maintain skills they have previously learned and preparing them for upcoming concepts. (See pg. 486)

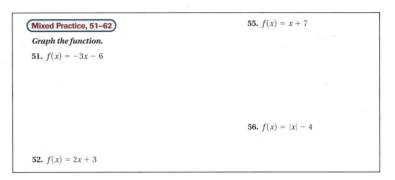

Find the slope–intercept form of the equation of a line that is parallel to the graphed line and that passes through the point plotted on the graph.

65. **66.**

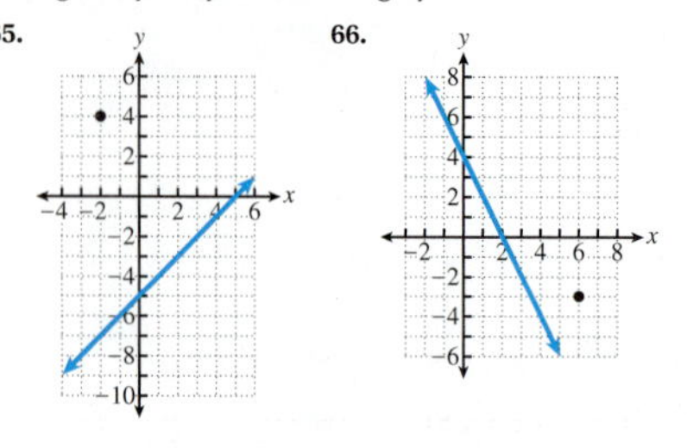

Find the slope–intercept form of the equation of a line that is perpendicular to the graphed line and that passes through the point plotted on the graph.

67. **68.**

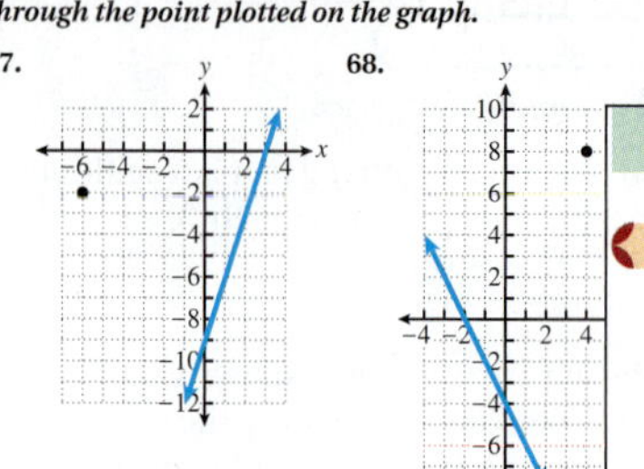

- **Community Exercises:** Building a community of learners within the classroom can make for a much more dynamic classroom and provide a sense of trust and collaboration among students. The Community Exercises, indicated in the Annotated Instructor's Edition with the CE icon, require thought, understanding, and communication. These optional exercises are provided as appropriate throughout the text. (See pp. 246)

- **Stretch Your Thinking Exercises:** Appearing at the end of each chapter, Stretch Your Thinking exercises ask students to go the extra mile by taking what they have learned in the chapter and applying it to a problem that requires some additional thought. These exercises can be done individually or in groups. (See pg. 278)

Stretch Your Thinking Chapter 2

Step 1: Form groups of no more than 4 or 5 students. Write the members' names in the chart below. For the best results, try to form groups in which all members are of the same sex. Each group should have a meter stick, ruler, or some type of measuring device.

Step 2: Measure the length of each member's foot (without shoes). Then measure and record the height of each member. Record these values in the appropriate column of the table.

Member's Name	Length of Foot	Height	Ordered Pair

Step 3: For each member, create an ordered pair in which the first coordinate is the length of the member's foot and the second coordinate is his/her height.

Step 4: Create an appropriate axis system using the horizontal axis for the foot length and the vertical axis for the heights.

Step 5: Plot your ordered pairs in the axis system you created. Try to be as accurate as possible. Starting from left to right, label your points A, B, C, D, etc. Visually, does it look like these points form a straight line? Would a straight line be an appropriate model for your data? Use a straight edge to check.

Step 6: Find the slope of the line that would connect point A and point B. Find the slope of the line that would connect point B and point C. Continue until you have used all of your points. Are these slopes the same? Should the slopes be the same if the points don't line up in a straight line? Have a group discussion and write down the groups' thoughts on these questions.

Step 7: Using a straight edge, try to find a straight line that best fits your ordered pairs. Calculate the equation of this line and draw it on your plane.

Step 8: Along with your completed chart, turn in the linear equation you created that your group feels best fits your data.

- **Classroom Examples:** Having in-class practice problems at your fingertips is extremely helpful to both new and experienced instructors. These instructor examples, called Classroom Examples, are included in the margins of the AIE as in-class practice problems. (See pg. 158)

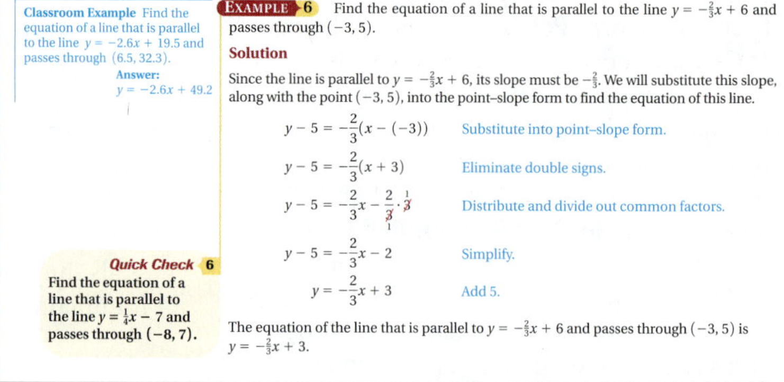

Classroom Example Find the equation of a line that is parallel to the line $y = -2.6x + 19.5$ and passes through $(6.5, 32.3)$.
Answer: $y = -2.6x + 49.2$

EXAMPLE 6 Find the equation of a line that is parallel to the line $y = -\frac{2}{3}x + 6$ and passes through $(-3, 5)$.

Solution

Since the line is parallel to $y = -\frac{2}{3}x + 6$, its slope must be $-\frac{2}{3}$. We will substitute this slope, along with the point $(-3, 5)$, into the point–slope form to find the equation of this line.

$$y - 5 = -\frac{2}{3}(x - (-3)) \quad \text{Substitute into point–slope form.}$$
$$y - 5 = -\frac{2}{3}(x + 3) \quad \text{Eliminate double signs.}$$
$$y - 5 = -\frac{2}{3}x - \frac{2}{3}\cdot 3 \quad \text{Distribute and divide out common factors.}$$
$$y - 5 = -\frac{2}{3}x - 2 \quad \text{Simplify.}$$
$$y = -\frac{2}{3}x + 3 \quad \text{Add 5.}$$

The equation of the line that is parallel to $y = -\frac{2}{3}x + 6$ and passes through $(-3, 5)$ is $y = -\frac{2}{3}x + 3$.

Quick Check 6
Find the equation of a line that is parallel to the line $y = \frac{1}{4}x - 7$ and passes through $(-8, 7)$.

Also Available

- **Worksheets for Classroom or Lab Practice** offer extra practice exercises for every section of the text with ample space for students to show their work. These lab- and classroom-friendly workbooks also list the learning objectives and key vocabulary terms for every text section, along with vocabulary practice problems.

Graphing Linear Equations

In this chapter we will examine linear equations in two variables. Equations involving two variables relate two unknown quantities to each other, such as a person's height and weight, or the number of items sold by a company and the profit that the company made. The primary focus is on graphing linear equations, which is the technique we use to display the solutions to an equation in two variables.

Study Strategy **Using Your Textbook Effectively** *In this chapter we will focus on how to get the most out of your textbook. Students who treat their books solely as a source of homework exercises are turning their backs on one of their best resources.*

Throughout this chapter we will revisit this study strategy and help you incorporate it into your study habits.

Building Your Study Strategy **Using Your Textbook Effectively, 1 Reading Ahead** One effective way to use your textbook is to read a section in the text before it is covered in class. This will give you an idea of the main concepts covered in the section, and these concepts can be clarified by your instructor in class.

When reading ahead you should scan the section. Look for definitions that are introduced in the section, as well as procedures developed in the section. Pay closer attention to the examples. If you find a step in the examples that you cannot understand, you can ask your instructor about it in class.

Building Your Study Strategies

Woodbury introduces a *Study Strategy* in each chapter opener. Each strategy is then revisited and expanded upon prior to each section's exercise set in *Building Your Study Strategy* boxes and then again at the end of the chapter. These helpful Study Strategies outline good study habits and ask students to apply these skills as they progress through the textbook. Study Strategy topics include Study Groups, Using Your Textbook Effectively, Test Taking, Overcoming Math Anxiety, and many more. (See pgs. 81, 96, 116, 159)

The ***Math Study Skills Workbook***, created specifically to complement the Woodbury text, gives students suggestions and activities that will help to make them better mathematics students. Students are given further guidance on better note-taking, time-management, test-taking skills, and other strategies.

Additional Student Support

Mathematicians in History

Mathematicians in History activities provide a structured opportunity for students to learn about the rich and diverse history of mathematics. These short research projects, which ask students to investigate the life of a prominent mathematician, can be assigned as independent work or used as a collaborative-learning activity. (See pg. 277)

Mathematicians in History

Ada Lovelace

Ada Lovelace, whose full name was Augusta Ada King, Countess of Lovelace, was an English mathematician who lived in the 19th century. Besides her mathematical contributions, she had a profound effect on the development of the computer and computer science.

Write a one-page summary (*or* make a poster) of the life of Ada Lovelace and her accomplishments.

Interesting issues:

- Where and when was Ada Lovelace born?
- Who was Ada Lovelace's famous father? What is he best known for?
- Ada's father had a mathematical nickname for her mother. What was it?
- One of Ada's early tutors was Augustus De Morgan, who became a famous logician and developed De Morgan's Laws. What are De Morgan's laws?
- Ada wrote a description of Charles Babbage's Analytical Engine. What was the Analytical Engine?
- In her description of Babbage's Analytical Engine, Ada wrote what is considered to be the first computer program. It calculated Bernoulli numbers. What are Bernoulli numbers?
- What did the U.S. Defense Department name after Ada Lovelace?
- Describe the circumstances of Ada's death.

Using Your Calculator

The optional *Using Your Calculator* feature is presented throughout the text, giving students guided calculator instruction with screen shots, as appropriate, to complement the material being covered. (See pg. 91)

Using Your Calculator When we graph a line using the TI-84, we will occasionally need to resize the window. In the previous example, the x-intercept is $(16, 0)$ and the y-intercept is $(0, 8000)$. Our graph should show these important points. After entering $8000 - 500x$ for Y_1, press the WINDOW key. Fill in the screen as shown, and press the GRAPH key to display the graph.

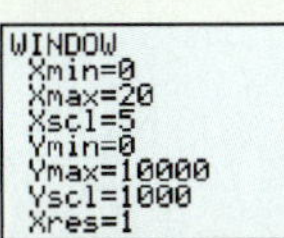

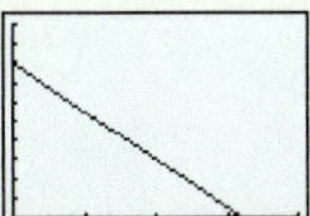

A Word of Caution

A Word of Caution tips, located throughout the text, help students avoid misconceptions by pointing out typical errors that students often make. (See pg. 108)

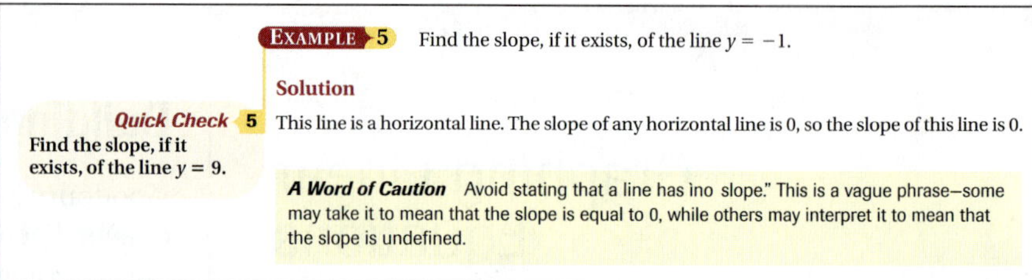

End-of-Chapter Content

Each chapter concludes with a **Chapter Summary,** a **summary** of the chapter's **Study Strategies, Chapter Review Exercises,** and a **Chapter Test.** Together, these are an excellent resource for extra practice and test preparation. Full solutions to highlighted chapter review exercises are provided in the back of the text as yet one more way for students to assess their understanding and check their work. A set of **Cumulative Review** exercises can be found after Chapters 4, 7, and 10. These exercises are strategically placed to help students review before midterm and final exams.

The **Pass the Test CD** provides students with video footage of an instructor working through the complete solution to all the exercises in all chapter tests in the textbook. The CD also contains interactive vocabulary flashcards, a Spanish glossary, and tips for time management.

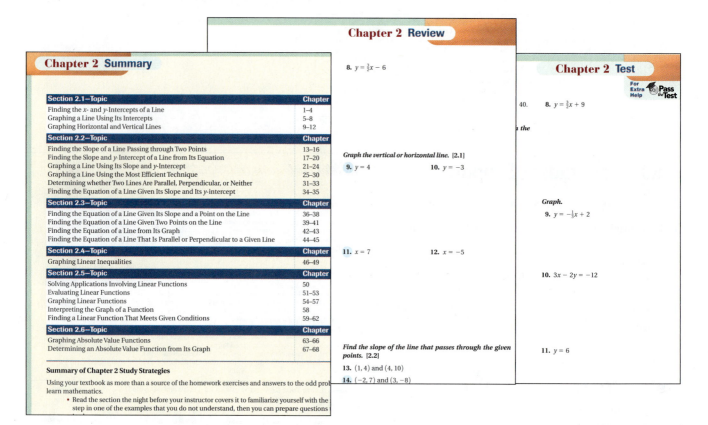

Chapter 2 Summary

Section 2.1–Topic	Chapter
Finding the x- and y-Intercepts of a Line	1–4
Graphing a Line Using Its Intercepts	5–8
Graphing Horizontal and Vertical Lines	9–12
Section 2.2–Topic	**Chapter**
Finding the Slope of a Line Passing through Two Points	13–16
Finding the Slope and y-Intercept of a Line from Its Equation	17–20
Graphing a Line Using Its Slope and y-Intercept	21–24
Graphing a Line Using the Most Efficient Technique	25–30
Determining whether Two Lines Are Parallel, Perpendicular, or Neither	31–33
Finding the Equation of a Line Given Its Slope and Its y-intercept	34–35
Section 2.3–Topic	**Chapter**
Finding the Equation of a Line Given Its Slope and a Point on the Line	36–38
Finding the Equation of a Line Given Two Points on the Line	39–41
Finding the Equation of a Line from Its Graph	42–43
Finding the Equation of a Line That Is Parallel or Perpendicular to a Given Line	44–45
Section 2.4–Topic	**Chapter**
Graphing Linear Inequalities	46–49
Section 2.5–Topic	**Chapter**
Solving Applications Involving Linear Functions	50
Evaluating Linear Functions	51–53
Graphing Linear Functions	54–57
Interpreting the Graph of a Function	58
Finding a Linear Function That Meets Given Conditions	59–62
Section 2.6–Topic	**Chapter**
Graphing Absolute Value Functions	63–66
Determining an Absolute Value Function from Its Graph	67–68

Summary of Chapter 2 Study Strategies

Using your textbook as more than a source of the homework exercises and answers to the odd prob
learn mathematics.

- Read the section the night before your instructor covers it to familiarize yourself with the step in one of the examples that you do not understand, then you can prepare questions

Chapter 2 Review

8. $y = \frac{3}{2}x - 6$

Graph the vertical or horizontal line. [2.1]

9. $y = 4$ **10.** $y = -3$

11. $x = 7$ **12.** $x = -5$

Find the slope of the line that passes through the given points. [2.2]

13. $(1, 4)$ and $(4, 10)$

14. $(-2, 7)$ and $(3, -8)$

Chapter 2 Test

For Extra Help Pass the Test

40.

the

8. $y = \frac{3}{2}x + 9$

Graph.

9. $y = -\frac{1}{3}x + 2$

10. $3x - 2y = -12$

11. $y = 6$

Overview of Supplements

The supplements available to students and instructors are designed to provide the extra support needed to help students be successful. As you can see from the list of supplements that follows, all areas of support are covered from guided solutions (video lectures and solutions manuals) to help in being a better math student (*Math Study Skills Workbook*). We hope that these additional supplements will help students master the skills, gain confidence in their mathematical abilities, and move on to the next course.

Student Supplements

Math Study Skills Workbook

- Created specifically to complement the Woodbury text
- Helps to make better mathematics students through study suggestions and activities to be completed by the student
- Further guidance on better note-taking, time-management, and test-taking skills, as well as other strategies
- Written by Alan Bass

ISBNs: 0-321-54306-8, 978-0-321-54306-6

Worksheets for Classroom or Lab Practice

- Extra practice exercises for every section of the text with ample space for students to show their work.
- These lab- and classroom-friendly workbooks also list the learning objectives and key vocabulary terms for every text section, along with vocabulary practice problems

ISBNs: 0-321-52315-6, 978-0-321-52315-0

Video Lectures on CD or DVD

- Short lectures for each section of the text presented by author George Woodbury and Mark Tom
- Complete set of digitized videos on CD-ROMs or DVDs for student use at home or on campus
- Ideal for distance learning or supplemental instruction
- Video lectures include optional English captions

CD ISBNs: 0-321-50655-3,
978-0-321-50655-9
DVD ISBNs: 0-321-55035-8
978-0-321-55035-4

Instructor Supplements

Annotated Instructor's Edition

- Contains answers to all exercises in the textbook and highlights special Community Exercises
- Provides Teaching Tips and Classroom Examples

ISBNs: 0-321-45955-5, 978-0-321-45955-8

Instructor and Adjunct Support Manual

- Includes resources to help both new and adjunct faculty with course preparation and classroom management by offering helpful teaching tips correlated to the sections of the text.

ISBNs: 0-321-46754-X, 978-0-321-46754-6

Printed Test Bank

- Three free-response tests per chapter and two multiple-choice tests per chapter
- Two free-response and two multiple-choice final exams
- Short quizzes of five to six questions for every section that can be used in class, for individual practice or for group work

ISBNs: 0-321-46539-3, 978-0-321-46539-9

TestGen®

- Enables instructors to build, edit, print, and administer tests
- Features a computerized bank of questions developed to cover all text objectives
- Algorithmically based, allowing instructors to create multiple, but equivalent, versions of the same question or test with the click of a button
- Instructors can modify questions or add new questions
- Tests can be printed or administered online
- The software and test bank are available for download from Pearson Education's online catalog.

Instructor's Solutions Manual

- Worked-out solutions to all section-level exercises
- Solutions to all Quick Check, Chapter Review, Chapter Test, and Cumulative Review exercises

ISBNs: 0-321-45959-8, 978-0-321-45959-6

Student Supplements

Pass the Test CD

- Provides students with a video of an instructor working through all exercises from all chapter tests in the textbook.
- Includes interactive vocabulary flashcards, a Spanish glossary, and a video with tips for time management.
- Optional English captioning
- Automatically included in every new copy of the textbook.

Student's Solutions Manual

- Worked-out solutions for the odd-numbered section-level exercises
- Solutions to all problems in the Chapter Review, Chapter Test, and Cumulative Review exercises

ISBNs: 0-321-46756-6, 978-0-321-46756-0

MathXL® Tutorials on CD

- Algorithmically generated practice exercises that correlate at the objective level to the content of the text
- Includes an example and a guided solution to accompany every exercise as well as video clips for selected exercises
- Recognizes student errors and provides appropriate feedback; generates printed summaries of student progress

ISBNs: 0-321-50669-3, 978-0-321-50669-6

Instructor Supplements

PowerPoint® Lecture Slides

- Present key concepts and definitions and examples from the text
- Available in MyMathLab or can be downloaded from Pearson Education's online catalog

Active Learning Questions

- Multiple-choice questions available for each chapter of the text
- Formatted in PowerPoint and can be used with classroom response systems
- Available in MyMathLab or can be downloaded from Pearson Education's online catalog

Pearson Adjunct Support Center

The Pearson Adjunct Support Center is staffed by qualified mathematics instructors with over 50 years of combined experience at both the community college and university level. Assistance is provided for faculty in the following areas:

- Suggested syllabus consultation
- Tips on using materials packaged with the book
- Book-specific content assistance
- Teaching suggestions including advice on classroom strategies

MathXL® www.mathxl.com

MathXL is a powerful online homework, tutorial, and assessment system that accompanies Pearson Education textbooks in mathematics or statistics. With MathXL, instructors can create, edit, and assign online homework and tests using algorithmically generated exercises correlated at the objective level to the textbook. They can also create and assign their own online exercises and import TestGen tests for added flexibility. All student work is tracked in MathXL's online gradebook. Students can take chapter tests in MathXL and receive personalized study plans based on their test results. The study plan diagnoses weaknesses and links students directly to tutorial exercises for the objectives they need to study and retest. Students can also access supplemental animations and video clips directly from selected exercises. MathXL is available to qualified adopters. For more information, visit our website at www.mathxl.com, or contact your Pearson Education sales representative.

MyMathLab® www.mymathlab.com

MyMathLab is a series of text-specific, easily customizable online courses for Pearson Education's textbooks in mathematics and statistics. Powered by CourseCompass™ (our online teaching and learning environment) and MathXL® (our online homework, tutorial, and assessment system), MyMathLab gives you the tools you need to deliver all or a portion of your course online, whether your students are in a lab setting or working from home. MyMathLab provides a rich and flexible set of course materials, featuring free-response exercises that are algorithmically generated for unlimited practice and mastery. Students can also use online tools, such as video lectures, animations, and a multimedia textbook, to independently improve their understanding and performance. Instructors can use MyMathLab's homework and test managers to select and assign online exercises correlated directly to the textbook, and they can also create and assign their own online exercises and import TestGen tests for added flexibility. MyMathLab's online gradebook—designed specifically for mathematics and statistics—automatically tracks students' homework and test results and gives the instructor control over how to calculate final grades. Instructors can also add offline (paper-and-pencil) grades to the gradebook. MyMathLab also includes access to the **Pearson Tutor Center,** which provides students with tutoring via toll-free phone, fax, email, and interactive Web sessions. MyMathLab is available to qualified adopters. For more information, visit our website at www.mymathlab.com or contact your sales representative.

InterAct Math® Tutorial Web site: www.interactmath.com

Get practice and tutorial help online! This interactive tutorial Web site provides algorithmically generated practice exercises that correlate directly to the exercises in the textbook. Students can retry an exercise as many times as they like with new values each time for unlimited practice and mastery. Every exercise is accompanied by an interactive guided solution that provides helpful feedback for incorrect answers, and students can also view a worked-out sample problem that steps them through an exercise similar to the one they're working on.

Acknowledgments

Writing a textbook such as this is a monumental task, and I would like to take this opportunity to thank everyone who has helped me along the way. The following reviewers provided thoughtful suggestions and were instrumental in the development of *Intermediate Algebra.*

Jerry Allison	*Black Hawk College*
Marie Aratari	*Oakland Community College*
Jan Archibald	*Ventura College*

O. Pauline Chow	*Harrisburg Area Community College*
Deborah Fries	*Wor-Wic Community College*
Shawna Haider	*Salt Lake Community College*
Pauline Hall	*Iowa State University*
Mary Beth Headlee	*Manatee Community College*
Michael Kaczmarski	*Honolulu Community College*
Mohammed Kazemi	*UNC–Charlotte*
Deanna Kindhart	*Illinois Central College*
Lonnie Larson	*Sacramento City College*
Mark Laurel	*MiraCosta College*
Mickey Levendusky	*Pima Community College*
Rhonda Macleod	*Tallahassee Community College*
Kathleen McDaniel	*Buena Vista University*
Robert Musselman	*California State University–Fresno*
Linda Pulsinelli	*Western Kentucky University*
Richard Rupp	*Del Mar College*
Jeanette Shea Shotwell	*Central Texas College*
Mary Ann Teel	*University of North Texas*
Timothy Thompson	*Oregon Institute of Technology*
Mary Lou Townsend	*Wor-Wic Community College*
Jackie Wing	*Angelina College*

I have truly enjoyed working with the team at Addison-Wesley. I do owe special thanks to my editor, Jenny Crum, as well as Lauren Morse, Antonio Arvelo, Ron Hampton, Jay Jenkins, Michelle Renda, Nathaniel Koven, Ceci Fleming, Carl Cottrell, and Sharon Tomasulo. Thanks are due to Greg Tobin and Maureen O'Connor for believing in my vision and taking a chance on me, and to Susan Winslow for getting this all started. Anne Scanlan-Rohrer is an outstanding developmental editor, and her hard work and suggestions have had a tremendous impact on this book. Stephanie Logan's assistance during the production process was invaluable, and Cheryl Davids and Sudhir Goel deserve credit for their help in accuracy checking the pages.

Thanks to Vineta Harper, Mark Tom, Don Rose, and Ross Rueger, my colleagues at College of the Sequoias, who have provided some great advice along the way and frequently listened to my ideas. I would also like to thank my students for keeping my fires burning. It truly is all about the students.

Most importantly, thanks to my wife Tina and our wonderful children Dylan and Alycia. They are truly my greatest blessing, and I love them more than words can say. The process of writing a textbook is long and difficult, and they have been supportive and understanding at every turn.

George Woodbury

CHAPTER One 1

Review of Real Numbers

This chapter reviews properties of real numbers and arithmetic that are important for students to know if they are to achieve success in algebra. The chapter also introduces several algebraic properties.

Study Strategy **Study Groups** *Throughout this book there are study tips to help you learn and succeed in this course. In this chapter we will focus on getting involved in a study group.*

Working with a study group is an excellent way to learn mathematics, improve your confidence and level of interest, and improve your performance on tests and quizzes. When working with a group, you will be able to work through questions about the material you are studying. Also, by explaining how to solve a particular type of problem to another person in your group, you will increase your ability to retain this knowledge.

We will revisit this study tip throughout this chapter and help you incorporate it into your study habits. See the end of Section 1.1 for tips on how to get a study group started.

1.1 Integers, Opposites, and Absolute Value

Objectives

1. **Understand sets of numbers.**
2. **Determine the greater of two integers.**
3. **Find the absolute value of an integer.**
4. **Add and subtract integers.**
5. **Multiply and divide integers.**
6. **Simplify exponents.**
7. **Use the order of operations to simplify arithmetic expressions.**

Objective 1 Understand sets of numbers. A **set** is a collection of objects, such as the set consisting of the numbers 1, 4, 9, and 16. This set can be written as $\{1, 4, 9, 16\}$. The braces, $\{\ \}$, are used to denote a set, and the values listed inside are said to be **elements** or members of the set. A set with no elements is called the **empty set**. A **subset** of a set is a collection of some or all of the elements of the set. For example, $\{1, 9\}$ is a subset of the set $\{1, 4, 9, 16\}$. The empty set is a subset of every set.

Real Numbers

This text, for the most part, deals with the set of real numbers.

Real Numbers

A number is a **real number** if it can be written as a fraction or as a decimal, including decimals that do not terminate or repeat.

One subset of the set of real numbers is the set of natural numbers.

Natural Numbers

The set of **natural numbers** is the set $\{1, 2, 3, \ldots\}$.

If we include the number 0 with the set of natural numbers, we have the set of **whole numbers**.

Whole Numbers

The set of **whole numbers** is the set $\{0, 1, 2, 3, \ldots\}$. This set can be displayed on a number line as follows.

The arrow on the right-hand side of the number line indicates that the values continue to increase in this direction. There is no largest whole number, but we say that the values approach infinity (∞).

Integers

Another important subset of the real numbers is the set of integers.

Integers

The set of **integers** is the set $\{\ldots, -3, -2, -1, 0, 1, 2, 3, \ldots\}$. We can display the set of integers on a number line as follows.

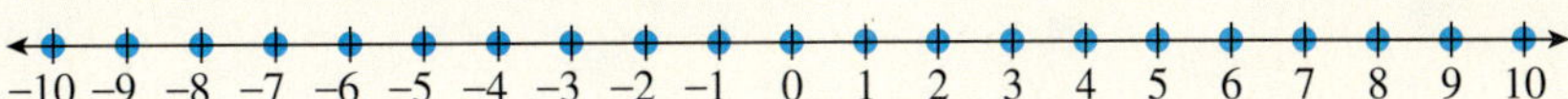

The arrow on the left side indicates that the values continue to decrease in this direction, and they are said to approach negative infinity $(-\infty)$.

Opposites

The set of integers is the set of whole numbers together with the opposites of the natural numbers. The **opposite** of a number is a number on the other side of 0 on the number line and is the same distance from 0 as that number. We denote the opposite of a real number a as $-a$. For example, -5 and 5 are opposites because they are both 5 units away from 0 and one is to the left of 0 while the other is to the right of 0.

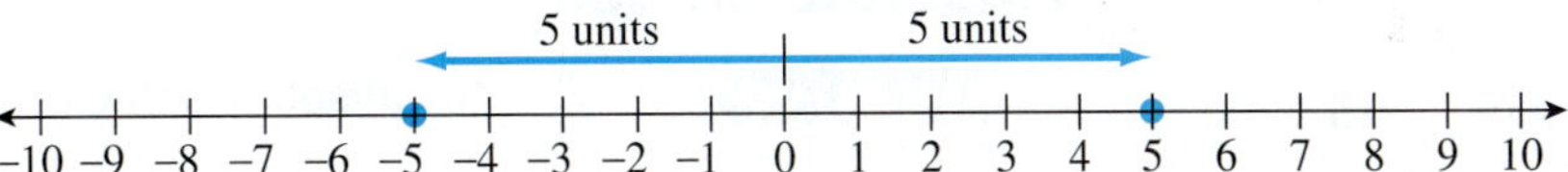

Numbers to the left of 0 on the number line are called **negative numbers.** Negative numbers represent a quantity less than 0. For example, if you have written checks that the balance in your checking account cannot cover, your balance would be a negative number. A temperature that is below 0°F, a golf score that is below par, and an elevation that is below sea level are other examples of quantities that can be represented by negative numbers.

Inequalities

Objective 2 **Determine the greater of two integers.** When comparing two whole numbers a and b, we say that a is **greater than** b, denoted $a > b$, if the number a is to the right of the number b on the number line. The number a is **less than** b, denoted $a < b$, if a is to the left of b on the number line. The statements $a > b$ and $a < b$ are called **inequalities.**

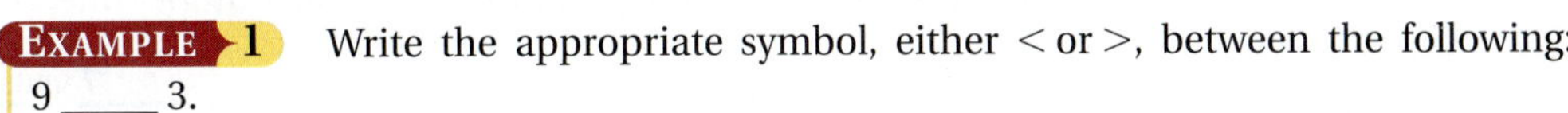

EXAMPLE 1 Write the appropriate symbol, either $<$ or $>$, between the following: 9 _____ 3.

Solution

Let's take a look at the two values graphed on a number line.

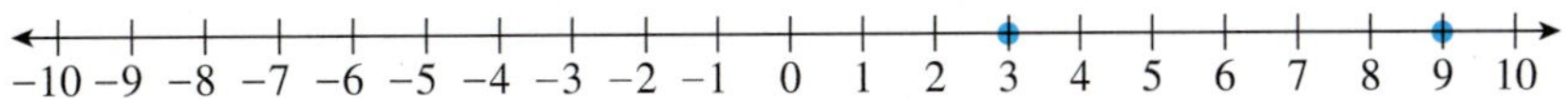

Since the number 9 is to the right of the number 3 on the number line, we say that 9 is greater than 3. So $9 > 3$.

EXAMPLE 2 Write the appropriate symbol, either $<$ or $>$, between the following: -6 ____ 2.

Solution

Looking at the number line, we can see that -6 is to the left of 2, so $-6 < 2$.

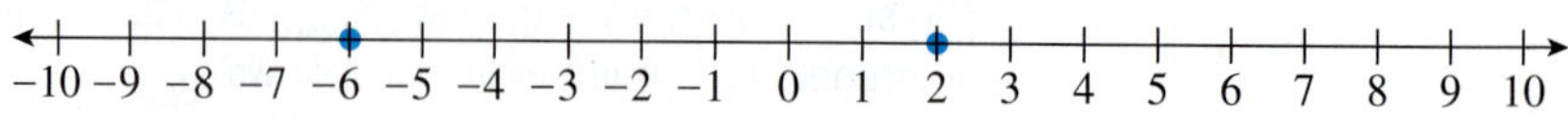

EXAMPLE 3 Write the appropriate symbol, either $<$ or $>$, between the following: -5 ____ -7.

Solution

On the number line, -5 is to the right of -7, so $-5 > -7$.

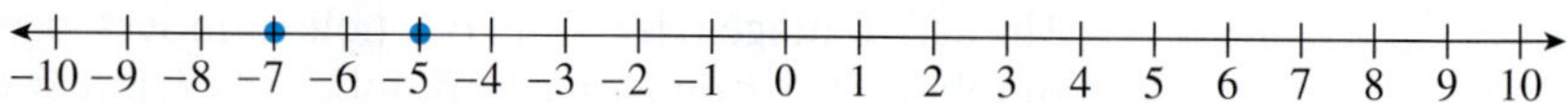

Quick Check 1

Write the appropriate symbol, either < or >, between the following:

a) 4 ____ 8

b) −10 ____ −15

c) 3 ____ −12.

Absolute Values

Objective 3 **Find the absolute value of an integer.**

Absolute Value

The **absolute value** of a number a, denoted $|a|$, is the distance between a and 0 on the number line.

Distance cannot be negative, so the absolute value of a number a will always be either 0 or positive.

EXAMPLE 4 Find the absolute value of 7.

Solution

The number 7 is 7 units away from 0 on the number line, so $|7| = 7$.

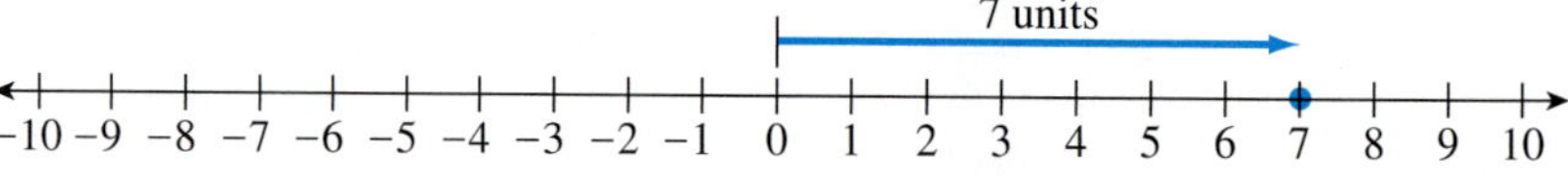

EXAMPLE 5 Find $|-3|$.

Solution

The number -3 is 3 units away from 0 on the number line, so $|-3| = 3$.

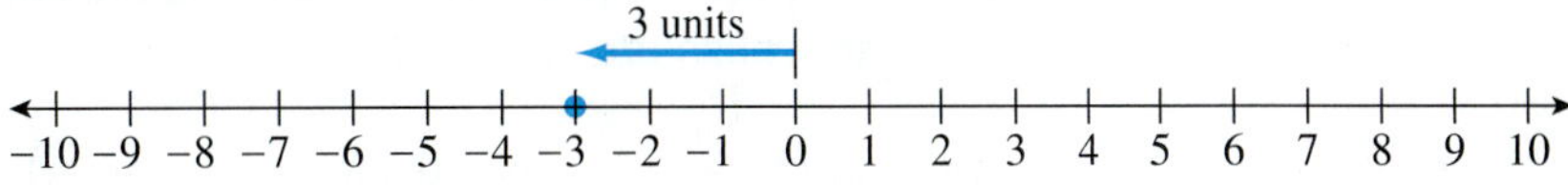

Quick Check 2
Find $|-8|$.

Addition and Subtraction of Integers

Objective 4 **Add and subtract integers.** We can use the number line to help us learn how to add and subtract integers. Suppose we were trying to add the integers 3 and -7, which could be written as $3 + (-7)$. On a number line, we will start at 0 and move 3 units in the positive, or right, direction. Adding -7 tells us to move 7 units in the negative, or left, direction.

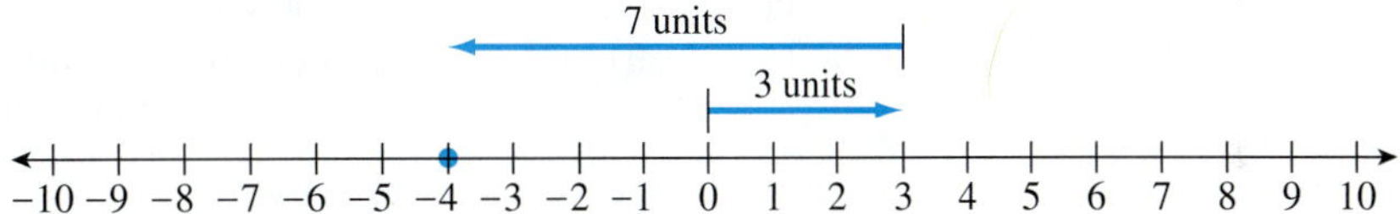

Since we end up at -4, this tells us that $3 + (-7) = -4$.

We can use a similar approach to verify an important property of opposites: the sum of two opposites is equal to 0.

Sum of Two Opposites

For any real number a, $a + (-a) = 0$.

Suppose that we wanted to add the opposites 4 and -4. Using the number line, we would begin at 0 and move 4 units to the right. We then move 4 units to the left, ending at 0. So, $4 + (-4) = 0$.

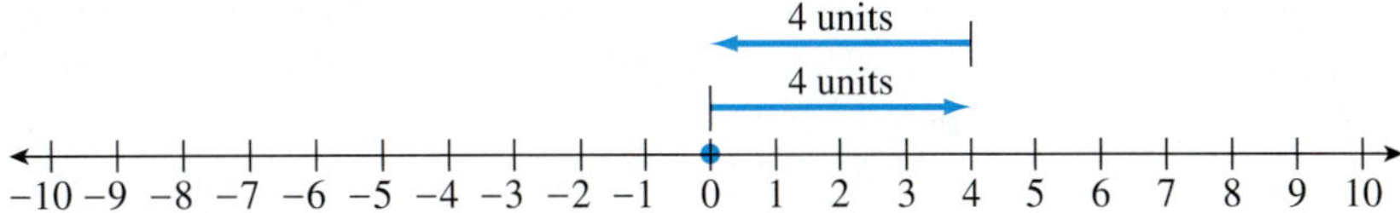

We now will examine another technique for finding the sum of a positive number and a negative number. In the sum $3 + (-7)$, the number 3 contributes to the sum in a positive fashion while the number -7 contributes to the sum in a negative fashion. The two numbers contribute to the sum in an opposite manner. We can think of the sum as the difference between these two contributions.

Adding a Positive Number and a Negative Number

1. **Take the absolute value of each number, and find the difference between these two absolute values.** This is the difference between the two numbers' contributions to the sum.
2. **The sign of the result is the same as the sign of the number that has the largest absolute value.**

For the sum $3 + (-7)$, we begin by taking the absolute value of each number: $|3| = 3$, $|-7| = 7$. The difference between the absolute values is 4. The sign of the sum is the same as the sign of the number that has the larger absolute value. In this case –7 has the larger absolute value, so our result is negative. Therefore, $3 + (-7) = -4$.

EXAMPLE 6 Find the sum $11 + (-9)$.

Solution

$\lvert 11\rvert = 11; \lvert -9\rvert = 9$	Find the absolute value of each number.
$11 - 9 = 2$	The difference between the absolute values is 2.
$11 + (-9) = 2$	Since the number with the larger absolute value is positive, the result is positive.

***Quick Check* 3**
Find the sum $8 + (-5)$.

Notice that in the previous example $11 + (-9)$ is equivalent to $11 - 9$, which also equals 2. What the two expressions share is that there is one number (11) that contributes to the total in a positive fashion and a second number (9) that contributes to the total in a negative fashion.

EXAMPLE 7 Find the sum $5 + (-16)$.

Solution

$\lvert 5\rvert = 5; \lvert -16\rvert = 16$	Find the absolute value of each number.
$16 - 5 = 11$	The difference between the absolute values is 11.
$5 + (-16) = -11$	Since the number with the larger absolute value is negative, the result is negative.

***Quick Check* 4**
Find the sum $9 + (-25)$.

Note that $-16 + 5$ is also equal to -11. The rules for adding a positive integer and a negative integer still apply when the first number is negative and the second number is positive.

To add two integers with the same sign, we add the absolute values of the integers and use the sign of the integers as the sign of the sum. Consider the sum $(-2) + (-5)$. On a number line we begin at 0 and move 2 units to the left. We then move 5 units further to the left, ending at -7.

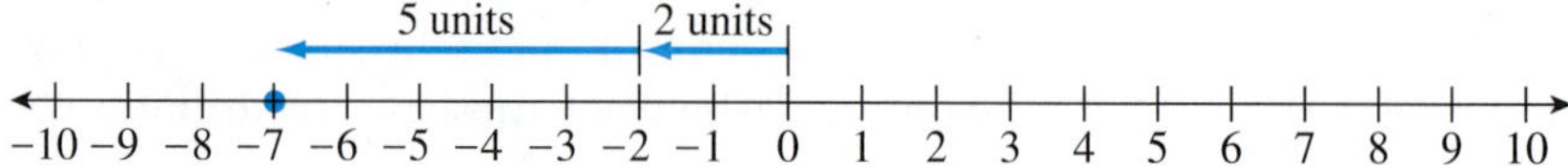

Since both contributions are negative, we can add the absolute values of the two numbers rather than finding their difference.

Adding Two (or More) Negative Numbers

1. Add the absolute values of each number.
2. The sign of the sum is negative.

EXAMPLE 8 Find the sum $-9 + (-8)$.

Solution

$\lvert -9 \rvert = 9; \lvert -8 \rvert = 8$	Find the absolute value of each number.
$9 + 8 = 17$	Add the absolute values.
$-9 + (-8) = -17$	Since both numbers were negative, the sum is negative.

Quick Check 5
Find the sum $-8 + (-13)$.

We could also begin the previous example by rewriting $-9 + (-8)$ as $-9 - 8$. After realizing that both values are contributing to the result in a negative fashion, we can add the negative contributions of 9 and 8 to produce a result of -17.

To subtract a negative integer from another integer, we will use the following property.

Subtraction of Real Numbers

For any real numbers a and b, $a - b = a + (-b)$.

This property tells us that we can add the opposite of b to a rather than subtracting b from a. Suppose that we were subtracting a negative integer, such as in the example $-8 - (-19)$. The property for subtraction of real numbers tells us that subtracting -19 is the same as adding its opposite, 19, so we would then convert this subtraction to $-8 + 19$. Remember that subtracting a negative number is equivalent to adding a positive number.

EXAMPLE 9 Find $13 - (-40)$.

Solution

$13 - (-40) = 13 + 40$	Subtracting -40 is the same as adding 40.
$= 53$	Add.

Quick Check 6
Find $19 - (-6)$.

General Strategy for Adding/Subtracting Integers

- Rewrite "double signs." Adding a negative number, for example $4 + (-5)$, can be rewritten as subtracting a positive number, $4 - 5$. Subtracting a negative number, for example $-2 - (-7)$, can be rewritten as adding a positive number, $-2 + 7$.
- Look at each integer and determine whether it is contributing positively or negatively to the result.
- Add up any integers contributing positively, if there are any, resulting in a single positive integer. Add up all integers that are contributing negatively, resulting in a single negative integer. Finish by finding the sum of these two integers.

Rather than being told to add or subtract, we may be asked to simplify a numerical expression. To **simplify** an expression, perform all arithmetic operations.

EXAMPLE 10 Simplify $6 - (-18) - 10 + (-25) - (-9) + 7$.

Solution

We begin by working on the double signs. This produces the following:

$$6 - (-18) - 10 + (-25) - (-9) + 7 = 6 + 18 - 10 - 25 + 9 + 7$$

Rewrite double signs.

$$= 40 - 35$$

The four integers that contribute in a positive fashion (6, 18, 9, and 7) have a sum of 40. The two integers that contribute in a negative fashion have a sum of -35.

$$= 5$$

Subtract.

Quick Check 7 Simplify $20 - 31 - (-17) - 8 + (-41) + 7$.

Using Your Calculator When you use your calculator, it is important for you to be able to distinguish between the subtraction key and the key for a negative number. On the TI-84, the subtraction key [-] is listed above the addition key on the right side of the calculator, while the negative key [(-)] is located to the left of the [ENTER] key at the bottom of the calculator. Here are two different ways to simplify the expression from the previous example on the TI-84.

```
6-(-18)-10+(-25)
-(-9)+7
```

```
6+18-10-25+9+7
```

In either case, pressing the [ENTER] key produces the result 5.

Multiplication and Division of Integers

Objective 5 **Multiply and divide integers.** The result obtained from multiplying two numbers is called the **product** of the two numbers. The numbers that are multiplied are called **factors**. When we multiply two positive integers, their product is also a positive integer. For example, the product of the two positive integers 4 and 7 is the positive integer 28. This can be written as $4 \cdot 7 = 28$. The product $4 \cdot 7$ can also be written as $4(7)$ or $(4)(7)$.

The product of a positive integer and a negative integer is a negative integer. For example, $4(-7) = -28$ and $(-9)(6) = -54$.

Products of Integers

$$(\text{Positive}) \cdot (\text{Positive}) = \text{Positive}$$
$$(\text{Positive}) \cdot (\text{Negative}) = \text{Negative}$$
$$(\text{Negative}) \cdot (\text{Positive}) = \text{Negative}$$

EXAMPLE 11 Find $7(-2)$.

Solution

We begin by multiplying 7 and 2, which equals 14. The next step is to determine the sign of our result. Whenever we multiply a positive integer by a negative integer the result is negative.

$$7(-2) = -14$$

***Quick Check* 8**
Find $5(-9)$.

A Word of Caution Note the difference between $7 - 2$ (a subtraction) and $7(-2)$ (a multiplication). A set of parentheses without a sign in front of it is used to indicate multiplication.

The product of two negative integers is a positive integer. For example, $(-2)(-5) = 10$.

Product of Two Negative Integers

$$(\text{Negative}) \cdot (\text{Negative}) = \text{Positive}$$

EXAMPLE 12 Find $(-8)(-7)$.

Solution

We begin by multiplying 8 and 7, which equals 56. The next step is to determine the sign of our result. Whenever we multiply a negative integer by a negative integer the result is positive.

$$(-8)(-7) = 56$$

***Quick Check* 9**
Find $-7(-13)$.

Products of Integers

- If a product contains an odd number of negative factors, then the result is negative.
- If a product contains an even number of negative factors, then the result is positive.

The main idea behind this principle is that every two negative factors multiply to be positive. If there are three negative factors, the product of the first two is a positive number. Multiplying this positive product by the third negative factor produces a negative product.

EXAMPLE 13 Find $6(-4)(-5)(-7)$.

Solution

Since there are three negative factors, our product will be negative.

$$6(-4)(-5)(-7) = -840$$

***Quick Check* 10**
Find $-3(-10)(6)(-8)$.

Before continuing on to division, we should mention multiplication by 0. Because any real number multiplied by 0 will be 0, we can state the multiplication property of 0.

Multiplication by 0

For any real number x,

$$0 \cdot x = 0$$
$$x \cdot 0 = 0.$$

When dividing one number, called the **dividend**, by another number, called the **divisor**, the result obtained is called the **quotient** of the two numbers.

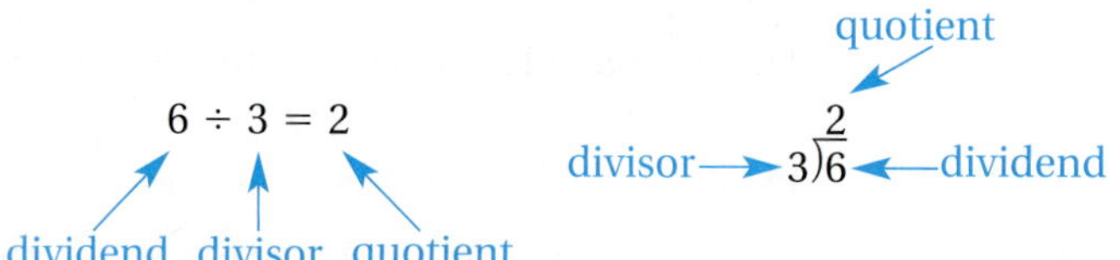

The statement "6 divided by 3 is equal to 2" is true because the product of the quotient and the divisor, $2 \cdot 3$, is equal to the dividend 6.

$$6 \div 3 = 2 \longleftrightarrow 2 \cdot 3 = 6$$

When we divide two integers that have the same sign (both positive or both negative), the quotient is positive. When we divide two integers that have different signs (one negative, one positive), the quotient is negative. Note that this is consistent with our rules for multiplication.

Quotients of Integers

(Positive) ÷ (Positive) = Positive
(Negative) ÷ (Negative) = Positive
(Positive) ÷ (Negative) = Negative
(Negative) ÷ (Positive) = Negative

EXAMPLE 14 Find $(-92) \div (-4)$.

Solution

When we divide a negative number by another negative number, the result is positive.

$$(-92) \div (-4) = 23 \qquad \text{Note that } -4 \cdot 23 = -92.$$

EXAMPLE 15 Find $(-91) \div 7$.

Solution

When we divide a negative number by a positive number, the result is negative.

$$(-91) \div 7 = -13$$

Quick Check 11
Find $128 \div (-8)$.

Whenever 0 is divided by any integer (except 0), the quotient is 0. For example, $0 \div 16 = 0$. We can check that this quotient is correct by multiplying the quotient by the divisor. Since $0 \cdot 16 = 0$, the quotient is correct.

Division by Zero

Whenever an integer is divided by 0, the quotient is said to be **undefined**.

We use the word **undefined** to state that an operation cannot be performed or is meaningless. For example, $41 \div 0$ is undefined. Suppose that $41 \div 0 = a$ for some real number a. In that case, the product $a \cdot 0$ would be equal to 41. Since we know that the product of 0 and any real number is equal to 0, there is no such number a.

Exponents

Objective 6 **Simplify exponents.** Using the same number repeatedly as a factor can be represented using **exponential notation**. For example, $3 \cdot 3 \cdot 3 \cdot 3 \cdot 3$ can be written as 3^5.

Base, Exponent

For the expression 3^5, the number being multiplied (3) is called the **base**. The **exponent** (5) tells us how many times that the base is used as a factor.

We read 3^5 as "*three raised to the fifth power*" or simply as "*three to the fifth power.*" When we raise a base to the second power, we usually say that the base is **squared**. When we raise a base to the third power, we usually say that the base is **cubed**. Exponents of 4 or higher do not have a special name.

EXAMPLE 16 Simplify 4^5.

Solution

In this example we want to use 4 as a factor 5 times.

$$\begin{aligned} 4^5 &= 4 \cdot 4 \cdot 4 \cdot 4 \cdot 4 && \text{Write 4 as a factor 5 times.} \\ &= 1024 && \text{Multiply.} \end{aligned}$$

Using Your Calculator Many calculators have a key that can be used to simplify expressions with exponents. On the TI-84, we use the [^] key. Other calculators may have a key that is labeled [y^x] or [x^y]. Here is the screen that you should see when using the TI-84 to simplify the expression in the previous example.

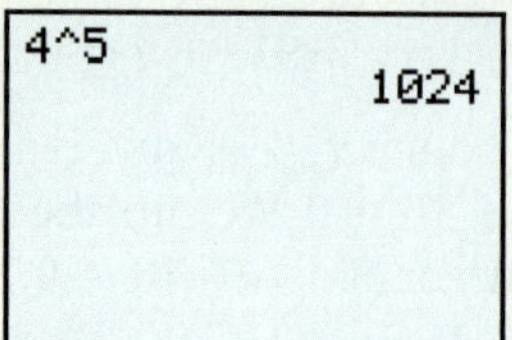

***Quick Check* 12**
Simplify 4^3.

EXAMPLE 17 Simplify $(\frac{2}{3})^4$.

Solution

When the base is a fraction, the same rules apply. We use $\frac{2}{3}$ as a factor four times.

$$\begin{aligned} \left(\frac{2}{3}\right)^4 &= \frac{2}{3} \cdot \frac{2}{3} \cdot \frac{2}{3} \cdot \frac{2}{3} && \text{Write as a product.} \\ &= \frac{16}{81} && \text{Multiply.} \end{aligned}$$

***Quick Check* 13**
Simplify $(\frac{1}{5})^3$.

Consider the expression $(-3)^2$. The base is -3, so we multiply $(-3) \cdot (-3)$, and the result is positive 9. However, in the expression -2^4, the negative sign is not included in a set of parentheses with the 2. So the base of this expression is 2 and not -2. We use 2 as a factor 4 times and then take the opposite of our result: $-2^4 = -(2 \cdot 2 \cdot 2 \cdot 2) = -16$.

Order of Operations

Objective 7 **Use the order of operations to simplify arithmetic expressions.**
Suppose we were asked to simplify the expression $2 + 4 \cdot 3$. We could obtain two different results depending on whether we performed the addition or the multiplication first.

Addition First	Multiplication First
$2 + 4 \cdot 3 = 6 \cdot 3$ $= 18$	$2 + 4 \cdot 3 = 2 + 12$ $= 14$

Only one result is correct. The **order of operations agreement** is a standard order in which arithmetic operations are performed, ensuring a single correct answer.

Order of Operations

1. **Remove grouping symbols.** We begin by simplifying all expressions that are inside parentheses, brackets, and absolute value bars. We also perform any operations in the numerator or denominator of a fraction. This is done by following steps 2 to 4 below.
2. **Perform any operations involving exponents.** After all grouping symbols have been removed from the expression, we simplify any exponential expressions.
3. **Multiply and divide.** These two operations have equal priority. We perform multiplications or divisions in the order they appear from left to right.
4. **Add and subtract.** At this point the only remaining operations should be additions and subtractions. Again these operations are of equal priority, and we perform them in the order that they appear from left to right.

Considering this information, the correct result when simplifying $2 + 4 \cdot 3$ is 14 because the multiplication takes precedence over the addition.

$$\begin{aligned} 2 + 4 \cdot 3 &= 2 + 12 && \text{Multiply } 4 \cdot 3. \\ &= 14 && \text{Add.} \end{aligned}$$

EXAMPLE 18 Simplify $\frac{2}{5} \cdot \frac{3}{8} - \frac{7}{4} \div \frac{2}{3}$.

Solution

The first operation we will perform is $\frac{2}{5} \cdot \frac{3}{8}$, since multiplication and division are performed left to right before subtraction and the multiplication appears before the division. To multiply two fractions, we begin by dividing out any factors that are common to a numerator and denominator, and then we multiply the remaining numerators and multiply the remaining denominators.

$$\begin{aligned} \frac{2}{5} \cdot \frac{3}{8} - \frac{7}{4} \div \frac{2}{3} &= \frac{\overset{1}{\cancel{2}}}{5} \cdot \frac{3}{\underset{4}{\cancel{8}}} - \frac{7}{4} \div \frac{2}{3} && \text{Divide out common factors.} \\ &= \frac{3}{20} - \frac{7}{4} \div \frac{2}{3} && \text{Multiply.} \\ &= \frac{3}{20} - \frac{7}{4} \cdot \frac{3}{2} && \text{Rewrite the division as a multiplication. Recall that to divide one fraction by another, we invert the second fraction and multiply.} \\ &= \frac{3}{20} - \frac{21}{8} && \text{Multiply.} \end{aligned}$$

$$= \frac{6}{40} - \frac{105}{40}$$

Rewrite each fraction with a common denominator of 40. Recall that two fractions must have the same denominator before we add or subtract them.

$$= -\frac{99}{40}$$

Subtract.

Quick Check 14
Simplify $\frac{3}{8} - \frac{4}{5} \div \frac{2}{9}$.

EXAMPLE 19 Simplify $9 - 2 \cdot 6 - 3^4$.

Solution

In this example the operation with the highest priority is 3^4, since simplifying exponents takes precedence over subtraction or multiplication. Be sure to note that the base is 3, and not -3, since the negative sign is not grouped with the 3 inside a set of parentheses.

$9 - 2 \cdot 6 - 3^4 = 9 - 2 \cdot 6 - 81$	Raise 3 to the 4th power.
$= 9 - 12 - 81$	Multiply $2 \cdot 6$.
$= -84$	Simplify.

Quick Check 15
Simplify
$-8 \cdot 7 + 15 - 2^3$.

EXAMPLE 20 Simplify $(-7)^2 - 4(-5)(6)$.

Solution

We begin by squaring negative 7, which equals 49.

$(-7)^2 - 4(-5)(6) = 49 - 4(-5)(6)$	Square -7.
$= 49 - (-120)$	Multiply $4(-5)(6)$.
$= 49 + 120$	Write as a sum, eliminating the double signs.
$= 169$	Add.

Quick Check 16
Simplify
$10^2 - 4(-5)(-8)$.

A Word of Caution When we square a negative number, such as $(-7)^2$ in the previous example, the result is a positive number.

EXAMPLE 21 Simplify $20 \div 4 + 5(9 - 3 \cdot 8)$.

Solution

The first step is to simplify the expression inside the set of parentheses. Here the multiplication takes precedence over the subtraction. Once we have simplified the expression inside the parentheses, we proceed to divide and multiply. We finish by subtracting.

$20 \div 4 + 5(9 - 3 \cdot 8) = 20 \div 4 + 5(9 - 24)$	Multiply $3 \cdot 8$.
$= 20 \div 4 + 5(-15)$	Subtract $9 - 24$.
$= 5 + 5(-15)$	Divide $20 \div 4$.
$= 5 - 75$	Multiply $5(-15)$.
$= -70$	Subtract.

Quick Check 17
Simplify
$36 \div 4 \cdot 2 - 3(6 \cdot 5 - 18)$.

Occasionally an expression will have one set of grouping symbols inside another set, such as the expression $12[8 - 3(7 - 2)] - 9(-6)$. We call this **nesting** grouping symbols. We begin by simplifying the innermost set of grouping symbols and work our way out from there.

EXAMPLE 22 Simplify $12[8 - 3(7 - 2)] - 9(-6)$.

Solution

We begin by simplifying the expression inside the set of parentheses. Once that has been done, we simplify the expression inside the square brackets.

$12[8 - 3(7 - 2)] - 9(-6) = 12[8 - 3(5)] - 9(-6)$ Subtract $7 - 2$, to simplify the expression inside the parentheses. Now we turn our attention to simplifying the expression inside the square brackets.

$= 12[8 - 15] - 9(-6)$ Multiply $3 \cdot 5$.

$= 12[-7] - 9(-6)$ Subtract $8 - 15$.

$= -84 - (-54)$ Multiply $12[-7]$ and $9(-6)$.

$= -84 + 54$ Rewrite without double signs.

$= -30$ Simplify.

Quick Check **18** Simplify $-7[3^2 + 4(6 - 13)]$.

Building Your Study Strategy **Study Groups, 1** **Who to Work With** To form a study group, you must begin with the question "Who do I want to work with?" Look for students who are serious about learning, such as students who seem prepared for each class or students who ask good questions during class. If a student does not appear to do their homework, or is constantly late for class, then this student may not be the best person to work with.

Look for students you feel you can get along with. You are about to spend a long period of time working with this group, sometimes under stressful conditions. Working with group members who get on your nerves will not help you learn.

If you take advantage of tutorial services provided by your college, keep an eye out for classmates who do the same. There is a strong chance that classmates who use the tutoring center are serious about learning mathematics and earning good grades.

EXERCISES 1.1

Vocabulary

1. A set with no elements is called the ________.

2. The arrow on the right side of a number line indicates that the values approach ________.

3. The sum of two negative integers is a(n) ________ integer.

4. Subtracting a negative integer can be rewritten as adding ________.

5. The product of a positive integer and a negative integer is a(n) ________ integer.

6. The product of a negative integer and a negative integer is a(n) ________ integer.

7. In a division problem, the number we divide by is called the ________.

8. Division by 0 is ________.

9. In exponential notation, the factor being multiplied repeatedly is called the ________.

10. In exponential notation, the exponent tells us ________.

11. When a base is ________, it is raised to the second power.

12. When a base is ________, it is raised to the third power.

Write the appropriate symbol, either < or >, between the following integers.

13. −6 ___ −11 **14.** −9 ___ −3

15. −16 ___ −14 **16.** −33 ___ −50

17. 0 ___ −20 **18.** −32 ___ 19

Find the following absolute values.

19. $|-19|$ **20.** $|7|$

21. $|21|$ **22.** $|-5|$

23. $-|16|$ **24.** $-|-9|$

Find the missing number, if possible. There may be more than one number that works, so find as many as possible. There may be no number that works.

25. $|?| = 6$

26. $|?| = 22$

27. $|?| + 8 = 33$

28. $2 \cdot |?| - 9 = 17$

Add.

29. $5 + (-14)$ **30.** $15 + (-7)$

31. $3 + (-25)$ **32.** $(-129) + 75$

33. $(-4) + (-26)$ **34.** $-29 + 16$

35. $-8.3 + 13.2$ **36.** $-14.6 + (-5.48)$

37. $-\frac{3}{4} + \frac{2}{5}$ **38.** $\frac{3}{10} + \left(-\frac{5}{8}\right)$

Subtract.

39. $8 - 16$ **40.** $21 - 39$

41. $(-17) - 32$ **42.** $(-4) - 13$

43. $12 - (-19)$ **44.** $(-14) - (-11)$

45. $6.4 - 13.7$ **46.** $-9.8 - (-24.3)$

47. $\frac{2}{9} - \left(-\frac{5}{6}\right)$ **48.** $-\frac{7}{12} - \frac{11}{20}$

Simplify.

49. $-15 + 28 - 19$

50. $7 - 16 - 35$

51. $15 + (-24) - 19 - (-7)$

52. $-28 + (-47) - (-9) - 12$

53. The temperature at 6 A.M. in Brookings, South Dakota, was −9°C. By 4 P.M., the temperature had risen by 15°C. What was the temperature at 4 P.M.?

FOR EXTRA HELP

Interactmath.com

MathXL Tutorials on CD

Video Lectures on CD

Addison-Wesley Math Tutor Center

Student's Solutions Manual

54. If a golfer completes a round at 3 strokes under par, her score is denoted -3. A pro golfer had rounds of -3, -5, 1, and -7 in a recent tournament. What was her total score for this tournament?

55. Dave drove from Death Valley, California, located 282 feet below sea level, to Denver, Colorado, located 5280 feet above sea level. What was the change in elevation that he experienced?

56. After withdrawing \$120 from her bank using an ATM card, Araceli had \$295 remaining in her savings account. How much money did Araceli have in the account prior to withdrawing the money?

Multiply.

57. $6(-9)$

58. $-2(13)$

59. $-4(-7)$

60. $-13 \cdot 8$

61. $5.4(-3.2)$

62. $-4.9(-4.18)$

63. $-\frac{3}{4} + \frac{2}{5}$

64. $\frac{3}{10} + \left(-\frac{5}{8}\right)$

65. $-7 \cdot 0$

66. $0(-29)$

67. $-4(-5)(11)$

68. $-3(-6)(-9)$

Divide, if possible.

69. $36 \div (-4)$

70. $-91 \div (-13)$

71. $-144 \div (-8)$

72. $-48 \div 3$

73. $-64.02 \div 6.6$

74. $-65.664 \div (-4.32)$

75. $-\frac{10}{27} \div \left(-\frac{8}{21}\right)$

76. $\frac{24}{55} \div \left(-\frac{78}{175}\right)$

77. $0 \div (-17)$

78. $0 \div 4$

79. $12 \div 0$

80. $-6 \div 0$

Simplify.

81. $-9(-17)$

82. $176 \div (-11)$

83. $8 - 19$

84. $8(-19)$

85. $-23 - (-7) - 36$

86. $10(-8)(-7)$

87. $-3(-9)(-13)(0)$

88. $-45 + (-26) - 87 - (-33)$

89. Carlos owns 12 bottles of wine, each worth \$1250. What is the total worth of these bottles of wine?

90. An Internet start-up company had an operating loss of \$28,600 last month. If the five partners decide to divide up the loss equally amongst themselves, what is the loss for each person?

91. Suze owns 1500 shares of a stock that dropped in value by \$8 per share last month. She owns 600 shares of a second stock that dropped in value by \$6 per share last month. She also owns 900 shares of a stock that went up by \$9 per share last month. What is Suze's net income on these three stocks for last month?

92. Donald's company lost \$32 million dollars in the year 2005. The company lost four times as much in 2006. The company went on to lose \$19 million more in 2007 than it had lost in 2006. How much money did Donald's company lose in 2007?

Simplify the given expression.

93. 4^3

94. 3^4

95. 10^5

96. 10^3

97. $\left(\frac{2}{5}\right)^2$

98. $\left(\frac{1}{8}\right)^5$

99. $(-5)^2$

100. -5^2

101. $-4^2 \cdot (-2)^3$

102. $-\left(\frac{2}{3}\right)^2 \cdot \left(\frac{3}{7}\right)^4$

Simplify the given expression.

103. $5 + 3 \cdot 4$

104. $9 - 2 \cdot 3$

105. $30 - 15 \div 5$

106. $8 + 7 \cdot 6 + 9$

107. $13 + 9^2 \div 3$

108. $5 \cdot 11 - 4 \cdot 20$

109. $5.7(-3.8) - 2.3^2$

110. $(-4.7)^2 - 4(13.8)(-5.9)$

111. $6^2 - 4(-3)(-7)$

112. $(-4)^2 - 4(2)(-9)$

113. $8 - 3(6 + 4 \cdot 9)$

114. $15 - 8(5 - 3 \cdot 6 - 4^2)$

115. $\frac{3}{4} + \frac{5}{4} \cdot \frac{6}{7}$

116. $12 \cdot \frac{5}{8} - \left(\frac{3}{4}\right)^2$

117. $\frac{4}{15}\left(\frac{23}{24} - \frac{1}{3}\right)$

118. $\frac{7}{45} \div \left(\frac{7}{12} - \frac{2}{3} + \frac{3}{4}\right) \cdot \frac{11}{14}$

119. $\frac{4 \cdot 6 - 3^2}{2 \cdot 7 \div 4^3}$

120. $\frac{7^2 + 24^2}{9^2 + 12^2}$

121. $10 - 4[6 - 2(9 - 7)]$

122. $-7[20 - 10 \div (3 - 8) - 5^2]$

123. $7 - 3 \cdot 8 - |6 - 9 \cdot 11 - 45|$

124. $-4 \cdot |6 \cdot 2 - 5 \cdot 7| - 3^4$

125. $20 - |5^2 - 7 \cdot 8 - 3^3|$

126. $|-3^2 + 34| - |(-5)^2 + 6(13)|$

Use arithmetic operation signs such as +, −, ·, and ÷ between the values to produce the desired results. (You may use parentheses as well.)

127. 5 6 13 7 = 24

128. 3 9 4 8 = 7

129. 9 7 3 6 = 5

130. 6 9 8 35 = −31

Find the missing number.

131. $3^? = 243$

132. $7^? = 2401$

133. $?^5 = 1024$

134. $?^5 = -32$

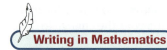

Writing in Mathematics

Answer in complete sentences.

135. Explain why subtracting a negative integer from another integer is the same as adding the opposite of that integer to it. Use the example $11 - (-5)$in your explanation.

136. Explain why a positive integer times a negative integer produces a negative integer.

137. Explain why division by 0 is undefined.

138. Explain the difference between $(-2)^6$ and -2^6.

*A **solutions manual page** is the solution to a problem as it would appear in a solutions manual for a textbook. Show and explain each step in solving the problem in such a way that a fellow student would be able to read and understand your work.*

139. ***Solutions Manual**** Write a solutions manual page for the following problem:

Simplify $3 - 4(20 \div 4 - 6 \cdot 8) - 5^2$.

*A **newsletter** assignment asks you to explain a certain mathematical topic. Your explanation should be in the form of a short, visually appealing article that might be published in a newsletter read by people who were interested in learning mathematics.*

140. ***Newsletter**** Write a newsletter explaining how to use the order of operations to simplify an expression.

***See Appendix B for details and sample answers.**

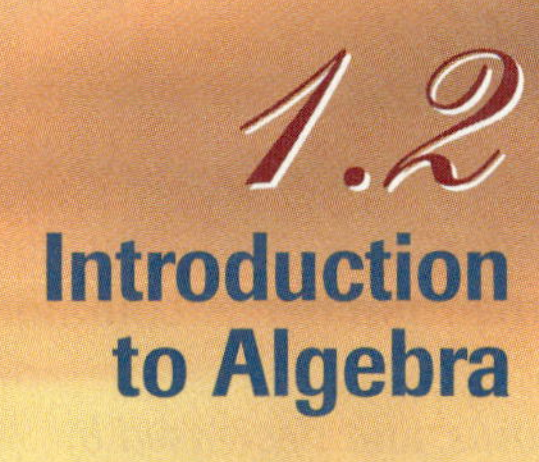

1.2 Introduction to Algebra

Objectives

1. **Build variable expressions.**
2. **Evaluate variable expressions.**
3. **Understand and use the commutative and associative properties of real numbers.**
4. **Understand and use the distributive property.**
5. **Identify terms and their coefficients.**
6. **Simplify variable expressions.**

Variables

The number of customers at a particular coffee shop changes from day to day, as does the closing price for a share of Google stock and the daily high temperature in Las Vegas, Nevada. Quantities that change, or vary, are often represented by variables. A **variable** is a letter or symbol that is used to represent a quantity which either changes or has an unknown value.

Variable Expressions

Objective 1 **Build variable expressions.** Suppose a movie theater charges \$8 admission per person. To determine the cost for a family to go to the movies, we multiply the number of people by 8 dollars. The number of people can change from family to family, so we can represent this quantity by a variable such as x. The cost for a family with x people can be written as $8 \cdot x$ or simply as $8x$. This expression for the cost, $8x$, is known as a variable expression. A **variable**, or **algebraic**, **expression** is a combination of one or more variables with numbers and/or arithmetic operations. When the operation is multiplication such as $8 \cdot x$, we often omit the multiplication dot. Here are some other examples of variable expressions.

$$3a + 5 \qquad\qquad x^2 + 3x - 10 \qquad\qquad \frac{y+5}{y-3}$$

EXAMPLE 1 Write an algebraic expression for "33 more than a number."

Solution

We must choose a variable to represent the unknown number. Let x represent the number. The expression is $x + 33$.

Phrases to look for that suggest addition are *more than, sum, plus, increased by,* and *total.*

***Quick Check* 1**

Write an algebraic expression for "the sum of a number and 62."

EXAMPLE 2 Write an algebraic expression for "13 less than a number."

Solution

Let x represent the number. The expression is $x - 13$.

Be careful with the order of subtraction when the phrase *less than* is used. Thirteen less than a number tells us that we need to subtract 13 from that number. A common error is to write the subtraction in the opposite order.

Phrases that suggest subtraction are *less than, difference, minus,* and *decreased by.*

Quick Check 2
Write an algebraic expression for "a number decreased by 6."

A Word of Caution Thirteen less than a number is written as $x - 13$, not $13 - x$.

EXAMPLE 3 Write an algebraic expression for "the product of 5 and two different numbers."

Quick Check 3
Write an algebraic expression for "3 times a number."

Solution

Since we are building an expression involving two different unknown numbers, we need to introduce two variables. Let x and y represent the two numbers. The expression is $5xy$.

Phrases that suggest multiplication are *product of, times, multiplied by, of,* and *twice.*

EXAMPLE 4 Five friends go out to dinner. If all of them decide to split evenly the cost of dinner, write a variable expression for the amount each friend would pay.

Solution

Let c represent the cost of the dinner. The expression is $c \div 5$ or $\frac{c}{5}$.

Phrases that suggest division are *quotient, divided by,* and *ratio.*

Quick Check 4
Write an algebraic expression for "the quotient of a number and 16."

Here are some different phrases that translate to $x + 5$, $x - 10$, $8x$, and $\frac{x}{6}$.

$x + 5$	• The sum of a number and 5 • A number plus 5 • A number increased by 5 • 5 more than a number • The total of a number and 5
$x - 10$	• 10 less than a number • The difference of a number and 10 • A number minus 10 • A number decreased by 10
$8x$	• The product of 8 and a number • 8 times a number • 8 multiplied by a number
$\frac{x}{6}$	• The quotient of a number and 6 • A number divided by 6 • The ratio of a number and 6

Evaluating Variable Expressions

Objective 2 **Evaluate variable expressions.** We will often have to evaluate variable expressions for particular values of variables. To do this we substitute the appropriate numerical value for each variable and then simplify the resulting expression using the order of operations.

EXAMPLE 5 Evaluate $5x - 3$ for $x = 8$.

Solution

The best first step when evaluating a variable expression is to rewrite the expression, replacing each variable with a set of parentheses. For example, rewrite $5x - 3$ as $5(\ ) - 3$. Then we can substitute the appropriate value for each variable and simplify.

$5x - 3$	
$5(8) - 3$	Substitute 8 for x.
$= 40 - 3$	Multiply.
$= 37$	Subtract.

The expression $5x - 3$ is equal to 37 for $x = 8$.

***Quick Check* 5**
Evaluate $2x + 17$ for $x = 38$.

EXAMPLE 6 Evaluate $x^2 - 10x - 18$ for $x = 4$.

Solution

$x^2 - 10x - 18$	
$(4)^2 - 10(4) - 18$	Substitute 4 for x.
$= 16 - 10(4) - 18$	Square 4.
$= 16 - 40 - 18$	Multiply.
$= -42$	Simplify.

***Quick Check* 6**
Evaluate $x^2 - 5x + 32$ for $x = -6$.

EXAMPLE 7 Evaluate $b^2 - 4ac$ for $a = 2$, $b = -5$, and $c = -9$.

Solution

$b^2 - 4ac$	
$(-5)^2 - 4(2)(-9)$	Substitute 2 for a, -5 for b, and -9 for c.
$= 25 - 4(2)(-9)$	Square negative 5.
$= 25 - (-72)$	Multiply.
$= 25 + 72$	Rewrite without double signs.
$= 97$	Add.

***Quick Check* 7**
Evaluate $b^2 - 4ac$ for $a = -3$, $b = -4$, and $c = 10$.

Properties of Real Numbers

Objective 3 **Understand and use the commutative and associative properties of real numbers.** We will now examine three properties of real numbers. The first is the **commutative property,** which is used for addition and multiplication.

Commutative Property

For all real numbers a and b,

$$a + b = b + a \text{ and } a \cdot b = b \cdot a.$$

This property states that changing the order of the numbers in a sum or in a product does not change the result. For example, $3 + 9 = 9 + 3$ and $7 \cdot 8 = 8 \cdot 7$. Note that this property does not work for subtraction or for division. Changing the order of the numbers in a difference or a quotient will generally change the result.

The second property that we will cover is the **associative property**, which also holds for addition and multiplication.

Associative Property

For all real numbers a, b, and c,

$$(a + b) + c = a + (b + c) \text{ and } (ab)c = a(bc).$$

This property states that changing the grouping of the numbers in a sum or in a product does not change the result. Check for yourself; does $(2 + 7) + 3$ equal $2 + (7 + 3)$? The first expression simplifies to $9 + 3$ while the second becomes $2 + 10$, both of which equal 12.

EXAMPLE 8 Simplify $3(14x)$ by applying the associative property.

Solution

By the associative property we know that $3(14x) = (3 \cdot 14)x$. We can now multiply $3 \cdot 14$ to produce a result of $42x$.

***Quick Check* 8** **Simplify $9(6a)$ by applying the associative property.**

EXAMPLE 9 Simplify $(2x + 9) + 16$ by applying the associative property.

***Quick Check* 9** **Simplify $(5x + 17) + 45$ by applying the associative property.**

Solution

By the associative property we know that $(2x + 9) + 16 = 2x + (9 + 16)$. This allows us to add 9 and 16. The simplified expression is $2x + 25$.

Objective 4 **Understand and use the distributive property.** The third property of real numbers is the **distributive property**.

Distributive Property

For all real numbers a, b, and c,

$$a(b + c) = ab + ac.$$

This property tells us that we can distribute the factor outside the parentheses to each number being added in the parentheses. We multiply the factor by each number inside the parentheses and then simplify. (This property holds true whether the operation inside the parentheses is addition or subtraction.) Consider the expression $3(5 + 4)$. The order of operations tells us that this is equal to $3 \cdot 9$ or 27. Here is how we simplify the expression using the distributive property.

$3(5 + 4) = 3 \cdot 5 + 3 \cdot 4$	Apply the distributive property.
$= 15 + 12$	Multiply $3 \cdot 5$ and $3 \cdot 4$.
$= 27$	Add.

Either way we get the same result.

EXAMPLE 10 Simplify $5(7x + 6)$ using the distributive property.

Solution

$5(7x + 6) = 5 \cdot 7x + 5 \cdot 6$	Distribute the 5.
$= 35x + 30$	Multiply. Recall from the example illustrating the associative property that $5 \cdot 7x$ equals $35x$.

***Quick Check* 10** **Simplify $10(3x - 8)$ using the distributive property.**

EXAMPLE 11 Simplify $6(7 + 2x - 3y)$ using the distributive property.

Solution

When there are more than two terms inside the parentheses, we distribute the factor to each term. Also, it is a good idea to distribute the factor and multiply mentally.

$$6(7 + 2x - 3y) = 42 + 12x - 18y$$

***Quick Check* 11**
Simplify $3(x - 5y - 8z)$ using the distributive property.

EXAMPLE 12 Simplify $-8(4x - 9)$ using the distributive property.

Solution

When the factor outside the parentheses is negative, the negative number must be distributed to each term inside the parentheses. This will change the sign of each term inside the parentheses.

$$
\begin{aligned}
-8(4x - 9) &= (-8) \cdot 4x - (-8) \cdot 9 && \text{Distribute the } -8. \\
&= -32x - (-72) && \text{Multiply.} \\
&= -32x + 72 && \text{Rewrite without double signs.}
\end{aligned}
$$

Quick Check 12
Simplify $-9(5x + 16)$ using the distributive property.

Simplifying Variable Expressions

Objective 5 **Identify terms and their coefficients.** In an algebraic expression a **term** is either a number, a variable, or a product of a number and variables. Terms in an algebraic expression are separated by addition or subtraction. In the expression $7x - 5y + 3$ there are three terms: $7x$, $-5y$, and 3. The numerical factor of a term is its **coefficient**. The coefficients of these terms are 7, -5, and 3. A term that does not contain a variable factor, such as the term 3 in the expression $7x - 5y + 3$, is called a **constant term**.

EXAMPLE 13 For the expression $9x^3 + x^2 - 7x - 30$, determine the number of terms, list them, and state the coefficient for each term.

Solution

This expression has four terms: $9x^3$, x^2, $-7x$, and -30.

What is the coefficient for the second term? Although it does not appear to have a coefficient, its coefficient is 1. This is because x^2 is the same as $1 \cdot x^2$. So the four coefficients are 9, 1, -7, and -30.

Quick Check 13 **For the expression $-x^2 + 8x - 37$, determine the number of terms, list them, and state the coefficient for each term.**

Objective 6 **Simplify variable expressions.** Two terms that have the same variable factors, or two terms that are constants, are called **like terms**. Consider the expression $9x + 8y + 6x - 3y + 8z - 5$. There are two sets of like terms in this expression: $9x$ and $6x$, as well as $8y$ and $-3y$. There are no like terms for $8z$, since no other term has z as its sole variable factor. Similarly, there are no like terms for the constant term -5.

Suppose we had to simplify the expression $8y - 3y$. This expression could be rewritten as $(8 - 3)y$ using the distributive property, or simply as $5y$.

Combining Like Terms

When we simplify variable expressions, we can combine like terms into a single term by adding or subtracting the coefficients of the like terms.

EXAMPLE 14 Simplify $3x + 14x$ by combining like terms.

Solution

These two terms are like terms, since they both have the same variable factors. We can simply add the two coefficients together to produce the expression $17x$.

$$3x + 14x = 17x$$

***Quick Check* 14**
Simplify $9a + 6b + a - 13b$ by combining like terms.

A general strategy for simplifying algebraic expressions is to begin by performing any distributive multiplications. After all multiplications have been performed, we can combine any like terms.

EXAMPLE 15 Simplify $6(3x - 11) - 2x + 31$.

Solution

The first step is to use the distributive property by distributing the 6 to each term inside the parentheses. Then we will be able to combine like terms.

$$\begin{aligned} 6(3x - 11) - 2x + 31 &= 18x - 66 - 2x + 31 && \text{Distribute the 6.} \\ &= 16x - 35 && \text{Combine like terms.} \end{aligned}$$

***Quick Check* 15**
Simplify $4(2x + 7) + 3x - 19$.

EXAMPLE 16 Simplify $3(5 - 8x) - 10(2x + 9)$.

Solution

We must be sure to distribute the *negative* 10 into the second set of parentheses.

$$\begin{aligned} 3(5 - 8x) - 10(2x + 9) &= 15 - 24x - 20x - 90 && \text{Distribute the 3 to each term in the first set of parentheses and distribute the } -10 \text{ to each term in the second set of parentheses.} \\ &= -44x - 75 && \text{Combine like terms.} \end{aligned}$$

Usually we will write our simplified result with variable terms preceding constant terms, but it would also be correct to write $-75 - 44x$, because of the commutative property.

***Quick Check* 16**
Simplify $5(2x - 11) - 4(x - 8)$.

A Word of Caution If a factor in front of a set of parentheses is negative or has a subtraction sign in front of it, we must distribute the negative sign along with the factor.

Building Your Study Strategy **Study Groups, 2 Where and When to Meet** Once you have formed a study group, you must determine where and when to meet. It is a good idea to meet at least twice a week, for at least an hour per session. Consider a location where quiet discussion is allowed, such as the library or tutorial center. Clearly, the group must meet at times when each person is available. Some groups like to meet during the hour before class, using the study group as a way to prepare for class. Other groups prefer to meet during the hour after class, allowing them to go over material while it is fresh in their minds. Another suggestion is to meet at a time when your instructor is holding office hours. This way, if the entire group is struggling with the same question, one or more members can visit the instructor for assistance. Some study groups prefer to meet off campus in the evening. One good place to meet would be at a coffee shop with tables large enough for everyone to work on, provided that the noise is not too distracting. The food court at a local mall would work as well. Other groups like to meet at each other's homes. This typically provides a comfortable, relaxing atmosphere in which to work. If your group is invited to meet at someone's house, try to be a perfect guest. Help set up and clean up if you can, or bring beverages and healthy snacks.

EXERCISES 1.2

Vocabulary

1. A(n) __________ is a letter or symbol used to represent a quantity that either changes or has an unknown value.

2. A combination of one or more variables with numbers and/or arithmetic operations is a(n) __________.

3. To __________ a variable expression, substitute the appropriate numerical value for each variable and then simplify the resulting expression using the order of operations.

4. For any real numbers a and b, the __________ property states that $a + b = b + a$ and $a \cdot b = b \cdot a$.

5. For any real numbers a, b, and c, the __________ property states that $(a + b) + c = a + (b + c)$ and $(a \cdot b) \cdot c = a \cdot (b \cdot c)$.

6. For any real numbers a, b, and c, the __________ property states that $a(b + c) = ab + ac$.

7. A(n) __________ is either a number, a variable, or a product of a number and variables.

8. Give the definition for a coefficient.

9. Two terms that have the same variable factors, or that are both constant terms, are called __________.

10. To combine two like terms, we add their __________.

Build a variable expression for the following phrases.

11. A number decreased by 5

12. The sum of a number and 27

13. A number divided by 7

14. Twice a number

15. Eight times a number

16. Eighteen minus a number

17. Three times a number, increased by 25

18. Nine less than six times a number

19. The quotient of fourteen and a number

20. The sum of two different numbers

21. Ten times the sum of two numbers

22. Half of the difference of a number and 16

23. B. J. McCay, a truck driver from the TV show *BJ & The Bear,* charged \$1.50 per mile to haul a load. If we let m represent the number of miles that McKay drove, build a variable expression for the amount of money he would charge to haul the load.

24. A college charges \$75/unit for tuition. If we let u represent the number of units that a student is taking, build a variable expression for the student's tuition.

25. Jim Rockford, a private investigator from the 1970s TV show *The Rockford Files,* charged \$200 per day plus expenses to take a case. If we let d represent the number of days that Rockford worked on a case, and if he had \$425 in expenses, build a variable expression for the amount of money he would charge for the case.

26. A computer consulting company charges \$1500 per day plus expenses to do field analysis of a crime scene. If we let d represent the number of days that the company consulted, and if the company had \$3750 in expenses, build a variable expression for the amount of money it would charge.

Evaluate the following algebraic expressions under the given conditions.

27. $5x - 3$ for $x = 7$

28. $7 - 4x$ for $x = 5$

29. $5a - 4b$ for $a = 8$ and $b = -13$

30. $-6x + 2y$ for $x = -9$ and $y = 4$

31. $\frac{3}{4}x - 7$ for $x = -12$

32. $\frac{2}{5}x + 13$ for $x = 15$

33. $6x - 22$ for $x = \frac{3}{2}$

34. $8x + 17$ for $x = -\frac{7}{4}$

35. $\frac{5}{6}x + \frac{11}{3}$ for $x = 14$

36. $\frac{7}{15}x - \frac{8}{5}$ for $x = 9$

37. $-2(x - 10)$ for $x = 4$

38. $(x - 7)(2x + 9)$ for $x = -6$

39. $x^2 + 7x - 30$ for $x = 6$

40. $x^2 - 9x - 18$ for $x = 3$

41. $a^2 - 8a - 11$ for $a = -2$

42. $y^2 + 10y - 13$ for $y = -3$

43. $m^2 - 36$ for $m = -7$

44. $7 - x^2$ for $x = -8$

45. $a^3 - 7a^2 - 25$ for $a = -3$

46. $x^9 - 4x^7 + 121x^2 - 329$ for $x = 0$

47. $b^2 - 4ac$ for $a = 1$, $b = 6$, and $c = -55$

48. $b^2 - 4ac$ for $a = -3$, $b = 8$, and $c = -2$

49. $b^2 - 4ac$ for $a = -8$, $b = -5$, and $c = 10$

50. $b^2 - 4ac$ for $a = 13$, $b = -10$, and $c = 7$

51. $-2(x + h) - 13$ for $x = 6$ and $h = 0.01$

52. $4(x + h) + 35$ for $x = -8$ and $h = 0.001$

53. $(x + h)^2 + 2(x + h) - 20$ for $x = 4$ and $h = 0.1$

54. $(x + h)^2 - 5(x + h) - 16$ for $x = -8$ and $h = 0.1$

55. The commutative property works for addition, but does not work in general for subtraction.

a) Give an example of two numbers a and b such that $a - b \neq b - a$.

b) Can you find two numbers a and b such that $a - b = b - a$?

56. The commutative property works for multiplication, but does not work in general for division.

a) Give an example of two numbers a and b such that $\frac{a}{b} \neq \frac{b}{a}$.

b) Can you find two numbers a and b such that $\frac{a}{b} = \frac{b}{a}$?

Simplify, where possible.

57. $8(x - 2)$

58. $10(3x + 11)$

59. $6(4 - 9x)$

60. $9(7x - 19)$

61. $-3(2x - 5)$

62. $-4(9x + 8)$

63. $12\left(\frac{3}{4}x - \frac{2}{3}\right)$

64. $16\left(\frac{5}{8}x + \frac{7}{4}\right)$

65. $2x + 11x$

66. $18x + 7x$

67. $3x - 12x$

68. $24a - 7a$

69. $2x - 19 + 4x + 6$

70. $7x + 25 - 19x - 86$

71. $55 - 32a + 65a - 88$

72. $86 + 21a - 45a + 79$

73. $\frac{1}{4}x - \frac{2}{3} + \frac{7}{2}x + \frac{1}{6}$

74. $\frac{7}{10}x + \frac{5}{8} + \frac{2}{5}x - \frac{19}{12}$

75. $8x - 9y$

76. $22 + 17m$

77. $8x + 15y - 3x - 26y$

78. $6m - 19n - 23m + 5n$

79. $2(4x - 9) + 3x + 20$

80. $7 - 25k + 3(5k - 8)$

81. $-5(4x - 13) - 24$

82. $9(7m - 6n) - 4n$

83. $11x - 8(2x - 15)$

84. $3 - 7(-12x + 7)$

85. $8(4z - 13) + 6(3z + 10)$

86. $-(2x - 17) - 3(6 - 5x)$

For the following expressions,
a) determine the number of terms;
b) write down each term; and
c) write down the coefficient for each.
(Be sure to simplify each expression before answering.)

87. $7x^2 - 10x - 35$

88. $-x^2 + 4x + 21$

89. $x - 34$

90. $12x^3 - 8x^2 - 17x + 43$

91. $4(2x^2 - 5x + 13) - 9x$

92. $5(x^2 - 4x - 9) - 6(7x - 2)$

93. $9(-4a + 8b + 7c) - 2(5b - 11c - 32)$

94. $6(-3x + 4y) - (2x + 13y)$

95. List four different algebraic expressions that simplify to $8x - 20$.

96. List four different algebraic expressions that evaluate to 27 for $x = 3$.

Writing in Mathematics

Answer in complete sentences.

97. Give an example of a real-world situation that can be described by the variable expression $45x$. Explain why the expression fits your situation.

98. Give an example of a real-world situation that can be described by the variable expression $10x + 30$. Explain why the expression fits your situation.

99. ***Solutions Manual**** Write a solutions manual page for the following problem:

Simplify $4(2x - 5) - 3(3x - 8) - 7x$.

100. ***Newsletter**** Write a newsletter explaining how to evaluate expressions.

***See Appendix B for details and sample answers.**

1.3 Linear Equations and Absolute Value Equations

Objectives

1. **Solve linear equations using the multiplication and addition properties of equality.**
2. **Solve linear equations using the five-step general strategy.**
3. **Identify linear equations that are identities or contradictions.**
4. **Solve a literal equation for a specified variable.**
5. **Solve absolute value equations.**

Linear Equations

Objective 1 Solve linear equations using the multiplication and addition properties of equality. An **equation** is a mathematical statement of equality between two expressions. It is a statement that asserts that the value of the expression on the left side of the equation is equal to the value of the expression on the right side. Here are a few examples of equations:

$$2x = 8 \qquad x + 17 = 20 \qquad 3x - 8 = 2x + 6 \qquad 5(2x - 9) + 3 = 7(3x + 16)$$

All of these are examples of linear equations. A **linear equation in one variable** has a single variable, and the exponent for that variable is 1. For example, if the variable in a linear equation is x, then the equation cannot have terms containing x^2 or x^3. The variable in a linear equation cannot appear in a denominator either. Here are some examples of equations that are not linear equations:

$x^2 - 5x - 6 = 0$	Variable is squared.
$m^3 - m^2 + 7m = 7$	Variable has exponents greater than 1.
$\dfrac{5x + 3}{x - 2} = -9$	Variable is in denominator.

A **solution** of an equation is a value that, when substituted for the variable in the equation, produces a true statement, such as $5 = 5$. For example, $x = 6$ is a solution of $7x - 5 = 37$, because when 6 is substituted for the variable x in the equation it results in a true statement.

$7x - 5 = 37$	
$7(6) - 5 = 37$	Substitute 6 for x.
$42 - 5 = 37$	Multiply.
$37 = 37$	Subtract.

However, $x = -2$ is not a solution of $5x + 8 = 2$, since a false statement results when -2 is substituted for x in the equation.

$5x + 8 = 2$	
$5(-2) + 8 = 2$	Substitute -2 for x.
$-10 + 8 = 2$	Multiply.
$-2 \neq 2$	Subtract.

The set of all solutions of an equation is called its **solution set**. The process of finding an equation's solution set is called **solving the equation**. When we find all the solutions to an equation, we write these values using set notation inside braces { }.

When we solve a linear equation, our goal is to convert it to an equivalent equation that has the variable isolated on one side with a constant on the other side, for example, $x = 3$ or $-5 = y$. The value that is on the opposite side of the equation from the variable after it has been isolated is the solution of the equation.

Multiplication Property of Equality

Our first tool for solving linear equations is the **multiplication property of equality**. It tells us that for any equation, if we multiply both sides of the equation by the same nonzero number, then both sides remain equal to each other.

Multiplication Property of Equality

For any algebraic expressions A and B, and any nonzero number n,

$$\text{if } A = B \text{ then } n \cdot A = n \cdot B.$$

Think of a scale that holds two weights in balance. If we double the amount of weight on each side of the scale, will the scale still be balanced? Of course it will.

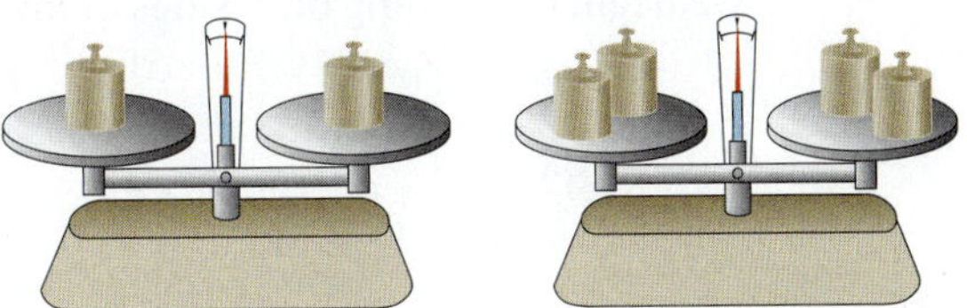

We can think of an equation as a scale, and the expressions on each side as the weights that are balanced. Multiplying both sides by a nonzero number leaves both sides still balanced and equal to each other.

EXAMPLE 1 Solve $\frac{x}{2} = 9$ using the multiplication property of equality.

Solution

The goal when solving this linear equation is to isolate the variable x on one side of the equation. We begin by multiplying both sides of this equation by 2. The expression $\frac{x}{2}$ is equivalent to $\frac{1}{2}x$, so when we multiply by 2 we are multiplying by the reciprocal of $\frac{1}{2}$. The product of reciprocals is equal to 1, so the resulting expression on the left side of the equation is $1x$ or x.

$$\frac{x}{2} = 9$$

$$2 \cdot \frac{x}{2} = 2 \cdot 9 \qquad \text{Multiply both sides by 2.}$$

$$\overset{1}{\cancel{2}} \cdot \frac{x}{\underset{1}{\cancel{2}}} = 2 \cdot 9 \qquad \text{Divide out common factors.}$$

$$x = 18 \qquad \text{Multiply.}$$

The solution that we have found is $x = 18$. Before moving on, we must check this value to be sure that we have made no mistakes and that it is actually a solution.

Check:

$$\frac{x}{2} = 9$$

$$\frac{18}{2} = 9 \qquad \text{Substitute 18 for } x.$$

$$9 = 9 \qquad \text{Divide.}$$

This is a true statement, so $x = 18$ is a solution and the solution set is $\{18\}$.

***Quick Check* 1**

Solve $\frac{x}{7} = -3$ using the multiplication property of equality.

The multiplication property of equality also allows us to divide both sides of an equation by the same nonzero number without affecting the equality of the two sides. This is because dividing both sides of an equation by a nonzero number, n, is equivalent to multiplying both sides of the equation by the reciprocal of the number, $\frac{1}{n}$.

EXAMPLE 2 Solve $3k = 51$ using the multiplication property of equality.

Solution

We begin by dividing both sides of the equation by 3, which will isolate the variable k.

$$3k = 51$$

$$\frac{3k}{3} = \frac{51}{3} \qquad \text{Divide both sides by 3.}$$

$$\frac{\overset{1}{\cancel{3}}k}{\underset{1}{\cancel{3}}} = \frac{51}{3} \qquad \text{Divide out common factors on the left side.}$$

$$k = 17 \qquad \text{Simplify.}$$

Again, we check our solution before writing it in solution set notation.

Check:

$$3k = 51$$

$$3(17) = 51 \qquad \text{Substitute 17 for } k.$$

$$51 = 51 \qquad \text{Multiply.}$$

$k = 17$ is indeed a solution and the solution set is $\{17\}$.

***Quick Check* 2**

Solve $9a = 54$ using the multiplication property of equality.

Addition Property of Equality

The **addition property of equality** tells us that we can add the same number to both sides of an equation, or subtract the same number from both sides of an equation, without affecting the equality of the two sides.

Addition Property of Equality

For any algebraic expressions A and B, and any number n,

$$\text{if } A = B \text{ then } A + n = B + n$$
$$\text{and } A - n = B - n.$$

This property helps us solve equations in which we have a number either added to or subtracted from a variable on one side of an equation.

EXAMPLE 3 Solve $x + 3 = 15$ using the addition property of equality.

Solution

In this example the number 3 is being added to the variable x. To isolate the variable we use the addition property of equality to subtract 3 from both sides of the equation.

$$x + 3 = 15$$
$$x + 3 - 3 = 15 - 3 \quad \text{Subtract 3 from both sides.}$$
$$x = 12 \quad \text{Simplify.}$$

Checking this solution, we will substitute 12 for x in the original equation.

Check:

$$x + 3 = 15$$
$$12 + 3 = 15 \quad \text{Substitute 12 for } x.$$
$$15 = 15 \quad \text{Add.}$$

We have a true statement, so the solution set is $\{12\}$.

***Quick Check* 3**
Solve $a + 11 = -7$ using the addition property of equality.

EXAMPLE 4 Solve $22 = y - 6$ using the addition property of equality.

Solution

When a value is subtracted from a variable, we isolate the variable by adding that value to both sides of the equation. In this example we will add 6 to both sides.

$$22 = y - 6$$
$$22 + 6 = y - 6 + 6 \quad \text{Add 6 to both sides.}$$
$$28 = y \quad \text{Add.}$$

Check:

$$22 = y - 6$$
$$22 = 28 - 6 \quad \text{Substitute 28 for } y.$$
$$22 = 22 \quad \text{Subtract.}$$

This is a true statement, so the solution set is $\{28\}$.

***Quick Check* 4**
Solve $x - 31 = -45$ using the addition property of equality.

Objective 2 **Solve linear equations using the five-step general strategy.** In the previous examples, we solved equations that required the use of only one operation to isolate the variable. Now we will learn how to solve equations requiring the use of both the multiplication and addition properties of equality.

Suppose we needed to solve the equation $2x - 6 = 10$. Should we divide both sides by 2 first? Should we add 6 to both sides first? We first determine which of the two numbers is most closely connected to the variable. The order of operations tells us that in the expression $2x - 6$, we first multiply 2 by x and then subtract 6 from the result. To isolate the variable x, we undo these operations in the opposite order. We will first add

6 to both sides to undo the subtraction, and then we will divide both sides by 2 to undo the multiplication.

EXAMPLE 5 Solve $2x - 6 = 10$.

Solution

We begin by isolating the variable term on the left side of the equation.

$$2x - 6 = 10$$
$$2x = 16 \quad \text{Add 6 to both sides to isolate } 2x.$$
$$x = 8 \quad \text{Divide both sides by 2.}$$

We will now check this solution.

Check:

$$2x - 6 = 10$$
$$2(8) - 6 = 10 \quad \text{Substitute 8 for } x.$$
$$16 - 6 = 10 \quad \text{Multiply.}$$
$$10 = 10 \quad \text{Subtract.}$$

Since both sides simplify to be 10, our solution is valid. The solution set is $\{8\}$.

***Quick Check* 5**
Solve $9x + 13 = -14$.

We will now examine a process that can be used to solve any linear equation. This five-step process not only works for types of equations we have already learned to solve, but also for more complicated equations such as $7x + 4 = 3x - 20$ or $5(2x - 9) + 3x = 7(3 - 8x)$.

Solving Linear Equations

1. **Simplify each side of the equation completely.**
   - Use the distributive property to clear any parentheses.
   - If there are fractions in the equation, multiply both sides of the equation by the least common multiple (LCM) of the denominators to clear the fractions from the equation. The LCM of two or more numbers is the smallest number that is a multiple of each of the numbers.
   - Combine any like terms that are on the same side of the equation. After completing this step, the equation should contain at most one variable term and at most one constant term on each side.
2. **Collect all variable terms on one side of the equation.** If there is a variable term on each side of the equation, use the addition property of equality to place both variable terms on the same side of the equation. It is a good idea to move the term with the smaller coefficient. For instance, if one side of the equation has a variable term $4x$ and the other side has a variable term $7x$, then we will subtract $4x$ from both sides of the equation as 4 is less than 7. This will prevent having to divide both sides of the equation by a negative number, which could lead to a careless sign error.
3. **Collect all constant terms on the other side of the equation.** If there is a constant term on each side of the equation, use the addition property of equality to isolate the variable term.

continued

4. **Divide both sides of the equation by the coefficient of the variable term.** At this point our equation should be in the form $ax = b$. We use the multiplication property of equality to find our solution.
5. **Check your solution.** Check that the value creates a true equation when substituted for the variable in the equation.

In the next example, we will solve equations that have variable terms and constant terms on both sides of the equation.

Example 6 Solve the equation $5x + 13 = 7x - 4$.

Solution

We have no fractions to clear in this equation, there are no multiplications to be performed, and there are no like terms to be combined. We begin by gathering the variable terms on one side of the equation.

$$5x + 13 = 7x - 4$$

$$5x + 13 - 5x = 7x - 4 - 5x$$ Subtract $5x$ from both sides to gather the variable terms on the right side of the equation.

$$13 = 2x - 4$$ Subtract.

$$13 + 4 = 2x - 4 + 4$$ Add 4 to both sides to isolate $2x$.

$$17 = 2x$$ Add.

$$\frac{17}{2} = \frac{2x}{2}$$ Divide both sides by 2 to isolate x.

$$\frac{17}{2} = x$$ Simplify.

The check of this solution is left to the reader. The solution set is $\{\frac{17}{2}\}$.

Quick Check 6
Solve the equation $8x + 21 = 3x - 9$.

Example 7 Solve $3(2x + 7) - 4x = 5x - 6$

Solution

We begin by using the distributive property of multiplication on the left side of the equation.

$$3(2x + 7) - 4x = 5x - 6$$

$$6x + 21 - 4x = 5x - 6$$ Distribute 3.

$$2x + 21 = 5x - 6$$ Combine like terms.

$$27 = 3x$$ Collect variable terms on the right side by subtracting $2x$. Add 6 to collect constants on the left side.

$$9 = x$$ Divide both sides by 3.

The check is left to the reader. The solution set is $\{9\}$.

Quick Check 7
Solve $3x - 2(4x - 5) = 2x - 11$.

EXAMPLE 8 Solve $\frac{4}{3}x + \frac{5}{6} = \frac{9}{4}x - 1$

Solution

This equation includes fractions. We can clear these fractions by multiplying each side of the equation by the LCM of the denominators, which is 12.

$$\frac{4}{3}x + \frac{5}{6} = \frac{9}{4}x - 1$$

$$12\left(\frac{4}{3}x + \frac{5}{6}\right) = 12\left(\frac{9}{4}x - 1\right) \quad \text{Multiply both sides by 12.}$$

$$\overset{4}{\cancel{12}} \cdot \frac{4}{\underset{1}{\cancel{3}}}x + \overset{2}{\cancel{12}} \cdot \frac{5}{\underset{1}{\cancel{6}}} = \overset{3}{\cancel{12}} \cdot \frac{9}{\underset{1}{\cancel{4}}}x - 12 \cdot 1 \quad \text{Distribute and divide out common factors.}$$

$$16x + 10 = 27x - 12 \quad \text{Multiply.}$$

$$10 = 11x - 12 \quad \text{Subtract } 16x.$$

$$22 = 11x \quad \text{Add 12.}$$

$$2 = x \quad \text{Divide both sides by 11.}$$

The check is left to the reader. The solution set is $\{2\}$.

Quick Check 8
Solve $\frac{2}{3}x - \frac{1}{5} = \frac{1}{2}x + \frac{5}{6}$.

Identities and Contradictions

Objective 3 **Identify linear equations that are identities or contradictions.**

In all the previous examples, the linear equations have had exactly one solution, but this will not always be the case. An equation that is always true regardless of the value substituted for the variable x is called an **identity**. The solution set for an identity is the set of all real numbers, denoted by $\mathbb{R}$

EXAMPLE 9 Solve $2(2x - 3) + 1 = 4x - 5$.

Solution

$$2(2x - 3) + 1 = 4x - 5$$

$$4x - 6 + 1 = 4x - 5 \quad \text{Distribute 2.}$$

$$4x - 5 = 4x - 5 \quad \text{Combine like terms.}$$

$$-5 = -5 \quad \text{Subtract } 4x \text{ from both sides.}$$

Quick Check 9
Solve
$(2x + 1) - (5 - 3x) = 2(3x - 2) - x$.

When we subtract $4x$ from each side in an attempt to collect all variable terms on one side of the equation, we eliminate the variable x from the resulting equation. Is the equation $-5 = -5$ a true statement? Yes, and it tells us that our equation is an identity and that our solution set is the set of all real numbers $\mathbb{R}$.

Notice that at one point in the previous example, the equation was $4x - 5 = 4x - 5$. If you reach a point where both sides of the equation are identical, then the equation is an identity.

Just as there are equations that are always true regardless of the value we choose for the variable, there are equations that are never true for any value of the variable.

These equations are called **contradictions** and have no solution. The solution set for these equations is the empty set or null set, and is denoted ∅. We can tell that an equation is a contradiction when we are solving it because the variable terms on each side of the equation will be eliminated, but in this case the resulting equation will be false, such as $1 = 2$.

EXAMPLE 10 Solve $4x + 3 = 2(2x + 3) - 1$.

Solution

$4x + 3 = 2(2x + 3) - 1$	
$4x + 3 = 4x + 6 - 1$	Distribute 2.
$4x + 3 = 4x + 5$	Combine like terms.
$3 = 5$	Subtract $4x$ from both sides.

After we subtract $4x$ from each side of the equation, the resulting equation $3 = 5$ is obviously false. This equation is a contradiction and its solution set is the empty set ∅.

Quick Check 10
Solve $5x - 7 = 6(x + 2) - x$.

Literal Equations

Objective 4 Solve a literal equation for a specified variable. A **literal equation** is an equation that contains two or more variables.

Perimeter of a Rectangle

The **perimeter** of a rectangle is a measure of the distance around the rectangle. The formula for the perimeter (P) of a rectangle with length L and width W is $P = 2L + 2W$.

$P = 2L + 2W$

W

L

The equation $P = 2L + 2W$ is a literal equation with three variables. Often we will be asked to solve literal equations for one variable in terms of the other variables in the equation. In the perimeter example, if we solved for the length L in terms of the width W and the perimeter P, then we would have a formula for the length of a rectangle if we knew its width and perimeter.

We solve literal equations by isolating the variable we wish to solve for. We will use the same general strategy that we used for solving linear equations. We treat the other variables in the equation as if they were constants.

EXAMPLE 11 Solve the literal equation $P = 2L + 2W$ (perimeter of a rectangle) for L.

Solution

We want to gather all terms containing the variable L on one side of the equation and gather all other terms on the other side. This can be done by subtracting $2W$ from both sides.

$$P = 2L + 2W$$
$$P - 2W = 2L + 2W - 2W \quad \text{Subtract } 2W \text{ to isolate } 2L.$$
$$P - 2W = 2L \quad \text{Subtract.}$$
$$\frac{P - 2W}{2} = \frac{2L}{2} \quad \text{Divide by 2 to isolate } L.$$
$$\frac{P - 2W}{2} = L$$

Quick Check 11
Solve the literal equation $3x + 5y = 4$ for y.

We usually rewrite the equation so that the variable we solved for appears on the left side: $L = \frac{P - 2W}{2}$.

Absolute Value Equations

Objective 5 **Solve absolute value equations.** We now introduce equations involving absolute values of linear expressions.

EXAMPLE 12 Solve $|x| = 5$.

Solution

Recall that the absolute value of a number x, denoted $|x|$, is a measure of the distance between 0 and that number x on a real number line. In this equation, we are looking for a number x that is 5 units away from 0.

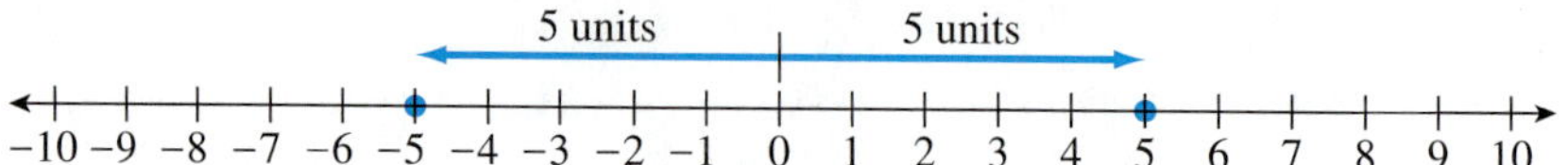

Quick Check 12
Solve $|x| = 7$.

There are two such numbers: 5 and -5. The solution set is $\{5, -5\}$.

The equation in the previous example is an **absolute value equation**. An absolute value equation relates the absolute value of an expression to a constant (e.g., $|2x - 3| = 7$) or to the absolute value of another expression (e.g., $|3x - 4| = |2x + 11|$).

Consider the absolute value equation $|2x - 3| = 7$. This equation involves a number, represented by $2x - 3$, whose absolute value is equal to 7. That number must either be 7 or -7. We begin to solve this equation by converting it into the following two equations.

$$2x - 3 = 7 \quad \text{or} \quad 2x - 3 = -7$$

In general, when we have an equation with the absolute value of an expression equal to a positive number, we set the expression (without the absolute value bars) equal to that number and its opposite. Solving those two equations will give us our solutions.

Solving Absolute Value Equations

For any expression X and any positive number a, the solutions to the equation

$$|X| = a$$

can be found by solving the two equations

$$X = a \text{ and } X = -a.$$

EXAMPLE 13 Solve $|2x - 3| = 7$.

Solution

We begin by converting the absolute value equation into two equations that do not involve absolute values.

$$|2x - 3| = 7$$

$$2x - 3 = 7 \quad \text{or} \quad 2x - 3 = -7 \qquad \text{Convert to two linear equations.}$$

$$2x = 10 \quad \text{or} \quad 2x = -4 \qquad \text{Add 3.}$$

$$x = 5 \quad \text{or} \quad x = -2 \qquad \text{Divide by 2.}$$

Check:

$x = 5$	$x = -2$
$\|2(5) - 3\| = 7$	$\|2(-2) - 3\| = 7$
$\|10 - 3\| = 7$	$\|-4 - 3\| = 7$
$\|7\| = 7$	$\|-7\| = 7$
$7 = 7$	$7 = 7$

Since both values check, our solution set is $\{-2, 5\}$.

***Quick Check* 13**
Solve $|3x + 8| = 5$.

A Word of Caution When solving an absolute value equation such as $|2x - 3| = 7$, be sure to rewrite the equation as two equations. Do not simply drop the absolute value bars and solve the resulting equation.

EXAMPLE 14 Solve $3|2x + 7| - 4 = 20$.

Solution

We must isolate the absolute value before converting the equation to two linear equations.

$$3|2x + 7| - 4 = 20$$

$$3|2x + 7| = 24 \qquad \text{Add 4 to both sides.}$$

$$|2x + 7| = 8 \qquad \text{Divide both sides by 3.}$$

$$2x + 7 = 8 \quad \text{or} \quad 2x + 7 = -8 \qquad \text{Convert to two linear equations.}$$

$$2x = 1 \quad \text{or} \quad 2x = -15 \qquad \text{Subtract 7 from both sides of each equation.}$$

$$x = \frac{1}{2} \quad \text{or} \quad x = -\frac{15}{2} \qquad \text{Divide both sides of each equation by 2.}$$

The solution set is $\{-\frac{15}{2}, \frac{1}{2}\}$. The check of these solutions is left to the reader.

***Quick Check* 14**
Solve $4|2x - 5| - 9 = 7$.

A Word of Caution When solving an absolute value equation, we must isolate the absolute value before rewriting the equation as two equations without absolute values.

Since the absolute value of a number is a measure of its distance from 0 on the number line, the absolute value of a number cannot be negative. Therefore an equation that has an absolute value equal to a negative number has no solution. For example, the equation $|x| = -3$ has no solution because there is no number whose absolute value is -3. We write $\varnothing$ for the solution set.

EXAMPLE 15 Solve $|3x - 4| + 8 = 6$.

Solution

We begin by isolating the absolute value.

$$|3x - 4| + 8 = 6$$
$$|3x - 4| = -2 \quad \text{Subtract 8 from both sides.}$$

Since an absolute value cannot equal -2, this equation has no solution. The solution set is $\varnothing$.

***Quick Check* 15**
Solve $|2x + 3| = -6$.

A Word of Caution An equation in which an absolute value is equal to a negative number has no solution. Do not try to rewrite the equation as two equivalent equations.

If two unknown numbers have the same absolute values, then either the two numbers are equal or they are opposites. We will use this idea to solve absolute equations such as $|3x - 4| = |2x + 11|$. If these two absolute values are equal, then either the two expressions inside the absolute value bars are equal, $3x - 4 = 2x + 11$, or the first expression is the opposite of the second expression, $3x - 4 = -(2x + 11)$. To find the solution to the original equation, we will solve these two resulting equations.

Solving Absolute Value Equations Involving Two Absolute Values

For any expressions X and Y, the solutions to the equation

$$|X| = |Y|$$

can be found by solving the two equations

$$X = Y \text{ and } X = -Y.$$

EXAMPLE 16 Solve $|3x - 4| = |2x + 11|$.

Solution

We begin by rewriting this equation as two equations that do not contain absolute values. For the first equation we set $3x - 4$ equal to $2x + 11$, and for the second equation we set $3x - 4$ equal to the opposite of $2x + 11$. We finish by solving each equation.

$$3x - 4 = 2x + 11 \quad \text{or} \quad 3x - 4 = -(2x + 11)$$

We will now solve each equation separately.

$$\begin{aligned} 3x - 4 &= 2x + 11 \\ x - 4 &= 11 && \text{Subtract } 2x. \\ x &= 15 && \text{Add 4.} \end{aligned}$$

$$\begin{aligned} 3x - 4 &= -(2x + 11) \\ 3x - 4 &= -2x - 11 && \text{Distribute.} \\ 5x - 4 &= -11 && \text{Add } 2x. \\ 5x &= -7 && \text{Add 4.} \\ x &= -\frac{7}{5} && \text{Divide both sides by 5.} \end{aligned}$$

The solution set is $\{-\frac{7}{5}, 15\}$. The check of these solutions is left to the reader.

Quick Check 16
Solve $|x - 9| = |2x + 13|$.

Building Your Study Strategy **Study Groups, 3** **Reviewing Homework** A study group can go over homework assignments together. It is important that each group member work on the assignment before arriving at the study session. If you struggled with a problem, or couldn't do it at all, ask for help or suggestions from your group members.

If there was a problem that you seem to understand better than others in your group, be sure to share your knowledge; explaining how to do a certain problem increases your chances of retaining that knowledge until the exam and beyond.

At some point, try several even-numbered problems from the text that are representative of the problems on the homework assignment. Check your solutions with other members of your group.

At the end of each session, quickly review what the group has accomplished. Finish the meeting by planning your next session, including the time and location of the meeting and what you plan to work on at that meeting.

EXERCISES 1.3

Vocabulary

1. A(n) __________ is a mathematical statement of equality between two expressions.

2. A(n) __________ of an equation is a value that, when substituted for the variable in the equation, produces a true statement.

3. Define a solution set of an equation.

4. Freebird's Pizza charges \$12 for a pizza. If the bill for an office pizza party is \$168, which equation can be used to determine how many pizzas were ordered?

   a) $x + 12 = 168$ b) $x - 12 = 168$

   c) $12x = 168$ d) $\frac{x}{12} = 168$

5. To clear fractions from an equation, multiply both sides of the equation by __________.

6. A contradiction is an equation that __________.

7. The solution set to a contradiction can be denoted $\varnothing$, which represents __________.

8. An equation that is always true is a(n) __________.

9. To solve the equation $2x - 9 = 7$, the best first step is to __________.

   a) divide both sides of the equation by 2

   b) add 9 to both sides of the equation

   c) subtract 7 from both sides of the equation

10. To solve the equation $3(4x + 5) = 11$, the best first step is to __________.

    a) distribute 3 on the left side of the equation

    b) divide both sides of the equation by 4

    c) subtract 5 from both sides of the equation

11. For any expression X and any positive number a, to solve the equation $|X| = a$ we rewrite the equation as the two equations __________ and __________.

12. The equation $|X| = a$ has no solutions if a is a __________ number.

Solve.

13. $x - 4 = 7$
14. $n + 2 = 6$
15. $m - 5 = -9$
16. $x + 7 = -3$
17. $2x = 10$
18. $-3x = 12$
19. $5t = -8$
20. $-2b = -28$
21. $2n - 5 = -11$
22. $3n + 4 = 19$
23. $-4x + 7 = -13$
24. $6x - 1 = 8$
25. $1.6x - 3.7 = 1.9$
26. $2.4x - 1.3 = 2.9$
27. $3x + 25 = -26$
28. $-5x + 33 = 48$
29. $2n - 5 = 5n + 11$
30. $8n - 27 = 5n + 12$
31. $4x + 18 = -x + 3$
32. $-6x + 29 = -2x + 5$
33. $-7b + 5 = 3b - 13$
34. $8t - 33 = 42 - 12t$
35. $2.3n + 2.17 = 4.7n - 3.11$

36. $3.5m + 6.21 = 1.2x + 16.56$

37. $2(2x - 4) - 7 = 6x - 11$

38. $5(3 - 2x) + 4x = 3(3x - 1) + 7$

39. $5x - 3(2x - 9) = 4(3x - 8) + 7x$

40. $10 - 6(2x + 1) = 13 - 9x$

41. $3(x + 4) - 2(x + 4) = 4(2x - 3) - 8(x - 2)$

42. $7x - 4(3x - 10) = 2(x - 8) + 6(3x + 1)$

43. $\frac{3}{4}t - 6 = \frac{1}{3}t - 1$

44. $\frac{1}{8}n - 7 = -\frac{1}{4}n - 13$

45. $\frac{1}{2}(2x - 7) - \frac{1}{3}x = \frac{4}{3}x - 5$

46. $\frac{2}{3}(x - 6) - \frac{1}{4}x = \frac{3}{8}(8 - x) + \frac{1}{2}x$

47. $2b - 9 = 2b - 9$

48. $4t + 11 = 4t - 11$

49. $3(2n - 1) + 5 = 2(3n + 2)$

50. $5(n - 3) - 3n = 2(n - 7) - 1$

51. $5x - 2(3 - 2x) = 3(3x - 2)$

52. $7x - (3x - 4) = 4x - 4$

Solve the following literal equations for the specified variable.

53. $-3x + y = 7$ for y

54. $-10x + y = -1$ for y

55. $6x + 2y = 15$ for y

56. $12x + 5y = -18$ for y

57. $8x - 4y = 11$ for y

58. $9x - y = -15$ for y

59. $P = a + b + c$ for a

60. $P = a + b + 2c$ for b

61. $A = L \cdot W$ for L

62. $A = L \cdot W$ for W

Solve.

63. $|x| = 2$

64. $|x| = 13$

65. $|x| = 929$

66. $|x| = 0$

67. $|x - 8| = 7$

68. $|2x - 11| = 5$

69. $|3x + 4| = 8$

70. $|4 - 5x| = 13$

71. $|x - 9| = -4$

72. $|x| - 9 = -4$

73. $|2x + 8 - 3x| = 6$

74. $|3(2 - x) + 7x| = 14$

75. $|x - 3| + 7 = 12$

76. $|x + 12| - 10 = 10$

77. $|4x - 11| - 19 = -8$

78. $|7x - 52| + 18 = 14$

79. $2|3x + 2| - 9 = 17$

80. $2|5x - 3| + 7 = 21$

81. $-|x + 3| - 10 = -16$

82. $-3|2x + 1| + 13 = -29$

83. $|4x - 5| = |3x - 23|$

84. $|x - 14| = |3x + 2|$

85. $|x + 12| = |2x - 9|$

86. $|2x + 1| = |x - 19|$

87. $|5x| = |2x - 21|$

88. $|7x - 11| = |4x + 6|$

89. $|x + 5| = 2x - 3$

90. $|x - 9| = 3x - 7$

91. $|2x - 7| = 3x + 8$

92. $|4x + 10| = 2x + 6$

93. Find an absolute value equation whose solution is $\{-2, 2\}$.

94. Find an absolute value equation whose solution is $\{-6, 10\}$.

95. Find an absolute value equation whose solution is $\{3, -5\}$.

96. Find an absolute value equation whose solution is $\{\frac{1}{2}, \frac{7}{2}\}$.

Writing in Mathematics

Answer in complete sentences.

97. Explain how to determine if a linear equation is an identity.

98. Explain how to determine if a linear equation is a contradiction.

99. Explain why the equation $|2x - 7| + 6 = 4$ has no solution.

100. Here is a student's work for solving the equation $|x - 3| = -5$.

$$|x - 3| = -5$$
$$x - 3 = -5 \quad \text{or} \quad x - 3 = 5$$
$$x = -2 \quad \text{or} \quad x = 8$$
$$\{-2, 8\}$$

Explain the student's error, and suggest how to avoid making a similar error in the future.

101. ***Solutions Manual**** Write a solutions manual page for the following problem:

Solve $|x - 8| + 3 = 7$.

102. ***Newsletter**** Write a newsletter explaining the steps for solving absolute value equations.

***See Appendix B for details and sample answers.**

1.4 Problem Solving: Applications of Linear Equations

Objectives

1. **Understand the six steps for solving applied problems.**
2. **Solve problems involving unknown numbers.**
3. **Solve problems involving geometric formulas.**
4. **Solve problems involving consecutive integers.**
5. **Solve problems involving motion.**
6. **Solve other applied problems.**

Introduction to Problem Solving

One of the most important skills developed in math classes is the skill of problem solving. Every day we are faced with making important decisions and predictions, and the thought process required in decision-making is quite similar to the process of solving mathematical problems.

When faced with a problem to solve in the real world, we first take inventory of the facts that we know. We also determine exactly what we are trying to figure out. Then we develop a plan for taking what we already know and using it to help us figure out how to solve the problem. Once we have solved the problem, we reflect on the route that we took, to make sure that it was a logical way to solve the problem. After examining our solution for correctness and practicality, we often finish by presenting our solution to others for their consideration, information, or approval.

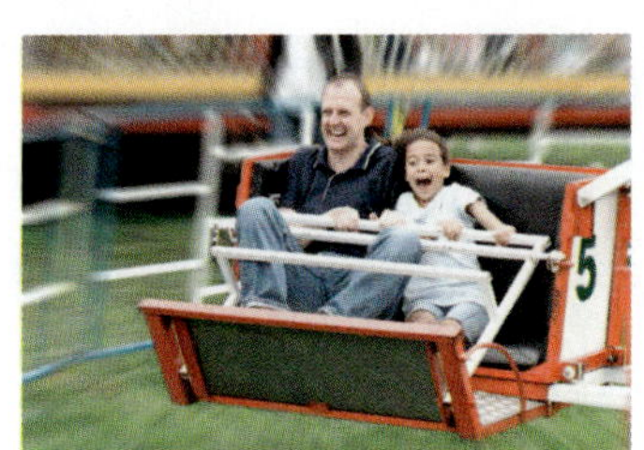

An example of such a real-world problem would be determining how much money you will need to bring on a trip to an amusement park. The problem to solve is figuring out how much money your trip will cost. Now think about the facts that you know: You know that admission to the park is \$40. You will also need to eat, and you estimate that lunch will cost you \$10, and it will cost another \$20 for snacks and drinks. The last time you were there, you spent \$15 on souvenirs. Based on inflation, you determine that you will need \$25 for souvenirs this time. This brings your estimate to \$95. To be safe, you add on another \$25 for emergencies. You decide to bring \$120 with you.

Objective 1 Understand the six steps for solving applied problems. In the previous example we identified the problem to be solved, gathered our facts, and used them to find a solution to the problem. Solving applied math problems will require a similar procedure. George Pólya was a leader in the study of problem solving. Here is a general plan for solving applied math problems, based on the work of Pólya's text *How to Solve It.*

Solving Applied Problems

1. **Read the problem.** This step is often overlooked, but misreading or misinterpreting the problem essentially guarantees an incorrect solution. Read the problem once quickly to get a rough idea of what is going on, and then read it more carefully a second time to gather all the important information.
2. **List all the important information.** Identify all known quantities presented in the problem, and determine which quantities we need to find. Creating a table to hold this information, or a drawing to represent the problem, is a good idea.

continued

3. **Assign a variable to the unknown quantity.** If there is more than one unknown quantity, express each one in terms of the same variable, if possible.
4. **Find an equation relating the known values to the variable expressions representing the unknown quantities.** Sometimes this equation can be translated directly from the wording of the problem. Other times the equation will depend on general facts that are known outside the statement of the problem, such as geometry formulas.
5. **Solve the equation.** Solving the equation is not the end of the problem, as the value of the variable is often not what we were originally looking for. If we were asked for the length and width of a rectangle and we reply "x equals 4," we have not answered the question. Check your solution to the equation.
6. **Present the solution.** Take the value of the variable from the solution to the equation and use it to figure out all unknown quantities. Check these values to be sure that they make sense in the context of the problem. For example, the length of a rectangle cannot be negative. Finally, present your solution in a complete sentence, using the proper units.

We will now put this strategy to use. All the applied problems in this section will lead to linear equations.

Objective 2 Solve problems involving unknown numbers.

EXAMPLE 1 Three less than four times a number is 89. Find the number.

Solution

This is not a very exciting problem, but it provides us with a good opportunity to practice setting up applied problems.

There is one unknown in this problem. Let's use n to represent the *unknown number.*

Unknown
Number: n

Now we translate the sentence "Three less than four times a number is 89" into an equation. In this case, four times a number is $4n$, so three less than four times a number is $4n - 3$.

$$\underbrace{\text{Three less than four times a number}}_{4n-3}\ \underbrace{\text{is}}_{=}\ \underbrace{89}_{89}$$

Now we solve the equation.

$$\begin{aligned} 4n - 3 &= 89 \\ 4n &= 92 && \text{Add 3 to both sides.} \\ n &= 23 && \text{Divide both sides by 4.} \end{aligned}$$

The solution to the equation is $n = 23$. Now we refer back to the table containing our unknown quantity. Since our unknown number was n, the solution to the problem is the same as the solution to the equation. The number is 23.

It is a good idea to check our solution. Four times our number is 92, and three less than that is 89.

***Quick Check* 1**
Five more than twice a number is 133. Find the number.

Geometry Problems

Objective 3 **Solve problems involving geometric formulas.** The following example involves the perimeter of a rectangle, which is a measure of the distance around the outside of the rectangle. The perimeter P of a rectangle whose length is L and whose width is W is given by the formula $P = 2L + 2W$.

EXAMPLE 2 Jorge has a rectangular swimming pool with a perimeter of 112 feet. If the length of his pool is 8 feet more than twice the width, find the dimensions of his pool.

Solution

The unknown quantities are the length and the width. Since the length is given in terms of the width, it is a good idea to pick a variable to represent the width. We can then write the length in terms of this variable. We will let w represent the width of the rectangle. The length is 8 feet more than twice the width, so we can express the length as $2w + 8$. One piece of information that we are given is that the perimeter is 112 feet. This information can be recorded in the following table or drawing.

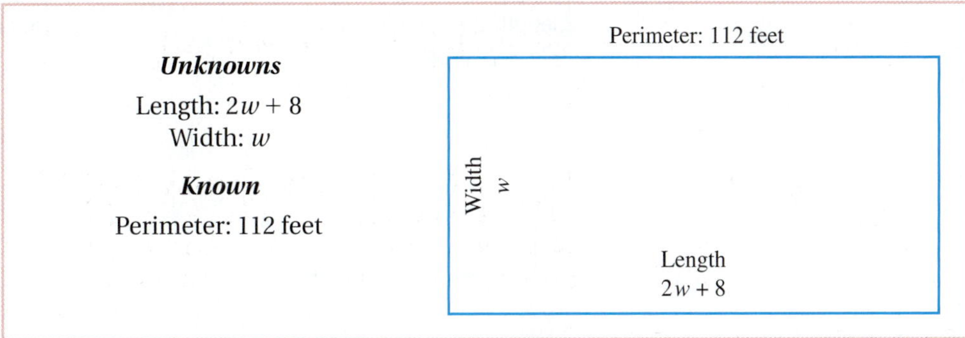

We start with the formula for perimeter, and substitute the appropriate values and expressions.

$P = 2L + 2W$	
$112 = 2(2w + 8) + 2w$	Substitute 112 for P, $2w + 8$ for L, and w for W.
$112 = 4w + 16 + 2w$	Distribute.
$112 = 6w + 16$	Combine like terms.
$96 = 6w$	Subtract 16 from both sides.
$16 = w$	Divide both sides by 6.

We now take this solution and substitute 16 for w in the expressions for length and width, which can be found in the table where we listed our unknowns.

$$\text{Length: } 2w + 8 = 2(16) + 8 = 40$$
$$\text{Width: } w = 16$$

The length of Jorge's pool is 40 feet, and the width is 16 feet. It checks that the perimeter of this rectangle is 112 feet.

Quick Check 2

A rectangular garden is surrounded by 140 feet of fencing. The length of the garden is 10 feet less than three times the width of the garden. Find the dimensions of the garden.

Here is a summary of useful formulas and definitions from geometry.

Geometry Formulas and Definitions

- Perimeter of a triangle with sides s_1, s_2, and s_3: $P = s_1 + s_2 + s_3$

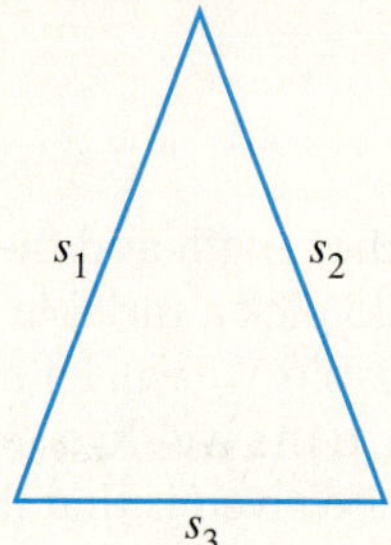

- Perimeter of a square with side s: $P = 4s$

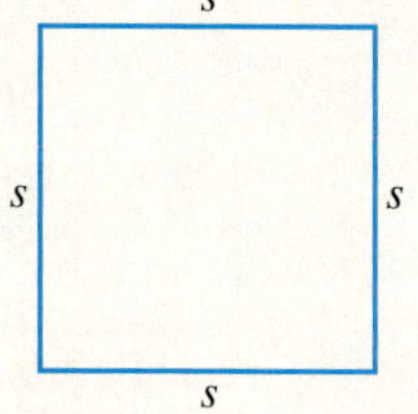

- Perimeter of a rectangle with length L and width W: $P = 2L + 2W$

- **Circumference** (distance around the outside) of a circle with radius r: $C = 2\pi r$

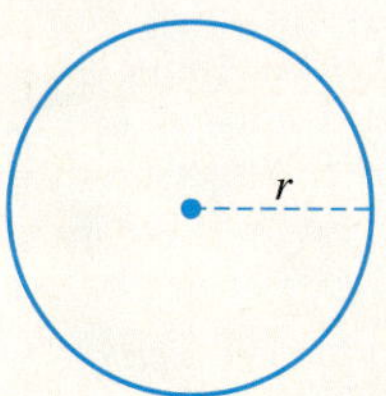

- Area of a triangle with base b and height h: $A = \frac{1}{2}bh$

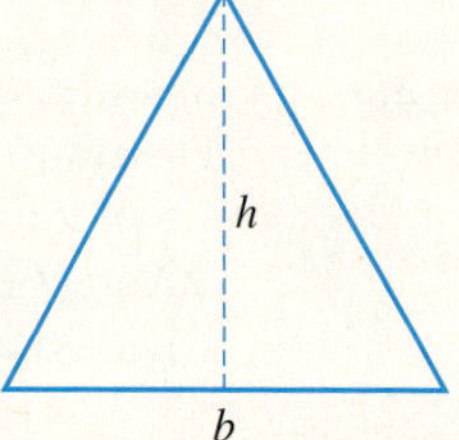

- Area of a square with side s: $A = s^2$

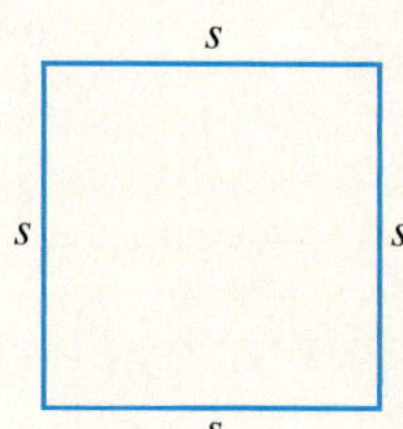

continued

- Area of a rectangle with length L and width W: $A = LW$

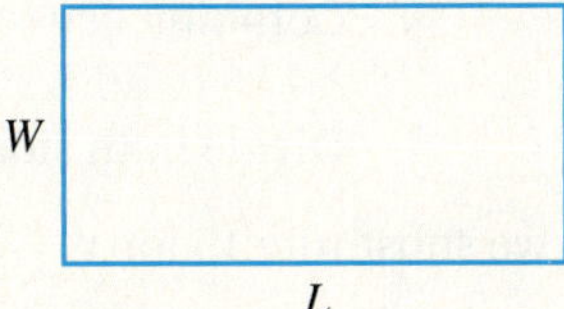

- Area of a circle with radius r: $A = \pi r^2$

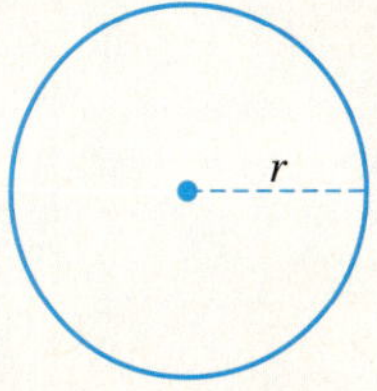

- An **equilateral triangle** is a triangle that has three equal sides.

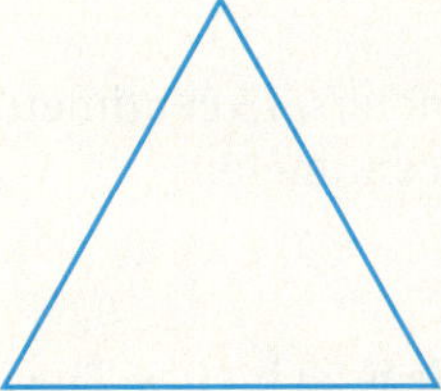

- An **isosceles triangle** is a triangle that has at least two equal sides.

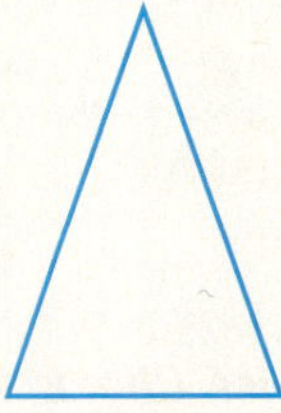

EXAMPLE 3 An isosceles triangle has a perimeter of 40 centimeters. The third side is 5 centimeters shorter than each of the two equal sides. Find the lengths of the three sides.

Solution

In this problem there are three unknowns: the lengths of the three sides. Because the triangle is an isosceles triangle, the first two sides have equal lengths. We can represent the length of each side by the variable x. The third side is 5 centimeters shorter than the other two sides and its length can be represented by $x - 5$. Here is a summary of this information, together with the known perimeter.

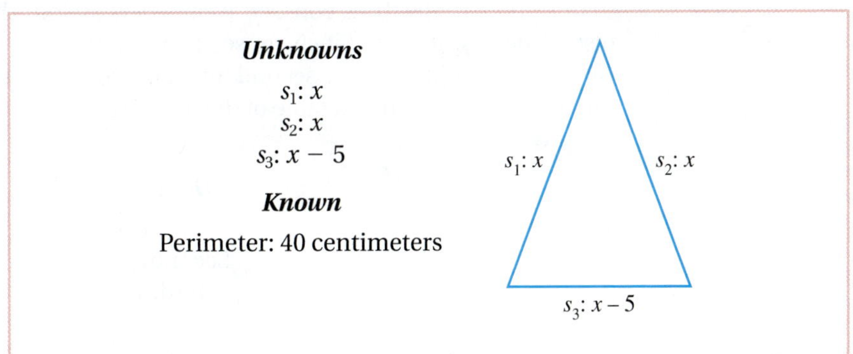

We now set up the equation and solve for x.

$$P = s_1 + s_2 + s_3$$
$$40 = x + x + (x - 5) \quad \text{Substitute 40 for } P,\ x \text{ for } s_1 \text{ and } s_2, \text{ and } x - 5 \text{ for } s_3.$$
$$40 = 3x - 5 \quad \text{Combine like terms.}$$
$$45 = 3x \quad \text{Add 5 to both sides.}$$
$$15 = x \quad \text{Divide both sides by 3.}$$

Looking back to the table, we substitute 15 for x.

$$s_1: x = 15$$
$$s_2: x = 15$$
$$s_3: x - 5 = 15 - 5 = 10$$

The three sides are 15 centimeters, 15 centimeters, and 10 centimeters. The perimeter of this triangle checks to be 40 centimeters.

Quick Check **3** **The perimeter of a triangle is 53 inches. The longest side is 1 inch longer than three times the shortest side. The other side of the triangle is 4 inches longer than twice the shortest side. Find the lengths of the three sides.**

Consecutive Integers

Objective 4 **Solve problems involving consecutive integers.** The following example involves consecutive integers, which are integers that are 1 unit apart from each other on the number line.

EXAMPLE 4 The sum of three consecutive integers is 141. Find them.

Solution

In this problem the three consecutive integers are our three unknowns. We will let x represent the first integer. Since consecutive integers are 1 unit apart from each other we can let $x + 1$ represent the second integer. Adding another 1, we can let $x + 2$ represent the third integer. Here is a table of the unknowns.

Unknowns

First: x
Second: $x + 1$
Third: $x + 2$

We know that the sum of these three integers is 141, which leads to our equation.

$$\begin{aligned} x + (x + 1) + (x + 2) &= 141 \\ 3x + 3 &= 141 && \text{Combine like terms.} \\ 3x &= 138 && \text{Subtract 3 from both sides.} \\ x &= 46 && \text{Divide both sides by 3.} \end{aligned}$$

We now substitute this solution to find our three unknowns.

First: $x = 46$
Second: $x + 1 = 46 + 1 = 47$
Third: $x + 2 = 46 + 2 = 48$

***Quick Check* 4**
The sum of three consecutive integers is 279. Find them.

The three integers are 46, 47, and 48. The three integers do add up to be 141.

Suppose that a problem involved consecutive *odd* integers rather than consecutive integers. Consider any string of consecutive odd integers, such as 15, 17, 19, 21, 23, ... How far apart are these consecutive odd integers? Each odd integer is 2 away from the previous one. When working with consecutive odd integers, or consecutive even integers for that matter, we will let x represent the first integer and then will add 2 to find the next consecutive integer of that type. Here is a table showing the pattern of unknowns for consecutive integers, consecutive odd integers, and consecutive even integers.

Consecutive Integers	Consecutive Odd Integers	Consecutive Even Integers
First: x	First: x	First: x
Second: $x + 1$	Second: $x + 2$	Second: $x + 2$
Third: $x + 2$	Third: $x + 4$	Third: $x + 4$
⋮	⋮	⋮

EXAMPLE 5 The largest of three consecutive odd integers is 87 less than twice the sum of the smaller two integers. Find them.

Solution

In this problem the three consecutive odd integers are our three unknowns. We will let x represent the first integer. Since consecutive odd integers are 2 units apart from each other, we can let $x + 2$ represent the second integer and $x + 4$ represent the third integer. Here is a table of the unknowns.

Unknowns
First: x
Second: $x + 2$
Third: $x + 4$

We know that the largest integer $(x + 4)$ is 87 less than twice the sum of the smaller two integers, which leads to our equation.

$$
\begin{aligned}
x + 4 &= 2(x + x + 2) - 87 && \\
x + 4 &= 2(2x + 2) - 87 && \text{Combine like terms.} \\
x + 4 &= 4x + 4 - 87 && \text{Distribute.} \\
x + 4 &= 4x - 83 && \text{Combine like terms.} \\
4 &= 3x - 83 && \text{Subtract } x \text{ from both sides.} \\
87 &= 3x && \text{Add 83 to both sides.} \\
x &= 29 && \text{Divide both sides by 3.}
\end{aligned}
$$

We now substitute this solution to find our three unknowns.

First: $x = 29$
Second: $x + 2 = 29 + 2 = 31$
Third: $x + 4 = 29 + 4 = 33$

The three integers are 29, 31, and 33.

Quick Check 5
If the largest of three consecutive even integers is subtracted from the sum of the two smaller integers, the result is 12. Find the integers.

Motion Problems

Objective 5 **Solve problems involving motion.**

Distance Formula

When an object, such as a car, moves at a constant speed or rate r for an amount of time t, then the distance traveled, d, is given by the formula $d = r \cdot t$.

Suppose that a person drove 297 miles in 4.5 hours. To determine the car's rate of speed for this trip, we can substitute 297 for d and 4.5 for t in the equation $d = r \cdot t$.

$$
\begin{aligned}
d &= r \cdot t && \\
297 &= r \cdot 4.5 && \text{Substitute 297 for } d \text{ and 4.5 for } t. \\
\frac{297}{4.5} &= \frac{r \cdot 4.5}{4.5} && \text{Divide both sides by 4.5.} \\
66 &= r && \text{Simplify.}
\end{aligned}
$$

The car's rate of speed for this trip was 66 mph.

EXAMPLE 6 Tina drove her car at a speed of 68 mph from her home to Las Vegas. Her mother Linda made the same trip at a rate of 85 mph, and it took her 1 hour less than Tina to make the trip. How far is it from Tina's home to Las Vegas?

Solution

For this problem, we can begin by setting up a table showing the relevant information. We will let t represent the time it took for Tina to make the trip. Since it took Linda 1 hour less to make the trip, we can represent her time by $t - 1$. We multiply each person's rate by the time she traveled to find the distance she traveled.

	Rate	Time	Distance ($d = r \cdot t$)
Tina	68	t	$68t$
Linda	85	$t - 1$	$85(t - 1)$

Since we know that both trips were exactly the same distance, we get our equation by setting the expression for Tina's distance equal to the expression for Linda's distance. We then solve for t.

$$68t = 85(t - 1)$$
$$68t = 85t - 85 \quad \text{Distribute.}$$
$$-17t = -85 \quad \text{Subtract } 85t \text{ from both sides.}$$
$$t = 5 \quad \text{Divide both sides by } -17.$$

At this point we must be careful to answer the appropriate question. We were asked to find the distance that was traveled. We can substitute 5 for t in either of the expressions for distance. Using the expression for the distance traveled by Tina, $68t = 68(5) = 340$. Using the expression for the distance traveled by Linda, $85(t - 1)$, produces the same result. It is 340 miles from Tina's home to Las Vegas.

Quick Check **6** **Jake drove at a speed of 60 mph for a certain amount of time. His brother Elwood then took over as the driver. Elwood drove at a speed of 55 mph for 2 hours longer than Jake had driven. If the two brothers drove a total of 455 miles, how long did Elwood drive?**

Other Problems

Objective 6 **Solve other applied problems.**

EXAMPLE 7 One number is 4 more than three times another number. If the sum of the two numbers is 72, find the numbers.

Solution

In this problem, there are two unknown numbers. We will let one of the numbers be x. Since we know that one number is 4 more than three times the other number, we can represent this number as $3x + 4$.

Unknowns

First: x

Second: $3x + 4$

Finally, we know that the sum of these two numbers is 72, which leads to the equation $x + 3x + 4 = 72$.

$$x + 3x + 4 = 72$$
$$4x + 4 = 72 \quad \text{Combine like terms.}$$
$$4x = 68 \quad \text{Subtract 4 from both sides.}$$
$$x = 17 \quad \text{Divide both sides by 4.}$$

We now substitute this solution to find the two unknown numbers.

First: $x = 17$

Second: $3x + 4 = 3(17) + 4 = 51 + 4 = 55$

The two numbers are 17 and 55. The sum of these two numbers is 72.

Quick Check 7 **A number is nine less than four times another number. If the sum of the two numbers is 101, find the numbers.**

We conclude this section with a problem involving coins.

EXAMPLE 8 Ernie has dimes and quarters in his pocket. He has 7 more dimes than quarters. If Ernie has \$4.55 in his pocket, how many dimes and quarters does he have?

Solution

We know the amount of money Ernie has in dimes and quarters totals \$4.55. Since the number of dimes is given in terms of the number of quarters, we will let q represent the number of quarters. Since we know that Ernie has 7 more dimes than quarters, we can represent the number of dimes by $q + 7$. A table can be quite helpful in organizing our information, as well as in finding our equation.

	Number of Coins	Value of Coin	Amount of Money
Dimes	$q + 7$	0.10	$0.10(q + 7)$
Quarters	q	0.25	$0.25q$

The amount of money that Ernie has in nickels and dimes is equal to \$4.55.

$$
\begin{aligned}
0.10(q + 7) + 0.25q &= 4.55 \\
0.10q + 0.70 + 0.25q &= 4.55 && \text{Distribute and multiply.} \\
0.35q + 0.70 &= 4.55 && \text{Combine like terms.} \\
0.35q &= 3.85 && \text{Subtract 0.70 from both sides.} \\
q &= 11 && \text{Divide both sides by 0.35.}
\end{aligned}
$$

The number of quarters is 11. The number of dimes $(q + 7)$ is $11 + 7$ or 18. Ernie has 18 dimes and 11 quarters. It is left to the reader to verify that these coins are worth a total of \$4.55.

Quick Check **8** **Bert has a collection of \$1 bills and \$5 bills in his wallet. He has 8 more \$1 bills than \$5 bills. If Bert has \$86 in his wallet, how many \$1 bills and \$5 bills does he have in his wallet?**

Building Your Study Strategy **Study Groups, 4** **Keeping in Touch** It is important for group members to share phone numbers and email addresses. This will allow you to contact other group members if you misplace the homework assignment for that day. You can also contact others in your group if you have to miss class. This allows you to find out what was covered in class and which problems were assigned for homework. The person you contact can also give you advice for certain problems.

Some members of study groups agree to call each other if they are having trouble with homework problems. Calling a group member for this purpose should be a last resort. Be sure that you have used all available resources (examples in the text, notes, etc.) and have given the problem your fullest effort. Otherwise, you may end up calling for help on all of the problems.

EXERCISES 1.4

Vocabulary

1. Which equation can be used to solve "9 less than twice a number is 35?"

   a) $2(n - 9) = 35$ b) $2n - 9 = 35$ c) $9 - 2n = 35$

2. State the formula for the perimeter of a rectangle.

3. If a rectangle has a perimeter of 110 feet and its width is 5 feet less than 3 times its length, which equation can be used to find the dimensions of the rectangle?

   a) $2(3L) + 2(L - 5) = 110$
   b) $2L + 2(3L - 5) = 110$
   c) $2(3W - 5) + 2W = 110$

4. Consecutive integers are integers that are ________ apart on a number line.

5. If we let x represent the first of three consecutive integers, then we can represent the other two integers by ________ and ________.

6. If we let x represent the first of three consecutive even integers, then we can represent the other two integers by ________ and ________.

7. If we let x represent the first of three consecutive odd integers, then we can represent the other two integers by ________ and ________.

8. Charles has 7 fewer \$5 bills than \$10 bills. The total value of these bills is \$265. Which equation can be used to determine how many of each type of bill Charles has?

   a) $5(n - 7) + 10n = 265$
   b) $5n - 7 + 10n = 265$
   c) $5n + 10(n - 7) = 265$
   d) $5n + 10n - 7 = 265$

9. Five more than twice a number is 27. Find the number.

10. Three more than nine times a number is 120. Find the number.

11. Six less than half a number is 15. Find the number.

12. Twelve less than one-third of a number is 17. Find the number.

13. Four times a number, minus 2.7, is 9.7. Find the number.

14. Three times a number, plus 9.2, is 59.3. Find the number.

15. The length of a rectangle is 6 meters more than its width, and the perimeter is 40 meters. Find the length and the width of the rectangle.

16. A rectangle has a perimeter of 98 feet. If its width is 11 feet less than its length, find the dimensions of the rectangle.

17. The length of a rectangle is 5 inches more than twice its width, and the perimeter is 106 inches. Find the length and the width of the rectangle.

18. The length of a rectangle is 7 feet less than three times its width, and the perimeter is 74 feet. Find the length and the width of the rectangle.

19. The width of a rectangle is 19.7 feet less than four times its length, and the perimeter is 22.6 feet. Find the length and the width of the rectangle.

20. The length of a rectangle is 3.2 meters less than six times its width, and the perimeter is 151.8 meters. Find the length and the width of the rectangle.

21. A rectangular garden is surrounded by 210 feet of fencing. If the length of the garden is 15 feet more than twice the width, find the dimensions of the garden.

22. A rectangular photograph has a perimeter of 27 inches. The length of the photograph is 3 inches less than twice its width. Find the dimensions of the photograph.

23. A rectangular concrete slab is poured for a basketball court, and the boards used around the outside of the rectangle used as forms for the slab have a total length of 150 feet. If the width of the basketball court is half its length, find the dimensions of the basketball court.

24. The length of a rectangular desk is 1.5 times as long as the width, and the perimeter of the desk is 156 inches. Find the dimensions of the desk.

25. A boxing ring is set up in the shape of a square, with a perimeter of 72 feet. What is the length of one of its sides?

26. The bases in a little league baseball infield are set out in the shape of a square. When a child hits a home run, she has to run 240 feet to make it all the way around the bases. How far is the distance from home plate to first base?

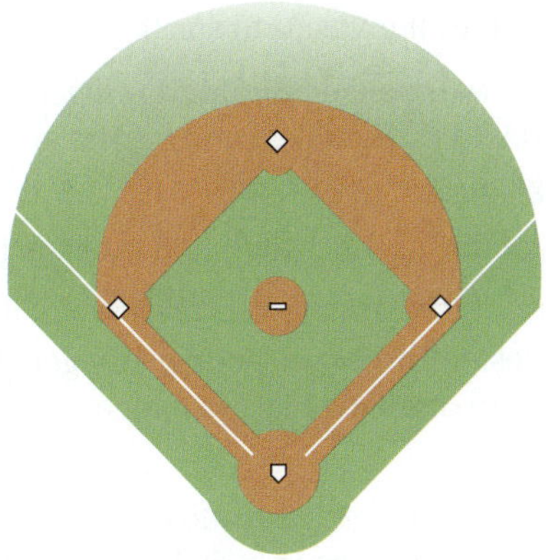

27. A yield sign is in the shape of an equilateral triangle with a perimeter of 108 centimeters. Find the length of each side of this triangle.

28. Darius buys a triangular sail shade that is in the shape of an isosceles triangle. The third side of the shade is 15 inches shorter than each of the other two sides. If the perimeter of this triangle is 411 inches, find the length of each side.

29. A yacht club sets up a race around three buoys forming a triangle. The second leg of the race is 10 miles longer than the first leg of the race. The third leg of the race is 10 miles less than twice the length of the first leg of the race. If the total length of the race is 160 miles, find the length of each leg of the race.

30. The cities of Boston, Chicago, and Memphis form a triangle. The distance from Boston to Chicago is 470 miles longer than the distance from Chicago to Memphis. The distance from Boston to Memphis is 260 miles longer than twice the distance from Chicago to Memphis. If the perimeter of the triangle connecting these three cities is 2850 miles, find the distance from Boston to Memphis.

31. A circle has a circumference of 94.2 centimeters. Use $\pi \approx 3.14$ to find the radius of the circle.

32. A circle has a circumference of 16π inches. Find its radius.

Two positive angles are said to be **complementary** if their measures add up to 90°.

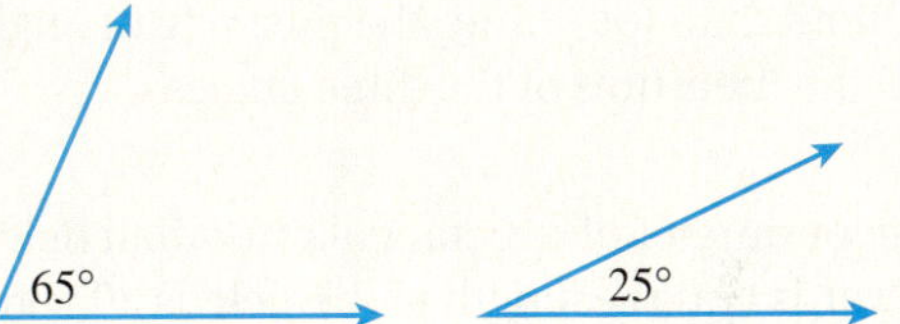

33. Angles A and B are complementary angles. If the measure of angle A is 60° less than the measure of angle B, find the measures of the two angles.

34. Angles A and B are complementary angles. Angle A is 22.8° more than angle B. Find the measures of the two angles.

35. An angle is 18° more than twice the measure of its complementary angle. Find the measures of the two angles.

36. An angle is four times the measure of its complementary angle. Find the measures of the two angles.

Two positive angles are said to be **supplementary** if their measures add up to 180°.

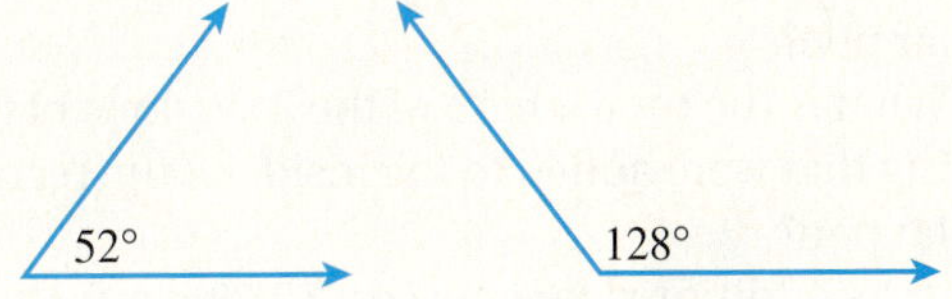

37. The measure of an angle is 35.2° more than the measure of its supplementary angle. Find the measures of the two angles.

38. The measure of an angle is 20° less than three times the measure of its supplementary angle. Find the measures of the two angles.

The measures of the three angles inside any triangle total 180°.

39. One angle in a triangle is 20° more than the smallest angle in the triangle, while the other angle is 40° more than the smaller angle. Find the measures of the three angles.

40. A triangle contains three angles: *A*, *B*, and *C*. Angle *B* is three times the measure of angle *A*, and angle *C* is 20° more than four times the measure of angle *A*. Find the measures of the three angles.

41. If the perimeter of a rectangular football field is 280 yards and the length of the field is 20 yards more than twice its width, find the *area* of the field.

42. If the perimeter of a rectangular back yard is 1080 feet and the length of the yard is 20 feet less than three times its width, find the *area* of the yard.

43. A rectangle has a perimeter of 54 inches. The length of the rectangle is 3 inches shorter than twice its width. If a square has a side as long as the length of this rectangle, find the perimeter of the square.

44. A rectangular pen with a perimeter of 160 feet is divided up into 3 square pens by building 2 fences inside the rectangular pen as shown.

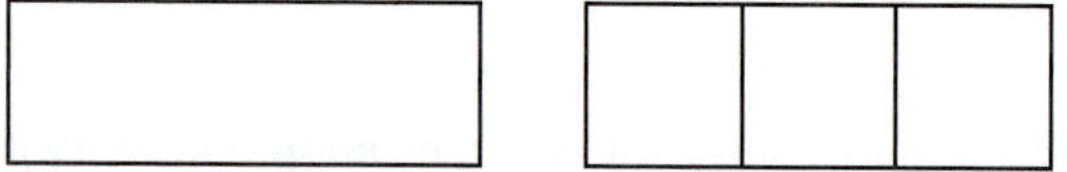

a) What are the original dimensions of the rectangular pen?
b) What is the total length of the 2 sections of fencing that were added to the inside of the rectangular pen?
c) If the additional fencing cost $3.79 per foot, find the cost of the 2 sections of fencing that were added.

45. The sum of four consecutive integers is 266. Find the four integers.

46. The sum of three consecutive integers is 159. Find the three integers.

47. The sum of three consecutive odd integers is 387. Find the three integers.

48. The sum of three consecutive odd integers is 111. Find the three integers.

49. The sum of three consecutive even integers is 144. Find the three integers.

50. The sum of five consecutive even integers is 430. Find the five integers.

51. Lou has a book open and the two page numbers that he is looking at have a sum of 537. What are the page numbers of the pages he is looking at?

52. Tyler is taking three consecutive days off in July for her vacation. The three dates add up to be 54. On what date does her vacation start?

53. The sum of four consecutive odd integers is 232. Find the product of the smallest and largest of the four odd integers.

54. The sum of five consecutive even integers is 640. Find the average of these five integers.

55. The smallest of three consecutive integers is 42 less than the sum of the two larger integers. Find the three integers.

56. There are three consecutive odd integers, and the sum of the smaller two integers is 60 less than 4 times the larger integer. Find the three odd integers.

57. Lance rode his bicycle at an average speed of 27 miles per hour for 3 hours. How far did he ride?

58. Danica made the 380-mile drive from San Francisco to Los Angeles in 5 hours. What was her average speed for the trip?

59. If Cheng drives at a speed of 72 miles per hour, how long will it take him to drive 450 miles?

60. A bullet train travels 240 kilometers per hour. How far can it travel in 1 hour and 15 minutes?

61. A swimmer crossed the English Channel in $7\frac{1}{2}$ hours. If the distance across the channel is 35 kilometers, find the swimmer's speed to the nearest tenth of a kilometer per hour.

62. In 1932, Amelia Earhart became the first woman to fly solo nonstop coast to coast. She flew 2447.8 miles in 19.1 hours. What was her average speed for the trip, to the nearest tenth of a mile per hour?

63. Susan drove 75 miles per hour on the way to Winslow, Arizona. On the way home, she drove 60 miles per hour and it took her 2 hours longer to drive home than it did to drive to Winslow.

a) What was the total driving time for Susan's trip?

b) How far does Susan live from Winslow?

64. Rodrigo rode his bike at a speed of 20 miles per hour to his vacation cabin. On his way back home, he rode his bike at a speed of 28 miles per hour through the countryside. The return trip took him 2 hours less time.

a) What was the total time for Rodrigo's trip?

b) How far does Rodrigo live from his vacation cabin?

65. Don started driving east at 7 A.M. at a speed of 70 miles per hour. If Rose leaves the same place heading east at 10 A.M. at a speed of 80 miles per hour, how long will it take her to catch up to Don?

66. A train left a city at 9 AM, heading directly north at a speed of 50 miles per hour. A bus left the same city at 11 AM, heading directly south at a speed of 65 miles per hour. At what time will the train and the bus be 560 miles apart?

67. One number is 19 more than another. If the sum of the two numbers is 85, find the numbers.

68. One number is 32 more than another. If the sum of the two numbers is 146, find the numbers.

69. One number is 36 less than another number. If the sum of the two numbers is 82, find the numbers.

70. One number is 55 less than another number. If the sum of the two numbers is 137, find the two numbers.

71. One number is 7 more than another number. If three times the smaller number is added to twice the larger number, the sum is 74. Find the two numbers.

72. One number is 14 more than another number. If four times the smaller number is added to five times the larger number, the sum is 367. Find the two numbers.

73. Amie has \$22.25 in quarters. How many quarters does she have?

74. A roll of dimes is worth \$5. How many dimes are in a roll?

75. Xochitl has a jar with dimes and quarters in it. The jar has 8 more quarters in it than it has dimes. If the total value of the coins is \$6.20, how many quarters are in the jar?

76. Jean's change purse has nickels and dimes in it. The number of nickels is 7 less than three times the number of dimes. If the total value of the coins is \$4.15, how many nickels are in the purse?

77. The film department held a fund-raiser by showing the movie *A Beautiful Mind.* Admission was \$3 for students and \$6 for non-students. The number of students who attended was 10 more than 3 times the number of non-students who attended. If the department raised \$780, how many students attended the movie?

78. A movie theater charges \$4.50 for children to see a matinee and \$7.25 for adults. At today's matinee there were 36 more children than adults, and the total receipts were \$890.50. How many children were at today's matinee?

Writing in Mathematics

Answer in complete sentences.

79. Write a word problem involving a rectangle with a length of 40 feet and a width of 25 feet. Explain how you created your problem.

80. Write a word problem whose solution is "There were 80 children and 30 adults in attendance." Explain how you created your problem.

81. Write a word problem for the given table and equation. Explain how you created your problem.

	Rate	Time	Distance ($d = r \cdot t$)
Steve	60	t	$60t$
Ross	70	$t + 2$	$70(t + 2)$

Equation: $60t + 70(t + 2) = 530$

82. Write a word problem for the given table and equation. Explain how you created your problem.

	Rate	Time	Distance ($d = r \cdot t$)
?	6	t	$6t$
?	10	$t - 1$	$10(t - 1)$

Equation: $10(t - 1) = 6t + 6$

83. ***Solutions Manual**** Write a solutions manual page for the following problem:

The width of a rectangle is 3 inches less than twice its length. If the perimeter of the rectangle is 36 inches, find its length and width.

84. ***Newsletter**** Write a newsletter explaining how to solve consecutive integer problems.

***See Appendix B for details and sample answers.**

Quick Review Exercises

Section 1.4

Solve.

1. $4x + 17 = 41$

2. $5x - 19 = 3x - 51$

3. $2x + 2(x + 8) = 68$

4. $x + 3(x + 2) = 2(x + 4) + 30$

1.5 Linear Inequalities and Absolute Value Inequalities

Objectives

1. **Graph the solutions of a linear inequality on a number line and express the solutions using interval notation.**
2. **Solve linear inequalities.**
3. **Solve compound linear inequalities.**
4. **Solve applied problems involving inequalities.**
5. **Solve absolute value inequalities.**

Suppose that you went to a doctor and she told you that your temperature was normal. This would mean that your temperature was 98.6°F. However, if the doctor told you that you had a fever, could you tell what your temperature was? No, all you would know is that it was above 98.6°F. If we let the variable t represent your temperature, this could be written as $t > 98.6$. This is an example of an inequality. An inequality is a mathematical statement comparing two quantities using the symbols $<$ (*less than*) or $>$ (*greater than*). It states that one quantity is smaller than (or larger than) the other quantity.

Two other symbols that may be used in an inequality are $\leq$ and $\geq$. The symbol $\leq$ is read as "*less than or equal to*" and the symbol $\geq$ is read as "*greater than or equal to.*" The inequality $x \leq 7$ is used to represent all real numbers that are either less than 7 or equal to 7. Inequalities involving the symbols $\leq$ or $\geq$ are often called **weak inequalities**, while inequalities involving the symbols $<$ or $>$ are called **strict inequalities** because they involve numbers that are strictly greater than (or less than) a given value.

Presenting Solutions to Inequalities

Objective 1 Graph the solutions of a linear inequality on a number line and express the solutions using interval notation.

Linear Inequality

A **linear inequality** is an inequality containing a linear expression.

A linear inequality often has infinitely many solutions, and it is not possible to list every solution. Consider the inequality $x < 3$. There are infinitely many solutions to this inequality, but one thing that all the solutions share is that they are all to the left of 3 on the number line. Values located to the right of 3 on the number line are greater than 3. The solutions can be represented on a number line as follows. All shaded numbers to the left are solutions.

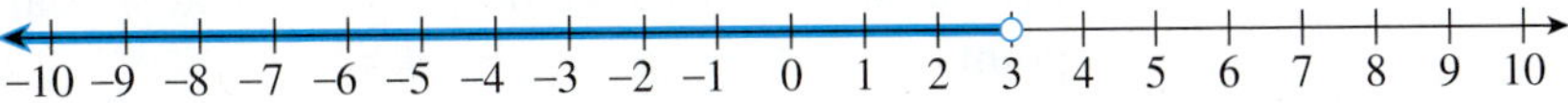

An **endpoint** of an inequality is a point on a number line that separates values that are solutions from values that are not. The open circle at $x = 3$ tells us that the endpoint is not included in the solution, because 3 is not less than 3. If the inequality was $x \leq 3$, then we would fill in the circle at $x = 3$.

Here are three more inequalities, with their solutions graphed on a number line.

Inequality	Solution
$x \le -2$	−10 −9 −8 −7 −6 −5 −4 −3 −2 −1 0 1 2 3 4 5 6 7 8 9 10
$x > 5$	−10 −9 −8 −7 −6 −5 −4 −3 −2 −1 0 1 2 3 4 5 6 7 8 9 10
$x \ge -4$	−10 −9 −8 −7 −6 −5 −4 −3 −2 −1 0 1 2 3 4 5 6 7 8 9 10

Saying that a is less than b is equivalent to saying that b is greater than a. In other words, the inequality $a < b$ is equivalent to the inequality $b > a$. This is an important tool, as we will want to rewrite an inequality so that the variable is on the left side of the inequality before graphing it on a number line.

Interval Notation

Another way to present the solutions to an inequality is by using **interval notation**. A range of values on a number line, such as the solutions to the inequality $x < -2$, is called an **interval**. Inequalities typically have one or more intervals as their solutions. Interval notation presents an interval that is a solution by listing its left and right endpoints. We use parentheses around the endpoints if the endpoints are not included as solutions, and we use brackets if the endpoints are included as solutions.

When an interval continues on indefinitely to the right on a number line, we will use the symbol ∞ (infinity) in the place of the right endpoint and enclose it with a parenthesis. Let's look again at the number line associated with the solutions to the inequality $x \ge -4$.

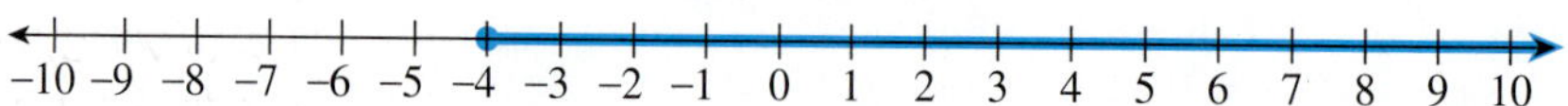

The solutions begin at -4 and include any number that is greater than -4. The interval is bounded by -4 on the left side and extends without bound on the right side, which can be written in interval notation as $[-4, \infty)$. The endpoint -4 is included in the interval, so we write a square bracket next to it, because it is a solution, while we will always write a parenthesis next to ∞, as it is not an actual number that ends the interval at that point.

If the inequality had been $x > -4$, instead of $x \ge -4$, then we would have written a parenthesis next to -4 rather than a square bracket. In other words, the solutions to this inequality can be expressed in interval notation $(-4, \infty)$.

For intervals that continue on indefinitely to the left on a number line, we will use $-\infty$ in the place of the left endpoint.

The following table shows the number line and interval notation associated with several types of simple linear inequalities.

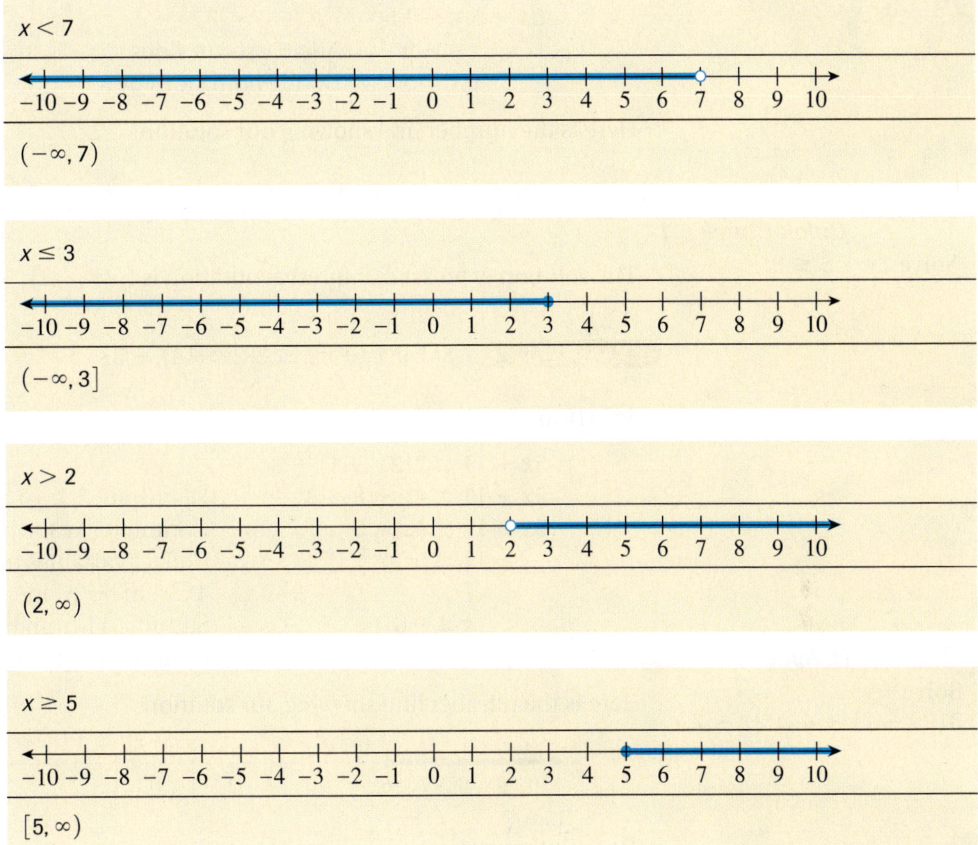

Recall that we use a closed circle whenever the endpoint is included in the interval, $\leq$ or $\geq$, and an open circle when the endpoint is not included, $<$ or $>$. When expressing the interval in interval notation, we use square brackets when the endpoint is included, and we use parentheses when the endpoint is not included.

Solving Linear Inequalities

Objective 2 **Solve linear inequalities.** Solving a linear inequality, such as $3x - 5 < -14$, is very similar to solving a linear equation. In fact, there is only one difference between the two procedures. Whenever we multiply both sides of an inequality by a negative number, or divide both sides by a negative number, the direction of the inequality sign changes. Why is this? Consider the inequality $3 < 5$, which is a true statement. If we multiply each side by -2, is the left side $(-2 \cdot 3 = -6)$ still less than the right side $(-2 \cdot 5 = -10)$? No, $-6 > -10$ because -6 is located to the right of -10 on the number line. Changing the direction of the inequality sign after multiplying or dividing by a negative number produces an inequality that is still a true statement.

EXAMPLE 1 Solve $3x - 5 < -14$.

Solution

$$\begin{aligned} 3x - 5 &< -14 \\ 3x &< -9 && \text{Add 5 to both sides.} \\ x &< -3 && \text{Divide both sides by 3.} \end{aligned}$$

Here is the number line showing our solution.

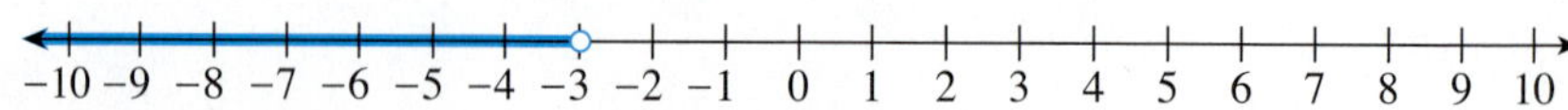

Quick Check 1
Solve $2x + 5 \leq 9$.

The solution expressed in interval notation is $(-\infty, -3)$.

EXAMPLE 2 Solve $5x + 11 \leq 2(2x + 4) - 3$.

Solution

$$\begin{aligned} 5x + 11 &\leq 2(2x + 4) - 3 \\ 5x + 11 &\leq 4x + 8 - 3 && \text{Distribute 2.} \\ 5x + 11 &\leq 4x + 5 && \text{Combine like terms.} \\ x + 11 &\leq 5 && \text{Collect variables on the left side by subtracting } 4x \text{ from both sides.} \\ x &\leq -6 && \text{Subtract 11 from both sides to collect constant terms on the right side.} \end{aligned}$$

Quick Check 2
Solve
$3(2x - 1) + 4 \geq 2x - 7$.

Here is the number line showing our solution.

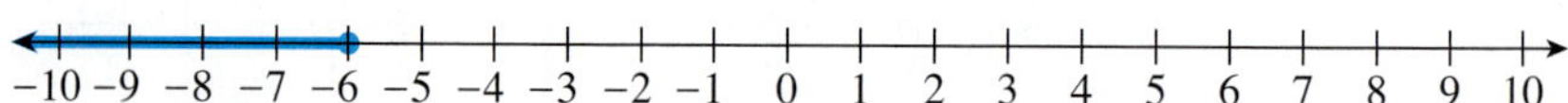

The solution expressed in interval notation is $(-\infty, -6]$.

EXAMPLE 3 Solve $-2x + 3 > 13$.

Solution

$$\begin{aligned} -2x + 3 &> 13 \\ -2x &> 10 && \text{Subtract 3 from both sides.} \\ \frac{-2x}{-2} &< \frac{10}{-2} && \text{Divide both sides by } -2. \text{ This changes the direction of the inequality sign from } > \text{ to } <. \\ x &< -5 && \text{Simplify.} \end{aligned}$$

Quick Check 3
Solve $5 - 2x > -5$.

Here is the number line showing our solution.

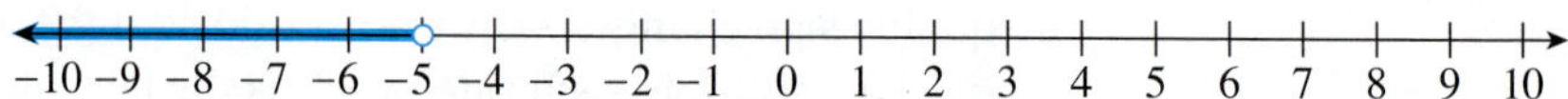

The solution expressed in interval notation is $(-\infty, -5)$.

A Word of Caution When we multiply or divide both sides of an inequality by a negative number, we must change the direction of the inequality sign.

Solving Compound Linear Inequalities

Objective 3 **Solve compound linear inequalities.** A compound inequality is made up of two simple inequalities. For instance, the compound inequality $3 \le x < 7$ represents numbers that are both greater than or equal to 3 and less than 7 at the same time. This can be represented graphically by the following number line.

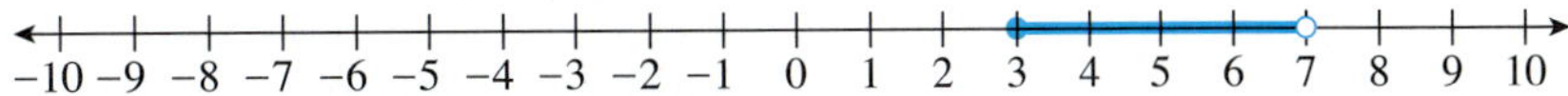

The interval notation for this interval is $[3, 7)$.

To solve a compound inequality of this type, we need to isolate the variable in the middle part of the inequality between two real numbers. A key aspect of our approach will be to work on all three parts of the inequality at once.

EXAMPLE 4 Solve $-3 < 2x - 1 < 5$.

Solution

We begin to solve this compound inequality by isolating the variable x in the middle part of the inequality.

$-3 < 2x - 1 < 5$

$-2 < 2x < 6$ Add 1 to each part of the inequality.

$-1 < x < 3$ Divide the three parts by 2.

Our solution consists of all real numbers greater than -1 and also less than 3, in other words, all real numbers that are between -1 and 3. Here is the number line showing the solution.

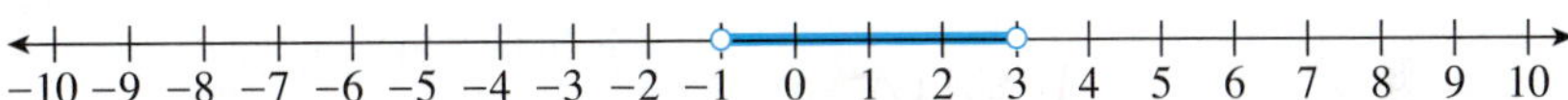

The solution expressed in interval notation is $(-1, 3)$.

***Quick Check* 4**
Solve $-8 \le 3x + 7 \le 1$.

Another type of compound inequality involves the word *or.* Consider the compound inequality $x < -2$ or $x > 1$. The solutions to this inequality are values that are solutions to either of the two inequalities. In other words, any number that is less than -2 $(x < -2)$ *or* is greater than 1 $(x > 1)$ is a solution to the compound inequality. Here are the solutions to this compound inequality, graphed on a number line.

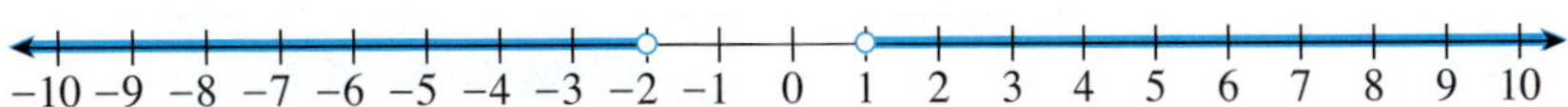

Notice that we have two different intervals that contain solutions. The solution expressed in interval notation is $(-\infty, -2) \cup (1, \infty)$. The symbol "$\cup$" represents a **union** in set theory, and combines the two intervals into one solution.

To solve a compound inequality involving "or," we simply solve each individual inequality separately.

EXAMPLE 5 Solve $x + 4 \le 7$ or $2x - 1 > 11$.

Solution

To solve a compound inequality of this type, we first solve each inequality separately. We begin by solving $x + 4 \le 7$.

$$x + 4 \le 7$$
$$x \le 3 \quad \text{Subtract 4 from both sides.}$$

Now we solve the other inequality.

$$2x - 1 > 11$$
$$2x > 12 \quad \text{Add 1 to both sides.}$$
$$x > 6 \quad \text{Divide both sides by 2.}$$

Here is a number line showing the solution.

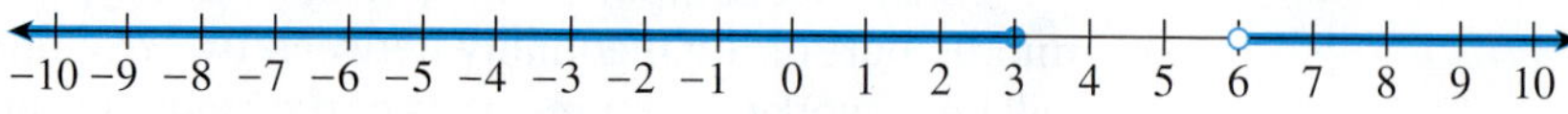

Quick Check 5
Solve $x + 3 < -3$ or $2x + 5 > 14$.

The solution expressed in interval notation is $(-\infty, 3] \cup (6, \infty)$.

Applications

Objective 4 **Solve applied problems involving inequalities.**

Many applied problems involve inequalities rather than equations. Here are some key phrases and their translations into inequalities.

Key Phrases for Inequalities

Phrase	Inequality
x is greater than *a* *x* is more than *a* *x* is higher than *a* *x* is above *a*	$x > a$
x is at least *a* *x* is *a* or higher	$x \ge a$
x is less than *a* *x* is lower than *a* *x* is below *a*	$x < a$
x is at most *a* *x* is *a* or lower	$x \le a$
x is between *a* and *b* *x* is more than *a* but less than *b*	$a < x < b$
x is between *a* and *b*, inclusive *x* is at least *a*, but no more than *b*	$a \le x \le b$

EXAMPLE 6 A sign at a movie theater states that you must be at least 17 years old in order to be admitted to an R-rated movie without a parent or guardian. Set up an inequality that shows the ages of people who are able to be admitted to an R-rated movie without a parent or guardian.

Solution

Let a represent the age of a person. If a person must be at least 17 years old, that means that the person's age must be 17 years old or older. In other words, the person's age must be greater than or equal to 17. The inequality is $a \geq 17$.

A Word of Caution The phrase "at least" means "greater than or equal to" ($\geq$) and does not mean "less than."

EXAMPLE 7 An instructor tells her class that a score of 90 or higher out of a possible 100 on the final exam is needed to earn an A. Set up an inequality that shows the scores that do not earn an A.

Quick Check 6
A student club has a bylaw that states a person must have an IQ of at least 120 in order to be admitted to the club. Set up an inequality that shows the IQ of people who are unable to be admitted to the club.

Solution

Let s represent a student's score. Since a score of 90 or higher will earn an A, any score lower than 90 will not. This inequality can be written as $s < 90$. If we happen to know that the lowest possible score is 0, we could also write this inequality as $0 \leq s < 90$.

We will now set up and solve applied problems involving inequalities.

EXAMPLE 8 If a student averages 70 or higher on the five tests given in a math class, then the student will pass the class. Gail's scores on the first four tests are 62, 57, 83, and 72. What scores on the fifth test will give Gail a passing average?

Solution

To find the average of five test scores, we add the five scores and divide by 5. Let x represent the score of the fifth test. The average can then be expressed as $\frac{62 + 57 + 83 + 72 + x}{5}$. We are interested in knowing which scores on the fifth test give an average that is 70 or higher.

Quick Check 7
If a student averages 90 or higher on the four tests given in a math class, then the student will earn an A in the class. Bobby's scores on the first three tests are 84, 99, and 80. What scores on the fourth test will give Bobby an A in the class?

$$\frac{62 + 57 + 83 + 72 + x}{5} \geq 70$$

$$\frac{274 + x}{5} \geq 70 \quad \text{Simplify the numerator.}$$

$$5 \cdot \frac{274 + x}{5} \geq 5 \cdot 70 \quad \text{Multiply both sides by 5 to clear the fraction.}$$

$$274 + x \geq 350 \quad \text{Simplify.}$$

$$x \geq 76 \quad \text{Subtract 274 from both sides.}$$

Gail must score at least 76 on the fifth test to pass the class.

Absolute Value Inequalities

Objective 5 **Solve absolute value inequalities.** As in the previous section, we will expand the topic to include absolute values. Consider the inequality $|x| < 4$. From the definition of absolute value, solutions to this inequality are values that are less than 4 units from 0 on the number line. Any real number in the following interval has an absolute value less than 4.

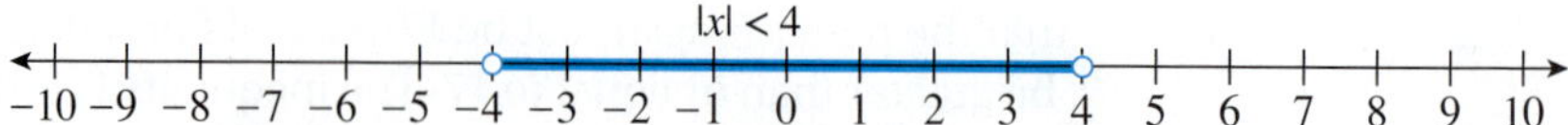

This can be expressed as the compound inequality $-4 < x < 4$. Using this idea, anytime we have an absolute value of an expression that is less than a positive number we will begin by "trapping" the expression between that positive number and its opposite, such as $-4 < x < 4$. We then proceed to solve the resulting compound inequality.

Solving Absolute Value Inequalities of the Form $|X| < a$ or $|X| \leq a$

For any expression X and any positive number a, the solutions to the inequality $|X| < a$ can be found by solving the compound inequality

$$-a < X < a.$$

Similarly, for any expression X and any positive number a, the solutions to the inequality $|X| \leq a$ can be found by solving the compound inequality

$$-a \leq X \leq a.$$

Example 9 Solve $|x - 4| < 2$.

Solution

The first step is to "trap" $x - 4$ between -2 and 2.

$|x - 4| < 2$

$-2 < x - 4 < 2$ "Trap" $x - 4$ between -2 and 2.

$2 < x < 6$ Add 4 to each part of the inequality.

Here is the number line showing our solution.

Quick Check 8
Solve $|x + 3| \leq 1$.

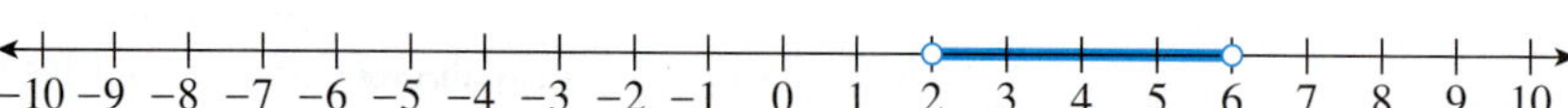

The solution expressed in interval notation is $(2, 6)$.

As with equations involving absolute values, we must first isolate the absolute value before rewriting the inequality as a compound inequality.

EXAMPLE 10 Solve $|3x + 7| - 5 \le 4$.

Solution

$\lvert 3x + 7\rvert - 5 \le 4$	
$\lvert 3x + 7\rvert \le 9$	Add 5 to both sides.
$-9 \le 3x + 7 \le 9$	Rewrite as a compound inequality.
$-16 \le 3x \le 2$	Subtract 7 from each part of the inequality.
$-\frac{16}{3} \le x \le \frac{2}{3}$	Divide each part of the inequality by 3.

Here is the number line showing our solution.

***Quick Check* 9**
Solve $|2x - 1| + 4 < 9$.

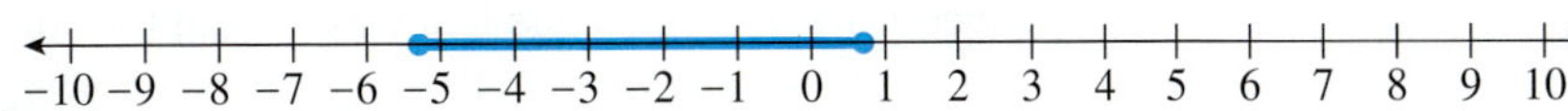

The solution expressed in interval notation is $\left[-\frac{16}{3}, \frac{2}{3}\right]$.

A Word of Caution When solving an absolute value inequality, we must isolate the absolute value before rewriting the inequality as a compound inequality.

If we have an inequality in which an absolute value is less than a negative number, such as $|x| < -2$, or less than 0, such as $|x| < 0$, then this inequality has no solution. Since an absolute value is always 0 or greater, it can never be *less than* a negative number.

EXAMPLE 11 Solve $|4x + 3| + 7 < 5$.

Solution

$\lvert 4x + 3\rvert + 7 < 5$	
$\lvert 4x + 3\rvert < -2$	Subtract 7.

Since an absolute value cannot be less than a negative number, this inequality has no solution: $\varnothing$.

***Quick Check* 10**
Solve $|x + 1| - 3 < -8$.

A Word of Caution An inequality which states that an absolute value is less than a negative number has no solution. Do not try to rewrite the inequality as a compound inequality.

Consider the inequality $|x| > 4$. Inequalities with an absolute value *greater* than a positive number must be approached in a manner different from the way one approaches absolute value inequalities with an absolute value *less* than a positive number. From the definition of absolute value, solutions to this inequality are values that are more than 4 units from 0 on the number line. Solutions to the inequality $|x| > 4$ are in one of two categories. First, any positive number greater than 4 will have an absolute value greater than 4 as well. Second, any negative number less than -4 will also have an absolute value greater than 4. The following number line shows where the solutions to this inequality can be found.

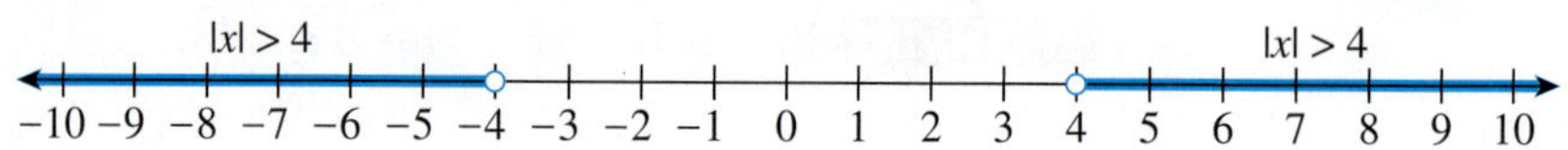

This can be expressed symbolically as $x < -4$ or $x > 4$.

Solving Absolute Value Inequalities of the Form $|X| > a$ or $|X| \geq a$

For any expression X and any positive number a, the solutions to the inequality $|X| > a$ can be found by solving the compound inequality

$$X < -a \quad \text{or} \quad X > a.$$

Similarly, for any expression X and any positive number a, the solutions to the inequality $|X| \geq a$ can be found by solving the compound inequality

$$X \leq -a \quad \text{or} \quad X \geq a.$$

EXAMPLE 12 Solve $|x + 3| > 2$.

Solution

We begin by converting this inequality into the compound inequality $x + 3 < -2$ or $x + 3 > 2$. We then solve each inequality.

$|x + 3| > 2$		
$x + 3 < -2$ or $x + 3 > 2$		Rewrite as a compound inequality.
$x < -5$ or $x > -1$		Subtract 3 to solve each inequality.

Here is the number line showing our solution.

The solution expressed in interval notation is $(-\infty, -5) \cup (-1, \infty)$.

Quick Check 11 Solve $|x - 2| > 4$.

Whether we are solving an absolute value equation or an absolute value inequality, the first step is always to isolate the absolute value. The following table shows the correct step to take when comparing an absolute value to a nonnegative number.

"Equal to"	$|2x + 9| = 5$	$2x + 9 = 5$ or $2x + 9 = -5$
"Less than" or "Less than or equal to"	$|x - 4| < 3$ or $|x - 4| \leq 3$	$-3 < x - 4 < 3$ or $-3 \leq x - 4 \leq 3$
"Greater than" or "Greater than or equal to"	$|3x + 5| > 7$ or $|3x + 5| \geq 7$	$3x + 5 < -7$ *or* $3x + 5 > 7$ or $3x + 5 \leq -7$ *or* $3x + 5 \geq 7$

EXAMPLE 13 Solve $|4x - 9| - 3 \geq 5$.

Solution

We begin by isolating the absolute value.

$$|4x - 9| - 3 \geq 5$$
$$|4x - 9| \geq 8 \quad \text{Add 3 to both sides.}$$

Since this inequality involves an absolute value that is *greater than* or equal to 8, we will rewrite this absolute value inequality as two linear inequalities.

$$4x - 9 \leq -8 \quad \text{or} \quad 4x - 9 \geq 8 \qquad \text{Convert to two linear inequalities.}$$
$$4x \leq 1 \quad \text{or} \quad 4x \geq 17 \qquad \text{Add 9 to both sides.}$$
$$x \leq \frac{1}{4} \quad \text{or} \quad x \geq \frac{17}{4} \qquad \text{Divide both sides by 4.}$$

Here is the number line showing our solution.

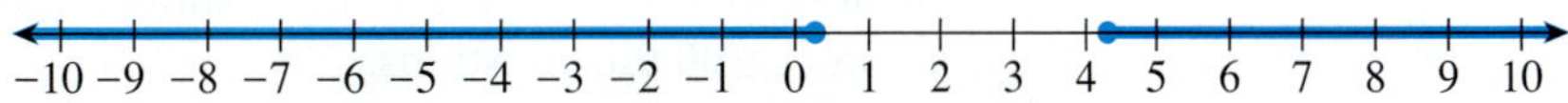

Quick Check 12
Solve $|2x + 7| + 5 > 10$.

The solution expressed in interval notation is $\left(-\infty, \frac{1}{4}\right] \cup \left[\frac{17}{4}, \infty\right)$.

Consider the inequality $|x| > -2$. Since the absolute value of any number must be 0 or greater, then $|x|$ must always be greater than -2 for any real number x. The solution set is the set of all real numbers $\mathbb{R}$.

EXAMPLE 14 Solve $|3x + 5| - 3 > -4$.

Solution

We begin by isolating the absolute value.

$$|3x + 5| - 3 > -4$$
$$|3x + 5| > -1 \quad \text{Add 3 to both sides to isolate the absolute value.}$$

This inequality must always be true, so our solution is the set of all real numbers $\mathbb{R}$. Here is the number line showing our solution.

−10 −9 −8 −7 −6 −5 −4 −3 −2 −1 0 1 2 3 4 5 6 7 8 9 10

The interval notation associated with the set of real numbers is $(-\infty, \infty)$.

Quick Check 13
Solve $|7x - 13| \geq -1$.

If an equation or inequality compares an absolute value to a negative number, the following table summarizes the solutions.

Equation or Inequality	Solution
$\|x + 3\| = -2$	$\varnothing$
$\|2x - 1\| < -4$	$\varnothing$
$\|3x + 4\| > -3$	All real numbers: $\mathbb{R}$

Building Your Study Strategy **Study Groups, 5 Doing Your Part** For any group or team to be effective, it must be made up of members who are committed to giving their fullest effort to success. Here are some pointers for being a productive study group member.

- **Arrive at each session fully prepared.** Be sure that you have completed all required homework assignments. Bring a list of any questions you have.
- **Stay focused during study sessions.** Do not spend your time socializing. It is all right to be friendly and begin each session with small talk, but be sure to shift the focus to math after a few minutes.
- **Be open minded.** During study sessions, you may be told that you are wrong. Keep in mind that the goal is to learn mathematics, not to always be correct initially.
- **Consider the feelings of others.** When a person has made a mistake, or does not know how to solve a particular problem, be supportive and encouraging.
- **Know when to speak and when to listen.** A study group works best when it is a collaborative body. One person should not be asking all the questions or answering all the questions.

EXERCISES 1.5

Vocabulary

1. A(n) __________ is an inequality containing linear expressions.

2. When graphing an inequality, we use a(n) __________ circle to indicate an endpoint that is not included as a solution.

3. In addition to a number line, we can express the solutions of an inequality using __________ notation.

4. When we solve a linear inequality, we must change the direction of the inequality whenever __________.

5. An inequality that is made of two or more individual inequalities is called a(n) __________ inequality.

6. Which inequality can be associated with the statement *The person's height is at least 80 inches?*

   a) $x \leq 80$ **b)** $x < 80$ **c)** $x \geq 80$ **d)** $x > 80$

Graph each inequality on a number line, and write it in interval notation.

7. $x < 4$

8. $x > -9$

9. $x \geq 6$

10. $x \leq 3$

11. $x \leq 2.4$

12. $x > 5.5$

13. $-5 < x < 2$

14. $1 < x \leq 9$

15. $x > 8$ or $x \le -5$

16. $x < -8$ or $x > 7$

Write an inequality associated with the given graph or interval notation.

17. 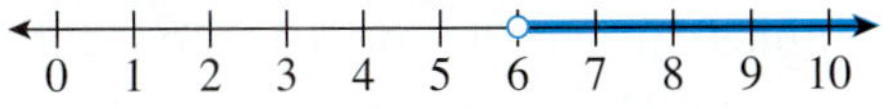

18. 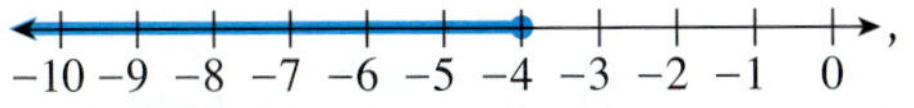

19. 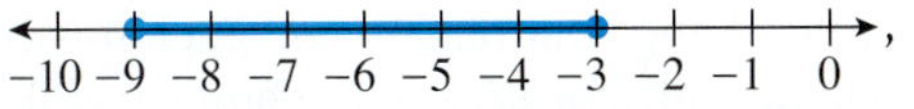

20. 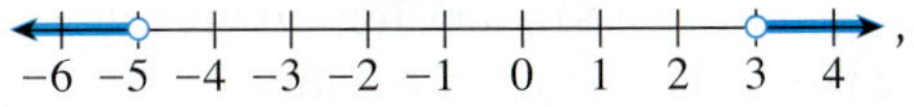

21. $(-\infty, 6) \cup (14, \infty)$

22. $(-6, 2)$

23. $(-\infty, -9)$

24. $[-16, \infty)$

Solve the inequality. Graph your solution on a number line and write it in interval notation.

25. $x - 4 < 3$

26. $x + 7 > 2$

27. $4x \le -20$

28. $-3x \le 24$

29. $3x \ge -10.5$

30. $2x \le 9.4$

31. $3x + 2 \ge -10$

32. $2x + 5 > 9$

33. $\frac{2}{3}x - \frac{5}{6} > -2$

34. $\frac{3}{5}x - \frac{1}{3} < \frac{2}{5}$

35. $5(2x + 1) - 7x \le 4x - 13$

36. $2x - 5(x - 3) < x + 9$

37. $8 - 3(6x - 4) > (2x + 15) - 7(3x + 4)$

38. $2(3x + 2) - (4x - 5) \ge -3(4 - 2x)$

39. $5 \le x - 3 \le 11$

40. $-2 < x + 5 < 4$

41. $-15 \le 5x \le 35$

42. $10 < 4x < 20$

43. $-9 < 2x + 5 < -4$

44. $-8 \le 3x + 7 \le 28$

45. $-6 < \frac{3}{5}x + \frac{2}{5} < -2$

46. $\frac{7}{4} \le \frac{1}{2}x - \frac{3}{8} \le \frac{5}{2}$

47. $x + 6 < 8$ or $x - 3 > 5$

48. $x - 4 \leq -7$ or $x - 8 \geq -1$

49. $3x + 4 \leq -17$ or $2x - 9 \geq 9$

50. $\frac{2}{3}x - 2 < -4$ or $x + \frac{5}{3} > \frac{17}{2}$

51. $5x - 11 < 3x - 3$ or $6x + 7 > 2x + 31$

52. $4x - 3 \leq -21$ or $4x - 3 \geq 21$

53. Curt needs to sell at least 7 computers today in order to receive a bonus. Write an inequality that shows the number of computers sold that will earn Curt the bonus.

54. If Gianluca gives up fewer that 3 goals in today's soccer game, he will set a record for the fewest goals allowed in a season by a goalie. Write an inequality that shows the number of goals that will give Gianluca the record.

55. A certain type of tree is tolerant to 20°F, which means that it can survive at temperatures down to 20°F. Write an inequality that shows the temperatures at which the tree cannot survive.

56. If Chad scores below 75% on the final exam, he will not pass the class. Write an inequality which shows the final exam scores that will allow Chad to pass the class.

57. The manager of a winery tells the owner that he has lost at least 5500 bottles of wine due to a forklift accident. Write an inequality for the number of bottles that have been lost.

58. A community college must have 8700 or more students this semester in order to meet its goals for growth. Write an inequality that shows the number of students for which the college will meet its goals.

59. Claudio is paid to get signatures on petitions. He gets paid \$25 per day, plus 5 cents for each signature he collects. How many signatures does he need to collect today in order to earn at least \$60?

60. Carolyn attends a charity wine-tasting festival. There is a \$20 admission fee, and a \$7.50 charge for each variety tasted. If Carolyn brings \$60 with her, how many varieties can she taste?

61. Angelica is going out of town on business. Her company will reimburse her up to \$75 for a rental car. She rents a car for \$24.95 plus 8 cents per mile. How many miles can Angelica drive without going over the amount that her company will reimburse?

62. An elementary school's booster club is holding a fundraiser by selling wrapping paper. Each roll of wrapping paper sells for \$7, and the booster club gets to keep half of the proceeds. If the booster club wants to raise at least \$10,000, how many rolls of wrapping paper does it need to sell?

63. Jay plans to give at least 15% of his annual earnings to charity. If Jay earned \$87,900 last year, how much should he have given to charity?

64. Deanna is looking for a formal dress to wear to a ball, and she has \$650 to spend. If the store charges 8% sales tax on each dress, what price range should Deanna be considering?

65. Students in a real estate class must have an average score of at least 85% on their exams in order to pass. If Traci has scored 82, 93, 81, and 90 on the first four exams, what scores on the fifth exam would allow her to pass the course? (Assume that the highest possible score is 100.)

66. Students with an average test score below 70 after the third test will be sent an Early Alert warning. Sam scored 57 and 63 on the first two tests. What scores on the third exam will save Sam from receiving an Early Alert warning? (Assume that the highest possible score is 100.)

Solve the inequality. Graph your solution on a number line and write your solution in interval notation.

67. $|x + 2| < 3$

68. $|x - 4| > 1$

69. $|2x + 6| > 4$

70. $|3x - 7| > 5$

71. $|3x| < 15$

72. $|2x - 9| < 7$

73. $|5x + 4| \geq -6$

74. $|4x - 6| < 10$

75. $|8x - 5| + 7 \geq 10$

76. $|x + 4| + 4 > 2$

77. $|4x + 9| - 7 \leq -3$

78. $|2x - 9| - 5 \geq 5$

79. $3|x - 7| \geq 9$

80. $4|3x - 5| \leq 16$

81. $|7x + 14| + 9 < 6$

82. $|3x - 1| + 5 \leq 9$

83. $|5x - 3| - 11 \leq 11$

84. $|6x - 11| - 5 \geq 8$

85. $|8 - x| \leq 5$

86. $|4 - 3x| \geq 10$

87. $2|3x + 2| + 7 > 19$

88. $3|x - 9| - 5 < 4$

89. $4|x + 5| - 11 > 17$

90. $2|2x + 10| + 19 \leq 11$

91. Find an absolute value inequality whose solution is $(-2, 2)$.

92. Find an absolute value inequality whose solution is $[-2, 6]$.

93. Find an absolute value inequality whose solution is $(-\infty, 1) \cup (9, \infty)$.

94. Find an absolute value inequality whose solution is $(-\infty, -4] \cup [-1, \infty)$.

Writing in Mathematics

Answer in complete sentences.

95. Explain why the inequality $|x + 5| < -3$ has no solutions while every real number is a solution to the inequality $|x + 5| > -3$.

96. Explain why the inequality $|x| - 7 < -4$ has solutions but the inequality $|x - 7| < -4$ does not.

97. ***Solutions Manual**** Write a solutions manual page for the following problem:

Solve $-10 < 3x + 2 \leq 29$.

98. ***Newsletter**** Write a newsletter explaining how to solve absolute value inequalities.

***See Appendix B for details and sample answers.**

Chapter 1 Summary

Section 1.1–Topic	Chapter Review Exercises
Determining Inequalities Involving Integers	1–2
Finding Absolute Values	3–4
Performing Arithmetic Operations with Integers	5–10
Simplifying Expressions Involving Exponents	11–14
Simplifying Expressions Using the Order of Operations	15–20
Section 1.2–Topic	**Chapter Review Exercises**
Building Variable Expressions	21–26
Evaluating Variable Expressions	27–30
Simplifying Variable Expressions	31–34
Identifying Terms and Their Coefficients	35–36
Section 1.3–Topic	**Chapter Review Exercises**
Determining if a Value Is a Solution of an Equation	37–38
Solving Linear Equations	39–48
Solving Literal Equations	49–52
Solving Absolute Value Equations	53–56
Section 1.4–Topic	**Chapter Review Exercises**
Solving Applied Problems Involving Linear Equations	57–61
Section 1.5–Topic	**Chapter Review Exercises**
Solving Linear Inequalities	62–65
Solving Compound Linear Inequalities	66–69
Solving Applied Problems Involving Inequalities	70–71
Solving Absolute Value Inequalities	72–75

Summary of Chapter 1 Study Strategies

- Creating a study group is one of the best ways to learn mathematics and improve your performance on quizzes and exams.
- Sometimes a concept will be easier for you to understand if it is explained to you by a peer. One of your group members may have some insight into a particular topic, and that person's explanation may be just the trick to turn on the "light bulb" in your head.
- When you explain a certain topic to a fellow student, or show a student how to solve a particular problem, you increase your chances of retaining this knowledge. Because you are forced to choose your words in such a way that the other student will completely understand, you have demonstrated that you truly understand.
- You will also find it helpful to have a support group: students who can lean on each other when times are bad. By supporting each other, you are increasing the chances that you will all learn and be successful in your class.

Write the appropriate symbol, either < or >, between the following integers. **[1.1]**

1. -8 ___ -13
2. -15 ___ 7

Find the following absolute values. **[1.1]**

3. $|5|$
4. $|-22|$

Simplify. **[1.1]**

5. $9 - 24$
6. $-13 - (-39)$
7. $6 - 17 - 42$
8. $4 - (-14) - 32$
9. $7(-8)$
10. $-144 \div (-6)$
11. 2^8
12. $\left(\frac{3}{4}\right)^3$
13. -5^2
14. $4^3 \cdot 3^2$
15. $9 + 3 \cdot 6$
16. $60 - 24 \div 6$
17. $11 - 9 \cdot 15 + 18$
18. $-4(7 - 3 \cdot 5) - 22$
19. $20 - 32 \div 4^2$
20. $\frac{2}{5} + \frac{3}{5} \cdot \frac{2}{21}$

Build a variable expression for the following phrases. **[1.2]**

21. A number decreased by 9
22. The sum of a number and 25
23. Seven less than three times a number
24. Fifteen more than twice a number
25. A coffee house charges $2.25 for a cup of coffee. If we let c represent the number of cups of coffee that the coffee house sells on a particular day, build a variable expression for the revenue from coffee sales. [1.2]
26. A rental company rents pickup trucks for $29.95, plus $0.50 per mile. If we let m represent the number of miles, build a variable expression for the cost of renting a pickup truck from this rental company. [1.2]

Worked-out solutions to Review Exercises marked with ● can be found on page AN-3.

Evaluate the following algebraic expressions under the given conditions. **[1.2]**

27. $2x + 9$ for $x = -7$
28. $13 - 4x$ for $x = 6$
29. $x^2 + 5x + 24$ for $x = -3$
30. $b^2 - 4ac$ for $a = 2$, $b = -7$, and $c = -4$

Simplify. **[1.2]**

31. $3(2x - 15)$
32. $7x - 19 - 3x - 16$
33. $4x - 2(3x - 11)$
34. $-3(4x + 9) + 2(7x - 15)$

For the following expressions,
a) determine the number of terms;
b) write down each term; and
c) write down the coefficient for each. **[1.2]**

35. $x^3 - 8x^2 + 7x - 15$
36. $3x^2 + x - 20$

Check to determine if the given value is a solution to the equation. **[1.3]**

37. $x = 3$, $2x - 7 = 13$
38. $x = -8$, $2(3x - 2) - (x + 7) = 3x - 27$

Solve. **[1.3]**

39. $x + 4 = 17$
40. $3x + 26 = -16$
41. $-23 = 2n - 39$
42. $51 - 6a = -9$
43. $\frac{x}{15} - \frac{3}{10} = \frac{1}{6}$
44. $\frac{2}{3}x + \frac{3}{4} = \frac{17}{6}$
45. $2x + 17 = 5x + 41$
46. $4x - 9 = -8x - 23$
47. $3n - 7 + n + 12 = 7n - 40$
48. $2(3m - 4) - (m - 13) = 2m - 13$

***Solve the following literal equations for the specified variable.* [1.3]**

49. $5x + y = 12$ for y

50. $4x + 3y = 24$ for y

51. $C = \pi \cdot d$ for d

52. $P = a + b + 2c$ for c

***Solve.* [1.3]**

53. $|x + 5| = 9$

54. $|2x - 3| = 17$

55. $|x - 4| - 7 = 6$

56. $|4x + 5| + 3 = 12$

57. One number is 35 more than another. If the sum of the two numbers is 117, find the numbers. [1.4]

58. The length of a rectangle is 7 feet longer than twice its width, and the perimeter is 92 feet. Find the length and the width of the rectangle. [1.4]

59. The sum of three consecutive odd integers is 171. Find the integers. [1.4]

60. If Chip drives at a speed of 66 miles per hour, how long will it take him to drive 165 miles? [1.4]

61. Alycia's piggy bank has nickels and dimes in it. The number of dimes is 2 more than five times the number of nickels. If the total value of the coins is $7.35, how many dimes are in the piggy bank? [1.4]

***Solve. Graph your solution on a number line, and express it in interval notation.* [1.5]**

62. $x + 4 < 9$

63. $x - 5 > -7$

64. $2x + 7 \geq 5$

65. $-3x + 7 \geq -17$

66. $4x < -16$ or $x + 6 > 5$

67. $x + 2 \leq -4$ or $5x - 19 \geq 11$

68. $-37 \leq 4x - 5 \leq 23$

69. $-9 < 2x + 7 < 9$

70. A baseball player is signing autographs at a convention for $26 each. How many autographs must he sign in order to earn at least $1000? [1.5]

71. If Dylan's test average is at least 80 but lower than 90, then his final grade will be a B. His scores on the first four tests were 88, 77, 78, and 90. What scores on the fifth exam will give him a B for the class? (Assume that the highest possible score is 100.) [1.5]

***Solve the inequality. Graph your solution on a number line and write your solution in interval notation.* [1.5]**

72. $|x + 3| < 6$

73. $|x - 5| > 4$

74. $|x - 6| + 7 \leq 8$

75. $|x - 4| - 6 \geq -3$

Chapter 1 Test

For Extra Help  Test solutions are found on the enclosed CD.

Write the appropriate symbol, either $<$ or $>$, between the following integers.

1. -9 ___ -7
2. Find the absolute value of $|-23|$

Simplify.

3. $9 - 24$
4. $6(-9)$
5. $20 - 9 \cdot 8$
6. $3 + 5 \cdot 14 - 8^2$
7. $10 - 7(4 - 2 \cdot 9 - 8)$
8. Build a variable expression for the phrase *nineteen less than twice a number.*
9. A tree trimmer charges \$100 to visit a house, plus \$80 per hour. If we let h represent the number of hours spent working at a particular house, build a variable expression for the tree trimmer's charge.
10. Evaluate $3x - 28$ for $x = -15$

Simplify.

11. $4(5x + 8) - 3x$
12. $2x - 6(3x - 16)$

Solve.

13. $4x - 19 = -43$
14. $\frac{x}{8} - \frac{5}{6} = \frac{7}{12}$
15. $6x + 20 = -3x + 2$
16. $3(x - 7) + 2(2x - 9) = 5x - 53$
17. Solve the literal equation $4x + 3y = 36$ for y.
18. Solve $|2x - 5| - 7 = 4$.
19. One number is 41 less than another. If the sum of the two numbers is 351, find the two numbers.
20. The length of a rectangle is 3 feet less than twice its width, and the perimeter is 36 feet. Find the length and the width of the rectangle.
21. The sum of three consecutive integers is 213. Find the three integers.

Solve. Graph your solution on a number line, and express it in interval notation.

22. $3x + 8 > -10$
23. $-13 \le 2x + 5 \le 19$
24. $|x + 7| < 2$
25. $|x + 2| + 5 \ge 10$

Mathematicians in History

Srinivasa Ramanujan

Srinivasa Ramanujan was a self-taught Indian mathematician viewed by many to be one of the greatest mathematical geniuses in history. As a student, he became so totally immersed in his work with mathematics that he ignored his other subjects and failed his college exams. Ramanujan's life is chronicled in the biography *The Man Who Knew Infinity: A Life of the Genius Ramanujan.*

Write a one-page summary (*or* make a poster) of the life of Srinivasa Ramanujan and his accomplishments.

Interesting issues:

- Where and when was Srinivasa Ramanujan born?
- At age 16, Ramanujan borrowed a mathematics book that strongly influenced his life as a mathematician. What was the title of the book?
- Ramanujan got married on July 14, 1909. The marriage was arranged by his mother. How old was his bride at the time?
- What jobs did Ramanujan hold in India?
- Which renowned mathematician invited Ramanujan to England in 1914?
- Ramanujan's health in England was poor. What was the cause of his poor health?
- What is the significance of the taxi cab number 1729?
- What were the circumstances that led to Ramanujan's death, and what was his age when he died?

Stretch Your Thinking Chapter 1

The workers at Smoothie Superb make a delicious orange juice and strawberry-banana smoothie called the "Berry Ana." They charge $3.78 for a regular size and give the customer one free nutritional boost. For each additional nutritional boost, they charge 43 cents. Let b represent the number of nutritional boosts in a drink.

a) Build a variable expression for the total cost of a "Berry Ana" smoothie with b nutritional boosts.

b) What would be the total cost for a customer to order a "Berry Ana" smoothie with two nutritional boosts?

During December, Smoothie Superb makes a specialty drink called the "Cold Terminator." The "Cold Terminator" is the exact same smoothie as the "Berry Ana," but it contains four nutritional boosts. A regular size "Cold Terminator" sells for $4.99.

c) If a customer wanted to order a "Berry Ana" smoothie with four nutritional boosts, would it be more cost effective to buy a "Cold Terminator"? Explain your answer.

d) Set up an inequality statement (using either $<$ or $<$) comparing the cost of a "Cold Terminator" with a "Berry Ana" with four boosts.

CHAPTER 2

Two

Graphing Linear Equations

In this chapter we will examine linear equations in two variables. Equations involving two variables relate two unknown quantities to each other, such as a person's height and weight, or the number of items sold by a company and the profit that the company made. The primary focus is on graphing linear equations, which is the technique we use to display the solutions to an equation in two variables.

Study Strategy **Using Your Textbook Effectively** *In this chapter we will focus on how to get the most out of your textbook. Students who treat their books solely as a source of homework exercises are turning their backs on one of their best resources.*

Throughout this chapter we will revisit this study strategy and help you incorporate it into your study habits.

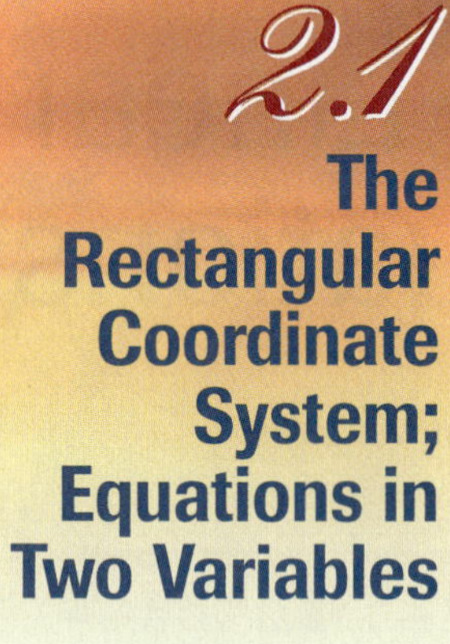

2.1 The Rectangular Coordinate System; Equations in Two Variables

Objectives

1. **Determine if an ordered pair is a solution to an equation in two variables.**
2. **Plot ordered pairs on a rectangular coordinate plane.**
3. **Graph linear equations in two variables.**
4. **Find the *x*- and *y*-intercepts of a line from its equation.**
5. **Graph linear equations using their intercepts.**
6. **Graph horizontal and vertical lines.**
7. **Interpret the graph of an applied linear equation.**

Equations in Two Variables and Their Solutions; Ordered Pairs

Objective 1 Determine if an ordered pair is a solution to an equation in two variables. In this section we will begin to examine equations in two variables, which relate two different quantities to each other. Specifically, we will be working with **linear equations in two variables.**

Standard Form of a Linear Equation

A linear equation in two variables is an equation that can be written in the form $Ax + By = C$, where A, B, and C are real numbers and A and B are not both 0. This is called the **standard form** of a linear equation in two variables.

A solution to an equation in two variables is a pair of values that, when substituted for the two variables, produces a true statement. Consider the linear equation $3x + 4y = 10$. One solution to this equation is $x = 2$ and $y = 1$. We see that this is a solution by substituting 2 for x and 1 for y into the equation.

$$3x + 4y = 10$$
$$3(2) + 4(1) = 10 \quad \text{Substitute 2 for } x \text{ and 1 for } y.$$
$$6 + 4 = 10 \quad \text{Multiply.}$$
$$10 = 10 \quad \text{Add.}$$

Since this produces a true statement, our pair of values together forms a solution. Solutions to this type of equation are written as **ordered pairs** (x, y). In our example the ordered pair $(2, 1)$ is a solution. We call it an ordered pair because the two values are listed in the specific order of x first and y second. The values are often referred to as **coordinates.**

EXAMPLE 1 Is the ordered pair $(-2, 1)$ a solution to the linear equation $y = 4x + 9$?

Solution

To determine whether this ordered pair is a solution we need to substitute -2 for x and 1 for y.

$$y = 4x + 9$$
$$(1) = 4(-2) + 9 \quad \text{Substitute } -2 \text{ for } x \text{ and 1 for } y.$$

$$1 = -8 + 9 \quad \text{Multiply.}$$
$$1 = 1 \quad \text{Simplify.}$$

Since this is a true statement, the ordered pair $(-2, 1)$ is a solution to the equation $y = 4x + 9$.

EXAMPLE 2 Is $(3, -1)$ a solution to the equation $2x - 3y = 3$?

Solution

Again, we check to see if this ordered pair is a solution by substituting the given values for x and y. We must be careful to substitute 3 for x and -1 for y, not vice versa.

$$2x - 3y = 3$$
$$2(3) - 3(-1) = 3 \quad \text{Substitute 3 for } x \text{ and } -1 \text{ for } y.$$
$$6 - (-3) = 3 \quad \text{Multiply.}$$
$$9 = 3 \quad \text{Simplify.}$$

Since this is a false statement, $(3, -1)$ is not a solution to the equation $2x - 3y = 3$.

Quick Check 1
Is $(-4, 0)$ a solution to the equation $3x - 4y = -12$?

The Rectangular Coordinate Plane

Objective 2 Plot ordered pairs on a rectangular coordinate plane. Ordered pairs can be displayed graphically using the **rectangular coordinate plane**. The rectangular coordinate plane is also known as the **Cartesian plane**, named after the famous French philosopher and mathematician René Descartes, who invented it.

The rectangular coordinate plane is made up of two number lines, known as **axes**. The axes are drawn at right angles to each other, intersecting at 0 on each axis. Here is how it looks.

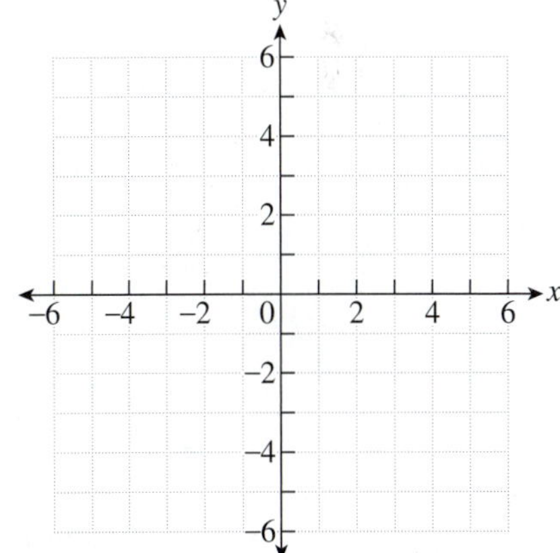

x-axis, y-axis, Origin

The **horizontal axis**, or ***x*-axis**, is used to show the first value of an ordered pair. The **vertical axis**, or ***y*-axis**, is used to show the second value of an ordered pair. The point where the two axes cross is called the **origin**, and represents the ordered pair $(0, 0)$.

Positive values of x are found to the right of the origin and negative values of x are found to the left of the origin. Positive values of y are found above the origin and negative values of y are found below the origin.

We can use the rectangular coordinate plane to represent ordered pairs that are solutions to an equation. To plot an ordered pair (x, y) on a rectangular coordinate plane we begin at the origin. The x-coordinate tells us how far to the left or right of the origin to move. The y-coordinate then tells us how far up or down to move from there.

Suppose we wanted to plot the ordered pair $(-3, 2)$ on a rectangular coordinate plane. The x-coordinate is -3 and that tells us to move 3 units to the left of the origin, as shown here. The y-coordinate is 2, so we move up 2 units to find the location of our ordered pair, and we place a point at this location. We will often refer to an ordered pair as a **point** when we plot it on a graph.

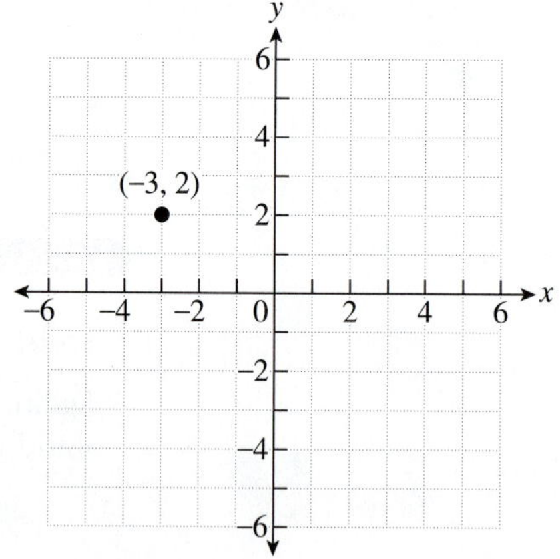

One final note about the rectangular coordinate plane is that the two axes divide the plane into four sections called **quadrants.** The four quadrants are labeled I, II, III, and IV. A point that lies on one of the axes is not considered to be in any of the quadrants.

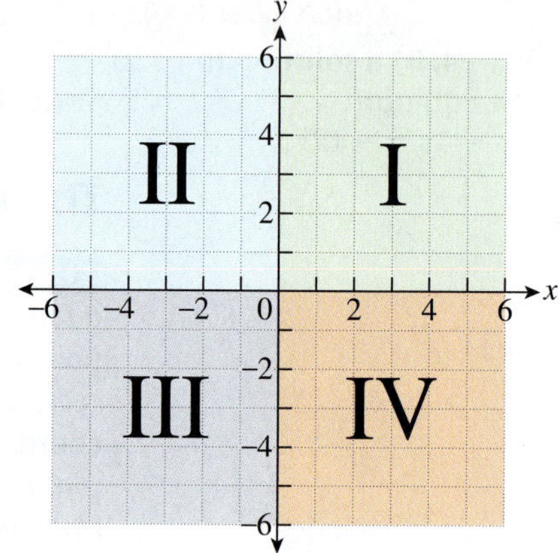

EXAMPLE 3 Plot the ordered pairs $(3, 5)$, $(2, -4)$, $(-5, -1)$, and $(-3, 0)$ on a rectangular coordinate plane.

Solution

The four ordered pairs are plotted on the following graph.

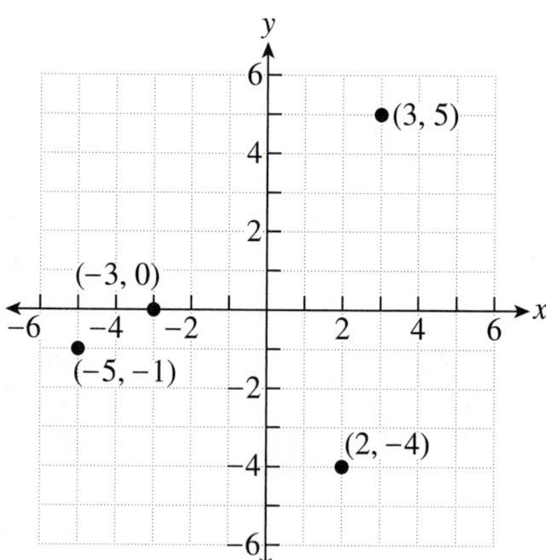

Quick Check **2** **Plot the ordered pairs $(6, 1)$, $(4, -1)$, $(-7, 2)$, and $(0, -3)$ on a rectangular coordinate plane.**

Choosing the appropriate scale for a graph is an important skill. To plot the ordered pair $(\frac{7}{2}, -4)$ on a rectangular coordinate plane, we could use a scale of $\frac{1}{2}$ on the axes. A rectangular plane with axes showing values ranging from -10 to 10 is not large enough to plot the ordered pair $(-45, 25)$. A good choice would be to label the axes showing values that range from -50 to 50. Since this is a wide range of values, it makes sense to increase the scale. Since each coordinate is a multiple of 5, we will use a scale of 5 units.

It is important to be able to read the coordinates of an ordered pair from a graph.

EXAMPLE 4 Find the coordinates of points A through G on the given graph.

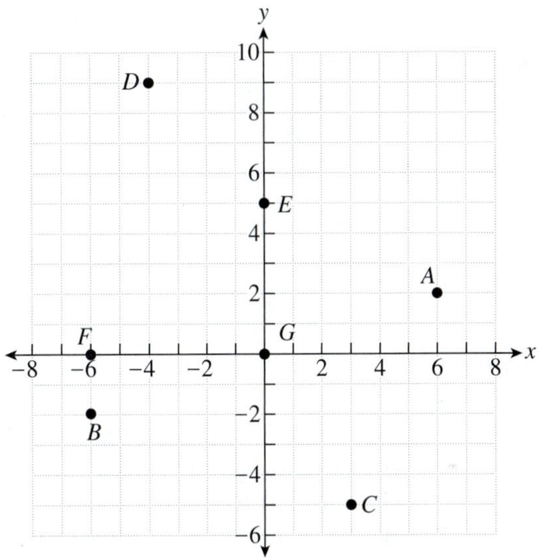

Solution

When determining coordinates for a point on a rectangular coordinate plane, we begin by finding the x-coordinate. We do this by determining how far to the left or right of the origin the point is. If the point is to the right, then the x-coordinate is positive. If the point is to the left, then the x-coordinate is negative. If the point is neither to the right nor to the left of the origin, then the x-coordinate is 0.

We next determine the y-coordinate of the point by measuring how far above or below the origin the point is. If the point is above the origin, then the y-coordinate is positive. If the point is below the origin, then the y-coordinate is negative. If the point is neither above nor below the origin, then the y-coordinate is 0.

Here is a table containing the coordinates of each point.

Point	*A*	*B*	*C*	*D*	*E*	*F*	*G*
Coordinates	(6, 2)	(−6, −2)	(3, −5)	(−4, 9)	(0, 5)	(−6, 0)	(0, 0)

Quick Check 3

Find the coordinates of points *A* through *E* on the given graph.

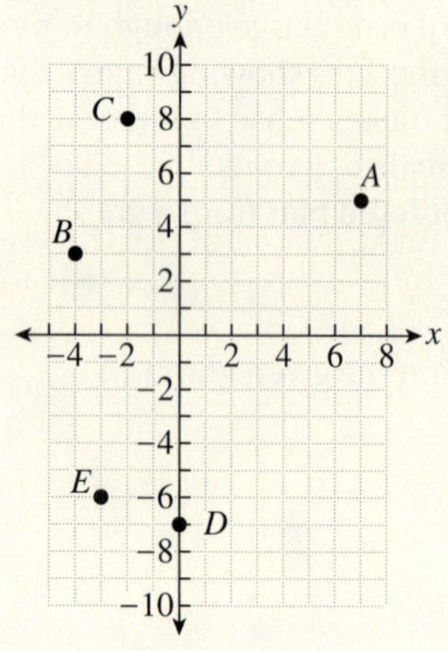

Graphing Linear Equations in Two Variables

Objective 3 **Graph linear equations in two variables.** The solutions to any linear equation in two variables follow a pattern. We will now begin to explore this pattern. Consider the linear equation $x + y = 7$. The reader can verify that the ordered pairs $(0, 7)$, $(3, 4)$, and $(5, 2)$ are all solutions to this equation.

When we plot these three points on a rectangular coordinate plane, it looks like this. Notice that the three points appear to fall on a straight line.

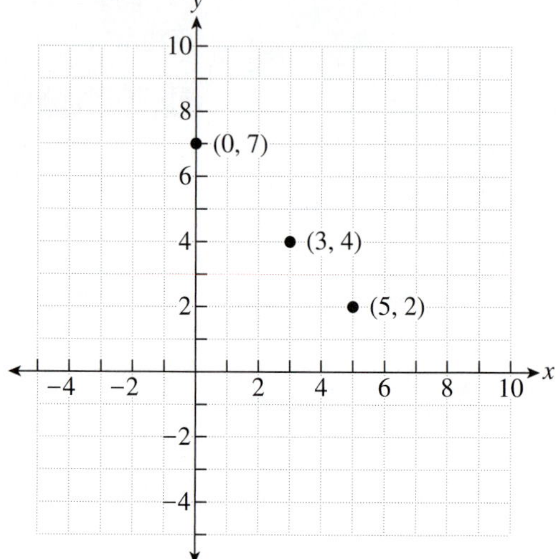

We can draw a straight line that passes through each of the three points. In addition to the three points on the graph, every other point which lies on the line is a solution to the equation. Every ordered pair that is a solution must lie on this line. The graph of the line is a way to present the solutions of the equation.

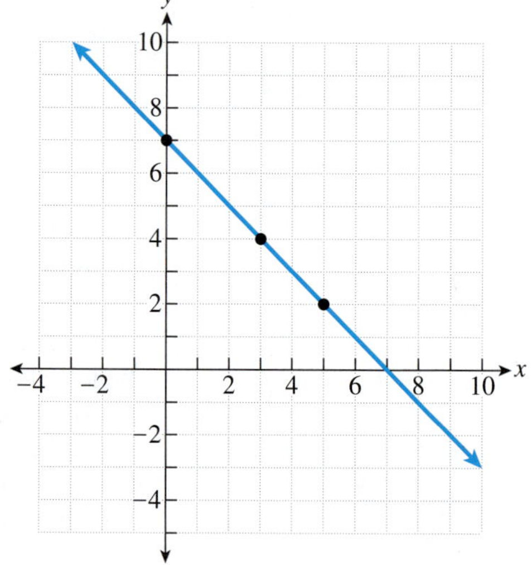

We can sketch a graph of the solutions to a linear equation by first plotting ordered pairs that are solutions and then drawing the straight line which goes through these points. The line we sketch continues indefinitely in both directions, so we draw arrows on both ends of the line. Although only two points are needed to determine a line, we will find three points. The third point is used as a check to make sure that the first two points are accurate. We will arbitrarily pick three values for one of the variables, and find the other coordinates as in the previous example.

EXAMPLE 5 Graph the linear equation $y = 3x - 4$.

Solution

When the equation has y isolated on one side of the equal sign, it is a good idea to select values for x and find the corresponding y-values. We find three points on the line by letting $x = 0$, $x = 1$, and $x = 2$. We did not need to use 0, 1, and 2 for x; they were chosen arbitrarily. If you start with other values, your ordered pairs will still be on the same line.

$x = 0$	$x = 1$	$x = 2$
$y = 3(0) - 4$ $y = 0 - 4$ $y = -4$	$y = 3(1) - 4$ $y = 3 - 4$ $y = -1$	$y = 3(2) - 4$ $y = 6 - 4$ $y = 2$

Our three ordered pairs are $(0, -4)$, $(1, -1)$, and $(2, 2)$. We plot these three points on a rectangular coordinate plane and draw a line through them.

Quick Check 4
Graph the linear equation $y = 2x + 1$.

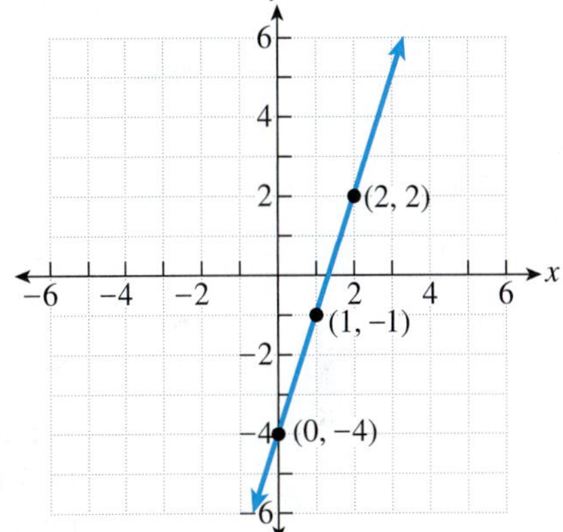

x-Intercepts and y-Intercepts

Objective 4 **Find the x- and y-intercepts of a line from its equation.** A line can be graphed provided that we know the coordinates of two ordered pairs that are on the line. We will now develop a systematic method for graphing linear equations in two variables, focusing on the **intercepts** of the line.

A point at which a graph crosses the x-axis is called an ***x*-intercept** and a point at which a graph crosses the y-axis is called a ***y*-intercept**.

Consider the following graph.

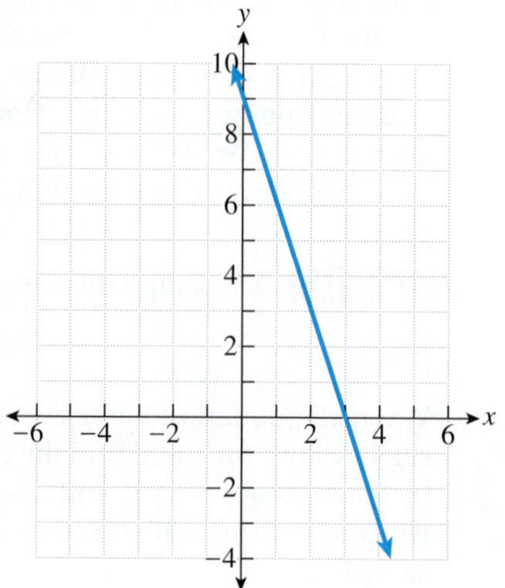

The x-intercept of this line is the point at which the graph crosses the x-axis. We see that this graph crosses the x-axis at the point $(3, 0)$, so the x-intercept is the point $(3, 0)$. This graph crosses the y-axis at the point $(0, 9)$, so its y-intercept is $(0, 9)$.

Some lines do not have both an x- and y-intercept. This horizontal line does not cross the x-axis, so it does not have an x-intercept. Its y-intercept is at the point $(0, -6)$.

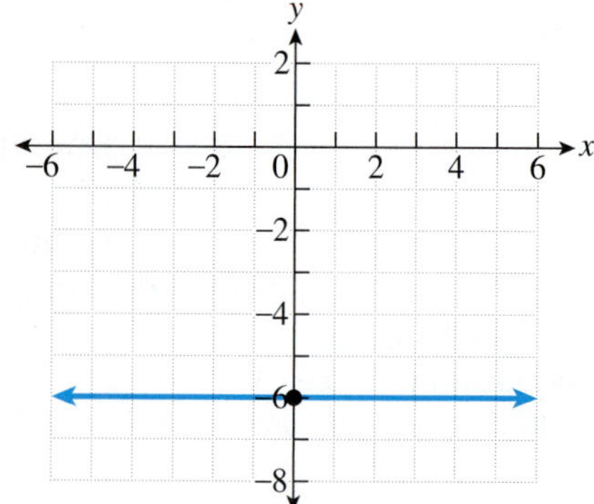

The x-intercept of this vertical line is the point $(4, 0)$. This graph does not cross the y-axis, so it does not have a y-intercept.

Any point that lies on the x-axis has a y-coordinate of 0. Similarly, any point that lies on the y-axis has an x-coordinate of 0. We can use these facts to find the intercepts of an equation's graph algebraically.

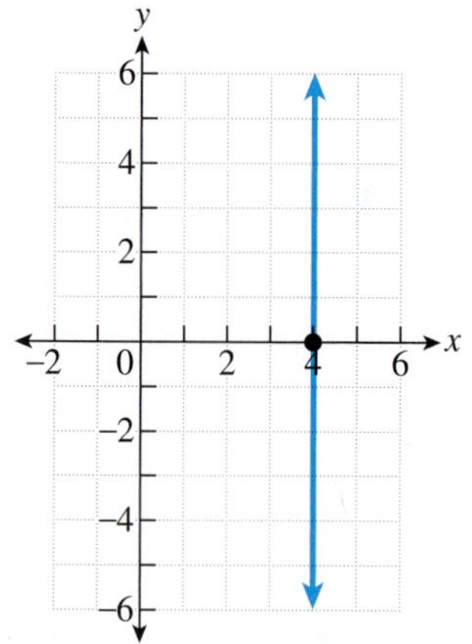

Finding x- and y-Intercepts

- Find the x-intercept by substituting 0 for y and solving for x.
- Find the y-intercept by substituting 0 for x and solving for y.

EXAMPLE 6 Find the x- and y-intercepts for the equation $4x - 3y = 12$.

Solution

Let's begin with the x-intercept by substituting 0 for y and solving for x.

$$\begin{aligned} 4x - 3y &= 12 \\ 4x - 3(0) &= 12 && \text{Substitute 0 for } y. \\ 4x &= 12 && \text{Simplify the left side of the equation.} \\ x &= 3 && \text{Divide both sides by 4.} \end{aligned}$$

The x-intercept is $(3, 0)$. To find the y-intercept we substitute 0 for x and solve for y.

$$\begin{aligned} 4x - 3y &= 12 \\ 4(0) - 3y &= 12 && \text{Substitute 0 for } x. \\ -3y &= 12 && \text{Simplify the left side of the equation.} \\ y &= -4 && \text{Divide both sides by } -3. \end{aligned}$$

The y-intercept is $(0, -4)$.

Quick Check 5
Find the x- and y-intercepts for the equation $x + 5y = 10$.

EXAMPLE 7 Find the x- and y-intercepts for the equation $y = 2x - 9$.

Solution

We will begin by finding the x-intercept. To do this we substitute 0 for y and solve for x.

$$\begin{aligned} y &= 2x - 9 \\ 0 &= 2x - 9 && \text{Substitute 0 for } y. \\ 9 &= 2x && \text{Add 9 to both sides.} \\ \tfrac{9}{2} &= x && \text{Divide both sides by 2.} \end{aligned}$$

The x-intercept is $(\frac{9}{2}, 0)$. To find the y-intercept we substitute 0 for x and solve for y.

$$\begin{aligned} y &= 2x - 9 \\ y &= 2(0) - 9 && \text{Substitute 0 for } x. \\ y &= -9 && \text{Simplify.} \end{aligned}$$

The y-intercept is at $(0, -9)$.

Quick Check 6
Find the x- and y-intercepts for the equation $y = 4x + 5$.

Graphing a Linear Equation Using Its Intercepts

Objective 5 **Graph linear equations using their intercepts.** To graph a linear equation, we will begin by finding any intercepts. This will often give us two points for our graph. Although only two points are necessary to graph a straight line, we will plot at least three points to be sure that our first two points are correct. In other words, we will use a third point as a "check" for the first two points. We find this third point by selecting a value for x, substituting it into the equation, and solving for y.

EXAMPLE 8 Graph $2x + y = 8$. Label any intercepts.

Solution

We begin by finding the x- and y-intercepts.

x-intercept

$$2x + (0) = 8 \quad \text{Substitute 0 for } y.$$
$$2x = 8 \quad \text{Simplify.}$$
$$x = 4 \quad \text{Divide both sides by 2.}$$

The x-intercept is $(4, 0)$.

y-intercept

$$2(0) + y = 8 \quad \text{Substitute 0 for } x.$$
$$y = 8 \quad \text{Simplify.}$$

The y-intercept is $(0, 8)$.

Before looking for a third point, let's plot the two intercepts on a graph. This can help us select a value to substitute for x to find a third point. It is a good idea to select a value for x that is somewhat near the x-coordinates of the other two points. For example, $x = 1$ would be a good choice for this example. If we choose a value of x that is too far away from the x-coordinates of the other two points, then we may have to change the scale of the axes or extend them to show this third point.

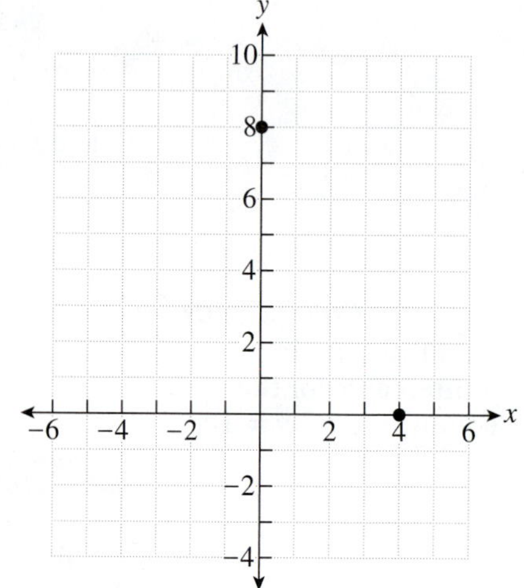

$$2(1) + y = 8 \quad \text{Substitute 1 for } x.$$
$$2 + y = 8 \quad \text{Simplify the left side of the equation.}$$
$$y = 6 \quad \text{Subtract 2.}$$

The point $(1, 6)$ is our third point. Keep in mind that the third point can vary depending on our choice for x, but the three points should still be on the same line. After plotting the third point, we finish by drawing a straight line that passes through all three points.

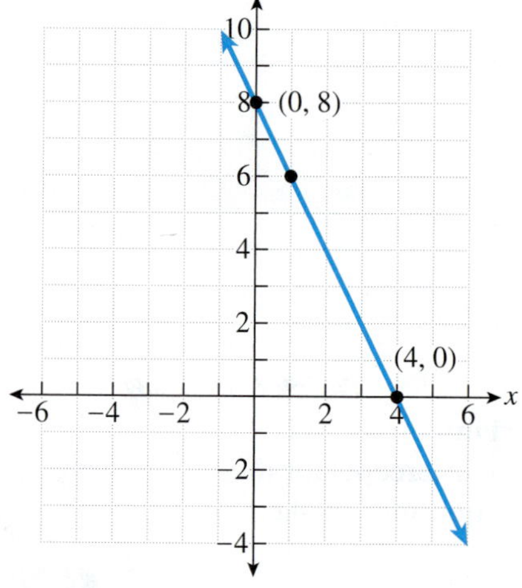

Quick Check 7
Graph $3x - 2y = -6$. Label any intercepts.

EXAMPLE 9 Graph $y = 3x - 9$. Label any intercepts.

Solution

We will begin by finding the x- and y-intercepts.

x-intercept $(y = 0)$	y-intercept $(x = 0)$
$0 = 3x - 9$ $9 = 3x$ $3 = x$ $(3, 0)$	$y = 3(0) - 9$ $y = -9$ $(0, -9)$

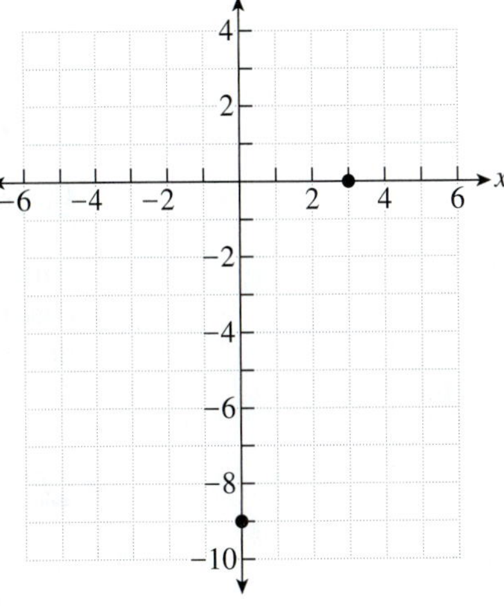

Here is a look at our two intercepts on a graph. Choosing $x = 1$ to find our third point is a good choice; it is close to the x-coordinates of the two intercepts and it will be easy to substitute 1 for x in the equation.

$y = 3(1) - 9$ Substitute 1 for x in the equation.

$y = -6$ Simplify.

Our third point is $(1, -6)$. Here is the graph.

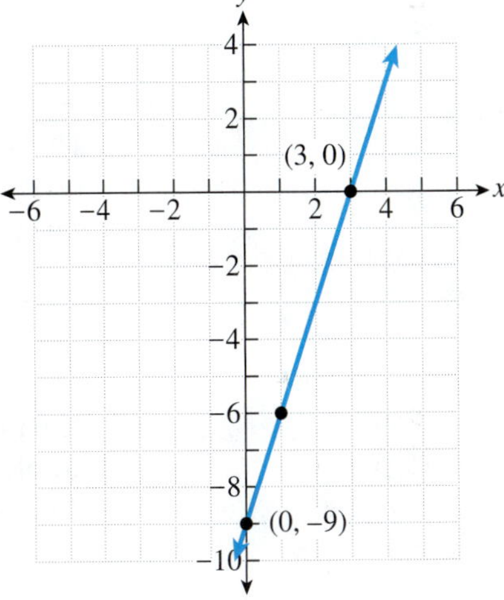

Quick Check 8
Graph $y = 2x - 10$.
Label any intercepts.

Using Your Calculator We can use the TI-84 to graph a line. To enter the equation of the line that we graphed in the previous example, press the [Y=] key. Enter $3x - 9$ next to Y_1, using the key labeled [X,T,Θ,n] to type the variable x. Now press the key labeled [GRAPH] to graph the line.

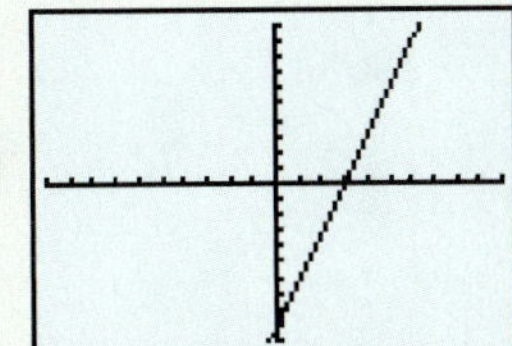

Graphing a Line Using Its Intercepts

- Find the x-intercept by substituting 0 for y and solving for x.
- Find the y-intercept by substituting 0 for x and solving for y.
- Plot these two intercepts on a coordinate system.
- Select a value of x that is appropriate based on the two points already plotted, substitute it into the equation, and solve for y to find the coordinates of a third point on the line.
- Graph the line that passes through these three points.

Graphing Linear Equations That Pass Through the Origin

Some lines do not have two intercepts. For example, a line that passes through the origin has its x- and y-intercepts at the same point. In this case, we will need to find two additional points.

EXAMPLE 10 Graph $x - 3y = 0$. Label any intercepts.

Solution

Let's start by finding the x-intercept.

$$x - 3(0) = 0 \quad \text{Substitute 0 for } y.$$
$$x = 0 \quad \text{Simplify.}$$

The x-intercept is at the origin $(0, 0)$. This point is also the y-intercept, which you can verify by substituting 0 for x and solving for y. We will need to find two more points before we graph the line. Suppose that we chose $x = 1$.

$$1 - 3y = 0 \quad \text{Substitute 1 for } x.$$
$$1 = 3y \quad \text{Add } 3y.$$
$$\tfrac{1}{3} = y \quad \text{Divide both sides by 3.}$$

The point $(1, \frac{1}{3})$ is on the line. However, it may not be easy for us to put this point on the graph because of its fractional y-coordinate. We can try to avoid this problem by solving the equation for y before choosing a value for x.

$$x - 3y = 0$$
$$x = 3y \quad \text{Add } 3y \text{ to both sides.}$$
$$\tfrac{1}{3}x = y \quad \text{or} \quad y = \tfrac{1}{3}x. \quad \text{Divide both sides by 3 and simplify.}$$

If we choose a value for x that is a multiple of 3, then we will have a y-coordinate that is an integer. We will use $x = 3$ and $x = -3$, although there are many other choices that would work.

$x = 3$	$x = -3$
$y = \frac{1}{3}(3)$	$y = \frac{1}{3}(-3)$
$y = 1$	$y = -1$
$(3, 1)$	$(-3, -1)$

Here is the graph.

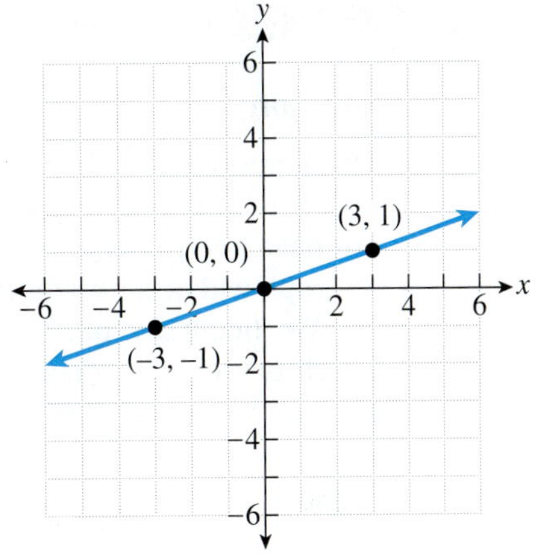

Quick Check **9** **Graph $y = -2x$. Label any intercepts.**

Graphing a Line That Passes through the Origin

- Determine that the x- and y-intercepts are both at the origin.
- Find a second point on the line by selecting a value of x and substituting it into the equation. Solve the equation for y.
- Find a third point on the line to serve as a check point by selecting another value of x and substituting it into the equation. Solve the equation for y.
- Graph the line that passes through these three points.

Horizontal Lines

Objective 6 **Graph horizontal and vertical lines.** Can we draw a line that does not cross the x-axis? Yes, a horizontal line can be drawn that does not cross the x-axis. Here's an example.

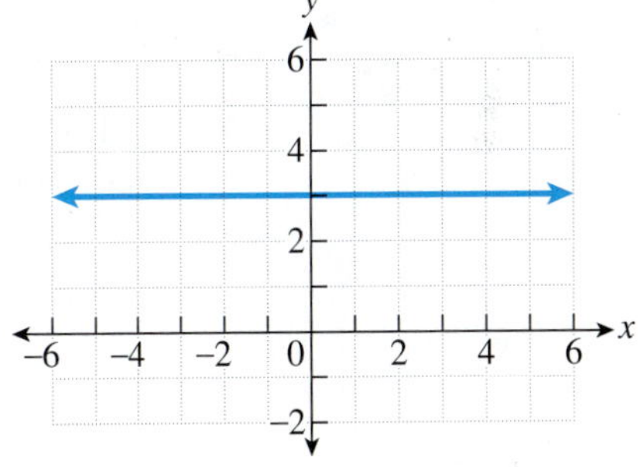

What is the equation of the line in this graph? Notice that each point has the same y-coordinate (3). The equation for the line is $y = 3$. The equation for a horizontal line is always of the form $y = b$, where b is a real number. To graph the line $y = b$, plot the y-intercept $(0, b)$ on the y-axis and draw a horizontal line through this point.

EXAMPLE 11 Graph $y = -5$. Label any intercepts.

Solution

This graph will be a horizontal line, as its equation is of the form $y = b$. If an equation does not have a term containing the variable x, then its graph will be a horizontal line.

We begin by plotting the point $(0, -5)$, which is the y-intercept. We then draw a horizontal line through this point.

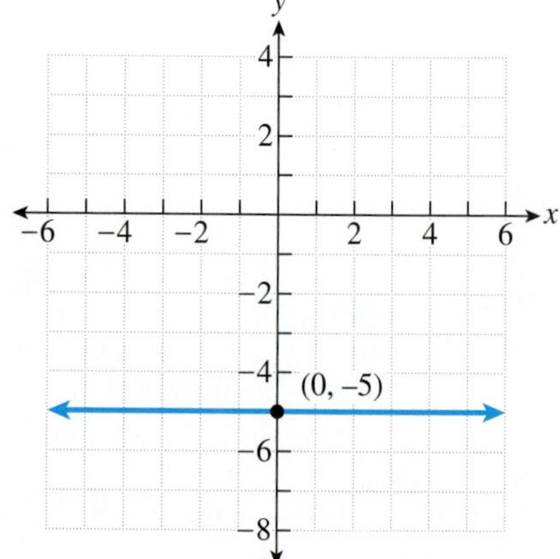

Quick Check 10
Graph $y = 4$. Label any intercepts.

Vertical Lines

Vertical lines have equations that are of the form $x = a$, where a is a real number. Just as the equation of a horizontal line does not have any terms containing the variable x, the equation of a vertical line does not have any terms containing the variable y. We begin to graph an equation of this form by plotting its x-intercept at $(a, 0)$. We then draw a vertical line through this point.

EXAMPLE 12 Graph $x = -2$. Label any intercepts.

Solution

The x-intercept for this line is $(-2, 0)$. We plot this point and then draw a vertical line through it.

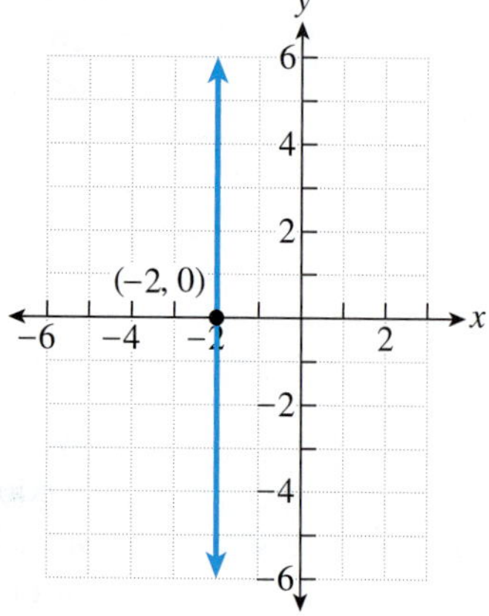

Quick Check **11** **Graph $x = 6$. Label any intercepts.**

Horizontal Lines	**Vertical Lines**
Equation: $y = b$	Equation: $x = a$
• Plot the y-intercept at $(0, b)$. • Graph the horizontal line passing through this point.	• Plot the x-intercept at $(a, 0)$. • Graph the vertical line passing through this point.

Applications of the Graphs of Linear Equations

Objective 7 **Interpret the graph of an applied linear equation.** The intercepts of a graph have important interpretations within the context of an applied problem. The y-intercept often gives us an initial condition, such as how much money is owed on a loan, or what the fixed costs are to attend college before adding in the number of units that a student is taking. The x-intercept often shows us a "break-even" point, such as a time when a business goes from a loss to a profit.

EXAMPLE 13 Mark borrowed \$8000 from his friend Tom, promising to pay him \$500 per month until he had paid off the loan. The amount of money (y) that Mark owes Tom after x months have passed is given by the equation $y = 8000 - 500x$.

Use the graph of this equation to answer the following questions. (Since the number of months cannot be negative, and since the amount of money owed will never be negative, the graph begins at $x = 0$ and at $y = 0$.)

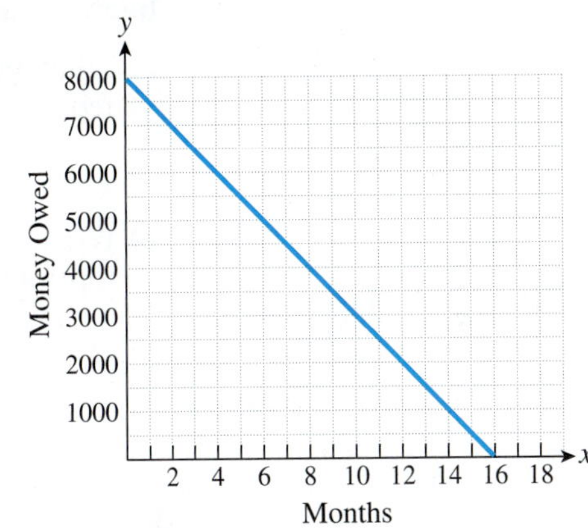

a) What does the y-intercept represent in this situation?

Solution

The y-intercept is $(0, 8000)$. It tells us that after borrowing the money from Tom, Mark owes Tom \$8000.

b) What does the x-intercept represent in this situation?

Solution

The x-intercept is $(16, 0)$. It tells us that Mark will pay off his debt in 16 months.

Using Your Calculator When we graph a line using the TI-84, we will occasionally need to resize the window. In the previous example, the x-intercept is $(16, 0)$ and the y-intercept is $(0, 8000)$. Our graph should show these important points. After entering $8000 - 500x$ for Y_1, press the WINDOW key. Fill in the screen as shown, and press the GRAPH key to display the graph.

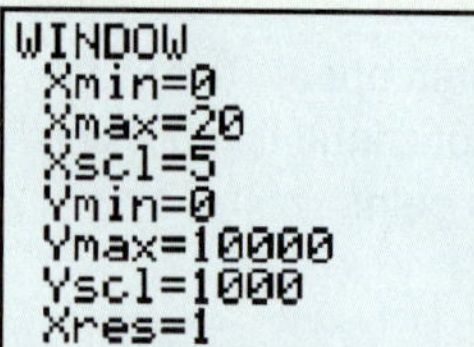

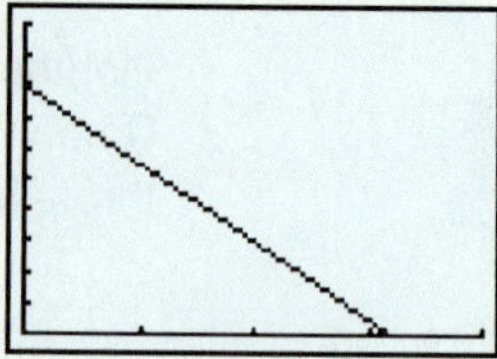

Quick Check 12 Yesenia invests $28,000 in a new business and expects to make a $4000 profit each month. Yesenia's net profit (y) after x months have passed is given by the equation $y = 4000x - 28{,}000$.

Use the graph of this equation to answer the following questions. (Since the number of months cannot be negative, the graph begins at $x = 0$.)

a) What does the y-intercept represent in this situation?

b) What does the x-intercept represent in this situation?

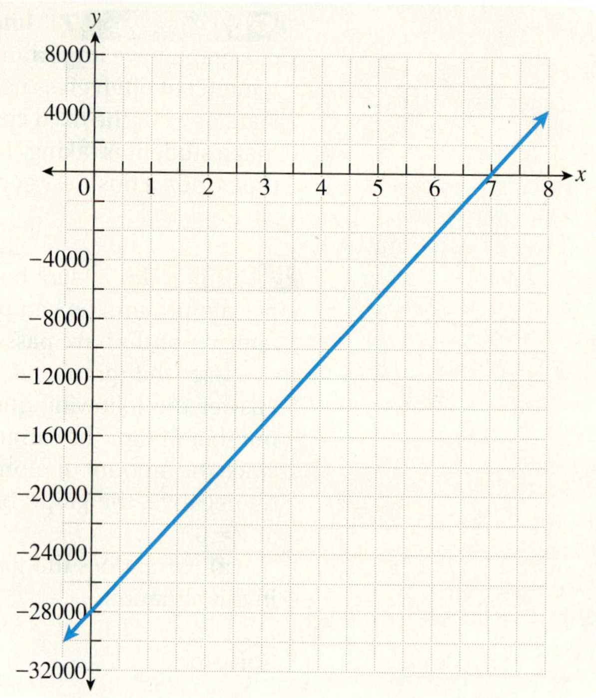

Building Your Study Strategy **Using Your Textbook Effectively, 1 Reading Ahead** One effective way to use your textbook is to read a section in the text before it is covered in class. This will give you an idea of the main concepts covered in the section, and these concepts can be clarified by your instructor in class.

When reading ahead you should scan the section. Look for definitions that are introduced in the section, as well as procedures developed in the section. Pay closer attention to the examples. If you find a step in the examples that you cannot understand, you can ask your instructor about it in class.

EXERCISES 2.1

Vocabulary

1. The standard form of a linear equation in two variables is ________.

2. Solutions to an equation in two variables are written as a(n) ________.

3. The point at $(0, 0)$ is called the ________.

4. A(n) ________ on a rectangular coordinate plane is used to show the solutions of a linear equation in two variables.

5. A point at which a graph crosses the x-axis is called a(n) ________.

6. A point at which a graph crosses the y-axis is called a(n) ________.

7. A ________ line has no y-intercept.

8. A ________ line has no x-intercept.

9. State the procedure for finding an x-intercept.

10. State the procedure for finding a y-intercept.

Is the ordered pair a solution to the given equation?

11. $(7, 2), 3x + 4y = 29$
12. $(5, -3), 6x - 2y = 24$
13. $(-6, 9), 2x + 3y = 15$
14. $(-5, -2), -3x + 8y = -1$
15. $(-8, 0), 3x - 7y = -5$
16. $(0, 4), 5x - 4y = -16$

Plot the points on a rectangular coordinate plane.

17. $A(3, 8)$ $B(-4, 2)$ $C(2, -5)$ $D(-6, -3)$

18. $A(-2, -1)$
$B(-5, 1)$
$C(1, -5)$
$D(-6, 0)$

19. $A(0, -3)$
$B(7, 0)$
$C(-3, 4)$
$D(4, -3)$

FOR EXTRA HELP

MyMathLab

MathXL

Interactmath.com

MathXL Tutorials on CD

Video Lectures on CD

Addison-Wesley Math Tutor Center

Student's Solutions Manual

20. **A** $(-6, 0)$

B $(0, -1)$

C $(0, -5)$

D $(4, 0)$

21. **A** $(\frac{5}{2}, 4)$

B $(-3, \frac{7}{2})$

C $(-\frac{9}{2}, -\frac{3}{2})$

D $(4, -\frac{1}{2})$

22. **A** $(-35, 25)$

B $(40, -15)$

C $(20, 30)$

D $(-15, -50)$

Find the coordinates of the labeled points A, B, C, and D.

23.

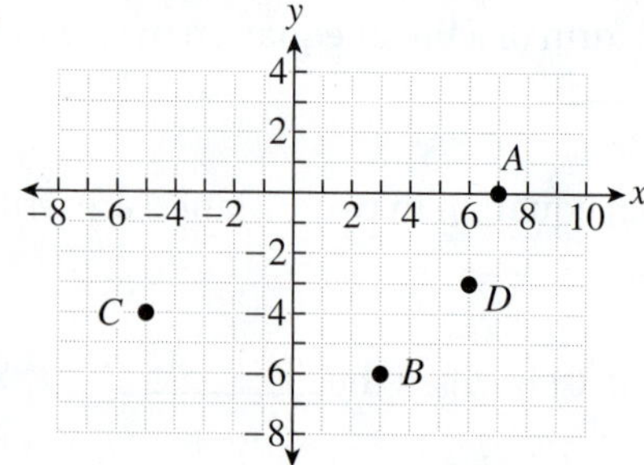

24.

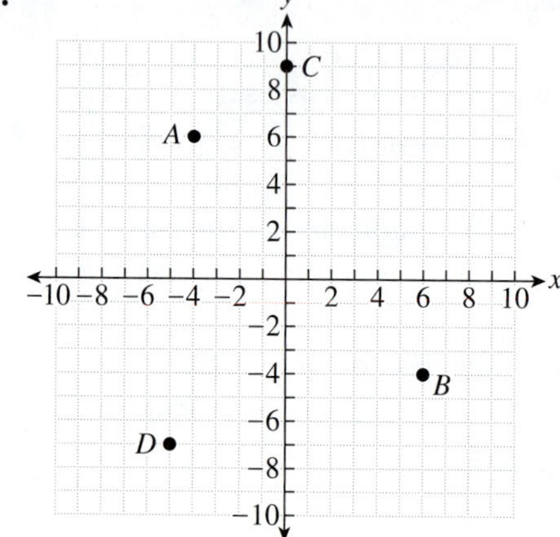

25.

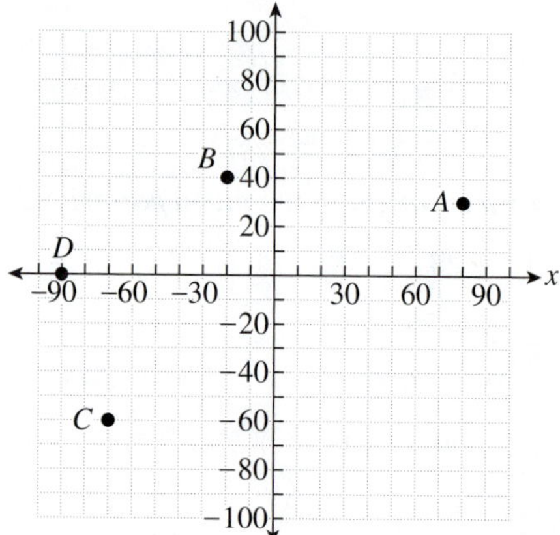

26.

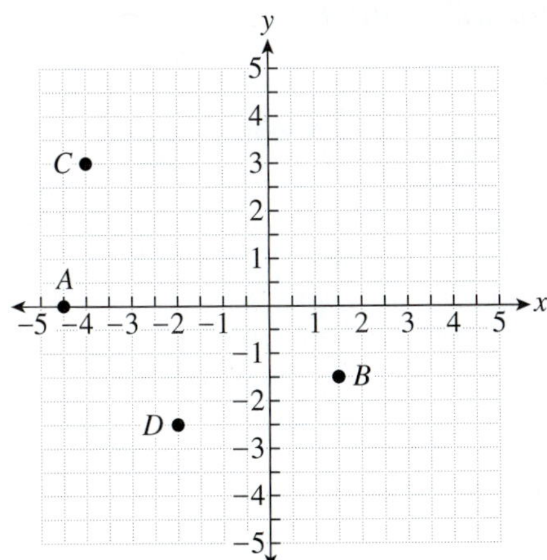

In which quadrant (I, II, III, or IV) is each given point?

27. $(-3, 8)$

28. $(6, -5)$

29. $(-4, -9)$

30. $(-10, 1)$

31. $\left(\frac{3}{2}, \frac{14}{3}\right)$

32. $(-87, -904)$

Using the given coordinate, find the other coordinate that makes the ordered pair a solution to the equation.

33. $6x - 3y = 30, (4, y)$

34. $2x + 7y = 11, (x, 3)$

35. $-x + 5y = -34, (x, -6)$

36. $8x + 4y = -20, (-7, y)$

37. $\frac{2}{3}x + 6y = 2, (12, y)$

38. $4x - \frac{3}{5}y = -14, (x, -10)$

Complete each table with ordered pairs that are solutions.

39. $3x + y = 10$

x	y
1	
	1
0	

40. $2x + 3y = 9$

x	y
−3	
	7
0	

41. $4x - 3y = 12$

x	y
−3	
0	
3	

42. $2x + 5y = -10$

x	y
−5	
0	
5	

43. $y = 4x - 7$

x	y
0	
1	
2	

44. $y = -\frac{2}{3}x + 8$

x	y
0	
3	
6	

Find the x- and y-intercepts from the graph.

45.

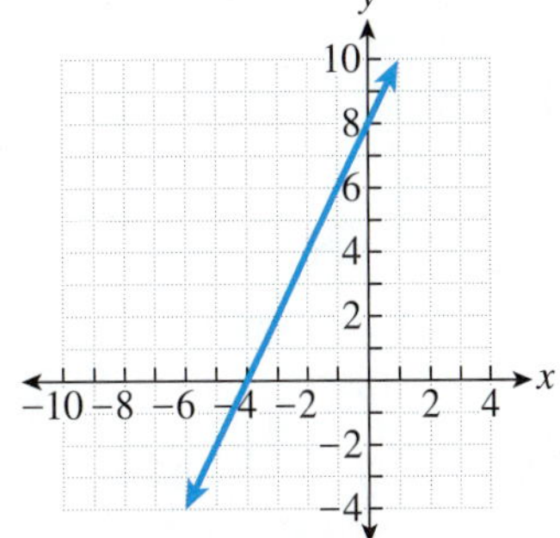

46.

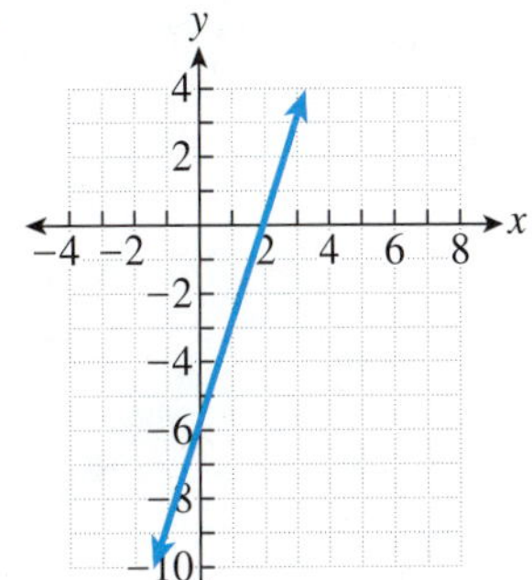

47.

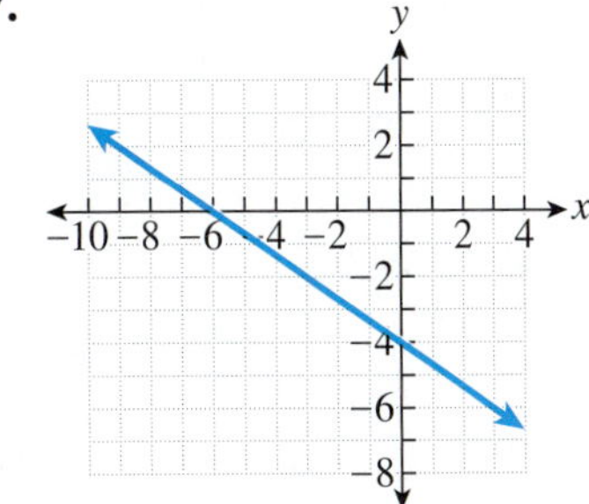

48.

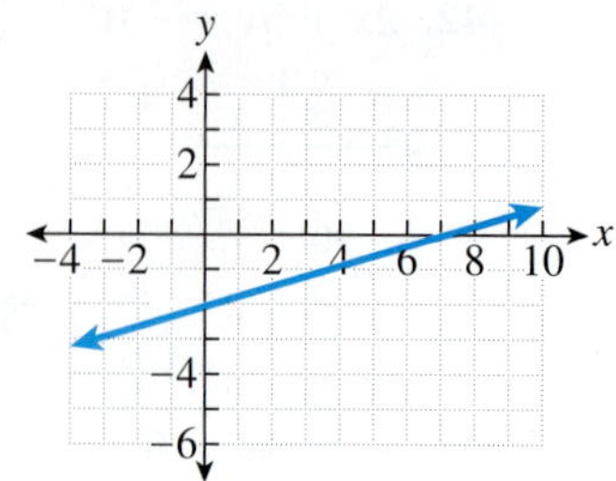

49.

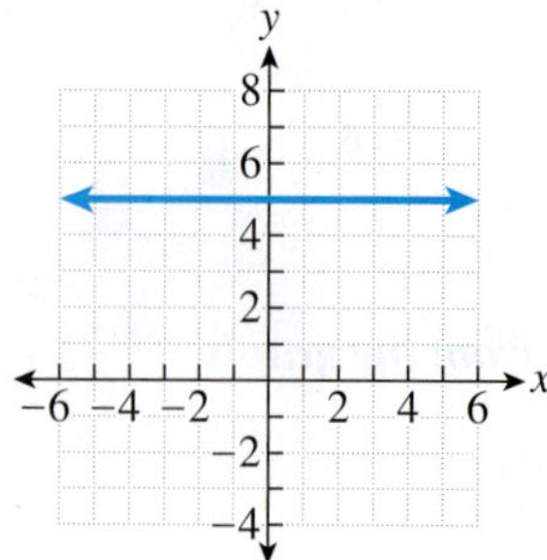

50.

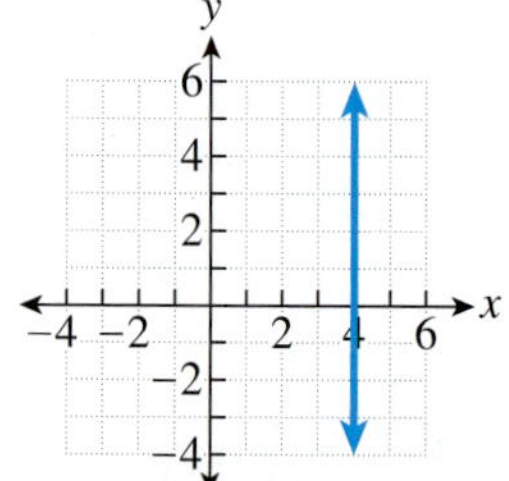

Find the x- and y-intercepts.

51. $x + y = 8$

52. $x - y = -5$

53. $3x - 4y = -24$

54. $5x + 6y = -30$

55. $-6x + 3y = -15$

56. $9x + 4y = 18$

57. $\frac{1}{2}x - \frac{3}{4}y = 3$

58. $\frac{1}{2}x - \frac{2}{3}y = -2$

59. $y = -2x - 32$

60. $y = 4x + 12$

61. $y = 5x$

62. $y = -4x$

Find the intercepts; then graph.

63. $x + y = 3$

64. $x - y = -6$

65. $x + 3y = 9$

66. $-5x + 2y = 10$

67. $2x + 3y = -18$

68. $4x - 7y = 28$

69. $2x - y = 8$

70. $x + 6y = 6$

71. $3x + 4y = 18$

72. $2x - 5y = 15$

73. $y = -2x$

74. $y = 3x$

75. $y = 4x + 6$

76. $y = 2x - 9$

77. $y = -8$

78. $y = 3$

79. $x = 7$

80. $x = -2$

81. Find three equations that have an x-intercept at $(6, 0)$.

82. Find three equations that have an x-intercept at $(-2, 0)$.

83. Find three equations that have a y-intercept at $(0, -5)$.

84. Find three equations that have a y-intercept at $(0, \frac{9}{2})$.

85. A farmer bought a new tractor in the year 2008, paying $16,000. For tax purposes, the tractor depreciates at $2000 per year. If we let x represent the number of years after 2008, then the value of the tractor (y) after x years is given by the equation $y = 16{,}000 - 2000x$.

a) Find the y-intercept of this equation. Explain, in your own words, what this intercept signifies.

b) Find the x-intercept of this equation. Explain, in your own words, what this intercept signifies.

86. Peter borrowed \$2000 from Leah, and agreed to pay her \$80 per month until he had paid back the entire \$2000. The equation that tells the amount of money (y) that Peter owes Leah after x months is $y = 2000 - 80x$.

a) Find the y-intercept of this equation. Explain, in your own words, what this intercept signifies.

b) Find the x-intercept of this equation. Explain, in your own words, what this intercept signifies.

87. At a country club, members pay an annual fee of \$2380. In addition, they must pay \$40 for each round of golf they play. The equation that gives the cost (y) to play x rounds of golf per year is $y = 2380 + 40x$.

a) Find the y-intercept of this equation. Explain, in your own words, what this intercept signifies.

b) In your own words, explain why this equation has no x-intercept within the context of this problem.

c) If a member plays two rounds of golf per month, how much will she pay to the country club in one year?

88. In addition to paying \$125 per unit, a community college student pays a registration fee of \$50 per semester. The equation that gives the cost (y) to take x units is $y = 125x + 50$.

a) Find the y-intercept of this equation. Within the context of this problem, is this cost possible? Explain why or why not.

b) In your own words, explain why this equation has no x-intercept within the context of this problem.

c) A full load for a student is 12 units. How much will a full-time student pay per semester?

Writing in Mathematics

Answer in complete sentences.

89. Explain, in your own words, why the order of the coordinates in an ordered pair is important. Is there a difference between plotting the ordered pair (a, b) and the ordered pair (b, a)?

90. If the ordered pair (a, b) is in quadrant II, in which quadrant is (b, a)? In which quadrant is $(-a, b)$? Explain how you determined your answers.

91. ***Solutions Manual**** Write a solutions manual page for the following problem:

Find the x- and y-intercepts, and then graph the equation $3x - 4y = 18$.

92. ***Newsletter**** Write a newsletter explaining how to graph a line using its intercepts.

***See Appendix B for details and sample answers.**

2.2 Slope of a Line

Objectives

1. **Understand the slope of a line.**
2. **Find the slope of a line passing through two points using the slope formula.**
3. **Find the slopes of horizontal and vertical lines.**
4. **Find the slope and *y*-intercept of a line from its equation.**
5. **Graph a line using its slope and *y*-intercept.**
6. **Determine whether two lines are parallel, perpendicular, or neither.**
7. **Interpret the slope and *y*-intercept in real-world applications.**

Slope of a Line

Objective 1 Understand the slope of a line. Here are four different lines that pass through the point $(0, 2)$:

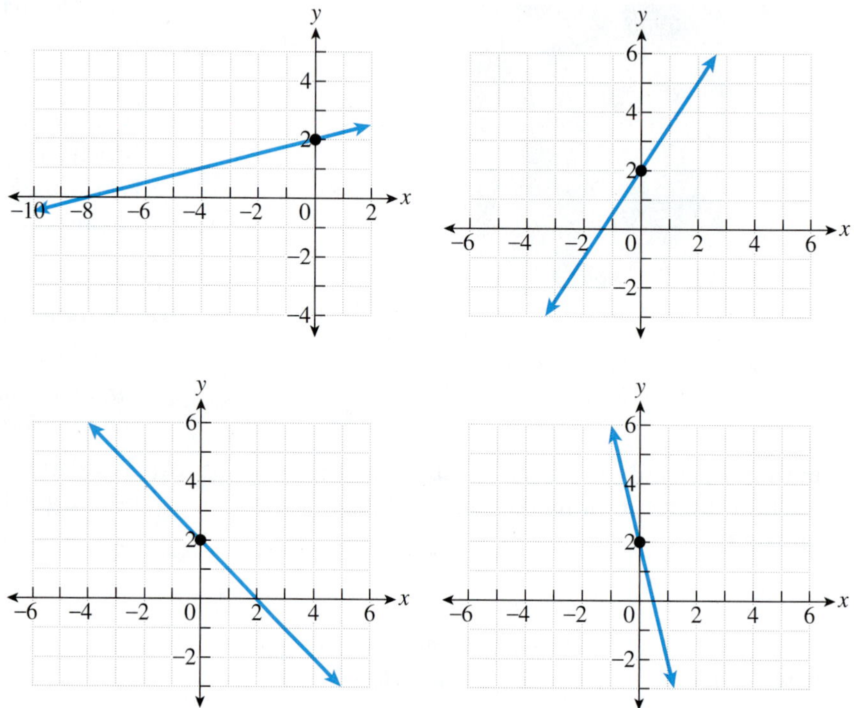

Notice that some of the lines are rising to the right, while others are falling to the right. Also, some are rising or falling more steeply than the others. The characteristic that distinguishes these lines is their slope.

The **slope** of a line is a measure of how steeply a line rises or falls from left to right. We use the letter m to stand for the slope of a line.

If a line rises from left to right, then its slope is positive. If a line falls from left to right, then its slope is negative.

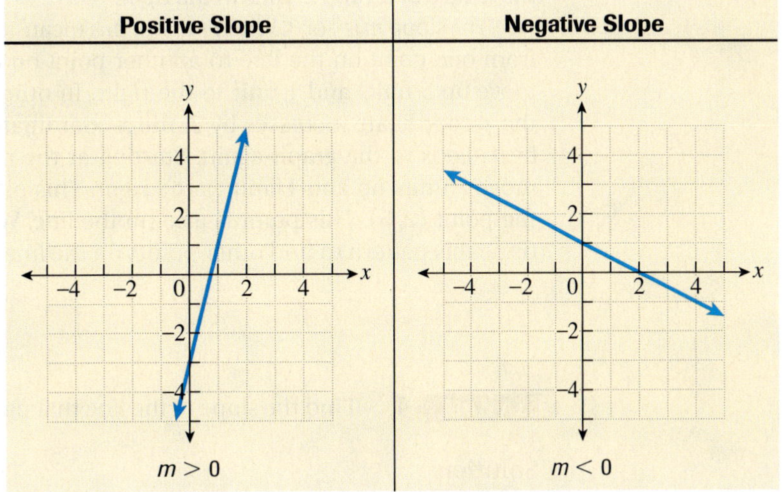

To find the slope of a line, we begin by selecting two points that are on the line. As we move from the point on the left to the point on the right, we measure how much the line rises or falls. The slope is equal to the distance of the rise or fall divided by the distance traveled from left to right. This is often referred to as "rise over run" or as the change in y divided by the change in x. Sometimes the change in y is written as Δy, and the change in x is written as Δx. The Greek letter Δ (delta) is often used to represent change in a quantity. The slope of a line is represented as $m = \dfrac{\text{rise}}{\text{run}}$ or $m = \dfrac{\Delta y}{\Delta x}$.

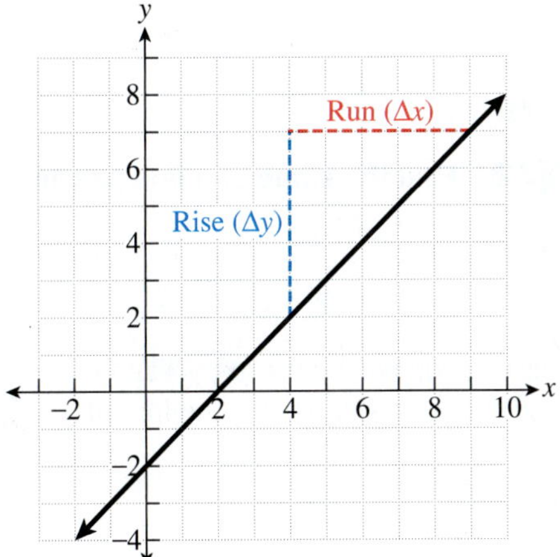

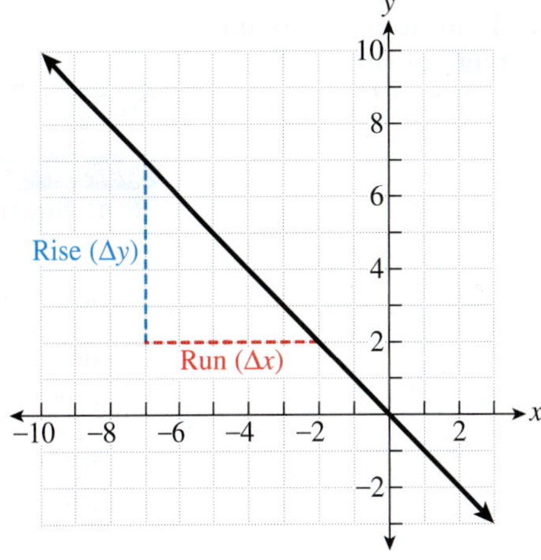

Consider the following line that passes through the points To move from the point on the left to the point on the right, we move up 6 units and 2 units to the right.

The slope m is $\frac{6}{2}$ or 3. When $m = 3$, this means that to move from one point on the line to another point on the line we move up 3 units and 1 unit to the right. In other words, as the y-coordinate increases by 3, the x-coordinate increases by 1. Look at the graph again. Starting at the point $(1, 2)$, move 3 units up and 1 unit to the right. This places you at the point $(2, 5)$. This point is also on the line. We can continue this pattern to find other points on the line.

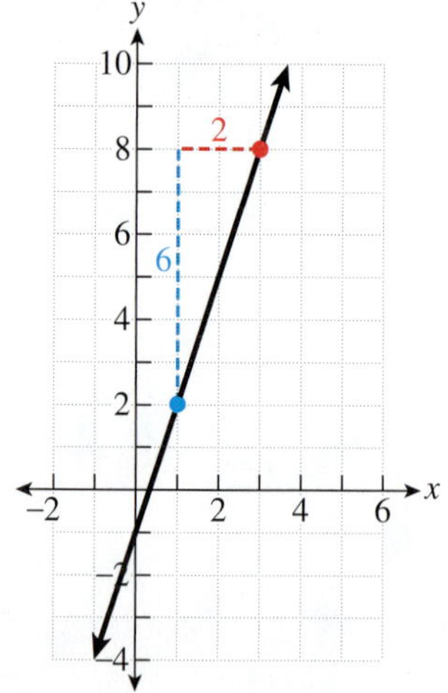

EXAMPLE 1 Find the slope of the line that passes through the points $(4, 9)$ and $(7, 3)$.

Solution

Again, we will begin by plotting the two points on a graph. Notice that this line falls from left to right, so its slope is negative. The line drops by 6 units as it moves 3 units to the right, so its slope is $\frac{-6}{3}$ or -2.

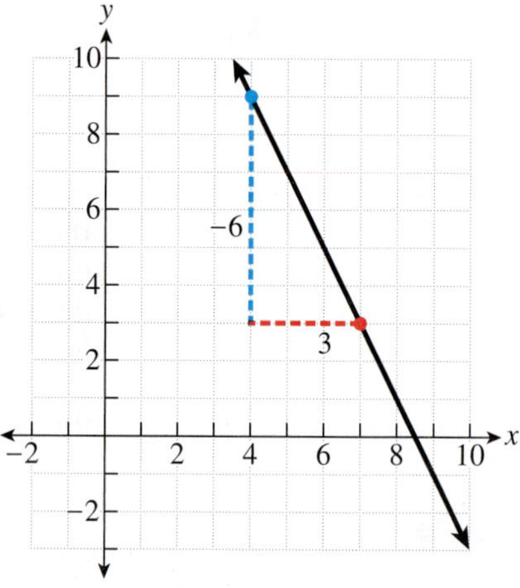

Quick Check 1
Find the slope of the line that passes through $(-3, 6)$ and $(1, -10)$.

Slope Formula

Objective 2 Find the slope of a line passing through two points using the slope formula.

Slope Formula

If a line passes through two points (x_1, y_1) and (x_2, y_2), then we can calculate its slope using the formula

$$m = \frac{y_2 - y_1}{x_2 - x_1}.$$

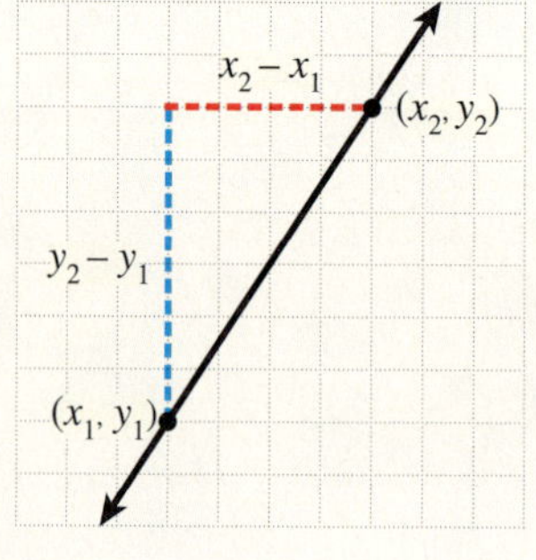

This is consistent with the technique that we used to find the slope in the previous example. The numerator $(y_2 - y_1)$ represents the vertical change, while the denominator $(x_2 - x_1)$ represents the horizontal change. Using this formula saves us from having to plot the points on a graph, but keep in mind that you can continue to use that technique.

We have previously seen that the slope of a line that passes through the points $(1, 2)$ and $(3, 8)$ is 3. We will now use the formula $m = \frac{y_2 - y_1}{x_2 - x_1}$ to find the slope of the line through these two points. We will use the point $(1, 2)$ as (x_1, y_1) and the point $(3, 8)$ as (x_2, y_2).

$$\begin{aligned} m &= \frac{8 - 2}{3 - 1} \\ &= \frac{6}{2} \\ &= 3 \end{aligned}$$

Notice that we get the same result as we did before. We could change the ordering of the points and still calculate the same slope. In other words, we could use the point $(3, 8)$ as (x_1, y_1) and the point $(1, 2)$ as (x_2, y_2), and still find that $m = 3$. We can subtract in either order as long as we are consistent. Whichever order we choose to subtract the y-coordinates in, we must subtract the x-coordinates in the same order.

EXAMPLE 2 Use the formula $m = \frac{y_2 - y_1}{x_2 - x_1}$ to find the slope of the line that passes through the two points $(1, 3)$ and $(5, 9)$.

Solution

One way to think of the numerator in this formula is "*the second y-coordinate minus the first y-coordinate,*" which in our example would be $9 - 3$. Using a similar approach for the x-coordinates in the denominator, we would begin with a denominator of $5 - 1$.

$$\begin{aligned} m &= \frac{9 - 3}{5 - 1} && \text{Substitute the coordinates of the points into the formula.} \\ &= \frac{6}{4} && \text{Subtract.} \\ &= \frac{3}{2} && \text{Simplify.} \end{aligned}$$

The slope of the line that passes through these two points is $\frac{3}{2}$. For every vertical change of 3 units upward, there is a horizontal change of 2 units to the right.

***Quick Check* 2** **Use the formula $m = \frac{y_2 - y_1}{x_2 - x_1}$ to find the slope of the line that passes through the two points $(8, 3)$ and $(2, 11)$.**

A Word of Caution It does *not* matter which point is labeled (x_1, y_1) and which point is labeled (x_2, y_2). It *is* important that the order in which the y-coordinates are subtracted in the numerator is the same as the order in which the x-coordinates are subtracted in the denominator.

EXAMPLE 3 Use the formula $m = \frac{y_2 - y_1}{x_2 - x_1}$ to find the slope of the line that passes through the two points $(-9, -4)$ and $(-5, 8)$.

Solution

In this example we learn to apply the formula to points that have negative x- or y-coordinates.

$$
\begin{aligned}
m &= \frac{8-(-4)}{-5-(-9)} && \text{Substitute into the formula.}\\
&= \frac{8+4}{-5+9} && \text{Eliminate double signs.}\\
&= \frac{12}{4} && \text{Simplify the numerator and denominator.}\\
&= 3 && \text{Simplify.}
\end{aligned}
$$

The slope of this line is 3. For every vertical change of 3 units upward, there is a horizontal change of 1 unit to the right.

Quick Check **3** **Use the formula $m = \frac{y_2 - y_1}{x_2 - x_1}$ to find the slope of the line that passes through the two points $(-1, 6)$ and $(3, -6)$.**

Horizontal and Vertical Lines

Objective 3 **Find the slopes of horizontal and vertical lines.** Let's look at the line that passes through the points $(1, 3)$ and $(4, 3)$.

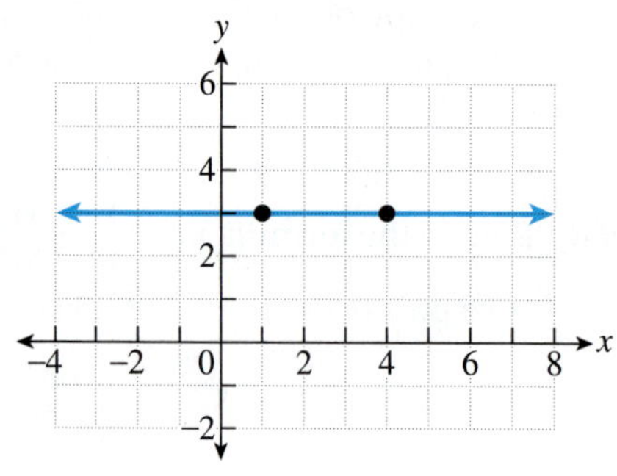

This is a horizontal line. What is the slope of this line?

$$\begin{aligned} m &= \frac{3 - 3}{4 - 1} \\ &= \frac{0}{3} \\ &= 0 \end{aligned}$$

The slope of this line is 0. The same is true for any horizontal line. The vertical change between any two points on a horizontal line is always equal to 0, and when we divide 0 by any nonzero number the result is equal to 0. Thus the slope of any horizontal line is equal to 0.

Here is a brief summary of the properties of horizontal lines.

Horizontal Lines

Equation: $y = b$, where b is a real number
y-intercept: $(0, b)$
Slope: $m = 0$

Look at the following vertical line.

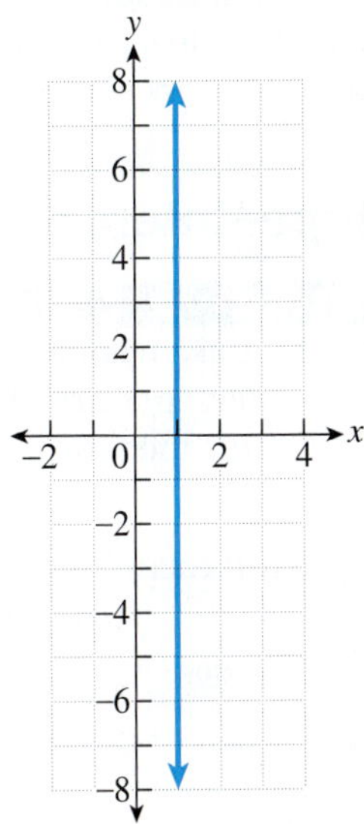

Notice that each point on this line has an x-coordinate of $x = 1$. We can attempt to find the slope of this line by selecting any two points on the line, such as $(1, 2)$ and $(1, 5)$, and using the slope formula.

$$\begin{aligned} m &= \frac{5 - 2}{1 - 1} \\ &= \frac{3}{0} \end{aligned}$$

The fraction $\frac{3}{0}$ is undefined. So the slope of a vertical line is said to be undefined.

Here is a brief summary of the properties of vertical lines.

Vertical Lines

Equation: $x = a$, where a is a real number
x-intercept: $(a, 0)$
Slope: Undefined

EXAMPLE 4 Find the slope, if it exists, of the line $x = 7$.

Solution

This line is a vertical line. The slope of a vertical line is undefined, so this line has undefined slope.

Quick Check 4
Find the slope, if it exists, of the line $x = -6$.

EXAMPLE 5 Find the slope, if it exists, of the line $y = -1$.

Solution

This line is a horizontal line. The slope of any horizontal line is 0, so the slope of this line is 0.

Quick Check 5
Find the slope, if it exists, of the line $y = 9$.

A Word of Caution Avoid stating that a line has "no slope." This is a vague phrase—some may take it to mean that the slope is equal to 0, while others may interpret it to mean that the slope is undefined.

Slope–Intercept Form of a Line

Objective 4 Find the slope and y-intercept of a line from its equation. If we solve a given equation for y so that it is in the form $y = mx + b$, this is called the **slope–intercept form** of a line. The number being multiplied by x is the slope m, while b represents the y-coordinate of the y-intercept.

Slope–Intercept Form of a Line

$y = mx + b$
m: Slope of the line
b: y-coordinate of the y-intercept

This means that if we were graphing the line $y = 2x + 5$, then the line would have a slope of 2 and a y-intercept of $(0, 5)$. Let's verify that this is true, beginning with the slope. Every time we increase the value of x by 1, y will increase by 2 because x is being multiplied by 2. If y increases by 2 (vertical change) every time x increases by 1 (horizontal change), this means that the slope of the line is equal to 2. As for the y-intercept, we can verify that this is true by substituting 0 for x in the equation.

$$y = 2x + 5$$
$$y = 2(0) + 5$$
$$y = 5$$

This shows that the y-intercept is at $(0, 5)$. Repeating this for the general form $y = mx + b$, we see that the y-coordinate of the y-intercept is b.

Example 6 Find the slope and y-intercept of the line $y = 4x - 11$.

Solution

Since this equation is already solved for y, we can read the slope and the y-intercept directly from the equation. The slope is 4, which is the coefficient of the term containing x in the equation. The y-intercept is $(0, -11)$ since -11 is the constant in the equation.

***Quick Check* 6**
Find the slope and the y-intercept of the line $y = -\frac{3}{5}x + 4$.

Example 7 Find the slope and y-intercept of the line $6x + 2y = 15$.

Solution

We begin by solving for y.

$$6x + 2y = 15$$
$$2y = -6x + 15 \qquad \text{Subtract } 6x \text{ from both sides.}$$
$$\frac{2y}{2} = \frac{-6x}{2} + \frac{15}{2} \qquad \text{Divide each term by 2.}$$
$$y = -3x + \frac{15}{2} \qquad \text{Simplify.}$$

The slope of this line is -3. The y-intercept is $\left(0, \frac{15}{2}\right)$.

***Quick Check* 7**
Find the slope and the y-intercept of the line $-3x + 3y = -18$.

A Word of Caution We cannot determine the slope and y-intercept of a line from its equation unless the equation has been solved for y first.

Graphing a Line Using Its Slope and y-Intercept

Objective 5 **Graph a line using its slope and y-intercept.** Once we have an equation in slope–intercept form, we can use this information to graph the line. We begin by plotting the y-intercept as our first point. We then use the slope of the line to find a second point. We could repeat this process to find additional points.

Example 8 Graph the line $y = -2x + 6$.

Solution

We will start at the y-intercept, which is $(0, 6)$. Since the slope is -2, this tells us that we can move down 2 units and 1 unit to the right to find another point on the line. This would give us a second point at $(1, 4)$. Repeating this process, we see that the point $(2, 2)$ is also on the line.

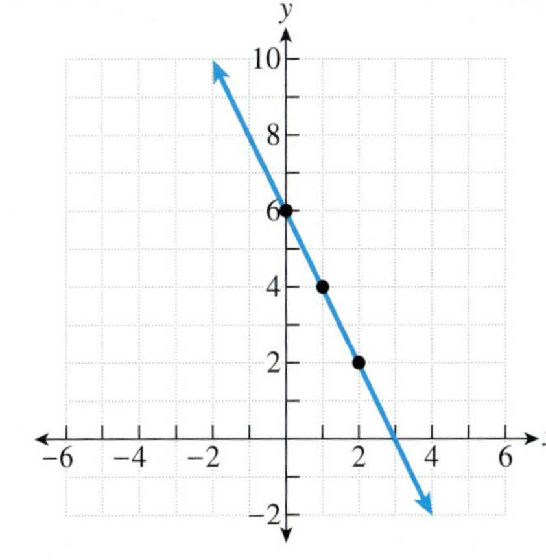

Quick Check 8 Graph the line $y = 3x - 9$.

EXAMPLE 9 Graph the line $y = \frac{2}{3}x - 4$.

Solution

We will start by plotting the y-intercept at $(0, -4)$. The slope is $\frac{2}{3}$ which tells us that we can move up by 2 units and 3 units to the right to find another point on the line. This will give us a second point at $(3, -2)$. Repeating this process, we can find another point at $(6, 0)$.

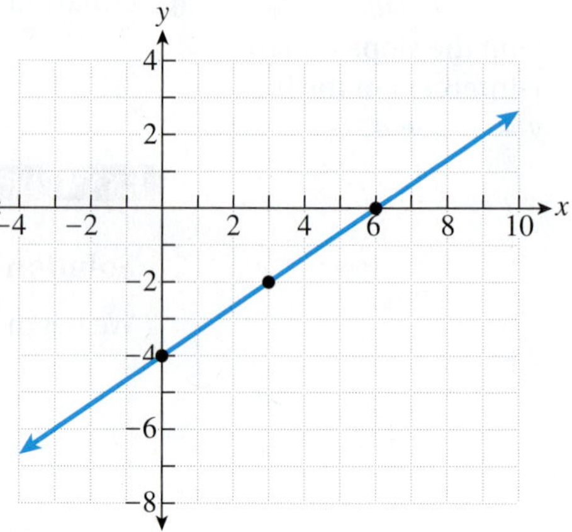

Quick Check 9
Graph the line $y = -\frac{1}{2}x - 3$.

Graphing a Line Using Its Slope and y-intercept

- Plot the y-intercept $(0, b)$.
- Use the slope m to find another point on the line.
- Graph the line that passes through these two points.

Parallel Lines

Objective 6 Determine whether two lines are parallel, perpendicular, or neither. We will now examine the relationship between two lines. Consider the following pair of lines.

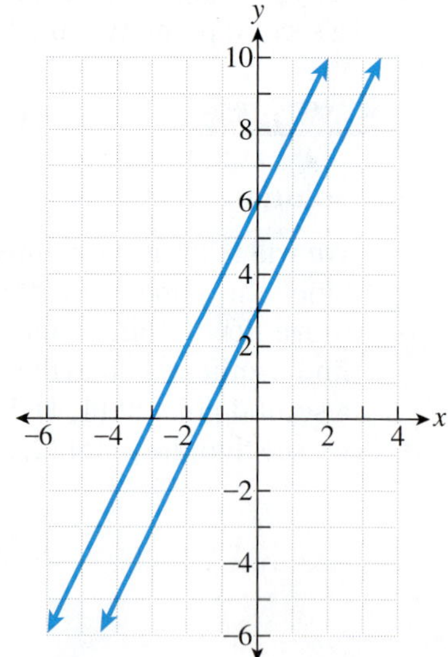

We notice that these two lines do not intersect. The lines have the same slope and are called **parallel lines.**

Parallel Lines

- Two nonvertical lines are **parallel** if they have the same slope. In other words, if we denote the slope of one line as m_1 and the slope of the other line as m_2, the two lines are parallel if $m_1 = m_2$.
- If two lines are both vertical, they are parallel.

Here are some examples of lines that are parallel.

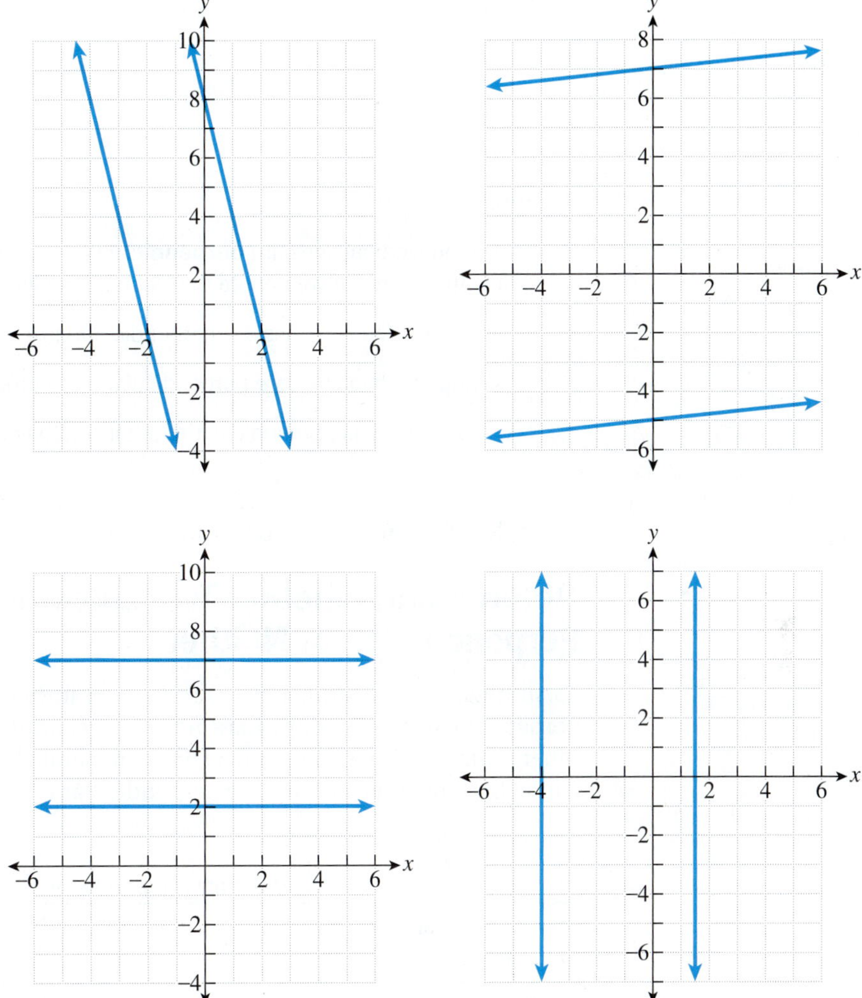

Any two horizontal lines, such as $y = 4$ and $y = 1$, are parallel to each other, since their slopes are both 0. Any two vertical lines, such as $x = 2$ and $x = -1$, are parallel to each other as well.

Perpendicular Lines

Two distinct lines that are not parallel intersect at one point. One special type of intersecting lines is **perpendicular lines**. Here is an example of two lines that are perpendicular. Perpendicular lines intersect at right angles. Notice that one of these lines has a positive slope and the other line has a negative slope.

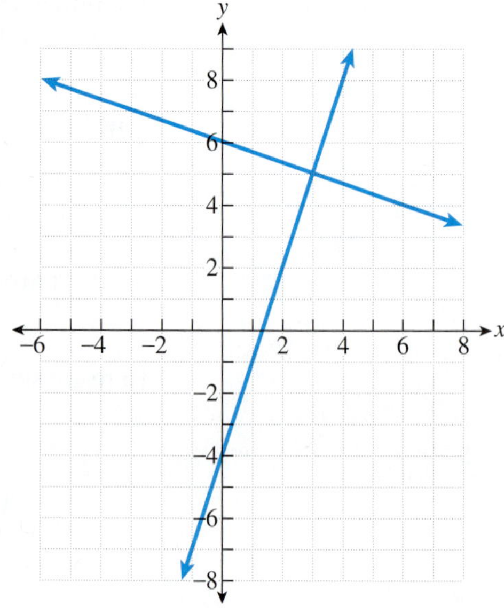

Perpendicular Lines

- Two nonvertical lines are **perpendicular** if their slopes are **negative reciprocals**. In other words, if we denote the slope of one line as m_1 and the slope of the other line as m_2, the two lines are perpendicular if $m_1 = -\dfrac{1}{m_2}$. (This is equivalent to saying that two lines are perpendicular if the product of their slopes is -1.)
- A vertical line is perpendicular to a horizontal line.

Examples of **negative reciprocals** are 5 and $-\frac{1}{5}$, -2 and $\frac{1}{2}$, and $-\frac{3}{10}$ and $\frac{10}{3}$.

Determining Whether Two Lines Are Parallel, Perpendicular, or Neither

To summarize, nonvertical lines are parallel if they have the same slope, and they are perpendicular if their slopes are negative reciprocals. Horizontal lines are parallel to other horizontal lines, and vertical lines are parallel to other vertical lines. Finally, a horizontal line and a vertical line are perpendicular to each other.

If . . .	the lines are parallel if . . .	the lines are perpendicular if . . .
two nonvertical lines have slopes m_1 and m_2	$m_1 = m_2$	$m_1 = -\dfrac{1}{m_2}$
one of the two lines is horizontal	the other line is horizontal	the other line is vertical
one of the two lines is vertical	the other line is vertical	the other line is horizontal

EXAMPLE 10 Are the two lines $y = 6x + 2$ and $y = \frac{1}{6}x - 3$ parallel, perpendicular, or neither?

Solution

The slope of the first line is 6 and the slope of the second line is $\frac{1}{6}$. The slopes are not equal, so the lines are not parallel. The slopes are not negative reciprocals, so the lines are not perpendicular either. The two lines are neither parallel nor perpendicular.

***Quick Check* 10**
Are the two lines $y = 2x - 5$ and $y = -2x$ parallel, perpendicular, or neither?

EXAMPLE 11 Are the two lines $-3x + y = 7$ and $-6x + 2y = -4$ parallel, perpendicular, or neither?

Solution

We find the slope of each line by solving each equation for y.

$$\begin{aligned} -3x + y &= 7 \\ y &= 3x + 7 && \text{Add } 3x \text{ to both sides.} \end{aligned}$$

The slope of the first line is 3. Now we find the slope of the second line.

$$\begin{aligned} -6x + 2y &= -4 \\ 2y &= 6x - 4 && \text{Add } 6x \text{ to both sides.} \\ y &= 3x - 2 && \text{Divide by 2.} \end{aligned}$$

The slope of the second line is also 3. Since the two slopes are equal, the lines are parallel.

***Quick Check* 11**
Are the two lines $y = 4x + 3$ and $8x - 2y = 12$ parallel, perpendicular, or neither?

EXAMPLE 12 Are the two lines $10x + 2y = 0$ and $x - 5y = 3$ parallel, perpendicular, or neither?

Solution

We begin by finding the slope of each line by solving for y.

$$\begin{aligned} 10x + 2y &= 0 \\ 2y &= -10x && \text{Subtract } 10x \text{ from both sides.} \\ y &= -5x && \text{Divide by 2.} \end{aligned}$$

The slope of the first line is -5. Now we find the slope of the second line.

$$\begin{aligned} x - 5y &= 3 \\ -5y &= -x + 3 && \text{Subtract } x \text{ from both sides.} \\ y &= \frac{1}{5}x - \frac{3}{5} && \text{Divide by } -5. \end{aligned}$$

The slope of the second line is $\frac{1}{5}$. Since the two slopes are negative reciprocals, the lines are perpendicular.

***Quick Check* 12**
Are the two lines $y = -\frac{2}{5}x + 9$ and $5x - 2y = -8$ parallel, perpendicular, or neither?

Applications

Objective 7 Interpret the slope and y-intercept in real-world applications. The concept of slope becomes much more interesting when we apply it to real-world situations.

Example 13 The number of Americans (y), in millions, who have a cellular phone can be approximated by the equation $y = 22.5x + 138.7$, where x is the number of years after 2002. Interpret the slope and y-intercept of this line.
(*Source:* Cellular Telecommunications Industry Association)

Solution

The slope of this line is 22.5, and this tells us that each year we can expect an additional 22.5 million Americans to have a cellular phone. (The number of Americans who have a cellular phone is increasing because the slope is positive.) The y-intercept at $(0, 138.7)$ tells us that approximately 138.7 million Americans had a cellular phone in 2002.

Quick Check **13 The number of women (y) enrolled in a graduate program in an American university in a given year can be approximated by the equation $y = 55{,}400x + 1{,}118{,}900$, where x is the number of years after 2001. Interpret the slope and y-intercept of this line.**
(*Source:* U.S. Department of Education, National Center for Education Statistics)

Building Your Study Strategy **Using Your Textbook Effectively, 2** **Quick Check Exercises** The Quick Check exercises following most examples in this text are designed to be similar to the examples in the book. After reading through, and possibly reworking an example, try the corresponding Quick Check exercise. After checking your answer in the back of the book, you will be able to determine whether you are ready to move forward in the section or whether you need further practice.

EXERCISES 2.2

Vocabulary

1. The __________ of a line is a measure of how steeply a line rises or falls from left to right.

2. A line that rises from left to right has __________ slope.

3. A line that falls from left to right has __________ slope.

4. State the formula for the slope of a line that passes through two points.

5. A(n) __________ line has a slope of 0, and the slope of a(n) __________ line is undefined.

6. The __________ form of the equation of a line is $y = mx + b$.

7. Two nonvertical lines are __________ if they have the same slope.

8. Two nonvertical lines are __________ if their slopes are negative reciprocals.

Determine whether the given line has a positive or a negative slope.

9.

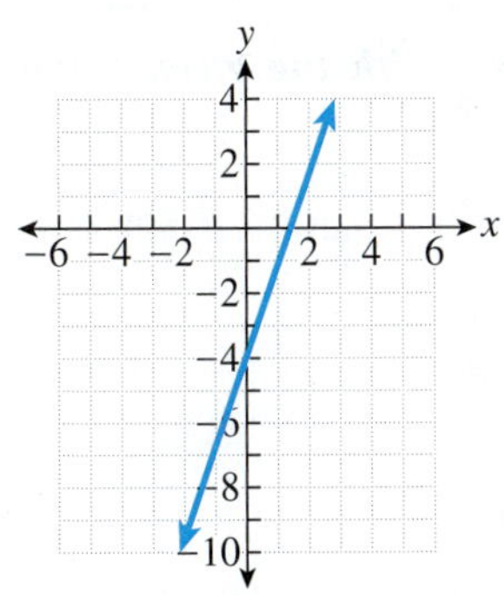

10.

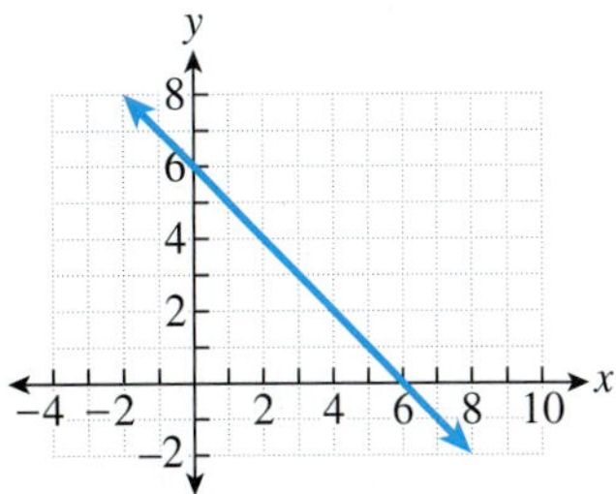

11.

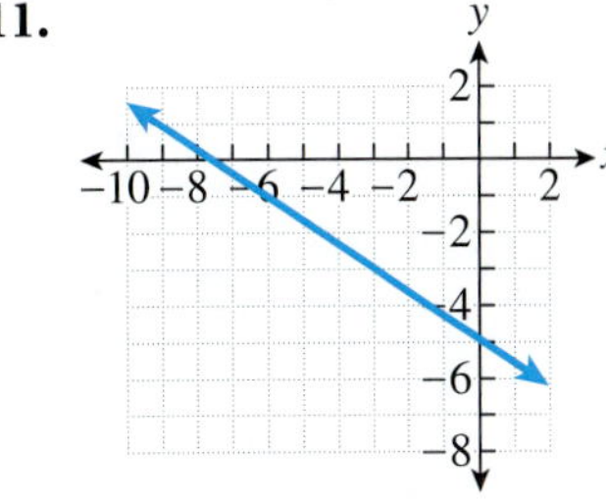

12.

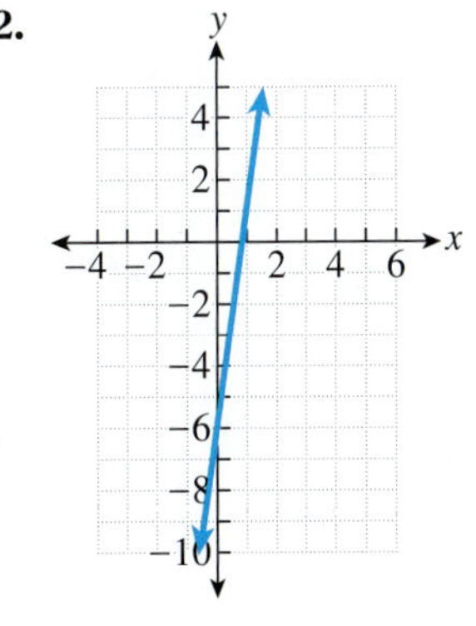

Find the slope of the given line. If the slope is undefined, state this.

13.

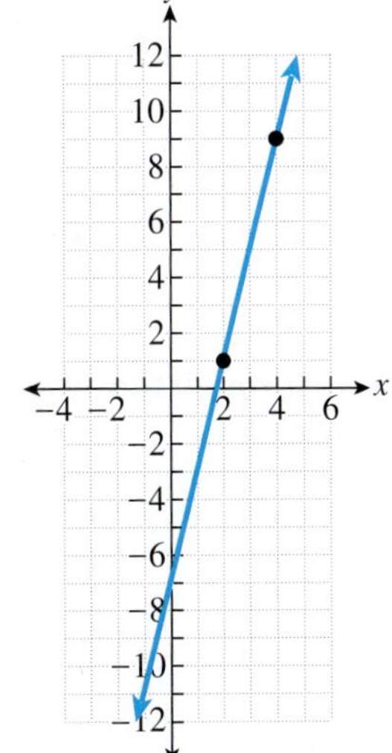

14.

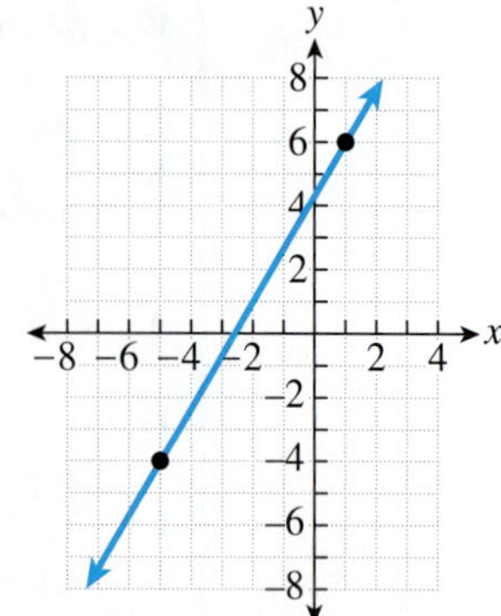

15.

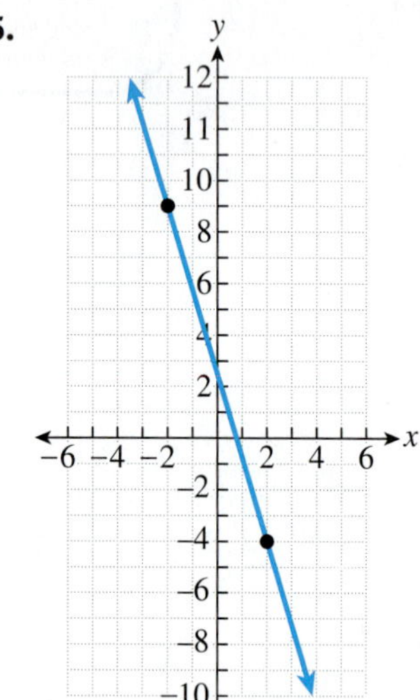

16.

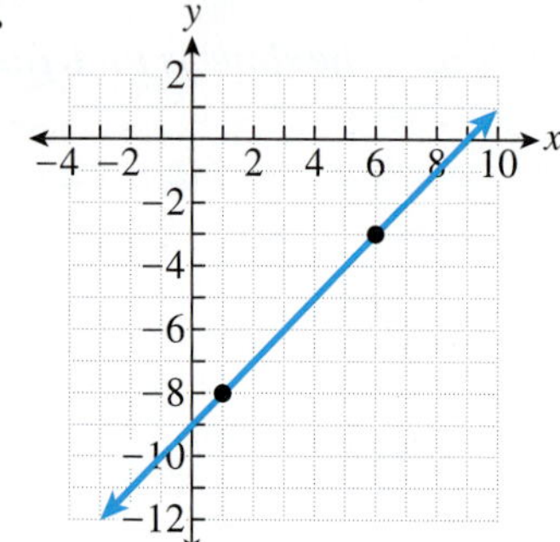

Find the slope of a line that passes through the given two points. If the slope is undefined, state this.

17. $(1, 2)$ and $(5, 18)$

18. $(9, -4)$ and $(-3, -12)$

19. $(-7, 8)$ and $(-1, -18)$

20. $(10, 0)$ and $(0, -8)$

21. $(5, 6)$ and $(-5, 6)$

22. $(-2, 3)$ and $(-2, 9)$

23. $(2, 102.7)$ and $(5, 98.2)$

24. $(5.2, 47.6)$ and $(7.6, 71)$

25. $(\frac{3}{4}, \frac{19}{3})$ and $(\frac{5}{2}, \frac{31}{6})$

26. $(\frac{1}{2}, 539{,}000)$ and $(\frac{3}{5}, 227{,}000)$

Determine the slope of the line. If the slope is undefined, state this.

27. $y = 5$

28. $x = 6$

29. $x = -3$

30. $y = -8$

Find the slope and the y-intercept of the given line.

31. $y = 5x - 8$

32. $y = -3x - 7$

33. $y = -\frac{2}{3}x + 6$

34. $y = x + 5$

35. $6x + 2y = 18$

36. $2x + 3y = -12$

37. $3x - 4y = 10$

38. $10x - 5y = -30$

39. $y = 2.3x - 42.9$

40. $3.6x + 4.8y = 15.6$

Find the equation of a line with the given slope and y-intercept.

41. Slope -7, y-intercept $(0, 8)$

42. Slope 6, y-intercept $(0, -9)$

43. Slope $\frac{2}{5}$, y-intercept $(0, -3)$

44. Slope $-\frac{1}{4}$, y-intercept $(0, \frac{5}{3})$

45. Slope 0, y-intercept $(0, 6)$

46. Slope 0, y-intercept $(0, -2)$

Graph using the slope and y-intercept.

47. $y = 2x + 6$

48. $y = 4x - 8$

55. $y = \frac{2}{3}x - 4$

56. $y = \frac{1}{4}x + 2$

49. $y = -3x + 3$

50. $y = -2x - 5$

57. $y = -\frac{3}{4}x - 6$

58. $y = -\frac{4}{3}x + 4$

51. $y = x - 6$

52. $y = -x - 2$

59. $y = 2.4x + 3.7$

60. $y = -3.5x + 6$

53. $y = -3x$

54. $y = 7x$

61. $y = 100x - 700$

62. $y = -500x + 2500$

63. $y = -20{,}000x + 120{,}000$

64. $y = 25{,}000x + 150{,}000$

Mixed Practice, 65–88

Graph the line using the most efficient technique.

65. $y = -x + 5$

66. $-5x + 4y = -20$

67. $y = \frac{2}{5}x - 2$

68. $y = x + 3$

69. $x = 4$

70. $y = \frac{3}{4}x$

71. $2x + 3y = 9$

72. $y = -4x + 10$

73. $y = -2x$

74. $4x - y = 8$

81. $y = -6x + 5$

82. $y = 5$

75. $y = -2x - 8$

76. $5x + 6y = 12$

83. $x + 3y = 6$

84. $y = -2x + 14$

77. $y = -3$

78. $x = -2$

85. $3x - 2y = 18$

86. $y = 2x + 7$

79. $y = 3x - 18$

80. $y = 8x - 4$

87. $y = 3x - 6$

88. $y = -\frac{1}{3}x + 3$

Graph the two lines and find the coordinates of the point of intersection.

89. $y = 3x - 9$ and $y = x - 1$

90. $x + 2y = 2$ and $y = \frac{2}{3}x - 6$

91. Over a 5-year period from 2000 to 2005, the value of a stock decreased by $15. If we let x represent the number of years after 2000, the value (y) of the stock is given by the equation $y = -3x + 68$.

a) Find the slope of the equation. In your own words, explain what this slope signifies.

b) Find the y-intercept of the equation. In your own words, explain what this y-intercept signifies.

c) Use this equation to predict the value of the stock in the year 2015.

92. The number of nursing students receiving associate's degrees (y) at a time x years after 2002 is approximated by the equation $y = 5400x + 40{,}400$. (Based on data from the years 2001–2004.) (*Source:* National Center for Education Statistics)

a) Find the slope of the equation. In your own words, explain what this slope signifies.

b) Find the y-intercept of the equation. In your own words, explain what this y-intercept signifies.

c) Use this equation to predict the number of nursing students receiving associate's degrees in the year 2009.

93. On a 700-foot stretch of highway through the mountains, the road drops by a total of 28 feet. Find the slope of the road.

94. Sahib walked 2 miles (10,560 feet) on a treadmill. The grade was set to 3%, which means that the slope of the treadmill is 3% or 0.03. Over the course of his workout, how many feet did Sahib climb (vertically)?

Are the two given lines parallel?

95. $y = 2x + 9, y = 2x - 5$

96. $y = 3x + 7, y = -3x + 4$

97. $6x + 2y = 12, 9x - 3y = 27$

98. $4x + 5y = 8, 12x + 15y = 30$

Are the two given lines perpendicular?

99. $y = 3x, y = \frac{1}{3}x + 6$

100. $y = \frac{5}{3}x - 7, y = -\frac{3}{5}x - 4$

101. $6x - 3y = 18, x + 2y = 4$

102. $x + y = 8, x - y = -5$

Are the two given lines parallel, perpendicular, or neither?

103. $y = 4x - 2, y = -4x + 8$

104. $y = -2x + 6, y = -\frac{1}{2}x - 6$

105. $y = \frac{2}{3}x + 9, y = -\frac{3}{2}x - 6$

106. $y = x - 7, y = x + 3$

107. $6x + 4y = -36, y = -\frac{3}{2}x + 1$

108. $10x + 6y = 15, 3x - 5y = 8$

109. Is the line that passes through $(-2, 5)$ and $(4, -7)$ parallel to the line $y = -2x + 7$?

110. Is the line that passes through $(2, 8)$ and $(11, -4)$ perpendicular to the line $y = \frac{3}{4}x - 6$?

Writing in Mathematics

111. In general, the larger a lawn is, the more time it takes to mow it. Would an equation relating the time needed to mow a lawn (y) to a lawn's area (x) have a positive or negative slope? Explain why.

112. The president of a community college predicts student enrollment figures will decrease for the next three years. Would an equation relating student enrollment (y) to the number of years after the statement was made (x) have a positive or negative slope for the next three years? Explain why.

113. Describe a real-world example of parallel lines.

114. Describe a real-world example of perpendicular lines.

115. Consider the equations $y = -\frac{2}{3}x - 5$ and $5x + 6y = -90$. For each equation, do you feel that graphing by finding the x- and y-intercepts or by using the slope and y-intercept is the most efficient way to graph the equation? Explain your choices.

116. ***Solutions Manual**** Write a solutions manual page for the following problem:

Find the slope of the line that passes through the points $(-2, 6)$ *and* $(4, -8)$.

117. ***Newsletter**** Write a newsletter explaining how to graph a line using its slope and y-intercept.

118. ***Newsletter**** Write a newsletter explaining how to find the slope of a line that passes through two given points.

***See Appendix B for details and sample answers.**

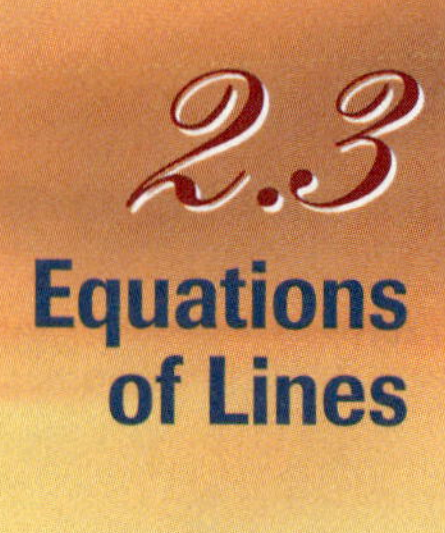

2.3 Equations of Lines

Objectives

1. **Find the equation of a line using the point–slope form.**
2. **Find the equation of a line given two points on the line.**
3. **Find a linear equation to describe real data.**
4. **Find the equation of a parallel or perpendicular line.**

We are already familiar with the slope–intercept form of the equation of a line: $y = mx + b$. When an equation is written in this form, we know both the slope of the line, m, and the y-coordinate of the y-intercept, b. This form is convenient for graphing lines. In this section we will look at another form of the equation of a line. We will also learn how to find the equation of a line if we know the slope of a line and any point through which the line passes.

Point–Slope Form of the Equation of a Line

Objective 1 **Find the equation of a line using the point–slope form.** In Section 2.2 we learned that if we know the slope of a line and its y-intercept, we can write the equation of the line using the slope–intercept form $y = mx + b$, where m is the slope of the line and b is the y-coordinate of the y-intercept. If we know the slope of a line and the coordinates of any point on that line, not just the y-intercept, we can write the equation of the line using the point–slope form of an equation.

Point–Slope Form of the Equation of a Line

The **point–slope form** of the equation of a line with slope m that passes through the point (x_1, y_1) is

$$y - y_1 = m(x - x_1).$$

This form can be derived directly from the slope formula $m = \frac{y_2 - y_1}{x_2 - x_1}$. If we let (x, y) represent an arbitrary point on the line, this formula becomes $m = \frac{y - y_1}{x - x_1}$. Multiplying both sides of that equation by $(x - x_1)$ produces the point–slope form of the equation of a line.

If a line has slope 2 and passes through the point (3, 1), we find its equation by substituting 2 for m, 3 for x_1, and 1 for y_1. Its equation is $y - 1 = 2(x - 3)$. This equation can be left in this form, but is usually converted to slope–intercept form or standard form.

EXAMPLE 1 Find the equation of a line whose slope is 2 and that passes through the point $(4, -3)$. Write the equation in slope–intercept form.

Solution

We will substitute 2 for m, 4 for x_1, and -3 for y_1 in the point–slope form. We will then solve the equation for y in order to write the equation in slope–intercept form.

$$y - (-3) = 2(x - 4) \quad \text{Substitute into point–slope form.}$$
$$y + 3 = 2(x - 4) \quad \text{Eliminate double signs.}$$
$$y + 3 = 2x - 8 \quad \text{Distribute 2.}$$
$$y = 2x - 11 \quad \text{Subtract 3 to isolate } y.$$

The equation of a line with slope 2 that passes through $(4, -3)$ is $y = 2x - 11$. Here is the graph of the line, showing that it passes through the point $(4, -3)$.

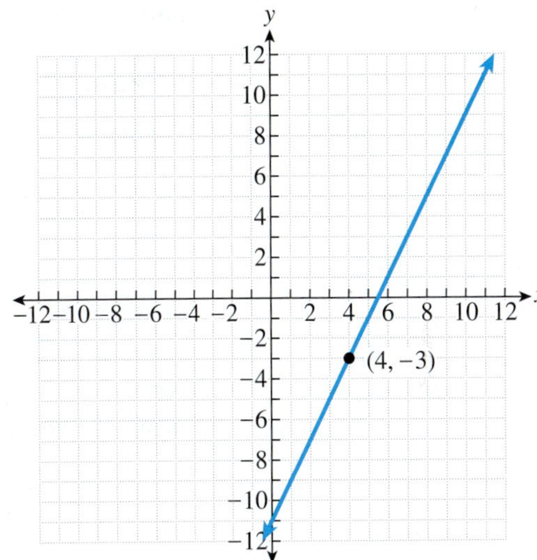

***Quick Check* 1**
Find the equation of a line whose slope is 3 and that passes through the point $(-2, 1)$. Write the equation in slope–intercept form.

EXAMPLE 2 Find the equation of a line whose slope is -1 and that passes through the point $(-7, 4)$. Write the equation in slope–intercept form.

Solution

We begin by substituting -1 for m, -7 for x_1, and 4 for y_1 in the point–slope form. After substituting, we solve the equation for y.

$$y - 4 = -1(x - (-7)) \quad \text{Substitute into point–slope form.}$$
$$y - 4 = -1(x + 7) \quad \text{Eliminate double signs.}$$
$$y - 4 = -x - 7 \quad \text{Distribute } -1.$$
$$y = -x - 3 \quad \text{Add 4 to isolate } y.$$

The equation of a line with slope -1 that passes through $(-7, 4)$ is $y = -x - 3$.

***Quick Check* 2**
Find the equation of a line whose slope is -2 and that passes through the point $(-1, -5)$. Write the equation in slope–intercept form.

In the previous examples, we converted the equation from point–slope form to slope–intercept form. This is a good idea in general. We use the point–slope form because it is a convenient form for finding the equation of a line if we know its slope and the coordinates of a point on the line. We convert the equation to slope–intercept form because it is easier to graph a line when its equation is in this form.

Finding the Equation of a Line Given Two Points on the Line

Objective 2 **Find the equation of a line given two points on the line.** Another use of the point–slope form is to find the equation of a line that passes through two given points. We begin by finding the slope of the line passing through those two points using the slope formula $m = \frac{y_2 - y_1}{x_2 - x_1}$. Then we use the point–slope form with this slope and either of the two points we were given. We finish by rewriting the equation in slope–intercept form.

EXAMPLE 3 Find the equation of a line that passes through the two points $(-3, -4)$ and $(-6, 2)$.

Solution

We begin by finding the slope of the line that passes through these two points.

$$m = \frac{2 - (-4)}{-6 - (-3)} \quad \text{Substitute into the formula } m = \frac{y_2 - y_1}{x_2 - x_1}.$$
$$= \frac{6}{-3} \quad \text{Simplify numerator and denominator.}$$
$$= -2 \quad \text{Simplify.}$$

We can now substitute into the point–slope form using either of the two points with $m = -2$. We will use $(-6, 2)$.

$$y - 2 = -2(x - (-6)) \quad \text{Substitute in point–slope form.}$$
$$y - 2 = -2(x + 6) \quad \text{Eliminate double signs.}$$
$$y - 2 = -2x - 12 \quad \text{Distribute } -2.$$
$$y = -2x - 10 \quad \text{Add 2 to isolate } y.$$

The equation of the line that passes through these two points is $y = -2x - 10$.

Here is the graph of the line, along with the points $(-3, -4)$ and $(-6, 2)$.

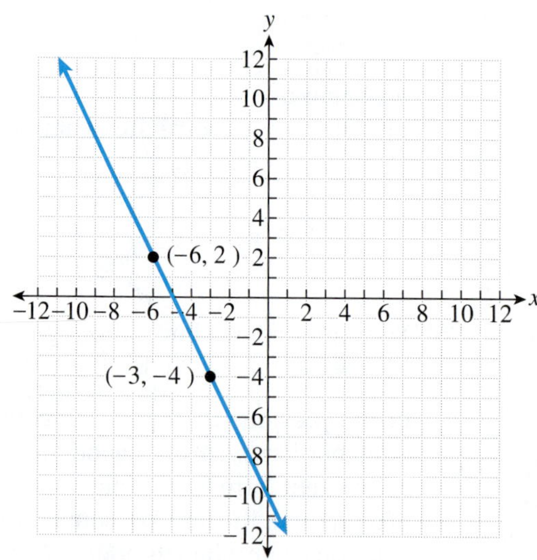

Quick Check 3

Find the equation of a line that passes through the two points $(6, -1)$ and $(-3, -7)$.

EXAMPLE 4 Find the equation of a line that passes through the two points $(5, -4)$ and $(5, 6)$.

Solution

We begin by attempting to find the slope of the line that passes through these two points.

$$m = \frac{6 - (-4)}{5 - 5}$$ Substitute into the slope formula.

$$= \frac{10}{0}$$ Simplify the numerator and denominator.

The slope is undefined, so this line is a vertical line. (We could have discovered this by plotting these two points on a graph.) The equation of this line is $x = 5$.

Quick Check 4

Find the equation of a line that passes through the two points $(-3, 8)$ and $(7, 8)$.

Finding a Linear Equation to Describe Real Data

Objective 3 **Find a linear equation to describe real data.**

EXAMPLE 5 In 1998 there were approximately 1,013,000 female graduate students in the United States. By 2003, this number had increased to approximately 1,233,000. Find a linear equation that describes the number (y) of female graduate students x years after 1998. (*Source:* U.S. Department of Education, National Center for Educational Statistics)

Solution

Since x represents the number of years after 1998, then $x = 0$ for the year 1998 and $x = 5$ for the year 2003. This tells us that two points on the line are $(0, 1{,}013{,}000)$ and $(5, 1{,}233{,}000)$. We begin by calculating the slope of the line.

$$m = \frac{1{,}233{,}000 - 1{,}013{,}000}{5 - 0}$$ Substitute into the slope formula.

$$= \frac{220{,}000}{5}$$ Simplify the numerator and denominator.

$$= 44{,}000$$ Divide.

The slope is 44,000, which tells us that the number of female graduate students increases at a rate of 44,000 students per year. In this example we know the y-intercept is $(0, 1{,}013{,}000)$, so we can write the equation directly in slope–intercept form. The equation is $y = 44{,}000x + 1{,}013{,}000$. (If we did not know the y-intercept of the line, we would find the equation of the line by substituting into the point–slope form.)

Quick Check 5 **In 2000 approximately 35.8 million people visited Las Vegas. By 2004, this number had increased to approximately 37.4 million people. Find a linear equation that tells the number (y) of people, in millions, who visited Las Vegas x years after 2000.** (*Source:* Las Vegas Convention and Visitors Authority)

Finding the Equation of a Parallel or Perpendicular Line

Objective 4 **Find the equation of a parallel or perpendicular line.** To find the equation of a line, we must know the slope of the line and the coordinates of a point on the line. In the previous examples, either we were given the slope or we calculated the slope using two points on the line. Sometimes the slope of the line is given in terms of another line. We could be given the equation of a line that is either parallel or perpendicular to the line for which we are trying to find the equation.

EXAMPLE 6 Find the equation of a line that is parallel to the line $y = -\frac{2}{3}x + 6$ and passes through $(-3, 5)$.

Solution

Since the line is parallel to $y = -\frac{2}{3}x + 6$, its slope must be $-\frac{2}{3}$. We will substitute this slope, along with the point $(-3, 5)$, into the point–slope form to find the equation of this line.

$$y - 5 = -\frac{2}{3}(x - (-3)) \quad \text{Substitute into point–slope form.}$$

$$y - 5 = -\frac{2}{3}(x + 3) \quad \text{Eliminate double signs.}$$

$$y - 5 = -\frac{2}{3}x - \frac{2}{\cancel{3}_1} \cdot \cancel{3}^1 \quad \text{Distribute and divide out common factors.}$$

$$y - 5 = -\frac{2}{3}x - 2 \quad \text{Simplify.}$$

$$y = -\frac{2}{3}x + 3 \quad \text{Add 5.}$$

Quick Check 6
Find the equation of a line that is parallel to the line $y = \frac{1}{4}x - 7$ and passes through $(-8, 7)$.

The equation of the line that is parallel to $y = -\frac{2}{3}x + 6$ and passes through $(-3, 5)$ is $y = -\frac{2}{3}x + 3$.

EXAMPLE 7 Find the equation of a line that is perpendicular to the line $x - 4y = 12$ and passes through $(-1, 8)$.

Solution

We must first find the slope of the line $x - 4y = 12$.

$$x - 4y = 12$$

$$-4y = -x + 12 \quad \text{Subtract } x \text{ from both sides.}$$

$$\frac{-4y}{-4} = \frac{-x}{-4} + \frac{12}{-4} \quad \text{Divide each term by } -4.$$

$$y = \frac{1}{4}x - 3 \quad \text{Simplify.}$$

The slope of the line is $\frac{1}{4}$, so the slope of a perpendicular line must be the negative reciprocal of $\frac{1}{4}$, or -4. We now can substitute into the point–slope form.

$$y - 8 = -4(x - (-1)) \quad \text{Substitute into point–slope form.}$$
$$y - 8 = -4(x + 1) \quad \text{Eliminate double signs.}$$
$$y - 8 = -4x - 4 \quad \text{Distribute.}$$
$$y = -4x + 4 \quad \text{Add 8.}$$

The equation of the line is $y = -4x + 4$.

***Quick Check* 7**

Find the equation of a line that is perpendicular to the line $2x + 3y = 15$ and passes through $(4, 11)$.

If we are given . . .	We find the equation by . . .
The slope and the y-intercept	Substituting m and b into the slope–intercept form $y = mx + b$
The slope and a point on the line	Substituting m and the coordinates of the point into the point–slope form $y - y_1 = m(x - x_1)$
Two points on the line	Calculating m using the slope formula $m = \frac{y_2 - y_1}{x_2 - x_1}$ and then substituting the slope and the coordinates of one of the points into the point-slope form $y - y_1 = m(x - x_1)$
A point on the line and the equation of a parallel line	Substituting the slope of that line and the coordinates of the point into the point-slope form $y - y_1 = m(x - x_1)$
A point on the line and the equation of a perpendicular line	Substituting the negative reciprocal of the slope of that line and the coordinates of the point into the point-slope form $y - y_1 = m(x - x_1)$

Keep in mind that if the slope of the line is undefined, then the line is vertical and its equation is of the form $x = a$, where a is the x-coordinate of the given point. We do not use the point–slope form or the slope–intercept form to find the equation of a vertical line.

***Building Your Study Strategy* Using Your Textbook Effectively, 3 Rewriting Your Notes** When you are rewriting your classroom notes, you may find your textbook to be helpful in supplementing your notes. If your instructor gave you a set of definitions in class, look up the corresponding definitions in the textbook. You can also look for examples similar to the ones provided by your instructor during class.

EXERCISES 2.3

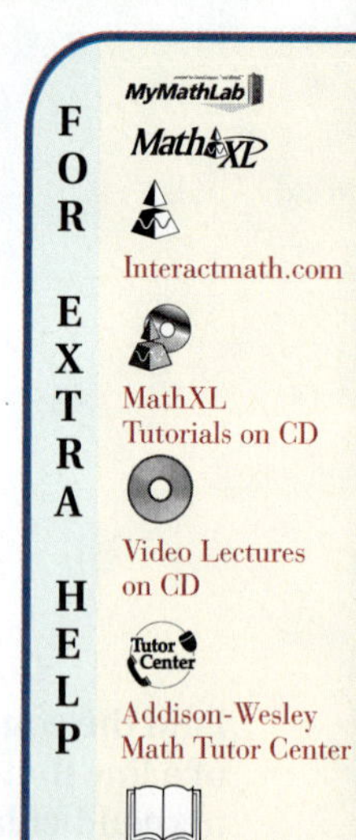

Vocabulary

1. The point–slope form of the equation of a line with slope m that passes through the point (x_1, y_1) is ________.

2. State the procedure for finding the equation of a line that passes through the points (x_1, y_1) and (x_2, y_2).

Write the following equations in slope–intercept form.

3. $-4x + 2y = -12$
4. $4x + 6y = 9$
5. $2x - 5y = 15$
6. $-8x - 6y = 20$
7. $8x - y = 9$
8. $10x + 8y = 0$

Find the equation of a line with the given slope and y-intercept.

9. Slope 5, y-intercept $(0, 3)$
10. Slope -1, y-intercept $(0, -6)$
11. Slope $\frac{3}{4}$, y-intercept $(0, -4)$
12. Slope 0, y-intercept $(0, 8)$
13. Slope $-\frac{2}{5}$, y-intercept $(0, 4)$
14. Slope $\frac{4}{7}$, y-intercept $(0, -9)$

Find the slope–intercept form of the equation of a line with the given slope that passes through the given point.

15. Slope 2, through $(3, -2)$
16. Slope 5, through $(-4, 2)$
17. Slope -3, through $(-5, -1)$
18. Slope -1, through $(2, -7)$
19. Slope $\frac{2}{3}$, through $(-6, 4)$
20. Slope $\frac{1}{5}$, through $(-5, 0)$
21. Slope $-\frac{1}{4}$, through $(8, 7)$
22. Slope $-\frac{3}{2}$, through $(-4, 9)$
23. Slope $\frac{5}{6}$, through $(4, -1)$
24. Slope $\frac{3}{8}$, through $(-2, -7)$
25. Slope -2.5, through $(3.2, -5.4)$
26. Slope 6.4, through $(3.5, 9.2)$

Find the slope–intercept form of the equation of a line with the given slope that passes through the given point. Graph the line and label the given point.

27. Slope 1, through $(5, 2)$

28. Slope -2, through $(3, -4)$

29. Slope $-\frac{4}{3}$, through $(-3, 8)$

30. Slope $\frac{1}{2}$, through $(-6, -2)$

31. Slope 0, through $(3, 4)$

32. Undefined slope, through $(3, 4)$

Find the slope–intercept form of the equation of a line that passes through the given points.

33. $(2, 1), (3, 5)$

34. $(1, -1), (3, -7)$

35. $(-2, 11), (4, -25)$

36. $(-5, -13), (3, -5)$

37. $(-9, -6), (18, 0)$

38. $(-8, 12), (4, 3)$

39. $(-6, 10), (6, 6)$

40. $(-2, -11), (6, 9)$

41. $(2.5, 37), (3.3, 81)$

42. $(4.2, 51.3), (6.6, 93.9)$

Find the slope–intercept form of the equation of a line that passes through the given points. Graph the line and label the given points.

43. $(1, -6), (5, 2)$

44. $(-2, 8), (7, -1)$

45. $(-4, 5), (8, -1)$

46. $(-2, -9), (10, 9)$

For Exercises 47–50, find the slope–intercept form of the equation of a line with the given x-intercept and y-intercept.

	47.	48.
x-intercept	(4, 0)	(−3, 0)
y-intercept	(0, 2)	(0, 6)

	49.	50.
x-intercept	(7, 0)	(−8, 0)
y-intercept	(0, −7)	(0, −6)

Find the slope–intercept form of the equation of the given line.

51.

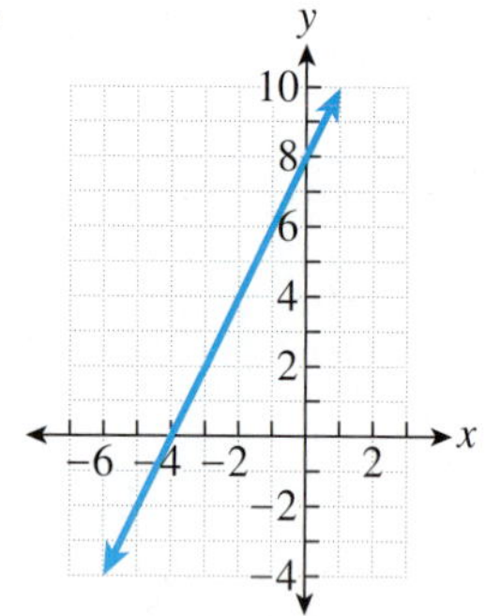

52.

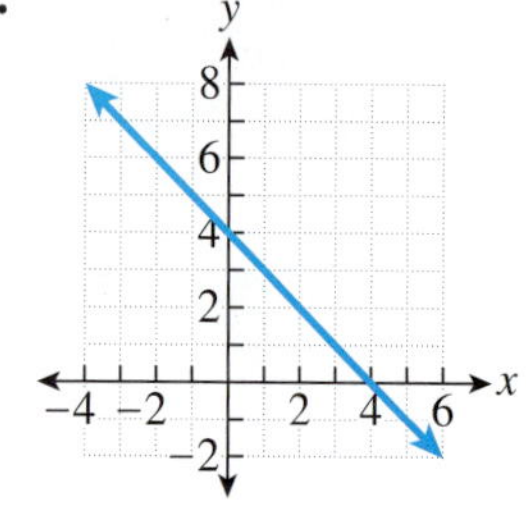

53.

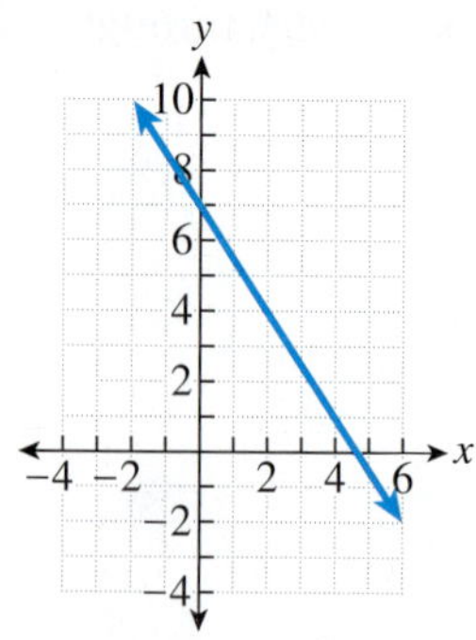

54.

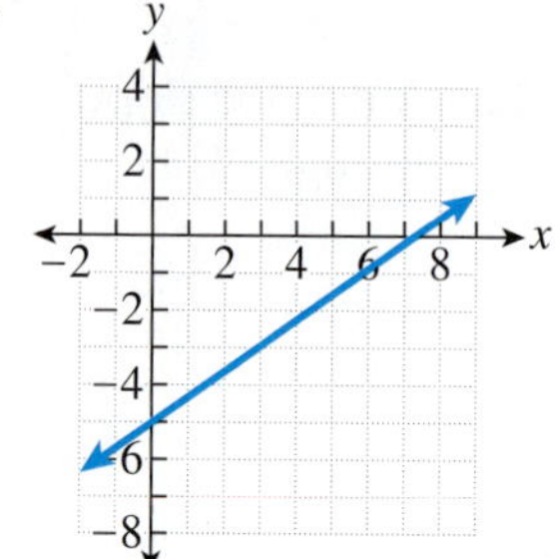

Find the slope–intercept form of the equation of a line that meets the given conditions.

55. Parallel to $y = -3x + 7$, through $(4, -2)$

56. Parallel to $y = 5x - 4$, through $(-2, 9)$

57. Parallel to $-4x + y = 8$, through $(5, 3)$

58. Parallel to $8x + 4y = 33$, through $(-3, -2)$

59. Perpendicular to $y = 3x - 8$, through $(6, -5)$

60. Perpendicular to $y = -\frac{1}{4}x + 9$, through $(-2, 7)$

61. Perpendicular to $2x - 5y = 20$, through $(-4, -6)$

62. Perpendicular to $3x + 4y = 24$, through $(-6, 3)$

63. Parallel to $y = 7$, through $(9, -6)$

64. Perpendicular to $x = -2$, through $(-4, -8)$

Find the slope–intercept form of the equation of a line that is parallel to the graphed line and that passes through the point plotted on the graph.

65.

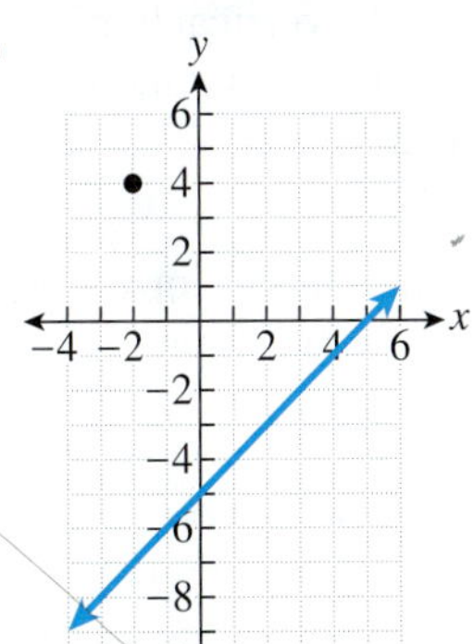

66.

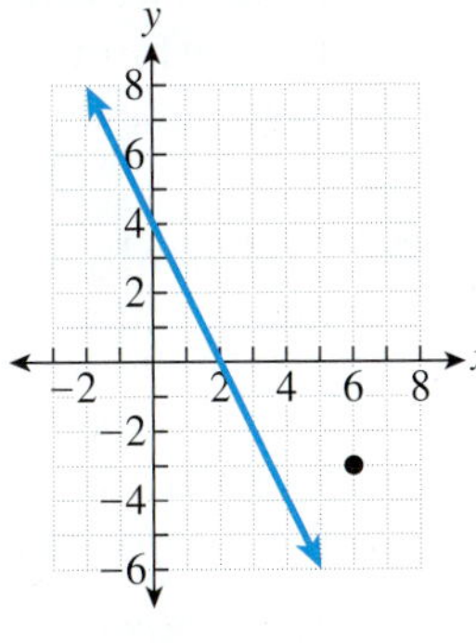

Find the slope–intercept form of the equation of a line that is perpendicular to the graphed line and that passes through the point plotted on the graph.

67.

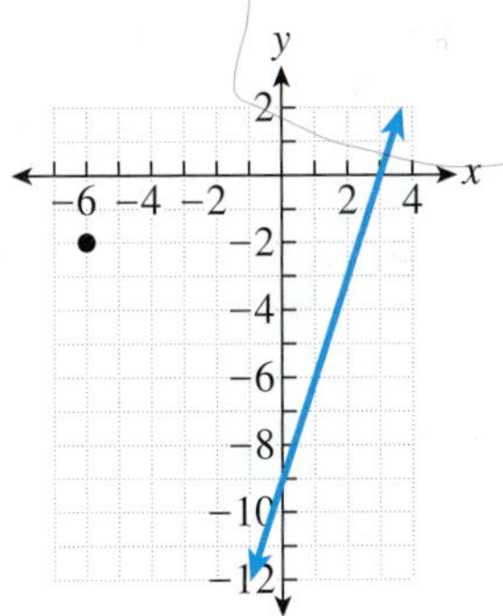

68.

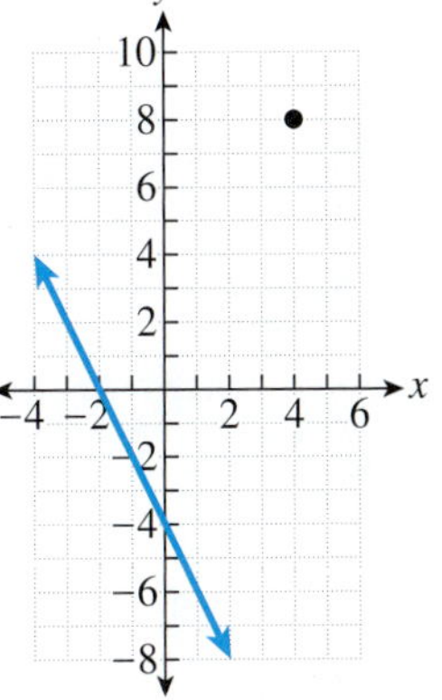

69. A car salesperson is paid a weekly salary, in addition to a commission for each car sold. Two weeks ago Jose sold 8 cars, and his paycheck for that week was \$2920. Last week Jose sold 3 cars, and his paycheck was \$1170.

a) Find a linear equation that calculates Jose's weekly pay in a week that he sells x cars.

b) What is Jose's weekly salary?

c) How much is Jose paid for each car that he sells?

70. John contracted to have a swimming pool built in his backyard. The contractor charged \$28,000. John paid the contractor a deposit, and now makes a fixed monthly payment until the balance has been paid in full. After 6 months John owed \$14,400. After 10 months John still owed \$12,000.

a) Find a linear equation that calculates John's balance (y) after x months.

b) How much was the deposit that John paid?

c) How many months will it take to pay off the entire balance?

71. A mechanic charges a service fee in addition to an hourly rate to fix a car. To fix Janessa's car, the mechanic took 3 hours and charged a total of \$525 (service fee plus hourly rate for 3 hours of work). Ronaldo's car took 5 hours to fix, and the charge was \$775.

a) Find a linear equation that calculates the charge (y) for a repair that takes x hours to perform.

b) How much is the repairman's service fee?

c) How much is the repairman's hourly rate?

72. Members of a country club pay an annual membership fee. In addition, they pay a greens fee each time they play a round of golf. Last year Mark played golf 75 times and paid the country club a total of \$13,250 (membership fee and greens fees). Last year Jared played golf 25 times and paid the country club \$8750.

a) Find a linear equation that calculates the amount owed (y) by a member who plays x rounds of golf in a year.

b) What is the annual membership fee at this country club?

c) What is the cost for a single round of golf?

73. An online book store charges a handling fee for each order placed, plus a shipping fee for each book. Don placed an order for 6 books, and he paid \$34.45 to ship his order. Rose placed an order for 9 books, and she paid \$48.70 to ship her order.

a) Find a linear equation that calculates the amount charged (y) to ship an order of x books.

b) What is the handling fee charged for each order?

c) What is the shipping fee for each book?

74. A teacher's association honors retirees by giving them a small bronze statue. When ordering, the association pays a fee to create the mold used to make all the statues, in addition to a fee for each statue created. An order of 6 statues would cost a total of \$729.65, while an order of 12 statues would cost \$1209.35.

a) Find a linear equation that calculates the charge (y) for an order of x statues.

b) How much is the fee to create the mold?

c) How much is the fee for each statue?

Writing in Mathematics

Answer in complete sentences.

75. Explain why the equation of a vertical line has the form $x = a$ and the equation of a horizontal line has the form $y = b$.

76. Explain why a horizontal line has a slope of 0 and a vertical line has undefined slope.

77. ***Solutions Manual**** Write a solutions manual page for the following problem:

Find the equation of the line with slope $m = -\frac{4}{3}$ *that passes through the point* $(6, -7)$.

78. ***Newsletter**** Write a newsletter explaining how to find the equation of a line that passes through two given points.

***See Appendix B for details and sample answers.**

Quick Review Exercises

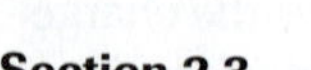

Section 2.3

Graph. Label any x- and y-intercepts.

1. $5x - 2y = 10$

2. $y = -3x - 8$

3. $y = \frac{2}{3}x - 4$

4. $y = -8$

2.4 Linear Inequalities

Objectives

1. **Determine whether an ordered pair is a solution of a linear inequality in two variables.**
2. **Graph a linear inequality in two variables.**
3. **Graph a linear inequality involving a horizontal or vertical line.**
4. **Graph linear inequalities associated with applied problems.**

In this section we will learn how to solve **linear inequalities** in two variables. Here are some examples of linear inequalities in two variables.

$$2x + 3y \leq 6 \qquad 5x - 4y \geq -8 \qquad -x + 9y < -18 \qquad -3x - 4y > 7$$

Solutions of Linear Inequalities in Two Variables

Objective 1 Determine whether an ordered pair is a solution of a linear inequality in two variables.

Solutions of Linear Inequalities

A solution of a linear inequality in two variables is an ordered pair (x, y) such that when the coordinates are substituted into the inequality, a true statement results.

For example, consider the linear inequality $2x + 3y \leq 6$. The ordered pair $(2, 0)$ is a solution because, when we substitute these coordinates into the inequality, it produces a true statement.

$$\begin{aligned} 2x + 3y &\leq 6 \\ 2(2) + 3(0) &\leq 6 \\ 4 + 0 &\leq 6 \\ 4 &\leq 6 \end{aligned}$$

Since this is a true statement, $(2, 0)$ is a solution. The ordered pair $(3, 4)$ is not a solution because $2(3) + 3(4)$ is not less than or equal to 6. Any ordered pair (x, y) for which $2x + 3y$ evaluates to be less than or equal to 6 is a solution to this inequality, and there are infinitely many solutions to this inequality. We will display our solutions on a graph.

Graphing Linear Inequalities in Two Variables

Objective 2 Graph a linear inequality in two variables. Ordered pairs that are solutions to the inequality $2x + 3y \leq 6$ are of one of two types: ordered pairs (x, y) that are solutions to $2x + 3y = 6$, and ordered pairs (x, y) that are solutions to $2x + 3y < 6$. Points satisfying $2x + 3y = 6$ lie on a line, so we begin by graphing this line. Since the equation is in standard form, a quick way to graph this line is by finding its x- and y-intercepts.

x-intercept ($y = 0$)	y-intercept ($x = 0$)
$2x + 3(0) = 6$	$2(0) + 3y = 6$
$2x + 0 = 6$	$0 + 3y = 6$
$2x = 6$	$3y = 6$
$x = 3$	$y = 2$
$(3, 0)$	$(0, 2)$

Here is the graph of the line.

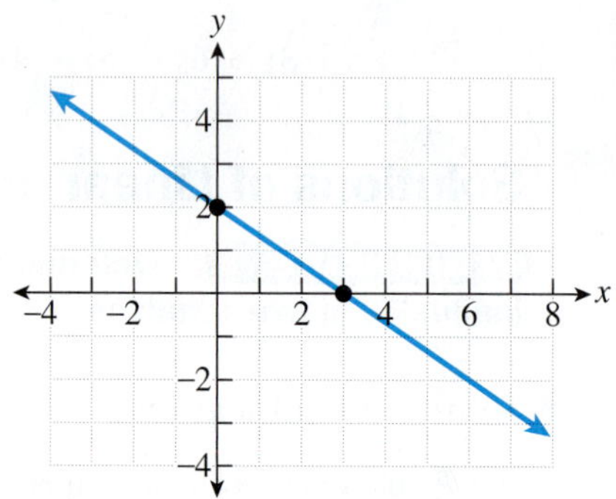

This line divides the plane into two half-planes and is the dividing line between ordered pairs for which $2x + 3y < 6$ and ordered pairs for which $2x + 3y > 6$. To finish graphing our solutions, we must determine which half-plane contains the ordered pairs for which $2x + 3y < 6$. To do this we will use a **test point**, which is a point that is not on the line. We substitute the test point's coordinates into the original inequality, and if the resulting inequality is true, this point and all other points on the same side of the line are solutions, and we will shade that half-plane. If the resulting inequality is false, then the solutions are on the other side of the line and we shade that half-plane instead. A wise choice for the test point is the origin $(0, 0)$ if it is not on the line which has been graphed, since its coordinates are easy to work with when substituting into the inequality. Since $(0, 0)$ is not on the line, we will use it as a test point.

$$\begin{aligned} \text{Test Point: } (0, 0)& \\ 2(0) + 3(0) &\le 6 \\ 0 + 0 &\le 6 \\ 0 &\le 6 \end{aligned}$$

Since this is a true inequality, $(0, 0)$ is a solution, and we will shade the half-plane containing this point.

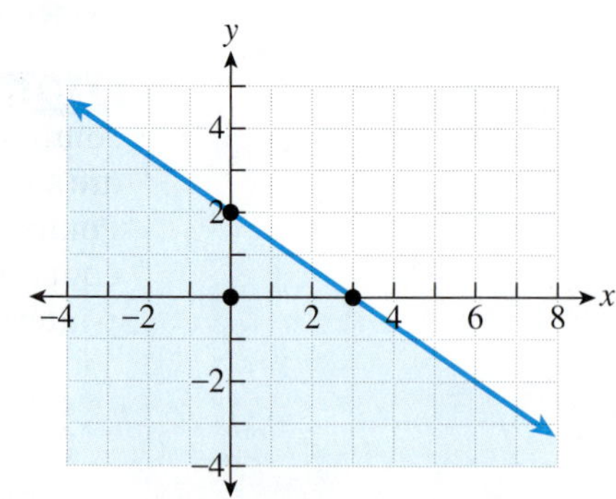

EXAMPLE 1 Graph $y \leq 2x - 6$.

Solution

We begin by graphing the line $y = 2x - 6$. This equation is in slope–intercept form, so we can graph it by plotting its y-intercept $(0, -6)$ and using the slope (up 2 units, 1 unit to the right) to find other points on the line.

Since the line does not pass through the origin, we can use $(0, 0)$ as a test point.

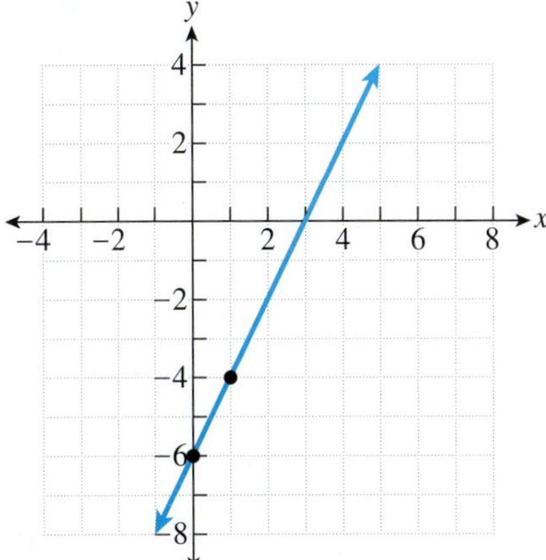

Test Point: $(0, 0)$
$0 \leq 2(0) - 6$
$0 \leq 0 - 6$
$0 \leq -6$

Quick Check 1
Graph $y \leq 3x + 9$.

This inequality is false, so the solutions are in the half-plane that does not contain the origin. Here is the graph of the inequality.

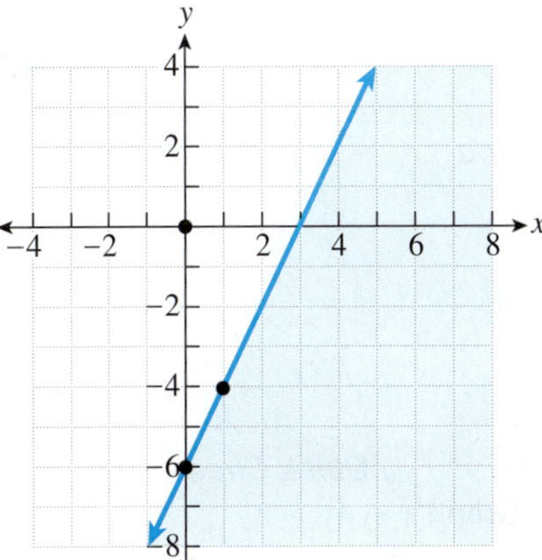

The first two inequalities we graphed were **weak linear inequalities**, which are inequalities involving the symbols $\leq$ or $\geq$. We now turn our attention to **strict linear inequalities**, which involve the symbols $<$ or $>$. An example of a strict linear inequality would be $4x - 3y < 12$. Ordered pairs for which $4x - 3y = 12$ are not solutions to this inequality, so the points on the line are not included as solutions. We denote this on the graph by graphing the line as a dashed or broken line. We still pick a test point and shade the appropriate half-plane in the same way that we did in the previous examples.

EXAMPLE 2 Graph $4x - 3y < 12$.

Solution

We start by graphing the line $4x - 3y = 12$ as a dashed line. This equation is in standard form, so we can graph the line by finding its intercepts.

x-intercept ($y = 0$)	y-intercept ($x = 0$)
$4x - 3(0) = 12$	$4(0) - 3y = 12$
$4x = 12$	$-3y = 12$
$x = 3$	$y = -4$
$(3, 0)$	$(0, -4)$

Here is the graph of the line, with an x-intercept at $(3, 0)$ and a y-intercept at $(0, -4)$. Notice that the line is a dashed line.

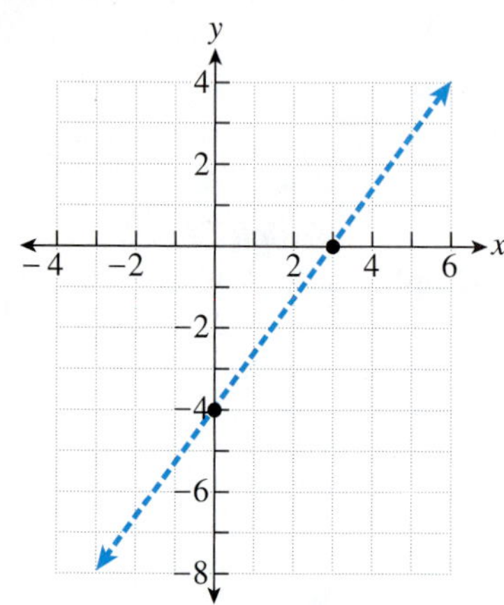

Since the line does not pass through the origin, we will use $(0, 0)$ as a test point.

$$\begin{aligned} &\text{Test Point: } (0, 0) \\ 4(0) - 3(0) &< 12 \\ 0 - 0 &< 12 \\ 0 &< 12 \end{aligned}$$

This inequality is true, so the origin is a solution. We shade the half-plane containing the origin.

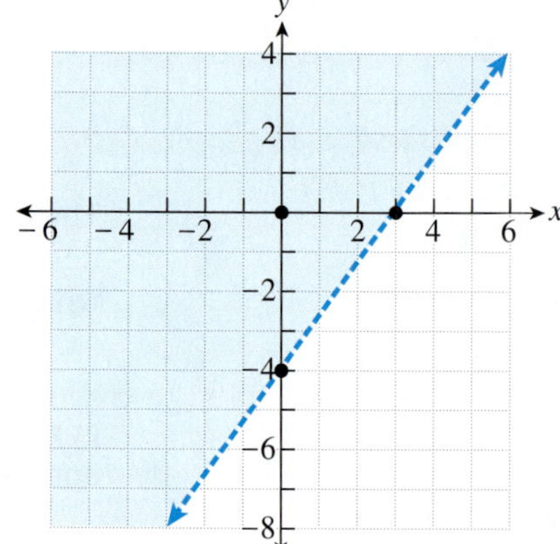

***Quick Check* 2**
Graph $x + 4y > 8$.

EXAMPLE 3 Graph $x > 2y - 9$.

Solution

This is a strict inequality, so we will graph the line $x = 2y - 9$ as a dashed line. We will begin by rewriting this equation in standard form as $x - 2y = -9$, so we can graph the line by finding its intercepts.

x-intercept ($y = 0$)	y-intercept ($x = 0$)
$x - 2(0) = -9$	$0 - 2y = -9$
$x = -9$	$-2y = -9$
	$y = \frac{9}{2}$
$(-9, 0)$	$(0, \frac{9}{2})$

Here is the graph of the dashed line, with an x-intercept at $(-9, 0)$ and a y-intercept at $(0, \frac{9}{2})$.

Since the line does not pass through the origin, we will use $(0, 0)$ as a test point.

Test Point: $(0, 0)$
$0 > 2(0) - 9$
$0 > -9$

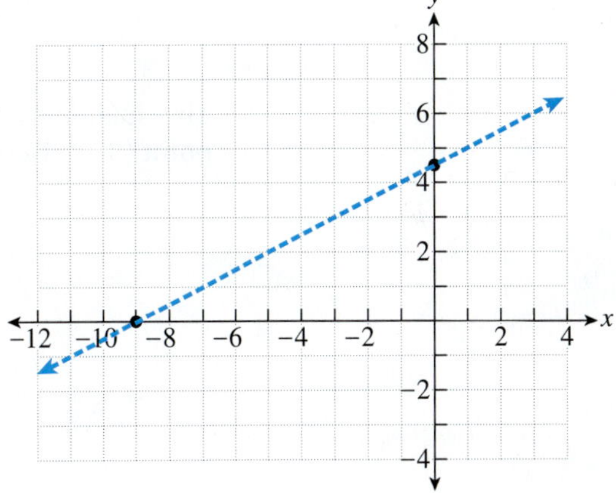

This inequality is true, so we will shade the half-plane that contains the test point $(0, 0)$. Here is the graph.

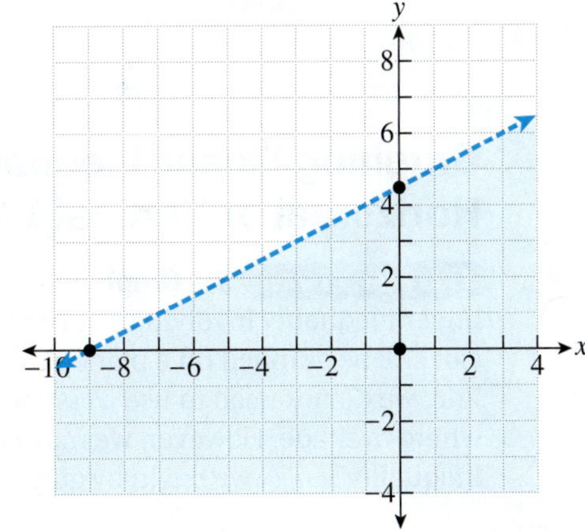

Quick Check 3
Graph $2x \leq 5y - 5$.

EXAMPLE 4 Graph $y \ge \frac{5}{3}x$.

Solution

This is a weak inequality, so we begin by graphing the line $y = \frac{5}{3}x$ as a solid line.

The equation is in slope–intercept form, with a slope of $\frac{5}{3}$ and a y-intercept at $(0, 0)$. After we plot the y-intercept, we can use the slope to find a second point at $(3, 5)$. Here is the graph of the line.

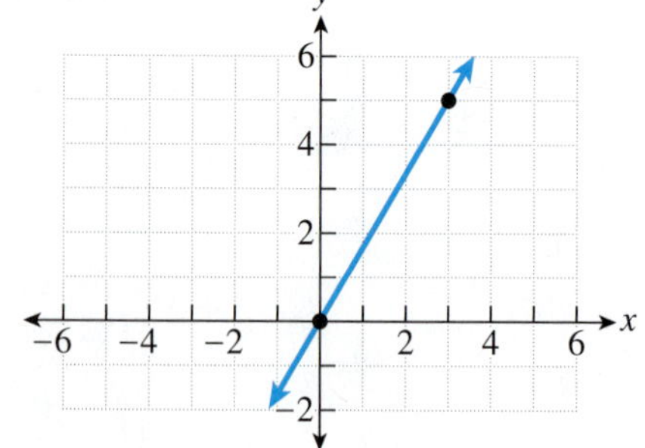

Since the line passes through the origin, we cannot use $(0, 0)$ as a test point. We will try to choose a point that is clearly not on the line, such as $(3, 0)$, which is to the right of the line.

Test Point: $(3, 0)$

$$0 \ge \frac{5}{3}(3)$$

$$0 \ge 5$$

This inequality is false, so we will shade the half-plane that does not contain the test point $(3, 0)$. Here is the graph.

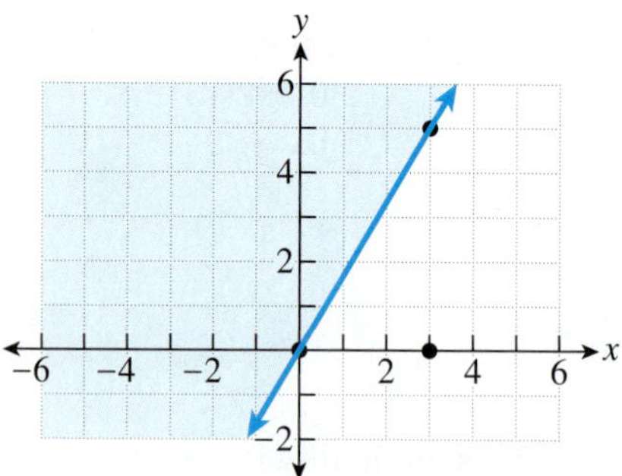

Quick Check 4

Graph $y < 5x$.

Graphing Linear Inequalities Involving Horizontal or Vertical Lines

Objective 3 **Graph a linear inequality involving a horizontal or vertical line.** A linear inequality involving a vertical line or a horizontal line has only one variable, but it can still be graphed on a plane rather than on a number line. After we graph the related line, we do not need to use a test point. Instead, we can use reasoning to determine where to shade. However, we may continue to use test points if we choose. To graph the inequality $x > 5$, we begin by graphing the vertical line $x = 5$ using a dashed line. The values of x that are greater than 5 are to the right of this line, so we shade the half-plane to the right of $x = 5$. If we had used the origin as a test point, we would have discovered that the resulting inequality $(0 > 5)$ is false, so we would have shaded the half-plane that does not contain the origin, producing the same graph.

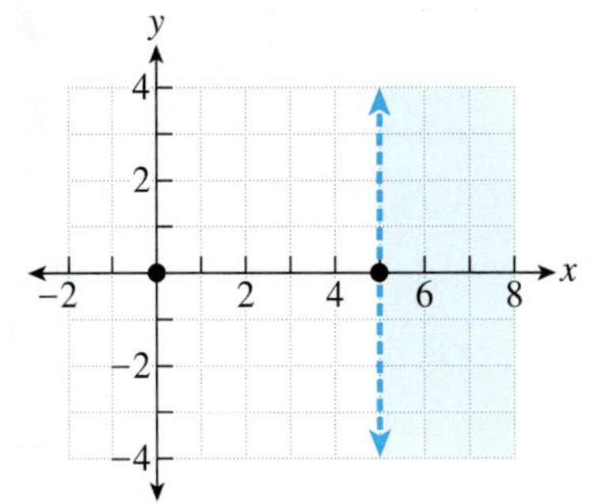

To graph the inequality $y \le -2$, we begin by graphing the horizontal line $y = -2$ using a solid line. The values of y that are less than -2 are below this line, so we shade the half-plane below the line.

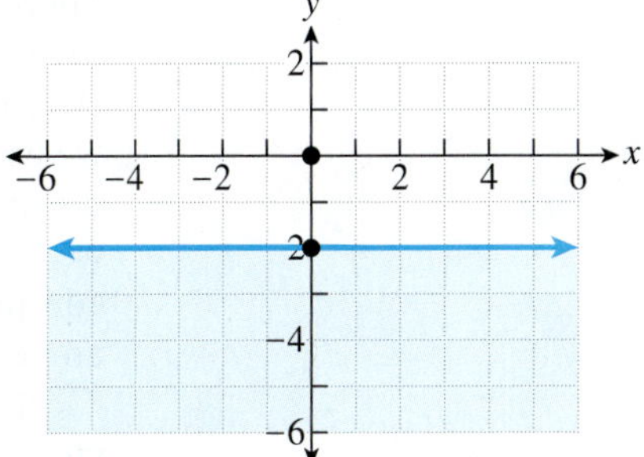

EXAMPLE 5 Graph $y > 4$.

Solution

This is a strict inequality, so we begin by graphing the horizontal line $y = 4$ as a dashed line. The values of y that are greater than 4 are above this line, so we shade the half-plane above the line.

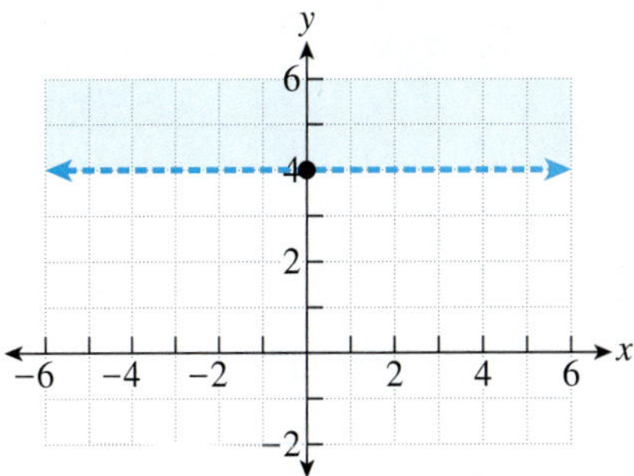

***Quick Check* 5**
Graph $x \le 3$.

Here is a brief summary of how to graph linear inequalities in two variables.

Graphing Linear Inequalities in Two Variables

- Graph the line related to the inequality, which is found by replacing the inequality symbol with an equal sign.
- If the inequality symbol includes equality ($\le$ or $\ge$), graph the line as a solid line.
- If the inequality symbol does not include equality ($<$ or $>$), graph the line as a dashed, or broken, line.
- Select a test point that is not on the line. Use the origin if possible. Substitute the coordinates of the point into the original inequality. If the resulting inequality is true, then shade the region on the side of the line that contains the test point. Otherwise, if the resulting inequality is false, shade the region on the side of the line that does not contain the test point.

Applications

Objective 4 Graph linear inequalities associated with applied problems.
We end the section with an applied problem.

EXAMPLE 6 A company is planning to buy some new desktop and laptop computers. The price quote for a desktop computer was \$900, while the price for a laptop computer was \$1200. The company has a budget of \$28,800. Set up and graph the appropriate inequality for the number of desktop computers and laptop computers that the company can buy.

Solution

There are two unknowns in this problem—the number of desktop computers that the company will buy and the number of laptop computers that the company will buy. We will let x represent the number of desktop computers that the company will buy and we will let y represent the number of laptop computers that the company will buy. Since the amount of money that the company will spend on computers cannot exceed \$28,800, we know that $900x + 1200y \leq 28{,}800$. (We also know that $x \geq 0$ and $y \geq 0$, since the company cannot buy a negative number of desktop or laptop computers. Therefore we are restricted to the first quadrant, the positive x-axis and the positive y-axis.)

To graph this inequality we begin by graphing $900x + 1200y = 28{,}800$ as a solid line. Since this equation is in standard form, we graph the line using its x-intercept $(32, 0)$ and its y-intercept $(0, 24)$.

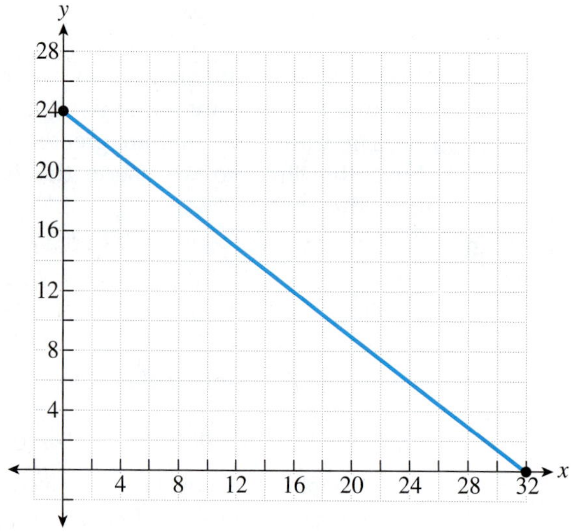

The origin is not on this line, so we can use $(0, 0)$ as a test point.

$$900x + 1200y \leq 28{,}800$$
$$900(0) + 1200(0) \leq 28{,}800 \quad \text{Substitute 0 for } x \text{ and 0 for } y.$$
$$0 \leq 28{,}800 \quad \text{True.}$$

This is a true statement, so we will shade on the side of the line that contains the origin.

Any ordered pair in the shaded region with integer coordinates is an actual solution to this problem. For example, the ordered pair $(20, 8)$ is in the shaded region and it tells us that the company can buy 20 desktop computers and 8 laptop computers and stay within its budget.

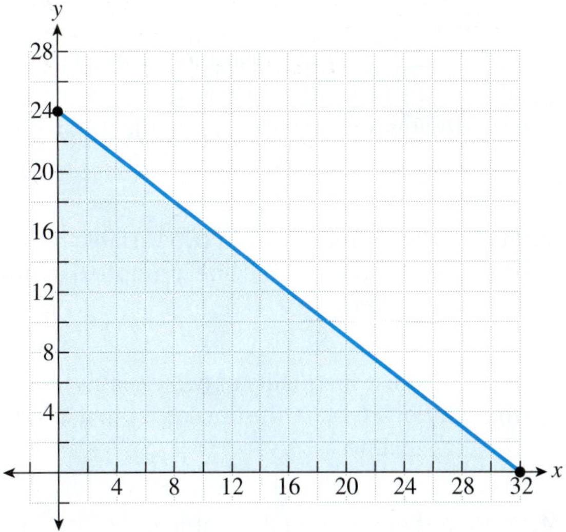

Quick Check **6** **A pharmaceutical company provides its sales representatives with vehicles; some receive vans and some receive sedans. This year the company will need to buy at least 30 vehicles. If the company purchases only vans, *x*, and sedans, *y*, set up and graph the appropriate inequality.**

Building Your Study Strategy **Using Your Textbook Effectively, 4** **Textbook Examples** You should keep your textbook nearby while you are working on your homework. If you are stuck on a particular problem, look through the section for an example that may give you an idea about how to proceed.

EXERCISES 2.4

FOR EXTRA HELP

MyMathLab
MathXL
Interactmath.com
MathXL Tutorials on CD
Video Lectures on CD
Addison-Wesley Math Tutor Center
Student's Solutions Manual

Vocabulary

1. A(n) __________ to a linear inequality in two variables is an ordered pair (x, y) such that when the coordinates are substituted into the inequality, a true statement results.

2. When a linear inequality in two variables involves the symbols $\le$ or $\ge$, the line is graphed as a(n) __________ line.

3. When a linear inequality in two variables involves the symbols $<$ or $>$, the line is graphed as a(n) __________ line.

4. A point that is not on the graph of the line and whose coordinates are used for determining which half-plane contains the solutions to the linear inequality is called a(n) __________.

5. Give an example of a strict linear inequality in two variables.

6. Give an example of a weak linear inequality in two variables.

Determine whether the ordered pair is a solution of the given linear inequality.

7. $4x - 3y \le 12$

a) $(0, 0)$ **b)** $(4, -2)$

8. $5x - 2y > 10$

a) $(2, 0)$ **b)** $(3, -2)$

9. $-7x + 4y < -15$

a) $(-2, 1)$ **b)** $(4, 3)$

10. $3x - 6y \ge -18$

a) $(-5, -1)$ **b)** $(-4, 1)$

11. $y \ge 3x - 8$

a) $(5, 9)$ **b)** $(-2, -10)$

12. $y < -2x + 7$

a) $(0, 0)$ **b)** $(8, -12)$

Complete the solution of the linear inequality by shading the appropriate region.

13. $4x + y \ge 7$

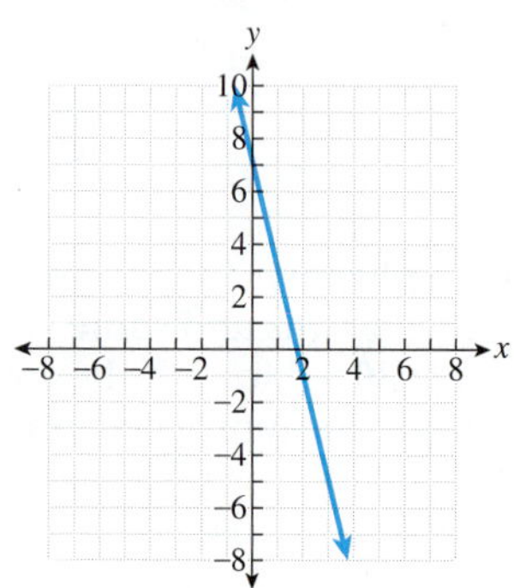

14. $-3x + 2y \le -6$

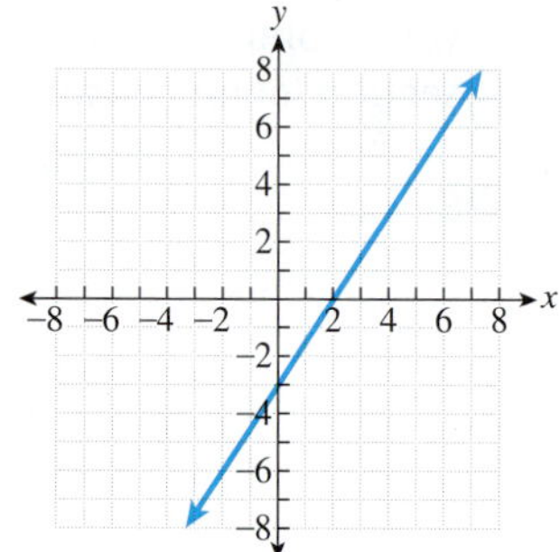

15. $2x + 8y < -4$

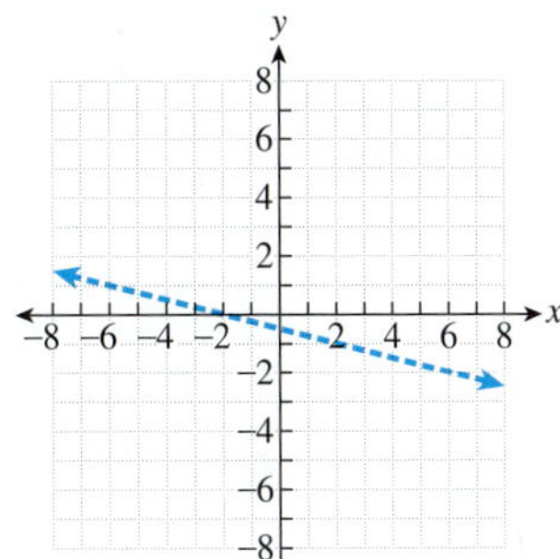

16. $10x - 2y > 0$

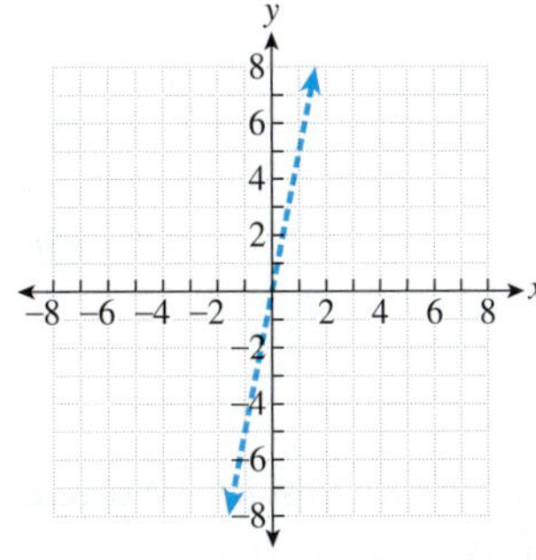

Graph the linear inequality on a plane.

17. $y \leq 3x - 9$

18. $y < -2x + 6$

19. $y > \frac{1}{3}x - 2$

20. $y \leq \frac{3}{2}x + 9$

21. $y < -x + 4$

22. $y > 4x - 2$

23. $y \geq 5x$

24. $y \leq -2x$

25. $y > 4x - 16$

26. $y \leq -x + 18$

27. $3x + 2y \leq 12$

28. $x + 4y \geq -8$

29. $4x - 5y > -20$

30. $7x - 3y < 21$

31. $5x + 2y > 0$

32. $x - y \leq 0$

33. $6x + 3y \geq 9$

34. $2x + 8y > -12$

35. $x < 3y - 5$

36. $x \leq -2y + 8$

37. $y > 4$

38. $y \leq -2$

39. $x \geq 5$

40. $x < 1$

Which graph, A or B, represents the solution of the linear inequality? Explain how you chose your answer.

41. $4x - 3y \leq 12$

A

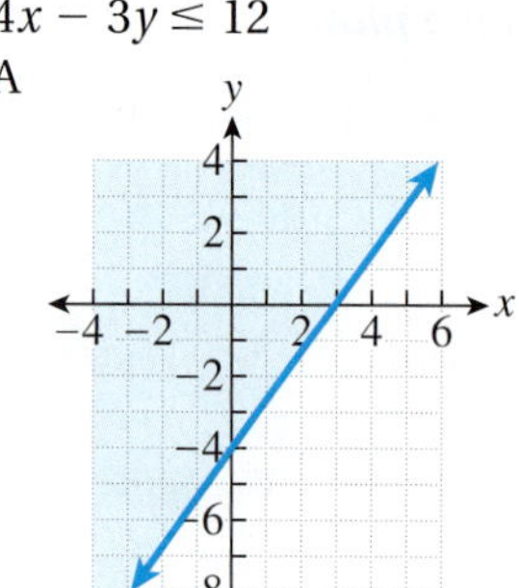

B

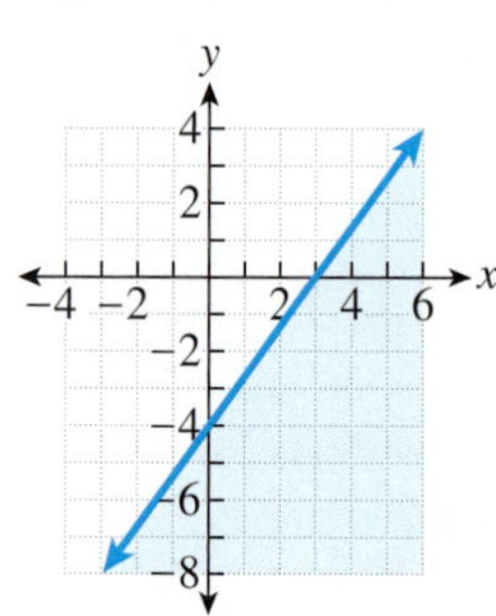

42. $y \geq 3x$

A

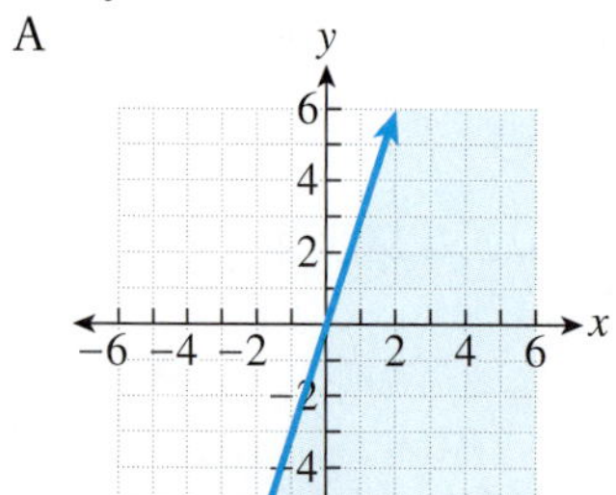

B

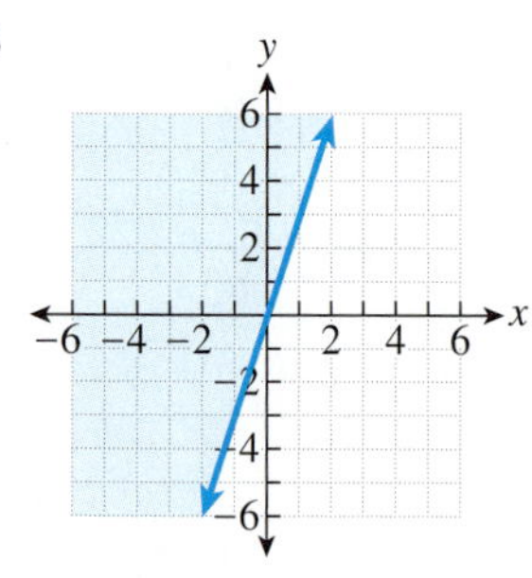

43. $5x + 2y < -16$

A

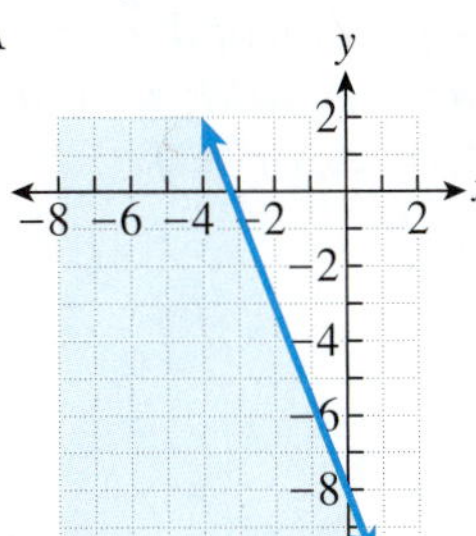

B

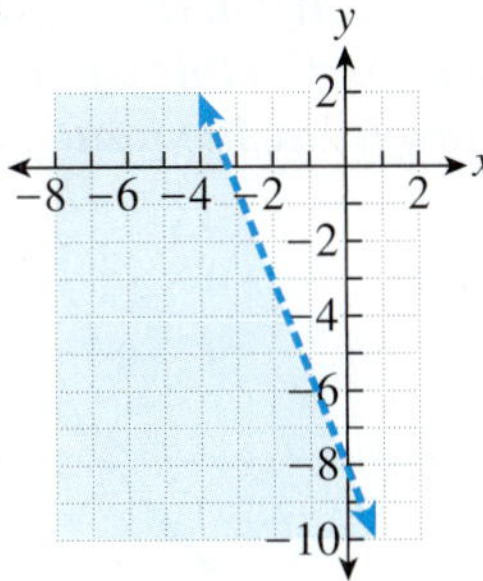

44. $x < 4y - 8$

A B

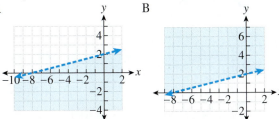

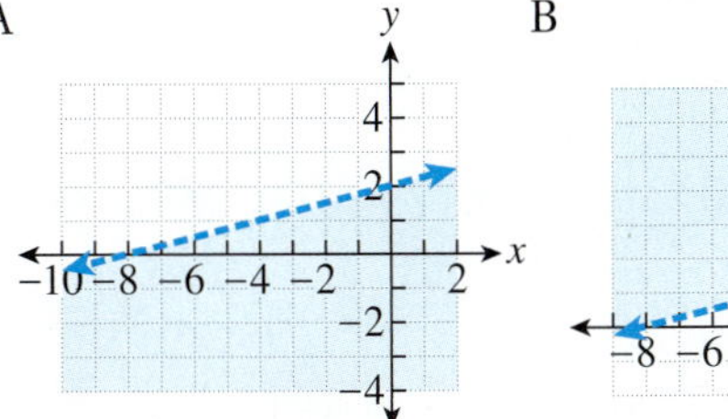

Determine the missing inequality sign ($<, >, \le$, or $\ge$) for the linear inequality based on the given graph.

45. $x + 3y$ ______ 6

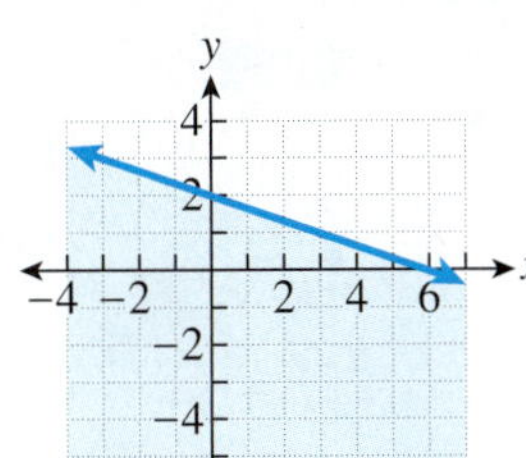

46. $4x + 5y$ ______ 20

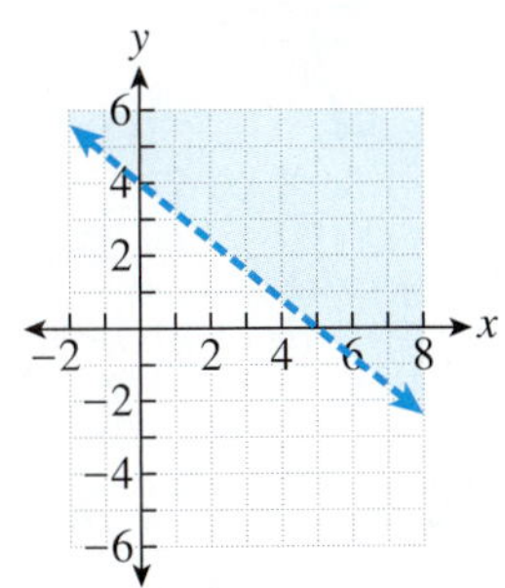

47. y ______ $-2x + 6$

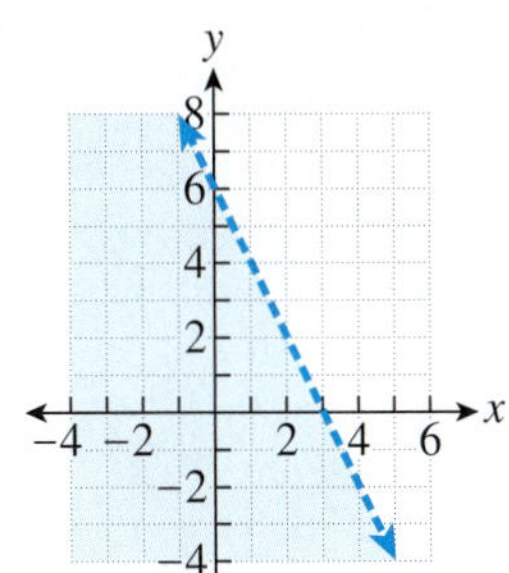

48. $3x - 2y$ ______ 18

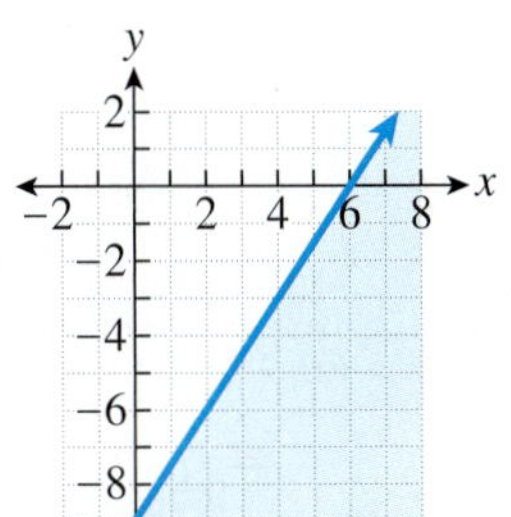

Mixed Practice, 49–60

Graph the given linear equation or linear inequality on a plane.

49. $x + 2y = -8$

50. $x < 2$

51. $4x - 3y < -24$

52. $y = -\frac{2}{3}x + 6$

53. $x = -7$

54. $y > -x + 6$

55. $y \leq 2x + 4$

56. $y = 3$

57. $y \geq -4$

58. $2x + 7y \geq -14$

59. $y = 3x - 6$

60. $5x - 3y = 15$

Determine the linear inequality associated with the given solution. (First, find the equation of the line. Then rewrite this equation as an inequality with the appropriate inequality sign: <, >, ≤, or ≥.)

61.

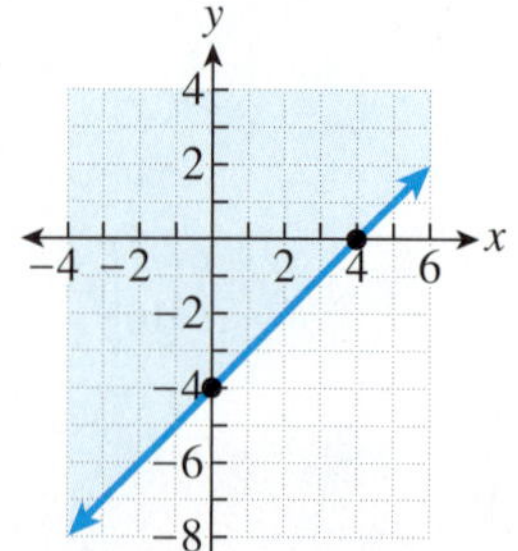

62.

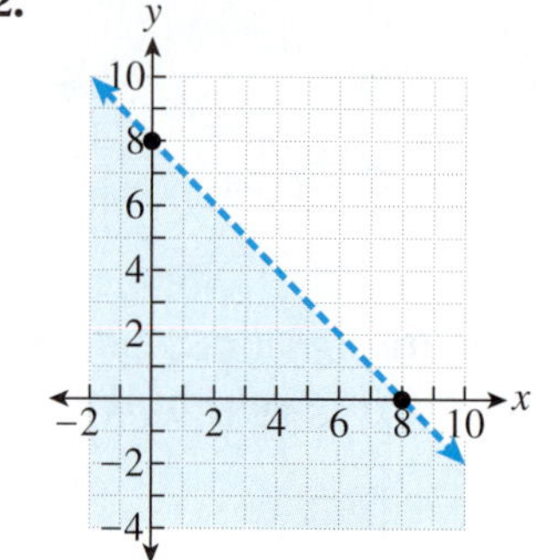

63.

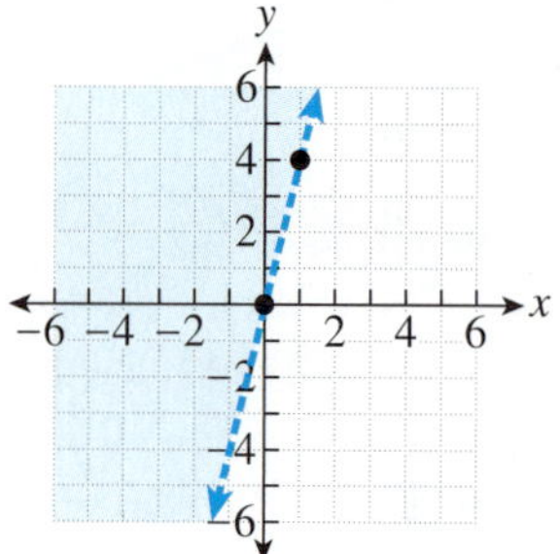

64.

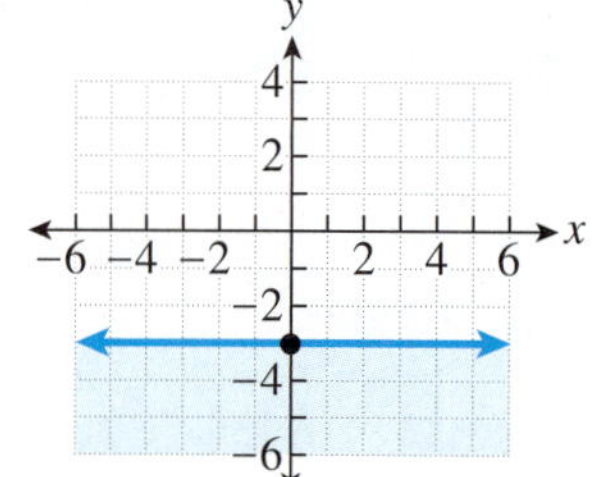

65. More than 40 people, some adults and the rest children, attended a movie. Set up and graph an inequality involving the number of adults (x) and children (y) in attendance.

66. An elevator has a warning posted inside the car that the maximum capacity is 1400 pounds. Suppose that the average weight of an adult is 175 pounds and the average weight of a child is 70 pounds.

a) Set up and graph an inequality involving the number of adults (x) and the number of children (y) who can safely ride on the elevator.

b) Using the graph from part a), can four adults and twelve children ride in the elevator at the same time?

67. A math instructor gives 4 points for each homework assignment and 10 points for each quiz. A student must earn at least 200 points from homework assignments and quizzes in order to pass the class. Set up and graph an inequality involving the number of homework assignments (x) and quizzes (y) required to pass the class.

68. An elementary school has a budget of less than \$30,000 to buy new computers. The school can purchase a desktop computer for \$600 and a laptop computer for \$750. Set up and graph an inequality involving the number of desktop computers (x) and laptop computers (y) the school can purchase.

69. A football team scores 3 points for each field goal and 7 points for each touchdown that they score. The team's coach has set a goal to score at least 550 points this season. Set up and graph an inequality involving the number of field goals (x) and touchdowns (y) the team must score in order to reach their goal.

70. A school club is selling boxes of chocolates and packages of popcorn as a fund-raiser. The club earns a profit of \$6 for each box of chocolates sold and \$4 for each package of popcorn sold. Set up and graph an inequality involving the number of boxes of chocolates (x) and packages of popcorn (y) the club must sell in order to raise at least \$2500.

Find the region that is the solution of both inequalities. First graph each inequality separately, and then shade the region that the two graphs have in common.

71. $y < x + 4$ and $y > -2x - 6$

72. $2x + 3y \leq 18$ and $3x - 4y \leq -12$

Writing in Mathematics

Answer in complete sentences.

73. True or false: every ordered pair is a solution to either $3x - 8y < 24$ or $3x - 8y > 24$. Explain your answer.

74. Explain why we use a dashed line when graphing inequalities involving the symbols $<$ and $>$, and why we use a solid line when graphing inequalities involving the symbols $\leq$ and $\geq$.

75. ***Solutions Manual**** Write a solutions manual page for the following problem:

Graph the linear inequality $2x + 3y < 12$.

76. ***Newsletter**** Write a newsletter explaining how to graph linear inequalities in two variables.

*See Appendix B for details and sample answers.

Objectives

1 **Define and understand function, domain, and range.**
2 **Evaluate functions.**
3 **Graph linear functions.**
4 **Interpret the graph of a linear function.**
5 **Determine the domain and range of a function from its graph.**
6 **Find a linear function that meets given conditions.**

A sandwich shop sells sandwiches for \$4 each. We know that it would cost \$4 to buy 1 sandwich, \$8 to buy 2 sandwiches, \$12 for 3 sandwiches, \$16 for 4 sandwiches, and so on. We also know that the general formula for the cost of x sandwiches in dollars is $4 \cdot x$. The cost depends on the number of sandwiches bought. We say that the cost is a function of the number of sandwiches bought.

Functions; Domain and Range

Objective 1 Define and understand function, domain, and range. A **relation** is a rule that takes an input value from one set and assigns a particular output value from another set to it. A relation for which each input value is assigned one and only one output value is called a **function**. In the previous example, the input value is the number of sandwiches bought, and the output value is the cost. For each number of sandwiches bought, there is only one possible cost. The rule for determining the cost is to multiply the number of sandwiches by \$4.

Sandwiches	1	2	3	4	...	x	...
	↓	↓	↓	↓	...	↓	
Cost	\$4	\$8	\$12	\$16		$\$4 \cdot x$	

The set of input values for a function is called the **domain** of the function. The domain for the previous example is the set of natural numbers $\{1, 2, 3, \ldots\}$. The set of output values for a function is called the **range** of the function. The range of the function in the sandwich shop example is $\{4, 8, 12, \ldots\}$.

Function, Domain, and Range

A **function** is a relation that takes an input value and assigns one and only one output value to it. The **domain** of the function is the set of input values, and the **range** is the set of output values.

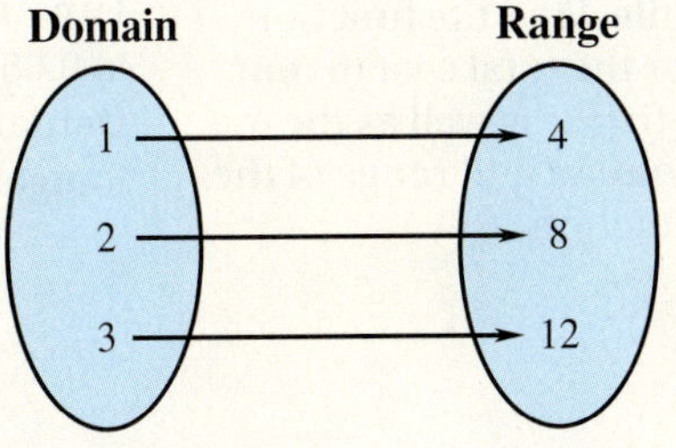

An input value for a function, along with its corresponding output value, can be written as an ordered pair. For example, if the cost for 5 sandwiches is $20, this can be represented by the ordered pair $(5, 20)$.

Some relations are not functions. If a relation associates one input value with more than one output value, then the relation is not a function.

EXAMPLE 1 **a)** If the input value for a relation is one of the 12 months and the output is the name of a student in your math class who has her birthday in that month, is this relation a function?

Solution

Since it is possible for two or more students to be born in a given month, this relation does not define a function. For example, if two or more students are born in October, then the input value of October is associated with more than one output value.

b) If the input value for a relation is one of the students in your math class and the output is the month of their birth, is the relation a function?

Solution

This relation is a function because each person can only be born in one month.

***Quick Check* 1**
If the input value for a relation is an address and the output value is the name of a person who lives at that address, is this relation a function? Explain your answer.

We can define functions between two finite sets. Consider the set of ordered pairs $\{(0, 0), (1, 1), (2, 4), (3, 9), (4, 16)\}$. A relation for which the x-coordinate of an ordered pair is the input value and the y-coordinate of that ordered pair is the output value is a function, because for this set there are no x-coordinates associated with more than one different y-coordinate. (If one x-coordinate was associated with two or more y-coordinates, such as in the ordered pairs $(5, 10)$ and $(5, 25)$, then the set would not represent a function.) The domain for this function would be the set of x-coordinates, or $\{0, 1, 2, 3, 4\}$. The range for this function would be the set of all y-coordinates, or $\{0, 1, 4, 9, 16\}$.

EXAMPLE 2 A telephone company has a long-distance plan that charges users a monthly rate of $5.95 per month, plus $0.05 per minute for individual calls. Find the function for the monthly long-distance bill, as well as the domain and range of the function.

***Quick Check* 2**
A rental agency rents large moving trucks for $39.95 plus $0.79 per mile. Find the function for the total cost to rent a truck, as well as the domain and range of the function.

Solution

To determine the monthly long-distance bill, we begin by multiplying the number of minutes by $0.05. To this we still need to add the $5.95 monthly rate.

Function: If m is the number of long-distance minutes in a month, then the cost is equal to $0.05m + 5.95$.

Domain: Set of possible number of long-distance minutes.

Range: Set of possible long-distance bills.

Function Notation

The perimeter of a square with side x can be found using the formula $P = 4x$. This is a function that can also be expressed using function notation as $P(x) = 4x$. **Function notation** is a way to present the output value of a function for the input x. The notation on the left side, $P(x)$, tells us the name of the function, P, as well as the input variable, x. $P(x)$ tells us that P is a function of x, and is read as "P of x." The parentheses are used to identify the input variable and are not used to indicate multiplication. We used P as the name of the function (P for perimeter), but we could have used any letter that we wanted. The letters f and g are frequently used for function names. The expression on the right side, $4x$, is the formula for the function.

$P(x) = 4x$	$P(x) = 4x$	$P(x) = 4x$
⇧	⇧	⇧
The variable inside the parentheses is the input variable for the function.	$P(x)$ is the output value of the function P when the input value is x.	The expression on the right side of the equation is the formula for this function.

Evaluating Functions

Objective 2 **Evaluate functions.** Suppose we wanted to find the perimeter of a square with a side of 3 inches. We are looking to evaluate our perimeter function, $P(x) = 4x$, for an input of 3, or, in other words, $P(3)$. When we find the output value of a function for a particular value of x, this is called **evaluating** the function. To evaluate a function for a particular value of the variable, we substitute that value for the variable in the function's formula, and then we simplify the resulting expression. To evaluate $P(3)$, we substitute 3 for x in the formula and simplify the resulting expression.

$$\begin{aligned} P(x) &= 4x \\ P(3) &= 4(3) \\ &= 12 \end{aligned}$$

The perimeter is 12 inches.

EXAMPLE 3 Let $f(x) = 3x + 17$. Find $f(9)$.

Solution

We need to replace x in the function's formula by 9 and simplify the resulting expression. It is a good idea to use parentheses when we substitute our input value.

$$\begin{aligned} f(9) &= 3(9) + 17 && \text{Substitute 9 for x.} \\ &= 27 + 17 && \text{Multiply.} \\ &= 44 && \text{Add.} \end{aligned}$$

$f(9) = 44$. This means that when the input is $x = 9$, the output of the function is 44.

EXAMPLE 4 Let $g(x) = 4x + 13$. Find $g(-8)$.

Solution

In this example we need to replace x by -8.

$$\begin{aligned} g(-8) &= 4(-8) + 13 && \text{Substitute } -8 \text{ for } x. \\ &= -32 + 13 && \text{Multiply.} \\ &= -19 && \text{Simplify.} \end{aligned}$$

Quick Check 3
Let $g(x) = -7x - 39$. Find $g(-4)$.

EXAMPLE 5 Let $f(x) = -\frac{3}{2}x + 5$. Find $f(6)$.

Solution

In this example we need to replace x by 6.

$$\begin{aligned} f(6) &= -\frac{3}{2}(6) + 5 && \text{Substitute 6 for } x. \\ &= -9 + 5 && \text{Multiply.} \\ &= -4 && \text{Simplify.} \end{aligned}$$

Quick Check 4
Let $f(x) = \frac{7}{3}x + 14$.
Find $f(9)$.

EXAMPLE 6 Let $g(x) = 2x - 8$. Find $g(0)$.

Solution

Often we will evaluate a function at $x = 0$. In this example the term $2x$ is equal to 0 when $x = 0$, leaving the function equal to -8.

$$\begin{aligned} g(0) &= 2(0) - 8 && \text{Substitute 0 for } x. \\ &= -8 && \text{Simplify.} \end{aligned}$$

Quick Check 5
Let $h(x) = -3x + 11$.
Find $h(0)$.

EXAMPLE 7 Let $g(x) = 8 - 3x$. Find $g(a + 3)$.

Solution

In this example we are substituting a variable expression for x in the function. After we replace x by $a + 3$, we will need to simplify the resulting variable expression.

$$\begin{aligned} g(a + 3) &= 8 - 3(a + 3) && \text{Replace } x \text{ by } a + 3. \\ &= 8 - 3a - 9 && \text{Distribute } -3. \\ &= -3a - 1 && \text{Collect like terms.} \end{aligned}$$

Quick Check 6
Let $g(x) = 7x - 12$. Find $g(a + 8)$.

Linear Functions and Their Graphs

Objective 3 **Graph linear functions.**

A **linear function** is a function of the form $f(x) = mx + b$, where m and b are real numbers.

Some examples of linear functions are $f(x) = x - 9$, $f(x) = 5x$, $f(x) = 3x + 11$, and $f(x) = 6$. We now turn our attention toward graphing linear functions. We graph any function $f(x)$ by plotting points of the form $(x, f(x))$. We treat the output value of the function $f(x)$ the way we treated the variable y when we were graphing linear equations in two variables. When we graph a function $f(x)$, the vertical axis is used to represent the output values of the function.

We can begin to graph a linear function by finding the y-intercept. As with a linear equation that is in slope–intercept form, the y-intercept for the graph of a linear function $f(x) = mx + b$ is the point $(0, b)$. For example, the y-intercept for the graph of the function $f(x) = 5x - 8$ is the point $(0, -8)$. In general, to find the y-intercept of any function $f(x)$, we can find $f(0)$. After plotting the y-intercept, we can then use the slope m to find other points.

EXAMPLE 8 Graph the linear function $f(x) = \frac{1}{2}x + 2$.

Solution

We can start with the y-intercept, which is $(0, 2)$. The slope of the line is $\frac{1}{2}$, so beginning at the point $(0, 2)$, we move up one unit and two units to the right. This leads to a second point at $(2, 3)$. Repeating this process, we see that another point on the line is $(4, 4)$. Here is the graph of the function $f(x) = \frac{1}{2}x + 2$.

Quick Check 7
Graph the linear function $f(x) = \frac{3}{4}x - 6$.

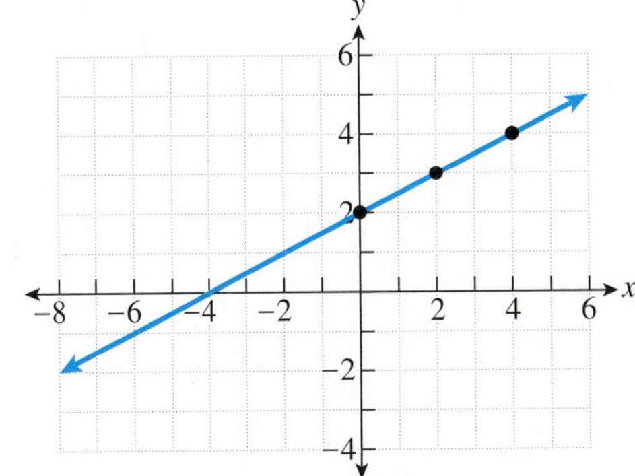

EXAMPLE 9 Graph the linear function $f(x) = -3$.

Solution

This function is known as a **constant function**. The function is constantly equal to -3 regardless of the input value x. Its graph is a horizontal line with a y-intercept at $(0, -3)$.

Quick Check 8
Graph the linear function $f(x) = 4$.

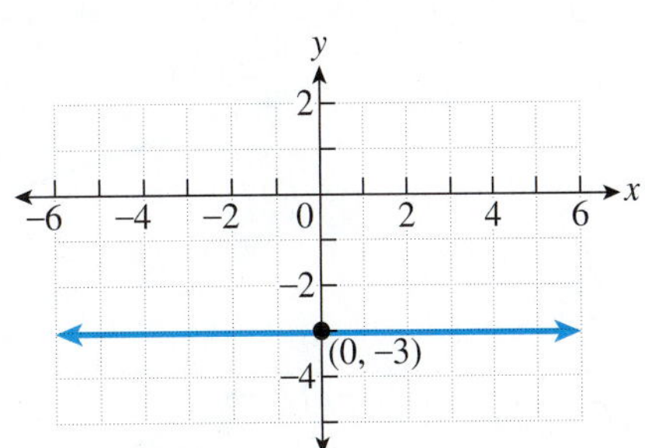

Interpreting the Graph of a Linear Function

Objective 4 **Interpret the graph of a linear function.** Here is the graph of a function $f(x)$.

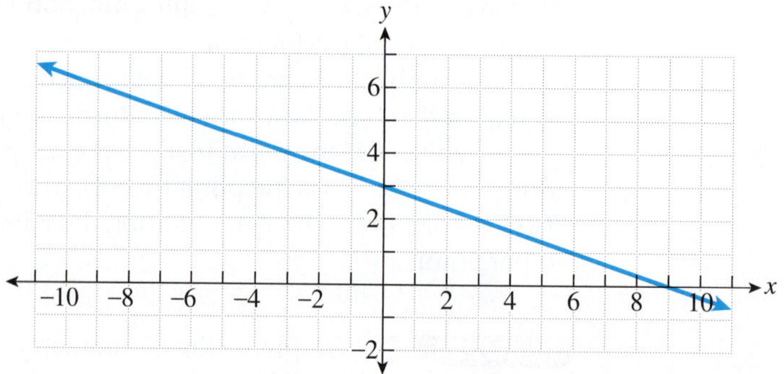

The ability to read and interpret a graph is an important skill. This line has an x-intercept at the point $(9, 0)$, so we know that $f(9) = 0$. The y-intercept is at the point $(0, 3)$, so $f(0) = 3$.

Suppose that we wanted to find $f(3)$ for this particular function. We can do this by finding a point on the line that has an x-coordinate of 3. The y-coordinate of this point is $f(3)$. In this case, $f(3) = 2$.

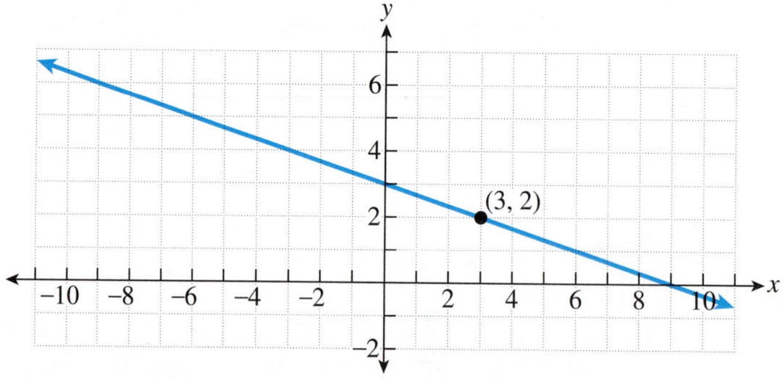

We can also use this graph to solve the equation $f(x) = 6$. Look for the point on the graph that has a y-coordinate of 6. The x-coordinate of this point is -9, so $x = -9$ is the solution to the equation $f(x) = 6$.

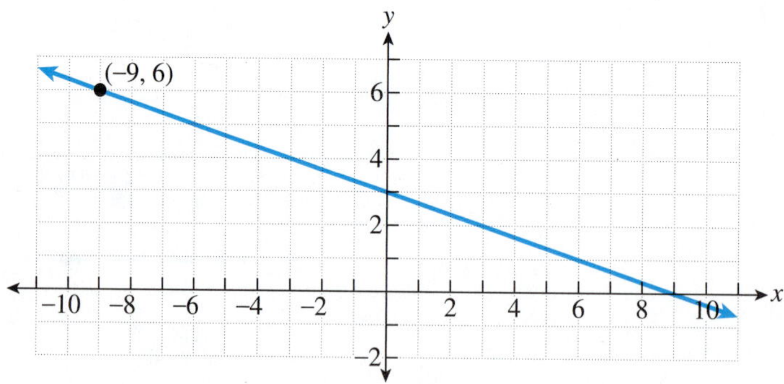

EXAMPLE 10 Consider the following graph of a function.

a) Find the x-intercept and the y-intercept.

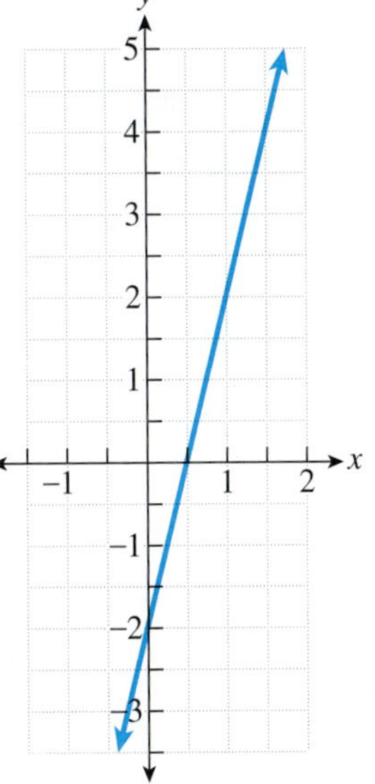

Solution

Note the scale on this graph, measured in units equal to $\frac{1}{2}$. The x-intercept is $(\frac{1}{2}, 0)$. The y-intercept is $(0, -2)$.

b) Find $f(1\frac{1}{2})$.

Solution

We are looking for a point on the line that has an x-coordinate of $1\frac{1}{2}$. The point is $(1\frac{1}{2}, 4)$, so $f(1\frac{1}{2}) = 4$.

c) Find a value x such that $f(x) = 2$.

Solution

We are looking for a point on the line that has a y-coordinate of 2, and this point is $(1, 2)$. The value that satisfies the equation $f(x) = 2$ is $x = 1$.

Quick Check **9** **Consider the following graph of a function.**

a) Find the x-intercept and the y-intercept.

b) Find $f(6)$.

c) Find a value x such that $f(x) = 6$.

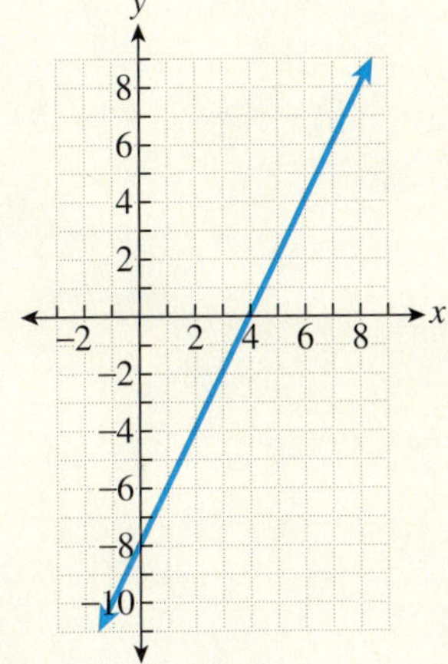

Objective 5 **Determine the domain and range of a function from its graph.**

The domain and range of a function can also be read from a graph. Recall that the domain of a function is the set of all input values. This corresponds to all of the x-coordinates of the points on the graph. The domain is the interval of values on the x-axis for which the graph exists. We read the domain from left to right on the graph.

The domain of a linear function is the set of all real numbers, which can be written in interval notation as $(-\infty, \infty)$. The graphs of linear functions continue on to the left and to the right. Look at the following three graphs of linear functions.

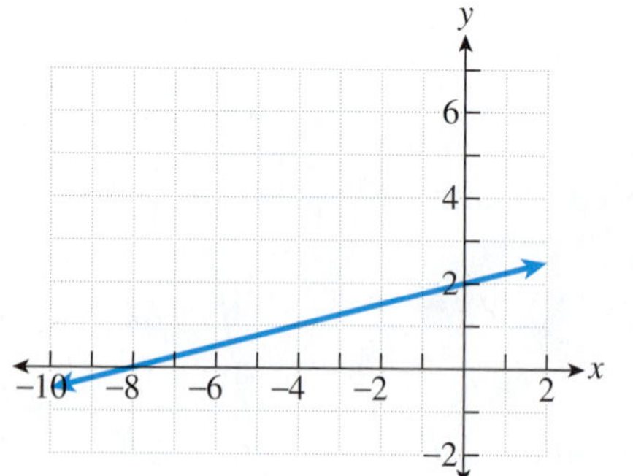

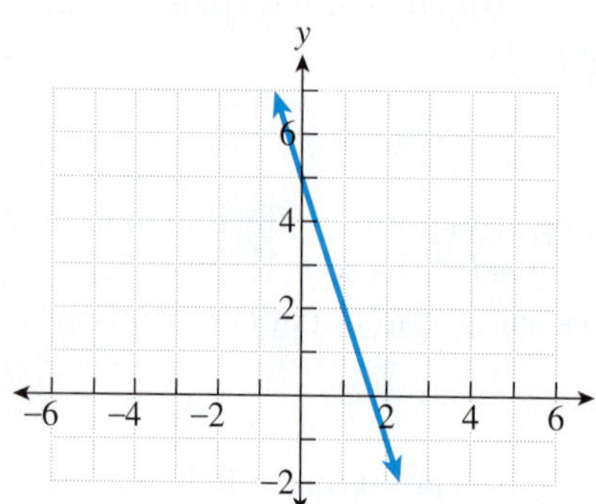

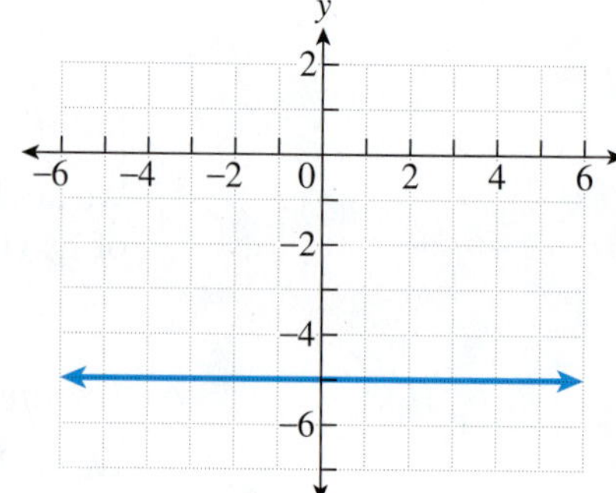

Each line continues to the left as well as to the right. This tells us that the graph exists for all values of x in the interval $(-\infty, \infty)$.

The range of a function can be read vertically from a graph. The range goes from the lowest point on the graph to the highest point. All linear functions have $(-\infty,\infty)$ as their range, except constant functions. Since a constant function of the form $f(x) = c$, where c is a real number, has only one possible output value, its range would be the set $\{c\}$. For example, the range of the constant function $f(x) = 3$ is $\{3\}$.

Finding a Linear Function That Meets Given Conditions

Objective 6 Find a linear function that meets given conditions. To find a linear function, $f(x) = mx + b$, that meets certain given conditions, we must find the values m and b. For example, if we are looking for a linear function whose graph has a slope of 3 and a y-intercept of $(0, 6)$, we know that $m = 3$ and $b = 6$. So, the function is $f(x) = 3x + 6$. Finding a linear function is similar to finding the equation of a line.

EXAMPLE 11 Find a linear function $f(x)$ whose graph has a slope of 2 that passes through the point $(3, 11)$.

Solution

Since the slope is 2, we know that $m = 2$ and that $f(x)$ is of the form $f(x) = 2x + b$. We also know that $f(3) = 11$ because the graph passes through the point $(3, 11)$. We will use this fact to help us find b.

$$f(3) = 11$$
$$2(3) + b = 11 \quad \text{Substitute 3 for } x \text{ in the function } f(x) = 2x + b.$$
$$6 + b = 11 \quad \text{Multiply.}$$
$$b = 5 \quad \text{Subtract 6 from both sides.}$$

Since the value of b is 5, the function is $f(x) = 2x + 5$.

Quick Check 10
Find a linear function $f(x)$ whose graph has a slope of -3 and passes through the point $(4, -7)$.

EXAMPLE 12 Find a linear function $f(x)$ whose graph passes through the points $(-3, 7)$ and $(6, -5)$.

Solution

We know that $f(-3) = 7$ and $f(6) = -5$ because the graph passes through the points $(-3, 7)$ and $(6, -5)$. We will begin by finding m.

$$m = \frac{-5 - 7}{6 - (-3)} \quad \text{Substitute into slope formula.}$$
$$= -\frac{4}{3} \quad \text{Simplify.}$$

The function is of the form $f(x) = -\frac{4}{3}x + b$. Now we will use the fact that $f(-3) = 7$.

$$f(-3) = 7$$
$$-\frac{4}{3}(-3) + b = 7 \quad \text{Substitute } -3 \text{ for } x \text{ in the function.}$$
$$4 + b = 7 \quad \text{Multiply.}$$
$$b = 3 \quad \text{Subtract 4 from both sides.}$$

The function is $f(x) = -\frac{4}{3}x + 3$.

***Quick Check* 11**
Find a linear function $f(x)$ whose graph passes through the points $(-4, -2)$ and $(2, -11)$.

Building Your Study Strategy **Using Your Textbook Effectively, 5 Note Cards**
Your textbook can be used to create a series of note cards for studying new terms and procedures or difficult problems. Each time you find a new term in the textbook, write the term on one side of a note card and its definition on the other side. Collect all of these cards and cycle through them, reading the term and trying to recite the definition. Check your definition on the other side of the card. You can do the same thing for each new procedure introduced in the textbook.

If there is a particular type of problem you are struggling with, find an example that has been worked out in the text. Write the problem on the front of the card and write the complete solution on the back of the card. Collect all of the cards containing difficult problems and select one at random. Try to solve the problem and then check the solution given on the back of the card. Repeat this for the other cards.

Some students only have trouble recalling the first step for solving particular problems. If you are in this group, you can write a series of problems on the front of some note cards and the first step on the back of the cards. You can then cycle through these cards in the same way. It is easy to carry a set of note cards with you. Any time you have 5 to 15 free minutes, cycle through them.

EXERCISES 2.5

Vocabulary

1. A(n) ______________ is a rule that takes one input value and assigns one and only one output value to it.

2. The ______________ of a function is the set of all possible input values and the ______________ of a function is the set of all possible output values.

3. Substituting a value for the function's variable and simplifying the resulting expression is called ______________ the function.

4. A(n) ______________ function is a function of the form $f(x) = mx + b$.

Here is a list showing the names of ten members of the Pro Football Hall of Fame, along with the NFL team for which they played most of their games. Use this list for exercises 5 and 6.

Player	NFL Team
Howie Long	Raiders
Ronnie Lott	49ers
Joe Montana	49ers
Lynn Swann	Steelers
Jim Kelly	Bills
Marcus Allen	Raiders
John Elway	Broncos
Barry Sanders	Lions
Dan Marino	Dolphins
Troy Aikman	Cowboys

5. Would a rule that took one of these ten players' names as an input and listed his NFL team as an output be a function? Explain why or why not.

6. Could a function be defined in the opposite direction, with the name of the NFL team as the input and the Hall of Fame player's name as the output? Explain why or why not.

For exercises 7–10 determine whether a function exists with
a) set A as the input and set B as the output; and
b) set B as the input and set A as the output.

7. High temperatures on July 10

Set A City	Set B Temperature
Boston, MA	88°F
Memphis, TN	94°F
Philadelphia, PA	88°F
Phoenix, AZ	111°F
Visalia, CA	102°F

8.

Set A Person	Set B Last 4 Digits of SSN
Jay Beckenstein	1234
Jenny Crum	5283
Jay Jenkins	6405
Lauren Morse	5555
Maureen O'Connor	9200

9.

Set A Person	Set B Birthday
Greg Erb	December 15
Karen Guardino	December 17
Jolene Lehr	October 15
Siméon Poisson	June 21
Lindsay Skay	May 28
Sharon Smith	June 21

10.

Set A Baseball Player	Set B Uniform Number
Josh Beckett	19
Manny Ramirez	24
Jason Varitek	33
Davis Ortiz	34
Curt Schilling	38

11. Could a function be defined with a man as an input and his child as an output? Why or why not?

12. Could a function be defined with a person as an input and that person's father as an output? Why or why not?

13. Could a function be defined with a person as an input and their favorite ice cream flavor as an output? Why or why not?

14. Could a function be defined with the last four digits of a phone number as an input and a person with those digits as the last four digits of their phone number as an output? Why or why not?

15. Could a function be defined with a student from your math class as an input and the student's math instructor as an output? Why or why not?

16. Could a function be defined with a person as an input and their favorite movie as an output? Why or why not?

For the given set of ordered pairs, determine whether a function could be defined for which the input would be an x-coordinate and the output would be the corresponding y-coordinate. If a function cannot be defined in this manner, explain why.

17. $\{(-2, 4), (-1, 1), (0, 0), (1, 1), (2, 4)\}$

18. $\{(4, -2), (1, -1), (0, 0), (1, 1), (4, 2)\}$

19. $\{(5, 1), (-2, -6), (-7, -6), (5, -2), (3, 4)\}$

20. $\{(-8, 5), (-4, 5), (0, 5), (4, 5), (8, 5)\}$

21. $\{(1, 1), (2, 2), (3, 3), (4, 4), (5, 5)\}$

22. $\{(4, -5), (4, -3), (4, 0), (4, 4), (4, 9)\}$

23. A Celsius temperature can be converted to a Fahrenheit temperature by multiplying it by 1.8 and then adding 32.

a) Create a function $F(x)$ that converts a Celsius temperature x to a Fahrenheit temperature.

b) Use the function $F(x)$ from part a) to convert the following Celsius temperatures to Fahrenheit temperatures.
0°C 100°C 25°C −5°C

24. To convert a Fahrenheit temperature to a Celsius temperature we first subtract 32 from it, and then multiply that difference by $\frac{5}{9}$.

a) Create a function $C(x)$ that converts a Fahrenheit temperature x into a Celsius temperature.

b) Use the function $C(x)$ from part a) to convert the following Fahrenheit temperatures to Celsius temperatures.
50°F 14°F 113°F −40°F

25. A restaurant hosts awards banquets. They charge $300 rental, plus $14 per person for food.

a) Create a function $f(x)$ for the amount the restaurant charges to host a banquet with x attendees.

b) Use the function $f(x)$ to determine how much the restaurant would charge to host a banquet with 150 attendees.

26. Admission to a county fair is $7, and each ride ticket is $0.50.

a) Create a function $f(x)$ for the amount that a person pays if he attends the fair and purchases x ride tickets.

b) Use the function $f(x)$ to determine how much a person pays if he attends the fair and purchases 45 ride tickets.

27. A major league baseball team sells tickets online. Tickets are sold for $18.75 each, and there is a $19.50 shipping charge for each order.

a) Create a function $f(x)$ for the total cost of ordering x tickets.

b) Use the function $f(x)$ to determine the total cost of ordering 6 tickets.

28. An online bookstore charges $2.50 per order for standard shipping, plus an additional $1.25 per book.

a) Create a function $f(x)$ for the shipping charges on an order of x books.

b) Use the function $f(x)$ to determine the shipping charges on an order of 7 books.

Evaluate the given function.

29. $f(x) = x + 7, f(-5)$

30. $f(x) = 3x, f(9)$

31. $g(x) = 5x - 8, g(-4)$

32. $h(x) = 6x + 1, h(11)$

33. $f(x) = -2x - 30, f(-8)$

34. $f(x) = 23 - 7x, f(6)$

35. $f(x) = 4x - 22, f(0)$

36. $f(x) = -7x + 16, f(0)$

37. $g(x) = 3x - 6, g(\frac{2}{3})$

38. $g(x) = 4x + 3, g(\frac{7}{4})$

39. $f(x) = 9x - 13.7, f(4.4)$

40. $f(x) = -14.2x + 37.6, f(5.8)$

41. $f(x) = 9x + 2, f(a)$

42. $f(x) = 3x - 11, f(b)$

43. $f(x) = 5x - 8, f(a + 4)$

44. $f(x) = 2x + 21, f(a - 6)$

45. $h(x) = -3x + 9, h(4a - 6)$

46. $h(x) = -5x - 22, h(3a + 7)$

47. $f(x) = 5.3x - 16.7, f(2n - 1)$

48. $f(x) = 3.4x + 11.9, f(3b - 8)$

49. $f(x) = \frac{2}{7}x - 3, f(14a + 35)$

50. $f(x) = -\frac{3}{4}x - 15, f(12a - 8)$

51. $f(x) = 7x + 8, f(x + h)$

52. $f(x) = 4x - 9, f(x + h)$

Graph the linear function.

53. $f(x) = -x + 5$

54. $f(x) = x + 3$

61. $f(x) = -\frac{3}{4}x + 6$

62. $f(x) = \frac{2}{5}x - 2$

55. $f(x) = 2x - 9$

56. $f(x) = -3x + 7$

63. $f(x) = 3x$

64. $f(x) = -2x$

57. $f(x) = -6x + 8$

58. $f(x) = 4x + 1$

65. $f(x) = -\frac{3}{2}x$

66. $f(x) = \frac{5}{8}x$

59. $f(x) = \frac{2}{3}x - 4$

60. $f(x) = -\frac{1}{2}x + 3$

67. $f(x) = 2.5x - 4.5$

68. $f(x) = -3.2x + 8.8$

69. $f(x) = -25x + 350$

70. $f(x) = 1000x + 7000$

71. $f(x) = 9$

72. $f(x) = 6$

73. $f(x) = -2$

74. $f(x) = -5$

75. Refer to the graph of the function $f(x)$.

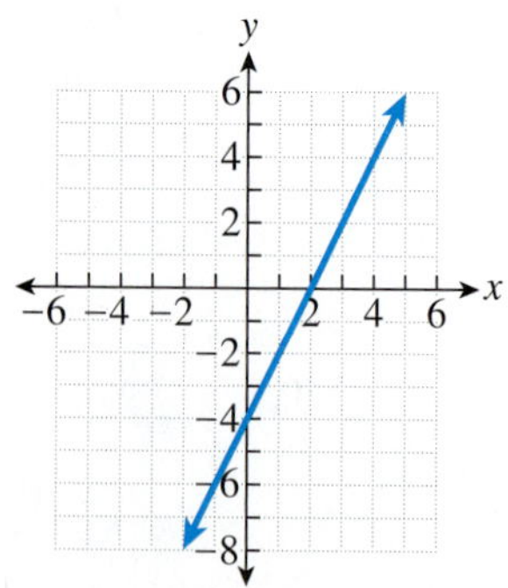

a) Find the x-intercept.

b) Find the y-intercept.

c) Find $f(-1)$.

d) Find a value a such that $f(a) = 6$.

76. Refer to the graph of the function $f(x)$.

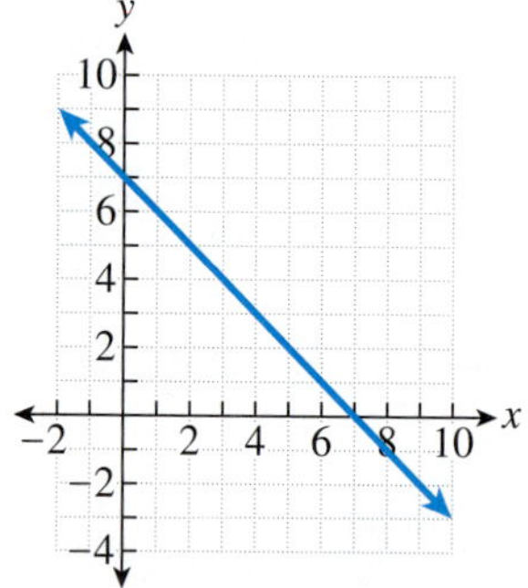

a) What is the x-intercept?

b) What is the y-intercept?

c) Find $f(3)$.

d) Find a value a such that $f(a) = -2$.

77. Refer to the graph of the function $f(x)$.

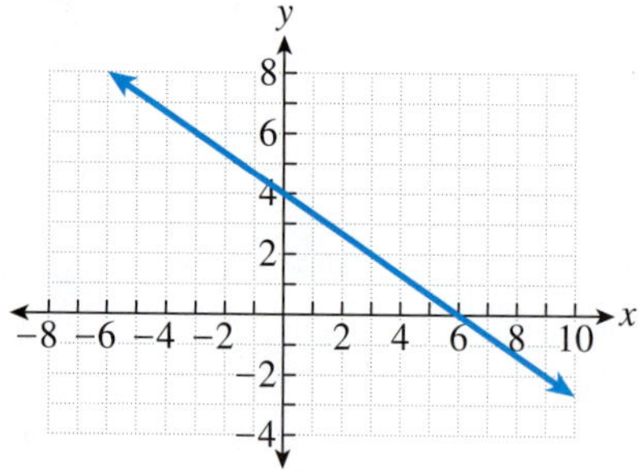

a) Find the x-intercept.

b) Find the y-intercept.

c) Find $f(9)$.

d) Find a value a such that $f(a) = 8$.

78. Refer to the graph of the function $f(x)$.

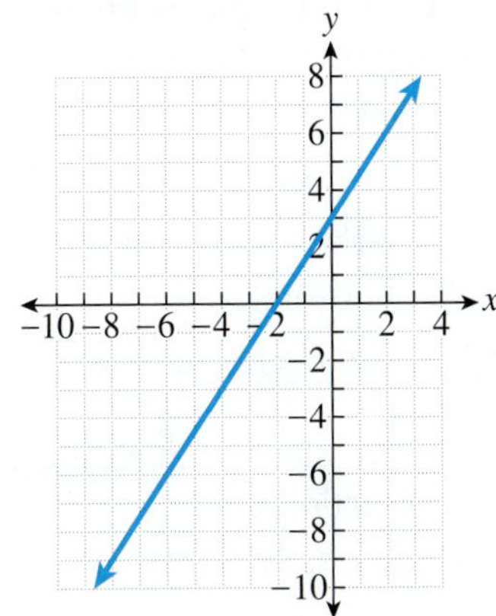

a) What is the x-intercept?

b) What is the y-intercept?

c) Find $f(2)$.

d) Find a value a such that $f(a) = -9$.

Find the domain and range of the given function.

79.

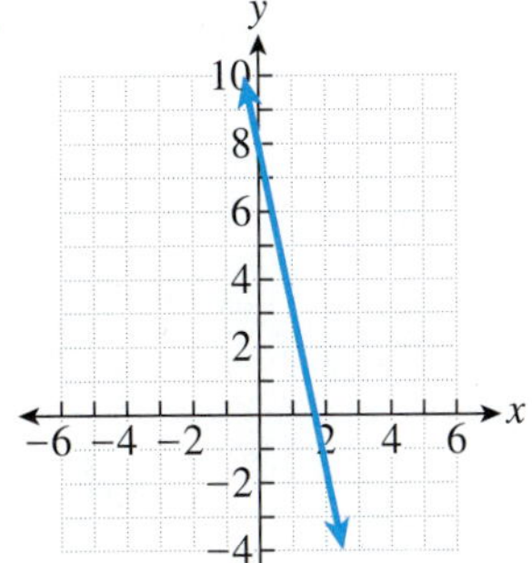

80.

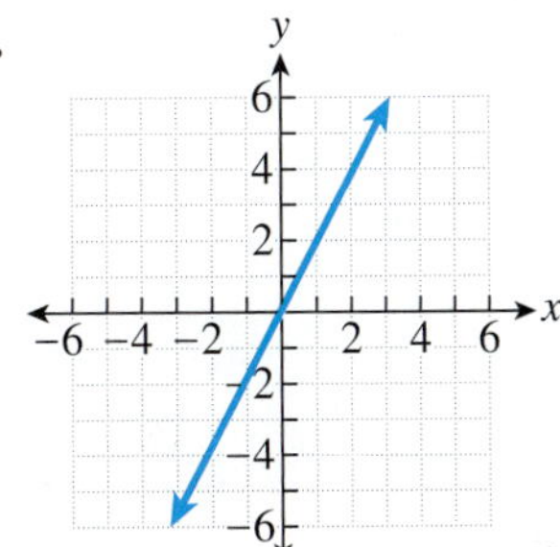

81.

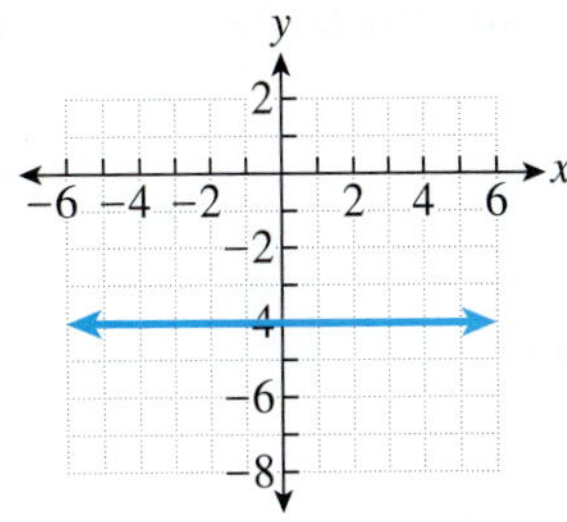

82.

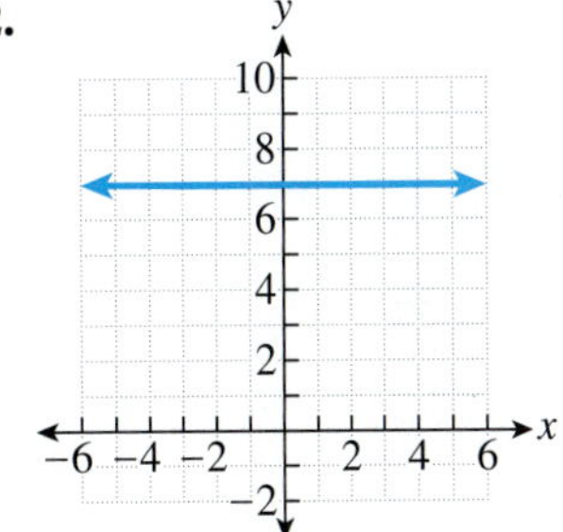

Create a linear function $f(x)$ whose graph has the given slope and y-intercept.

83. Slope of 2, y-intercept at $(0, 7)$

84. Slope of 4, y-intercept at $(0, -10)$

85. Slope of -1, y-intercept at $(0, -6)$

86. Slope of $(-\frac{5}{4})$, y-intercept at $(0, \frac{3}{4})$

Create a linear function $f(x)$, whose graph has the given slope, that meets the given condition.

87. Slope of 4, $f(2) = 5$

88. Slope of 2, $f(7) = -9$

89. Slope of -3, $f(-1) = 8$

90. Slope of -1, $f(-5) = -13$

91. Slope of $\frac{2}{3}$, $f(6) = -4$

92. Slope of $-\frac{3}{5}$, $f(-10) = -9$

93. Slope of 6.4, $f(4) = 14.4$

94. Slope of 15.5, $f(228.4) = 2867.3$

Create a linear function $f(x)$ that meets the given conditions.

95. $f(1) = 3, f(4) = 9$

96. $f(2) = 7, f(6) = -9$

97. $f(-3) = -6, f(5) = -14$

98. $f(-4) = 1, f(2) = -17$

99. $f(-3) = -10, f(6) = 2$

100. $f(-8) = 22, f(-2) = 13$

101. $f(3.2) = -30.4, f(-1.5) = -2.2$

102. $f(-5.4) = -29.1, f(13.8) = 47.7$

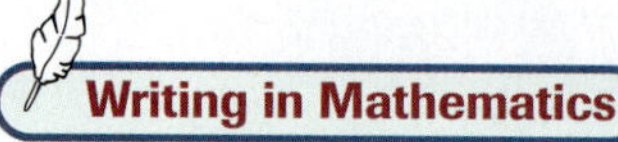

Answer in complete sentences.

103. Give an example of two sets A and B and a relation that is a function from set A to set B. Explain why your relation meets the definition of a function. Give another relation that would not be a function from set B to set A. Explain why that relation does not meet the definition of a function.

104. ***Solutions Manual**** Write a solutions manual page for the following problem:

Graph the linear function $f(x) = \frac{2}{3}x - 4$.

105. ***Newsletter**** Write a newsletter explaining how to graph linear functions.

***See Appendix B for details and sample answers.**

2.6 Absolute Value Functions

Objectives

1. **Graph absolute value functions.**
2. **Determine an absolute value function from its graph.**

Absolute Value Functions

Objective 1 **Graph absolute value functions.** We will now graph our first nonlinear function: the **absolute value function** $f(x) = |x|$. We will begin to graph this function by creating a table of function values for various values of x.

x	$f(x) = \|x\|$	$(x, f(x))$
-2	$f(-2) = \|-2\| = 2$	$(-2, 2)$
-1	$f(-1) = \|-1\| = 1$	$(-1, 1)$
0	$f(0) = \|0\| = 0$	$(0, 0)$
1	$f(1) = \|1\| = 1$	$(1, 1)$
2	$f(2) = \|2\| = 2$	$(2, 2)$

Here is the graph. Notice that the graph is not a straight line, but instead is V-shaped. The most important point for graphing an absolute value function is the point of the V, where the graph changes from falling to rising. The turning point is the origin for this particular function.

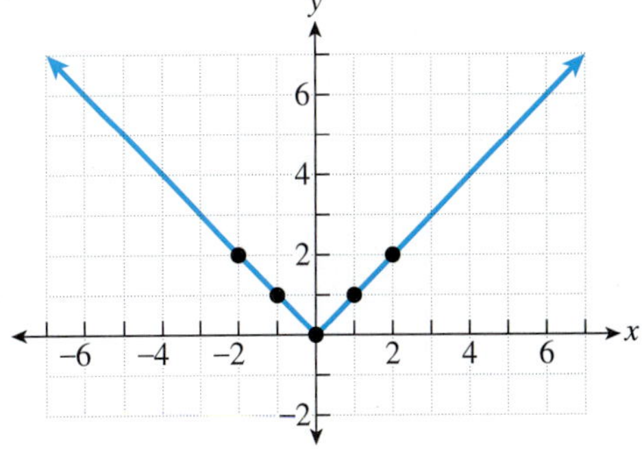

To graph an absolute value function, we find the x-coordinate of the turning point, select two values to the left and right of this value, and create a table of values. Choosing values for x that are both to the left and to the right of the turning point is crucial. Suppose that we had chosen only positive values for x when graphing $f(x) = |x|$. All of our points would have been on the straight line located to the right of the origin, and our graph would have been a straight line rather than being V-shaped.

The turning point occurs at the value of x that makes the expression inside the absolute value bars equal to 0. This is because the minimum output of an absolute value occurs when we take the absolute value of 0.

Graphing an Absolute Value Function

- Determine the value of x for which the expression inside the absolute value bars is equal to 0.
- In addition to this value, select two values that are less than this value and two that are greater.
- Create a table of function values for these values of x.
- Place the points $(x, f(x))$ on the graph and draw the V-shaped graph that passes through these points.

EXAMPLE 1 Graph $f(x) = |x - 3|$.

Solution

We begin by finding the value of x that makes the expression inside the absolute value bars equal to 0.

$x - 3 = 0$ Set $x - 3$ equal to 0.

$x = 3$ Add 3.

Now we select two values that are less than 3, such as 1 and 2, and two values that are greater than 3, such as 4 and 5. Next we create a table of values.

x	$f(x) = \|x - 3\|$	$(x, f(x))$
1	$f(1) = \|1 - 3\| = \|-2\| = 2$	$(1, 2)$
2	$f(2) = \|2 - 3\| = \|-1\| = 1$	$(2, 1)$
3	$f(3) = \|3 - 3\| = \|0\| = 0$	$(3, 0)$
4	$f(4) = \|4 - 3\| = \|1\| = 1$	$(4, 1)$
5	$f(5) = \|5 - 3\| = \|2\| = 2$	$(5, 2)$

Here is the graph, based on the points from our table.

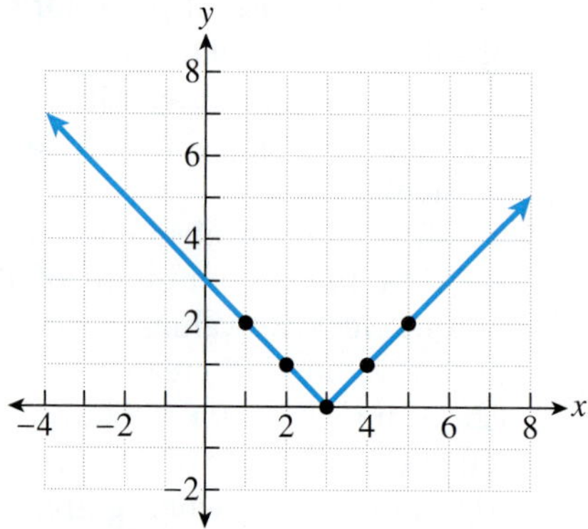

Quick Check 1
Graph the function $f(x) = |x + 1|$.

Using Your Calculator We can graph absolute value functions using the TI-84. (To access the absolute value function on the TI-84, press the MATH key and you will find it under the **NUM** menu.) To enter the equation of the function from Example 1, $f(x) = |x-3|$, press the Y= key. Enter $|x - 3|$ next to Y_1. Now press the key labeled GRAPH to graph the function.

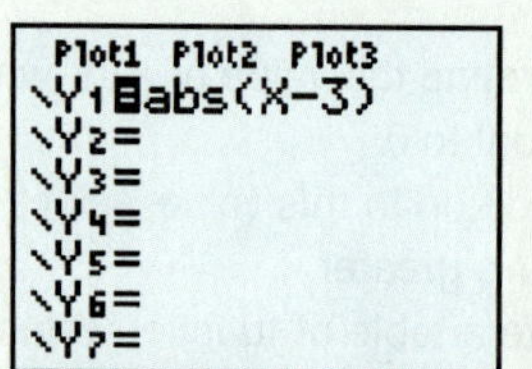

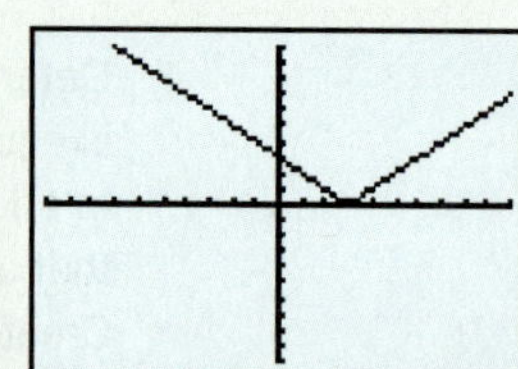

EXAMPLE 2 Graph $f(x) = |x + 2| - 1$.

Solution

We begin by finding the value of x that makes the expression inside the absolute value bars equal to 0.

$$x + 2 = 0 \quad \text{Set } x + 2 \text{ equal to 0.}$$
$$x = -2 \quad \text{Subtract 2.}$$

Now we select two values that are less than -2, such as -4 and -3, and two values that are greater than -2, such as -1 and 0. Next we create a table of values.

x	$f(x) = \lvert x + 2\rvert - 1$	$(x, f(x))$
-4	$f(-4) = \lvert -4 + 2\rvert - 1 = 2 - 1 = 1$	$(-4, 1)$
-3	$f(-3) = \lvert -3 + 2\rvert - 1 = 1 - 1 = 0$	$(-3, 0)$
-2	$f(-2) = \lvert -2 + 2\rvert - 1 = 0 - 1 = -1$	$(-2, -1)$
-1	$f(-1) = \lvert -1 + 2\rvert - 1 = 1 - 1 = 0$	$(-1, 0)$
0	$f(0) = \lvert -0 + 2\rvert - 1 = 2 - 1 = 1$	$(0, 1)$

Here is the graph, based on the points from our table.

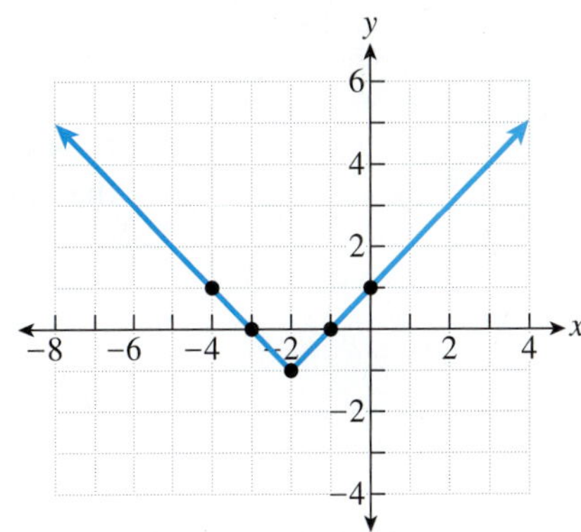

Quick Check 2
Graph the function $f(x) = |x + 3| + 4$.

Recall that the domain of a function is the set of all possible input values for the function and the range of a function is the set of all possible output values for the function. From the graph of a function, the domain can be read from left to right and the range can be read from bottom to top. Let's look at the graph of $f(x) = |x|$ once again.

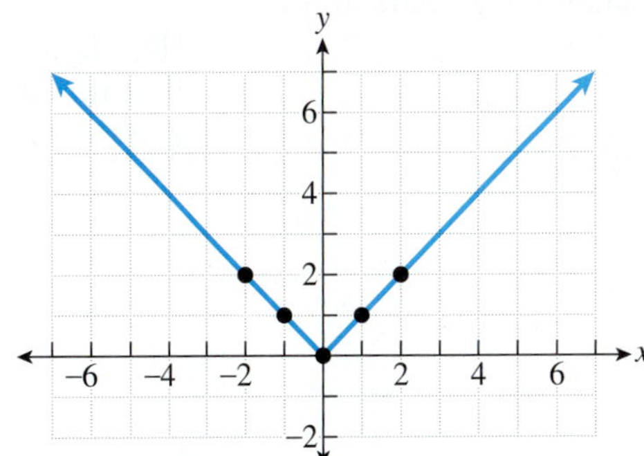

Notice that the graph extends out to infinity on the left and to infinity on the right. Its domain is the set of real numbers $\mathbb{R}$, which is expressed in interval notation as $(-\infty, \infty)$. The lowest value of this function is 0, but it has no upper bound. The range of this function is $[0, \infty)$.

Let's reexamine the graph of $f(x) = |x + 2| - 1$. The domain of this function is also the set of real numbers. What is the range? The lowest function value for this function is -1, so the range is $[-1, \infty)$.

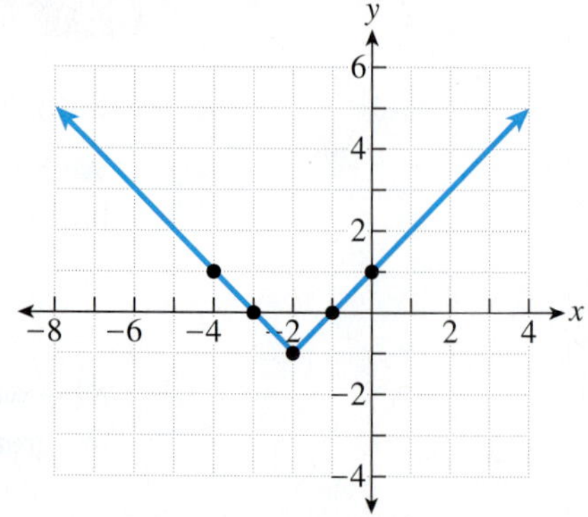

EXAMPLE 3 Graph the function $f(x) = |x - 3| + 5$, and find the domain and range of the function.

Solution

We begin by finding the x-coordinate of the turning point of the graph. To do this we set the expression inside the absolute value bars equal to 0 and solve for x.

$x - 3 = 0$ Set $x - 3$ equal to 0.
$x = 3$ Add 3.

Now we can create a table of values, using two values for x that are less than 3 and two that are greater than 3.

x	$f(x) = \|x - 3\| + 5$	$(x, f(x))$
1	$f(1) = \|1 - 3\| + 5 = 7$	(1, 7)
2	$f(2) = \|2 - 3\| + 5 = 6$	(2, 6)
3	$f(3) = \|3 - 3\| + 5 = 5$	(3, 5)
4	$f(4) = \|4 - 3\| + 5 = 6$	(4, 6)
5	$f(5) = \|5 - 3\| + 5 = 7$	(5, 7)

Quick Check 3
Graph the function $f(x) = |x - 4| - 2$, and state the domain and range.

Here is the graph. From this graph we can see that the domain is the set of all real numbers, and the range is $[5, \infty)$.

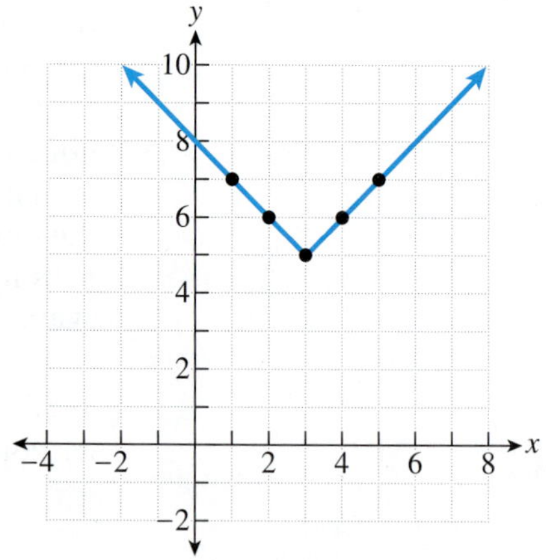

A Word of Caution When we graph an absolute value function, choosing values for x wisely will result in a series of points that forms a "V." If all of our points appear to lie on a straight line, then we have either chosen values for x that are all on one side of the point where the graph changes from decreasing to increasing, or we have made some mistakes when evaluating the function. In either case, we should go back and check our work.

EXAMPLE 4 Graph the function $f(x) = |2x + 3| - 1$, and find the domain and range of the function.

Solution

We begin by finding the x-coordinate of the turning point of the graph.

$$2x + 3 = 0 \qquad \text{Set } 2x + 3 \text{ equal to 0.}$$

$$x = -\frac{3}{2} \qquad \text{Solve for } x.$$

Now we can create a table of values, using two values for x that are less than $-\frac{3}{2}$ and two that are greater than $-\frac{3}{2}$. With the exception of the turning point, we will use integer values for x.

x	$f(x) = \lvert 2x + 3\rvert - 1$	$(x, f(x))$
-3	$f(-3) = \lvert 2(-3) + 3\rvert - 1 = 2$	$(-3, 2)$
-2	$f(-2) = \lvert 2(-2) + 3\rvert - 1 = 0$	$(-2, 0)$
$-\frac{3}{2}$	$f(-\frac{3}{2}) = \lvert 2(-\frac{3}{2}) + 3\rvert - 1 = -1$	$(-\frac{3}{2}, -1)$
-1	$f(-1) = \lvert 2(-1) + 3\rvert - 1 = 0$	$(-1, 0)$
0	$f(0) = \lvert 2(0) + 3\rvert - 1 = 2$	$(0, 2)$

***Quick Check* 4**
Graph the function $f(x) = |5x - 2| - 8$, and state the domain and range.

Here is the graph. From this graph we can see that the domain is the set of all real numbers, and the range is $[-1, \infty)$.

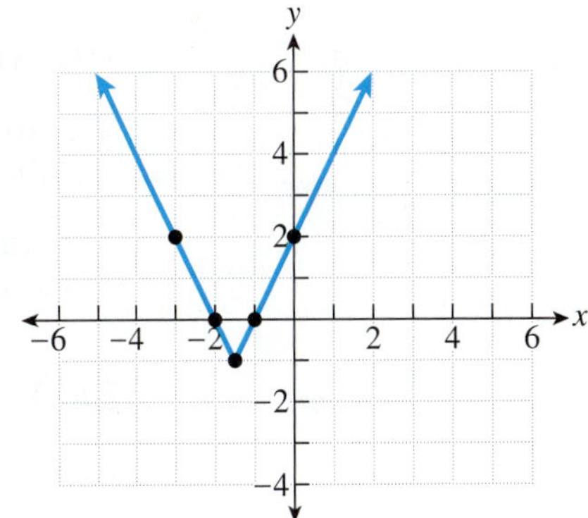

The graphs of some absolute value functions open downward, rather than upward. If an absolute value function has a negative sign in front of the absolute value bars, then its graph opens downward.

EXAMPLE 5 Graph $f(x) = -|x + 4| + 2$, and state the domain and range.

Solution

Although there is a negative sign in front of the absolute value bars, we will use the same approach for graphing this function as for the previous absolute value functions. We begin by finding the x-coordinate of the turning point by setting $x + 4$ equal to 0 and solving for x. In this case, $x = -4$ is the x-coordinate of the turning point. Now we can create a table of values, using two values for x that are less than -4 and two that are greater than -4.

x	$f(x) = -\|x + 4\| + 2$	$(x, f(x))$
-6	$f(-6) = -\|-6 + 4\| + 2 = -2 + 2 = 0$	$(-6, 0)$
-5	$f(-5) = -\|-5 + 4\| + 2 = -1 + 2 = 1$	$(-5, 1)$
-4	$f(-4) = -\|-4 + 4\| + 2 = 0 + 2 = 2$	$(-4, 2)$
-3	$f(-3) = -\|-3 + 4\| + 2 = -1 + 2 = 1$	$(-3, 1)$
-2	$f(-2) = -\|-2 + 4\| + 2 = -2 + 2 = 0$	$(-2, 0)$

Quick Check 5
Graph the function $f(x) = -|x - 1| - 3$, and state the domain and range.

Here is the graph. From this graph, we can see that the domain is the set of all real numbers. This time the function has a maximum value of 2 with no lower bound, so the range is $(-\infty, 2]$.

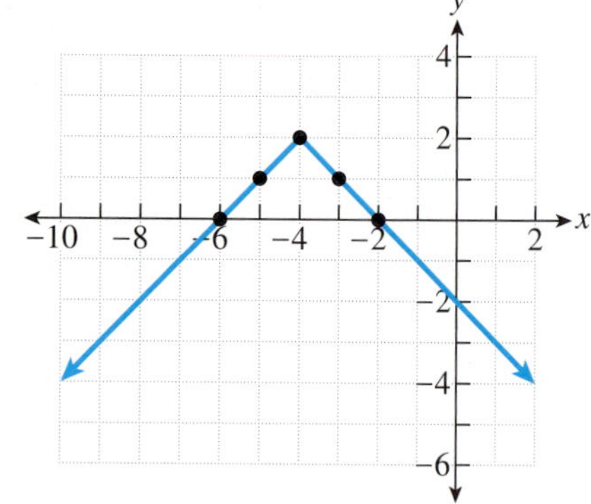

Intercepts of an Absolute Value Function

We can find the intercepts of an absolute value function $f(x)$ the same way that we found the intercepts of a linear function.

- x-intercept: Set the function $f(x)$ equal to 0 and solve for x.
- y-intercept: Find $f(0)$.

EXAMPLE 6 Find any intercepts of the absolute value function $f(x) = |x - 5| - 8$.

Solution

To find the x-intercept, we set the function equal to 0 and solve for x.

$$f(x) = 0 \quad \text{Set } f(x) \text{ equal to 0.}$$
$$|x - 5| - 8 = 0 \quad \text{Replace } f(x) \text{ by its formula.}$$
$$|x - 5| = 8 \quad \text{Add 8.}$$
$$x - 5 = 8 \quad \text{or} \quad x - 5 = -8 \quad \text{Rewrite as two linear equations.}$$
$$x = 13 \quad \text{or} \quad x = -3 \quad \text{Add 5 to both sides of each equation.}$$

So, the x-intercepts are $(13, 0)$ and $(-3, 0)$. To find the y-intercept, we evaluate the function at $x = 0$.

Quick Check 6
Find any intercepts of the function $f(x) = |x + 2| - 6$.

$$\begin{aligned} f(0) &= |0 - 5| - 8 && \text{Substitute 0 for } x. \\ &= |-5| - 8 && \text{Simplify } 0 - 5. \\ &= 5 - 8 && \text{Simplify the absolute value.} \\ &= -3 && \text{Subtract.} \end{aligned}$$

The y-intercept is at $(0,-3)$.

Finding an Absolute Value Function from Its Graph

Objective 2 **Determine an absolute value function from its graph.** To find an absolute value function from its graph, we will focus on the turning point of the graph.

EXAMPLE 7 Find the absolute value function $f(x)$ on the graph.

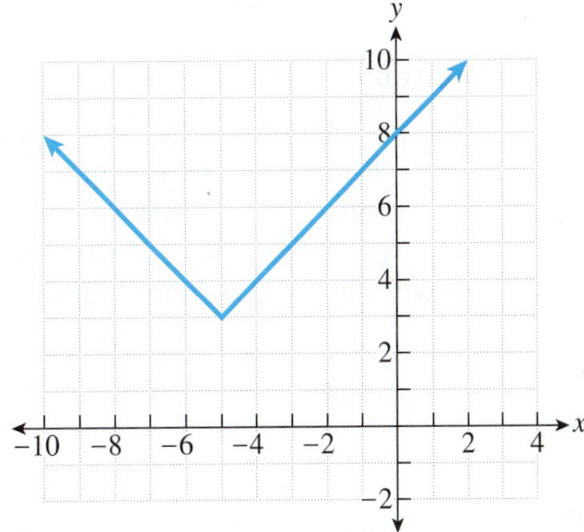

Solution

To find the function, let's focus on the turning point of this graph, which is at $(-5, 3)$. Since the x-coordinate is -5, the expression $x + 5$ must be inside the absolute value bars. The y-coordinate of the turning point is 3, which tells us that 3 is added to the absolute value. The function is of the form $f(x) = a|x + 5| + 3$, where a is a real number. To determine the value of a, we need to use the coordinates of another point on the graph, such as the y-intercept $(0, 8)$. This point tells us that $f(0) = 8$. We will use this fact to find a.

Quick Check 7
Find the absolute value function on the graph.

$$\begin{aligned} f(0) &= 8 && \\ a|0 + 5| + 3 &= 8 && \text{Substitute 0 for } x \text{ in the function } f(x). \\ 5a + 3 &= 8 && \text{Simplify the absolute value.} \\ 5a &= 5 && \text{Subtract 3.} \\ a &= 1 && \text{Divide both sides by 5.} \end{aligned}$$

Substituting 1 for a, the function is $f(x) = |x + 5| + 3$.

In the next example we will try to find an absolute value function for a graph that opens downward. This tells us that there will be a negative sign in front of the absolute value bars. We will proceed exactly as we did in the previous example.

EXAMPLE 8 Find the absolute value function $f(x)$ on the graph.

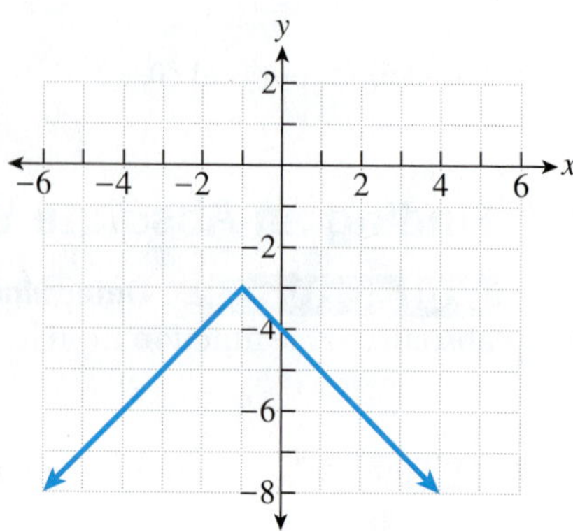

Solution

The turning point of this graph is at $(-1, -3)$. Since the x-coordinate is -1, the expression $x + 1$ must be inside the absolute value bars. The y-coordinate of the turning point is -3, which tells us that 3 is subtracted from the absolute value. The function is of the form $f(x) = a|x + 1| - 3$, where a is a negative real number. To determine the value of a, we will use the coordinates of the y-intercept $(0, -4)$. This point tells us that $f(0) = -4$.

$$f(0) = -4$$
$$a|0 + 1| - 3 = -4 \quad \text{Substitute 0 for } x \text{ in the function } f(x).$$
$$a - 3 = -4 \quad \text{Simplify the absolute value.}$$
$$a = -1 \quad \text{Add 3.}$$

Substituting -1 for a, the function is $f(x) = -|x + 1| - 3$.

Quick Check 8
Find the absolute value function on the graph.

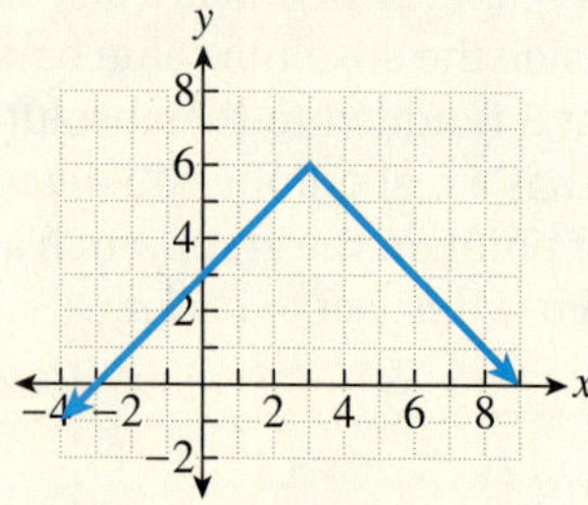

Building Your Study Strategy **Using Your Textbook Effectively, 6 More Features** Two features of this textbook that you may find helpful are labeled "A Word of Caution" and "Using Your Calculator." Each "Word of Caution" warns you about common student errors for certain problems and tells you how to avoid repeating the same mistake. After a section is covered in class, look through the section for this feature before attempting the homework exercises.

"Using Your Calculator" shows how you can use the Texas Instruments TI-84 calculator to help with selected examples in the text. If you own one of these calculators, look for this feature before beginning your homework. If you own a different calculator, you should be able to interpret the directions for your calculator.

EXERCISES 2.6

Vocabulary

1. A(n) ________ is a function of the form $f(x) = |x|$.

2. The graph of an absolute value function is ________ -shaped.

3. Find $f(0)$ to find the ________ of a function.

4. Set $f(x) = 0$ to find the ________ of a function.

Graph the absolute value function. State the domain and range of the function.

5. $f(x) = |x + 3|$

6. $f(x) = |x + 4|$

7. $f(x) = |x - 2|$

8. $f(x) = |x - 1|$

9. $f(x) = |x| + 2$

10. $f(x) = |x| + 5$

11. $f(x) = |x| - 3$

12. $f(x) = |x| - 1$

13. $f(x) = |x + 1| - 3$

14. $f(x) = |x - 3| - 4$

15. $f(x) = |x - 4| + 2$

16. $f(x) = |x + 2| + 5$

17. $f(x) = |3x - 2|$

18. $f(x) = |2x + 7|$

19. $f(x) = |2x + 5| - 3$

20. $f(x) = |4x - 6| + 2$

21. $f(x) = |x - 2.5| + 3.5$

22. $f(x) = |x + 1.4| - 6.3$

23. $f(x) = |x + \frac{9}{4}| - \frac{3}{2}$

24. $f(x) = |x - \frac{2}{5}| + \frac{13}{4}$

25. $f(x) = 3|x - 2| - 2$

26. $f(x) = 2|x + 1| + 4$

27. $f(x) = -|x - 3|$

28. $f(x) = -|x| - 3$

29. $f(x) = -|x + 2| - 7$

30. $f(x) = -|x - 5| + 4$

31. $f(x) = -2|x - 3| - 2$

32. $f(x) = -3|x + 2| + 5$

Find the intercepts of the absolute value function.

33. $f(x) = |x + 7| - 2$

34. $f(x) = |x - 4| - 5$

35. $f(x) = |x - 6|$

36. $f(x) = |x| - 3$

37. $f(x) = |x - 2| + 3$

38. $f(x) = |x + 5| + 6$

Use the graph of the absolute function $f(x)$ to find values of x that are solutions of the inequality. Express your solution in interval notation.

39. $f(x) \leq 0$

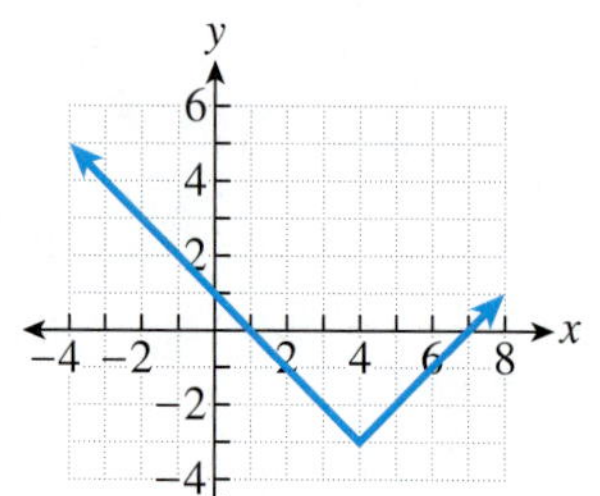

40. $f(x) < 0$

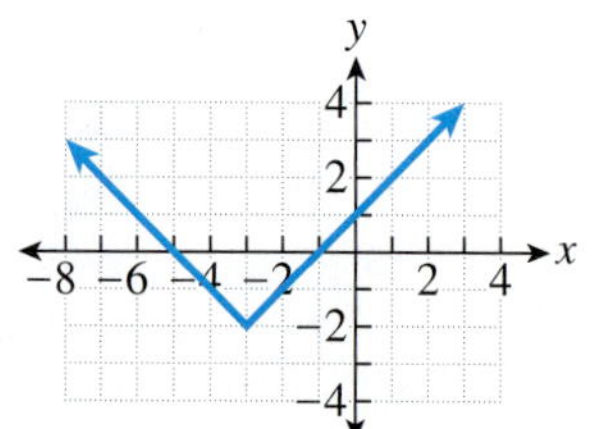

41. $f(x) > 0$

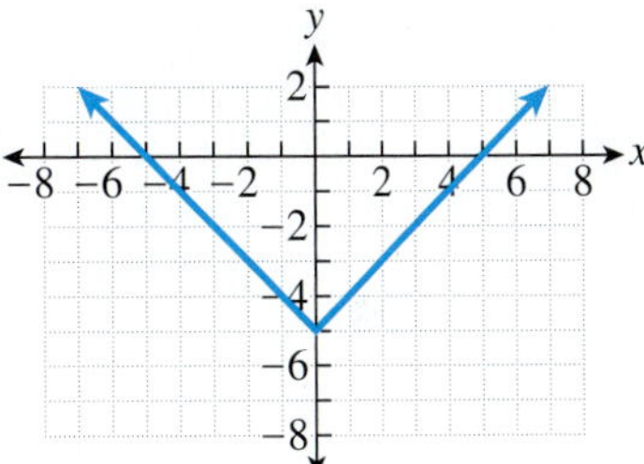

42. $f(x) \geq 0$

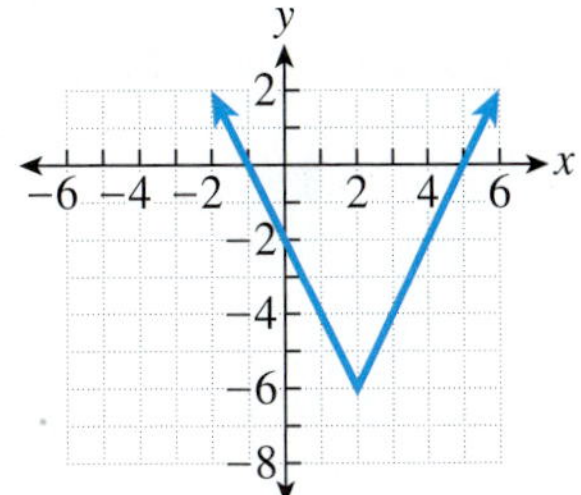

Determine the absolute value function $f(x)$ that has been graphed.

43.

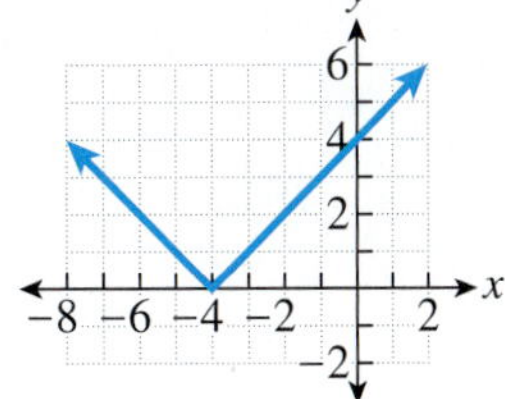

44.

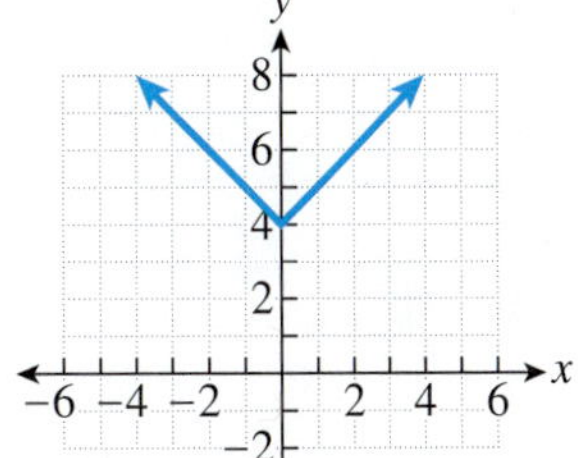

45.

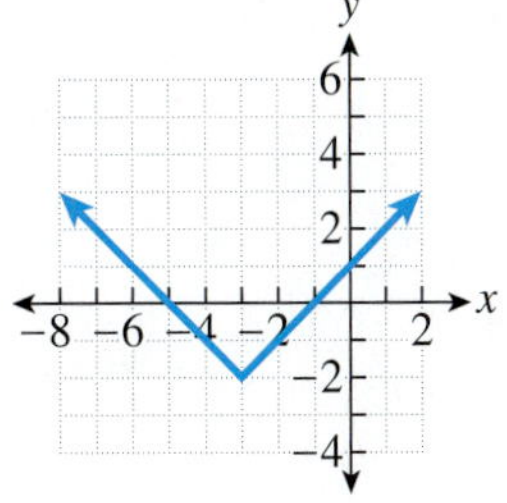

46.

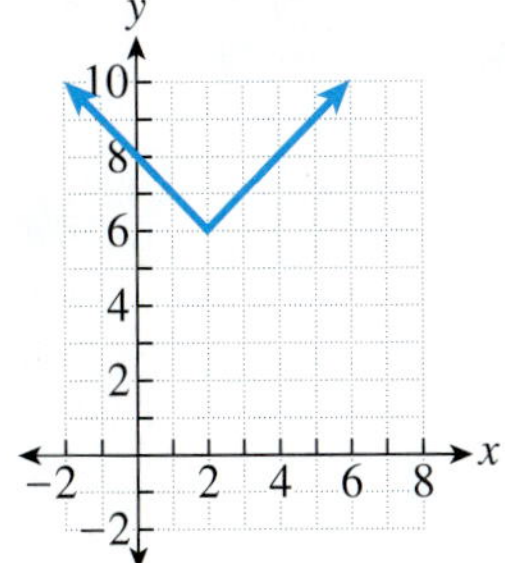

47.

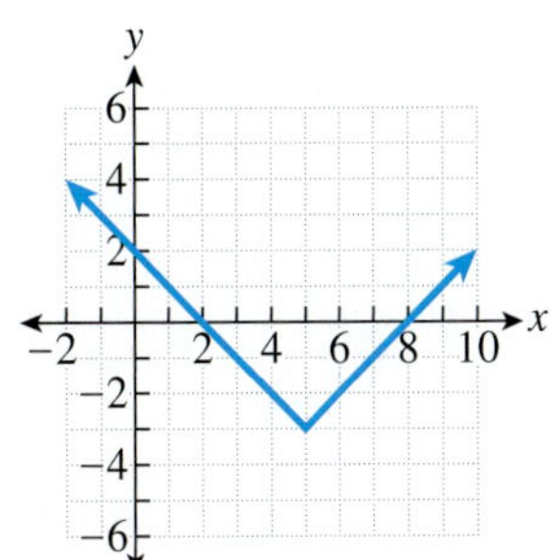

48.

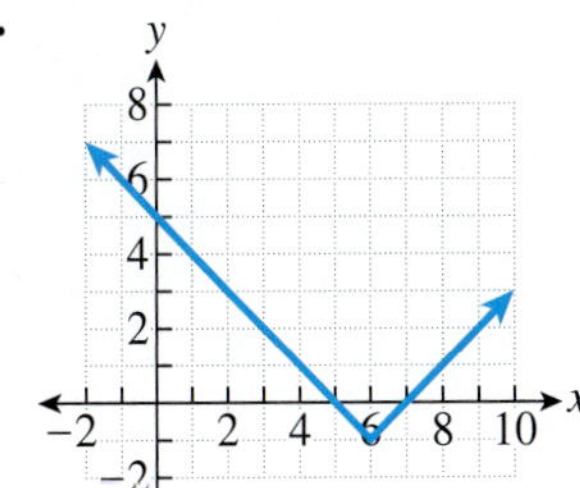

49.

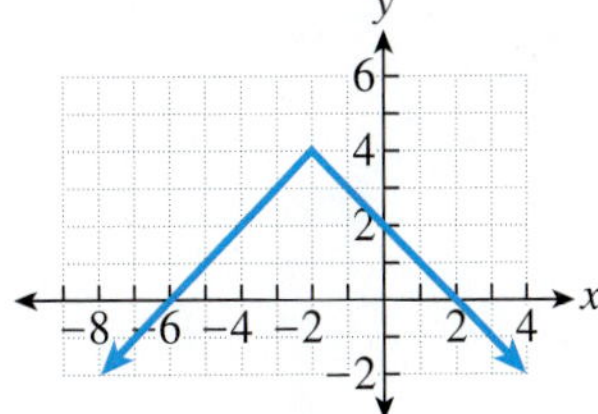

50.

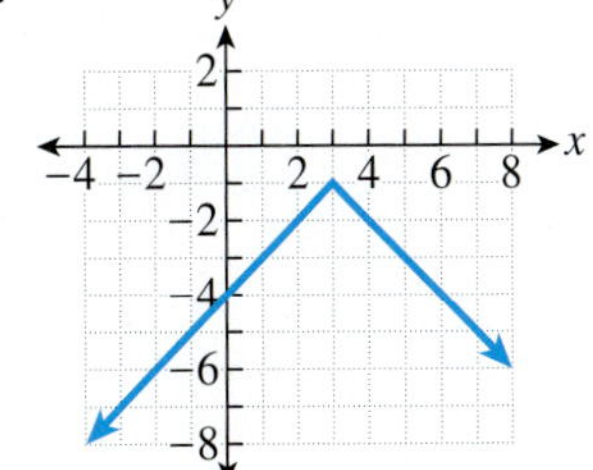

Mixed Practice, 51–62

Graph the function.

51. $f(x) = -3x - 6$

52. $f(x) = 2x + 3$

53. $f(x) = |x - 2| - 5$

54. $f(x) = |x - 6| + 3$

55. $f(x) = x + 7$

56. $f(x) = |x| - 4$

57. $f(x) = |x + 2|$

58. $f(x) = \frac{7}{3}x - 4$

59. $f(x) = |x + 3| - 4$

60. $f(x) = -5$

61. $f(x) = \frac{4}{9}x$

62. $f(x) = -|x + 8| + 5$

Writing in Mathematics

Answer in complete sentences.

63. Explain the differences between graphing a linear function and graphing an absolute value function.

64. A student is trying to graph the function $f(x) = |x - 5| + 1$, and evaluates the function for the following values of x: $-2, -1, 0, 1, 2$. Will the student's graph be correct? If not, explain what the error is and how to avoid that error.

65. ***Solutions Manual**** Write a solutions manual page for the following problem:

Graph the absolute value function $f(x) = |x - 2| - 5$.

66. ***Newsletter**** Write a newsletter explaining how to graph absolute value functions.

***See Appendix B for details and sample answers.**

Chapter 2 Summary

Section 2.1—Topic	Chapter Review Exercises
Finding the x- and y-Intercepts of a Line	1–4
Graphing a Line Using Its Intercepts	5–8
Graphing Horizontal and Vertical Lines	9–12
Section 2.2—Topic	**Chapter Review Exercises**
Finding the Slope of a Line Passing through Two Points	13–16
Finding the Slope and y-Intercept of a Line from Its Equation	17–20
Graphing a Line Using Its Slope and y-Intercept	21–24
Graphing a Line Using the Most Efficient Technique	25–30
Determining whether Two Lines Are Parallel, Perpendicular, or Neither	31–33
Finding the Equation of a Line Given Its Slope and Its y-intercept	34–35
Section 2.3—Topic	**Chapter Review Exercises**
Finding the Equation of a Line Given Its Slope and a Point on the Line	36–38
Finding the Equation of a Line Given Two Points on the Line	39–41
Finding the Equation of a Line from Its Graph	42–43
Finding the Equation of a Line That Is Parallel or Perpendicular to a Given Line	44–45
Section 2.4—Topic	**Chapter Review Exercises**
Graphing Linear Inequalities	46–49
Section 2.5—Topic	**Chapter Review Exercises**
Solving Applications Involving Linear Functions	50
Evaluating Linear Functions	51–53
Graphing Linear Functions	54–57
Interpreting the Graph of a Function	58
Finding a Linear Function That Meets Given Conditions	59–62
Section 2.6—Topic	**Chapter Review Exercises**
Graphing Absolute Value Functions	63–66
Determining an Absolute Value Function from Its Graph	67–68

Summary of Chapter 2 Study Strategies

Using your textbook as more than a source of the homework exercises and answers to the odd problems will help you to learn mathematics.

- Read the section the night before your instructor covers it to familiarize yourself with the material. If you find a step in one of the examples that you do not understand, then you can prepare questions to ask your instructor in class.
- Reread the section after it is covered in class. After you read through an example, attempt the "Quick Check" exercise that follows it. The answers to these exercises can be found at the back of the textbook. As you work through the homework exercises, refer back to the examples found in that section for assistance.
- Use the textbook to create a series of note cards for each section. Note cards can be used to memorize new terms or procedures and can help you remember how to solve particular problems.
- In each section, look for the feature labeled "A Word of Caution." These will point out common student mistakes and offer advice for not making the same mistakes yourself.
- Finally, the feature labeled "Using Your Calculator" will show you how to use the TI-84 calculator to help with the problems in that section.

Find the x-intercept and y-intercept, if possible. [2.1]

1. $2x - 3y = -24$
2. $3x + 4y = 18$
3. $6x - 9y = 0$
4. $y = 3x - 4$

Find the x- and y-intercepts, and use them to graph the equation. [2.1]

5. $4x + 3y = 24$

6. $x - 2y = -6$

7. $5x + y = 0$

8. $y = \frac{3}{2}x - 6$

Graph the vertical or horizontal line. [2.1]

9. $y = 4$

10. $y = -3$

11. $x = 7$

12. $x = -5$

Find the slope of the line that passes through the given points. [2.2]

13. $(1, 4)$ and $(4, 10)$
14. $(-2, 7)$ and $(3, -8)$
15. $(-9, 8)$ and $(-1, 2)$
16. $(-6, -7)$ and $(0, 8)$

Find the slope and the y-intercept of the given line. [2.2]

17. $y = -3x + 2$
18. $y = \frac{2}{5}x - 9$
19. $8x - 2y = 10$
20. $14x + 6y = 24$

Worked-out solutions to Review Exercises marked with ⬤ can be found on page AN-12.

Graph using the slope and y-intercept. **[2.2]**

21. $y = 2x - 8$

22. $y = -4x + 4$

23. $y = \frac{3}{4}x - 6$

24. $y = -2x$

Graph. **[2.1, 2.2]**

25. $y = x - 5$

26. $3x - 2y = -18$

27. $y = 7$

28. $8x + 3y = 24$

29. $x = -8$

30. $y = \frac{4}{3}x + 8$

Are the two given lines parallel, perpendicular, or neither? **[2.2]**

31. $2x + y = 7$
$2x - y = -3$

32. $3x + 2y = 8$
$y = \frac{2}{3}x + 6$

33. $12x + 9y = 18$
$8x + 6y = -12$

Find the equation of a line with the given slope and y-intercept. **[2.2]**

34. Slope -3, y-intercept $(0, 6)$

35. Slope $\frac{4}{3}$, y-intercept $(0, -8)$

Find the slope–intercept form of the equation of a line with the given slope that passes through the given point. **[2.3]**

36. Slope 3, through $(4, 7)$

37. Slope -2, through $(-6, -1)$

38. Slope $\frac{3}{5}$, through $(10, -8)$

Find the slope–intercept form of the equation of a line that passes through the two given points. **[2.3]**

39. $(-1, 6)$ and $(3, -2)$

40. $(-6, 1)$ and $(3, 13)$

41. $(4, -2)$ and $(-10, 5)$

***Find the equation of the line that has been graphed.* [2.3]**

42.

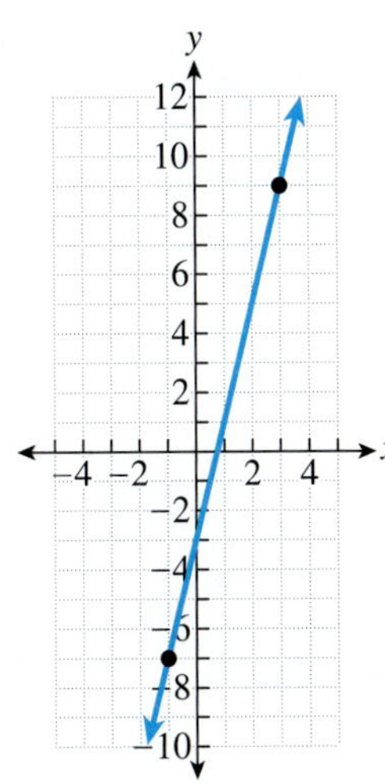

43.

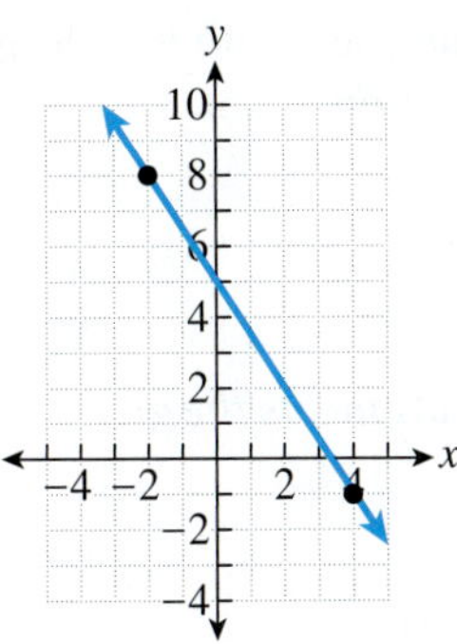

44. Find the equation of a line that is parallel to $y = -3x + 10$ that passes through the point $(5, -6)$. **[2.3]**

45. Find the equation of a line that is perpendicular to $4x + 3y = -8$ that passes through the point $(-8, 1)$. **[2.3]**

***Graph the inequality on a plane.* [2.4]**

46. $y > x + 6$

47. $3x + 4y \leq 12$

48. $5x - 2y \geq 10$

49. $y < 6$

50. A major league baseball pitcher signed a contract to earn $3,000,000 for the upcoming season. He also earns a bonus of $50,000 for each game he pitches. **[2.5]**

a) Create a function $f(x)$ for the amount of money the pitcher will earn if he pitches in x games.

b) Use the function from part a) to determine how much money he will earn if he pitches in 27 games.

c) Use the function from part a) to determine how many games he will need to pitch in to earn $5,000,000.

***Evaluate the given function.* [2.5]**

51. $f(x) = 4x - 27, f(3)$

52. $f(x) = 21 - 6x, f(-5)$

53. $f(x) = 5x + 19, f(3a - 2)$

Graph the linear function. **[2.5]**

54. $f(x) = 3x - 6$

55. $f(x) = \frac{2}{3}x + 4$

56. $f(x) = -x + 7$

57. $f(x) = \frac{4}{5}x$

58. Consider the following graph of a function $f(x)$. **[2.5]**

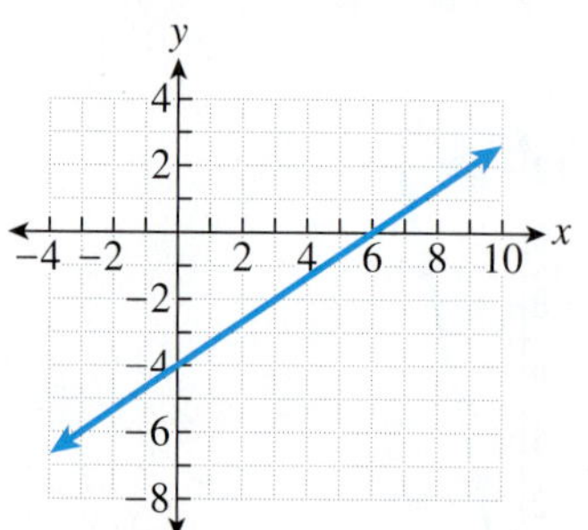

a) Find $f(-3)$.

b) Find all values x for which $f(x) = 2$.

c) Find the domain of $f(x)$.

d) Find the range of $f(x)$.

Create a linear function $f(x)$, whose graph has the given slope, that meets the given condition. **[2.5]**

59. Slope of -3, $f(4) = 1$

60. Slope of $\frac{2}{3}$, $f(9) = -17$

Create a linear function $f(x)$ that meets the given conditions. **[2.5]**

61. $f(-2) = -5, f(3) = 15$

62. $f(-8) = -2, f(20) = -23$

Graph the absolute value function. State the domain and range of the function. **[2.6]**

63. $f(x) = |x + 2|$

64. $f(x) = |x - 3| + 1$

65. $f(x) = |x - 1| - 5$

66. $f(x) = -|x + 2| + 4$

Determine the absolute value function $f(x)$ that has been graphed. **[2.6]**

67.

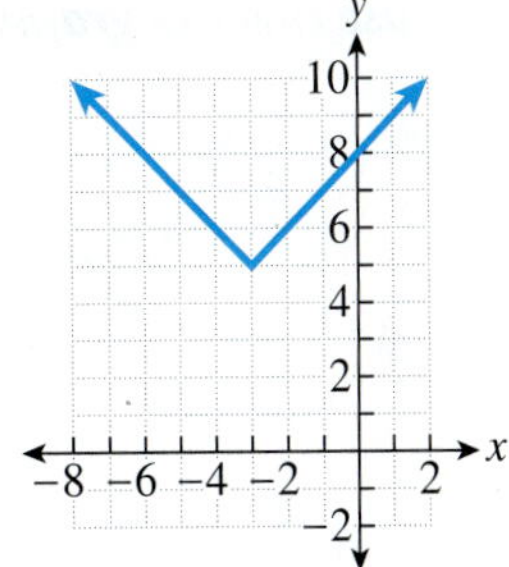

68.

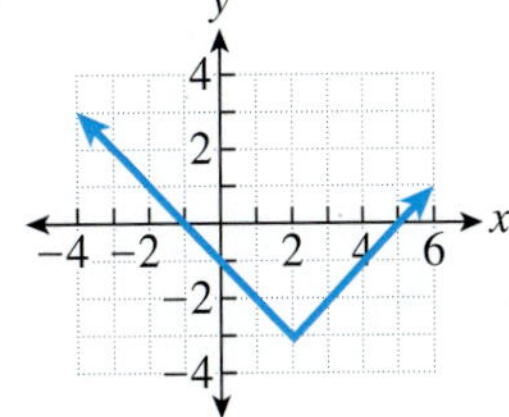

Chapter 2 Test

For Extra Help — Pass the Test — Test solutions are found on the enclosed CD.

1. Find the x-intercept and y-intercept of $5x - 4y = 40$.

Find the x- and y-intercepts, and use them to graph the equation.

2. $2x - y = -8$

3. $5x + 2y = 15$

4. Find the slope of the line that passes through the points $(-9, 4)$ and $(-1, -8)$.

Find the slope and the y-intercept of the given line.

5. $y = -\frac{3}{4}x + 10$
6. $3x - 5y = 45$

Graph using the slope and y-intercept.

7. $y = -4x + 8$

8. $y = \frac{3}{2}x + 9$

Graph.

9. $y = -\frac{1}{3}x + 2$

10. $3x - 2y = -12$

11. $y = 6$

12. Are the two lines $8x + 6y = 48$ and $9x - 12y = 36$ parallel, perpendicular, or neither?

13. Find the equation of a line whose slope is -7 and y-intercept is $(0, 12)$.

14. Find the slope–intercept form of the equation of a line with a slope of $\frac{4}{3}$ that passes through the point $(6, 3)$.

15. Find the slope–intercept form of the equation of a line that passes through the points $(-2, 19)$ and $(4, -5)$.

16. Evaluate the given function. $f(x) = 5x - 32, f(-4)$

17. Graph the inequality on a plane. $y \geq 2x + 5$

18. Graph the function $f(x) = -\frac{3}{4}x + 6$.

19. Graph the function $f(x) = |x + 2| - 1$. State the domain and the range.

20. Determine the absolute value function $f(x)$ that has been graphed.

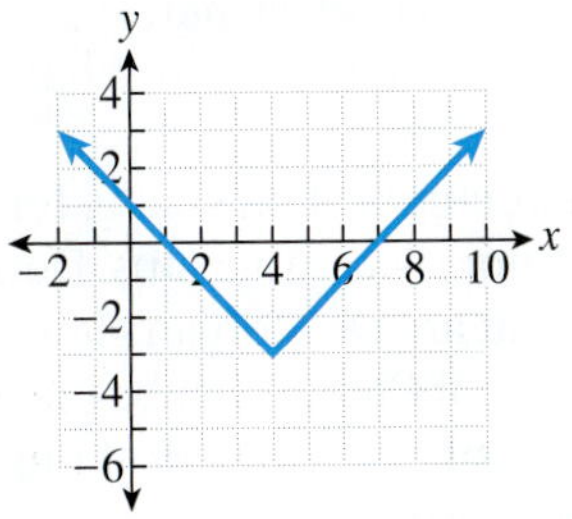

Mathematicians in History

Ada Lovelace

Ada Lovelace, whose full name was Augusta Ada King, Countess of Lovelace, was an English mathematician who lived in the 19th century. Besides her mathematical contributions, she had a profound effect on the development of the computer and computer science.

Write a one-page summary (*or* make a poster) of the life of Ada Lovelace and her accomplishments.

Interesting issues:

- Where and when was Ada Lovelace born?
- Who was Ada Lovelace's famous father? What is he best known for?
- Ada's father had a mathematical nickname for her mother. What was it?
- One of Ada's early tutors was Augustus De Morgan, who became a famous logician and developed De Morgan's Laws. What are De Morgan's laws?
- Ada wrote a description of Charles Babbage's Analytical Engine. What was the Analytical Engine?
- In her description of Babbage's Analytical Engine, Ada wrote what is considered to be the first computer program. It calculated Bernoulli numbers. What are Bernoulli numbers?
- What did the U.S. Defense Department name after Ada Lovelace?
- Describe the circumstances of Ada's death.

Stretch Your Thinking ❯ Chapter 2

Step 1: Form groups of no more than 4 or 5 students. Write the members' names in the chart below. For the best results, try to form groups in which all members are of the same sex. Each group should have a meter stick, ruler, or some type of measuring device.

Step 2: Measure the length of each member's foot (without shoes). Then measure and record the height of each member. Record these values in the appropriate column of the table.

Member's Name	Length of Foot	Height	Ordered Pair

Step 3: For each member, create an ordered pair in which the first coordinate is the length of the member's foot and the second coordinate is his/her height.

Step 4: Create an appropriate axis system using the horizontal axis for the foot length and the vertical axis for the heights.

Step 5: Plot your ordered pairs in the axis system you created. Try to be as accurate as possible. Starting from left to right, label your points A, B, C, D, etc. Visually, does it look like these points form a straight line? Would a straight line be an appropriate model for your data? Use a straight edge to check.

Step 6: Find the slope of the line that would connect point A and point B. Find the slope of the line that would connect point B and point C. Continue until you have used all of your points. Are these slopes the same? Should the slopes be the same if the points don't line up in a straight line? Have a group discussion and write down the groups' thoughts on these questions.

Step 7: Using a straight edge, try to find a straight line that best fits your ordered pairs. Calculate the equation of this line and draw it on your plane.

Step 8: Along with your completed chart, turn in the linear equation you created that your group feels best fits your data.

CHAPTER Three 3

Systems of Equations

In this chapter we will learn to set up and solve systems of equations. A system of equations is a set of two or more associated equations containing two or more variables. We will review solving systems of linear equations in two variables by three methods: graphing, the substitution method, and the addition method. We will also learn to set up systems of equations to solve applied problems. After learning to solve systems of linear inequalities in two variables, we will move on to solving systems of three linear equations in three variables. We will finish the chapter by learning how to solve systems of linear equations by using matrices and also by using Cramer's rule.

Study Strategy **Making the Best Use of Your Resources** *In this chapter we will focus on how to get the most out of your available resources. What works best for one student may not work for another student. Consider all the different resources presented in this chapter, and use the ones you feel could help you.*

Throughout this chapter we will revisit this study tip and help you incorporate it into your study habits.

3.1 Systems of Two Equations in Two Unknowns

Objectives

1. **Determine whether an ordered pair is a solution to a system of two equations in two unknowns.**
2. **Solve a system of two linear equations in two unknowns graphically.**
3. **Solve a system of two linear equations using the substitution method.**
4. **Solve a system of two linear equations using the addition method.**

Systems of Linear Equations and Their Solutions

Objective 1 Determine whether an ordered pair is a solution to a system of two equations in two unknowns. A **system of linear equations** consists of two or more linear equations. Here are some examples.

$$\begin{aligned} 5x + 4y &= 20 \\ 2x - y &= 6 \end{aligned} \qquad \begin{aligned} y &= \tfrac{3}{5}x \\ y &= -4x + 3 \end{aligned} \qquad \begin{aligned} x + y &= 11 \\ y &= 5 \end{aligned}$$

We examine the equations in a system together to find the solution(s) that the equations have in common.

Solution of a System of Linear Equations

A **solution of a system of linear equations** is an ordered pair (x, y) that is a solution to each equation in the system.

EXAMPLE 1 Is the ordered pair $(4, -3)$ a solution to the given system of equations?

$$\begin{aligned} 3x + y &= 9 \\ 2x - 5y &= 23 \end{aligned}$$

Solution

To determine if the ordered pair $(4, -3)$ is a solution, we need to substitute 4 for x and -3 for y into each equation. If the resulting equations are both true, then the ordered pair is a solution to the system. Otherwise, it is not.

$$\begin{aligned} 3x + y &= 9 \\ 3(4) + (-3) &= 9 \\ 12 - 3 &= 9 \\ 9 &= 9 \end{aligned} \qquad \begin{aligned} 2x - 5y &= 23 \\ 2(4) - 5(-3) &= 23 \\ 8 + 15 &= 23 \\ 23 &= 23 \end{aligned}$$

Since $(4, -3)$ is a solution to each equation, it is a solution to the system of equations.

EXAMPLE 2 Is the ordered pair $(-4, -1)$ a solution to the given system of equations?

$$y = 2x + 7$$
$$y = 11 - 3x$$

Solution

Again we begin by substituting -4 for x and -1 for y in each equation.

$y = 2x + 7$	$y = 11 - 3x$
$(-1) = 2(-4) + 7$	$(-1) = 11 - 3(-4)$
$-1 = -8 + 7$	$-1 = 11 + 12$
$-1 = -1$	$-1 = 23$

Although $(-4, -1)$ is a solution to the equation $y = 2x + 7$, it is not a solution to the equation $y = 11 - 3x$. Therefore $(-4, -1)$ is not a solution to the system of equations.

Quick Check 1

Is the ordered pair $(-8, 7)$ a solution to the given system of equations?

$$6x - 3y = -69$$
$$-x + 2y = 22$$

Systems of Linear Equations and Their Solutions

Objective 2 **Solve a system of two linear equations in two unknowns graphically.** Systems of two linear equations can be solved graphically. Since an ordered pair that is a solution to a system of two equations must be a solution to each equation, we know that this ordered pair is on the graph of each equation. To solve a system of two linear equations, we graph each line and look for points of intersection. Often there is only a single solution. Here are some graphical examples of systems of equations that have the ordered pair $(3, 1)$ as a solution.

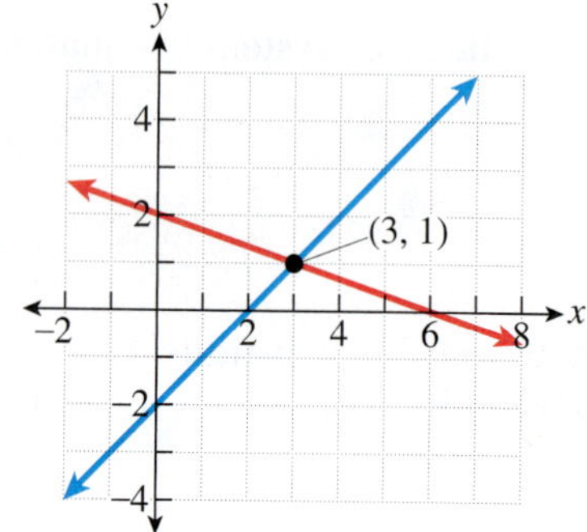

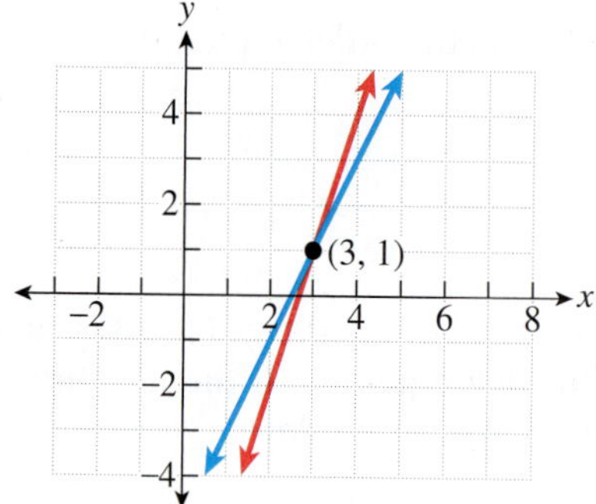

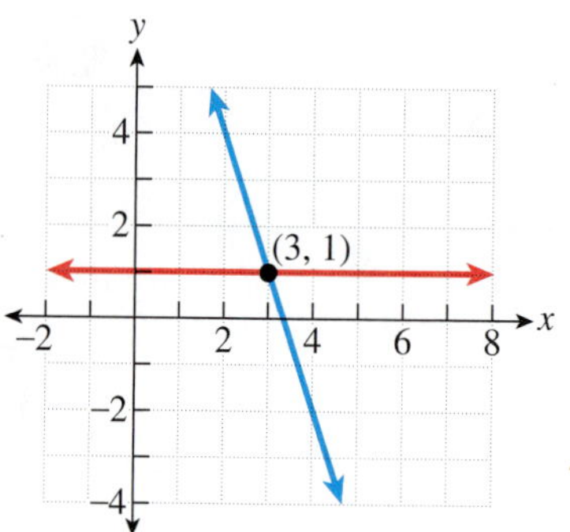

The graph of each equation in a system of linear equations is a line, and if these two lines cross at exactly one point, then this ordered pair is the only solution to the system. In this case, the system is said to be **independent**.

EXAMPLE 3 Solve the system by graphing.

$$y = 2x + 5$$
$$y = -x - 1$$

Solution

We begin by graphing each line.

To graph $y = 2x + 5$, we first plot the y-intercept $(0, 5)$. We then use the slope of the line, $m = 2$, to find a second point on the line.

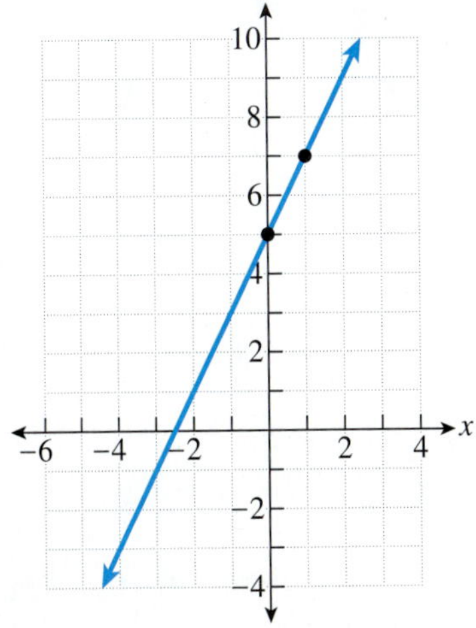

To graph $y = -x - 1$, we first plot the y-intercept $(0, -1)$. We then use the slope of the line, $m = -1$, to find a second point on the line.

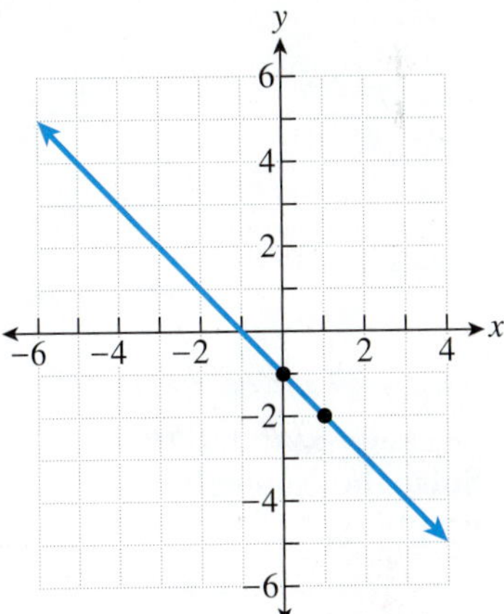

The graph at the right shows both lines. These two lines intersect at the point $(-2, 1)$, and this point of intersection is the solution to this independent system of equations.

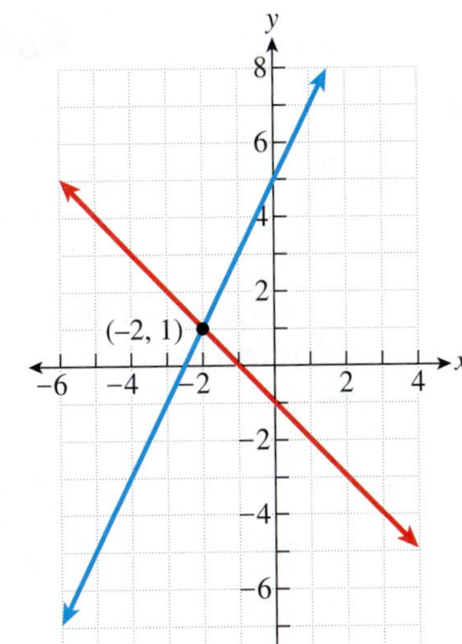

***Quick Check* 2**

Solve the system by graphing.

$$y = 3x - 6$$
$$y = x + 2$$

If the two lines associated with a system of equations are parallel lines that do not cross, then the system has no solution and is said to be **inconsistent**. An inconsistent system has no solution and we use the empty set, $\varnothing$, to represent this.

EXAMPLE 4 Solve the system by graphing.

$$y = 3x - 9$$
$$y = 3x + 6$$

Solution

We begin by graphing each line.

The graph of $y = 3x - 9$ has a y-intercept of $(0, -9)$ and a slope of 3. The graph of $y = 3x + 6$ has a y-intercept of $(0, 6)$ and a slope of 3. Here are both lines on the same set of axes.

These lines are parallel, so this system of equations is inconsistent and has no solution.

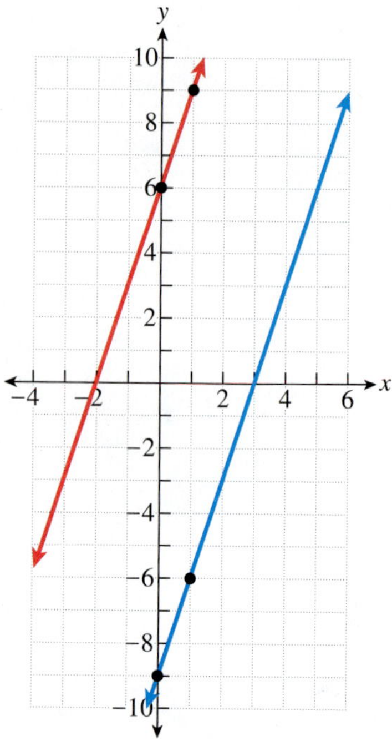

Quick Check 3
Solve the system by graphing.

$$y = -5x + 8$$
$$y = -5x + 2$$

If the two lines associated with a system of equations are actually the same line, then the system is said to be a **dependent** system.

EXAMPLE 5 Solve the system by graphing.

$$x + y = 5$$
$$2x + 2y = 10$$

Solution

We begin by graphing each line.

The graph of $x + y = 5$ has an x-intercept of $(5, 0)$ and a y-intercept of $(0, 5)$. The graph of $2x + 2y = 10$ has the same intercepts. Both lines are identical, so this system of equations is a dependent system. Each ordered pair that is on the line is a solution. We write the solution in the form $(x, 5 - x)$ by solving one of the equations for y in terms of x.

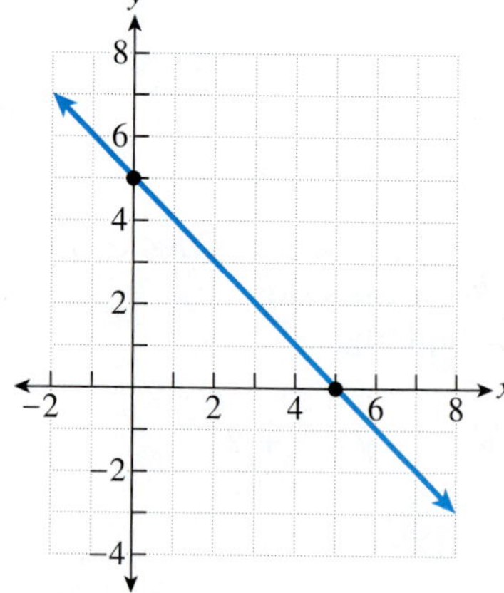

Quick Check 4
Solve the system by graphing.

$$y = 3x - 12$$
$$-6x + 2y = -24$$

Solving systems of equations by graphing does have a major drawback. If the solution coordinates are not integers, we will not be able to determine them accurately from a graph. Consider the graph of the system

$$4x - 3y = 12$$
$$2x + 5y = 10$$

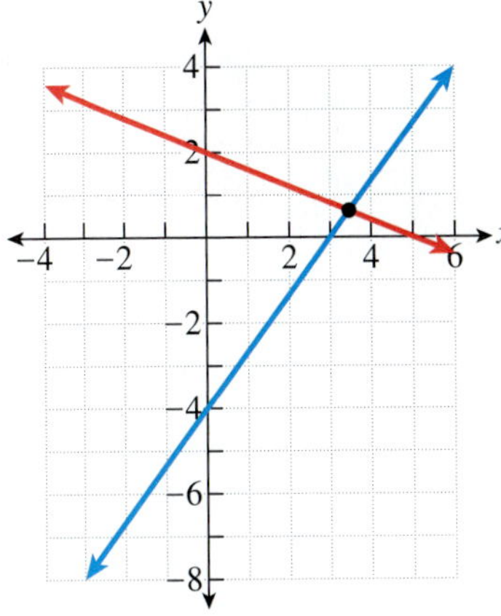

We can see that the x-coordinate of this solution is between 3 and 4, and the y-coordinate of this solution is between 0 and 1. However, we cannot determine the exact coordinates of this solution from the graph. We will develop two algebraic techniques for finding solutions of linear systems of equations, to use instead of graphing.

The Substitution Method

Objective 3 Solve a system of two linear equations using the substitution method. The substitution technique requires us to solve one of the equations for one of the variables in terms of the other variable. The expression that we find is then *substituted* for that variable in the other equation, giving us one equation with one variable. After solving that equation, we substitute this value into the first expression.

The Substitution Method

To solve a system of equations by using the substitution method, we solve one of the equations for one of the variables and then substitute that expression for the variable in the other equation.

Consider the following system of equations.

$$y = x - 5$$
$$2x + 3y = 20$$

Looking at the first equation, $y = x - 5$, we see that y is equal to the expression $x - 5$. We can use this fact to replace y by $x - 5$ in the second equation, $2x + 3y = 20$. We are substituting the expression $x - 5$ for y since the two expressions are equivalent.

EXAMPLE 6 Solve the system by substitution.

$$y = x - 5$$
$$2x + 3y = 20$$

Solution

Since the first equation is solved for y in terms of x, we can substitute $x - 5$ for y in the second equation.

$$y = x - 5$$
$$\downarrow$$
$$2x + 3y = 20$$

This will produce a single equation with only one variable, x.

$2x + 3y = 20$	
$2x + 3(x - 5) = 20$	Substitute $x - 5$ for y.
$2x + 3x - 15 = 20$	Distribute.
$5x - 15 = 20$	Combine like terms.
$5x = 35$	Add 15.
$x = 7$	Divide both sides by 5.

We are now halfway to our solution. We know that the x-coordinate of the solution is 7 and we now need to find the y-coordinate.

A Word of Caution The solution to a system of equations in two variables is an ordered pair, not a single value.

We can find the y-coordinate by substituting 7 for x in either of the original equations. It is easier to substitute the value into the equation $y = x - 5$, since y is already isolated in this equation. We could have chosen the other equation; it would produce the same solution. In general, it is a good idea to substitute the first value we find back into the equation that was used to make the original substitution.

$y = x - 5$	
$y = (7) - 5$	Substitute 7 for x.
$y = 2$	Subtract.

The solution to this system is $(7, 2)$.

This solution can be checked by substituting 7 for x and 2 for y in the equation $2x + 3y = 20$ as shown below.

Check

$2(7) + 3(2) = 20$	Substitute 7 for x and 2 for y.
$14 + 6 = 20$	Multiply.
$20 = 20$	Add.

This equation is true, so our solution checks.

***Quick Check* 5** Solve the system by substitution.

$$y = 3x - 2$$
$$2x + 5y = 58$$

In the next example we must solve one of the equations for x or y before applying the substitution method.

EXAMPLE 7 Solve by substitution.

$$x + 3y = 1$$
$$3x - 2y = 14$$

Solution

If there is a variable with a coefficient of 1, then we should try to solve the equation for that variable. Using this strategy will help us avoid the use of fractions. Solving the first equation for x gives us $x = 1 - 3y$. We can then substitute the expression $1 - 3y$ into the equation $3x - 2y = 14$ for x.

$3x - 2y = 14$	
$3(1 - 3y) - 2y = 14$	Substitute $1 - 3y$ for x.
$3 - 9y - 2y = 14$	Distribute.
$3 - 11y = 14$	Combine like terms.
$-11y = 11$	Subtract 3.
$y = -1$	Divide both sides by -11.

Next we substitute this value for y into the equation $x = 1 - 3y$.

$x = 1 - 3(-1)$	Substitute -1 for y.
$x = 1 + 3$	Multiply.
$x = 4$	Divide.

The ordered pair solution is $(4, -1)$.

This solution can be checked by substituting 4 for x and -1 for y in the equation $3x - 2y = 14$. The check is left to the reader.

***Quick Check* 6**
Solve by substitution.

$$2x - 5y = -24$$
$$4x + y = -4$$

EXAMPLE 8 Solve by substitution.

$$2x - 3y = -22$$
$$6x + 5y = 18$$

Solution

Neither equation has a variable term with a coefficient of 1, so we must choose a variable for which to solve one equation. Solving the first equation for x gives us $x = \frac{3}{2}y - 11$. We then substitute the expression $\frac{3}{2}y - 11$ into the equation $6x + 5y = 18$ for x.

$6x + 5y = 18$	
$6(\frac{3}{2}y - 11) + 5y = 18$	Substitute $\frac{3}{2}y - 11$ for x.
$9y - 66 + 5y = 18$	Distribute.
$14y - 66 = 18$	Combine like terms.
$14y = 84$	Add 66.
$y = 6$	Divide both sides by 14.

Next, we substitute this value for y into the equation $x = \frac{3}{2}y - 11$.

$x = \frac{3}{2}(6) - 11$	Substitute 6 for y.
$x = 9 - 11$	Multiply.
$x = -2$	Subtract.

The ordered pair solution is $(-2, 6)$.

***Quick Check* 7**
Solve by substitution.

$$2x + 7y = -11$$
$$-4x + 9y = -47$$

We finish our investigation of the substitution method with a system that is dependent.

EXAMPLE 9 Solve by substitution.

$$y = 4x + 5$$
$$8x - 2y = -10$$

Solution

Since the first equation is already solved for y, we can substitute $4x + 5$ into the equation $8x - 2y = -10$ for y.

$$\begin{aligned} 8x - 2y &= -10 & \\ 8x - 2(4x + 5) &= -10 & \text{Substitute } 4x + 5 \text{ for } y. \\ 8x - 8x - 10 &= -10 & \text{Distribute.} \\ -10 &= -10 & \text{Combine like terms.} \end{aligned}$$

This equation is an identity that is true for all values of x. This system of equations is a dependent system with infinitely many solutions. Since we have already shown that $y = 4x + 5$, the ordered pairs that are solutions have the form $(x, 4x + 5)$.

Quick Check 8
Solve by substitution.

$$y = -2x + 3$$
$$8x + 4y = 12$$

The Addition Method

Objective 4 Solve a system of two linear equations using the addition method.

The Addition Method

The **addition method** is an algebraic alternative to the substitution method. To use it, we combine the two equations in a system of linear equations into a single equation with a single variable by adding them together.

Solving a system of equations by substitution is fairly easy when one of the equations contains a variable term with a coefficient of 1 or -1. However, the substitution method can become quite tedious for solving a system when this is not the case.

As with the substitution method, the goal of the addition method is to combine the two given equations into a single equation with only one variable. This method is based on the addition property of equality from Chapter 1, which tells us that when we add the same number to both sides of an equation, the equation remains true. (For any real numbers a, b, and c, if $a = b$ then $a + c = b + c$.) This property can be extended to cover adding equal expressions to both sides of an equation: if $a = b$ and $c = d$ then $a + c = b + d$.

Consider the system

$$3x + 2y = 20$$
$$5x - 2y = -4$$

If we add the left side of the first equation $(3x + 2y)$ to the left side of the second equation $(5x - 2y)$, this will be equal to the sum of the two numbers on the right side of the equations. Here is the result of this addition.

$$\begin{array}{rcr} 3x + 2y &=& 20 \\ 5x - 2y &=& -4 \\ \hline 8x \qquad &=& 16 \end{array}$$

Notice that when we added these two equations, the two terms containing y canceled and this left an equation containing only the variable x. We next solve this equation for x and substitute this value for x into either of the two original equations to find y. Our next example will walk through this entire process.

EXAMPLE 10 Solve the system by addition.

$$\begin{aligned} 3x + 2y &= 20 \\ 5x - 2y &= -4 \end{aligned}$$

Solution

Since the two terms containing y are opposites, we can use addition to eliminate y.

$$\begin{array}{rcrl} 3x + 2y &=& 20 & \\ 5x - 2y &=& -4 & \text{Add the two equations.} \\ \hline 8x \qquad &=& 16 & \\ x &=& 2 & \text{Divide both sides by 8.} \end{array}$$

The x-coordinate of our solution is 2. We now substitute this value for x in the first original equation. (We could have chosen the other equation; it will produce exactly the same solution.)

$$\begin{aligned} 3x + 2y &= 20 & & \\ 3(2) + 2y &= 20 & & \text{Substitute 2 for } x. \\ 6 + 2y &= 20 & & \text{Multiply.} \\ 2y &= 14 & & \text{Subtract 6.} \\ y &= 7 & & \text{Divide both sides by 2.} \end{aligned}$$

The solution to this system is the ordered pair $(2, 7)$. We could check this solution by substituting 2 for x and 7 for y in the equation $5x - 2y = -4$ as shown below.

Check

$$\begin{aligned} 5(2) - 2(7) &= -4 & & \text{Substitute 2 for } x \text{ and 7 for } y. \\ 10 - 14 &= -4 & & \text{Multiply.} \\ -4 &= -4 & & \text{Subtract.} \end{aligned}$$

This equation is true, so our solution checks.

***Quick Check* 9**

Solve the system by addition.

$$\begin{aligned} 4x + 3y &= 31 \\ -4x + 5y &= -23 \end{aligned}$$

If the two equations in a system do not contain a pair of opposite variable terms, we can still use the addition method. We must first multiply both sides of one or both equations by a constant in such a way that two of the variable terms become opposites. Then we proceed as usual.

EXAMPLE 11 Solve by addition.

$$2x + 5y = 18$$
$$4x + 7y = 24$$

Solution

If we multiply the first equation by -2, the coefficient for the x term will be -4, which is the opposite of the coefficient of the x term in the second equation. This will then allow us to add the equations and eliminate the variable x.

$$\begin{aligned} 2x + 5y &= 18 \\ 4x + 7y &= 24 \end{aligned} \xrightarrow{\text{Multiply by } -2.} \begin{aligned} -4x - 10y &= -36 \\ 4x + 7y &= 24 \end{aligned}$$

$$\begin{aligned} -4x - 10y &= -36 \\ 4x + 7y &= 24 \quad \text{Add.} \\ \hline -3y &= -12 \\ y &= 4 \quad \text{Divide both sides by } -3. \end{aligned}$$

We now substitute 4 for y into either of the original equations to find x.

$$\begin{aligned} 2x + 5y &= 18 \\ 2x + 5(4) &= 18 \quad \text{Substitute 4 for } y. \\ 2x + 20 &= 18 \quad \text{Multiply.} \\ 2x &= -2 \quad \text{Subtract 20.} \\ x &= -1 \quad \text{Divide both sides by 2.} \end{aligned}$$

Our solution is $(-1, 4)$.

This solution can be checked by substituting -1 for x and 4 for y in the equation $4x + 7y = 24$. The check is left to the reader.

Quick Check 10
Solve by addition.

$$6x - 5y = 74$$
$$7x + 10y = 23$$

EXAMPLE 12 Solve by addition.

$$x + y = 3000$$
$$0.03x + 0.05y = 126$$

Solution

We will begin by clearing the second equation of decimals. This can be accomplished by multiplying both sides of that equation by 100.

$$\begin{aligned} x + y &= 3000 \\ 0.03x + 0.05y &= 126 \end{aligned} \xrightarrow{\text{Multiply by 100.}} \begin{aligned} x + y &= 3000 \\ 3x + 5y &= 12{,}600 \end{aligned}$$

If we multiply the first equation by -3, the coefficient for the x term will be -3, which is the opposite of the coefficient of the x term in the second equation.

$$\begin{aligned} x + y &= 3000 \\ 3x + 5y &= 12{,}600 \end{aligned} \xrightarrow{\text{Multiply by } -3.} \begin{aligned} -3x - 3y &= -9000 \\ 3x + 5y &= 12{,}600 \end{aligned}$$

$$\begin{aligned} -3x - 3y &= -9000 \\ 3x + 5y &= 12{,}600 \quad \text{Add.} \\ \hline 2y &= 3600 \\ y &= 1800 \quad \text{Divide both sides by 2.} \end{aligned}$$

We now substitute 1800 for y into either of the original equations to find x.

$$x + y = 3000$$
$$x + (1800) = 3000 \quad \text{Substitute 1800 for } y.$$
$$x = 1200 \quad \text{Subtract 1800.}$$

Our solution is (1200, 1800).

Quick Check 11
Solve by addition.

$$x + y = 4400$$
$$0.09x - 0.06y = 171$$

If at least one equation in a system of equations contains fractions, we can begin to solve the system by clearing the equations of fraction(s). This can be accomplished by multiplying both sides of the equation by the LCM of the denominators.

EXAMPLE 13 Solve by addition.

$$4x - 2y = 20$$
$$\tfrac{1}{2}x + \tfrac{2}{3}y = 8$$

Solution

We will begin by clearing the second equation of fractions. This can be accomplished by multiplying both sides of that equation by the LCM of the denominators 2 and 3, which is 6.

$$4x - 2y = 20 \qquad\qquad 4x - 2y = 20$$
$$\tfrac{1}{2}x + \tfrac{2}{3}y = 8 \xrightarrow{\text{Multiply by 6.}} 3x + 4y = 48$$

If we multiply the first equation by 2, the coefficient for the y term will be -4, which is the opposite of the coefficient of the y term in the second equation.

$$4x - 2y = 20 \xrightarrow{\text{Multiply by 2.}} 8x - 4y = 40$$
$$3x + 4y = 48 \qquad\qquad 3x + 4y = 48$$

$$8x - 4y = 40$$
$$3x + 4y = 48 \quad \text{Add.}$$
$$11x \quad = 88$$

$$x = 8 \quad \text{Divide both sides by 11.}$$

We now substitute 8 for x into either of the original equations to find y.

$$4x - 2y = 20$$
$$4(8) - 2y = 20 \quad \text{Substitute 8 for } x.$$
$$32 - 2y = 20 \quad \text{Multiply.}$$
$$-2y = -12 \quad \text{Subtract 32.}$$
$$y = 6 \quad \text{Divide by } -2.$$

Our solution is (8, 6).

Quick Check 12
Solve by addition.

$$6x - y = -33$$
$$\tfrac{3}{4}x + \tfrac{5}{3}y = 12$$

In the next example we will need to multiply both equations in order to apply the addition method.

EXAMPLE 14 Solve by addition.

$$2x + 3y = 22$$
$$-3x + 7y = 13$$

Solution

If we multiply the first equation by 3 and the second equation by 2, the coefficients for the x term will be opposites. This will allow us to add the equations and eliminate the variable x.

$$2x + 3y = 22 \xrightarrow{\text{Multiply by 3.}} 6x + 9y = 66$$
$$-3x + 7y = 13 \xrightarrow{\text{Multiply by 2.}} -6x + 14y = 26$$

$$\begin{array}{r} 6x + 9y = 66 \\ \underline{-6x + 14y = 26} \\ 23y = 92 \end{array} \quad \text{Add.}$$

$$y = 4 \quad \text{Divide both sides by 23.}$$

We now substitute 4 for y into either of the original equations to find x.

$$2x + 3y = 22$$
$$2x + 3(4) = 22 \quad \text{Substitute 4 for } y.$$
$$2x + 12 = 22 \quad \text{Multiply.}$$
$$2x = 10 \quad \text{Subtract 12.}$$
$$x = 5 \quad \text{Divide both sides by 2.}$$

Our solution is $(5, 4)$.

***Quick Check* 13**
Solve by addition.

$$11x - 4y = 112$$
$$7x + 6y = 20$$

We will finish with an example of an inconsistent system of equations.

EXAMPLE 15 Solve.

$$x - 2y = 12$$
$$-3x + 6y = 36$$

Solution

If we multiply the first equation by 3, the coefficient for the x term will be 3, which is the opposite of the coefficient of the x term in the second equation.

$$x - 2y = 12 \xrightarrow{\text{Multiply by 3.}} 3x - 6y = 36$$
$$-3x + 6y = 36 \qquad -3x + 6y = 36$$

$$\begin{array}{r} 3x - 6y = 36 \\ \underline{-3x + 6y = 36} \\ 0 = 72 \end{array} \quad \text{Add.}$$

This equation is a contradiction and has no solution. The system is an inconsistent system.

***Quick Check* 14**
Solve by addition.

$$3x - 4y = 9$$
$$6x - 8y = -18$$

The substitution and addition methods for solving a system of equations will solve any linear system of equations. There are certain times when one method is easier to apply than the other, and knowing which method to choose for a particular system of equations can save time and prevent errors. In general, if one of the equations in the system is already solved for one of the variables, such as $y = 3x$ or $x = -4y + 5$, then using the substitution method is a good choice. The substitution method also works well when one of the equations contains a variable term with a coefficient of 1 or -1 (such as $x + 6y = 17$ or $7x - y = 14$). Otherwise, consider using the addition method.

Building Your Study Strategy **Making the Best Use of Your Resources, 1 Office Hours** One of the most important resources available to you is your instructor. Take advantage of your instructor's experience and knowledge. If you have a question during the lecture, be sure to ask. The question that you have is probably on the minds of several other students as well.

Learn where your instructor's office is located and when office hours are held. Visiting your instructor during office hours is a great way to get one-on-one help. There are some questions that are difficult for the instructor to answer completely during the class period that are more easily handled in your instructor's office. Your instructor may also be able to give you extra problems to work on. In addition to answering your questions during class, some instructors are available for questions either right before or right after class.

EXERCISES 3.1

Vocabulary

1. A(n) ________ consists of two or more linear equations.

2. An ordered pair (x, y) is a solution to a system of two linear equations if ________.

3. A system of two linear equations is a(n) ________ system if it has exactly one solution.

4. A system of two linear equations is a(n) ________ system if it has no solution.

5. A system of two linear equations is a(n) ________ system if it has infinitely many solutions.

6. To solve a system of equations by substitution, begin by solving ________.

7. To solve the system of equations $\begin{aligned} 3x + 2y &= 9 \\ 4x - y &= 13 \end{aligned}$ by substitution, which equation should be solved for which variable first?

a) first equation for x

b) first equation for y

c) second equation for x

d) second equation for y

8. If an equation in a system of equations contains fractions, we can clear the fractions by multiplying both sides of the equation by ________.

Is the ordered pair a solution to the given system of equations?

9. $(-4, 20)$, $y = -3x + 8$
$7x + 2y = 12$

10. $(5, -4)$, $3x + 2y = 7$
$x - 4y = -11$

11. $(3, 2)$, $6x + 5y = 28$
$-3x - 8y = 7$

12. $(-3, -6)$, $9x + y = -33$
$2x + 4y = -30$

13. $(8, -12)$, $\frac{3}{4}x + \frac{5}{3}y = -14$
$\frac{1}{2}x - \frac{1}{6}y = 6$

14. $(6000, 2500)$, $x + y = 8500$
$0.08x + 0.05y = 605$

Solve the system by graphing. If the system is inconsistent and has no solution, state this. If the system is dependent, write the form of the solution for any real number x.

15. $y = x + 6$
$y = -2x - 3$

16. $y = -3x + 8$
$y = 2x - 7$

17. $2x + 3y = 12$
$x + y = 2$

18. $2x + 5y = 10$
$x - y = -9$

19. $y = \frac{4}{3}x - 8$
$8x - 6y = 24$

20. $7x - 3y = 21$
$y = \frac{1}{2}x + 4$

21. $-4x + 5y = 20$
$y = \frac{2}{5}x + 6$

22. $y = 3x - 6$
$6x - 2y = 12$

Solve the system by substitution. If the system is inconsistent and has no solution, state this. If the system is dependent, write the form of the solution for any real number x.

23. $y = 2x - 3$
$3x - 2y = 2$

24. $y = 3x + 8$
$4x - y = -14$

25. $y = \frac{5}{3}x - 8$
$-10x + 6y = -48$

26. $x = \frac{3}{4}y + 7$
$2x - 3y = 2$

27. $x = 2y - 11$
$3x - 8y = -29$

28. $x = -\frac{3}{4}y + 4$
$4x + 3y = -12$

29. $y = \frac{4}{9}x + 17$
$3x - 2y = -15$

30. $y = 5000 - x$
$0.03x + 0.07y = 276$

Solve the system by addition. If the system is inconsistent and has no solution, state this. If the system is dependent, write the form of the solution for any real number x.

31. $5x + y = 39$
$3x - y = 17$

32. $-2x - 7y = 41$
$2x + 3y = -21$

33. $x - y = 8$
$-3x + 3y = -24$

34. $3x + y = 17$
$5x - 4y = 34$

35. $\frac{2}{3}x + \frac{1}{4}y = \frac{19}{6}$
$\frac{3}{4}x - \frac{1}{2}y = 2$

36. $2x + 5y = 1$
$3x - 4y = 36$

37. $2.5x + 3.5y = 47$
$1.5x + 1.4y = 23.3$

38. $4x - 6y = 10$
$6x - 9y = 20$

Mixed Practice, 39–64

Solve by addition or substitution.

39. $x + 3y = 1$
$-x + 2y = -11$

40. $8x + y = 92$
$5x - y = 25$

41. $10x + 7y = -5$
$4x + 7y = -23$

42. $2x - 9y = 6$
$2x + 4y = 32$

43. $x + 3y = 17$
$-2x + 2y = 6$

44. $3x + 7y = 14$
$5x - 4y = -8$

45. $5x + 9y = 33$
$y = 2$

46. $-2x + 3y = -1$
$5x = 25$

47. $y = \frac{1}{2}x - 3$
$3x + 2y = 10$

48. $x = 6y + 9$
$\frac{1}{3}x + 2y = -9$

49. $y = 3x - 4$
$9x - 3y = 12$

50. $y = \frac{3}{5}x - 2$
$6x - 10y = -20$

51. $2x - 3y = 10$
$4x - 2y = 28$

52. $x + 2y = -12$
$3x - 4y = 26$

53. $x - 4y = 7$
$3x + 2y = 14$

54. $5x + y = -2$
$3x - 7y = -43$

55. $4x + 3y = -28$
$3x - 2y = -4$

56. $-5x + 4y = 22$
$2x + 5y = 11$

57. $6x - 4y = 11$
$9x - 6y = 17$

58. $4x + y = 13$
$12x + 3y = 39$

59. $\frac{3}{4}x + \frac{1}{3}y = 1$
$\frac{2}{5}x - \frac{1}{4}y = \frac{31}{10}$

60. $\frac{1}{6}x + \frac{1}{2}y = -3$
$\frac{2}{3}x - \frac{3}{4}y = \frac{7}{4}$

61. $x + y = 80$
$0.5x + 0.3y = 36$

62. $x + y = 2000$
$0.04x - 0.09y = 15$

63. $0.2x + 1.5y = 16$
$3x - 4y = 28$

64. $0.15x + 0.25y = 26$
$0.03x + 0.06y = 6$

65. Draw the graph of a linear equation that, along with the given line, forms an inconsistent system.

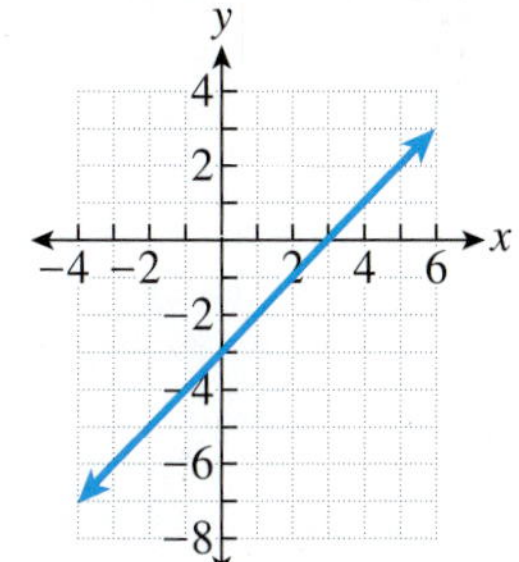

66. Draw the graph of a linear equation that, along with the given line, forms a dependent system.

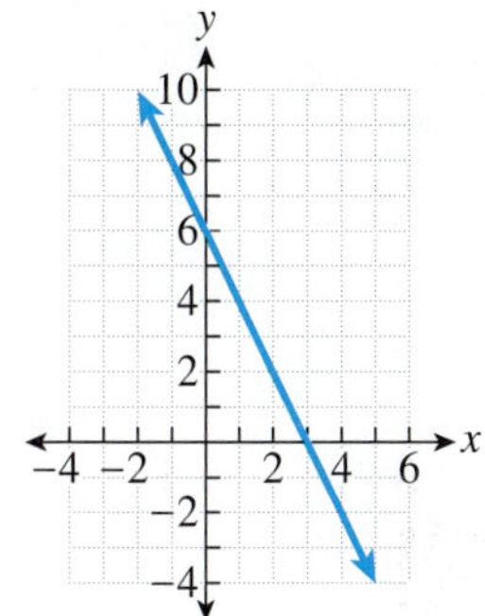

67. Draw the graph of a linear equation that, along with the given line, forms a system of equations with the ordered pair $(4, -3)$ as the only solution.

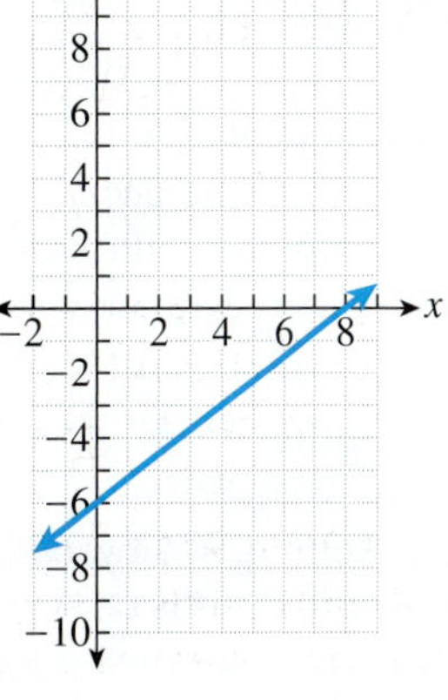

68. Draw the graphs of two linear equations that form a system of equations with the ordered pair $(-5, 2)$ as the only solution.

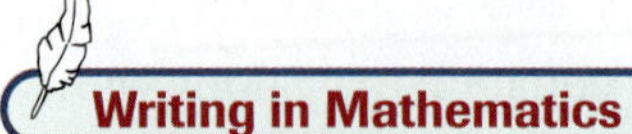

Answer in complete sentences.

69. Explain why the solution to an independent system of equations is the ordered pair that is the point of intersection for the two lines.

70. When we are solving a system of equations by graphing, we may not be able to accurately determine the coordinates of the solution to an independent system. Explain why.

71. When you use the substitution method, how can you tell that a system of equations is inconsistent and has no solutions? How can you tell that a system of equations is dependent and has infinitely many solutions?

72. Explain the differences between the addition method and the substitution method, and how to determine which method to use for a particular system of equations.

73. ***Solutions Manual**** Write a solutions manual page for the following problem:

Solve the following system of equations by any method.

$$4x + y = 23$$
$$3x - 4y = 3$$

74. ***Newsletter**** Write a newsletter explaining how to solve a system of two linear equations by the substitution method.

75. ***Newsletter**** Write a newsletter explaining how to solve a system of two linear equations by the addition method.

***See Appendix B for details and sample answers.**

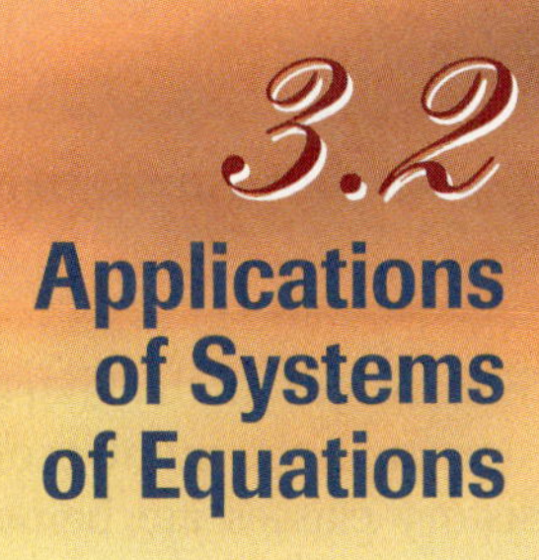

3.2 Applications of Systems of Equations

Objectives

1. **Solve applied problems involving systems of equations using the substitution method or the addition method.**
2. **Solve geometry problems using a system of equations.**
3. **Solve interest problems using a system of equations.**
4. **Solve mixture problems using a system of equations.**
5. **Solve motion problems using a system of equations.**

Objective 1 Solve applied problems involving systems of equations using the substitution method or the addition method. This section focuses on applications involving systems of linear equations with two variables. We will use both the substitution and addition methods to solve these systems. When you are trying to solve a system, use the method you feel will be more efficient.

We begin to solve an applied problem by identifying the two unknown quantities and choosing a variable to represent each quantity. We then need to find two equations that relate these quantities.

EXAMPLE 1 The sum of two numbers is 90, and their difference is 18. What are the two numbers?

Solution

The two unknown quantities in this problem are the two numbers. We begin with a table of the unknown quantities.

Unknowns:
First Number: x
Second Number: y

Since the sum of the two numbers is 90, we know that one of our equations is $x + y = 90$. We need a second equation, and this can be found from the fact that the difference of the two numbers is 18. This translates to the equation $x - y = 18$. Here is the system of equations that we need to solve.

$$\begin{aligned} x + y &= 90 \\ x - y &= 18 \end{aligned}$$

We will solve this system of equations using the addition method, since the terms containing y are already opposites.

$$\begin{array}{r} x + y = 90 \\ \underline{x - y = 18} \\ 2x = 108 \end{array} \quad \text{Add to eliminate } y.$$

$$x = 54 \quad \text{Divide both sides by 2.}$$

To find the other number, we need to substitute 54 for x in one of the original equations and solve for y. We will use the first equation.

Quick Check 1
The sum of two numbers is 124, and their difference is 20. What are the two numbers?

$$\begin{aligned} x + y &= 90 \\ 54 + y &= 90 && \text{Substitute 54 for } x. \\ y &= 36 && \text{Subtract 54.} \end{aligned}$$

The second number is 36, so the two numbers are 54 and 36. A quick check shows that the two numbers have a sum of 90 and a difference of 18.

EXAMPLE 2 A community college is offering a total of 112 history and math classes this semester. The number of math classes being offered is 48 more than the number of history classes being offered. How many math classes and history classes are being offered this semester?

Solution

The two unknown quantities in this problem are the number of history classes and the number of math classes being offered this semester. We begin with a table of the unknown quantities.

Unknowns:
Number of history classes: h
Number of math classes: m

We choose the variables h and m because it will remind us that they stand for the number of ***h***istory classes (h) and ***m***ath classes (m) being offered this semester. The problem tells us the total number of history and math classes being offered is 112. This translates to the equation $h + m = 112$. We need a second equation, and this can be found from the sentence that tells us that the number of math classes being offered is 48 more than the number of history classes being offered. This translates to the equation $m = h + 48$. Here is the system of equations that we need to solve.

$$\begin{aligned} h + m &= 112 \\ m &= h + 48 \end{aligned}$$

We will solve this system of equations using the substitution method, since the second equation has already been solved for m. We will substitute the expression $h + 48$ for the variable m in the second equation.

$$\begin{aligned} h + m &= 112 \\ h + (h + 48) &= 112 && \text{Substitute } h + 48 \text{ for } m. \\ 2h + 48 &= 112 && \text{Combine like terms.} \\ 2h &= 64 && \text{Subtract 48.} \\ h &= 32 && \text{Divide both sides by 2.} \end{aligned}$$

Quick Check 2
Jessica and Vanessa own a total of 160 books. Jessica owns 30 more books than Vanessa does. How many books does each woman own?

There are 32 history classes being offered this semester. To find out how many math classes are being offered this semester, we can substitute 32 for h in the equation $m = h + 48$.

$$\begin{aligned} m &= 32 + 48 && \text{Substitute 32 for } h. \\ m &= 80 && \text{Add.} \end{aligned}$$

There are 80 math classes being offered this semester, and 32 history classes. A quick check shows that the total number of these two types of classes is 112 and that the number of math classes is 48 more than the number of history classes.

EXAMPLE 3 Tina's purse contains \$5 bills and \$10 bills. If the purse contains 39 bills worth a total of \$355, how many \$5 bills are there?

Solution

The two unknown quantities in this problem are the number of \$5 bills and the number of \$10 bills. We start with a table of unknowns.

Unknowns:

Number of \$5 bills: f

Number of \$10 bills: t

The first equation comes from the fact that there are 39 bills in the purse. Since all of these bills are either \$5 bills or \$10 bills, the number of \$5 bills, f, plus the number of \$10 bills, t, must equal 39. The first equation is $f + t = 39$.

The second equation comes from the amount of money in the purse. The amount of money from \$5 bills, $\$5f$, plus the amount of money from \$10 bills, $\$10t$, must equal the amount of money in the purse. The second equation is $5f + 10t = 355$. Here is the system of equations that we need to solve.

$$\begin{aligned} f + t &= 39 \\ 5f + 10t &= 355 \end{aligned}$$

Before we solve this system of equations, let's look at a table that contains all the pertinent information for this problem.

	Number of Bills	Value/Bill (\$)	Money in Purse (\$)
\$5 Bills	f	5	$5f$
\$10 Bills	t	10	$10t$
Total	39		355

Notice that the first equation in the system, $f + t = 39$, can be found in the first column of the table (labeled "Number of Bills"). The second equation in the system, $5f + 10t = 355$, can be found in the last column in the table (labeled "Money in Purse (\$)").

Since neither equation is already solved for f or t, we will use the addition method to solve for t. (Note that you could solve this system by the substitution method if you prefer, as the first equation can easily be solved for f or t.)

$$\begin{aligned} f + t &= 39 \\ 5f + 10t &= 355 \end{aligned} \xrightarrow{\text{Multiply by } -5.} \begin{aligned} -5f - 5t &= -195 \\ 5f + 10t &= 355 \end{aligned}$$

$$\begin{aligned} -5f - 5t &= -195 \\ 5f + 10t &= 355 \\ \hline 5t &= 160 \end{aligned} \quad \text{Add to eliminate } f.$$

$$t = 32 \quad \text{Divide by 5 to solve for } t.$$

There were 32 \$10 bills. We now substitute this value for t in the equation $f + t = 39$.

$$\begin{aligned} f + 32 &= 39 && \text{Substitute 32 for } t. \\ f &= 7 && \text{Subtract 32.} \end{aligned}$$

Quick Check 3
A movie was attended by 84 people. If admission was \$7 for adults and \$4 for children, and the total box office receipts were \$417, how many adults and how many children attended the movie?

There were 7 \$5 bills and 32 \$10 bills in the purse.

We should check our solution for accuracy. If there were 7 \$5 bills and 32 \$10 bills in the purse, then that is a total of 39 bills. Also, the 7 \$5 bills are worth \$35, while the 32 \$10 bills are worth \$320. The total amount of money is \$355, so our solution checks.

Geometry Problems

Objective 2 **Solve geometry problems using a system of equations.** In Section 1.4 we solved problems involving the perimeter of a rectangle. We will now use a system of two equations in two variables to solve the same types of problems.

Example 4 The length of a rectangle is 19 feet more than its width. If the perimeter of the rectangle is 102 feet, find the length and the width of the rectangle.

Solution

The two unknowns are the length and the width of the rectangle. We will use the variables l and w for the length and width, respectively.

Unknowns:
Length: l
Width: w

We are told that the length is 19 feet more than the width of the rectangle. We express this relationship in an equation as $l = w + 19$. The second equation in the system comes from the fact that the perimeter is 102 feet. Since the perimeter of a rectangle is equal to twice the length plus twice the width, our second equation can be written as $2l + 2w = 102$. Here is the system we are solving:

$$\begin{aligned} l &= w + 19 \\ 2l + 2w &= 102 \end{aligned}$$

We will use the substitution method to solve this system, as the first equation is already solved for l. We begin by substituting the expression $w + 19$ for l in the equation $2l + 2w = 102$.

$$\begin{aligned} 2(w + 19) + 2w &= 102 && \text{Substitute } w + 19 \text{ for } l. \\ 2w + 38 + 2w &= 102 && \text{Distribute.} \\ 4w + 38 &= 102 && \text{Combine like terms.} \\ 4w &= 64 && \text{Subtract 38.} \\ w &= 16 && \text{Divide both sides by 4.} \end{aligned}$$

The width of the rectangle is 16 feet. To solve for the length, we substitute 16 for w in the equation $l = w + 19$.

$$\begin{aligned} l &= 16 + 19 && \text{Substitute 16 for } w. \\ l &= 35 && \text{Add.} \end{aligned}$$

The length of the rectangle is 35 feet, and the width is 16 feet. This length is 19 feet more than this width, and the perimeter of this rectangle $(2 \cdot 35 + 2 \cdot 16)$ is 102 feet, so our solution checks.

Quick Check 4
A rectangular concrete slab is poured for a new patio. The perimeter of the slab is 70 feet and the length of the slab is 15 feet longer than the width. Find the length and width of the new patio.

EXAMPLE 5 A fruit packing house is in the shape of a rectangle. The length of the packing house is twice its width. If the perimeter of the packing house is 1050 feet, find the length and the width of the packing house.

Solution

The two unknowns are the length and the width of the rectangle. Again, we will use the variables l and w for the length and width, respectively.

Unknowns:

Length: l

Width: w

We are told that the length is twice the width of the rectangle. We express this relationship in an equation as $l = 2w$. The second equation in the system comes from the fact that the perimeter is 1050 feet. Since the perimeter of a rectangle is equal to twice the length plus twice the width, our second equation can be written as $2l + 2w = 1050$. Here is the system we are solving.

$$l = 2w$$
$$2l + 2w = 1050$$

We will use the substitution method to solve this system, as the first equation is already solved for l. We begin by substituting the expression $2w$ for l in the equation $2l + 2w = 1050$.

$$2(2w) + 2w = 1050 \quad \text{Substitute } 2w \text{ for } l.$$
$$4w + 2w = 1050 \quad \text{Multiply.}$$
$$6w = 1050 \quad \text{Combine like terms.}$$
$$w = 175 \quad \text{Divide both sides by 6.}$$

The width of the packing house is 175 feet. To solve for the length, we substitute 175 for w in the equation $l = 2w$.

$$l = 2(175) \quad \text{Substitute 175 for } w.$$
$$l = 350 \quad \text{Multiply.}$$

The width of the packing house is 175 feet, and the length is 350 feet. The check is left to the reader.

***Quick Check* 5**
The perimeter of a rectangle is 80 inches and the length of the rectangle is three times the width. Find the length and width of the rectangle.

Interest Problems

Objective 3 **Solve interest problems using a system of equations.** For certain problems, it is easier to set up a system of equations in two variables than to express both unknown quantities in terms of the same variable. For example, the mixture and interest problems are often easier to solve when working with a system of equations in two variables. We will begin with a problem involving interest.

EXAMPLE 6 Omar invested \$4000 in two certificates of deposit (CDs). He deposited some of the money in a CD that paid 4% annual interest, and the rest in a CD that paid 3% annual interest. If Omar earned \$145 in interest in the first year, how much did he invest in each CD?

Solution

The two unknowns are the amount invested at 4% and the amount invested at 3%. We will let x represent the amount invested at 4% interest, and y represent the amount invested at 3% interest. To determine the interest earned, we multiply the principal by the interest rate. If the first account earns 4% interest, then we can represent the interest earned by $0.04x$. In a similar fashion, we can represent the interest earned from the investment at 3% interest as $0.03y$. The following table summarizes this information.

Account	Principal	Interest Rate	Interest Earned
CD 1	x	0.04	$0.04x$
CD 2	y	0.03	$0.03y$
Total	4000		145

The first equation in our system comes from the fact that the amount invested at 4% interest plus the amount invested at 3% is equal to the total amount invested. As an equation, this can be represented as $x + y = 4000$. This equation can be found in our table in the column labeled "Principal." The second equation comes from the amount of interest earned, and can be found in the column labeled "Interest Earned." We know that Omar earned $0.04x$ in interest from the first CD and he earned $0.03y$ in interest from the second CD. Since the total amount of interest earned comes from these two CDs, the second equation in our system is $0.04x + 0.03y = 145$. Here is the system we must solve.

$$\begin{aligned} x + y &= 4000 \\ 0.04x + 0.03y &= 145 \end{aligned}$$

We will clear the second equation of decimals by multiplying both sides of the equation by 100.

$$\begin{aligned} x + y &= 4000 \\ 0.04x + 0.03y &= 145 \end{aligned} \xrightarrow{\text{Multiply by 100.}} \begin{aligned} x + y &= 4000 \\ 4x + 3y &= 14{,}500 \end{aligned}$$

Quick Check 6 Carolyn invested \$2800 in two certificates of deposit (CDs). She deposited some of the money in a CD that paid 5% annual interest, and the rest in a CD that paid 3% annual interest. If Carolyn earned \$119 in interest in the first year, how much did she invest in each CD?

To solve this system of equations, we can use the addition method. (The substitution method would work just as well.) If we multiply the first equation by -3, then the variable terms containing y will be opposites.

$$\begin{aligned} x + y &= 4000 \\ 4x + 3y &= 14{,}500 \end{aligned} \xrightarrow{\text{Multiply by } -3.} \begin{aligned} -3x - 3y &= -12{,}000 \\ 4x + 3y &= 14{,}500 \end{aligned}$$

$$\begin{aligned} -3x - 3y &= -12{,}000 \\ 4x + 3y &= 14{,}500 \\ \hline x \quad\quad &= 2500 \end{aligned} \qquad \text{Add to eliminate } y.$$

We know that Omar invested \$2500 at 4% interest. To find the amount invested at 3%, we substitute 2500 for x in the original equation $x + y = 4000$.

$$\begin{aligned} 2500 + y &= 4000 && \text{Substitute 2500 for } x. \\ y &= 1500 && \text{Subtract 2500.} \end{aligned}$$

Omar invested \$2500 at 4% interest and \$1500 at 3% interest.

Mixture Problems

Objective 4 **Solve mixture problems using a system of equations.**

EXAMPLE 7 Leah works at a coffee shop that sells Kona coffee beans for \$40 per pound and Jamaican Blue Mountain coffee beans for \$50 per pound. The coffee shop also sells a blend made up of Kona coffee and Jamaican Blue Mountain coffee, for \$43.50 per pound. If Leah's boss asks her to make 20 pounds of this blend, how much Kona coffee and Jamaican Blue Mountain coffee should she mix together?

Solution

The two unknowns are the number of pounds of Kona coffee (k) and the number of pounds of Jamaican coffee (j) that need to be mixed together to form the blend. Since Leah needs to make 20 pounds of coffee with a value of \$43.50 per pound, the total value of this mixture can be found by multiplying 20 by 43.50. The total cost of this blend is \$870. The following table summarizes this information.

Coffee	Pounds	Cost per Pound	Total Cost
Kona	k	40	$40k$
Jamaican	j	50	$50j$
Total	20	43.50	870

The first equation in our system comes from the fact that the weight of the Kona coffee plus the weight of the Jamaican coffee must equal 20 pounds. As an equation, this can be represented by $k + j = 20$. This equation can be found in our table in the column labeled "Pounds." The second equation comes from the total cost of the coffee. We know that the total cost is \$870 (20 pounds multiplied by \$43.50/pound), so the second equation in our system is $40k + 50j = 870$. This equation can be found in the column labeled "Total Cost." Here is the system we must solve.

$$\begin{aligned} k + j &= 20 \\ 40k + 50j &= 870 \end{aligned}$$

To solve this system of equations, we can use the addition method. If we multiply the first equation by -40, then the variable terms containing k will be opposites.

$$\begin{aligned} k + \quad j &= 20 \\ 40k + 50j &= 870 \end{aligned} \xrightarrow{\text{Multiply by } -40.} \begin{aligned} -40k - 40j &= -800 \\ 40k + 50j &= 870 \end{aligned}$$

$$\begin{aligned} -40k - 40j &= -800 \\ 40k + 50j &= 870 \\ \hline 10j &= 70 \end{aligned} \qquad \text{Add to eliminate } k.$$

$$j = 7 \qquad \text{Divide both sides by 10.}$$

Quick Check 7
Scott is making 36 pounds of a mixture of pecans and cashews that his shop will sell for \$4.50 per pound. If the shop sells pecans for \$5.50 per pound and cashews for \$4 per pound, how many pounds of pecans and cashews should Scott mix together?

We know that Leah must use 7 pounds of Jamaican coffee. To determine how much Kona coffee Leah must use, we substitute 7 for j in the equation $k + j = 20$.

$$k + 7 = 20 \quad \text{Substitute 7 for } j.$$
$$k = 13 \quad \text{Subtract 7.}$$

Leah must use 13 pounds of Kona coffee and 7 pounds of Jamaican coffee.

The previous example is a mixture problem. In this mixture problem, two items of different values or costs are combined into a mixture with a specific cost. In other mixture problems, two different solutions of different concentration levels are mixed together to form a single solution with a given strength, as in the next example.

EXAMPLE 8 Julio has one solution that is 30% alcohol and a second solution that is 45% alcohol. He wants to combine these solutions to make 66 liters of a solution that is 35% alcohol. How many liters of each original solution need to be used?

Solution

The two unknowns are the volume of the 30% alcohol solution and the volume of the 45% alcohol solution that need to be mixed together in order to make 66 liters of a solution that is 35% alcohol.

Solution	Volume of Solution (liters)	% Concentration of Alcohol	Volume of Alcohol (liters)
30%	x	0.3	$0.3x$
45%	y	0.45	$0.45y$
Mixture (35%)	66	0.35	23.1

The first equation in our system comes from the fact that the volume of the two solutions must equal 66 liters. As an equation, this can be represented by $x + y = 66$. This equation can be found in our table in the column labeled "Volume of Solution." The second equation comes from the total volume of alcohol, under the column labeled "Volume of Alcohol." We know that the total volume of alcohol is 23.1 liters (35% of the 66 liters in the mixture is alcohol), so the second equation in our system is $0.3x + 0.45y = 23.1$. Here is the system we must solve.

$$x + y = 66$$
$$0.3x + 0.45y = 23.1$$

We begin by clearing the second equation of decimals. This can be done by multiplying both sides of the second equation by 100.

$$\begin{aligned} x + y &= 66 \\ 0.3x + 0.45y &= 23.1 \end{aligned} \quad \xrightarrow{\text{Multiply by 100.}} \quad \begin{aligned} x + y &= 66 \\ 30x + 45y &= 2310 \end{aligned}$$

To solve this system of equations, we can use the addition method. If we multiply the first equation by -30, then the variable terms containing x will be opposites.

$$\begin{aligned} x + y &= 66 \\ 30x + 45y &= 2310 \end{aligned} \quad \xrightarrow{\text{Multiply by } -30.} \quad \begin{aligned} -30x - 30y &= -1980 \\ 30x + 45y &= 2310 \end{aligned}$$

$$\begin{aligned} -30x - 30y &= -1980 \\ 30x + 45y &= 2310 \\ \hline 15y &= 330 \end{aligned} \quad \text{Add to eliminate } x.$$

$$y = 22 \quad \text{Divide both sides by 15.}$$

Quick Check 8
A chemist has one solution that is 75% acid and a second solution that is 55% acid. She wants to combine the solutions to make a new solution that is 59% acid. If she needs to make 600 milliliters of this new solution, how many milliliters of the two solutions should she mix?

We know that Julio must use 22 liters of the solution that is 45% alcohol. To determine how much of the 30% alcohol solution Julio must use, we substitute 22 for y in the equation $x + y = 66$.

$$\begin{aligned} x + 22 &= 66 \quad \text{Substitute 22 for } y. \\ x &= 44 \quad \text{Subtract 22.} \end{aligned}$$

Julio must use 44 liters of the 30% alcohol solution and 22 liters of the 45% alcohol solution.

Motion Problems

Objective 5 Solve motion problems using a system of equations. We will now turn our attention to motion problems, which were introduced in Section 1.4. Recall that the equation for motion problems is $d = r \cdot t$, where r is the rate of speed, t is the time, and d is the distance traveled. For example, if a car is traveling at a speed of 50 miles per hour for 3 hours, then the distance traveled is $50 \cdot 3$ or 150 miles.

EXAMPLE 9 Joanna is training for a triathlon. Yesterday she focused on running and biking, spending a total of 6 hours on the two activities. Joanna runs at a speed of 8 miles per hour and rides her bike at a speed of 25 miles per hour. If Joanna covered a total of 116 miles yesterday, how much time did she spend running and how much time did she spend riding?

Solution

The two unknowns in this problem are the amount of time Joanna ran and the amount of time she rode her bike. We will let x represent the time that Joanna ran and y represent the amount of time that she rode her bike. All of our information is displayed in the following table.

	Rate	Time	Distance ($d = r \cdot t$)
Running	8	x	$8x$
Biking	25	y	$25y$
Total		6	116

We know that Joanna spent a total of 6 hours on these two activities, so $x + y = 6$. We also know that the total distance traveled was 116 miles. This leads to the equation $8x + 25y = 116$. Here is the system of equations we must solve.

$$\begin{aligned} x + y &= 6 \\ 8x + 25y &= 116 \end{aligned}$$

We can solve the first equation for x ($x = 6 - y$), so we will use the substitution method to solve this system of equations. (Note that we could have used the addition method instead.) We substitute the expression $6 - y$ for x in the equation $8x + 25y = 116$.

$$\begin{aligned} 8(6 - y) + 25y &= 116 && \text{Substitute } 6 - y \text{ for } x. \\ 48 - 8y + 25y &= 116 && \text{Distribute.} \\ 48 + 17y &= 116 && \text{Combine like terms.} \\ 17y &= 68 && \text{Subtract 48.} \\ y &= 4 && \text{Divide both sides by 17.} \end{aligned}$$

Joanna spent 4 hours riding her bike. To find out how long she was running, we substitute 4 for y in the equation $x = 6 - y$ and solve for x.

$$\begin{aligned} x &= 6 - 4 && \text{Substitute 4 for } y. \\ x &= 2 && \text{Subtract.} \end{aligned}$$

Joanna spent 2 hours running and 4 hours riding her bike.

Quick Check 9

While hauling a load, Marge drove at an average speed of 65 miles per hour, except for a period of time when she was driving through a construction zone. In the construction zone, Marge's average speed was 35 miles per hour. If it took Marge 8 hours to drive 475 miles, how long was she in the construction zone?

Suppose that a boat travels at a speed of 15 miles per hour in still water. The boat will travel faster than 15 miles per hour while heading downstream, because the speed of the current is pushing the boat forward. The boat will travel slower than 15 miles per hour while heading upstream, because the speed of the current is pushing the boat back. If the speed of the current is 3 miles per hour, this boat would travel 18 miles per hour downstream $(15 + 3)$ and 12 miles per hour upstream $(15 - 3)$. The speed of an airplane is affected in a similar fashion depending on whether it is flying with the wind or against it.

EXAMPLE 10 Eva can take her kayak from her campsite to the nearest store, which is 6 miles downstream, in 1 hour. Returning upstream from the store to the campsite takes 3 hours. How fast can Eva paddle her kayak in still water, and what is the speed of the current?

Solution

The two unknowns in this problem are the speed of the kayak in still water and the speed of the current. We will let k represent the speed of the kayak in still water, and we will let c represent the speed of the current. The information for this problem can be summarized in the following table.

	Rate	Time	Distance ($d = r \cdot t$)
Downstream	$k + c$	1	$k + c$
Upstream	$k - c$	3	$3(k - c)$

Since the distance in each direction is 6 miles, we know that both $k + c$ and $3(k - c)$ equal 6. Here is the system we need to solve.

$$\begin{aligned} k + c &= 6 \\ 3k - 3c &= 6 \end{aligned}$$

We will solve this system of equations by using the addition method. Multiplying both sides of the first equation by 3 will help us eliminate the variable c.

$$k + c = 6 \xrightarrow{\text{Multiply by 3.}} 3k + 3c = 18$$
$$3k - 3c = 6 \qquad\qquad 3k - 3c = 6$$

$$\begin{aligned} 3k + 3c &= 18 \\ 3k - 3c &= 6 \\ \hline 6k \quad &= 24 \end{aligned}$$ Add to eliminate c.

$k = 4$ Divide both sides by 6.

The speed of the kayak in still water is 4 miles per hour. To find the speed of the current we substitute 4 for k in the original equation $k + c = 6$.

$4 + c = 6$ Substitute 4 for k.
$c = 2$ Subtract 4.

The speed of the current is 2 miles per hour and the speed of the kayak is 4 miles per hour.

Quick Check 10
An airplane traveling with a tailwind can make a 1500-mile trip in 3 hours. However, traveling into the same wind, the plane would take $3\frac{3}{4}$ hours to fly 1500 miles. What is the speed of the plane in calm air, and what is the speed of the wind?

Building Your Study Strategy **Making the Best Use of Your Resources, 2**
Tutoring Most college campuses have a tutorial center providing free tutoring in math and other subjects. Some tutorial centers offer walk-in tutoring, others schedule appointments for one-on-one or group tutoring, and some combine these approaches.

Walk-in tutoring allows you to come in whenever you want to and ask questions as they arise. One-on-one tutoring is designed for students who will require more extensive assistance. If you do not completely understand a tutor's answer or directions, do not be afraid to ask for further explanation.

While it is a good idea to ask a tutor if you are doing a problem correctly or to give you an idea of a starting point for a problem, it is not a good idea to ask a tutor to do a problem for you. We learn by doing, not by watching. It may look easy while the tutor is working a problem, but this will not help you to do the next problem.

EXERCISES 3.2

FOR EXTRA HELP
MyMathLab
MathXL
Interactmath.com
MathXL Tutorials on CD
Video Lectures on CD
Addison-Wesley Math Tutor Center
Student's Solutions Manual

Vocabulary

1. State the formula for the perimeter of a rectangle.
2. State the formula for calculating simple interest.
3. State the formula for calculating distance.
4. A chemist is mixing a solution that is 35% alcohol with a second solution that is 55% alcohol to create 80 ml of a solution that is 40% alcohol. Which system of equations can we use to solve this problem?

   a) $x + y = 40$
   $0.35x + 0.55y = 80$

   b) $x + y = 80$
   $0.35x + 0.55y = 0.40$

   c) $x + y = 80$
   $0.35x + 0.55y = 0.40(80)$

5. A town has two youth baseball leagues, with a total of 1200 children participating. The northern league has 300 more children than the southern league. How many children are in each league?
6. The University of California, Los Angeles, (UCLA) has 7914 more students than the University of Southern California (USC). If the two schools have a total of 41,708 students, how many students attend each university?
7. Don's age is 5 more than 3 times Rose's age. If we added Don's age to Rose's age, the total would be 57. How old is Don? How old is Rose?

8. Mark was 32 when his son Tom was born. Today the total of their ages is 74. How old are Mark and Tom?

9. Two baseball players hit 95 home runs combined last season. The first player hit 16 fewer home runs than twice the number of home runs hit by the second player. How many home runs did each player hit?

10. A piece of wire 300 centimeters long was cut into two pieces. The longer of the two pieces is 40 centimeters longer than 3 times the length of the shorter piece. How long is each piece?

11. On a history exam, each multiple-choice question is worth 5 points, and each true/false question is worth 2 points. Deb answered 25 questions correctly on the exam and earned a score of 89 points. How many multiple-choice questions did she answer correctly, and how many true/false questions did she answer correctly?

12. On a sociology exam students earn 5 points for each correct response and lose 2 points for each incorrect response or unanswered question. The exam contained 20 questions, and Phil earned a score of 72 points. How many questions did Phil answer correctly?

13. Juliana has 28 coins in her purse. Some are dimes and the rest are quarters. The value of the coins is $5.95. How many dimes does she have?

14. Monique has a total of 37 $10 and $20 bills in her purse. The value of the bills is $610. How many $10 bills does she have?

15. The cost to attend a movie is $8.50 for adults, while children are admitted for $4.75. If a movie was attended by 158 people who paid a total of $1106.75, how many adults and how many children attended the movie?

16. The cost to attend a college play is $6 for the general public, while students are admitted for $2. If the play was attended by 389 people who paid a total of $1090, how many non-students and how many students attended the play?

17. Last week a softball team ordered 5 pizzas and 3 pitchers of soda after a game, and the bill was $88. This week they ordered 4 pizzas and 4 pitchers of soda and the bill was $80. What is the price of a single pizza and what is the price of a pitcher of soda?

18. A sandwich shop sells small sandwiches and large sandwiches. An order of 6 small sandwiches and 3 large sandwiches costs $38.25, while an order of 4 small sandwiches and 7 large sandwiches costs $54.25. What is the cost for a small sandwich? What is the cost for a large sandwich?

19. An office manager occasionally brings coffee for her workers. One day she brought 4 regular coffees and 8 lattes, and it cost her $37.20. Another day she brought 9 coffees and 3 lattes, and it cost her $28.20. How much does a coffee cost? How much does a latte cost?

20. A florist sold 6 carnations and 6 roses for $20.70 to one of his customers. Another customer purchased 30 carnations and 15 roses for $67.50. What is the cost for a single carnation and a single rose?

Recall from chapter 1 that two positive angles are said to be complementary if their measures add up to be 90°, and supplementary if their measures add up to be 180°.

21. Two angles are complementary. The measure of one angle is 16° less than the measure of the other angle. Find the measure of each angle.

22. Two angles are complementary. The measure of one angle is 4° more than 3 times the measure of the other angle. Find the measure of each angle.

23. Two angles are supplementary. The measure of one angle is 12° less than 5 times the measure of the other angle. Find the measure of each angle.

24. Two angles are supplementary. The measure of one angle is 15° more than twice the measure of the other angle. Find the measure of each angle.

25. The perimeter of a rectangle is 80 inches. The width of the rectangle is 4 inches less than the length of the rectangle. Find the dimensions of the rectangle.

26. A rectangular vegetable garden is surrounded by 148 feet of fence. The length of the garden is 2 feet more than twice the width of the garden. Find the dimensions of this garden.

27. A rectangular oil painting has a perimeter of 101.4 inches. The length of the painting is 1.6 times the width of the painting. Find the dimensions of the painting.

28. Tina needs 195 inches of fabric to use as binding around a rectangular quilt she is making. The width of the quilt is 7.5 inches less than the length of the quilt. Find the dimensions of the quilt.

29. The perimeter of a rectangular table is 200 inches. The length of the table is 14 inches less than twice the width of the table. Find the dimensions of the table.

30. A rectangular tennis court is surrounded by 380 feet of fencing. The length of the court is 5 feet less than twice the width of the court. Find the dimensions of the court.

31. A home is being built on a rectangular lot. The perimeter of the lot is 1160 feet, and the width of the lot is 20 feet less than half the length of the lot. Find the dimensions of this lot.

32. A rectangular door is surrounded by 224 inches of weather stripping, which runs around the entire door. The height of the door is 2.5 times the width of the door. Find the dimensions of the door.

33. A swimming pool is in the shape of a rectangle. The length of the pool is 15 feet more than the width. A concrete deck 6 feet wide is added around the pool, and the perimeter of the deck is 198 feet.

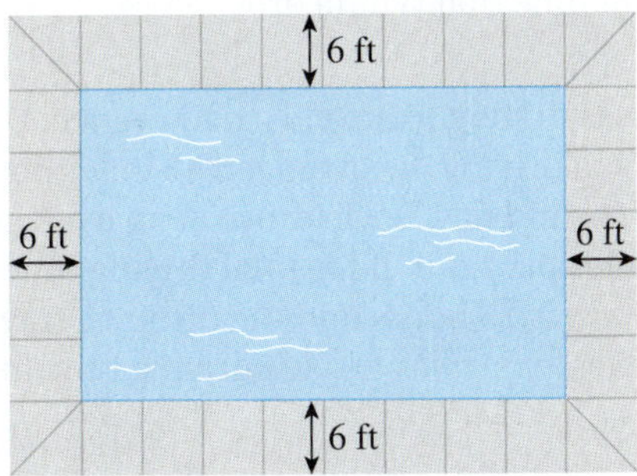

Find the dimensions of the swimming pool.

34. A farmer fences in a rectangular area as a pasture for her horses next to a building, using the existing building as the fourth side of the rectangle as shown.

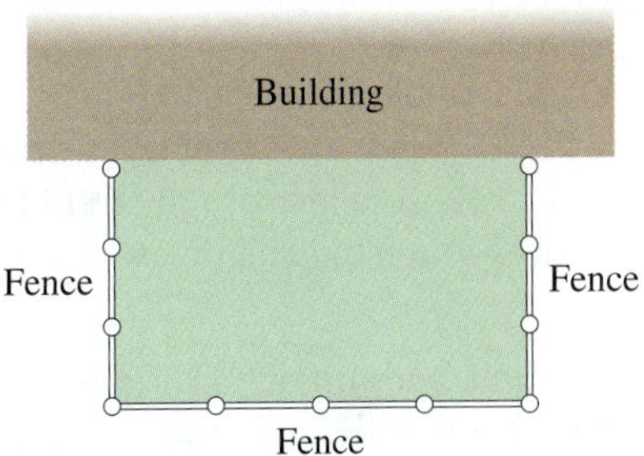

The side of the fence parallel to the building is 48 feet longer than either of the other two sides. If the farmer used 120 feet of fencing, find the dimensions of the pasture.

35. Randy invested a total of \$2000 in two mutual funds. One fund earned a 7% profit, while the other earned an 8% profit. If Randy's total profit was \$152, how much did he invest in each mutual fund?

36. Yolanda invested a total of \$3000 in two mutual funds. One fund earned a 10% profit, while the other earned a 12% profit. If Yolanda's total profit was \$320, how much did she invest in each mutual fund?

37. Lucy deposited a total of \$2200 in two different certificates of deposit (CDs). One CD paid 5% annual interest and the other paid 3% annual interest. At the end of one year she had earned \$96 in interest from the two CDs. How much did Lucy invest in each CD?

38. Angela deposited a total of \$5800 in two different certificates of deposit (CDs). One CD paid 3% annual interest, and the other paid 4% annual interest. At the end of one year she had earned \$213.50 in interest from the two CDs. How much did Angela invest in each CD?

39. Mark deposited a total of \$32,000 in two different CDs. One CD paid 4.5% annual interest, and the other paid 3.75% annual interest. At the end of one year he had earned \$1387.50 in interest from the two CDs. How much did Mark invest in each CD?

40. Tom invested $8650 in two mutual funds. One fund earned a 9.5% profit, while the other earned a 12% profit. Between the two accounts Tom made a profit of $888. How much did he invest in each account?

41. Alma invested a total of $12,000 in two mutual funds. While one fund earned a 7% profit, the other fund had a loss of 20%. Alma lost a total of $1320 from the two funds. How much did she invest in each fund?

42. Dennis has had a bad run of luck in the stock market. He invested a total of $6500 in two mutual funds. The first fund had a loss of 9%, and the second fund had a loss of 42%. Between the two funds, Dennis lost $1410. How much did he invest in each fund?

43. A coffee shop sells Kona coffee beans for $40 per pound and South American coffee beans for $12 per pound. If the manager wants to make 20 pounds of a mixture of these two types of beans that can be sold for $19 per pound, how many pounds of each type of bean should she use?

44. Nancy runs a winery in Napa, California. Her vineyards produce a Merlot wine that is worth $40 per liter, as well as a Cabernet Sauvignon wine that is worth $65 per liter. She wants to blend these two wines to create 2000 liters of wine that is worth $60 per liter. How many liters of each type of wine does Nancy need to blend?

45. Jorge sells dried fruits and nuts at his store. He sells a mixture of dried fruit for $3.90 per pound, and he sells mixed nuts for $6.40 per pound. He wants to mix the dried fruit and nuts to create 50 pounds of a mixture that he can sell for $4.72 per pound. How many pounds of dried fruit and how many pounds of mixed nuts should Jorge mix?

46. A farmer sells almonds for $6.99 per pound and walnuts for $5.79 per pound. If she wants to create 30 pounds of a mixture of these two types of nuts that she can sell for $6.24 per pound, how many of each type should she mix?

47. A chemist has one solution that is 60% acid and a second solution that is 36% acid. She wants to combine the solutions to make a new solution that is 42% acid. If she needs to make 440 milliliters of this new solution, how many milliliters of the two solutions should she mix?

48. A metallurgist has one brass alloy that is 69% copper and a second brass alloy that is 42% copper. He needs to melt and combine these alloys in such a way to make 75 grams of an alloy that is 60% copper. How much of each alloy should he use?

49. A bartender has two liquors that are 20% and 60% alcohol respectively. She needs to mix these liquors together to create 10 ounces of a mixture that is 30% alcohol. How much of each type of liquor does she need to mix?

50. A farmer has two types of feed that are 10% protein and 18% protein respectively. How many pounds of each need to be combined to make a mixture of 40 pounds of feed that is 14.8% protein?

51. A bartender mixes a batch of rum and cola, which is made of rum (40% alcohol) and cola (no alcohol). How much of each needs to be mixed to make 2 liters of rum and cola that is 25% alcohol?

52. An auto mechanic has an antifreeze solution that is 10% antifreeze. He needs to mix this solution with pure antifreeze to produce 9 liters of a solution that is 25% antifreeze. How much of each does he need to mix?

53. Fred and Wilma were driving to the coast for a vacation. During the first part of the trip, Fred drove at a speed of 70 miles per hour. They switched drivers and Wilma drove the rest of the way at a speed of 80 miles per hour. It took them 8 hours to reach the coast, which was 585 miles away from their home. Find the length of time that Fred drove, as well as the length of time that Wilma drove.

54. Doyle was driving to attend a conference. During the first day of driving, he averaged 66 miles per hour. On the second day Doyle drove at an average speed of 81 miles per hour. If his total driving time for the two days was 15 hours and the distance traveled was 1080 miles, what was Doyle's driving time for each day?

55. A racecar driver averaged 150 miles per hour over the first part of a 500-mile race, but engine troubles dropped her average speed to 125 miles per hour for the rest of the race. If the race took the driver

3.5 hours to complete, after what period of time did her engine trouble start?

56. A truck driver started for his destination at an average speed of 68 miles per hour. Once snow began to fall, he had to drop his speed to 48 miles per hour. It took the truck driver 5 hours to reach his destination, which was 285 miles away. How many hours after the driver started did the snow begin to fall?

57. A kayak can travel 15 miles downstream in 3 hours, while it would take 5 hours to make the same trip upstream. Find the speed of the kayak in still water, as well as the speed of the current.

58. A rowboat would take 2 hours to travel 8 miles upstream, while it would only take 1 hour to travel the same distance downstream. Find the speed of the rowboat in still water, as well as the speed of the current.

59. An airplane traveling with a tailwind can make a 2000-mile trip in 4 hours. However, traveling into the same wind, the plane would take 5 hours to fly 2000 miles. What is the speed of the plane in calm air, and what is the speed of the wind?

60. A small plane can fly 300 miles with a tailwind in 3 hours. Flying the same distance into the same wind would take 5 hours. Find the speed of the plane in calm air and the speed of the wind.

Writing in Mathematics

Answer in complete sentences.

61. Write a word problem for which the solution is "The length of the rectangle is 15 m and the width is 8 m."

62. Write a word problem for which the solution is "$5000 at 6% and $3000 at 5%."

63. Write a word problem for which the solution is "80 ml of 40% acid solution and 10 ml of 58% acid solution."

64. Write a word problem for which the solution is "Bill drove for 3 hours and Margie drove for 4 hours."

65. ***Solutions Manual**** Write a solutions manual page for the following problem:

A play was attended by 300 people. Admission was $8 for adults and $3 for children. If the total receipts for the play were $2150, how many children attended the play?

66. ***Newsletter**** Write a newsletter explaining how to solve interest problems using a system of two equations.

*See Appendix B for details and sample answers.

Quick Review Exercises

Section 3.2

Graph. Label any x- and y-intercepts.

1. $x - 4y = 6$
2. $y = 2x - 5$
3. $y = -\frac{1}{4}x - 2$
4. $y = 3$

3.3 Systems of Linear Inequalities

Objectives

1. **Solve a system of linear inequalities by graphing.**
2. **Solve a system of linear inequalities with more than two inequalities.**
3. **Graph systems of linear inequalities associated with applied problems.**

Just as there are systems of linear equations, there are also **systems of linear inequalities**. A system of linear inequalities is made up of two or more linear inequalities. As with systems of equations, an ordered pair is a solution to a system of linear inequalities if it is a solution to each linear inequality in the system.

Solving a System of Linear Inequalities by Graphing

Objective 1 **Solve a system of linear inequalities by graphing.** To find the solution set to a system of linear inequalities, we begin by graphing each inequality separately. If the solutions to the individual inequalities intersect, the region of intersection is the solution set to the system of inequalities.

EXAMPLE 1 Graph the system of inequalities.

$$y \geq x - 4$$
$$y < -3x + 9$$

Solution

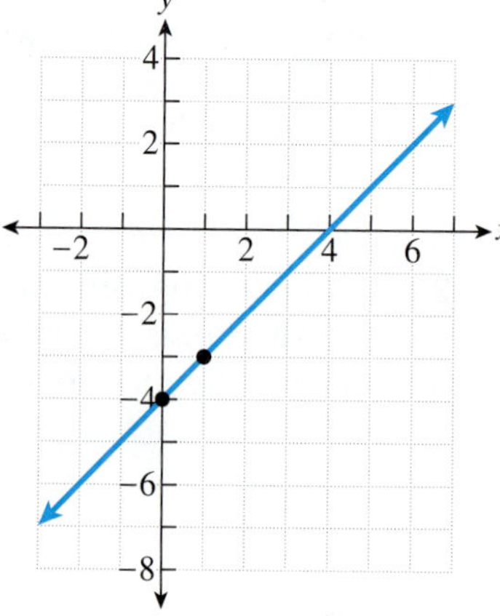

We begin by graphing the inequality $y \geq x - 4$. To do this we graph the line $y = x - 4$ as a solid line. Recall that we use a solid line when the points on the line are solutions to the inequality. Since this equation is in slope–intercept form, an efficient way to graph it is by plotting the y-intercept at $(0, -4)$ and then using the slope of 1 to find additional points on the line.

Since the origin is not on this line, we can use $(0, 0)$ as a test point. Substitute 0 for x and 0 for y in the inequality $y \geq x - 4$.

$$0 \geq 0 - 4$$
$$0 \geq -4$$

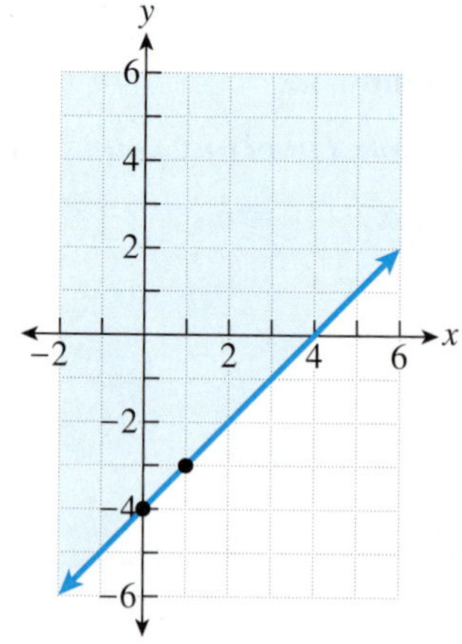

Substituting these coordinates into the inequality $y \geq x - 4$ produces a true statement, so we shade the half-plane containing $(0, 0)$.

We now turn our attention to the second inequality in the system.

To graph the inequality $y < -3x + 9$, we use a dashed line to graph the equation $y = -3x + 9$. Recall that the dashed line is used to signify that no point on the line is a solution to the inequality. Again, this equation is in slope–intercept form, so we graph the line by plotting the y-intercept at $(0, 9)$ and use the slope of -3 to find additional points on the graph.

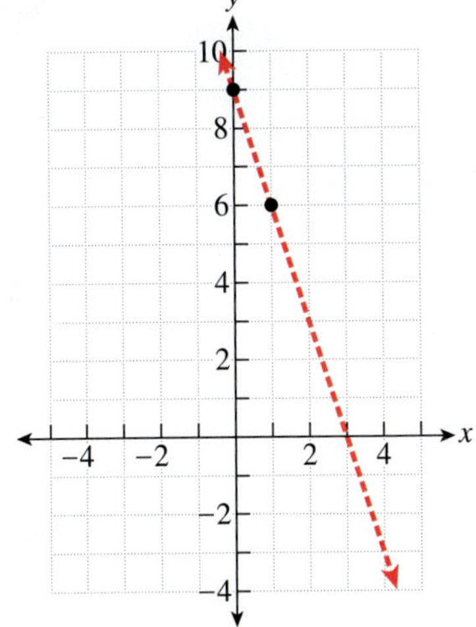

We can use the origin as a test point for this inequality as well. Substituting 0 for x and 0 for y produces a true statement.

$$0 < -3(0) + 9$$
$$0 < 0 + 9$$
$$0 < 9$$

We shade the half plane on the same side of the line as the origin.

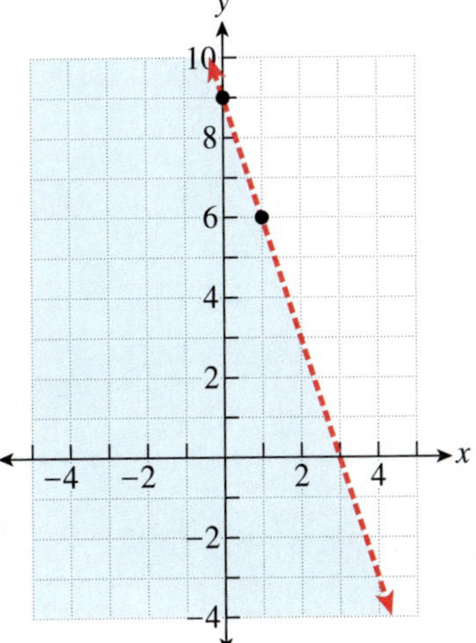

Quick Check 1

Graph the system of inequalities.

$y \geq x - 6$

$y \leq -\frac{2}{3}x + 4$

The solution to the system of linear inequalities is the region where the two solutions intersect.

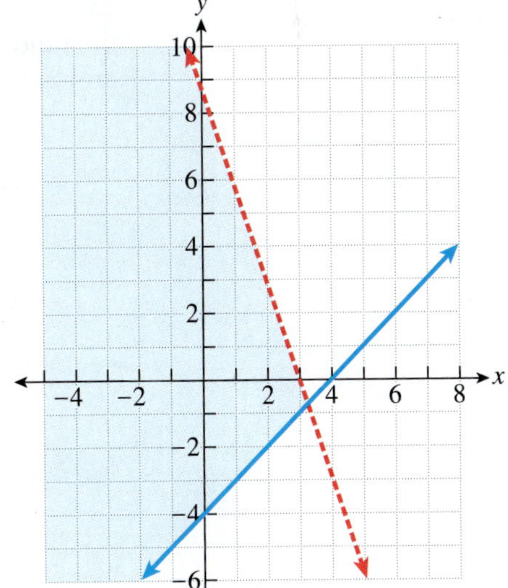

EXAMPLE 2 Graph the system of inequalities.

$$3x + 4y \leq 12$$
$$2x - y < -8$$

Solution

We begin by graphing the line $3x + 4y = 12$ using a solid line. This equation is in standard form, so we next find the x-intercept and the y-intercept.

x-intercept ($y = 0$)	y-intercept ($x = 0$)
$3x + 4(0) = 12$	$3(0) + 4y = 12$
$3x = 12$	$4y = 12$
$x = 4$	$y = 3$
$(4, 0)$	$(0, 3)$

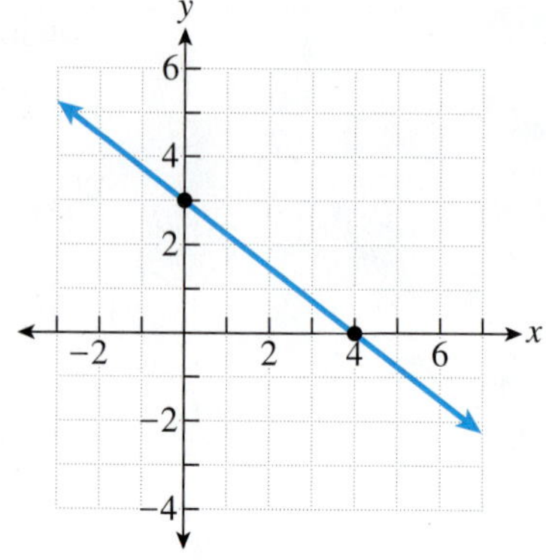

Now we determine which half-plane to shade. Since the line does not pass through the origin, we can choose $(0, 0)$ as a test point.

$$3x + 4y \leq 12$$
$$3(0) + 4(0) \leq 12$$
$$0 \leq 12$$

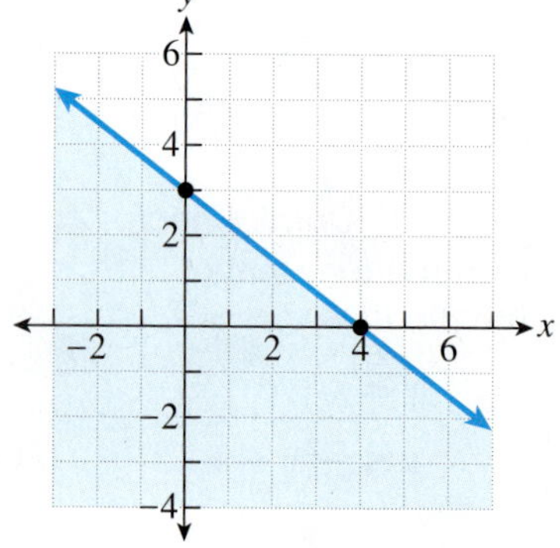

Substituting this ordered pair into the inequality produces a true statement. Thus, we shade the half-plane containing the origin.

Now we turn our attention to the inequality $2x - y < -8$. We begin by graphing the line $2x - y = -8$ using a dashed line. This equation is in standard form, so we next find the x-intercept and the y-intercept.

x-intercept ($y = 0$)	y-intercept ($x = 0$)
$2x - (0) = -8$	$2(0) - y = -8$
$2x = -8$	$-y = -8$
$x = -4$	$y = 8$
$(-4, 0)$	$(0, 8)$

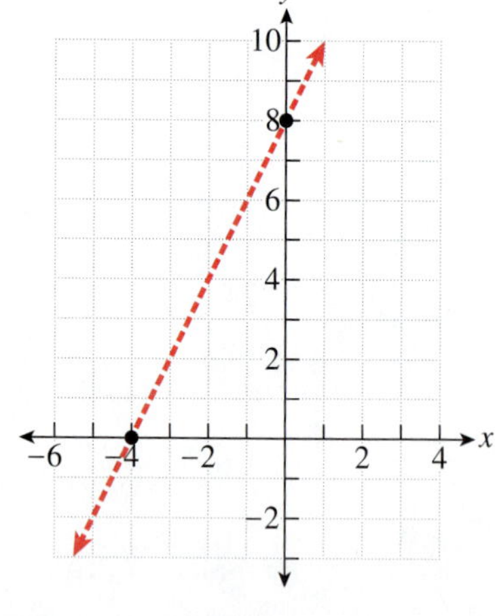

Now we determine which half-plane to shade. Since the line does not pass through the origin, we can choose $(0, 0)$ as a test point.

$$\begin{aligned} 2x - y &< -8 \\ 2(0) - (0) &< -8 \\ 0 &< -8 \end{aligned}$$

Substituting this ordered pair into the inequality produces a false statement. Thus, we shade the half-plane on the side of the line that does not contain the origin.

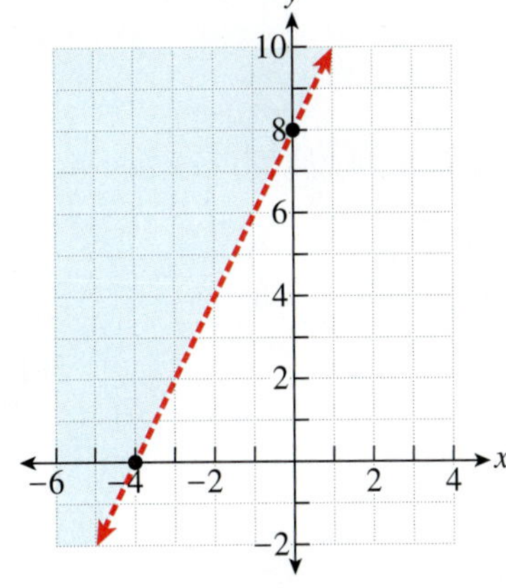

Quick Check 2

Graph the system of inequalities.

$$\begin{aligned} 2x + 5y &> 10 \\ x - 3y &> -9 \end{aligned}$$

The solution to the system of equations is the region where the two shaded areas intersect.

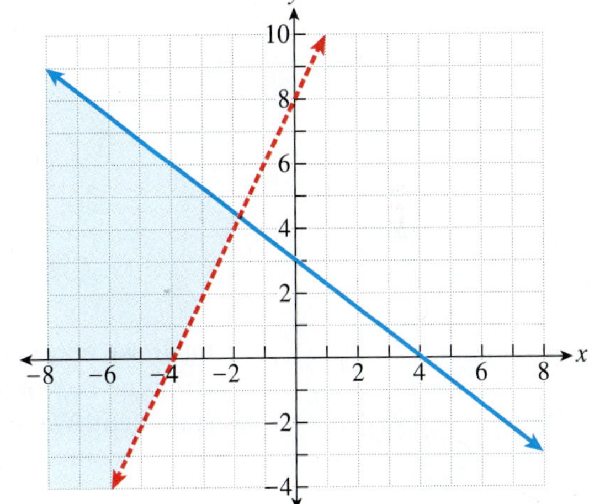

EXAMPLE 3 Graph the system of inequalities.

$$\begin{aligned} y &< -2x + 5 \\ y &\geq -3 \end{aligned}$$

Solution

We begin by graphing the line $y = -2x + 5$ using a dashed line. This equation is in slope–intercept form, so we plot the y-intercept $(0, 5)$ and then we use the slope of -2 to find additional points on the line.

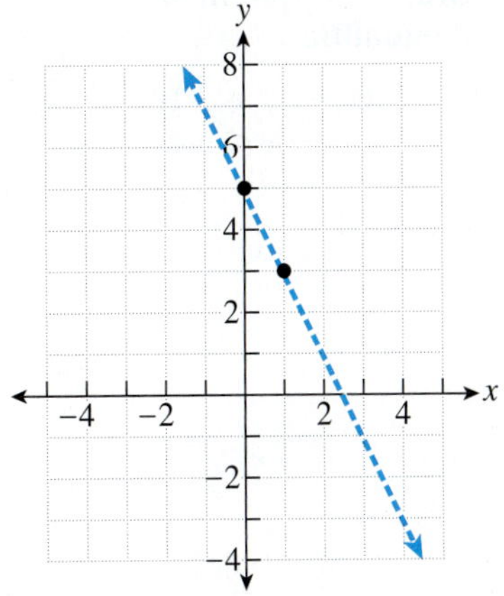

Now we determine which half-plane to shade. Since the line does not pass through the origin, we can choose $(0, 0)$ as a test point.

$$y < -2x + 5$$
$$0 < -2(0) + 5$$
$$0 < 5$$

Substituting this ordered pair into the inequality produces a true statement. Thus, we shade the half-plane containing the origin.

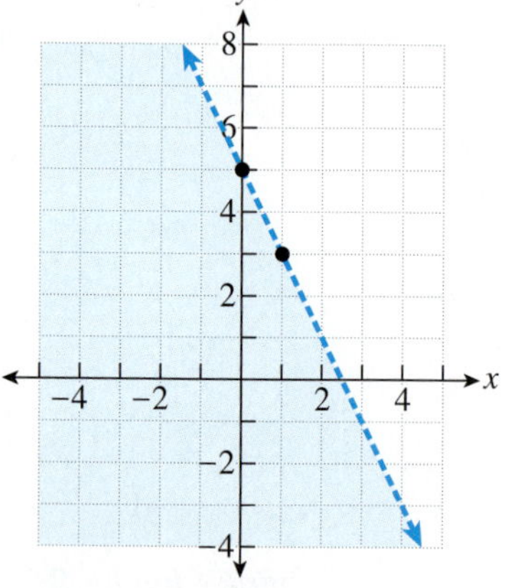

Now we turn our attention to the inequality $y \geq -3$. We begin by graphing the line $y = -3$ using a solid line. This line is a horizontal line with a y-intercept at $(0, -3)$.

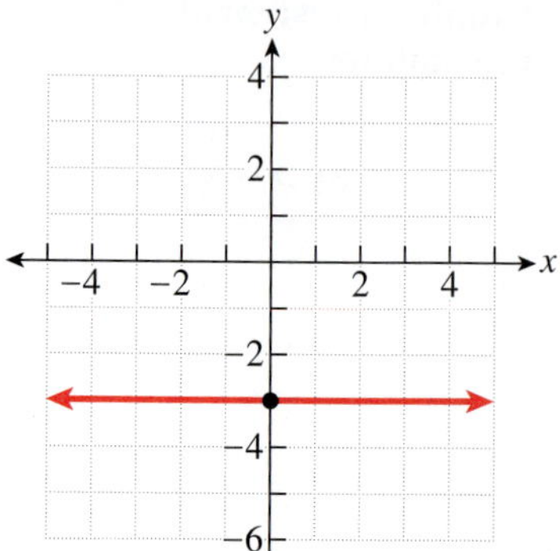

Again we may choose the origin as a test point. Substituting the ordered pair $(0, 0)$ produces a true statement. We shade the half-plane above the line $y = -3$ containing the origin.

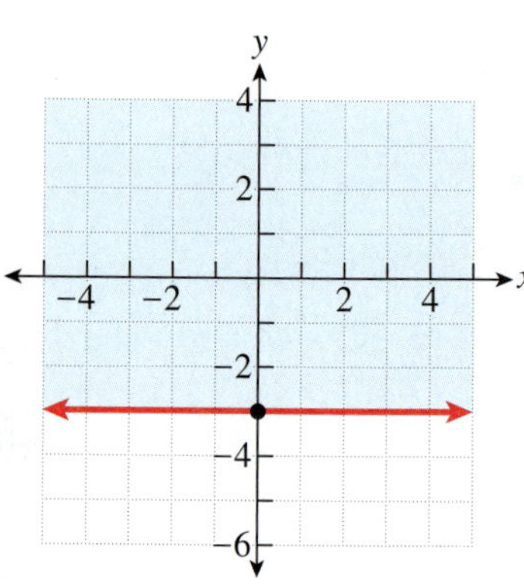

Quick Check 3

Graph the system of inequalities.

$$-3x + 2y \leq -18$$
$$y < -4$$

The solution to the system of equations is the region where the two shaded areas intersect.

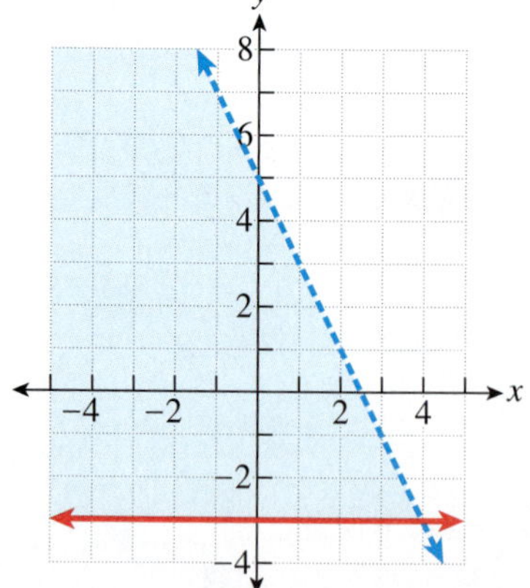

Solving a System of Linear Inequalities with More Than Two Inequalities

Objective 2 Solve a system of linear inequalities with more than two inequalities. The next example involves a system of three linear inequalities. Regardless of the number of inequalities in a system, we still find the solution in the same manner. Graph each inequality individually, and then find the region where the shaded regions all intersect.

EXAMPLE 4 Graph the system of inequalities.

$$\begin{aligned} 2x - 3y &\geq -6 \\ x &\geq 2 \\ y &\geq -6 \end{aligned}$$

Solution

We begin by graphing the equation $2x - 3y = -6$ with a solid line. Since the equation is in general form, we can graph it by finding its x- and y-intercepts.

$2x - 3y = -6$	
x-intercept $(y = 0)$	y-intercept $(x = 0)$
$2x - 3(0) = -6$	$2(0) - 3y = -6$
$2x = -6$	$-3y = -6$
$x = -3$	$y = 2$
$(-3, 0)$	$(0, 2)$

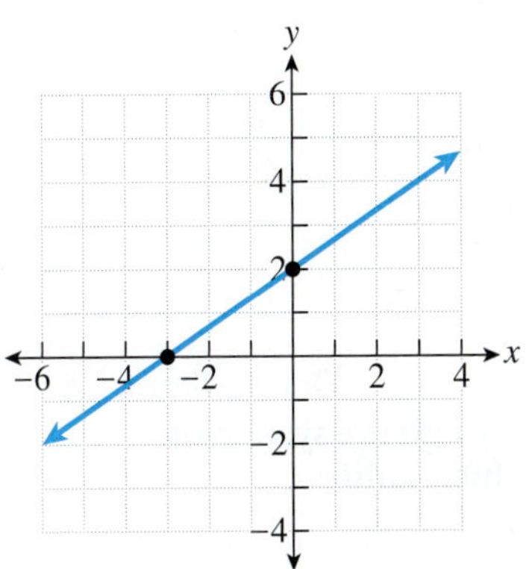

Since the line does not pass through the origin, we can use $(0, 0)$ as a test point. When we substitute 0 for x and 0 for y in the inequality $2x - 3y \geq -6$, we obtain a true statement.

$$\begin{aligned} 2(0) - 3(0) &\geq -6 \\ 0 &\geq -6 \end{aligned}$$

Since this statement is true, we shade the half-plane on the side of the line containing the point $(0, 0)$.

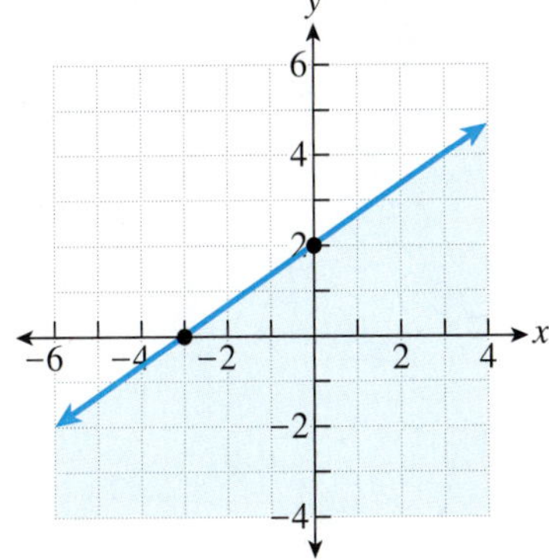

To graph the second inequality, $x \geq 2$, we begin with the graph of the vertical line $x = 2$. This line is also solid. We shade to the right of the vertical line because this is the region where we find values of x that are greater than 2.

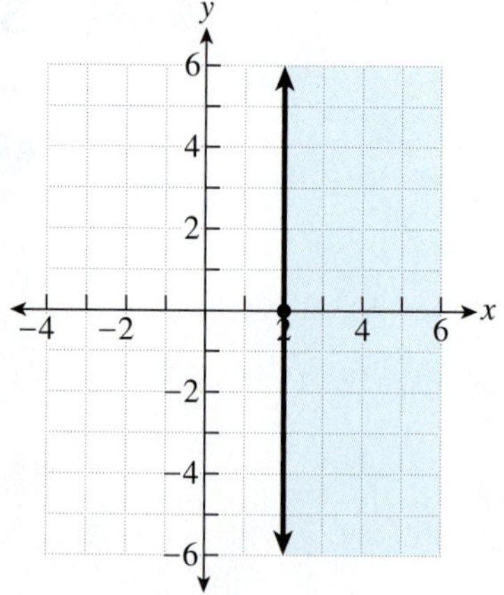

For the third inequality, $y \geq -6$, we graph the solid horizontal line $y = -6$. We shade above the horizontal line $y = -6$ because this is the region where we find values of y that are greater than -6.

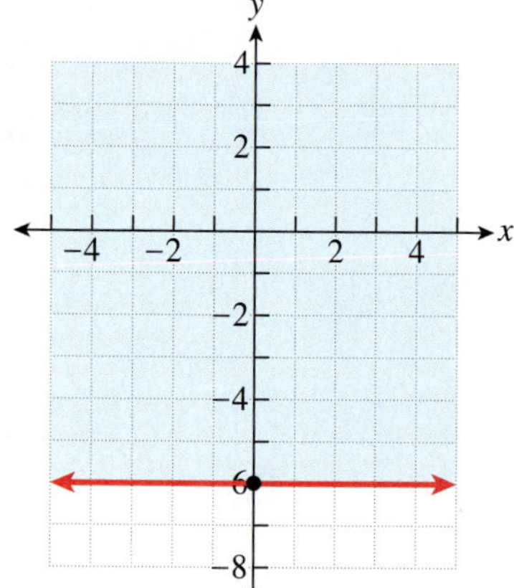

Quick Check 4

Graph the system of inequalities.

$$\begin{aligned} x + y &\leq 4 \\ x &\geq -3 \\ y &\geq 1 \end{aligned}$$

We now display the solutions to the system of inequalities by finding the region of intersection for these three inequalities.

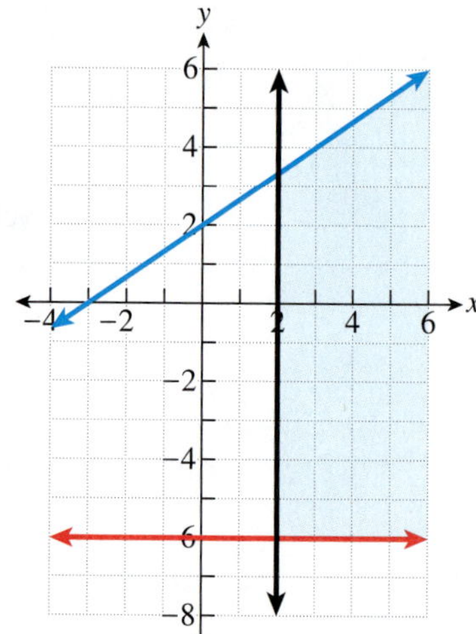

Graphing Systems of Linear Inequalities Associated with Applied Problems

Objective 3 **Graph systems of linear inequalities associated with applied problems.**

EXAMPLE 5 A politician wants to poll registered voters about their feelings on education. She tells her campaign manager to contact at least 300 voters and to poll at least twice as many women as men. Set up and graph a system of inequalities for the politician's wishes.

Solution

We will let x represent the number of men and y represent the number of women. Since the politician wants to contact at least 300 voters, we know that our first inequality is $x + y \geq 300$. She also wants the number of women to be at least twice the number of men, so the second inequality is $y \geq 2x$. The system of inequalities is $\begin{matrix} x + y \geq 300 \\ y \geq 2x \end{matrix}$.

To graph the first inequality, we graph the solid line $x + y = 300$. This line has an x-intercept at $(300, 0)$ and a y-intercept at $(0, 300)$. Using a test point of $(0, 0)$ in the inequality $x + y \geq 300$ produces a false inequality, so we shade the half-plane that does not contain the origin.

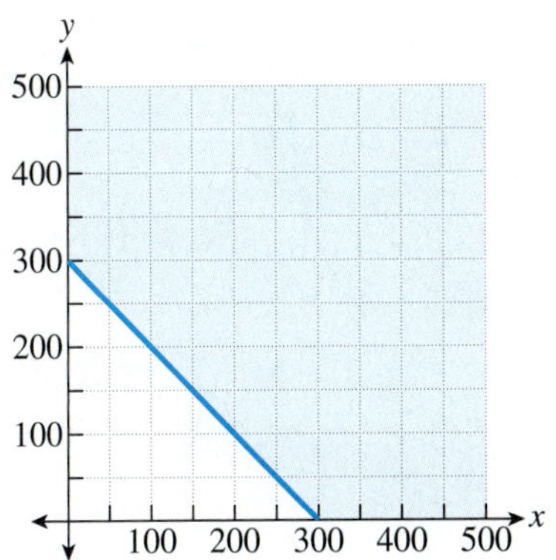

To graph the second inequality, we graph the solid line $y = 2x$. This line has a y-intercept at the origin, and a slope of 2. Using a test point of $(100, 0)$ in the inequality $y \geq 2x$ produces a false inequality, so we shade the half-plane that does not contain the point $(100, 0)$.

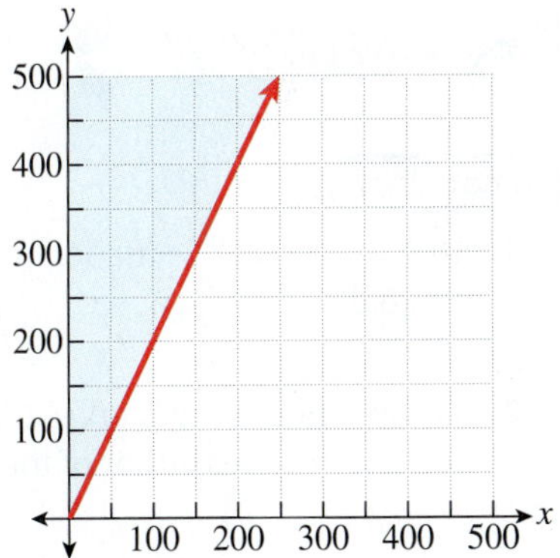

We now display the solutions to the system of inequalities by finding the region of intersection for these two inequalities.

Any point in the shaded region with integer coefficients, such as $(100, 400)$, is a solution to the problem. If 100 men and 400 women were polled, the politician's requests are satisfied.

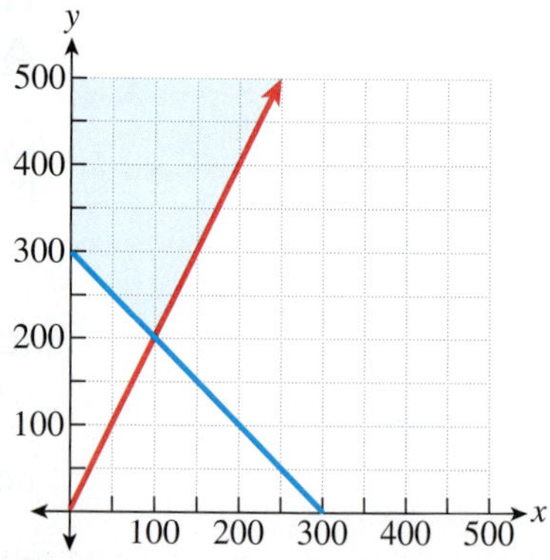

Quick Check **5** **A salesperson has set up a meeting to sell vacation time-shares. He has booked a room that has only 250 seats, so he can invite no more than 250 people. Ideally, he wants there to be at least 4 times as many retirees as people who are still working. Set up and graph a system of inequalities for this situation, letting *x* represent the number of people who are still working and *y* represent the number of retirees.**

Building Your Study Strategy **Making the Best Use of Your Resources, 3** **Study Skills Courses** Some colleges run short-term study skills courses or seminars, and these courses can be quite helpful to students who are learning mathematics. These courses cover everything from note-taking and test-taking to overcoming math anxiety. The tips and suggestions that you pick up can only help to improve your understanding of mathematics. To find out if there are any such courses or seminars at your school, ask an academic counselor.

EXERCISES 3.3

Vocabulary

1. A(n) __________ is made up of two or more linear inequalities.

2. A linear inequality is graphed with a(n) __________ line if it involves either of the symbols $<$ or $>$.

3. A linear inequality is graphed with a(n) __________ line if it involves either of the symbols $\leq$ or $\geq$.

4. When solving a system of linear inequalities, shade the region where the solutions of the individual inequalities __________.

Graph the system of inequalities.

5. $y \geq 4x - 7$
$y \leq -x + 6$

6. $y \geq 2x - 2$
$y \leq x + 1$

7. $y < x - 4$
$y > -2x - 7$

8. $y < -\frac{2}{3}x + 8$
$y > \frac{3}{2}x - 3$

9. $y < 2x + 6$
$y > 2x - 5$

10. $y \geq 4x - 8$
$y < -\frac{1}{3}x$

11. $y \geq 4$
$y \leq 5x - 2$

12. $y \leq -x + 2$
$y < -3$

13. $y \geq \frac{1}{2}x - 3$
$x > 5$

14. $y \geq -3x + 5$
$x \geq 1$

15. $x + 3y \leq 6$
$4x - y < 8$

16. $5x + 2y < 10$
$-2x + y > 6$

17. $6x - 2y \geq -12$
$3x + 6y \leq 24$

18. $-3x + y \leq 3$
$4x + 3y \geq -24$

19. $x + 8y < 8$
$5x - 3y \geq 15$

20. $5x + 8y \leq 40$
$-4x + 7y > 28$

21. $y \geq 2x + 5$
$2x + 3y > -9$

22. $y < 4x + 6$
$4x - y < 8$

29. $y \geq -2$
$y < \frac{1}{5}x - 1$

30. $x \leq 7$
$2x + 6y \leq 12$

23. $y \geq 2x$
$9x + 3y < 27$

24. $y < -3x$
$6x - 5y \geq 30$

31. $x < -4$
$y > 2$

32. $x \geq -8$
$y \leq 5$

25. $y > \frac{1}{4}x + 6$
$7x + 2y \leq -14$

26. $y \geq -\frac{2}{5}x - 7$
$x - y > 9$

33. $y < 3x$
$y < 9$
$x > 1$

34. $x - y < 5$
$y < -2x + 6$
$y > -4$

27. $x > 5$
$y \geq \frac{1}{2}x - 3$

28. $-3x + 5y \geq 15$
$y < 4$

35. $3x - 2y \leq -10$
$y < -4x - 6$
$-x + 3y < 12$

36. $4x + y < 4$
$y < 3x - 2$
$2x - 3y \leq 12$

37. $x \geq -7$
$x \leq 6$
$-x + 2y > -8$

38. $y < 8$
$y > -3$
$2x + y < 3$

39. The dean of a college is setting up a workshop to discuss Student Learning Outcomes for staff and faculty members, and she wants at least 100 people to attend. She also wants the number of faculty members in attendance to be at least twice the number of staff members in attendance. Set up and graph a system of inequalities for this situation, letting x represent the number of staff members and y represent the number of faculty members.

40. The dean of a community college plans to give a student satisfaction survey to at least 500 students at his college. Because most of the students at his college plan to transfer, he wants to be sure that the number of transfer students who take the survey is at least 4 times the number of non-transfer students who take the survey. Set up and graph a system of inequalities for this situation, letting x represent the number of non-transfer students and y represent the number of transfer students.

41. The director of financial aid at a college has the authority to grant up to 60 scholarships. He wants at least as many female students to receive scholarships as male students. Set up and graph a system of inequalities for this situation, letting x represent the number of male students and y represent the number of female students.

42. A couple's wedding reception has room for at most 200 people. The number of people at the reception who were friends of the bride was at least 60 more than the number of people who were friends of the groom. Set up and graph a system of inequalities for this situation, letting x represent the number of people who were friends of the groom and y represent the number of people who were friends of the bride.

43. Write a system of linear inequalities whose solution is the entire second quadrant, as shown.

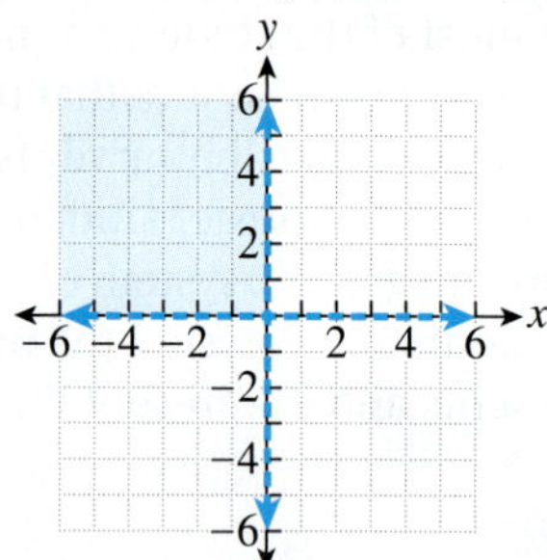

44. Write a system of linear inequalities whose solution is the entire fourth quadrant.

45. Find the area of the region defined by the system of linear inequalities.

$x \geq 1$

$y \geq 3$

$y \leq -x + 8$

46. Write a system of linear inequalities whose solution set is shown on the given graph.

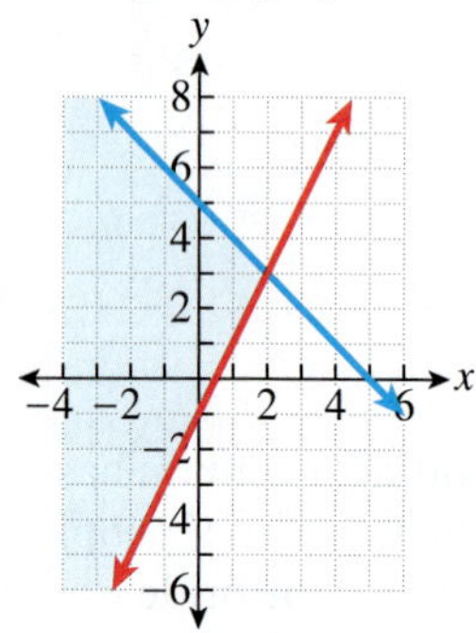

Answer in complete sentences.

47. ***Solutions Manual**** Write a solutions manual page for the following problem:

Solve the system of linear inequalities.

$y < 3x + 5$

$2x + 3y \geq 6$

48. ***Newsletter**** Write a newsletter explaining how to solve a system of linear inequalities.

***See Appendix B for details and sample answers.**

3.4 Systems of Three Equations in Three Unknowns

Objectives

1. **Determine whether an ordered triple is a solution of a system of three equations in three unknowns.**
2. **Solve systems of three equations in three unknowns.**
3. **Solve applied problems using a system of three equations in three unknowns.**

Solutions to a System of Three Equations in Three Unknowns

Objective 1 Determine whether an ordered triple is a solution of a system of three equations in three unknowns. We now turn our attention toward **systems of three equations in three unknowns**. Each of the three equations will be of the form $ax + by + cz = d$, where a, b, c, and d are real numbers. If there is a unique solution to such a system, it will be an **ordered triple** (x, y, z).

EXAMPLE 1 Is $(3, 2, 4)$ a solution to the following system of equations?

$$2x - 3y + z = 4$$
$$x + 2y - z = 3$$
$$2x - y - 2z = -4$$

Solution

To determine whether the ordered triple is a solution to the system of equations, we will substitute 3 for x, 2 for y, and 4 for z in each equation. If each equation is true for these three values, then the ordered pair is a solution to the system of equations.

$2x - 3y + z = 4$	$x + 2y - z = 3$	$2x - y - 2z = -4$
$2(3) - 3(2) + (4) = 4$	$(3) + 2(2) - (4) = 3$	$2(3) - (2) - 2(4) = -4$
$6 - 6 + 4 = 4$	$3 + 4 - 4 = 3$	$6 - 2 - 8 = -4$
$4 = 4$	$3 = 3$	$-4 = -4$

Since the ordered triple makes each equation in the system true, the ordered triple $(3, 2, 4)$ is a solution to the system of equations.

***Quick Check* 1**

Is $(3, -4, 8)$ a solution to the following system of equations?

$$x + y + z = 7$$
$$3x + 2y - z = -7$$
$$2x - y + z = 18$$

EXAMPLE 2 Is $(5, -2, 3)$ a solution to the system of equations?

$$x + y + z = 6$$
$$3x - 2y + z = 14$$
$$2x + 5y - z = -3$$

Solution

We will begin by substituting 5 for x, -2 for y, and 3 for z in each equation. If each equation is true for these three values, then the ordered pair is a solution to the system of equations.

$x + y + z = 6$	$3x - 2y + z = 14$	$2x + 5y - z = -3$
$(5) + (-2) + (3) = 6$	$3(5) - 2(-2) + (3) = 14$	$2(5) + 5(-2) - (3) = -3$
$6 = 6$	$15 + 4 + 3 = 14$	$10 - 10 - 3 = -3$
	$22 = 14$	$-3 = -3$

Quick Check 2
Is $(4, -2, 5)$ a solution to the following system of equations?

$$x - 2y - z = 3$$
$$3x + 2y + 2z = 26$$
$$5x - 3y - 4z = 6$$

Since $(5, -2, 3)$ is not a solution to the equation $3x - 2y + z = 14$, the ordered triple $(5, -2, 3)$ is not a solution to the system of equations.

Solving a System of Three Equations in Three Unknowns

Objective 2 Solve systems of three equations in three unknowns. The graph of a linear equation with three variables is a two-dimensional **plane**.

For a system of three linear equations in three unknowns, if the planes associated with the equations intersect at a single point, this ordered triple (x, y, z) is the solution to the system of equations.

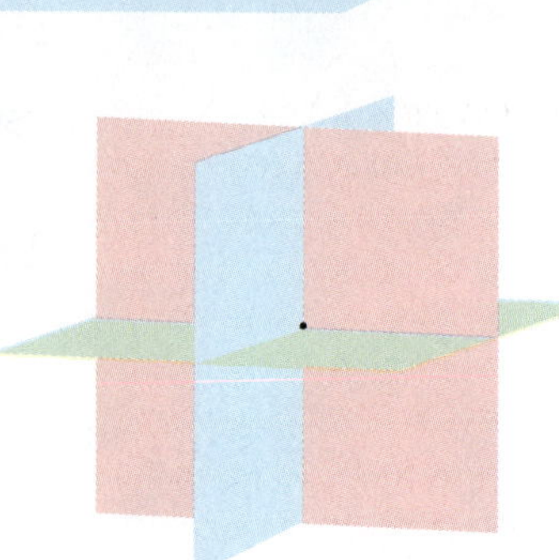

Some systems of three equations in three unknowns have no solution. Such a system is called an inconsistent system. Trying to solve an inconsistent system will produce an equation that is a contradiction, such as $0 = 1$. Graphically, the following systems are inconsistent, as the three planes do not have a point in common.

Some systems of three equations in three unknowns have infinitely many solutions. Such a system is called a dependent system of equations. Trying to solve a dependent system of equations will produce an equation that is an identity, such as $0 = 0$. Graphically, if at least two of the planes are identical, or if the intersection of the three planes is a line, then the system is a dependent system.

Since graphing equations in three dimensions is beyond the scope of this text, we will focus on solving systems of three linear equations in three unknowns algebraically.

Solving a System of Three Equations in Three Unknowns

- Select two of the equations and use the addition method to eliminate one of the variables.
- Select a different pair of equations and use the addition method to eliminate the same variable, leaving us with two equations in two unknowns.
- Solve the system of two equations for the two unknowns.
- Substitute these two values in any of the original three equations and solve for the unknown variable.

We will now apply this technique to the system of equations from the first example.

EXAMPLE 3 Solve the system.

$$\begin{aligned} 2x - 3y + z &= 4 \quad &\text{(Eq. 1)} \\ x + 2y - z &= 3 \quad &\text{(Eq. 2)} \\ 2x - y - 2z &= -4 \quad &\text{(Eq. 3)} \end{aligned}$$

Solution

Notice that the three equations have been labeled. This will help to keep track of the equations as we proceed to solve the system. We must choose a variable to eliminate first, and in this example we will eliminate z first. We may begin by adding Equations 1 and 2 together, as the coefficients of the z-terms are opposites.

$$\begin{array}{rcl} 2x - 3y + z = 4 & & \text{(Eq. 1)} \\ \underline{x + 2y - z = 3} & & \underline{\text{(Eq. 2)}} \\ 3x - y \quad\quad = 7 & & \text{(Eq. 4)} \end{array}$$

We need to select a different pair of equations and eliminate z. By multiplying Equation 1 by 2, and adding it to Equation 3, we can eliminate z once again.

$$\begin{array}{rcl} 4x - 6y + 2z = 8 & & (2 \cdot \text{Eq. 1}) \\ \underline{2x - y - 2z = -4} & & \underline{\text{(Eq. 3)}} \\ 6x - 7y \quad\quad = 4 & & \text{(Eq. 5)} \end{array}$$

We now work to solve the system of equations formed by Equations 4 and 5, which is a system of two equations in two unknowns. We will solve this system using the addition method as well, although we could also choose the substitution method if it was convenient. We will eliminate the variable y by multiplying Equation 4 by -7.

$$\begin{aligned} 3x - y &= 7 \\ 6x - 7y &= 4 \end{aligned} \quad \xrightarrow{\text{Multiply by } -7.} \quad \begin{aligned} -21x + 7y &= -49 \\ 6x - 7y &= 4 \end{aligned}$$

$$\begin{array}{rcr} -21x + 7y &=& -49 \\ 6x - 7y &=& 4 \\ \hline -15x \quad &=& -45 \end{array}$$

$x = 3$ Divide both sides by -15.

We now substitute 3 for x in either of the equations that contained only two variables. We will use Equation 4.

$3x - y = 7$
$3(3) - y = 7$ Substitute 3 for x in Equation 4.
$9 - y = 7$ Multiply.
$-y = -2$ Subtract 9.
$y = 2$ Divide both sides by -1.

To find z, we will substitute 3 for x and 2 for y in either of the original equations that contained three variables. We will use Equation 1.

$2x - 3y + z = 4$
$2(3) - 3(2) + z = 4$ Substitute 3 for x and 2 for y in Equation 1.
$6 - 6 + z = 4$ Multiply.
$0 + z = 4$ Combine like terms.
$z = 4$

The solution to this system is the ordered triple $(3, 2, 4)$.

Quick Check 3

Solve the system.

$$\begin{aligned} x + y + z &= 0 \\ 2x - y + 3z &= -9 \\ -3x + 2y - 4z &= 13 \end{aligned}$$

EXAMPLE 4 Solve the system.

$$\begin{aligned} 3x + 2y + 2z &= 7 && \text{(Eq. 1)} \\ 4x + 4y - 3z &= -17 && \text{(Eq. 2)} \\ -x - 6y + 4z &= 13 && \text{(Eq. 3)} \end{aligned}$$

Solution

In this example we will eliminate x first. It is a good idea to choose x because the coefficient of the x-term in Equation 3 can easily be multiplied to become the opposite of the coefficients of the x-terms in Equations 1 and 2. We begin by multiplying Equation 3 by 3 and adding it to Equation 1.

$$\begin{array}{rcll} 3x + 2y + 2z &=& 7 & \text{(Eq. 1)} \\ -3x - 18y + 12z &=& 39 & (3 \cdot \text{Eq. 3}) \\ \hline -16y + 14z &=& 46 & \text{(Eq. 4)} \end{array}$$

We need to select a different pair of equations and eliminate x. By multiplying Equation 3 by 4 and adding it to Equation 2, we can eliminate x once again.

$$\begin{array}{rll} 4x + 4y - 3z &= -17 & \text{(Eq. 2)} \\ -4x - 24y + 16z &= 52 & (4 \cdot \text{Eq. 3}) \\ \hline -20y + 13z &= 35 & \text{(Eq. 5)} \end{array}$$

We now solve the system of equations formed by Equations 4 and 5. We will use the addition method to eliminate the variable y. To determine the constant by which to multiply each equation, we should find the least common multiple (LCM) of 16 and 20, which is 80. If we multiply Equation 4 by 5, then the coefficient of the y-term will be -80. If we multiply Equation 5 by -4, then the coefficient of the y-term will be 80. This will make the coefficients of the y-terms opposites, allowing us to add the equations together and eliminate y.

$$\begin{array}{rcr} -16y + 14z = 46 & \xrightarrow{\text{Multiply by 5.}} & -80y + 70z = 230 \\ -20y + 13z = 35 & \xrightarrow{\text{Multiply by } -4.} & 80y - 52z = -140 \end{array}$$

$$\begin{array}{rr} -80y + 70z = & 230 \\ 80y - 52z = & -140 \\ \hline 18z = & 90 \end{array}$$

$z = 5$ Divide both sides by 18.

We now substitute 5 for z in either of the equations that contained only two variables. We will use Equation 4.

$$\begin{array}{rl} -16y + 14z = 46 & \\ -16y + 14(5) = 46 & \text{Substitute 5 for } z \text{ in Equation 4.} \\ -16y + 70 = 46 & \text{Multiply.} \\ -16y = -24 & \text{Subtract 70.} \\ y = \frac{3}{2} & \text{Divide both sides by } -16 \text{ and simplify.} \end{array}$$

To find x, we will substitute 5 for z and $\frac{3}{2}$ for y in either of the original equations that contained three variables. We will use Equation 1.

$$\begin{array}{rl} 3x + 2y + 2z = 7 & \\ 3x + 2\left(\frac{3}{2}\right) + 2(5) = 7 & \text{Substitute } \frac{3}{2} \text{ for } y \text{ and 5 for } z \text{ in Equation 1.} \\ 3x + 3 + 10 = 7 & \text{Multiply.} \\ 3x + 13 = 7 & \text{Combine like terms.} \\ 3x = -6 & \text{Subtract 13.} \\ x = -2 & \text{Divide both sides by 3.} \end{array}$$

The solution to this system is the ordered triple $\left(-2, \frac{3}{2}, 5\right)$.

***Quick Check* 4**

Solve the system.

$$\begin{aligned} 2x + y - z &= -8 \\ -4x - 3y + 3z &= 25 \\ 6x + 5y + 4z &= 3 \end{aligned}$$

Example 5 Solve the system.

$$\begin{aligned} x + 2y + 3z &= 10 \quad &\text{(Eq. 1)} \\ 4x + 3y - 2z &= -29 \quad &\text{(Eq. 2)} \\ x + y &= -5 \quad &\text{(Eq. 3)} \end{aligned}$$

Solution

In this example, notice that Equation 3 has only two variables: x and y. If we can combine Equation 1 and Equation 2 to eliminate z, then we will have two equations in two unknowns. Multiplying Equation 1 by 2 produces a z-term with a coefficient of 6, and multiplying Equation 2 by 3 produces a z-term with a coefficient of -6. Adding the two equations at this point will eliminate the variable z.

$$\begin{aligned} 2x + 4y + 6z &= 20 \quad &(2 \cdot \text{Eq. 1}) \\ 12x + 9y - 6z &= -87 \quad &(3 \cdot \text{Eq. 2}) \\ \hline 14x + 13y &= -67 \quad &\text{(Eq. 4)} \end{aligned}$$

We now solve the system of equations formed by Equations 3 and 4. We will use the addition method to eliminate the variable y. If we multiply Equation 3 by -13 and add it to Equation 4, then we will eliminate the variable y.

$$\begin{aligned} x + y &= -5 \quad \xrightarrow{\text{Multiply by } -13.} \quad -13x - 13y = 65 \\ 14x + 13y &= -67 \qquad\qquad\qquad\quad 14x + 13y = -67 \end{aligned}$$

$$\begin{aligned} -13x - 13y &= 65 \\ 14x + 13y &= -67 \\ \hline x &= -2 \end{aligned}$$

We now substitute -2 for x in either of the equations that contained only two variables. We will use Equation 3.

$$\begin{aligned} x + y &= -5 \\ (-2) + y &= -5 \quad &\text{Substitute } -2 \text{ for } x \text{ in Equation 3.} \\ y &= -3 \quad &\text{Add 2.} \end{aligned}$$

To find z, we will substitute -2 for x and -3 for y in either of the original equations that contained three variables. We will use Equation 1.

$$\begin{aligned} x + 2y + 3z &= 10 \\ (-2) + 2(-3) + 3z &= 10 \quad &\text{Substitute } -2 \text{ for } x \text{ and } -3 \text{ for } z \text{ in Equation 1.} \\ -2 - 6 + 3z &= 10 \quad &\text{Multiply.} \\ 3z - 8 &= 10 \quad &\text{Combine like terms.} \\ 3z &= 18 \quad &\text{Add 8.} \\ z &= 6 \quad &\text{Divide both sides by 3.} \end{aligned}$$

The solution to this system is the ordered triple $(-2, -3, 6)$.

Quick Check 5

Solve the system.

$$\begin{aligned} x - 3y + 5z &= 34 \\ 4x + y + 2z &= 17 \\ x - z &= 1 \end{aligned}$$

Applications of Systems of Three Equations in Three Unknowns

Objective 3 Solve applied problems using a system of three equations in three unknowns. We finish the section with an application involving a system of three equations in three unknowns.

EXAMPLE 6 A minor league baseball team has three prices for admission to their games. Adults are charged \$7, senior citizens are charged \$5, and children are charged \$4. At last night's game there were 1000 people in attendance, generating \$5600 in receipts. If we know that there were 300 more children than senior citizens at the game, find how many adult tickets, senior citizen tickets, and child tickets were sold.

Solution

There are three unknowns in this problem: the number of adult tickets sold, the number of senior citizen tickets sold, and the number of child tickets sold. We will let a, s, and c represent these three quantities, respectively. Here is a table of the unknowns.

Unknowns:

Number of adult tickets sold: a
Number of senior citizen tickets sold: s
Number of child tickets sold: c

Solving a problem with three unknowns requires that we establish a system of three equations. Since we know that there were 1000 people at last night's game, the first equation in our system is $a + s + c = 1000$.

We also know that the revenue from all of the tickets was \$5600, which will lead to our second equation. The amount received from the adult tickets was \$7a (a tickets at \$7 each). Likewise, the amount received from senior citizen tickets was \$5s, and the amount received from child tickets was \$4c. The second equation in our system is $7a + 5s + 4c = 5600$.

Our third equation comes from knowing that there were 300 more children than senior citizens at the game. In other words, the number of children at the game was equal to the number of senior citizens plus 300. Our third equation is $c = s + 300$, which can be rewritten as $-s + c = 300$.

Here is our system of equations:

$$\begin{aligned} a + s + c &= 1000 \quad &\text{(Eq. 1)} \\ 7a + 5s + 4c &= 5600 \quad &\text{(Eq. 2)} \\ -s + c &= 300 \quad &\text{(Eq. 3)} \end{aligned}$$

Since Equation 3 has only two variables, we will use the addition method to combine Equations 1 and 2, eliminating the variable a. We begin by adding -7 times Equation 1 to Equation 2.

$$\begin{aligned} -7a - 7s - 7c &= -7000 \quad &(-7 \cdot \text{Eq. 1}) \\ 7a + 5s + 4c &= 5600 \quad &\text{(Eq. 2)} \\ \hline -2s - 3c &= -1400 \quad &\text{(Eq. 4)} \end{aligned}$$

We now have two equations that contain only the two variables s and c. We can solve the system of equations formed by Equations 3 and 4. We will use the addition method to eliminate the variable c by adding three times Equation 3 to Equation 4.

$$\begin{aligned} -s + c &= 300 \\ -2s - 3c &= -1400 \end{aligned} \xrightarrow{\text{Multiply by 3.}} \begin{aligned} -3s + 3c &= 900 \\ -2s - 3c &= -1400 \end{aligned}$$

$$\begin{aligned} -3s + 3c &= 900 \\ \underline{-2s - 3c} &\underline{= -1400} \\ -5s \qquad &= -500 \end{aligned}$$

$$s = 100 \qquad \text{Divide both sides by } -5.$$

There were 100 senior citizens at the game. To find the number of children at the game, we will substitute 100 for s in Equation 3.

$$\begin{aligned} -s + c &= 300 \\ -100 + c &= 300 && \text{Substitute 100 for } s \text{ in Equation 3.} \\ c &= 400 && \text{Add 100.} \end{aligned}$$

There were 400 children at the game. (Since we knew that there were 300 more children than senior citizens at the game, we could have just added 300 to 100.) To find the number of adults at the game, we will substitute 100 for s and 400 for c in Equation 1.

$$\begin{aligned} a + s + c &= 1000 \\ a + (100) + (400) &= 1000 && \text{Substitute 100 for } s \text{ and 400 for } c \text{ in Equation 1.} \\ a &= 500 && \text{Solve for } a. \end{aligned}$$

There were 500 adults, 100 senior citizens, and 400 children at last night's game. The reader can verify that the total revenue for 500 adults, 100 senior citizens, and 400 children is $5600.

Quick Check 6

Gilbert looks into his wallet and finds $1, $5, and $10 bills totaling $111. There are 30 bills in all, and Gilbert has 4 more $5 bills than he has $10 bills. How many bills of each type does Gilbert have in his wallet?

Building Your Study Strategy **Making the Best Use of Your Resources, 4**

Student Solutions Manual The student solutions manual for a mathematics textbook can be a valuable resource when used properly, but can be detrimental to learning if used improperly. You should refer to the solutions manual to check your work or to see where you went wrong with your solution. The detailed solution can be helpful in this case.

Some students do their homework with the solutions manual sitting open in front of them, looking at what they should write for the next line. If you do nothing except essentially copy the solutions manual onto your own paper, you will most likely have tremendous difficulty on the next exam.

EXERCISES 3.4

Vocabulary

1. The solution of a system of three equations in three unknowns is a(n) __________.

2. To solve a system of three equations in three unknowns, begin by eliminating one of the variables to create a system of __________ equations in __________ unknowns.

Is the ordered triple a solution to the given system of equations?

3. $(3, 2, 6)$
$x + y + z = 11$
$x - y + 2z = 17$
$4x - 3y - 2z = -6$

4. $(2, -4, -1)$
$x + y - z = -1$
$5x + 2y + 2z = 0$
$\frac{1}{2}x + 2y - 6z = -1$

5. $(-5, 2, 0)$
$x + 2y + 3z = -1$
$3x - y - 6z = -17$
$4x - 3y + z = -26$

6. $(1, -1, -6)$
$x - y - z = 8$
$5x + 7y + z = -8$
$-2x + 5y - 4z = 21$

7. $(7500, 3500, 1000)$
$x + y + z = 12{,}000$
$0.08x + 0.04y + 0.03z = 770$
$x - 2y = 500$

8. $(50, 30, 20)$
$x + y + z = 100$
$x + 0.3y + 0.6z = 71$
$0.2y - 0.3z = 0$

Solve.

9. $x + 3y - 2z = -3$
$-2x - 2y + 3z = -2$
$-3x + 4y - 5z = -17$

10. $3x - y + 2z = -10$
$4x + 2y + 3z = 2$
$2x + 3y + 2z = 9$

11. $x + y + z = 1$
$3x + 2y + z = 7$
$4x + 4y - z = 19$

12. $x + y + z = 6$
$-2x - y + 3z = 43$
$3x + 2y + 2z = 3$

13. $x - 4y + 7z = -5$
$3x - 6y + 5z = 4$
$-5x + 10y + 4z = -19$

14. $x + y + 4z = -6$
$-2x + y - 4z = 6$
$3x + 2y - 8z = -2$

15. $2x + 3y + z = 5$
$-2x - y + 4z = 60$
$x + 2y - 8z = -30$

16. $x - y - z = 16$
$2x + 3y + 2z = -5$
$4x + 5y + 4z = -5$

17. $2x - 3y - z = -6$
$-3x + 2y + 4z = 24$
$5x - 4y + 3z = -18$

18. $x + 4y + 2z = -3$
$2x - 3y - z = 19$
$4x + 9y + 5z = 3$

19. $-x + y + 2z = -20$
$3x - 2y - 6z = 52$
$5x - 4y + 8z = 2$

20. $x + y - z = 39$
$2x + y + 3z = -10$
$3x - 2y - 2z = 37$

21. $4x + 2y + z = -5$
$2x - y - 6z = 0$
$-8x + 6y + 3z = -30$

22. $x + 3y - 6z = 8$
$3x + 4y - 2z = 9$
$5x + 6y - 12z = 22$

23. $\frac{1}{2}x + \frac{1}{3}y + z = 3$
$\frac{2}{3}x - 4y - 3z = 13$
$3x + 4y + 5z = 11$

24. $\frac{3}{4}x - \frac{5}{2}y - \frac{1}{3}z = -14$
$x + \frac{3}{4}y + \frac{7}{2}z = -26$
$2x - 3y - 4z = -4$

25. $x + 2y + z = 8$
$3x + 4y + 2z = 9$
$3x - 4y = -5$

26. $2x + 3y - z = 10$
$-x + 4y + 2z = 200$
$5x + 2y = 88$

27. $6x + y + 6z = 8$
$3x + 5y - 2z = -44$
$2x + 7z = 34$

28. $-x + 10y - 9z = -26$
$4x + 2y + z = 202$
$2y + z = 106$

29. $3x + 2y - 3z = 5$
$-2x + 3y + 5z = 32$
$5y - 4z = -4$

30. $3x - 2y - 2z = -43$
$-x + 5y + z = 49$
$x + z = -9$

31. $x + y + z = 10{,}000$
$0.06x + 0.07y + 0.05z = 620$
$x - 2y = 0$

32. $x + y + z = 20{,}000$
$-0.02x - 0.07y + 0.2z = 2505$
$6x - z = 1000$

33. $x + 2y + 3z = 14$
$3y - z = 3$
$z = 3$

34. $3x + 8y - 2z = 3$
$5x + 4z = 49$
$3x = 15$

35. $7x + 5y - 4z = 14$
$3y - 2z = 1$
$2y = -10$

36. $-4x - 3y + 12z = -188$
$7x + 6y = 104$
$-6x = -48$

37. Create a system of three equations whose solution is $(1, 2, 3)$.

38. Create a system of three equations whose solution is $(5, -4, 8)$.

39. Create a system of three equations whose solution is $(9, 0, -4)$.

40. Create a system of three equations whose solution is $(1000, 2500, 3500)$.

41. Jonah opens his wallet and finds that all of the bills are either $1 bills, $5 bills, or $10 bills. There are 35 bills in his wallet. There are four more $5 bills than $1 bills. If the total value of the bills is $246, how many $10 bills does Jonah have?

42. A cash register has no $1 bills in it. All of the bills are either $5 bills, $10 bills, or $20 bills. There are 37 bills in the cash register, and the total value of these bills is $310. If the number of $10 bills in the register is twice the number of $20 bills, how many $5 bills are in the register?

43. Carlos has a coin jar that contains nickels, dimes, and quarters. There are 83 coins in the jar. There are nine more nickels than dimes. If the total value of the coins in the jar is $10.90, how many dimes does the jar contain?

44. Annie has $1905 in poker chips in front of her. All the chips are either $5 chips, $25 chips, or $100 chips. There are 19 more $5 chips than $25 chips, and altogether Annie has 66 chips. How many $25 chips does Annie have?

45. A campus club held a bake sale as a fund-raiser, selling coffee, muffins, and bacon-and-egg sandwiches. They charged $1 for a cup of coffee, $2 for a muffin, and $3 for a bacon-and-egg sandwich. They sold a total of 50 items, raising $85 dollars. If they sold 5 more muffins than cups of coffee, how many bacon-and-egg sandwiches did they sell?

46. A grocery store was supporting a local charity by having shoppers donate $1, $5, or $20 at the checkout stand. A total of 110 shoppers donated $430. The amount of money raised from $5 donations was $50 more than the amount of money raised from $20 donations. How many shoppers donated $20?

47. The college book store sells t-shirts for \$15, golf shirts for \$20, and sweatshirts for \$30. This semester they sold 124 shirts for a total of \$2760. The amount of money raised from t-shirt sales was \$240 higher than the amount of money raised from golf shirt sales. How many sweatshirts did they sell?

48. A minor league baseball team sells children's tickets for \$5, senior citizen tickets for \$6, and adult tickets for \$7. At last night's game there were 901 fans, and they paid a total of \$5991. If there were 92 more children at the game than there were senior citizens, how many adult tickets did the team sell?

49. The measures of the three angles in a triangle must total 180°. The measure of angle A is 20° less than the measure of angle C. The measure of angle C is twice the measure of angle B. Find the measure of each angle.

50. The measures of the three angles in a triangle must total 180°. The measure of angle A is 15° more than the measure of angle B. The measure of angle B is 15° greater than four times the measure of angle C. Find the measure of each angle.

51. Mary invested a total of \$40,000 in three different bank accounts. One account pays an annual interest rate of 3%, the second account pays 5% annual interest, and the third account pays 6% annual interest. In one year, Mary earned a total of \$1960 in interest from these three accounts. If Mary invested \$8000 more in the account that pays 5% interest than she did in the account that pays 6% interest, find the amount she invested in each account. (Recall the introduction to problems about interest in Section 3.2.)

52. Dennis invested a total of \$100,000 in three different mutual funds. After one year the first fund had shown a profit of 8%, the second fund had shown a loss of 5%, and the third fund had shown a profit of 2%. In the first year Dennis had made a profit of \$2900 from the three funds. If Dennis invested \$10,000 more in the fund that lost 5% than he invested in the fund that made a 2% profit, how much did he initially invest in each fund?

Writing in Mathematics

Answer in complete sentences.

53. Write a word problem whose solution is "There were 40 children, 200 adults, and 60 senior citizens in attendance." Your problem must lead to a system of three equations in three unknowns. Explain how you created your problem.

54. Write a word problem for the given system of equations.

$$\begin{aligned} x + y + z &= 350{,}000 \\ 0.05x + 0.07y - 0.04z &= 17{,}000 \\ y &= 2x \end{aligned}$$

Explain how you created your problem.

55. ***Solutions Manual**** Write a solutions manual page for the following problem:

Solve the system of equations.

$$\begin{aligned} 3x + 2y - z &= 18 \\ 2x - 5y + z &= 27 \\ 4x + 3y + 2z &= -11 \end{aligned}$$

56. ***Newsletter**** Write a newsletter explaining how to solve a system of three linear equations in three unknowns.

***See Appendix B for details and sample answers.**

3.5 Using Matrices to Solve Systems of Equations

Objectives

1. Solve systems of two equations in two unknowns using matrices.
2. Solve systems of three equations in three unknowns using matrices.
3. Solve applied problems using matrices.

Solving a System of Two Equations in Two Unknowns Using Matrices

Objective 1 Solve systems of two equations in two unknowns using matrices.

Matrix

A **matrix** (plural **matrices**) is a rectangular array of numbers, such as $\begin{bmatrix} 1 & 5 & 7 & 20 \\ 0 & 4 & -3 & 5 \\ 0 & 0 & 6 & 6 \end{bmatrix}$.
The numbers in a matrix are called **elements**. A matrix is often described by the number of **rows** (horizontal) and number of **columns** (vertical) it contains. The matrix in this example is a three by four, or 3×4, matrix, because it has 3 rows and 4 columns. We refer to the first row of a matrix as R_1, the second row as R_2, and so on.

We can use matrices to solve systems of equations in a manner that is similar to using the addition method. For example, to solve the system of equations $\begin{aligned} 2x - 5y &= -9 \\ 4x + 3y &= -5 \end{aligned}$, we begin by creating the **augmented matrix** $\left[\begin{array}{cc:c} 2 & -5 & -9 \\ 4 & 3 & -5 \end{array}\right]$. The first column of the matrix contains the coefficients of the variable x from the two equations. The second column contains the coefficients of the variable y. The dashed vertical line separates these coefficients from the constants in each equation.

We will use **row operations** to convert a matrix into a form in which we can easily solve the system of equations.

Rules of Row Operations

1. Any two rows of a matrix can be interchanged.
2. The elements in any row can be multiplied by any nonzero number.
3. The elements in any row can be changed by adding a nonzero multiple of another row's elements to them.

To solve a system of two equations in two unknowns, we must transform the augmented matrix associated with that system into one of the form $\left[\begin{array}{cc:c} a & b & c \\ 0 & d & e \end{array}\right]$, in other words, to transform the matrix into one whose first element in the second row is 0. When the element below the main diagonal is 0, the matrix is **in upper triangular form**.

Upper Triangular Form of a 2 × 3 Matrix

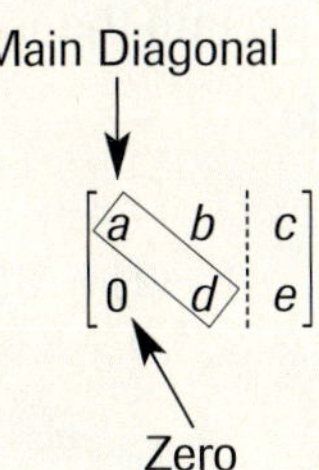

The matrix $\begin{bmatrix} a & b & c \\ 0 & d & e \end{bmatrix}$ corresponds to the system of equations $\begin{aligned} ax + by &= c \\ dy &= e \end{aligned}$. We can easily solve the second equation for y, and substitute that value in the first equation to solve for x. Substituting the solution from one row into an equation from another row is called **back substituting**.

EXAMPLE 1 Solve $\begin{aligned} 2x - 5y &= -9 \\ 4x + 3y &= -5 \end{aligned}$ using a matrix.

Solution

The augmented matrix associated with this system is $\begin{bmatrix} 2 & -5 & -9 \\ 4 & 3 & -5 \end{bmatrix}$. We need to change the first element in row 2 to 0, and this is done by adding some multiple of the first row to the second row. Consider the first element in the first row, 2, and the first element in the second row, 4. The question that we should ask is "What number, multiplied by 2, is equal to -4?" The answer to the question is -2. We will multiply the first row by -2 and add the result to the second row. Here are the operations for each column.

Column 1	Column 2	Column 3	
$-2(2)$	$-2(-5)$	$-2(-9)$	-2(Row 1)
$+\ 4$	$+\ 3$	$+\ -5$	+ Row 2
0	13	13	New Row 2

$$\begin{bmatrix} 2 & -5 & -9 \\ 4 & 3 & -5 \end{bmatrix}$$

$$\begin{bmatrix} 2 & -5 & -9 \\ 0 & 13 & 13 \end{bmatrix} \quad \text{Multiply } R_1 \text{ by } -2 \text{ and add to } R_2.$$

The second row of this matrix is equivalent to the equation $13y = 13$, which can be solved by dividing both sides of the equation by 13. The solution to this equation is $y = 1$.

The first row of this matrix is equivalent to the equation $2x - 5y = -9$. If we substitute 1 for y, then we can solve this equation for x.

$$\begin{aligned} 2x - 5(1) &= -9 && \text{Substitute 1 for } y. \\ 2x - 5 &= -9 && \text{Multiply.} \\ 2x &= -4 && \text{Add 5.} \\ x &= -2 && \text{Divide by 2.} \end{aligned}$$

The solution to the system of equations is $(-2, 1)$. This solution can be checked by substituting into the second equation in the original system, $4x + 3y = -5$.

Check

$$\begin{aligned} 4x + 3y &= -5 \\ 4(-2) + 3(1) &= -5 && \text{Substitute } -2 \text{ for } x \text{ and } 1 \text{ for } y. \\ -8 + 3 &= -5 && \text{Multiply.} \\ -5 &= -5 && \text{Simplify} \end{aligned}$$

The solution checks.

Quick Check 1
Solve $\begin{aligned} -2x + 7y &= 39 \\ 8x + 3y &= -1 \end{aligned}$ using a matrix.

EXAMPLE 2 Solve $\begin{aligned} 6x - 5y &= 24 \\ -3x + 4y &= -3 \end{aligned}$ using a matrix.

Solution

The augmented matrix associated with this system is $\left[\begin{array}{cc:c} 6 & -5 & 24 \\ -3 & 4 & -3 \end{array}\right]$. Notice that the first element R_1, 6, is a multiple of the first element of R_2, -3. The system would be easier to solve if the rows were switched so the first element in R_2 was a multiple of the first element on R_1. We will begin by switching the rows.

$$\left[\begin{array}{cc:c} 6 & -5 & 24 \\ -3 & 4 & -3 \end{array}\right]$$

$$\left[\begin{array}{cc:c} -3 & 4 & -3 \\ 6 & -5 & 24 \end{array}\right] \quad \text{Switch } R_1 \text{ and } R_2.$$

$$\left[\begin{array}{cc:c} -3 & 4 & -3 \\ 0 & 3 & 18 \end{array}\right] \quad \begin{array}{l} \text{Multiply } R_1 \text{ by 2 and add to } R_2. \\ 2(-3) + 6, 2(4) + (-5), 2(-3) + 24 \end{array}$$

The second row of this matrix is equivalent to the equation $3y = 18$, which can be solved by dividing both sides of the equation by 3. The solution to this equation is $y = 6$.

The first row of this matrix is equivalent to the equation $-3x + 4y = -3$. If we substitute 6 for y, then we can solve this equation for x.

$$\begin{aligned} -3x + 4(6) &= -3 && \text{Substitute 6 for } y. \\ -3x + 24 &= -3 && \text{Multiply.} \\ -3x &= -27 && \text{Subtract 24.} \\ x &= 9 && \text{Divide by } -3. \end{aligned}$$

The solution to the system of equations is $(9, 6)$.

Quick Check 2
Solve $\begin{aligned} 20x - 9y &= -78 \\ 5x + 11y &= 7 \end{aligned}$ using a matrix.

In the first two examples, one of the elements in the first column was a multiple of the other element in the first column. When this is not the case, we will have to multiply each row by a number in such a way that the two elements in the first column become opposites.

EXAMPLE 3 Solve $\begin{aligned} 3x + 4y &= 6 \\ 2x - y &= -7 \end{aligned}$.

Solution

The augmented matrix associated with this system is $\left[\begin{array}{cc|c} 3 & 4 & 6 \\ 2 & -1 & -7 \end{array}\right]$. Neither element in the first column is a multiple of the other element in the first column. The two elements in the first column, 3 and 2, have an LCM of 6, so we will multiply row 2 by 3 to make the first element in row 2 into a multiple of the first element in row 1.

$$\left[\begin{array}{cc|c} 3 & 4 & 6 \\ 2 & -1 & -7 \end{array}\right]$$

$$\left[\begin{array}{cc|c} 3 & 4 & 6 \\ 6 & -3 & -21 \end{array}\right] \quad \text{Multiply } R_2 \text{ by 3.}$$

$$\left[\begin{array}{cc|c} 3 & 4 & 6 \\ 0 & -11 & -33 \end{array}\right] \quad \text{Multiply } R_1 \text{ by } -2 \text{ and add to } R_2.$$

The second row of this matrix is equivalent to the equation $-11y = -33$. We will solve the second equation for y.

$$\begin{aligned} -11y &= -33 \\ y &= 3 \quad && \text{Divide both sides by } -11. \end{aligned}$$

Substitute 3 for y in the equation $3x + 4y = 6$ and solve for x.

$$\begin{aligned} 3x + 4(3) &= 6 && \text{Substitute 3 for } y. \\ 3x + 12 &= 6 && \text{Multiply.} \\ 3x &= -6 && \text{Subtract 12.} \\ x &= -2 && \text{Divide by 3.} \end{aligned}$$

The solution to the system of equations is $(-2, 3)$.

Quick Check 3
Solve $\begin{aligned} 4x + 7y &= -39 \\ -3x + 8y &= 16 \end{aligned}$ using a matrix.

Solving a System of Three Equations in Three Unknowns Using Matrices

Objective 2 Solve systems of three equations in three unknowns using matrices. To solve a system of three equations in three unknowns, we begin by setting up a 3×4 augmented matrix. Our goal is to use row operations to transform the matrix to one of the form $\left[\begin{array}{ccc|c} a & b & c & d \\ 0 & e & f & g \\ 0 & 0 & h & i \end{array}\right]$. This is the upper triangular form of a 3×4 matrix, as all of the elements below the main diagonal are 0.

Upper Triangular Form of a 3 × 4 Matrix

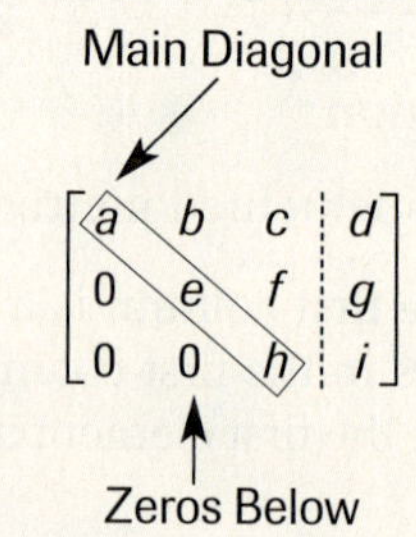

We will begin by "zeroing out" the first column under row 1. This can be done by adding some multiple of row 1 to row 2, and some other multiple of row 1 to row 3. Once this has been accomplished, we need to do the same for the second column under row 2.

At this point the third row will be equivalent to an equation that only contains the variable z. Once that equation has been solved, we can substitute that value for z in the equation that is equivalent to row 2, allowing us to solve for y. These two values can be substituted for y and z in the equation that is equivalent to row 1, allowing us to solve for x.

EXAMPLE 4 Solve the system
$$\begin{aligned} x + y + z &= 1 \\ 2x - 3y + 4z &= 29. \\ -x - 2y + 3z &= 8 \end{aligned}$$

Solution

The augmented matrix associated with this system is $\left[\begin{array}{rrr|r} 1 & 1 & 1 & 1 \\ 2 & -3 & 4 & 29 \\ -1 & -2 & 3 & 8 \end{array}\right]$. We will begin by changing the first element in row 2 and row 3 to 0.

$$\left[\begin{array}{rrr|r} 1 & 1 & 1 & 1 \\ 2 & -3 & 4 & 29 \\ -1 & -2 & 3 & 8 \end{array}\right]$$

$$\text{New } R_2 \rightarrow \left[\begin{array}{rrr|r} 1 & 1 & 1 & 1 \\ 0 & -5 & 2 & 27 \\ -1 & -2 & 3 & 8 \end{array}\right]$$

Multiply R_1 by -2 and add to R_2.
$-2(1) + 2, -2(1) + (-3),$
$-2(1) + 4, -2(1) + 29$

$$\text{New } R_3 \rightarrow \left[\begin{array}{rrr|r} 1 & 1 & 1 & 1 \\ 0 & -5 & 2 & 27 \\ 0 & -1 & 4 & 9 \end{array}\right]$$

Add R_1 to R_3.
$1 + (-1), 1 + (-2), 1 + 3, 1 + 8$

The elements in the first column under row 1 are now both 0. Next we need to change the second element of row 3, -1, to a 0. We begin to do that by multiplying row 3 by -5, so adding row 2 to row 3 will zero out the second column of the third row. In general, we multiply row 2 and row 3 by constants such that the elements in column 2 are opposites.

$$\text{New } R_3 \rightarrow \left[\begin{array}{ccc|c} 1 & 1 & 1 & 1 \\ 0 & -5 & 2 & 27 \\ 0 & 5 & -20 & -45 \end{array}\right]$$ Multiply R_3 by -5. $-5(0), -5(-1), -5(4), -5(9)$

$$\text{New } R_3 \rightarrow \left[\begin{array}{ccc|c} 1 & 1 & 1 & 1 \\ 0 & -5 & 2 & 27 \\ 0 & 0 & -18 & -18 \end{array}\right]$$ Add R_2 to R_3. $0 + 0, -5 + 5,$

The last row is equivalent to the equation $-18z = -18$, whose solution is $z = 1$. We substitute this value for z in the equation that is equivalent to row 2, $-5y + 2z = 27$.

$$\begin{aligned} -5y + 2(1) &= 27 && \text{Substitute 1 for } z. \\ -5y + 2 &= 27 && \text{Multiply.} \\ -5y &= 25 && \text{Subtract 2.} \\ y &= -5 && \text{Divide by } -5. \end{aligned}$$

Now we substitute -5 for y and 1 for z in the equation that is equivalent to row 1, $x + y + z = 1$.

$$\begin{aligned} x + (-5) + (1) &= 1 && \text{Substitute } -5 \text{ for } y \text{ and 1 for } z. \\ x - 4 &= 1 && \text{Combine like terms.} \\ x &= 5 && \text{Add 4.} \end{aligned}$$

The solution to this system is $(5, -5, 1)$.

Quick Check 4
Solve
$$\begin{aligned} x - y + 2z &= 6 \\ 4x - 2y + 3z &= 19 \\ -7x + 4y - 8z &= -33 \end{aligned}$$
using a matrix.

Solving Applied Problems Using Matrices

Objective 3 **Solve applied problems using matrices.** Once we have found a system of equations associated with an applied problem, we can solve it using matrices rather than previous techniques.

EXAMPLE 5 The perimeter of a rectangle is 260 feet. If the length of the rectangle is 16 feet longer than its width, find the dimensions of the rectangle.

Solution

The two unknowns are the length and the width of the rectangle. We will use the variables l and w for the length and width, respectively.

Unknowns:

Length: l

Width: w

We are told that the perimeter of the rectangle is 260 feet. Since the perimeter of a rectangle is equal to twice the length plus twice the width, our first equation can be written as $2l + 2w = 260$. The second equation in the system comes from the fact that the length of the rectangle is 16 feet more than its width. We express this relationship in an equation as $l = w + 16$. Here is the system we are solving.

$$\begin{aligned} 2l + 2w &= 260 \\ l &= w + 16 \end{aligned}$$

We must rewrite the second equation in general form as $\begin{aligned} 2l + 2w &= 260 \\ l - w &= 16 \end{aligned}$ before creating an augmented matrix. The augmented matrix associated with this system is $\left[\begin{array}{rr|r} 2 & 2 & 260 \\ 1 & -1 & 16 \end{array}\right]$. Notice that the first element of R_1 is a multiple of the first element of R_2. The system would be easier to solve if the rows were switched so the first element in R_2 was a multiple of the first element in R_1. We will begin by switching R_1 and R_2.

$$\left[\begin{array}{rr|r} 2 & 2 & 260 \\ 1 & -1 & 16 \end{array}\right]$$

$$\left[\begin{array}{rr|r} 1 & -1 & 16 \\ 2 & 2 & 260 \end{array}\right] \qquad \text{Switch } R_1 \text{ and } R_2.$$

$$\text{New } R_2 \rightarrow \left[\begin{array}{rr|r} 1 & -1 & 16 \\ 0 & 4 & 228 \end{array}\right] \qquad \begin{array}{l} \text{Multiply } R_1 \text{ by } -2 \text{ and add to } R_2. \\ -2(1) + 2, -2(-1) + 2, -2(16) + 260 \end{array}$$

Quick Check 5
The perimeter of a rectangle is 42 inches. If the length of the rectangle is 5 inches longer than its width, find the dimensions of the rectangle.

The second row of this matrix is equivalent to the equation $4w = 228$, which can be solved by dividing both sides of the equation by 4. The solution to this equation is $w = 57$.

The first row of this matrix is equivalent to the equation $l - w = 16$. If we substitute 57 for w, then we can solve this equation for l.

$$\begin{aligned} l - 57 &= 16 && \text{Substitute 57 for } w. \\ l &= 73 && \text{Add 57.} \end{aligned}$$

The length of the rectangle is 73 feet and the width is 57 feet. A quick check confirms that the length of the rectangle is 16 feet more than its width, and that the perimeter $(2(73) + 2(57))$ is 260 feet.

EXAMPLE 6 Luz invested \$7500 in three mutual funds. During the first year, one of the mutual funds earned an 8% profit. Another of the funds earned a 13% profit in the first year. The third fund showed a loss of 4% in the first year. She invested \$500 more in the fund that earned a 13% profit than she invested in the fund that lost 4%. If Luz made a net profit of \$555 from these three mutual funds in the first year, how much did she invest in each?

Solution

This problem is quite similar to the interest problems that were covered in Section 3.2. The three unknowns are the amount invested (x) in the fund that earned an 8% profit, the amount invested (y) in the fund that earned a 13% profit, and the amount (z) invested in the fund that lost 4%. The following table summarizes the important information.

Mutual Fund	Principal	Profit Rate	Profit
Fund 1	x	0.08	$0.08x$
Fund 2	y	0.13	$0.13y$
Fund 3	z	-0.04	$-0.04z$
Total	7500		555

The first equation in our system comes from the information that the amounts invested in the funds total \$7500. As an equation, this can be represented as $x + y + z = 7500$. This equation can be found in our table in the column labeled "Principal."

Since we know that she invested \$500 more in the second fund than she did in the third fund, the second equation in our system is $y = z + 500$. In standard form, this equation can be rewritten as $y - z = 500$.

The third equation in our system comes from the amount of profit earned and can be found in the last column. The total profit is \$555, so the equation is $0.08x + 0.13y - 0.04z = 555$. Here is the system we must solve.

$$x + y + z = 7500$$
$$y - z = 500$$
$$0.08x + 0.13y - 0.04z = 555$$

We will clear the third equation of decimals by multiplying both sides of the equation by 100.

$$0.08x + 0.13y - 0.04z = 555 \xrightarrow{\text{Multiply by 100.}} 8x + 13y - 4z = 55{,}500$$

The augmented matrix associated with this system is $\left[\begin{array}{ccc|c} 1 & 1 & 1 & 7500 \\ 0 & 1 & -1 & 500 \\ 8 & 13 & -4 & 55{,}500 \end{array}\right]$. We will begin by zeroing out the rest of the first column.

$$\left[\begin{array}{ccc|c} 1 & 1 & 1 & 7500 \\ 0 & 1 & -1 & 500 \\ 8 & 13 & -4 & 55{,}500 \end{array}\right]$$

$$\begin{array}{r} \\ \\ \text{New } R_3 \rightarrow \end{array} \left[\begin{array}{ccc|c} 1 & 1 & 1 & 7500 \\ 0 & 1 & -1 & 500 \\ 0 & 5 & -12 & -4500 \end{array}\right]$$

Multiply R_1 by -8 and add to R_3.
$-8(1) + 8, -8(1) + 13, -8(1) + (-4),$
$-8(7500) + 55{,}500$

The elements in the first column under row 1 are now both 0. Next we need to change the second element of row 3 to a 0. This can be accomplished by multiplying row 2 by -5 and adding it to row 3.

$$\begin{array}{r} \\ \\ \text{New } R_3 \rightarrow \end{array} \left[\begin{array}{ccc|c} 1 & 1 & 1 & 7500 \\ 0 & 1 & -1 & 500 \\ 0 & 0 & -7 & -7000 \end{array}\right]$$

Multiply R_2 *by* -5 and add to R_3.
$-5(0) + 0, -5(1) + 5, -5(-1) + (-12),$
$-5(500) + (-4500)$

The last row is equivalent to the equation $-7z = -7000$, whose solution is $z = 1000$. We substitute this value for z in the equation that is equivalent to row 2, $y - z = 500$.

$$y - (1000) = 500 \quad \text{Substitute 1000 for } z.$$
$$y = 1500 \quad \text{Add 1000.}$$

Now we substitute 1500 for y and 1000 for z in the equation that is equivalent to row 1, $x + y + z = 7500$.

$$x + (1500) + (1000) = 7500 \quad \text{Substitute 1500 for } y \text{ and 1000 for } z.$$
$$x + 2500 = 7500 \quad \text{Combine like terms.}$$
$$x = 5000 \quad \text{Subtract 2500.}$$

Luz invested \$5000 in the fund that earned an 8% profit, \$1500 in the fund that earned a 13% profit, and \$1000 in the fund that lost 4%.

A quick check will verify that the total amount invested $(5000 + 1500 + 1000)$ was 7500, that the amount invested in the fund that earned a 13% profit (\$1500) was \$500 more than the amount invested in the fund that lost 4% (\$1000), and that the total profit $(5000(0.08) + 1500(0.13) + 1000(-0.04))$ was \$555.

Quick Check 6

Javier invested \$10,000 in three bank accounts, which paid 3%, 4%, and 6% annual interest, respectively. He invested \$500 more in the account that paid 6% interest than he invested in the account that paid 4% interest. If Javier earned \$455 from these three accounts in the first year, how much did he invest in each?

Building Your Study Strategy **Making the Best Use of Your Resources, 5**

Classmates Some frequently overlooked resources are the other students who are taking the class with you.

- If you have a quick question about a particular problem, a classmate can help you.
- If your notes are incomplete for a certain day, a classmate may allow you to use their notes to fill in the holes in your own notes.
- If you are forced to miss class, a quick call to a classmate can help you find out what was covered in class and what the homework assignment is.
- You can form a study group with your fellow classmates.
- A classmate can also help to motivate you when you are feeling low.

EXERCISES 3.5

Vocabulary

1. A(n) ________ is a rectangular array of numbers.

2. The numbers in a matrix are called ________.

3. A 3×4 matrix has 3 ________ and 4 ________.

4. The ________ begins with the first element in the first row and moves diagonally down the matrix to the right.

5. An array in which every element below the main diagonal is 0 is in ________ form.

6. Substituting the solution from one row into an equation from another row is called ________.

Solve the system of equations using matrices.

7. $x - 4y = 2$
$-3x + 2y = -16$

8. $x - 2y = -12$
$-2x + 5y = 31$

9. $-x + 6y = 22$
$3x + 7y = 9$

10. $-x + 10y = -12$
$6x - 5y = -38$

11. $2x + 3y = 32$
$4x + 5y = 58$

12. $5x + 2y = 17$
$10x - 3y = -43$

13. $3x + 4y = -1$
$12x - 11y = 104$

14. $6x - 11y = 87$
$18x + 7y = -99$

15. $-4x + y = 22$
$8x - 3y = -50$

16. $-3x + 4y = -21$
$15x + 17y = 105$

17. $8x + 9y = 6$
$-16x - 3y = -22$

18. $6x - 16y = -2$
$-18x + 8y = -24$

19. $2x + 3y = 13$
$5x - 4y = -2$

20. $3x - 2y = 16$
$7x + 5y = 18$

21. $-4x + 9y = 75$
$3x - 7y = -58$

22. $x + 2y = -23$
$6x + y = -39$

23. $11x + 7y = 42$
$19x - 13y = -78$

24. $15x + 8y = 9$
$-10x + 9y = -6$

25. $x + 3y + 2z = 11$
$3x - 2y - z = 4$
$-2x + 4y + 3z = 5$

26. $x - y + z = 1$
$4x + 3y + 2z = 10$
$4x - 2y - 5z = 31$

27. $x + y + z = 3$
$2x + 3y + 4z = 16$
$5x + 4y + 6z = 23$

28. $x + y + z = 2$
$-2x - 3y + 8z = -26$
$5x - 4y - 9z = 20$

29. $2x + 3y - 5z = 27$
$6x - 4y + 7z = 7$
$-4x + 3y + 8z = -16$

30. $3x + 2y + 3z = -21$
$-9x + 4y - 4z = 103$
$6x + y + 5z = -56$

31. $4x + 3y + 8z = -10$
$x - 4y - 7z = 17$
$-2x + 9y + 3z = -51$

32. $5x + 3y + 6z = 131$
$7x - 4y - z = 72$
$-x + 2y - 3z = -10$

33. $4x + 3y + 5z = -40$
$-3x - 6y + 7z = -35$
$5x + 15y + z = 10$

34. $3x + 7y - 8z = 60$
$5x - 4y + 7z = -44$
$-4x + y + 10z = -36$

35. $x + 2y + 2z = 22$
$\frac{1}{2}x - \frac{2}{3}y + \frac{7}{4}z = -7$
$\frac{5}{6}x + \frac{8}{9}y - \frac{3}{2}z = 24$

36. $6x + y + 3z = -40$
$-5x + \frac{3}{2}y - z = 20$
$\frac{9}{2}x + \frac{1}{5}y + \frac{3}{4}z = 19$

37. $x + y + z = 10{,}000$
$x - y = 1000$
$0.04x + 0.05y + 0.06z = 460$

38. $x + y + z = 9500$
$-y + z = 1000$
$0.08x + 0.1y - 0.07z = 155$

Solve the system of four equations in four unknowns using matrices.

39. $a + b + c + d = 7$
$2a - 3b - 3c + 4d = -20$
$-3a + b - 2c - 5d = -5$
$4a - 2b - 5c - 7d = 2$

40. $a + b + c + d = 3$
$-a + 2b + 4c - 3d = -14$
$a - 3b - 8c + 6d = 20$
$2a - b - 2c + 4d = 17$

41. $a + 2b + 3c + 4d = -10$
$3a + 4b - 3c - 2d = 76$
$2a - 3b + 6c - 3d = -3$
$-5a - 6b + c + 2d = -102$

42. $a - 5b + 7c + 9d = 67$
$-a + 2b + 5c + 10d = 75$
$8a + 3b - 9c - d = -41$
$-6a + 5b + c + d = 33$

Solve using matrices.

43. The perimeter of a rectangle is 74 inches. The length of the rectangle is 11 inches more than the width of the rectangle. Find the dimensions of the rectangle.

44. A tennis court is surrounded by a rectangular fence. The perimeter of the fence is 250 feet. The width of the court is 100 feet less than twice its length. Find the dimensions of this court.

45. The cost to attend a movie is $8 for adults, while children are admitted for $5. If a movie was attended by 63 people who paid a total of $366, how many adults and how many children attended the movie?

46. The cost to attend a lecture is $10 for the general public, while college students are admitted for only $3. If the lecture was attended by 245 people who paid a total of $1344, how many college students attended the lecture?

47. Simon invested a total of $4000 in two mutual funds. One fund earned a 4% profit while the other earned a 15% profit. If Simon's total profit was $468, how much did he invest in each mutual fund?

48. Paula invested a total of $1500 in two mutual funds. One fund earned a 12% profit while the other earned a 32% profit. If Paula's total profit was $290, how much did she invest in each mutual fund?

49. Gordon sells dried fruits and nuts at his store. He sells a mixture of dried fruit for $4 per pound, and he sells mixed nuts for $6 per pound. He wants to mix the dried fruit and nuts to create 40 pounds of a mixture that he can sell for $4.75 per pound. How many pounds of dried fruit and how many of mixed nuts should Gordon mix?

50. A farmer sells almonds for $8 per pound and walnuts for $5 per pound. If she wants to create 60 pounds of a mixture of these two types of nuts that she can sell for $6.25 per pound, how many of each type should she use?

51. Tran has a coin jar that contains nickels, dimes, and quarters. There are 194 coins in the jar. There are 5 more dimes than quarters. If the total value of the coins in the jar is $26.70, how many quarters does the jar contain?

52. Marisela's deposit bag contains $5 bills, $10 bills, and $20 bills. There are 280 bills in the bag, and their value is $3780. There are 36 more $20 bills than $10 bills. How many $20 bills are in Marisela's deposit bag?

53. Martha invested a total of $32,000 in three different mutual funds. One fund earned a profit of 10%, while the other two funds had losses of 3% and 2%, respectively. Martha lost a total of $230 from these three funds. If Martha invested $14,000 more in the fund that lost 2% than she did in the fund that lost 3%, find the amount she invested in each fund.

54. Warren invested a total of $40,000 in three different mutual funds. After 1 year the first fund had shown a profit of 15%, the second fund had shown a loss of 12%, and the third fund had shown a profit of 9%. In the first year Warren had made a profit of $3090 from the three funds. If Warren invested $1000 more in the fund that lost 12% than he invested in the fund that made a 15% profit, how much did he initially invest in each fund?

Write the system of equations associated with the given augmented matrix and then make up a word problem associated with the system of equations.

55. $\left[\begin{array}{cc|c} 2 & 2 & 48 \\ 1 & -2 & 3 \end{array}\right]$

56. $\left[\begin{array}{cc|c} 1 & 1 & 92 \\ 8 & 5 & 640 \end{array}\right]$

57. $\left[\begin{array}{ccc|c} 1 & 1 & 1 & 6000 \\ 0.04 & 0.05 & 0.06 & 290 \\ 1 & -1 & 0 & -1000 \end{array}\right]$

58. $\left[\begin{array}{ccc|c} 1 & 1 & 1 & 250 \\ 9 & 8 & 6 & 1970 \\ 0 & -2 & 1 & 0 \end{array}\right]$

Writing in Mathematics

59. ***Solutions Manual**** Write a solutions manual page for the following problem:

Solve the system of equations using matrices.
$3x + 5y = 41$
$4x - 3y = 16$

60. ***Newsletter**** Write a newsletter explaining how to solve a system of three equations in three unknowns using matrices.

***See Appendix B for details and sample answers.**

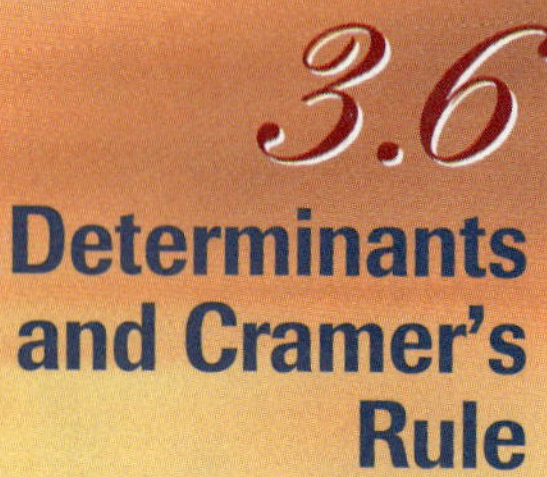

3.6 Determinants and Cramer's Rule

Objectives

1. **Find the determinant of a 2 × 2 matrix.**
2. **Use Cramer's rule to solve systems of two linear equations in two unknowns.**
3. **Find the determinant of a 3 × 3 matrix.**
4. **Use Cramer's rule to solve systems of three linear equations in three unknowns.**
5. **Use Cramer's rule to solve applied problems.**

Determinant of a 2 × 2 Matrix

Objective 1 **Find the determinant of a 2 × 2 matrix.** A matrix that has an equal number of rows and columns is called a **square matrix**. For example, the 2×2 matrix $\begin{bmatrix} 5 & 9 \\ 3 & 2 \end{bmatrix}$ is a square matrix.

Determinant

The **determinant** of a 2×2 square matrix $\begin{bmatrix} a & b \\ c & d \end{bmatrix}$ is a real number that is denoted $\begin{vmatrix} a & b \\ c & d \end{vmatrix}$. The calculation of the determinant of a 2×2 matrix is defined as follows.

$$\begin{vmatrix} a & b \\ c & d \end{vmatrix} = ad - bc$$

The following graphic may be helpful when calculating the determinant of a 2×2 matrix.

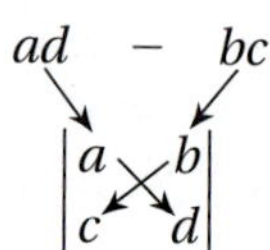

EXAMPLE 1 Evaluate the determinant $\begin{vmatrix} 4 & 7 \\ 13 & 9 \end{vmatrix}$.

Solution

We use the formula $\begin{vmatrix} a & b \\ c & d \end{vmatrix} = ad - bc$ to evaluate the determinant of a 2×2 matrix.

$$\begin{aligned} \begin{vmatrix} 4 & 7 \\ 13 & 9 \end{vmatrix} &= 4 \cdot 9 - 7 \cdot 13 && \text{Apply the formula for the determinant.} \\ &= 36 - 91 && \text{Multiply.} \\ &= -55 && \text{Subtract.} \end{aligned}$$

The determinant is -55.

Quick Check 1
Evaluate the determinant $\begin{vmatrix} 8 & 5 \\ 11 & 12 \end{vmatrix}$.

EXAMPLE 2 Evaluate the determinant $\begin{vmatrix} 6 & -2 \\ 7 & 3 \end{vmatrix}$.

Solution

In this example, one of the elements is a negative number so we must be careful with our signs.

$$\begin{aligned} \begin{vmatrix} 6 & -2 \\ 7 & 3 \end{vmatrix} &= 6 \cdot 3 - (-2) \cdot 7 && \text{Apply the formula for the determinant.} \\ &= 18 + 14 && \text{Multiply.} \\ &= 32 && \text{Add.} \end{aligned}$$

The determinant is 32.

Quick Check 2
Evaluate the determinant $\begin{vmatrix} 2 & -7 \\ 9 & -8 \end{vmatrix}$.

Using Cramer's Rule to Solve a System of Two Linear Equations in Two Unknowns

Objective 2 **Use Cramer's rule to solve systems of two linear equations in two unknowns.** One major application of determinants is in solving systems of linear equations using a technique known as **Cramer's rule**.

Cramer's Rule for Two Linear Equations in Two Unknowns

For the system of equations $\begin{aligned} a_1x + b_1y &= c_1 \\ a_2x + b_2y &= c_2 \end{aligned}$, we define

$$D = \begin{vmatrix} a_1 & b_1 \\ a_2 & b_2 \end{vmatrix}, D_x = \begin{vmatrix} c_1 & b_1 \\ c_2 & b_2 \end{vmatrix}, \text{ and } D_y = \begin{vmatrix} a_1 & c_1 \\ a_2 & c_2 \end{vmatrix}.$$

The solutions to the system of equations are given by

$$x = \frac{D_x}{D} \text{ and } y = \frac{D_y}{D},$$

provided that $D \neq 0$.

Notice that D is the determinant of the matrix containing the coefficients of x and y, one equation per row. To calculate D_x, replace the coefficients of the x-terms in D with the constant terms from the two equations. To calculate D_y, replace the coefficients of the y-terms in D with the constant terms from the two equations.

EXAMPLE 3 Solve the system $\begin{aligned} 3x + 4y &= 5 \\ 2x - 3y &= 26 \end{aligned}$ using Cramer's rule.

Solution

We will begin by calculating D, which is the determinant of the matrix containing the coefficients of the variable terms.

$$D = \begin{vmatrix} 3 & 4 \\ 2 & -3 \end{vmatrix}$$ Substitute 3 for a_1, 4 for b_1, 2 for a_2, and -3 for b_2.

$= 3(-3) - 4(2)$ Evaluate the determinant.

$= -17$ Simplify.

Since $D \neq 0$, we can continue on to calculate D_x. (If $D = 0$, the system is either inconsistent or dependent and must be solved by other methods.) We replace the first column in D with the constant terms from the two equations.

$$D_x = \begin{vmatrix} 5 & 4 \\ 26 & -3 \end{vmatrix}$$ Substitute 5 for c_1, 4 for b_1, 26 for c_2, and -3 for b_2.

$= 5(-3) - 4(26)$ Evaluate the determinant.

$= -119$ Simplify.

Now we calculate D_y by replacing the second column in D by the constant terms from the two equations.

$$D_y = \begin{vmatrix} 3 & 5 \\ 2 & 26 \end{vmatrix}$$ Substitute 3 for a_1, 5 for c_1, 2 for a_2, and 26 for c_2.

$= 3(26) - 5(2)$ Evaluate the determinant.

$= 68$ Simplify.

Once all three determinants have been calculated, we solve for x and y.

$$x = \frac{D_x}{D} = \frac{-119}{-17} = 7 \qquad y = \frac{D_y}{D} = \frac{68}{-17} = -4$$

The solution to this system is $(7, -4)$. We can check our solution by substituting into the two original equations.

Check

$3x + 4y = 5$	$2x - 3y = 26$
$3(7) + 4(-4) = 5$	$2(7) - 3(-4) = 26$
$21 - 16 = 5$	$14 + 12 = 26$
$5 = 5$	$26 = 26$

***Quick Check* 3**
Solve the system
$8x + 3y = 60$
$7x - 6y = 18$
using Cramer's rule.

EXAMPLE 4 Solve the system $\begin{aligned} 6x - 4y &= 18 \\ -x + 6y &= 1 \end{aligned}$ using Cramer's rule.

Solution

We will begin by calculating D, which is the determinant of the matrix containing the coefficients of the variable terms.

$$D = \begin{vmatrix} 6 & -4 \\ -1 & 6 \end{vmatrix} = 6(6) - (-4)(-1) = 32$$

Since $D \neq 0$, we can continue on to calculate D_x and D_y.

$$D_x = \begin{vmatrix} 18 & -4 \\ 1 & 6 \end{vmatrix} = 18(6) - (-4)(1) = 112$$

$$D_y = \begin{vmatrix} 6 & 18 \\ -1 & 1 \end{vmatrix} = 6(1) - 18(-1) = 24$$

Once all three determinants have been calculated, we solve for x and y.

$$x = \frac{D_x}{D} = \frac{112}{32} = \frac{7}{2} \qquad y = \frac{D_y}{D} = \frac{24}{32} = \frac{3}{4}$$

The solution to this system is $\left(\frac{7}{2}, \frac{3}{4}\right)$. The check is left to the reader.

Quick Check 4
Solve the system
$3x + 10y = 6$
$-5x + 8y = 27$
using Cramer's rule.

Determinants of 3 × 3 Matrices

Objective 3 **Find the determinant of a 3 × 3 matrix.**

Determinant of a 3 × 3 Matrix

The determinant of a 3 × 3 matrix is defined in terms of three 2 × 2 determinants.

$$\begin{vmatrix} a_1 & b_1 & c_1 \\ a_2 & b_2 & c_2 \\ a_3 & b_3 & c_3 \end{vmatrix} = a_1 \cdot \begin{vmatrix} b_2 & c_2 \\ b_3 & c_3 \end{vmatrix} - a_2 \cdot \begin{vmatrix} b_1 & c_1 \\ b_3 & c_3 \end{vmatrix} + a_3 \cdot \begin{vmatrix} b_1 & c_1 \\ b_2 & c_2 \end{vmatrix}$$

The determinant $\begin{vmatrix} b_2 & c_2 \\ b_3 & c_3 \end{vmatrix}$ is said to be the **minor** of the element a_1. The minor of an element can be found by crossing out the row and column of the original determinants that contain that element.

$$\begin{vmatrix} a_1 & b_1 & c_1 \\ a_2 & b_2 & c_2 \\ a_3 & b_3 & c_3 \end{vmatrix}$$

Using the formula for the determinant of a 3 × 3 matrix is called **expanding** by the minors of the first column. However, we can expand by the minors of any column or row, by multiplying that row's or that column's elements by their respective minors. To determine whether to add or subtract the product of an element and its minor, begin by adding the number of that element's row to the number of that element's column. If the total is even we add the product, and if the total is odd we subtract the product.

For example, the element a_1 is in the first row and first column. Since $1 + 1 = 2$ is even, we add the product of a_1 and its minor. The element a_2 is in the second row and first column. Since $2 + 1 = 3$ is odd, we subtract the product of a_2 and its minor. For a 3×3 matrix, the following matrix can help you determine whether to add or subtract based on the location of the element you are working with.

$$\begin{bmatrix} + & - & + \\ - & + & - \\ + & - & + \end{bmatrix}$$

In the remaining examples, we will evaluate the determinant of a 3×3 matrix by expanding by the minors of the first column, so we may use the previously listed formula.

EXAMPLE 5 Evaluate the determinant $\begin{vmatrix} 6 & 3 & 5 \\ 2 & -4 & -3 \\ 9 & -7 & 8 \end{vmatrix}$.

Solution

We will apply the formula $\begin{vmatrix} a_1 & b_1 & c_1 \\ a_2 & b_2 & c_2 \\ a_3 & b_3 & c_3 \end{vmatrix} = a_1 \cdot \begin{vmatrix} b_2 & c_2 \\ b_3 & c_3 \end{vmatrix} - a_2 \cdot \begin{vmatrix} b_1 & c_1 \\ b_3 & c_3 \end{vmatrix} + a_3 \cdot \begin{vmatrix} b_1 & c_1 \\ b_2 & c_2 \end{vmatrix}$.

$$\begin{vmatrix} 6 & 3 & 5 \\ 2 & -4 & -3 \\ 9 & -7 & 8 \end{vmatrix} = 6 \cdot \begin{vmatrix} -4 & -3 \\ -7 & 8 \end{vmatrix} - 2 \cdot \begin{vmatrix} 3 & 5 \\ -7 & 8 \end{vmatrix} + 9 \cdot \begin{vmatrix} 3 & 5 \\ -4 & -3 \end{vmatrix}$$

Rewrite in terms of products of elements of the first column and their minors.

$$= 6[(-4)(8) - (-3)(-7)] - 2[3(8) - 5(-7)] + 9[3(-3) - 5(-4)]$$

Evaluate the three minors.

$$= 6(-53) - 2(59) + 9(11)$$

Simplify each minor.

$$= -318 - 118 + 99$$

Multiply.

$$= -337$$

Simplify.

So, $\begin{vmatrix} 6 & 3 & 5 \\ 2 & -4 & -3 \\ 9 & -7 & 8 \end{vmatrix} = -337$.

***Quick Check* 5**

Evaluate the determinant $\begin{vmatrix} 3 & 10 & 12 \\ -4 & -6 & 1 \\ 2 & 5 & 7 \end{vmatrix}$.

Using Cramer's Rule to Solve a System of Three Linear Equations in Three Unknowns

Objective 4 Use Cramer's rule to solve systems of three linear equations in three unknowns. We now advance to Cramer's rule for a system of three linear equations in three unknowns.

Cramer's Rule for Three Linear Equations in Three Unknowns

For the system of equations $\begin{aligned} a_1x + b_1y + c_1z &= d_1 \\ a_2x + b_2y + c_2z &= d_2 \\ a_3x + b_3y + c_3z &= d_3 \end{aligned}$, we define

$$D = \begin{vmatrix} a_1 & b_1 & c_1 \\ a_2 & b_2 & c_2 \\ a_3 & b_3 & c_3 \end{vmatrix},\ D_x = \begin{vmatrix} d_1 & b_1 & c_1 \\ d_2 & b_2 & c_2 \\ d_3 & b_3 & c_3 \end{vmatrix},\ D_y = \begin{vmatrix} a_1 & d_1 & c_1 \\ a_2 & d_2 & c_2 \\ a_3 & d_3 & c_3 \end{vmatrix},\ \text{and } D_z = \begin{vmatrix} a_1 & b_1 & d_1 \\ a_2 & b_2 & d_2 \\ a_3 & b_3 & d_3 \end{vmatrix}.$$

The solutions to the system of equations are given by

$$x = \frac{D_x}{D},\ y = \frac{D_y}{D},\ \text{and } z = \frac{D_z}{D},$$

provided that $D \neq 0$.

Notice that D is the determinant of the matrix containing the coefficients of x, y, and z, one equation per row. To calculate D_x, replace the coefficients of the x-terms in D with the constant terms from the three equations. To calculate D_y, replace the coefficients of the y-terms in D with the constant terms from the three equations. To calculate D_z, replace the coefficients of the z-terms in D with the constant terms from the three equations.

EXAMPLE 6 Solve the system $\begin{aligned} 2x + 4y - z &= 7 \\ 3x + y + 6z &= 43 \\ 4x - 3y - 2z &= 3 \end{aligned}$ using Cramer's rule.

Solution

We will begin by calculating D, which is the determinant of the matrix containing the coefficients of the variable terms.

$$D = \begin{vmatrix} 2 & 4 & -1 \\ 3 & 1 & 6 \\ 4 & -3 & -2 \end{vmatrix}$$ Substitute coefficients.

$$= 2 \cdot \begin{vmatrix} 1 & 6 \\ -3 & -2 \end{vmatrix} - 3 \cdot \begin{vmatrix} 4 & -1 \\ -3 & -2 \end{vmatrix} + 4 \cdot \begin{vmatrix} 4 & -1 \\ 1 & 6 \end{vmatrix}$$ Rewrite in terms of products of elements of the first column and their minors.

$$= 2(16) - 3(-11) + 4(25)$$ Evaluate each minor.

$$= 165$$ Simplify.

Since $D \neq 0$, we can continue on to calculate D_x. We replace the first column in D with the constant terms from the three equations.

$$D_x = \begin{vmatrix} 7 & 4 & -1 \\ 43 & 1 & 6 \\ 3 & -3 & -2 \end{vmatrix}$$ Substitute coefficients.

$$= 7 \cdot \begin{vmatrix} 1 & 6 \\ -3 & -2 \end{vmatrix} - 43 \cdot \begin{vmatrix} 4 & -1 \\ -3 & -2 \end{vmatrix} + 3 \cdot \begin{vmatrix} 4 & -1 \\ 1 & 6 \end{vmatrix}$$ Rewrite in terms of products of elements of the first column and their minors.

$$= 7(16) - 43(-11) + 3(25)$$ Evaluate each minor.

$$= 660$$ Simplify.

Now we calculate D_y by replacing the second column in D by the constant terms from the three equations.

$$D_y = \begin{vmatrix} 2 & 7 & -1 \\ 3 & 43 & 6 \\ 4 & 3 & -2 \end{vmatrix}$$ Substitute coefficients.

$$= 2 \cdot \begin{vmatrix} 43 & 6 \\ 3 & -2 \end{vmatrix} - 3 \cdot \begin{vmatrix} 7 & -1 \\ 3 & -2 \end{vmatrix} + 4 \cdot \begin{vmatrix} 7 & -1 \\ 43 & 6 \end{vmatrix}$$ Rewrite in terms of products of elements of the first column and their minors.

$$= 2(-104) - 3(-11) + 4(85)$$ Evaluate each minor.

$$= 165$$ Simplify.

Now we calculate D_z by replacing the third column in D by the constant terms from the three equations.

$$D_z = \begin{vmatrix} 2 & 4 & 7 \\ 3 & 1 & 43 \\ 4 & -3 & 3 \end{vmatrix}$$ Substitute coefficients.

$$= 2 \cdot \begin{vmatrix} 1 & 43 \\ -3 & 3 \end{vmatrix} - 3 \cdot \begin{vmatrix} 4 & 7 \\ -3 & 3 \end{vmatrix} + 4 \cdot \begin{vmatrix} 4 & 7 \\ 1 & 43 \end{vmatrix}$$ Rewrite in terms of products of elements of the first column and their minors.

$$= 2(132) - 3(33) + 4(165)$$ Evaluate each minor.

$$= 825$$ Simplify.

Once all three determinants have been calculated, we solve for x, y, and z.

$$x = \frac{D_x}{D} = \frac{660}{165} = 4 \qquad y = \frac{D_y}{D} = \frac{165}{165} = 1 \qquad z = \frac{D_z}{D} = \frac{825}{165} = 5$$

The solution to this system is (4, 1, 5). We can check our solution by substituting into the three original equations.

Quick Check 6

Solve the system

$$x + 3y + 2z = -5$$
$$-4x - y + 3z = -46$$
$$6x + 5y + 4z = 28$$

using Cramer's rule.

Check

$2x + 4y - z = 7$	$3x + y + 6z = 43$	$4x - 3y - 2z = 3$
$2(4) + 4(1) - (5) = 7$	$3(4) + 1 + 6(5) = 43$	$4(4) - 3(1) - 2(5) = 3$
$8 + 4 - 5 = 7$	$12 + 1 + 30 = 43$	$16 - 3 - 10 = 3$
$7 = 7$	$43 = 43$	$3 = 3$

EXAMPLE 7 Solve the system
$$\begin{aligned} 3x + 2y - 3z &= -12 \\ 2x - 5y + 4z &= 35 \\ x - y - 8z &= -27 \end{aligned}$$
using Cramer's rule.

Solution

We will begin by calculating D, which is the determinant of the matrix containing the coefficients of the variable terms.

$$D = \begin{vmatrix} 3 & 2 & -3 \\ 2 & -5 & 4 \\ 1 & -1 & -8 \end{vmatrix} = 3(44) - 2(-19) + 1(-7) = 163$$

Since $D \neq 0$, we can continue on to calculate D_x, D_y, and D_z.

$$D_x = \begin{vmatrix} -12 & 2 & -3 \\ 35 & -5 & 4 \\ -27 & -1 & -8 \end{vmatrix} = (-12)(44) - 35(-19) + (-27)(-7) = 326$$

$$D_y = \begin{vmatrix} 3 & -12 & -3 \\ 2 & 35 & 4 \\ 1 & -27 & -8 \end{vmatrix} = 3(-172) - 2(15) + 1(57) = -489$$

$$D_z = \begin{vmatrix} 3 & 2 & -12 \\ 2 & -5 & 35 \\ 1 & -1 & -27 \end{vmatrix} = 3(170) - 2(-66) + 1(10) = 652$$

Once all four determinants have been calculated, we solve for x, y, and z.

$$x = \frac{D_x}{D} = \frac{326}{163} = 2 \qquad y = \frac{D_y}{D} = \frac{-489}{163} = -3 \qquad z = \frac{D_z}{D} = \frac{652}{163} = 4$$

The solution to this system is $(2, -3, 4)$. The check is left to the reader.

Quick Check 7
Solve the system
$$\begin{aligned} x + 2y + 3z &= 6 \\ 2x - y + z &= -3 \\ 4x + 3y - 5z &= -15 \end{aligned}$$
using Cramer's rule.

Cramer's rule, as we've seen, can be quite tedious for solving systems of three linear equations in three unknowns. It can be used to solve quickly for one of the variables without solving for the others. Computer software packages or calculators that calculate determinants can help speed the process up as well.

Using Cramer's Rule to Solve Applied Problems

Objective 5 **Use Cramer's rule to solve applied problems.** We will finish the section by using Cramer's rule to solve an applied problem.

EXAMPLE 8 Amber has \$10 bills and \$20 bills in her pocket. If she has 31 bills that are worth a total of \$530, how many bills of each type does she have?

Solution

Before we solve this system of equations, let's look at a table that contains all of the pertinent information for this problem.

	Number of Bills	Value/Bill (\$)	Money in Pocket (\$)
\$10 Bills	x	10	$10x$
\$20 Bills	y	20	$20y$
Total	31		530

The number of \$10 bills plus the number of \$20 bills must be 31, so the first equation in the system is $x + y = 31$. The amount of money in \$10 bills plus the amount of money in \$20 bills must total \$530, so the second equation in the system is $10x + 20y = 530$.

To solve the system $\begin{matrix} x + y = 31 \\ 10x + 20y = 530 \end{matrix}$, we will begin by calculating D, which is the determinant of the matrix containing the coefficients of the variable terms.

$$D = \begin{vmatrix} 1 & 1 \\ 10 & 20 \end{vmatrix} = 1(20) - 1(10) = 10$$

Since $D \neq 0$, we can continue on to calculate D_x and D_y. To calculate D_x, we replace the first column in D with the constants from the two equations. To calculate D_y, we replace the second column in D with the constants from the two equations.

$$D_x = \begin{vmatrix} 31 & 1 \\ 530 & 20 \end{vmatrix} = 31(20) - 1(530) = 90$$

$$D_y = \begin{vmatrix} 1 & 31 \\ 10 & 530 \end{vmatrix} = 1(530) - 31(10) = 220$$

Once all three determinants have been calculated, we solve for x and y.

$$x = \frac{D_x}{D} = \frac{90}{10} = 9 \qquad y = \frac{D_y}{D} = \frac{220}{10} = 22$$

Amber has 9 \$10 bills and 22 \$20 bills. A quick check by the reader will verify that these 31 bills are worth \$530.

***Quick Check* 8**
Linda sold 43 items at a craft fair. Some of the items were priced at \$48, and the rest were priced at \$20. If Linda took in \$1280 at the craft fair, how many of each type of item did she sell?

***Building Your Study Strategy* Making the Best Use of Your Resources, 6**
Internet Resources Consider using the Internet as a resource that will allow you to conduct further research on topics you are learning. There are a great number of Web sites dedicated to mathematics. These sites have alternate explanations, examples, and practice problems. If you find a Web site that helps you with one particular topic, check that site when you are researching another topic. Keep in mind that anyone can create a Web site, so there is no guarantee that the information is correct. If you have any questions, be sure to ask your instructor.

EXERCISES 3.6

Vocabulary

1. A matrix that has an equal number of rows and columns is called a(n) ________.

2. For a 2 × 2 matrix, the formula for the determinant $\begin{vmatrix} a & b \\ c & d \end{vmatrix}$ is ________.

3. A technique for solving systems of linear equations using determinants is ________.

4. When using Cramer's rule to solve a system of linear equations, the determinant D contains the coefficients of the ________ terms.

Evaluate the determinant.

5. $\begin{vmatrix} 5 & 8 \\ 3 & 7 \end{vmatrix}$

6. $\begin{vmatrix} 12 & 7 \\ 6 & 9 \end{vmatrix}$

7. $\begin{vmatrix} 6 & -5 \\ 4 & 8 \end{vmatrix}$

8. $\begin{vmatrix} 10 & 9 \\ 8 & -11 \end{vmatrix}$

9. $\begin{vmatrix} 4 & 6 \\ -7 & -13 \end{vmatrix}$

10. $\begin{vmatrix} 9 & -8 \\ -5 & -15 \end{vmatrix}$

Solve the system of equations using Cramer's rule.

11. $x + 2y = 8$
$4x + 3y = 17$

12. $x + y = 9$
$5x + 4y = 41$

13. $3x + y = 2$
$2x - 7y = -60$

14. $2x + 5y = -1$
$3x - 2y = 27$

15. $2x - 3y = 0$
$3x - 2y = -15$

16. $3x - 2y = 0$
$4x + 5y = -92$

17. $7x + 9y = 54$
$6x - 5y = -30$

18. $3x - 4y = -12$
$-5x + 8y = 20$

19. $5x + 3y = -5$
$-4x + 7y = 192$

20. $7x - 6y = -13$
$8x + 7y = 290$

21. $4x + 2y = 10$
$6x + 5y = 18$

22. $9x - 12y = 8$
$7x - 2y = 5$

23. $3x + 4y = 5$
$-9x + 8y = 50$

24. $3x - 8y = 2$
$5x + 6y = -45$

25. $x + y = 5000$
$0.2x + 0.4y = 1400$

26. $x + y = 40{,}000$
$0.03x + 0.05y = 1700$

27. $x + 3y = 4.4$
$-0.2x + 0.6y = 2.6$

28. $6x + 5y = -0.4$
$3x - 4y = 20.6$

Evaluate the determinant.

29. $\begin{vmatrix} 1 & 2 & 3 \\ 2 & 5 & 1 \\ 3 & 6 & 4 \end{vmatrix}$

30. $\begin{vmatrix} 1 & 4 & 2 \\ 3 & 4 & 6 \\ 5 & 7 & 3 \end{vmatrix}$

31. $\begin{vmatrix} 2 & 3 & -2 \\ 5 & 5 & 1 \\ 4 & -4 & -3 \end{vmatrix}$

32. $\begin{vmatrix} 3 & -6 & -2 \\ 3 & 4 & -5 \\ 8 & 2 & -7 \end{vmatrix}$

33. $\begin{vmatrix} 1 & 3 & 2 \\ -3 & 4 & -7 \\ -2 & -6 & 9 \end{vmatrix}$

34. $\begin{vmatrix} 4 & -9 & 0 \\ -2 & 5 & -7 \\ 5 & -1 & 3 \end{vmatrix}$

Solve the system of equations using Cramer's rule.

35. $3x + 2y + z = 15$
$x - 3y + 2z = 13$
$4x - 5y - z = -4$

36. $2x + 3y + 4z = 16$
$3x - 5y - 6z = 5$
$4x - 5y - z = 10$

37. $x + y + z = -5$
$x - 3y + 5z = 7$
$2x - y - 3z = -9$

38. $2x + y - z = 2$
$-3x - 2y + 4z = 8$
$-2x + 5y + 10z = -17$

FOR EXTRA HELP

MyMathLab

MathXL

Interactmath.com

MathXL Tutorials on CD

Video Lectures on CD

Addison-Wesley Math Tutor Center

Student's Solutions Manual

39. $5x + 4y + 3z = 2$
$3x - 2y - 8z = -40$
$-2x + 3y + 6z = 24$

40. $x + y + z = 0$
$3x - 2y + 4z = -40$
$5x + 6y + 7z = 8$

41. $2x + y + 4z = 0$
$4x - 5y - 8z = 10$
$x - 3y + 6z = 8$

42. $x + 3y + 6z = -4$
$-2x - 9y + 12z = -18$
$5x + 2y + 4z = 19$

43. $x + y + z = 6000$
$-x + y - z = 0$
$0.03x + 0.02y + 0.05z = 170$

44. $x + y + z = 2000$
$3x - y + 2z = 3300$
$0.1x + 0.12y + 0.2z = 280$

45. $x + 2y + 3z = 7$
$\frac{1}{2}x + \frac{3}{4}y - \frac{1}{3}z = -1$
$\frac{2}{3}x + \frac{5}{2}y + 2z = 0$

46. $\frac{1}{2}x + \frac{2}{3}y - \frac{5}{6}z = 13$
$3x + \frac{4}{3}y + \frac{7}{2}z = -32$
$-\frac{5}{2}x + 4y - \frac{3}{4}z = 16$

A **parallelogram** is a four-sided plane figure that has two sets of opposite parallel sides. The area of a parallelogram with vertices at $(0, 0)$, (a, c), (b, d), and $(a + b, c + d)$ is equal to the absolute value of the determinant $\begin{vmatrix} a & b \\ c & d \end{vmatrix}$. Use this fact to find the area of the given parallelogram.

47.

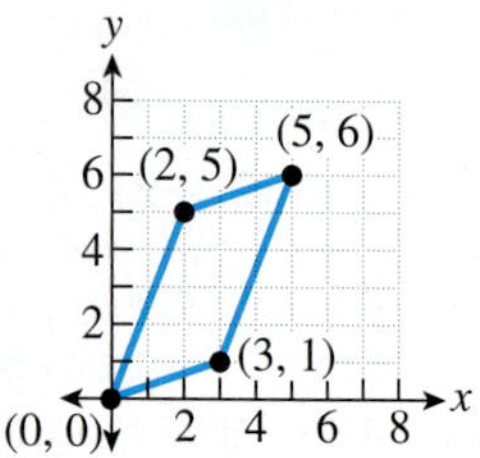

48.

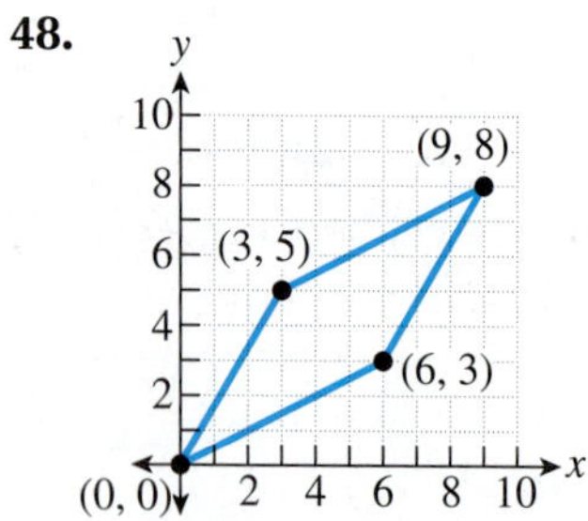

49.

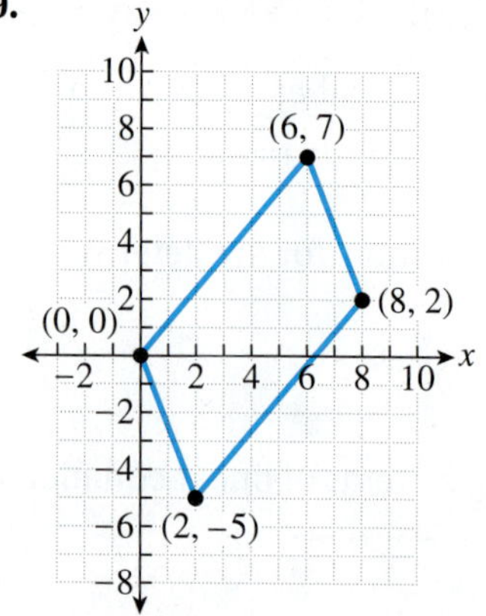

50.

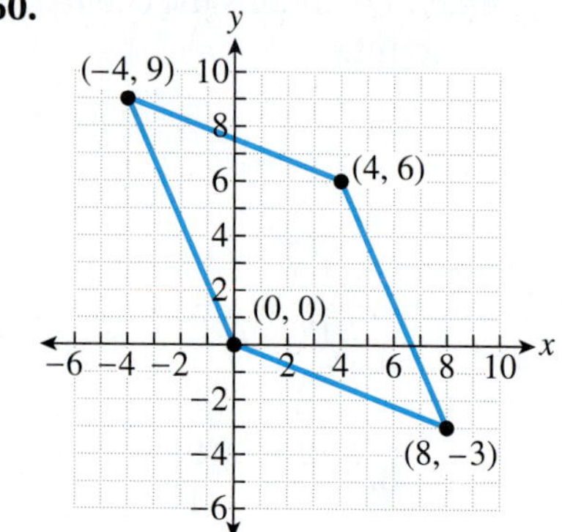

Solve using Cramer's rule.

51. The perimeter of a rectangular yard is 220 feet. The length of the yard is 30 feet more than its width. Find the dimensions of the yard.

52. A family has installed a rectangular swimming pool in their back yard and surrounded it with 130 feet of fencing. The length of the pool area is 10 feet less than twice its width. Find the dimensions of the pool area.

53. The Math Club at a college is selling hats and T-shirts to raise money for a field trip. The club is charging \$10 for a hat and \$8 for a T-shirt. In total, they sold 52 items, raising \$458. How many T-shirts did they sell?

54. Rooms at a hotel cost \$100 per night. The hotel gives a \$20 discount to members of its Preferred Guest club. Last night 73 rooms were occupied, and the hotel took in \$7060. How many Preferred Guest members stayed at the hotel last night?

55. Dan invested a total of $2500 in two bank accounts. The first account paid 3.5% annual interest while the other account paid 4% interest. If Dan earned $97.50 in interest during the first year, how much did he invest in each account?

56. Roseanne invested a total of $5000 in two bank accounts. The first account paid 4% annual interest while the other account paid 6% interest. If Roseanne earned $270 in interest during the first year, how much did she invest in each account?

57. If a coffee shop sells Kona coffee beans for $40 per pound and Latin American coffee beans for $12 per pound, how many pounds of each type of coffee beans should be mixed together to create 70 pounds of a blend that can be sold for $15 per pound?

58. A farmer sells apple cider for $15 per gallon and pomegranate juice for $20 per gallon. If she wants to create 50 gallons of a mixture of these two drinks that she can sell for $18 per gallon, how much of each type should she use?

59. Alycia has a coin jar that contains pennies, nickels, and dimes. There are 210 coins in the jar. There are 15 more nickels than pennies. If the total value of the coins in the jar is $14.65, how many pennies does the jar contain?

60. Dylan's wallet has $1 bills, $5 bills, and $10 bills inside. There are 29 bills in the wallet, and their value is $209. There are 5 more $5 bills than $1 bills. How many $10 bills are in Dylan's wallet?

61. Stephanie invested a total of $2800 in three different mutual funds. The funds earned profits of 8%, 9%, and 15%, respectively. Stephanie earned a profit of $318 from these three funds. If Stephanie invested $200 more in the fund that earned 15% than she did in the fund that earned 9%, find the amount she invested in each fund.

62. Ellen invested a total of $19,000 in three different mutual funds. After one year the first fund had shown a profit of 7%, the second fund had shown a loss of 6%, and the third fund had shown a profit of 11%. In the first year Ellen made a profit of $1180 from the three funds. If Ellen invested twice as much in the fund that made an 11% profit than she invested in the fund that lost 6%, how much did she initially invest in each fund?

Writing in Mathematics

63. ***Solutions Manual**** Write a solutions manual page for the following problem:

Evaluate the determinant. $\begin{vmatrix} 5 & 2 & 3 \\ 4 & -2 & -7 \\ -3 & 1 & 4 \end{vmatrix}$

64. ***Newsletter**** Write a newsletter explaining how to solve a system of two linear equations in two unknowns using Cramer's rule.

***See Appendix B for details and sample answers.**

Chapter 3 Summary

Section 3.1—Topic	Chapter Review Exercises
Solving Systems of Linear Equations by Graphing	1–4
Solving Systems of Linear Equations Using the Substitution Method	5–8
Solving Systems of Linear Equations Using the Addition Method	9–12
Solving Systems of Linear Equations by Any Method	13–18
Section 3.2—Topic	**Chapter Review Exercises**
Solving Applications Using Systems of Linear Equations	19–24
Section 3.3—Topic	**Chapter Review Exercises**
Solving Systems of Linear Inequalities	25–28
Section 3.4—Topic	**Chapter Review Exercises**
Solving Systems of Three Linear Equations in Three Variables Using the Addition Method	29–32
Solving Applied Problems Using a System of Three Linear Equations in Three Variables	33–35
Section 3.5—Topic	**Chapter Review Exercises**
Solving Systems of Linear Equations Using Matrices	36–41
Section 3.6—Topic	**Chapter Review Exercises**
Finding the Determinant of a Matrix	42–47
Solving Systems of Linear Equations Using Cramer's Rule	48–53

Summary of Chapter 3 Study Strategies

You have a great number of resources at your disposal to help you learn mathematics, and your success may depend on how well you use them.

- Your instructor is a valuable resource who can answer your questions and provide advice. In addition to being available during the class period and office hours, some instructors take questions from students right before class begins or just after class ends.
- The tutorial center is another valuable resource. Keep in mind that the goal is for you to understand the material and to be able to solve the problems yourself.
- Many colleges offer short courses or seminars in study skills. Ask an academic counselor about these courses and whether they would be helpful.
- Some other resources discussed in this chapter were the student solutions manual, your classmates, and the Internet.

Chapter 3 Review

Solve the system by graphing. If the system is inconsistent and has no solution, state this. If the system is dependent, write the form of the solution for any real number x. **[3.1]**

1. $y = x + 1$
$y = -3x + 9$

2. $y = \frac{1}{2}x - 2$
$y = -\frac{1}{2}x + 6$

3. $y = -x + 3$
$y = 4x - 7$

4. $2x + 4y = -4$
$2x + 3y = -6$

Solve the system by substitution. If the system is inconsistent and has no solution, state this. If the system is dependent, write the form of the solution for any real number x. **[3.1]**

5. $x = 3y - 4$
$4x + 5y = 35$

6. $y = \frac{1}{2}x - 3$
$-3x + 2y = 10$

7. $x = 2y + 7$
$-3x + 6y = 21$

8. $-2x + y = -9$
$4x - 3y = 19$

Solve the system by addition. If the system is inconsistent and has no solution, state this. If the system is dependent, write the form of the solution for any real number x. **[3.1]**

9. $3x - 4y = -22$
$-3x + 5y = 23$

10. $5x + 2y = 18$
$3x - 4y = -10$

11. $4x + 7y = 39$
$6x - 5y = -50$

12. $2x - y = 1$
$-6x + 3y = -3$

Worked-out solutions to Review Exercises marked with ◯ can be found on page AN-17.

Solve the system by substitution or addition. If the system is inconsistent and has no solution, state this. If the system is dependent, write the form of the solution for any real number x. **[3.1]**

13. $3x - 2y = 23$
$-x + 4y = -21$

14. $x + y = -4$
$3x - 5y = -36$

15. $y = \frac{3}{2}x - 6$
$5x - 4y = 32$

16. $x + y = 3000$
$0.04x + 0.07y = 174$

17. $y = 4x - 3$
$-12x + 3y = -9$

18. $\frac{2}{3}x + \frac{1}{4}y = -4$
$\frac{4}{9}x + \frac{5}{6}y = \frac{8}{3}$

19. The sum of two numbers is 105. One number is 6 less than twice the other number. Find the two numbers. [3.2]

20. A nightclub has a \$10 cover charge. On Wednesday nights, the club has a "Ladies' Night" promotion and women are admitted for only \$1. If the total attendance on the last "Ladies' Night" was 149 patrons and the club collected \$491 in cover charges, how many women were at the club on the last "Ladies' Night"? [3.2]

21. The length of a rectangular room is 8 feet more than its width. Shirley has installed a baseboard around the outside of the room, skipping the doorway which is 3 feet wide. If she installed 81 feet of baseboard, find the length and width of the room. [3.2]

22. Donald deposited \$4800 in two bank accounts, one paying 5% annual interest and the other paying 7% annual interest. In the first year Donald earned \$310 in interest. How much did he deposit in each account? [3.2]

23. One juice blend contains 19% fruit juice, while a second blend is 35% fruit juice. How many milliliters of each type must be mixed together to make 1000 milliliters of a solution that is 25% fruit juice? [3.2]

24. A kayak can travel 18,000 meters upstream in 15 minutes. It only takes 12 minutes for the kayak to travel the same distance downstream. Find the speed of the kayak, in meters per minute, in still water. [3.2]

Solve the system of inequalities. **[3.3]**

25. $x + 3y \le 6$
$-3x + 2y \ge -18$

26. $3x + 4y < 24$
$y \le \frac{1}{4}x + 2$

27. $y > 2x$
$y < -4x + 8$

28. $y \le 6$
$3x - 4y < -12$

Solve the system by addition. **[3.4]**

29. $3x + 2y + z = 13$
$2x + 3y - 2z = 0$
$4x - 7y - 3z = 28$

30. $x + y + z = 5$
$3x + 2y + 4z = 24$
$4x - y + 2z = -5$

31. $-2x + 3y + 5z = -58$
$\frac{1}{2}x + \frac{2}{3}y + \frac{5}{6}z = -2$
$2x + 2y + z = 16$

32. $x + y + z = -2$
$3x - 2y + 6z = 31$
$3y + 4z = 1$

33. The measures of the three angles, A, B, and C, inside a triangle add up to be 180°. The measure of angle C is 6° less than the sum of the measures of angles A and B. If two times the measure of angle A is added to the measure of angle B, the result is 168°. Find the measure of each angle. [3.4]

34. A quilter set up a booth at a craft show. She sold twin-sized quilts for $100, queen-sized quilts for $175, and king-sized quilts for $325. At the show she sold 25 quilts, for a total of $3550. The amount of money she raised from selling twin-sized quilts was $100 more than the amount of money she raised from selling queen-sized quilts. How many twin-sized quilts did she sell? [3.4]

35. Amir invested a total of $50,000 in three different mutual funds. After 1 year the first fund had shown a profit of 10%, the second fund had shown a profit of 16%, and the third fund had shown a loss of 5%. Despite the poor performance of the third fund, in the first year Amir made a profit of $1430. If Amir invested $7000 more in the fund that earned a 10% profit than he invested in the fund that made a 16% profit, how much did he initially invest in each fund? [3.4]

Solve the system of equations using matrices. **[3.5]**

36. $x + 3y = -11$
$-4x - 5y = 9$

37. $x + y = 13$
$3x + 2y = 29$

38. $4x - 5y = 46$
$6x + y = 1$

39. $3x - 2y = 18$
$10x + 7y = -104$

40. $x - y + z = 5$
$-3x + 2y - 4z = -22$
$2x + y + 3z = 25$

41. $2x + y - z = -1$
$3x + 2y + 3z = -40$
$-4x + 4y - 5z = 89$

Evaluate the determinant. **[3.6]**

42. $\begin{vmatrix} 6 & 3 \\ -8 & 7 \end{vmatrix}$

43. $\begin{vmatrix} 2 & 5 \\ 13 & -8 \end{vmatrix}$

44. $\begin{vmatrix} 1 & -7 \\ -10 & -3 \end{vmatrix}$

45. $\begin{vmatrix} 9 & 10 \\ 0 & -13 \end{vmatrix}$

46. $\begin{vmatrix} 2 & 1 & -1 \\ 3 & 2 & 3 \\ -4 & 4 & -5 \end{vmatrix}$

47. $\begin{vmatrix} 1 & -4 & 19 \\ -3 & 7 & -12 \\ 2 & 6 & 13 \end{vmatrix}$

Solve the system of equations using Cramer's rule. **[3.6]**

48. $3x + 2y = -1$
$4x - 7y = 76$

49. $x - y = 5$
$-3x + 2y = -8$

50. $5x - 2y = 0$
$4x + 9y = 159$

51. $7x + 6y = -30$
$5x + 8y = -1$

52. $x + y + z = -5$
$3x + 2y + 3z = -9$
$4x - y + 5z = 7$

53. $x + 2y + 3z = 0$
$3x - y - 2z = 15$
$6x + 9y - 4z = -80$

Chapter 3 Test

Test solutions are found on the enclosed CD.

If the system is inconsistent and has no solution, state this. If the system is dependent, write the form of the solution for any real number x.

1. Solve the system by graphing.
$y = -3x + 6$
$y = -2x + 5$

2. Solve the system by substitution.
$y = 4x - 9$
$3x - 4y = -29$

3. Solve the system by addition.
$3x + 8y = 27$
$7x - 6y = -85$

Solve the system by the method of your choice.

4. $-\frac{1}{2}x + \frac{3}{4}y = 15$
$\frac{2}{3}x - \frac{5}{8}y = -14$

5. $10x + 2y = 11$
$y = -5x + 17$

6. $4x + 7y = 22$
$-6x + 8y = -70$

7. Shannon has a rectangular garden that is surrounded by 80 feet of fencing. The length is 5 feet less than twice its width. Find the dimensions of Shannon's garden.

8. Gail bought shares in two mutual funds, investing a total of \$5000. One fund's shares went up by 8%, while the other fund went up in value by 22%. This was a total profit of \$792 for Gail. How much did she invest in each mutual fund?

9. Solve the system of inequalities.
$y \geq 2x - 7$
$3x + 2y < 6$

10. Solve the system by addition.
$x + 2y + z = -13$
$2x + 5y - 4z = -8$
$-4x + 3y + 2z = -38$

11. A coin jar contains nickels, dimes, and quarters. There are 59 coins in the jar, worth a total of \$7.80. If there are 5 more nickels than dimes in the jar, how many quarters are in the jar?

Solve the system of equations using matrices.

12. $2x + 3y = -15$
$4x - 9y = 75$

13. $x - y + 2z = -21$
$-3x + y + 5z = -43$
$6x - 3y - 4z = 29$

Evaluate the determinant.

14. $\begin{vmatrix} 9 & -8 \\ 4 & 11 \end{vmatrix}$

15. $\begin{vmatrix} 3 & 2 & 6 \\ -2 & 8 & 7 \\ -4 & -9 & 10 \end{vmatrix}$

Solve the system of equations using Cramer's rule.

16. $2x + y = 25$
$-5x + 6y = -3$

Mathematicians in History

Andrew Wiles

Andrew Wiles is a British-American research mathematician who gained worldwide fame with his 1995 proof of Fermat's Last Theorem, which had been unproven for over 350 years. On his pursuit of the proof of this theorem, Wiles said "*I had this very rare privilege of being able to pursue in my adult life what had been my childhood dream.*"

Write a one-page summary (*or* make a poster) of the life of Andrew Wiles and his accomplishments.

Interesting issues:

- Wiles was the first mathematician to prove Fermat's Last Theorem. What is Fermat's Last Theorem?
- Under what circumstances was Wiles introduced to Fermat's Last Theorem?
- In 1982, Wiles became a professor at what U.S. university?
- Fermat wrote that he had a "truly marvelous" proof of Fermat's Last Theorem. What was his reason for not writing his proof?
- What did Wiles say after he finished writing the proof to Fermat's Last Theorem?
- According to Wiles, now that Fermat's Last Theorem has been proved, what is now the greatest problem for mathematicians?
- As a mathematician, Wiles has an Erdös number of 3. What does that mean?

Stretch Your Thinking Chapter 3

Have you ever wondered how math problems are created? Well, today you will be able to explore the answer to that question by creating your very own math application. You have studied various applications of systems of equation (Section 3.2), and now it's your turn to create an application problem that can be solved using systems of equations.

The first step is to decide upon the context of your problem. You need to make sure to have two unknowns for which you will be solving. Will you be solving for the measure of two angles? Will you be looking for the quantity of two types of coffee beans being mixed together? Will you be looking for the dimensions of a rectangular object? Try to be as creative as possible when thinking about the context.

Once you have decided upon the context, you will need to decide on the values of your two unknowns. Do some research; your problem should contain real-world values. You will then need to create two distinct equations that relate these values to each other. Think about the equations you saw in Section 3.2 and how you can make your own equations.

After you have your two equations, solve the system using one of the methods you learned in this chapter. Did you arrive at the correct values of your unknowns? Have a friend solve your system and see if he/she was able to correctly solve the system. You may need to talk to your instructor if you are having difficulties in making sure that your equations correctly relate your unknowns to each other.

Once you have finalized your equations, you need a well-formed paragraph that describes the context of your problem and that gives your equations in mathematical language. Again, refer to Section 3.2 for examples of wording and paragraph structure. After you feel that your paragraph correctly states the problem you intended, switch with someone in class and see if he/she can solve your application.

You should turn your paragraph, the two equations, and your solution in to your instructor. Have fun!

CHAPTER Four 4

Exponents and Polynomials

The expressions and equations we have examined to this point in the text have been linear. In this chapter we begin to examine expressions and functions that are nonlinear, including polynomials such as $-16t^2 + 32t + 128$. We will learn to add, subtract, and multiply polynomials. After we learn how to simplify expressions containing exponents, we will learn about scientific notation and use it to perform arithmetic calculations.

In the remainder of the chapter we will begin to learn how to solve quadratic equations. A* quadratic equation *is an equation of the form $ax^2 + bx + c = 0$ $(a \neq 0)$, where a, b, and c are real numbers. To solve a quadratic equation, we must be able to factor the quadratic expression $ax^2 + bx + c$. Sections 4.4 through 4.7 focus on factoring techniques.

We finish the chapter by learning to solve quadratic equations and their applications.

Study Strategy **Math Anxiety** *In this chapter we will focus on how to overcome math anxiety, a condition shared by many students. Math anxiety can prevent a student from learning mathematics, and it can seriously affect a student's performance on quizzes and exams.*

4.1 Exponents

Objectives

1. **Use the properties of exponents.**
2. **Use two or more properties of exponents to simplify an expression.**
3. **Evaluate functions using the rules for exponents.**

We have used exponents to represent repeated multiplication of a particular factor. For example, the expression 3^7 is used to represent a product in which 3 is a factor 7 times.

$$3^7 = 3 \cdot 3 \cdot 3 \cdot 3 \cdot 3 \cdot 3 \cdot 3$$

Recall that in the expression 3^7, the number 3 is called the base and the number 7 is called the exponent.

Objective 1 Use the properties of exponents. In this section we will introduce several properties of exponents and learn how to apply these properties to simplify expressions involving exponents.

Product Rule

Product Rule

For any base x, $x^m \cdot x^n = x^{m+n}$.

When multiplying two expressions with the same base, we keep the base and add the exponents. Consider the example $x^3 \cdot x^7$. We know that $x^3 = x \cdot x \cdot x$ and $x^7 = x \cdot x \cdot x \cdot x \cdot x \cdot x \cdot x$. So $x^3 \cdot x^7$ can be rewritten as $(x \cdot x \cdot x) \cdot (x \cdot x \cdot x \cdot x \cdot x \cdot x \cdot x)$. Since x is being repeated as a factor 10 times, this expression can be rewritten as x^{10}.

$$x^3 \cdot x^7 = x^{3+7} = x^{10}$$

A Word of Caution When multiplying two expressions with the same base, we keep the base and add the exponents. Do not multiply the exponents.

$$x^3 \cdot x^7 = x^{10}, \text{ not } x^{21}$$

EXAMPLE 1 Simplify $x^4 \cdot x^6$.

Solution

Since we are multiplying and the two bases are the same, we keep the base and add the exponents.

$$x^4 \cdot x^6 = x^{4+6} \quad \text{Keep the base, add the exponents.}$$
$$= x^{10} \quad \text{Add.}$$

***Quick Check* 1**
Simplify $y^{13} \cdot y^8$.

EXAMPLE 2 Simplify $(x + 5)^3 \cdot (x + 5)^6$.

Solution

The base for these expressions is the sum $x + 5$.

Quick Check 2
Simplify $(a - b)^8 \cdot (a - b)^9$.

$$\begin{aligned}(x + 5)^3 \cdot (x + 5)^6 &= (x + 5)^{3+6} && \text{Keep the base, add the exponents.}\\ &= (x + 5)^9 && \text{Add.}\end{aligned}$$

EXAMPLE 3 Simplify $(x^4y^7)(x^5y)$.

Solution

When we have a product involving more than one base, we can use the associative and commutative properties of multiplication to simplify the product.

$$\begin{aligned}(x^4y^7)(x^5y) &= x^4y^7x^5y && \text{Write without parentheses.}\\ &= x^4x^5y^7y && \text{Group like bases together, applying the commutative property of multiplication.}\\ &= x^9y^8 && \text{For each base, add the exponents.}\end{aligned}$$

In future problems it is not necessary to show the application of the associative and commutative properties of multiplication. Look for bases that are the same, add their exponents, and write each base to the calculated power.

Quick Check 3
Simplify $(x^{11}y^8)(x^{10}y^{20})$.

Power Rule

The second property of exponents involves raising an expression with an exponent to a power.

Power Rule

For any base x, $(x^m)^n = x^{m \cdot n}$.

In essence, this property tells us that when we raise a power to a power, we keep the base and multiply the exponents. To show why this is true, consider the expression $(x^3)^7$. Whenever we raise an expression to the seventh power, this means that the expression is repeated as a factor seven times. So $(x^3)^7 = x^3 \cdot x^3 \cdot x^3 \cdot x^3 \cdot x^3 \cdot x^3 \cdot x^3$. Applying the first property from this section, $x^3 \cdot x^3 \cdot x^3 \cdot x^3 \cdot x^3 \cdot x^3 \cdot x^3 = x^{3+3+3+3+3+3+3}$ or x^{21}. This exponent could be found by multiplying 3 by 7.

$$(x^3)^7 = x^{3 \cdot 7} = x^{21}$$

A Word of Caution When raising an expression with an exponent to another power, we keep the base and multiply the exponents. Do not add the exponents.

$$(x^3)^7 = x^{21}, \text{ not } x^{10}$$

EXAMPLE 4 Simplify $(x^4)^9$.

Solution

In this example we are raising a power to another power, so we keep the base and multiply the exponents.

$$(x^4)^9 = x^{4 \cdot 9} \quad \text{Keep the base, multiply the exponents.}$$
$$= x^{36} \quad \text{Multiply.}$$

Quick Check 4
Simplify $(x^8)^6$.

Power of a Product Rule

The third property introduced in this section involves raising a product to a power.

Power of a Product Rule

For any bases x and y, $(xy)^n = x^n y^n$.

This property tells us that when we raise a product to a power, we should raise each factor to that power. Consider the expression $(xy)^5$. We know that this can be rewritten as $xy \cdot xy \cdot xy \cdot xy \cdot xy$. Using the associative and commutative properties of multiplication, we can rewrite this expression as $x \cdot x \cdot x \cdot x \cdot x \cdot y \cdot y \cdot y \cdot y \cdot y$, which simplifies to x^5y^5.

EXAMPLE 5 Simplify $(3a)^6$.

Solution

We begin by raising each factor to the sixth power.

$$(3a)^6 = 3^6a^6 \quad \text{Raise each factor to the sixth power.}$$
$$= 729a^6 \quad \text{Simplify } 3^6.$$

Quick Check 5
Simplify $(5x)^3$.

A Word of Caution This property only applies when we raise a product to a power, not a sum or a difference. Although $(xy)^2 = x^2y^2$, we cannot use the same approach to simplify $(x + y)^2$ or $(x - y)^2$.

$$(x + y)^2 \neq x^2 + y^2$$
$$(x - y)^2 \neq x^2 - y^2$$

For example, let $x = 12$ and $y = 5$. In this case $(x + y)^2$ is equal to 17^2 or 289, but $x^2 + y^2$ is equal to $12^2 + 5^2$ or 169. This shows that, in general, $(x + y)^2$ is not equal to $x^2 + y^2$.

Quotient Rule

The next property involves fractions that contain the same base in their numerators and denominators. It also applies to dividing exponential expressions with the same base.

Quotient Rule

For any base x, $\dfrac{x^m}{x^n} = x^{m-n}$ $(x \neq 0)$.

Note that this property has a restriction; it does not apply when the base is 0. This is because division by 0 is undefined. This property tells us that when we are dividing two expressions that have the same base, we subtract the exponent of the denominator from the exponent of the numerator. Consider the expression $\frac{x^7}{x^3}$. This can be rewritten as $\frac{x \cdot x \cdot x \cdot x \cdot x \cdot x \cdot x}{x \cdot x \cdot x}$. From our previous work, we know that this fraction can be simplified by dividing out common factors in the numerator and denominator.

$$\frac{\overset{1}{\cancel{x}} \cdot \overset{1}{\cancel{x}} \cdot \overset{1}{\cancel{x}} \cdot x \cdot x \cdot x \cdot x}{\underset{1}{\cancel{x}} \cdot \underset{1}{\cancel{x}} \cdot \underset{1}{\cancel{x}}} = x^4$$

One way to think of this is that the number of factors in the numerator has been reduced by 3, because there were three factors in the denominator. The property tells us that we can simplify this expression by subtracting the exponents, which produces the same result.

$$\frac{x^7}{x^3} = x^{7-3} = x^4$$

Example 6 Simplify $\frac{y^{11}}{y^6}$. (Assume $y \neq 0$.)

Solution

Since the bases are the same, we keep the base and subtract the exponents.

$$\frac{y^{11}}{y^6} = y^{11-6} \quad \text{Keep the base, subtract the exponents.}$$
$$= y^5 \quad \text{Subtract.}$$

Quick Check 6
Simplify $\frac{x^{28}}{x^7}$. (Assume $x \neq 0$.)

Example 7 Simplify $a^{18} \div a^3$. (Assume $a \neq 0$.)

Solution

Although it is in a different form than the previous example, $a^{18} \div a^3$ is equivalent to $\frac{a^{18}}{a^3}$. We keep the base and subtract the exponents.

$$a^{18} \div a^3 = a^{18-3} \quad \text{Keep the base, subtract the exponents.}$$
$$= a^{15} \quad \text{Subtract.}$$

Quick Check 7
Simplify $x^{17} \div x^7$. (Assume $x \neq 0$.)

A Word of Caution When dividing two expressions with the same base, keep the base and subtract the exponents. Do not divide the exponents.

$$a^{18} \div a^3 = a^{15}, \text{ not } a^6$$

Example 8 Simplify $\frac{12x^{15}y^{11}}{3x^6y}$. (Assume $x, y \neq 0$.)

Solution

Although we will subtract the exponents for the bases x and y, we will not subtract the numerical factors 12 and 3. We simplify numerical factors to lowest terms.

$$\frac{12x^{15}y^{11}}{3x^6y} = 4x^9y^{10}$$ Simplify $\frac{12}{3}$. Keep the variable bases, subtract the exponents.

Quick Check 8
Simplify $\frac{54a^{12}b^7c^{19}}{6ab^6c^{10}}$.
(Assume $a, b, c \neq 0$.)

Zero Exponent Rule

Consider the expression $\frac{2^7}{2^7}$. We know that any number (except 0) divided by itself is equal to 1. The quotient rule also tells us that we can subtract the exponents in this expression, so $\frac{2^7}{2^7} = 2^{7-7} = 2^0$. Since we have shown that $\frac{2^7}{2^7}$ is equal to both 2^0 and 1, then the expression 2^0 must be equal to 1. The following property involves raising a base to the power of zero.

Zero Exponent Rule

For any base x, $x^0 = 1$ $(x \neq 0)$.

Example 9 Simplify 6^0.

Solution

Applying the zero exponent rule, we see that $6^0 = 1$.

A Word of Caution When raising a nonzero base to the power of zero, the result is 1, not 0.

$$6^0 = 1, \text{ not } 0$$

Quick Check 9
Simplify. (Assume $x \neq 0$.)

a) 5^0
b) $(7x)^0$
c) $7x^0$

The expressions $(4x)^0$ and $4x^0$ are different $(x \neq 0)$. In the first expression, the base that is being raised to the power of zero is $4x$. So $(4x)^0 = 1$, since any nonzero base raised to the power of zero is equal to 1. In the second expression, the base that is being raised to the power of zero is x, not $4x$. The expression $4x^0$ simplifies to be $4 \cdot 1$ or 4.

Power of a Quotient Rule

The concluding property in this section deals with raising a quotient to a power. It is similar to the property involving raising a product to a power.

Power of a Quotient Rule

For any bases x and y, $\left(\frac{x}{y}\right)^n = \frac{x^n}{y^n}$ $(y \neq 0)$.

To raise a quotient to a power, we raise both the numerator and the denominator to that power. The restriction $y \neq 0$ is again due to the fact that division by 0 is undefined. Consider the expression $\left(\frac{r}{s}\right)^3$, where $s \neq 0$. The property tells us that this is equivalent to $\frac{r^3}{s^3}$, and here's why.

$$\left(\frac{r}{s}\right)^3 = \frac{r}{s} \cdot \frac{r}{s} \cdot \frac{r}{s}$$ Repeat $\frac{r}{s}$ as a factor three times.

$$= \frac{r \cdot r \cdot r}{s \cdot s \cdot s}$$ Use the definition of multiplication for fractions.

$$= \frac{r^3}{s^3}$$ Rewrite the numerator and denominator using exponents.

A Word of Caution When raising a quotient to a power, raise both the numerator and the denominator to that power. Do not just raise the numerator to that power.

$$\left(\frac{r}{s}\right)^3 = \frac{r^3}{s^3}, \text{ not } \frac{r^3}{s} \quad (\text{Assume } s \neq 0.)$$

EXAMPLE 10 Simplify $\left(\frac{5}{x}\right)^3$. (Assume $x \neq 0$.)

Solution

We begin by raising each factor in the numerator and denominator to the third power, and then we simplify the resulting expression.

$$\left(\frac{5}{x}\right)^3 = \frac{5^3}{x^3}$$ Raise each factor in the numerator and denominator to the third power.

$$= \frac{125}{x^3}$$ Simplify the numerator.

***Quick Check* 10**

Simplify $\left(\frac{7}{c}\right)^2$.

(Assume $c \neq 0$.)

Here is a brief summary of the properties introduced in this section.

Properties of Exponents

1. **Product Rule:** For any base x, $x^m \cdot x^n = x^{m+n}$.
2. **Power Rule:** For any base x, $(x^m)^n = x^{m \cdot n}$.
3. **Power of a Product Rule:** For any bases x and y, $(xy)^n = x^n y^n$.
4. **Quotient Rule:** For any base x, $\frac{x^m}{x^n} = x^{m-n}$ $(x \neq 0)$.
5. **Zero Exponent Rule:** For any base x, $x^0 = 1$ $(x \neq 0)$.
6. **Power of a Quotient Rule:** For any bases x and y, $\left(\frac{x}{y}\right)^n = \frac{x^n}{y^n}$ $(y \neq 0)$.

Objective 2 Use two or more properties of exponents to simplify an expression. To simplify some expressions, we will have to apply two or more properties of exponents.

EXAMPLE 11 Simplify $(x^2)^6(x^5)^4$.

Solution

To simplify this expression, we must use both the power and the product rules.

$$\begin{aligned}(x^2)^6(x^5)^4 &= x^{2\cdot 6}x^{5\cdot 4} && \text{Apply the power rule.}\\ &= x^{12}x^{20} && \text{Multiply.}\\ &= x^{32} && \text{Keep the base, add the exponents.}\end{aligned}$$

***Quick Check* 11**
Simplify $(x^{10})^3(x^4)^8$.

EXAMPLE 12 Simplify $(x^5y^4)^7$.

Solution

When we raise each factor to the seventh power, we are raising a power to a power and therefore multiply the exponents.

$$\begin{aligned}(x^5y^4)^7 &= (x^5)^7(y^4)^7 && \text{Raise each factor to the seventh power.}\\ &= x^{35}y^{28} && \text{Keep the base, multiply the exponents.}\end{aligned}$$

***Quick Check* 12**
Simplify $(a^9b^{12})^7$.

EXAMPLE 13 Simplify $(2x^6y^4)^3(3x^5y^6)^2$.

Solution

In this example we begin by raising each factor to the appropriate power. Then we multiply expressions with the same base.

$$\begin{aligned}(2x^6y^4)^3(3x^5y^6)^2 &= 2^3x^{18}y^{12}\cdot 3^2x^{10}y^{12} && \text{Raise each base to the appropriate power.}\\ &= 8x^{18}y^{12}\cdot 9x^{10}y^{12} && \text{Simplify } 2^3 \text{ and } 3^2.\\ &= 72x^{28}y^{24} && \text{Simplify.}\end{aligned}$$

***Quick Check* 13**
Simplify $(7x^4y)^2(2x^5y^8)^4$.

EXAMPLE 14 Simplify $\left(\dfrac{x^3y^2}{2z^5}\right)^7$. (Assume $z \neq 0$.)

Solution

We begin by raising each factor in the numerator and denominator to the seventh power, and then we simplify the resulting expression.

$$\begin{aligned}\left(\frac{x^3y^2}{2z^5}\right)^7 &= \frac{(x^3)^7(y^2)^7}{(2)^7(z^5)^7} && \text{Raise each factor in the numerator and denominator to the seventh power.}\\ &= \frac{x^{21}y^{14}}{128z^{35}} && \text{Simplify.}\end{aligned}$$

***Quick Check* 14**
Simplify $\left(\dfrac{3a^7}{b^5c^8}\right)^4$.
(Assume $b, c \neq 0$.)

Objective 3 **Evaluate functions using the rules for exponents.** We conclude this section by investigating functions containing exponents.

EXAMPLE 15 For the function $f(x) = x^3$, evaluate $f(-5)$.

Solution

We substitute -5 for x and simplify the resulting expression.

$$f(-5) = (-5)^3 \quad \text{Substitute } -5 \text{ for } x.$$
$$= -125 \quad \text{Simplify.}$$

Quick Check 15
For the function $f(x) = x^4$, evaluate $f(-3)$.

EXAMPLE 16 For the function $f(x) = 6x^7$, evaluate $f(a^5)$.

Solution

We substitute a^5 for x and simplify.

$$f(a^5) = 6(a^5)^7 \quad \text{Substitute } a^5 \text{ for } x.$$
$$= 6a^{5 \cdot 7} \quad \text{Simplify using the power rule.}$$
$$= 6a^{35} \quad \text{Multiply.}$$

Quick Check 16
For the function $f(x) = 10x^3$, evaluate $f(b^9)$.

EXAMPLE 17 The number of feet traveled by a free-falling object in t seconds is given by the function $f(t) = 16t^2$. If a bowling ball is dropped from a helicopter, how far would it fall in 5 seconds?

Solution

To solve this problem we need to find $f(5)$.

$$f(5) = 16 \cdot (5)^2 \quad \text{Substitute 5 for } t.$$
$$= 400 \quad \text{Simplify.}$$

The bowling ball would fall 400 feet in 5 seconds.

Quick Check 17
The volume of a cube, in cubic units, whose edge is x is given by the function $V(x) = x^3$. Use the function to find the volume of a cube whose edge is 6 cm.

Building Your Study Strategy **Math Anxiety, 1** **Cause of Math Anxiety** If you have been avoiding taking a math class, if you panic when asked a mathematical question, or if you feel that you cannot learn mathematics, then you may have a condition known as math anxiety. Like many other anxiety-related conditions, math anxiety may be traced back to an event that first triggered the negative feelings toward mathematics. If you are going to overcome your anxiety, the first step is to understand the cause of your anxiety.

Many students can recall one embarrassing moment in their life that may be the initial cause of their anxiety. Were you ridiculed as a child by a teacher or another student when you couldn't solve a math problem? Were you expected to be a mathematical genius like a parent or older sibling, but were not able to measure up to their reputation? One negative event can start math anxiety, and the journey to overcome math anxiety can begin with a single success. If you are able to complete a homework assignment, or show improvement on a quiz or exam, then celebrate your success. Let this be your vindication and consider your slate to have been wiped clean.

Exercises 4.1

Vocabulary

1. The product rule states that for any base x, ________.

2. The power rule states that for any base x, ________.

3. The power of a product rule states that for any bases x and y, ________.

4. The quotient rule states that for any nonzero base x, ________.

5. The zero exponent rule states that for any nonzero base x, ________.

6. The power of a quotient rule states that for any bases x and y $(y \neq 0)$, ________.

Simplify.

7. $2^3 \cdot 2^4$
8. $6^3 \cdot 6$
9. $x^3 \cdot x^{10}$
10. $a^7 \cdot a^8$
11. $m^{15} \cdot m^{18}$
12. $n^{17} \cdot n^{40}$
13. $x^5 \cdot x^7 \cdot x^2$
14. $x^6 \cdot x^9 \cdot x$
15. $(x - y)^6 \cdot (x - y)^5$
16. $(x + y)^2 \cdot (x + y)^5$
17. $(a + 2)^8 \cdot (a + 2)^{11}$
18. $(b - 8) \cdot (b - 8)^8$
19. $(x^3y^4)(x^2y^6)$
20. $(a^7b^{11})(a^{19}b^3)$
21. $(7m^5n^2)(8m^3n^9)$
22. $(5x^{12}y^9)(14x^{11}y^{23})$
23. $(2^3)^2$
24. $(3^2)^4$
25. $(x^7)^4$
26. $(x^5)^{12}$
27. $(a^9)^8$
28. $(b^{13})^7$
29. $(x^{10})^{12}$
30. $(x^{15})^{15}$
31. $(2x)^4$
32. $(3x)^5$
33. $(4x^3)^2$
34. $(7x^8)^2$
35. $(x^8y^{11})^3$
36. $(x^9y^7)^6$
37. $(-9x^7y)^2$
38. $(5x^{14}y^9)^3$

Simplify. (Assume all variables are nonzero.)

39. $\dfrac{x^{13}}{x^6}$
40. $\dfrac{x^{18}}{x^9}$
41. $d^{24} \div d^8$
42. $c^{18} \div c^{12}$
43. $\dfrac{40x^{10}}{5x^2}$
44. $\dfrac{78y^{22}}{6y^{15}}$
45. $\dfrac{(x-9)^{16}}{(x-9)^5}$
46. $\dfrac{(b+7)^{25}}{(b+7)^9}$
47. $\dfrac{a^{30}b^{21}}{a^5b^9}$
48. $\dfrac{m^{13}n^{18}}{m^{12}n^{12}}$
49. $\dfrac{20x^9y^{15}z^{21}}{4x^3y^{13}z}$
50. $\dfrac{36a^{16}b^{25}c^{36}}{3a^4b^5c^6}$
51. 4^0
52. 10^0
53. -6^0
54. $(-6)^0$
55. $\left(\dfrac{9}{2}\right)^0$
56. $\dfrac{9}{2^0}$
57. $(12x)^0$
58. $(15a^{11}b^{23}c^{45})^0$
59. $\left(\dfrac{3}{4}\right)^3$
60. $\left(\dfrac{x}{5}\right)^4$
61. $\left(\dfrac{a}{b}\right)^{12}$
62. $\left(\dfrac{m}{n}\right)^8$
63. $\left(\dfrac{x^2}{y^6}\right)^9$
64. $\left(\dfrac{y^{11}}{x^7}\right)^4$

Find the missing exponent(s). (Assume all variables are nonzero.)

65. $x^4 \cdot x^? = x^{13}$
66. $a^? \cdot a^7 = a^{23}$
67. $(x^9)^? = x^{54}$
68. $(x^6)^?(x^9)^{13} = x^{165}$
69. $(x^4y^7)^? = x^{48}y^{84}$
70. $(m^?n^?)^8 = m^{24}n^{104}$
71. $\dfrac{x^{21}}{x^?} = x^6$
72. $\dfrac{y^?}{y^5} = y^{19}$

FOR EXTRA HELP

MyMathLab
MathXL
Interactmath.com
MathXL Tutorials on CD
Video Lectures on CD
Tutor Center
Addison-Wesley Math Tutor Center
Student's Solutions Manual

Simplify. (Assume all variables are nonzero.)

73. $(x^5)^4(x^6)^3$

74. $(x^9)^7(x^8)^{12}$

75. $\dfrac{(x^7)^8}{(x^9)^4}$

76. $\dfrac{(x^{11})^{12}}{(x^{13})^4}$

77. $(x^4yz^5)^6(y^9z^4)^7$

78. $(a^7b^{14}c^{21})^8(a^9b^6c^3)^2$

79. $\dfrac{(a^5b^4c^7)^5}{(ab^2c^3)^6}$

80. $\dfrac{(x^{11}y^6z^8)^{10}}{(x^{13}y^7z^2)^4}$

81. $\left(\dfrac{-5a^4}{b^{12}}\right)^2$

82. $\left(\dfrac{x^{16}}{8y^{11}}\right)^3$

83. $\left(\dfrac{a^8b^{15}}{b^6c^7}\right)^8$

84. $\left(\dfrac{x^{10}y^7}{y^3z^7}\right)^6$

Mixed Practice, 85–98

Simplify.

85. $(a^8)^3$

86. $(20y^4)^0$

87. $\left(\dfrac{3x^2}{4y^7}\right)^4$

88. $(x^{16}y^{15}z^3)^5$

89. $\dfrac{x^{18}}{x}$

90. $\dfrac{x^{37}}{x^{37}}$

91. $\left(\dfrac{a^5b^{10}}{3c^7}\right)^5$

92. $c^{17} \cdot c^{17}$

93. $(b^{15})^4$

94. $\dfrac{x^{18}y^{11}}{xy^8}$

95. $(x^{14})^7$

96. $(-2x^9y^8z)^4$

97. $\left(\dfrac{6a^8b^6}{5c^5d^7}\right)^4$

98. $(-7x^4y^5z^6)^3$

Evaluate the given function.

99. $f(x) = x^2, f(7)$

100. $g(x) = x^3, g(4)$

101. $g(x) = x^4, g(-3)$

102. $f(x) = x^8, f(-1)$

103. $f(x) = 5x^6, f(2)$

104. $f(x) = -8x^3, f(4)$

105. $f(x) = x^6, f(a^8)$

106. $f(x) = x^5, f(b^3)$

107. $f(x) = x^3, f(2a^7)$

108. $f(x) = x^4, f(5b^9)$

The number of feet traveled by a free-falling object in t seconds is given by the function $f(t) = 16t^2$.

109. If a ball is dropped from the top of a building, how far would it fall in 4 seconds?

110. How far would a skydiver fall in 9 seconds?

The area of a square, in square units, with side x is given by the function $A(x) = x^2$.

111. Use the function to find the area of a square that is 60 feet long on a side.

112. Use the function to find the area of a square that is 23 meters long on a side.

The area of a circle, in square units, with radius r is given by the function $A(r) = \pi r^2$.

113. Use the function to find the area of the base of a circular storage tank whose radius is 18 feet. Use $\pi \approx 3.14$.

114. Columbus Circle in New York City is a traffic circle, or roundabout. The radius of the inner circle is 107 feet. Use the function to find the area of the inner circle. Use $\pi \approx 3.14$.

Writing in Mathematics

Answer in complete sentences.

115. Which expression is equal to x^9: $(x^5)^4$ or $x^5 \cdot x^4$? Explain your answer.

116. Explain why the power of a product rule, $(xy)^n = x^ny^n$, does not apply to the expression $(x + y)^2$. In other words, why is $(x + y)^2$ not equal to $x^2 + y^2$?

117. Explain, in your own words, why $8^0 = 1$.

118. Explain the difference between the expressions $\dfrac{a^3}{b}$ and $\left(\dfrac{a}{b}\right)^3$.

119. ***Solutions Manual**** Write a solutions manual page for the following problem:

Simplify $(3x^2y^3z^5)^3(4x^5yz^3)^2$.

120. ***Newsletter**** Write a newsletter explaining how to use the six properties of exponents: product rule, power rule, power of a product rule, quotient rule, zero exponent rule, and power of a quotient rule.

***See Appendix B for details and sample answers.**

4.2 Negative Exponents; Scientific Notation

Objectives

1. **Understand negative exponents.**
2. **Use the rules of exponents to simplify expressions containing negative exponents.**
3. **Convert numbers from standard notation to scientific notation.**
4. **Convert numbers from scientific notation to standard notation.**
5. **Perform arithmetic operations with numbers in scientific notation.**
6. **Use scientific notation to solve applied problems.**

Negative Exponents

Objective 1 Understand negative exponents. In this section we introduce the concept of a **negative exponent.**

Negative Exponents

For any nonzero base x, $x^{-n} = \dfrac{1}{x^n}$.

For example, $2^{-3} = \dfrac{1}{2^3}$ or $\dfrac{1}{8}$.

Let's examine this definition. Consider the expression $\dfrac{2^4}{2^7}$. If we apply the property $\dfrac{x^m}{x^n} = x^{m-n}$ from the previous section, we see that $\dfrac{2^4}{2^7}$ simplifies to equal 2^{-3}.

$$\frac{2^4}{2^7} = 2^{4-7} = 2^{-3}$$

If we simplify $\dfrac{2^4}{2^7}$ by dividing out common factors, the result is $\dfrac{1}{2^3}$.

$$\frac{2^4}{2^7} = \frac{\overset{1}{\cancel{2}} \cdot \overset{1}{\cancel{2}} \cdot \overset{1}{\cancel{2}} \cdot \overset{1}{\cancel{2}}}{\underset{1}{\cancel{2}} \cdot \underset{1}{\cancel{2}} \cdot \underset{1}{\cancel{2}} \cdot \underset{1}{\cancel{2}} \cdot 2 \cdot 2 \cdot 2} = \frac{1}{2^3}$$

Since $\dfrac{2^4}{2^7}$ is equal to both 2^{-3} and $\dfrac{1}{2^3}$, $2^{-3} = \dfrac{1}{2^3}$.

A Word of Caution Raising a positive base to a negative exponent is not the same as raising the opposite of that base to a positive power.

$$2^{-3} = \frac{1}{2^3}, \text{ not } (-2)^3$$

EXAMPLE 1 Rewrite 5^{-2} without using negative exponents and simplify.

Solution

When raising a number to a negative exponent, we begin by rewriting the expression without negative exponents. We finish by raising the base to the appropriate positive power.

$$5^{-2} = \frac{1}{5^2} \quad \text{Rewrite without negative exponents.}$$
$$= \frac{1}{25} \quad \text{Raise 5 to the second power.}$$

Quick Check 1
Rewrite 2^{-5} without using negative exponents and simplify.

EXAMPLE 2 Rewrite the expression $7x^{-8}$ without using negative exponents. (Assume $x \neq 0$.)

Solution

The base in this example is x. The exponent does not apply to the 7 since there are no parentheses. When we rewrite the expression without a negative exponent, the number 7 is unaffected.

$$7x^{-8} = 7\left(\frac{1}{x^8}\right) = \frac{7}{x^8}$$

Note the difference between the expressions $7x^{-8}$ and $(7x)^{-8}$. In the first expression, the base x is being raised to the -8th power, leaving the 7 in the numerator when the expression is rewritten without negative exponents. The base in the second expression is $7x$, so we rewrite $(7x)^{-8}$ as $\frac{1}{(7x)^8}$.

Quick Check 2
Rewrite the expression x^3y^{-6} without using negative exponents. (Assume $y \neq 0$.)

EXAMPLE 3 Rewrite the expression $\frac{a^6}{b^{-2}}$ without using negative exponents. (Assume $b \neq 0$.)

Solution

In this expression, a factor in the denominator has a negative exponent.

$$\frac{a^6}{b^{-2}} = \frac{a^6}{\left(\frac{1}{b^2}\right)} \quad \text{Rewrite } b^{-2} \text{ as } \frac{1}{b^2}.$$
$$= a^6 \cdot \frac{b^2}{1} \quad \text{To divide by } \frac{1}{b^2}\text{, multiply by its reciprocal.}$$
$$= a^6b^2 \quad \text{Simplify.}$$

If a base with a negative exponent is a factor in the denominator of a fraction, we move it to the numerator to join the other factor and change its exponent to a positive exponent.

Quick Check 3
Rewrite $\frac{x^6}{y^{-9}}$ without using negative exponents. (Assume $y \neq 0$.)

EXAMPLE 4 Rewrite the expression $\frac{a^5b^{-4}}{c^{-6}d^3}$ without using negative exponents. (Assume all variables are nonzero.)

Solution

Of the four factors in this example, two of them (b^{-4} and c^{-6}) need to be rewritten without using negative exponents. This can be done by changing the sign of their exponents and moving the factors to the other side of the fraction bar.

$$\frac{a^5b^{-4}}{c^{-6}d^3} = \frac{a^5c^6}{b^4d^3}$$

Quick Check 4

Rewrite $\frac{x^{-5}y^{-2}}{z^{-1}w^7}$ without using negative exponents. (Assume all variables are nonzero.)

Using the Rules of Exponents with Negative Exponents

Objective 2 **Use the rules of exponents to simplify expressions containing negative exponents.** All the properties of exponents developed in the last section hold true for negative exponents. When we are simplifying expressions involving negative exponents, there are two general routes we can take. We can choose to apply the appropriate property first, and then we can rewrite the expression without negative exponents. Or, in some circumstances, it will be more convenient to rewrite the expression without negative exponents before attempting to apply the appropriate property.

EXAMPLE 5 Simplify the expression $x^8 \cdot x^{-3}$. Write the result without using negative exponents. (Assume $x \neq 0$.)

Solution

This example uses the product rule $x^m \cdot x^n = x^{m+n}$.

$$\begin{aligned} x^8 \cdot x^{-3} &= x^{8+(-3)} && \text{Keep the base, add the exponents.} \\ &= x^5 && \text{Simplify.} \end{aligned}$$

Quick Check 5

Simplify the expression $x^{-20} \cdot x^8$. Write the result without using negative exponents. (Assume $x \neq 0$.)

An alternate approach would be to rewrite the expression without negative exponents first.

$$\begin{aligned} x^8 \cdot x^{-3} &= \frac{x^8}{x^3} && \text{Rewrite without negative exponents.} \\ &= x^5 && \text{Keep the base, subtract the exponents.} \end{aligned}$$

Use the approach that seems clearer to you.

EXAMPLE 6 Simplify the expression $(y^7)^{-5}$. Write the result without using negative exponents. (Assume $y \neq 0$.)

Quick Check 6

Simplify the expression $(x^{-3})^{-8}$. Write the result without using negative exponents. (Assume $x \neq 0$.)

Solution

This example uses the power rule $(x^m)^n = x^{m \cdot n}$.

$$\begin{aligned} (y^7)^{-5} &= y^{-35} && \text{Keep the base, multiply the exponents.} \\ &= \frac{1}{y^{35}} && \text{Rewrite without using negative exponents.} \end{aligned}$$

EXAMPLE 7 Simplify the expression $(5a^{-6}b^7)^{-2}$. Write the result without using negative exponents. (Assume $a, b \neq 0$.)

Solution

This example uses the power of a product rule $(xy)^n = x^n y^n$.

$$\begin{aligned}(5a^{-6}b^7)^{-2} &= 5^{-2}a^{-6(-2)}b^{7(-2)} && \text{Raise each factor to the power of } -2.\\ &= 5^{-2}a^{12}b^{-14} && \text{Simplify each exponent.}\\ &= \frac{a^{12}}{5^2b^{14}} && \text{Rewrite without using negative exponents.}\\ &= \frac{a^{12}}{25b^{14}} && \text{Simplify } 5^2.\end{aligned}$$

Quick Check 7

Simplify the expression $(2a^5b^{-2}c)^{-4}$. Write the result without using negative exponents. (Assume $a, b, c \neq 0$.)

EXAMPLE 8 Simplify the expression $\frac{x^4y^9z^3}{x^{18}y^5z^6}$. Write the result without using negative exponents. (Assume $x, y, z \neq 0$.)

Solution

This example uses the quotient rule $\frac{x^m}{x^n} = x^{m-n}$.

$$\begin{aligned}\frac{x^4y^9z^3}{x^{18}y^5z^6} &= x^{-14}y^4z^{-3} && \text{Keep the base, subtract the exponents.}\\ &= \frac{y^4}{x^{14}z^3} && \text{Rewrite without using negative exponents.}\end{aligned}$$

Quick Check 8

Simplify the expression $\frac{x^{13}y^{12}z^{11}}{x^8y^{16}z^{24}}$. Write the result without using negative exponents. (Assume $x, y, z \neq 0$.)

EXAMPLE 9 Simplify the expression $\left(\frac{x^2y^3}{z^4}\right)^{-2}$. Write the result without using negative exponents. (Assume $x, y, z \neq 0$.)

Solution

This example uses the power of a quotient rule $\left(\frac{x}{y}\right)^n = \frac{x^n}{y^n}$.

$$\begin{aligned}\left(\frac{x^2y^3}{z^4}\right)^{-2} &= \frac{x^{-4}y^{-6}}{z^{-8}} && \text{Raise each factor in the numerator and denominator to the power of } -2.\\ &= \frac{z^8}{x^4y^6} && \text{Rewrite without using negative exponents.}\end{aligned}$$

Quick Check 9

Simplify the expression $\left(\frac{a^8}{b^2c^7}\right)^{-5}$. Write the result without using negative exponents. (Assume $a, b, c \neq 0$.)

Notice in the previous example that $\left(\frac{x^2y^3}{z^4}\right)^{-2} = \frac{z^8}{x^4y^6}$, which is equivalent to $\left(\frac{z^4}{x^2y^3}\right)^2$. This leads to an important property involving quotients that are raised to a negative power.

For any $a, b \neq 0$, $\left(\frac{a}{b}\right)^{-n} = \left(\frac{b}{a}\right)^{n}$.

Scientific Notation

Objective 3 Convert numbers from standard notation to scientific notation.
Scientific notation is used to represent numbers that are very large, such as 93,000,000, or very small, such as 0.0000324.

Rewriting a Number in Scientific Notation

To convert a number to scientific notation, we rewrite it in the form $a \times 10^{b}$ where $1 \leq a < 10$ and b is an integer.

First, consider the following table listing several powers of 10.

10^5	10^4	10^3	10^2	10^1	10^0	10^{-1}	10^{-2}	10^{-3}	10^{-4}	10^{-5}
100,000	10,000	1000	100	10	1	0.1	0.01	0.001	0.0001	0.00001

Notice that all of the positive powers of 10 are numbers that are 10 or higher. All of the negative powers of 10 are numbers between 0 and 1.

To convert a number to scientific notation, we first move the decimal point so that it immediately follows the first nonzero digit in the number. Count the number of decimal places that the decimal point moves. This will give us the power of 10 when the number is written in scientific notation. If the original number was 10 or larger, then the exponent is positive, but if the original number was less than 1, then the exponent is negative.

EXAMPLE 10 Convert 0.000043 to scientific notation.

Solution

Move the decimal point so that it follows the first nonzero digit, which is 4.

0 000043

To do this, we must move the decimal point 5 places to the right. Since 0.000043 is less than 1, this exponent must be negative.

$0.000043 = 4.3 \times 10^{-5}$

EXAMPLE 11 Convert 67,000,000 to scientific notation.

Quick Check 10
Convert to scientific notation.

a) 0.00965
b) 35,200,000,000

Solution

Move the decimal point so that it follows the digit 6. To do this, we must move the decimal point 7 places to the left. Since 67,000,000 is 10 or larger, the exponent will be a positive 7.

$67{,}000{,}000 = 6.7 \times 10^{7}$

Objective 4 **Convert numbers from scientific notation to standard notation.**

EXAMPLE 12 Convert 1.92×10^4 from scientific notation to standard notation.

Solution

$$\begin{aligned} 1.92 \times 10^4 &= 1.92 \times 10{,}000 && \text{Rewrite } 10^4 \text{ as } 10{,}000. \\ &= 19{,}200 && \text{Multiply.} \end{aligned}$$

To multiply a number by 10^4 or 10,000, we move the decimal point 4 places to the right. Note that the power of 10 is positive in this example.

EXAMPLE 13 Convert 2.04×10^{-7} from scientific notation to standard notation.

Solution

$$\begin{aligned} 2.04 \times 10^{-7} &= 2.04 \times 0.0000001 && \text{Rewrite } 10^{-7} \text{ as } 0.0000001. \\ &= 0.000000204 && \text{Multiply.} \end{aligned}$$

To multiply a decimal number by 10^{-7} or 0.0000001, we move the decimal point 7 places to the left. Note that the power of 10 is negative in this example.

If you are unsure about which direction to move the decimal point, try thinking about whether we are making the number larger or smaller. Multiplying by positive powers of 10 makes the number larger, so move the decimal point to the right. Multiplying by negative powers of 10 makes the number smaller, so move the decimal point to the left.

***Quick Check* 11**
Convert from scientific notation to standard notation.

a) 8.7×10^{-3}
b) 5.56×10^9

Objective 5 **Perform arithmetic operations with numbers in scientific notation.** When we are performing calculations involving very large and/or very small numbers, using scientific notation can be very convenient.

EXAMPLE 14 Multiply $(2.2 \times 10^7)(2.8 \times 10^{13})$. Express your answer using scientific notation.

Solution

When we are multiplying two numbers that are written in scientific notation, we may first multiply the two decimal numbers. We then multiply the powers of 10 using the product rule $x^m \cdot x^n = x^{m+n}$.

$$\begin{aligned} (2.2 \times 10^7)(2.8 \times 10^{13}) &= (2.2)(2.8)(10^7)(10^{13}) && \text{Reorder the factors.} \\ &= 6.16 \times 10^{20} && \text{Multiply decimal numbers. Add the exponents for the base 10.} \end{aligned}$$

***Quick Check* 12**
Multiply $(3.2 \times 10^6)(2.6 \times 10^{10})$. Express your answer using scientific notation.

Using Your Calculator The TI-84 can help us perform calculations with numbers in scientific notation. To enter a number in scientific notation, we use the second function **EE** above the key labeled [,]. For example, to enter the number 2.2×10^7, type 2.2 [2nd] [,] 7. Here is the screen shot showing how to multiply $(2.2 \times 10^7)(2.8 \times 10^{13})$.

```
2.2E7*2.8E13
          6.16E20
```

EXAMPLE 15 Multiply $(4.3 \times 10^{11})(6.2 \times 10^{20})$. Express your answer using scientific notation.

Solution

$$(4.3 \times 10^{11})(6.2 \times 10^{20}) = (4.3)(6.2)(10^{11})(10^{20})$$ Reorder the factors.

$$= 26.66 \times 10^{31}$$ Multiply decimal numbers. Add the exponents for the base 10.

Although this answer is correct, it is not in scientific notation, as the number 26.66 has two digits to the left of the decimal point. In order to rewrite 26.66 as 2.666, we move the decimal point one place to the left, which will increase the power of 10 from 31 to 32. Here is another way to think of it.

$$26.66 \times 10^{31} = (2.666 \times 10^1) \times 10^{31} = 2.666 \times 10^{32}$$

So, $(4.3 \times 10^{11})(6.2 \times 10^{20}) = 2.666 \times 10^{32}$.

***Quick Check* 13**
Multiply $(5.5 \times 10^3)(4.9 \times 10^{-19})$. Express your answer using scientific notation.

EXAMPLE 16 Divide $(1.2 \times 10^{-3}) \div (4.8 \times 10^8)$. Express your answer using scientific notation.

Solution

To divide numbers that are written in scientific notation, we begin by dividing the decimal numbers. We then divide the powers of 10 separately, using the property $\frac{x^m}{x^n} = x^{m-n}$.

$$(1.2 \times 10^{-3}) \div (4.8 \times 10^8) = \frac{1.2 \times 10^{-3}}{4.8 \times 10^8}$$ Rewrite as a fraction.

$$= \frac{1.2}{4.8} \times 10^{-3-8}$$ Divide the decimal numbers. Subtract the exponents for the powers of 10.

$$= 0.25 \times 10^{-11}$$ Divide 1.2 by 4.8. Simplify the exponent.

Our answer is not in scientific notation, as the number 0.25 is less than one. In order to rewrite 0.25 as 2.5, we move the decimal point one place to the right. This will decrease the exponent by one from -11 to -12.

$$0.25 \times 10^{-11} = (2.5 \times 10^{-1}) \times 10^{-11} = 2.5 \times 10^{-12}$$

So, $(1.2 \times 10^{-3}) \div (4.8 \times 10^{8}) = 2.5 \times 10^{-10}$.

***Quick Check* 14**
Find $(1.3 \times 10^{-4}) \div (6.5 \times 10^{-21})$. Express your answer using scientific notation.

Applications

Objective 6 **Use scientific notation to solve applied problems.** We finish the section with an applied problem using scientific notation.

EXAMPLE 17 At its farthest point, the planet Jupiter is 510 million miles from the Sun. How many seconds does it take for light from the sun to reach Jupiter, if the speed of light is 1.86×10^5 miles per second?

Solution

To determine the length of time that the light takes, we divide the distance traveled by the rate of speed. To solve this problem, we need to divide the distance of 510 million miles by the speed of light. Written in scientific notation, the distance is 5.1×10^8.

$$\frac{5.1 \times 10^8}{1.86 \times 10^5} \approx 2.7 \times 10^3$$

Divide the decimal numbers, approximate with a calculator. Subtract the exponents for the powers of 10.

The light will reach Jupiter in approximately 2.7×10^3 or 2700 seconds, or 45 minutes.

***Quick Check* 15**
How many seconds does it take for light to travel 651,000,000 miles? (The speed of light is 1.86×10^5 miles per second.)

***Building Your Study Strategy* Math Anxiety, 2 Mathematical Autobiography**
One effective tool for understanding your past difficulties in mathematics, and how those past difficulties could hinder your ability to learn mathematics today, is to write a mathematical autobiography. Write down your successes and your failures, as well as the reasons why you think you succeeded or failed. Go back into your mathematical past as far as you can remember. Write down how friends, relatives, and teachers affected you. Let your autobiography sit for a few days and then read it over. Read it on an analytical level as if another person had written it and detach yourself from it personally. Look for patterns, and think about strategies to reverse the bad patterns and continue any good patterns that you can find.

EXERCISES 4.2

Vocabulary

1. For any nonzero base x, $x^{-n} =$ __________.

2. A number is written in __________ if it is in the form $a \times 10^b$ where $1 \le a < 10$ and b is an integer.

Rewrite the expression without using negative exponents. (Assume all variables represent nonzero real numbers.)

3. 5^{-2}

4. 6^{-2}

5. 2^{-5}

6. 4^{-3}

7. -3^{-3}

8. -2^{-8}

9. a^{-10}

10. x^{-12}

11. $6x^{-2}$

12. $3x^{-8}$

13. $-4m^{-3}$

14. $-10m^{-6}$

15. $\dfrac{1}{x^{-4}}$

16. $\dfrac{1}{x^{-6}}$

17. $\dfrac{9}{x^{-2}}$

18. $\dfrac{x^4}{y^{-8}}$

19. $\dfrac{x^{-5}}{y^{-7}}$

20. $\dfrac{x^{-8}}{y^6}$

21. $\dfrac{-3a^{-2}b^4}{c^6}$

22. $\dfrac{-5x^2y^{-8}}{z^{-5}}$

23. $\dfrac{a^{-7}b^{-6}}{c^{-2}d}$

24. $\dfrac{a^4b^{-5}}{c^2d^{-10}}$

Simplify the expression. Write the result without using negative exponents. (Assume all variables represent nonzero real numbers.)

25. $x^{-4} \cdot x^{12}$

26. $x^9 \cdot x^{-7}$

27. $a^{-3} \cdot a^{-8}$

28. $b^{-13} \cdot b^{-12}$

29. $m^9 \cdot m^{-13}$

30. $n^{-16} \cdot n^5$

31. $(x^3)^{-6}$

32. $(x^9)^{-4}$

33. $(x^{-5})^{-8}$

34. $(x^{-7})^{-3}$

35. $(x^7y^3z^{-4})^{-3}$

36. $(x^{-5}y^{-4}z^3)^{-6}$

37. $(2a^{-4}b^{-10}c^3)^{-5}$

38. $(-3a^2b^{-4}c^{-9})^4$

39. $\dfrac{x^{-3}}{x^{11}}$

40. $\dfrac{x^9}{x^{-6}}$

41. $\dfrac{a^2}{a^{-7}}$

42. $\dfrac{a^{-6}}{a^5}$

43. $\dfrac{y^{-12}}{y^{-5}}$

44. $\dfrac{x^{-2}}{x^{-8}}$

45. $\dfrac{x^{15}}{x^{33}}$

46. $\dfrac{x^5}{x^{19}}$

47. $\dfrac{x^4 \cdot x^{-12}}{x^{-6}}$

48. $\dfrac{x^{-6}}{x^{-13} \cdot x^2}$

49. $\left(\dfrac{x^6}{y^3}\right)^{-4}$

50. $\left(\dfrac{3a^4}{b^7}\right)^{-5}$

51. $\left(\dfrac{a^5b^{-6}}{2c^{-9}d^4}\right)^{-3}$

52. $\left(\dfrac{x^{-6}y^{11}}{wz^{-3}}\right)^{-7}$

Convert the given number to scientific notation.

53. 0.00039

54. 67,000

55. 395,000,000

56. 0.00000261

57. 0.00000000009

58. 4,500,000

Convert the given number to standard notation.

59. 9.2×10^{-5}

60. 7.5×10^5

FOR EXTRA HELP

Interactmath.com

MathXL Tutorials on CD

Video Lectures on CD

Addison-Wesley Math Tutor Center

Student's Solutions Manual

61. 5.927×10^9

62. 3.3×10^{-8}

63. 9.03×10^{-3}

64. 8.0×10^{11}

Perform the following calculations. Express your answer using scientific notation.

65. $(1.6 \times 10^6)(4.4 \times 10^8)$

66. $(3.7 \times 10^{-6})(2.6 \times 10^{-11})$

67. $(1.8 \times 10^{13}) \div (3.6 \times 10^{-7})$

68. $(1.504 \times 10^{-10}) \div (4.7 \times 10^8)$

69. $(6.8 \times 10^{-11})(3.7 \times 10^6)$

70. $(6.3 \times 10^{16}) \div (8.4 \times 10^{-17})$

71. $(4.12 \times 10^{16})(9.3 \times 10^8)$

72. $(7.5 \times 10^{-23})(8.2 \times 10^{-19})$

73. (5,900,000,000,000)(0.0000003)

74. 0.000000000018 ÷ 3,000,000,000,000

75. If a computer can perform a calculation in 0.0000000008 seconds, how long would it take the computer to perform 5,000,000,000 calculations?

76. If a computer can perform a calculation in 0.0000000008 seconds, how many calculations can it perform in 1 minute?

77. The speed of light is 1.86×10^5 miles per second. How far does light from the Sun travel in 1 hour?

78. The speed of light is 1.86×10^5 miles per second. At the farthest point in its orbit, the planet Venus is 67,689,000 miles from the Sun. How many seconds does it take for light from the Sun to reach Venus?

79. If the mass of a typical grain of salt is 5.85×10^{-5} grams, and a container has 737 grams of salt in it, how many grains of salt are in the container?

80. The speed of light is 1.86×10^5 miles per second. How far can light from the Sun travel in 1 year? (This distance is often referred to as a "light-year.")

81. For the quarter ending in June 2006, Microsoft showed total revenues of $\$1.18 \times 10^{10}$ and Apple showed total revenues of $\$4.37 \times 10^9$. What was the combined revenue of the two companies for the quarter that ended in June 2006? (*Source:* Stargeek.com)

82. McDonald's restaurants serve nearly 5.0×10^7 customers daily. How many customers do McDonald's restaurants serve in one year? (*Source:* McDonald's Corporation)

83. A new supercomputer performs 2.8×10^{13} calculations per second. How many calculations can it perform in one hour (3600 seconds)? (*Source:* RIKEN)

84. A new supercomputer performs 2.8×10^{13} calculations per second. How many calculations can it perform in one year (31,536,000 seconds)? (*Source:* RIKEN)

85. The pie chart shows the percent of Starbucks Corporation revenues that come from the United States, as well international revenues. If Starbucks' revenues totaled $\$6.4 \times 10^9$ in 2005, what were the U.S. revenues? (*Source:* Starbucks Corporation)

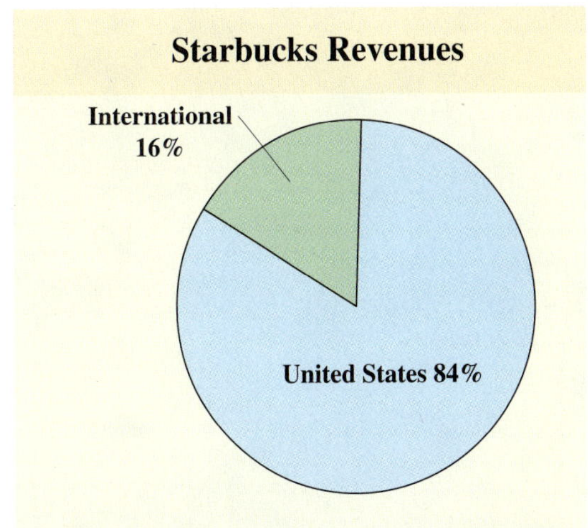

86. The pie chart shows the percent of U.S. college students who received some type of financial aid in 2003–04. If there were approximately 1.91×10^7 U.S. college students in 2003–04, how many of them received some type of financial aid? (*Source:* U.S. Department of Education, National Center for Educational Statistics, National Postsecondary Student Aid Study, 2003–04)

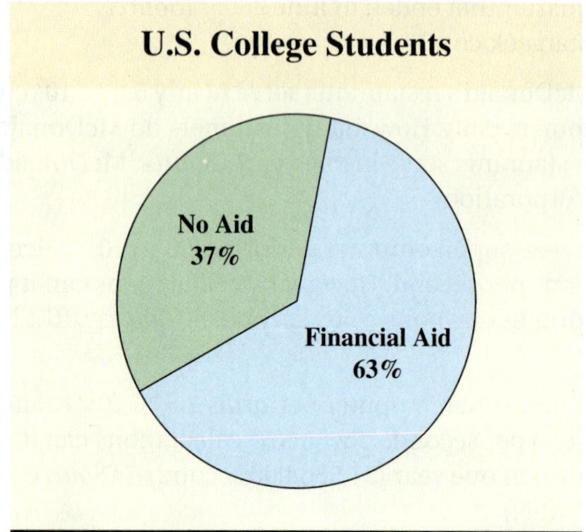

Answer in complete sentences.

87. Explain the difference between the expressions 4^{-2} and -4^2.

88. Describe three real-world examples of numbers that use scientific notation.

89. Explain the statement "When you are performing calculations involving very large and/or very small numbers, it can be quite helpful to use scientific notation."

90. ***Solutions Manual**** Write a solutions manual page for the following problem:

Simplify $(6a^{-3}b^5c^{-4})^{-2}$.

91. ***Newsletter**** Write a newsletter explaining how to convert to and from scientific notation.

***See Appendix B for details and sample answers.**

4.3 Polynomials; Addition, Subtraction, and Multiplication of Polynomials

Objectives

1 **Identify polynomials and understand the vocabulary used to describe them.**
2 **Evaluate polynomials.**
3 **Add and subtract polynomials.**
4 **Multiply polynomials.**
5 **Find special products.**
6 **Understand polynomials in several variables.**

Polynomials in a Single Variable

Objective 1 Identify polynomials and understand the vocabulary used to describe them. Many real-world phenomena cannot be described by linear expressions and linear functions. For example, if a car is traveling at x miles per hour on dry pavement, the distance in feet required for the car to come to a complete stop can be approximated by the expression $0.06x^2 + 1.1x + 0.02$. This nonlinear expression is an example of a **polynomial.**

Polynomials

A **polynomial in a single variable *x*** is a sum of **terms** of the form ax^n, where a is a real number and n is a whole number.

Here are some examples of a polynomial in a single variable x.

$$x^2 - 9x + 14 \qquad 4x^5 - 13 \qquad x^7 - 4x^5 + 8x^2 + 13x$$

An expression is *not* a polynomial if it contains a term with a variable that is raised to a power other than a whole number (such as $x^{3/4}$ or x^{-2}), or if it has a term with a variable in the denominator $\left(\text{such as } \frac{5}{x^3}\right)$.

Degree of a Term

For a polynomial in a single variable, the **degree of a term** is equal to the variable's exponent.

For example, the degree of the term $8x^9$ is 9. A **constant term** is a term that does not contain a variable. The degree of a constant term is 0, because a constant term such as -15 can be rewritten as $-15x^0$. The polynomial $4x^5 - 9x^4 + 2x^3 - 7x - 11$ has five terms, and their degrees are 5, 4, 3, 1, and 0, respectively.

Term	$4x^5$	$-9x^4$	$2x^3$	$-7x$	-11
Degree	5	4	3	1	0

A polynomial with only one term is called a **monomial**. A **binomial** is a polynomial that has two terms, while a **trinomial** is a polynomial that has three terms. (We do not have special names to describe polynomials with four or more terms.) Here are some examples.

Monomial	Binomial	Trinomial
$2x$	$x^2 - 25$	$x^2 - 15x - 76$
$9x^2$	$7x + 3$	$4x^2 + 12x + 9$
$-4x^3$	$5x^3 - 135$	$x^6 + 11x^5 - 26x^4$

EXAMPLE 1 Classify the polynomial as a monomial, binomial, or trinomial. List the degree of each term.

a) $9x^2 - 49$ **b)** $-15x^8$ **c)** $x^3 - 7x^2 + 15x$

Solution

a) $9x^2 - 49$ has two terms, so it is a binomial. The degrees of its two terms are 2 and 0.
b) $-15x^8$ has only one term, so it is a monomial. The degree of this term is 8.
c) $x^3 - 7x^2 + 15x$ is a trinomial as it has three terms. The degrees of those terms are 3, 2, and 1.

***Quick Check* 1**
Classify each polynomial as a monomial, binomial, or trinomial. List the degree of each term.
a) $x^2 + 7x - 30$
b) $16x^4 - 81$
c) $18x^4$

The **coefficient** of a term is the numerical part of a term. When you are determining the coefficient of a term, be sure to include its sign. In the polynomial $8x^3 - 6x^2 - 5x + 13$, the four terms have coefficients 8, −6, −5, and 13 respectively.

Term	$8x^3$	$-6x^2$	$-5x$	13
Coefficient	8	−6	−5	13

EXAMPLE 2 For the trinomial $x^2 - 3x - 12$, determine the coefficient of each term.

Solution

The first term of this trinomial, x^2, has a coefficient of 1. Even though we do not see a coefficient in front of the term, the coefficient is 1 because $x^2 = 1 \cdot x^2$. The second term, $-3x$, has a coefficient of −3. The coefficient of the constant term is −12.

***Quick Check* 2**
For the polynomial $x^3 + 2x^2 - x + 8$, determine the coefficient of each term.

We will write polynomials in **descending order**, writing the term of highest degree first, followed by the term of next highest degree, and so on. For example, the polynomial $3x^2 - 9 - 5x^5 + 7x$ is written in descending order as $-5x^5 + 3x^2 + 7x - 9$. The term of highest degree is called the **leading term**, and its coefficient is called the **leading coefficient**.

Degree of a Polynomial

The degree of the leading term is also called the **degree of the polynomial**.

EXAMPLE 3 Rewrite the polynomial $x^3 + 10 + 2x^4 - x$ in descending order and identify the leading term, the leading coefficient, and the degree of the polynomial.

Quick Check 3
Rewrite the polynomial $3x^2 + 25 - 7x - 6x^3$ in descending order and identify the leading term, the leading coefficient, and the degree of the polynomial.

Solution

To write a polynomial in descending order, we write the terms according to their degree from highest to lowest. In descending order this polynomial is $2x^4 + x^3 - x + 10$. The leading term is $2x^4$, and the leading coefficient is 2. The leading term has degree 4, so this polynomial has degree 4.

Evaluating Polynomials

Objective 2 **Evaluate polynomials.** To **evaluate a polynomial** for a particular value of a variable, we substitute the value for the variable in the polynomial and simplify the resulting expression.

EXAMPLE 4 Evaluate $x^3 - 5x^2 - 10x - 30$ for $x = -3$.

Solution

$$\begin{aligned}
&(-3)^3 - 5(-3)^2 - 10(-3) - 30 && \text{Substitute } -3 \text{ for } x.\\
&= -27 - 5(9) - 10(-3) - 30 && \text{Simplify terms involving exponents.}\\
&= -27 - 45 + 30 - 30 && \text{Multiply.}\\
&= -72 && \text{Simplify.}
\end{aligned}$$

Quick Check 4
Evaluate $2x^4 - 3x^2 + 12x - 28$ for $x = -2$.

A **polynomial function** is a function that is described by a polynomial, such as $f(x) = x^3 - 5x^2 - 5x + 7$. Linear functions of the form $f(x) = mx + b$ are first-degree polynomial functions.

EXAMPLE 5 For the polynomial function $f(x) = x^2 + 6x - 16$, find $f(9)$.

Solution

Recall that the notation $f(9)$ tells us to substitute 9 for x in the function. After substituting we simplify the resulting expression.

$$\begin{aligned}
f(9) &= (9)^2 + 6(9) - 16 && \text{Substitute 9 for } x.\\
&= 81 + 6(9) - 16 && \text{Simplify terms involving exponents.}\\
&= 81 + 54 - 16 && \text{Multiply.}\\
&= 119 && \text{Simplify.}
\end{aligned}$$

Quick Check 5
For the polynomial function $f(x) = x^4 + 3x^3 - 21x$, find $f(4)$.

Adding and Subtracting Polynomials

Objective 3 **Add and subtract polynomials.** Just as we can perform arithmetic operations combining two numbers into one number, we can perform operations combining two polynomials into one polynomial, such as adding and subtracting.

Adding Polynomials

To add two polynomials together, we combine their like terms.

Recall that two terms are like terms if they have the same variables with the same exponents.

EXAMPLE 6 Add $(x^2 - 3x - 28) + (2x^2 - 9x + 16)$.

Solution

To add these polynomials we drop the parentheses and combine like terms.

$(x^2 - 3x - 28) + (2x^2 - 9x + 16) = x^2 - 3x - 28 + 2x^2 - 9x + 16$ Rewrite without parentheses.

$= 3x^2 - 12x - 12$ Combine like terms. $x^2 + 2x^2 = 3x^2$, $-3x - 9x = -12x$, $-28 + 16 = -12$

A Word of Caution The terms $3x^2$ *and* $-12x$ are not like terms, as the variables do not have the same exponents.

Quick Check **6** Add $(x^3 - 5x^2 - 6x + 12) + (x^2 + 16x + 60)$.

To subtract one polynomial from another, such as $(x^2 + 2x - 8) - (3x^2 - 8x + 26)$, we change the sign of each term that is being subtracted and then combine like terms. Changing the sign of each term in the polynomial that is being subtracted is equivalent to applying the distributive property with -1.

EXAMPLE 7 Subtract $(x^2 + 2x - 8) - (3x^2 - 8x + 26)$.

Solution

We remove the parentheses by changing the sign of each term in the polynomial that is being subtracted. We then can combine like terms.

$(x^2 + 2x - 8) - (3x^2 - 8x + 26) = x^2 + 2x - 8 - 3x^2 + 8x - 26$ Distribute to remove parentheses.

$= -2x^2 + 10x - 34$ Combine like terms.

A Word of Caution When subtracting a polynomial from a second polynomial, be sure to change the sign of each term in the polynomial that is being subtracted.

Quick Check **7** Subtract $(4x^2 + 3x + 15) - (6x^2 - 11x + 40)$.

EXAMPLE 8 Given the polynomial functions $f(x) = x^2 + 8x - 20$ and $g(x) = 3x^2 - 10x + 32$, find $f(x) - g(x)$.

Quick Check 8
Given the polynomial functions $f(x) = 3x^2 + 25x - 49$ and $g(x) = -2x^2 - 6x + 34$, find $f(x) - g(x)$.

Solution

We will substitute the appropriate expressions for $f(x)$ and $g(x)$, and simplify the resulting expression.

$$f(x) - g(x) = (x^2 + 8x - 20) - (3x^2 - 10x + 32) \quad \text{Substitute.}$$
$$= x^2 + 8x - 20 - 3x^2 + 10x - 32 \quad \text{Distribute to remove parentheses.}$$
$$= -2x^2 + 18x - 52 \quad \text{Combine like terms.}$$

Objective 4 **Multiply polynomials.** Now that we have learned how to add and subtract polynomials, we will explore multiplication of polynomials. We begin by learning how to multiply monomials.

Multiplying Monomials

To multiply monomials, we begin by multiplying their coefficients, and then we multiply the variable factors using the property of exponents that states for any real number x, $x^m \cdot x^n = x^{m+n}$.

EXAMPLE 9 Find $6x^3 \cdot 5x^2$.

Solution

We multiply the coefficients first, and then we multiply variables.

$$6x^3 \cdot 5x^2 = 30x^{3+2} \quad \text{Multiply coefficients. Multiply variables using the property } x^m \cdot x^n = x^{m+n}.$$
$$= 30x^5 \quad \text{Simplify the exponent.}$$

Quick Check 9
Find $4x \cdot 11x^2$.

Multiplying a Monomial by a Polynomial

We now advance to multiplying a monomial by a polynomial containing two or more terms, such as $4x(3x^2 - x + 8)$. To do this, we use the distributive property $a(b + c) = ab + ac$. To find the product $4x(3x^2 - x + 8)$, we multiply the monomial $4x$ by each term of the polynomial $3x^2 - x + 8$.

Multiplying a Monomial by a Polynomial

To multiply a monomial by a polynomial containing two or more terms, multiply the monomial by each term of the polynomial.

EXAMPLE 10 Multiply $4x(3x^2 - x + 8)$.

Solution

We begin by distributing the monomial $4x$ to each term of the polynomial.

$$\begin{aligned} 4x(3x^2 - x + 8) &= 4x \cdot 3x^2 - 4x \cdot x + 4x \cdot 8 && \text{Distribute } 4x. \\ &= 12x^3 - 4x^2 + 32x && \text{Multiply.} \end{aligned}$$

Although we will continue to show the distribution of the monomial, your goal should be to perform this task mentally.

***Quick Check* 10** Multiply $7x^2(x^2 - 6x - 27)$.

EXAMPLE 11 Multiply $-2x^3(x^4 + 2x^3 - 9x)$.

Solution

Notice that the coefficient of the monomial being distributed is negative. We must distribute $-2x^3$, and multiplying by this term changes the sign of each term in the polynomial.

$$\begin{aligned} -2x^3(x^4 + 2x^3 - 9x) &= (-2x^3)(x^4) + (-2x^3)(2x^3) - (-2x^3)(9x) && \text{Distribute } -2x^3. \\ &= -2x^7 - 4x^6 + 18x^4 && \text{Simplify.} \end{aligned}$$

A Word of Caution When multiplying a polynomial by a term with a negative coefficient, be sure to change the sign of each term in the polynomial.

***Quick Check* 11** Multiply $-6x(x^3 - x^2 - 12x + 60)$.

Multiplying Polynomials

Multiplying Two Polynomials

To multiply two polynomials when each contains two or more terms, we multiply each term in the first polynomial by each term in the second polynomial.

Suppose that we wanted to multiply $(x + 9)(x + 7)$. We could distribute the factor $(x + 9)$ to each term in the second polynomial as follows.

$$(x + 9)(x + 7) = (x + 9) \cdot x + (x + 9) \cdot 7$$

We could then perform the two multiplications by first distributing x in the first product and then distributing 7 in the second product.

$$\begin{aligned} (x + 9)(x + 7) &= (x + 9) \cdot x + (x + 9) \cdot 7 \\ &= x \cdot x + 9 \cdot x + x \cdot 7 + 9 \cdot 7 \\ &= x^2 + 9x + 7x + 63 \\ &= x^2 + 16x + 63 \end{aligned}$$

Each term in the first polynomial is multiplied by each term in the second polynomial.

EXAMPLE 12 Multiply $(x + 6)(x + 4)$.

Solution

We begin by taking the first term, x, in the first polynomial and multiplying it by each term in the second polynomial. We then repeat this for the second term, 6, in the first polynomial.

$(x + 6)(x + 4) = x \cdot x + x \cdot 4 + 6 \cdot x + 6 \cdot 4$	Distribute the term x from the first polynomial, then distribute the term 6.
$= x^2 + 4x + 6x + 24$	Multiply.
$= x^2 + 10x + 24$	Combine like terms. ($4x$ and $6x$)

Quick Check 12
Multiply $(x + 8)(x + 5)$.

We often refer to the process of multiplying a binomial by another binomial as "FOIL." FOIL is an acronym for **F**irst, **O**uter, **I**nner, **L**ast, which describes the four multiplications that occur when we multiply two binomials. Here are the four multiplications that were performed in the previous example.

First	Outer	Inner	Last
$x \cdot x$ $(x + 6)(x + 4)$	$x \cdot 4$ $(x + 6)(x + 4)$	$(x + 6)(x + 4)$ $6 \cdot x$	$(x + 6)(x + 4)$ $6 \cdot 4$

EXAMPLE 13 Multiply $(x - 7)(3x + 4)$.

Solution

When multiplying polynomials we must be careful with our signs. When we distribute the second term of $x - 7$ to the second polynomial, we must distribute a negative 7.

$(x - 7)(3x + 4) = x \cdot 3x + x \cdot 4 - 7 \cdot 3x - 7 \cdot 4$	Distribute using FOIL.
$= 3x^2 + 4x - 21x - 28$	Multiply.
$= 3x^2 - 17x - 28$	Combine like terms.

Quick Check 13
Multiply $(2x - 7)(5x - 8)$.

The term FOIL applies only when we multiply a binomial by another binomial. If we are multiplying a binomial by a trinomial, there are six multiplications that must be performed and the term FOIL cannot be used. Keep in mind that each term in the first polynomial must be multiplied by each term in the second polynomial.

EXAMPLE 14 Given the functions $f(x) = 4x + 5$ and $g(x) = x^2 - 4x + 25$, find $f(x) \cdot g(x)$.

Solution

We will substitute the appropriate expressions for $f(x)$ and $g(x)$ in parentheses, and simplify the resulting expression. We need to multiply each term in the first polynomial by each term in the second polynomial.

$$f(x) \cdot g(x) = (4x + 5)(x^2 - 4x + 25)$$ Substitute for $f(x)$ and $g(x)$.

$$= 4x \cdot x^2 - 4x \cdot 4x + 4x \cdot 25 + 5 \cdot x^2 - 5 \cdot 4x + 5 \cdot 25$$ Distribute.

$$= 4x^3 - 16x^2 + 100x + 5x^2 - 20x + 125$$ Multiply.

$$= 4x^3 - 11x^2 + 80x + 125$$ Combine like terms.

Quick Check 14
Given the functions $f(x) = x - 7$ and $g(x) = x^2 + 9x - 10$, find $f(x) \cdot g(x)$.

In the previous example, we may choose to align our products vertically to collect like terms together.

$$(4x + 5)(x^2 - 4x + 25) = \begin{array}{rrrr} 4x^3 & -16x^2 & +100x & \\ & +5x^2 & -20x & +125 \\ \hline 4x^3 & -11x^2 & +80x & +125 \end{array}$$

Special Products

Objective 5 **Find special products.** We finish this section by examining some special products. The first special product is of the form $(a + b)(a - b)$. In words, this is the product of the sum and the difference of two terms. Here is the multiplication.

$$(a + b)(a - b) = a^2 - ab + ab - b^2$$ Distribute.

$$= a^2 - b^2$$ Combine like terms.

Notice that when we were combining like terms, two of the terms were opposites. This left us with only two terms. We can use this result to assist us anytime we multiply two binomials of the form $(a + b)(a - b)$.

$$(a + b)(a - b) = a^2 - b^2$$

EXAMPLE 15 Multiply $(x + 8)(x - 8)$.

Solution

$$(x + 8)(x - 8) = x^2 - 8^2$$ Multiply using the pattern $(a + b)(a - b) = a^2 - b^2$.

$$= x^2 - 64$$ Simplify.

Quick Check 15
Multiply $(x + 7)(x - 7)$.

EXAMPLE 16 Multiply $(2x - 9)(2x + 9)$.

Solution

Although the difference is listed first, we can still multiply using the same pattern.

$$(2x - 9)(2x + 9) = (2x)^2 - 9^2$$ Multiply using the pattern $(a + b)(a - b) = a^2 - b^2$.

$$= 4x^2 - 81$$ Simplify.

***Quick Check* 16**
Multiply $(3x - 2)(3x + 2)$.

The other special product we will examine is the square of a binomial, such as $(x - 5)^2$ or $(2x + 7)^2$. Here are the patterns for squaring binomials of the form $a + b$ and $a - b$.

$$(a + b)^2 = a^2 + 2ab + b^2$$
$$(a - b)^2 = a^2 - 2ab + b^2$$

Let's derive the first of these patterns.

$$(a + b)^2 = (a + b)(a + b)$$ To square a binomial, we multiply it by itself.

$$= a^2 + ab + ab + b^2$$ Distribute (FOIL).

$$= a^2 + 2ab + b^2$$ Combine like terms.

The second pattern can be derived in the same fashion.

EXAMPLE 17 Multiply $(x - 5)^2$.

Solution

We can use the pattern $(a - b)^2 = a^2 - 2ab + b^2$, substituting x for a and 5 for b.

$$a^2 - 2ab + b^2$$

$$x^2 - 2 \cdot x \cdot 5 + 5^2$$ Substitute x for a and 5 for b.

$$= x^2 - 10x + 25$$ Multiply.

***Quick Check* 17**
Multiply $(2x - 3)^2$.

EXAMPLE 18 Multiply $(2x + 7)^2$.

Solution

We can use the pattern $(a + b)^2 = a^2 + 2ab + b^2$, substituting $2x$ for a and 7 for b.

$$a^2 + 2ab + b^2$$

$$(2x)^2 + 2 \cdot (2x) \cdot 7 + 7^2$$ Substitute $2x$ for a and 7 for b.

$$= 4x^2 + 28x + 49$$ Multiply.

***Quick Check* 18**
Multiply $(x + 9)^2$.

Although the patterns developed for these special products may help to save time when multiplying, keep in mind that we can find products of these types by multiplying as we did earlier in this section. When we square a binomial, we can start by rewriting the expression as the product of the binomial and itself. For example, we could rewrite $(8x - 7)^2$ as $(8x - 7)(8x - 7)$ and then multiply.

Polynomials in Several Variables

Objective 6 Understand polynomials in several variables. While the polynomials that we have examined to this point contained a single variable, some polynomials contain two or more variables. Here are some examples.

$$x^2y^2 + 3xy - 10 \qquad 7a^3 - 8a^2b + 3ab^2 - 15b^3 \qquad x^2 + 2xh + h^2 + 5x + 5h$$

We evaluate polynomials in several variables by substituting values for each variable and simplifying the resulting expression.

EXAMPLE 19 Evaluate the polynomial $x^2 + 4xy - 3y^2$ for $x = 2$ and $y = -7$.

Solution

We will substitute 2 for x and -7 for y. Be careful to substitute the right value for each variable.

$x^2 + 4xy - 3y^2$	
$(2)^2 + 4(2)(-7) - 3(-7)^2$	Substitute 2 for x and -7 for y.
$= 4 + 4(2)(-7) - 3(49)$	Simplify terms involving exponents.
$= 4 - 56 - 147$	Multiply.
$= -199$	Simplify.

Quick Check 19
Evaluate the polynomial $b^2 - 4ac$ for $a = 1$, $b = -9$, and $c = -52$.

For a polynomial in several variables, the degree of a term is equal to the sum of the exponents of its variable factors. For example, the degree of the term $3x^2y^5$ is $2 + 5$ or 7.

EXAMPLE 20 Find the degree of each term in the polynomial $a^6b^2 - 5a^5b^4 + 30a^4b^6$. In addition, find the degree of the polynomial.

Solution

The first term, a^6b^2, has degree 8. The second term, $-5a^5b^4$, has degree 9. The third term, $30a^4b^6$, has degree 10. The degree of a polynomial is equal to the highest degree of any of its terms, so the degree of this polynomial is 10.

Quick Check 20
Find the degree of each term in the polynomial $x^4 - 8x^3y^3 + 21x^2y^6$. In addition, find the degree of the polynomial.

Building Your Study Strategy **Math Anxiety, 3 Test Anxiety** Many students confuse test anxiety with math anxiety. If you feel that you can learn mathematics and you understand the material but "freeze up" on tests, then you may have test anxiety, especially if this happens in other classes besides your math class. Speak to your academic counselor about this. Many colleges offer seminars or short-term classes about how to reduce test anxiety; ask your counselor whether such a course would be appropriate for you.

EXERCISES 4.3

Vocabulary

1. A(n) __________ in a single variable x is a sum of terms of the form ax^n, where a is a real number and n is a whole number.

2. For a polynomial in a single variable x, the __________ of a term is equal to its exponent.

3. A polynomial with only one term is called a(n) __________.

4. A polynomial with two terms is called a(n) __________.

5. A polynomial with three terms is called a(n) __________.

6. The __________ of a term is the numerical part of a term.

7. When the terms of a polynomial are written from highest degree to lowest degree, the polynomial is said to be in __________.

8. The __________ of a polynomial is the term that has the greatest degree.

9. To multiply a polynomial by another polynomial, __________ each term in the first polynomial by each term in the second polynomial.

10. The acronym __________ can remind us how to multiply a binomial by another binomial.

List the degree and coefficient of each term in the given polynomial.

11. $6x^3 - 5x^2 + 8x - 19$

12. $-x^5 + 12x^3 + 7x - 6$

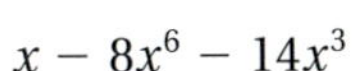

13. $x - 8x^6 - 14x^3$

14. $9x^4 - x^3 + x^2 + 3x + 10$

Identify the given polynomial as a monomial, binomial or trinomial, and state its degree.

15. $x^2 - 7x - 44$

16. $3x^5$

17. $5x - 30$

18. $8x^4 - 32x^2$

19. $-16x^2$

20. $x^4 - 12x^2 - 45$

Evaluate the polynomial for the given value of the variable.

21. $x^2 + 3x + 18$ for $x = 4$

22. $x^3 - 7x^2 - 10x$ for $x = 6$

23. $x^2 + 6x - 16$ for $x = -5$

24. $x^4 + x^3 - 2x^2 + 9x + 40$ for $x = -3$

Evaluate the given polynomial function.

25. $f(x) = x^2 - 10x + 32, f(7)$

26. $f(x) = x^2 + 5x - 36, f(-10)$

27. $g(x) = 2x^3 - 3x^2 + 6x + 24, g(-4)$

28. $g(x) = -x^3 + 3x^2 - 9x + 18, g(6)$

Add or subtract.

29. $(4x^2 + 10x - 15) + (x^2 - 17x + 17)$

30. $(3x^2 - 6x - 41) + (-8x^2 - 6x + 26)$

31. $(x^3 + 2x^2 - 4x - 8) + (7x^2 - 6x + 35)$

32. $(-6x^2 + 17x - 38) + (-2x^3 + 12x^2 - 31x)$

33. $(3x^2 - 8x + 11) - (2x^2 + 7x - 32)$

FOR EXTRA HELP

MyMathLab
MathXL
Interactmath.com
MathXL Tutorials on CD
Video Lectures on CD
Tutor Center
Addison-Wesley Math Tutor Center
Student's Solutions Manual

34. $(x^2 + x - 56) - (5x^2 - 6x + 40)$

35. $(9x^5 - 4x^3 + 5x^2) - (14x^5 - 4x^3 - 5x^2 + 1)$

36. $(x^3 - 6x^2 + 25x) - (9x^2 - 5x + 16)$

37. $(2x^2 + 7x - 24) + (2x^2 + 7x - 24)$

38. $(x^2 - 3x - 18) - (x^2 - 3x - 18)$

Multiply.

39. $4x \cdot 9x^2$

40. $8x^2 \cdot 7x^2$

41. $-2x^4 \cdot 10x^6$

42. $-7x^5(-13x^{10})$

43. $x^3 \cdot 4x^2 \cdot 11x$

44. $-15x^7 \cdot 5x^{12} \cdot 8x^{10}$

Multiply.

45. $3(4x - 9)$

46. $8(2x + 11)$

47. $-4(3x - 7)$

48. $-2(6x + 19)$

49. $5x(8x + 9)$

50. $-2x(7x - 15)$

51. $-x^3(4x^2 - 6x + 9)$

52. $3x^3(2x^4 + 8x^2 - 35)$

Multiply.

53. $(x + 5)(x + 4)$

54. $(x + 11)(x - 3)$

55. $(x - 9)(x + 2)$

56. $(x - 6)(x - 7)$

57. $(3x + 2)(x - 6)$

58. $(5x - 4)(3x + 8)$

59. $(2x + 5)(x^2 - 3x - 10)$

60. $(3x - 4)(9x^2 + 12x + 16)$

61. $(x^2 - 2x + 8)(x^2 + 5x - 24)$

62. $(2x^2 - 3x - 4)(x^2 + 7x + 12)$

For the given functions $f(x)$ and $g(x)$, find $f(x) + g(x)$ and $f(x) - g(x)$.

63. $f(x) = 8x^2 + 9x - 16$, $g(x) = 3x^2 - 9x + 10$

64. $f(x) = 2x^2 - 10x - 23$, $g(x) = -x^2 - 6x + 40$

For the given functions $f(x)$ and $g(x)$, find $f(x) \cdot g(x)$.

65. $f(x) = x - 4$, $g(x) = x + 10$

66. $f(x) = -2x^3$, $g(x) = x^2 - 7x - 30$

67. $f(x) = 3x^6$, $g(x) = -8x^3$

68. $f(x) = 5x - 2$, $g(x) = x^2 + 4x - 16$

Find the special product using the appropriate formula.

69. $(x + 9)(x - 9)$

70. $(x - 5)(x + 5)$

71. $(3x - 2)(3x + 2)$

72. $(4x + 7)(4x - 7)$

73. $(x + 8)^2$

74. $(x - 2)^2$

75. $(3x - 5)^2$

76. $(2x + 13)^2$

Find the missing polynomial.

77. $(2x^2 + 6x - 19) + ? = 8x^2 + 3x - 7$

78. $(5x^2 - 4x - 16) - ? = -4x^2 + 9x - 33$

Find the missing factor or term.

79. $?(3x^2 - 4x - 28) = 12x^4 - 16x^3 - 112x^2$

80. $-2x^3(?) = -12x^{12} + 18x^9 - 38x^5$

81. $(x + ?)(x + 6) = x^2 + 10x + 24$

82. $(x + 8)(x - ?) = x^2 + 3x - 40$

83. $(x + 12) \cdot ? = x^2 - 144$

84. $(7x - 3) \cdot ? = 49x^2 - 9$

Mixed Practice, 85–96

Simplify.

85. $(x + 6)(x - 5)$

86. $-9x^4 \cdot 6x^3$

87. $3x^2(x^2 - 5x - 36)$

88. $(x + 10)(x - 10)$

89. $(x^2 - 3x - 18) + (5x^2 + 2x - 24)$

90. $(x - 11)^2$

91. $(3x^2 - 4x + 32) - (4x^2 + 7x - 50)$

92. $(5x + 2)(4x + 9)$

93. $(x^2 + 8x + 7) - (-x^2 - 5x + 22)$

94. $-6x(x^2 - 6x - 8)$

95. $(3x^2 + 25x - 200) + (7x^2 - 11x + 96)$

96. $(x - 3)(2x^2 - 9x - 12)$

Evaluate the polynomial.

97. $x^2 + 4xy - 5y^2$ for $x = -6$ and $y = 3$

98. $2x^2 + 14xy + y^2$ for $x = 5$ and $y = -8$

99. $b^2 - 4ac$ for $a = -4$, $b = -3$, and $c = 6$

100. $b^2 - 4ac$ for $a = 8$, $b = -4$, and $c = 3$

For each polynomial, list the degree of each term and the degree of the polynomial.

101. $4x^7y - 11x^5y^4 + 6x^3y^7$

102. $-x^9y^6 + 4x^6y^4 + 5x^3y^2$

103. $xy^7z - 2x^3y^4z^5 + 9x^4y^2z^2$

104. $5a^9b^{10}c^{12} + 7a^{16}b^6c^{11} + a^3b^{10}c^{10}$

Simplify.

105. $(6x^2 + 9xy - 7y^2) + (-3x^2 + 4xy + 5y^2)$

106. $(x^2y^2 - 8xy - 15) - (3x^2y^2 + 11xy - 21)$

107. $(4a^3b - 9a^2b^2 + 13ab^3) - (7ab^3 + 14a^2b^2 - 8a^3b)$

108. $(8m^4n^2 - 21m^3n^3 + 20m^2n^4) + (-9m^2n^4 - m^4n^2)$

109. $7x^5y^4 \cdot 8x^9y^3$

110. $-6x^4y \cdot 15x^6y^5$

111. $-2xy^2(5x^2 - 4xy + 4y^2)$

112. $3x^5y^4(3xy - 8x^3y^2 - 19x)$

113. $(x + 7y)(x - 3y)$

114. $(xy - 9)(xy - 8)$

115. Ellen is the yearbook advisor at a high school, and she determines that the revenue generated from selling yearbooks can be approximated by the function $R(x) = -6x^2 + 240x$, where x represents the price of a yearbook in dollars. How much money will be generated if the yearbook is sold for $20?

116. The average cost per shirt, in dollars, to produce x T-shirts is given by the function $f(x) = 0.00015x^2 - 0.06x + 10.125$. What is the average cost per shirt to produce 350 T-shirts?

117. The number of students earning master's degrees in mathematics in the United States in a particular year can be approximated by the function $g(x) = -11x^2 + 477x - 1104$, where x represents the number of years after 1970. Use the function to estimate the number of students who will have earned a master's degree in mathematics in the United States in the year 2008. (*Source:* U.S. Department of Education, National Center for Education Statistics)

118. The number of births, in thousands, in the United States in a particular year can be approximated by the function $f(x) = 6x^2 - 72x + 4137$, where x represents the number of years after 1990. Use the function to predict the number of births in the United States in the year 2015. (*Source:* National Center for Health Statistics, U.S. Department of Health and Human Services)

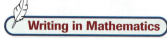

Writing in Mathematics

Answer in complete sentences.

119. Explain why it is a good idea to use parentheses when evaluating a polynomial for negative values of a variable. Do you feel that using parentheses is a good idea for evaluating any polynomial?

120. A classmate made the following error when subtracting two polynomials.

$$(3x^2 + 2x - 7) - (x^2 - x + 9)$$
$$= 3x^2 + 2x - 7 - x^2 - x + 9$$

Explain the error to your classmate, and give advice on how to avoid making this error.

121. Explain the difference between simplifying the expression $8x^2 + 3x^2$ and simplifying the expression $(8x^2)(3x^2)$. Discuss how the coefficients and exponents are handled differently.

122. To simplify the expression $(x + 8)^2$, we can either multiply $(x + 8)(x + 8)$ or use the special product formula $(a + b)^2 = a^2 + 2ab + b^2$. Which method do you prefer? Which method do you feel you will be able to remember? Explain your answer.

123. ***Solutions Manual**** Write a solutions manual page for the following problem:

Simplify $(3x^2 + 4x - 7) - (2x^2 - 6x + 19)$.

124. ***Solutions Manual**** Write a solutions manual page for the following problem:

Simplify $(2x - 3)(5x^2 + 4x - 9)$.

125. ***Newsletter**** Write a newsletter that covers the definitions related to polynomials.

126. ***Newsletter**** Write a newsletter explaining how to multiply two binomials.

***See Appendix B for details and sample answers.**

4.4 An Introduction to Factoring; The Greatest Common Factor; Factoring by Grouping

Objectives

1. **Find the greatest common factor (GCF) of two or more integers.**
2. **Find the GCF of two or more variable terms.**
3. **Factor the GCF out of each term of a polynomial.**
4. **Factor out a common binomial factor from a polynomial.**
5. **Factor a polynomial by grouping.**

In this section we will begin to learn how to **factor** a polynomial. A polynomial has been **factored** when it is represented as the product of two or more polynomials.

Greatest Common Factor

Objective 1 **Find the greatest common factor (GCF) of two or more integers.** Before we begin to learn how to factor polynomials, we will go over the procedure for finding the **greatest common factor (GCF)** of two or more integers. The greatest common factor of two or more integers is the largest whole number that is a factor of each integer.

Finding the GCF of Two or More Integers

Find the prime factorization of each integer by expressing each integer as the product of prime numbers.

Write the prime factors that are common to each integer; the GCF is the product of these prime factors.

Recall that a number greater than 1 is prime if its only factors are 1 and itself. The first eight prime numbers are 2, 3, 5, 7, 11, 13, 17, and 19.

Example 1 Find the GCF of 60 and 126.

Solution

The prime factorization of 60 is $2 \cdot 2 \cdot 3 \cdot 5$ and the prime factorization of 126 is $2 \cdot 3 \cdot 3 \cdot 7$. The prime factors that they share in common are 2 and 3, so the GCF is $2 \cdot 3$ or 6. (This means that 6 is the greatest number that divides evenly into both 60 and 126.)

***Quick Check* 1**
Find the GCF of 12 and 28.

Example 2 Find the GCF of 108, 270, and 504.

Solution

We begin with the prime factorizations.

$$108 = 2^2 \cdot 3^3 \qquad 270 = 2 \cdot 3^3 \cdot 5 \qquad 504 = 2^3 \cdot 3^2 \cdot 7$$

The only primes that are factors of all three numbers are 2 and 3. To determine what power of each factor to use in the GCF, we select the smallest power that we can find in either number. We choose the smallest power because this is what the integers have in common. The smallest power of 2 that is a factor of any of the three numbers is 2^1, and the smallest power of 3 that is a factor is 3^2. The GCF is $2 \cdot 3^2$ or 18.

Quick Check 2
Find the GCF of 48, 72, and 84.

Note that if two or more integers do not have any common prime factors, then their GCF is 1. For example, the GCF of 8, 12, and 15 is 1, since there are no prime numbers that are factors of all three numbers.

Objective 2 **Find the GCF of two or more variable terms.** We can also find the GCF of two or more variable terms. For a variable factor to be included in the GCF, it must be a factor of each term. As with prime factors, the exponent for a variable factor used in the GCF is the smallest exponent that can be found for that variable in any one term.

EXAMPLE 3 Find the GCF of $10x^3$ and $15x^4$.

Solution

The GCF of the two coefficients is 5. The variable x is a factor of each term, and its smallest exponent is 3. The GCF of these two terms is $5x^3$.

EXAMPLE 4 Find the GCF of $a^2b^4c^8$, a^5b^3, and $a^8b^6c^5$.

Solution

Only the variables a and b are factors of each term. (The variable c is not a factor of the second term.) The smallest power of a in any one term is 2, and the smallest power of b is 3. The GCF is a^2b^3.

Quick Check 3
Find the GCF.

a) $30x^4$ **and** $75x^8$
b) $x^8y^6z^4$, $x^3y^5z^7$, **and** x^5z^2

Factoring Out the Greatest Common Factor

Objective 3 **Factor the GCF out of each term of a polynomial.** The first step in factoring any polynomial is to factor out the GCF from all of the terms. This process uses the distributive property, and you may think of it as "undistributing" the GCF from each term. Consider the polynomial $4x^2 + 8x + 20$. The GCF of these three terms is 4, and the polynomial can be rewritten as the product $4(x^2 + 2x + 5)$. Notice that the GCF has been factored out of each term, and the polynomial inside the parentheses is the polynomial that we would multiply 4 by in order to equal $4x^2 + 8x + 20$.

$$\begin{aligned} 4(x^2 + 2x + 5) &= 4 \cdot x^2 + 4 \cdot 2x + 4 \cdot 5 && \text{Distribute.} \\ &= 4x^2 + 8x + 20 && \text{Multiply.} \end{aligned}$$

EXAMPLE 5 Factor $6x^3 - 18x^2 + 42x$ by factoring out the GCF.

Solution

We begin by finding the GCF, which is $6x$. The next task is to fill in the missing terms of the polynomial inside the parentheses so that the product of $6x$ and that polynomial is $6x^3 - 18x^2 + 42x$.

$$6x(? - ? + ?)$$

To find the missing terms, we can divide each term of the original polynomial by $6x$.

$$\frac{6x^3}{6x} = x^2 \qquad -\frac{18x^2}{6x} = -3x \qquad \frac{42x}{6x} = 7$$

So, $6x^3 - 18x^2 + 42x = 6x(x^2 - 3x + 7)$.

We can check our answer by multiplying $6x(x^2 - 3x + 7)$, which should equal $6x^3 - 18x^2 + 42x$. The check is left to the reader.

***Quick Check* 4**
Factor $5x^4 - 30x^3 - 75x^2$ by factoring out the GCF.

EXAMPLE 6 Factor $6x^4 + 18x^3 + 3x^2$ by factoring out the GCF.

Solution

The GCF for these three terms is $3x^2$. Notice that this is also the third term.

$$6x^4 + 18x^3 + 3x^2 = 3x^2(2x^2 + 6x + 1) \qquad \text{Factor out the GCF.}$$

When a term in a polynomial is the GCF of the polynomial, factoring out the GCF leaves a 1 in that term's place. Why? Because we need to determine what we multiply $3x^2$ by in order to equal $3x^2$, and $3x^2 \cdot 1 = 3x^2$.

***Quick Check* 5**
Factor $14x^2 - 21x - 7$ by factoring out the GCF.

A Word of Caution Be sure to write a 1 in the place of a term that was the GCF when factoring out the GCF from a polynomial.

Objective 4 Factor out a common binomial factor from a polynomial.
Occasionally the GCF of two terms will be a binomial or some other polynomial with more than one term. Consider the expression $8x(x - 5) + 3(x - 5)$, which contains the two terms $8x(x - 5)$ and $3(x - 5)$. Each term has the binomial $x - 5$ as a factor. This common factor can be factored out of this expression.

EXAMPLE 7 Factor $8x(x - 5) + 3(x - 5)$ by factoring out the GCF.

Solution

The binomial $x - 5$ is a common factor for these two terms. We begin by factoring out this common factor.

$$8x\underbrace{(x - 5)} + 3\underbrace{(x - 5)}$$

$$= (x - 5)(? + ?)$$

Quick Check 6
Factor
$2x^2(x + 7) + 9(x + 7)$
by factoring out the GCF.

After factoring out $x - 5$, what factors remain in the first term? The only factor that remains is $8x$. In the second term, the only factor remaining is 3.

$$8x(x - 5) + 3(x - 5) = (x - 5)(8x + 3)$$

EXAMPLE 8 Factor $5x(4x + 3) - 6(4x + 3)$ by factoring out the GCF.

Solution

The GCF of these two terms is $4x + 3$, which can be factored out as follows.

$$5x(4x + 3) - 6(4x + 3) = (4x + 3)(5x - 6)$$ Factor out the common factor $4x + 3$.

Quick Check 7
Factor
$9x(3x - 4) - 2(3x - 4)$
by factoring out the GCF.

Notice that the second term was negative, which led to $5x - 6$, rather than $5x + 6$, as the other factor.

Again, factoring out the GCF will be our first step in factoring any polynomial. Often this will make our factoring easier, and sometimes we cannot factor the expression without first factoring out the GCF.

Factoring by Grouping

Objective 5 **Factor a polynomial by grouping.** We now turn our attention to a factoring technique known as **factoring by grouping**. Consider the polynomial $3x^3 + 33x^2 + 7x + 77$. The GCF for the four terms is 1, so we cannot factor the polynomial by factoring out the GCF. However, the first pair of terms has a common factor of $3x^2$ and the second pair of terms has a common factor of 7. If we factor $3x^2$ out of the first two terms and 7 out of the last two terms, we produce the following.

$$3x^2(x + 11) + 7(x + 11)$$

Notice that the two resulting terms have a common factor of $x + 11$. We can factor out this common binomial factor.

$$3x^3 + 33x^2 + 7x + 77$$
$$= 3x^2(x + 11) + 7(x + 11)$$ Factor $3x^2$ from the first two terms and 7 from the last two terms.
$$= (x + 11)(3x^2 + 7)$$ Factor out the common binomial factor $x + 11$.

Factoring a Polynomial with Four Terms by Grouping

Factor a common factor out of the first two terms and another common factor out of the last two terms. If the two "groups" share a common binomial factor, factor out this binomial to complete the factoring of the polynomial.

If the two "groups" do not share a common binomial factor, we can try rearranging the terms of the polynomial in a different order. If we cannot find two "groups" that share a common factor, then we cannot factor the polynomial by this method.

EXAMPLE 9 Factor $4x^3 - 28x^2 + 5x - 35$ by grouping.

Solution

We first check for a factor that is common to each of the four terms. Since there is no common factor other than 1, we proceed to factoring by grouping.

$$4x^3 - 28x^2 + 5x - 35 = 4x^2(x - 7) + 5x - 35$$ Factor $4x^2$ out of first two terms.

$$= 4x^2(x - 7) + 5(x - 7)$$ Factor 5 out of last two terms.

$$= (x - 7)(4x^2 + 5)$$ Factor out the common factor $x - 7$.

Quick Check 8

Factor $x^3 + 4x^2 + 7x + 28$ by grouping.

We can check our factoring by multiplying the two factors together. The check is left to the reader.

EXAMPLE 10 Factor $9x^2 + 72x - 8x - 64$ by grouping.

Solution

Since the four terms have no common factors other than 1, we factor by grouping. After factoring $9x$ out of the first two terms, we must factor a *negative* 8 out of the last two terms. If we factored a positive 8 out of the last two terms, rather than a negative 8, the two binomial factors would be different.

$$9x^2 + 72x - 8x - 64 = 9x(x + 8) - 8(x + 8)$$ Factor the common factor $9x$ out of the first two terms and the common factor -8 out of the last two terms.

$$= (x + 8)(9x - 8)$$ Factor out the common factor $x + 8$.

Quick Check 9

Factor $2x^2 - 22x - 3x + 33$ by grouping.

When the third of the four terms is negative, this often indicates that a negative common factor will need to be factored from the last two terms.

A Word of Caution It is often necessary to factor a negative common factor from two terms when factoring by grouping.

EXAMPLE 11 Factor $5x^3 - 30x^2 + x - 6$ by grouping.

Solution

Again, the four terms have no common factor other than 1, so we may proceed to factor this polynomial by grouping. Notice that the last two terms have no common factor other than 1. When this is the case, we will actually factor out the common factor of 1 from the two terms.

Quick Check 10

Factor $7x^3 + 21x^2 + x + 3$ by grouping.

$$5x^3 - 30x^2 + x - 6 = 5x^2(x - 6) + 1(x - 6)$$ Factor the common factors of $5x^2$ and 1 from the first two terms and the last two terms, respectively.

$$= (x - 6)(5x^2 + 1)$$ Factor out the common factor $x - 6$.

In the next example, the four terms have a common factor other than 1. We will factor out the GCF from the polynomial before attempting to use factoring by grouping.

EXAMPLE 12 Factor $2x^3 - 10x^2 - 6x + 30$ completely.

Solution

The four terms share a common factor of 2, so we will factor this out before proceeding with factoring by grouping.

$$2x^3 - 10x^2 - 6x + 30 = 2(x^3 - 5x^2 - 3x + 15)$$ Factor out the common factor 2.

$$= 2[x^2(x - 5) - 3(x - 5)]$$ Factor out the common factor x^2 from the first two terms. Factor out the common factor -3 from the last two terms.

$$= 2(x - 5)(x^2 - 3)$$ Factor out the common factor $x - 5$.

If we had not factored out the common factor 2 before factoring this polynomial by grouping, we would have factored $2x^3 - 10x^2 - 6x + 30$ to be $(x - 5)(2x^2 - 6)$. If we stop here, we have not factored this polynomial completely, as $2x^2 - 6$ has a common factor of 2.

Quick Check 11

Factor $5x^2 + 35x - 10x - 70$ by grouping.

Occasionally, despite our best efforts, a polynomial *cannot* be factored. Here is an example of just such a polynomial. Consider the polynomial $x^2 - 5x + 3x + 15$. The first two terms have a common factor of x and the last two terms have a common factor of 3.

$$x^2 - 5x + 3x + 15 = x(x - 5) + 3(x + 5)$$

Since the two binomials are not the same, we cannot factor a common factor out of the two terms. Reordering the terms $-5x$ and $3x$ leads to the same problem. The polynomial cannot be factored.

There are instances when factoring by grouping fails, yet the polynomial can be factored by other techniques. For example, the polynomial $x^3 + 2x^2 - 7x - 24$ cannot be factored by grouping. However, it can be shown that $x^3 + 2x^2 - 7x - 24 = (x - 3)(x^2 + 5x + 8)$.

Building Your Study Strategy **Math Anxiety, 4** **Proper Placement** Many students think they have math anxiety because they constantly feel that they are in over their heads, when in fact they may not be prepared for that particular class. Taking a course for which you do not have the prerequisite skills is not a good idea, and success will be difficult to achieve. For example, if you are taking an intermediate algebra course but are struggling with performing arithmetic operations with positive and negative numbers, then perhaps you would be better off taking an elementary algebra class.

If this situation applies to you, talk to your instructor. He or she will be able to advise whether you should be taking another course or whether you have the skills to be successful in this course.

EXERCISES 4.4

FOR EXTRA HELP

MyMathLab

MathXL

Interactmath.com

MathXL Tutorials on CD

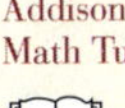

Video Lectures on CD

Tutor Center

Addison-Wesley Math Tutor Center

Student's Solutions Manual

Vocabulary

1. A polynomial has been __________ when it is represented as the product of two or more polynomials.
2. The __________ of two or more integers is the largest whole number that is a factor of each integer.
3. The exponent for a variable factor used in the GCF is the __________ exponent that can be found for that variable in any one term.
4. The process of factoring a polynomial by first breaking the polynomial into two sets of terms is called factoring by __________.

Find the greatest common factor (GCF).

5. 10, 12
6. 15, 33
7. 40, 56
8. 9, 27
9. 36, 48, 72
10. 12, 16, 45
11. x^5, x^4
12. y^6, y^9
13. a^5b^7, a^2b^9
14. $a^5b^4c^3, a^4bc^8, a^2b^6c^7$
15. $6x^7, 10x^5$
16. $28x^8, 42x^{10}$
17. $12a^5b^6c^3, 24a^3c^4, 30a^4b^7c$
18. $36x^7y^2z^5, 72x^4y^3z^2, 90y^5z^5$

Factor the GCF out of the given expression.

19. $4x - 24$
20. $5a + 60$
21. $7x^2 + 5x$
22. $8x^2 - 25x$
23. $12x^3 + 18x^2$
24. $15x^6 - 35x^4$
25. $28x^4 + 7x^2$
26. $30x^6 - 3x^5$
27. $4x^2 + 20x + 76$
28. $6x^2 + 18x - 39$
29. $12x^5 - 24x^4 - 32x^3$
30. $9x^4 - 15x^3 - 33x^2$
31. $42x^4 - 54x^3 + 6x^2$
32. $40x^5 + 150x^4 - 10x^3$
33. $m^2n^3 - m^7n^2 + m^6n^8$
34. $x^9y^2 + x^6y^4 - x^3y^6$
35. $36a^7b^5 - 60a^4b^4 - 48a^2b^8$
36. $25r^{10}s^4 + 15r^4s^6 - 45r^5s$
37. $-6x^3 + 15x^2 - 36x$
38. $-8a^5 - 16a^4 + 40a^2$
39. $10x^4 + 12x^2 + 2x$
40. $24x^6 - 32x^4 - 8x^2$
41. $4x(3x + 8) + 7(3x + 8)$
42. $x(2x - 9) + 5(2x - 9)$
43. $x(4x - 9) - 8(4x - 9)$
44. $2x(2x - 3) - 5(2x - 3)$
45. $5x(x - 6) + (x - 6)$
46. $4x(3x - 7) - (3x - 7)$

Factor by grouping.

47. $x^2 + 3x + 6x + 18$
48. $x^2 + 8x + 2x + 16$
49. $x^2 - 9x + 7x - 63$
50. $x^2 - 8x + 3x - 24$
51. $x^2 + 10x - 12x - 120$
52. $x^2 - 7x - 11x + 77$
53. $3x^2 + 12x + 10x + 40$
54. $2x^2 - 6x + 7x - 21$
55. $5x^2 + 20x - 2x - 8$
56. $3x^2 - 18x - 8x + 48$
57. $2x^2 + 16x + x + 8$
58. $6x^2 - 90x + x - 15$

59. $3x^2 - 12x - x + 4$

60. $21x^2 - 56x - 3x + 8$

61. $x^3 + 6x^2 + 5x + 30$

62. $3x^3 - 6x^2 + 2x - 4$

63. $5x^3 + 40x + 3x^2 + 24$

64. $11x^3 + 33x + 6x^2 + 18$

65. $2x^2 + 8x + 12x + 48$

66. $3x^2 + 6x + 27x + 54$

67. $5x^2 - 15x + 40x - 120$

68. $4x^2 + 24x - 4x - 24$

Find the polynomial that factors to be the given product by grouping.

69. $(x - 5)(x^2 + 7)$

70. $(x + 3)(x^2 - 11)$

71. $(2x - 9)(5x^2 - 3)$

72. $3(x + 4)(x^2 + 6)$

Writing in Mathematics

Answer in complete sentences.

73. When factoring a polynomial, explain how to check your work.

74. Give an example of a four-term polynomial that can be factored by grouping. Use this example to explain the process of factoring by grouping.

75. Explain why it is necessary to factor a negative common factor from the last two terms of the polynomial $x^3 - 5x^2 - 7x + 35$ in order to factor it by grouping.

76. ***Solutions Manual**** Write a solutions manual page for the following problem:

Factor $x^3 - 6x^2 - 7x + 42$.

77. ***Newsletter**** Write a newsletter explaining how to factor out a GCF from a polynomial.

***See Appendix B for details and sample answers.**

Quick Review Exercises

Section 4.4

Multiply.

1. $(x + 7)(x + 4)$

2. $(x - 9)(x - 11)$

3. $(x + 5)(x - 8)$

4. $(x - 10)(x + 6)$

5. $(2x + 5)(x + 4)$

6. $(3x - 8)(2x - 7)$

7. $(x + 6)(3x - 4)$

8. $(4x - 3)(2x + 9)$

4.5 Factoring Trinomials of Degree 2

Objectives

1. **Factor a trinomial of the form $x^2 + bx + c$.**
2. **Factor a trinomial by first factoring out a common factor.**
3. **Factor a trinomial of the form $ax^2 + bx + c$ $(a \neq 1)$ by grouping.**
4. **Factor a trinomial of the form $ax^2 + bx + c$ $(a \neq 1)$ by trial and error.**
5. **Factor a perfect square trinomial.**
6. **Factor a trinomial in several variables.**

Objective 1 Factor a trinomial of the form $x^2 + bx + c$. In this section we will learn how to factor trinomials that have degree 2. We will begin with trinomials of this type that have a leading coefficient of 1. Some examples of this type of polynomial are $x^2 + 12x + 32$, $x^2 - 9x + 14$, $x^2 + 8x - 20$, and $x^2 - x - 12$.

Factoring Polynomials of the Form $x^2 + bx + c$

If $x^2 + bx + c$, where b and c are integers, is factorable, it can be factored to be the product of two binomials of the form $(x + m)(x + n)$, where m and n are integers.

We will explore this concept by multiplying two binomials of the form $(x + m)(x + n)$, and using the result to help us learn to factor trinomials of the form $x^2 + bx + c$.

Multiply the two binomials $x + 4$ and $x + 8$.

$$(x + 4)(x + 8) = x^2 + 8x + 4x + 32 \quad \text{Distribute.}$$
$$= x^2 + 12x + 32 \quad \text{Combine like terms.}$$

The two binomials have a product that is a trinomial of the form $x^2 + bx + c$ with $b = 12$ and $c = 32$. Notice that the two numbers 4 and 8 have a product of 32 and a sum of 12; in other words, their product is equal to c and their sum is equal to b.

$$4 \cdot 8 = 32$$
$$4 + 8 = 12$$

We will use this pattern to help us factor trinomials of the form $x^2 + bx + c$. We will look for two integers m and n with a product equal to c and a sum equal to b. If we can find two such integers, then the trinomial factors to be $(x + m)(x + n)$.

Factoring $x^2 + bx + c$ When c Is Positive

EXAMPLE 1 Factor $x^2 + 9x + 20$.

Solution

We begin, as always, by looking for common factors. The terms in this trinomial have no common factors other than 1. We are looking for two integers m and n whose product is 20 and whose sum is 9. Here are the factors of 20.

Factors	$1 \cdot 20$	$2 \cdot 10$	$4 \cdot 5$
Sum	21	12	9

The pair of factors that have a sum of 9 are 4 and 5. The trinomial $x^2 + 9x + 20$ factors to be $(x + 4)(x + 5)$. Be aware that this could also be written as $(x + 5)(x + 4)$.

We can check our work by multiplying $x + 4$ by $x + 5$. If the product equals $x^2 + 9x + 20$, then we have factored correctly. The check is left to the reader.

EXAMPLE 2 Factor $x^2 - 11x + 28$.

Solution

The major difference between this trinomial and the previous one is that the x term has a negative coefficient. We are looking for two integers m and n whose product is 28 and whose sum is -11. Since the product of these two integers is positive, they must have the same sign. The sum of these two integers is negative, so both integers must be negative. Here are the negative factors of 28.

Factors	$(-1)(-28)$	$(-2)(-14)$	$(-4)(-7)$
Sum	-29	-16	-11

***Quick Check* 1**

Factor.

a) $x^2 + 10x + 16$

b) $x^2 - 8x + 15$

The pair that has a sum of -11 is -4 and -7, so $x^2 - 11x + 28 = (x - 4)(x - 7)$.

From the previous examples, we see that when the constant term c is positive, as in $x^2 + 9x + 20$ and $x^2 - 11x + 28$, then the two numbers we are looking for, m and n, will both have the same sign. Both numbers will be positive if b is positive, and both numbers will be negative if b is negative.

Factoring $x^2 + bx + c$ When c Is Negative

If the product of two numbers is negative, one of the numbers must be negative and the other number must be positive. So if the constant term c is negative, then m and n must have opposite signs.

EXAMPLE 3 Factor $x^2 + 4x - 32$.

Solution

These three terms do not have any common factors, so we look for two integers m and n such that their product is -32 and their sum is 4. Since the product is negative, we are looking for one negative number and one positive number. Here are the factors of -32, along with their sums.

Factors	$-1 \cdot 32$	$-2 \cdot 16$	$-4 \cdot 8$	$-8 \cdot 4$	$-16 \cdot 2$	$-32 \cdot 1$
Sum	31	14	4	-4	-14	-31

The pair of integers that adds up to 4 is -4 and 8, so $x^2 + 4x - 32 = (x - 4)(x + 8)$.

Example 4 Factor $x^2 - 5x - 24$.

Solution

Since there are no common factors to factor out, we start looking for two integers m and n that have a product of -24 and a sum of -5. Here are the factors of -24.

Factors	$-1 \cdot 24$	$-2 \cdot 12$	$-3 \cdot 8$	$-4 \cdot 6$	$-6 \cdot 4$	$-8 \cdot 3$	$-12 \cdot 2$	$-24 \cdot 1$
Sum	23	10	5	2	-2	-5	-10	-23

Quick Check 2

Factor.

a) $x^2 + 5x - 14$

b) $x^2 - 2x - 48$

Our integers m and n are -8 and 3. $x^2 - 5x - 24 = (x - 8)(x + 3)$.

It is not necessary to list each set of factors as we did in the previous examples. We can often find the two integers m and n quickly through trial and error.

Example 5 Factor.

a) $x^2 - 10x - 24$

Solution

The two integers whose product is -24 and sum is -10 are -12 and 2.

$$x^2 - 10x - 24 = (x - 12)(x + 2)$$

b) $x^2 + 10x + 24$

Solution

This example is similar to the last example, but we are looking for two integers that have a product of *positive* 24 instead of -24. Also, the sum of these two integers is 10 instead of -10. The two integers are 4 and 6.

$$x^2 + 10x + 24 = (x + 4)(x + 6)$$

c) $x^2 + 10x - 24$

Solution

The two integers that have a product of -24 and a sum of 10 are -2 and 12.

$$x^2 + 10x - 24 = (x - 2)(x + 12)$$

Quick Check 3

Factor.

a) $x^2 + 5x - 6$

b) $x^2 + 5x + 6$

c) $x^2 - 5x + 6$

d) $x^2 - 5x - 6$

d) $x^2 - 10x + 24$

Solution

The integers -4 and -6 have a product of 24 and a sum of -10,

$$x^2 - 10x + 24 = (x - 4)(x - 6)$$

Trinomials That Are Prime

Not every trinomial of the form $x^2 + bx + c$ can be factored. For example, there may not be a pair of integers that have a product of c and a sum of b. In this case we say that the trinomial cannot be factored; it is **prime.** For example, $x^2 + 9x + 12$ is prime because there do not exist two numbers whose product is 12 and whose sum is 9.

EXAMPLE 6 Factor $x^2 + 11x - 18$.

Solution

As there are no common factors to factor out, we start looking for two integers m and n that have a product of -18 and a sum of 11. Here are the factors of -18.

Factors	$-1 \cdot 18$	$-2 \cdot 9$	$-3 \cdot 6$	$-6 \cdot 3$	$-9 \cdot 2$	$-18 \cdot 1$
Sum	17	7	3	-3	-7	-17

There are no integers that have a product of -18 and a sum of 11, so the trinomial $x^2 + 11x - 18$ is prime.

***Quick Check* 4**
Factor $x^2 + 9x - 20$.

Factoring a Trinomial Whose Terms Have a Common Factor

Objective 2 **Factor a trinomial by first factoring out a common factor.** We now turn our attention to factoring trinomials whose terms contain common factors other than 1. After we factor out a common factor, we will attempt to factor the remaining polynomial factor.

EXAMPLE 7 Factor $3x^2 + 42x + 144$.

Solution

The GCF of these three terms is 3.

A Word of Caution When factoring a polynomial, begin by factoring out the GCF of all of the terms.

After we have factored out the GCF, the expression is $3(x^2 + 14x + 48)$. We now try to factor the trinomial $x^2 + 14x + 48$ by finding two integers that have a product of 48 and a sum of 14. The integers that satisfy these conditions are 6 and 8.

$$3x^2 + 42x + 144 = 3(x^2 + 14x + 48) \quad \text{Factor out the GCF (3).}$$
$$= 3(x + 6)(x + 8) \quad \text{Factor } x^2 + 14x + 48.$$

***Quick Check* 5**
Factor $5x^2 - 5x - 60$.

A Word of Caution When we factor out a common factor from a polynomial, that common factor must be written in all following stages of factoring. Don't "lose" the common factor.

EXAMPLE 8 Factor $-x^2 - 2x + 63$.

Solution

The leading coefficient for this trinomial is -1, but to factor the trinomial the leading coefficient must be a *positive* 1. We begin by factoring out a -1 from each term, which will change the sign of each term.

$$-x^2 - 2x + 63 = -(x^2 + 2x - 63) \quad \text{Factor out } -1\text{, leaving the leading coefficient 1.}$$

$$= -(x + 9)(x - 7) \quad \text{Factor } x^2 + 2x - 63 \text{ by finding two integers whose product is } -63 \text{ and whose sum is 2.}$$

***Quick Check* 6**
Factor $-x^2 + 12x - 27$.

Factoring Trinomials of the Form $ax^2 + bx + c$ $(a \neq 1)$

If a trinomial's leading coefficient is not 1, or if the degree of the trinomial is not 2, then we will attempt to factor out a common factor so that the trinomial factor is of the form $x^2 + bx + c$. There will be times when this is not possible, so we will now learn to factor trinomials of the form $ax^2 + bx + c$, where $a \neq 1$. Some examples of this type of trinomial are $2x^2 + 13x + 20$, $8x^2 - 22x - 21$, and $3x^2 - 11x + 8$.

There are two methods for factoring this type of trinomial that we will examine: factoring by grouping and factoring by trial and error. Your instructor may prefer one of these methods to the other, or an altogether different method, and may ask you to use one method only. However, if you are allowed to use either method, use the method that you feel more comfortable using.

Factoring by Grouping

Objective 3 **Factor a trinomial of the form $ax^2 + bx + c$ $(a \neq 1)$ by grouping.** We begin with factoring by grouping. Consider the product $(3x + 2)(2x + 5)$.

$$(3x + 2)(2x + 5) = 6x^2 + 15x + 4x + 10 \quad \text{Distribute.}$$

$$= 6x^2 + 19x + 10 \quad \text{Combine like terms.}$$

Since $(3x + 2)(2x + 5) = 6x^2 + 19x + 10$, we know that $6x^2 + 19x + 10$ can be factored as $(3x + 2)(2x + 5)$. By rewriting the middle term $19x$ as $15x + 4x$, we can factor by grouping. The important skill is being able to determine that $15x + 4x$ is the right way to rewrite $19x$, instead of $16x + 3x$, $9x + 10x$, $22x - 3x$, or any other two terms that simplify to $19x$.

Factoring $ax^2 + bx + c$ ($a \neq 1$) by Grouping

1. Multiply $a \cdot c$.
2. Find two integers whose product is $a \cdot c$ and whose sum is b.
3. Rewrite the term bx as two terms, using the two integers found in step 2.
4. Factor the resulting polynomial by grouping.

To factor the polynomial $6x^2 + 19x + 10$ by grouping, we begin by multiplying $6 \cdot 10$, which equals 60. Next we look for two integers whose product is 60 and whose sum is 19, and those two integers are 15 and 4. We can then rewrite $19x$ as $15x + 4x$. (These terms could be written in the opposite order and would still lead to the correct factoring.) We then factor the polynomial $6x^2 + 15x + 4x + 10$ by grouping.

$$6x^2 + 19x + 10 = 6x^2 + 15x + 4x + 10$$ Rewrite $19x$ as $15x + 4x$.

$$= 3x(2x + 5) + 2(2x + 5)$$ Factor the common factor $3x$ from the first two terms and the common factor 2 from the last two terms.

$$= (2x + 5)(3x + 2)$$ Factor out the common factor $2x + 5$.

Now we will examine several examples using this technique.

EXAMPLE 9 Factor $2x^2 + 7x + 6$ by grouping.

Solution

We first check to see if there are any common factors; in this case, there are not. We proceed with factoring by grouping. Since $2 \cdot 6 = 12$, we look for two integers with a product of 12 and a sum of 7, which is the coefficient of the middle term. The integers 3 and 4 meet these criteria, so we rewrite $7x$ as $3x + 4x$.

$$2x^2 + 7x + 6 = 2x^2 + 3x + 4x + 6$$ Find two integers whose product is $2 \cdot 6 = 12$ and whose sum is 7. Rewrite $7x$ as $3x + 4x$.

$$= x(2x + 3) + 2(2x + 3)$$ Factor the common factor x from the first two terms and the common factor 2 from the last two terms.

$$= (2x + 3)(x + 2)$$ Factor out the common factor $2x + 3$.

We check that this factoring is correct by multiplying $2x + 3$ by $x + 2$, which should equal $2x^2 + 7x + 6$. The check is left to the reader.

Quick Check 7
Factor $3x^2 + 14x + 8$ by grouping.

EXAMPLE 10 Factor $3x^2 - 10x + 8$ by grouping.

Solution

We begin by multiplying $3 \cdot 8$ as there are no common factors other than 1. We look for two integers whose product is 24 and whose sum is -10. The integers are -4 and -6, so we rewrite $-10x$ as $-4x - 6x$ and then factor by grouping.

$3x^2 - 10x + 8 = 3x^2 - 4x - 6x + 8$	Find two integers with a product of 24 and a sum of -10. Rewrite $-10x$ as $-4x - 6x$.
$= x(3x - 4) - 2(3x - 4)$	Factor out the common factor x from the first two terms and the common factor -2 from the last two terms.
$= (3x - 4)(x - 2)$	Factor out the common factor $3x - 4$.

Quick Check 8
Factor $4x^2 - 25x + 36$ by grouping.

EXAMPLE 11 Factor $12x^2 + 11x - 15$ by grouping.

Solution

The terms have no common factor other than 1, so we begin by multiplying $12(-15)$, which equals -180. We need to find two integers with a product of -180 and a sum of 11. We can start by listing the different factors of 180. Since we know that one of the integers will be positive and one will be negative, we look for two factors in this list that have a difference of 11.

$1 \cdot 180 \quad 2 \cdot 90 \quad 3 \cdot 60 \quad 4 \cdot 45 \quad 5 \cdot 36 \quad 6 \cdot 30 \quad 9 \cdot 20 \quad 10 \cdot 18 \quad 12 \cdot 15$

The pair 9 and 20 has a difference of 11 and a product of -180. We rewrite $11x$ as $-9x + 20x$ and then factor by grouping.

$12x^2 + 11x - 15 = 12x^2 - 9x + 20x - 15$	Find two integers whose product is -180 and whose sum is 11. Rewrite $11x$ as $-9x + 20x$.
$= 3x(4x - 3) + 5(4x - 3)$	Factor the common factor $3x$ out of the first two terms and the common factor 5 out of the last two terms.
$= (4x - 3)(3x + 5)$	Factor out the common factor $4x - 3$.

Quick Check 9
Factor $8x^2 - 26x + 15$ by grouping.

Factoring by Trial and Error

Objective 4 Factor a trinomial of the form $ax^2 + bx + c$ ($a \neq 1$) by trial and error. We now turn our attention to the method of trial and error. We need to factor a trinomial $ax^2 + bx + c$, and once factored it will have the following form.

(variable term + constant)(variable term + constant)

The product of the variable terms will be ax^2, and the product of the constants will equal c. We will use these facts to get started by listing all of the factors of ax^2 and c.

Consider the trinomial $2x^2 + 7x + 6$, which has no common factors other than 1. We begin by listing all of the factors of $2x^2$ and 6. Factors of $2x^2$ are $x \cdot 2x$. Factors of 6 are $1 \cdot 6$ and $2 \cdot 3$.

Since there is only one pair of factors whose product is $2x^2$, we know that if this trinomial is factorable, it will be of the form $(2x + ?)(x + ?)$. We will substitute the different factors of 6 in all possible orders in place of the question marks, satisfying the following products.

$$\overset{2x^2}{\overline{(2x}} + ?)(\overline{x} + ?) \quad \text{with } (2x)(x) = 2x^2 \text{ and } (?)(?) = 6$$

We will keep substituting factors of 6 until the middle term of the product of the two binomials is $7x$. Rather than fully distributing for each trial, we need only look at the two products shown in the following graphic.

$$(2x + ?)(x + ?)$$

When the sum of these two products is $7x$, we have found the correct factors.

We begin by using the factors 1 and 6, which leads to the binomial factors $(2x + 1)(x + 6)$. Since the middle term for these factors is $13x$, we have not found the correct factors. We then switch the positions of 1 and 6 and try again. This time, when we multiply out $(2x + 6)(x + 1)$, the middle term equals $8x$, not $7x$. We need to try another pairing, so we try the other factors of 6, which are 2 and 3. When we multiply out $(2x + 3)(x + 2)$, the middle term is $7x$, so these are the correct factors.

$$(2x + 3)(x + 2) \quad \text{inner: } 3x, \quad \text{outer: } 4x$$

$2x^2 + 7x + 6 = (2x + 3)(x + 2)$.

EXAMPLE 12 Factor $6x^2 - 7x - 3$ by trial and error.

Solution

There are no common factors other than 1, so we begin by listing the factors of $6x^2$ and -3.

Factors of $6x^2$ are $x \cdot 6x$ and $2x \cdot 3x$
Factors of -3 are $-1 \cdot 3$ and $-3 \cdot 1$

We are looking for two factors that produce the middle term $-7x$ when multiplied. We will begin by using x and $6x$ for the variable terms and -1 and 3 for the constants.

Factors	Middle Term
$(x - 1)(6x + 3)$	$-3x$
$(x + 3)(6x - 1)$	$17x$

Neither of these middle terms is $-7x$, so we have not found the correct factors. Also, since neither middle term is the opposite of $-7x$, switching the signs of the constants 1 and 3 will not lead to the correct factors either. So $(x + 1)(6x - 3)$ and $(x - 3)(6x + 1)$ are not correct. We switch to the pair $2x$ and $3x$.

Factors	Middle Term
$(2x - 1)(3x + 3)$	$3x$
$(2x + 3)(3x - 1)$	$7x$

Since the second of these products produces a middle term that is the opposite of $-7x$, we can get the correct factoring by changing the signs of the constant terms. $6x^2 - 7x - 3 = (2x - 3)(3x + 1)$.

***Quick Check* 10**
Factor $2x^2 + 9x - 18$ by trial and error.

In the previous example, there are certain pairings that we could have skipped, such as $(x - 1)(6x + 3)$. Notice that the second binomial factor has a common factor of 3. If this were the correct factoring of $6x^2 - 7x - 3$, then the trinomial itself would have a common factor of 3 as well, which it does not. As it turns out, of the eight possible trial-and-error combinations, we could skip the following four of them because one of the binomial factors has a common factor of 3:

$$(x - 1)(6x + 3) \quad (x + 1)(6x - 3) \quad (2x - 1)(3x + 3) \quad (2x + 1)(3x - 3)$$

This leaves only four combinations to check:

$$(x + 3)(6x - 1) \quad (x - 3)(6x + 1) \quad (2x - 3)(3x + 1) \quad (2x + 3)(3x - 1)$$

Taking advantage of this shortcut will allow us to be more efficient in our factoring, which will be helpful when we try to factor a trinomial whose terms have more factors. In the next example we will again factor $12x^2 + 11x - 15$, which we factored by grouping in Example 11; this time we will use trial and error.

EXAMPLE 13 Factor $12x^2 + 11x - 15$.

Solution

Since there are no common factors to factor out, we begin by listing the factors of $12x^2$ and -15.

Factors of $12x^2$ are $x \cdot 12x$, $2x \cdot 6x$, and $3x \cdot 4x$
Factors of -15 are $-1 \cdot 15$, $-15 \cdot 1$, $-3 \cdot 5$, and $-5 \cdot 3$

At first glance this may seem like it will take a while, but there are some ways that we can shorten the process. For example, we may skip over any pairings resulting in a binomial factor in which both terms have a common factor other than 1, such as $3x - 3$. If the terms of a binomial factor had a common factor other than 1, then the original polynomial would also have a common factor, but we know that $12x^2 + 11x - 15$ does not have a common factor that can be factored out.

If the pair of constants -1 and 15 does not produce the desired middle term and does not produce the opposite of the desired middle term, then we do not need to use the pair of constants -15 and 1. Instead we can move on to the next pair of constants.

Eventually, through trial and error, we find that $12x^2 + 11x - 15 = (3x + 5)(4x - 3)$.

***Quick Check* 11**
Factor $12x^2 + 5x - 72$ by trial and error.

EXAMPLE 14 Factor $3x^2 + 33x + 54$.

Solution

The three terms have a common factor of 3. After we factor out this common factor, the trinomial factor will be $x^2 + 11x + 18$, which is a trinomial with a leading coefficient of 1. We look for two integers whose product is 18 and whose sum is 11. Those two integers are 2 and 9.

$$3x^2 + 33x + 54 = 3(x^2 + 11x + 18)$$ Factor out the common factor 3.

$$= 3(x + 2)(x + 9)$$ To factor $x^2 + 11x + 18$ we need to find two integers whose product is 18 and whose sum is 11. The two integers are 2 and 9.

Quick Check 12
Factor $8x^2 + 40x - 112$.

Factoring out the common factor in the previous example makes factoring the trinomial much more manageable. If we were to try factoring without factoring out the common factor, our task would be much more tedious.

Factoring a Perfect Square Trinomial

Objective 5 **Factor a perfect square trinomial.**

EXAMPLE 15 Factor $x^2 + 10x + 25$.

Solution

Since the leading coefficient is 1, we begin by looking for two integers that have a product of 25 and a sum of 10. Since the product of 5 and 5 is 25 and their sum is 10, $x^2 + 10x + 25 = (x + 5)(x + 5)$. Notice that the same factor is listed twice. We can rewrite this as $(x + 5)^2$.

$$x^2 + 10x + 25 = (x + 5)^2$$

Quick Check 13
Factor $x^2 - 16x + 64$.

When a trinomial factors to equal the square of a binomial, we call it a **perfect square trinomial.**

Factoring Trinomials in Several Variables

Objective 6 **Factor a trinomial in several variables.** Trinomials in several variables can also be factored. In the next example we will explore the similarities and the differences between factoring trinomials in two variables and trinomials in one variable.

EXAMPLE 16 Factor $x^2 + 12xy + 27y^2$.

Solution

These three terms have no common factors, and the leading coefficient is 1. Suppose that the variable y were not included; in other words, suppose that we were asked to factor $x^2 + 12x + 27$. After looking for two integers whose product is 27 and whose sum is 12, we would see that $x^2 + 12x + 27 = (x + 3)(x + 9)$. Now we can examine the changes that are necessary because of the second variable y. Note that $3 \cdot 9 = 27$, not $27y^2$. However, $3y \cdot 9y$ does equal $27y^2$, and $x \cdot 9y + x \cdot 3y$ does equal $12xy$. This trinomial factors to be $(x + 3y)(x + 9y)$.

***Quick Check* 14**
Factor $x^2 - 6xy - 40y^2$.

We can factor a trinomial in two or more variables by first ignoring all of the variables except for the first one. After determining how the trinomial factors for the first variable, we can then determine where the rest of the variables appear in the factored form of the trinomial. We can check that our factoring is correct by multiplying, making sure that the product is equal to the trinomial.

EXAMPLE 17 Factor $p^2r^2 - 2pr - 48$.

Solution

If the variable r did not appear in the trinomial, and we were factoring $p^2 - 2p - 48$, we would look for two integers that have a product of -48 and a sum of -2. The two integers are -8 and 6, so the polynomial $p^2 - 2p - 48 = (p - 8)(p + 6)$. Now we work on the variable r. Since the first term in the trinomial is p^2r^2, the first term in each binomial must be pr. $p^2r^2 - 2pr - 48 = (pr - 8)(pr + 6)$.

***Quick Check* 15**
Factor $x^2y^2 + xy - 42$.

Building Your Study Strategy **Math Anxiety, 5 Giving a Full Effort** Many students who perform poorly in their math class attribute their performance to math anxiety, when, in reality, poor study skills are to blame. Be sure that you are giving math your fullest possible effort, including

- working with a study group,
- reading the text before the material is covered in class,
- rereading the text after the material is covered in class,
- completing each homework assignment,
- making note cards for particularly difficult problems or procedures,
- getting help if there is a problem that you do not understand,
- asking your instructor questions when you do not understand, and
- seeing a tutor.

Exercises 4.5

For Extra Help: MyMathLab, MathXL, Interactmath.com, MathXL Tutorials on CD, Video Lectures on CD, Addison-Wesley Math Tutor Center, Student's Solutions Manual

Vocabulary

1. A polynomial of the form $x^2 + bx + c$ is a second-degree trinomial with a(n) ________ of 1.

2. A(n) ________ trinomial is a trinomial with two identical binomial factors.

3. A polynomial is ________ if it cannot be factored.

4. Before attempting to factor a second-degree trinomial, it is wise to factor out any ________.

5. A second-degree trinomial with a leading coefficient not equal to 1 can be written in the form ________.

6. The two techniques for factoring trinomials of the form $ax^2 + bx + c$, $(a \neq 1)$ are called factoring by ________ and by ________.

Factor completely. If the polynomial cannot be factored, write "prime."

7. $x^2 + 8x + 12$
8. $x^2 + 13x + 40$
9. $x^2 + 12x + 27$
10. $x^2 + 13x + 42$
11. $x^2 - 14x + 48$
12. $x^2 - 9x + 14$
13. $x^2 - 3x + 10$
14. $x^2 - 7x + 12$
15. $x^2 + 3x - 18$
16. $x^2 + 9x - 14$
17. $x^2 + 4x - 45$
18. $x^2 + 3x - 4$
19. $x^2 - 4x - 21$
20. $x^2 - 5x - 50$
21. $x^2 - x - 72$
22. $x^2 - 11x - 30$
23. $x^2 + 14x + 45$
24. $x^2 + 6x - 16$
25. $x^2 + 3x + 40$
26. $x^2 - 15x + 54$
27. $x^2 - 3x - 28$
28. $x^2 + 2x - 24$
29. $x^2 + 19x - 90$
30. $x^2 + 18x + 80$
31. $x^2 + x - 30$
32. $x^2 + 10x + 8$
33. $x^2 + 6x - 27$
34. $x^2 - 12x + 27$
35. $x^2 - x - 6$
36. $x^2 + 7x - 18$
37. $3x^2 - 30x + 48$
38. $4x^2 + 32x + 60$
39. $2x^2 - 4x - 126$
40. $10x^2 + 10x - 20$
41. $5x^3 - 30x^2 + 25x$
42. $2x^4 - 2x^3 - 60x^2$
43. $-x^2 - 7x + 30$
44. $-x^2 - 11x - 28$
45. $-3x^2 + 39x - 120$
46. $-2x^2 + 8x + 64$
47. $2x^2 + 13x + 15$
48. $3x^2 + 10x + 3$
49. $2x^2 + x - 36$
50. $5x^2 + 6x - 8$
51. $7x^2 + 41x - 6$
52. $3x^2 - 28x - 20$
53. $2x^2 - 19x + 24$
54. $3x^2 - 19x + 20$
55. $4x^2 + 16x + 15$
56. $12x^2 + 17x + 6$
57. $9x^2 + 15x - 14$
58. $8x^2 + 6x - 5$
59. $15x^2 - 34x + 15$
60. $12x^2 - 43x + 36$
61. $12x^2 + 41x + 15$
62. $12x^2 - 52x - 9$

63. $12x^2 + 23x - 24$

64. $12x^2 - 28x + 15$

65. $12x^2 + 35x + 18$

66. $20x^2 + 91x + 60$

67. $x^2 + 8x + 16$

68. $x^2 + 12x + 36$

69. $x^2 - 4x + 4$

70. $x^2 - 14x + 49$

71. $x^2 - 2x + 1$

72. $x^2 + 18x + 81$

73. $9x^2 - 12x + 4$

74. $4x^2 + 20x + 25$

75. $x^2 + 16xy + 63y^2$

76. $x^2 - xy - 42y^2$

77. $x^2y^2 - 14xy + 45$

78. $x^2y^2 + 14xy + 40$

79. $x^2 - 6xy + 9y^2$

80. $x^2y^2 + 10xy + 25$

81. $2x^2 - 7xy + 3y^2$

82. $4x^2 + 8xy + 3y^2$

83. $6x^2y^2 - 5xy - 4$

84. $5x^2y^2 + 7xy - 24$

Mixed Practice, 85–108

Factor completely, using the appropriate technique.

85. $x^2 - 7x$

86. $x^3 + 6x^2 - 10x - 60$

87. $x^2 - 2x - 24$

88. $x^2 + 11x + 30$

89. $x^3 + 6x^2 + 8x + 48$

90. $3x^3 - 21x^2 + 4x - 28$

91. $x^2 + 11x + 24$

92. $6x^2 + x - 2$

93. $3x^2 + 6x - 105$

94. $x^2 - 12x + 35$

95. $3x^2 - 14x + 16$

96. $x^2 + 5x - 24$

97. $4x^2 + 9x - 9$

98. $x^2 - 8x - 20$

99. $5x^3 - 15x^2 - 6x + 18$

100. $15x^3 + 18x^2$

101. $x^2 + 3x - 54$

102. $4x^2 + 20x + 21$

103. $x^2 + 14x + 36$

104. $x^2 - 8x + 16$

105. $x^2 - 9x + 8$

106. $x^2 + 13x - 42$

107. $x^2 + 14x + 49$

108. $4x^2 - 24x - 64$

Writing in Mathematics

Answer in complete sentences.

109. Explain why it is a good idea to factor out a common factor of -1 from the expression $-x^2 + 7x + 30$ before attempting to factor the trinomial.

110. Explain why it is a good idea to factor out a common factor from a trinomial of the form $ax^2 + bx + c$ before trying other factoring techniques.

111. ***Solutions Manual**** Write a solutions manual page for the following problem:

Factor $3x^2 - 9x - 84$.

112. ***Solutions Manual**** Write a solutions manual page for the following problem:

Factor $6x^2 - 35x + 36$.

113. ***Newsletter**** Write a newsletter explaining how to factor a trinomial of the form $x^2 + bx + c$.

114. ***Newsletter**** Write a newsletter explaining how to factor a trinomial of the form $ax^2 + bx + c$, $(a \neq 1)$, either by grouping or by trial and error.

***See Appendix B for details and sample answers.**

4.6 Factoring Special Binomials

Objectives

1. **Factor a difference of squares.**
2. **Factor the factors of a difference of squares.**
3. **Factor a difference of cubes.**
4. **Factor a sum of cubes.**

Difference of Squares

Objective 1 Factor a difference of squares. Recall the special product $(a + b)(a - b) = a^2 - b^2$ from Section 4.3. The binomial $a^2 - b^2$ is a **difference of squares**. Based on this special product, a difference of squares factors in the following manner.

Difference of Squares

$$a^2 - b^2 = (a + b)(a - b)$$

To identify a binomial as a difference of squares, we must verify that each term is a perfect square. Variable factors must have exponents that are multiples of 2 such as x^2, y^4, and a^6. Any constant must also be a perfect square. Here are the first ten perfect squares.

1, 4, 9, 16, 25, 36, 49, 64, 81, 100

EXAMPLE 1 Factor $x^2 - 49$.

Solution

This binomial is a difference of squares, as it can be rewritten $(x)^2 - (7)^2$. Rewriting the binomial in this form helps us to use the formula $a^2 - b^2 = (a + b)(a - b)$.

$$\begin{aligned} x^2 - 49 &= (x)^2 - (7)^2 && \text{Rewrite each term as a square.} \\ &= (x + 7)(x - 7) && \text{Factor using the formula for a difference of squares, } a^2 - b^2 = (a+b)(a-b). \end{aligned}$$

We can check our work by multiplying $x + 7$ by $x - 7$, which should equal $x^2 - 49$. The check is left to the reader.

Quick Check 1
Factor $x^2 - 25$.

EXAMPLE 2 Factor $4x^2 - 81$.

Solution

This binomial is a difference of squares, as it can be rewritten $(2x)^2 - (9)^2$.

$$\begin{aligned} 4x^2 - 81 &= (2x)^2 - (9)^2 && \text{Rewrite each term as a square.} \\ &= (2x + 9)(2x - 9) && \text{Factor using the formula for a difference of squares, } a^2 - b^2 = (a+b)(a-b). \end{aligned}$$

Quick Check 2
Factor $9x^2 - 16$.

Although we will continue to write each term as a perfect square, if you can factor a difference of squares without writing each term as a perfect square, then do not feel that you *must* rewrite each term as a perfect square before proceeding.

EXAMPLE 3 Factor $36a^6 - 25b^4$.

Solution

The first term of this binomial is a square, as 36 is a square and the exponent for the variable factor a is even. The first term can be rewritten $(6a^3)^2$. In a similar fashion, $25b^4$ can be rewritten $(5b^2)^2$. This binomial is a difference of squares.

$$\begin{aligned} 36a^6 - 25b^4 &= (6a^3)^2 - (5b^2)^2 && \text{Rewrite each term as a square.} \\ &= (6a^3 + 5b^2)(6a^3 - 5b^2) && \text{Factor as a difference of squares.} \end{aligned}$$

Quick Check 3
Factor $49x^{12} - 25y^{16}$.

The binomial $8x^2 - 81$ is not a difference of squares because the coefficient of the first term (8) is not a square. Since there are no common factors other than 1, this binomial cannot be factored.

EXAMPLE 4 Factor $5x^2 - 320$.

Solution

Although this binomial is not a difference of squares (5 and 320 are not squares), the common factor of 5 can be factored out. After we factor out the common factor 5, we have $5(x^2 - 64)$. The binomial in parentheses is a difference of squares and can be factored accordingly.

$$\begin{aligned} 5x^2 - 320 &= 5(x^2 - 64) && \text{Factor out the common factor 5.} \\ &= 5\left[(x)^2 - (8)^2\right] && \text{Rewrite each term in the binomial factor as a square.} \\ &= 5(x + 8)(x - 8) && \text{Factor as a difference of squares.} \end{aligned}$$

Quick Check 4
Factor $3x^2 - 75$.

If a binomial is a **sum of squares,** then it cannot be factored unless the two terms have a common factor. Some examples of a sum of squares are $x^2 + 49$, $4a^2 + 9b^2$, and $64a^{10} + 25b^4$.

Sum of Squares

A sum of squares, $a^2 + b^2$, cannot be factored.

Objective 2 **Factor the factors of a difference of squares.** Occasionally, after we have factored a difference of squares, one or more of the factors may still be factorable. For example, consider the binomial $x^4 - 16$, which factors to be $(x^2 + 4)(x^2 - 4)$. The first factor, $x^2 + 4$, is a sum of squares and cannot be factored. However, the second factor, $x^2 - 4$, is a difference of squares and can be factored further.

EXAMPLE 5 Factor $x^4 - 16$ completely.

Solution

This is a difference of squares and we factor it accordingly. After we factor the binomial as a difference of squares, we see that one of the binomial factors $(x^2 - 4)$ is a difference of squares and must be factored as well.

$$
\begin{aligned}
x^4 - 16 &= (x^2)^2 - (4)^2 && \text{Rewrite each term as a perfect square.}\\
&= (x^2 + 4)(x^2 - 4) && \text{Factor as a difference of squares.}\\
&= (x^2 + 4)[(x)^2 - (2)^2] && \text{Rewrite each term in the binomial } x^2 - 4 \text{ as a square.}\\
&= (x^2 + 4)(x + 2)(x - 2) && \text{Factor } x^2 - 4 \text{ as a difference of squares.}
\end{aligned}
$$

***Quick Check* 5**
Factor $a^4 - 81$.

A Word of Caution When you are factoring a difference of squares, check the binomial factor containing a difference to see if it can be factored.

Difference of Cubes

Objective 3 **Factor a difference of cubes.** Another special binomial that can be factored is a **difference of cubes**. In this case, both terms are perfect cubes rather than squares. Here is the formula for factoring a difference of cubes.

Difference of Cubes

$$a^3 - b^3 = (a - b)(a^2 + ab + b^2)$$

Before proceeding, let's multiply out $(a - b)(a^2 + ab + b^2)$ to show that it actually equals $a^3 - b^3$.

$$
\begin{aligned}
(a - b)(a^2 + ab + b^2) &= a \cdot a^2 + a \cdot ab + a \cdot b^2 - b \cdot a^2 - b \cdot ab - b \cdot b^2 && \text{Distribute.}\\
&= a^3 + a^2b + ab^2 - a^2b - ab^2 - b^3 && \text{Multiply.}\\
&= a^3 - b^3 && \text{Combine like terms.}
\end{aligned}
$$

We see that the formula is correct.

In order for a term to be a cube, its variable factors must have exponents that are multiples of 3, such as x^3, y^6, and z^9. Also each constant factor must be a perfect cube. Here are the first ten perfect cubes.

1, 8, 27, 64, 125, 216, 343, 512, 729, 1000

We need to memorize the formula for factoring a difference of cubes. There are some patterns that can help us to remember the formula. A difference of cubes, $a^3 - b^3$, has two factors: a binomial $(a - b)$ and a trinomial $(a^2 + ab + b^2)$. The binomial looks just like the difference of cubes without the cubes, including the sign between the terms. The signs between the three terms in the trinomial are both addition signs. To find the actual terms in the trinomial factor, the following diagram may be helpful.

$$a^3 - b^3 = (\overset{\text{1st}}{a} - \overset{\text{2nd}}{b})\;(\underset{\text{1st}\cdot\text{1st}}{a^2} + \overset{\text{1st}\cdot\text{2nd}}{ab} + \underset{\text{2nd}\cdot\text{2nd}}{b^2})$$

EXAMPLE 6 Factor $x^3 - 8$.

Solution

We begin by looking for common factors other than 1 that the two terms share, but there are none. This binomial is not a difference of squares, as the exponent of the variable term is not a multiple of 2 and the constant 8 is not a square. The binomial is, however, a difference of cubes. We can rewrite the term x^3 as $(x)^3$ and we can rewrite 8 as $(2)^3$. By rewriting $x^3 - 8$ as $(x)^3 - (2)^3$, we will be better able to identify the terms that are in the binomial and trinomial factors.

$x^3 - 8 = (x)^3 - (2)^3$ Rewrite each term as a perfect cube.

$= (x - 2)(x \cdot x + x \cdot 2 + 2 \cdot 2)$ Factor as a difference of cubes. The binomial factor is the same as $(x)^3 - (2)^3$ without the cubes. Treating x as the "1st" term in the binomial factor and 2 as the "2nd" term, the terms in the trinomial are "1st · 1st" + "1st · 2nd" + "2nd · 2nd."

$= (x - 2)(x^2 + 2x + 4)$ Simplify each term in the trinomial factor.

Quick Check 6
Factor $x^3 - 216$.

A Word of Caution A difference of cubes $x^3 - y^3$ cannot be factored as $(x - y)^3$.

EXAMPLE 7 Factor $m^9 - 64$.

Solution

Although the number 64 is a square, this binomial is not a difference of squares because m^9 is not a square. In order for a term to be a square, its variable factors must have exponents that are multiples of 2. This binomial is a difference of cubes. The exponent in the first term is a multiple of 3, and the number 64 is equal to 4^3.

$m^9 - 64 = (m^3)^3 - (4)^3$ Rewrite each term as a perfect cube.

$= (m^3 - 4)(m^3 \cdot m^3 + m^3 \cdot 4 + 4 \cdot 4)$ Factor as a difference of cubes.

$= (m^3 - 4)(m^6 + 4m^3 + 16)$ Simplify each term in the trinomial factor.

Quick Check 7
Factor $x^{18} - 8$.

A Word of Caution When factoring a difference of cubes $a^3 - b^3$, do not factor the trinomial factor $a^2 + ab + b^2$.

Sum of Cubes

Objective 4 **Factor a sum of cubes.** Unlike a sum of squares, a **sum of cubes** can be factored. Here is the formula.

Sum of Cubes

$$a^3 + b^3 = (a + b)(a^2 - ab + b^2)$$

Notice that the terms in the factors are the same as the factors in a *difference* of cubes, with the exception of some of their signs. When we factor a sum of cubes, the binomial factor is a sum rather than a difference. The sign of the middle term in the trinomial is negative rather than positive. The last term in the trinomial factor is positive, just as it was in the formula for a difference of cubes. The diagram shows the differences between the formula for a difference of cubes and a sum of cubes.

$$a^3 - b^3 = (a - b)(a^2 + ab + b^2)$$

$$a^3 + b^3 = (a + b)(a^2 - ab + b^2)$$

EXAMPLE 8 Factor $z^3 + 27$.

Solution

This binomial is a sum of cubes and can be rewritten as $(z)^3 + (3)^3$. We then can factor the sum of cubes using the formula.

$$\begin{aligned} z^3 + 27 &= (z)^3 + (3)^3 && \text{Rewrite each term as a perfect cube.} \\ &= (z + 3)(z \cdot z - z \cdot 3 + 3 \cdot 3) && \text{Factor as a sum of cubes.} \\ &= (z + 3)(z^2 - 3z + 9) && \text{Simplify each term in the trinomial factor.} \end{aligned}$$

Quick Check 8
Factor $x^3 + 343$.

A Word of Caution A sum of cubes $x^3 + y^3$ cannot be factored as $(x + y)^3$.

EXAMPLE 9 Factor $64p^3 + 729r^3$.

Solution

This binomial is a sum of cubes and can be rewritten as $(4p)^3 + (9r)^3$.

$64p^3 + 729r^3 = (4p)^3 + (9r)^3$ Rewrite each term as a perfect cube.

$= (4p + 9r)(4p \cdot 4p - 4p \cdot 9r + 9r \cdot 9r)$ Factor as a sum of cubes.

$= (4p + 9r)(16p^2 - 36pr + 81r^2)$ Simplify each term in the trinomial factor.

Quick Check 9
Factor $125n^3 + 216$.

We finish this section by summarizing the strategies for factoring binomials.

Factoring Binomials

- Factor out the GCF, if there is a common factor other than 1.
- Determine whether both terms are perfect squares.
  If the binomial is a sum of squares, it cannot be factored.
  If the binomial is a difference of squares, we can factor it by using the formula $a^2 - b^2 = (a + b)(a - b)$. Keep in mind that some of the resulting factors can be differences of squares as well, which will need to be further factored.
- If both terms are not perfect squares, determine whether they both are perfect cubes.
  If the binomial is a difference of cubes, we factor using the formula $a^3 - b^3 = (a - b)(a^2 + ab + b^2)$.
  If the binomial is a sum of cubes, we factor using the formula $a^3 + b^3 = (a + b)(a^2 - ab + b^2)$.

Building Your Study Strategy **Math Anxiety, 6 Relax** Learning to relax in stressful situations will help you keep your anxiety under control. By being able to relax, you will be able to suppress the physical symptoms associated with math anxiety, such as feeling tense and uncomfortable, sweating, having shortness of breath, and having an accelerated heartbeat.

Wanting to relax and being able to relax are often two different things. Many students use deep breathing exercises. Close your eyes and take a deep breath. Try to clear your mind and relax your entire body. Slowly exhale, and then repeat the process until you feel relaxed.

There are other relaxation methods that may work for you, but you must make sure that whatever technique you use, it is something you can practice in class. For instance, primal scream therapy would be inappropriate during an exam.

EXERCISES 4.6

Vocabulary

1. A binomial of the form $a^2 - b^2$ is a(n) ________.

2. A numerical factor is a(n) ________ if it can be written as n^2 for some integer n.

3. A variable factor is a perfect square if its exponents are ________.

4. A binomial of the form $a^2 + b^2$ is a(n) ________.

5. A binomial of the form $a^3 - b^3$ is a(n) ________.

6. A binomial of the form $a^3 + b^3$ is a(n) ________.

Factor completely. If the polynomial cannot be factored, write "prime."

7. $x^2 - 16$
8. $x^2 - 9$
9. $a^2 - 25$
10. $b^2 - 36$
11. $49x^2 - 9$
12. $64x^2 - 81$
13. $x^2 - 6$
14. $x^2 - 24$
15. $a^2 - 64b^2$
16. $m^2 - 16n^2$
17. $25x^2 - 64y^2$
18. $36x^2 - 121y^2$
19. $25 - x^2$
20. $121 - x^2$
21. $3x^2 - 108$
22. $8x^2 - 200$
23. $x^2 + 16$
24. $x^2 + 9$
25. $4x^2 + 16$
26. $9x^2 + 225y^2$
27. $x^4 - 16$
28. $x^4 - 1$
29. $x^4 - 49y^2$
30. $x^8 - 36y^4$

31. $x^3 - 1$
32. $x^3 - y^3$
33. $x^3 - 27$
34. $a^3 - 216$
35. $125 - y^3$
36. $64 - x^3$
37. $x^3 - 8y^3$
38. $y^3 - 27x^3$
39. $125a^3 - 729b^3$
40. $27r^3 - 8s^3$

41. $x^3 + y^3$
42. $x^3 + 125$
43. $x^3 + 8$
44. $b^3 + 64$
45. $x^6 + 216$
46. $y^{12} + 27$
47. $8m^3 + n^3$
48. $x^3 + 125y^3$

49. $5x^3 - 40$
50. $7x^3 + 189$
51. $16x^3 + 250y^3$
52. $108a^4 - 32ab^3$

Mixed Practice, 53–88

Factor completely, using the appropriate technique. If the polynomial cannot be factored, write "prime."

53. $x^2 - 10x + 25$
54. $x^3 + 8x^2 + 5x + 40$

55. $x^2 + 9x + 14$

56. $3x^2 - 6x - 45$

57. $x^2 - 9x$

58. $x^2 - 3x + 28$

59. $6x^2 - 5x - 25$

60. $-4x^2 - 10x + 66$

61. $x^4 + 2x^3 - 48x^2$

62. $x^2 + 22x + 112$

63. $-7x^2 - 175$

64. $36 - x^2$

65. $-3x^2 + 51x - 216$

66. $2x^2 + 50x$

67. $x^2 - 3x - 54$

68. $x^2 + 4x - 60$

69. $49x^2 - 36y^2$

70. $x^2 + 4x - 5$

71. $3x^2 + 8x - 4$

72. $x^3 - 216y^3$

73. $x^2 - 64y^6$

74. $x^3 - 64y^6$

75. $3x^4 + 21x^3 + 45x^2$

76. $20x^2 - 61x + 36$

77. $6x^2 + 216$

78. $36x^2 + 6x - 210$

79. $x^2 + 10x + 21$

80. $216x^3 - y^3z^6$

81. $3x^2 + 5x - 28$

82. $2x^2 + 11x + 5$

83. $x^3 + 9x^2 - 5x - 45$

84. $4a^2 + 9b^2$

85. $x^3 + 125y^3$

86. $27x^3 - 64$

87. $4x^5 - 12x^4 + 9x^3$

88. $x^3 - 5x^2 - x + 5$

Determine which factoring technique is appropriate for the given polynomial.

89. ____ $x^2 - 11x + 40$	**a.** factor out GCF
90. ____ $x^2 + 81$	**b.** factor by grouping
91. ____ $5a^4b^5 - 15a^3b^6 + 20a^8b^2$	**c.** trinomial, leading coefficient of 1
92. ____ $3x^2 + 16x - 12$	**d.** trinomial, leading coefficient other than 1
93. ____ $27x^3 - 1000$	**e.** difference of squares
94. ____ $x^3 - 5x^2 - 6x + 30$	**f.** sum of squares
95. ____ $x^3 + 216y^3$	**g.** difference of cubes
96. ____ $x^2 - 121$	**h.** sum of cubes

Writing in Mathematics

Answer in complete sentences.

97. Explain how to identify a binomial as a difference of squares.

98. A student was asked to factor $4 - x^2$. The student's answer was $(x + 2)(x - 2)$. Was the student correct? If not, what was the student's error? Explain fully.

99. Explain how to identify a binomial as a difference of cubes or as a sum of cubes.

100. Explain, in your own words, how to remember the formula for factoring a difference of cubes.

101. ***Solutions Manual**** Write a solutions manual page for the following problem:

Factor $8x^3 - 27y^3$.

102. ***Newsletter**** Write a newsletter explaining how to factor a difference of squares.

***See Appendix B for details and sample answers.**

4.7 Factoring Polynomials: A General Strategy

Objective

1 Understand the strategy for factoring a general polynomial.

Objective 1 Understand the strategy for factoring a general polynomial. In this section we will review the different factoring techniques introduced in this chapter and develop a general strategy for factoring any polynomial. Keep in mind that a polynomial must be factored completely, meaning that none of its polynomial factors can be factored further. For example, if we factored the trinomial $4x^2 + 24x + 32$ to be $(2x + 4)(2x + 8)$, this would not be factored completely because the two terms in each binomial factor have a common factor of 2 that must be factored out as follows.

$$\begin{aligned} 4x^2 + 24x + 32 &= (2x + 4)(2x + 8) \\ &= 2(x + 2) \cdot 2(x + 4) \\ &= 4(x + 2)(x + 4) \end{aligned}$$

If we factor out all common factors as our first step when factoring a polynomial, we can avoid having to factor out common factors at the end of the process. We could have factored out the common factor of 4 from the polynomial $4x^2 + 24x + 32$ at the very beginning. In addition to ensuring that the polynomial has been factored completely, factoring out common factors will often make our work easier, and will occasionally allow us to factor polynomials that we would not be able to factor otherwise.

Here is a general strategy for factoring any polynomial.

Factoring Polynomials

1. Factor out any common factors.
2. Determine the number of terms in the polynomial.
   a. If there are only **2 terms**, check to see if the binomial is one of the special binomials discussed in Section 4.6.
      - Difference of Squares: $a^2 - b^2 = (a + b)(a - b)$
      - Sum of Squares: $a^2 + b^2$ is not factorable
      - Difference of Cubes: $a^3 - b^3 = (a - b)(a^2 + ab + b^2)$
      - Sum of Cubes: $a^3 + b^3 = (a + b)(a^2 - ab + b^2)$

   b. If there are **3 terms**, try to factor the trinomial using the techniques discussed in Section 4.5.
      - $x^2 + bx + c = (x + m)(x + n)$: Find two integers m and n whose product is c and whose sum is b.
      - $ax^2 + bx + c\ (a \neq 1)$: We have two methods for factoring this type of trinomial: grouping, and trial and error.

   c. If there are **4 terms**, try factoring by grouping, discussed in Section 4.4.
3. After the polynomial has been factored, be sure that any factor with two or more terms does not have any common factors other than 1. If there are common factors, factor them out.
4. Check your factoring through multiplication.

We will now factor several polynomials of various forms. Some of the examples will have twists that we did not see in the previous sections. The focus will be on identifying the best technique.

Example 1 Factor $8x^2 + 32x - 96$ completely.

Solution

These three terms have a common factor of 8, and after we factor it out, we have $8(x^2 + 4x - 12)$. This is a quadratic trinomial with a leading coefficient of 1, so we factor it using the method introduced in Section 4.5.

Quick Check 1
Factor $3x^2 - 21x - 54$ completely.

$$8x^2 + 32x - 96 = 8(x^2 + 4x - 12)$$ Factor out the GCF 8.

$$= 8(x + 6)(x - 2)$$ Find two integers whose product is -12 and whose sum is 4. The integers are 6 and -2.

Example 2 Factor $-4x^2 + 12x + 40$ completely.

Solution

These three terms have a common factor of -4, and after we factor it out, we have $-4(x^2 - 3x - 10)$. This remaining polynomial is a quadratic trinomial with a leading coefficient of 1.

Quick Check 2
Factor $-8x^2 - 80x + 192$ completely.

$$-4x^2 + 12x + 40 = -4(x^2 - 3x - 10)$$ Factor out the GCF -4.

$$= -4(x - 5)(x + 2)$$ Find two integers whose product is -10 and whose sum is -3. The integers are -5 and 2.

Example 3 Factor $6x^2 - 150y^2$ completely.

Solution

These two terms have a common factor of 6, and after we factor it out, we have $6(x^2 - 25y^2)$. This is a difference of squares, so we factor it using the method introduced in Section 4.6.

Quick Check 3
Factor $12x^2 - 300y^4$ completely.

$$6x^2 - 150y^2 = 6(x^2 - 25y^2)$$ Factor out the GCF 6.

$$= 6(x + 5y)(x - 5y)$$ Factor the difference of squares.

Example 4 Factor $27r^9 - 64s^3t^6$ completely.

Solution

There are no common factors other than 1, so we begin by noticing that this is a binomial. This is not a difference of squares, as $27r^9$ cannot be rewritten as a perfect square. It is, however, a difference of cubes. The variable factors all have exponents that are multiples of 3, and the coefficients are perfect cubes. (Recall the list of the first 10 perfect cubes given in Section 4.6: $27 = 3^3$ and $64 = 4^3$.) We begin by rewriting each term as a perfect cube.

$27r^9 - 64s^3t^6 = (3r^3)^3 - (4st^2)^3$ Rewrite each term as a perfect cube.

$= (3r^3 - 4st^2)(3r^3 \cdot 3r^3 + 3r^3 \cdot 4st^2 + 4st^2 \cdot 4st^2)$ Factor as a difference of cubes. (Recall the pattern for the trinomial factor: "1st · 1st" + "1st · 2nd" + "2nd · 2nd.")

$= (3r^3 - 4st^2)(9r^6 + 12r^3st^2 + 16s^2t^4)$ Simplify each term in the trinomial factor.

For a difference or sum of cubes, check to be sure that the binomial factor cannot be factored further as one of our special binomials.

***Quick Check* 4** **Factor $8x^9y^{15} - 125z^3$ completely.** $(2x^3y^5 - 5z)(4x^6y^{10} + 10x^3y^5z + 25z^2)$

EXAMPLE 5 Factor $12x^3 + 8x^2 - 27x - 18$ completely.

Solution

This polynomial has four terms, so we will try to factor by grouping. If we have trouble factoring the polynomial as written, we can always rearrange the order of its terms.

$12x^3 + 8x^2 - 27x - 18 = 4x^2(3x + 2) - 9(3x + 2)$ Factor the common factor $4x^2$ from the first two terms and factor the common factor -9 from the last two terms.

$= (3x + 2)(4x^2 - 9)$ Factor out the common factor $3x + 2$.

Notice that the factor $4x^2 - 9$ is a difference of squares and must be factored further.

$$12x^3 + 8x^2 - 27x - 18 = 4x^2(3x + 2) - 9(3x + 2)$$
$$= (3x + 2)(4x^2 - 9)$$
$$= (3x + 2)(2x + 3)(2x - 3)$$
Factor the difference of squares. $4x^2 - 9 = (2x + 3)(2x - 3)$

***Quick Check* 5** **Factor $x^4 + 6x^3 + 27x + 162$ completely.**

If a binomial is both a difference of squares and a difference of cubes, such as $x^6 - 64$, to factor it completely we must start by factoring it as a difference of squares. The next example illustrates this.

EXAMPLE 6 Factor $x^6 - 64$ completely.

Solution

There are no common factors to factor out, so we begin by factoring this binomial as a difference of squares.

$x^6 - 64 = (x^3)^2 - (8)^2$ Rewrite each term as a perfect square.

$= (x^3 + 8)(x^3 - 8)$ Factor as a difference of squares.

Notice that each factor can be factored further, as $x^3 + 8$ is a sum of cubes and $x^3 - 8$ is a difference of cubes. We use the fact that $x^3 + 8 = (x + 2)(x^2 - 2x + 4)$ and $x^3 - 8 = (x - 2)(x^2 + 2x + 4)$ to complete the factoring.

$$\begin{aligned} x^6 - 64 &= (x^3)^2 - (8)^2 \\ &= (x^3 + 8)(x^3 - 8) \\ &= (x + 2)(x^2 - 2x + 4)(x - 2)(x^2 + 2x + 4). \end{aligned}$$

Note what would have occurred if we had first factored $x^6 - 64$ as a difference of cubes.

$$x^6 - 64 = (x^2)^3 - (4)^3$$ Rewrite each term as a perfect cube.

$$= (x^2 - 4)(x^4 + 4x^2 + 16)$$ Factor the difference of cubes.

$$= (x + 2)(x - 2)(x^4 + 4x^2 + 16)$$ Factor the difference of squares.

Quick Check 6
Factor $a^6 - b^6$ completely.

At this point, we do not know how to factor $x^4 + 4x^2 + 16$, so the technique of Example 6, factoring first as a difference of squares, has factored the binomial $x^6 - 64$ more completely.

EXAMPLE 7 Factor $8x^6 + 1$ completely.

Solution

The two terms do not have a common factor other than 1, so we need to determine whether this binomial matches one of our special forms. The binomial can be rewritten as a sum of cubes, $(2x^2)^3 + (1)^3$.

Quick Check 7
Factor $x^6 + 27y^3$ completely.

$$8x^6 + 1 = (2x^2)^3 + (1)^3$$ Rewrite each term as a perfect cube.

$$= (2x^2 + 1)(4x^4 - 2x^2 + 1)$$ Factor as a sum of cubes.

EXAMPLE 8 Factor $20x^6 + 105x^5 - 90x^4$ completely.

Solution

The three terms have a common factor of $5x^4$, so we begin by factoring this out of the trinomial. After doing this, we can factor the trinomial factor $4x^2 + 21x - 18$ by using either of the techniques for factoring a trinomial of the form $ax^2 + bx + c\,(a \neq 0)$ that were introduced in Section 4.5. We will factor it by grouping, although you may prefer to use trial and error.

$$20x^6 + 105x^5 - 90x^4 = 5x^4(4x^2 + 21x - 18)$$ Factor out the common factor $5x^4$.

$$= 5x^4(4x^2 - 3x + 24x - 18)$$ To factor by grouping we need to find two integers whose product is equal to $4(-18) = -72$ and whose sum is 21. The two integers are -3 and 24, so we rewrite the term $21x$ as $-3x + 24x$.

$$= 5x^4[x(4x - 3) + 6(4x - 3)]$$ Factor out the common factor x from the first two terms and factor out the common factor 6 from the last two terms.

$$= 5x^4(4x - 3)(x + 6)$$ Factor out the common factor $4x - 3$.

Quick Check 8
Factor $12x^2 + 52x + 16$ completely.

Building Your Study Strategy **Math Anxiety, 7** **Positive Attitude** Math anxiety causes many students to think negatively about math. A positive attitude will lead to more success than a negative attitude. Jot down some of the negative thoughts that you have about math and your ability to learn and understand mathematics. Write each one of these thoughts on a note card, then on the other side of the note card, write the exact opposite statement on the other side. For example, if you write "I am not smart enough to learn mathematics," on the other side of the card, you should write "I am smart. I do well in all of my classes, and I can learn mathematics also!" Once you have finished your note cards, cycle through the positive thoughts on a regular basis.

A little confidence will go a long way toward improving your performance in class. Have the confidence to sit in the front row and ask questions during class, knowing that having your questions answered will increase your understanding and your chances for a better grade on the next exam. Finally, each time you have a success, no matter how small, reward yourself.

EXERCISES 4.7

Vocabulary

1. The first step for factoring a trinomial is to factor out any __________.
2. A binomial of the form $a^2 - b^2$ is a(n) __________.
3. A binomial of the form $a^2 + b^2$ is a(n) __________.
4. A binomial of the form $a^3 - b^3$ is a(n) __________.
5. A binomial of the form $a^3 + b^3$ is a(n) __________.
6. Factoring can be checked by __________ the factors.

Factor completely. If the polynomial cannot be factored, write "prime."

7. $27x^3 - 8y^3$
8. $x^2 - 3x - 108$
9. $x^3 - 8x^2 - x + 8$
10. $x^2 - 4x + 32$
11. $x^6 - 1$
12. $-4x^2 - 4x + 80$
13. $x^2 - 20x + 91$
14. $x^2 - 36$
15. $x^6 + 64y^3$
16. $x^2 + 13x - 40$
17. $5x^2 - 50x + 105$
18. $6x^2 + 33x + 36$
19. $7x^2 + 31x - 20$
20. $x^3 - 1000y^9$
21. $x^2 + 13x + 36$
22. $x^4 + 25$
23. $x^4 + 5x^3 + 8x + 40$
24. $9x^2 + 63x + 90$
25. $9x^2 - 24x + 16$
26. $a^3b^4 + 2a^2b^5 - 4a^5b$
27. $m^2 + 16mn + 64n^2$
28. $a^2b^2 - 3ab - 40$
29. $x^2 + 3x - 208$

30. $x^5 - 13x^4 + 42x^3$

31. $18x^2 - 30x + 6$

32. $x^3 + 7x^2 - 9x - 63$

33. $x^2 + 17x - 18$

34. $x^6 - 16y^4$

35. $4x^2 - 31x - 8$

36. $16x^4 + 49y^2$

37. $18x^2 - 12x + 66$

38. $x^2y^2 + 13xy + 12$

39. $10x^2 + 7x + 1$

40. $9 + 64x^2y^2$

41. $4x^3 + 12x^2 - 9x - 27$

42. $5x^3 + 320$

43. $x^2 - 21x - 100$

44. $x^8 - 81$

45. $3x^2 + 9x - 84$

46. $8x^2 - 34x + 35$

47. $x^6 - y^6$

48. $x^9 + 8$

49. $-x^2 + 8x - 12$

50. $3x^2 - 5x - 15$

51. $12x^2 + 7x - 12$

52. $x^2 - 19x + 70$

53. $x^6 + 216$

54. $9x^2 - 22x + 8$

55. $125x^6 + 27y^3$

56. $x^4 - 25$

57. $x^2 + x - 72$

58. $8x^2 + 47x - 6$

59. $x^2 + 16x + 64$

60. $x^{15} - 1$

61. $6x^2 + 33x + 36$

62. $x^5 - 49x^3$

Find the missing term(s).

63. $x^2 + 15x - 54 = (x - 3)(x + ?)$

64. $8x^3 + 343y^3z^{12} = (2x + 7yz^4)(4x^2 - 14xyz^4 + ?)$

65. $x^2 - 529 = (x + ?)(x - ?)$

66. $12x^2 + 35x - 187 = (4x - 11)(?)$

67. $108x^3y^6 - 1372z^9 = 4(?)(9x^2y^4 + 21xy^2z^3 + 49z^6)$

68. $x^6 - 4096 = (x + 4)(x^2 - 4x + 16)(?)(x^2 + 4x + 16)$

Writing in Mathematics

Answer in complete sentences.

69. ***Solutions Manual**** Write a solutions manual page for the following problem:

Factor.

- $x^3 + 3x^2 - 10x - 30$
- $x^2 - 5x - 36$
- $3x^2 + 7x - 20$
- $9x^2 - 25$
- $x^3 - 343$
- $x^3 + 125$

70. ***Newsletter**** Write a newsletter explaining the following strategies for factoring polynomials:

- Factoring out the GCF,
- Factoring by grouping,
- Factoring trinomials of the form $x^2 + bx + c$,
- Factoring polynomials of the form $ax^2 + bx + c\ (a \neq 1)$ by grouping,
- Factoring polynomials of the form $ax^2 + bx + c\ (a \neq 1)$ by trial and error,
- Factoring a difference of squares,
- Factoring a difference of cubes, and
- Factoring a sum of cubes.

***See Appendix B for details and sample answers.**

4.8 Solving Quadratic Equations by Factoring

Objectives

1. **Solve an equation using the zero-factor property of real numbers.**
2. **Solve a quadratic equation by factoring.**
3. **Solve a quadratic equation that is not in standard form.**
4. **Solve a quadratic equation that has coefficients that are fractions.**
5. **Find a quadratic equation, given its solutions.**
6. **Solve applied problems using quadratic equations.**

Quadratic Equations

A **quadratic equation** is an equation that can be written as $ax^2 + bx + c = 0$, where a, b, and c are real numbers and $a \neq 0$. This form is referred to as the **standard form of a quadratic equation.**

We have already learned to solve linear equations $(ax + b = 0)$. The difference between these two types of equations is that a quadratic equation has a second-degree term, ax^2. Because of the second-degree term, the techniques used to solve linear equations will not work for quadratic equations.

The Zero-Factor Property of Real Numbers

Objective 1 **Solve an equation using the zero-factor property of real numbers.**
To solve a quadratic equation, we will use the **zero-factor property of real numbers.**

Zero-Factor Property of Real Numbers

If $a \cdot b = 0$, then $a = 0$ or $b = 0$.

The principle behind this property is that if two or more unknown numbers have a product of zero, then at least one of the numbers must be zero. This property only holds true when the product is equal to 0, not for any other numbers.

EXAMPLE 1 Use the zero-factor property to solve the equation $x(x - 6) = 0$.

Solution

In this example, the product of two unknown numbers, x and $x - 6$, is equal to 0. The zero-factor property tells us that either $x = 0$ or $x - 6 = 0$. Essentially, we have taken an equation that we did not know how to solve yet, $x(x - 6) = 0$, and rewritten it as two linear equations that we do know how to solve.

$x(x - 6) = 0$

$x = 0$ or $x - 6 = 0$ Set each factor equal to 0.

$x = 0$ or $x = 6$ Solve each linear equation.

There are two solutions to this equation, 0 and 6. We write these solutions in a solution set as $\{0, 6\}$.

EXAMPLE 2 Use the zero-factor property to solve the equation $(x + 2)(5x - 4) = 0$.

Solution

Applying the zero-factor property gives us the equations $x + 2 = 0$ and $5x - 4 = 0$. We solve each equation to find the solutions to the original equation.

$$(x + 2)(5x - 4) = 0$$

$x + 2 = 0$ or $5x - 4 = 0$ — Set each factor equal to 0.

$x = -2$ or $5x = 4$ — Solve each linear equation.

$x = -2$ or $x = \frac{4}{5}$

The solution set for this equation is $\{-2, \frac{4}{5}\}$.

***Quick Check* 1**

Use the zero-factor property to solve the equation.

a) $(x + 6)(x - 5) = 0$
b) $x(3x - 16) = 0$

If an equation has the product of more than two factors equal to 0, such as $(x + 4)(x + 1)(x - 2) = 0$, we set each factor equal to 0 and solve. The solution set to this equation is $\{-4, -1, 2\}$.

Solving Quadratic Equations by Factoring

Objective 2 **Solve a quadratic equation by factoring.** To solve a quadratic equation by factoring, we will use the following procedure.

Solving Quadratic Equations by Factoring

1. **Write the equation in standard form: $ax^2 + bx + c = 0$.** *We need to collect all the terms on one side of the equation. It helps to collect all the terms on the side that makes the coefficient of the squared term positive.*
2. **Factor the expression $ax^2 + bx + c$ completely.** *If you are struggling with factoring, refer back to Sections 4.4 through 4.7.*
3. **Set each factor equal to 0 and solve the resulting equations.** *Each of these equations should be a linear equation.*
4. **Finish by checking the solutions.** *This is an excellent opportunity to catch mistakes in factoring.*

EXAMPLE 3 Solve $x^2 + 2x - 48 = 0$.

Solution

This equation is already in standard form, so we begin by factoring the expression $x^2 + 2x - 48$.

$$x^2 + 2x - 48 = 0$$

$(x + 8)(x - 6) = 0$ — Factor.

$x + 8 = 0$ or $x - 6 = 0$ — Set each factor equal to 0.

$x = -8$ or $x = 6$ — Solve each equation.

The solution set is $\{-8, 6\}$. Now we will check our solutions.

Check $(x = -8)$

$$\begin{aligned} x^2 + 2x - 48 &= 0 \\ (-8)^2 + 2(-8) - 48 &= 0 \\ 64 + 2(-8) - 48 &= 0 \\ 64 - 16 - 48 &= 0 \\ 0 &= 0 \end{aligned}$$

Check $(x = 6)$

$$\begin{aligned} x^2 + 2x - 48 &= 0 \\ (6)^2 + 2(6) - 48 &= 0 \\ 36 + 2(6) - 48 &= 0 \\ 36 + 12 - 48 &= 0 \\ 0 &= 0 \end{aligned}$$

The check shows our solutions are correct.

***Quick Check* 2**
Solve $x^2 - 6x - 40 = 0$.

A Word of Caution Pay close attention to the directions of a problem. If you are asked to solve a quadratic equation, do not just factor the quadratic expression and stop. If you are asked to factor a quadratic expression, do not set each factor equal to 0 and solve the resulting equations.

EXAMPLE 4 Solve $3x^2 - 39x + 126 = 0$.

Solution

The equation is in standard form. To factor $3x^2 - 39x + 126$, we begin by factoring out the common factor 3.

$$\begin{aligned} 3x^2 - 39x + 126 &= 0 \\ 3(x^2 - 13x + 42) &= 0 && \text{Factor out the GCF.} \\ 3(x - 6)(x - 7) &= 0 && \text{Factor } x^2 - 13x + 42. \\ x - 6 = 0 \quad \text{or} \quad x - 7 &= 0 && \text{Set each factor containing a variable equal to 0.} \\ & && \text{We can ignore the numerical factor 3.} \\ x = 6 \quad \text{or} \quad x &= 7 && \text{Solve each equation.} \end{aligned}$$

The solution set is $\{6, 7\}$. The check is left to the reader.

We do not need to set a numerical factor equal to 0, because such an equation will not have a solution. For example, if we had set the common factor 3 equal to 0, we would have realized that the equation $3 = 0$ has no solution. The zero-factor property tells us that at least one of the three factors must equal 0; we just know that it cannot be the factor 3 that is equal to 0.

***Quick Check* 3**
Solve
$4x^2 + 36x + 32 = 0$.

A Word of Caution A numerical common factor does not affect the solutions of a quadratic equation.

EXAMPLE 5 Solve $2x^2 + x - 45 = 0$.

Solution

When we factor the expression $2x^2 + x - 45$, there is no common factor other than 1 that can be factored out. So the leading coefficient of this trinomial is not 1. We can factor by trial and error or by grouping. We will use factoring by grouping. We look for two integers whose product is equal to $2(-45)$ or -90 and whose sum is 1. The two integers are -9 and 10, so we can rewrite the term x as $-9x + 10x$ and then factor by grouping. (Refer to Section 4.4 to review this technique.)

$$2x^2 + x - 45 = 0$$
$$2x^2 - 9x + 10x - 45 = 0 \quad \text{Rewrite } x \text{ as } -9x + 10x.$$
$$x(2x - 9) + 5(2x - 9) = 0 \quad \text{Factor out the common factor } x \text{ from the first two terms and factor out the common factor 5 from the last two terms.}$$
$$(2x - 9)(x + 5) = 0 \quad \text{Factor out the common factor } 2x - 9.$$
$$2x - 9 = 0 \quad \text{or} \quad x + 5 = 0 \quad \text{Set each factor equal to 0.}$$
$$2x = 9 \quad \text{or} \quad x = -5 \quad \text{Solve each equation.}$$
$$x = \tfrac{9}{2} \quad \text{or} \quad x = -5$$

The solution set is $\{-5, \frac{9}{2}\}$. The check is left to the reader.

Quick Check 4
Solve $6x^2 - 17x + 12 = 0$.

Objective 3 **Solve a quadratic equation that is not in standard form.** We now turn our attention to equations that are not already in standard form. In each of the following examples, the check of the solutions is left to the reader.

EXAMPLE 6 Solve $x^2 = 36$.

Solution

We begin by rewriting this equation in standard form. We can do this by subtracting 36 from each side of the equation. Once this has been done, the expression to be factored in this example is a difference of squares.

$$x^2 = 36$$
$$x^2 - 36 = 0 \quad \text{Subtract 36.}$$
$$(x + 6)(x - 6) = 0 \quad \text{Factor.}$$
$$x + 6 = 0 \quad \text{or} \quad x - 6 = 0 \quad \text{Set each factor equal to 0.}$$
$$x = -6 \quad \text{or} \quad x = 6 \quad \text{Solve each equation.}$$

The solution set is $\{-6, 6\}$.

Quick Check 5
Solve $x^2 + 18 = 99$.

EXAMPLE 7 Solve $x^2 + 9 = -6x$.

Solution

To rewrite this equation in standard form, we need to add $6x$ so that all terms will be on the left side of the equation. When we add $6x$, we must be sure to write the terms in descending order.

$$x^2 + 9 = -6x$$
$$x^2 + 6x + 9 = 0 \quad \text{Add } 6x.$$
$$(x + 3)(x + 3) = 0 \quad \text{Factor.}$$
$$x + 3 = 0 \quad \text{or} \quad x + 3 = 0 \quad \text{Set each factor equal to 0.}$$
$$x = -3 \quad \text{or} \quad x = -3 \quad \text{Solve each equation.}$$

Quick Check 6
Solve $x^2 + 8x = 20$.

Notice that both solutions are identical. In this case we only need to write the repeated solution once. The solution set is $\{-3\}$.

Occasionally we will need to simplify one or both sides of an equation in order to rewrite the equation in standard form. For instance, to solve the equation $x(x + 8) = -15$, we must first multiply x by $x + 8$. You may be wondering why we would want to multiply out the left side, since it is already factored. Although it is factored, the product is equal to -15, not 0, and we need to set it equal to 0 before factoring.

Example 8 Solve $x(x + 8) = -15$.

Solution

As mentioned, we must first multiply x by $x + 8$. Then we can rewrite the equation in standard form.

$x(x + 8) = -15$	
$x^2 + 8x = -15$	Multiply.
$x^2 + 8x + 15 = 0$	Add 15.
$(x + 5)(x + 3) = 0$	Factor.
$x + 5 = 0 \quad \text{or} \quad x + 3 = 0$	Set each factor equal to 0.
$x = -5 \quad \text{or} \quad x = -3$	Solve each equation.

The solution set is $\{-5, -3\}$.

Quick Check 7
Solve
$(x + 4)(x - 3) = 18$.

A Word of Caution Be sure that the equation you are solving is written as a product equal to 0 before setting each factor equal to 0 and solving.

Objective 4 Solve a quadratic equation that has coefficients that are fractions. If an equation contains fractions, clearing those fractions makes it easier to factor the quadratic expression. We can clear the fractions by multiplying both sides of the equation by the LCM of the denominators.

Example 9 Solve $\frac{1}{6}x^2 - \frac{5}{3}x + 4 = 0$.

Solution

The LCM for these two denominators is 6, so we can clear the fractions by multiplying both sides of the equation by 6.

$\frac{1}{6}x^2 - \frac{5}{3}x + 4 = 0$	
$6(\frac{1}{6}x^2 - \frac{5}{3}x + 4) = 6 \cdot 0$	Multiply both sides by 6, which is the LCM of the denominators.
$\overset{1}{\cancel{6}} \cdot \frac{1}{\underset{1}{\cancel{6}}}x^2 - \overset{2}{\cancel{6}} \cdot \frac{5}{\underset{1}{\cancel{3}}}x + 6 \cdot 4 = 6 \cdot 0$	Distribute and divide out common factors.

$x^2 - 10x + 24 = 0$	Multiply.
$(x-4)(x-6) = 0$	Factor.
$x - 4 = 0$ or $x - 6 = 0$	Set each factor equal to 0.
$x = 4$ or $x = 6$	Solve each equation.

The solution set is $\{4, 6\}$.

Quick Check 8
Solve $\frac{1}{4}x^2 + \frac{3}{2}x - 4 = 0$.

Finding a Quadratic Equation, Given Its Solutions

Objective 5 **Find a quadratic equation, given its solutions.** If we know the two solutions to a quadratic equation, then we can find an equation with these solutions. The next example illustrates this process.

EXAMPLE 10 Find a quadratic equation in standard form, with integer coefficients, that has the solution set $\{-3, 7\}$.

Solution

We know that $x = -3$ is a solution to the equation. This tells us that $x + 3$ is a factor of the quadratic expression. Similarly, knowing that $x = 7$ is a solution tells us that $x - 7$ is a factor of the quadratic expression. Multiplying these two factors will give us a quadratic equation with these two solutions.

$x = -3$ or $x = 7$	Begin with the solutions.
$x + 3 = 0$ or $x - 7 = 0$	Rewrite each equation so the right side is equal to 0.
$(x+3)(x-7) = 0$	Write an equation that has these two expressions as factors.
$x^2 - 4x - 21 = 0$	Multiply.

A quadratic equation that has the solution set $\{-3, 7\}$ is $x^2 - 4x - 21 = 0$.

Quick Check 9
Find a quadratic equation in standard form, with integer coefficients, that has the solution set $\{-10, -3\}$.

Notice that we say *a* quadratic equation rather than *the* quadratic equation. There are infinitely many quadratic equations with integer coefficients that have this solution set. For example, multiplying both sides of our equation by 2 gives us the equation $2x^2 - 8x - 42 = 0$, which has the same solution set.

Applications

Objective 6 **Solve applied problems using quadratic equations.**

Consecutive Integer Problems

Recall that consecutive integers follow the pattern $x, x + 1, x + 2$, and so on. Consecutive even integers and consecutive odd integers follow the pattern $x, x + 2, x + 4$, and so on.

EXAMPLE 11 The product of two consecutive positive integers is 132. Find the two integers.

Solution

We begin by creating a table of unknowns. It is important to represent each unknown in terms of the same variable. In this problem we are looking for two consecutive positive integers. We can let x represent the first integer. Since the integers are consecutive, we can represent the second integer by $x + 1$.

Unknowns
First: x
Second: $x + 1$

We know the product is 132, which leads to the equation $x(x + 1) = 132$.

$$x(x + 1) = 132$$
$$x^2 + x = 132 \quad \text{Multiply } x \text{ by } x + 1.$$
$$x^2 + x - 132 = 0 \quad \text{Subtract 132.}$$
$$(x + 12)(x - 11) = 0 \quad \text{Factor.}$$
$$x + 12 = 0 \quad \text{or} \quad x - 11 = 0 \quad \text{Set each factor equal to 0.}$$
$$x = -12 \quad \text{or} \quad x = 11 \quad \text{Solve each equation.}$$

Since we know the integers are positive, we omit the solution $x = -12$. We use the table of our unknowns with the solution $x = 11$ to find the solution to our problem.

Unknowns
First: $x = 11$
Second: $x + 1 = 11 + 1 = 12$

The two integers are 11 and 12. The product of these two integers is 132.

***Quick Check* 10**
The product of two consecutive positive integers is 72. Find the two integers.

A Word of Caution Be sure to check the practicality of your answers to an applied problem. In the previous example, the integers that we were looking for were positive, so we omitted solutions that led to negative integers.

EXAMPLE 12 The product of two consecutive odd positive integers is 63. Find the two integers.

Solution

We begin by creating a table of unknowns. We let x represent the first integer. Since the integers are odd consecutive integers, we can represent the second integer by $x + 2$.

Unknowns

First: x
Second: $x + 2$

We know the product is 63, which leads to the equation $x(x + 2) = 63$.

$$\begin{aligned} x(x+2) &= 63 \\ x^2 + 2x &= 63 && \text{Multiply } x \text{ by } x + 2. \\ x^2 + 2x - 63 &= 0 && \text{Subtract 63.} \\ (x+9)(x-7) &= 0 && \text{Factor.} \\ x + 9 = 0 \quad \text{or} \quad x - 7 &= 0 && \text{Set each factor equal to 0.} \\ x = -9 \quad \text{or} \quad x &= 7 && \text{Solve each equation.} \end{aligned}$$

Since we know the integers are positive, we omit the solution $x = -9$. We use the table of our unknowns with the solution $x = 7$ to find the solution to our problem.

Unknowns

First: $x = 7$
Second: $x + 2 = 7 + 2 = 9$

The two integers are 7 and 9. The product of these two integers is 63.

***Quick Check* 11**
The product of two consecutive even positive integers is 440. Find the two integers.

Geometry Problems

The next example involves the area of a rectangle. Recall that the area of a rectangle is equal to the product of its length and width.

Area = Length · Width.

EXAMPLE 13 The area of a rectangle is 105 square meters. If the length of the rectangle is 8 meters more than its width, find the dimensions of the rectangle.

Solution

For this problem the unknowns are the length and the width of the rectangle. Since the length is defined in terms of the width, we let w represent the width. The length can be represented by the expression $w + 8$ as it is 8 meters longer than the width.

Unknowns

Length: $w + 8$
Width: w

Since the product of the length and the width is equal to the area of the rectangle, the equation we need to solve is $w(w + 8) = 105$.

$$\begin{aligned} w(w+8) &= 105 \\ w^2 + 8w &= 105 && \text{Multiply.} \\ w^2 + 8w - 105 &= 0 && \text{Subtract 105.} \\ (w+15)(w-7) &= 0 && \text{Factor.} \\ w + 15 = 0 \quad \text{or} \quad w - 7 &= 0 && \text{Set each factor equal to 0.} \\ w = -15 \quad \text{or} \quad w &= 7 && \text{Solve each equation.} \end{aligned}$$

We omit the negative solution $w = -15$, as a rectangle cannot have a negative width or length. We can use the solution $w = 7$ and the table of unknowns to find the length and the width of the rectangle.

Quick Check 12

The area of a rectangle is 60 square feet. If the length of the rectangle is 4 feet more than its width, find the dimensions of the rectangle.

Unknowns

Length: $w + 8 = 7 + 8 = 15$
Width: $w = 7$

The length of the rectangle is 15 meters and the width is 7 meters. We can verify that the area of a rectangle with these dimensions is 105 square meters.

EXAMPLE 14 A homeowner has installed an in-ground spa in the backyard. The spa is rectangular in shape, with a length that is 4 feet more than its width. The homeowner put a 1-foot-wide concrete border around the spa. If the area covered by the spa and the border is 96 square feet, find the dimensions of the spa.

Solution

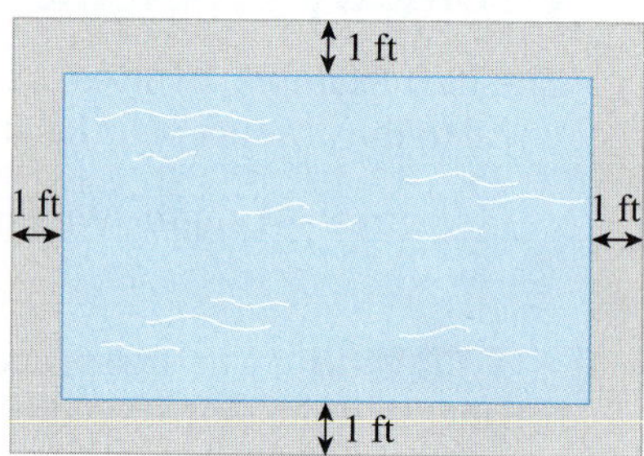

The unknowns are the length and width of the rectangular spa. We will let x represent the width of the spa.

Unknowns

Length: $x + 4$
Width: x

Since we know the area covered by the spa and the border, our equation must involve the outer rectangle in the picture. The width of the border can be represented by $x + 2$ because we need to add 2 feet to the width of the spa (x). The width of the spa is increased by 1 foot on 2 sides. In a similar fashion, the length of the border is $x + 6$ because we add 2 feet to the length of the spa.

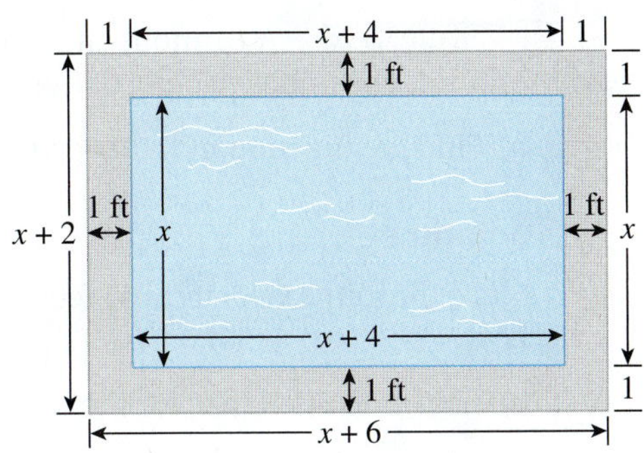

The equation we will solve is $(x + 6)(x + 2) = 96$, since the area covered is 96 square feet.

$$
\begin{aligned}
(x + 6)(x + 2) &= 96 \\
x^2 + 2x + 6x + 12 &= 96 && \text{Multiply.} \\
x^2 + 8x + 12 &= 96 && \text{Combine like terms.} \\
x^2 + 8x - 84 &= 0 && \text{Subtract 96.} \\
(x + 14)(x - 6) &= 0 && \text{Factor.} \\
x + 14 = 0 \quad \text{or} \quad x - 6 &= 0 && \text{Set each factor equal to 0.} \\
x = -14 \quad \text{or} \quad x &= 6 && \text{Solve each equation.}
\end{aligned}
$$

Quick Check 13

A homeowner has poured a rectangular concrete slab in her backyard to use as a barbecue area. The length is 3 feet more than its width. There is a 2-foot–wide flower bed around the barbecue area. If the area covered by the barbecue area and the flower bed is 270 square feet, find the dimensions of the barbecue area.

We will omit the negative solution $x = -14$, as the dimensions of the spa cannot be negative. We use the solution $x = 6$ and the table of unknowns to find the length and the width of the rectangle.

Unknowns

Length: $x + 4 = 6 + 4 = 10$

Width: $x = 6$

The length of the spa is 10 feet and the width of the spa is 6 feet. We can verify that the area covered by the spa and border is indeed 96 square feet.

Projectile Problems

The height, in feet, of a projectile after t seconds can be found using the function $h(t) = -16t^2 + v_0t + s$, where v_0 is the initial velocity of the projectile and s is the initial height.

$h(t) = -16t^2 + v_0t + s$

t: Time (seconds)

v_0: Initial Velocity (feet/second)

s: Initial Height (feet)

This formula does not cover projectiles that continue to propel themselves after launch, such as a rocket with an engine.

Example 15 A cannonball is fired upward from a platform that is 96 feet above the ground. The initial velocity of the object is 80 feet per second.

a) Find the function $h(t)$ that gives the height of the cannonball in feet after t seconds.

Solution

Since the initial velocity is 80 feet per second and the initial height is 96 feet, the function is $h(t) = -16t^2 + 80t + 96$.

b) How long will it take for the cannonball to land on the ground?

Solution

The cannonball's height when it lands on the ground is 0 feet, so we set the function equal to 0 and solve for the time t in seconds.

$-16t^2 + 80t + 96 = 0$	Set $h(t)$ equal to 0.
$0 = 16t^2 - 80t - 96$	Collect all terms on the right side of the equation so the leading coefficient is positive.
$0 = 16(t^2 - 5t - 6)$	Factor out the GCF (16).
$0 = 16(t - 6)(t + 1)$	Factor $t^2 - 5t - 6$.
$t - 6 = 0$ or $t + 1 = 0$	Set each variable factor equal to 0.
$t = 6$ or $t = -1$	Solve each equation.

We will omit the negative solution $t = -1$, as the time must be positive. It takes the cannonball 6 seconds to land on the ground.

Quick Check 14

A model rocket is launched upward from the ground with an initial velocity of 128 feet per second.

a) Find the function $h(t)$ that gives the height of the rocket in feet after t seconds.

b) How long will it take for the rocket to land on the ground?

***Building Your Study Strategy* Math Anxiety, 8 Avoiding Procrastination** Some students put off doing their math homework or studying due to negative feelings about mathematics. Procrastination is your enemy. Convince yourself that you can do it and get started. If you wait until you are tired, you will not be able to give your best effort. If you do not do the homework assignment at all, you will have trouble following the next day's material. Try scheduling a time to devote to mathematics each day, and stick to your schedule.

EXERCISES 4.8

Vocabulary

1. A(n) __________ is an equation that can be written as $ax^2 + bx + c = 0$, where a, b, and c are real numbers and $a \neq 0$.

2. A quadratic equation is in standard form if it is in the form __________.

3. The __________ property of real numbers states that if $a \cdot b = 0$, then $a = 0$ or $b = 0$.

4. Once a quadratic expression has been set equal to 0 and factored, we __________ and solve.

Solve.

5. $(x + 5)(x - 18) = 0$
6. $(x - 4)(x - 9) = 0$
7. $x(x - 15) = 0$
8. $x(2x + 21) = 0$
9. $5(x - 6)(x + 2) = 0$
10. $-4(x + 7)(x - 5) = 0$
11. $(3x + 20)(x + 7) = 0$
12. $(x - 3)(5x - 8) = 0$
13. $x^2 - 5x + 6 = 0$
14. $x^2 - 14x + 45 = 0$
15. $x^2 + 8x + 7 = 0$
16. $x^2 + 10x + 16 = 0$
17. $x^2 + 7x - 30 = 0$
18. $x^2 - x - 72 = 0$
19. $x^2 - 13x - 48 = 0$
20. $x^2 + 5x - 36 = 0$
21. $x^2 - 10x + 25 = 0$
22. $x^2 + 18x + 81 = 0$
23. $x^2 - 25 = 0$
24. $x^2 - 4 = 0$
25. $x^2 + 8x = 0$
26. $x^2 - 16x = 0$
27. $3x^2 - 8x = 0$
28. $12x^2 + 7x = 0$
29. $2x^2 + 6x - 108 = 0$
30. $5x^2 - 5x - 150 = 0$
31. $-x^2 + 5x + 24 = 0$
32. $-x^2 + 11x - 18 = 0$
33. $-3x^2 + 6x + 105 = 0$
34. $-6x^2 + 66x - 60 = 0$
35. $5x^2 + 18x - 8 = 0$
36. $6x^2 + 23x + 15 = 0$
37. $x^2 - x = 12$
38. $x^2 + 11x = -18$
39. $x^2 = 10x - 25$
40. $x^2 - 5x = 84$
41. $x^2 + 15x = 4x - 30$
42. $x^2 + 5x = 2x + 28$
43. $2x^2 - 7x - 27 = x^2 - 6x + 45$
44. $x^2 + 6x = 2x^2 + 11x - 6$
45. $x^2 = 100$
46. $x^2 = 144$
47. $x^2 + 8x = 8x + 49$
48. $x^2 - 3x = -3x + 81$
49. $x(x + 2) = 2(x + 32)$
50. $x(x + 13) = (7x + 8) + (6x + 17)$
51. $x(x + 11) = -28$
52. $x(x - 6) = 72$
53. $(x - 8)(x - 5) = 70$
54. $(x + 9)(x - 7) = -55$
55. $(x + 3)(x + 8) = (x + 7)(x + 2)$
56. $(x - 3)(x + 4) = (x + 10)(x - 6)$
57. $\frac{1}{2}x^2 - \frac{1}{3}x - \frac{4}{3} = 0$
58. $\frac{1}{6}x^2 - \frac{3}{4}x + \frac{1}{3} = 0$
59. $\frac{4}{3}x^2 + \frac{4}{3}x - \frac{7}{4} = 0$
60. $\frac{3}{2}x^2 - 3x - \frac{8}{3} = 0$

FOR EXTRA HELP

MyMathLab

MathXL

Interactmath.com

MathXL Tutorials on CD

Video Lectures on CD

Addison-Wesley Math Tutor Center

Student's Solutions Manual

Find a quadratic equation with integer coefficients that has the given solution set.

61. $\{2, 8\}$

62. $\{-6, 3\}$

63. $\{-9, 0\}$

64. $\{6\}$

65. $\{-10, 10\}$

66. $\{-2, \frac{2}{3}\}$

67. $\{-\frac{3}{5}, \frac{5}{3}\}$

68. $\{-\frac{7}{2}, -\frac{3}{4}\}$

Use the given solution for each quadratic equation to find the other solution.

69. $x^2 + 13x - 378 = 0, x = 14$

70. $x^2 - 73x + 1302 = 0, x = 31$

71. $6x^2 + 5x - 300 = 0, x = -\frac{15}{2}$

72. $48x^2 + 10x - 875 = 0, x = \frac{25}{6}$

73. Two consecutive positive integers have a product of 90. Find the integers.

74. Two consecutive positive integers have a product of 240. Find the integers.

75. Two consecutive even positive integers have a product of 168. Find the integers.

76. Two consecutive even negative integers have a product of 48. Find the integers.

77. Two consecutive odd negative integers have a product of 195. Find the integers.

78. Two consecutive odd positive integers have a product of 575. Find the integers.

79. The width of a rectangular rug is 4 feet less than its length. If the area of the rug is 60 square feet, find the length and width of the rug.

80. The width of a rectangular swimming pool is half its length, and the area covered by the pool is 450 square feet. Find the length and width of the swimming pool.

81. The length of a rectangular classroom is 3 times its width. If the area covered by the classroom is 300 square feet, find the length and width of the classroom.

82. The length of a rectangular photo frame is 4 centimeters less than twice its width. If the area of this rectangle is 160 square centimeters, find the length and width of the frame.

83. Jared has a rectangular garden in his back yard that covers 300 square meters. The length of the garden is 10 meters less than twice its width. Find the dimensions of the garden.

84. The height of a doorway is 1 foot less than 3 times its width. If the area of the doorway is 24 square feet, find the dimensions of the doorway.

85. A photo's width is 2 inches less than its length. A border of 1 inch is placed around the photo, and the area covered by the photo and its border is 120 square inches. Find the dimensions of the photo itself.

86. The width of a rectangular quilt is 20 inches less than its length. After a 5-inch border is placed around the quilt, its area is 2400 square inches. Find the original dimensions of the quilt.

For problems 87–90, use the function $h(t) = -16t^2 + v_0t + s$.

87. A projectile was launched upward from a building 288 feet tall. If the initial velocity of the projectile was 112 feet per second, how long will it take the projectile to land on the ground?

88. Standing on a platform 64 feet high, Marquel throws a softball upward at a speed of 48 feet per second. How long will it take for the softball to land on the ground?

89. A projectile is launched upward from the ground with an initial velocity of 192 feet per second. How long will it take for the projectile to land on the ground?

90. A projectile is launched upward from the ground with an initial velocity of 320 feet per second. How long will it take for the projectile to land on the ground?

Writing in Mathematics

Answer in complete sentences.

91. Explain the zero–product property of real numbers. Describe how this property is used when solving quadratic equations by factoring.

92. When solving a quadratic equation by factoring, explain why we should not factor until one side of the quadratic equation is set equal to 0.

93. Explain why the common factor of 2 has no effect on the solutions of the equation $2(x - 7)(x + 5) = 0$.

94. Write a word problem involving a rectangle with a length of 24 feet and width of 16 feet. Your problem must lead to a quadratic equation.

95. ***Solutions Manual**** Write a solutions manual page for the following problem:

Solve $\frac{1}{6}x^2 - \frac{1}{3}x - 4 = 0$.

96. ***Newsletter**** Write a newsletter explaining how to solve an equation of the form $ax^2 + bx + c = 0$.

***See Appendix B for details and sample answers.**

Chapter 4 Summary

Section 4.1—Topic	Chapter Review Exercises
Simplifying Expressions Using the Rules of Exponents	1–8
Section 4.2—Topic	**Chapter Review Exercises**
Simplifying Expressions with Negative Exponents	9–16
Rewriting Numbers Using Scientific Notation	17–18
Rewriting Numbers That Are in Scientific Notation	19–20
Performing Calculations Involving Scientific Notation	21–22
Solving Applied Problems Using Scientific Notation	23–24
Section 4.3—Topic	**Chapter Review Exercises**
Evaluating Polynomials	25–28
Adding and Subtracting Polynomials	29–32
Evaluating Polynomial Functions	33–34
Adding and Subtracting Polynomial Functions	35–36
Multiplying Polynomials	37–46
Multiplying Polynomial Functions	47–48
Section 4.4—Topic	**Chapter Review Exercises**
Factoring Out the GCF	49
Factoring by Grouping	50–51
Section 4.5—Topic	**Chapter Review Exercises**
Factoring Trinomials of the Form $x^2 + bx + c$	52–57
Factoring Trinomials of the Form $ax^2 + bx + c\ (a \neq 0)$	58–59
Section 4.6—Topic	**Chapter Review Exercises**
Factoring Binomials: Difference of Squares, Difference of Cubes, and Sum of Cubes	60–64
Section 4.7—Topic	**Chapter Review Exercises**
Factoring Polynomials	49–64
Section 4.8—Topic	**Chapter Review Exercises**
Solving Quadratic Equations by Factoring	65–74
Finding a Quadratic Equation Given Its Solution Set	75–76
Solving Applications of Quadratic Equations	77–80

Summary of Chapter 4 Study Strategies

Math anxiety can hinder your success in a mathematics class, but it can be overcome.

- The first step is to understand what has caused your anxiety to begin with.
- Use relaxation techniques to overcome the physical symptoms of math anxiety, and give your best effort.
- Develop a positive attitude and confidence in your abilities.
- Avoid procrastinating.
- Many students believe that they have math anxiety, but poor performance can be caused by other factors as well. Do you completely understand the material, but "freeze up" on the tests? Does this happen in other classes as well? If so, you may have test anxiety, not math anxiety.
- Some students do poorly in their class because they are taking a class that they are not prepared for. Discuss your correct placement with your instructor and your academic counselor if you feel that this may be the case for you.
- Finally, be sure that poor study skills are not the cause of your difficulties. If you are not giving your fullest effort, you cannot expect to learn mathematics.

Chapter 4 Review

Simplify the expression. Write the result without using negative exponents. (Assume all variables represent nonzero real numbers.) **[4.1, 4.2]**

1. $\frac{x^7}{x^5}$

2. $(3x^6)^4$

3. $12x^0$

4. $\left(\frac{2x^5}{5y^3}\right)^3$

5. $9x^6 \cdot 3x^4$

6. $-7a^8b^3 \cdot 4ab^{12}$

7. $(-8x^{10}y^8z^7)^2$

8. $(a^{13}b^{10}c^{19})^5$

9. 8^{-2}

10. $3x^{-6}$

11. $(6x^{-10})^{-2}$

12. $\frac{x^{12}}{x^{-7}}$

13. $m^{14} \cdot m^{-21}$

14. $(3x^{-2}y^{-3}z^6)^{-3}$

15. $a^{-18} \cdot a^{-8}$

16. $\frac{x^{-16}}{x^{-5}}$

Rewrite in scientific notation. **[4.2]**

17. 3,900,000

18. 0.0000065

Convert to standard notation. **[4.2]**

19. 7.2×10^{-8}

20. 4.09×10^{11}

Perform the following calculations. Express your answer using scientific notation. **[4.2]**

21. $(2.5 \times 10^8)(8.4 \times 10^7)$

22. $(4.8 \times 10^{-6}) \div (1.6 \times 10^3)$

23. The mass of a hydrogen atom is 1.66×10^{-24} grams. What is the mass of 2.0×10^{16} hydrogen atoms? **[4.2]**

24. If a computer can perform a calculation in 5.0×10^{-12} seconds, how long would it take the computer to perform 4.0×10^{15} calculations? **[4.2]**

Evaluate the polynomial for the given value of the variable. **[4.3]**

25. $x^2 + 7x - 16$ for $x = -4$

26. $3x^2 - 8x + 6$ for $x = 3$

27. $-x^2 + 10x + 15$ for $x = 8$

28. $-2x^2 + 5x - 19$ for $x = -5$

Add or subtract. **[4.3]**

29. $(x^2 + 3x - 18) + (x^2 - 5x + 24)$

30. $(5x^2 - 19) + (2x^2 + 3x + 6)$

31. $(x^2 + 9x - 10) - (3x^2 - 5x + 42)$

32. $(3x^3 + 2x + 24) - (5x^2 - 7x + 15)$

Evaluate the given polynomial function. **[4.3]**

33. $f(x) = x^2 - 9x + 20, f(-8)$

34. $f(x) = x^2 + 5x - 12, f(2b^2)$

Given the functions $f(x)$ and $g(x)$, find $f(x) + g(x)$ and $f(x) - g(x)$. **[4.3]**

35. $f(x) = x^2 + 4x - 8, g(x) = x^2 - 5x - 32$

36. $f(x) = 2x^2 - 8x - 25, g(x) = 3x^2 + 7x - 6$

Multiply. **[4.3]**

37. $(x - 10)^2$

38. $(x - 7)(x + 12)$

39. $7x(2x^2 - 5x - 9)$

40. $-6x^5 \cdot 9x^4$

41. $(x - 9)(x + 9)$

42. $(4x - 7)(2x + 9)$

43. $(5x + 4)(5x - 4)$

44. $(3x + 8)^2$

45. $-5x^2(7x^2 - 9x - 10)$

46. $(x - 6)(2x^2 - 3x - 8)$

Given the functions $f(x)$ and $g(x)$, find $f(x) \cdot g(x)$. **[4.3]**

47. $f(x) = x + 5, g(x) = 3x + 10$

48. $f(x) = -4x^3, g(x) = -3x^2 + 5x - 12$

Worked-out solutions to Review Exercises marked with ● can be found on page AN-23.

Factor completely. **[4.4, 4.5, 4.6, 4.7]**

49. $2x^3y^4 - 8x^2y^5 + 4x^4y^6$

50. $x^3 - 10x^2 + 3x - 30$

51. $x^3 - 7x^2 - 4x + 28$

52. $x^2 + 6x + 9$

53. $x^2 + 6x - 27$

54. $-4x^2 + 28x + 32$

55. $x^2 - 6x + 40$

56. $x^2 + 15x + 56$

57. $x^2 - 9x - 10$

58. $8x^2 + 27x - 20$

59. $4x^2 - 4x - 15$

60. $6x^2 - 384$

61. $9x^2 - 49y^2$

62. $x^3 + 125y^3$

63. $x^2 + 144$

64. $x^3 - 512$

Solve. **[4.8]**

65. $(x - 4)(2x + 3) = 0$

66. $x^2 - 36 = 0$

67. $x^2 - 12x + 35 = 0$

68. $x^2 + 8x - 48 = 0$

69. $x^2 - 2x - 63 = 0$

70. $x^2 - 15x = 0$

71. $x^2 + 18x + 80 = 0$

72. $3x^2 - 17x + 10 = 0$

73. $x(x + 6) = 55$

74. $(x + 3)(x + 7) = 5$

Find a quadratic equation with integer coefficients that has the given solution set. **[4.8]**

75. $\{-4, 3\}$

76. $\{-\frac{2}{3}, 8\}$

77. Two consecutive odd positive integers have a product of 323. Find the two integers. [4.8]

78. The length of a rectangle is 1 meter less than twice its width. The area of the rectangle is 66 square meters. Find the length and the width of the rectangle. [4.8]

For Exercises 79 and 80, use the function $h(t) = -16t^2 + v_0t + s$.

79. A ball is thrown upward with an initial velocity of 96 feet per second from the edge of a cliff that is 112 feet above a river. How long will it take for the ball to land in the river? [4.8]

80. A projectile is launched upward with an initial velocity of 80 feet per second. How long will it take for the projectile to land on the ground? [4.8]

Chapter 4 Test

For Extra Help — Pass the Test — Test solutions are found on the enclosed CD.

Simplify the expression. Write the result without using negative exponents. (Assume all variables represent nonzero real numbers.)

1. $\dfrac{x^9}{x^4}$
2. $(x^5y^6z^{10})^4$
3. 4^{-3}
4. $\dfrac{x^{-8}}{x^{20}}$
5. $x^{26} \cdot x^{-19}$
6. $(4xy^{-2}z^3)^{-2}$

Rewrite in scientific notation.

7. 490,000,000

Convert to standard notation.

8. 3.6×10^{-5}

Perform the following calculations. Express your answer using scientific notation.

9. $(7.2 \times 10^8)(3.4 \times 10^{-3})$

Add or subtract.

10. $(3x^2 + 4x - 12) - (5x^2 - 7x + 20)$
11. $(x^2 + 9x + 16) + (x^2 - 12x - 8)$

Multiply.

12. $(x + 9)(x - 9)$
13. $(3x + 2)(2x - 7)$

Factor completely.

14. $x^3 + 4x^2 - 6x - 24$
15. $x^2 - 13x + 30$
16. $x^2 - 6x - 72$
17. $6x^2 - 23x - 4$
18. $4x^2 - 25y^2$
19. $8x^3 - 125$

Solve.

20. $x^2 - 49 = 0$
21. $x^2 - 13x + 40 = 0$
22. $x^2 + 3x - 54 = 0$

Find a quadratic equation with integer coefficients that has the given solution set.

23. $\{-4, 7\}$
24. The length of a rectangle is 2 feet more than twice its width. The area of the rectangle is 144 square feet. Find the length and the width of the rectangle.
25. A ball is thrown upward with an initial velocity of 32 feet per second from the edge of a cliff 128 feet above a beach. How long will it take for the ball to land on the beach? Use the function $h(t) = -16t^2 + v_0t + s$.

Mathematicians in History

Sir Isaac Newton

Sir Isaac Newton was an English mathematician and scientist who lived in the 17th and 18th centuries. His work with motion and gravity helped us to better understand our world and our solar system. Alexander Pope once said "Nature and Nature's laws lay hid in the night; God said, Let Newton be! And all was light."

Write a one-page summary (*or* make a poster) of the life of Sir Isaac Newton and his accomplishments.

Interesting issues:

Where and when was Sir Isaac Newton born?

- Describe Newton's upbringing and his relationship with his mother and stepfather.
- It has been said that an apple was the inspiration for Newton's ideas about the force of gravity. Explain how the apple is believed to have inspired Newton's ideas.
- In a letter to Robert Hooke, Newton wrote "If I have been able to see further, it was only because I stood on the shoulders of giants." Explain what this statement means.
- Newton is often referred to as the "Father of Calculus." What is calculus?
- What are Newton's three laws of motion? Explain what they mean in your own words.
- Where was Newton buried?
- What did Newton invent for his pets?

At a recent school carnival, George agreed to be launched out of a cannon to become the world's first flying mathematician. He landed after 3.4625 seconds. George's height, in feet, after t seconds is given by the function $h(t) = -16t^2 + 55.4t$.

a. Complete the following chart listing George's height at 0.5 second intervals.

Time (in seconds)	Height (in feet)
0	
0.5	
1	
1.5	
2	
2.5	
3	
3.5	

b. Create a coordinate system in which the horizontal axis represents time (in seconds) and the vertical axis represents height (in feet). Plot these eight points and connect them to see George's course of flight through the air.

Use the table and graph to answer the following questions:

c. What was George's height when he was launched out of the cannon?

d. What was George's height after 0.5 seconds?

e. How long did it take George to reach a height of 39.4 feet?

f. What was George's height when he landed on the ground? How long did it take (from the time of being launched) for George to land on the ground?

g. When did George reach his maximum height? What was his maximum height?

Cumulative Review Chapters 1–4

Simplify the given expression.

1. $7 - 3(9 - 2 \cdot 6) - 8^2$

2. Evaluate $x^2 + 4x - 45$ for $x = -7$

Simplify.

3. $9(2x - 7) - 3(4x + 5)$

Solve

4. $6x + 25 = 10$

5. $4(3x - 5) - 7x = 2x - 8$

6. Solve the literal equation $-9x + 4y = 23$ for y.

7. Solve $|x - 8| - 3 = 10$.

8. The length of a rectangle is 5 feet less than three times its width, and the perimeter is 62 feet. Find the length and the width of the rectangle.

Solve. Graph your solution on a number line, and express it in interval notation.

9. $5x + 2 > -28$

Solve the inequality. Graph your solution on a number line and write your solution in interval notation.

10. $|x + 4| < 5$

Find the x- and y-intercepts, and use them to graph the equation.

11. $-4x + 3y = 12$

12. Find the slope of the line that passes through the points $(-2, 7)$ and $(-6, -3)$.

Find the slope and y-intercept of the given line.

13. $6x - 2y = 24$

Graph using the slope and y-intercept.

14. $y = -\frac{3}{2}x + 9$

15. Are the two lines $10x + 4y = 25$ and $y = \frac{5}{2}x - 8$ parallel, perpendicular, or neither?

16. Find the slope–intercept form of the equation of a line with a slope of -4 that passes through the point $(2, -9)$.

17. Find the slope–intercept form of the equation of a line that passes through the points $(-3, 8)$ and $(6, -10)$.

18. Graph the inequality $y \geq \frac{3}{4}x - 6$.

19. Graph the function $f(x) = 4x - 1$.

20. Graph the function $f(x) = |x - 1| - 6$. State the domain and the range.

21. Solve the system by graphing.

$y = -2x + 2$
$y = \frac{1}{3}x - 5$

22. Solve the system by substitution.

$y = 3x - 10$
$2x - 5y = -54$

23. Solve the system by addition.

$5x - 4y = 41$
$2x + 10y = -30$

24. Marisela bought shares in two mutual funds, investing a total of \$8000. One fund's shares went up by 7%, while the other fund went up in value by 16%. This was a total profit of \$1055 for Marisela. How much was invested in each mutual fund?

25. Solve the system of inequalities.

$y \geq x - 8$
$x + 5y < 5$

26. Solve the system by addition.

$3x + 2y + z = 17$
$2x + 3y - 2z = 2$
$-4x + 5y + 3z = -49$

Solve the system of equations using martices.

27. $7x + 4y = 22$
$x - 3y = -29$

28. $x - y + 2z = -11$
$-2x + y + 7z = -37$
$4x - 5y - 5z = 17$

Evaluate the determinant.

29. $\begin{vmatrix} 5 & 2 & -3 \\ 7 & -1 & 4 \\ -2 & 9 & 10 \end{vmatrix}$

Solve the system of equations using Cramer's rule.

30. $5x + 2y = -14$
$-4x + 3y = 48$

Simplify the expression. Write the result without using negative exponents. (Assume all variables represent nonzero numbers.)

31. $(3x^4y^7z^8)^4$

32. $\dfrac{a^2b^{-6}}{a^{-9}b^{-2}}$

33. $x^{22} \cdot x^{-31}$

Rewrite in scientific notation.

34. 32,400,000

Add or subtract.

35. $(x^2 + 7x - 16) - (3x^2 - 8x - 41)$

36. $(x^2 - 3x - 19) + (x^2 + 11x - 8)$

Multiply.

37. $(5x - 4)(2x - 9)$

Factor completely.

38. $x^2 - 5x - 84$

39. $12x^2 - 8x - 15$

40. $18x^2 - 98y^2$

Solve.

41. $x^2 + 10x - 21 = 4x + 19$

42. A projectile is launched upward with an initial velocity of 48 feet per second from a platform 160 feet above the ground. How long will it take for the ball to land on the beach? Use the function $h(t) = -16t^2 + v_o t + s$.

CHAPTER 5

Rational Expressions and Equations

In this chapter, we will examine rational expressions, which are fractions whose numerator and denominator are polynomials. We will learn to simplify rational expressions and how to add, subtract, multiply, and divide two rational expressions. Rational expressions have many applications, such as determining the maximum load that a wooden beam can support, finding the illumination from a light source, and solving work–rate problems.

***Study Strategy* Test Taking** *To be successful in a math class, as well as understanding the material, you must be a good test taker. In this chapter, we will discuss the test-taking skills necessary for success in a math class.*

5.1 Rational Expressions and Functions

Objectives

1. Evaluate rational expressions.
2. Find the values for which a rational expression is undefined.
3. Simplify rational expressions to lowest terms.
4. Identify factors in the numerator and denominator that are opposites.
5. Evaluate rational functions.
6. Find the domain of rational functions.

A **rational expression** is a quotient of two polynomials, such as $\frac{x^2 + 15x + 44}{x^2 - 16}$. The denominator of a rational expression must not be zero, as division by zero is undefined. A major difference between rational expressions and linear or quadratic expressions is that a rational expression has one or more variables in the denominator.

Evaluating Rational Expressions

Objective 1 **Evaluate rational expressions.** We can evaluate a rational expression for a particular value of the variable just as we evaluated polynomials. We substitute the value for the variable in the expression and then simplify. When simplifying, we evaluate the numerator and denominator separately, and then simplify the resulting fraction.

EXAMPLE 1 Evaluate the rational expression $\frac{x^2 + 3x - 20}{x^2 - 5x - 8}$ for $x = -4$.

Solution

We begin by substituting -4 for x.

$$\frac{(-4)^2 + 3(-4) - 20}{(-4)^2 - 5(-4) - 8}$$ Substitute -4 for x.

$$= \frac{16 - 12 - 20}{16 + 20 - 8}$$ Simplify each term in the numerator and denominator. Note: $(-4)^2 = 16$

$$= \frac{-16}{28}$$ Simplify the numerator and denominator.

$$= -\frac{4}{7}$$ Simplify.

Quick Check 1

Evaluate the rational expression $\frac{x^2 - 6x - 1}{x^2 + 7x + 4}$ for $x = -2$.

Finding Values for Which a Rational Expression Is Undefined

Objective 2 **Find values for which a rational expression is undefined.** Rational expressions are undefined for values of the variable that cause the denominator to equal 0, as division by 0 is undefined. In general, to find the values for which a rational expression is undefined, we set the denominator equal to 0, ignoring the numerator, and solve the resulting equation.

Example 2 Find the values for which the rational expression $\frac{8}{2x - 3}$ is undefined.

Solution

We begin by setting the denominator, $2x - 3$, equal to 0 and then we solve for x.

$$2x - 3 = 0 \quad \text{Set the denominator equal to 0.}$$
$$2x = 3 \quad \text{Add 3 to both sides of the equation.}$$
$$x = \frac{3}{2} \quad \text{Divide both sides by 2.}$$

The expression $\frac{8}{2x - 3}$ is undefined for $x = \frac{3}{2}$.

Quick Check 2
Find the values for which $\frac{6}{5x + 9}$ is undefined.

Example 3 Find the values for which $\frac{x^2 + 9x}{x^2 - 3x - 40}$ is undefined.

Solution

We begin by setting the denominator $x^2 - 3x - 40$ equal to 0, ignoring the numerator. Notice that the resulting equation is quadratic and can be solved by factoring.

$$x^2 - 3x - 40 = 0 \quad \text{Set the denominator equal to 0.}$$
$$(x + 5)(x - 8) = 0 \quad \text{Factor } x^2 - 3x - 40.$$
$$x + 5 = 0 \quad \text{or} \quad x - 8 = 0 \quad \text{Set each factor equal to 0.}$$
$$x = -5 \quad \text{or} \quad x = 8 \quad \text{Solve.}$$

The expression $\frac{x^2 + 9x}{x^2 - 3x - 40}$ is undefined when $x = -5$ or $x = 8$.

Quick Check 3
Find the values for which $\frac{4x - 7}{x^2 - 10x + 21}$ is undefined.

Simplifying Rational Expressions to Lowest Terms

Objective 3 **Simplify rational expressions to lowest terms.** Rational expressions are often referred to as ***algebraic fractions***. As with numerical fractions, we will learn to simplify rational expressions to lowest terms. In later sections, we will learn to add, subtract, multiply, and divide rational expressions.

We simplified a numerical fraction to lowest terms by dividing out factors that were common to the numerator and denominator. For example, consider the fraction $\frac{30}{84}$. To simplify this fraction, we could begin by factoring the numerator and denominator.

$$\frac{30}{84} = \frac{2 \cdot 3 \cdot 5}{2 \cdot 2 \cdot 3 \cdot 7}$$

The numerator and denominator have common factors of 2 and 3, which are divided out to simplify the fraction to lowest terms.

$$\frac{\overset{1}{\cancel{2}} \cdot \overset{1}{\cancel{3}} \cdot 5}{\underset{1}{\cancel{2}} \cdot 2 \cdot \underset{1}{\cancel{3}} \cdot 7} = \frac{5}{2 \cdot 7} \quad \text{or} \quad \frac{5}{14}$$

Simplifying Rational Expressions

To simplify a rational expression to lowest terms, we first factor the numerator and denominator completely. Then we divide out common factors in the numerator and denominator.

If P, Q, and R are polynomials, $Q \neq 0$, and $R \neq 0$, then $\frac{PR}{QR} = \frac{P}{Q}$.

EXAMPLE 4 Simplify the rational expression $\frac{15x^4}{6x^7}$. (Assume $x \neq 0$.)

Solution

This rational expression has a numerator and denominator that are monomials. In this case, we can simplify the expression using the properties of exponents developed in Chapter 4.

$$\frac{15x^4}{6x^7} = \frac{\overset{5}{\cancel{15}}x^4}{\underset{2}{\cancel{6}}x^7} \quad \text{Divide out the common factor 3.}$$

$$= \frac{5}{2x^3} \quad \text{Divide numerator and denominator by } x^4.$$

Quick Check 4 Simplify the rational expression $\frac{12x^8}{8x^2}$. (Assume $x \neq 0$.)

EXAMPLE 5 Simplify $\frac{(x-4)(x+6)}{(x+6)(x-1)(x-4)}$. (Assume the denominator is nonzero.)

Solution

In this example the numerator and denominator have already been factored. Notice that they share the common factors $x - 4$ and $x + 6$.

$$\frac{(x-4)(x+6)}{(x+6)(x-1)(x-4)} = \frac{\overset{1}{\cancel{(x-4)}}\overset{1}{\cancel{(x+6)}}}{\underset{1}{\cancel{(x+6)}}(x-1)\underset{1}{\cancel{(x-4)}}} \quad \text{Divide out common factors.}$$

$$= \frac{1}{x-1} \quad \text{Simplify.}$$

Quick Check 5 Simplify $\frac{(x-5)(x-3)(x-1)}{(x+1)(x-5)(x-3)}$. (Assume the denominator is nonzero.)

Notice that both factors were divided out of the numerator. Be careful to note that the numerator is equal to 1 in such a situation.

A Word of Caution If each factor in the numerator of a rational expression is divided out when simplifying the expression, be sure to write a 1 in the numerator.

EXAMPLE 6 Simplify $\frac{x^2 + 11x + 24}{x^2 + 4x - 32}$. (Assume the denominator is nonzero.)

Quick Check 6 Simplify $\frac{x^2 - 3x - 18}{x^2 + 3x - 54}$. (Assume the denominator is nonzero.)

Solution

The trinomials in the numerator and denominator must be factored before we can simplify this expression. (For a review of factoring techniques, you may refer to Sections 4.4 through 4.7.)

$$\frac{x^2 + 11x + 24}{x^2 + 4x - 32} = \frac{(x+3)(x+8)}{(x+8)(x-4)} \quad \text{Factor numerator and denominator.}$$

$$= \frac{(x+3)\overset{1}{\cancel{(x+8)}}}{\underset{1}{\cancel{(x+8)}}(x-4)} \quad \text{Divide out common factors.}$$

$$= \frac{(x+3)}{(x-4)} \quad \text{Simplify.}$$

A Word of Caution When simplifying a rational expression, be very careful to divide out only expressions that are common factors of the numerator and denominator. You cannot *reduce* individual terms in the numerator and denominator as in the following examples:

$$\frac{x+8}{x-6} \neq \frac{x+\overset{4}{\cancel{8}}}{x-\underset{3}{\cancel{6}}}$$

$$\frac{x^2-25}{x^2-36} \neq \frac{\overset{1}{\cancel{x^2}}-25}{\underset{1}{\cancel{x^2}}-36}$$

To avoid this, remember to factor the numerator and denominator completely before attempting to divide out common factors.

EXAMPLE 7 Simplify $\frac{2x^2 - 5x - 12}{2x^2 + 2x - 40}$. (Assume the denominator is nonzero.)

Solution

The trinomials in the numerator and denominator must be factored before we can simplify this expression. The numerator $2x^2 - 5x - 12$ is a trinomial with a leading coefficient that is not equal to 1 and can be factored by grouping or by trial and error. (For a review of these factoring techniques, you may refer to Section 4.5.)

$$2x^2 - 5x - 12 = (2x+3)(x-4)$$

The denominator $2x^2 + 2x - 40$ has a common factor of 2 that must be factored out first.

$$2x^2 + 2x - 40 = 2(x^2 + x - 20)$$
$$= 2(x+5)(x-4)$$

Now we can simplify the rational expression.

$$\frac{2x^2 - 5x - 12}{2x^2 + 2x - 40} = \frac{(2x+3)(x-4)}{2(x+5)(x-4)} \quad \text{Factor numerator and denominator.}$$

$$= \frac{(2x+3)\overset{1}{\cancel{(x-4)}}}{2(x+5)\underset{1}{\cancel{(x-4)}}} \quad \text{Divide out common factors.}$$

$$= \frac{2x+3}{2(x+5)} \quad \text{Simplify.}$$

***Quick Check* 7**

Simplify $\frac{3x^2 - 10x - 8}{x^2 - 16}$. (Assume the denominator is nonzero.)

Identifying Factors in the Numerator and Denominator That Are Opposites

Objective 4 Identify factors in the numerator and denominator that are opposites. Two expressions of the form $a - b$ and $b - a$ are **opposites**. Subtraction in the opposite order produces the opposite result. Consider the expressions $a - b$ and $b - a$ when $a = 10$ and $b = 4$. In this case, $a - b = 10 - 4$ or 6, and $b - a = 4 - 10$ or -6. We can also see that $a - b$ is the opposite of $b - a$ by noting that $-(b - a) = -b + a = a - b$.

This is a useful result when we are simplifying rational expressions. The rational expression $\frac{a - b}{b - a}$ simplifies to -1, as any fraction whose numerator is the opposite of its denominator is equal to -1. If a rational expression has a factor in the numerator that is the opposite of a factor in the denominator, these two factors can be divided out to equal -1, as in the following example. We write the -1 in the numerator.

EXAMPLE 8 Simplify $\frac{x^2 - 7x + 12}{9 - x^2}$. (Assume that the denominator is nonzero.)

Solution

We begin by factoring the numerator and denominator completely.

$$\frac{x^2 - 7x + 12}{9 - x^2} = \frac{(x - 3)(x - 4)}{(3 + x)(3 - x)} \quad \text{Factor the numerator and denominator.}$$

$$= \frac{\overset{-1}{\cancel{(x - 3)}}(x - 4)}{(3 + x)\underset{1}{\cancel{(3 - x)}}} \quad \text{Divide out the opposite factors.}$$

$$= -\frac{x - 4}{3 + x} \quad \text{Simplify, writing the negative sign in front of the fraction.}$$

Quick Check 8

Simplify $\frac{7x - x^2}{x^2 - 13x + 42}$. (Assume the denominator is nonzero.)

A Word of Caution Two expressions of the form $a + b$ and $b + a$ are *not* opposites but are equal to each other. Addition in the opposite order produces the same result. When we divide two expressions of the form $a + b$ and $b + a$, the result is 1, not -1. For example, $\frac{x + 2}{2 + x} = 1$.

Rational Functions

Objective 5 Evaluate rational functions.

A **rational function** $r(x)$ is a function of the form $r(x) = \frac{f(x)}{g(x)}$, where $f(x)$ and $g(x)$ are polynomials and $g(x) \neq 0$.

We begin our investigation of rational functions by learning to evaluate them.

EXAMPLE 9 For $r(x) = \dfrac{x^2 + 9x - 20}{x^2 - 3x + 16}$, find $r(6)$.

Solution

We begin by substituting 6 for x in the function.

$$r(6) = \frac{(6)^2 + 9(6) - 20}{(6)^2 - 3(6) + 16}$$ Substitute 6 for x.

$$= \frac{36 + 54 - 20}{36 - 18 + 16}$$ Simplify each term in the numerator and denominator.

$$= \frac{70}{34}$$ Simplify the numerator and denominator.

$$= \frac{35}{17}$$ Simplify to lowest terms.

Quick Check 9

For $r(x) = \dfrac{x^2 - 5x + 6}{x^2 + 4x + 12}$, find $r(-6)$.

Finding the Domain of a Rational Function

Objective 6 **Find the domain of rational functions.** Rational functions differ from linear functions and quadratic functions in that the domain of a rational function is not always the set of real numbers. We have to exclude any value that causes the function to be undefined, namely any value for which the denominator is equal to 0. Suppose that the function $r(x)$ was undefined for $x = 6$. Then the domain of $r(x)$ is the set of all real numbers except 6. This can be expressed in interval notation as $(-\infty, 6) \cup (6, \infty)$, which is the union of the set of all real numbers that are less than 6 with the set of all real numbers that are greater than 6.

EXAMPLE 10 Find the domain of $r(x) = \dfrac{x^2 + 25}{x^2 - 14x + 45}$.

Solution

We begin by setting the denominator equal to 0 and solving for x.

$x^2 - 14x + 45 = 0$ Set the denominator equal to 0.

$(x - 5)(x - 9) = 0$ Factor $x^2 - 14x + 45$.

$x - 5 = 0$ or $x - 9 = 0$ Set each factor equal to 0.

$x = 5$ or $x = 9$ Solve each equation.

The domain of the function is the set of all real numbers except 5 and 9. In interval notation this can be written as $(-\infty, 5) \cup (5, 9) \cup (9, \infty)$.

Quick Check 10

Find the domain of $r(x) = \dfrac{7x + 24}{x^2 + 6x - 16}$.

We must determine the domain of a rational function before simplifying it. If we divide a common factor out of the numerator before finding the domain of the function, we will miss a value of x that must be excluded from the domain of the function.

Building Your Study Strategy **Test Taking, 1** **Preparing Yourself** The first test-taking skill we will discuss is preparing yourself completely. As legendary basketball coach John Wooden once said, "Failing to prepare is preparing to fail." Start preparing for the exam well in advance; do not plan to study (or cram) on the night before the exam. To adequately prepare, you should know what the format of the exam will be. What topics will be covered on the test? How long will the test be?

Review your old homework assignments, notes, and note cards, spending more time on problems or concepts that you struggled with before. Work through the chapter review and chapter test in the text, to identify any areas of weakness for you. If you are having trouble with a particular type of problem, go back to that section in the text and review.

Make a practice test for yourself, and take it under test conditions without using your text or notes. Allow yourself the same amount of time that you will be allowed for the actual test, so you will know if you are working fast enough.

EXERCISES 5.1

Vocabulary

1. A(n) __________ is a quotient of two polynomials.
2. Rational expressions are undefined for values of the variable that cause the __________ to equal 0.
3. A rational expression is said to be __________ if its numerator and denominator do not have any common factors.
4. Two expressions of the form $a - b$ and $b - a$ are __________.
5. A __________ $r(x)$ is a function of the form $r(x) = \dfrac{f(x)}{g(x)}$, where $f(x)$ and $g(x)$ are polynomials and $g(x) \neq 0$.
6. The __________ of a rational function excludes all values for which the function is undefined.

Evaluate the rational expression for the given value of the variable.

7. $\dfrac{9}{x+8}$ for $x = 4$
8. $\dfrac{20}{3x-4}$ for $x = -8$
9. $\dfrac{5x-6}{x+10}$ for $x = -6$
10. $\dfrac{3x-10}{2x+4}$ for $x = 18$
11. $\dfrac{x^2-3x-2}{x^2+4x+3}$ for $x = 5$
12. $\dfrac{x^2-7x-18}{x^2+2x+3}$ for $x = 3$
13. $\dfrac{x^2+10x-9}{x^2+6}$ for $x = -2$
14. $\dfrac{x^2+5x+6}{x^2-3x-20}$ for $x = -7$

Find all values of the variable for which the rational expression is undefined.

15. $\dfrac{x+8}{x-6}$

16. $\dfrac{x-4}{2x-9}$

17. $\dfrac{x-9}{x(x+5)}$

18. $\dfrac{x+3}{(x+3)(x+7)}$

19. $\dfrac{3x+4}{x^2+15x+54}$

20. $\dfrac{x-10}{x^2-4x-32}$

21. $\dfrac{x}{x^2-25}$

22. $\dfrac{x^2+8}{x^2-3x}$

Simplify the given rational expression. (Assume that the denominator in each case is nonzero.)

23. $\dfrac{6x^3}{21x^7}$

24. $\dfrac{25x^9}{20x^{15}}$

25. $\dfrac{x-2}{(x+7)(x-2)}$

26. $\dfrac{(x-9)(x+8)}{x+8}$

27. $\dfrac{(x+6)(x-4)}{(x-4)(x-6)}$

28. $\dfrac{(x-7)(x-5)}{x(x-7)}$

29. $\dfrac{x^2+x-30}{x^2-7x+10}$

30. $\dfrac{x^2+6x}{x^2+10x+24}$

31. $\dfrac{x^2-11x+24}{x^2+x-72}$

32. $\dfrac{x^2+5x-14}{x^2+x-42}$

33. $\dfrac{x^2-12x+20}{x^2-4x+4}$

34. $\dfrac{x^2+3x-28}{x^2+14x+49}$

35. $\dfrac{x^2-36}{2x^2+15x+18}$

36. $\dfrac{3x^2-13x+4}{x^2-16}$

37. $\dfrac{4x^2-40x+96}{x^2-3x-4}$

38. $\dfrac{2x^2+14x+20}{3x^2-3x-90}$

39. $\dfrac{2x^2+5x-12}{x^2-3x-28}$

40. $\dfrac{3x^2-20x+12}{2x^2-7x-30}$

Determine whether the two given binomials are opposites or not.

41. $x+6$ and $6+x$

42. $x-8$ and $8-x$

43. $12-x$ and $x-12$

44. $5x-4$ and $4x-5$

45. $x+10$ and $-x-10$

46. $3x+5$ and $3x-5$

Simplify the given rational expression. (Assume that the denominator in each case is nonzero.)

47. $\dfrac{x-6}{(2+x)(6-x)}$

48. $\dfrac{(9+x)(9-x)}{x^2-2x-63}$

49. $\dfrac{x^2-11x+28}{16-x^2}$

50. $\dfrac{2x-x^2}{x^2+x-6}$

51. $\dfrac{36-x^2}{x^2+13x+42}$

52. $\dfrac{x^2-5x-14}{35-5x}$

Evaluate the given rational function.

53. $r(x) = \dfrac{3x - 6}{x^2 + 3x + 10}$, $r(-2)$

54. $r(x) = \dfrac{4x + 2}{x^2 + 5x - 14}$, $r(7)$

55. $r(x) = \dfrac{x^2 + 2x - 8}{x^2 - x - 8}$, $r(5)$

56. $r(x) = \dfrac{x^2 - 25}{x^2 + 10x + 24}$, $r(-4)$

57. $r(x) = \dfrac{x^3 - 2x^2 + 4x + 5}{3x^2 + 4x - 11}$, $r(3)$

58. $r(x) = \dfrac{x^3 - 6x^2 + 7x + 8}{x^3 + 5x^2 - 8x - 9}$, $r(2)$

Find the domain of the given rational function.

59. $r(x) = \dfrac{x^2 + 13x + 20}{x^2 - 6x}$

60. $r(x) = \dfrac{x^2 - 5x - 19}{x^2 + 4x + 3}$

61. $r(x) = \dfrac{x^2 - 7x - 10}{x^2 + 4x - 45}$

62. $r(x) = \dfrac{x^2 + 2x - 15}{x^2 - 25}$

63. $r(x) = \dfrac{x^2 + 48}{x^2 + 4x - 21}$

64. $r(x) = \dfrac{x^2 + 20x - 9}{x^2 - 17x + 72}$

Identify the given function as a linear function, a quadratic function, or a rational function.

65. $f(x) = x^2 - 8x + 16$

66. $f(x) = 5x + 15$

67. $f(x) = \dfrac{3x^2 + 2x + 12}{x}$

68. $f(x) = \dfrac{x^2 + 6x + 25}{2x^2 - 3x + 10}$

69. $f(x) = \dfrac{3}{4}x - 6$

70. $f(x) = \dfrac{2}{5}x^2 - \dfrac{3}{2}x + 5$

Use the given graph of a rational function $r(x)$ to determine the following.

71.

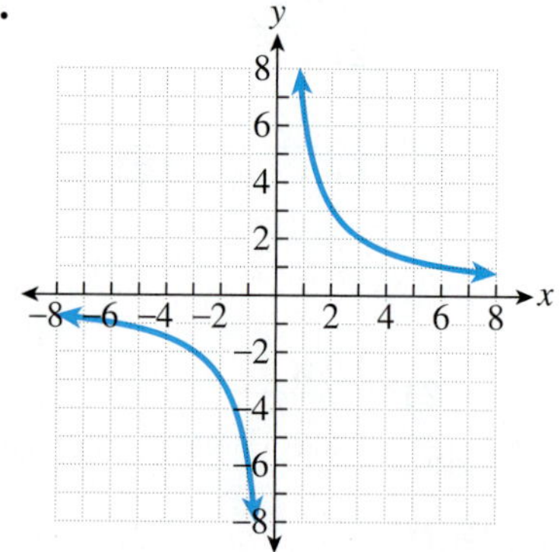

a) Find $r(-1)$.

b) Find all values x such that $r(x) = -3$.

72.

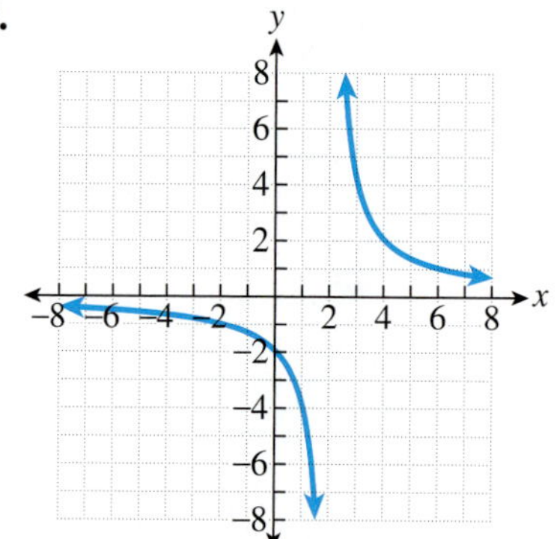

a) Find $r(3)$.

b) Find all values x such that $r(x) = -2$.

73.

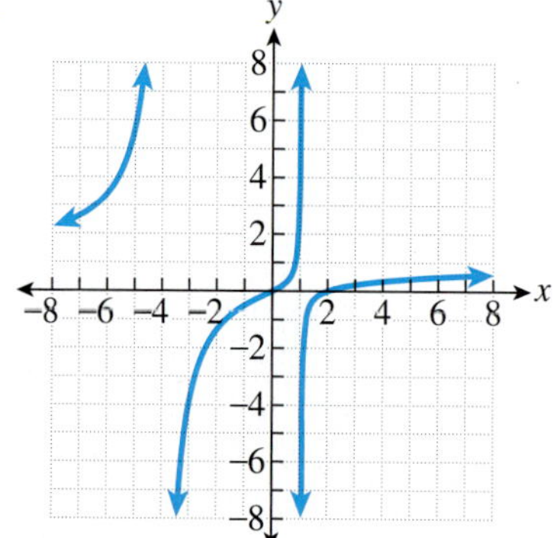

a) Find $r(-5)$.

b) Find all values x such that $r(x) = 0$.

74.

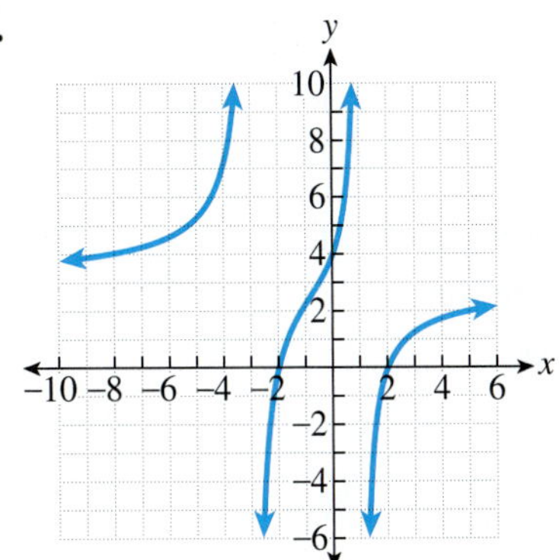

a) Find $r(0)$.

b) Find all values x such that $r(x) = 0$.

Writing in Mathematics

Answer in complete sentences.

75. Explain how to find the values for which a rational expression is undefined.

76. Is the rational expression $\frac{(x+2)(x-7)}{(x-5)(x+2)}$ undefined for the value $x = -2$? Explain your answer.

77. Explain how to determine if two factors are opposites. Use examples to illustrate your explanation.

78. ***Solutions Manual**** Write a solutions manual page for the following problem:
Find the values for which the rational expression $\frac{x^2 + 3x - 28}{x^2 - 16}$ *is undefined.*

79. ***Newsletter**** Write a newsletter that explains how to simplify a rational expression.

***See Appendix B for details and sample answers.**

5.2 Multiplication and Division of Rational Expressions

Objectives

1. **Multiply two rational expressions.**
2. **Multiply two rational functions.**
3. **Divide a rational expression by another rational expression.**
4. **Divide a rational function by another rational function.**

Multiplying Rational Expressions

Objective 1 Multiply two rational expressions. In this section, we will learn how to multiply and divide rational expressions. Multiplying rational expressions is similar to multiplying numerical fractions. Suppose that we needed to multiply $\frac{4}{9} \cdot \frac{21}{10}$. Before multiplying, we can divide out factors common to one of the numerators and one of the denominators. For example, the first numerator (4) and the second denominator (10) have a common factor of 2 that can be divided out of each. The second numerator (21) and the first denominator (9) have a common factor of 3 that can be divided out as well.

$$\frac{4}{9} \cdot \frac{21}{10} = \frac{2 \cdot 2}{3 \cdot 3} \cdot \frac{3 \cdot 7}{2 \cdot 5} \quad \text{Factor each numerator and denominator.}$$

$$= \frac{\overset{1}{\cancel{2}} \cdot 2}{\underset{1}{\cancel{3}} \cdot 3} \cdot \frac{\overset{1}{\cancel{3}} \cdot 7}{\underset{1}{\cancel{2}} \cdot 5} \quad \text{Divide out common factors.}$$

$$= \frac{2 \cdot 7}{3 \cdot 5} \quad \text{Multiply remaining factors.}$$

$$= \frac{14}{15}$$

Multiplying Rational Expressions

$$\frac{A}{B} \cdot \frac{C}{D} = \frac{AC}{BD} \qquad B \neq 0 \text{ and } D \neq 0$$

To multiply two rational expressions, we will begin by factoring each numerator and denominator completely. After dividing out factors common to a numerator and a denominator, we will express the product of the two rational expressions as a single rational expression, leaving the numerator and denominator in factored form.

Example 1 Multiply $\frac{x^2 - 6x - 27}{x^2 - 11x + 18} \cdot \frac{x^2 - 4}{x^2 + 9x + 18}$.

Solution

We begin by factoring each numerator and denominator. Then we proceed to divide out common factors.

$$\frac{x^2 - 6x - 27}{x^2 - 11x + 18} \cdot \frac{x^2 - 4}{x^2 + 9x + 18}$$

$$= \frac{(x-9)(x+3)}{(x-2)(x-9)} \cdot \frac{(x+2)(x-2)}{(x+3)(x+6)} \quad \text{Factor numerators and denominators completely.}$$

$$= \frac{\overset{1}{\cancel{(x-9)}}\overset{1}{\cancel{(x+3)}}}{\underset{1}{\cancel{(x-2)}}\underset{1}{\cancel{(x-9)}}} \cdot \frac{(x+2)\overset{1}{\cancel{(x-2)}}}{\underset{1}{\cancel{(x+3)}}(x+6)}$$ Divide out common factors.

$$= \frac{x+2}{x+6}$$ Multiply.

Quick Check 1 Multiply $\frac{x^2+11x+18}{x^2+2x-15} \cdot \frac{x^2+13x+40}{x^2+5x-36}$.

EXAMPLE 2 Multiply $\frac{x^2-2x-35}{3x-x^2} \cdot \frac{x^2-13x+30}{2x^2-9x-35}$.

Solution

Again, we begin by completely factoring both numerators and denominators.

$$\frac{x^2-2x-35}{3x-x^2} \cdot \frac{x^2-13x+30}{2x^2-9x-35}$$

$$= \frac{(x+5)(x-7)}{x(3-x)} \cdot \frac{(x-3)(x-10)}{(2x+5)(x-7)}$$ Factor completely.

$$= \frac{(x+5)\overset{1}{\cancel{(x-7)}}}{x\underset{1}{\cancel{(3-x)}}} \cdot \frac{\overset{-1}{\cancel{(x-3)}}(x-10)}{(2x+5)\underset{1}{\cancel{(x-7)}}}$$ Divide out common factors. Notice that the factors $x-3$ and $3-x$ are opposites and divide out to equal -1.

$$= -\frac{(x+5)(x-10)}{x(2x+5)}$$ Multiply remaining factors. Write the negative sign in the numerator in front of the fraction.

Quick Check 2 Multiply $\frac{x^2-7x-30}{x^2-2x-8} \cdot \frac{x^2-9x+20}{100-x^2}$.

Here is a summary of the procedure for multiplying rational expressions.

Multiplying Rational Expressions

- Completely factor each numerator and each denominator.
- Divide out factors that are common to a numerator and a denominator, and divide out factors in a numerator and denominator that are opposites.
- Multiply the remaining factors, leaving the numerator and denominator in factored form.

Multiplying Rational Functions

Objective 2 **Multiply two rational functions.**

EXAMPLE 3 For $f(x) = \dfrac{x^2 + 4x - 12}{x^2 - 13x + 40}$ and $g(x) = \dfrac{x^2 - 6x - 16}{x^2 - 36}$, find $f(x) \cdot g(x)$.

Solution

We replace $f(x)$ and $g(x)$ by their formulas and proceed to multiply.

$$f(x) \cdot g(x) = \frac{x^2 + 4x - 12}{x^2 - 13x + 40} \cdot \frac{x^2 - 6x - 16}{x^2 - 36}$$ Replace $f(x)$ and $g(x)$ with their formulas.

$$= \frac{(x+6)(x-2)}{(x-5)(x-8)} \cdot \frac{(x+2)(x-8)}{(x+6)(x-6)}$$ Factor completely.

$$= \frac{\overset{1}{\cancel{(x+6)}}(x-2)}{(x-5)\underset{1}{\cancel{(x-8)}}} \cdot \frac{(x+2)\overset{1}{\cancel{(x-8)}}}{\underset{1}{\cancel{(x+6)}}(x-6)}$$ Divide out common factors.

$$= \frac{(x-2)(x+2)}{(x-5)(x-6)}$$ Multiply remaining factors.

***Quick Check* 3** For $f(x) = \dfrac{x + 8}{x^2 - x - 20}$ and $g(x) = \dfrac{x^2 - 15x + 50}{x^2 + 8x}$, find $f(x) \cdot g(x)$.

Dividing a Rational Expression by Another Rational Expression

Objective 3 **Divide a rational expression by another rational expression.**

Dividing a rational expression by another rational expression is similar to dividing a numerical fraction by another numerical fraction. We replace the divisor, which is the rational expression we are dividing by, by its reciprocal and then multiply. Replacing the divisor by its reciprocal is also called **inverting** the divisor.

Dividing Rational Expressions

$$\frac{A}{B} \div \frac{C}{D} = \frac{A}{B} \cdot \frac{D}{C} \qquad B \neq 0,\ C \neq 0,\ \text{and } D \neq 0$$

EXAMPLE 4 Divide $\dfrac{x^2 + 12x + 32}{x^2 - 11x + 30} \div \dfrac{x^2 + x - 12}{x^2 - 5x}$.

Solution

We begin by inverting the divisor and multiplying. In this example we must factor each numerator and denominator completely.

$$\frac{x^2 + 12x + 32}{x^2 - 11x + 30} \div \frac{x^2 + x - 12}{x^2 - 5x}$$

$$= \frac{x^2 + 12x + 32}{x^2 - 11x + 30} \cdot \frac{x^2 - 5x}{x^2 + x - 12}$$ Invert the divisor and multiply.

$$= \frac{(x+4)(x+8)}{(x-5)(x-6)} \cdot \frac{x(x-5)}{(x+4)(x-3)}$$ Factor completely.

$$= \frac{\overset{1}{\cancel{(x+4)}}(x+8)}{\underset{1}{\cancel{(x-5)}}(x-6)} \cdot \frac{x\overset{1}{\cancel{(x-5)}}}{\underset{1}{\cancel{(x+4)}}(x-3)}$$ Divide out common factors.

$$= \frac{x(x+8)}{(x-6)(x-3)}$$ Multiply remaining factors.

Quick Check 4 Divide $\frac{x^2 - 49}{x^2 - 14x + 49} \div \frac{x^2 + 10x + 21}{x^2 - 6x - 7}$.

EXAMPLE 5 Divide $\frac{x^2 - 9x - 10}{x^2 - 6x + 8} \div \frac{x^2 - 1}{2 - x}$.

Solution

We rewrite the problem as a multiplication problem by inverting the divisor. We then factor each numerator and denominator completely before dividing out common factors. (You may want to factor at the same time you invert the divisor.)

$$\frac{x^2 - 9x - 10}{x^2 - 6x + 8} \div \frac{x^2 - 1}{2 - x}$$

$$= \frac{x^2 - 9x - 10}{x^2 - 6x + 8} \cdot \frac{2 - x}{x^2 - 1}$$ Invert the divisor and multiply.

$$= \frac{(x-10)(x+1)}{(x-2)(x-4)} \cdot \frac{2-x}{(x+1)(x-1)}$$ Factor completely.

$$= \frac{(x-10)\overset{1}{\cancel{(x+1)}}}{\underset{1}{\cancel{(x-2)}}(x-4)} \cdot \frac{\overset{-1}{\cancel{(2-x)}}}{\underset{1}{\cancel{(x+1)}}(x-1)}$$ Divide out common factors. Note that $2 - x$ and $x - 2$ are opposites.

$$= -\frac{x-10}{(x-4)(x-1)}$$ Multiply remaining factors.

Quick Check 5 Divide $\frac{x^2 - 4x - 12}{x^2 + 13x + 22} \div \frac{6 - x}{x^2 + 15x + 44}$.

Here is a summary of the procedure for dividing rational expressions.

Dividing Rational Expressions

- Invert the divisor and change the operation from division to multiplication.
- Completely factor each numerator and each denominator.
- Divide out factors that are common to a numerator and a denominator, and divide out factors in a numerator and denominator that are opposites.
- Multiply the remaining factors, leaving the numerator and denominator in factored form.

Dividing a Rational Function by Another Rational Function

Objective 4 **Divide a rational function by another rational function.**

EXAMPLE 6 For $f(x) = \frac{x^2 - 4x - 21}{x + 5}$ and $g(x) = x^2 + 8x + 15$, find $f(x) \div g(x)$.

Solution

Replace $f(x)$ and $g(x)$ by their formulas and divide. Treat $g(x)$ as a rational function with a denominator of 1. We can see that its reciprocal is $\frac{1}{x^2 + 8x + 15}$.

$$f(x) \div g(x) = \frac{x^2 - 4x - 21}{x + 5} \div (x^2 + 8x + 15)$$ Replace $f(x)$ and $g(x)$ by their formulas.

$$= \frac{x^2 - 4x - 21}{x + 5} \cdot \frac{1}{x^2 + 8x + 15}$$ Invert the divisor and multiply.

$$= \frac{(x - 7)(x + 3)}{x + 5} \cdot \frac{1}{(x + 3)(x + 5)}$$ Factor completely.

$$= \frac{(x - 7)\overset{1}{\cancel{(x + 3)}}}{x + 5} \cdot \frac{1}{\underset{1}{\cancel{(x + 3)}}(x + 5)}$$ Divide out common factors.

$$= \frac{x - 7}{(x + 5)^2}$$ Multiply remaining factors.

Quick Check 6

For $f(x) = \frac{x^2 + 16x + 64}{x^2 + x - 56}$ and $g(x) = x^2 + 7x - 8$, find $f(x) \div g(x)$.

Building Your Study Strategy **Test Taking, 2 A Good Night's Sleep** Get a good night's sleep on the night before the exam. Tired students do not think as well as students who are rested. Some studies suggest that good-quality sleep on the two nights prior to an exam can have a positive effect on test scores. Also be sure to eat properly before an exam. Hungry students can be distracted during an exam.

EXERCISES 5.2

Vocabulary

1. To multiply two rational expressions, begin by __________ each numerator and denominator completely.

2. When multiplying two rational expressions, __________ out factors common to a numerator and a denominator.

3. When multiplying two rational expressions, once the numerator and the denominator do not share any common factors, express the product as a single rational expression, leaving the numerator and denominator in __________ form.

4. When dividing by a rational expression, replace the divisor by its __________ and then multiply.

Multiply.

5. $\frac{x+3}{(x-4)(x-6)} \cdot \frac{(x-4)(x+1)}{(x-1)(x+3)}$

6. $\frac{x+7}{x+8} \cdot \frac{(x+8)(x-2)}{(x+7)(x+2)}$

7. $\frac{x-1}{x-3} \cdot \frac{x^2+6x-27}{x^2-9x+8}$

8. $\frac{x^2+3x-10}{x^2+2x-35} \cdot \frac{x-5}{x-2}$

9. $\frac{x^2+4x}{x^2-2x-15} \cdot \frac{x^2-6x+5}{x^2+14x+40}$

10. $\frac{x^2+7x-18}{x^2-2x-48} \cdot \frac{x^2+5x-6}{x^2-2x}$

11. $\frac{8x-x^2}{x^2-5x+6} \cdot \frac{x^2+7x-30}{x^2+2x-80}$

12. $\frac{x^2-x-72}{x^2+12x+35} \cdot \frac{x^2+8x+7}{9x-x^2}$

13. $\frac{x^2+2x-8}{x^2-15x+54} \cdot \frac{81-x^2}{x^2-2x-24}$

14. $\frac{x^2+x-20}{64-x^2} \cdot \frac{x^2-15x+56}{x^2+9x+20}$

15. $\frac{x^2-18x+81}{x^2+10x+16} \cdot \frac{2x^2+13x-24}{x^2-5x-36}$

16. $\frac{x^2-4x-12}{x^2+3x-4} \cdot \frac{5x^2+23x+12}{x^2-14x+48}$

17. $\frac{x^2-144}{3x^2+10x+3} \cdot \frac{x^2+18x+45}{x^2-21x+108}$

18. $\frac{x^2-4x-77}{x^2-20x} \cdot \frac{x^2-15x-100}{x^2-6x-55}$

19. $\frac{2x^2-14x-36}{x^2-x-42} \cdot \frac{x^2+2x-24}{3x^2-12}$

20. $\frac{x^2+3x-40}{4x^2+36x+56} \cdot \frac{5x^2+25x+30}{x^2-4x-5}$

21. $\frac{2x^2+x-3}{x^2-12x+32} \cdot \frac{x^2-x-56}{2x^2+13x+15}$

22. $\frac{3x^2-4x}{x^2-3x-54} \cdot \frac{x^2+8x+12}{3x^2+2x-8}$

23. $\frac{x^3-5x^2+7x-35}{x^2+9x+18} \cdot \frac{x^2+5x+6}{x^2-6x+5}$

24. $\frac{x^3+6x^2-3x-18}{x^2-49} \cdot \frac{x^2+7x}{x^2-4x-60}$

25. $\frac{x^3-8}{x^2-5x-36} \cdot \frac{x^2-13x+36}{x^2-x-2}$

FOR EXTRA HELP

Interactmath.com

MathXL Tutorials on CD

Video Lectures on CD

Addison-Wesley Math Tutor Center

Student's Solutions Manual

26. $\dfrac{x^3 - 512}{x^2 + 4x + 4} \cdot \dfrac{x^2 + 10x + 16}{x^2 - 64}$

27. $\dfrac{x^3 + 1}{x^2 + 5x} \cdot \dfrac{x^2 + 3x - 10}{x^2 + 13x + 12}$

28. $\dfrac{x^3 + 1000}{x + 3} \cdot \dfrac{x^2 + 10x + 21}{x^2 + 5x - 50}$

For the given functions f(x) and g(x), find f(x) · g(x).

29. $f(x) = \dfrac{7}{4 - x}, g(x) = \dfrac{x^2 - x - 12}{x^2 + 12x + 27}$

30. $f(x) = \dfrac{x^2 - 3x - 10}{x^2 + 2x - 3}, g(x) = \dfrac{x^2 + 8x - 9}{x^2 - 25}$

31. $f(x) = \dfrac{x^2 - 6x - 72}{x^2 - x - 6}, g(x) = \dfrac{x^2 - 6x + 9}{x^2 + 3x - 18}$

32. $f(x) = \dfrac{x^2 - 18x + 80}{x^2 + 2x - 63}, g(x) = \dfrac{x^2 + 13x + 36}{64 - x^2}$

33. $f(x) = \dfrac{x^2 + 5x - 24}{x^2 - 2x - 3}, g(x) = x^2 - 7x - 8$

34. $f(x) = x - 6, g(x) = \dfrac{x^2 + 5x - 66}{x^2 - 12x + 36}$

Find the missing numerator and denominator.

35. $\dfrac{(x + 5)(x - 2)}{(2x + 1)(x - 7)} \cdot \dfrac{?}{?} = \dfrac{(x + 5)}{(2x + 1)}$

36. $\dfrac{x^2 - 5x - 6}{x^2 + 15x + 54} \cdot \dfrac{?}{?} = \dfrac{(x + 1)(x + 5)}{(x + 6)^2}$

37. $\dfrac{x^2 - 10x + 16}{x^2 + 4x - 77} \cdot \dfrac{?}{?} = -\dfrac{x^2 - 11x + 24}{x^2 - 5x - 14}$

38. $\dfrac{x^2 + 4x}{x^2 - 81} \cdot \dfrac{?}{?} = \dfrac{x^2 - 6x}{x^2 - 7x - 18}$

Divide.

39. $\dfrac{(x + 7)(x - 5)}{(x + 2)(x + 15)} \div \dfrac{x + 7}{x + 15}$

40. $\dfrac{(x + 6)(x + 4)}{x - 5} \div \dfrac{x + 6}{(x - 2)(x - 5)}$

41. $\dfrac{x^2 + 15x + 56}{x^2 - 9x + 18} \div \dfrac{x^2 + 9x + 8}{x^2 - 7x + 6}$

42. $\dfrac{x^2 - 4x - 45}{x^2 + 4x + 4} \div \dfrac{x^2 - 16x + 63}{x^2 - 8x - 20}$

43. $\dfrac{x^2 + 16x + 63}{x^2 - 5x - 6} \div \dfrac{x^2 + 17x + 70}{x^2 + 11x + 10}$

44. $\dfrac{x^2 - 10x + 16}{x^2 + 4x - 5} \div \dfrac{x^2 - 5x - 24}{x^2 - 5x + 4}$

45. $\dfrac{x^2 + 14x + 45}{x^2 + 7x + 12} \div \dfrac{x^2 + 18x + 81}{3x^2 + 8x - 16}$

46. $\dfrac{x^2 - 12x + 35}{x^2 - 3x - 10} \div \dfrac{x^2 - 8x + 7}{x^2 + x - 2}$

47. $\dfrac{x^2 - 8x - 9}{x^2 - 10x + 9} \div \dfrac{x^2 - 2x - 3}{x^2 - 4x + 3}$

48. $\dfrac{6x^2 + 7x + 1}{x^2 - 8x + 16} \div \dfrac{x^2 + 5x + 4}{x^2 - 16}$

49. $\dfrac{x^2 - 8x + 12}{x^2 - 8x + 15} \div \dfrac{4 - x^2}{x^2 + 5x - 50}$

50. $\dfrac{9 - x^2}{x^2 + 3x + 2} \div \dfrac{x^2 - 12x + 27}{x^2 + 10x + 9}$

51. $\dfrac{4x - x^2}{x^2 - 2x + 1} \div \dfrac{x^2 - 14x + 40}{x^2 + 6x - 7}$

52. $\dfrac{x - x^2}{x^2 + 11x + 30} \div \dfrac{x^2 - 10x + 9}{x^2 + 12x + 36}$

53. $\dfrac{3x^2 - 3x - 6}{2x^2 - 32x + 128} \div \dfrac{x^2 + 8x - 20}{x^2 - 64}$

54. $\dfrac{4x^2 + 24x + 20}{x^2 + 3x - 54} \div \dfrac{5x^2 + 105x + 100}{x^2 + 18x + 81}$

55. $\dfrac{x^2 + 3x}{2x^2 + 7x - 4} \div \dfrac{x^2 + 7x}{2x^2 + 3x - 2}$

56. $\dfrac{3x^2 - 13x - 10}{x^2 - 3x + 2} \div \dfrac{3x^2 + 11x + 6}{x^2 - 11x + 10}$

57. $\frac{x^3 - 5x^2 - 8x + 40}{x^2 + 3x} \div \frac{x^2 - x - 20}{x^2 + 5x + 6}$

58. $\frac{3x^3 + 7x^2 + 36x + 84}{x^2 - 16} \div \frac{3x + 7}{4x - x^2}$

59. $\frac{x^3 - 27}{x^2 - 8x + 12} \div \frac{9 - x^2}{x^2 + x - 42}$

60. $\frac{x^3 - 125}{x^2 - 4x - 5} \div \frac{x^2 + 15x + 54}{x^2 + 10x + 9}$

61. $\frac{x^3 + 216}{x + 3} \div \frac{x^2 + 21x + 90}{x^2 + 6x + 9}$

62. $\frac{x^3 + 64}{x^2 + 13x + 30} \div \frac{x^2 + 14x + 40}{x^2 + 10x + 21}$

For the given functions $f(x)$ and $g(x)$, find $f(x) \div g(x)$.

63. $f(x) = \frac{x^2 + 13x + 42}{x^2 - 10x + 25}, g(x) = x + 6$

64. $f(x) = 3x + 15, g(x) = \frac{x^2 + 6x + 5}{x^2 + 17x + 72}$

65. $f(x) = \frac{x^2 + 6x + 9}{x^2 + 20x + 96}, g(x) = \frac{x^2 - 9}{x^2 + 8x}$

66. $f(x) = \frac{x^2 + 2x - 80}{x^2 + 10x + 25}, g(x) = \frac{x^2 + 16x + 60}{x^2 - 25}$

67. $f(x) = \frac{2x^2 + 3x}{x^2 - 16}, g(x) = \frac{2x^2 - 9x - 18}{x^2 - 19x + 60}$

68. $f(x) = \frac{x^2 + 20x + 100}{3x^2 - 10x + 3}, g(x) = \frac{x^2 - 100}{3x^2 + 11x - 4}$

Find the missing numerator and denominator.

69. $\frac{(x + 9)(x - 6)}{(x - 3)(x - 7)} \div \frac{?}{?} = \frac{x + 9}{x - 7}$

70. $\frac{x^2 - 5x + 4}{x^2 - 14x + 45} \div \frac{?}{?} = \frac{(x - 1)(x - 6)}{(x - 9)^2}$

71. $\frac{x^2 - 10x + 9}{x^2 + 7x} \div \frac{?}{?} = \frac{x^2 - 13x + 12}{x^2 + 11x + 28}$

72. $\frac{x^2 + 12x + 36}{x^2 - 100} \div \frac{?}{?} = \frac{x^2 - 36}{x^2 + 10x}$

Writing in Mathematics

Answer in complete sentences.

73. Explain the similarities between dividing numerical fractions and dividing rational expressions. Are there any differences?

74. Explain why the restrictions $B \neq 0$, $C \neq 0$, and $D \neq 0$ are necessary when dividing $\frac{A}{B} \div \frac{C}{D}$.

75. ***Solutions Manual**** Write a solutions manual page for the following problem:

Divide $\frac{x^2 - 6x - 27}{x^2 + 4x - 5} \div \frac{9x - x^2}{x^2 - 25}$.

76. ***Newsletter**** Write a newsletter that explains how to multiply two rational expressions.

***See Appendix B for details and sample answers.**

5.3 Addition and Subtraction of Rational Expressions

Objectives

1. **Add rational expressions with the same denominator.**
2. **Subtract rational expressions with the same denominator.**
3. **Add or subtract rational expressions with opposite denominators.**
4. **Find the least common denominator (LCD) of two or more rational expressions.**
5. **Add or subtract rational expressions with unlike denominators.**

Now that we have learned how to multiply and divide rational expressions, we move on to addition and subtraction. We know from our work with numerical fractions that two fractions must have the same denominator before we can add or subtract them. The same holds true for rational expressions. We will begin with rational expressions that already have the same denominator.

Adding Rational Expressions with the Same Denominator

Objective 1 **Add rational expressions with the same denominator.** To add fractions that have the same denominator, we add the numerators and place the result over the common denominator. We will follow the same procedure when adding two rational expressions. Of course, we should check that our result is in simplest terms.

Adding Rational Expressions with the Same Denominator

$$\frac{A}{C} + \frac{B}{C} = \frac{A+B}{C} \qquad C \neq 0$$

EXAMPLE 1 Add $\frac{9}{x+2} + \frac{7}{x+2}$.

Solution

These two fractions have the same denominator, so we add the two numerators and place the result over the common denominator $x + 2$.

$$\frac{9}{x+2} + \frac{7}{x+2} = \frac{9+7}{x+2} \quad \text{Add numerators, placing the sum over the common denominator.}$$

$$= \frac{16}{x+2} \quad \text{Simplify the numerator.}$$

The numerator and denominator do not have any common factors, so this is our final result. A common error is to attempt to divide a common factor out of 16 in the numerator and 2 in the denominator, but the number 2 is a term of the denominator and not a factor.

Quick Check 1

Add $\frac{3}{2x-5} + \frac{7}{2x-5}$.

EXAMPLE 2 Add $\dfrac{3x-8}{x^2-5x-24}+\dfrac{x+20}{x^2-5x-24}$.

Solution

The two denominators are the same, so we may add.

$$\frac{3x-8}{x^2-5x-24}+\frac{x+20}{x^2-5x-24}$$

$$=\frac{(3x-8)+(x+20)}{x^2-5x-24}$$ Add numerators.

$$=\frac{4x+12}{x^2-5x-24}$$ Combine like terms.

$$=\frac{4\overset{1}{\cancel{(x+3)}}}{(x-8)\underset{1}{\cancel{(x+3)}}}$$ Factor numerator and denominator and divide out the common factor.

$$=\frac{4}{x-8}$$ Simplify.

Quick Check 2 Add $\dfrac{2x+25}{x^2+10x+24}+\dfrac{3x-5}{x^2+10x+24}$.

EXAMPLE 3 Add $\dfrac{x^2+2x-9}{x^2-5x+4}+\dfrac{2x+4}{x^2-5x+4}$.

Solution

The denominators are the same, so we add the numerators and then simplify.

$$\frac{x^2+2x-9}{x^2-5x+4}+\frac{2x+4}{x^2-5x+4}$$

$$=\frac{(x^2+2x-9)+(2x+4)}{x^2-5x+4}$$ Add numerators.

$$=\frac{x^2+4x-5}{x^2-5x+4}$$ Combine like terms.

$$=\frac{(x+5)\overset{1}{\cancel{(x-1)}}}{\underset{1}{\cancel{(x-1)}}(x-4)}$$ Factor numerator and denominator and divide out the common factor.

$$=\frac{x+5}{x-4}$$ Simplify.

Quick Check 3 Add $\dfrac{x^2+4x+20}{x^2+7x-18}+\dfrac{8x+7}{x^2+7x-18}$.

Subtracting Rational Expressions with the Same Denominator

Objective 2 **Subtract rational expressions with the same denominator.** Subtracting two rational expressions with the same denominator is just like adding them, except that we subtract the two numerators rather than adding them.

Subtracting Rational Expressions with the Same Denominator

$$\frac{A}{C} - \frac{B}{C} = \frac{A - B}{C} \qquad C \neq 0$$

EXAMPLE 4 Subtract $\frac{4x - 3}{x - 4} - \frac{x + 9}{x - 4}$.

Solution

The two denominators are the same, so we may subtract these two rational expressions. When the numerator of the second fraction has more than one term we must remember that we are subtracting the whole numerator and not just the first term. When we subtract the numerators and place the difference over the common denominator, it is a good idea to write each numerator in a set of parentheses. This will remind us to subtract each term in the second numerator.

$$\frac{4x - 3}{x - 4} - \frac{x + 9}{x - 4} = \frac{(4x - 3) - (x + 9)}{x - 4} \qquad \text{Subtract numerators.}$$

$$= \frac{4x - 3 - x - 9}{x - 4} \qquad \text{Distribute.}$$

$$= \frac{3x - 12}{x - 4} \qquad \text{Combine like terms.}$$

$$= \frac{3\overset{1}{\cancel{(x - 4)}}}{\underset{1}{\cancel{x - 4}}} \qquad \text{Factor numerator and divide out the common factor.}$$

$$= 3 \qquad \text{Simplify.}$$

Quick Check 4

Subtract $\frac{3x + 4}{x + 8} - \frac{x - 12}{x + 8}$.

Notice that the denominator $x - 4$ was also a factor of the numerator, leaving a denominator of 1.

EXAMPLE 5 Subtract $\frac{2x^2 - x - 15}{6x - x^2} - \frac{x^2 + 3x - 3}{6x - x^2}$.

Solution

The denominators are the same, so we can subtract these two rational expressions.

$$\frac{2x^2 - x - 15}{6x - x^2} - \frac{x^2 + 3x - 3}{6x - x^2}$$

$$= \frac{(2x^2 - x - 15) - (x^2 + 3x - 3)}{6x - x^2} \qquad \text{Subtract numerators.}$$

$= \frac{2x^2 - x - 15 - x^2 - 3x + 3}{6x - x^2}$ Change the sign of each term in the second set of parentheses by distributing -1.

$= \frac{x^2 - 4x - 12}{6x - x^2}$ Combine like terms.

$= \frac{\overset{-1}{\cancel{(x - 6)}}(x + 2)}{x\underset{1}{\cancel{(6 - x)}}}$ Factor numerator and denominator and divide out the common factor. The -1 results from the fact that $x - 6$ and $6 - x$ are opposites.

$= -\frac{x + 2}{x}$ Simplify.

Quick Check 5 Subtract $\frac{3x^2 - 6x - 14}{25 - x^2} - \frac{2x^2 - 8x + 21}{25 - x^2}$.

A Word of Caution When subtracting a rational expression whose numerator contains more than one term, be sure to subtract the entire numerator and not just the first term. One way to remember this is by placing the numerators inside sets of parentheses.

Adding or Subtracting Rational Expressions with Opposite Denominators

Objective 3 Add or subtract rational expressions with opposite denominators.

Consider the expression $\frac{10}{x - 2} + \frac{3}{2 - x}$. Are the two denominators the same? No, but they are opposites. We can rewrite the denominator $2 - x$ as its opposite $x - 2$ if we also rewrite the operation (addition) as its opposite (subtraction). In other words, we can rewrite the expression $\frac{10}{x - 2} + \frac{3}{2 - x}$ as $\frac{10}{x - 2} - \frac{3}{x - 2}$. Once the denominators are the same, we can subtract the numerators.

Example 6 Add $\frac{8x}{3x - 15} + \frac{40}{15 - 3x}$.

Solution

The two denominators are opposites, so we may change the second denominator to $3x - 15$ by changing the operation from addition to subtraction.

$\frac{8x}{3x - 15} + \frac{40}{15 - 3x}$

$= \frac{8x}{3x - 15} - \frac{40}{3x - 15}$ Rewrite the second denominator as $3x - 15$ by changing the operation from addition to subtraction.

$$= \frac{8x - 40}{3x - 15}$$ Subtract the numerators.

$$= \frac{8\cancel{(x-5)}^{1}}{3\cancel{(x-5)}_{1}}$$ Factor the numerator and denominator and divide out the common factor.

$$= \frac{8}{3}$$ Simplify.

Quick Check 6
Add $\dfrac{2x}{3x - 18} + \dfrac{12}{18 - 3x}$.

EXAMPLE 7 Subtract $\dfrac{x^2 + x + 5}{x^2 - 9} - \dfrac{6x + 7}{9 - x^2}$.

Solution

These two denominators are opposites, so we begin by rewriting the second rational expression in such a way that the two rational expressions have the same denominator.

$$\frac{x^2 + x + 5}{x^2 - 9} - \frac{6x + 7}{9 - x^2}$$

$$= \frac{x^2 + x + 5}{x^2 - 9} + \frac{6x + 7}{x^2 - 9}$$ Rewrite the second denominator as $x^2 - 9$ by changing the operation from subtraction to addition.

$$= \frac{(x^2 + x + 5) + (6x + 7)}{x^2 - 9}$$ Add the numerators.

$$= \frac{x^2 + 7x + 12}{x^2 - 9}$$ Combine like terms.

$$= \frac{\cancel{(x+3)}^{1}(x + 4)}{\cancel{(x+3)}_{1}(x - 3)}$$ Factor the numerator and denominator and divide out the common factor.

$$= \frac{x + 4}{x - 3}$$ Simplify.

Quick Check 7
Subtract
$\dfrac{x^2 - 7x + 10}{x^2 - 16} - \dfrac{x - 2}{16 - x^2}$.

The Least Common Denominator of Two or More Rational Expressions

Objective 4 **Find the least common denominator (LCD) of two or more rational expressions.** If two numerical fractions do not have the same denominator, we cannot add or subtract the fractions until we rewrite them as equivalent fractions with a common denominator. The same holds true for rational expressions with unlike denominators. We will begin by learning how to find the **least common denominator (LCD)** for two or more rational expressions.

Finding the LCD of Two Rational Expressions

Begin by completely factoring each denominator, and then identify each expression that is a factor of one or both denominators. The LCD is equal to the product of these factors.

If an expression is a repeated factor of one or more of the denominators, then we repeat it as a factor in the LCD as well. The exponent used for this factor is equal to the greatest power that the factor is raised to in any one denominator.

EXAMPLE 8 Find the LCD of $\frac{3}{10a^2b}$ and $\frac{5}{12a^5}$.

Solution

We begin with the coefficients 10 and 12. The smallest number that both divide into evenly is 60. Moving on to variable factors in the denominator, we see that the variables a and b are factors of one or both denominators. Note that the variable a is raised to the fifth power in the second denominator, so the LCD must contain a factor of a^5. The LCD is $60a^5b$.

Quick Check 8 Find the LCD of $\frac{7}{6x^3y^2}$ and $\frac{3}{8x^6y}$.

EXAMPLE 9 Find the LCD of $\frac{x+7}{x^2+x-30}$ and $\frac{x-4}{x^2-36}$.

Solution

We begin by factoring each denominator.

$$\frac{x+7}{x^2+x-30}=\frac{x+7}{(x+6)(x-5)} \qquad \frac{x-4}{x^2-36}=\frac{x-4}{(x+6)(x-6)}$$

The factors in the denominators are $x+6$, $x-5$, and $x-6$. Since no expression is repeated as a factor in any one denominator, the LCD is $(x+6)(x-5)(x-6)$.

Quick Check 9 Find the LCD of $\frac{x+5}{x^2+11x+24}$ and $\frac{x-3}{x^2-3x-18}$.

EXAMPLE 10 Find the LCD of $\frac{x+9}{x^2+4x-21}$ and $\frac{x}{x^2-6x+9}$.

Solution

Again, we begin by factoring each denominator.

$$\frac{x+9}{x^2+4x-21}=\frac{x+9}{(x+7)(x-3)} \qquad \frac{x}{x^2-6x+9}=\frac{x}{(x-3)(x-3)}$$

The two expressions that are factors are $x+7$ and $x-3$; the factor $x-3$ is repeated twice in the second denominator. So the LCD must have $x-3$ as a factor twice as well. The LCD is $(x+7)(x-3)(x-3)$ or $(x+7)(x-3)^2$.

Quick Check 10 Find the LCD of $\frac{2x+5}{x^2-36}$ and $\frac{x-9}{x^2+12x+36}$.

Adding or Subtracting Rational Expressions with Unlike Denominators

Objective 5 Add or subtract rational expressions with the same denominators. To add or subtract two rational expressions that do not have the same denominator, we begin by finding the LCD. We then convert each rational expression to an equivalent rational expression that has the LCD as its denominator. We can then add or subtract as we did in the previous section. As always, we should attempt to simplify the resulting rational expression.

EXAMPLE 11 Add $\frac{5}{x+4} + \frac{3}{x+6}$.

Solution

The two denominators are not the same, so we begin by finding the LCD for these two rational expressions. Each denominator has a single factor, and the LCD is the product of these two denominators. The LCD is $(x + 4)(x + 6)$. We will multiply $\frac{5}{x+4}$ by $\frac{x+6}{x+6}$ to write it as an equivalent fraction whose denominator is the LCD. We need to multiply $\frac{3}{x+6}$ by $\frac{x+4}{x+4}$ to write it as an equivalent fraction whose denominator is the LCD.

$$\frac{5}{x+4} + \frac{3}{x+6}$$

$$= \frac{5}{x+4} \cdot \frac{x+6}{x+6} + \frac{3}{x+6} \cdot \frac{x+4}{x+4}$$ Multiply to rewrite each expression as an equivalent rational expression that has the LCD as its denominator.

$$= \frac{5x+30}{(x+4)(x+6)} + \frac{3x+12}{(x+4)(x+6)}$$ Distribute in each numerator, but do not distribute in the denominators.

$$= \frac{(5x+30)+(3x+12)}{(x+4)(x+6)}$$ Add the numerators, writing the sum over the common denominator.

$$= \frac{8x+42}{(x+4)(x+6)}$$ Combine like terms.

$$= \frac{2(4x+21)}{(x+4)(x+6)}$$ Factor the numerator.

Since the numerator and denominator do not have any common factors, this rational expression cannot be simplified any further.

***Quick Check* 11**
Add $\frac{7}{x-5} + \frac{2}{x+8}$.

A Word of Caution When adding two rational expressions, we cannot simply add the two numerators together and place their sum over the sum of the two denominators.

$$\frac{5}{x+4} + \frac{3}{x+6} \neq \frac{5+3}{(x+4)+(x+6)}$$

We must first find a common denominator and rewrite each rational expression as an equivalent expression whose denominator is equal to the common denominator.

When adding or subtracting rational expressions, leave the denominator in factored form. After we simplify the numerator, we factor the numerator if possible and check the denominator for common factors that can be divided out.

EXAMPLE 12 Add $\dfrac{x}{x^2 + 6x + 8} + \dfrac{4}{x^2 + 8x + 12}$.

Solution

In this example we must factor each denominator to find the LCD.

$$\frac{x}{x^2 + 6x + 8} + \frac{4}{x^2 + 8x + 12}$$

$$= \frac{x}{(x + 2)(x + 4)} + \frac{4}{(x + 2)(x + 6)}$$ Factor each denominator. The LCD is $(x + 2)(x + 4)(x + 6)$.

$$= \frac{x}{(x + 2)(x + 4)} \cdot \frac{x + 6}{x + 6} + \frac{4}{(x + 2)(x + 6)} \cdot \frac{x + 4}{x + 4}$$ Multiply to rewrite each expression as an equivalent rational expression that has the LCD as its denominator.

$$= \frac{x^2 + 6x}{(x + 2)(x + 4)(x + 6)} + \frac{4x + 16}{(x + 2)(x + 4)(x + 6)}$$ Distribute in each numerator.

$$= \frac{(x^2 + 6x) + (4x + 16)}{(x + 2)(x + 4)(x + 6)}$$ Add the numerators, writing the sum over the LCD.

$$= \frac{x^2 + 10x + 16}{(x + 2)(x + 4)(x + 6)}$$ Combine like terms.

$$= \frac{\overset{1}{\cancel{(x + 2)}}(x + 8)}{\underset{1}{\cancel{(x + 2)}}(x + 4)(x + 6)}$$ Factor the numerator and divide out the common factor.

$$= \frac{x + 8}{(x + 4)(x + 6)}$$ Simplify.

There is another method for creating two equivalent rational expressions that have the same denominators. In this example, once the two denominators had been factored, we had the expression $\dfrac{x}{(x + 2)(x + 4)} + \dfrac{4}{(x + 2)(x + 6)}$. Notice that the first denominator is missing the factor $x + 6$ that appears in the second denominator, so we can multiply the numerator and denominator of the first rational expression by $x + 6$. In a similar fashion, the second denominator is missing the factor $x + 4$ that appears in the first denominator, so we can multiply the numerator and denominator of the second rational expression by $x + 4$.

$$\frac{x}{(x + 2)(x + 4)} + \frac{4}{(x + 2)(x + 6)} = \frac{x}{(x + 2)(x + 4)} \cdot \frac{x + 6}{x + 6} + \frac{4}{(x + 2)(x + 6)} \cdot \frac{x + 4}{x + 4}$$

At this point, both rational expressions share the same denominator, $(x + 2)(x + 4)(x + 6)$, so we may proceed with the addition problem.

Quick Check 12 Add $\frac{x}{x^2 - 4x + 3} + \frac{3}{x^2 + 4x - 5}$.

EXAMPLE 13 Subtract $\frac{x}{x^2 - 64} - \frac{3}{x^2 - 10x + 16}$.

Solution

We begin by factoring each denominator to find the LCD. We must be careful that we subtract the entire second numerator, not just the first term. In other words, the subtraction must change the sign of each term in the second numerator before we combine like terms. Using parentheses around the two numerators before starting to subtract them will help with this.

$$\frac{x}{x^2 - 64} - \frac{3}{x^2 - 10x + 16}$$

$$= \frac{x}{(x+8)(x-8)} - \frac{3}{(x-2)(x-8)}$$ Factor each denominator. The LCD is $(x + 8)(x - 8)(x - 2)$.

$$= \frac{x}{(x+8)(x-8)} \cdot \frac{x-2}{x-2} - \frac{3}{(x-2)(x-8)} \cdot \frac{x+8}{x+8}$$ Multiply to rewrite each expression as an equivalent rational expression that has the LCD as its denominator.

$$= \frac{x^2 - 2x}{(x+8)(x-8)(x-2)} - \frac{3x + 24}{(x+8)(x-8)(x-2)}$$ Distribute in each numerator.

$$= \frac{(x^2 - 2x) - (3x + 24)}{(x+8)(x-8)(x-2)}$$ Subtract the numerators, writing the difference over the LCD.

$$= \frac{x^2 - 2x - 3x - 24}{(x+8)(x-8)(x-2)}$$ Distribute.

$$= \frac{x^2 - 5x - 24}{(x+8)(x-8)(x-2)}$$ Combine like terms.

$$= \frac{\overset{1}{\cancel{(x-8)}}(x+3)}{(x+8)\underset{1}{\cancel{(x-8)}}(x-2)}$$ Factor the numerator and divide out the common factor.

$$= \frac{x+3}{(x+8)(x-2)}$$ Simplify.

Quick Check 13 Subtract $\frac{x}{x^2 + x - 2} - \frac{2}{x^2 + 7x + 10}$.

EXAMPLE 14 Add $\frac{x-5}{x^2+14x+33}+\frac{10}{x^2+17x+66}$.

Solution

Notice that the first numerator contains a binomial. We must be careful when multiplying to create equivalent rational expressions with the same denominator. We begin by factoring each denominator to find the LCD.

$$\frac{x-5}{x^2+14x+33}+\frac{10}{x^2+17x+66}$$

$$=\frac{x-5}{(x+3)(x+11)}+\frac{10}{(x+6)(x+11)}$$

Factor each denominator. The LCD is $(x+3)(x+11)(x+6)$.

$$=\frac{x-5}{(x+3)(x+11)}\cdot\frac{x+6}{x+6}+\frac{10}{(x+6)(x+11)}\cdot\frac{x+3}{x+3}$$

Multiply to rewrite each expression as an equivalent rational expression with the LCD as its denominator.

$$=\frac{x^2+x-30}{(x+3)(x+11)(x+6)}+\frac{10x+30}{(x+3)(x+11)(x+6)}$$

Distribute in each numerator.

$$=\frac{(x^2+x-30)+(10x+30)}{(x+3)(x+11)(x+6)}$$

Add the numerators, writing the sum over the LCD.

$$=\frac{x^2+11x}{(x+3)(x+11)(x+6)}$$

Combine like terms.

$$=\frac{x\overset{1}{\cancel{(x+11)}}}{(x+3)\underset{1}{\cancel{(x+11)}}(x+6)}$$

Factor the numerator and divide out the common factor.

$$=\frac{x}{(x+3)(x+6)}$$

Simplify.

Quick Check **14** Add $\frac{x+6}{x^2+9x+14}+\frac{2}{x^2+4x-21}$.

Building Your Study Strategy **Test Taking, 3** **Write Down Important Information.** As soon as you receive your test, write down any formulas, rules, or procedures that will help you during the exam, such as a table or formula that you use to solve a particular type of word problem. Once you have written these down, you can refer to them as you work through the test. Write it down on the test while it is still fresh in your memory. That way, when you reach the appropriate problem on the test, you will not need to worry about not being able to remember the table or formula. By writing all of this information on your test, you will eliminate memorization difficulties that arise when taking a test. However, if you do not understand the material or how to use what you have written down, then you will struggle on the test. There is no substitute for understanding what you are doing.

Exercises 5.3

Vocabulary

1. To add fractions that have the same denominator, we add the __________ and place the result over the common denominator.

2. To subtract fractions that have the same denominator, we __________ the numerators and place the result over the common denominator.

3. When subtracting a rational expression whose numerator contains more than one term, subtract the entire numerator and not just the __________.

4. When adding two rational expressions with opposite denominators, we can replace the second denominator with its opposite by changing the addition to __________.

5. The __________ of two rational expressions is an expression that is a product of all the factors of the two denominators.

6. To add two rational expressions that have unlike denominators, we begin by converting each rational expression to a(n) __________ rational expression that has the LCD as its denominator.

Add.

7. $\dfrac{7}{x-3} + \dfrac{11}{x-3}$

8. $\dfrac{13}{x+5} + \dfrac{4}{x+5}$

9. $\dfrac{3x}{x+4} + \dfrac{12}{x+4}$

10. $\dfrac{5x}{4x+24} + \dfrac{30}{4x+24}$

11. $\dfrac{x}{x^2+3x-18} + \dfrac{6}{x^2+3x-18}$

12. $\dfrac{x}{x^2+14x+45} + \dfrac{9}{x^2+14x+45}$

13. $\dfrac{x^2+x-8}{x^2+13x+42} + \dfrac{3x-13}{x^2+13x+42}$

14. $\dfrac{x^2-7x+9}{x^2-6x-16} + \dfrac{4x-49}{x^2-6x-16}$

Subtract.

15. $\dfrac{16}{x+6} - \dfrac{7}{x+6}$

16. $\dfrac{4}{x-5} - \dfrac{14}{x-5}$

17. $\dfrac{6x}{x-7} - \dfrac{42}{x-7}$

18. $\dfrac{3x}{2x-22} - \dfrac{33}{2x-22}$

19. $\dfrac{x}{x^2-13x+36} - \dfrac{4}{x^2-13x+36}$

20. $\dfrac{x}{x^2+7x-30} - \dfrac{3}{x^2+7x-30}$

21. $\dfrac{5x-7}{x^2-36} - \dfrac{2x+11}{x^2-36}$

22. $\dfrac{12x-5}{x^2-5x-6} - \dfrac{3x-14}{x^2-5x-6}$

23. $\dfrac{x^2+8x+7}{x^2+2x-35} - \dfrac{3x+21}{x^2+2x-35}$

24. $\dfrac{x^2+4x-4}{x^2-11x+24} - \dfrac{4x+5}{x^2-11x+24}$

25. $\dfrac{2x^2+17x+23}{x^2+6x} - \dfrac{x^2+5x-13}{x^2+6x}$

26. $\dfrac{3x^2+8x-2}{x^2-25} - \dfrac{2x^2+11x+8}{x^2-25}$

Add or subtract.

27. $\dfrac{7x}{x-8} + \dfrac{56}{8-x}$

28. $\dfrac{5x}{2x-10} + \dfrac{25}{10-2x}$

FOR EXTRA HELP

MyMathLab

MathXL

Interactmath.com

MathXL Tutorials on CD

Video Lectures on CD

Addison-Wesley Math Tutor Center

Student's Solutions Manual

29. $\dfrac{x^2 + 4x}{x - 3} - \dfrac{x - 24}{3 - x}$

30. $\dfrac{x^2 - 12x}{x - 4} - \dfrac{3x + 20}{4 - x}$

31. $\dfrac{x^2 - 9x + 33}{x^2 - 64} + \dfrac{6x - 23}{64 - x^2}$

32. $\dfrac{x^2 - 18x + 10}{x^2 - 16} - \dfrac{3x + 34}{16 - x^2}$

Find the missing numerator.

33. $\dfrac{?}{x + 7} + \dfrac{28}{x + 7} = 4$

34. $\dfrac{x^2 + 7x - 9}{x - 4} - \dfrac{?}{x - 4} = x + 9$

35. $\dfrac{x^2 - 6x - 12}{(x - 5)(x - 2)} + \dfrac{?}{(x - 5)(x - 2)} = \dfrac{x + 6}{x - 2}$

36. $\dfrac{x^2 + 3x + 10}{x^2 + 4x - 32} - \dfrac{?}{x^2 + 4x - 32} = \dfrac{x - 4}{x + 8}$

Find the LCD of the given rational expressions.

37. $\dfrac{9}{4x}, \dfrac{5}{6x}$

38. $\dfrac{3}{10x^3}, \dfrac{7}{8x^5}$

39. $\dfrac{11}{x - 8}, \dfrac{1}{x + 4}$

40. $\dfrac{x - 3}{x - 4}, \dfrac{x + 6}{2x + 5}$

41. $\dfrac{x + 4}{x^2 + 9x + 14}, \dfrac{x}{x^2 - 4}$

42. $\dfrac{x - 5}{x^2 + 3x - 54}, \dfrac{x + 4}{x^2 - x - 30}$

43. $\dfrac{3x - 8}{x^2 - 6x + 9}, \dfrac{x + 1}{x^2 - 7x + 12}$

44. $\dfrac{x - 6}{x^2 + 14x + 49}, \dfrac{9}{x^2 + 7x}$

Add or subtract.

45. $\dfrac{5x}{6} + \dfrac{2x}{9}$

46. $\dfrac{7x}{12} - \dfrac{x}{20}$

47. $\dfrac{9}{8x} - \dfrac{11}{6x}$

48. $\dfrac{7}{9x} + \dfrac{3}{10x}$

49. $\dfrac{2}{x + 5} + \dfrac{3}{x + 4}$

50. $\dfrac{3}{x - 6} - \dfrac{7}{x + 1}$

51. $\dfrac{5}{(x - 10)(x - 5)} - \dfrac{4}{(x - 5)(x - 9)}$

52. $\dfrac{8}{(x + 2)(x + 6)} + \dfrac{2}{(x + 6)(x + 7)}$

53. $\dfrac{9}{x^2 + x - 2} + \dfrac{6}{x^2 - 4x + 3}$

54. $\dfrac{3}{x^2 + 7x + 10} + \dfrac{5}{x^2 - x - 6}$

55. $\dfrac{5}{x^2 - 16x + 63} - \dfrac{2}{x^2 - 15x + 56}$

56. $\dfrac{7}{x^2 + 8x + 15} - \dfrac{4}{x^2 + 7x + 12}$

57. $\dfrac{1}{x^2 - 15x + 54} - \dfrac{6}{x^2 - 81}$

58. $\dfrac{4}{x^2 - 4} - \dfrac{7}{x^2 - 3x - 10}$

59. $\dfrac{5}{2x^2 + 3x - 2} + \dfrac{7}{2x^2 - 9x + 4}$

60. $\dfrac{7}{2x^2 - 15x - 27} - \dfrac{3}{2x^2 - 3x - 9}$

61. $\dfrac{x + 3}{x^2 + 11x + 30} + \dfrac{4}{x^2 + 12x + 35}$

62. $\dfrac{x - 1}{x^2 + 13x + 40} + \dfrac{3}{x^2 + 15x + 56}$

63. $\dfrac{x + 2}{x^2 - 36} - \dfrac{2}{x^2 - 9x + 18}$

64. $\dfrac{x-4}{x^2-8x+15}-\dfrac{6}{x^2+2x-35}$

65. $\dfrac{x+2}{x^2+6x-16}+\dfrac{3}{x^2+11x+24}$

66. $\dfrac{x+5}{x^2-2x-3}+\dfrac{1}{x^2+3x+2}$

67. $\dfrac{x-1}{x^2+7x+10}-\dfrac{3}{x^2+x-2}$

68. $\dfrac{x+7}{x^2+5x+4}-\dfrac{4}{x^2+4x+3}$

69. $\dfrac{x+10}{x^2+15x+54}+\dfrac{7}{x^2+6x}$

70. $\dfrac{x+3}{x^2-7x}-\dfrac{5}{x^2-49}$

71. $\dfrac{x+1}{x^2-5x+6}+\dfrac{x-8}{x^2-6x+8}$

72. $\dfrac{x+1}{x^2+9x+20}+\dfrac{x+10}{x^2+10x+24}$

For the given rational functions $f(x)$ and $g(x)$, find $f(x)+g(x)$.

73. $f(x)=\dfrac{3x}{4x+28},\ g(x)=\dfrac{21}{4x+28}$

74. $f(x)=\dfrac{x^2+7x+11}{x^2-9},\ g(x)=\dfrac{2x+5}{9-x^2}$

75. $f(x)=\dfrac{3}{x^2-3x-18},\ g(x)=\dfrac{2}{x^2+12x+27}$

76. $f(x)=\dfrac{x+5}{x^2+x},\ g(x)=\dfrac{5}{x^2-x}$

For the given rational functions $f(x)$ and $g(x)$, find $f(x)-g(x)$.

77. $f(x)=\dfrac{13}{x-3},\ g(x)=\dfrac{4}{x-3}$

78. $f(x)=\dfrac{x^2+4x}{x-6},\ g(x)=\dfrac{x+6}{6-x}$

79. $f(x)=\dfrac{6}{x^2+12x+27},\ g(x)=\dfrac{2}{x^2+16x+63}$

80. $f(x)=\dfrac{x+6}{x^2+18x+80},\ g(x)=\dfrac{2}{x^2+14x+48}$

Mixed Practice, 81–98

Add or subtract.

81. $\dfrac{x-2}{15}+\dfrac{x+3}{18}$

82. $\dfrac{x-6}{x^2+9x-36}+\dfrac{2}{x^2+4x-21}$

83. $\dfrac{4}{x^2-10x+21}+\dfrac{5}{x^2-x-42}$

84. $\dfrac{3x+7}{x-7}-\dfrac{x-35}{7-x}$

85. $\dfrac{x^2-10x+2}{x^2+9x-10}+\dfrac{3x+4}{x^2+9x-10}$

86. $\dfrac{8}{x^2-x-20}-\dfrac{6}{x^2-16}$

87. $\dfrac{x^2+5x+21}{x^2-81}-\dfrac{10x+33}{81-x^2}$

88. $\dfrac{11}{6a}-\dfrac{7}{10a}$

89. $\dfrac{3x-8}{x^2-25}-\dfrac{x+2}{x^2-25}$

90. $\dfrac{x^2-2x+26}{x^2+x-56}-\dfrac{8x+5}{x^2+x-56}$

91. $\dfrac{x+5}{x^2-17x+70}-\dfrac{4}{x^2-15x+56}$

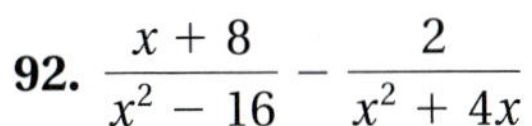

92. $\dfrac{x+8}{x^2-16} - \dfrac{2}{x^2+4x}$

93. $\dfrac{5}{3x-4} + \dfrac{2}{4-3x}$

94. $\dfrac{3}{x^2+9x+18} + \dfrac{7}{x^2-x-12}$

95. $\dfrac{x-5}{x^2+5x+4} + \dfrac{6}{x^2+6x+8}$

96. $\dfrac{x^2-6x+24}{x^2-8x} + \dfrac{7x-16}{8x-x^2}$

97. $\dfrac{2}{x^2+6x+8} - \dfrac{3}{x^2+7x+10}$

98. $\dfrac{x^2-7x-9}{x^2+8x+12} + \dfrac{11x+13}{x^2+8x+12}$

Writing in Mathematics

Answer in complete sentences.

99. Explain how to determine that two rational expressions have opposite denominators.

100. Here is a student's solution to a problem on an exam. Describe the student's error, and provide the correct solution. Assuming that the problem was worth 10 points, how many points would you give to the student for his solution? Explain your reasoning.

$$\frac{5}{(x+3)(x+2)} + \frac{7}{(x+3)(x+6)}$$
$$= \frac{5}{(x+3)(x+2)} \cdot \frac{x+6}{x+6} + \frac{7}{(x+3)(x+6)} \cdot \frac{x+2}{x+2}$$
$$= \frac{5(x+6)+7(x+2)}{(x+3)(x+2)(x+6)}$$
$$= \frac{5\overset{1}{\cancel{(x+6)}} + 7\overset{1}{\cancel{(x+2)}}}{(x+3)\underset{1}{\cancel{(x+2)}}\underset{1}{\cancel{(x+6)}}}$$
$$= \frac{12}{x+3}$$

101. ***Solutions Manual**** Write a solutions manual page for the following problem:

Subtract $\dfrac{x^2+3x+3}{x^2+6x-16} - \dfrac{12x-11}{x^2+6x-16}$.

102. ***Solutions Manual**** Write a solutions manual page for the following problem:

Add $\dfrac{x+6}{x^2+9x+20} + \dfrac{4}{x^2+6x+8}$.

103. ***Newsletter**** Write a newsletter that explains how to add two rational expressions with the same denominator.

104. ***Newsletter**** Write a newsletter that explains how to subtract two rational expressions with unlike denominators.

***See Appendix B for details and sample answers.**

5.4 Complex Fractions

Objectives

1. **Simplify complex numerical fractions.**
2. **Simplify complex fractions containing variables.**

Complex Fractions

A **complex fraction** is a fraction or rational expression containing one or more fractions in its numerator or denominator. Here are some examples.

$$\frac{\frac{1}{2}+\frac{5}{3}}{\frac{10}{3}-\frac{7}{4}} \qquad \frac{\frac{x+5}{x-9}}{1-\frac{5}{x}} \qquad \frac{\frac{1}{2}+\frac{1}{x}}{\frac{1}{4}-\frac{1}{x^2}} \qquad \frac{1+\frac{3}{x}-\frac{28}{x^2}}{1+\frac{7}{x}}$$

Objective 1 **Simplify complex numerical fractions.** To simplify a complex fraction, we must rewrite it in such a way that its numerator and denominator do not contain fractions. This can be done by finding the LCD of all fractions within the complex fraction and then multiplying the numerator and denominator by this LCD. This will clear the fractions within the complex fraction. We finish by simplifying the resulting rational expression, if possible.

We begin with a complex fraction made up of numerical fractions.

EXAMPLE 1 Simplify the complex fraction $\dfrac{3+\frac{5}{6}}{\frac{2}{3}+\frac{3}{4}}$.

Solution

The LCD of the three denominators (6, 3, and 4) is 12, so we begin by multiplying the complex fraction by $\frac{12}{12}$. Notice that when we multiply by $\frac{12}{12}$, we are really multiplying by 1, which does not change the value of the original expression.

$$\frac{3+\frac{5}{6}}{\frac{2}{3}+\frac{3}{4}} = \frac{12}{12}\cdot\frac{3+\frac{5}{6}}{\frac{2}{3}+\frac{3}{4}}$$ Multiply the numerator and denominator by the LCD.

$$= \frac{12\cdot 3+\overset{2}{\cancel{12}}\cdot\frac{5}{\underset{1}{\cancel{6}}}}{\overset{4}{\cancel{12}}\cdot\frac{2}{\underset{1}{\cancel{3}}}+\overset{3}{\cancel{12}}\cdot\frac{3}{\underset{1}{\cancel{4}}}}$$ Distribute and divide out common factors.

$$= \frac{36 + 10}{8 + 9} \qquad \text{Multiply.}$$

$$= \frac{46}{17} \qquad \text{Simplify the numerator and denominator.}$$

There is another method for simplifying complex fractions. We can rewrite the numerator as a single fraction by adding $3 + \frac{5}{6}$, which would equal $\frac{23}{6}$.

$$3 + \frac{5}{6} = \frac{18}{6} + \frac{5}{6} = \frac{23}{6}$$

We can also rewrite the denominator as a single fraction by adding $\frac{2}{3} + \frac{3}{4}$, which would equal $\frac{17}{12}$.

$$\frac{2}{3} + \frac{3}{4} = \frac{8}{12} + \frac{9}{12} = \frac{17}{12}$$

Once the numerator and denominator are single fractions, we can rewrite $\dfrac{\frac{23}{6}}{\frac{17}{12}}$ as a division problem $\left(\frac{23}{6} \div \frac{17}{12}\right)$ and simplify from there. This method produces the same result, $\frac{46}{17}$.

$$\frac{23}{6} \div \frac{17}{12} = \frac{23}{\cancel{6}_{1}} \cdot \frac{\cancel{12}^{2}}{17} = \frac{46}{17}$$

In most of the remaining examples we will use the LCD method, as the technique is similar to the technique used to solve rational equations in the next section.

Quick Check 1
Simplify the complex fraction $\dfrac{\frac{5}{4} + \frac{8}{5}}{\frac{3}{10} - \frac{1}{8}}$.

Simplifying Complex Fractions Containing Variables

Objective 2 Simplify complex fractions containing variables.

Example 2 Simplify $\dfrac{1 + \frac{5}{x}}{1 - \frac{25}{x^2}}$.

Solution

The LCD for the two simple fractions with denominators x and x^2 is x^2, so we will begin by multiplying the complex fraction by $\frac{x^2}{x^2}$.

$$\frac{1 + \frac{5}{x}}{1 - \frac{25}{x^2}} = \frac{x^2}{x^2} \cdot \frac{1 + \frac{5}{x}}{1 - \frac{25}{x^2}} \qquad \text{Multiply the numerator and denominator by the LCD.}$$

$$= \frac{x^2 \cdot 1 + \overset{x}{\cancel{x^2}} \cdot \frac{5}{\underset{1}{\cancel{x}}}}{x^2 \cdot 1 - \overset{1}{\cancel{x^2}} \cdot \frac{25}{\underset{1}{\cancel{x^2}}}}$$
Distribute and divide out common factors, clearing the fractions.

$$= \frac{x^2 + 5x}{x^2 - 25}$$
Multiply.

$$= \frac{x\overset{1}{\cancel{(x + 5)}}}{\underset{1}{\cancel{(x + 5)}}(x - 5)}$$
Factor the numerator and denominator and divide out the common factor.

$$= \frac{x}{x - 5}$$
Simplify.

Once we have cleared the fractions, resulting in a rational expression such as $\frac{x^2 + 5x}{x^2 - 25}$, then we simplify the rational expression by factoring and dividing out common factors.

Quick Check 2
Simplify $\dfrac{\frac{1}{49} - \frac{1}{x^2}}{\frac{1}{7} + \frac{1}{x}}$.

Example 3 Simplify $\dfrac{1 + \frac{6}{x} - \frac{16}{x^2}}{1 - \frac{5}{x} + \frac{6}{x^2}}$.

Solution

We begin by multiplying the numerator and denominator by the LCD of all denominators. In this case the LCD is x^2.

$$\frac{1 + \frac{6}{x} - \frac{16}{x^2}}{1 - \frac{5}{x} + \frac{6}{x^2}} = \frac{x^2}{x^2} \cdot \frac{1 + \frac{6}{x} - \frac{16}{x^2}}{1 - \frac{5}{x} + \frac{6}{x^2}}$$
Multiply the numerator and denominator by the LCD.

$$= \frac{x^2 \cdot 1 + \overset{x}{\cancel{x^2}} \frac{6}{\underset{1}{\cancel{x}}} - \overset{1}{\cancel{x^2}} \frac{16}{\underset{1}{\cancel{x^2}}}}{x^2 \cdot 1 - \overset{x}{\cancel{x^2}} \cdot \frac{5}{\underset{1}{\cancel{x}}} + \overset{1}{\cancel{x^2}} \cdot \frac{6}{\underset{1}{\cancel{x^2}}}}$$
Distribute and divide out common factors.

$$= \frac{x^2 + 6x - 16}{x^2 - 5x + 6}$$
Multiply.

$$= \frac{(x + 8)\overset{1}{\cancel{(x - 2)}}}{\underset{1}{\cancel{(x - 2)}}(x - 3)}$$
Factor the numerator and denominator and divide out the common factor.

$$= \frac{x + 8}{x - 3}$$
Simplify.

Quick Check 3
Simplify $\dfrac{1 - \frac{8}{x} - \frac{9}{x^2}}{1 - \frac{81}{x^2}}$.

Example 4 Simplify $\dfrac{\dfrac{6}{x+5}-\dfrac{3}{x+3}}{\dfrac{x}{x+3}+\dfrac{2}{x+5}}$.

Solution

The LCD for the four simple fractions is $(x+3)(x+5)$.

$$\frac{\frac{6}{x+5}-\frac{3}{x+3}}{\frac{x}{x+3}+\frac{2}{x+5}}$$

$$=\frac{(x+3)(x+5)}{(x+3)(x+5)}\cdot\frac{\frac{6}{x+5}-\frac{3}{x+3}}{\frac{x}{x+3}+\frac{2}{x+5}}$$ Multiply the numerator and denominator by the LCD

$$=\frac{(x+3)\cancel{(x+5)}^{1}\cdot\frac{6}{\cancel{x+5}_{1}}-\cancel{(x+3)}^{1}(x+5)\cdot\frac{3}{\cancel{x+3}_{1}}}{\cancel{(x+3)}^{1}(x+5)\cdot\frac{x}{\cancel{x+3}_{1}}+(x+3)\cancel{(x+5)}^{1}\cdot\frac{2}{\cancel{x+5}_{1}}}$$ Distribute and divide out common factors, clearing the fractions.

$$=\frac{6(x+3)-3(x+5)}{x(x+5)+2(x+3)}$$ Multiply.

$$=\frac{6x+18-3x-15}{x^2+5x+2x+6}$$ Distribute.

$$=\frac{3x+3}{x^2+7x+6}$$ Combine like terms.

$$=\frac{3\cancel{(x+1)}^{1}}{\cancel{(x+1)}_{1}(x+6)}$$ Factor the numerator and denominator and divide out the common factor.

$$=\frac{3}{x+6}$$ Simplify.

Quick Check 4

Simplify $\dfrac{\dfrac{1}{x+6}+\dfrac{1}{x-2}}{\dfrac{x}{x+6}-\dfrac{2}{x-2}}$.

Example 5 Simplify $\dfrac{\dfrac{x^2+10x+21}{x^2-5x-36}}{\dfrac{x^2-9}{x^2+12x+32}}$.

Solution

In this case it will be easier to rewrite the complex fraction as a division problem rather than multiplying the numerator and denominator by the LCD. This is a wise idea when we have a complex fraction with a single rational expression in its numerator and a single rational expression in its denominator. We have used this method when dividing rational expressions.

$$\frac{\frac{x^2+10x+21}{x^2-5x-36}}{\frac{x^2-9}{x^2+12x+32}} = \frac{x^2+10x+21}{x^2-5x-36} \div \frac{x^2-9}{x^2+12x+32}$$ Rewrite as a division problem.

$$= \frac{x^2+10x+21}{x^2-5x-36} \cdot \frac{x^2+12x+32}{x^2-9}$$ Invert the divisor and multiply.

$$= \frac{(x+3)(x+7)}{(x+4)(x-9)} \cdot \frac{(x+4)(x+8)}{(x+3)(x-3)}$$ Factor each numerator and denominator.

$$= \frac{\overset{1}{\cancel{(x+3)}}(x+7)}{\underset{1}{\cancel{(x+4)}}(x-9)} \cdot \frac{\overset{1}{\cancel{(x+4)}}(x+8)}{\underset{1}{\cancel{(x+3)}}(x-3)}$$ Divide out common factors.

$$= \frac{(x+7)(x+8)}{(x-9)(x-3)}$$ Simplify.

***Quick Check* 5**

Simplify $\dfrac{\frac{x^2-8x+16}{x^2+8x-9}}{\frac{x^2+x-20}{x^2+2x-63}}$.

Building Your Study Strategy **Test Taking, 4** **Read the Test** In the same way you would begin to solve a word problem, you should begin to take a test by briefly reading through it. This will give you an idea of how many problems you have to solve, how many word problems there are, and roughly how much time you can devote to each problem. It is a good idea to establish a schedule, such as "I need to be done with 12 problems by the time half of the class period is over." This way you will know whether you need to speed up during the second half of the exam or if you have plenty of time.

EXERCISES 5.4

Vocabulary

1. A(n) ________ is a fraction or rational expression containing one or more fractions in its numerator or denominator.

2. To simplify a complex fraction, multiply the numerator and denominator by the ________ of all fractions within the complex fraction.

Simplify the complex fraction.

3. $\dfrac{\frac{3}{5}-\frac{2}{3}}{\frac{1}{6}+\frac{3}{10}}$

4. $\dfrac{\frac{5}{6}+\frac{11}{8}}{\frac{3}{4}-\frac{1}{12}}$

5. $\dfrac{3-\frac{3}{4}}{\frac{7}{8}-\frac{1}{6}}$

6. $\dfrac{\frac{7}{10}-3}{\frac{3}{5}+\frac{5}{4}}$

7. $\dfrac{x+\frac{2}{9}}{x+\frac{3}{2}}$

8. $\dfrac{x-\frac{4}{5}}{x+\frac{3}{8}}$

9. $\dfrac{10-\frac{15}{x}}{x-\frac{3}{2}}$

10. $\dfrac{3+\frac{4}{x}}{\frac{x}{2}+\frac{2}{3}}$

11. $\dfrac{4+\frac{12}{x}}{1-\frac{9}{x^2}}$

12. $\dfrac{3-\frac{108}{x^2}}{1+\frac{6}{x}}$

13. $\dfrac{\frac{1}{9}-\frac{1}{x^2}}{\frac{1}{3}-\frac{1}{x}}$

14. $\dfrac{\frac{1}{x}+\frac{1}{5}}{\frac{1}{x^2}-\frac{1}{25}}$

15. $\dfrac{\frac{7}{x-5}-\frac{4}{x-2}}{\frac{x+2}{x-5}}$

16. $\dfrac{\frac{5}{x-9}+\frac{3}{x-1}}{2-\frac{x+2}{x-1}}$

17. $\dfrac{\frac{6}{x+7}+\frac{5}{x-4}}{2-\frac{x+13}{x+7}}$

18. $\dfrac{\frac{4}{x+4}-\frac{5}{x+3}}{5+\frac{x+33}{x+3}}$

19. $\dfrac{1+\frac{3}{x}-\frac{10}{x^2}}{1-\frac{1}{x}-\frac{30}{x^2}}$

20. $\dfrac{1+\frac{8}{x}+\frac{16}{x^2}}{1+\frac{13}{x}+\frac{36}{x^2}}$

21. $\dfrac{1-\frac{5}{x}-\frac{24}{x^2}}{1-\frac{64}{x^2}}$

22. $\dfrac{1-\frac{4}{x^2}}{1+\frac{1}{x}-\frac{6}{x^2}}$

23. $\dfrac{x+7+\frac{6}{x}}{1-\frac{4}{x}-\frac{5}{x^2}}$

24. $\dfrac{1-\frac{14}{x}+\frac{40}{x^2}}{x-7+\frac{12}{x}}$

25. $\dfrac{\frac{x^2+12x+35}{x^2-3x-40}}{\frac{x^2+9x+14}{x^2-64}}$

26. $\dfrac{\frac{x^2-7x+6}{x^2-16}}{\frac{x^2-6x}{x^2+13x+36}}$

27. $\dfrac{\frac{x^2+17x+72}{x^2+5x+6}}{\frac{x^2-81}{x^2-2x-15}}$

28. $\dfrac{\frac{x^2-2x}{x^2-10x+16}}{\frac{x^2+4x-60}{x^2-14x+48}}$

29. $\dfrac{\frac{1}{x^2+15x+36}}{\frac{1}{x^2-7x-30}}$

30. $\dfrac{\frac{1}{x^2-25x+100}}{\frac{1}{x^2+15x-100}}$

Mixed Practice, 31–54

Simplify the given rational expression, using the techniques developed in Sections 5.1 through 5.4.

31. $\dfrac{x^2+3x+5}{x^2+2x-48}+\dfrac{2x-29}{x^2+2x-48}$

32. $\dfrac{x^2+5x+6}{x^2+5x-6}\div\dfrac{x^2-9}{x^2-x-42}$

33. $\dfrac{x+3}{x^2-2x-24}+\dfrac{2}{x^2-8x+12}$

34. $\dfrac{5}{x^2+4x-96}-\dfrac{1}{x^2-12x+32}$

35. $\dfrac{x^2+3x}{x^2-9x+18}\div\dfrac{x^2-25}{x^2-11x+30}$

36. $\dfrac{x^2-64}{24x-3x^2}$

37. $\frac{x-1}{x^2-3x} - \frac{6}{x^2+3x-18}$

38. $\frac{x^2+12x+20}{x^2+11x+30} \cdot \frac{x^2+x-30}{x^2+13x+30}$

39. $\frac{9x}{2x-6} + \frac{27}{6-2x}$

40. $\dfrac{1-\frac{18}{x}+\frac{81}{x^2}}{1-\frac{1}{x}-\frac{72}{x^2}}$

41. $\dfrac{\frac{3}{x-3}+\frac{2}{x+12}}{\frac{x+6}{x-3}}$

42. $\frac{3}{x^2+7x+12} + \frac{9}{x^2+11x+28}$

43. $\frac{9}{x^2+13x+22} - \frac{4}{x^2+20x+99}$

44. $\frac{4x-x^2}{x^2+13x+42} \cdot \frac{49-x^2}{x^2-x-12}$

45. $\dfrac{\frac{1}{3}-\frac{1}{x}}{\frac{1}{9}-\frac{1}{x^2}}$

46. $\frac{x^2+5x+20}{x^2-4} - \frac{5x-4}{4-x^2}$

47. $\frac{5}{x^2+7x+12} + \frac{6}{x^2+10x+24}$

48. $\frac{6x}{2x-9} + \frac{27}{9-2x}$

49. $\frac{x^2+13x+30}{x^2+13x-30} \cdot \frac{x^2-5x+6}{x^2-9}$

50. $\frac{x+6}{x^2+9x+20} + \frac{4}{x^2+6x+8}$

51. $\frac{x^2+4x-10}{x^2-5x-14} - \frac{5x+32}{x^2-5x-14}$

52. $\dfrac{x+\frac{3}{10}}{x-\frac{2}{5}}$

53. $\frac{x^2+8x}{x^2+15x-16} \div \frac{x^2-64}{x^2-10x+9}$

54. $\frac{x-5}{x^2-4x+3} - \frac{6}{x^2+x-2}$

Writing in Mathematics

Answer in complete sentences.

55. Explain what a complex fraction is. Compare and contrast complex fractions and the rational expressions found in Section 5.1.

56. One method for simplifying complex fractions is to rewrite the complex fraction as one rational expression divided by another rational expression. When is this method the most efficient way to simplify a complex fraction?

57. ***Solutions Manual**** Write a solutions manual page for the following problem:

Simplify $\dfrac{\frac{3}{x-9}+\frac{6}{x-3}}{3-\frac{x+5}{x-3}}$.

58. ***Newsletter**** Write a newsletter that explains how to simplify complex fractions.

***See Appendix B for details and sample answers.**

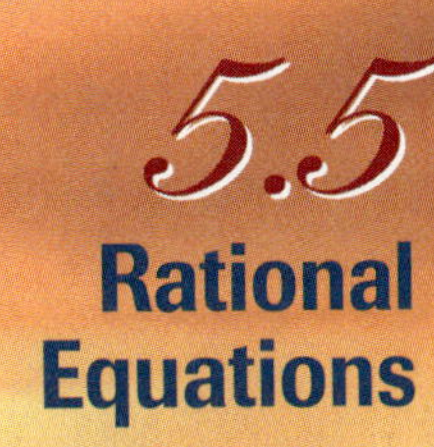

5.5 Rational Equations

Objectives

1. Solve rational equations.
2. Solve literal equations.

Solving Rational Equations

Objective 1 Solve rational equations. In this section, we learn how to solve **rational equations**, which are equations containing at least one rational expression. The main goal is to rewrite the equation as an equivalent equation that does not contain a rational expression. We then solve the equation using methods developed in earlier chapters.

In Chapter 1, we learned how to solve an equation containing fractions such as the equation $\frac{1}{4}x - \frac{3}{5} = \frac{9}{10}$. We began by finding the LCD of all fractions, and then multiplied both sides of the equation by that LCD to clear the equation of fractions. We will employ the same technique in this section. There is a major difference, though, when solving equations containing a variable in a denominator. Occasionally we will find a solution that causes one of the rational expressions in the equation to be undefined. If a denominator of a rational expression is equal to 0 when the value of a solution is substituted for the variable, then that solution must be omitted from the answer set and is called an **extraneous solution**. We must check each solution that we find to make sure that it is not an extraneous solution.

Here is a summary of the method for solving rational equations:

Solving Rational Equations

1. Find the LCD of all denominators in the equation.
2. Multiply both sides of the equation by the LCD to clear the equation of fractions.
3. Solve the resulting equation.
4. Check for extraneous solutions.

Example 1 Solve $\frac{9}{x} + \frac{2}{3} = \frac{17}{12}$.

Solution

We begin by finding the LCD of these three fractions, which is $12x$. Now we multiply both sides of the equation by the LCD to clear the equation of fractions. Once this has been done, we can solve the resulting equation.

$$\frac{9}{x} + \frac{2}{3} = \frac{17}{12}$$

$$12x \cdot \left(\frac{9}{x} + \frac{2}{3}\right) = 12x \cdot \frac{17}{12} \quad \text{Multiply both sides of the equation by the LCD.}$$

$$12\overset{1}{\cancel{x}} \cdot \frac{9}{\underset{1}{\cancel{x}}} + \overset{4}{\cancel{12}}x \cdot \frac{2}{\underset{1}{\cancel{3}}} = \overset{1}{\cancel{12}}x \cdot \frac{17}{\underset{1}{\cancel{12}}} \quad \text{Distribute and divide out common factors.}$$

$$108 + 8x = 17x \quad \text{Multiply. The resulting equation is linear.}$$

$$108 = 9x$$ Subtract $8x$ to collect all variable terms on one side of the equation.

$$12 = x$$ Divide both sides by 9.

Check:

$$\frac{9}{(12)} + \frac{2}{3} = \frac{17}{12}$$ Substitute 12 for x.

$$\frac{3}{4} + \frac{2}{3} = \frac{17}{12}$$ Simplify the fraction $\frac{9}{12}$. The LCD of these fractions is 12.

$$\frac{9}{12} + \frac{8}{12} = \frac{17}{12}$$ Write each fraction with a common denominator of 12.

$$\frac{17}{12} = \frac{17}{12}$$ Add.

Quick Check 1
Solve $\frac{38}{x} - \frac{3}{5} = \frac{2}{3}$.

Since $x = 12$ does not make any rational expression in the original equation undefined, this value is a solution. The solution set is $\{12\}$.

When checking whether a solution is an extraneous solution, we need only determine whether the solution causes the LCD to equal 0. If the LCD is equal to 0 for this solution, then one or more rational expressions are undefined and the solution is an extraneous solution. Also, if the LCD is equal to 0, then we have multiplied both sides of the equation by 0. The multiplication property of equality says we can multiply both sides of an equation by any *nonzero* number without affecting the equality of both sides. In the previous example, the only solution that could possibly be an extraneous solution is $x = 0$, because that is the only value of x for which the LCD is equal to 0.

EXAMPLE 2 Solve $x + 6 - \frac{27}{x} = 0$.

Solution

The LCD in this example is x. The LCD is equal to 0 only if $x = 0$. If we find that $x = 0$ is a solution, then we must omit that solution as an extraneous solution.

$$x + 6 - \frac{27}{x} = 0$$

$$x \cdot \left(x + 6 - \frac{27}{x}\right) = x \cdot 0$$ Multiply each side of the equation by the LCD, x.

$$x \cdot x + x \cdot 6 - \overset{1}{\cancel{x}} \cdot \frac{27}{\underset{1}{\cancel{x}}} = x \cdot 0$$ Distribute and divide out common factors.

$$x^2 + 6x - 27 = 0$$ Multiply. The resulting equation is quadratic.

$$(x + 9)(x - 3) = 0$$ Factor.

$$x = -9 \quad \text{or} \quad x = 3$$ Set each factor equal to 0 and solve.

Quick Check 2
Solve $1 = \frac{4}{x} + \frac{32}{x^2}$.

You may verify that neither solution causes the LCD to equal 0. The solution set is $\{-9, 3\}$.

A Word of Caution When solving a rational equation, the use of the LCD is completely different than when we are adding or subtracting rational expressions. We use the LCD to clear the denominators of the rational expressions when solving a rational equation. When adding or subtracting rational expressions, we rewrite each expression as an equivalent expression whose denominator is the LCD.

EXAMPLE 3 Solve $\frac{x+5}{x-4} + 3 = \frac{2x+1}{x-4}$.

Solution

The LCD is $x - 4$, so we will begin to solve this equation by multiplying both sides of the equation by $x - 4$.

$$\frac{x+5}{x-4} + 3 = \frac{2x+1}{x-4} \quad \text{The LCD is } x-4.$$

$$(x-4)\cdot\left(\frac{x+5}{x-4} + 3\right) = (x-4)\cdot\frac{2x+1}{x-4} \quad \text{Multiply both sides by the LCD.}$$

$$\overset{1}{\cancel{(x-4)}}\cdot\frac{x+5}{\underset{1}{\cancel{x-4}}} + (x-4)3 = \overset{1}{\cancel{(x-4)}}\cdot\frac{2x+1}{\underset{1}{\cancel{x-4}}} \quad \text{Distribute and divide out common factors.}$$

$$x + 5 + 3x - 12 = 2x + 1 \quad \text{Multiply. The resulting equation is linear.}$$

$$4x - 7 = 2x + 1 \quad \text{Combine like terms.}$$

$$2x - 7 = 1 \quad \text{Subtract } 2x.$$

$$2x = 8 \quad \text{Add 7.}$$

$$x = 4 \quad \text{Divide both sides by 2.}$$

The LCD is equal to 0 when $x = 4$, and two rational expressions in the original equation are undefined when $x = 4$.

Check:

$$\frac{(4)+5}{(4)-4} + 3 = \frac{2(4)+1}{(4)-4} \quad \text{Substitute 4 for } x.$$

$$\frac{9}{0} + 3 = \frac{9}{0} \quad \text{Simplify each numerator and denominator.}$$

So, although $x = 4$ is a solution to the equation obtained once the fractions have been cleared, it is not a solution to the original equation. This solution is an extraneous solution and since there are no other solutions, this equation has no solution. Recall that we write the solution set as $\varnothing$ when there is no solution.

***Quick Check* 3**

Solve $\frac{x+3}{x+2} - 4 = \frac{3x+7}{x+2}$.

EXAMPLE 4 Solve $\frac{x-1}{x^2+4x+3} = \frac{5}{x^2-3x-4}$.

Solution

We begin by factoring the denominators to find the LCD.

$$\frac{x-1}{x^2+4x+3} = \frac{5}{x^2-3x-4}$$

$$\frac{x-1}{(x+1)(x+3)}=\frac{5}{(x-4)(x+1)}$$

The LCD is $(x+1)(x+3)(x-4)$.

$$\cancel{(x+1)}\cancel{(x+3)}(x-4)\cdot\frac{x-1}{\cancel{(x+1)}\cancel{(x+3)}}=\cancel{(x+1)}(x+3)\cancel{(x-4)}\cdot\frac{5}{\cancel{(x-4)}\cancel{(x+1)}}$$

Multiply by the LCD. Divide out common factors.

$$(x-4)(x-1)=5(x+3)$$ Multiply remaining factors.

$$x^2-5x+4=5x+15$$ Multiply.

$$x^2-10x-11=0$$ Collect all terms on the left side. The resulting equation is quadratic.

$$(x+1)(x-11)=0$$ Factor.

$$x=-1 \quad \text{or} \quad x=11$$ Set each factor equal to 0 and solve.

The solution $x=-1$ is an extraneous solution as it makes the LCD equal to 0. It is left to the reader to verify that the solution $x=11$ checks. The solution set for this equation is $\{11\}$.

Quick Check 4 Solve $\dfrac{x+8}{x^2-10x+24}=\dfrac{6}{x^2-9x+20}$.

EXAMPLE 5 Solve $\dfrac{x}{x+5}+\dfrac{3}{x+8}=\dfrac{7x+20}{x^2+13x+40}$.

Solution

$$\frac{x}{x+5}+\frac{3}{x+8}=\frac{7x+20}{x^2+13x+40}$$ Factor the denominator.

$$\frac{x}{x+5}+\frac{3}{x+8}=\frac{7x+20}{(x+5)(x+8)}$$ The LCD is $(x+5)(x+8)$.

$$(x+5)(x+8)\cdot\left(\frac{x}{x+5}+\frac{3}{x+8}\right)=(x+5)(x+8)\cdot\frac{7x+20}{(x+5)(x+8)}$$ Multiply by the LCD.

$$\cancel{(x+5)}(x+8)\cdot\frac{x}{\cancel{(x+5)}}+(x+5)\cancel{(x+8)}\cdot\frac{3}{\cancel{(x+8)}}=\cancel{(x+5)}\cancel{(x+8)}\cdot\frac{7x+20}{\cancel{(x+5)}\cancel{(x+8)}}$$

Distribute and divide out common factors.

$$x(x+8)+3(x+5)=7x+20$$ Multiply remaining factors.

$$x^2+8x+3x+15=7x+20$$ Distribute.

$$x^2 + 11x + 15 = 7x + 20$$ Combine like terms.

$$x^2 + 4x - 5 = 0$$ Collect all terms on the left side. The resulting equation is quadratic.

$$(x + 5)(x - 1) = 0$$ Factor.

$$x = -5 \quad \text{or} \quad x = 1$$ Set each factor equal to 0 and solve.

The solution $x = -5$ is an extraneous solution as it makes the LCD equal to 0. It is left to the reader to verify that the solution $x = 1$ checks. The solution set for this equation is $\{1\}$.

Quick Check 5 Solve $\dfrac{x}{x-3} - \dfrac{6}{x-9} = \dfrac{x-45}{x^2 - 12x + 27}$.

Literal Equations

Objective 2 Solve literal equations. Recall that a literal equation is an equation containing two or more variables, and we solve the equation for one of the variables by isolating that variable on one side of the equation. In this section, we will learn how to solve literal equations containing one or more rational expressions.

EXAMPLE 6 Solve the literal equation $\dfrac{2}{x} + \dfrac{5}{y} = \dfrac{3}{4}$ for x.

Solution

We begin by multiplying by the LCD $(4xy)$ to clear the equation of fractions.

$$\frac{2}{x} + \frac{5}{y} = \frac{3}{4}$$

$$4xy \cdot \left(\frac{2}{x} + \frac{5}{y}\right) = 4xy \cdot \frac{3}{4}$$ Multiply by the LCD, $4xy$.

$$4\overset{1}{\cancel{x}}y \cdot \frac{2}{\underset{1}{\cancel{x}}} + 4x\overset{1}{\cancel{y}} \cdot \frac{5}{\underset{1}{\cancel{y}}} = \overset{1}{\cancel{4}}xy \cdot \frac{3}{\underset{1}{\cancel{4}}}$$ Distribute and divide out common factors.

$$8y + 20x = 3xy$$ Multiply remaining factors.

Notice that there are two terms that contain the variable we are solving for. We need to collect both of these terms on the same side of the equation and then factor x out of those terms. This will allow us to divide and isolate x.

$$8y + 20x = 3xy$$

$$8y = 3xy - 20x$$ Subtract $20x$ to collect all terms with x on the right side of the equation.

$$8y = x(3y - 20)$$ Factor out the common factor x.

$$\frac{8y}{3y - 20} = \frac{x\overset{1}{\cancel{(3y - 20)}}}{\underset{1}{\cancel{3y - 20}}}$$ Divide both sides by $3y - 20$ to isolate x.

$$\frac{8y}{3y-20} = x \qquad \text{Simplify.}$$

$$x = \frac{8y}{3y-20} \qquad \text{Rewrite with the variable we are solving for on the left side of the equation.}$$

As in Chapter 1, we will rewrite our solution so that the variable we are solving for is on the left side.

***Quick Check* 6**
Solve the literal equation $\frac{4}{x} + \frac{2}{y} = \frac{3}{z}$ for x.

EXAMPLE 7 Solve the literal equation $\frac{1}{x+y} = \frac{1}{b} + 3$ for y.

Solution

We begin by multiplying both sides of the equation by the LCD, which is $b(x+y)$.

$$\frac{1}{x+y} = \frac{1}{b} + 3$$

$$b \cdot (x+y) \cdot \frac{1}{x+y} = b \cdot (x+y) \cdot \left(\frac{1}{b} + 3\right) \qquad \text{Multiply both sides by the LCD.}$$

$$b\overset{1}{\cancel{(x+y)}} \cdot \frac{1}{\underset{1}{\cancel{x+y}}} = \overset{1}{\cancel{b}}(x+y) \cdot \frac{1}{\underset{1}{\cancel{b}}} + b(x+y) \cdot 3 \qquad \text{Distribute and divide out common factors.}$$

$$b = x + y + 3b(x+y) \qquad \text{Multiply remaining factors.}$$

$$b = x + y + 3bx + 3by \qquad \text{Distribute.}$$

$$b - x - 3bx = y + 3by \qquad \text{Subtract } x \text{ and } 3bx \text{ so only the terms containing } y \text{ are on the right side.}$$

$$b - x - 3bx = y(1+3b) \qquad \text{Factor out the common factor } y.$$

$$\frac{b-x-3bx}{1+3b} = \frac{y\overset{1}{\cancel{(1+3b)}}}{\underset{1}{\cancel{1+3b}}} \qquad \text{Divide both sides by } 1+3b \text{ to isolate } y.$$

$$y = \frac{b-x-3bx}{1+3b} \qquad \text{Rewrite with } y \text{ on the left side.}$$

***Quick Check* 7**
Solve the literal equation $\frac{1}{x-2} + \frac{2}{y} = 5$ for x.

Building Your Study Strategy **Test Taking, 5 Check Point Values** Not all problems on a test are assigned the same point value. Identify which problems are worth more points than others. It is important not to be in a position of having to rush through the problems that have higher point values because you didn't notice that they were worth more until you got to them at the end of the test.

EXERCISES 5.5

Vocabulary

1. A(n) __________ is an equation containing at least one rational expression.

2. To solve a rational equation, begin by multiplying both sides of the equation by the __________ of the denominators in the equation.

3. A(n) __________ of a rational equation is a solution that causes one or more of the rational expressions in the equation to be undefined.

4. A(n) __________ equation is an equation containing two or more variables.

Solve.

5. $\frac{x}{4} + \frac{7}{5} = \frac{3}{2}$

6. $\frac{x}{8} + \frac{5}{12} = \frac{17}{6}$

7. $\frac{2x}{3} + 4 = \frac{x}{8} + \frac{5}{6}$

8. $\frac{x}{4} - \frac{13}{4} = \frac{x+6}{6} - \frac{7}{2}$

9. $\frac{22}{x} - \frac{3}{4} = \frac{5}{8}$

10. $2 + \frac{27}{x} = \frac{17}{4}$

11. $\frac{9}{x} = \frac{15}{20}$

12. $5 - \frac{1}{4x} = \frac{1}{12x} - \frac{1}{3x}$

13. $x - 4 + \frac{16}{x} = 6$

14. $x - \frac{18}{x} = 3 + \frac{10}{x}$

15. $\frac{x}{9} = \frac{2}{x} - \frac{1}{3}$

16. $\frac{x}{6} + \frac{4}{x} = \frac{5}{3}$

17. $\frac{2x+1}{x-5} = \frac{x-8}{x-5}$

18. $\frac{7x+6}{x+4} = \frac{3x-10}{x+4}$

19. $\frac{5x-3}{4x-3} = \frac{13x-9}{4x-3}$

20. $\frac{x^2+8x}{x+6} = \frac{9x+42}{x+6}$

21. $3 + \frac{15x-4}{x+4} = \frac{x^2+20x}{x+4}$

22. $5 + \frac{3x-8}{x-2} = \frac{x^2+8x-54}{x-2}$

23. $\frac{3x-1}{x-7} - 4 = \frac{2x+3}{x-7}$

24. $\frac{4x-11}{3x-4} - \frac{x+24}{3x-4} = 2$

25. $x + \frac{2x+31}{x-5} = \frac{14x-39}{x-5}$

26. $x + \frac{x-25}{x+8} = \frac{6x+15}{x+8}$

27. $\frac{4}{x+7} = \frac{5}{2x-1}$

28. $\frac{8}{x+2} = \frac{18}{2x+10}$

29. $\frac{6}{x+14} = \frac{15}{8-2x}$

30. $\frac{4}{x-3} = \frac{14}{8x+3}$

31. $\frac{x+3}{4x+9} = \frac{3}{x+5}$

32. $\frac{x-6}{x+6} = \frac{3}{x+2}$

33. $\frac{x+4}{x+8} = \frac{2}{x-1}$

34. $\frac{x+7}{x+10} = \frac{5}{x-2}$

35. $\frac{x-4}{x^2+5x-24} = \frac{6}{x^2+15x+56}$

36. $\frac{x-3}{x^2+3x-4} = \frac{7}{x^2+13x+36}$

37. $\frac{x+4}{x^2-x-2} = \frac{6}{x^2-4x-5}$

38. $\frac{x-4}{x^2-3x-54} = \frac{2}{x^2-12x+27}$

39. $\frac{x+5}{x^2-10x-11} = \frac{4}{x^2-19x+88}$

40. $\frac{x+6}{x^2+8x+15} = \frac{4}{x^2+2x-15}$

FOR EXTRA HELP

Interactmath.com

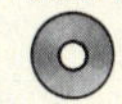
MathXL Tutorials on CD

Video Lectures on CD

Addison-Wesley Math Tutor Center

Student's Solutions Manual

41. $\frac{2}{x+5} + \frac{3}{x+3} = \frac{x-3}{x^2+8x+15}$

42. $\frac{5}{x} + \frac{6}{x-4} = \frac{3x+4}{x^2-4x}$

43. $\frac{3}{x-2} - \frac{7}{x-8} = \frac{x-20}{x^2-10x+16}$

44. $\frac{x}{x+2} + \frac{3}{x-4} = \frac{4x+12}{x^2-2x-8}$

45. $\frac{x+4}{x-6} + \frac{3}{x+2} = \frac{x+23}{x^2-4x-12}$

46. $\frac{x+6}{x+7} - \frac{5}{x-7} = \frac{3x-7}{x^2-49}$

47. $\frac{x-3}{x+1} - \frac{7}{x-1} = \frac{x+9}{x^2-1}$

48. $\frac{x-6}{x} + \frac{9}{x-5} = \frac{11x}{x^2-5x}$

49. $\frac{x+3}{x^2+9x+20} + \frac{1}{x^2-4x-32} = \frac{2}{x^2-3x-40}$

50. $\frac{x+5}{x^2+5x-14} - \frac{4}{x^2+10x+21} = \frac{5}{x^2+x-6}$

51. $\frac{x-2}{x^2-x-20} - \frac{7}{x^2+x-12} = \frac{2}{x^2-8x+15}$

52. $\frac{x+5}{x^2-9} + \frac{8}{x^2+5x-24} = \frac{2}{x^2+11x+24}$

53. If $x = 2$ is a solution to the equation $\frac{x+1}{x+6} + \frac{1}{x+4} = \frac{?}{x^2+10x+24}$,

 a) Find the constant in the missing numerator.

 b) Find the other solution to the equation.

54. If $x = 5$ is a solution to the equation $\frac{x-3}{x+4} + \frac{?}{x+1} = \frac{5x+32}{x^2+5x+4}$,

 a) Find the constant in the missing numerator.

 b) Find the other solution to the equation.

Solve for the specified variable.

55. $W = \frac{A}{L}$ for L

56. $h = \frac{2A}{b}$ for b

57. $y = \frac{x}{3x-2}$ for x

58. $y = \frac{5x+4}{2x}$ for x

59. $\frac{x}{3r} + \frac{y}{4r} = 1$ for r

60. $\frac{1}{a} - \frac{2}{b} = 3$ for b

61. $\frac{1}{a} + \frac{1}{b} = \frac{1}{c}$ for c

62. $1 + \frac{a}{b} = \frac{c}{d}$ for b

63. $x = \frac{a+b}{a-b}$ for b

64. $\frac{1}{R} = \frac{1}{R_1} + \frac{1}{R_2}$ for R

65. $n = \frac{n_1+n_2}{n_1 \cdot n_2}$ for n_1

66. $p = \frac{x_1+x_2}{n_1+n_2}$ for n_1

Writing in Mathematics

Answer in complete sentences.

67. Explain how to determine whether a solution is an extraneous solution.

68. We use the LCD when we add two rational expressions, as well as when we solve a rational equation. Explain how the LCD is used differently for these two types of problems.

69. ***Solutions Manual**** Write a solutions manual page for the following problem:

 Solve $\frac{x-6}{x^2-9} + \frac{5}{x^2+4x-21} = \frac{2}{x^2+10x+21}$.

70. ***Newsletter**** Write a newsletter that explains how to solve rational equations.

***See Appendix B for details and sample answers.**

5.6 Applications of Rational Equations

Objectives

1. **Solve applied problems involving the reciprocal of a number.**
2. **Solve applied work-rate problems.**
3. **Solve applied uniform motion problems.**
4. **Solve variation problems.**

In this section, we will look at applied problems requiring the use of rational equations to solve them. We begin with problems involving reciprocals.

Solving Applied Problems Involving the Reciprocal of a Number

Objective 1 **Solve applied problems involving the reciprocal of a number.**

EXAMPLE 1 The sum of the reciprocal of a number and $\frac{1}{4}$ is $\frac{1}{3}$. Find the number.

Solution

There is only one unknown in this problem, and we will let x represent it.

Unknown
Number: x

The reciprocal of this number can be written as $\frac{1}{x}$. We are told that the sum of this reciprocal and $\frac{1}{4}$ is $\frac{1}{3}$, which leads to the equation $\frac{1}{x} + \frac{1}{4} = \frac{1}{3}$.

$$\frac{1}{x} + \frac{1}{4} = \frac{1}{3}$$ The LCD is $12x$.

$$12x \cdot \left(\frac{1}{x} + \frac{1}{4}\right) = 12x \cdot \frac{1}{3}$$ Multiply both sides by the LCD.

$$\overset{1}{12\cancel{x}} \cdot \frac{1}{\underset{1}{\cancel{x}}} + \overset{3}{\cancel{12}}x \cdot \frac{1}{\underset{1}{\cancel{4}}} = \overset{4}{\cancel{12}}x \cdot \frac{1}{\underset{1}{\cancel{3}}}$$ Distribute and divide out common factors.

$$12 + 3x = 4x$$ Multiply remaining factors. The resulting equation is linear.

$$12 = x$$ Subtract $3x$.

The unknown number is 12. The reader should verify that $\frac{1}{12} + \frac{1}{4} = \frac{1}{3}$.

Quick Check 1
The sum of the reciprocal of a number and $\frac{2}{5}$ is $\frac{1}{2}$. Find the number.

EXAMPLE 2 One positive number is five larger than another positive number. If the reciprocal of the smaller number is added to eight times the reciprocal of the larger number, the sum is equal to 1. Find the two numbers.

Solution

In this problem there are two unknown numbers. If we let x represent the smaller number, then we can write the larger number as $x + 5$.

Unknowns
Smaller Number: x
Larger Number: $x + 5$

Since the unknown numbers are both positive, we must omit any solutions for which either the smaller number or larger number are not positive.

The reciprocal of the smaller number is $\frac{1}{x}$ and eight times the reciprocal of the larger number is $8 \cdot \frac{1}{x+5}$ or $\frac{8}{x+5}$. This leads to the equation $\frac{1}{x} + \frac{8}{x+5} = 1$.

$$\frac{1}{x} + \frac{8}{x+5} = 1$$ The LCD is $x(x + 5)$.

$$x \cdot (x+5) \cdot \left(\frac{1}{x} + \frac{8}{x+5}\right) = x(x+5) \cdot 1$$ Multiply both sides by the LCD.

$$\overset{1}{\cancel{x}}(x+5) \cdot \frac{1}{\underset{1}{\cancel{x}}} + x\overset{1}{\cancel{(x+5)}} \cdot \frac{8}{\underset{1}{\cancel{x+5}}} = x(x+5) \cdot 1$$ Distribute and divide out common factors.

$$x + 5 + 8x = x(x+5)$$ Multiply remaining factors.

$$x + 5 + 8x = x^2 + 5x$$ Distribute. The resulting equation is quadratic.

$$9x + 5 = x^2 + 5x$$ Combine like terms.

$$0 = x^2 - 4x - 5$$ Collect all terms on the right side of the equation.

$$0 = (x+1)(x-5)$$ Factor.

$$x = -1 \quad \text{or} \quad x = 5$$ Set each factor equal to 0 and solve.

Since we were told that the numbers must be positive, we can omit the solution $x = -1$. We now return to the table of unknowns to find the two numbers.

Smaller Number: $x = 5$
Larger Number: $x + 5 = 5 + 5 = 10$

The two numbers are 5 and 10. The reader should verify that $\frac{1}{5} + 8 \cdot \frac{1}{10} = 1$.

***Quick Check* 2**
One positive number is 10 less than another positive number. If seven times the reciprocal of the smaller number is added to six times the reciprocal of the larger number, the sum is equal to 1. Find the two numbers.

Work–Rate Problems

Objective 2 **Solve applied work–rate problems.** Now we turn our attention to problems known as **work–rate problems**. These problems usually involve two or more people or objects working together to perform a job, such as Kathy and Leah painting a room together or two copy machines processing an exam. In general, the equation we will be solving corresponds to the following:

Portion of the job completed by person 1	+	Portion of the job completed by person 2	=	1 (Completed job)

To determine the portion of the job completed by each person, we must know the **rate** at which the person works. If it takes Kathy 5 hours to paint a room, how much of the room could she paint in one hour? She could paint $\frac{1}{5}$ of the room in 1 hour, and this is her working rate. In general, the work rate for a person is equal to the reciprocal of the time it takes for that person to complete the whole job. If we multiply the work rate for a person by the amount of time that person has been working, then this tells us the portion of the job that the person has completed.

EXAMPLE 3 Working alone, Kathy can paint a room in 5 hours. Leah can paint the same room in only 3 hours. How long would it take the two of them to paint the room if they work together?

Solution

A good approach to any work–rate problem is to start with the following table and fill in the information.

	Time to complete the job alone	Work rate	Time Working	Portion of the Job Completed
Person 1				
Person 2				

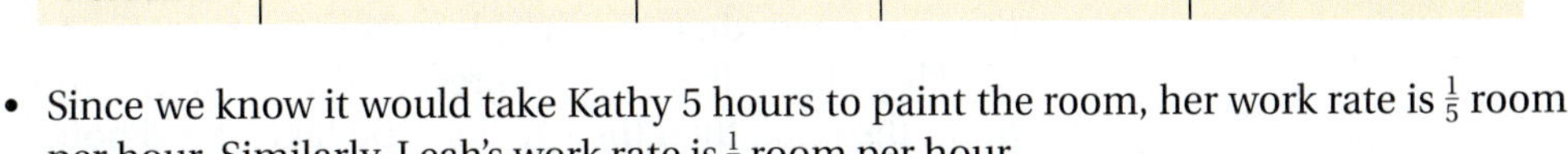

- Since we know it would take Kathy 5 hours to paint the room, her work rate is $\frac{1}{5}$ room per hour. Similarly, Leah's work rate is $\frac{1}{3}$ room per hour.
- The unknown in this problem is the amount of time they will be working together, which we will represent by the variable t.
- Finally, to determine the portion of the job completed by each person, we multiply the person's work rate by the time they have been working.

	Time to complete the job alone	Work rate	Time Working	Portion of the Job Completed
Kathy	5 hours	$\frac{1}{5}$	t	$\frac{t}{5}$
Leah	3 hours	$\frac{1}{3}$	t	$\frac{t}{3}$

To find the equation we need to solve, we add the portion of the room painted by Kathy to the portion of the room painted by Leah and set this sum equal to 1. The equation for this problem is $\frac{t}{5} + \frac{t}{3} = 1$.

Quick Check 3

Marco's new printer, working alone, can print a complete set of brochures in 15 minutes. His old printer can print the set of brochures in 40 minutes. How long would it take the two printers to print the set of brochures if they work together?

$$\frac{t}{5} + \frac{t}{3} = 1 \qquad \text{The LCD is 15.}$$

$$15 \cdot \left(\frac{t}{5} + \frac{t}{3}\right) = 15 \cdot 1 \qquad \text{Multiply both sides by the LCD.}$$

$$\overset{3}{\cancel{15}} \cdot \frac{t}{\underset{1}{\cancel{5}}} + \overset{5}{\cancel{15}} \cdot \frac{t}{\underset{1}{\cancel{3}}} = 15 \cdot 1 \qquad \text{Distribute and divide out common factors.}$$

$$3t + 5t = 15 \qquad \text{Multiply remaining factors.}$$

$$8t = 15 \qquad \text{Combine like terms.}$$

$$t = \frac{15}{8} \quad \text{or} \quad 1\frac{7}{8} \qquad \text{Divide both sides by 8.}$$

It would take them $1\frac{7}{8}$ hours to paint the room if they worked together. To convert the answer to hours and minutes we multiply $\frac{7}{8}$ of an hour by 60 minutes per hour, which is $52\frac{1}{2}$ minutes. It would take them approximately one hour and $52\frac{1}{2}$ minutes. The check of this solution is left to the reader.

Example 4 Two drainpipes, working together, can drain a pool in 3 hours. Working alone, it would take the smaller pipe 8 hours longer than it would take the larger pipe to drain the pool. How long would it take the smaller pipe alone to drain the pool?

Solution

In this example we know the amount of time it would take the pipes to drain the pool if they were working together, but we do not know how long it would take each pipe working alone. If we let t represent the time, in hours, that it takes the larger pipe to drain the pool, then the smaller pipe takes $t + 8$ hours to drain the pool. We will begin with a table.

	Time to Complete the Job Alone	Work Rate	Time Working	Portion of the Job Completed
Smaller pipe	$t + 8$ hours	$\frac{1}{t+8}$	3 hours	$\frac{3}{t+8}$
Larger pipe	t hours	$\frac{1}{t}$	3 hours	$\frac{3}{t}$

When we add the portion of the pool drained by the smaller pipe in 3 hours to the portion of the pool drained by the larger pipe, the sum will equal 1. The equation is $\frac{3}{t+8} + \frac{3}{t} = 1$.

$\frac{3}{t+8} + \frac{3}{t} = 1$	The LCD is $t(t+8)$.
$t(t+8) \cdot \left(\frac{3}{t+8} + \frac{3}{t}\right) = t(t+8) \cdot 1$	Multiply both sides by the LCD.
$\overset{1}{t(\cancel{t+8})} \cdot \frac{3}{\underset{1}{\cancel{t+8}}} + \overset{1}{\cancel{t}}(t+8) \cdot \frac{3}{\underset{1}{\cancel{t}}} = t(t+8) \cdot 1$	Distribute and divide out common factors.
$3t + 3(t+8) = t(t+8)$	Multiply remaining factors.
$3t + 3t + 24 = t^2 + 8t$	Distribute. The resulting equation is quadratic.
$6t + 24 = t^2 + 8t$	Combine like terms.
$0 = t^2 + 2t - 24$	Collect all terms on the right side of the equation.
$0 = (t+6)(t-4)$	Factor.
$t = -6$ or $t = 4$	Set each factor equal to 0 and solve.

Since the time spent by each pipe must be positive, we may immediately omit the solution $t = -6$. The reader may check the solution $t = 4$ by verifying that $\frac{3}{4+8} + \frac{3}{4} = 1$. We now use $t = 4$ to find the amount of time it would take the smaller pipe to drain the pool. Since $t + 8$ represents the amount of time it would take for the smaller pipe to drain the pool, it would take the smaller pipe $4 + 8$ or 12 hours to drain the pool.

Quick Check 4

Two drainpipes, working together, can drain a tank in 4 hours. Working alone, it would take the smaller pipe 6 hours longer than it would take the larger pipe to drain the tank. How long would it take the smaller pipe alone to drain the tank?

Uniform Motion Problems

Objective 3 **Solve applied uniform motion problems.** We now turn our attention to uniform motion problems. Recall that if an object is moving at a constant rate of speed for a certain amount of time, then the distance traveled by the object is equal to the product of its rate of speed and the length of time it traveled. The formula we used earlier in the text is rate · time = distance or $r \cdot t = d$. If we solve this formula for the time traveled, we have

$$\text{time} = \frac{\text{distance}}{\text{rate}} \quad \text{or} \quad t = \frac{d}{r}$$

EXAMPLE 5 Drake drove 30 miles, one way, to make a sales call. His second sales call was 35 miles away from the first sales call, and he drove 10 miles per hour faster than he did on his way to his first sales call. If the driving time was 1 hour for the entire trip, find Drake's driving speed on the way to his second sales call.

Solution

We know that the distance traveled to the first sales call is 30 miles, and the distance traveled to the second sales call is 35 miles. Drake's speed on the way to the second sales call was 10 miles per hour faster than it was on the way to the first sales call. We will let r represent his rate of speed on the way to the first sales call, and we can represent his rate of speed on the way to the second sales call as $r + 10$. To find an expression for the time spent on each part of the trip, we divide the distance by the rate.

We can summarize this information in a table:

	Distance (d)	Rate (r)	Time (t)
To First Sales Call	30 miles	r	$\frac{30}{r}$
To Second Sales Call	35 miles	$r + 10$	$\frac{35}{r+10}$

The equation we need to solve comes from the fact that the driving time on the way to the first sales call plus the driving time on the way to the second sales call is equal to 1 hour. If we add the time spent on the way to the first sales call $\left(\frac{30}{r}\right)$ to the time spent on the way the second sales call $\left(\frac{35}{r+10}\right)$, this will be equal to 1. The equation we need to solve is $\frac{30}{r} + \frac{35}{r+10} = 1$.

$$\frac{30}{r} + \frac{35}{r+10} = 1 \quad \text{The LCD is } r(r+10).$$

$$r(r+10) \cdot \left(\frac{30}{r} + \frac{35}{r+10}\right) = r(r+10) \cdot 1 \quad \text{Multiply both sides by the LCD.}$$

$$\overset{1}{\cancel{r}}(r+10) \cdot \frac{30}{\underset{1}{\cancel{r}}} + r(\overset{1}{\cancel{r+10}}) \cdot \frac{35}{\underset{1}{\cancel{r+10}}} = r(r+10) \cdot 1 \quad \text{Distribute and divide out common factors.}$$

$$30(r+10) + 35r = r(r+10) \quad \text{Multiply remaining factors.}$$

$$30r + 300 + 35r = r^2 + 10r \quad \text{Distribute.}$$

$$65r + 300 = r^2 + 10r \quad \text{Combine like terms.}$$

$$0 = r^2 - 55r - 300 \quad \text{Collect all terms on the right side of the equation.}$$

$$0 = (r-60)(r+5) \quad \text{Factor.}$$

$$r = 60 \quad \text{or} \quad r = -5 \quad \text{Set each factor equal to 0 and solve.}$$

Quick Check 5

Liana drove 240 miles to pick up a friend and then returned home. On the way home, she drove 20 miles per hour faster than she did on her way to pick up her friend. If the total driving time for the trip was 7 hours, find Liana's driving speed on the way home.

We omit the solution $r = -5$, as Drake's speed cannot be negative. The expression for Drake's driving speed on the way to the second sales call is $r + 10$, so his speed on the way to the second sales call was $60 + 10$ or 70 miles per hour. The reader can check the solution by verifying that $\frac{30}{60} + \frac{35}{70} = 1$.

EXAMPLE 6 Stephanie took her kayak to the Kaweah River, which flows downstream at a rate of 2 kilometers per hour. She paddled 15 km upstream, and then paddled downstream to her starting point. If this round-trip took a total of 4 hours, find the speed that Stephanie can paddle in still water.

Solution

We will let r represent the speed that Stephanie can paddle in still water. Since the current of the river is 2 kilometers per hour, Stephanie's kayak travels at a speed of $r - 2$ kilometers per hour when she is paddling upstream. This is because the current is pushing against the kayak. Stephanie travels at a speed of $r + 2$ kilometers per hour when she is paddling downstream, as the current is flowing in the same direction as the kayak. The equation we will solve includes the time spent paddling upstream and downstream. To find expressions in terms of r for the time spent in each direction, we divide the distance (15 km) by the rate of speed. Here is a table containing the relevant information.

	Distance	Rate	Time
Upstream	15 km	$r - 2$	$\frac{15}{r-2}$
Downstream	15 km	$r + 2$	$\frac{15}{r+2}$

We are told that the time needed to make the round-trip is 4 hours. In other words, the time spent paddling upstream plus the time spent paddling downstream is equal to 4 hours, or $\frac{15}{r-2} + \frac{15}{r+2} = 4$.

$$\frac{15}{r-2} + \frac{15}{r+2} = 4 \qquad \text{The LCD is } (r-2)(r+2).$$

$$(r-2)(r+2) \cdot \left(\frac{15}{r-2} + \frac{15}{r+2}\right) = (r-2)(r+2) \cdot 4 \qquad \text{Multiply both sides by the LCD.}$$

$$\overset{1}{\cancel{(r-2)}}(r+2) \cdot \frac{15}{\underset{1}{\cancel{r-2}}} + (r-2)\overset{1}{\cancel{(r+2)}} \cdot \frac{15}{\underset{1}{\cancel{r+2}}} = (r-2)(r+2) \cdot 4 \qquad \text{Distribute and divide out common factors.}$$

$$15(r+2) + 15(r-2) = 4(r-2)(r+2) \qquad \text{Multiply remaining factors.}$$

$$15r + 30 + 15r - 30 = 4r^2 - 16 \qquad \text{Multiply.}$$

$$30r = 4r^2 - 16 \qquad \text{Combine like terms.}$$

$$0 = 4r^2 - 30r - 16 \qquad \text{Collect all terms on the right side of the equation.}$$

$$0 = 2(2r^2 - 15r - 8) \qquad \text{Factor out the common factor 2.}$$

$$0 = 2(2r+1)(r-8) \qquad \text{Factor.}$$

$$r = -\frac{1}{2} \quad \text{or} \quad r = 8 \qquad \text{Set each variable factor equal to 0 and solve.}$$

Quick Check 6

Jacob took his canoe to a river that flows downstream at a rate of 2 miles per hour. He paddled 22.5 miles downstream, and then returned back to the camp he started from. If the round-trip took him 7 hours, find the speed that Jacob can paddle in still water.

We omit the negative solution, as the speed of the kayak in still water must be positive. Stephanie's kayak travels at a speed of 8 kilometers per hour in still water. The reader can check this solution by verifying that $\frac{15}{8-2} + \frac{15}{8+2} = 4$.

Variation

Objective 4 **Solve applied variation problems.** In the remaining examples, we will investigate the concept of **variation** between two or more quantities. Two quantities are said to **vary directly** if an increase in one quantity produces a proportional increase in the other quantity, and a decrease in one quantity produces a proportional decrease in the other quantity. For example, suppose that you have a part-time job that pays by the hour. The hours you work in a week and the amount of money you earn (before taxes) vary directly. As the hours you work increase, the amount of money you earn increases by the same factor. If there is a decrease in the number of hours you work, the amount of money that you earn decreases by the same factor.

Direct Variation

If a quantity y varies directly as a quantity x, then the two quantities are related by the equation

$$y = kx$$

where k is called the **constant of variation**.

For example, if you are paid \$12 per hour at your part-time job, then the amount of money you earn (y) and the number of hours you work (x) are related by the equation $y = 12x$. The value of k in this situation is 12, and it tells us that each time x increases by 1 hour, y increases by \$12.

EXAMPLE 7 y varies directly as x. If $y = 54$ when $x = 9$, find y when $x = 17$.

Solution

We begin by finding k. Using $y = 54$ when $x = 9$, we can use the equation $54 = k \cdot 9$ to find k.

$54 = k \cdot 9$ Substitute 54 for y and 9 for x into $y = kx$.

$6 = k$ Divide both sides by 9.

Now we use this value of k to find y when $x = 17$.

$y = 6 \cdot 17$ Substitute 6 for k and 17 for x into $y = kx$.

$y = 102$ Multiply.

EXAMPLE 8 The distance a train travels varies directly as the time it is traveling. If a train can travel 399 miles in 7 hours, how far can it travel in 12 hours?

Solution

The first step in a variation problem is to find k. We will let y represent the distance traveled and x represent the time. To find the variation constant, we will substitute the related information (399 miles in 7 hours) into the equation for direct variation.

$399 = k \cdot 7$ Substitute 399 for y and 7 for x into $y = kx$.

$57 = k$ Divide both sides by 7.

Now we will use this variation constant to find the distance traveled in 12 hours.

$y = 57 \cdot 12$ Substitute 57 for k and 12 for x into $y = kx$.

$y = 684$ Multiply.

The train travels 684 miles in 12 hours.

Quick Check **7** **The tuition a college student pays varies directly as the number of units the student is taking. If a student pays \$720 to take 15 units in a semester, how much would a student pay to take 12 units?**

In some cases, an increase in one quantity produces a *decrease* in another quantity. In this case we say that the quantities **vary inversely**. For example, suppose that you and some of your friends are going to buy a birthday gift for someone, splitting the cost equally. As the number of people who are contributing increases, the cost for each person decreases. The cost for each person varies inversely as the number of people contributing.

Inverse Variation

If a quantity y varies inversely as a quantity x, then the two quantities are related by the equation

$$y = \frac{k}{x}$$

where k is the constant of variation.

EXAMPLE 9 y varies inversely as x. If $y = 24$ when $x = 5$, find y when $x = 8$.

Solution

We begin by finding k.

$24 = \frac{k}{5}$ Substitute 24 for y and 5 for x into $y = \frac{k}{x}$.

$120 = k$ Multiply both sides by 5.

Now we use this value of k to find y when $x = 8$.

$y = \frac{120}{8}$ Substitute 120 for k and 8 for x into $y = \frac{k}{x}$.

$y = 15$ Divide.

EXAMPLE 10 The time required to drive from Cincinnati to Cleveland varies inversely as the average speed of the car. If it takes 4 hours to make the drive at 60 miles per hour, how long would it take to make the drive at 80 miles per hour?

Solution

Again, we begin by finding k. We will let y represent the time required to drive from Cincinnati to Cleveland and x represent the average speed of the car. To find k, we will substitute the related information (4 hours to make the trip at 60 miles per hour) into the equation for inverse variation.

$4 = \frac{k}{60}$ Substitute 4 for y and 60 for x into $y = \frac{k}{x}$.

$240 = k$ Multiply both sides by 60.

Now we will use this variation constant to find the time required to drive from Cincinnati to Cleveland at 80 miles per hour.

$y = \frac{240}{80}$ Substitute 240 for k and 80 for x into $y = \frac{k}{x}$.

$y = 3$ Divide.

It would take 3 hours to drive from Cincinnati to Cleveland at 80 miles per hour.

Quick Check 8 **The time required to complete a cycling race varies inversely as the average speed of the cyclist. If it takes 15 minutes to complete the race at an average speed of 40 kilometers per hour, how long would it take to complete the race at an average speed of 30 kilometers per hour?**

Often a quantity varies depending on two or more variables. A quantity y is said to **vary jointly** as two quantities x and z if it varies directly as the product of these two quantities. The equation in such a case is $y = kxz$.

EXAMPLE 11 y varies jointly as x and the square of z. If $y = 1800$ when $x = 9$ and $z = 5$, find y when $x = 15$ and $z = 10$.

Solution

We begin by finding k, using the equation $y = kxz^2$.

$1800 = k \cdot 9 \cdot 5^2$ Substitute 1800 for y, 9 for x, and 5 for z into $y = kxz^2$.

$1800 = k \cdot 225$ Simplify.

$8 = k$ Divide both sides by 225.

Now we use this value of k to find y when $x = 15$ and $z = 10$.

$y = 8 \cdot 15 \cdot 10^2$ Substitute 8 for k, 15 for x, and 10 for z into $y = kxz^2$.

$y = 12{,}000$ Simplify.

Quick Check 9 **y varies jointly as x and the square root of z. If $y = 600$ when $x = 40$ and $z = 36$, find y when $x = 200$ and $z = 25$.**

Building Your Study Strategy **Test Taking, 6** **Easy Problems First?** When taking a test, work on the easiest problems first, saving the more difficult problems for later. One benefit to this approach is that you will gain confidence as you progress through the test, so you are confident when you attempt to solve a difficult problem. Students who struggle on a problem on the test will lose confidence, and this may cause them to miss later problems that they know how to do. Also, if you complete the easier problems quickly first, you will save time to work on the few difficult problems. One exception to this practice is to begin the test by first attempting the most difficult type of problem for you. For example, if a particular type of word problem has been difficult for you to solve throughout your preparation for the exam, you may want to focus on how to solve that type of problem just before arriving for the test. With the solution of this type of problem committed to your short-term memory, your best opportunity for success comes at the beginning of the test prior to solving any other problems.

EXERCISES 5.6

Vocabulary

1. The __________ of a number x is $\frac{1}{x}$.

2. The __________ of a person is the amount of work he or she does in one unit of time.

3. If an object is moving at a constant rate of speed for a certain amount of time, then the length of time it traveled is equal to the distance traveled by the object __________ by its rate of speed.

4. If a canoe can travel at a rate of r miles per hour in still water and a river's current is c miles per hour, then the rate of speed that the canoe travels at while moving upstream is given by the expression __________.

5. Two quantities are said to __________ if an increase in one quantity produces a proportional increase in the other quantity.

6. Two quantities are said to __________ if an increase in one quantity produces a proportional decrease in the other quantity.

7. The sum of the reciprocal of a number and $\frac{2}{3}$ is $\frac{11}{12}$. Find the number.

8. The sum of the reciprocal of a number and $\frac{5}{6}$ is $\frac{23}{24}$. Find the number.

9. The sum of 5 times the reciprocal of a number and $\frac{7}{2}$ is $\frac{31}{6}$. Find the number.

10. The sum of 7 times the reciprocal of a number and $\frac{3}{4}$ is $\frac{4}{3}$. Find the number.

11. The difference of the reciprocal of a number and $\frac{3}{8}$ is $-\frac{7}{40}$. Find the number.

12. The difference of the reciprocal of a number and $\frac{3}{14}$ is $\frac{2}{7}$. Find the number.

13. One positive number is four larger than another positive number. If three times the reciprocal of the smaller number is added to two times the reciprocal of the larger number, the sum is equal to 1. Find the two numbers.

14. One positive number is six less than another positive number. If eight times the reciprocal of the smaller number is added to six times the reciprocal of the larger number, the sum is equal to 1. Find the two numbers.

15. One positive number is four less than another positive number. If the reciprocal of the smaller number is added to five times the reciprocal of the larger number, the sum is equal to $\frac{13}{24}$. Find the two numbers.

16. One positive number is four larger than another positive number. If five times the reciprocal of the smaller number is added to three times the reciprocal of the larger number, the sum is equal to $\frac{17}{15}$. Find the two numbers.

17. One copy machine can run off a set of copies in 30 minutes. A newer machine can do the same job in 20 minutes. How long would it take the two machines, working together, to make all of the necessary copies?

18. Dylan can mow a lawn in 30 minutes, while Alycia takes 45 minutes to mow the same lawn. If Dylan and Alycia work together, using two lawn mowers, how long would it take them to mow the lawn?

19. Dianne can completely weed her vegetable garden in 75 minutes. Her friend Gabby can do the same task in 90 minutes. If they worked together, how long would it take Dianne and Gabby to weed the vegetable garden?

20. One hose can fill a 40,000-gallon swimming pool in 16 hours. A hose from the neighbor's house can fill a swimming pool of that size in 20 hours. If the two hoses run at the same time, how long would they take to fill the swimming pool?

21. Anita can paint a room in 2 hours. Barbara can do the same job in 3 hours, while it would take Carol 4 hours to paint the entire room. If the three friends work together, how long would it take them to paint an entire room?

22. A company has printed out 1000 surveys to put in preaddressed envelopes and mail out. Bill can fill all of the envelopes in 8 hours, Terrell can do the job in 10 hours, and Jerry would take 15 hours to do the job. If all three work together, how long will it take them to stuff the 1000 envelopes?

23. Two drainpipes, working together, can drain a pool in 10 hours. Working alone, it would take the smaller pipe 15 hours longer than it would take the larger pipe to drain the pool. How long would it take the smaller pipe alone to drain the pool?

24. Jacqui and Jill, working together, can paint the interior of a house in 6 hours. If Jill were working alone, it would take her 5 hours longer than it would take Jacqui to paint the interior of a house. How long would it take Jacqui to paint the interior of a house by herself?

25. An electrician and his assistant can wire a house in 4 hours. If they worked alone, it would take the assistant 6 hours longer than the electrician to wire the house. How long would it take the assistant to wire the house?

26. Wayne and Ed can clean an entire building in 2 hours. Wayne can clean the entire building by himself in 3 hours less time than Ed can. How long would it take Wayne to clean the building by himself?

27. Two pipes, working together, can drain a tank in 6 minutes. The smaller pipe, working alone, would take 3 times as long as the larger pipe to drain the tank. How long would it take the smaller pipe, working alone, to drain the tank?

28. Carolyn and Laura can muck out their stable in 9 minutes. Working alone, it would take Laura three times as long as it would take Carolyn to muck out the stable. How long would it take Laura, working alone, to muck out the stable?

29. A college has two printers available to print out class rosters for the first day of classes. The older printer takes 10 hours longer than the newer printer to print out a complete set of class rosters, so the dean of registration decides to use the newer printer. After 3 hours the newer printer breaks down, so the dean must switch to the older printer. It takes the older printer an additional 8 hours to complete the job. How long would it have taken the older printer to print out a complete set of class rosters?

30. A bricklayer's apprentice takes 10 hours longer than the bricklayer to make a fireplace. The apprentice worked alone on a fireplace for 5 hours, after which the bricklayer began to help. It took 2 more hours for the pair to finish the fireplace. How long would it take the apprentice to make a fireplace on his own?

31. Ross is training for a triathlon. He rides his bicycle at a speed that is 20 miles per hour faster than his running speed. If Ross can cycle 42 miles in the same amount of time that it takes him to run 12 miles, what is Ross's running speed?

32. Ariel rides her bicycle at a speed that is 10 miles per hour faster than the speed that Sharon rides her bicycle. Ariel can ride 75 miles in the same amount of time that it takes for Sharon to ride 50 miles. How fast does Ariel ride her bicycle?

33. Jared is training to run a marathon. Today he ran 14 miles in 2 hours. After running the first 8 miles at a certain speed, he increased his speed by 3 miles per hour for the remaining 6 miles. Find the speed at which Jared ran the last 6 miles.

34. Cecilia drove 195 miles home from college to spend the weekend with her family. After the first 105 miles, she increased her driving speed by 12 miles per hour. If it took Cecilia 3 hours to get home, find her speed for the first 105 miles of the trip.

35. A salmon is swimming in a river that is flowing downstream at a speed of 8 kilometers per hour. The salmon can swim 8 kilometers upstream in the same time that it would take to swim 12 kilometers downstream. What is the speed of the salmon in still water?

36. Janet is swimming in a river that flows downstream at a speed of 0.5 meters per second. It takes her the same amount of time to swim 300 meters upstream as it does to swim 500 meters downstream. Find Janet's swimming speed (in meters per second) in still water.

37. An airplane flies 300 miles with a 40 mile per hour tailwind, and then flies 300 miles back into the 40 mile per hour wind. If the time for the round-trip was 4 hours, find the speed of the airplane in calm air.

38. Selma is kayaking in a river that flows downstream at a rate of 1 mile per hour. Selma paddles 9 miles downstream and then turns around and paddles 10 miles upstream, and the trip takes 4 hours.

a) How fast can Selma paddle in still water?

b) Selma is now 1 mile upstream of her starting point. How many minutes will it take her to paddle back to her starting point?

39. y varies directly as x. If $y = 54$ when $x = 9$, find y when $x = 21$.

40. y varies directly as x. If $y = 91$ when $x = 7$, find y when $x = 18$.

41. y varies directly as x. If $y = 40$ when $x = 15$, find y when $x = 27$.

42. y varies directly as x. If $y = 48$ when $x = 84$, find y when $x = 35$.

43. y varies inversely as x. If $y = 12$ when $x = 14$, find y when $x = 8$.

44. y varies inversely as x. If $y = 20$ when $x = 24$, find y when $x = 30$.

45. y varies inversely as x. If $y = 10$ when $x = 9$, find y when $x = 20$.

46. y varies inversely as x. If $y = 50$ when $x = 24$, find y when $x = 250$.

47. y varies directly as the square of x. If $y = 72$ when $x = 3$, find y when $x = 6$.

48. y varies inversely as the square of x. If $y = 10$ when $x = 8$, find y when $x = 2$.

49. y varies jointly as x and z. If $y = 630$ when $x = 6$ and $z = 7$, find y when $x = 5$ and $z = 12$.

50. y varies directly as x and inversely as z. If $y = 18$ when $x = 12$ and $z = 8$, find y when $x = 55$ and $z = 15$.

51. Stuart's gross pay varies directly as the number of hours he works. In a week that he worked 24 hours, his gross pay was $258. What would Stuart's gross pay be if he worked 36 hours?

52. The height that a ball bounces varies directly as the height it is dropped from. If a ball dropped from a height of 48 inches bounces 20 inches, how high would the ball bounce if it were dropped from a height of 72 inches?

53. **Ohm's law.** In a circuit the electric current (in amperes) varies directly as the voltage. If the current is 8 amperes when the voltage is 24 volts, find the current when the voltage is 9 volts.

54. **Hooke's law.** The distance that a hanging object stretches a spring varies directly as the mass of the object. If a 5-kilogram weight stretches a spring by 32 centimeters, how far would a 2-kilogram weight stretch the spring?

55. The amount of money that each person must contribute to buy a retirement gift for a coworker varies inversely as the number of people contributing. If 16 people would each have to contribute $15, how much would each person have to contribute if there were 24 people?

56. The maximum load that a wooden beam can support varies inversely as its length. If a beam that is 6 feet long can support 900 pounds, what is the maximum load that can be supported by a beam that is 8 feet long?

57. In a circuit the electric current (in amperes) varies inversely as the resistance (in ohms). If the current is 30 amperes when the resistance is 3 ohms, find the current when the resistance is 5 ohms.

58. **Boyle's law.** The volume of a gas varies inversely as the pressure upon it. The volume of a gas is 128 cubic centimeters when it is under a pressure of 50 kilograms per square centimeter. Find the volume when the pressure is reduced to 40 kilograms per square centimeter.

59. The illumination of an object varies inversely as the square of its distance from the source of light. If a light source provides an illumination of 36 foot-candles at a distance of 8 feet, find the illumination at a distance of 24 feet. (One foot-candle is the amount of illumination produced by a standard candle at a distance of one foot.)

60. The illumination of an object varies inversely as the square of its distance from the source of light. If a light source provides an illumination of 75 foot-candles at a distance of 6 feet, find the illumination at a distance of 30 feet.

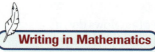

Writing in Mathematics

Answer in complete sentences.

61. Write a work–rate word problem that can be solved by the equation $\frac{3}{t+6} + \frac{10}{t} = 1$. Explain how you created your problem.

62. Write a uniform motion word problem that can be solved by the equation $\frac{3}{r-4} + \frac{3}{r+4} = 1$. Explain how you created your problem.

63. ***Solutions Manual**** Write a solutions manual page for the following problem:

James drives 150 miles to his job each day. Due to a rainstorm on his trip home, he had to drive 25 miles per hour slower than he drove on his way to work, and it took him 1 hour longer to get home than it did to get to work. How fast was James driving on his way home?

64. ***Newsletter**** Write a newsletter that explains how to solve work–rate problems.

***See Appendix B for details and sample answers.**

Quick Review Exercises

Section 5.6

Simplify.

1. $\frac{x^{12}}{x^3}$

2. $\frac{-54x^9}{6x^4}$

3. $(x^3 - 5x^2 - 7x + 10) - (x^3 + 8x^2)$

4. $(-4x^2 - 9x + 13) - (2x^2 - 11x + 21)$

5.7 Division of Polynomials

Objectives

1. **Divide a monomial by a monomial.**
2. **Divide a polynomial by a monomial.**
3. **Divide a polynomial by a polynomial using long division.**
4. **Use placeholders when dividing a polynomial by a polynomial.**

A rational expression is a quotient of two polynomials. In this section, we will explore another method for simplifying a quotient of two polynomials by using division. This technique is particularly useful when the numerator and denominator do not contain common factors.

Dividing Monomials by Monomials

Objective 1 Divide a monomial by a monomial. We will begin by reviewing how to divide a monomial by another monomial, such as $\frac{4x^5}{2x^2}$ or $\frac{30a^5b^4}{6ab^2}$.

Dividing a Monomial by a Monomial

To divide a monomial by another monomial, we divide the coefficients first. Then we divide the variables using the quotient rule $\frac{x^m}{x^n} = x^{m-n}$, assuming that no variable in the denominator is equal to 0.

EXAMPLE 1 Divide $\frac{20x^8}{5x^3}$. (Assume $x \neq 0$.)

Solution

$$\frac{20x^8}{5x^3} = 4x^{8-3} \quad \text{Divide coefficients. Subtract exponents for } x.$$

$$= 4x^5 \quad \text{Simplify the exponent.}$$

We can check our quotients by using multiplication. If $\frac{20x^8}{5x^3} = 4x^5$, then we know that $5x^3 \cdot 4x^5$ should be equal to $20x^8$.

Check:

$$5x^3 \cdot 4x^5 = 20x^{3+5} \quad \text{Multiply coefficients. Keep the base } x \text{, add the exponents.}$$

$$= 20x^8 \quad \text{Simplify the exponent.}$$

Our quotient of $4x^5$ checks.

Quick Check **1** Divide $\frac{42x^{10}}{7x^2}$. (Assume $x \neq 0$.)

Example 2 Divide $\frac{32x^7y^6}{4xy^6}$. (Assume $x, y \neq 0$.)

Solution

When there is more than one variable, as in this example, we divide the coefficients and then divide the variables one at a time.

$$\frac{32x^7y^6}{4xy^6} = 8x^6y^0 \quad \text{Divide coefficients, subtract exponents.}$$

$$= 8x^6 \quad \text{Rewrite without } y \text{ as a factor. } (y^0 = 1)$$

Quick Check 2

Divide $\frac{56x^8y^5}{4x^3y^2}$. (Assume $x, y \neq 0$.)

Dividing Polynomials by Monomials

Objective 2 **Divide a polynomial by a monomial.** Now we move on to dividing a polynomial by a monomial, such as $\frac{20x^5 - 30x^3 - 35x^2}{5x^2}$.

Dividing a Polynomial by a Monomial

To divide a polynomial by a monomial, divide each term of the polynomial by the monomial.

Example 3 Divide $\frac{32x^2 + 40x - 4}{4}$.

Solution

We will divide each term in the numerator by 4.

$$\frac{32x^2 + 40x - 4}{4} = \frac{32x^2}{4} + \frac{40x}{4} - \frac{4}{4} \quad \text{Divide each term in the numerator by 4.}$$

$$= 8x^2 + 10x - 1 \quad \text{Divide.}$$

If $\frac{32x^2 + 40x - 4}{4} = 8x^2 + 10x - 1$, then $4(8x^2 + 10x - 1)$ should equal $32x^2 + 40x - 4$. We can use this to check our work.

Check:

$$4(8x^2 + 10x - 1) = 4 \cdot 8x^2 + 4 \cdot 10x - 4 \cdot 1 \quad \text{Distribute.}$$

$$= 32x^2 + 40x - 4 \quad \text{Multiply.}$$

Our quotient of $8x^2 + 10x - 1$ checks.

Quick Check 3

Divide $\frac{6x^2 - 30x - 18}{6}$.

EXAMPLE 4 Divide $\dfrac{20x^5 - 30x^3 - 35x^2}{5x^2}$. (Assume $x \neq 0$.)

Solution

In this example we will divide each term in the numerator by $5x^2$.

$$\frac{20x^5 - 30x^3 - 35x^2}{5x^2} = \frac{20x^5}{5x^2} - \frac{30x^3}{5x^2} - \frac{35x^2}{5x^2}$$ Divide each term in the numerator by $5x^2$.

$$= 4x^3 - 6x - 7$$ Divide.

Quick Check 4

Divide $\dfrac{8x^9 - 24x^7 + 120x^4}{8x^3}$. (Assume $x \neq 0$.)

Dividing a Polynomial by a Polynomial (Long Division)

Objective 3 Divide a polynomial by another polynomial using long division. To divide a polynomial by another polynomial containing at least two terms, we use a procedure similar to long division. Before outlining this procedure, let's review some of the terms associated with long division.

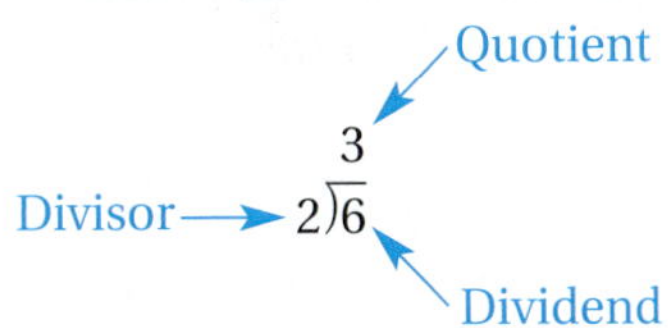

Suppose that we were asked to divide $\dfrac{x^2 - 14x + 48}{x - 8}$. The polynomial in the numerator is the dividend and the polynomial in the denominator is the divisor. We may rewrite this division as $x - 8\overline{)x^2 - 14x + 48}$. We must be sure to write both the divisor and the dividend in descending order. (The term with the highest degree goes first, then the term with the next highest degree, and so on.) We perform the division using the following steps.

Division by a Polynomial

1. Divide the term in the dividend with the highest degree by the term in the divisor with the highest degree. Add this result to the quotient.
2. Multiply the monomial obtained in step 1 by the divisor, writing the result underneath the dividend. (Align like terms vertically.)
3. Subtract the product obtained in step 2 from the dividend. (Recall that to subtract a polynomial from another polynomial, we change the signs of its terms and then combine like terms.)
4. Repeat steps 1–3 with the result of step 3 as the new dividend. Keep repeating this procedure until the degree of the new dividend is less than the degree of the divisor.

EXAMPLE 5 Divide $\frac{x^2 - 14x + 48}{x - 8}$.

Solution

We begin by writing $x - 8\overline{)x^2 - 14x + 48}$. We divide the term in the dividend with the highest degree (x^2) by the term in the divisor with the highest degree (x). Since $\frac{x^2}{x} = x$, we will write x in the quotient and multiply x by the divisor $x - 8$, writing this product underneath the dividend.

$$\begin{array}{r}
x \\
x - 8\overline{)x^2 - 14x + 48} \\
x^2 - 8x
\end{array}$$

Multiply $x(x - 8)$.

To subtract, we may change the signs of the second polynomial and then combine like terms.

$$\begin{array}{r}
x \\
x - 8\overline{)x^2 - 14x + 48} \\
\underline{\overset{-}{x^2} \overset{+}{\not{-}} 8x \downarrow} \\
- 6x + 48
\end{array}$$

Change the signs and combine like terms.

We now begin the process again by dividing $-6x$ by x, which equals -6. Multiply -6 by $x - 8$ and subtract.

$$\begin{array}{r}
x - 6 \\
x - 8\overline{)x^2 - 14x + 48} \\
\underline{\overset{-}{x^2} \overset{+}{\not{-}} 8x \downarrow} \\
- 6x + 48 \\
\underline{\overset{+}{\not{-}} 6x \overset{-}{\not{+}} 48} \\
0
\end{array}$$

Multiply -6 by $x - 8$. Subtract the product.

The remainder of 0 tells us that we are through, since its degree is less than the degree of the divisor $x - 8$. The expression written above the division box $(x - 6)$ is the quotient.

$$\frac{x^2 - 14x + 48}{x - 8} = x - 6.$$

We can check our work by multiplying the quotient $(x - 6)$ by the divisor $(x - 8)$, which should equal the dividend $(x^2 - 14x + 48)$.

Quick Check 5

Divide $\frac{x^2 + 4x - 21}{x + 7}$.

Check:

$$\begin{aligned}
(x - 6)(x - 8) &= x^2 - 8x - 6x + 48 \\
&= x^2 - 14x + 48
\end{aligned}$$

Our division checks; the quotient is $x - 6$.

In the previous example, the remainder of 0 also tells us that $x - 8$ divides into $x^2 - 14x + 48$ evenly, so $x - 8$ is a **factor** of $x^2 - 14x + 48$. The quotient, $x - 6$, is also a factor of $x^2 - 14x + 48$.

In the next example, we will learn how to write a quotient when there is a nonzero remainder.

EXAMPLE 6 Divide $x^2 - 13x + 33$ by $x - 4$.

Solution

Since the divisor and dividend are already written in descending order, we may begin to divide.

$$\begin{array}{r l}
 & x - 9 \\
x - 4 \overline{)} & x^2 - 13x + 33 \\
 & \underline{-x^2 + 4x \quad \downarrow} \quad \text{Multiply } x \text{ by } x - 4 \text{ and subtract.} \\
 & -9x + 33 \\
 & \underline{+9x - 36} \quad \text{Multiply } -9 \text{ by } x - 4 \text{ and subtract.} \\
 & -3
\end{array}$$

Quick Check 6
Divide $\dfrac{x^2 + 5x - 5}{x + 7}$.

The remainder is -3. After the quotient, we write a fraction with the remainder in the numerator and the divisor in the denominator. Since the remainder is negative, we subtract this fraction from the quotient. If the remainder had been positive, we would have added this fraction to the quotient.

$$\frac{x^2 - 13x + 33}{x - 4} = x - 9 - \frac{3}{x - 4}$$

EXAMPLE 7 Divide $6x^2 - 13x + 13$ by $3x + 4$.

Solution

Since the divisor and dividend are already written in descending order, we may begin to divide.

$$\begin{array}{r l}
 & 2x - 7 \\
3x + 4 \overline{)} & 6x^2 - 13x + 13 \\
 & \underline{-6x^2 - 8x \quad \downarrow} \quad \text{Multiply } 2x \text{ by } 3x + 4 \text{ and subtract.} \\
 & -21x + 13 \\
 & \underline{+21x + 28} \quad \text{Multiply } -7 \text{ by } 3x + 4 \text{ and subtract.} \\
 & 41
\end{array}$$

Quick Check 7
Divide $\dfrac{8x^2 + 2x - 77}{4x - 9}$.

The remainder is 41. We write the quotient, and add a fraction with the remainder in the numerator and the divisor in the denominator.

$$\frac{6x^2 - 13x + 13}{3x + 4} = 2x - 7 + \frac{41}{3x + 4}$$

Using Placeholders When Dividing a Polynomial by a Polynomial

Objective 4 **Use placeholders when dividing a polynomial by a polynomial.** Suppose that we wanted to divide $x^3 - 19x - 8$ by $x - 6$. Notice that the dividend is missing an x^2 term. When this is the case we add the term $0x^2$ as a **placeholder**. We add placeholders to dividends that are missing terms of a particular degree.

Example 8 Divide $(x^3 - 19x - 8) \div (x - 6)$.

Solution

The degree of the dividend is 3, and each degree lower than 3 must be represented in the dividend. We will add the term $0x^2$ as a placeholder. The divisor does not have any missing terms.

$$\begin{array}{r|l}
 & x^2 + 6x + 17 \\
\hline
x - 6 & x^3 + 0x^2 - 19x - 8 \\
 & \underline{-x^3 \overset{+}{\not -} 6x^2 \quad \downarrow \quad \downarrow} \\
 & 6x^2 - 19x - 8 \\
 & \underline{-6x^2 \overset{+}{\not -} 36x \quad \downarrow} \\
 & 17x - 8 \\
 & \underline{\overset{-}{\not +} 17x \overset{+}{\not -} 102} \\
 & 94
\end{array}$$

Multiply x^2 by $x - 6$ and subtract.

Multiply $6x$ by $x - 6$ and subtract.

Multiply 17 by $x - 6$ and subtract.

$$(x^3 - 19x - 8) \div (x - 6) = x^2 + 6x + 17 + \frac{94}{x - 6}.$$

Quick Check 8

Divide $\dfrac{x^3 + 6x^2 + 43}{x + 9}$.

Building Your Study Strategy **Test Taking, 7 Review Your Test** Try to leave yourself enough time to review the test at the end of the test period. Check for careless errors, which can cost you a significant number of points. Also be sure that your answers make sense within the context of the problem. For example, if the question asks how tall a person is and your answer is 68 feet tall, then chances are that something has gone astray.

Check for problems, or parts of problems, that you may have skipped and left blank. There is no worse feeling in a math class than getting a test back and realizing that you simply forgot to do one or more of the problems.

Take all of the allotted time to review your test. There is no reward for turning in a test early, and the longer that you spend on the test, the more likely it is that you will find a mistake or a problem where you can gain points. Keep in mind that a majority of the students who turn in tests early do so because they were not prepared for the test and cannot do several of the problems.

EXERCISES 5.7

Vocabulary

1. To divide a monomial by another monomial, divide the coefficients first and then divide the variables by __________ the exponents of like bases.

2. To divide a polynomial by a monomial, divide each __________ of the polynomial by the monomial.

3. If one polynomial divides evenly into another polynomial, with no remainder, then the first polynomial is said to be a(n) __________ of the second polynomial.

4. __________ are added to dividends that are missing terms of a particular degree.

Divide. (Assume all variables are nonzero.)

5. $\dfrac{42x^8}{6x^2}$

6. $\dfrac{56x^7}{7x^3}$

7. $\dfrac{-60x^8}{5x}$

8. $\dfrac{32x^{12}}{8x^{11}}$

9. $\dfrac{26x^7y^6}{2xy^5}$

10. $\dfrac{24x^5y^{12}}{4x^5y^4}$

11. $(120x^5) \div (10x^2)$

12. $(39x^{10}) \div (3x^5)$

13. $\dfrac{45x^9}{25x^7}$

14. $\dfrac{-56x^{15}}{12x^8}$

15. $\dfrac{3x^{14}}{36x^3}$

16. $\dfrac{7x^{21}}{42x^8}$

Find the missing monomial. (Assume all variables are nonzero.)

17. $\dfrac{?}{4x^3} = 3x^8$

18. $\dfrac{?}{7x^6} = -9x^{11}$

19. $\dfrac{32x^{10}}{?} = 8x^3$

20. $\dfrac{36x^{16}}{?} = \dfrac{6x^2}{5}$

Divide. (Assume all variables are nonzero.)

21. $\dfrac{10x^2 - 22x - 2}{2}$

22. $\dfrac{5x^2 + 45x}{5}$

23. $\dfrac{3x^4 + 39x^3 - 21x^2 + 9x}{3x}$

24. $\dfrac{28x^6 - 40x^3 - 4x^2}{4x}$

25. $(14x^8 + 21x^6 - 35x^3) \div (7x^2)$

26. $(-18x^{10} + 30x^9 - 6x^8) \div (6x^2)$

27. $\dfrac{48x^6 - 56x^5}{-8x^3}$

28. $\dfrac{-12x^5 + 8x^4 + 16x^3 - 4x^2}{-4x^2}$

29. $\dfrac{9x^5y^6 - 27x^2y^5 + 3x^3y^4}{3xy^4}$

30. $\dfrac{2a^8b^6 - 18a^5b^4 - 30a^2b^2}{2a^2b}$

Find the missing dividend or divisor. (Assume all variables are nonzero.)

31. $\dfrac{27x^3 - 45x^2 - 12x}{?} = 9x^2 - 15x - 4$

32. $\dfrac{4x^7 - 20x^5 + 36x^4}{?} = x^4 - 5x^2 + 9x$

33. $\dfrac{?}{6x^4} = x^2 - 7x - 19$

34. $\dfrac{?}{7x^2} = -x^4 - 8x^2 + 2$

Divide using long division.

35. $\dfrac{x^2 + 13x + 40}{x + 5}$

36. $\dfrac{x^2 + 15x + 36}{x + 3}$

37. $\dfrac{x^2 + x - 72}{x - 8}$

38. $\dfrac{x^2 + 5x - 14}{x + 7}$

39. $(x^2 - 21x + 90) \div (x - 6)$

40. $(x^2 - 15x + 77) \div (x - 9)$

41. $\frac{x^2 + 10x - 32}{x - 3}$

42. $\frac{x^2 + 14x + 37}{x + 5}$

43. $\frac{x^3 - 8x^2 - 21x + 9}{x + 2}$

44. $\frac{x^3 + 5x^2 - 13x - 32}{x + 4}$

45. $\frac{2x^2 + 19x + 30}{x + 8}$

46. $\frac{3x^2 - 13x - 75}{x - 7}$

47. $\frac{6x^2 - 17x - 75}{2x - 9}$

48. $\frac{4x^2 - 34x + 92}{2x - 5}$

49. $\frac{x^2 - 49}{x - 7}$

50. $(9x^2 - 169) \div (3x + 13)$

51. $\frac{x^4 + 2x^2 + 8x - 27}{x - 5}$

52. $\frac{x^5 - 9x^4 - 10x^2 + x + 17}{x + 2}$

53. $\frac{x^3 - 64}{x - 4}$

54. $\frac{x^3 + 512}{x + 8}$

Find the missing dividend or divisor.

55. $\frac{?}{x + 7} = x - 6$

56. $\frac{?}{x - 9} = x - 15$

57. $\frac{x^2 - 14x - 72}{?} = x + 4$

58. $\frac{6x^2 - 5x - 99}{?} = 2x - 9$

59. Is $x + 11$ a factor of $x^2 + 34x + 253$?

60. Is $x - 3$ a factor of $x^2 + 11x - 39$?

61. Is $2x + 7$ a factor of $6x^2 + x - 63$?

62. Is $5x + 8$ a factor of $45x^2 + 37x - 56$?

Mixed Practice, 63–80

Divide. (Assume all variables are nonzero.)

63. $\frac{x^2 - 11x + 20}{x - 4}$

64. $\frac{x^2 + 6x - 7}{x + 9}$

65. $\frac{x^3 + 2x^2 - 7}{x + 6}$

66. $\frac{24x^{10}}{4x^2}$

67. $\frac{9x^6 - 12x^5 + 21x^4 - 3x^3}{3x^3}$

68. $\frac{x^2 - x - 210}{x + 14}$

69. $\frac{4x^2 + 19x - 48}{x + 7}$

70. $\frac{3x^2 - 9x - 51}{x - 5}$

71. $\frac{20x^2 - 37x + 7}{4x - 1}$

72. $\frac{8x^2 - 26x + 27}{2x - 3}$

73. $\frac{8x^8 - 6x^6 - 18x^5 + 4x^3}{-2x^3}$

74. $\dfrac{6x^2 + 5x - 391}{3x - 23}$

75. $\dfrac{8x^3 - 26x - 9}{4x + 6}$

76. $\dfrac{72a^9b^{10}c^{11}}{6ab^3c^{11}}$

77. $\dfrac{6x^3 - 37x^2 + 36x + 70}{2x - 7}$

78. $\dfrac{30x^{10} - 42x^7 - 57x^4}{3x^4}$

79. $\dfrac{-54x^{10}y^4z^8}{6x^7y^3z}$

80. $\dfrac{x^5 + 242}{x + 3}$

Writing in Mathematics

Answer in complete sentences.

81. Explain the use of placeholders when dividing a polynomial by another polynomial.

82. ***Solutions Manual**** Write a solutions manual page for the following problem:

Divide $\dfrac{6x^2 - 17x - 19}{2x - 7}$.

83. ***Newsletter**** Write a newsletter that explains how to divide a polynomial by another polynomial using long division.

***See Appendix B for details and sample answers.**

Chapter 5 Summary

Section 5.1—Topic	Chapter Review Exercises
Evaluating Rational Expressions	1–4
Finding Values for Which a Rational Expression Is Undefined	5–6
Simplifying Rational Expressions to Lowest Terms	7–10
Evaluating Rational Functions	11–12
Finding the Domain of a Rational Function	13–14
Section 5.2—Topic	**Chapter Review Exercises**
Multiplying Two Rational Expressions	15–18
Multiplying Two Rational Functions	19–20
Dividing a Rational Expression by Another Rational Expression	21–24
Dividing a Rational Function by Another Rational Function	25–26
Section 5.3—Topic	**Chapter Review Exercises**
Adding or Subtracting Rational Expressions with the Same Denominator	27–29
Adding or Subtracting Rational Expressions with Opposite Denominators	30–32
Adding or Subtracting Rational Expressions with Unlike Denominators	33–37
Adding or Subtracting Rational Functions	38–39
Section 5.4—Topic	**Chapter Review Exercises**
Simplifying Complex Numerical Fractions	40–43
Section 5.5—Topic	**Chapter Review Exercises**
Solving Rational Equations	44–51
Solving Literal Equations	52–54
Section 5.6—Topic	**Chapter Review Exercises**
Solving Applied Problems Involving Rational Equations	55–64
Section 5.7—Topic	**Chapter Review Exercises**
Dividing Polynomials	65–72

Summary of Chapter 5 Study Strategies

Attending class each day and doing all of your homework does not guarantee a good grade when taking an exam. Through careful preparation and the adoption of the test-taking strategies introduced in this chapter, you can maximize your grade on a math exam. Here is a summary of the points that have been introduced.

- Make the most of your time before the exam.
- Write down important facts as soon as you get your test.
- Briefly read through the test.
- Begin by solving the easier problems first.
- Review your test as thoroughly as possible before turning it in.

Evaluate the rational expression for the given value of the variable. [5.1]

1. $\dfrac{9}{x-5}$ for $x = -7$

2. $\dfrac{x-4}{x+20}$ for $x = 16$

3. $\dfrac{x^2+7x-8}{x^2-2x+6}$ for $x = 6$

4. $\dfrac{x^2-3x+12}{x^2+5x-4}$ for $x = -8$

Find all values for which the rational expression is undefined. [5.1]

5. $\dfrac{9}{2x-3}$

6. $\dfrac{x^2+15x+56}{x^2+3x-28}$

Simplify the given rational expression. (Assume that all denominators are nonzero.) [5.1]

7. $\dfrac{x-8}{x^2-64}$

8. $\dfrac{x^2+7x-18}{x^2-7x+10}$

9. $\dfrac{4x-x^2}{x^2-11x+28}$

10. $\dfrac{2x^2+11x-6}{x^2+2x-24}$

Evaluate the given rational function. [5.1]

11. $r(x) = \dfrac{x-9}{x^2-5x+16}$, $r(-3)$

12. $r(x) = \dfrac{x^2+3x+14}{x^2-5x}$, $r(7)$

Find the domain of the given rational function. [5.1]

13. $r(x) = \dfrac{x^2+11x+32}{x^2-9x}$

14. $r(x) = \dfrac{x^2+3x-25}{x^2-5x-6}$

Multiply. [5.2]

15. $\dfrac{x+8}{x-3} \cdot \dfrac{x^2-13x+30}{x^2+14x+48}$

16. $\dfrac{x^2+16x+63}{x^2+x-12} \cdot \dfrac{x^2-16}{x^2+3x-54}$

17. $\dfrac{x^2+5x}{3x^2+4x-4} \cdot \dfrac{x^2+4x+4}{x^2-4x-45}$

18. $\dfrac{49-x^2}{x+6} \cdot \dfrac{x^2+5x-6}{x^2-4x-21}$

For the given functions f(x) and g(x), find f(x) · g(x). [5.2]

19. $f(x) = \dfrac{x+8}{x^2+2x}$, $g(x) = \dfrac{x^2+12x+20}{x^2+2x-80}$

20. $f(x) = \dfrac{x^2+x-30}{x^2+2x-99}$, $g(x) = \dfrac{x^2-5x-36}{x^2-x-20}$

Divide. [5.2]

21. $\dfrac{x^2+8x+7}{x^2-3x-40} \div \dfrac{x+1}{x+5}$

22. $\dfrac{x^2+10x-24}{x^2+10x+24} \div \dfrac{x^2-5x+6}{x^2+13x+42}$

23. $\dfrac{5x^2+7x-6}{x^2+8x} \div \dfrac{x^2-3x-10}{x^2-64}$

24. $\dfrac{100-x^2}{x-7} \div \dfrac{x^2-13x+30}{x^2-x-42}$

For the given functions f(x) and g(x), find f(x) ÷ g(x). [5.2]

25. $f(x) = \dfrac{x^2+3x}{x^2+8x+16}$, $g(x) = \dfrac{x^2+10x+21}{x^2-3x-28}$

26. $f(x) = \dfrac{x^2+15x-16}{x^2+9x+14}$, $g(x) = \dfrac{x^2-6x+5}{x^2-4x-12}$

Add or subtract. [5.3]

27. $\dfrac{15}{x+4} + \dfrac{9}{x+4}$

28. $\dfrac{x^2+4x}{x^2+11x+24} - \dfrac{7x+18}{x^2+11x+24}$

29. $\dfrac{x^2-6x-8}{x^2+4x-45} + \dfrac{6x-17}{x^2+4x-45}$

30. $\dfrac{4x+9}{x-3} - \dfrac{2x-27}{3-x}$

Worked-out solutions to Review Exercises marked with ● can be found on page AN-26.

31. $\dfrac{2x^2 + 3x + 21}{x - 2} + \dfrac{x^2 + 13x + 5}{2 - x}$

32. $\dfrac{x^2 + 6x + 10}{x^2 - 16} - \dfrac{8x + 30}{16 - x^2}$

33. $\dfrac{5}{x^2 - 3x - 4} + \dfrac{4}{x^2 - 12x + 32}$

34. $\dfrac{9}{x^2 + 13x + 22} - \dfrac{2}{x^2 + 20x + 99}$

35. $\dfrac{x}{x^2 - 16} - \dfrac{5}{x^2 + 2x - 24}$

36. $\dfrac{x + 6}{x^2 + 18x + 80} - \dfrac{2}{x^2 + 14x + 48}$

37. $\dfrac{x + 5}{x^2 + 4x + 3} + \dfrac{x - 5}{x^2 + 5x + 4}$

For the given rational functions f(x) and g(x), find f(x) + g(x). **[5.3]**

38. $f(x) = \dfrac{3}{x^2 - 4x - 5}, g(x) = \dfrac{1}{x^2 - 12x + 35}$

For the given rational functions f(x) and g(x), find f(x) − g(x). **[5.3]**

39. $f(x) = \dfrac{x - 9}{x^2 - 17x + 70}, g(x) = \dfrac{4}{x^2 - 8x - 20}$

Simplify the complex fraction. **[5.4]**

40. $\dfrac{1 + \dfrac{6}{x}}{1 - \dfrac{36}{x^2}}$

41. $\dfrac{\dfrac{x + 8}{x - 4} - \dfrac{2}{x}}{\dfrac{x^2 + 9x + 20}{x^2 - 4x}}$

42. $\dfrac{1 + \dfrac{19}{x} + \dfrac{90}{x^2}}{1 + \dfrac{8}{x} - \dfrac{9}{x^2}}$

43. $\dfrac{\dfrac{x^2 + 5x - 14}{x^2 - 7x + 12}}{\dfrac{x^2 + 13x + 42}{x^2 - 16}}$

Solve. **[5.5]**

44. $\dfrac{10}{x} - \dfrac{1}{6} = \dfrac{11}{24}$

45. $x + 6 - \dfrac{48}{x} = 4$

46. $\dfrac{2}{x - 10} = \dfrac{7}{x + 15}$

47. $\dfrac{2x + 3}{x + 3} = \dfrac{x - 2}{x - 5}$

48. $\dfrac{5}{x - 4} + \dfrac{4}{x + 3} = \dfrac{x^2 - 3x + 31}{x^2 - x - 12}$

49. $\dfrac{x}{x + 6} + \dfrac{7}{x - 8} = \dfrac{84}{x^2 - 2x - 48}$

50. $\dfrac{x - 3}{x^2 - x - 30} + \dfrac{4}{x^2 + x - 20} = \dfrac{3}{x^2 - 10x + 24}$

51. $\dfrac{6}{x^2 + 7x - 8} - \dfrac{2}{x^2 + 2x - 3} = \dfrac{x + 4}{x^2 + 11x + 24}$

Solve for the specified variable. **[5.5]**

52. $\dfrac{b}{2} = \dfrac{A}{h}$ for h

53. $y = \dfrac{5x}{2x - 3}$ for x

54. $\dfrac{x}{3r} - \dfrac{y}{4r} = \dfrac{1}{6}$ for r

55. The sum of the reciprocal of a number and $\frac{7}{12}$ is $\frac{3}{4}$. Find the number. [5.6]

56. One positive number is 12 more than another positive number. If four times the reciprocal of the smaller number is added to six times the reciprocal of the larger number, the sum is equal to 1. Find the two numbers. [5.6]

57. Jerry can mow a lawn in 40 minutes, while George takes 50 minutes to mow the same lawn. If Jerry and George work together, using two lawn mowers, how long would it take them to mow the lawn? [5.6]

58. Two pipes, working together, can fill a tank in 10 hours. Working alone, it would take the smaller pipe 15 hours longer than it would take the larger pipe to fill the tank. How long would it take the smaller pipe alone to fill the tank? [5.6]

59. Margaret had to make a 305-mile drive to Atlanta. After driving the first 110 miles, she increased her speed by 10 miles per hour. If the drive took her exactly 5 hours, find the speed at which she was driving for the first 110 miles. [5.6]

60. Sarah is kayaking in a river that is flowing downstream at a speed of 2 miles per hour. Sarah paddled 4 miles upstream, then turned around and paddled 8 miles downstream. This took Sarah a total of 2 hours. What is the speed that Sarah can paddle in still water? [5.6]

61. The number of calories in a glass of soda varies directly as the amount of soda. If a 12-ounce serving of soda has 180 calories, how many calories are there in an 8-ounce glass of soda? [5.6]

62. The distance required for a car to stop after applying the brakes varies directly as the square of the speed of the car. If it takes 125 feet for a car traveling at 50 miles per hour to stop, how far would it take for a car traveling 80 miles per hour to come to a stop? [5.6]

63. The maximum load that a wooden beam can support varies inversely as its length. If a beam that is 8 feet long can support 725 pounds, what is the maximum load that can be supported by a beam that is 10 feet long? [5.6]

64. The illumination of an object varies inversely as the square of its distance from the source of light. If a light source provides an illumination of 30 foot-candles at a distance of 10 feet, find the illumination at a distance of 5 feet. [5.6]

Divide. **[5.7]**

65. $\dfrac{42x^9}{7x^7}$

66. $\dfrac{24x^5 - 40x^4 - 88x^3 + 8x^2}{8x^2}$

67. $\dfrac{x^2 + 21x - 130}{x - 5}$

68. $\dfrac{x^2 - 5x - 113}{x + 8}$

69. $\dfrac{6x^2 + 11x - 1}{x + 3}$

70. $\dfrac{8x^3 - 38x^2 + 13x + 80}{2x - 5}$

71. $\dfrac{x^3 + 7x - 20}{x - 6}$

72. $\dfrac{x^3 - 515}{x - 8}$

Chapter 5 Test

Test solutions are found on the enclosed CD.

Evaluate the rational expression for the given value of the variable.

1. $\frac{8}{x - 6}$ for $x = -12$

Find all values for which the rational expression is undefined.

2. $\frac{x^2 + 13x + 42}{x^2 - 13x + 40}$

Simplify the given rational expression. (Assume that all denominators are nonzero.)

3. $\frac{x^2 - 13x + 36}{x^2 - 2x - 63}$

Evaluate the rational function.

4. $r(x) = \frac{x^2 + 6x - 20}{x^2 - 2x + 24}$, $r(-4)$

Multiply.

5. $\frac{x^2 + 7x - 8}{x^2 + 7x + 12} \cdot \frac{x^2 - 7x - 30}{x^2 - 6x + 5}$

Divide.

6. $\frac{x^2 - 6x + 9}{x^2 + 6x + 8} \div \frac{9 - x^2}{x^2 - 8x - 20}$

For the given functions $f(x)$ and $g(x)$, find $f(x) \cdot g(x)$.

7. $f(x) = \frac{x^2 + x - 42}{x^2 - 2x}$, $g(x) = \frac{x - 2}{x^2 + 2x - 35}$

Add or subtract.

8. $\frac{5x - 3}{2x + 8} - \frac{2x - 15}{2x + 8}$

9. $\frac{2x^2 + 8x + 13}{x - 5} + \frac{x^2 + 4x + 58}{5 - x}$

10. $\frac{x - 5}{x^2 - 4x + 3} - \frac{6}{x^2 + x - 2}$

11. $\frac{3}{x^2 - 16x + 63} + \frac{6}{x^2 - 10x + 21}$

For the given rational functions $f(x)$ and $g(x)$, find $f(x) - g(x)$.

12. $f(x) = \frac{x}{x^2 + 9x + 18}$, $g(x) = \frac{5}{x^2 + x - 6}$

Simplify the complex fraction.

13. $\dfrac{1 - \frac{4}{x} - \frac{12}{x^2}}{1 - \frac{15}{x} + \frac{54}{x^2}}$

Solve.

14. $\frac{14}{x} + \frac{4}{15} = \frac{29}{30}$

15. $\frac{x + 6}{x^2 + 6x + 8} - \frac{4}{x^2 - 6x - 16} = \frac{3}{x^2 - 4x - 32}$

Solve for the specified variable.

16. $x = \frac{4y}{y + 8}$ for y

17. Two pipes, working together, can fill a tank in 60 minutes. Working alone, it would take the smaller pipe 90 minutes longer than it would take the larger pipe to fill the tank. How long would it take the smaller pipe alone to fill the tank?

18. Preparing for a race, Lance rode his bicycle 23 miles. After the first 8 miles, he increased his speed by 5 miles per hour. If the ride took him exactly 1 hour, find the speed at which he was riding for the first 8 miles.

Divide.

19. $\frac{2x^2 - 13x - 99}{x + 4}$

20. $\frac{2x^3 - 9x^2 - 109}{x - 7}$

Mathematicians in History

John Nash

John Nash is an American mathematician whose research has greatly affected mathematics and economics, as well as many other fields. When Nash was applying to graduate school at Princeton, one of his math professors said simply, "This man is a genius."

Write a one-page summary (*or* make a poster) of the life of John Nash and his accomplishments.

Interesting issues:

- Where and when was John Nash born?
- Describe Nash's childhood, as well as his life as a college student.
- What mental illness struck Nash in the late 1950's?
- Nash was an influential figure in the field of game theory. What is game theory?
- In 1994, Nash won the Nobel Prize for Economics. Exactly what did Nash win the prize for?
- One of Nash's nicknames is "The Phantom of Fine Hall." Why was this nickname chosen for him?
- What color sneakers did Nash wear?
- Sylvia Nasar wrote a biography of Nash's life, which was made into an Academy Award–winning movie. What was the title of the book and movie?
- What actor played John Nash in the movie?

Stretch Your Thinking › Chapter 5

The weight W of an object varies inversely as the square of the distance d from the center of the Earth. At sea level (3978 miles from the center of the Earth) a person weighs 150 pounds. The formula used to compute the weight of this person at different distances from the center of the Earth is $W = \frac{2{,}373{,}672{,}600}{d^2}$.

a) Use the given formula to calculate the weight of this person at different distances from the center of the Earth. Round to the nearest tenth of a pound.

Location	Distance from center of Earth	Weight
On top of the world's tallest building, Taipei 101 (Taipei, Taiwan)	3978.316 miles	
On top of the world's tallest structure, KVLY-TV mast (Mayville, ND)	3978.391 miles	
In an airplane	3984.629 miles	
In the space station	4201.402 miles	
Halfway to the Moon	123,406 miles	
Halfway to Mars	4,925,728 miles	

b) What do you notice about the weight of the person as the distance from the center of the Earth increases?

c) What do you think is the reason for what you've observed in b)?

d) As accurately as possible, plot these points on an axis system in which the horizontal axis represents the distance and the vertical axis represents the weight. Do the points support your observations from b)?

e) How far from the center of the Earth would this 150-pound person have to travel to weigh 100 pounds?

f) How far from the center of the Earth would this 150-pound person have to travel to weigh 75 pounds?

g) How far from the center of the Earth would this 150-pound person have to travel to weigh 0 pounds?

CHAPTER 6

Six

Radical Expressions and Equations

In this chapter, we will investigate radical expressions and equations and their applications. Among the applications is the method for finding the distance between any two objects and a way to determine the speed a car was traveling by measuring the skid marks left by its tires.

Study Strategy **Doing Your Homework** *Doing a homework assignment should not be viewed as just some requirement. Homework exercises are assigned to help you to learn mathematics. In this chapter, we will discuss how to do your homework and get the most out of your effort.*

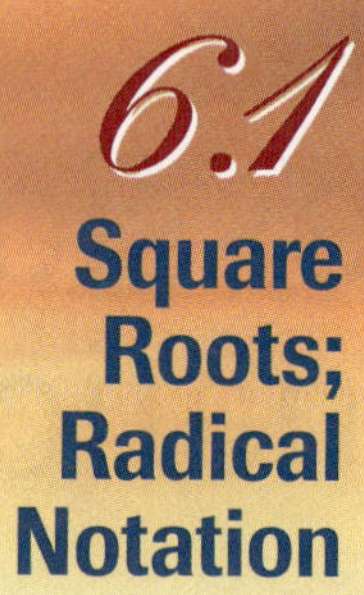

6.1 Square Roots; Radical Notation

Objectives

1. **Find the square root of a number.**
2. **Simplify the square root of a variable expression.**
3. **Approximate the square root of a number by using a calculator.**
4. **Find *n*th roots.**
5. **Multiply radical expressions.**
6. **Divide radical expressions.**
7. **Evaluate radical functions.**
8. **Find the domain of a radical function.**

Square Roots

Consider the equation $x^2 = 36$. There are two solutions of this equation: $x = 6$ and $x = -6$.

Square Root

A number a is a **square root** of a number b if $a^2 = b$.

The numbers 6 and -6 are square roots of 36, since $6^2 = 36$ and $(-6)^2 = 36$. The number 6 is the positive square root of 36, while the number -6 is the negative square root of 36.

Objective 1 **Find the square root of a number.**

Principal Square Root

The **principal square root** of b, denoted $\sqrt{b}$, for $b > 0$, is the positive number a such that $a^2 = b$.

The expression $\sqrt{b}$ is called a **radical expression**. The sign $\sqrt{\ }$ is called a **radical sign**, while the expression contained inside the radical sign is called the **radicand**.

EXAMPLE 1 Simplify $\sqrt{25}$.

Solution

We are looking for a positive number a such that $a^2 = 25$. The number is 5, so $\sqrt{25} = 5$.

EXAMPLE 2 Simplify $\sqrt{\frac{1}{9}}$.

Solution

Since $\left(\frac{1}{3}\right)^2 = \frac{1}{3} \cdot \frac{1}{3} = \frac{1}{9}$, $\sqrt{\frac{1}{9}} = \frac{1}{3}$.

EXAMPLE 3 Simplify $-\sqrt{49}$.

Solution

In this example, we are looking for the negative square root of 49, which is -7. So $-\sqrt{49} = -7$. We can first find the principal square root of 49 and then make it negative.

Quick Check 1

Simplify.

a) $\sqrt{49}$ b) $\sqrt{\frac{4}{25}}$

c) $-\sqrt{36}$

The principal square root of a negative number, such as $\sqrt{-49}$, is not a real number because there is no real number a such that $a^2 = -49$. We will learn in Section 6.6 that $\sqrt{-49}$ is called an *imaginary* number.

Square Roots of Variable Expressions

Objective 2 Simplify the square root of a variable expression. Now we turn our attention toward simplifying radical expressions containing variables, such as $\sqrt{x^6}$.

Simplifying $\sqrt{a^2}$

For any real number a, $\sqrt{a^2} = |a|$.

You may be wondering why the absolute value bars are necessary. For any nonnegative number a, $\sqrt{a^2} = a$. For example, $\sqrt{8^2} = 8$. The absolute value bars are necessary to include negative values of a. Suppose that $a = -10$. Then $\sqrt{a^2} = \sqrt{(-10)^2}$, which is equal to $\sqrt{100}$, or 10. So $\sqrt{a^2}$ equals the opposite of a. In either case, the principal square root of a^2 will be a positive number.

$$\sqrt{a^2} = \begin{cases} a & \text{if } a \geq 0 \\ -a & \text{if } a < 0 \end{cases}$$

The absolute value bars address both cases. We must use absolute values when dealing with variables, because we do not know whether the variable is negative.

EXAMPLE 4 Simplify $\sqrt{x^6}$.

Solution

We begin by rewriting the radicand as a square. Note that $x^6 = (x^3)^2$.

$$\begin{aligned} \sqrt{x^6} &= \sqrt{(x^3)^2} && \text{Rewrite radicand as a square.} \\ &= |x^3| && \text{Simplify.} \end{aligned}$$

Since we do not know whether x is negative, the absolute value bars are necessary.

EXAMPLE 5 Simplify $\sqrt{64y^{10}}$.

Solution

Again, we rewrite the radicand as a square.

$$\begin{aligned} \sqrt{64y^{10}} &= \sqrt{(8y^5)^2} && \text{Rewrite radicand as a square.} \\ &= |8y^5| && \text{Simplify.} \end{aligned}$$

Quick Check 2
Simplify.
a) $\sqrt{a^{14}}$ b) $\sqrt{25b^{22}}$

Since we know that 8 is a positive number, we can remove it from the absolute value bars. This allows us to write the expression as $8|y^5|$.

EXAMPLE 6 Simplify $\sqrt{36x^4}$.

Solution

We rewrite the radicand as a square.

$$\begin{aligned}\sqrt{36x^4} &= \sqrt{(6x^2)^2} && \text{Rewrite radicand as a square.}\\ &= |6x^2| && \text{Simplify.}\end{aligned}$$

Since we know that 6 is a positive number, it can be removed from the absolute value bars. Because x^2 cannot be negative either, it may be removed from the absolute value bars as well.

$$\sqrt{36x^4} = 6x^2$$

Quick Check 3
Simplify $\sqrt{64x^{24}}$.

From this point on, we will assume that all variable factors in a radicand represent nonnegative real numbers. This eliminates the need to use absolute value bars when simplifying radical expressions whose radicand contains variables.

Approximating Square Roots Using a Calculator

Objective 3 **Approximate the square root of a number by using a calculator.**
Consider the expression $\sqrt{12}$. There is no positive integer that is the square root of 12. In such a case, we can use a calculator to approximate the radical expression. All calculators have a function for calculating square roots. Rounding to the nearest thousandth, we see that $\sqrt{12} \approx 3.464$, rounded to the nearest thousandth. The symbol $\approx$ is read as *is approximately equal to,* so the principal square root of 12 is approximately equal to 3.464. If we square 3.464, it is equal to 11.999296, which is very close to 12.

EXAMPLE 7 Approximate $\sqrt{42}$ to the nearest thousandth, using a calculator.

Solution

Since $6^2 = 36$ and $7^2 = 49$, we know that $\sqrt{42}$ must be a number between 6 and 7.

$$\sqrt{42} \approx 6.481$$

Quick Check 4
Approximate $\sqrt{109}$ to the nearest thousandth, using a calculator.

Using Your Calculator We can approximate square roots by using the TI-84.

```
√(42)
        6.480740698
```

*n*th Roots

Objective 4 **Find *n*th roots.** Now we move on to discuss roots other than square roots.

Principal *n*th Root

For any positive integer $n > 1$ and any number b, if $a^n = b$ and a and b both have the same sign, then a is the **principal *n*th root** of b, denoted $a = \sqrt[n]{b}$. The number n is called the **index** of the radical.

A square root has an index of 2, and the radical is written without the index. If the index is 3, this is called a **cube root**. If n is even, then the principal nth root is a nonnegative number, but if n is odd, the principal nth root is of the same sign as the radicand. As with square roots, an even root of a negative number is not a real number. For example, $\sqrt[6]{-64}$ is not a real number. However, an odd root of a negative number, such as $\sqrt[3]{-64}$, is a negative real number.

EXAMPLE 8 Simplify $\sqrt[3]{125}$.

Solution

Look for a number that, when cubed, is equal to 125. Since $125 = 5^3$, $\sqrt[3]{125} = \sqrt[3]{(5)^3} = 5$.

EXAMPLE 9 Simplify $\sqrt[4]{81}$.

Solution

In this example, we are looking for a number that, when raised to the fourth power, is equal to 81. (We can use a factor tree to factor 81.) Since $81 = 3^4$, $\sqrt[4]{81} = \sqrt[4]{(3)^4} = 3$.

Using Your Calculator Find the function for calculating *n*th roots on the TI-84 by pressing the MATH key and selecting option 5 under the MATH menu.

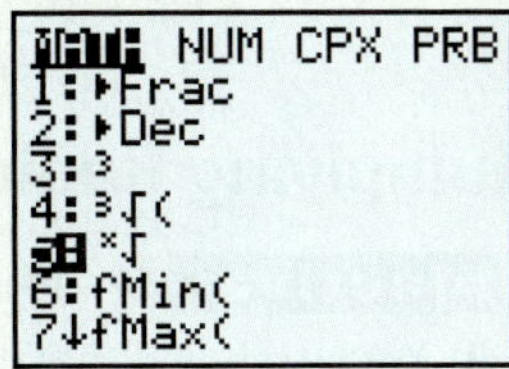

Press the index for the radical, then the *n*th root function, and then the radicand inside a set of parentheses. Here is the screen shot of the calculation of $\sqrt[4]{81}$:

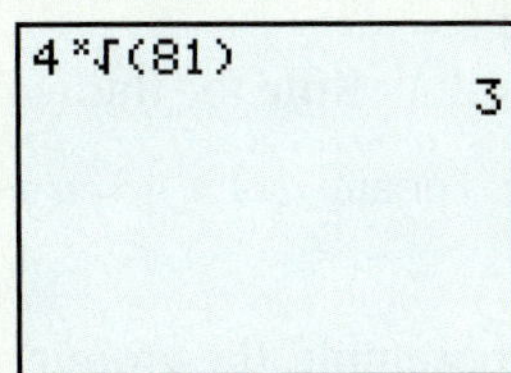

EXAMPLE 10 Simplify $\sqrt[5]{-32}$.

Solution

Notice that the radicand is negative. So we are looking for a negative number that, when raised to the fifth power, is equal to -32. (If the index were even, then this expression would not be a real number.) Since $(-2)^5 = -32$, $\sqrt[5]{-32} = \sqrt[5]{(-2)^5} = -2$.

Quick Check 5

Simplify.

a) $\sqrt[3]{27}$ b) $\sqrt[4]{256}$
c) $\sqrt[3]{-343}$

For any nonnegative number x, $\sqrt[n]{x^n} = x$.

EXAMPLE 11 Simplify $\sqrt[4]{x^{20}}$. (Assume x is nonnegative.)

Solution

We begin by rewriting the radicand as an expression raised to the fourth power. We can rewrite x^{20} as $(x^5)^4$.

$$\begin{aligned} \sqrt[4]{x^{20}} &= \sqrt[4]{(x^5)^4} && \text{Rewrite radicand as an expression raised to the fourth power.} \\ &= x^5 && \text{Simplify.} \end{aligned}$$

Quick Check 6

Simplify $\sqrt[6]{x^{42}}$. (Assume x is nonnegative.)

EXAMPLE 12 Simplify $\sqrt[3]{64a^3b^9c^{21}}$. (Assume a, b, and c are nonnegative.)

Solution

We begin by rewriting the radicand as a cube. We can rewrite $64a^3b^9c^{21}$ as $(4ab^3c^7)^3$.

$$\begin{aligned} \sqrt[3]{64a^3b^9c^{21}} &= \sqrt[3]{(4ab^3c^7)^3} && \text{Rewrite radicand as a cube.} \\ &= 4ab^3c^7 && \text{Simplify.} \end{aligned}$$

Quick Check 7

Simplify $\sqrt[5]{-32x^{35}y^{40}}$. (Assume x and y are nonnegative.)

EXAMPLE 13 Simplify $\sqrt{x^2 - 8x + 16}$. (Assume $x \geq 4$.)

Solution

This is a square root, so we must begin by rewriting the radicand as a square. If we factor the radicand, we see that it can be expressed as $(x - 4)(x - 4)$, or $(x - 4)^2$.

$$\begin{aligned} \sqrt{x^2 - 8x + 16} &= \sqrt{(x - 4)^2} && \text{Rewrite the radicand as a square.} \\ &= x - 4 && \text{Simplify. (Since } x \geq 4,\ x - 4 \text{ is nonnegative.)} \end{aligned}$$

Quick Check 8

Simplify $\sqrt{x^2 + 10x + 25}$. (Assume $x \geq -5$.)

Multiplying Radical Expressions

Objective 5 **Multiply radical expressions.** We know that $\sqrt{9} \cdot \sqrt{100} = 3 \cdot 10$, or 30. We also know that $\sqrt{9 \cdot 100} = \sqrt{900}$, or 30. In this case, we see that $\sqrt{9} \cdot \sqrt{100} = \sqrt{9 \cdot 100}$. If two radical expressions with nonnegative radicands have the same index, then we can multiply the two expressions by multiplying the two radicands and writing the product inside the same radical.

Product Rule for Radicals

For any root n, if $\sqrt[n]{a}$ and $\sqrt[n]{b}$ are real numbers, then $\sqrt[n]{a} \cdot \sqrt[n]{b} = \sqrt[n]{ab}$.

For example, the product of $\sqrt{2}$ and $\sqrt{8}$ is $\sqrt{16}$, which simplifies to equal 4.

EXAMPLE 14 Multiply $\sqrt{45n} \cdot \sqrt{5n}$. (Assume n is nonnegative.)

Quick Check 9
Multiply $\sqrt{6b^5} \cdot \sqrt{150b^3}$. (Assume b is nonnegative.)

Solution

Since both radicals are square roots, we can multiply the radicands.

$$\sqrt{45n} \cdot \sqrt{5n} = \sqrt{225n^2} \quad \text{Multiply the radicands.}$$
$$= 15n \quad \text{Simplify the square root.}$$

EXAMPLE 15 Multiply $7\sqrt{2} \cdot 9\sqrt{2}$.

Solution

We multiply the factors in front of the radicals by each other and multiply the radicands by each other.

Quick Check 10
Multiply $3\sqrt{6} \cdot 8\sqrt{6}$.

$$7\sqrt{2} \cdot 9\sqrt{2} = 63\sqrt{4} \quad \text{Multiply the factors in front of the radicals } (7 \cdot 9). \text{ Multiply the radicands.}$$
$$= 63 \cdot 2 \quad \text{Simplify the square root.}$$
$$= 126 \quad \text{Multiply.}$$

Dividing Radical Expressions

Objective 6 **Divide radical expressions.** We can rewrite the quotient of two radical expressions that have the same index as the quotient of the two radicands inside the same radical.

Quotient Rule for Radicals

For any root n, if $\sqrt[n]{a}$ and $\sqrt[n]{b}$ are real numbers, then $b \neq 0$, $\dfrac{\sqrt[n]{a}}{\sqrt[n]{b}} = \sqrt[n]{\dfrac{a}{b}}$.

EXAMPLE 16 Simplify $\dfrac{\sqrt{108}}{\sqrt{3}}$.

Solution

Since both radicals are square roots, we begin by dividing the radicands. We write the quotient of the radicands under a single square root.

$$\frac{\sqrt{108}}{\sqrt{3}} = \sqrt{\frac{108}{3}} \quad \text{Rewrite as the square root of the quotient of the radicands.}$$
$$= \sqrt{36} \quad \text{Divide.}$$
$$= 6 \quad \text{Simplify the square root.}$$

EXAMPLE 17 Simplify $\frac{\sqrt[3]{40b^7}}{\sqrt[3]{5b^4}}$. (Assume b is nonnegative.)

Solution

Since the index of both radicals is the same, we begin by dividing the radicands.

$$\frac{\sqrt[3]{40b^7}}{\sqrt[3]{5b^4}} = \sqrt[3]{\frac{40b^7}{5b^4}} \qquad \text{Divide the radicands.}$$
$$= \sqrt[3]{8b^3} \qquad \text{Simplify the radicand.}$$
$$= 2b \qquad \text{Simplify the radical.}$$

***Quick Check* 11** **Simplify.**

a) $\frac{\sqrt{350}}{\sqrt{14}}$ **b)** $\frac{\sqrt[5]{2916a^{18}}}{\sqrt[5]{12a^3}}$ **(Assume a is nonnegative.)**

Radical Functions

Objective 7 **Evaluate radical functions.** A **radical function** is a function that involves radicals, such as $f(x) = \sqrt{x-4} + 3$.

EXAMPLE 18 For the radical function $f(x) = \sqrt{x+5} - 2$, find $f(-1)$.

Solution

To evaluate this function, we substitute -1 for x and simplify.

$$f(-1) = \sqrt{(-1)+5} - 2 \qquad \text{Substitute } -1 \text{ for } x.$$
$$= \sqrt{4} - 2 \qquad \text{Simplify radicand.}$$
$$= 2 - 2 \qquad \text{Take the square root of 4.}$$
$$= 0 \qquad \text{Subtract.}$$

***Quick Check* 12**
For the radical function $f(x) = \sqrt{3x+13} + 33$, find $f(-3)$.

Finding the Domain of Radical Functions

Objective 8 **Find the domain of a radical function.** Radical functions involving even roots are different than the functions we have seen to this point in that their domain is restricted. To find the domain of a radical function involving an even root, we need to find the values of the variable that make the radicand nonnegative. In other words, set the radicand greater than or equal to zero and solve. The domain of a radical function involving odd roots is the set of all real numbers. Remember that square roots are considered to be even roots with an index of 2.

EXAMPLE 19 Find the domain of the radical function $f(x) = \sqrt{x - 9} + 7$. Express your answer in interval notation.

Solution

Since the radical has an even index, we begin by setting the radicand $(x - 9)$ greater than or equal to zero. We solve this inequality to find the domain.

$$x - 9 \geq 0 \quad \text{Set the radicand greater than or equal to 0.}$$
$$x \geq 9 \quad \text{Add 9.}$$

The domain of the function is $[9, \infty)$.

***Quick Check* 13**
Find the domain of the radical function $f(x) = \sqrt[6]{x + 18} - 30$. Express your answer in interval notation.

EXAMPLE 20 Find the domain of the radical function $f(x) = \sqrt[5]{14x - 9} + 21$. Express your answer in interval notation.

Solution

Since this radical function involves an odd root, its domain is the set of all real numbers $\mathbb{R}$. This can be expressed in interval notation as $(-\infty, \infty)$.

***Quick Check* 14** **Find the domain of the radical function $f(x) = \sqrt[3]{16x - 409} + 38$. Express your answer in interval notation.**

Building Your Study Strategy **Doing Your Homework, 1** **Review First** Before beginning any homework assignment, it is a good idea to review first. Start by going over your notes from class to remind you of the type of problems covered by your instructor and how your instructor solved them. Your notes may contain the steps for certain procedures and advice from your instructor about typical errors to avoid. After reviewing your notes, keep them handy for further reference as you proceed through the homework assignment.

You should then review the appropriate section in the text. Pay particular attention to the examples, as they show the solution to problems similar to the problems you are about to solve in the homework exercises. Look for the feature labeled "A Word of Caution" for advice on how to avoid making common errors. Finally, look for summaries of procedures introduced in the section. As you proceed through the homework assignment, refer back to the section as needed.

EXERCISES 6.1

Vocabulary

1. A number a is a(n) _________ of a number b if $a^2 = b$.

2. The _________ square root of b, denoted $\sqrt{b}$, for $b > 0$, is the positive number a such that $a^2 = b$.

3. For any positive integer $n > 1$ and any number b, if $a^n = b$ and a and b both have the same sign, then a is the principal _________ of b, denoted $a = \sqrt[n]{b}$.

4. For the expression $a = \sqrt[n]{b}$, n is called the _________ of the radical.

5. The expression contained inside a radical is called the _________.

6. A radical with an index of 3 is also known as a(n) _________ root.

7. A(n) _________ is a function that involves radicals.

8. The domain of a radical function involving an even root consists of values of the variable for which the radicand is _________.

Simplify the radical expression. Indicate if the expression is not a real number.

9. $\sqrt{36}$
10. $\sqrt{64}$
11. $\sqrt{4}$
12. $\sqrt{100}$
13. $\sqrt{\frac{1}{16}}$
14. $\sqrt{\frac{1}{25}}$
15. $\sqrt{\frac{36}{25}}$
16. $\sqrt{\frac{4}{81}}$
17. $-\sqrt{16}$
18. $-\sqrt{36}$
19. $\sqrt{-144}$
20. $\sqrt{-25}$

Simplify the radical expression. Where appropriate, include absolute values.

21. $\sqrt{a^{16}}$
22. $\sqrt{b^{12}}$
23. $\sqrt{x^{20}}$
24. $\sqrt{x^8}$
25. $\sqrt{9x^6}$
26. $\sqrt{16x^2}$
27. $\sqrt{\frac{1}{49}x^4}$
28. $\sqrt{\frac{1}{100}x^{24}}$
29. $\sqrt{m^{14}n^{10}}$
30. $\sqrt{a^{18}b^6}$
31. $\sqrt{x^4y^8z^{32}}$
32. $\sqrt{x^{20}y^2z^{10}}$

Find the missing number or expression. Assume all variables represent nonnegative real numbers.

33. $\sqrt{?} = 9$
34. $\sqrt{?} = 16$
35. $\sqrt{?} = 3x$
36. $\sqrt{?} = 5a^3$

Approximate to the nearest thousandth, using a calculator.

37. $\sqrt{55}$
38. $\sqrt{98}$
39. $\sqrt{326}$
40. $\sqrt{409}$
41. $\sqrt{0.53}$
42. $\sqrt{0.06}$

Simplify the radical expression. Assume all variables represent nonnegative real numbers.

43. $\sqrt[3]{8}$
44. $\sqrt[3]{-27}$
45. $\sqrt[4]{625}$
46. $\sqrt[4]{256}$
47. $\sqrt[3]{a^{18}}$
48. $\sqrt[4]{b^{44}}$
49. $\sqrt[3]{343m^{21}}$
50. $\sqrt[3]{216n^{42}}$
51. $\sqrt[4]{81x^{12}y^{20}}$
52. $\sqrt[6]{64s^{36}t^{54}}$
53. $\sqrt{x^2 + 6x + 9}$, $x \geq -3$
54. $\sqrt{x^2 + 12x + 36}$, $x \geq -6$

Simplify.

55. $\sqrt{27} \cdot \sqrt{3}$
56. $\sqrt{8} \cdot \sqrt{8}$
57. $\sqrt{10} \cdot \sqrt{90}$
58. $\sqrt{2} \cdot \sqrt{72}$
59. $\frac{\sqrt{180}}{\sqrt{5}}$
60. $\frac{\sqrt{63}}{\sqrt{7}}$
61. $\frac{\sqrt{800}}{\sqrt{8}}$
62. $\frac{\sqrt{1872}}{\sqrt{13}}$

63. $\sqrt[3]{6} \cdot \sqrt[3]{36}$

64. $\sqrt[3]{12} \cdot \sqrt[3]{18}$

65. $\sqrt[4]{48} \cdot \sqrt[4]{27}$

66. $\sqrt[5]{16} \cdot \sqrt[5]{64}$

67. $\dfrac{\sqrt[3]{297}}{\sqrt[3]{11}}$

68. $\dfrac{\sqrt[4]{144}}{\sqrt[4]{9}}$

69. $4\sqrt{5} \cdot 10\sqrt{5}$

70. $16\sqrt{3} \cdot 3\sqrt{3}$

71. $10\sqrt{80} \cdot \sqrt{45}$

72. $2\sqrt{49} \cdot 15\sqrt{121}$

Find the missing number.

73. $\sqrt{20} \cdot \sqrt{?} = 10$

74. $\sqrt{21} \cdot \sqrt{?} = 42$

75. $\dfrac{\sqrt{?}}{\sqrt{8}} = 7$

76. $\dfrac{\sqrt{?}}{\sqrt{6}} = 6$

Evaluate the radical function. Round to the nearest thousandth if necessary.

77. $f(x) = \sqrt{x - 5}$; find $f(21)$.

78. $f(x) = \sqrt{3x + 4}$; find $f(20)$.

79. $f(x) = \sqrt{2x + 19} - 2$; find $f(3)$.

80. $f(x) = \sqrt{4x - 11} - 10$; find $f(5)$.

81. $f(x) = \sqrt{x} + 6$; find $f(8)$.

82. $f(x) = \sqrt{3x + 1} + 5$; find $f(4)$.

83. $f(x) = \sqrt{x^2 - 8x + 16}$; find $f(9)$.

84. $f(x) = \sqrt{x^2 - 7x + 19}$; find $f(10)$.

Find the domain of the radical function. Express your answer in interval notation.

85. $f(x) = \sqrt{x - 8} + 10$

86. $f(x) = \sqrt{4x + 3} - 9$

87. $f(x) = \sqrt{2x - 15} - 8$

88. $f(x) = \sqrt{5x - 95}$

89. $f(x) = \sqrt[4]{4x + 10} + 2$

90. $f(x) = \sqrt[6]{8x - 50} - 11$

91. $f(x) = \sqrt[3]{15x - 42} + 6$

92. $f(x) = \sqrt[5]{10x + 115} - 45$

Writing in Mathematics

Answer in complete sentences.

93. Which of the following are real numbers, and which are not real numbers: $-\sqrt{64}$, $\sqrt{-64}$, $-\sqrt[3]{64}$, $\sqrt[3]{-64}$? Explain your reasoning.

94. If we do not know whether x is a nonnegative number, explain why $\sqrt{x^2} = |x|$ rather than x.

95. Explain how to find the domain of a radical function. Use examples.

96. True or False: For any real number x, $\sqrt{x} \geq \sqrt[3]{x}$. Explain your reasoning.

97. ***Solutions Manual**** Write a solutions manual page for the following problem:

Simplify $\sqrt[3]{125x^9y^6}$.

98. ***Newsletter**** Write a newsletter that explains how to simplify the principal square root of a variable expression.

***See Appendix B for details and sample answers.**

6.2 Rational Exponents

Objectives

1. **Simplify expressions containing exponents of the form $1/n$.**
2. **Simplify expressions containing exponents of the form m/n.**
3. **Simplify expressions containing rational exponents.**
4. **Simplify expressions containing negative rational exponents.**
5. **Use rational exponents to simplify radical expressions.**

Rational Exponents of the Form $1/n$

Objective 1 Simplify expressions containing exponents of the form $1/n$. We have used exponents to represent repeated multiplication. For example, x^n tells us that the base x is a factor n times.

$$x^3 = x \cdot x \cdot x$$
$$x^6 = x \cdot x \cdot x \cdot x \cdot x \cdot x$$

We run into a problem with this definition when we encounter an expression with a fractional exponent, such as $x^{1/2}$. Saying that the base x is a factor $\frac{1}{2}$ times does not make any sense. Using the properties of exponents introduced in Chapter 4, we know that $x^{1/2} \cdot x^{1/2} = x^{1/2+1/2}$, or x. If we multiply $x^{1/2}$ by itself, the result is x. The same is true when we multiply $\sqrt{x}$ by itself, suggesting that $x^{1/2} = \sqrt{x}$.

> For any integer $n > 1$, we define $a^{1/n}$ to be the nth root of a, or $\sqrt[n]{a}$.

EXAMPLE 1 Rewrite $81^{1/4}$ as a radical expression and simplify if possible.

Solution

$81^{1/4} = \sqrt[4]{81}$ Rewrite as a radical expression. An exponent of $1/4$ is equivalent to a fourth root.

$= \sqrt[4]{3^4}$ Rewrite radicand as a number to the fourth power.

$= 3$ Simplify.

EXAMPLE 2 Rewrite $(-125x^6)^{1/3}$ as a radical expression and simplify if possible.

Solution

$(-125x^6)^{1/3} = \sqrt[3]{(-5x^2)^3}$ Rewrite as a radical expression. Rewrite the radicand as a cube.

$= -5x^2$ Simplify. Keep in mind that an odd root of a negative number is negative.

***Quick Check* 1** Rewrite as a radical expression and simplify if possible.

a) $36^{1/2}$ b) $(-32x^{15})^{1/5}$

EXAMPLE 3 Rewrite $(a^{10}b^5c^{20})^{1/5}$ as a radical expression and simplify if possible.

Solution

$$\begin{aligned}(a^{10}b^5c^{20})^{1/5} &= \sqrt[5]{a^{10}b^5c^{20}} && \text{Rewrite as a radical expression.}\\ &= \sqrt[5]{(a^2bc^4)^5} && \text{Rewrite radicand.}\\ &= a^2bc^4 && \text{Simplify.}\end{aligned}$$

Quick Check **2** **Rewrite $(x^{32}y^4z^{24})^{1/4}$ as a radical expression and simplify if possible. Assume all variables represent nonnegative values.**

For any negative number x and even integer n, $x^{1/n}$ is not a real number. For example, $(-64)^{1/6}$ is not a real number, because $(-64)^{1/6} = \sqrt[6]{-64}$ and an even root of a negative number is not a real number.

EXAMPLE 4 Rewrite $\sqrt[7]{x}$ by using rational exponents.

Solution

$$\sqrt[7]{x} = x^{1/7} \qquad \text{Rewrite, using the definition } \sqrt[n]{x} = x^{1/n}.$$

Quick Check **3**
Rewrite $\sqrt[9]{a}$, using rational exponents.

Rational Exponents of the Form *m/n*

Objective 2 **Simplify expressions containing exponents of the form *m/n*.**

We now turn our attention to rational exponents of the form m/n, for any integers m and $n > 1$. The expression $a^{m/n}$ can be rewritten as $(a^{1/n})^m$, which is equivalent to $(\sqrt[n]{a})^m$.

For any integers m and n, $n > 1$, we define $a^{m/n}$ to be $(\sqrt[n]{a})^m$. This is also equivalent to $\sqrt[n]{a^m}$. The denominator in the exponent, n, is the root we are taking. The numerator in the exponent, m, is the power to which we raise this radical.

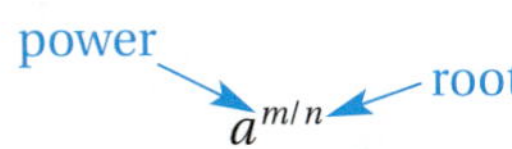

EXAMPLE 5 Rewrite $(243x^{10})^{3/5}$ as a radical expression and simplify if possible.

Quick Check **4**
Rewrite $(4096x^{18})^{5/6}$ as a radical expression and simplify if possible. (Assume x is nonnegative.)

Solution

$$\begin{aligned}(243x^{10})^{3/5} &= (\sqrt[5]{243x^{10}})^3 && \text{Rewrite as a radical expression.}\\ &= \left(\sqrt[5]{(3x^2)^5}\right)^3 && \text{Rewrite radicand.}\\ &= (3x^2)^3 && \text{Simplify the radical.}\\ &= 27x^6 && \text{Raise } 3x^2 \text{ to the third power.}\end{aligned}$$

EXAMPLE 6 Rewrite $\sqrt[5]{x^3}$ by using rational exponents.

Solution

$\sqrt[5]{x^3} = x^{3/5}$ Rewrite, using the definition $\sqrt[n]{x^m} = x^{m/n}$.

***Quick Check* 5**

Rewrite $\sqrt[12]{x^5}$ by using rational exponents. (Assume x is nonnegative.)

Simplifying Expressions Containing Rational Exponents

Objective 3 **Simplify expressions containing rational exponents.** The properties of exponents developed in Chapter 4 for integer exponents are true for fractional exponents as well. Here is a summary of those properties:

Properties of Exponents

For any bases x and y:

1. $x^m \cdot x^n = x^{m+n}$
2. $(x^m)^n = x^{m \cdot n}$
3. $(xy)^n = x^n y^n$
4. $\dfrac{x^m}{x^n} = x^{m-n}\ (x \neq 0)$
5. $x^0 = 1\ (x \neq 0)$
6. $\left(\dfrac{x}{y}\right)^n = \dfrac{x^n}{y^n}\ (y \neq 0)$
7. $x^{-n} = \dfrac{1}{x^n}\ (x \neq 0)$

EXAMPLE 7 Simplify the expression $x^{4/3} \cdot x^{5/6}$. (Assume x is nonnegative.)

Solution

When multiplying two expressions with the same base, we add the exponents and keep the base.

$x^{4/3} \cdot x^{5/6} = x^{\frac{4}{3}+\frac{5}{6}}$ Add the exponents; keep the base.

$= x^{\frac{8}{6}+\frac{5}{6}}$ Rewrite the fractions with a common denominator.

$= x^{13/6}$ Add.

***Quick Check* 6**

Simplify the expression $x^{5/8} \cdot x^{7/12}$. (Assume x is nonnegative.)

EXAMPLE 8 Simplify the expression $(x^{3/4})^{2/5}$. (Assume x is nonnegative.)

Solution

When raising an exponential expression to another power, we multiply the exponents and keep the base.

$(x^{3/4})^{2/5} = x^{\frac{3}{4}\cdot\frac{2}{5}}$ Multiply the exponents; keep the base.

$= x^{3/10}$ Multiply.

***Quick Check* 7**

Simplify the expression $(x^{7/10})^{4/9}$. (Assume x is nonnegative.)

EXAMPLE 9 Simplify the expression $\left(\frac{b^4}{c^2d^6}\right)^{3/2}$, where $c \neq 0$, $d \neq 0$. (Assume all variables represent nonnegative values.)

Solution

We begin by raising each factor to the $\frac{3}{2}$ power.

$$\left(\frac{b^4}{c^2d^6}\right)^{3/2} = \frac{b^{4\cdot\frac{3}{2}}}{c^{2\cdot\frac{3}{2}}d^{6\cdot\frac{3}{2}}}$$ Raise each factor to the $\frac{3}{2}$ power.

$$= \frac{b^6}{c^3d^9}$$ Multiply exponents.

Quick Check 8
Simplify the expression $\left(\frac{x^9}{y^3z^{12}}\right)^{5/3}$, where $y \neq 0, z \neq 0$.

Simplifying Expressions Containing Negative Rational Exponents

Objective 4 **Simplify expressions containing negative rational exponents.**

EXAMPLE 10 Simplify the expression $64^{-5/6}$.

Solution

We begin by rewriting the expression with a positive exponent.

$$64^{-5/6} = \frac{1}{64^{5/6}}$$ Rewrite the expression with a positive exponent.

$$= \frac{1}{(\sqrt[6]{64})^5}$$ Rewrite in radical notation.

$$= \frac{1}{2^5}$$ Simplify the radical.

$$= \frac{1}{32}$$ Raise 2 to the fifth power.

Quick Check 9
Simplify the expression $81^{-3/2}$.

Using Rational Exponents to Simplify Radical Expressions

Objective 5 **Use rational exponents to simplify radical expressions.** In the previous section, we learned that we may multiply two radicals if they have the same index. In other words, $\sqrt[n]{a} \cdot \sqrt[n]{b} = \sqrt[n]{ab}$ if $\sqrt[n]{a}$ and $\sqrt[n]{b}$ are real numbers. If the two indices are not the same, we can use fractional exponents to multiply the radicals.

EXAMPLE 11 Simplify the expression $\sqrt{a} \cdot \sqrt[5]{a}$. (Assume a is nonnegative.) Express your answer in radical notation.

Solution

We will begin by rewriting the radicals using fractional exponents.

$$\begin{aligned} \sqrt{a} \cdot \sqrt[5]{a} &= a^{1/2} \cdot a^{1/5} && \text{Rewrite both radicals using fractional exponents.} \\ &= a^{\frac{1}{2}+\frac{1}{5}} && \text{Add the exponents, keeping the base.} \\ &= a^{\frac{5}{10}+\frac{2}{10}} && \text{Rewrite each fraction with a common denominator.} \\ &= a^{7/10} && \text{Add.} \\ &= \sqrt[10]{a^7} && \text{Rewrite in radical notation.} \end{aligned}$$

***Quick Check* 10**

Simplify the expression $\sqrt[3]{x} \cdot \sqrt[4]{x}$. (Assume x is nonnegative.) Express your answer in radical notation.

***Building Your Study Strategy* Doing Your Homework, 2 Neat and Complete** Two important words to keep in mind when working on your homework exercises are *neat* and *complete*. When your homework is neat, it is easier to check your work, it helps an instructor spot an error if you need help with a particular problem, and it will be easier to understand when you review an old homework assignment prior to an exam.

Be as complete as you can when working on the homework exercises. By listing each step, you are increasing your chances of being able to remember all of the steps necessary to solve a similar problem on an exam or quiz. When you review an old homework assignment prior to an exam, you may have difficulty remembering how to solve a problem if you did not write down all of the necessary steps. If you make a mistake while working on a particular exercise, make note of the mistake that you made, to avoid making a similar mistake later.

EXERCISES 6.2

Vocabulary

1. For any integer $n > 1$, $a^{1/n} =$ __________.

2. For any integers m and n, $n > 1$, $a^{m/n} =$ __________.

Rewrite each radical expression using rational exponents.

3. $\sqrt[4]{x}$
4. $\sqrt[5]{a}$
5. $\sqrt{7}$
6. $\sqrt[3]{10}$
7. $9\sqrt[4]{d}$
8. $\sqrt[4]{9d}$

Rewrite as a radical expression and simplify if possible. Assume all variables represent nonnegative real numbers.

9. $64^{1/2}$
10. $25^{1/2}$
11. $64^{1/3}$
12. $1296^{1/4}$
13. $(-343)^{1/3}$
14. $(-243)^{1/5}$
15. $(x^{12})^{1/4}$
16. $(y^{15})^{1/3}$
17. $(x^{40}y^{35})^{1/5}$
18. $(x^{22}y^{36})^{1/2}$
19. $(16x^{32}y^{60})^{1/4}$
20. $(27x^{15}y^{30}z^{45})^{1/3}$
21. $(-216x^{18}y^{15}z^{21})^{1/3}$
22. $(-32x^{5}y^{25}z^{125})^{1/5}$

Rewrite each radical expression, using rational exponents. Assume all variables represent nonnegative real numbers.

23. $(\sqrt[5]{x})^3$
24. $(\sqrt[4]{x})^7$
25. $(\sqrt[9]{y})^8$
26. $(\sqrt[12]{a})^{17}$
27. $\sqrt[8]{x^5}$
28. $\sqrt[4]{a^{15}}$
29. $(\sqrt[3]{3x^2})^7$
30. $(\sqrt[9]{2x^4})^2$
31. $(\sqrt[8]{10x^4y^5})^3$
32. $(\sqrt[5]{a^6b^7c^8})^4$

Rewrite as a radical expression and simplify if possible. Assume all variables represent nonnegative real numbers.

33. $25^{3/2}$
34. $16^{3/4}$
35. $32^{2/5}$
36. $1000^{7/3}$
37. $(x^{12})^{2/3}$
38. $(x^{20})^{6/5}$
39. $(256a^8b^{24})^{3/4}$
40. $(4x^{16}y^{32}z^{64})^{5/2}$
41. $(-125x^9y^{15}z^3)^{4/3}$
42. $(256a^4b^8c^{20})^{7/4}$

Simplify the expression. Assume all variables represent nonnegative real numbers.

43. $x^{7/10} \cdot x^{1/10}$
44. $x^{3/4} \cdot x^{11/4}$
45. $x^{2/5} \cdot x^{1/4}$
46. $x^{5/8} \cdot x^{1/6}$
47. $x^{4/3} \cdot x^{1/2}$
48. $x^{10/7} \cdot x^{1/14}$
49. $a^{5/3} \cdot a^{3/4} \cdot a^{1/6}$
50. $m^{4/5} \cdot m^{5/2} \cdot m^{7/10}$
51. $x^{1/8}y^{2/5} \cdot x^{3/8}y^{1/10}$
52. $x^{5/7}y^{2/9} \cdot x^{3/2}y^{2/3}$
53. $(x^{1/6})^{3/5}$
54. $(x^{5/4})^{2/3}$
55. $(b^{2/3})^{2/3}$
56. $(n^{3/7})^{5/9}$
57. $(x^4)^{7/8}$
58. $(x^{7/12})^3$
59. $\dfrac{x^{4/5}}{x^{3/10}}\ (x \neq 0)$
60. $\dfrac{x^{7/6}}{x^{11/12}}\ (x \neq 0)$
61. $\dfrac{x^{5/8}}{x^{1/3}}\ (x \neq 0)$
62. $\dfrac{x^{9/10}}{x^{21/40}}\ (x \neq 0)$
63. $\dfrac{x^{13/14}}{x^{13/14}}\ (x \neq 0)$
64. $\dfrac{x^{9/8}}{x^{9/8}}\ (x \neq 0)$
65. $(x^{5/4})^0\ (x \neq 0)$
66. $(x^{2/9})^0\ (x \neq 0)$
67. $8^{-2/3}$
68. $81^{-1/4}$
69. $49^{-1/2}$
70. $25^{-3/2}$
71. $32^{-3/5} \cdot 32^{-4/5}$
72. $27^{-5/3} \cdot 27^{-4/3}$
73. $6^{-1/4} \cdot 6^{-3/4}$
74. $10^{-3/2} \cdot 10^{-5/2}$
75. $\dfrac{125^{2/3}}{125^{7/3}}$
76. $\dfrac{16^{3/4}}{16^{10/4}}$
77. $\dfrac{216^{7/3}}{216^{8/3}}$
78. $\dfrac{4^{3/2}}{4^5}$

FOR EXTRA HELP

Interactmath.com

MathXL Tutorials on CD

Video Lectures on CD

Addison-Wesley Math Tutor Center

Student's Solutions Manual

Simplify each expression. Assume all variables represent nonnegative real numbers. Express your answer in radical notation.

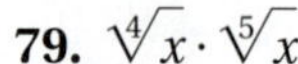

79. $\sqrt[4]{x} \cdot \sqrt[5]{x}$

80. $\sqrt[9]{x} \cdot \sqrt[18]{x}$

81. $\sqrt[12]{a} \cdot \sqrt[4]{a}$

82. $\sqrt[3]{b} \cdot \sqrt[8]{b}$

83. $\dfrac{\sqrt[4]{m}}{\sqrt[12]{m}} \; (m \neq 0)$

84. $\dfrac{\sqrt{n}}{\sqrt[3]{n}} \; (n \neq 0)$

85. $\dfrac{\sqrt[30]{x}}{\sqrt[5]{x}} \; (x \neq 0)$

86. $\dfrac{\sqrt[10]{x}}{\sqrt[5]{x}} \; (x \neq 0)$

87. List four expressions containing rational exponents of the form $\dfrac{1}{n}$ that are equivalent to 4.

88. List four expressions containing rational exponents of the form $\dfrac{m}{n}$ that are equivalent to 9.

89. List four expressions containing rational exponents that are equivalent to $5x^3$.

90. List four expressions containing rational exponents that are equivalent to $a^4b^8c^{12}$.

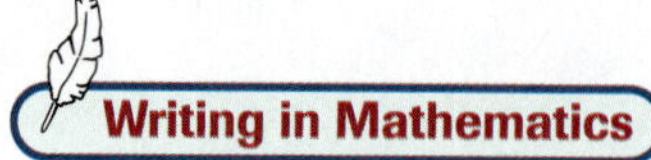

Writing in Mathematics

Answer in complete sentences.

91. Is $-16^{1/2}$ a real number? Explain your answer.

92. Is $16^{-1/2}$ a real number? Explain your answer.

93. ***Solutions Manual**** Write a solutions manual page for the following problem:
Simplify $(16x^8)^{3/4}$.

94. ***Newsletter**** Write a newsletter that explains how to rewrite an expression with rational exponents in radical form.

***See Appendix B for details and sample answers.**

6.3 Simplifying, Adding, and Subtracting Radical Expressions

Objectives

1 **Simplify radical expressions by using the product property.**

2 **Add or subtract radical expressions containing like radicals.**

3 **Simplify radical expressions before adding or subtracting.**

Simplifying Radical Expressions Using the Product Property

Objective 1 **Simplify radical expressions by using the product property.** A radical expression is considered simplified if the radicand contains no factors with exponents greater than or equal to the index of the radical. For example, $\sqrt[3]{x^5}$ is not simplified because there is an exponent inside the radical that is greater than the index of the radical. The goal for simplifying radical expressions is to remove as many factors as possible from the radicand. We will use the product property for radical expressions to help us with this.

We could rewrite $\sqrt[3]{x^5}$ as $\sqrt[3]{x^3 \cdot x^2}$, and then rewrite this radical expression as the product of two radicals. Using the product property for radicals, $\sqrt[n]{a} \cdot \sqrt[n]{b} = \sqrt[n]{ab}$, we know that $\sqrt[3]{x^3 \cdot x^2} = \sqrt[3]{x^3} \cdot \sqrt[3]{x^2}$. The reason for rewriting $\sqrt[3]{x^5}$ as $\sqrt[3]{x^3} \cdot \sqrt[3]{x^2}$ is that the radical $\sqrt[3]{x^3}$ equals x.

$$\begin{aligned} \sqrt[3]{x^5} &= \sqrt[3]{x^3 \cdot x^2} \\ &= \sqrt[3]{x^3} \cdot \sqrt[3]{x^2} \\ &= x\sqrt[3]{x^2} \end{aligned}$$

Now the radicand contains no factors with an exponent that is greater than or equal to the index 3 and is simplified.

Simplifying Radical Expressions

- Completely factor any numerical factors in the radicand.
- Rewrite each factor as a product of two factors. The exponent for the first factor should be the largest multiple of the radical's index that is less than or equal to the factor's original exponent.
- Use the product property to remove factors from the radicand.

Example 1 Simplify $\sqrt[4]{a^{23}}$. (Assume a is nonnegative.)

Solution

We begin by rewriting a^{23} as a product of two factors. The largest multiple of the index (4) that is less than or equal to the exponent for this factor (23) is 20, so we will rewrite a^{23} as $a^{20} \cdot a^3$.

$$\begin{aligned} \sqrt[4]{a^{23}} &= \sqrt[4]{a^{20} \cdot a^3} && \text{Rewrite } a^{23} \text{ as the product of two factors.} \\ &= \sqrt[4]{a^{20}} \cdot \sqrt[4]{a^3} && \text{Use the product property of radicals to rewrite the radical as the product of two radicals.} \\ &= a^5\sqrt[4]{a^3} && \text{Simplify the radical.} \end{aligned}$$

Quick Check 1

Simplify $\sqrt[5]{x^{17}}$. (Assume x is nonnegative.)

EXAMPLE 2 Simplify $\sqrt{x^{11}y^{10}z^5}$. (Assume all variables represent nonnegative values.)

Solution

Again, we begin by rewriting factors as a product of two factors. In this example, the exponent of the factor y is a multiple of the index 2. We do not need to rewrite this factor as the product of two factors.

Quick Check 2
Simplify $\sqrt{a^8b^{15}c^7d}$. (Assume all variables represent nonnegative values.)

$$\begin{aligned}\sqrt{x^{11}y^{10}z^5} &= \sqrt{(x^{10}\cdot x)y^{10}(z^4\cdot z)} && \text{Rewrite factors.}\\ &= \sqrt{x^{10}y^{10}z^4}\cdot\sqrt{xz} && \text{Rewrite as the product of two radicals.}\\ &= x^5y^5z^2\sqrt{xz} && \text{Simplify the radical.}\end{aligned}$$

EXAMPLE 3 Simplify $\sqrt{24}$.

Solution

We begin by rewriting 24, using its prime factorization $(2^3\cdot 3)$.

$$\begin{aligned}\sqrt{24} &= \sqrt{2^3\cdot 3} && \text{Factor 24.}\\ &= \sqrt{(2^2\cdot 2)\cdot 3} && \text{Rewrite } 2^3 \text{ as } 2^2\cdot 2.\\ &= \sqrt{2^2}\cdot\sqrt{2\cdot 3} && \text{Rewrite as the product of two radicals.}\\ &= 2\sqrt{6} && \text{Simplify.}\end{aligned}$$

We could have used a different tactic to simplify this square root. The largest factor of 24 that is a perfect square is 4, so we could begin by rewriting 24 as $4\cdot 6$. Since we know that the square root of 4 is 2, we can factor 4 out of the radicand and write it as 2 in front of the radical.

Quick Check 3
Simplify $\sqrt{90}$.

$$\begin{aligned}\sqrt{24} &= \sqrt{4\cdot 6}\\ &= 2\sqrt{6}\end{aligned}$$

EXAMPLE 4 Simplify $\sqrt[3]{324}$.

Solution

We begin by factoring 324 to be $2^2\cdot 3^4$.

Quick Check 4
Simplify $\sqrt[3]{280}$.

$$\begin{aligned}\sqrt[3]{324} &= \sqrt[3]{2^2\cdot 3^4} && \text{Factor 324.}\\ &= \sqrt[3]{2^2\cdot(3^3\cdot 3)} && \text{Rewrite } 3^4 \text{ as } 3^3\cdot 3.\\ &= \sqrt[3]{3^3}\cdot\sqrt[3]{2^2\cdot 3} && \text{Rewrite as the product of two radicals.}\\ &= 3\sqrt[3]{12} && \text{Simplify.}\end{aligned}$$

There is an alternative approach for simplifying radical expressions. Suppose we were trying to simplify $\sqrt[5]{a^{48}}$. Using the previous method, we would arrive at the answer $a^9\sqrt[5]{a^3}$.

$$\begin{aligned}\sqrt[5]{a^{48}} &= \sqrt[5]{a^{45}\cdot a^3} && \text{Rewrite } a^{48} \text{ as } a^{45}\cdot a^3 \text{, since 45 is the highest multiple of 5 that is less than or equal to 48.}\\ &= \sqrt[5]{a^{45}}\cdot\sqrt[5]{a^3} && \text{Rewrite as the product of two radicals.}\\ &= a^9\sqrt[5]{a^3} && \text{Simplify.}\end{aligned}$$

We know that for every five times that a is repeated as a factor in the radicand, we can take a^5 out of the radicand and write it as a in front of the radical. We need to determine how many groups of five can be removed from the radicand, using division. Notice that if we divide the exponent 48 by the index 5, the quotient is 9 with a remainder of 3. When we divide the exponent of a factor in the radicand by the index of the radical, the quotient tells us the exponent of the factor removed from the radicand and the remainder tells us the exponent of the factor remaining in the radicand.

An Alternative Approach for Simplifying $\sqrt[n]{x^p}$

- Divide p by n: $\frac{p}{n} = q + \frac{r}{n}$
- The quotient q tells us how many times x will be a factor in front of the radical.
- The remainder r tells us how many times x will remain as a factor in the radicand.

$$\sqrt[n]{x^p} = x^q\sqrt[n]{x^r}$$

EXAMPLE 5 Simplify $\sqrt[6]{a^{31}b^{18}c^5d^{53}}$. (Assume all variables represent nonnegative values.)

Solution

We will work with one factor at a time, beginning with a. The index, 6, divides into the exponent, 31, five times with a remainder of one. This tells us we can write a^5 as a factor in front of the radical and can write a^1, or a, in the radicand.

$$\sqrt[6]{a^{31}b^{18}c^5d^{53}} = a^5\sqrt[6]{ab^{18}c^5d^{53}}$$

For the factor b, $18 \div 6 = 3$ with a remainder of 0. We will write b^3 as a factor in front of the radical, and since the remainder is 0, we will not write b as a factor in the radicand. For the factor c, the index does not divide into 5, so c^5 remains as a factor in the radicand. Finally, for the factor d, $53 \div 6 = 8$ with a remainder of 5. We will write d^8 as a factor in front of the radical and d^5 as a factor in the radicand.

$$\sqrt[6]{a^{31}b^{18}c^5d^{53}} = a^5b^3d^8\sqrt[6]{ac^5d^5}$$

Quick Check 5

Simplify $\sqrt[5]{x^{33}y^6z^{50}w^{18}}$.

Adding and Subtracting Radical Expressions Containing Like Radicals

Objective 2 Add or subtract radical expressions containing like radicals.

Like Radicals

Two radical expressions are called **like radicals** if they have the same index and the same radicand.

The radical expressions $5\sqrt[3]{4x}$ and $9\sqrt[3]{4x}$ are like radicals because they have the same index (3) and the same radicand $(4x)$. Here are some examples of radical expressions that are not like radicals:

$\sqrt{5}$ and $\sqrt[3]{5}$ The two radicals have different indices.

$\sqrt[4]{7x^2y^3}$ and $\sqrt[4]{7x^3y^2}$ The two radicands are different.

We can add and subtract radical expressions by combining like radicals similar to the way we combine like terms. We add or subtract the coefficients in front of the like radicals.

$$6\sqrt{2} + 3\sqrt{2} = 9\sqrt{2}$$

EXAMPLE 6 Simplify $5\sqrt[3]{4x} - 15\sqrt[3]{4x}$.

Solution

Notice that the radicals are like radicals. We can combine these two expressions by subtracting the coefficients.

$$5\sqrt[3]{4x} - 15\sqrt[3]{4x} = -10\sqrt[3]{4x}$$ Combine the like radicals by subtracting the coefficients and keeping the radical.

Quick Check 6 Simplify $7\sqrt[5]{2x^2} + 14\sqrt[5]{2x^2}$.

EXAMPLE 7 Simplify $12\sqrt{13} + \sqrt{3} + \sqrt{3} - 6\sqrt{13}$.

Solution

There are two pairs of like radicals in this example. There are two radical expressions containing $\sqrt{13}$ and two radical expressions containing $\sqrt{3}$.

$12\sqrt{13} + \sqrt{3} + \sqrt{3} - 6\sqrt{13}$

$= 6\sqrt{13} + \sqrt{3} + \sqrt{3}$ Subtract $12\sqrt{13} - 6\sqrt{13}$.

$= 6\sqrt{13} + 2\sqrt{3}$ Add $\sqrt{3} + \sqrt{3}$.

Quick Check 7 Simplify $9\sqrt{10} - 13\sqrt{5} + 6\sqrt{10} + 8\sqrt{5}$.

EXAMPLE 8 Simplify $18\sqrt[5]{x^3y^2} - 2\sqrt[5]{x^2y^3} - 8\sqrt[5]{x^3y^2}$.

Solution

Of the three terms, only the first and third contain like radicals.

$$18\sqrt[5]{x^3y^2} - 2\sqrt[5]{x^2y^3} - 8\sqrt[5]{x^3y^2} = 10\sqrt[5]{x^3y^2} - 2\sqrt[5]{x^2y^3}$$

Subtract $18\sqrt[5]{x^3y^2} - 8\sqrt[5]{x^3y^2}$.

Quick Check 8 Simplify $16\sqrt[5]{a^4b^3} - 11\sqrt[7]{a^4b^3} - 5\sqrt[5]{a^4b^3}$.

Objective 3 **Simplify radical expressions before adding or subtracting.** Are the expressions $\sqrt{24}$ and $\sqrt{54}$ like radicals? We must simplify each radical completely before we can determine whether the two expressions are like radicals. In this case, $\sqrt{24} = 2\sqrt{6}$ and $\sqrt{54} = 3\sqrt{6}$, so $\sqrt{24}$ and $\sqrt{54}$ are like radicals.

EXAMPLE 9 Simplify $\sqrt{12} + \sqrt{3}$.

Solution

We begin by simplifying each radical completely. Since 12 can be written as $4 \cdot 3$, $\sqrt{12}$ can be simplified to be $2\sqrt{3}$. We could also use the prime factorization of 12 $(2^2 \cdot 3)$ to simplify $\sqrt{12}$.

$$\begin{aligned} \sqrt{12} + \sqrt{3} &= \sqrt{4 \cdot 3} + \sqrt{3} && \text{Factor 12.} \\ &= 2\sqrt{3} + \sqrt{3} && \text{Simplify } \sqrt{4 \cdot 3}. \\ &= 3\sqrt{3} && \text{Add.} \end{aligned}$$

***Quick Check* 9**
Simplify $\sqrt{63} + \sqrt{7}$.

EXAMPLE 10 Simplify $\sqrt{45} - \sqrt{80} - \sqrt{20}$.

Solution

In this example, we must simplify all three radicals before proceeding.

$$\begin{aligned} \sqrt{45} - \sqrt{80} - \sqrt{20} &= \sqrt{9 \cdot 5} - \sqrt{16 \cdot 5} - \sqrt{4 \cdot 5} && \text{Factor each radicand.} \\ &= 3\sqrt{5} - 4\sqrt{5} - 2\sqrt{5} && \text{Simplify each radical.} \\ &= -3\sqrt{5} && \text{Combine like radicals.} \end{aligned}$$

***Quick Check* 10**
Simplify
$\sqrt{18} - \sqrt{32} + \sqrt{98}$.

Building Your Study Strategy **Doing Your Homework, 3 When You Are Stuck**
Eventually, there will be a homework exercise that you cannot answer correctly. Rather than giving up, here are some options to consider:

- Review your class notes. There may be a similar problem that was discussed in class. If so, you can use the solution of this problem to help you figure out the homework exercise that you were unable to do.
- Review the related section in the text. You may be able to find a similar example, or there may be a "Word of Caution" warning you about typical errors on that type of problem.
- Call someone in your study group. A member of your study group may have already completed the problem and can help you figure out what to do.

If you still cannot solve the problem, move on and try the next problem. Be sure to ask your instructor the very next day about your unsolved problem.

EXERCISES 6.3

Vocabulary

1. Two radical expressions are called __________ if they have the same index and the same radicand.
2. To add radical expressions with like radicals, add the __________ of the radicals and place the sum in front of the like radical.

Simplify.

3. $\sqrt{12}$
4. $\sqrt{18}$
5. $\sqrt{80}$
6. $\sqrt{175}$
7. $\sqrt{363}$
8. $\sqrt{288}$
9. $\sqrt[3]{81}$
10. $\sqrt[3]{48}$
11. $\sqrt[3]{750}$
12. $\sqrt[3]{1029}$
13. $\sqrt[4]{1200}$
14. $\sqrt[5]{448}$

Simplify the radical expression. Assume all variables represent nonnegative real numbers.

15. $\sqrt{x^9}$
16. $\sqrt{a^{15}}$
17. $\sqrt[3]{m^{13}}$
18. $\sqrt[3]{x^{40}}$
19. $\sqrt[5]{x^{74}}$
20. $\sqrt[4]{b^{82}}$
21. $\sqrt{x^{15}y^{12}}$
22. $\sqrt{a^{23}b^3}$
23. $\sqrt{xy^7}$
24. $\sqrt{x^{20}y^{11}}$
25. $\sqrt{x^{33}y^{17}z^{16}}$
26. $\sqrt{r^{19}s^{18}t^{28}}$
27. $\sqrt[3]{a^{12}b^2c^{19}}$
28. $\sqrt[3]{x^{30}y^{21}z^{25}}$
29. $\sqrt[6]{a^{69}b^{35}c^5}$
30. $\sqrt[5]{x^{80}y^{40}z^{11}}$
31. $\sqrt{8a^9b^8}$
32. $\sqrt{180x^{25}y^{21}}$
33. $\sqrt[4]{112a^3b^{13}c^{23}}$
34. $\sqrt[5]{64x^{72}y^{54}z^{104}}$

Add or subtract. Assume all variables represent nonnegative real numbers.

35. $10\sqrt{5} + 17\sqrt{5}$
36. $8\sqrt{11} - 19\sqrt{11}$
37. $9\sqrt[3]{4} - 15\sqrt[3]{4}$
38. $14\sqrt[3]{20} + 6\sqrt[3]{20}$
39. $40\sqrt{21} - 12\sqrt{21} - 33\sqrt{21}$
40. $6\sqrt{7} - 44\sqrt{7} + 16\sqrt{7}$
41. $7\sqrt{10} + 6\sqrt{15} - 12\sqrt{15} + 9\sqrt{10}$
42. $\sqrt{19} + 18\sqrt{6} - 13\sqrt{19} + 24\sqrt{6}$
43. $(9\sqrt{2} + 5\sqrt{3}) - (6\sqrt{2} - 5\sqrt{3})$
44. $(4\sqrt{5} - 3\sqrt{14}) - (8\sqrt{14} - 16\sqrt{5})$
45. $5\sqrt{x} - 7\sqrt{x}$
46. $13\sqrt{y} + 12\sqrt{y}$
47. $3\sqrt[5]{a} + 8\sqrt[5]{a}$
48. $11\sqrt[4]{x} - 2\sqrt[4]{x}$
49. $16\sqrt{x} - 9\sqrt{x} + 15\sqrt{x}$
50. $-24\sqrt{x} - 17\sqrt{x} + 3\sqrt{x}$
51. $4\sqrt{x} - 11\sqrt{y} - 6\sqrt{y} + \sqrt{x}$
52. $-2\sqrt{a} + 9\sqrt{b} - 18\sqrt{a} - 5\sqrt{b}$
53. $8\sqrt[3]{x} + 7\sqrt[4]{x} - 10\sqrt[4]{x} + 13\sqrt[3]{x}$
54. $\sqrt[4]{ab^3} + 12\sqrt[4]{a^2b^3} + 23\sqrt[4]{ab^3} - 25\sqrt[4]{a^2b^3}$
55. $8\sqrt{2} + \sqrt{50}$
56. $2\sqrt{48} + 5\sqrt{3}$
57. $7\sqrt{28} - 3\sqrt{63} + 16\sqrt{7}$
58. $6\sqrt{8} - 13\sqrt{18} + 9\sqrt{200}$
59. $2\sqrt[3]{81} - 5\sqrt[3]{192} - 10\sqrt[3]{3}$
60. $\sqrt[4]{2} + 6\sqrt[4]{32} + 9\sqrt[4]{512}$

FOR EXTRA HELP

MyMathLab
MathXL
Interactmath.com
MathXL Tutorials on CD
Video Lectures on CD
Tutor Center
Addison-Wesley Math Tutor Center
Student's Solutions Manual

61. $10\sqrt{75} - 3\sqrt{216} + 6\sqrt{96} + 14\sqrt{192}$

62. $13\sqrt{128} + 6\sqrt{169} - 7\sqrt{121} - 11\sqrt{162}$

Find the missing radical expression.

63. $(8\sqrt{3} + 7\sqrt{2}) + (?) = 15\sqrt{3} - 4\sqrt{2}$

64. $(3\sqrt{200} - 6\sqrt{108}) + (?) = 19\sqrt{2} - 50\sqrt{3}$

65. $(3\sqrt{50} - \sqrt{405}) - (?) = 3\sqrt{98} - 13\sqrt{20}$

66. $(4\sqrt{224} - 2\sqrt{360}) - (?) = \sqrt{640} + 4\sqrt{350}$

67. List four expressions involving a sum or difference of radicals that are equivalent to $7\sqrt{2}$.

68. List four expressions involving a sum or difference of radicals that are equivalent to $-6\sqrt{3}$.

69. List four expressions involving a sum or difference of radicals that are equivalent to $9\sqrt{5} + 4\sqrt{3}$. At least one of your terms must contain $\sqrt{20}$.

70. List four expressions involving a sum or difference of radicals that are equivalent to $20\sqrt{2} - 19\sqrt{7}$. At least one of your terms must contain $\sqrt{18}$, and another term must contain $\sqrt{28}$.

Writing in Mathematics

Answer in complete sentences.

71. Explain how to simplify $\sqrt{360}$.

72. Explain how to determine whether radical expressions are like radicals.

73. ***Solutions Manual**** Write a solutions manual page for the following problem:
Add $9\sqrt{12} + 6\sqrt{75}$.

74. ***Newsletter**** Write a newsletter that explains how to simplify a radical of the form $\sqrt[n]{x^p}$.

***See Appendix B for details and sample answers.**

6.4 Multiplying and Dividing Radical Expressions

Objectives

1. **Multiply radical expressions.**
2. **Use the distributive property to multiply radical expressions.**
3. **Multiply radical expressions that have two or more terms.**
4. **Multiply radical expressions that are conjugates.**
5. **Rationalize a denominator that has one term.**
6. **Rationalize a denominator that has two terms.**

Multiplying Two Radical Expressions

Objective 1 Multiply radical expressions. In Section 6.1, we learned how to multiply one radical by another, as well as how to divide one radical by another.

Multiplying Radicals with the Same Index

For any positive integer $n > 1$, if $\sqrt[n]{a}$ and $\sqrt[n]{b}$ are real numbers then

$$\sqrt[n]{x} \cdot \sqrt[n]{y} = \sqrt[n]{xy} \qquad \text{and} \qquad \frac{\sqrt[n]{x}}{\sqrt[n]{y}} = \sqrt[n]{\frac{x}{y}}.$$

In this section, we will build on that knowledge and learn how to multiply and divide expressions containing two or more radicals. We begin with a review of multiplying radicals.

Example 1 Multiply $\sqrt{18} \cdot \sqrt{8}$.

Solution

Since the index of each radical is the same and neither radicand is a perfect square, we begin by multiplying the two radicands.

$$\sqrt{18} \cdot \sqrt{8} = \sqrt{144} \qquad \text{Multiply the radicands.}$$
$$= 12 \qquad \text{Simplify the radical.}$$

Example 2 Multiply $\sqrt[3]{a^{13}b^7c^2} \cdot \sqrt[3]{a^{10}b^8c^{17}}$.

Solution

Since there are powers in each radical greater than the index of that radical, we could simplify each radical first. However, we would then have to multiply and simplify the radical again. A more efficient approach is to multiply first and then simplify only once.

$$\sqrt[3]{a^{13}b^7c^2} \cdot \sqrt[3]{a^{10}b^8c^{17}} = \sqrt[3]{a^{23}b^{15}c^{19}} \qquad \text{Multiply the radicands by adding the exponents for each factor.}$$
$$= a^7b^5c^6\sqrt[3]{a^2c} \qquad \text{Simplify the radical. For each factor, we take out as many groups of 3 as possible.}$$

EXAMPLE 3 Multiply $9\sqrt{6} \cdot 7\sqrt{10}$.

Solution

Since the index is the same for each radical, we can multiply the radicands together. The factors in front of each radical, 9 and 7, will be multiplied by each other as well. After multiplying, we finish by simplifying the radical completely.

$$9\sqrt{6} \cdot 7\sqrt{10} = 63\sqrt{60}$$ Multiply factors in front of the radicals and multiply the radicands.

$$= 63\sqrt{2^2 \cdot 3 \cdot 5}$$ Factor the radicand.

$$= 63 \cdot 2\sqrt{3 \cdot 5}$$ Simplify the radical.

$$= 126\sqrt{15}$$ Multiply.

Quick Check 1

Multiply.

a) $\sqrt{45} \cdot \sqrt{80}$

b) $\sqrt[3]{x^4y^2z} \cdot \sqrt[3]{x^8yz^4}$

c) $4\sqrt{8} \cdot 9\sqrt{6}$

It is important to note that whenever we multiply the square root of an expression by the square root of the same expression, the product is equal to the expression itself, as long as the expression is nonnegative.

Multiplying a Square Root by Itself

For any nonnegative x, $\sqrt{x} \cdot \sqrt{x} = x$.

Using the Distributive Property with Radical Expressions

Objective 2 **Use the distributive property to multiply radical expressions.**

Now we will use the distributive property to multiply radical expressions.

EXAMPLE 4 Multiply $\sqrt{5}(\sqrt{10} - \sqrt{5})$.

Solution

We begin by distributing $\sqrt{5}$ to each term in the parentheses, and then we multiply the radicals as in the previous examples.

$$\sqrt{5}(\sqrt{10} - \sqrt{5}) = \sqrt{5} \cdot \sqrt{10} - \sqrt{5} \cdot \sqrt{5}$$ Distribute $\sqrt{5}$.

$$= \sqrt{50} - 5$$ Multiply. Recall that $\sqrt{5} \cdot \sqrt{5} = 5$.

$$= 5\sqrt{2} - 5$$ Simplify the radical.

Quick Check 2

Multiply $\sqrt{12}(\sqrt{3} + \sqrt{15})$.

EXAMPLE 5 Multiply $\sqrt[4]{x^{11}y^6}(\sqrt[4]{x^5y^{19}} + \sqrt[4]{x^{11}y^2})$. (Assume x and y are nonnegative.)

Solution

We begin by using the distributive property. Since each radical has an index of 4, we can then multiply the radicals.

$$\sqrt[4]{x^{11}y^6}\left(\sqrt[4]{x^5y^{19}} + \sqrt[4]{x^{11}y^2}\right) = \sqrt[4]{x^{11}y^6} \cdot \sqrt[4]{x^5y^{19}} + \sqrt[4]{x^{11}y^6} \cdot \sqrt[4]{x^{11}y^2}$$
Distribute $\sqrt[4]{x^{11}y^6}$.

$$= \sqrt[4]{x^{16}y^{25}} + \sqrt[4]{x^{22}y^8}$$
Multiply by adding exponents for each factor.

$$= x^4y^6\sqrt[4]{y} + x^5y^2\sqrt[4]{x^2}$$
Simplify each radical.

Since the radicals are not like radicals, we cannot simplify this expression any further.

Quick Check 3
Multiply
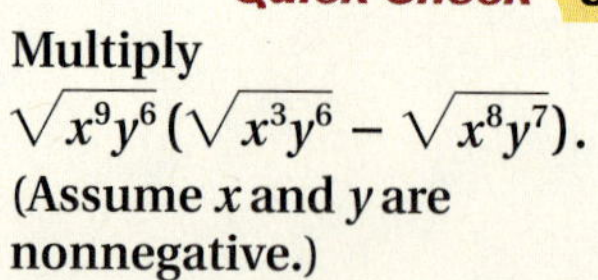
$\sqrt{x^9y^6}\left(\sqrt{x^3y^6} - \sqrt{x^8y^7}\right)$.
(Assume x and y are nonnegative.)

Multiplying Radical Expressions That Have at Least Two Terms

Objective 3 **Multiply radical expressions that have two or more terms.**

EXAMPLE 6 Multiply $(8\sqrt{6} + \sqrt{2})(3\sqrt{12} - 4\sqrt{3})$.

Solution

We begin by multiplying each term in the first set of parentheses by each term in the second set of parentheses, using the distributive property. Since there are two terms in each set of parentheses, we can use the FOIL technique. Multiply factors outside a radical by factors outside a radical, and multiply radicands by radicands.

$$(8\sqrt{6} + \sqrt{2})(3\sqrt{12} - 4\sqrt{3})$$

$$= 8 \cdot 3\sqrt{6 \cdot 12} - 8 \cdot 4\sqrt{6 \cdot 3} + 3\sqrt{2 \cdot 12} - 4\sqrt{2 \cdot 3}$$
Distribute.

$$= 24\sqrt{72} - 32\sqrt{18} + 3\sqrt{24} - 4\sqrt{6}$$
Multiply.

$$= 24 \cdot 6\sqrt{2} - 32 \cdot 3\sqrt{2} + 3 \cdot 2\sqrt{6} - 4\sqrt{6}$$
Simplify each radical.

$$= 144\sqrt{2} - 96\sqrt{2} + 6\sqrt{6} - 4\sqrt{6}$$
Multiply.

$$= 48\sqrt{2} + 2\sqrt{6}$$
Combine like radicals.

EXAMPLE 7 Multiply $(\sqrt{5} + \sqrt{6})^2$.

Solution

To square any binomial, we multiply it by itself.

$$(\sqrt{5} + \sqrt{6})^2 = (\sqrt{5} + \sqrt{6})(\sqrt{5} + \sqrt{6})$$
Square the binomial $\sqrt{5} + \sqrt{6}$ by multiplying it by itself.

$$= \sqrt{5} \cdot \sqrt{5} + \sqrt{5} \cdot \sqrt{6} + \sqrt{6} \cdot \sqrt{5} + \sqrt{6} \cdot \sqrt{6}$$
Distribute.

$$= 5 + \sqrt{30} + \sqrt{30} + 6$$
Multiply.

$$= 11 + 2\sqrt{30}$$
Combine like terms.

Quick Check 4 Multiply. a) $(5\sqrt{3} + 4\sqrt{2})(7\sqrt{3} - 6\sqrt{2})$ b) $(\sqrt{7} - \sqrt{10})^2$

A Word of Caution Whenever we square a binomial, such as $(\sqrt{5} + \sqrt{6})^2$, we must multiply the binomial by itself. We cannot simply square each term.

$$(a + b)^2 \neq a^2 + b^2$$

Multiplying Conjugates

Objective 4 **Multiply radical expressions that are conjugates.** The expressions $\sqrt{13} + \sqrt{5}$ and $\sqrt{13} - \sqrt{5}$ are called **conjugates**. Two expressions are conjugates if they are of the form $x + y$ and $x - y$. Notice that the two terms are the same, with the exception of the sign of the second term.

The multiplication of two conjugates follows a pattern. Let's look at the product $(\sqrt{x} + \sqrt{y})(\sqrt{x} - \sqrt{y})$.

$$(\sqrt{x} + \sqrt{y})(\sqrt{x} - \sqrt{y}) = x - \sqrt{xy} + \sqrt{xy} - y$$ Distribute. Note that $\sqrt{x} \cdot \sqrt{x} = x$ and $\sqrt{y} \cdot \sqrt{y} = y$.

$$= x - y$$ Combine the two opposite terms $-\sqrt{xy}$ and $\sqrt{xy}$.

Whenever we multiply conjugates, the two middle terms will be opposites of each other and therefore their sum is 0. We can multiply the first term in the first set of parentheses by the first term in the second set of parentheses; then multiply the second term in the first set of parentheses by the second term in the second set of parentheses; and, finally, place a minus sign between two products.

Multiplication of Two Conjugates

$$(a + b)(a - b) = a^2 - b^2$$

Example 8 Multiply $(\sqrt{17} + \sqrt{23})(\sqrt{17} - \sqrt{23})$.

Solution

These two expressions are conjugates, so we multiply them accordingly.

$$(\sqrt{17} + \sqrt{23})(\sqrt{17} - \sqrt{23}) = (\sqrt{17})^2 - (\sqrt{23})^2$$ Multiply, using the rule for multiplying conjugates.

$$= 17 - 23$$ Square each square root.

$$= -6$$ Subtract.

EXAMPLE 9 Multiply $(2\sqrt{6} - 8\sqrt{5})(2\sqrt{6} + 8\sqrt{5})$.

Solution

When we multiply two conjugates, we must remember to multiply the factors in front of the radicals by each other and to multiply the radicands by each other.

$$(2\sqrt{6} - 8\sqrt{5})(2\sqrt{6} + 8\sqrt{5}) = 2\sqrt{6} \cdot 2\sqrt{6} - 8\sqrt{5} \cdot 8\sqrt{5}$$ Multiply first term by first term, second term by second term.

$$= 4 \cdot 6 - 64 \cdot 5$$ Multiply.

$$= 24 - 320$$ Multiply.

$$= -296$$ Subtract.

Quick Check 5 Multiply. a) $(\sqrt{38} - \sqrt{29})(\sqrt{38} + \sqrt{29})$ b) $(8\sqrt{11} - 5\sqrt{7})(8\sqrt{11} + 5\sqrt{7})$

Rationalizing the Denominator

Objective 5 **Rationalize a denominator that has one term.** Earlier in this chapter, we introduced a criterion for determining whether a radical was simplified. We stated that for a radical to be simplified, its index must be greater than any power within the radical. There are two other rules that we now add:

- There can be no fractions in a radicand.
- There can be no radicals in the denominator of a fraction.

For example, we would not consider the following expressions simplified: $\sqrt{\frac{3}{10}}$, $\frac{9}{\sqrt{2}}$, $\frac{6}{\sqrt{4} - \sqrt{3}}$, and $\frac{\sqrt{6} + \sqrt{12}}{\sqrt{3} - 8}$. The process of rewriting an expression without a radical in its denominator is called **rationalizing the denominator**.

The rational expression $\frac{\sqrt{16}}{\sqrt{49}}$ is not simplified, as there is a radical in the denominator. However, we know that $\sqrt{49} = 7$, so we can simplify the denominator in such a way that it no longer contains a radical.

$$\frac{\sqrt{16}}{\sqrt{49}} = \frac{4}{7}$$ Simplify the numerator and denominator.

The radical expression $\sqrt{\frac{75}{3}}$ is not simplified, as there is a fraction inside the radical. We can simplify $\frac{75}{3}$ to be 25, rewriting the radical without a fraction inside.

$$\sqrt{\frac{75}{3}} = \sqrt{25} = 5$$ Simplify the fraction first, then $\sqrt{25}$.

Suppose that we needed to simplify $\frac{\sqrt{15}}{\sqrt{2}}$. We cannot simplify $\sqrt{2}$, and the fraction itself cannot be simplified either. In such a case, we will multiply both the numerator and denominator by an expression that will allow us to rewrite the denominator without a radical. Then we simplify.

EXAMPLE 10 Rationalize the denominator: $\frac{\sqrt{15}}{\sqrt{2}}$.

Solution

If we multiply the denominator by $\sqrt{2}$, the denominator would equal 2 and will be rationalized.

$$\frac{\sqrt{15}}{\sqrt{2}} = \frac{\sqrt{15}}{\sqrt{2}} \cdot \frac{\sqrt{2}}{\sqrt{2}}$$ Multiply by $\frac{\sqrt{2}}{\sqrt{2}}$, which makes the denominator equal to 2. Multiplying by $\frac{\sqrt{2}}{\sqrt{2}}$ is equivalent to multiplying by 1.

$$= \frac{\sqrt{30}}{2}$$ Multiply.

Since $\sqrt{30}$ cannot be simplified, this expression cannot be simplified further.

***Quick Check* 6**
Rationalize the denominator: $\frac{\sqrt{70}}{\sqrt{3}}$.

EXAMPLE 11 Rationalize the denominator: $\frac{11}{\sqrt{12}}$.

Solution

At first glance, we might think that multiplying the numerator and denominator by $\sqrt{12}$ is the correct way to proceed. However, if we multiply the numerator and denominator by $\sqrt{3}$, the radicand in the denominator will be 36, which is a perfect square.

$$\frac{11}{\sqrt{12}} = \frac{11}{\sqrt{12}} \cdot \frac{\sqrt{3}}{\sqrt{3}}$$ Multiply by $\frac{\sqrt{3}}{\sqrt{3}}$ to make the radicand in the denominator a perfect square.

$$= \frac{11\sqrt{3}}{\sqrt{36}}$$ Multiply.

$$= \frac{11\sqrt{3}}{6}$$ Simplify the radical in the denominator.

Multiplying by $\frac{\sqrt{12}}{\sqrt{12}}$ would also be valid, but the subsequent process of simplifying would be difficult. One way to determine the best expression by which to multiply is to completely factor the radicand in the denominator. In this example, $12 = 2^2 \cdot 3$. The factor 2 is already a perfect square, but the factor 3 is not. Multiplying by $\sqrt{3}$ makes the factor 3 a perfect square as well.

***Quick Check* 7**
Rationalize the denominator: $\frac{2}{\sqrt{20}}$.

EXAMPLE 12 Rationalize the denominator: $\sqrt{\frac{5a^3b^{14}c^6}{80a^9b^7c^{11}}}$. (Assume all variables represent nonnegative values.)

Solution

Notice that the numerator and denominator have common factors. We will begin by simplifying the fraction to lowest terms.

$$\sqrt{\frac{5a^3b^{14}c^6}{80a^9b^7c^{11}}} = \sqrt{\frac{b^7}{16a^6c^5}}$$ Divide out common factors and simplify.

$$= \frac{\sqrt{b^7}}{\sqrt{16a^6c^5}}$$ Rewrite as the quotient of two square roots. Notice that the factors 16 and a^6 are already perfect squares, but c^5 is not.

$$= \frac{\sqrt{b^7}}{\sqrt{16a^6c^5}} \cdot \frac{\sqrt{c}}{\sqrt{c}}$$ Multiply by $\frac{\sqrt{c}}{\sqrt{c}}$.

$$= \frac{\sqrt{b^7c}}{\sqrt{16a^6c^6}}$$ Multiply.

$$= \frac{b^3\sqrt{bc}}{4a^3c^3}$$ Simplify both radicals.

***Quick Check* 8**
Rationalize the denominator: $\sqrt{\frac{12x^2y^8z^5}{75x^5y^3z^{15}}}$. (Assume all variables represent nonnegative values.)

Rationalizing a Denominator That Has Two Terms

Objective 6 **Rationalize a denominator that has two terms.** In the previous examples, each denominator had only one term. If a denominator is a binomial that contains one or two square roots, then we rationalize the denominator by multiplying the numerator and denominator by the conjugate of the denominator. For example, consider the expression $\frac{6}{\sqrt{11}+\sqrt{7}}$. We know from earlier in the section that multiplying $\sqrt{11}+\sqrt{7}$ by its conjugate $\sqrt{11}-\sqrt{7}$ will produce a product that does not contain a radical.

EXAMPLE 13 Rationalize the denominator: $\frac{6}{\sqrt{11}+\sqrt{7}}$.

Solution

Since this denominator is a binomial, we will multiply the numerator and denominator by the conjugate of the denominator.

$$\frac{6}{\sqrt{11}+\sqrt{7}} = \frac{6}{\sqrt{11}+\sqrt{7}} \cdot \frac{\sqrt{11}-\sqrt{7}}{\sqrt{11}-\sqrt{7}}$$ Multiply the numerator and denominator by the conjugate of the denominator $(\sqrt{11}-\sqrt{7})$.

$$= \frac{6\sqrt{11}-6\sqrt{7}}{\sqrt{11}\cdot\sqrt{11}-\sqrt{7}\cdot\sqrt{7}}$$ Multiply. Since $\sqrt{11}+\sqrt{7}$ and $\sqrt{11}-\sqrt{7}$ are conjugates, we only need to multiply $\sqrt{11}$ by $\sqrt{11}$ and $\sqrt{7}$ by $\sqrt{7}$.

$$= \frac{6\sqrt{11}-6\sqrt{7}}{11-7}$$ Simplify the products in the denominator.

$$= \frac{6\sqrt{11} - 6\sqrt{7}}{4} \quad \text{Subtract.}$$

$$= \frac{6(\sqrt{11} - \sqrt{7})}{4} \quad \text{Factor the numerator.}$$

$$= \frac{\overset{3}{\cancel{6}}(\sqrt{11} - \sqrt{7})}{\underset{2}{\cancel{4}}} \quad \text{Divide out the common factor 2.}$$

$$= \frac{3(\sqrt{11} - \sqrt{7})}{2} \quad \text{Simplify.}$$

***Quick Check* 9**
Rationalize the denominator: $\frac{\sqrt{15}}{\sqrt{5} + \sqrt{3}}$.

EXAMPLE 14 Rationalize the denominator: $\frac{4\sqrt{3} - 3\sqrt{5}}{2\sqrt{3} - \sqrt{5}}$.

Solution

Since the denominator is a binomial, we begin by multiplying the numerator and denominator by the conjugate of the denominator, which is $2\sqrt{3} + \sqrt{5}$.

$$\frac{4\sqrt{3} - 3\sqrt{5}}{2\sqrt{3} - \sqrt{5}} = \frac{4\sqrt{3} - 3\sqrt{5}}{2\sqrt{3} - \sqrt{5}} \cdot \frac{2\sqrt{3} + \sqrt{5}}{2\sqrt{3} + \sqrt{5}}$$

Multiply the numerator and denominator by the conjugate of the denominator.

$$= \frac{4\sqrt{3} \cdot 2\sqrt{3} + 4\sqrt{3} \cdot \sqrt{5} - 3\sqrt{5} \cdot 2\sqrt{3} - 3\sqrt{5} \cdot \sqrt{5}}{2\sqrt{3} \cdot 2\sqrt{3} - \sqrt{5} \cdot \sqrt{5}}$$

Multiply. We must multiply each term in the first numerator by each term in the second numerator. To multiply the denominators, we can take advantage of the fact that the two denominators are conjugates.

$$= \frac{8 \cdot 3 + 4\sqrt{15} - 6\sqrt{15} - 3 \cdot 5}{4 \cdot 3 - 5} \quad \text{Simplify each product.}$$

$$= \frac{24 + 4\sqrt{15} - 6\sqrt{15} - 15}{12 - 5} \quad \text{Multiply.}$$

$$= \frac{9 - 2\sqrt{15}}{7} \quad \text{Combine like terms and like radicals.}$$

***Quick Check* 10**
Rationalize the denominator: $\frac{2\sqrt{6} - 7\sqrt{2}}{2\sqrt{6} + 3\sqrt{2}}$.

Building Your Study Strategy **Doing Your Homework, 4 Homework Diary**
Once you finish the last exercise of a homework assignment, it is not necessarily time to stop. Try summarizing what you have just accomplished in a homework diary. In this diary you can list the types of problems you have solved, the types of problems you struggled with, and important notes about these problems. When you are ready to begin preparing for a quiz or exam, this diary will help you to decide where to focus your attention. If you are using note cards as a study aide, this would be a good time to create a set of study cards for this section. You now know which problems are more difficult for you, as well as which procedures you will need to know in the future.

Summarizing your homework efforts in these ways will help you to retain what you learned from the assignment.

Exercises 6.4

Vocabulary

1. For any nonnegative x, $\sqrt{x} \cdot \sqrt{x} =$ ________.

2. Two expressions of the form $x + y$ and ________ are called conjugates.

3. The process of rewriting an expression without a radical in its denominator is called ________ the denominator.

4. To rationalize a denominator containing two terms and at least one square root, multiply the numerator and denominator by the ________ of the denominator.

Multiply. Assume all variables represent nonnegative real numbers.

5. $\sqrt{4} \cdot \sqrt{9}$
6. $\sqrt{25} \cdot \sqrt{81}$
7. $\sqrt{2} \cdot \sqrt{18}$
8. $\sqrt{3} \cdot \sqrt{48}$
9. $\sqrt[3]{5} \cdot \sqrt[3]{25}$
10. $\sqrt[3]{12} \cdot \sqrt[3]{42}$
11. $5\sqrt{8} \cdot 3\sqrt{2}$
12. $6\sqrt{12} \cdot 3\sqrt{75}$
13. $\sqrt{x^{11}} \cdot \sqrt{x^5}$
14. $\sqrt{x^{21}} \cdot \sqrt{x}$
15. $\sqrt{m^{10}n^7} \cdot \sqrt{m^9n^9}$
16. $\sqrt{s^3t} \cdot \sqrt{s^5t^{19}}$
17. $\sqrt[4]{x^8y^{13}} \cdot \sqrt[4]{x^{33}y^{15}}$
18. $\sqrt[7]{a^{22}b^{20}} \cdot \sqrt[7]{a^{34}b^{19}}$

Multiply. Assume all variables represent nonnegative real numbers.

19. $\sqrt{2}(\sqrt{8} - \sqrt{6})$
20. $\sqrt{6}(\sqrt{2} + \sqrt{3})$
21. $\sqrt{98}(\sqrt{18} + \sqrt{10})$
22. $\sqrt{15}(\sqrt{30} - \sqrt{35})$
23. $\sqrt{10}(3\sqrt{6} - 8\sqrt{15})$
24. $12\sqrt{15}(7\sqrt{3} - 11\sqrt{5})$
25. $\sqrt{m^5n^9}(\sqrt{m^3n^4} + \sqrt{m^8n^7})$
26. $\sqrt{xy^{11}}(\sqrt{x^7y^3} - \sqrt{x^{39}y^{15}})$
27. $\sqrt[3]{a^7b^{10}c^{13}}(\sqrt[3]{a^{11}b^{35}} + \sqrt[3]{a^{23}c^8})$
28. $\sqrt[3]{x^4y^7z^{10}}(\sqrt[3]{x^8y^{16}z^{24}} - \sqrt[3]{x^{16}y^{12}z^2})$

Multiply.

29. $(\sqrt{2} + \sqrt{5})(\sqrt{2} - \sqrt{3})$
30. $(\sqrt{3} + \sqrt{8})(\sqrt{6} - \sqrt{2})$
31. $(\sqrt{10} - \sqrt{5})(\sqrt{10} + \sqrt{5})$
32. $(\sqrt{14} + \sqrt{6})(\sqrt{7} - \sqrt{54})$
33. $(5\sqrt{3} + \sqrt{2})(6\sqrt{2} - 11\sqrt{3})$
34. $(5\sqrt{12} + \sqrt{5})(3\sqrt{20} + 2\sqrt{21})$
35. $(9 + 3\sqrt{8})(\sqrt{6} - 7\sqrt{50})$
36. $(3 + 4\sqrt{7})(5\sqrt{14} - 8)$
37. $(4\sqrt{5} - 3\sqrt{2})^2$
38. $(6 + 7\sqrt{3})^2$

Multiply the conjugates.

39. $(\sqrt{7} - \sqrt{10})(\sqrt{7} + \sqrt{10})$
40. $(\sqrt{26} + \sqrt{3})(\sqrt{26} - \sqrt{3})$
41. $(9\sqrt{6} + 3\sqrt{8})(9\sqrt{6} - 3\sqrt{8})$
42. $(5\sqrt{18} - 2\sqrt{24})(5\sqrt{18} + 2\sqrt{24})$
43. $(8 - 11\sqrt{2})(8 + 11\sqrt{2})$
44. $(4\sqrt{3} + 15)(4\sqrt{3} - 15)$

Simplify. Assume all variables represent nonnegative real numbers.

45. $\dfrac{\sqrt{36}}{\sqrt{25}}$
46. $\dfrac{\sqrt{49}}{\sqrt{100}}$
47. $\sqrt{\dfrac{12}{x^4y^6}}$
48. $\sqrt{\dfrac{b^7}{a^{10}c^2}}$

FOR EXTRA HELP

MyMathLab

MathXL

Interactmath.com

MathXL Tutorials on CD

Video Lectures on CD

Tutor Center

Addison-Wesley Math Tutor Center

Student's Solutions Manual

49. $\sqrt{\dfrac{126}{7}}$

50. $\sqrt{\dfrac{100}{5}}$

51. $\sqrt{\dfrac{135}{20}}$

52. $\sqrt{\dfrac{14}{18}}$

Rationalize the denominator and simplify. Assume all variables represent nonnegative real numbers.

53. $\dfrac{\sqrt{8}}{\sqrt{3}}$

54. $\dfrac{\sqrt{24}}{\sqrt{7}}$

55. $\dfrac{1}{\sqrt{2}}$

56. $\dfrac{5}{\sqrt{5}}$

57. $\sqrt{\dfrac{27}{11a^7}}$

58. $\sqrt{\dfrac{80}{3n^{10}}}$

59. $\sqrt[3]{\dfrac{s^8}{2r^2t}}$

60. $\sqrt[3]{\dfrac{2x^{10}}{7y^{16}z^{22}}}$

61. $\dfrac{15}{\sqrt{18}}$

62. $\dfrac{2\sqrt{3}}{\sqrt{32}}$

63. $\dfrac{ab^6}{\sqrt{a^5b^4c^7}}$

64. $\dfrac{2x^8}{\sqrt{10xy^3}}$

Rationalize the denominator.

65. $\dfrac{9}{\sqrt{13}-\sqrt{7}}$

66. $\dfrac{10}{\sqrt{20}+\sqrt{2}}$

67. $\dfrac{4\sqrt{3}}{\sqrt{3}+\sqrt{11}}$

68. $\dfrac{5\sqrt{5}}{\sqrt{15}-\sqrt{5}}$

69. $\dfrac{6\sqrt{3}}{\sqrt{13}-3}$

70. $\dfrac{8\sqrt{2}}{4-\sqrt{6}}$

71. $\dfrac{\sqrt{3}+\sqrt{8}}{\sqrt{6}-\sqrt{2}}$

72. $\dfrac{\sqrt{5}+\sqrt{2}}{\sqrt{10}-\sqrt{2}}$

73. $\dfrac{\sqrt{8}+\sqrt{32}}{\sqrt{8}-\sqrt{32}}$

74. $\dfrac{\sqrt{13}-\sqrt{11}}{\sqrt{11}-\sqrt{13}}$

75. $\dfrac{4\sqrt{2}-\sqrt{3}}{2\sqrt{2}-3\sqrt{3}}$

76. $\dfrac{\sqrt{12}-9\sqrt{10}}{6\sqrt{3}-4\sqrt{5}}$

Mixed Practice, 77–94

Simplify. Assume all variables represent nonnegative real numbers.

77. $\sqrt{\dfrac{90}{98}}$

78. $\sqrt{10x^5}\cdot\sqrt{18x^9}$

79. $(8\sqrt{7}+5\sqrt{5})^2$

80. $\dfrac{28}{\sqrt{32}}$

81. $\sqrt[3]{25a^8bc^7}\cdot\sqrt[3]{25ab^5c^7}$

82. $\sqrt[3]{\dfrac{2x^4y^{13}}{54x^{10}y^2}}$

83. $\dfrac{10\sqrt{5}-7\sqrt{7}}{\sqrt{5}+\sqrt{7}}$

84. $\sqrt{60}(7\sqrt{3}-2\sqrt{15})$

85. $(3\sqrt{10}-7\sqrt{2})(4\sqrt{10}+9\sqrt{5})$

86. $\dfrac{8\sqrt{2}}{\sqrt{10}}$

87. $-\sqrt{7x^9}(2\sqrt{14x}-\sqrt{7x^{13}})$

88. $(9\sqrt{7}-8\sqrt{8})(9\sqrt{7}+8\sqrt{8})$

89. $\sqrt{\dfrac{a^5b^8}{12c^9}}$

90. $\dfrac{13\sqrt{5}+6}{10-7\sqrt{5}}$

91. $\dfrac{10x^2}{\sqrt[3]{36x^8}}$

92. $(20\sqrt{3} - 7\sqrt{10})(8\sqrt{3} + 15\sqrt{2})$

93. $(18\sqrt{7} - \sqrt{11})(18\sqrt{7} + \sqrt{11})$

94. $(4\sqrt{14} - 3\sqrt{7})^2$

95. Develop a general formula for the product $(\sqrt{a} + \sqrt{b})^2$.

96. Develop a general formula for the product $(a\sqrt{b} + c\sqrt{d})^2$.

97. List four different pairs of conjugates whose product is 20.

98. List four different pairs of conjugates whose product is 38.

99. List a sum of two radical expressions whose square is equivalent to $98 + 40\sqrt{6}$.

100. List a sum of two radical expressions whose square is equivalent to $23 - 6\sqrt{10}$.

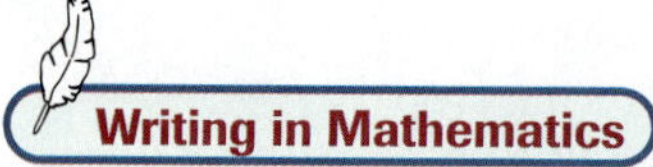

Answer in complete sentences.

101. Explain how to determine whether two radical expressions are conjugates.

102. Explain how to multiply conjugates. Use an example to illustrate the process.

103. ***Solutions Manual**** Write a solutions manual page for the following problem:

Simplify $(5\sqrt{2} + 4\sqrt{3})(8\sqrt{2} - 3\sqrt{3})$.

104. ***Newsletter**** Write a newsletter that explains how to rationalize a denominator containing two terms and at least one square root.

*See Appendix B for details and sample answers.

Quick Review Exercises

Section 6.4

Find the prime factorization of the given number.

1. 144
2. 1400
3. 1024
4. 29,106

6.5 Radical Equations and Applications of Radical Equations

Objectives

1. **Solve radical equations.**
2. **Solve equations containing radical functions.**
3. **Solve equations containing rational exponents.**
4. **Solve equations in which a radical is equal to a variable expression.**
5. **Solve equations containing two radicals.**
6. **Solve applied problems involving a pendulum and its period.**
7. **Solve other applied problems involving radicals.**

Solving Radical Equations

Objective 1 **Solve radical equations.** A **radical equation** is an equation containing one or more radicals. Here are some examples of radical equations:

$$\sqrt{x} = 9 \qquad \sqrt[3]{2x-5} = 3 \qquad x + \sqrt{x} = 20$$
$$\sqrt[4]{3x-8} = \sqrt[4]{2x+11} \qquad \sqrt{x-4} + \sqrt{x+8} = 6$$

In this section, we will learn how to solve radical equations. We will find a way to convert a radical equation to an equivalent equation that we already know how to solve.

Raising Equal Numbers to the Same Power

If two numbers a and b are equal, then for any n, $a^n = b^n$.

If we raise two equal numbers to the same power, they remain equal to each other. We will use this fact to solve radical equations.

Solving Radical Equations

- Isolate a radical term containing the variable on one side of the equation.
- Raise both sides of the equation to the nth power, where n is the index of the radical. *For any nonnegative number x and any integer $n > 1$,* $(\sqrt[n]{x})^n = x$.
- If the resulting equation does not contain a radical, solve the equation. If the resulting equation does contain a radical, begin the process again by isolating the radical on one side of the equation.
- Check the solution(s).

It is crucial that we check all solutions when solving a radical equation, as raising both sides of an equation to an even power can introduce **extraneous solutions**. An extraneous solution is a solution of the equation that results when we raise both sides to a certain power, but it is not a solution of the original equation. A value may be a solution of the equation we obtain after raising both sides of the equation to the nth power, without being a solution of the original equation. If both sides of an equation are raised to an even power, the potential solutions must be checked in the original equation.

EXAMPLE 1 Solve $\sqrt{x-5} = 3$.

Solution

Since the radical $\sqrt{x-5}$ is already isolated on the left side of the equation, we may begin by squaring both sides of the equation.

$$\sqrt{x-5} = 3$$
$$(\sqrt{x-5})^2 = 3^2 \quad \text{Square both sides.}$$
$$x - 5 = 9 \quad \text{Simplify.}$$
$$x = 14 \quad \text{Add 5.}$$

We need to check this solution against the original equation.

$$\sqrt{14-5} = 3 \quad \text{Substitute 14 for } x.$$
$$\sqrt{9} = 3 \quad \text{Subtract.}$$
$$3 = 3 \quad \text{Simplify the square root.}$$

The solution $x = 14$ checks, so the solution set is $\{14\}$.

EXAMPLE 2 Solve $\sqrt[3]{6x+4} + 7 = 11$.

Solution

In this example, we begin by isolating the radical.

$$\sqrt[3]{6x+4} + 7 = 11$$
$$\sqrt[3]{6x+4} = 4 \quad \text{Subtract 7 to isolate the radical.}$$
$$(\sqrt[3]{6x+4})^3 = 4^3 \quad \text{Raise both sides of the equation to the third power.}$$
$$6x + 4 = 64 \quad \text{Simplify.}$$
$$6x = 60 \quad \text{Subtract 4.}$$
$$x = 10 \quad \text{Divide both sides by 6.}$$

Now we check the solution, using the original equation.

$$\sqrt[3]{6(10)+4} + 7 = 11 \quad \text{Substitute 10 for } x.$$
$$\sqrt[3]{64} + 7 = 11 \quad \text{Simplify the radicand.}$$
$$4 + 7 = 11 \quad \text{Simplify the cube root.}$$
$$11 = 11 \quad \text{Add. The solution checks.}$$

The solution set is $\{10\}$. Because we raised both sides of the equation to an odd power, we did not introduce any extraneous solutions.

***Quick Check* 1**

Solve.

a) $\sqrt{x+2} = 7$

b) $\sqrt[3]{x+9} + 10 = 6$

EXAMPLE 3 Solve $\sqrt{x} + 8 = 5$.

Solution

We begin by isolating $\sqrt{x}$ on the left side of the equation.

$$\sqrt{x} + 8 = 5$$
$$\sqrt{x} = -3 \quad \text{Subtract 8 to isolate the square root.}$$
$$(\sqrt{x})^2 = (-3)^2 \quad \text{Square both sides of the equation.}$$
$$x = 9 \quad \text{Simplify.}$$

Now we check this solution, using the original equation.

$\sqrt{9} + 8 = 5$	Substitute 9 for x.
$3 + 8 = 5$	Simplify the square root. The principal square root of 9 is 3, not -3.
$11 = 5$	Add.

This solution does not check, so it is an extraneous solution. The equation has no solution; the solution set is $\varnothing$.

If we obtain an equation in which an even root is equal to a negative number, such as $\sqrt{x} = -3$, this equation will not have any solutions. This is because the principal even root of a number, if it exists, cannot be negative.

Quick Check 2

Solve $\sqrt{2x-9} - 8 = -11$.

Solving Equations Involving Radical Functions

Objective 2 Solve equations containing radical functions.

EXAMPLE 4 For $f(x) = \sqrt{2x+11} - 6$, find all values x for which $f(x) = 3$.

Solution

We begin by setting the function equal to 3.

$f(x) = 3$	
$\sqrt{2x+11} - 6 = 3$	Replace $f(x)$ by its formula.
$\sqrt{2x+11} = 9$	Add 6 to isolate the radical.
$(\sqrt{2x+11})^2 = 9^2$	Square both sides.
$2x + 11 = 81$	Simplify.
$2x = 70$	Subtract 11.
$x = 35$	Divide both sides by 2.

It is left to the reader to verify that the solution is not an extraneous solution: $f(35) = 3$.

Quick Check 3

For $f(x) = \sqrt{3x-8} + 4$, find all values x for which $f(x) = 9$.

Solving Equations with Rational Exponents

Objective 3 Solve equations containing rational exponents. The next examples involve equations containing fractional exponents. Recall that $x^{1/n} = \sqrt[n]{x}$.

EXAMPLE 5 Solve $x^{1/3} + 7 = -3$.

Solution

We begin by rewriting $x^{1/3}$ as $\sqrt[3]{x}$.

$x^{1/3} + 7 = -3$	
$\sqrt[3]{x} + 7 = -3$	Rewrite $x^{1/3}$ using radical notation as $\sqrt[3]{x}$.
$\sqrt[3]{x} = -10$	Subtract 7 to isolate the radical.
$(\sqrt[3]{x})^3 = (-10)^3$	Raise both sides to the third power.
$x = -1000$	Simplify.

It is left to the reader to verify that the solution is not an extraneous solution. The solution set is $\{-1000\}$.

Example 6 Solve $(1 - 5x)^{1/2} - 3 = 3$.

Solution

We begin by rewriting the equation using a radical.

$$(1 - 5x)^{1/2} - 3 = 3$$
$$\sqrt{1 - 5x} - 3 = 3 \qquad \text{Rewrite, using radical notation.}$$
$$\sqrt{1 - 5x} = 6 \qquad \text{Add 3 to isolate the radical.}$$
$$(\sqrt{1 - 5x})^2 = 6^2 \qquad \text{Square both sides.}$$
$$1 - 5x = 36 \qquad \text{Simplify.}$$
$$-5x = 35 \qquad \text{Subtract 1.}$$
$$x = -7 \qquad \text{Divide both sides by } -5.$$

It is left to the reader to verify that the solution is not an extraneous solution. The solution set is $\{-7\}$.

Quick Check 4

Solve.

a) $x^{1/2} - 10 = -7$
b) $(x + 4)^{1/3} + 8 = 2$

Solving Equations in Which a Radical Is Equal to a Variable Expression

Objective 4 **Solve equations in which a radical is equal to a variable expression.** After we isolated the radical in all of the previous examples, the resulting equation had a radical expression equal to a constant. In the next example, we will learn how to solve equations that result in a radical equal to a variable expression.

Example 7 Solve $\sqrt{6x + 16} = x$.

Solution

Since the radical is already isolated, we begin by squaring both sides. This will result in a quadratic equation, which we solve by collecting all terms on one side of the equation and factoring.

$$\sqrt{6x + 16} = x$$
$$(\sqrt{6x + 16})^2 = x^2 \qquad \text{Square both sides.}$$
$$6x + 16 = x^2 \qquad \text{Simplify.}$$
$$0 = x^2 - 6x - 16 \qquad \text{Collect all terms on the right side of the equation by subtracting } 6x \text{ and } 16.$$
$$0 = (x - 8)(x + 2) \qquad \text{Factor.}$$
$$x = 8 \quad \text{or} \quad x = -2 \qquad \text{Set each factor equal to 0 and solve.}$$

Now we check both solutions.

$x = 8$	$x = -2$
$\sqrt{6(8) + 16} = 8$	$\sqrt{6(-2) + 16} = -2$
$\sqrt{48 + 16} = 8$	$\sqrt{-12 + 16} = -2$
$\sqrt{64} = 8$	$\sqrt{4} = -2$
$8 = 8$	$2 = -2$
True	False

The solution $x = -2$ is an extraneous solution. The solution set is $\{8\}$.

Quick Check 5

Solve $\sqrt{12x - 20} = x$.

EXAMPLE 8 Solve $\sqrt{x} + 6 = x$.

Solution

We begin by isolating the radical.

$\sqrt{x} + 6 = x$	
$\sqrt{x} = x - 6$	Subtract 6 to isolate the radical.
$(\sqrt{x})^2 = (x - 6)^2$	Square both sides.
$x = (x - 6)(x - 6)$	Square the binomial by multiplying it by itself.
$x = x^2 - 12x + 36$	Multiply. The resulting equation is quadratic.
$0 = x^2 - 13x + 36$	Subtract x to collect all terms on the right side of the equation.
$0 = (x - 4)(x - 9)$	Factor.
$x = 4$ or $x = 9$	Set each factor equal to 0 and solve.

Now we check both solutions.

$x = 4$	$x = 9$
$\sqrt{4} + 6 = 4$	$\sqrt{9} + 6 = 9$
$2 + 6 = 4$	$3 + 6 = 9$
$8 = 4$	$9 = 9$
False	True

Quick Check 6
Solve $\sqrt{2x + 4} = x$.

The solution $x = 4$ is an extraneous solution. The solution set is $\{9\}$.

Solving Radical Equations Containing Two Radicals

Objective 5 **Solve equations containing two radicals.**

EXAMPLE 9 Solve $\sqrt[5]{6x + 5} = \sqrt[5]{4x - 3}$.

Solution

We will raise both sides to the fifth power. Since both radicals have the same index, this will result in an equation that does not contain a radical.

$\sqrt[5]{6x + 5} = \sqrt[5]{4x - 3}$	
$(\sqrt[5]{6x + 5})^5 = (\sqrt[5]{4x - 3})^5$	Raise both sides to the fifth power.
$6x + 5 = 4x - 3$	Simplify.
$2x + 5 = -3$	Subtract $4x$ from both sides.
$2x = -8$	Subtract 5.
$x = -4$	Divide both sides by 2.

Quick Check 7
Solve
$\sqrt[3]{5x - 11} = \sqrt[3]{7x + 33}$.

It is left to the reader to verify that the solution is not an extraneous solution. This solution checks, and the solution set is $\{-4\}$.

Occasionally, equations containing two square roots will still contain a square root after we have squared both sides. This will require us to square both sides a second time.

EXAMPLE 10 Solve $\sqrt{x+6} - \sqrt{x-1} = 1$.

Solution

We must begin by isolating one of the two radicals on the left side of the equation. We will isolate $\sqrt{x+6}$, as it is positive.

$$\sqrt{x+6} - \sqrt{x-1} = 1$$

$$\sqrt{x+6} = 1 + \sqrt{x-1}$$ Add $\sqrt{x-1}$ to isolate the radical $\sqrt{x+6}$ on the left side.

$$(\sqrt{x+6})^2 = (1 + \sqrt{x-1})^2$$ Square both sides.

$$x + 6 = (1 + \sqrt{x-1})(1 + \sqrt{x-1})$$ Square the binomial on the right side by multiplying it by itself.

$$x + 6 = 1 \cdot 1 + 1 \cdot \sqrt{x-1} + 1 \cdot \sqrt{x-1} + \sqrt{x-1} \cdot \sqrt{x-1}$$ Distribute.

$$x + 6 = 1 + 2\sqrt{x-1} + x - 1$$ Simplify.

$$x + 6 = 2\sqrt{x-1} + x$$ Combine like terms.

$$6 = 2\sqrt{x-1}$$ Subtract x to isolate the radical.

$$3 = \sqrt{x-1}$$ Divide both sides by 2.

$$3^2 = (\sqrt{x-1})^2$$ Square both sides.

$$9 = x - 1$$ Simplify.

$$10 = x$$ Add 1.

It is left to the reader to verify that the solution is not an extraneous solution. The solution set is $\{10\}$.

Quick Check 8
Solve.
$\sqrt{x+3} - \sqrt{x-2} = 1.$

A Pendulum and Its Period

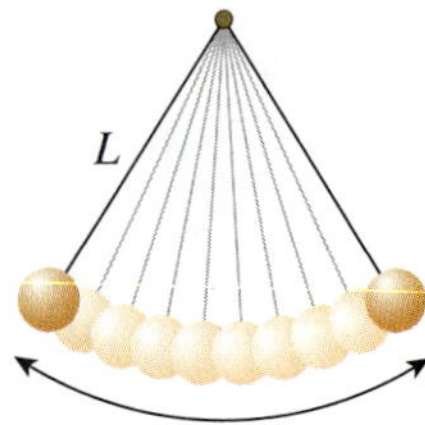

Objective 6 **Solve applied problems involving a pendulum and its period.**

The **period** of a pendulum is the amount of time it takes to swing from one extreme to the other and then back again. The period T of a pendulum in seconds can be found by the formula $T = 2\pi\sqrt{\frac{L}{32}}$, where L is the length of the pendulum in feet.

EXAMPLE 11 A pendulum has a length of 3 feet. Find its period, rounded to the nearest hundredth of a second.

Solution

We substitute 3 for L in the formula and simplify to find the period T.

$$T = 2\pi\sqrt{\frac{L}{32}}$$

$$T = 2\pi\sqrt{\frac{3}{32}}$$ Substitute 3 for L.

$$T \approx 1.92$$ Approximate, using a calculator.

The period of a pendulum that is 3 feet long is approximately 1.92 seconds.

Quick Check 9
A pendulum has a length of 6 feet. Find its period, rounded to the nearest hundredth of a second.

EXAMPLE 12 If a pendulum has a period of 1 second, find its length in feet. Round to the nearest hundredth of a foot.

Solution

In this example, we substitute 1 for T and solve for L. To solve this equation for L, we isolate the radical and then square both sides.

$$T = 2\pi\sqrt{\frac{L}{32}}$$

$$1 = 2\pi\sqrt{\frac{L}{32}} \qquad \text{Substitute 1 for } T.$$

$$\frac{1}{2\pi} = \frac{2\pi\sqrt{\frac{L}{32}}}{2\pi} \qquad \text{Divide by } 2\pi \text{ to isolate the radical.}$$

$$\frac{1}{2\pi} = \sqrt{\frac{L}{32}} \qquad \text{Simplify.}$$

$$\left(\frac{1}{2\pi}\right)^2 = \left(\sqrt{\frac{L}{32}}\right)^2 \qquad \text{Square both sides.}$$

$$\frac{1}{4\pi^2} = \frac{L}{32} \qquad \text{Simplify.}$$

$$32 \cdot \frac{1}{4\pi^2} = 32 \cdot \frac{L}{32} \qquad \text{Multiply by 32 to isolate } L.$$

$$\overset{8}{\cancel{32}} \cdot \frac{1}{\underset{1}{\cancel{4}}\pi^2} = \overset{1}{\cancel{32}} \cdot \frac{L}{\underset{1}{\cancel{32}}} \qquad \text{Divide out common factors.}$$

$$\frac{8}{\pi^2} = L \qquad \text{Simplify.}$$

$$L \approx 0.81 \qquad \text{Approximate, using a calculator.}$$

The length of the pendulum is approximately 0.81 feet.

***Quick Check* 10**

If a pendulum has a period of 3 seconds, find its length in feet. Round to the nearest hundredth of a foot.

Other Applications Involving Radicals

Objective 7 **Solve other applied problems involving radicals.**

EXAMPLE 13 A vehicle made 150 feet of skid marks on the asphalt before crashing. The speed, s, in miles per hour, that the vehicle was traveling when it started skidding can be approximated by the formula $s = \sqrt{30df}$, where d represents the length of the skid marks in feet and f represents the drag factor of the road. If the drag factor for asphalt is 0.75, find the speed the car was traveling. Round to the nearest mile per hour.

Solution

We begin by substituting 150 for d and 0.75 for f.

$s = \sqrt{30df}$

$s = \sqrt{30(150)(0.75)}$ Substitute 150 for d and 0.75 for f.

$s = \sqrt{3375}$ Simplify the radicand.

$s \approx 58$ Approximate, using a calculator.

The car was traveling at approximately 58 miles per hour when it started skidding.

Quick Check **11** **A vehicle made 215 feet of skid marks on the asphalt before crashing. The speed, s, that the vehicle was traveling in miles per hour when it started skidding can be approximated by the formula $s = \sqrt{30df}$, where d represents the length of the skid marks in feet and f represents the drag factor of the road. If the drag factor for asphalt is 0.75, find the speed that the car was traveling when it started skidding. Round to the nearest mile per hour.**

Building Your Study Strategy **Doing Your Homework, 5 Regular Schedule** Try to establish a regular schedule for doing your homework. Some students find it advantageous to work on their homework as soon as possible after class while the material is still fresh in their minds. You may be able to find other students in your class who can work with you in the library. If you have to wait until the evening to do your homework, try to complete your assignment before you get tired. If you have more than one subject to work on, begin with math.

EXERCISES 6.5

Vocabulary

1. A(n) __________ equation is an equation containing one or more radicals.

2. If two numbers a and b are equal, then for any n, $a^n =$ __________.

3. To solve a radical equation, first __________ one radical containing the variable on one side of the equation.

4. A(n) __________ solution is a solution to the equation that results when we raise both sides to a certain power, but it is not a solution to the original equation.

5. An object suspended from a support so that it swings freely back and forth under the influence of gravity is called a(n) __________.

6. The __________ of a pendulum is the amount of time it takes the pendulum to swing from one extreme to the other and then back again.

Solve. Check for extraneous solutions.

7. $\sqrt{x + 2} = 5$
8. $\sqrt{x - 4} = 4$
9. $\sqrt[3]{2x - 5} = 3$
10. $\sqrt[4]{3x - 8} = 2$
11. $\sqrt{5x - 19} = -6$
12. $\sqrt[3]{2x + 15} = -1$
13. $\sqrt{4x - 3} - 9 = -4$
14. $\sqrt{7x + 13} + 19 = 11$
15. $\sqrt[6]{x - 6} - 9 = -10$
16. $\sqrt[3]{x + 7} - 5 = -3$
17. $\sqrt{x^2 + 3x - 3} = 5$
18. $\sqrt{x^2 - 9} = 4$
19. For the function $f(x) = \sqrt{3x + 9}$, find all values x for which $f(x) = 12$.
20. For the function $f(x) = \sqrt{x - 8} + 7$, find all values x for which $f(x) = 13$.
21. For the function $f(x) = \sqrt[3]{5x - 1} + 5$, find all values x for which $f(x) = 9$.
22. For the function $f(x) = \sqrt[4]{2x - 3} - 7$, find all values x for which $f(x) = -4$.
23. For the function $f(x) = \sqrt{x^2 - 5x + 2} + 10$, find all values x for which $f(x) = 14$.
24. For the function $f(x) = \sqrt{x^2 + 8x + 40} - 3$, find all values x for which $f(x) = 2$.

Solve. Check for extraneous solutions.

25. $x^{1/2} + 8 = 10$
26. $x^{1/3} - 11 = -20$
27. $(x + 10)^{1/2} - 4 = 3$
28. $(2x - 35)^{1/2} + 6 = 17$
29. $(5x + 6)^{1/3} - 8 = -2$
30. $(x^2 - 10x + 49)^{1/2} + 2 = 7$
31. $x = \sqrt{2x + 48}$
32. $\sqrt{3x + 10} = x$
33. $\sqrt{4x + 13} = x - 2$
34. $x + 9 = \sqrt{6x + 46}$
35. $\sqrt{2x - 5} + 4 = x$
36. $\sqrt{3x + 13} - 3 = x$
37. $x = \sqrt{49 - 8x} + 7$
38. $x = \sqrt{2x + 9} - 5$
39. $2x - 3 = \sqrt{30 - 7x}$
40. $3x + 5 = \sqrt{27x + 27}$
41. $3x = 1 + \sqrt{4x^2 + x + 7}$
42. $x = \sqrt{54 + 5x - x^2} - 3$
43. $\sqrt{4x - 15} = \sqrt{3x + 11}$
44. $\sqrt[3]{6x + 7} = \sqrt[3]{x - 5}$
45. $\sqrt[4]{x^2 - 8x + 4} = \sqrt[4]{3x - 14}$
46. $\sqrt{5x^2 + 3x - 11} = \sqrt{4x^2 - 6x - 25}$
47. $\sqrt{3x^2 + 6x + 10} = \sqrt{x^2 + 5x + 55}$
48. $\sqrt{8x^2 - 3x - 15} = \sqrt{2x^2 - 8x - 11}$
49. $\sqrt{x + 4} = \sqrt{x - 1} + 1$
50. $\sqrt{x + 14} - \sqrt{x - 10} = 2$
51. $\sqrt{x - 9} + \sqrt{5x - 14} = 7$

52. $\sqrt{4x - 15} + 3 = \sqrt{6x}$

53. $\sqrt{3x - 5} = 5 + \sqrt{2x + 3}$

54. $\sqrt{5x + 1} = 2 - \sqrt{5x - 1}$

For Exercises 55–60, use the formula $T = 2\pi\sqrt{\frac{L}{32}}$.

55. A pendulum has a length of 4 feet. Find its period, rounded to the nearest hundredth of a second.

56. A pendulum has a length of 6 feet. Find its period, rounded to the nearest hundredth of a second.

57. A pendulum has a length of 2.5 feet. Find its period, rounded to the nearest hundredth of a second.

58. A pendulum has a length of 1.4 feet. Find its period, rounded to the nearest hundredth of a second.

59. If a pendulum has a period of 0.6 second, find its length in feet. Round to the nearest tenth of a foot.

60. If a pendulum has a period of 1.3 seconds, find its length in feet. Round to the nearest tenth of a foot.

If a pendulum has a length of L inches, then its period in seconds, T, can be found by the formula $T = 2\pi\sqrt{\frac{L}{384}}$.

61. A pendulum has a length of 17 inches. Find its period, rounded to the nearest hundredth of a second.

62. If a pendulum has a period of 0.45 second, find its length in inches. Round to the nearest tenth of an inch.

If a pendulum has a length of L meters, then its period in seconds, T, can be found by the formula $T = 2\pi\sqrt{\frac{L}{9.8}}$.

63. A pendulum has a length of 3 meters. Find its period, rounded to the nearest hundredth of a second.

64. If a pendulum has a period of 1.2 seconds, find its length in meters. Round to the nearest tenth of a meter.

Skid-mark analysis is one way to estimate the speed a car was traveling prior to an accident. The speed, s, that the vehicle was traveling in miles per hour can be approximated by the formula $s = \sqrt{30df}$, ***where d represents the length of the skid marks in feet and f represents the drag factor of the road.***

65. A vehicle that was involved in an accident made 70 feet of skid marks on the asphalt before crashing. If the drag factor for asphalt is 0.75, find the speed the car was traveling when it started skidding. Round to the nearest mile per hour.

66. A vehicle that was involved in an accident made 240 feet of skid marks on the asphalt before crashing. If the drag factor for asphalt is 0.75, find the speed the car was traveling when it started skidding. Round to the nearest mile per hour.

67. A vehicle that was involved in an accident made 185 feet of skid marks on a concrete road before crashing. If the drag factor for concrete is 0.95, find the speed the car was traveling when it started skidding. Round to the nearest mile per hour.

68. A vehicle that was involved in an accident made 60 feet of skid marks on a concrete road before crashing. If the drag factor for concrete is 0.95, find the speed the car was traveling when it started skidding. Round to the nearest mile per hour.

A water tank has a hole at the bottom, and the rate r at which water flows out of the hole in gallons per minute can be found by the formula $r = 19.8\sqrt{d}$, ***where d represents the depth of the water in the tank, in feet.***

69. Find the rate of water flow if the depth of water in the tank is 49 feet.

70. Find the rate of water flow if the depth of water in the tank is 9 feet.

71. Find the rate of water flow if the depth of water in the tank is 13 feet. Round to the nearest tenth of a gallon per minute.

72. Find the rate of water flow if the depth of water in the tank is 22 feet. Round to the nearest tenth of a gallon per minute.

73. If water is flowing out of the tank at a rate of 30 gallons per minute, find the depth of water in the tank. Round to the nearest tenth of a foot.

74. If water is flowing out of the tank at a rate of 100 gallons per minute, find the depth of water in the tank. Round to the nearest tenth of a foot.

The hull speed of a sailboat is the maximum speed that the hull can attain from wind power. The hull speed, h, in knots can be calculated by the formula $h = 1.34\sqrt{L}$, where L is the length of the water line in feet.

75. Find the hull speed of a sailboat with a waterline of 36 feet.

76. Find the hull speed of a sailboat with a waterline of 28 feet. Round to the nearest hundredth of a knot.

77. If the hull speed of a sailboat is 9.5 knots, find the length of its waterline. Round to the nearest tenth of a foot.

78. If the hull speed of a sailboat is 6.4 knots, find the length of its waterline. Round to the nearest tenth of a foot.

A tsunami is a great sea wave caused by a natural phenomenon such as an earthquake or volcanic activity. The speed of a tsunami, s, in miles per hour at any particular point can be found from the formula $s = 308.29\sqrt{d}$, where d is the depth of the ocean in miles at that point.

79. Find the speed of a tsunami if the depth of the water is 3 miles. Round to the nearest mile per hour.

80. Find the speed of a tsunami if the depth of the water is 0.4 mile. Round to the nearest mile per hour.

81. If a tsunami is traveling at 600 miles per hour, what is the depth of the ocean at that location? Round to the nearest tenth of a mile.

82. If a tsunami is traveling at 450 miles per hour, what is the depth of the ocean at that location? Round to the nearest tenth of a mile.

83. List four different radical equations that each have $x = 3$ as a solution.

84. List four different radical equations that each have $x = -9$ as a solution.

Writing in Mathematics

Answer in complete sentences.

85. Write a real-world word problem that involves finding the period of a pendulum. Solve your problem, explaining each step of the process.

86. Write a real-world word problem that involves estimating the speed a car was traveling when it started skidding, based on the length of its skid marks. Solve your problem, explaining each step of the process.

87. What is an extraneous solution to an equation? Explain how to determine that a solution to a radical equation is actually an extraneous solution.

88. ***Solutions Manual**** Write a solutions manual page for the following problem:

Solve $\sqrt{3x - 2} = x - 2$.

89. ***Newsletter**** Write a newsletter that explains how to solve a radical equation.

***See Appendix B for details and sample answers.**

6.6 The Complex Numbers

Objectives

1. **Rewrite square roots of negative numbers as imaginary numbers.**
2. **Add and subtract complex numbers.**
3. **Multiply imaginary numbers.**
4. **Multiply complex numbers.**
5. **Divide by a complex number.**
6. **Divide by an imaginary number.**
7. **Simplify expressions containing powers of *i*.**

Imaginary Numbers

Objective 1 Rewrite square roots of negative numbers as imaginary numbers.

Section 6.1 stated that the square root of a negative number, such as $\sqrt{-1}$ or $\sqrt{-25}$, was not a real number. This is because there is no real number that equals a negative number when it is squared. The square root of a negative number is an **imaginary number**.

We define the **imaginary unit** i to be a number that is equal to $\sqrt{-1}$. The number i has the property that $i^2 = -1$.

Imaginary Unit i

$$i = \sqrt{-1}$$
$$i^2 = -1$$

All imaginary numbers can be expressed in terms of i because $\sqrt{-1}$ is a factor of every imaginary number.

EXAMPLE 1 Express $\sqrt{-25}$ in terms of i.

Solution

Whenever we have a square root with a negative radicand, we begin by factoring out i. Then we simplify the resulting square root.

$\sqrt{-25} = \sqrt{25(-1)}$ Rewrite -25 as $25(-1)$.
$= \sqrt{25} \cdot \sqrt{-1}$ Rewrite as the product of two square roots.
$= \sqrt{25}\, i$ Rewrite $\sqrt{-1}$ as i.
$= 5i$ Simplify the square root.

As you become more experienced at working with imaginary numbers, you may wish to combine a few of the previous steps into one step.

EXAMPLE 2 Express $\sqrt{-40}$ in terms of i.

Solution

$\sqrt{-40} = \sqrt{40} \cdot \sqrt{-1}$ Rewrite as the product of two square roots.
$= 2\sqrt{10} \cdot i$ Simplify the square root. Rewrite $\sqrt{-1}$ as i.
$= 2i\sqrt{10}$ Rewrite with i in front of the radical.

A Word of Caution After rewriting the square root of a negative number, such as $\sqrt{-40}$, as an imaginary number, be sure that i does not appear in the radicand. Instead, i should be written in front of the radical.

EXAMPLE 3 Express $-\sqrt{-12}$ in terms of i.

Solution

Notice that in this example there is a negative sign in front of the square root as well as in the radicand. We simplify the square root first, making the result negative.

$$-\sqrt{-12} = -\sqrt{12} \cdot \sqrt{-1} \quad \text{Rewrite as the product of two square roots.}$$
$$= -2i\sqrt{3} \quad \text{Simplify the square root. Rewrite } \sqrt{-1} \text{ as } i.$$

***Quick Check* 1**

Express in terms of i.

a) $\sqrt{-36}$

b) $\sqrt{-63}$

c) $-\sqrt{-54}$

Complex Numbers

The set of imaginary numbers and the set of real numbers are subsets of the set of **complex numbers**.

Complex Numbers

A **complex number** is a number of the form $a + bi$, where a and b are real numbers.

Real numbers are complex numbers for which $b = 0$, while imaginary numbers are complex numbers for which $a = 0$ but $b \neq 0$. We often associate the word *complex* with something that is difficult, but here *complex* refers to the fact that these numbers are made up of two parts.

Real and Imaginary Parts of a Complex Number

For the complex number $a + bi$, the number a is the **real part** and the number b is the **imaginary part.**

Addition and Subtraction of Complex Numbers

Objective 2 **Add or subtract complex numbers.** We now focus on operations involving complex numbers. We add two complex numbers by adding the two real parts and adding the two imaginary parts. We can follow a similar technique for subtraction.

EXAMPLE 4 Simplify $(6 + 5i) + (7 - 2i)$.

Solution

We begin by removing the parentheses, and then we combine the two real parts and the two imaginary parts of these complex numbers.

$$(6 + 5i) + (7 - 2i) = 6 + 5i + 7 - 2i \quad \text{Remove parentheses.}$$
$$= 13 + 3i \quad \text{Combine the two real parts. Combine the two imaginary parts.}$$

Notice that the process of adding these two complex numbers is similar to simplifying the expression $(6 + 5x) + (7 - 2x)$.

EXAMPLE 5 Simplify $(-2 + 9i) - (8 - 5i)$.

Solution

We must distribute the negative sign to both parts of the second complex number, just as we distributed negative signs when we subtracted variable expressions.

$$(-2 + 9i) - (8 - 5i) = -2 + 9i - 8 + 5i \quad \text{Distribute.}$$
$$= -10 + 14i \quad \text{Combine the two real parts, and combine the two imaginary parts.}$$

Quick Check 2 Simplify.

a) $(3 + 12i) + (-8 + 15i)$
b) $(14 - 6i) - (3 + 22i)$

Multiplying Imaginary Numbers

Objective 3 **Multiply imaginary numbers.** Before learning to multiply two complex numbers, we will discuss the multiplication of two imaginary numbers. Suppose that we wanted to multiply $6i$ by $9i$. Just as $6x \cdot 9x = 54x^2$, the product $6i \cdot 9i$ is equal to $54i^2$. However, recall that $i^2 = -1$. So this product is $54(-1)$, or -54. When multiplying two imaginary numbers, we substitute -1 for i^2.

EXAMPLE 6 Multiply $8i \cdot 13i$.

Solution

$$8i \cdot 13i = 104i^2 \quad \text{Multiply.}$$
$$= 104(-1) \quad \text{Rewrite } i^2 \text{ as } -1.$$
$$= -104 \quad \text{Multiply.}$$

Quick Check 3 Multiply: $4i \cdot 15i$.

EXAMPLE 7 Multiply: $\sqrt{-24} \cdot \sqrt{-45}$.

Solution

Although it may be tempting to multiply -24 by -45 and combine the two square roots into one, we cannot do this. The property $\sqrt{x} \cdot \sqrt{y} = \sqrt{xy}$ holds true only if either x or y is nonnegative. We must rewrite each square root as an imaginary number before multiplying.

$$\sqrt{-24} \cdot \sqrt{-45} = \sqrt{24}\,i \cdot \sqrt{45}\,i \quad \text{Rewrite each radical as an imaginary number.}$$
$$= \sqrt{2^3 \cdot 3}\,i \cdot \sqrt{3^2 \cdot 5}\,i \quad \text{Factor each radicand.}$$

$= \sqrt{2^3 \cdot 3^3 \cdot 5\, i^2}$ Multiply the two radicands.
$= 2 \cdot 3\sqrt{2 \cdot 3 \cdot 5\, i^2}$ Simplify the square root.
$= -6\sqrt{30}$ Rewrite i^2 as -1 and simplify.

When we multiply $\sqrt{24}$ by $\sqrt{45}$, our work will be easier if we factor 24 and 45 before multiplying, rather than trying to simplify $\sqrt{1080}$.

Quick Check 4
Multiply $\sqrt{-5} \cdot \sqrt{-120}$.

Multiplying Complex Numbers

Objective 4 **Multiply complex numbers.** We multiply two complex numbers by using the distributive property. Often, a product of two complex numbers will contain a term with i^2, and we will rewrite i^2 as -1.

EXAMPLE 8 Multiply $3i(4 - 5i)$.

Solution

We begin by multiplying $3i$ by both terms in the parentheses.

$3i(4 - 5i) = 3i \cdot 4 - 3i \cdot 5i$ Distribute.
$= 12i - 15i^2$ Multiply.
$= 12i + 15$ Rewrite i^2 as -1 and simplify.
$= 15 + 12i$ Rewrite in the form $a + bi$.

Quick Check 5
Multiply $-6i(7 + 8i)$.

EXAMPLE 9 Multiply $(9 + i)(2 + 7i)$.

Solution

In this example, we must multiply each term in the first set of parentheses by each term in the second set.

$(9 + i)(2 + 7i) = 9 \cdot 2 + 9 \cdot 7i + i \cdot 2 + i \cdot 7i$ Distribute (FOIL).
$= 18 + 63i + 2i + 7i^2$ Multiply.
$= 18 + 65i - 7$ Add $63i + 2i$. Rewrite i^2 as -1 and simplify.
$= 11 + 65i$ Combine like terms.

Quick Check 6
Multiply $(3 - 2i)(8 + i)$.

EXAMPLE 10 Multiply $(7 + 2i)^2$.

Solution

Recall that we square a binomial by multiplying it by itself.

$(7 + 2i)^2 = (7 + 2i)(7 + 2i)$ Multiply $7 + 2i$ by itself.
$= 7 \cdot 7 + 7 \cdot 2i + 2i \cdot 7 + 2i \cdot 2i$ Distribute.
$= 49 + 14i + 14i + 4i^2$ Multiply.
$= 49 + 28i - 4$ Add $14i + 14i$. Rewrite i^2 as -1 and simplify.
$= 45 + 28i$ Combine like terms.

Quick Check 7
Multiply $(9 - 4i)^2$.

EXAMPLE 11 Multiply $(5 + 6i)(5 - 6i)$.

Solution

$$\begin{aligned}(5 + 6i)(5 - 6i) &= 5 \cdot 5 - 5 \cdot 6i + 6i \cdot 5 - 6i \cdot 6i && \text{Distribute.}\\ &= 25 - 30i + 30i - 36i^2 && \text{Multiply.}\\ &= 25 - 36i^2 && \text{Combine like terms.}\\ &= 25 + 36 && \text{Rewrite } i^2 \text{ as } -1 \text{ and simplify.}\\ &= 61 && \text{Add.}\end{aligned}$$

***Quick Check* 8**
Multiply $(7 + 4i)(7 - 4i)$.

In the previous example, the two complex numbers that were multiplied were conjugates. Their product was a real number and did not have an imaginary part. Two complex numbers of the form $a + bi$ and $a - bi$ are conjugates, and their product will always be equal to $a^2 + b^2$.

$$\begin{aligned}(a + bi)(a - bi) &= a^2 - abi + abi - b^2i^2 && \text{Distribute.}\\ &= a^2 - b^2i^2 && \text{Combine like terms.}\\ &= a^2 + b^2 && \text{Rewrite } i^2 \text{ as } -1 \text{ and simplify.}\end{aligned}$$

EXAMPLE 12 Multiply $(10 - 3i)(10 + 3i)$.

Solution

We will use the fact that $(a + bi)(a - bi) = a^2 + b^2$ for two complex numbers that are conjugates.

$$\begin{aligned}(10 - 3i)(10 + 3i) &= 10^2 + 3^2 && \text{The product equals } a^2 + b^2.\\ &= 100 + 9 && \text{Square 10 and 3.}\\ &= 109 && \text{Add.}\end{aligned}$$

***Quick Check* 9**
Multiply $(11 - 4i)(11 + 4i)$.

Dividing by a Complex Number

Objective 5 **Divide by a complex number.** Since the imaginary number i is a square root $(\sqrt{-1})$, a simplified expression cannot contain i in its denominator. We will use conjugates to rewrite the expression without i in the denominator, in a procedure similar to rationalizing a denominator (Section 6.4).

EXAMPLE 13 Simplify $\dfrac{2}{3 + i}$.

Solution

We begin by multiplying the numerator and denominator by the conjugate of the denominator, which is $3 - i$.

$$\frac{2}{3+i} = \frac{2}{3+i} \cdot \frac{3-i}{3-i}$$ Multiply the numerator and denominator by the conjugate of the denominator.

$$= \frac{6-2i}{3^2+1^2}$$ Multiply the numerators by distributing 2 to both terms in the second numerator. Multiply the denominators by using the fact that $(a+bi)(a-bi) = a^2 + b^2$.

$$= \frac{6-2i}{10}$$ Simplify the denominator.

$$= \frac{\overset{1}{\cancel{2}}(3-i)}{\underset{5}{\cancel{10}}}$$ Factor the numerator. Divide out factors common to the numerator and denominator.

$$= \frac{3-i}{5}$$ Simplify.

$$= \frac{3}{5} - \frac{1}{5}i$$ Rewrite in the form $a + bi$.

Quick Check 10

Simplify $\frac{15}{8+6i}$.

Answers in the back of the text will be written in both forms: as a single fraction and in $a + bi$ form. Your instructor will let you know which form is preferred for your class.

EXAMPLE 14 Simplify $\frac{1+6i}{2+5i}$.

Solution

In this example, the numerator has two terms, and we must multiply by using the distributive property.

$$\frac{1+6i}{2+5i} = \frac{1+6i}{2+5i} \cdot \frac{2-5i}{2-5i}$$ Multiply the numerator and denominator by the conjugate of the denominator.

$$= \frac{2-5i+12i-30i^2}{2^2+5^2}$$ Multiply.

$$= \frac{2+7i-30i^2}{4+25}$$ Combine like terms in the numerator. Square 2 and 5 in the denominator.

$$= \frac{2+7i+30}{29}$$ Rewrite i^2 as -1 and simplify. Simplify the denominator.

$$= \frac{32+7i}{29}$$ Combine like terms.

$$= \frac{32}{29} + \frac{7}{29}i$$ Rewrite in the form $a + bi$.

Quick Check 11

Simplify $\frac{9+2i}{7-3i}$.

Note that the expression $\frac{1+6i}{2+5i}$ is equivalent to $(1+6i) \div (2+5i)$. If we are asked to divide a complex number by another complex number, we begin by rewriting the expression as a fraction and then proceed as in the previous example.

Dividing by an Imaginary Number

Objective 6 **Divide by an imaginary number.** When a denominator is an imaginary number, we multiply the numerator and denominator by i to rewrite the fraction without i in the denominator.

Example 15 Simplify $\dfrac{5+2i}{3i}$.

Solution

Since the denominator is an imaginary number, we will begin by multiplying the fraction by $\frac{i}{i}$.

$$\frac{5+2i}{3i} = \frac{5+2i}{3i} \cdot \frac{i}{i} \quad \text{Multiply the numerator and denominator by } i.$$
$$= \frac{5i+2i^2}{3i^2} \quad \text{Multiply.}$$
$$= \frac{5i-2}{-3} \quad \text{Rewrite } i^2 \text{ as } -1 \text{ and simplify.}$$
$$= -\frac{5i-2}{3} \quad \text{Factor } -1 \text{ out of the denominator.}$$
$$= \frac{-5i+2}{3} \quad \text{Distribute.}$$
$$= \frac{2}{3} - \frac{5}{3}i \quad \text{Rewrite in } a - bi \text{ form.}$$

Quick Check 12

Simplify $\dfrac{15-8i}{12i}$.

Powers of *i*

Objective 7 **Simplify expressions containing powers of *i*.** Occasionally, i may be raised to a power greater than 2. We finish this section by learning to simplify such expressions. The expression i^3 can be rewritten as $i^2 \cdot i$, which is equivalent to $-1 \cdot i$, or $-i$. The expression i^4 can be rewritten as $i^2 \cdot i^2$, which is equivalent to $-1(-1)$, or 1. The following table shows the first four positive powers of i:

i	i^2	i^3	i^4
↓	↓	↓	↓
i	-1	$-i$	1

We can use the fact that $i^4 = 1$ to simplify greater powers of i.

$$i^5 = i^4 \cdot i = 1 \cdot i = i \qquad i^6 = i^4 \cdot i^2 = 1(-1) = -1$$
$$i^7 = i^4 \cdot i^3 = 1 \cdot i^3 = 1(-i) = -i \qquad i^8 = i^4 \cdot i^4 = 1 \cdot 1 = 1$$

We can now see that there is a pattern.

i	i^2	i^3	i^4	i^5	i^6	i^7	i^8
↓	↓	↓	↓	↓	↓	↓	↓
i	-1	$-i$	1	i	-1	$-i$	1

In general, to simplify i^n, where n is a whole number, we divide n by 4. If the remainder is equal to r, then $i^n = i^r$.

Remainder	0	1	2	3
i^n	1	i	-1	$-i$

EXAMPLE 16 Simplify i^{27}.

Solution

If we divide the exponent 27 by 4, the remainder is 3. So, $i^{27} = i^3 = -i$.

***Quick Check* 13**
Simplify i^{30}.

***Building Your Study Strategy* Doing Your Homework, 6 Complete Each Assignment** It is crucial that you are able to complete an assignment before the next class session. Since math is a sequential topic, if you do not understand the material from the previous day, then you will have difficulty understanding the next day's material.

EXERCISES 6.6

Vocabulary

1. The square root of a negative number is a(n) ________ number.
2. The imaginary unit i is a number that is equal to ________.
3. The number i has the property that $i^2 =$ ________.
4. A(n) ________ number is a number of the form $a + bi$, where a and b are real numbers.
5. For the complex number $a + bi$, the number a is the ________ part.
6. For the complex number $a + bi$, the number b is the ________ part.

Express in terms of i.

7. $\sqrt{-4}$
8. $\sqrt{-16}$
9. $\sqrt{-81}$
10. $\sqrt{-100}$
11. $-\sqrt{-121}$
12. $-\sqrt{-196}$
13. $\sqrt{-27}$
14. $\sqrt{-80}$
15. $\sqrt{-32}$
16. $-\sqrt{-735}$

Add or subtract the complex numbers.

17. $(8 + 9i) + (3 + 5i)$
18. $(13 + 11i) - (6 + 2i)$

FOR EXTRA HELP

MyMathLab
MathXL
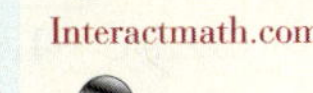
Interactmath.com

MathXL Tutorials on CD

Video Lectures on CD

Addison-Wesley Math Tutor Center
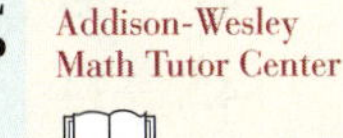

Student's Solutions Manual

19. $(6 - 7i) - (2 + 10i)$

20. $(1 + 6i) + (5 - 12i)$

21. $(-3 + 8i) + (3 - 4i)$

22. $(5 - 9i) + (5 + 9i)$

23. $(12 + 13i) - (12 - 13i)$

24. $(14 - i) - (21 + 11i)$

25. $(1 - 12i) - (8 + 8i) + (5 - 4i)$

26. $(-9 + 2i) - (5 + 6i) - (10 - 20i)$

Find the missing complex number.

27. $(4 - 7i) + ? = 10 + 2i$

28. $(11 + 10i) + ? = 20 - 3i$

29. $(3 + 4i) - ? = 1 - 5i$

30. $? - (-9 + 7i) = -15 - 3i$

Multiply.

31. $7i \cdot 14i$

32. $3i \cdot 8i$

33. $-4i \cdot 5i$

34. $i(-9i)$

35. $\sqrt{-49} \cdot \sqrt{-64}$

36. $\sqrt{-25} \cdot \sqrt{-25}$

37. $\sqrt{-12} \cdot \sqrt{-18}$

38. $\sqrt{-28} \cdot \sqrt{-7}$

39. $\sqrt{-20} \cdot \sqrt{-45}$

40. $\sqrt{-27} \cdot \sqrt{-50}$

Find two imaginary numbers with the given product. (Answers may vary.)

41. -65

42. -14

43. 40

44. 12

Multiply.

45. $6i(8 - 3i)$

46. $4i(2 + 5i)$

47. $-7i(4 - 7i)$

48. $-3i(9 + 8i)$

49. $(8 + i)(3 - 5i)$

50. $(7 + 6i)(7 - 6i)$

51. $(6 - 4i)^2$

52. $(5 - 2i)(8 - 10i)$

53. $(9 - i)(9 + i)$

54. $(10 + 3i)^2$

55. $(1 + 7i)^2$

56. $(5i + 4)(4i - 5)$

Multiply the conjugates, using the fact that $(a + bi)(a - bi) = a^2 + b^2$.

57. $(5 + 2i)(5 - 2i)$

58. $(7 + 4i)(7 - 4i)$

59. $(12 - 8i)(12 + 8i)$

60. $(6 - i)(6 + i)$

Rationalize the denominator.

61. $\dfrac{5}{2 + i}$

62. $\dfrac{4}{5 - 3i}$

63. $\dfrac{15i}{7 - 4i}$

64. $\dfrac{6i}{9 + 5i}$

65. $\dfrac{4 - 9i}{3 + 2i}$

66. $\dfrac{1 + 3i}{6 - i}$

67. $\dfrac{5 + 6i}{5 - 6i}$

68. $\dfrac{7 + 5i}{5 + 7i}$

Rationalize the denominator.

69. $\dfrac{8}{i}$

70. $\dfrac{5}{6i}$

71. $\dfrac{5 - 7i}{3i}$

72. $\dfrac{3 + 10i}{5i}$

73. $\dfrac{6 + i}{-4i}$

74. $\dfrac{2 - 9i}{-8i}$

Divide.

75. $4i \div (7 + 3i)$

76. $(9 + 6i) \div (3i)$

77. $(6 - i) \div (3 + 4i)$

78. $(5 - 10i) \div (12i - 5)$

Find the missing complex number.

79. $(4 + i)(?) = 4 + 35i$

80. $(7 - 5i)(?) = 31 - i$

81. $\dfrac{?}{1 + 9i} = 4 - 3i$

82. $\dfrac{?}{10 - 3i} = 6 - 4i$

Simplify.

83. i^{15}

84. i^{34}

85. i^{200}

86. i^{65}

87. $i^{13} \cdot i^{17}$

88. $\dfrac{i^{51}}{i^{20}}$

89. $i^{39} + i^{29}$

90. $i^{103} - i^{161}$

Mixed Practice, 91–114

Simplify.

91. $(7 + 2i)(6 - 3i)$

92. $(4 - 6i)(4 + 6i)$

93. $\dfrac{9 + 4i}{5i}$

94. $(18 - 13i) - (8 - 5i)$

95. $12i \div (4 - 2i)$

96. $\dfrac{10 - 9i}{5 - i}$

97. $(2 - 5i)(2 + 5i)$

98. $(14 - 11i) + (-23 + 6i)$

99. $\sqrt{-24} \cdot \sqrt{-27}$

100. $(6 - 4i) \div (8i)$

101. $\dfrac{15i}{9 - 2i}$

102. $(16 - 9i)^2$

103. $\sqrt{-252}$

104. $(11 + 4i)(6 - 13i)$

105. $(9 - 3i)^2$

106. $7i \cdot 12i$

107. $(8 - 11i) + (15 + 21i)$

108. i^{207}

109. $-14i \cdot 9i$

110. $\sqrt{-20} \cdot \sqrt{-15}$

111. i^{323}

112. $\dfrac{7 - 6i}{-2i}$

113. $(30 - 17i) - (54 - 33i)$

114. $\sqrt{-2673}$

Writing in Mathematics

Answer in complete sentences.

115. Explain how adding two complex numbers is similar to adding two variable expressions.

116. ***Solutions Manual**** Write a solutions manual page for the following problem: *Simplify* $\sqrt{-84}$.

117. ***Newsletter**** Write a newsletter that explains how to multiply complex numbers.

*See Appendix B for details and sample answers.

Chapter 6 Summary

Section 6.1–Topic	Chapter Review Exercises
Simplifying Radicals	1–6
Approximating Square Roots	7–8
Evaluating Radical Functions	9–10
Finding the Domain of Radical Functions	11–12
Section 6.2–Topic	**Chapter Review Exercises**
Rewriting Radical Expressions by Using Rational Exponents	13–16
Simplifying Expressions with Rational Exponents	17–24
Section 6.3–Topic	**Chapter Review Exercises**
Simplifying Radical Expressions	25–30
Adding and Subtracting Radical Expressions	31–34
Section 6.4–Topic	**Chapter Review Exercises**
Multiplying Radical Expressions	35–40
Rationalizing the Denominator of Radical Expressions	41–50
Section 6.5–Topic	**Chapter Review Exercises**
Solving Radical Equations	51–56
Solving Equations Involving Radical Functions	57–58
Solving Applied Problems Involving Radical Equations	59–60
Section 6.6–Topic	**Chapter Review Exercises**
Simplifying Square Roots of Negative Numbers	61–64
Adding and Subtracting Complex Numbers	65–68
Multiplying Complex Numbers	69–74
Rationalizing Denominators Containing Complex Numbers	75–78
Simplifying Expressions Containing Powers of i	79–80

Summary of Chapter 6 Study Strategies

A homework assignment should not be viewed as a chore to be completed, but rather as an opportunity to increase your understanding of mathematics. Here is a summary of the ideas presented in this chapter:

- Review your notes and the text before beginning a homework assignment.
- Be neat and complete when doing a homework assignment. *If your work is neat, it will be easier for you to spot errors. When you are reviewing for an exam, your homework will be easier to review if it is neat. Do not skip any steps, even if you feel that they are not necessary.*
- Strategies for homework problems you cannot answer:
  - *Look in your notes and the text for similar problems or suggestions about this type of problem.*
  - *Call another student in the class for help.*
  - *Get help from a tutor at the tutorial center on your campus.*
  - *Ask your instructor for help at the first opportunity.*
- Complete a homework session by summarizing your work. *Prepare notes on difficult problems or concepts. These hints will be helpful when you begin to review for an exam.*
- Keep up-to-date with your homework assignments. *Falling behind will make it difficult for you to learn the new material presented in class.*

Simplify the radical expression. Assume all variables represent nonnegative real numbers. **[6.1]**

1. $\sqrt{25}$

2. $\sqrt{169}$

3. $\sqrt[3]{-64}$

4. $\sqrt[3]{x^9}$

5. $\sqrt{81x^{12}}$

6. $\sqrt[5]{243x^{15}y^{10}z^{25}}$

Approximate to the nearest thousandth, using a calculator. **[6.1]**

7. $\sqrt{17}$

8. $\sqrt{41}$

Evaluate the radical function. Round to the nearest thousandth if necessary. **[6.1]**

9. $f(x) = \sqrt{x - 10}, f(91)$

10. $f(x) = \sqrt{x^2 + 6x - 19}, f(3)$

Find the domain of the radical function. Express your answer in interval notation. **[6.1]**

11. $f(x) = \sqrt{2x + 6}$

12. $f(x) = \sqrt[4]{3x + 25}$

Rewrite each radical expression by using rational exponents. Assume all variables represent nonnegative real numbers. **[6.2]**

13. $\sqrt[5]{a}$

14. $(\sqrt[4]{x})^3$

15. $(\sqrt{x})^{13}$

16. $\sqrt[5]{n^8}$

Rewrite as a radical expression and simplify if possible. Assume all variables represent nonnegative real numbers. **[6.2]**

17. $4^{1/2}$

18. $(343x^{21})^{2/3}$

Simplify the expression. Assume all variables represent nonnegative real numbers. **[6.2]**

19. $x^{3/2} \cdot x^{1/6}$

20. $(x^{3/8})^{7/6}$

Worked-out solutions to Review Exercises marked with can be found on page AN-29.

21. $\dfrac{x^{2/3}}{x^{5/12}}$

22. $49^{-1/2}$

23. $64^{-5/3}$

24. $\dfrac{27^{4/3}}{27^{8/3}}$

Simplify the radical expression. Assume all variables represent nonnegative real numbers. **[6.3]**

25. $\sqrt{28}$

26. $\sqrt[3]{x^{10}}$

27. $\sqrt[5]{x^{11}y^8z^{15}}$

28. $\sqrt[3]{405a^{25}b^{12}c^{16}}$

29. $\sqrt{180r^3s^{11}t^{12}}$

30. $\sqrt{175r^6s^8t^4}$

Add or subtract. Assume all variables represent nonnegative real numbers. **[6.3]**

31. $4\sqrt{2} + 8\sqrt{2}$

32. $7\sqrt[3]{x} - \sqrt[3]{x}$

33. $5\sqrt{20} + 3\sqrt{125}$

34. $6\sqrt{12} - 8\sqrt{50} + 9\sqrt{75}$

Multiply. Assume all variables represent nonnegative real numbers. **[6.4]**

35. $\sqrt[3]{4x^2} \cdot \sqrt[3]{18x^4}$

36. $5\sqrt{3}(2\sqrt{6} - 4\sqrt{3})$

37. $(\sqrt{3} + \sqrt{8})(\sqrt{6} - \sqrt{2})$

38. $(10\sqrt{5} - 7\sqrt{11})(9\sqrt{5} + 2\sqrt{11})$

39. $(2\sqrt{15} - 9\sqrt{3})^2$

40. $(2\sqrt{13} - 10\sqrt{7})(2\sqrt{13} + 10\sqrt{7})$

Simplify. Assume all variables represent nonnegative real numbers. **[6.4]**

41. $\dfrac{\sqrt{224x^5}}{\sqrt{2x}}$

42. $\sqrt{\dfrac{50a^9b^8c^3}{8a^3bc^{13}}}$

Rationalize the denominator and simplify. Assume all variables represent nonnegative real numbers. **[6.4]**

43. $\sqrt[3]{\dfrac{16}{5}}$

44. $\dfrac{8}{\sqrt{14}}$

45. $\sqrt{\dfrac{11}{18}}$

46. $\dfrac{a^5bc^3}{\sqrt{a^4b^5c}}$

47. $\dfrac{8}{\sqrt{11}-\sqrt{5}}$

48. $\dfrac{2\sqrt{3}}{10\sqrt{15}-3\sqrt{12}}$

49. $\dfrac{\sqrt{7}+5\sqrt{2}}{2\sqrt{7}-3\sqrt{2}}$

50. $\dfrac{3-\sqrt{5}}{6+7\sqrt{5}}$

Solve. **[6.5]**

51. $\sqrt{4x+1}=7$

52. $\sqrt[3]{7x-15}=5$

53. $\sqrt{3x+22}+2=x$

54. $\sqrt{17-4x}-x=1$

55. $\sqrt{x+5}=\sqrt{3x-13}$

56. $\sqrt{x+8}=2-\sqrt{x-12}$

57. For the function $f(x)=\sqrt{3x+4}$, find all values of x for which $f(x)=4$. [6.5]

58. For the function $f(x)=\sqrt[3]{x^2-6x+11}+8$, find all values of x for which $f(x)=11$. [6.5]

59. A pendulum has a length of 5 feet. Find its period, rounded to the nearest hundredth of a second. Use the formula $T=2\pi\sqrt{\dfrac{L}{32}}$. [6.5]

60. A vehicle involved in an accident made 200 feet of skid marks on the asphalt. The speed, s, that the vehicle was traveling in miles per hour when it started skidding can be approximated by the formula $s=\sqrt{30df}$, where d represents the length of the skid marks in feet and f represents the drag factor of the road. If the drag factor for asphalt is 0.75, find the speed the car was traveling. Round to the nearest mile per hour. [6.5]

Express in terms of i. **[6.6]**

61. $\sqrt{-49}$

62. $\sqrt{-40}$

63. $\sqrt{-252}$

64. $-\sqrt{-675}$

Add or subtract the complex numbers. **[6.6]**

65. $(4+2i)+(7+3i)$

66. $(9-4i)+(7+2i)$

67. $(11-8i)-(10-13i)$

68. $(6+12i)-(5+5i)$

Multiply. **[6.6]**

69. $2i\cdot 8i$

70. $\sqrt{-6}\cdot\sqrt{-50}$

71. $7i(3-4i)$

72. $(10+2i)(13+8i)$

73. $(9-3i)^2$

74. $(3+14i)(3-14i)$

Rationalize the denominator. **[6.6]**

75. $\dfrac{6}{7-3i}$

76. $\dfrac{15i}{6+8i}$

77. $\dfrac{2+i}{9-5i}$

78. $\dfrac{16}{i}$

Simplify. **[6.6]**

79. i^{53}

80. i^{22}

Chapter 6 Test

For Extra Help — Pass the Test — Test solutions are found on the enclosed CD.

Simplify the radical expression. Assume all variables represent nonnegative real numbers.

1. $\sqrt{25x^{10}y^{14}}$

2. For $f(x) = \sqrt{x^2 - 9x + 6}$, evaluate $f(-5)$. Round to the nearest thousandth.

Simplify the expression. Assume all variables represent nonnegative real numbers.

3. $a^{3/10} \cdot a^{1/5}$

4. $\dfrac{b^{7/8}}{b^{5/12}}$

5. $125^{-2/3}$

Simplify the radical expression. Assume all variables represent nonnegative real numbers.

6. $\sqrt{72}$

7. $\sqrt[4]{a^{21}b^{42}c^4}$

Add or subtract. Assume all variables represent nonnegative real numbers.

8. $8\sqrt{18} + 11\sqrt{2} - 3\sqrt{200}$

Multiply. Assume all variables represent nonnegative real numbers.

9. $4\sqrt{5}(7\sqrt{10} - 2\sqrt{35})$

10. $(3\sqrt{6} - 4\sqrt{8})(9\sqrt{6} - 2\sqrt{8})$

Rationalize the denominator and simplify. Assume all variables represent nonnegative real numbers.

11. $\dfrac{x^3y^2}{\sqrt{x^3y^{11}}}$

12. $\dfrac{15\sqrt{7} - 8\sqrt{2}}{4\sqrt{7} + 9\sqrt{2}}$

Solve.

13. $\sqrt[3]{6x + 5} = 5$

14. $\sqrt{x + 28} + 2 = x$

15. If a pendulum has a period of 4 seconds, find its length, rounded to the nearest tenth of a foot. Use the formula $T = 2\pi\sqrt{\dfrac{L}{32}}$.

16. Express $\sqrt{-99}$ in terms of i.

Add or subtract the complex numbers.

17. $(2 - 3i) - (20 + 8i)$

Multiply.

18. $(6 + 7i)(6 - 7i)$

Rationalize the denominator.

19. $\dfrac{4 + 9i}{3 - 5i}$

Mathematicians in History

Pythagoras of Samos

Pythagoras of Samos, often referred to simply as Pythagoras, was the leader of a society known as the Pythagoreans. This society was partially religious and partially scientific in nature. One of their mathematical achievements is the first proof of a theorem that related the lengths of the sides of a right triangle. This theorem is known as the Pythagorean theorem.

Write a one-page summary (*or* make a poster) of the life of Pythagoras, the mathematical achievements of Pythagoras and his society, and the beliefs of the Pythagoreans.

Interesting issues:

- What was the "semicircle"?
- Who were the *mathematikoi*, and by what rules did they lead their lives?
- What number did the Pythagoreans consider to be the "best" number, and why?
- Which society was the first to know of the Pythagorean theorem?
- The Pythagoreans believed that all relationships could be expressed numerically as a ratio of two integers. They discovered, however, that the diagonal of a square whose side has length 1 is an irrational number. This number is $\sqrt{2}$. This discovery rocked the foundation of their system of beliefs, and they swore each other to secrecy regarding this discovery. A Pythagorean named Hippasus told others outside of the society of this irrational number. What was the fate of Hippasus?

Have you ever watched a child swing back and forth on a playground swing? Do you remember seeing in the movies a person swinging a pocket-watch back and forth in order to put another person into a trance? This motion of the swing and pocket-watch is called a pendulum. Today, you will be creating your own pendulum and using the data you collect to discover a mathematical model.

Step 1: Form groups of no more than four or five students. Each group should have a yardstick or ruler, a piece of string (approximately 2 yards long), a weight tied to the end of the string (hardware washers work well), and a stopwatch (or some method for counting seconds).

Step 2: Measure a length for your pendulum (in inches) and fasten your string to the edge of a deck or table at this length. For example, you might want to start with a 36-inch pendulum. Write the length of your pendulum in the following table:

Length of Pendulum (Inches)	Number of Seconds for 10 Swings					Average Time
	Trial 1	Trial 2	Trial 3	Trial 4	Trial 5	

Step 3: Pull the pendulum back and release. Calculate the number of seconds it takes for your pendulum to make 10 swings. (Note that a swing is one complete back-and-forth motion.) Record the time in the chart. Repeat this process three to five times for the same length. Calculate the average time needed for this given length.

Step 4: Change the length of your pendulum and record the length in the chart. Record the time needed for 10 swings. Repeat this process three to five times for the same length. Calculate the average time needed for this given length.

Step 5: Repeat step 4 a few more times for different lengths of the pendulum.

Step 6: What did you notice about the length of the pendulum in relation to the length of time needed for 10 swings? Did the time increase as the length increased? Or did the time decrease as the length increased?

Step 7: Fill in the following table:

Length of Pendulum	Square Root Length of Pendulum	Average Time for 10 Swings	Average Time for 1 Swing	Average Time for 1 Swing Divided by Square Root of Length of Pendulum

Step 8: What do you notice about the last column? As it turns out, these values should be relatively the same and should be around 0.32. There is actually a formula that uses square roots and that relates the time it takes for a pendulum to make one swing and the length of the pendulum. This formula is $T = 2\pi\sqrt{\frac{L}{384}}$, where T is the time for one swing and L is the length of the pendulum in inches.

Step 9: See how close your times were to the times the formula would predict. Were your actual times close to the predicted times?

CHAPTER Seven 7

Quadratic Equations

In this chapter, we will learn how to apply techniques other than factoring to solve quadratic equations. **Quadratic equations** ***are equations that can be written in the form $ax^2 + bx + c = 0$, where $a \neq 0$. One important goal of this chapter is determining which technique will provide the most efficient solution to a particular equation. We will also apply these techniques to new application problems that previously could not be solved because they produce equations which are not factorable. In this chapter, we will also examine the graphs of quadratic equations and functions. The graph of a quadratic equation is called a parabola and is U shaped. The techniques used to graph linear equations will also be applied in graphing quadratic equations, along with some new techniques. After examining inequalities that involve quadratic and rational expressions, the chapter ends by examining the graphs of other functions.***

Study Strategy **Time Management** *When asked why they are having difficulties in a particular class, many students claim that they simply do not have enough time. However, a great number of these students have enough time, but lack the time-management skills to make the most of their time. It seems as if they fritter and waste the hours in an offhand way. In this chapter, we will discuss effective time-management strategies, focusing on how to make more efficient use of available time.*

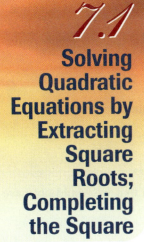

7.1 Solving Quadratic Equations by Extracting Square Roots; Completing the Square

Objectives

1. **Solve quadratic equations by factoring.**
2. **Solve quadratic equations by extracting square roots.**
3. **Solve quadratic equations by extracting square roots involving a linear expression that is squared.**
4. **Solve applied problems by extracting square roots.**
5. **Solve quadratic equations by completing the square.**
6. **Find a quadratic equation, given its solutions.**

Solving Quadratic Equations by Factoring

Objective 1 Solve quadratic equations by factoring. In Chapter 4, we learned how to solve quadratic equations through the use of factoring. We begin this section with a review of this technique. We start by simplifying both sides of the equation completely. This includes any distributing that must be done and combining like terms when possible. After that, we need to collect all of the terms on one side of the equation with 0 on the other side of the equation. Recall that it is a good idea to move the terms to the side of the equation that will produce a positive second-degree term. After the equation contains an expression equal to 0, we need to factor the expression. Finally, we set each factor equal to 0 and solve the resulting equation.

EXAMPLE 1 Solve $x^2 - 7x = -10$.

Solution

$$x^2 - 7x = -10$$

$x^2 - 7x + 10 = 0$ — Add 10 to collect all terms on the left side.

$(x - 2)(x - 5) = 0$ — Factor.

$x - 2 = 0$ or $x - 5 = 0$ — Set each factor equal to 0.

$x = 2$ or $x = 5$ — Solve the resulting equations.

The solution set is $\{2, 5\}$.

EXAMPLE 2 Solve $x^2 = 49$.

Solution

$$x^2 = 49$$

$x^2 - 49 = 0$ — Set the equation equal to 0.

$(x + 7)(x - 7) = 0$ — Factor. (Difference of squares)

$x + 7 = 0$ or $x - 7 = 0$ — Set each factor equal to 0.

$x = -7$ or $x = 7$ — Solve the resulting equations.

The solution set is $\{-7, 7\}$.

***Quick Check* 1**

Solve.

a) $x^2 = 2x + 15$
b) $x^2 = 121$

Solving Quadratic Equations by Extracting Square Roots

Objective 2 Solve quadratic equations by extracting square roots. In the preceding example, we were trying to find a number x that, when squared, equals 49. The principal square root of 49 is 7, which gives us one of the two solutions. In an equation where we have a squared term equal to a constant, we can take the square root of both sides of the equation to find the solution, as long as we take both the positive and negative square root of the constant. The symbol $\pm$ is used to represent both the positive and negative square root and is read as "plus or minus." For example, $x = \pm 7$ means $x = 7$ or $x = -7$. This technique for solving quadratic equations is called **extracting square roots**.

Extracting Square Roots

1. Isolate the squared term.
2. Take the square root of each side. (Remember to take both the *positive* and *negative* $(\pm)$ square root of the constant.)
3. Simplify the square root.
4. Solve by isolating the variable.

Now we will use extracting square roots to solve an equation that we would not have been able to solve by factoring.

EXAMPLE 3 Solve $x^2 - 50 = 0$.

Solution

The expression $x^2 - 50$ cannot be factored because 50 is not a perfect square. We proceed to solve this equation by extracting square roots.

$$x^2 - 50 = 0$$

$$x^2 = 50 \quad \text{Add 50.}$$

$$\sqrt{x^2} = \pm\sqrt{50} \quad \text{Take the square root of each side.}$$

$$x = \pm 5\sqrt{2} \quad \text{Simplify the square root. } \sqrt{50} = \sqrt{25 \cdot 2} = 5\sqrt{2}$$

The solution set is $\{-5\sqrt{2}, 5\sqrt{2}\}$. Using a calculator, we find that the solutions are approximately ± 7.07.

A Word of Caution When taking the square root of both sides of an equation, do not forget to use the symbol $\pm$.

This technique can also be used to find complex solutions to equations, as well as real solutions.

EXAMPLE 4 Solve $x^2 = -16$.

Solution

If we had added the 16 to the left side of the equation, then the resulting equation would have been $x^2 + 16 = 0$. The expression $x^2 + 16$ is not factorable. (Recall that the sum of two squares is not factorable.) The only way for us to find a solution would be to take the square roots of both sides of the original equation. We proceed to solve this equation by extracting square roots.

$$x^2 = -16$$

$$\sqrt{x^2} = \pm\sqrt{-16} \quad \text{Take the square root of each side.}$$

$$x = \pm 4i \quad \text{Simplify the square root.}$$

The solution set is $\{-4i, 4i\}$. Note that these two solutions are not real numbers.

Quick Check 2

Solve.

a) $x^2 = 28$

b) $x^2 + 32 = 0$

Objective 3 Solve quadratic equations by extracting square roots involving a linear expression that is squared. Extracting square roots is an excellent technique to solve quadratic equations whenever our equation is made up of a squared term and a constant term. Consider the equation $(2x - 7)^2 = 25$. To use factoring to solve this equation, we would first have to square $2x - 7$, then collect all terms on the left side of the equation by subtracting 25, and finally hope that the resulting expression could be factored. Extracting square roots is a more efficient way to solve this equation.

EXAMPLE 5 Solve $(2x - 7)^2 = 25$.

Solution

$$(2x - 7)^2 = 25$$

$$\sqrt{(2x - 7)^2} = \pm\sqrt{25} \quad \text{Take the square root of each side.}$$

$$2x - 7 = \pm 5 \quad \text{Simplify the square root. Note that the square root of } (2x - 7)^2 \text{ is } 2x - 7.$$

$$2x = 7 \pm 5 \quad \text{Add 7.}$$

$$x = \frac{7 \pm 5}{2} \quad \text{Divide by 2.}$$

$\frac{7 + 5}{2} = 6$ and $\frac{7 - 5}{2} = 1$, so the solution set is $\{1, 6\}$.

EXAMPLE 6 Solve $(3x + 4)^2 + 35 = 15$.

Solution

$$(3x + 4)^2 + 35 = 15$$

$$(3x + 4)^2 = -20 \quad \text{Subtract 35 to isolate the squared term.}$$

$$\sqrt{(3x + 4)^2} = \pm\sqrt{-20} \quad \text{Take the square root of each side.}$$

$$3x + 4 = \pm 2i\sqrt{5} \quad \text{Simplify the square root.}$$

$$3x = -4 \pm 2i\sqrt{5} \quad \text{Subtract 4.}$$

$$x = \frac{-4 \pm 2i\sqrt{5}}{3} \quad \text{Divide both sides by 3.}$$

Since we cannot simplify the numerator in this case, these are our solutions. The solution set is $\left\{\frac{-4 - 2i\sqrt{5}}{3}, \frac{-4 + 2i\sqrt{5}}{3}\right\}$.

***Quick Check* 3**
Solve.

a) $(3x + 1)^2 = 100$
b) $2(2x - 3)^2 + 17 = -1$

Solving Applied Problems by Extracting Square Roots

Objective 4 Solve applied problems by extracting square roots. Problems involving the area of a geometric figure often lead to quadratic equations, as area is measured in square units. Here are two useful area formulas which can lead to quadratic equations that can be solved by extracting square roots.

Figure	Square	Circle
	s	r
Dimensions	Side s^2	Radius r
Area	$A = s^2$	$A = \pi r^2$

EXAMPLE 7 A square has an area of 40 square centimeters. Find the length of its side, rounded to the nearest tenth of a centimeter.

Solution

In this problem, the unknown quantity is the length of the side of the square, while we know that the area is 40 cm^2. We will let x represent the length of the side. Since the area of a square is equal to its side squared, the equation we need to solve is $x^2 = 40$.

$$x^2 = 40$$

$$\sqrt{x^2} = \pm\sqrt{40} \quad \text{Take the square root of each side.}$$

$$x = \pm 2\sqrt{10} \quad \text{Simplify the square roots.}$$

Since x represents the length of the side of the square, it must be positive. We may omit the solution $x = -2\sqrt{10}$. The solution is $x = 2\sqrt{10}$ cm. Rounding this to the nearest tenth, we know that the length of the side of the square is 6.3 cm.

***Quick Check* 4**
A square has an area of 150 square inches. Find the length of its side, rounded to the nearest tenth of an inch.

EXAMPLE 8 A circle has an area of 100 square feet. Find the radius of the circle, rounded to the nearest tenth of a foot.

Solution

In this problem, the unknown quantity is the radius of the circle, and we will let r represent the radius. Since the area of a circle is given by the formula $A = \pi r^2$, the equation we need to solve is $\pi r^2 = 100$.

$$\pi r^2 = 100$$

$$\frac{\cancel{\pi}^{1} r^2}{\cancel{\pi}_{1}} = \frac{100}{\pi} \quad \text{Divide both sides by } \pi.$$

$$\sqrt{r^2} = \pm\sqrt{\frac{100}{\pi}} \quad \text{Take the square root of each side.}$$

$$r = \pm\sqrt{\frac{100}{\pi}} \quad \text{Simplify.}$$

Again, we may omit the negative solution. The solution is $r = \sqrt{\frac{100}{\pi}}$ ft. Rounding this to the nearest tenth, the radius is 5.6 ft.

***Quick Check* 5**
A circle has an area of 20 square centimeters. Find the radius of the circle, rounded to the nearest tenth of a centimeter.

Solving Quadratic Equations by Completing the Square

Objective 5 **Solve quadratic equations by completing the square.** We can solve any quadratic equation by converting it to an equation with a squared term equal to a constant. Rewriting the equation in this form allows us to solve it by extracting square roots. We will now examine a procedure for doing this called **completing the square**. This procedure is used for an equation such as $x^2 - 6x - 16 = 0$, which has both a second-degree term (x^2) and a first-degree term $(-6x)$.

The first step is to isolate the variable terms on one side of the equation with the constant term on the other side. This can be done by adding 16 to both sides of the equation. The resulting equation is $x^2 - 6x = 16$.

The next step is to add a number to both sides of the equation that makes the side of the equation containing the variable terms into a perfect square trinomial. To do this, we take half of the coefficient of the first-degree term, which in this case is -6, and square it.

$$\left(\frac{-6}{2}\right)^2 = (-3)^2 = 9$$

After adding 9 to both sides of the equation, we obtain the equation $x^2 - 6x + 9 = 25$. The expression on the left side of the equation, $x^2 - 6x + 9$, is a perfect square trinomial and can be factored as $(x - 3)^2$. The resulting equation, $(x - 3)^2 = 25$, is in the correct form for extracting square roots. (This equation will be solved completely in the next example.)

Here is the procedure for solving a quadratic equation by completing the square, provided that the coefficient of the squared term is 1.

Completing the Square

1. Isolate all variable terms on one side of the equation, with the constant term on the other side of the equation.
2. Identify the coefficient of the first-degree term. Take half of that number, square it, and add that to both sides of the equation.
3. Factor the resulting perfect square trinomial.
4. Take the square root of each side of the equation. Be sure to include $\pm$ on the side where the constant is.
5. Solve the resulting equation.

In the following example, we will complete the solution of $x^2 - 6x - 16 = 0$ by completing the square.

EXAMPLE 9 Solve $x^2 - 6x - 16 = 0$ by completing the square.

Solution

$$x^2 - 6x - 16 = 0$$
$$x^2 - 6x = 16 \qquad \text{Add 16.}$$

$$x^2 - 6x + 9 = 16 + 9$$ $(\frac{-6}{2})^2 = (-3)^2 = 9$. Add 9 to complete the square.

$$(x-3)^2 = 25$$ Simplify. Factor the left side.

$$\sqrt{(x-3)^2} = \pm\sqrt{25}$$ Take the square root of each side.

$$x - 3 = \pm 5$$ Simplify the square root.

$$x = 3 \pm 5$$ Add 3.

$3 + 5 = 8$ and $3 - 5 = -2$, so the solution set is $\{-2, 8\}$.

Note that the equation in the previous example could have been solved with less work by factoring $x^2 - 6x - 16$. Always use factoring whenever possible, because it often leads to the quickest and most direct solutions.

Example 10 Solve $x^2 + 8x + 20 = 0$ by completing the square.

Solution

$$x^2 + 8x + 20 = 0$$

$$x^2 + 8x = -20$$ Subtract 20.

$$x^2 + 8x + 16 = -20 + 16$$ Half of 8 is 4, which equals 16 when squared: $(\frac{8}{2})^2 = (4)^2 = 16$. Add 16.

$$x^2 + 8x + 16 = -4$$ Simplify.

$$(x+4)^2 = -4$$ Factor the left side.

$$\sqrt{(x+4)^2} = \pm\sqrt{-4}$$ Take the square root of each side.

$$x + 4 = \pm 2i$$ Simplify the square root.

$$x = -4 \pm 2i$$ Subtract 4.

The solution set is $\{-4 - 2i, -4 + 2i\}$.

Quick Check 6

Solve by completing the square.

a) $x^2 + 8x + 12 = 0$
b) $x^2 - 6x + 10 = 0$

Note that the expression $x^2 + 8x + 20$ does not factor, so completing the square is the only technique we have for solving this equation. In the first two examples of completing the square, the coefficient of the first-degree term was an even integer. If this is not the case, we must use fractions to complete the square.

Example 11 Solve $x^2 - 5x - 5 = 0$ by completing the square.

Solution

The expression $x^2 - 5x - 5$ does not factor, so we proceed with completing the square.

$$x^2 - 5x - 5 = 0$$

$$x^2 - 5x = 5$$ Add 5.

$$x^2 - 5x + \frac{25}{4} = 5 + \frac{25}{4}$$ Half of -5 is $-\frac{5}{2}$, which equals $\frac{25}{4}$ when squared: $(-\frac{5}{2})^2 = \frac{25}{4}$. Add $\frac{25}{4}$.

$$x^2 - 5x + \frac{25}{4} = \frac{45}{4}$$ Add 5 and $\frac{25}{4}$ by rewriting 5 as a fraction whose denominator is 4: $5 + \frac{25}{4} = \frac{20}{4} + \frac{25}{4}$.

$$\left(x - \frac{5}{2}\right)^2 = \frac{45}{4}$$ Factor.

$$\sqrt{\left(x-\frac{5}{2}\right)^2} = \pm\sqrt{\frac{45}{4}}$$ Take the square root of each side.

$$x - \frac{5}{2} = \pm\frac{3\sqrt{5}}{2}$$ Simplify the square root.

$$x = \frac{5}{2} \pm \frac{3\sqrt{5}}{2}$$ Add $\frac{5}{2}$.

The solution set is $\left\{\frac{5-3\sqrt{5}}{2}, \frac{5+3\sqrt{5}}{2}\right\}$.

Quick Check 7
Solve $x^2 + 7x - 18 = 0$ by completing the square.

A Word of Caution To solve a quadratic equation by completing the square, we must be sure that the leading coefficient is positive 1. If the leading coefficient is not equal to 1, then we divide both sides of the equation by the leading coefficient.

EXAMPLE 12 Solve $3x^2 + 18x + 30 = 0$ by completing the square.

Solution

The leading coefficient is not equal to 1, so we begin by dividing each term on both sides of the equation by 3. We are then able to solve this equation by completing the square.

$$3x^2 + 18x + 30 = 0$$

$$\frac{3x^2 + 18x + 30}{3} = \frac{0}{3}$$ Divide both sides of the equation by 3, so that the leading coefficient is equal to 1.

$$x^2 + 6x + 10 = 0$$ Simplify.

$$x^2 + 6x = -10$$ Subtract 10.

$$x^2 + 6x + 9 = -10 + 9$$ Half of 6 is 3, which equals 9 when squared: $\left(\frac{6}{2}\right)^2 = (3)^2 = 9$.

$$x^2 + 6x + 9 = -1$$ Simplify.

$$(x+3)^2 = -1$$ Factor.

$$\sqrt{(x+3)^2} = \pm\sqrt{-1}$$ Take the square root of each side.

$$x + 3 = \pm i$$ Simplify each square root.

$$x = -3 \pm i$$ Subtract 3.

The solution set is $\{-3 - i, -3 + i\}$.

Quick Check 8
Solve $2x^2 - 32x + 92 = 0$ by completing the square.

Finding a Quadratic Equation, Given Its Solutions

Objective 6 **Find a quadratic equation, given its solutions.**

EXAMPLE 13 Find a quadratic equation in standard form, with integer coefficients, that has the solution set $\{-6, \frac{1}{2}\}$.

Solution

We know that $x = -6$ is a solution to the equation. This tells us that $x + 6$ is a factor of the quadratic expression. Similarly, knowing that $x = \frac{1}{2}$ is a solution tells us that $2x - 1$ is a factor of the quadratic expression.

$$x = \frac{1}{2}$$

$$2 \cdot x = \frac{1}{2} \cdot 2$$ Multiply both sides by 2 to clear the fractions.

$$2x = 1$$ Simplify.

$$2x - 1 = 0$$ Subtract 1.

Multiplying these two factors will give us a quadratic equation with these two solutions.

$x = -6$ or $x = \frac{1}{2}$ Begin with the solutions.

$x + 6 = 0$ or $2x - 1 = 0$ Rewrite each equation so that the right side is equal to 0.

$(x + 6)(2x - 1) = 0$ Write an equation that has these two expressions as factors.

$2x^2 - x + 12x - 6 = 0$ Multiply.

$2x^2 + 11x - 6 = 0$ Simplify.

Quick Check 9
Find a quadratic equation in standard form, with integer coefficients, that has the solution set $\{\frac{5}{3}, -2\}$.

A quadratic equation that has the solution set $\{-6, \frac{1}{2}\}$ is $2x^2 + 11x - 6 = 0$.

Notice that we say *a* quadratic equation rather than *the* quadratic equation. There are infinitely many quadratic equations with integer coefficients that have this solution set. For example, multiplying both sides of our equation by 3 gives us the equation $6x^2 + 33x - 18 = 0$, which has the same solution set.

EXAMPLE 14 Find a quadratic equation in standard form, with integer coefficients, that has the solution set $\{-6i, 6i\}$.

Solution

We know that $x = -6i$ is a solution to the equation. This tells us that $x + 6i$ is a factor of the quadratic expression. Similarly, knowing that $x = 6i$ is a solution tells us that $x - 6i$ is a factor of the quadratic expression. Multiplying these two factors will give us a quadratic equation with these two solutions.

$x = -6i$ or $x = 6i$ Begin with the solutions.

$x + 6i = 0$ or $x - 6i = 0$ Rewrite each equation so that the right side is equal to 0.

$(x + 6i)(x - 6i) = 0$ Write an equation with one side having these two expressions as factors.

$x^2 - 6ix + 6ix - 36i^2 = 0$ Multiply.

$x^2 + 36 = 0$ Simplify. Recall that $i^2 = -1$.

Quick Check 10
Find a quadratic equation in standard form, with integer coefficients, that has the solution set $\{-10i, 10i\}$.

A quadratic equation that has the solution set $\{-6i, 6i\}$ is $x^2 + 36 = 0$.

Building Your Study Strategy **Time Management, 1** **Keeping Track** The first step in achieving effective time management is keeping track of your time over a one-week period. Mark down the times that you spend in class, working, cooking, watching TV, eating, sleeping, being with family, socializing, getting tutorial help, preparing for school, and, most importantly, studying. This will give you a good idea of how much extra time you have to devote to studying, as well as how much time you devote to other activities. Most students do not realize how much time they waste in a day.

Once you have made any changes to your current schedule, try to schedule more time for studying math. You should be spending between two and four hours studying per week for each hour that you spend in the classroom. Try to schedule some time each day, rather than cramming it all into the weekend. Leave some blank periods for flexibility; these periods can be used for emergencies for any of your classes.

EXERCISES 7.1

Vocabulary

1. The method of solving an equation by taking the square root of both sides is called solving by ________.
2. In order to solve an equation by extracting square roots, first isolate the expression that is ________.
3. The method of solving an equation by rewriting the variable expression as a square of a binomial is called solving by ________.
4. To solve the equation $x^2 + bx = c$ by completing the square, first add ________ to both sides of the equation.

Solve by factoring.

5. $x^2 - 3x - 10 = 0$
6. $x^2 + 9x + 14 = 0$
7. $x^2 + 11x + 24 = 0$
8. $x^2 + 7x - 18 = 0$
9. $x^2 + 6x + 8 = 0$
10. $x^2 + 15x + 56 = 0$
11. $x^2 - 2x = 24$
12. $x^2 - 45 = -4x$
13. $x^2 - 64 = 0$
14. $x^2 - 25 = 0$
15. $x^2 - 4x = 7x + 26$
16. $x^2 + 8x + 13 = 5x + 41$

Solve by extracting square roots.

17. $x^2 = 36$
18. $x^2 = 4$
19. $x^2 - 98 = 0$
20. $x^2 - 48 = 0$
21. $x^2 = -25$
22. $x^2 = -12$
23. $x^2 + 14 = 42$
24. $x^2 + 75 = 30$
25. $(x - 5)^2 = 36$
26. $(x - 9)^2 = 54$
27. $(x + 3)^2 = -27$
28. $(x + 14)^2 = -144$

FOR EXTRA HELP

MyMathLab
MathXL
Interactmath.com
MathXL Tutorials on CD
Video Lectures on CD
Tutor Center
Addison-Wesley Math Tutor Center
Student's Solutions Manual

29. $(x + 5)^2 + 33 = 15$

30. $(x + 3)^2 - 11 = 38$

31. $2(x - 9)^2 - 17 = 83$

32. $3(x + 1)^2 + 11 = 83$

33. $\left(x - \frac{2}{5}\right)^2 = \frac{49}{25}$

34. $\left(x + \frac{3}{4}\right)^2 = -\frac{25}{16}$

Fill in the missing term that makes the expression a perfect square trinomial. Factor the resulting expression.

35. $x^2 + 12x +$ ___

36. $x^2 - 2x +$ ___

37. $x^2 - 5x +$ ___

38. $x^2 + 13x +$ ___

39. $x^2 - 8x +$ ___

40. $x^2 - 16x +$ ___

41. $x^2 + \frac{1}{4}x +$ ___

42. $x^2 + x +$ ___

For Exercises 43–52, approximate to the nearest tenth when necessary.

43. The area of a square is 81 square meters. Find the length of a side of the square.

44. The area of a square is 144 square feet. Find the length of a side of the square.

45. The area of a square is 128 square inches. Find the length of a side of the square.

46. The area of a square is 200 square centimeters. Find the length of a side of the square.

47. The area of a circle is 24 square inches. Find the radius of the circle.

48. The area of a circle is 60 square centimeters. Find the radius of this circle.

49. The area of a circle is 250 square meters. Find the radius of the circle.

50. The area of a circle is 48 square feet. Find the radius of this circle.

51. The area of a circle is 32π square feet. Find the radius of the circle.

52. The area of a circle is 49π square meters. Find the radius of the circle.

Solve by completing the square.

53. $x^2 - 8x - 33 = 0$

54. $x^2 + 6x - 7 = 0$

55. $x^2 + 4x + 7 = 0$

56. $x^2 - 10x + 18 = 0$

57. $x^2 + 8x = 9$

58. $x^2 + 12x = -27$

59. $x^2 - 14x + 30 = 0$

60. $x^2 + 20x + 125 = 0$

61. $x^2 + 8x + 44 = 0$

62. $x^2 + 26x - 84 = 0$

63. $x^2 - 16 = -6x$

64. $x^2 - 48 = 2x$

65. $x^2 - 9x - 32 = 0$

66. $x^2 + 5x + 16 = 0$

67. $x^2 + 7x + 6 = 0$

68. $x^2 + 13x - 30 = 0$

69. $x^2 + 3x = 5$

70. $x^2 - x = 7$

71. $x^2 + \frac{5}{2}x + 1 = 0$

72. $x^2 - \frac{23}{6}x + \frac{7}{2} = 0$

73. $2x^2 - 2x - 144 = 0$

74. $4x^2 - 56x + 172 = 0$

75. $5x^2 + 20x + 200 = 0$

76. $3x^2 + 54x + 258 = 0$

Find a quadratic equation with integer coefficients that has the following solution set:

77. $\{-5, 2\}$

78. $\{-4, -3\}$

79. $\{6, 8\}$

80. $\{0, 5\}$

81. $\left\{-2, \frac{3}{4}\right\}$

82. $\left\{\frac{1}{2}, \frac{11}{4}\right\}$

83. $\{-3, 3\}$

84. $\{-6, 6\}$

85. $\{-5i, 5i\}$

86. $\{-i, i\}$

Mixed Practice, 87–110

Solve by any method (factoring, extracting square roots, or completing the square).

87. $x^2 + 13x + 30 = 0$

88. $x^2 + 12 = 0$

89. $x^2 + 6x - 17 = 0$

90. $(x - 5)^2 = 1$

91. $x^2 - 6x - 7 = 0$

92. $x^2 - 6x = -10$

93. $(x - 4)^2 - 11 = 16$

94. $x^2 - 5x - 36 = 0$

95. $x^2 - 4x = 20$

96. $x^2 + 85 = 18x$

97. $x^2 - 5x - 6 = 0$

98. $(2x + 8)^2 - 5 = 11$

99. $x^2 - 8x + 19 = 0$

100. $x^2 - 20x + 91 = 0$

101. $3(x + 3)^2 + 13 = -11$

102. $x^2 - 9 = 0$

103. $x^2 + 3x + 9 = 0$

104. $x^2 - 2x + 50 = 0$

105. $x(x + 5) - 7(x + 5) = 0$

106. $(4x - 3)(4x - 3) = 25$

107. $-16x^2 + 64x + 80 = 0$

108. $\left(x + \frac{b}{2a}\right)^2 = \frac{b^2 - 4ac}{4a^2}$, where a, b, and c are constants and $a \neq 0$.

109. $\frac{2}{3}x^2 + \frac{8}{3}x - \frac{10}{3} = 0$

110. $x^2 - 10x + 11 = 0$

For each rational expression, list the values that are excluded from the domain. (Recall that a real number is excluded from the domain of a rational expression if it causes the denominator to equal 0 when it is substituted into the expression.)

111. $\frac{7}{x^2 - 7x + 12}$

112. $\frac{5}{x^2 - 3x}$

113. $\frac{x + 2}{x^2 - 40}$

114. $\frac{x - 6}{x^2 + 25}$

115. $\frac{x^2 - 11x + 28}{(x - 5)^2 - 16}$

116. $\frac{x^2 + 49}{3(4x - 3)^2 - 24}$

117. Fill in the blank in the equation $x^2 - 5x + 7 =$ ________ so that the equation has

a) $x = 8$ as a solution

b) $x = -6$ as a solution

118. Fill in the blank in the equation $x^2 + 2x - 13 =$ ________ so that the equation has

a) $x = 7$ as a solution

b) $x = -3$ as a solution

119. Fill in the blank in the equation $x^2 + 8x + 10 =$ ________ so that the equation has

a) two real solutions

b) two nonreal solutions

120. Fill in the blank in the equation $x^2 - 6x + 34 =$ ________ so that the equation has

a) two real solutions

b) two nonreal solutions

Writing in Mathematics

Answer in complete sentences.

121. Explain why we use the symbol ± when taking the square root of each side of an equation.

122. If the coefficient of the second-degree term is not 1, we divide both sides of the equation by that coefficient before attempting to complete the square. Explain why dividing both sides of the equation by this coefficient does not affect the solutions of the equation.

123. ***Solutions Manual**** Write a solutions manual page for the following problem:

Solve $(x-6)^2+20=-28$ *by extracting square roots.*

124. ***Newsletter**** Write a newsletter explaining how to solve a quadratic equation by completing the square.

***See Appendix B for details and sample answers.**

7.2 The Quadratic Formula

Objectives

1. **Derive the quadratic formula.**
2. **Identify the coefficients *a*, *b*, and *c* of a quadratic equation.**
3. **Solve quadratic equations by the quadratic formula.**
4. **Use the discriminant to determine the number and type of solutions of a quadratic equation.**
5. **Use the discriminant to determine whether a quadratic expression is factorable.**
6. **Solve projectile motion problems.**

The Quadratic Formula

Objective 1 **Derive the quadratic formula.** Completing the square to solve a quadratic equation can be tedious. As an alternative, in this section we will develop and use the **quadratic formula**, which is a numerical formula that gives the solution(s) of *any* quadratic equation.

We derive the quadratic formula by solving the general equation $ax^2 + bx + c = 0$ (where a, b, and c are real numbers and $a \neq 0$) by completing the square. Recall that the coefficient of the second-degree term must be equal to 1, so the first step is to divide both sides of the equation $ax^2 + bx + c = 0$ by a.

$$ax^2 + bx + c = 0$$

$$x^2 + \frac{b}{a}x + \frac{c}{a} = 0$$ Divide both sides by a.

$$x^2 + \frac{b}{a}x = -\frac{c}{a}$$ Subtract the constant term $\frac{c}{a}$.

$$x^2 + \frac{b}{a}x + \frac{b^2}{4a^2} = -\frac{c}{a} + \frac{b^2}{4a^2}$$ Half of $\frac{b}{a}$ is $\frac{b}{2a}$. $\left(\frac{b}{2a}\right)^2 = \frac{b^2}{4a^2}$. Add $\frac{b^2}{4a^2}$ to both sides of the equation.

$$x^2 + \frac{b}{a}x + \frac{b^2}{4a^2} = \frac{b^2 - 4ac}{4a^2}$$ Simplify the right side of the equation as a single fraction by rewriting $-\frac{c}{a}$ as $-\frac{4ac}{4a^2}$.

$$\left(x + \frac{b}{2a}\right)^2 = \frac{b^2 - 4ac}{4a^2}$$ Factor the left side of the equation.

$$\sqrt{\left(x + \frac{b}{2a}\right)^2} = \pm\sqrt{\frac{b^2 - 4ac}{4a^2}}$$ Take the square root of both sides.

$$x + \frac{b}{2a} = \pm\frac{\sqrt{b^2 - 4ac}}{\sqrt{4a^2}}$$ Simplify the square root on the left side. Rewrite the right side as the quotient of two square roots.

$$x + \frac{b}{2a} = \pm\frac{\sqrt{b^2 - 4ac}}{2a}$$ Simplify the square root in the denominator.

$$x = -\frac{b}{2a} \pm \frac{\sqrt{b^2 - 4ac}}{2a}$$ Subtract $\frac{b}{2a}$.

$$x = \frac{-b \pm \sqrt{b^2 - 4ac}}{2a}$$ Rewrite as a single fraction.

The Quadratic Formula

For a general quadratic equation $ax^2 + bx + c = 0$ (where a, b, and c are real numbers and $a \neq 0$), the quadratic formula tells us that the solutions of this equation are given by

$$x = \frac{-b \pm \sqrt{b^2 - 4ac}}{2a}.$$

If we can identify the coefficients a, b, and c in a quadratic equation, then we can find the solutions of the equation by substituting these values for a, b, and c in the quadratic formula.

Identifying the Coefficients to Be Used in the Quadratic Formula

Objective 2 **Identify the coefficients *a*, *b*, and *c* of a quadratic equation.** The first step in using the quadratic formula is to identify the coefficients a, b, and c. We can do this only after our equation is in standard form: $ax^2 + bx + c = 0$.

EXAMPLE 1 For the quadratic equation, identify a, b, and c.

a) $x^2 - 6x + 8 = 0$

Solution

$a = 1$, $b = -6$, and $c = 8$. Be sure to include the negative sign when identifying coefficients that are negative.

b) $3x^2 - 5x = 7$

Solution

Before identifying our coefficients, we must convert this equation to standard form: $3x^2 - 5x - 7 = 0$. So $a = 3$, $b = -5$, and $c = -7$.

c) $x^2 + 8 = 0$

Solution

In this example, we do not see a first-degree term. In a case like this, $b = 0$. The coefficients that we do see tell us that $a = 1$ and $c = 8$.

***Quick Check* 1**

For the given quadratic equation, identify *a*, *b*, and *c*.

a) $x^2 + 11x - 13 = 0$
b) $5x^2 - 11 = 9x$
c) $x^2 = 30$

Solving Quadratic Equations by the Quadratic Formula

Objective 3 **Solve quadratic equations by using the quadratic formula.** Now we will solve several equations, using the quadratic formula.

EXAMPLE 2 Solve $x^2 - 6x + 8 = 0$.

Solution

We can use the quadratic formula with $a = 1$, $b = -6$, and $c = 8$.

$$x = \frac{-b \pm \sqrt{b^2 - 4ac}}{2a}$$

$$x = \frac{6 \pm \sqrt{(-6)^2 - 4(1)(8)}}{2(1)}$$

Substitute 1 for a, -6 for b, and 8 for c. When you use the formula, it is a good idea to think of the $-b$ in the numerator as the "opposite of b." The opposite of -6 is 6.

$$x = \frac{6 \pm \sqrt{36 - 32}}{2}$$

Simplify each term in the radicand.

$$x = \frac{6 \pm \sqrt{4}}{2}$$

Subtract.

$$x = \frac{6 \pm 2}{2}$$

Simplify the square root.

$\frac{6 + 2}{2} = 4$ and $\frac{6 - 2}{2} = 2$; so the solution set is $\{2, 4\}$.

Note that the equation in the previous example could have been solved much more quickly by factoring $x^2 - 6x + 8$ to be $(x - 4)(x - 2)$ and then solving. Use factoring whenever you can, treating the quadratic formula as an alternative.

EXAMPLE 3 Solve $3x^2 + 8x = -3$.

Solution

To solve this equation, we must rewrite the equation in standard form by collecting all terms on the left side of the equation. In other words, we will rewrite the equation as $3x^2 + 8x + 3 = 0$.

If an equation does not quickly factor, we are wise to proceed directly to the quadratic formula rather than trying to factor an expression that may not be factorable. Here, $a = 3$, $b = 8$, and $c = 3$.

$$x = \frac{-8 \pm \sqrt{(8)^2 - 4(3)(3)}}{2(3)}$$

Substitute 3 for a, 8 for b, and 3 for c in the quadratic formula.

$$x = \frac{-8 \pm \sqrt{28}}{6}$$

Simplify the radicand.

$$x = \frac{-8 \pm 2\sqrt{7}}{6}$$

Simplify the square root.

$$x = \frac{\overset{1}{\cancel{2}}(-4 \pm \sqrt{7})}{\underset{3}{\cancel{6}}}$$

Factor the numerator and divide out the common factor.

$$x = \frac{-4 \pm \sqrt{7}}{3}$$

Simplify.

The solution set is $\left\{\frac{-4-\sqrt{7}}{3}, \frac{-4+\sqrt{7}}{3}\right\}$. These solutions are approximately -2.22 and -0.45, respectively.

Using Your Calculator Here is the screen that shows how to approximate $\frac{-4+\sqrt{7}}{3}$ and $\frac{-4-\sqrt{7}}{3}$:

```
(-4+√(7))/3
       -.4514162296
(-4-√(7))/3
       -2.215250437
```

Quick Check 2 Solve by using the quadratic formula.

a) $x^2 + 7x - 30 = 0$

b) $7x^2 + 21x = -9$

EXAMPLE 4 Solve $x^2 - 4x = -5$.

Solution

We begin by rewriting the equation in standard form: $x^2 - 4x + 5 = 0$. The quadratic expression in this equation does not factor, so we use the quadratic formula with $a = 1$, $b = -4$, and $c = 5$.

$$x = \frac{4 \pm \sqrt{(-4)^2 - 4(1)(5)}}{2(1)}$$ Substitute 1 for *a*, −4 for *b*, and 5 for *c* in the quadratic formula.

$$x = \frac{4 \pm \sqrt{-4}}{2}$$ Simplify the radicand.

$$x = \frac{4 \pm 2i}{2}$$ Simplify the square root. Be sure to include "*i*", as we are taking the square root of a negative number.

$$x = \frac{\overset{1}{\cancel{2}}(2 \pm i)}{\underset{1}{\cancel{2}}}$$ Factor the numerator and divide out the common factor.

$$x = 2 \pm i$$ Simplify.

The solution set is $\{2 - i, 2 + i\}$.

Quick Check 3 Solve $x^2 - 7x = -19$.

If an equation has coefficients that are fractions, multiply each side of the equation by the LCD to clear the equation of fractions. If we can solve the equation by factoring, it will be easier to factor without the fractions involved. If we cannot solve by factoring, we will find the quadratic formula easier to simplify by using integers rather than fractions.

EXAMPLE 5 Solve $\frac{1}{2}x^2 - x + \frac{1}{3} = 0$.

Solution

The first step is to clear the fractions by multiplying each side of the equation by the LCD, which in this example is 6.

$$6 \cdot \left(\frac{1}{2}x^2 - x + \frac{1}{3}\right) = 6 \cdot 0$$ Multiply both sides of the equation by the LCD.

$$3x^2 - 6x + 2 = 0$$ Distribute and simplify.

Now we can use the quadratic formula with $a = 3$, $b = -6$, and $c = 2$.

$$x = \frac{6 \pm \sqrt{(-6)^2 - 4(3)(2)}}{2(3)}$$ Substitute 3 for a, -6 for b, and 2 for c in the quadratic formula.

$$x = \frac{6 \pm \sqrt{12}}{6}$$ Simplify the radicand.

$$x = \frac{6 \pm 2\sqrt{3}}{6}$$ Simplify the square root.

$$x = \frac{\overset{1}{\cancel{2}}(3 \pm \sqrt{3})}{\underset{3}{\cancel{6}}}$$ Factor the numerator and divide out common factors.

$$x = \frac{3 \pm \sqrt{3}}{3}$$ Simplify.

The solution set is $\left\{\frac{3 - \sqrt{3}}{3}, \frac{3 + \sqrt{3}}{3}\right\}$. These solutions are approximately 0.42 and 1.58, respectively.

Quick Check 4

Solve $\frac{1}{4}x^2 + \frac{1}{3}x + \frac{1}{2} = 0$.

In addition to clearing any fractions, make sure that the coefficient (a) of the second-degree term is positive. If this term is negative, you can collect all terms on the other side of the equation or multiply each side of the equation by negative 1.

EXAMPLE 6 Solve $-2x^2 + 11x + 6 = 0$.

Solution

We begin by rewriting the equation so that a is positive, because factoring a quadratic expression is more convenient when the leading coefficient is positive. Also, you may find that simplifying the quadratic formula is easier when the leading coefficient, a, is positive.

$$0 = 2x^2 - 11x - 6$$ Collect all terms on the right side of the equation.

Often, when the leading coefficient is not 1, factoring can be difficult or time consuming—if the expression is factorable at all. In such a situation, go right to the quadratic formula. In this example, we can use the quadratic formula with $a = 2$, $b = -11$, and $c = -6$.

$$x = \frac{11 \pm \sqrt{(-11)^2 - 4(2)(-6)}}{2(2)}$$ Substitute 2 for a, -11 for b, and -6 for c in the quadratic formula.

$$x = \frac{11 \pm \sqrt{169}}{4}$$ Simplify the radicand.

$$x = \frac{11 \pm 13}{4}$$ Simplify the square root.

$\frac{11+13}{4} = 6$ and $\frac{11-13}{4} = -\frac{1}{2}$; so the solution set is $\{-\frac{1}{2}, 6\}$.

Quick Check 5
Solve $-6x^2 - 7x + 20 = 0$.

Although the quadratic formula can be used to solve *any* quadratic equation, it does not always provide the most efficient way to solve a particular equation. We should always check to see whether factoring or extracting square roots can be used before using the quadratic formula.

Using the Discriminant to Determine the Number and Type of Solutions of a Quadratic Equation

Objective 4 **Use the discriminant to determine the number and type of solutions of a quadratic equation.** In the quadratic formula, the expression $b^2 - 4ac$ is called the **discriminant**. The discriminant can be used to give us some information about our solutions. If the discriminant is negative $(b^2 - 4ac < 0)$, then the equation has two nonreal complex solutions. This is because we take the square root of a negative number in the quadratic formula. If the discriminant is zero, $b^2 - 4ac = 0$, then the equation has one real solution. In this case, the quadratic formula simplifies to be $x = \frac{-b \pm \sqrt{0}}{2a}$, or simply, $x = \frac{-b}{2a}$. If the discriminant is positive, $b^2 - 4ac > 0$, then the equation has two real solutions. The square root of a positive discriminant is a real number, so the quadratic formula produces two real solutions. The following chart summarizes what the discriminant tells us about the number and type of solutions of an equation:

Solutions of a Quadratic Equation Based on the Discriminant

$b^2 - 4ac$	Number and Type of Solutions
Negative	Two Nonreal Complex Solutions
Zero	One Real Solution
Positive	Two Real Solutions

It is important to know that an equation has no solutions that are real numbers, when working on an applied problem. (It will also be important when we are graphing quadratic equations, which is covered later in this chapter.)

Example 7 For the quadratic equation, use the discriminant to determine the number and type of solutions.

a) $x^2 - 6x - 16 = 0$

Solution

$$(-6)^2 - 4(1)(-16) = 100$$ Substitute 1 for a, -6 for b, and -16 for c.

Since the discriminant is positive, this equation has two real solutions.

b) $x^2 + 36 = 0$

Solution

$$(0)^2 - 4(1)(36) = -144 \qquad \text{Substitute 1 for } a\text{, 0 for } b\text{, and 36 for } c.$$

The discriminant is negative. This equation has two nonreal complex solutions.

c) $x^2 + 10x + 25 = 0$

Solution

$$(10)^2 - 4(1)(25) = 0 \qquad \text{Substitute 1 for } a\text{, 10 for } b\text{, and 25 for } c.$$

Since the discriminant equals 0, this equation has one real solution.

***Quick Check* 6**

For each given quadratic equation, use the discriminant to determine the number and type of solutions.

a) $x^2 + 18x - 63 = 0$
b) $x^2 + 5x + 42 = 0$
c) $x^2 - 20x + 100 = 0$

Using the Discriminant to Determine Whether a Quadratic Expression Is Factorable

Objective 5 **Use the discriminant to determine whether a quadratic expression is factorable.** The discriminant can also be used to tell us whether a quadratic expression is factorable. If the discriminant is equal to 0 or a positive number that is a perfect square (1, 4, 9, etc.), then the expression is factorable.

EXAMPLE 8 For the quadratic expression, use the discriminant to determine whether the expression can be factored.

a) $x^2 - 6x - 27$

Solution

$$(-6)^2 - 4(1)(-27) = 144 \qquad \text{Substitute 1 for } a\text{, } -6 \text{ for } b\text{, and } -27 \text{ for } c.$$

The discriminant is a perfect square ($\sqrt{144} = 12$), so the expression can be factored.

$$x^2 - 6x - 27 = (x - 9)(x + 3)$$

b) $2x^2 + 7x + 4$

Solution

$$(7)^2 - 4(2)(4) = 17 \qquad \text{Substitute 2 for } a\text{, 7 for } b\text{, and 4 for } c.$$

The discriminant is not a perfect square, so the expression is not factorable.

***Quick Check* 7**

For each given quadratic expression, use the discriminant to determine whether the expression can be factored.

a) $x^2 + 14x + 12$
b) $5x^2 - 36x - 32$

Projectile Motion Problems

Objective 6 **Solve projectile motion problems.** An application that leads to a quadratic equation involves the height in feet of an object propelled into the air after t seconds. Recall that the height, in feet, of a projectile after t seconds can be found by the function $h(t) = -16t^2 + v_0t + s$, where v_0 is the initial velocity of the projectile and s is the initial height.

Height of a Projectile

$h(t) = -16t^2 + v_0t + s$
t: Time (seconds)
v_0: Initial velocity (feet/second)
s: Initial height (feet)

EXAMPLE 9 A rock is thrown from ground level at a speed of 48 feet/second.

a) How long will it take until the rock lands on the ground?

Solution

The initial velocity of the rock is 48 feet per second, so $v_0 = 48$. Since the rock is thrown from ground level, the initial height is 0 feet. The function for the height of the rock after t seconds is $h(t) = -16t^2 + 48t$.

The rock's height when it lands on the ground is 0 feet, so we set the function equal to 0 and solve for the time t in seconds.

$-16t^2 + 48t = 0$	Set the function equal to 0.
$0 = 16t^2 - 48t$	Collect all terms on the right side of the equation so that the leading coefficient is positive.
$0 = 16t(t - 3)$	Factor out the GCF $(16t)$.
$16t = 0$ or $t - 3 = 0$	Set each variable factor equal to 0.
$t = 0$ or $t = 3$	Solve each equation.

The time of 0 seconds corresponds to the precise moment that the rock was thrown and does not represent the time required to land on the ground. Our solution is 3 seconds.

b) When will the rock be at a height of 32 feet?

Solution

This problem is similar to part (a), except that we will set the function $h(t)$ equal to 32 rather than 0.

$-16t^2 + 48t = 32$	Set the function equal to 32.
$0 = 16t^2 - 48t + 32$	Collect all terms on the right side of the equation.
$0 = 16(t^2 - 3t + 2)$	Factor out the GCF (16).
$0 = 16(t - 1)(t - 2)$	Factor the trinomial.
$t - 1 = 0$ or $t - 2 = 0$	Set each variable factor equal to 0.
$t = 1$ or $t = 2$	Solve each equation.

Do both of these answers make sense? The rock first passes a height of 32 feet after 1 second while it is on its way up. It is at a height of 32 feet after 2 seconds while it is on its way down. The rock is 32 feet above the ground after 1 second and again after 2 seconds.

Quick Check 8

A rock is thrown upward from ground level at an initial velocity of 80 feet per second.

a) After how many seconds will the rock land on the ground?

b) When will the rock be at a height of 96 feet?

EXAMPLE 10 A golf ball is launched by a slingshot from a platform that is 95 feet high at an initial velocity of 88 feet per second.

a) When will the golf ball be at a height of 175 feet? (Round to the nearest hundredth of a second.)

Solution

The initial velocity of the golf ball is 88 feet per second, so $v_0 = 88$. Since the golf ball was launched from a platform 95 feet high, the initial height $s = 95$. The function for the height of the golf ball after t seconds is $h(t) = -16t^2 + 88t + 95$.

To find when the golf ball is at a height of 175 feet, we set the function equal to 175 and solve for the time t in seconds.

$-16t^2 + 88t + 95 = 175$	Set the function equal to 175.
$0 = 16t^2 - 88t + 80$	Collect all terms on the right side of the equation.
$0 = 8(2t^2 - 11t + 10)$	Factor out the GCF (8).
$0 = 2t^2 - 11t + 10$	Divide both sides by 8.

This trinomial does not factor, so we will use the quadratic formula with $a = 2$, $b = -11$, and $c = 10$ to solve this equation.

$t = \frac{11 \pm \sqrt{(-11)^2 - 4(2)(10)}}{2(2)}$	Substitute 2 for a, -11 for b, and 10 for c.
$t = \frac{11 \pm \sqrt{41}}{4}$	Simplify the radicand and the denominator.

We use a calculator to approximate these solutions.

$$\frac{11 + \sqrt{41}}{4} \approx 4.35 \qquad \frac{11 - \sqrt{41}}{4} \approx 1.15$$

The golf ball is at a height of 175 feet after approximately 1.15 seconds and again after approximately 4.35 seconds.

Quick Check 9

A projectile is launched from the top of a building 40 feet high at an initial velocity of 36 feet per second.

a) When will the projectile be at a height of 50 feet? (Round to the nearest hundredth of a second.)

b) Will the projectile ever reach a height of 80 feet? If so, when will it be at this height?

b) Will the golf ball ever reach a height of 250 feet? If so, when will it be at this height?

Solution

We begin by setting the function equal to 250 and solving for t.

$-16t^2 + 88t + 95 = 250$	Set the function equal to 250.
$0 = 16t^2 - 88t + 155$	Collect all terms on the right side of the equation.

This trinomial does not have any common factors other than 1, so we will use the quadratic formula with $a = 16$, $b = -88$, and $c = 155$ to solve this equation.

$t = \frac{88 \pm \sqrt{(-88)^2 - 4(16)(155)}}{2(16)}$	Substitute 16 for a, -88 for b, and 155 for c.
$t = \frac{88 \pm \sqrt{-2176}}{32}$	Simplify the radicand and the denominator.

Since the radicand is negative, this equation has no real-number solutions. The golf ball does not reach a height of 250 feet.

A Word of Caution When the quadratic formula is used to solve an applied problem, a negative radicand indicates that there are no real solutions to this problem.

Building Your Study Strategy **Time Management 2, Study After Class** The best time to study new material is as soon as possible after the class period. After you have made a schedule of your current commitments, look for a block of time that is as close to your class period as possible. You can study the new material on campus if necessary. Try to study each day at the same time. It is ideal to begin each study session by reworking your notes; then move on to attempting the homework exercises.

If possible, establish a second study period during the day that will be used for review purposes. This second study period should take place later in the day and can be used to review homework or notes or even to read ahead for the next class period.

Exercises 7.2

Vocabulary

1. The __________ formula is a formula for calculating the solutions of a quadratic equation.
2. In the quadratic formula, the expression $b^2 - 4ac$ is called the __________.
3. If the discriminant is negative, then the quadratic equation has __________ real solutions.
4. If the discriminant is zero, then the quadratic equation has __________ unique real solution.
5. If the discriminant is positive, then the quadratic equation has __________ real solutions.
6. If the discriminant is 0 or a positive perfect square, then the quadratic expression in the equation is __________.

Solve by using the quadratic formula.

7. $x^2 - 5x - 36 = 0$
8. $x^2 + 4x - 45 = 0$
9. $x^2 - 4x + 2 = 0$
10. $x^2 + 10x + 13 = 0$
11. $x^2 + x + 7 = 0$
12. $x^2 - 3x + 18 = 0$
13. $x^2 + 6x = 0$
14. $x^2 + 8x + 10 = 0$
15. $x^2 + 3x - 10 = 0$
16. $x^2 + 5x + 12 = 0$
17. $x^2 + 5x + 6 = 0$
18. $x^2 + 11x + 28 = 0$
19. $x^2 - 4 = 0$
20. $2x^2 + 3x - 35 = 0$
21. $x^2 - 15 = 2x$

22. $x^2 + 8x = 12$

23. $x^2 - 3x = 9$

24. $x^2 = 7x$

25. $-4 = 19x - 5x^2$

26. $4x - x^2 = 3$

27. $x^2 - 6x + 9 = 0$

28. $x^2 + 10x + 25 = 0$

29. $x^2 - 24 = 0$

30. $x^2 + 49 = 0$

31. $x(x - 4) + 3x = 20$

32. $(2x + 1)(x - 3) = -9$

33. $x^2 - \frac{1}{5}x + \frac{3}{4} = 0$

34. $\frac{2}{3}x^2 - \frac{3}{5}x + \frac{1}{4} = 0$

35. $-x^2 + 7x - 12 = 0$

36. $-2x^2 + 15x = 8$

For each of the following quadratic equations, use the discriminant to determine the number and type of solutions.

37. $x^2 + 12x - 30 = 0$

38. $x^2 - 9x + 21 = 0$

39. $2x^2 - 3x + 5 = 0$

40. $25x^2 - 20x + 4 = 0$

41. $x^2 + \frac{2}{5}x + \frac{5}{6} = 0$

42. $x^2 - 5x - 9 = 0$

43. $9x^2 - 12x + 4 = 0$

44. $x^2 - \frac{2}{3}x + \frac{1}{9} = 0$

Use the discriminant to determine whether each of the given quadratic expressions is factorable. If the expression can be factored, write "factorable." Otherwise, write "prime."

45. $x^2 - 8x + 19$

46. $x^2 + 6x - 40$

47. $x^2 + 15x + 54$

48. $x^2 - 11x + 27$

49. $3x^2 - 5x - 8$

50. $2x^2 - 15x - 21$

51. $5x^2 + 29x + 20$

52. $10x^2 + 11x + 3$

Mixed Practice, 53–88

Solve each of the following quadratic equations, using the most efficient technique (factoring, extracting square roots, completing the square, or quadratic formula):

53. $x^2 - 5x - 15 = 0$

54. $x^2 - 68 = 0$

55. $3x^2 + 2x - 1 = 0$

56. $x^2 - 20x + 91 = 0$

57. $(5x - 4)^2 = 36$

58. $7x(8x - 3) + 4(8x - 3) = 0$

59. $2x^2 + 8x = -9$

60. $x^2 + 179x = 0$

61. $x^2 + 324 = 0$

62. $x^2 + \frac{3}{5}x - \frac{1}{12} = 0$

63. $6x^2 - 29x + 28 = 0$

64. $x^2 + 8x - 9 = 0$

65. $3(2x + 1)^2 - 7 = 23$

66. $5x^2 + 11x - 9 = 0$

67. $x^2 - 9x - 21 = 0$

68. $4x^2 - 25 = 0$

69. $16x^2 - 24x + 9 = 0$

70. $x^2 - 15x + 50 = 0$

71. $x^2 + x + 20 = 0$

72. $x^2 + 13x + 36 = 0$

73. $x^2 - 4x - 2 = 0$

74. $3x^2 - 2x - 16 = 0$

75. $x^2 - 6x + 10 = 0$

76. $(x - 8)^2 + 13 = 134$

77. $\frac{3}{4}x^2 + \frac{2}{3}x - \frac{1}{2} = 0$

78. $x^2 + 3x - 18 = 0$

79. $2x(3x + 7) - 5(3x + 7) = 0$

80. $(2x + 3)^2 - 10 = 71$

81. $2(2x - 9)^2 + 13 = 77$

82. $x^2 + x - 72 = 0$

83. $x^2 + 3x - 4 = 0$

84. $x^2 + 10x + 21 = 0$

85. $x^2 - 10x + 18 = 0$

86. $x^2 + 2x + 4 = 0$

87. $x^2 - 16x + 63 = 0$

88. $4x^2 - 12x - 11 = 0$

For Exercises 89 through 98, use the function $h(t) = -16t^2 + v_0t + s$.

89. An object is launched upward from the ground with an initial speed of 128 feet per second. How long will it take until the object lands on the ground?

90. An object is launched upward from a platform 160 feet above the ground with an initial speed of 48 feet per second. How long will it take until it lands on the ground?

91. Nick is standing on a cliff above a beach. He throws a rock upward from a height 90 feet above the beach at a speed of 70 feet per second. How long will it take until the rock lands on the beach? Round to the nearest tenth of a second.

92. Jan is standing on the roof of a building. She launches a water balloon upward from a height of 20 feet at an initial speed of 44 feet per second. How long will it take until it lands on the ground? Round to the nearest tenth of a second.

93. An object is launched upward from ground level with an initial velocity of 64 feet per second. At what time(s) is the object 48 feet above the ground?

94. An object is launched upward from ground level with an initial speed of 75 feet per second. At what time(s) is the object 70 feet above the ground? Round to the nearest tenth of a second.

95. An object is launched upward from the top of a building 80 feet high, with an initial velocity of 100 feet per second. At what time(s) is the object 200 feet above the ground? Round to the nearest tenth of a second.

96. An object is launched upward from a cliff 240 feet above a beach, with an initial velocity of 80 feet per second. At what time(s) is the object 300 feet above the ground? Round to the nearest tenth of a second.

97. A boy throws a rock upward, with an initial velocity of 16 feet per second, at a streetlight that is 25 feet above the ground. Does the rock ever reach the height of the streetlight? Explain why, in your own words.

98. A football player kicks a football straight upward. If it takes 4 seconds for the ball to land on the ground, find the initial velocity of the football.

99. Derive a formula that would provide the general solution of the linear equation $ax + b = 0$, where a and b are real numbers and $a \neq 0$.

Writing in Mathematics

Answer in complete sentences.

100. When each method is possible, which method of solving quadratic equations do you prefer: solving by factoring, completing the square, or the quadratic formula? Explain your answer.

101. Explain how the discriminant tells us whether an equation has two real solutions, one real solution, or two complex solutions.

102. ***Solutions Manual**** Write a solutions manual page for the following problem:

Solve $\frac{1}{2}x^2 - \frac{5}{6}x + 1 = 0$ *by using the quadratic formula.*

103. ***Newsletter**** Write a newsletter explaining how to solve quadratic equations by using the quadratic formula.

***See Appendix B for details and sample answers.**

7.3 Equations That Are Quadratic in Form

Objectives

1. **Solve equations by making a *u*-substitution.**
2. **Solve radical equations.**
3. **Solve rational equations.**
4. **Solve work-rate problems.**

In this section, we will learn how to solve several types of equations that are **quadratic in form**. For example, $x^4 - 13x^2 + 36 = 0$ is not a quadratic equation, but if we rewrite it as $(x^2)^2 - 13(x^2) + 36 = 0$, then we can see that it looks like a quadratic equation.

Solving Equations by Making a *u*-Substitution

Objective 1 Solve equations by making a *u*-substitution. One approach to solving equations that are quadratic in form is to use a ***u*-substitution**. We substitute the variable u for an expression, such as x^2, so that the resulting equation is a quadratic equation in u. In other words, the equation can be rewritten in the form $au^2 + bu + c = 0$. We can then solve this quadratic equation by the methods of Sections 7.1 and 7.2 (factoring, extracting square roots, completing the square, and quadratic formula). After solving this equation for u, we replace u by the expression it previously substituted for and then solve the resulting equations for the original variable.

EXAMPLE 1 Solve $x^4 - 13x^2 + 36 = 0$.

Solution

Let $u = x^2$. We can then replace x^2 in the original equation by u, and we can replace x^4 by u^2. The resulting equation will be $u^2 - 13u + 36 = 0$, which is quadratic.

$$x^4 - 13x^2 + 36 = 0$$

$$u^2 - 13u + 36 = 0 \qquad \text{Substitute } u \text{ for } x^2.$$

$$(u - 4)(u - 9) = 0 \qquad \text{Factor.}$$

$$u - 4 = 0 \quad \text{or} \quad u - 9 = 0 \qquad \text{Set each factor equal to 0.}$$

$$u = 4 \quad \text{or} \quad u = 9 \qquad \text{Solve.}$$

Now we replace u by x^2 and solve the resulting equations for x.

$$u = 4 \quad \text{or} \quad u = 9$$

$$x^2 = 4 \quad \text{or} \quad x^2 = 9 \qquad \text{Substitute } x^2 \text{ for } u.$$

$$\sqrt{x^2} = \pm\sqrt{4} \quad \text{or} \quad \sqrt{x^2} = \pm\sqrt{9} \qquad \text{Solve by taking the square roots of both sides of the equation.}$$

$$x = \pm 2 \quad \text{or} \quad x = \pm 3 \qquad \text{Simplify the square root.}$$

The solution set is $\{-2, 2, -3, 3\}$.

Quick Check 1

Solve $x^4 - x^2 - 12 = 0$.

A Word of Caution When solving an equation by using a *u*-substitution, *do not stop after solving for u.* You must solve for the variable in the original equation.

The challenge is determining when a *u*-substitution will be helpful and determining what to let u represent. Look for an equation in which the variable part of the first term is the square of the variable part of a second term; in other words, its exponent is twice the exponent of the second term. Then let u represent the variable part with the smaller exponent.

EXAMPLE 2 Find the u-substitution that will convert the equation to a quadratic equation.

a) $x - 7\sqrt{x} - 30 = 0$

Solution

Let $u = \sqrt{x}$. This will allow us to replace $\sqrt{x}$ by u and x by u^2, since $u^2 = (\sqrt{x})^2 = x$. The resulting equation will be $u^2 - 7u - 30 = 0$, which is quadratic.

b) $x^{2/3} + 9x^{1/3} + 8 = 0$

Solution

Let $u = x^{1/3}$. We can then replace $x^{2/3}$ by u^2, since $(x^{1/3})^2 = x^{2/3}$. The resulting equation will be $u^2 + 9u + 8 = 0$.

c) $(x^2 + 6x)^2 + 13(x^2 + 6x) + 40 = 0$

Solution

Let $u = x^2 + 6x$ as this is the expression that is being squared. The resulting equation will be $u^2 + 13u + 40 = 0$.

Quick Check 2 **Find the u-substitution that will convert the given equation to a quadratic equation.**

a) $x + 11\sqrt{x} - 26 = 0$
b) $2x^{2/3} - 17x^{1/3} + 8 = 0$
c) $(x^2 - 4x)^2 - 9(x^2 - 4x) - 36 = 0$

EXAMPLE 3 Solve $x + 3\sqrt{x} - 10 = 0$.

Solution

Let $u = \sqrt{x}$. The resulting equation is $u^2 + 3u - 10 = 0$, which is a quadratic equation solvable by factoring. If we could not use factoring, then we would have to use the quadratic formula to solve for u.

$x + 3\sqrt{x} - 10 = 0$			
$u^2 + 3u - 10 = 0$			Replace $\sqrt{x}$ by u.
$(u - 2)(u + 5) = 0$			Factor.
$u - 2 = 0$	or	$u + 5 = 0$	Set each factor equal to 0.
$u = 2$	or	$u = -5$	Solve for u.
$\sqrt{x} = 2$	or	$\sqrt{x} = -5$	Replace u by
$(\sqrt{x})^2 = 2^2$	or	$(\sqrt{x})^2 = (-5)^2$	Square both sides of the equation.
$x = 4$	or	$x = 25$	Simplify.

Recall that any time we square each side of an equation, we must check for extraneous roots.

Check $(x = 25)$:

$$(25) + 3\sqrt{(25)} - 10 = 0 \quad \text{Substitute 25 for } x \text{ in the original equation.}$$
$$25 + 3 \cdot 5 - 10 = 0 \quad \text{Simplify the square root.}$$
$$30 = 0 \quad \text{Simplify.}$$

So $x = 25$ is not a solution of the equation. We could have seen this before squaring each side of the equation $\sqrt{x} = -5$. The square root of x cannot be negative, so the equation $\sqrt{x} = -5$ cannot have a solution. The check whether $x = 4$ is actually a solution is left to the reader. The solution set is $\{4\}$.

A Word of Caution Whenever we square both sides of an equation, such as in the previous example, we must check our solutions for extraneous roots.

EXAMPLE 4 Solve $x^{2/3} - 6x^{1/3} - 7 = 0$.

Solution

Let $u = x^{1/3}$. The resulting equation is $u^2 - 6u - 7 = 0$, which can be solved by factoring.

$$x^{2/3} - 6x^{1/3} - 7 = 0$$
$$u^2 - 6u - 7 = 0 \quad \text{Replace } x^{1/3} \text{ by } u.$$
$$(u - 7)(u + 1) = 0 \quad \text{Factor.}$$
$$u - 7 = 0 \quad \text{or} \quad u + 1 = 0 \quad \text{Set each factor equal to 0.}$$
$$u = 7 \quad \text{or} \quad u = -1 \quad \text{Solve for } u.$$
$$x^{1/3} = 7 \quad \text{or} \quad x^{1/3} = -1 \quad \text{Replace } u \text{ by } x^{1/3}.$$
$$(x^{1/3})^3 = 7^3 \quad \text{or} \quad (x^{1/3})^3 = (-1)^3 \quad \text{Raise each side to the third power.}$$
$$x = 343 \quad \text{or} \quad x = -1 \quad \text{Simplify.}$$

The solution set is $\{-1, 343\}$.

***Quick Check* 3**

Solve.

a) $x - 6x^{1/2} + 5 = 0$

b) $x^{2/3} - 4x^{1/3} + 3 = 0.$

Solving Radical Equations

Objective 2 **Solve radical equations.** Some equations that contain square roots cannot be solved by a u-substitution. When this happens, we will use the techniques developed in Section 6.5. We begin by isolating the radical, and then we proceed to square both sides of the equation. This can lead to an equation that is quadratic.

EXAMPLE 5 Solve $\sqrt{x + 7} + 5 = x$.

Solution

We begin by isolating the radical so that we may square each side of the equation.

$$\sqrt{x + 7} + 5 = x$$
$$\sqrt{x + 7} = x - 5 \quad \text{Subtract 5 to isolate the radical.}$$
$$(\sqrt{x + 7})^2 = (x - 5)^2 \quad \text{Square both sides.}$$
$$x + 7 = (x - 5)(x - 5) \quad \text{Square the binomial by multiplying it by itself.}$$

$$x + 7 = x^2 - 10x + 25 \quad \text{Multiply.}$$
$$0 = x^2 - 11x + 18 \quad \text{Subtract } x \text{ and 7 to collect all terms on the right side of the equation.}$$
$$0 = (x - 2)(x - 9) \quad \text{Factor.}$$
$$x - 2 = 0 \quad \text{or} \quad x - 9 = 0 \quad \text{Set each factor equal to 0.}$$
$$x = 2 \quad \text{or} \quad x = 9 \quad \text{Solve.}$$

Since we have squared each side of the equation, we must check for extraneous roots.

$x = 2$	$x = 9$
$\sqrt{(2) + 7} + 5 = (2)$	$\sqrt{(9) + 7} + 5 = (9)$
$\sqrt{9} + 5 = 2$	$\sqrt{16} + 5 = 9$
$3 + 5 = 2$	$4 + 5 = 9$
$8 = 2$	$9 = 9$
False	True

Quick Check 4 Solve $x + 7 = \sqrt{x + 9}$.

The solution $x = 2$ is an extraneous solution and must be omitted. The solution set is $\{9\}$.

Solving Rational Equations

Objective 3 Solve rational equations. Solving rational equations, which were covered in Chapter 5, often requires that we solve a quadratic equation. We begin to solve a rational equation by finding the LCD and multiplying each side of the equation by it to clear the equation of fractions. The resulting equation could be quadratic, as demonstrated in the next example. Once we solve the resulting equation, any solution that causes a denominator in the original equation to be equal to 0 must be omitted.

EXAMPLE 6 Solve $\dfrac{x}{x-4} + \dfrac{2}{x+3} = \dfrac{6}{x^2 - x - 12}$.

Solution

We begin by factoring the denominators to find the LCD. The LCD is $(x - 4)(x + 3)$, and solutions of $x = 4$ and $x = -3$ must be omitted, since either would result in a denominator of 0.

$$\frac{x}{x-4} + \frac{2}{x+3} = \frac{6}{x^2 - x - 12}$$
$$\frac{x}{x-4} + \frac{2}{x+3} = \frac{6}{(x-4)(x+3)} \quad \text{The LCD is } (x-4)(x+3).$$
$$(x-4)(x+3)\cdot\left(\frac{x}{x-4} + \frac{2}{x+3}\right) = (x-4)(x+3)\cdot\frac{6}{(x-4)(x+3)}$$

Multiply by the LCD.

$$\overset{1}{\cancel{(x-4)}}(x+3)\cdot\frac{x}{\underset{1}{\cancel{(x-4)}}} + (x-4)\overset{1}{\cancel{(x+3)}}\cdot\frac{2}{\underset{1}{\cancel{(x+3)}}} = \overset{1}{\cancel{(x-4)}}\overset{1}{\cancel{(x+3)}}\cdot\frac{6}{\underset{1}{\cancel{(x-4)}}\underset{1}{\cancel{(x+3)}}}$$

Distribute and divide out common factors.

$x(x+3)+2(x-4)=6$	Multiply remaining factors.
$x^2+3x+2x-8=6$	Multiply.
$x^2+5x-8=6$	Combine like terms.
$x^2+5x-14=0$	Collect all terms on the left side by subtracting 6.
$(x+7)(x-2)=0$	Factor.
$x+7=0$ or $x-2=0$	Set each factor equal to 0.
$x=-7$ or $x=2$	Solve.

Since neither solution causes a denominator to equal 0, we do not need to omit either solution. The solution set is $\{-7, 2\}$.

***Quick Check* 5**
Solve $\dfrac{2}{x+1}+\dfrac{1}{x-1}=1.$

Solving Work-Rate Problems

Objective 4 Solve work-rate problems. The last example of the section is a work-rate problem. Work-rate problems involve rational equations and were introduced in Chapter 5.

EXAMPLE 7 A water tower has two drainpipes attached to it. Working alone, the smaller pipe would take 15 minutes longer than the larger pipe to empty the tower. If both drainpipes work together, the tower can be drained in 30 minutes. How long would it take the small pipe, working alone, to drain the tower? (Round your answer to the nearest tenth of a minute.)

Solution

If we let t represent the amount of time that it takes for the larger pipe to drain the tower, then the time required for the small pipe to drain the tower can be represented by $t+15$. Recall that the work-rate is the reciprocal of the time required to complete the entire job. So the work-rate for the smaller pipe is $\frac{1}{t+15}$, and the work-rate for the large pipe is $\frac{1}{t}$. To determine the portion of the job completed by each pipe when they work together, we multiply the work-rate for each pipe by the amount of time that it takes for the two pipes to drain the tower while working together. Here is a table showing the important information:

Pipe	Time to Complete the Job Alone	Work-Rate	Time Working	Portion of the Job Completed
Smaller	$t+15$ minutes	$\frac{1}{t+15}$	30	$\frac{30}{t+15}$
Larger	t minutes	$\frac{1}{t}$	30	$\frac{30}{t}$

After adding the portion of the tower drained by the smaller pipe in 30 minutes to the portion of the tower drained by the larger pipe, the sum will equal 1, which represents finishing the entire job. The equation is $\frac{30}{t+15}+\frac{30}{t}=1$.

$$\frac{30}{t+15} + \frac{30}{t} = 1$$ The LCD is $t(t+15)$.

$$t(t+15)\cdot\left(\frac{30}{t+15} + \frac{30}{t}\right) = t(t+15)\cdot 1$$ Multiply both sides by the LCD.

$$\overset{1}{t(\cancel{t+15})}\cdot\frac{30}{\underset{1}{\cancel{(t+15)}}} + \overset{1}{\cancel{t}}(t+15)\cdot\frac{30}{\underset{1}{\cancel{t}}} = t(t+15)\cdot 1$$ Distribute and divide out common factors.

$$30t + 30(t+15) = t(t+15)$$ Multiply remaining factors.

$$30t + 30t + 450 = t^2 + 15t$$ Multiply. The resulting equation is quadratic.

$$60t + 450 = t^2 + 15t$$ Combine like terms.

$$0 = t^2 - 45t - 450$$ Collect all terms on the right side of the equation.

The quadratic expression does not factor, so we will use the quadratic formula.

$$t = \frac{45 \pm \sqrt{(-45)^2 - 4(1)(-450)}}{2(1)}$$ Substitute 1 for a, -45 for b, and -450 for c.

$$t = \frac{45 \pm \sqrt{3825}}{2}$$ Simplify the radicand.

At this point, we must use a calculator to approximate the solutions for t.

$$\frac{45+\sqrt{3825}}{2} \approx 53.4 \qquad \frac{45-\sqrt{3825}}{2} \approx -8.4$$

We omit the negative solution, so $t \approx 53.4$. The amount of time required by the small pipe is represented by $t + 15$, so it would take the small pipe approximately $53.4 + 15$, or 68.4, minutes to drain the tank.

***Quick Check* 6**

Working alone, Gabe can clean the gymnasium floor in 50 minutes less time than it takes Rob. If both janitors work together, it takes them 45 minutes to clean the gymnasium floor. How long would it take Gabe, working alone, to clean the gymnasium floor? (Round your answer to the nearest tenth of a minute.)

***Building Your Study Strategy* Time Management, 3 When to Study** When should you study math? Try to schedule your math study sessions for the time of day when you feel that you are at your sharpest. If you feel that you are most alert in the mornings, then you should reserve as much time as possible in the mornings to study math. Other students will find that it is better for them to study math in the middle of the day, in the evening, or at night. If you constantly get tired while studying late at night, change your study schedule so that you can study math before you get tired.

EXERCISES 7.3

FOR EXTRA HELP

MyMathLab

MathXL

Interactmath.com

MathXL Tutorials on CD

Video Lectures on CD

Addison-Wesley Math Tutor Center

Student's Solutions Manual

Vocabulary

1. Replacing a variable expression by the variable u in order to rewrite an equation as a quadratic equation is called solving by ________.

2. Whenever both sides of an equation are squared, it is necessary to check for ________ solutions.

Solve by making a u-substitution.

3. $x^4 - 5x^2 + 4 = 0$
4. $x^4 + 5x^2 - 36 = 0$
5. $x^4 - 6x^2 + 9 = 0$
6. $x^4 + 12x^2 + 32 = 0$
7. $x^4 + 7x^2 - 18 = 0$
8. $x^4 + x^2 - 2 = 0$
9. $x^4 - 13x^2 + 36 = 0$
10. $x^4 + 8x^2 + 16 = 0$
11. $x - 9\sqrt{x} + 8 = 0$
12. $x + 3\sqrt{x} - 40 = 0$
13. $x - 20\sqrt{x} + 64 = 0$
14. $x + 13\sqrt{x} - 30 = 0$
15. $x - 2x^{1/2} - 3 = 0$
16. $x + 11x^{1/2} + 18 = 0$
17. $x - 7x^{1/2} + 10 = 0$
18. $x + 5x^{1/2} - 14 = 0$
19. $x^6 - 28x^3 + 27 = 0$
20. $x^6 - 19x^3 - 216 = 0$
21. $x^6 + 16x^3 + 64 = 0$
22. $x^6 - 17x^3 + 16 = 0$
23. $x^6 + 1001x^3 + 1000 = 0$
24. $x^6 - 2x^3 + 1 = 0$
25. $x^{2/3} + 5x^{1/3} - 6 = 0$
26. $x^{2/3} - 12x^{1/3} + 20 = 0$
27. $x^{2/3} + 9x^{1/3} + 20 = 0$
28. $x^{2/3} - 4x^{1/3} - 45 = 0$
29. $(x - 3)^2 + 5(x - 3) + 4 = 0$
30. $(x + 7)^2 + 2(x + 7) - 24 = 0$
31. $(2x - 9)^2 - 6(2x - 9) - 27 = 0$
32. $(4x + 2)^2 - 4(4x + 2) - 60 = 0$

Solve.

33. $\sqrt{x^2 + 6x} = 4$
34. $\sqrt{x^2 - 8x} = 3$
35. $\sqrt{3x - 6} = x - 2$
36. $\sqrt{4x + 52} = x + 5$
37. $\sqrt{x + 15} - x = 3$
38. $\sqrt{3x^2 + 8x + 5} - 5 = 2x$
39. $\sqrt{x - 1} + 2 = \sqrt{2x + 5}$
40. $\sqrt{2x + 6} = 5 - \sqrt{x - 4}$

Solve.

41. $x = \dfrac{10}{x - 3}$
42. $x + 13 = \dfrac{30}{x}$
43. $1 + \dfrac{7}{x} + \dfrac{6}{x^2} = 0$
44. $1 + \dfrac{5}{x} - \dfrac{24}{x^2} = 0$
45. $\dfrac{1}{x} + \dfrac{7}{x + 2} = \dfrac{10}{x(x + 2)}$
46. $\dfrac{4}{x + 3} + \dfrac{1}{x - 5} = \dfrac{3}{x^2 - 2x - 15}$
47. $\dfrac{2}{x + 3} + \dfrac{x + 7}{x + 1} = \dfrac{9}{4}$
48. $\dfrac{x}{x + 7} - \dfrac{4}{x + 2} = \dfrac{7}{x^2 + 9x + 14}$
49. One small pipe takes twice as long to fill a tank as a larger pipe does. If it takes the two pipes 40 minutes to fill the tank when working together, how long will it take each pipe individually to fill the tank?

50. Rob takes 5 hours longer than Genevieve to paint a room. If they work together, they can paint a room in 6 hours. How long does it take Rob to paint a room by himself?

51. Daisuke takes 1 hour more than Hideki to rake the leaves from their yard in the fall. If the two work together, they can rake the leaves in 2 hours. How long does it take Daisuke, working alone, to rake the leaves? Round to the nearest tenth of an hour.

52. A water tank has two drainpipes attached to it. The larger pipe can drain the tank in 2 hours less than the smaller pipe. If both pipes are being used, they can drain the tank in 4 hours. How long would it take the smaller pipe to drain the tank working alone? Round to the nearest tenth of an hour.

53. A new copy machine can print a set of brochures in 20 minutes less than an older machine. If both machines work simultaneously, they can print the set of brochures in 30 minutes. How long would it take the older machine to print the set of brochures if it worked alone? Round to the nearest tenth of a minute.

54. It takes Leo 10 minutes more than Joyce to plant a flat of marigolds. If the two can plant a flat of marigolds in 16 minutes, how long does it take Leo to plant a flat of marigolds? Round to the nearest tenth of a minute.

Mixed Practice, 55–78

Solve, using the technique of your choice.

55. $x^2 - 8x - 13 = 0$

56. $(7x - 11)^2 - 6(7x - 11) = 0$

57. $x - 2\sqrt{x} - 48 = 0$

58. $x^2 + 15x + 54 = 0$

59. $(3x - 2)^2 = 32$

60. $x^{2/3} + 3x^{1/3} - 4 = 0$

61. $x^2 - 5x + 14 = 0$

62. $x^4 - 5x^2 - 36 = 0$

63. $x^2 - 6x - 91 = 0$

64. $(2x + 15)^2 = 18$

65. $x^6 - 7x^3 - 8 = 0$

66. $3x^2 + x - 8 = 0$

67. $\sqrt{3x - 2} = x - 2$

68. $x + \sqrt{x} - 12 = 0$

69. $(5x + 3)^2 + 2(5x + 3) - 15 = 0$

70. $x = \sqrt{6x - 27} + 3$

71. $(x + 8)^2 = -144$

72. $\dfrac{x + 3}{x - 4} + \dfrac{9}{x + 2} = \dfrac{2}{x^2 - 2x - 8}$

73. $x - 3\sqrt{x} - 4 = 0$

74. $x^2 + 13x + 50 = 0$

75. $x^{2/3} + 12x^{1/3} + 35 = 0$

76. $x^2 - 15x + 56 = 0$

77. $x^2 + 10x + 25 = 0$

78. $(x + 4)^2 = -108$

Writing in Mathematics

Answer in complete sentences.

79. Suppose that you are solving an equation that is quadratic in form by using the substitution $u = x^2$. Once you have solved the equation for u, explain how you would solve for x.

80. Suppose that you are solving an equation that is quadratic in form by using the substitution $u = \sqrt{x}$. Once you have solved the equation for u, explain how you would solve for x.

81. Give an example of an equation containing a square root that would be best solved by a u-substitution. Explain why you feel that using a u-substitution is a more efficient method for solving this equation than isolating the square root and squaring both sides of the equation.

82. Give an example of an equation containing a square root that would be best solved by isolating the square root and squaring both sides of the equation. Explain why you feel that isolating the square root and squaring both sides of the equation is a more efficient method for solving this equation than using a u-substitution.

83. ***Solutions Manual**** Write a solutions manual page for the following problem:

Two pipes can drain a tank in 7 minutes. It takes the smaller pipe 12 minutes longer to drain the tank alone than it takes the larger pipe to drain the tank alone. How long would it take each pipe, working alone, to drain the tank?

84. ***Newsletter**** Write a newsletter explaining how to solve equations with u-substitution.

***See Appendix B for details and sample answers.**

Quick Review Exercises

Section 7.3

Solve.

1. $(x - 3)^2 = 49$

2. $x^2 - 9x - 36 = 0$

3. $x^2 - 6x - 13 = 0$

4. $x^2 + 11x + 40 = 0$

7.4 Graphing Quadratic Equations and Quadratic Functions

Objectives

1. **Graph quadratic equations.**
2. **Graph parabolas that open downward.**
3. **Graph quadratic functions of the form $f(x) = ax^2 + bx + c$.**
4. **Find the vertex of a parabola by completing the square.**
5. **Graph quadratic equations of the form $y = a(x - h)^2 + k$.**
6. **Graph quadratic functions of the form $f(x) = a(x - h)^2 + k$.**

Graphing Quadratic Equations

Objective 1 **Graph quadratic equations.** The graphs of quadratic equations are not lines like the graphs of linear equations, or V-shaped like the graphs of absolute value equations. The graphs of quadratic equations are U-shaped and are called parabolas. Let's consider the graph of the most basic quadratic equation: $y = x^2$. We will first create a table of ordered pairs that will represent points on our graph.

x	$y = x^2$
−2	4
−1	1
0	0
1	1
2	4

The figure that follows shows these points and the graph of $y = x^2$. Notice that the shape, called a parabola, is not a straight line, but instead is U-shaped.

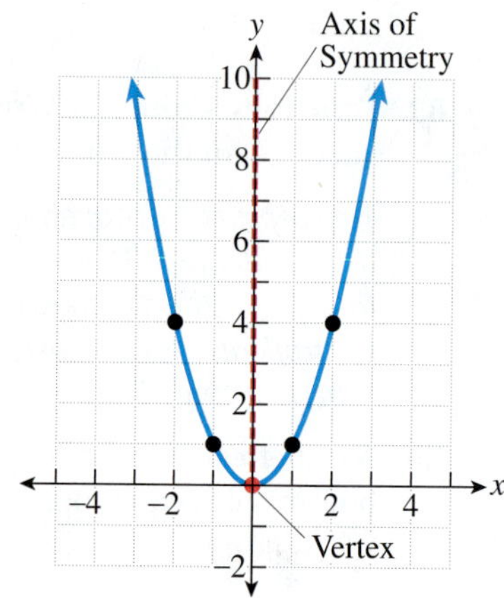

The point where the graph changes from decreasing to increasing is called the vertex. In the figure, the vertex is located at the bottom of the parabola. Notice that if we draw a vertical line through the vertex of the parabola, the left and right sides become mirror images. The graph of an equation is said to be **symmetric** if we can fold the graph along a line and the two sides of the graph coincide; in other words, a graph is symmetric if one side of the graph is a mirror image of the other side. A parabola is always symmetric, and we call the vertical line through the vertex the **axis of symmetry**. For this parabola, the equation of the axis of symmetry is $x = 0$.

We graph parabolas by plotting points, and the choice of our points is very important. We look for the y-intercept, the x-intercept(s) if there are any, and the vertex. We also use the axis of symmetry to help us find "mirror" points that are symmetric to points we have already graphed.

As before, we find the y-intercept by substituting 0 for x and solving for y. We will notice that the y-intercept of a quadratic equation in standard form $(y = ax^2 + bx + c)$ is always the point $(0, c)$. We find the x-intercepts, if there are any, by substituting 0 for y and solving for x. This equation will be quadratic, and we solve it by previous techniques.

A parabola opens upward if $a > 0$, and the vertex will be at the lowest point of the parabola. A parabola opens downward if $a < 0$, and the vertex will be at the highest point of the parabola. Parabolas that open downward will be covered later in this section. To learn how to find the coordinates of the vertex, we begin by completing the square for the equation $y = ax^2 + bx + c$.

$$y = ax^2 + bx + c$$

$$y - c = a\left(x^2 + \frac{b}{a}x\right)$$ Subtract c from both sides. Factor a from the two terms containing x.

$$y - c + \frac{b^2}{4a} = a\left(x^2 + \frac{b}{a}x + \frac{b^2}{4a^2}\right)$$ Half of $\frac{b}{a}$ is $\frac{b}{2a}$. Add $\left(\frac{b}{2a}\right)^2$ or $\frac{b^2}{4a^2}$ to the terms inside the parentheses. Since there is a factor in front of the parentheses, we add $a \cdot \frac{b^2}{4a^2}$ or $\frac{b^2}{4a}$ to the left side.

$$y = a\left(x + \frac{b}{2a}\right)^2 + \frac{4ac - b^2}{4a}$$ Factor the trinomial inside the parentheses. Collect all terms on the right side of the equation. $c - \frac{b^2}{4a} = \frac{4ac - b^2}{4a}$.

Since a squared expression cannot be negative, the minimum value of y occurs when $x + \frac{b}{2a} = 0$, or in other words, when $x = \frac{-b}{2a}$.

If the equation is in standard form, $y = ax^2 + bx + c$, then we can find the x-coordinate of the vertex by the formula $x = \frac{-b}{2a}$. We then find the y-coordinate of the vertex by substituting this value for x in the original equation.

EXAMPLE 1 Graph $y = x^2 + 6x + 8$.

Solution

We begin by setting $x = 0$ and solving for y to find the y-intercept.

$$y = (0)^2 + 6(0) + 8 = 8$$ Substitute 0 for x in the original equation.

The y-intercept is $(0, 8)$. To find the x-intercepts, we substitute 0 for y and attempt to solve for x.

$$0 = x^2 + 6x + 8$$ Substitute 0 for y in the original equation.

$$0 = (x + 2)(x + 4)$$ Factor the trinomial.

$$x = -2 \quad \text{or} \quad x = -4$$ Set each factor equal to 0 and solve.

The x-intercepts are $(-2, 0)$ and $(-4, 0)$. Finally, we find the vertex. We begin by finding the x-coordinate of the vertex.

$$x = \frac{-6}{2(1)} = -3 \qquad \text{Substitute 1 for } a \text{ and 6 for } b \text{ into } x = \tfrac{-b}{2a}.$$

Now we substitute this value for x into the original equation and solve for y.

$$y = (-3)^2 + 6(-3) + 8 \qquad \text{Substitute } -3 \text{ for } x.$$
$$y = -1 \qquad \text{Simplify.}$$

The vertex is at $(-3, -1)$.

The next figure on the left is a sketch showing the locations of the four points we have found. If we add the axis of symmetry, we see that the y-intercept $(0, 8)$ is three units to the right of the axis of symmetry. There is a mirror point with the same y-coordinate located three units to the left of the axis of symmetry.

This produces the graph shown at the right:

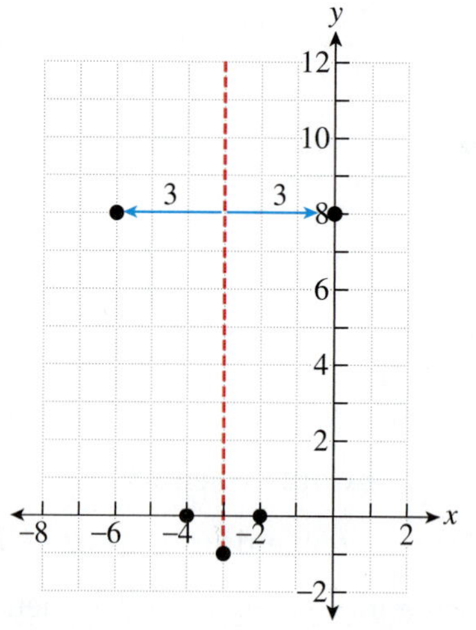

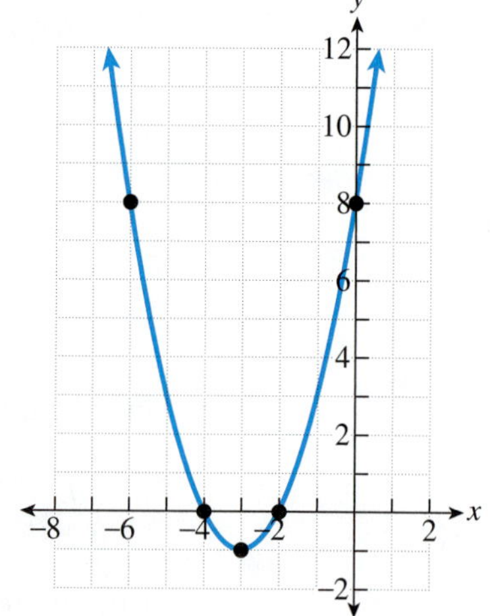

Quick Check 1

Graph $y = x^2 - 8x + 7$.

Using Your Calculator We can use the TI-84 to graph parabolas. To graph $y = x^2 + 6x + 8$, begin by pushing the [Y=] key and typing $x^2 + 6x + 8$ next to Y_1 as shown in the screen shot on the left.

To graph the parabola, press the [GRAPH] key. On the right is the screen that you should see in the standard viewing window:

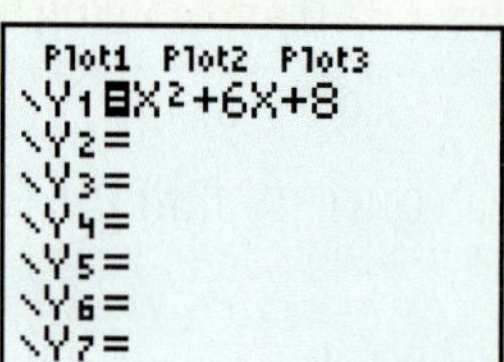

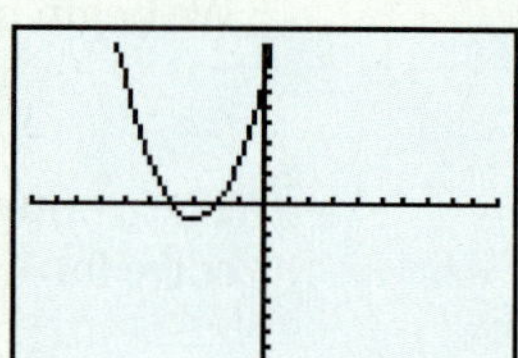

EXAMPLE 2 Graph $y = x^2 - 4x - 7$.

Solution

We begin by substituting 0 for x in the original equation and solving for y to find the y-intercept.

$$y = (0)^2 - 4(0) - 7 = -7 \qquad \text{Substitute 0 for } x.$$

The y-intercept is $(0, -7)$. Again, the y-intercept is the point $(0, c)$. To find the x-intercepts, we substitute 0 for y in the original equation and attempt to solve for x.

$$0 = x^2 - 4x - 7 \qquad \text{Substitute 0 for } y.$$

Since we cannot factor this expression, we must use the quadratic formula to find the x-intercepts.

$$x = \frac{4 \pm \sqrt{(-4)^2 - 4(1)(-7)}}{2(1)} \qquad \text{Substitute 1 for } a, -4 \text{ for } b, \text{ and } -7 \text{ for } c.$$

$$x = \frac{4 \pm \sqrt{44}}{2} \qquad \text{Simplify the radicand and denominator.}$$

$$x = \frac{4 \pm 2\sqrt{11}}{2} \qquad \text{Simplify the square root.}$$

$$x = \frac{\overset{1}{\cancel{2}}(2 \pm \sqrt{11})}{\underset{1}{\cancel{2}}} \qquad \text{Divide out common factors.}$$

$$x = 2 \pm \sqrt{11} \qquad \text{Simplify.}$$

Since $2 + \sqrt{11} \approx 5.3$ and $2 - \sqrt{11} \approx -1.3$, the x-intercepts are approximately $(5.3, 0)$ and $(-1.3, 0)$. Now we find the vertex, using the formula $x = \frac{-b}{2a}$.

Quick Check 2
Graph $y = x^2 + 6x - 9$.

$$x = \frac{4}{2(1)} = 2 \qquad \text{Substitute 1 for } a \text{ and } -4 \text{ for } b.$$

We substitute 2 for x in the original equation and solve for y to find the y-coordinate of the vertex.

$$y = (2)^2 - 4(2) - 7 \qquad \text{Substitute 2 for } x.$$
$$y = -11 \qquad \text{Simplify.}$$

The vertex is at $(2, -11)$.

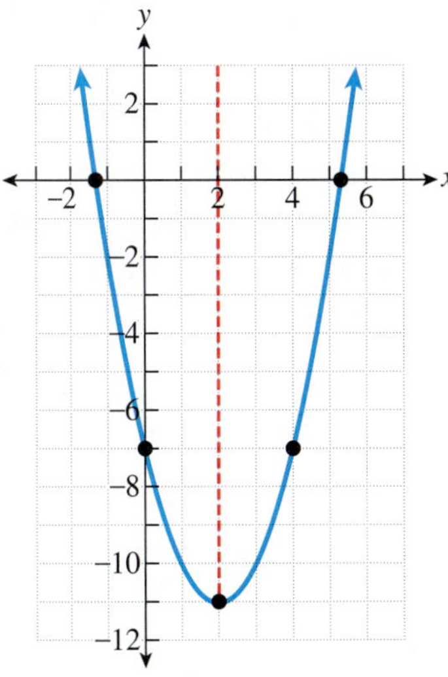

At the right is a sketch of the parabola showing the vertex, x-intercepts, y-intercept, and the axis of symmetry $(x = 2)$. Also shown is the point $(4, -7)$, which is symmetric to the y-intercept.

Occasionally, a parabola will not have any x-intercepts. In this case, we will have a negative discriminant $(b^2 - 4ac)$ when using the quadratic formula. However, we can determine that a parabola does not have any x-intercepts by altering the order we have been using to find our intercepts and vertex. If the vertex is above the x-axis, then the parabola does not have x-intercepts. If the vertex is on the x-axis, then the vertex is the only x-intercept of the parabola.

Example 3 Graph $y = x^2 - 2x + 2$.

Solution

In this example, we begin by finding the vertex rather than the intercepts. The x-coordinate of the vertex is found by the formula $x = \frac{-b}{2a}$.

$$x = \frac{2}{2(1)} = 1 \quad \text{Substitute 1 for } a \text{ and } -2 \text{ for } b.$$

Now we substitute 1 for x in the original equation and solve for y.

$$y = (1)^2 - 2(1) + 2 \quad \text{Substitute 1 for } x.$$
$$y = 1 \quad \text{Simplify.}$$

The vertex is at $(1, 1)$. To find the y-intercept, we substitute 0 for x in the original equation.

$$y = (0)^2 - 2(0) + 2 = 2 \quad \text{Substitute 0 for } x.$$

Quick Check 3
Graph $y = x^2 + 6x + 12$.

The y-intercept is $(0, 2)$.

If we plot the vertex and the y-intercept on the graph, we see there are no x-intercepts. The vertex is above the x-axis, and the parabola only moves in an upward direction from there. Therefore, this parabola does not have any x-intercepts. If we chose to use the quadratic formula to find the x-intercepts, we would have ended up with $x = \frac{2 \pm \sqrt{-4}}{2}$. Since the discriminant is negative, the equation $0 = x^2 - 2x + 2$ does not have any real solutions and the parabola does not have any x-intercepts.

At the right is the graph, showing the axis of symmetry $(x = 1)$ and a third point, $(2, 2)$, that is symmetric to the y-intercept.

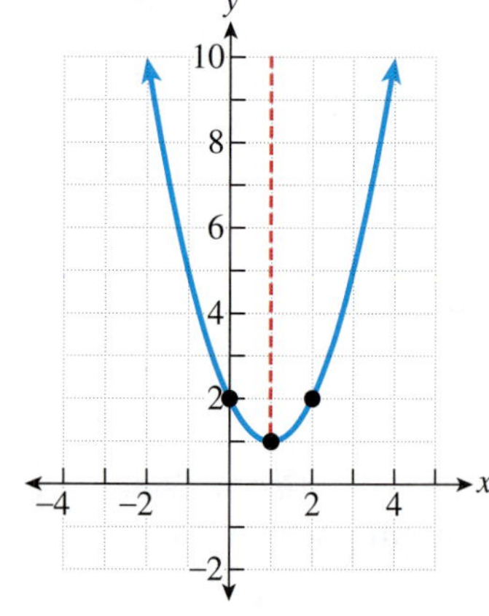

A Word of Caution If a parabola that opens upward has its vertex above the x-axis, then there are no x-intercepts.

Graphing Parabolas That Open Downward

Objective 2 **Graph parabolas that open downward.** Some parabolas open downward rather than upward. The way to determine which way a parabola will open is by writing the equation in standard form: $y = ax^2 + bx + c$. If a is positive, as it was in our previous examples, then the parabola will open upward. If a is negative, then the parabola will open downward. For instance, the graph of $y = -3x^2 + 5x - 7$ would open downward because the coefficient of the second-degree term is negative. The following example shows how to graph a parabola that opens downward:

Example 4 Graph $y = -x^2 + 4x + 12$.

Solution

We begin by finding the x-coordinate of the vertex.

$$x = \frac{-4}{2(-1)} = 2 \qquad \text{Substitute } -1 \text{ for } a \text{ and } 4 \text{ for } b.$$

Now substitute 2 for x in the original equation and solve for y.

$$y = -(2)^2 + 4(2) + 12 \qquad \text{Substitute 2 for } x.$$
$$y = 16 \qquad \text{Simplify.}$$

The vertex is at $(2, 16)$. To find the y-intercept, substitute 0 for x and solve for y.

$$y = -(0)^2 + 4(0) + 12 = 12 \qquad \text{Substitute 0 for } x.$$

The y-intercept is $(0, 12)$. If we plot the vertex and the y-intercept on the graph, we will see that there must be two x-intercepts. The vertex is above the x-axis, and the parabola opens downward, so the graph must cross the x-axis. To find the x-intercepts, we substitute 0 for y in the original equation and solve for x.

$$0 = -x^2 + 4x + 12 \qquad \text{Substitute 0 for } y.$$
$$x^2 - 4x - 12 = 0 \qquad \text{Collect all terms on the left side of the equation so that the coefficient of the squared term is positive.}$$
$$(x-6)(x+2) = 0 \qquad \text{Factor.}$$
$$x = 6 \quad \text{or} \quad x = -2 \qquad \text{Set each factor equal to 0 and solve.}$$

Quick Check 4
Graph $y = -x^2 + 9x + 10$.

The x-intercepts are $(6, 0)$ and $(-2, 0)$.

Following is the graph, showing the axis of symmetry $(x = 2)$ and the point $(4, 12)$ that is symmetric to the y-intercept:

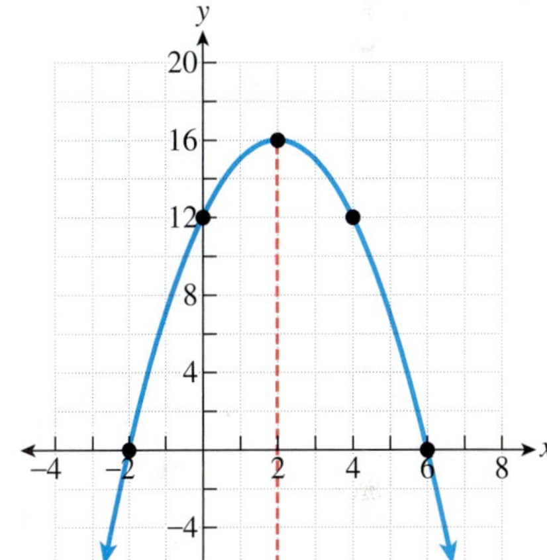

EXAMPLE 5 Graph $y = -\frac{1}{2}x^2 - x + 4$.

Solution

This parabola will open downward, because the second-degree term has a negative coefficient. We begin by finding the x-coordinate of the vertex.

$$x = \frac{1}{2\left(-\frac{1}{2}\right)} = -1 \qquad \text{Substitute } -\tfrac{1}{2} \text{ for } a \text{ and } -1 \text{ for } b.$$

Now we substitute -1 for x in the original equation and solve for y.

$$y = -\frac{1}{2}(-1)^2 - (-1) + 4 \quad \text{Substitute } -1 \text{ for } x.$$

$$y = -\frac{1}{2} + 1 + 4 \quad \text{Simplify each term.}$$

$$y = -\frac{1}{2} + \frac{2}{2} + \frac{8}{2} \quad \text{Rewrite each fraction with a common denominator of 2.}$$

$$y = \frac{9}{2} \quad \text{Simplify.}$$

The vertex is at $(-1, \frac{9}{2})$. For an equation of the form $y = ax^2 + bx + c$, the y-intercept is at the point $(0, c)$. The y-intercept of this parabola is $(0, 4)$.

If we put the vertex and the y-intercept on the graph, we will be able to see that there must be two x-intercepts. The vertex is above the x-axis, and the parabola opens downward, so the graph must cross the x-axis. To find the x-intercepts, we substitute 0 for y in the original equation and attempt to solve for x.

Quick Check 5

Graph $y = -\frac{1}{4}x^2 - \frac{5}{2}x - 4$.

$$0 = -\frac{1}{2}x^2 - x + 4 \quad \text{Substitute 0 for } y.$$

$$\frac{1}{2}x^2 + x - 4 = 0 \quad \text{Collect all terms on the left side of the equation so that the coefficient of the squared term is positive.}$$

$$2 \cdot \left(\frac{1}{2}x^2 + x - 4\right) = 2 \cdot 0 \quad \text{Multiply both sides of the equation by the LCD (2) to clear the equation of fractions.}$$

$$x^2 + 2x - 8 = 0 \quad \text{Distribute and simplify.}$$

$$(x - 2)(x + 4) = 0 \quad \text{Factor.}$$

$$x = 2 \quad \text{or} \quad x = -4 \quad \text{Set each factor equal to 0 and solve.}$$

The x-intercepts are $(2, 0)$ and $(-4, 0)$.

Following is the graph, with the axis of symmetry plotted to find a fifth point that is symmetric to the y-intercept:

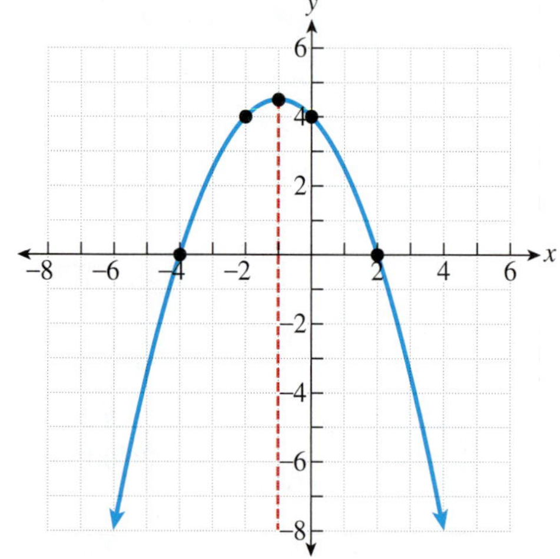

If a parabola that opens downward has its vertex below the x-axis, then it has no x-intercepts.

Here is a summary for graphing parabolas.

Graphing Parabolas

- Determine whether the parabola opens upward or downward. For an equation in standard form $(y = ax^2 + bx + c)$, the parabola will open upward if a is positive. If a is negative, the parabola will open downward.
- Find the vertex of the parabola. Use the formula $x = \dfrac{-b}{2a}$ to find the x-coordinate of the vertex. Substitute this result for x in the original equation to find the y-coordinate.
- Find the y-intercept of the parabola. We find the y-intercept by letting $x = 0$.
- After plotting the vertex and the y-intercept, determine whether there are any x-intercepts. If there are any x-intercepts, we find them by letting $y = 0$ and solving the resulting quadratic equation for x. If we use the quadratic formula to solve this equation and the discriminant, $b^2 - 4ac$, is negative, then there are no x-intercepts.

Graphing Quadratic Functions of the Form $f(x) = ax^2 + bx + c$

Objective 3 **Graph quadratic functions of the form $f(x) = ax^2 + bx + c$.** We now move on to graphing quadratic functions of the form $f(x) = ax^2 + bx + c$, $a \neq 0$. These graphs are also parabolas, and the x-coordinate of the vertex of the graph of a quadratic function $f(x) = ax^2 + bx + c$ can be found by the formula $x = \dfrac{-b}{2a}$. Once we know the x-coordinate of the vertex, we evaluate the function for that value to find the y-coordinate of the vertex. In other words, the y-coordinate of the vertex is $f\left(\dfrac{-b}{2a}\right)$. Thus, the vertex of the parabola is at the point $\left(\dfrac{-b}{2a}, f\left(\dfrac{-b}{2a}\right)\right)$.

To find the y-intercept of a parabola, determine $f(0)$. If the function is of the form $f(x) = ax^2 + bx + c$, then the y-intercept will be located at the point $(0, c)$.

To find the x-intercepts of a parabola, we set the function $f(x)$ equal to 0 and solve for x.

EXAMPLE 6 Graph $f(x) = x^2 - 8x + 12$.

Solution

In this quadratic function, $a = 1$, $b = -8$, and $c = 12$. Since a is positive, this parabola opens upward. We begin by finding the coordinates of the vertex. We find the x-coordinate of the vertex by using the formula $x = \dfrac{-b}{2a}$.

$$x = \frac{-(-8)}{2(1)} = 4 \qquad \text{Substitute 1 for } a \text{ and } -8 \text{ for } b \text{ and simplify.}$$

To find the y-coordinate of the vertex, we evaluate the function when $x = 4$.

$$f(4) = (4)^2 - 8(4) + 12 \quad \text{Substitute 4 for } x.$$
$$= -4 \quad \text{Simplify.}$$

The vertex is $(4, -4)$. Because the vertex is below the x-axis and the parabola opens upward, the graph must have two x-intercepts.

Since the y-intercept is the point $(0, c)$, the y-intercept is $(0, 12)$.

To find the x-intercepts, we set $f(x)$ equal to 0 and solve for x.

$$x^2 - 8x + 12 = 0 \quad \text{Set the function equal to 0.}$$
$$(x - 2)(x - 6) = 0 \quad \text{Factor the quadratic expression.}$$
$$x = 2 \quad \text{or} \quad x = 6 \quad \text{Set each factor equal to 0 and solve.}$$

Quick Check 6

Graph $f(x) = x^2 + 2x - 24$.

The x-intercepts are $(2, 0)$ and $(6, 0)$.

We now can use the axis of symmetry, $x = 4$, to find the point that is symmetric to the y-intercept.

Since the point $(0, 12)$ is 4 units to the left of the axis of symmetry, there must be a symmetric point on the parabola with the same y-coordinate that is 4 units to the right of the axis of symmetry. This point is $(8, 12)$. Once all five points have been plotted, we finish by drawing a smooth, U-shaped curve through them.

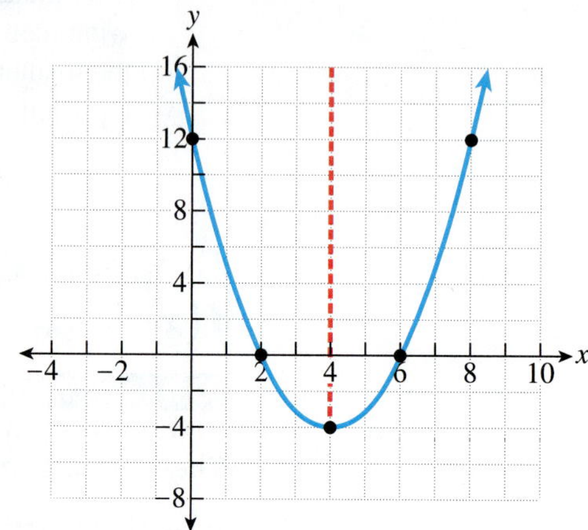

In the next example, we will graph a parabola that opens downward. In addition, we will have to use the quadratic formula to find the x-intercepts.

EXAMPLE 7 Graph $f(x) = -x^2 + 6x + 5$.

Solution

In this quadratic function, $a = -1$. Since a is negative, this parabola opens downward. We begin by finding the coordinates of the vertex. We find the x-coordinate of the vertex by using the formula $x = \frac{-b}{2a}$.

$$x = \frac{-6}{2(-1)} = 3 \quad \text{Substitute } -1 \text{ for } a \text{ and 6 for } b \text{ and simplify.}$$

To find the y-coordinate of the vertex, we evaluate the function when $x = 3$.

$$f(3) = -(3)^2 + 6(3) + 5 \quad \text{Substitute 3 for } x.$$
$$= 14 \quad \text{Simplify.}$$

The vertex is $(3, 14)$. Since the vertex is above the x-axis and the parabola opens downward, the graph must have two x-intercepts.

The y-intercept is $(0, 5)$.

To find the x-intercepts, we set $f(x)$ equal to 0 and solve for x.

$-x^2 + 6x + 5 = 0$ Set the function equal to 0.

$0 = x^2 - 6x - 5$ Collect all terms on the right side of the equation so that the coefficient of the second-degree term is positive.

Since the expression $x^2 - 6x - 5$ cannot be factored, we will use the quadratic formula to find the x-intercepts.

$$x = \frac{6 \pm \sqrt{(-6)^2 - 4(1)(-5)}}{2(1)}$$ Substitute 1 for a, -6 for b, and -5 for c in the quadratic formula.

$$x = \frac{6 \pm \sqrt{56}}{2}$$ Simplify the radicand and denominator.

$$x = \frac{6 \pm 2\sqrt{14}}{2}$$ Simplify the radical.

$$x = \frac{\overset{1}{\cancel{2}}(3 \pm \sqrt{14})}{\underset{1}{\cancel{2}}}$$ Divide out common factors.

$$x = 3 \pm \sqrt{14}$$ Simplify.

Quick Check 7

Graph $f(x) = -x^2 + 4x + 3$.

Using a calculator, we find that $3 + \sqrt{14} \approx 6.7$ and $3 - \sqrt{14} \approx -0.7$, so the x-intercepts are located at approximately $(6.7, 0)$ and $(-0.7, 0)$.

Here is the graph. We can find the point $(6, 5)$ by using the axis of symmetry $(x = 3)$ and the y-intercept. Once all five points have been plotted, we draw an inverted U-shaped curve through them.

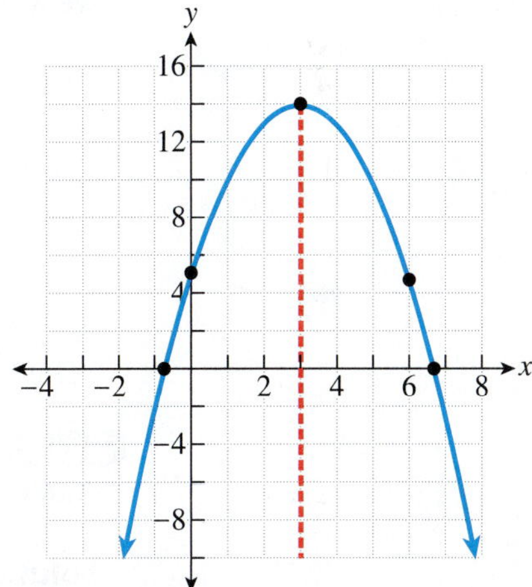

Finding the Vertex of a Parabola by Completing the Square

Objective 4 **Find the vertex of a parabola by completing the square.** As an alternative to using the formula $x = \frac{-b}{2a}$, we could find the coordinates of the vertex by completing the square.

Finding the Vertex of $y = a(x - h)^2 + k$

The graph of the quadratic equation $y = a(x - h)^2 + k$ is a parabola with axis of symmetry $x = h$ and vertex (h, k). The parabola opens upward if a is positive and opens downward if a is negative.

Example 8 Find the vertex and axis of symmetry for the parabola.

a) $y = (x - 4)^2 + 3$

Solution

This equation is in the form $y = a(x - h)^2 + k$. The axis of symmetry is $x = 4$, and the vertex is $(4, 3)$.

b) $y = -(x + 1)^2 - 4$

Solution

This parabola opens downward, but this does not affect how we find the axis of symmetry or the vertex. The axis of symmetry is $x = -1$, and the vertex is $(-1, -4)$.

***Quick Check* 8**
Find the vertex and axis of symmetry for the parabola.

a) $y = (x + 2)^2 - 8$
b) $y = -2(x - 8)^2 + 7$

Example 9 Find the vertex of the parabola $y = x^2 - 14x + 33$ by completing the square.

Solution

We will begin by isolating the terms containing x on the right side of the equation. Then we will complete the square.

$$y = x^2 - 14x + 33$$

$$y - 33 = x^2 - 14x$$ Subtract 33 to isolate the terms containing x.

$$y - 33 + 49 = x^2 - 14x + 49$$ Take half of the coefficient of the first-degree term, square it, and add it to both sides of the equation. $(\frac{-14}{2})^2 = (-7)^2 = 49$.

$$y + 16 = (x - 7)^2$$ Factor the trinomial as a perfect square.

$$y = (x - 7)^2 - 16$$ Subtract 16 to isolate y.

The vertex of the parabola is $(7, -16)$.

Example 10 Find the vertex of the parabola $y = -2x^2 - 12x + 21$ by completing the square.

Solution

After isolating the terms containing x, we must factor out -2 so that the coefficient of the squared term is 1.

$$y = -2x^2 - 12x + 21$$

$$y - 21 = -2x^2 - 12x$$ Subtract 21 to isolate the terms containing x.

$$y - 21 = -2(x^2 + 6x)$$ Factor out the common factor -2.

$$y - 21 - 18 = -2(x^2 + 6x + 9)$$ $(\frac{6}{2})^2 = 3^2 = 9$. Add 9 in the parentheses. This is equivalent to subtracting 18 on the right side because of the factor -2 that is multiplied by 9. Subtract 18 from the left side.

$$y - 39 = -2(x + 3)^2$$ Factor the trinomial as a perfect square.

$$y = -2(x + 3)^2 + 39$$ Add 39 to isolate y.

The vertex is $(-3, 39)$.

***Quick Check* 9**
Find the vertex of the parabola by completing the square.

a) $y = x^2 + 6x - 40$
b) $y = -x^2 + 12x + 45$

Graphing Quadratic Equations of the Form $y = a(x - h)^2 + k$

Objective 5 **Graph quadratic equations of the form $y = a(x - h)^2 + k$.**

EXAMPLE 11 Graph $y = (x + 4)^2 - 9$.

Solution

This parabola opens upward, and the vertex is $(-4, -9)$. Next, we find the y-intercept.

$$y = (0 + 4)^2 - 9 \quad \text{Substitute 0 for } x.$$
$$y = 7 \quad \text{Simplify.}$$

The y-intercept is $(0, 7)$. Since the parabola opens upward and the vertex is below the x-axis, the parabola has two x-intercepts. We find the coordinates of the x-intercepts by substituting 0 for y and solving for x by extracting square roots.

$$0 = (x + 4)^2 - 9 \quad \text{Substitute 0 for } y.$$
$$9 = (x + 4)^2 \quad \text{Add 9 to isolate } (x + 4)^2.$$
$$\pm\sqrt{9} = \sqrt{(x + 4)^2} \quad \text{Take the square root of each side.}$$
$$\pm 3 = x + 4 \quad \text{Simplify each square root.}$$
$$-4 \pm 3 = x \quad \text{Subtract 4 to isolate } x.$$

The x-coordinates of the x-intercepts are $-4 + 3 = -1$ and $-4 - 3 = -7$. The x-intercepts are $(-1, 0)$ and $(-7, 0)$. The axis of symmetry is $x = -4$, and the point $(-8, 7)$ is symmetric to the y-intercept.

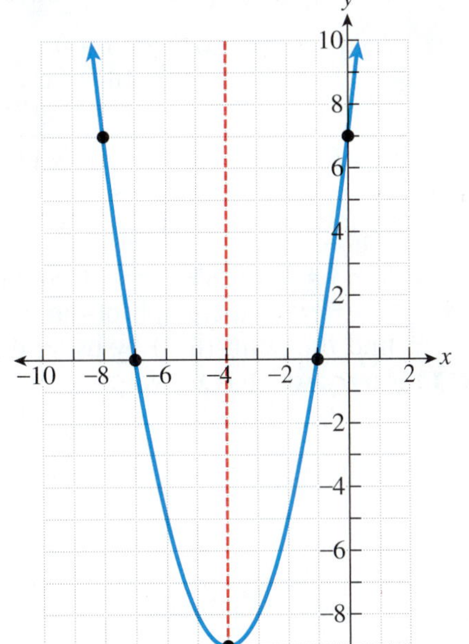

Quick Check 10

Graph $y = -(x - 2)^2 + 1$.

Graphing Quadratic Functions of the Form $f(x) = a(x - h)^2 + k$

Objective 6 **Graph quadratic functions of the form $f(x) = a(x - h)^2 + k$.** We will now graph quadratic functions of the form $f(x) = a(x - h)^2 + k$. The graph of the quadratic function $f(x) = a(x - h)^2 + k$ is a parabola with axis of symmetry $x = h$ and vertex (h, k). The parabola opens upward if a is positive and opens downward if a is negative. We still find the y-intercept of the function by finding $f(0)$. We also find the x-intercepts, if there are any, by setting the function $f(x)$ equal to 0 and solving for x.

EXAMPLE 12 Graph $f(x) = -(x - 3)^2 + 16$.

Solution

For this quadratic function, $a = -1$. Since a is negative, this parabola opens downward and its vertex is $(3, 16)$. Since the vertex is above the x-axis and the parabola opens downward, this parabola has two x-intercepts.

We next find the y-intercepts by evaluating the function at $x = 0$.

$$\begin{aligned} f(0) &= -(0 - 3)^2 + 16 && \text{Substitute 0 for } x. \\ &= 7 && \text{Simplify.} \end{aligned}$$

The y-intercept is $(0, 7)$.

To find the x-intercepts, we set the function equal to 0 and solve for x. We will solve the resulting equation by extracting square roots.

$$\begin{aligned} -(x - 3)^2 + 16 &= 0 && \text{Set the function equal to 0.} \\ 16 &= (x - 3)^2 && \text{Add } (x-3)^2 \text{ to isolate the squared expression.} \\ \pm\sqrt{16} &= \sqrt{(x - 3)^2} && \text{Take the square root of both sides.} \\ \pm 4 &= x - 3 && \text{Simplify each square root.} \\ 3 \pm 4 &= x && \text{Add 3 to isolate } x. \end{aligned}$$

$3 + 4 = 7$ and $3 - 4 = -1$; so the x-intercepts are $(7, 0)$ and $(-1, 0)$.

Here is the graph of the function. We can find the point $(6, 7)$ by using the axis of symmetry, which is $x = 3$.

Quick Check 11

Graph $f(x) = (x + 1)^2 - 8$.

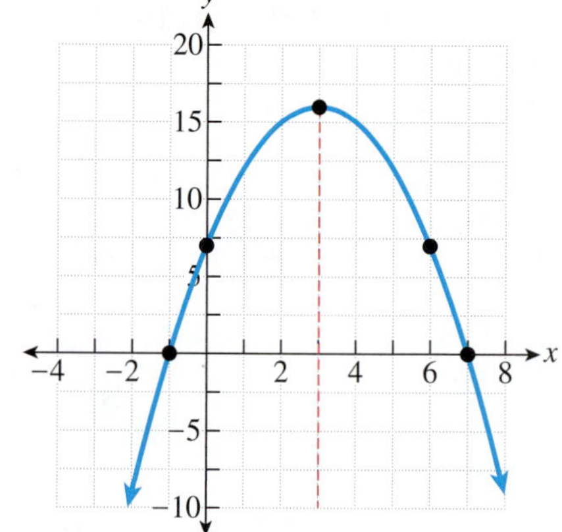

Building Your Study Strategy **Time Management, 4** **Setting Goals** One way to get the most from a study session is to establish a set of goals to accomplish for each session. For example, during a 90-minute session, you may set a goal of completing the homework assignment, in addition to creating a set of note cards for the most recent section in the textbook. Setting goals will encourage you to work quickly and efficiently. Many students set a goal of studying for a certain amount of time, but time alone is not a worthy goal. Create a "to-do" list each time you start a study session, and you will find that you will have a much greater chance of reaching your goals.

EXERCISES 7.4

FOR EXTRA HELP

MyMathLab

MathXL

Interactmath.com

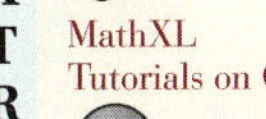

MathXL Tutorials on CD

Video Lectures on CD

Addison-Wesley Math Tutor Center

Student's Solutions Manual

Vocabulary

1. The graph of a quadratic equation is a U-shaped graph called a ________.
2. A parabola opens upward if the leading coefficient is ________.
3. A parabola opens downward if the leading coefficient is ________.
4. The turning point of a parabola is called its ________.
5. To find the ________ of a parabola, substitute 0 for x and solve for y.
6. To find the ________ of a parabola, substitute 0 for y and solve for x.

Find the vertex of the parabola associated with each of the following quadratic equations, as well as the equation of the axis of symmetry.

7. $y = x^2 - 6x - 21$
8. $y = x^2 + 8x - 15$
9. $y = x^2 + 14x + 40$
10. $y = x^2 - 4x - 45$
11. $y = -x^2 - 10x + 32$
12. $y = -x^2 + 4x + 13$
13. $y = x^2 - 9x + 14$
14. $y = x^2 + 7x + 7$
15. $y = 3x^2 + 12x - 20$
16. $y = -2x^2 - 6x + 17$
17. $y = x^2 - 7$
18. $y = x^2 - 8x$
19. $y = (x - 3)^2 - 4$
20. $y = (x + 5)^2 + 1$
21. $y = -(x + 4)^2 + 4$
22. $y = -(x - 10)^2 + 15$

Find the x- and y-intercepts of the parabola associated with the given quadratic equations. If necessary, round to the nearest tenth. If the parabola does not have any x-intercepts, state "no x-intercepts."

23. $y = x^2 + 6x + 5$
24. $y = x^2 - 8x - 6$
25. $y = -x^2 + 4x - 7$
26. $y = x^2 - 8x + 7$
27. $y = 2x^2 - 7x - 13$
28. $y = -x^2 + x + 72$
29. $y = x^2 + 4x$

30. $y = -x^2 + 12x - 43$

31. $y = x^2 - 6x + 9$

32. $y = -x^2 + 16x - 64$

33. $y = (x + 2)^2 - 1$

34. $y = (x - 7)^2 - 8$

35. $y = -(x - 5)^2 - 16$

36. $y = -(x + 6)^2 + 9$

Graph the given parabolas. Label the vertex and all intercepts.

37. $y = x^2 - 2x - 3$

38. $y = x^2 - 7x + 5$

39. $y = -x^2 + 4x$

40. $y = x^2 - 8x - 20$

41. $y = -x^2 + 3x - 4$

42. $y = -x^2 + 4$

43. $y = 2x^2 + 8x - 25$

44. $y = x^2 - 2x - 1$

45. $y = x^2 + 6x + 10$

50. $y = x^2 + x - 6$

51. $y = x^2 - 12$

46. $y = x^2 + 8x + 18$

47. $y = -x^2 + 6x - 6$

52. $y = -x^2 + 4x - 4$

53. $y = \frac{1}{2}x^2 - 2x + \frac{3}{2}$

48. $y = x^2 - 5x$

49. $y = x^2 + 8x + 16$

54. $y = \frac{1}{3}x^2 + 2x - 9$

55. $y = (x - 2)^2 + 3$

56. $y = (x - 5)^2 + 1$

57. $y = -(x + 1)^2 + 3$

58. $y = -(x + 3)^2 + 8$

Graph the function. Label the vertex, y-intercept, and any x-intercepts.

59. $f(x) = x^2 - 4x + 3$

60. $f(x) = x^2 - 2x - 8$

61. $f(x) = x^2 + 4x - 21$

62. $f(x) = x^2 + 10x + 16$

63. $f(x) = -x^2 + 6x + 7$

64. $f(x) = -x^2 + 2x + 15$

65. $f(x) = x^2 + 6x - 5$

66. $f(x) = x^2 - 3x - 8$

67. $f(x) = x^2 - 6x + 10$

68. $f(x) = x^2 + 4x + 8$

69. $f(x) = -x^2 - 4x - 6$

70. $f(x) = -x^2 - 2x - 7$

71. $f(x) = (x + 4)^2 - 9$

72. $f(x) = (x + 1)^2 - 4$

73. $f(x) = -(x + 1)^2 + 16$

74. $f(x) = -(x - 3)^2 + 9$

75. $f(x) = (x - 4)^2 - 5$

76. $f(x) = (x + 2)^2 - 3$

77. $f(x) = (x + 2)^2 + 10$

78. $f(x) = -(x + 3)^2 - 16$

Find an equation of a parabola that meets the given conditions.

79. Axis of symmetry: $x = 4$, range: $[-9, \infty)$

80. Axis of symmetry: $x = 2$, range: $(-\infty, 6]$

81. x-intercepts: $(-5, 0)$ and $(3, 0)$

82. Axis of symmetry: $x = -5$, no x-intercepts

Writing in Mathematics

Answer in complete sentences.

83. Explain how to determine that a parabola does not have any x-intercepts.

84. Explain how to find the vertex of a parabola by completing the square. Use an example as an illustration of the process.

85. ***Solutions Manual**** Write a solutions manual page for the following problem:

Graph $y = (x - 3)^2 - 4$.

86. ***Newsletter**** Write a newsletter explaining how to graph an equation of the form $y = ax^2 + bx + c$.

***See Appendix B for details and sample answers.**

7.5 Applications Using Quadratic Equations

Objectives

1. **Solve applied geometric problems.**
2. **Solve problems by using the Pythagorean theorem.**
3. **Solve applied problems by using the Pythagorean theorem.**
4. **Find the maximum or minimum value of a quadratic function.**
5. **Solve applied maximum/minimum problems.**

In this section, we will learn how to solve applied problems resulting in quadratic equations.

Solving Applied Geometric Problems

Objective 1 **Solve applied geometric problems.** Problems involving the area of a geometric figure often lead to quadratic equations, as area is measured in square units. Here are some useful area formulas:

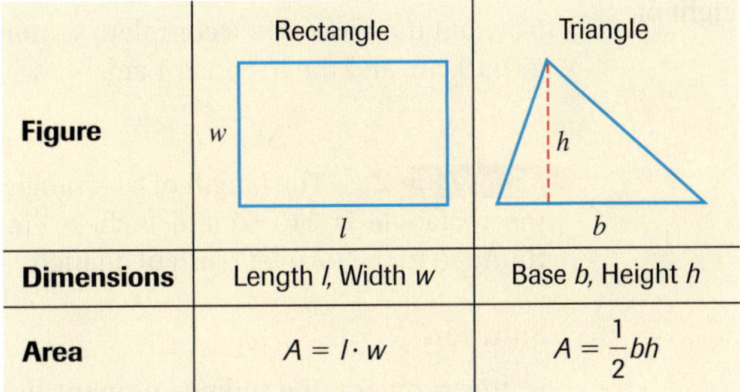

	Rectangle	Triangle
Figure	w, l	h, b
Dimensions	Length l, Width w	Base b, Height h
Area	$A = l \cdot w$	$A = \frac{1}{2}bh$

EXAMPLE 1 The height of a triangle is 5 centimeters less than its base. The area is 18 square centimeters. Find the base and the height of the triangle.

Solution

In this problem, the unknown quantities are the base and the height, while we know that the area is 18 square centimeters. Since the height is given in terms of the base, a wise choice is to represent the base of the triangle by x. Since the length is 5 cm less than the base, it can be represented by $x - 5$. This information is summarized in the following table:

Unknowns	***Known***	
Base: x Height: $x - 5$	Area: 18 cm^2	$x - 5$ x

Since the area of a triangle is given by the formula $A = \frac{1}{2}bh$, the equation we need to solve is $\frac{1}{2}x(x - 5) = 18$.

$$\frac{1}{2}x(x-5) = 18$$

$$\overset{1}{\cancel{2}} \cdot \frac{1}{\underset{1}{\cancel{2}}}x(x-5) = 2 \cdot 18 \qquad \text{Multiply both sides by 2.}$$

$$x^2 - 5x = 36 \qquad \text{Simplify each side of the equation.}$$

$$x^2 - 5x - 36 = 0 \qquad \text{Rewrite in standard form.}$$

$$(x-9)(x+4) = 0 \qquad \text{Factor.}$$

$$x - 9 = 0 \quad \text{or} \quad x + 4 = 0 \qquad \text{Set each factor equal to 0.}$$

$$x = 9 \quad \text{or} \quad x = -4 \qquad \text{Solve.}$$

> **Quick Check 1**
> **The base of a triangle is 2 inches longer than 3 times its height. If the area of the triangle is 60 square inches, find the base and height of the triangle.**

Look back to the table of unknowns. If $x = -4$, then the length is -4 cm and the height is -9 cm, which is not possible. The solutions derived from $x = -4$ are omitted. If $x = 9$, then the base is 9 cm, while the width is $9 - 5 = 4$ cm.

Base: $x = 9$
Height: $x - 5 = 9 - 5 = 4$

Now put the answer in a complete sentence with the proper units. The base of the triangle is 9 cm, and the height is 4 cm.

EXAMPLE 2 The length of a rectangle is 7 inches less than twice its width. The area of the rectangle is 240 square inches. Find the length and the width of the rectangle, rounded to the nearest tenth of an inch.

Solution

In this problem, the unknown quantities are the length and the width, while we know that the area is 240 square inches. Since the length is given in terms of the width, a wise choice is to represent the width of the rectangle by x. Since the length is 7 inches less than twice the width, it can be represented by $2x - 7$. This information is summarized in the following table:

Unknowns	***Known***	
Length: $2x - 7$ Width: x	Area: 240 in.^{2}	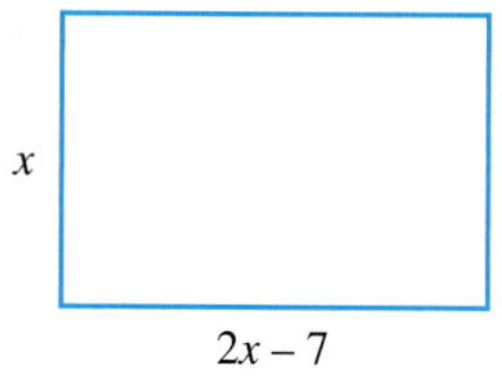

Since the area of a rectangle is equal to its length times its width, the equation we need to solve is $x(2x - 7) = 240$.

$$x(2x-7) = 240$$

$$2x^2 - 7x = 240 \qquad \text{Distribute.}$$

$$2x^2 - 7x - 240 = 0 \qquad \text{Rewrite in standard form.}$$

This trinomial does not factor, so we will use the quadratic formula with $a = 2$, $b = -7$, and $c = -240$ to solve this equation.

$$x = \frac{7 \pm \sqrt{(-7)^2 - 4(2)(-240)}}{2(2)}$$ Substitute 2 for a, -7 for b, and -240 for c.

$$x = \frac{7 \pm \sqrt{1969}}{4}$$ Simplify the radicand and the denominator.

We use a calculator to approximate these solutions.

$$\frac{7 + \sqrt{1969}}{4} \approx 12.8 \qquad \frac{7 - \sqrt{1969}}{4} \approx -9.3$$

We may omit the negative solution, as the length and width of the rectangle would both be negative. If $x \approx 12.8$, then the width is approximately 12.8 inches, while the length is approximately $2(12.8) - 7 = 18.6$ inches.

Width: $x \approx 12.8$
Length: $2x - 7 \approx 2(12.8) - 7 = 18.6$

The width of the rectangle is approximately 12.8 inches, and the length is approximately 18.6 inches.

Quick Check 2

The length of a rectangle is 8 inches more than its width. The area of the rectangle is 350 square inches. Find the length and the width of the rectangle, rounded to the nearest tenth of an inch.

Solving Problems by the Pythagorean Theorem

Objective 2 **Solve problems by using the Pythagorean theorem.** Other applications of geometry that lead to quadratic equations involve the Pythagorean theorem, which is an equation that relates the length of the three sides of a right triangle.

The side opposing the right angle is called the **hypotenuse** and is labeled c in the figure at the left. The other two sides that form the right angle are called the **legs** of the triangle and are labeled a and b. (It makes no difference which is a and which is b.)

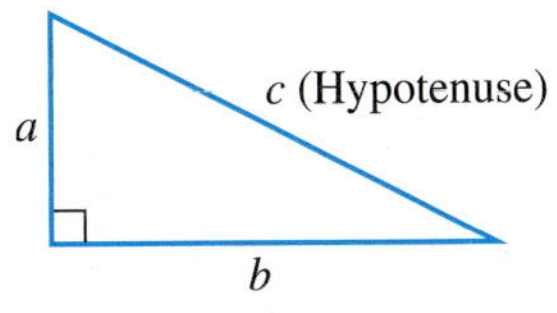

Pythagorean Theorem

For any right triangle whose hypotenuse has length c and whose legs have lengths a and b, respectively,

$$a^2 + b^2 = c^2.$$

Example 3 A right triangle has a hypotenuse that measures 18 inches, and one of its legs is 6 inches long. Find the length of the other leg, to the nearest tenth of an inch.

Solution

In this problem, the length of one of the legs is unknown. We can label the unknown leg as either a or b.

Unknowns	*Known*
a	b: 6 in. Hypotenuse (c): 18 in.

$$
\begin{aligned}
a^2 + 6^2 &= 18^2 && \text{Substitute 6 for } b \text{ and 18 for } c \text{ in } a^2 + b^2 = c^2.\\
a^2 + 36 &= 324 && \text{Square 6 and 18.}\\
a^2 &= 288 && \text{Subtract 36.}\\
\sqrt{a^2} &= \pm\sqrt{288} && \text{Solve by extracting square roots.}\\
a &= \pm 12\sqrt{2} && \text{Simplify the square root.}
\end{aligned}
$$

Since the length of a leg must be a positive number, we are concerned only about $12\sqrt{2}$, which rounds to be 17.0 inches. The length of the other leg is approximately 17.0 inches.

Quick Check 3
One leg of a right triangle measures 5 inches, while the hypotenuse measures 11 inches. Find, to the nearest hundredth of an inch, the length of the other leg of the triangle.

Applications of the Pythagorean Theorem

Objective 3 Solve applied problems by using the Pythagorean theorem. Now we turn our attention to solving applied problems by the Pythagorean theorem. In these problems, we begin by drawing a picture of the situation. We must be able to identify a right triangle in our figure in order to apply the Pythagorean theorem.

EXAMPLE 4 A 5-foot ladder is leaning against a wall. If the bottom of the ladder is 3 feet from the base of the wall, how high up the wall is the top of the ladder?

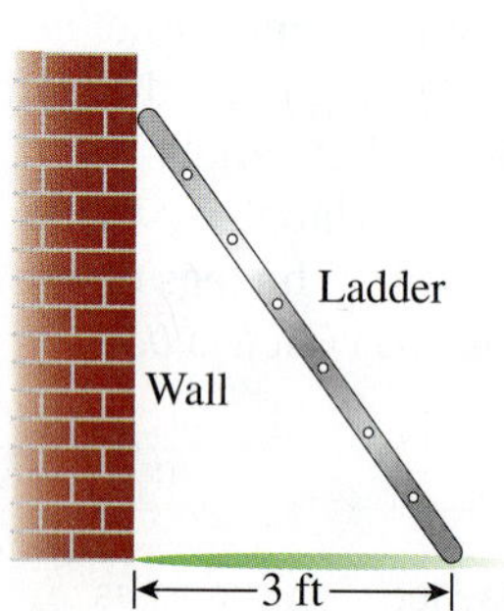

Solution

The ladder, the wall, and the ground form a right triangle, with the ladder being the hypotenuse. In this problem the height of the wall, which is the length of one of the legs in the right triangle, is unknown.

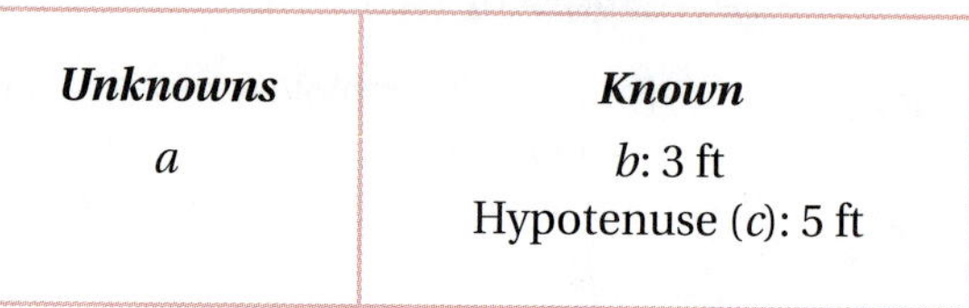

Unknowns	*Known*
a	b: 3 ft Hypotenuse (c): 5 ft

Quick Check 4
An 8-foot ladder is leaning against a wall. If the bottom of the ladder is 2 feet from the base of the wall, how high up the wall is the top of the ladder? (Round to the nearest tenth of a foot.)

$$
\begin{aligned}
a^2 + 3^2 &= 5^2 && \text{Substitute 3 for } b \text{ and 5 for } c \text{ in } a^2 + b^2 = c^2.\\
a^2 + 9 &= 25 && \text{Square 3 and 5.}\\
a^2 &= 16 && \text{Subtract 9.}\\
\sqrt{a^2} &= \pm\sqrt{16} && \text{Solve by extracting square roots.}\\
a &= \pm 4 && \text{Simplify the square root.}
\end{aligned}
$$

Again, the negative solution does not make sense in this problem. The ladder is resting at a point on the wall that is 4 feet above the ground.

EXAMPLE 5 The Modesto airport is located 120 miles north and 50 miles west of the Visalia airport. If a plane flies directly from Visalia to Modesto, how many miles is the flight?

50 miles
Modesto
N
W
E
S
120 miles
Visalia

Solution

The directions of north and west form a 90-degree angle, so our picture shows us that we have a right triangle whose hypotenuse (the direct distance from Visalia to Modesto) is unknown.

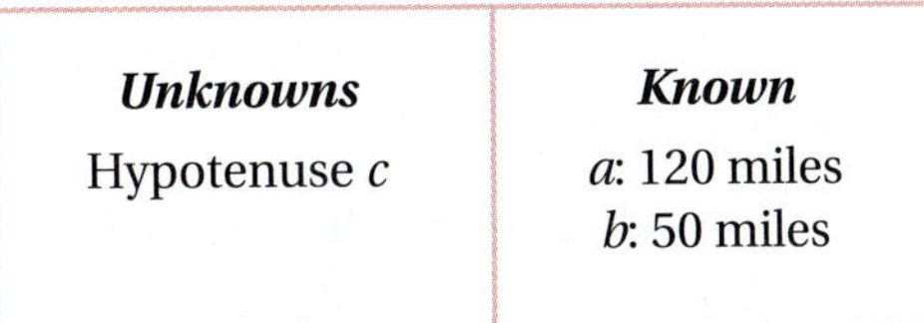

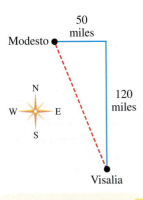

Unknowns	*Known*
Hypotenuse c	a: 120 miles b: 50 miles

$$120^2 + 50^2 = c^2 \quad \text{Substitute 120 for } a \text{ and 50 for } b \text{ in } a^2 + b^2 = c^2.$$
$$14{,}400 + 2500 = c^2 \quad \text{Square 120 and 50.}$$
$$16{,}900 = c^2 \quad \text{Simplify.}$$
$$\pm\sqrt{16{,}900} = \sqrt{c^2} \quad \text{Solve by extracting square roots.}$$
$$\pm 130 = c \quad \text{Simplify the square root.}$$

The negative solution does not make sense in this problem. The direct distance from the Visalia airport to the Modesto airport is 130 miles.

***Quick Check* 5**
Cassie's backyard is in the shape of a rectangle whose dimensions are 70 feet by 240 feet. She needs a hose that will extend from one corner of her yard to the corner that is diagonally opposite to it. How long does the hose have to be?

Maximum and Minimum Values of Quadratic Functions

Objective 4 **Find the maximum or minimum value of a quadratic function.**

Maximum and Minimum Values of a Function

The greatest possible output of a function is called the **maximum value** of the function. The least possible output of a function is called the **minimum value** of the function.

Each parabola that opens upward has a minimum value at its vertex.

We can see that this parabola does not go below the line $y = -4$, which is the y-coordinate of the vertex. The minimum value of this parabola is -4.

To find the minimum value of a quadratic function whose graph is a parabola that opens upward, we need to find the y-coordinate of its vertex.

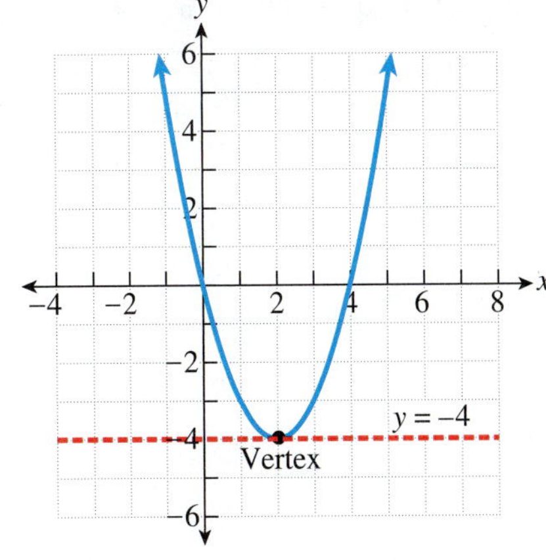

EXAMPLE 6 Find the minimum value of the function $f(x) = 3x^2 - 12x + 17$.

Solution

Since the graph of this function is a parabola that opens upward $(a > 0)$, the function has a minimum value at its vertex. To find the vertex, we begin by finding the x-coordinate by using the formula $x = \frac{-b}{2a}$.

$$x = \frac{-(-12)}{2(3)} = 2$$ Substitute 3 for a and -12 for b and simplify.

The minimum value of this function occurs when $x = 2$. Now, to find the minimum value, we evaluate the function when $x = 2$.

$$f(2) = 3(2)^2 - 12(2) + 17$$ Substitute 2 for x.
$$= 5$$ Simplify.

The minimum value for this function is 5.

Do all quadratic functions have a minimum value? No. If the graph of a quadratic function is a parabola that opens downward, then the function does not have a minimum value. Such a function does have a maximum value, and this maximum value can be found at the vertex.

EXAMPLE 7 Find the maximum or minimum value of the function $f(x) = -2x^2 - 8x - 13$.

Solution

Since this parabola will open downward, the function has a maximum value at the vertex.

$$x = \frac{-(-8)}{2(-2)} = -2$$ Substitute -2 for a and -8 for b and simplify.

The maximum value occurs when $x = -2$. Now we evaluate the function for this value of x to find the maximum value of the function.

$$f(-2) = -2(-2)^2 - 8(-2) - 13$$ Substitute -2 for x.
$$= -2(4) - 8(-2) - 13$$ Square -2.
$$= -8 + 16 - 13$$ Multiply.
$$= -5$$ Simplify.

The maximum value of this function is -5 and occurs when $x = -2$.

Quick Check 6

Find the maximum or minimum value of the quadratic function.

a) $f(x) = -x^2 + 10x - 35$
b) $f(x) = 5x^2 - 20x + 67$

Applied Maximum/Minimum Problems

Objective 5 **Solve applied maximum/minimum problems.**

EXAMPLE 8 What is the maximum product of two numbers whose sum is 40?

Solution

There are two unknown numbers in this problem. If we let x represent the first number, then we can represent the second number by $40 - x$.

Unknowns
1: x
2: $40 - x$

The product of these two numbers is given by the function $f(x) = x(40 - x)$. This function simplifies to be $f(x) = -x^2 + 40x$. Since this function is quadratic and its graph is a parabola that opens downward $(a < 0)$, it has a maximum value. Its maximum value can be found at the vertex. The maximum product occurs when $x = \frac{-b}{2a}$.

$$x = \frac{-40}{2(-1)} = 20$$ Substitute -1 for a and 40 for b and simplify.

The maximum product occurs when $x = 20$.

1: $x = 20$
2: $40 - x = 40 - 20 = 20$

Quick Check 7
What is the maximum product of two numbers whose sum is 62?

The two numbers whose sum is 40 that have the greatest product are 20 and 20. The maximum product is 400.

EXAMPLE 9 A farmer wants to fence off a rectangular pen, adjacent to his barn, for some pigs. If he has 120 feet of fencing, what dimensions would give the pen the largest possible area?

Solution

There are two unknowns in this problem, the length and the width of the rectangle.

Suppose that we let x represent the width of the rectangle. Since the perimeter is equal to 120 feet, the length of the remaining side is $120 - 2x$.

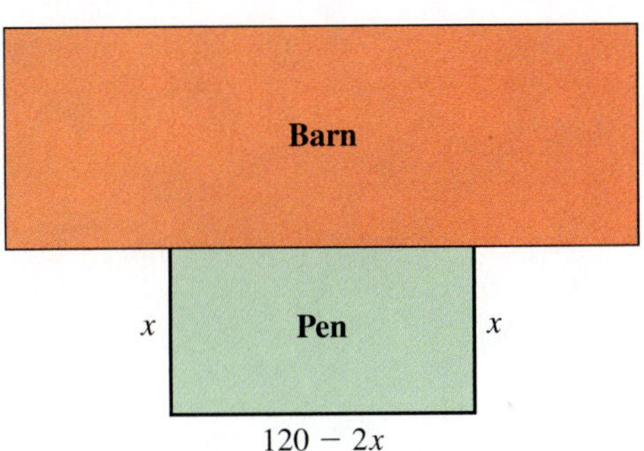

Unknowns

Length: $120 - 2x$
Width: x

The area of a rectangle is equal to its length times its width. The area of this pen is given by the function $A(x) = x(120 - 2x)$. This function simplifies to be $A(x) = -2x^2 + 120x$. Since this function is quadratic and its graph is a parabola that opens downward (a is negative), it has a maximum value that can be found at the vertex. The maximum area occurs when $x = \frac{-b}{2a}$.

$$x = \frac{-120}{2(-2)} = 30$$ Substitute -2 for a and 120 for b, and simplify.

The maximum area occurs when $x = 30$.

Length: $120 - 2x = 120 - 2(30) = 60$
Width: $x = 30$

Quick Check 8

A rancher wants to fence in a rectangular corral that is adjacent to a stable area, so only three sides of the rectangle will need to be fenced in. What is the largest possible rectangular area that can be fenced in with 264 meters of fencing?

The maximum area occurs when the length is 60 feet and the width (adjacent to the barn) is 30 feet. The maximum area that can be enclosed is 1800 square feet.

Building Your Study Strategy **Time Management, 5 Difficult Subjects First**
If you have more than one subject to study, as most students do, then the order in which you study the subjects is important. A wise idea is to arrange your study schedule so that you will be studying the most difficult subject first, while you are most alert. If possible, avoid studying two difficult subjects back to back. If you cannot arrange a study break between these two topics, try to study a less demanding subject between two difficult subjects.

EXERCISES 7.5

Vocabulary

1. In a right triangle, the side opposite the right angle is called the ________.

2. In a right triangle, the sides that are adjacent to the right angle are called ________.

3. The ________ theorem states that for any right triangle whose hypotenuse has length c and whose two legs have lengths a and b, respectively, $a^2 + b^2 = c^2$.

4. The maximum or minimum value of a quadratic function occurs at the ________.

For all problems, approximate to the nearest tenth when necessary.

5. The length of a rectangle is 2 inches more than its width. If the area of the rectangle is 80 square inches, find the length and width of the rectangle.

6. The length of a rectangle is 7 inches less than 3 times the width. If the area of the rectangle is 40 square inches, find the length and width of the rectangle.

7. The length of a rectangle is twice its width. If the area of the rectangle is 98 square inches, find the length and width of the rectangle.

8. The length of a rectangle is 1 foot more than 5 times its width. If the area of the rectangle is 328 square feet, find the length and width of the rectangle.

9. The length of a rectangle is 1 inch less than 7 times its width. If the area of the rectangle is 50 square inches, find the dimensions of the rectangle. Round to the nearest tenth of an inch.

10. The width of a rectangle is half of its length. If the area of the rectangle is 300 square inches, find the dimensions of the rectangle. Round to the nearest tenth of an inch.

11. The base of a triangle is 5 inches more than its height. If the area of the triangle is 42 square inches, find the base and height of the triangle.

12. The height of a triangle is 1 foot less than 3 times its base. If the area of the triangle is 22 square feet, find the base and height of the triangle.

13. The height of a triangle is 1 inch more than twice its base. If the area of the triangle is 15 square inches, find the base and height of the triangle. Round to the nearest tenth of an inch.

14. The height of a triangle is 7 inches less than its base. If the area of the triangle is 32 square inches, find the base and height of the triangle. Round to the nearest tenth of an inch.

15. A rectangular photograph has an area of 40 square inches. If the width of the photograph is 3 inches more than its height, find the dimensions of the photograph.

16. The area of a rectangular poster is 1400 square centimeters. If the width of the poster is 5 centimeters less than its height, find the dimensions of the poster.

17. The length of a rectangular quilt is twice its width, and the area of the quilt is 12.5 square feet. Find the dimensions of the quilt.

18. The length of a rectangular room is 3 feet more than twice its width. If the area of the room is 160 square feet, find the dimensions of the room. Round to the nearest tenth of a foot.

19. The width of a rectangular table is 16 inches less than its length. If the area of the table is 540 square inches, find the dimensions of the table. Round to the nearest tenth of an inch.

20. Steve has a rectangular lawn, and the length of the lawn is 25 feet more than the width of the lawn. If the area of the lawn is 7000 square feet, find the dimensions of the lawn. Round to the nearest tenth of a foot.

21. A kite is in the shape of a triangle. The base of the kite is twice its height. If the area of the kite is 256 square inches, find the base and height of the kite.

22. A hang glider is triangular in shape. If the base of the hang glider is 2 feet more than twice its height, and the area of the hang glider is 30 square feet, find the hang glider's base and height.

23. The sail on a sailboat is shaped like a triangle, with an area of 46 square feet. The height of the sail is 8 feet more than the base of the sail. Find the base and height of the sail. Round to the nearest tenth of a foot.

24. A kite is in the shape of a triangle and is made with 250 square inches of material. The base of the kite is 20 inches longer than the height of the kite. Find the base and the height of the kite. Round to the nearest tenth of an inch.

25. The two legs of a right triangle are 7 inches and 24 inches, respectively. Find the hypotenuse of the triangle.

26. The two legs of a right triangle are 6 inches and 11 inches, respectively. Find the hypotenuse of the triangle, rounded to the nearest tenth of an inch.

27. A right triangle with a hypotenuse of 10 centimeters has a leg that measures 5 centimeters. Find the length of the other leg, rounded to the nearest tenth of a centimeter.

28. A right triangle has a leg that measures 15 inches and a hypotenuse that measures 30 inches. Find the length of the other leg, rounded to the nearest tenth of an inch.

29. Patti drove 80 miles to the west and then drove 60 miles south. How far is she from her starting location?

30. Victor flew to a city that was 700 miles north and 2400 miles east of his starting point. How far did he fly to reach this city?

31. George is casting a shadow on the ground. If George is 2 feet shorter than the length of the shadow on the ground and the tip of the shadow is 10 feet from the top of George's head, how tall is George?

32. The base of a 15-foot ladder is leaning against the wall. If the distance between the base of the ladder and the wall is 3 feet less than the height of the top of the ladder on the wall, how high up the wall does the ladder reach?

33. A guy wire 40 feet long runs from the top of a pole to a spot on the ground. If the height of the pole is 5 feet more than the distance from the base of the pole to the spot where the guy wire is anchored, how tall is the pole? Round to the nearest tenth of a foot.

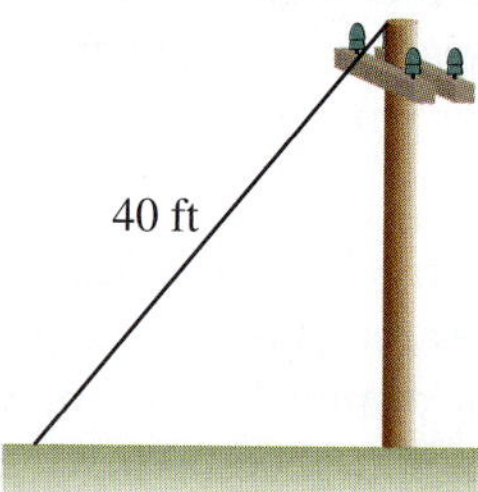

34. A 12-foot ramp leads to a doorway. The height of the doorway is 8 feet less than the horizontal distance covered by the ramp. How high is the doorway? Round to the nearest tenth of a foot.

35. A rectangular computer screen is 13 inches wide and 10 inches high. Find the length of its diagonal. Round to the nearest tenth of an inch.

36. The bases on a baseball diamond form a square whose side is 90 feet. How far is it from home plate to second base? Round to the nearest tenth of a foot.

37. The length of a rectangular quilt is 1 foot more than its width. If the diagonal of the quilt is 6.5 feet, find the length and width of the quilt. Round to the nearest tenth of a foot.

38. The diagonal of a rectangular table is 9 feet, and the length of the table is 5 feet more than its width. Find the length and width of the table. Round to the nearest tenth of a foot.

Use the following fact about right triangles for Exercises 39–42: The legs of a right triangle represent the base and height of that triangle.

39. The height of a right triangle is 3 feet less than the base of the triangle. The area of the triangle is 54 square feet.

a) Use this information to find the base and height of this triangle.

b) Find the hypotenuse of the triangle.

40. The base of a right triangle is 4 feet less than twice the height of the triangle. The area of the triangle is 24 square feet.

a) Use this information to find the base and height of this triangle.

b) Then use that information to find the hypotenuse of the triangle.

41. A road sign indicating falling rocks is in the shape of an equilateral triangle, with each side measuring 80 centimeters

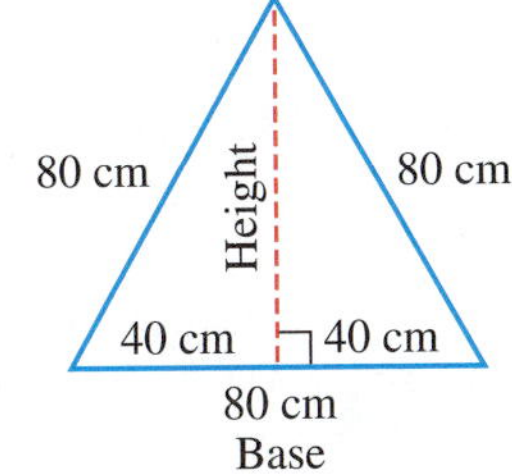

a) Use the Pythagorean theorem to find the height of the triangle. Round to the nearest tenth of a centimeter.

b) Find the area of the sign.

42. A farmer fenced in a corral in the shape of an equilateral triangle, with each side measuring 30 feet.

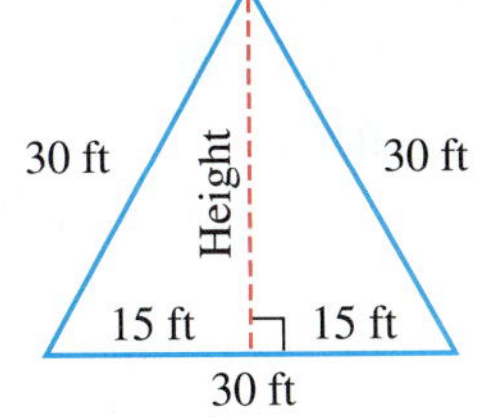

a) Use the Pythagorean theorem to find the height of the triangle. Round to the nearest tenth of a foot.

b) Find the area of the corral.

Determine whether the given quadratic function has a maximum value or a minimum value. Then find that maximum or minimum value.

43. $f(x) = x^2 - 8x + 20$

44. $f(x) = x^2 - 6x - 35$

45. $f(x) = x^2 + 12x - 42$

46. $f(x) = x^2 + 9x + 60$

47. $f(x) = -x^2 + 14x - 76$

48. $f(x) = -x^2 - 20x - 57$

49. $f(x) = (x - 7)^2 + 18$

50. $f(x) = (x + 3)^2 - 19$

51. $f(x) = -(x + 16)^2 - 33$

52. $f(x) = -(x - 14)^2 + 47$

53. $f(x) = 7(x - 4)^2 + 46$

54. $f(x) = -8(x + 11)^2 - 50$

55. A girl throws a rock upward with an initial velocity of 32 feet/second. The height of the rock (in feet) after t seconds is given by the function $h(t) = -16t^2 + 32t + 5$. What is the maximum height that the rock reaches?

56. A pilot flying at a height of 5000 feet determines that she must eject from her plane. The ejection seat launches with an initial velocity of 144 feet/second. The height of the pilot (in feet) t seconds after ejection is given by the function $h(t) = -16t^2 + 144t + 5000$. What is the maximum height that the pilot reaches?

57. At the start of a college football game, a referee tosses a coin to determine which team will receive the opening kickoff. The height of the coin (in meters) after t seconds is given by the function $h(t) = -4.9t^2 + 2.1t + 1.5$. What is the maximum height that the coin reaches?

58. If a man on the moon throws a rock upward with an initial velocity of 78 feet per second, the height of the rock (in feet) after t seconds is given by the function $h(t) = -2.6t^2 + 78t + 5$. What is the maximum height that the rock reaches?

59. What is the maximum product of two numbers whose sum is 24?

60. What is the maximum product of two numbers whose sum is 100?

61. A farmer has 300 feet of fencing to make a rectangular corral. What dimensions will make a corral that has the maximum area? What is the maximum area possible?

62. Emeril wants to fence off a rectangular herb garden adjacent to his house, using the house to form the fourth side of the rectangle as shown.

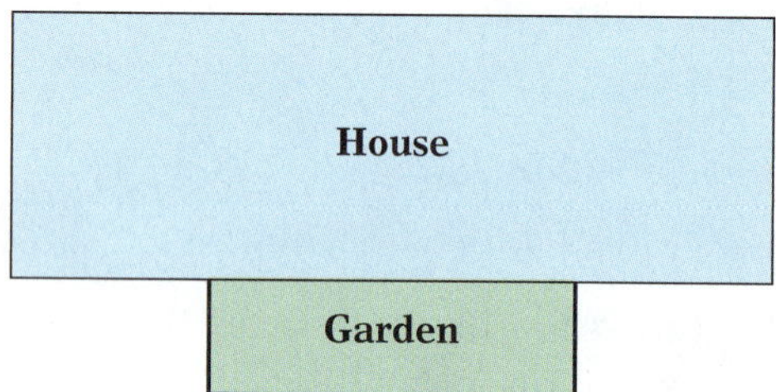

If Emeril has 80 feet of fencing, what dimensions of the herb garden will produce the maximum area? What is the maximum possible area?

63. A company produces small frying pans. The average cost per pan in dollars is given by the function $C(x) = 0.000625x^2 - 0.0875x + 6.5$, where x is the number of pans produced (in thousands). For what number of pans is the average cost per pan minimized? What is the minimum cost per pan that is possible?

64. A company produces mini CD players/radios. The average cost per radio in dollars is given by the function $C(x) = 0.009x^2 - 0.432x + 9.3$, where x is the number of radios produced (in thousands). For what number of radios is the average cost per radio minimized? What is the minimum cost per radio that is possible?

Writing in Mathematics

Answer in complete sentences.

65. Write a word problem whose solution is "The length of the rectangle is 14 feet, and the width is 9 feet." The problem must lead to a quadratic equation.

66. Write a word problem associated with the equation $60^2 + b^2 = 90^2$. Explain how you created the problem. Solve your problem, explaining each step.

67. ***Solutions Manual**** Write a solutions manual page for the following problem:

The length of a rectangle is 8 centimeters longer than twice its width. If the area of the rectangle is 150 square centimeters, find the length and the width of the rectangle. (Round to the nearest tenth of a centimeter.)

68. ***Newsletter**** Write a newsletter explaining how to solve applied maximum/minimum problems involving quadratic functions.

***See Appendix B for details and sample answers.**

7.6 Quadratic and Rational Inequalities

Objectives

1. **Solve quadratic inequalities.**
2. **Solve quadratic inequalities graphically.**
3. **Solve rational inequalities.**
4. **Solve inequalities involving functions.**
5. **Solve applied problems involving inequalities.**

Quadratic Inequalities

Objective 1 Solve quadratic inequalities. In this section, we will expand our knowledge of inequalities to include two different types of inequalities: quadratic inequalities and rational inequalities.

Quadratic Inequalities

A **quadratic inequality** is an inequality that can be rewritten as $ax^2 + bx + c < 0$, $ax^2 + bx + c \le 0$, $ax^2 + bx + c > 0$, or $ax^2 + bx + c \ge 0$.

The first step to solving a quadratic inequality is to find its **zeros**, which are values of x for which $ax^2 + bx + c = 0$. For example, to find the zeros for the inequality $x^2 + 9x - 22 > 0$, we set $x^2 + 9x - 22$ equal to 0 and solve for x.

$$x^2 + 9x - 22 = 0 \quad \text{Set the quadratic expression equal to 0.}$$
$$(x + 11)(x - 2) = 0 \quad \text{Factor.}$$
$$x = -11 \quad \text{or} \quad x = 2 \quad \text{Set each factor equal to 0 and solve.}$$

The zeros for this inequality are -11 and 2.

The zeros of an inequality divide the number line into intervals and often act as boundary points. In any particular interval created by the zeros of an inequality, either each real number in the interval is a solution of the inequality or each real number in the interval is not a solution of the inequality. By picking one value in each interval and testing it, we can determine which intervals contain solutions and which do not.

When solving a strict inequality involving the symbol $<$ or $>$, the zeros are not included in the solutions of the inequality, and we place an open circle on each zero on the number line. When solving a weak inequality involving the symbol $\le$ or $\ge$, the zeros are included in the solutions of the inequality, and we place a closed circle on each zero.

EXAMPLE 1 Solve $x^2 + x - 20 \le 0$.

Solution

We begin by finding the zeros.

$$x^2 + x - 20 = 0 \quad \text{Set the quadratic expression equal to 0.}$$
$$(x + 5)(x - 4) = 0 \quad \text{Factor.}$$
$$x = -5 \quad \text{or} \quad x = 4 \quad \text{Set each factor equal to 0 and solve.}$$

The zeros for this inequality are -5 and 4. We plot these zeros on a number line and create boundaries between the three intervals, which have been labeled as I, II, and III in the figure that follows. The zeros are included as solutions, and we place closed circles on the number line at $x = -5$ and $x = 4$.

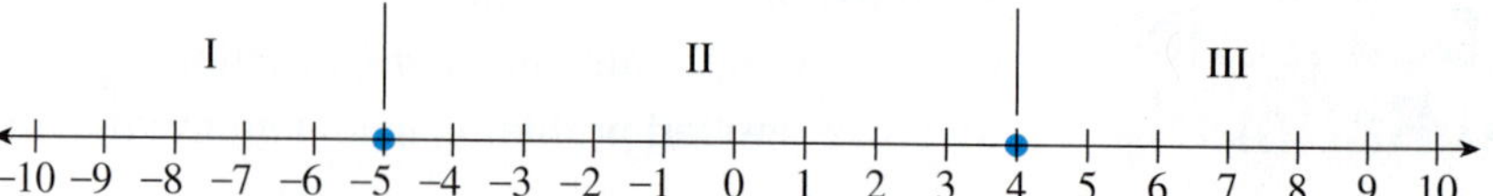

Now we choose a value from each of the three intervals we have created to be our test points. A **test point** is a value of the variable x that we use to evaluate the expression in the inequality, allowing us to determine which intervals contain solutions of the inequality. We will use $x = -6$, $x = 0$, and $x = 5$. Because we are looking for intervals where $x^2 + x - 20 \leq 0$, our solution will be made up of the intervals whose test points produce a negative result when substituted into the expression $x^2 + x - 20$.

Test Point (x)	-6	0	5
$x^2 + x - 20$	$(-6)^2 + (-6) - 20$ $= 36 - 6 - 20$ $= 10$	$(0)^2 + (0) - 20$ $= 0 + 0 - 20$ $= -20$	$(5)^2 + (5) - 20$ $= 25 + 5 - 20$ $= 10$

The only test point for which the expression $x^2 + x - 20$ is negative is $x = 0$, which is in interval II; so the interval $[-5, 4]$ is the solution of this inequality.

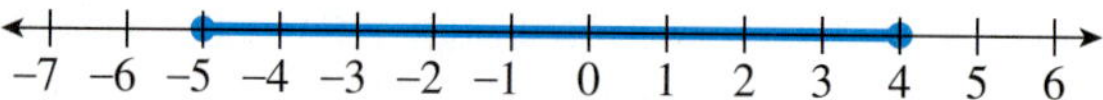

When the expression we are working with can be factored, we can use a sign chart to determine which intervals are solutions to the inequality. A **sign chart** is a chart that can be used to determine whether an expression is positive or negative for certain intervals of real numbers. A sign chart focuses on whether a factor is positive or negative for each interval, by the use of a test point. If we know the sign of each factor, then we can easily find the sign of the product, allowing us to solve our inequality. The initial sign chart should look like the following:

Test Point	$x + 5$	$x - 4$	$(x + 5)(x - 4)$
-6			
0			
5			

We fill in the first row by determining the sign of each factor when $x = -6$. Since $-6 + 5$ is negative, we put a negative sign in the first row underneath $x + 5$. In the same way, $x - 4$ is negative when $x = -6$, so we put another negative sign in the first row underneath $x - 4$. The product of two negative factors is positive, so we put a positive sign in the first row under $(x + 5)(x - 4)$. Here is the sign chart after it has been completed:

Test Point	$x + 5$	$x - 4$	$(x + 5)(x - 4)$
−6	−	−	+
0	+	−	−
5	+	+	+

Quick Check 1
Solve
$x^2 - 14x + 48 \le 0$.

Notice that the product is negative in the interval containing the test point $x = 0$. Therefore, our solution is the interval $[-5, 4]$.

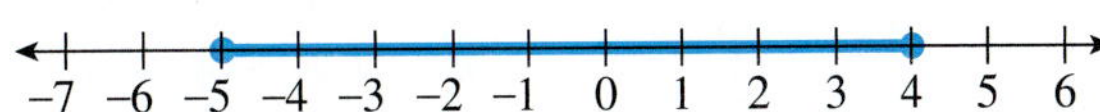

If the quadratic expression does not factor, then we cannot use a sign chart. In such a case, we will find the zeros of the expression by using the quadratic formula. We will then substitute the values for our test points directly into the quadratic expression.

EXAMPLE 2 Solve $x^2 + 6x + 4 > 0$.

Solution

We find the zeros by solving the equation $x^2 + 6x + 4 = 0$. Since $x^2 + 6x + 4$ does not factor, we will use the quadratic formula to solve the equation.

$$x = \frac{-6 \pm \sqrt{6^2 - 4(1)(4)}}{2(1)} \quad \text{Substitute 1 for } a \text{, 6 for } b \text{, and 4 for } c.$$

$$x = \frac{-6 \pm \sqrt{20}}{2} \quad \text{Simplify the radicand.}$$

$$x = \frac{-6 \pm 2\sqrt{5}}{2} \quad \text{Simplify the square root.}$$

$$x = \frac{\overset{1}{\cancel{2}}(-3 \pm \sqrt{5})}{\underset{1}{\cancel{2}}} \quad \text{Divide out common factors.}$$

$$x = -3 \pm \sqrt{5} \quad \text{Simplify.}$$

The zeros for this inequality are $-3 + \sqrt{5}$ and $-3 - \sqrt{5}$. We need an approximate value for each zero to determine where each zero belongs on the number line. These two zeros are approximately equal to -0.8 and -5.2. The zeros are not included as soutions, so we place open circles at the zeros.

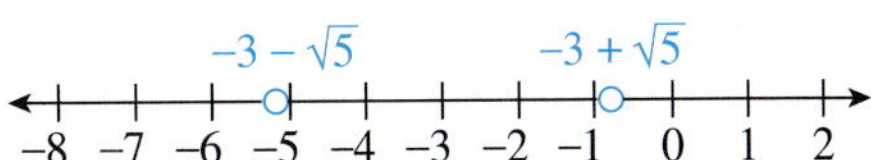

We will use $x = -6$, $x = -1$, and $x = 0$ as test points. We are looking for intervals where the expression $x^2 + 6x + 4$ is greater than 0, and our solution will be made up of the intervals whose test points produce a positive result when substituted into $x^2 + 6x + 4$.

Test Point	$x^2 + 6x + 4$	Sign
-6	$(-6)^2 + 6(-6) + 4 = 4$	+
-1	$(-1)^2 + 6(-1) + 4 = -1$	−
0	$(0)^2 + 6(0) + 4 = 4$	+

Notice that the product is positive in the intervals containing the test points $x = -6$ and $x = 0$. Here is our solution:

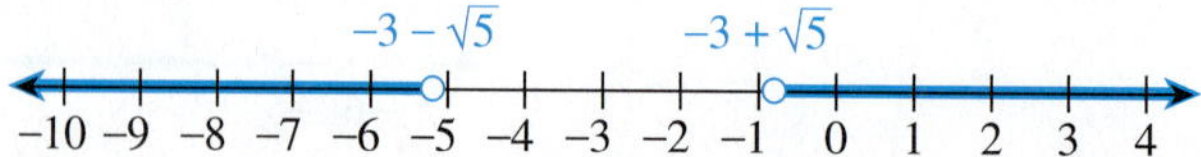

In interval notation, we can express this solution as $(-\infty, -3 - \sqrt{5}) \cup (-3 + \sqrt{5}, \infty)$. (Notice that we used the exact values in our solution, not the approximate values.)

***Quick Check* 2**

Solve $x^2 + 3x - 15 \geq 0$.

If the expression in our inequality has no zeros, then either every real number is a solution of the inequality or it has no solutions at all.

EXAMPLE 3 Solve $x^2 - 5x + 7 \leq 0$.

Solution

We begin by looking for the zeros of $x^2 - 5x + 7$. Since this expression does not factor, we will use the quadratic formula.

$$x = \frac{5 \pm \sqrt{(-5)^2 - 4(1)(7)}}{2(1)} \qquad \text{Substitute 1 for } a, -5 \text{ for } b, \text{ and 7 for } c.$$

$$x = \frac{5 \pm \sqrt{-3}}{2} \qquad \text{Simplify the radicand and denominator.}$$

The discriminant is negative, so this expression has no real zeros. In this case, we can pick any real number and use it as our only test point. When $x = 0$, we can see that $x^2 - 5x + 7$ is equal to 7, which is not less than or equal to 0. This inequality has no solution.

***Quick Check* 3**

Solve $x^2 - 5x + 20 > 0$.

Solving Quadratic Inequalities from the Graph of a Quadratic Equation

Objective 2 **Solve quadratic inequalities graphically.** Consider the graph of $y = x^2 - 6x + 8$. The graph of the equation drops below the x-axis in the interval $(2, 4)$. This tells us that the interval $(2, 4)$ is the solution of the inequality $x^2 - 6x + 8 < 0$. For any value of x that is less than 2 or greater than 4, the expression $x^2 - 6x + 8$ is greater than 0.

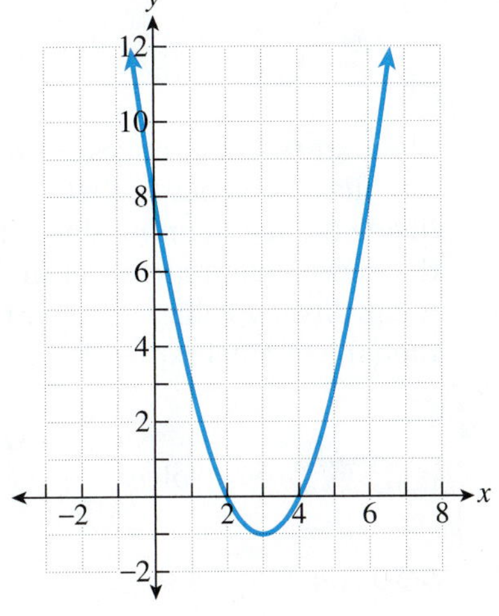

EXAMPLE 4 Use the graph of $y = -x^2 + 3x + 10$ to solve the inequality $-x^2 + 3x + 10 > 0$.

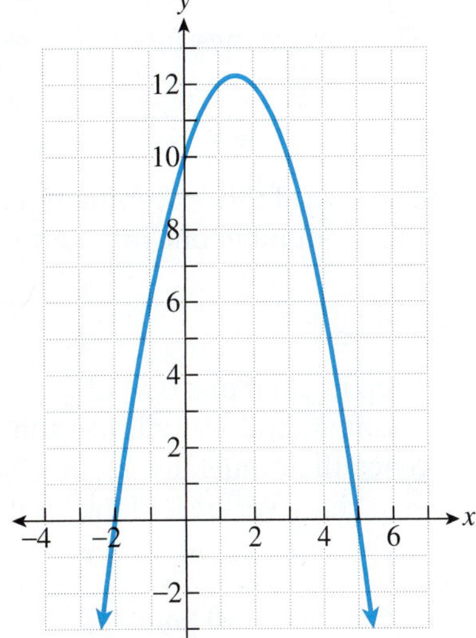

Solution

We are looking for intervals for which the graph is above the x-axis. Our solution in this case is the interval $(-2, 5)$.

Quick Check 4

Use the graph of $y = x^2 - 5x - 14$ to solve the inequality $x^2 - 5x - 14 > 0$.

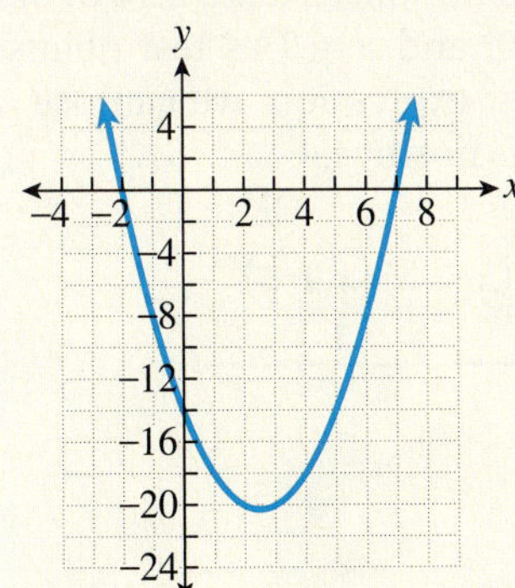

Rational Inequalities

Objective 3 **Solve rational inequalities.**

Rational Inequalities

A **rational inequality** is an inequality that involves a rational expression, such as

$$\frac{x+2}{x-4} < 0.$$

There are two differences between solving a rational inequality and a quadratic inequality. The first difference is that we use the zeros of the numerator and the denominator to divide the real number line into intervals. The second difference is that the zeros from the denominator are never included in the solution, even when the inequality is a weak inequality involving $\leq$ or $\geq$. This is because if a value causes a denominator to equal 0, then the rational expression is undefined for that value.

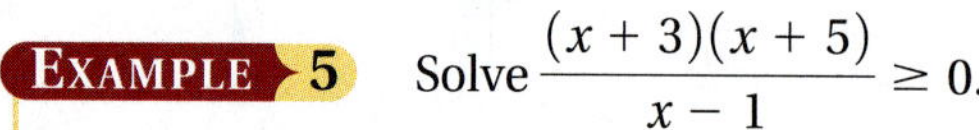

EXAMPLE 5 Solve $\dfrac{(x+3)(x+5)}{x-1} \geq 0.$

Solution

We begin by finding the zeros of the numerator.

$(x+3)(x+5) = 0$ Set the numerator equal to 0.

$x = -3$ or $x = -5$ Set each factor equal to 0 and solve.

The zeros of the numerator are -3 and -5. Now we find the zeros of the denominator.

$x - 1 = 0$ Set the denominator equal to 0.

$x = 1$ Solve.

The zero of the denominator is 1. Here are all of the zeros on a single number line, dividing the number line into four intervals:

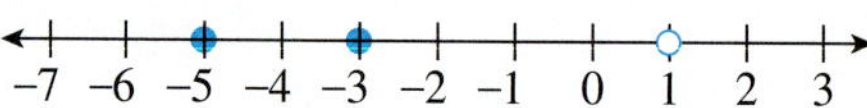

Keep in mind that the value $x = 1$ cannot be included in any solution, as it is a zero of the denominator. We will use the values $x = -6$, $x = -4$, $x = 0$, and $x = 2$ as test points. Since the numerator factors and the denominator is a linear expression, we can use a sign chart. We are looking for intervals where the expression is positive.

Test Point	$x+3$	$x+5$	$x-1$	$\dfrac{(x+3)(x+5)}{x-1}$
−6	−	−	−	−
−4	−	+	−	+
0	+	+	−	−
2	+	+	+	+

The intervals containing $x = -4$ and $x = 2$ are solutions of this inequality, as represented on the following number line:

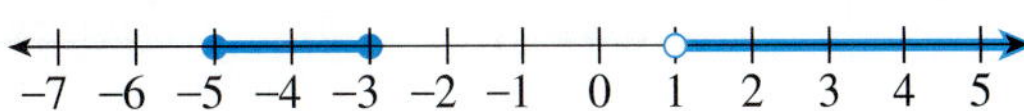

This can be represented in interval notation by $[-5, -3] \cup (1, \infty)$.

***Quick Check* 5**

Solve $\dfrac{(x+2)(x-6)}{(x-8)(x+7)} \leq 0.$

A Word of Caution When we solve a rational inequality, the zeros of the denominator are always excluded as solutions because the rational expression is undefined for those values.

EXAMPLE 6 Solve $\dfrac{x^2 - 4x - 12}{x^2 - 9} < 0.$

Solution

We begin by finding the zeros of the numerator.

$x^2 - 4x - 12 = 0$			Set the numerator equal to 0.
$(x-6)(x+2) = 0$			Factor.
$x - 6 = 0$	or	$x + 2 = 0$	Set each factor equal to 0.
$x = 6$	or	$x = -2$	Solve.

The zeros of the numerator are 6 and -2. Now we find the zeros of the denominator.

$x^2 - 9 = 0$			Set the denominator equal to 0.
$(x+3)(x-3) = 0$			Factor.
$x + 3 = 0$	or	$x - 3 = 0$	Set each factor equal to 0.
$x = -3$	or	$x = 3$	Solve.

The zeros of the denominator are -3 and 3. Here are all of the zeros on a single number line, dividing the number line into five intervals:

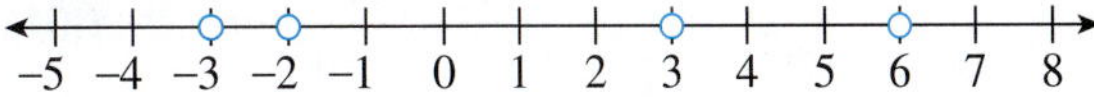

In this example, none of the zeros can be included in any solution, as the inequality was strictly less than 0 and not less than or equal to 0. We will use the values $x = -4$, $x = -2.5$, $x = 0$, $x = 4$, and $x = 7$ as test points. Here is the sign chart; keep in mind that we are looking for intervals in which this expression is negative:

Test Point	$x - 6$	$x + 2$	$x + 3$	$x - 3$	$\frac{(x-6)(x+2)}{(x+3)(x-3)}$
-4	$-$	$-$	$-$	$-$	$+$
-2.5	$-$	$-$	$+$	$-$	$-$
0	$-$	$+$	$+$	$-$	$+$
4	$-$	$+$	$+$	$+$	$-$
7	$+$	$+$	$+$	$+$	$+$

The intervals containing $x = -2.5$ and $x = 4$ are solutions of this inequality, as represented on the following number line:

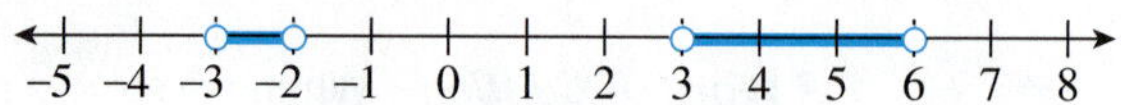

This can be represented in interval notation by $(-3, -2) \cup (3, 6)$.

Quick Check 6

Solve $\frac{x^2 - 6x}{x^2 + 5x + 4} > 0$.

When solving rational inequalities, if the inequality contains expressions on both sides, then we must rewrite the inequality in such a way that there is a single rational expression on one side of the inequality and 0 on the other side. We then solve the inequality in the same way as the previous examples.

EXAMPLE 7 Solve $\frac{x^2 + 4x + 9}{x + 6} \geq 2$.

Solution

We begin by subtracting 2 from both sides of the inequality, and then we combine the left side of the inequality as a single rational expression.

$$\frac{x^2 + 4x + 9}{x + 6} \geq 2$$

$$\frac{x^2 + 4x + 9}{x + 6} - 2 \geq 0$$ Subtract 2 from both sides.

$$\frac{x^2 + 4x + 9}{x + 6} - 2 \cdot \frac{x + 6}{x + 6} \geq 0$$ Multiply 2 by $\frac{x + 6}{x + 6}$ so that both expressions have the same denominator.

$$\frac{x^2 + 4x + 9 - 2x - 12}{x + 6} \geq 0$$ Distribute and combine numerators.

$$\frac{x^2 + 2x - 3}{x + 6} \geq 0$$ Simplify the numerator.

Now we can find the zeros of the numerator.

$$x^2 + 2x - 3 = 0 \quad \text{Set the numerator equal to 0.}$$
$$(x - 1)(x + 3) = 0 \quad \text{Factor.}$$
$$x - 1 = 0 \quad \text{or} \quad x + 3 = 0 \quad \text{Set each factor equal to 0.}$$
$$x = 1 \quad \text{or} \quad x = -3 \quad \text{Solve.}$$

The zeros of the numerator are 1 and -3. Now we find the zeros of the denominator.

$$x + 6 = 0 \quad \text{Set the denominator equal to 0.}$$
$$x = -6 \quad \text{Solve.}$$

The only zero of the denominator is -6. Here are all of the zeros on a single number line, dividing the number line into four intervals:

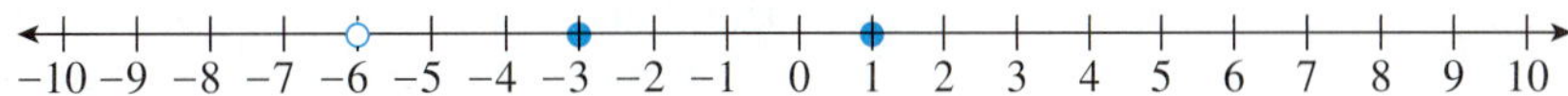

In this example, the zeros from the numerator are included as solutions, while the zeros from the denominator are excluded as solutions. We will use the values $x = -7$, $x = -4$, $x = 0$, and $x = 2$ as test points. Here is the sign chart; keep in mind that we are looking for intervals in which the expression $\frac{(x+3)(x-1)}{x+6}$ is positive:

Test Point	$x + 3$	$x - 1$	$x + 6$	$\frac{(x+3)(x-1)}{x+6}$
-7	$-$	$-$	$-$	$-$
-4	$-$	$-$	$+$	$+$
0	$+$	$-$	$+$	$-$
2	$+$	$+$	$+$	$+$

Quick Check 7

Solve $\frac{x^2 - 17}{x - 3} < 4.$

The intervals containing $x = -4$ and $x = 2$ are solutions to this inequality, as represented on the following number line:

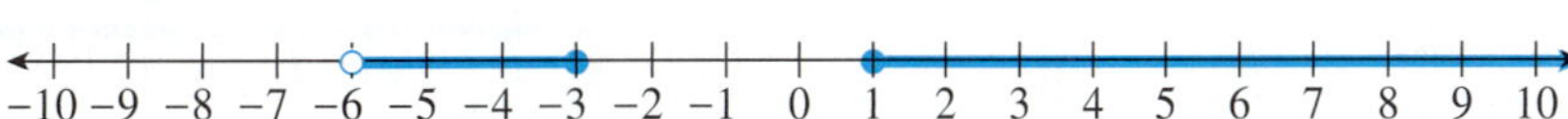

This can be represented in interval notation by $(-6, -3] \cup [1, \infty)$.

Solving Inequalities Involving Functions

Objective 4 **Solve inequalities involving functions.** We move on to examining inequalities involving functions.

EXAMPLE 8 Given $f(x) = x^2 - 2x - 16$, find all values x for which $f(x) \le 8$.

Solution

We begin by setting the function less than or equal to 8.

$$f(x) \le 8$$
$$x^2 - 2x - 16 \le 8 \quad \text{Set the function less than or equal to 8.}$$
$$x^2 - 2x - 24 \le 0 \quad \text{Subtract 8 to write the inequality in general form.}$$

We now find the zeros for this inequality.

$$x^2 - 2x - 24 = 0 \quad \text{Set the quadratic expression equal to 0.}$$
$$(x-6)(x+4) = 0 \quad \text{Factor.}$$
$$x - 6 = 0 \quad \text{or} \quad x + 4 = 0 \quad \text{Set each factor equal to 0.}$$
$$x = 6 \quad \text{or} \quad x = -4 \quad \text{Solve.}$$

The zeros for this inequality are 6 and -4.

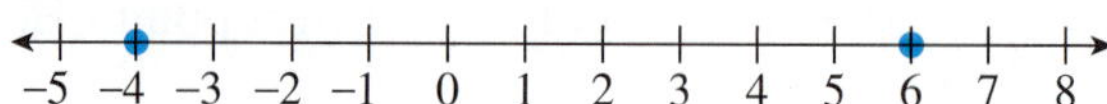

We will use $x = -5$, $x = 0$, and $x = 7$ as our test points. Our solution will be made up of the intervals whose test points produce a negative result when substituted into the expression $x^2 - 2x - 24$. Here is the sign chart after it has been completed:

Test Point	$x - 6$	$x + 4$	$(x-6)(x+4)$
-5	$-$	$-$	$+$
0	$-$	$+$	$-$
7	$+$	$+$	$+$

Quick Check 8

Given $f(x) = x^2 + 10x + 35$, find all values x for which $f(x) < 14$.

The product is negative in the interval containing the test point $x = 0$. Here is our solution represented on a number line:

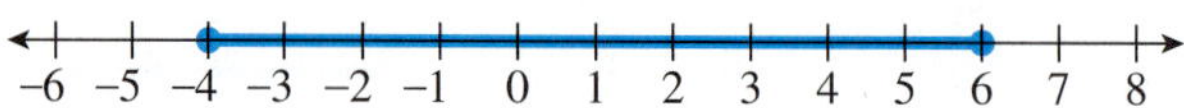

In interval notation, the solution is expressed as $[-4, 6]$.

Solving Applied Problems Involving Inequalities

Objective 5 **Solve applied problems involving inequalities.**

EXAMPLE 9 A projectile is fired from the roof of a building 72 feet tall, with an initial velocity of 96 feet per second. The height of the projectile, in feet, after t seconds is given by the function $h(t) = -16t^2 + 96t + 72$. For what length of time is the projectile at least 200 feet above the ground?

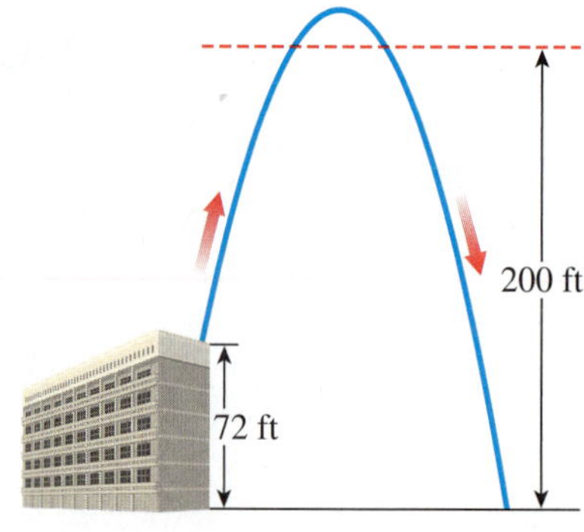

Solution

To find the time interval that the projectile is at least 200 feet above the ground, we must solve the inequality $h(t) \geq 200$.

$$h(t) \geq 200$$

$$-16t^2 + 96t + 72 \geq 200 \quad \text{Replace } h(t) \text{ by } -16t^2 + 96t + 72.$$

$$-16t^2 + 96t - 128 \geq 0 \quad \text{Subtract 200 to rewrite the inequality in standard form.}$$

$$16t^2 - 96t + 128 \leq 0 \quad \text{Multiply both sides of the inequality by } -1. \text{ Change the direction of the inequality.}$$

Now we find the zeros of the inequality.

$$16t^2 - 96t + 128 = 0$$

$$16(t^2 - 6t + 8) = 0 \quad \text{Factor out the common factor 16.}$$

$$16(t-2)(t-4) = 0 \quad \text{Factor the trinomial.}$$

$$t = 2 \quad \text{or} \quad t = 4 \quad \text{Set each variable factor equal to 0 and solve.}$$

The zeros of the inequality are $t = 2$ and $t = 4$.

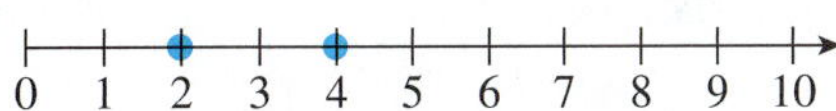

We will use $t = 1$, $t = 3$, and $t = 5$ as our test points. (Note that t must be greater than or equal to 0, since it represents the amount of time since the projectile was fired.) In this case, we will substitute the values for our test points into the function $h(t)$. If the function's output for one of the test points is 200 or higher, then each point in the interval containing the test point is a solution to the inequality $h(t) \geq 200$.

$t = 1$	$t = 3$	$t = 5$
$h(1) = -16(1)^2 + 96(1) + 72$	$h(3) = -16(3)^2 + 96(3) + 72$	$h(5) = -16(5)^2 + 96(5) + 72$
$= -16(1) + 96 + 72$	$= -16(9) + 288 + 72$	$= -16(25) + 480 + 72$
$= -16 + 96 + 72$	$= -144 + 288 + 72$	$= -400 + 480 + 72$
$= 152$	$= 216$	$= 152$

The inequality $h(t) \geq 200$ is true only in the interval containing the test point $t = 3$, so the projectile's height is at least 200 feet from 2 seconds after launch until 4 seconds after launch.

Quick Check 9

A projectile is fired from ground level with an initial velocity of 272 feet per second. The height of the projectile, in feet, after t seconds is given by the function $h(t) = -16t^2 + 272t$. For what length of time is the projectile at least 960 feet above the ground?

Building Your Study Strategy **Time Management, 6** **Brief Breaks** One suggestion to keep your mental energy at its highest while you are studying is to take brief 10-minute study breaks. You can use this time to take a short, brisk walk, to make a healthy snack, or to make a quick phone call. One study break per hour will help to keep you from feeling fatigued.

EXERCISES 7.6

Vocabulary

1. A ________ inequality is an inequality involving a quadratic expression.

2. The ________ of a quadratic inequality in standard form are values of the variable for which the quadratic expression is equal to 0.

3. The zeros of a ________ inequality are not included in the solutions of that inequality.

4. The zeros from the ________ of a rational inequality are never included as a solution.

Solve each quadratic inequality. Express your solution on a number line, using interval notation.

5. $(x - 5)(x - 1) < 0$

6. $(x + 3)(x - 2) > 0$

7. $-2(x - 6)(x + 3) \leq 0$

8. $x(x - 4) > 0$

9. $x^2 - 13x + 30 \leq 0$

10. $x^2 + 2x - 63 < 0$

11. $x^2 + 8x + 15 > 0$

12. $x^2 - 7x + 10 \leq 0$

13. $x^2 - 7x < 0$

14. $x^2 - 16 > 0$

15. $x^2 + 6x + 3 > 0$

16. $x^2 - 5x - 11 < 0$

17. $3x^2 - 5x - 1 \geq 0$

18. $2x^2 + 7x + 4 \leq 0$

19. $x^2 + 3x + 5 > 0$

20. $x^2 + x + 8 < 0$

21. $-x^2 + 10x - 5 > 0$

22. $x^2 + 6x \leq 27$

23. $x^2 - 4x \geq 45$

24. $-x^2 + 6x + 1 < 0$

Solve each polynomial inequality. Express your solution on a number line, using interval notation.

25. $(x - 1)(x + 2)(x - 4) \geq 0$

26. $(x + 8)(x - 5)(x - 2) \leq 0$

27. $(x + 6)(x + 2)(x + 3) < 0$

28. $x(x - 1)(x + 8)(x + 2) > 0$

29. $(x - 7)^2(x + 5)(x + 9) \leq 0$

30. $(x - 10)^3(x + 4)^2(x - 4) \geq 0$

Solve each rational inequality. Express your solution on a number line, using interval notation.

31. $\dfrac{x - 3}{x + 4} < 0$

32. $\dfrac{x + 8}{x - 6} > 0$

33. $\dfrac{(x + 2)(x - 2)}{x + 1} \leq 0$

34. $\dfrac{(x + 7)(x - 4)}{(x - 5)(x + 10)} < 0$

35. $\dfrac{x^2 - 4x - 5}{x + 2} < 0$

36. $\dfrac{x + 7}{x^2 - 14x + 48} \leq 0$

37. $\dfrac{x^2 + 7x - 8}{x^2 - 49} \geq 0$

38. $\dfrac{x^2 - 13x + 36}{x^2 + 12x + 36} > 0$

39. $\dfrac{x^2 - 12x + 35}{x^2 + 6x - 16} \leq 0$

40. $\dfrac{x^2 + 15x + 44}{x^2 - 7x - 60} \geq 0$

41. $\dfrac{x^2 - 4x + 3}{x^2 + 4x - 32} < 0$

42. $\dfrac{x^2 - 81}{x^2 - 15x + 54} \leq 0$

43. $\dfrac{4x - 7}{x - 4} > 3$

44. $\dfrac{6x + 18}{x + 2} \leq 5$

45. $\dfrac{x^2 + 9x + 12}{x + 3} \geq 2$

46. $\dfrac{x^2 - 3x - 46}{x - 7} < 4$

47. $\dfrac{x^2 + 3x + 10}{x - 2} \leq x$

48. $\dfrac{x^2 + 7x + 21}{x + 4} \geq x$

49. A projectile is fired upwards from the roof of a building 27 feet tall with an initial velocity of 64 feet per second. The height of the projectile, in feet, after t seconds is given by the function $h(t) = -16t^2 + 64t + 27$. For what length of time is the projectile at least 75 feet above the ground?

50. A projectile is fired upwards from the roof of a building 98 feet tall with an initial velocity of 208 feet per second. The height of the projectile, in feet, after t seconds is given by the function $h(t) = -16t^2 + 208t + 98$. For what length of time is the projectile at least 450 feet above the ground?

51. A projectile is fired upwards from ground level with an initial velocity of 32 feet per second. The height of the projectile, in feet, after t seconds is given by the function $h(t) = -16t^2 + 32t$. For what length of time is the projectile at least 12 feet above the ground?

52. A projectile is fired upwards from ground level with an initial velocity of 48 feet per second. The height of the projectile, in feet, after t seconds is given by the function $h(t) = -16t^2 + 48t$. For what length of time is the projectile at least 20 feet above the ground?

53. A projectile is fired upwards from the roof of a building 276 feet tall with an initial velocity of 144 feet per second. The height of the projectile, in feet, after t seconds is given by the function $h(t) = -16t^2 + 144t + 276$. For how long is the projectile at least 500 feet above the ground?

54. A projectile is fired upwards from the roof of a building 20 feet tall with an initial velocity of 84 feet per second. The height of the projectile, in feet, after t seconds, is given by the function $h(t) = -16t^2 + 84t + 20$. For how long is the projectile at least 100 feet above the ground?

55. Given $f(x) = x^2 - 8x + 12$, find all values x for which $f(x) \le 0$.

56. Given $f(x) = x^2 + 13x - 5$, find all values x for which $f(x) \ge 0$.

57. Given $f(x) = x^2 - 5x + 13$, find all values x for which $f(x) \ge 7$.

58. Given $f(x) = x^2 + 4x - 31$, find all values x for which $f(x) < 14$.

59. Given $f(x) = x^2 - 8x + 24$, find all values x for which $f(x) \le -9$.

60. Given $f(x) = x^2 + 6x - 9$, find all values x for which $f(x) < -49$.

61. Given $f(x) = \dfrac{x+9}{x-6}$, find all values x for which $f(x) \le 0$.

62. Given $f(x) = \dfrac{x^2 - 17x + 70}{x^2 + 2x - 48}$, find all values x for which $f(x) < 0$.

63. Given $f(x) = \dfrac{x^2 - 8x - 33}{x^2 + 13x + 42}$, find all values x for which $f(x) > 0$.

64. Given $f(x) = \dfrac{x^2 + 14x + 24}{x^2 + 6x + 5}$, find all values x for which $f(x) \ge 0$.

Find each quadratic inequality whose solution is given.

65.
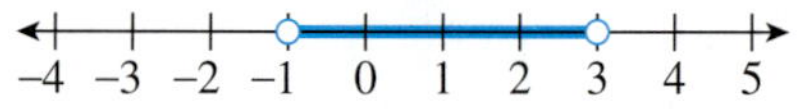

66.
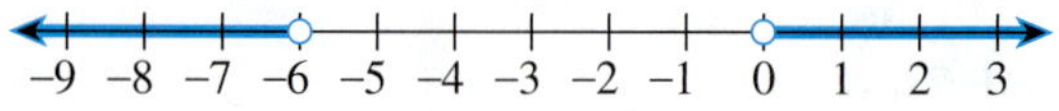

67.
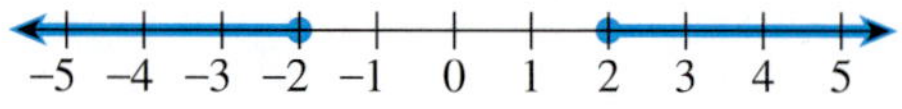

68.
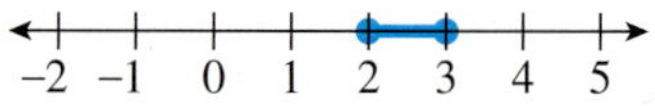

Find each rational inequality whose solution is given.

69.
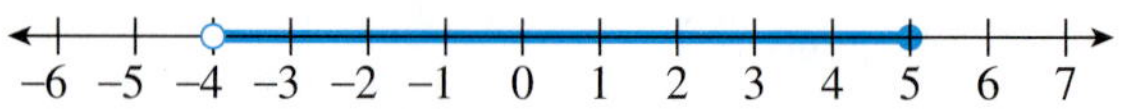

70.
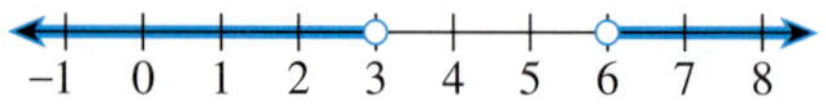

71. 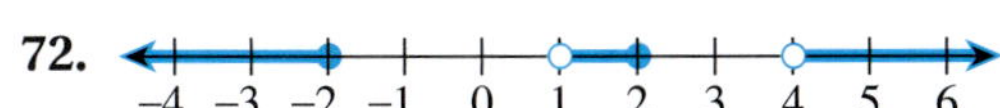

72.

73. Use the graph of $y = x^2 + 6x - 7$ to solve $x^2 + 6x - 7 < 0$.

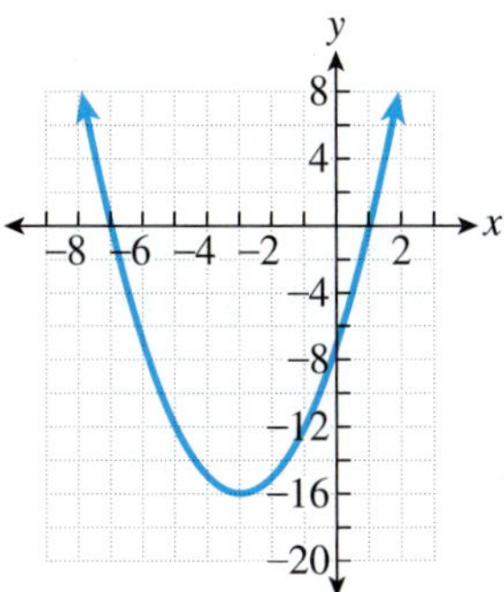

74. Use the graph of $y = -x^2 + 3x + 4$ to solve $-x^2 + 3x + 4 \geq 0$.

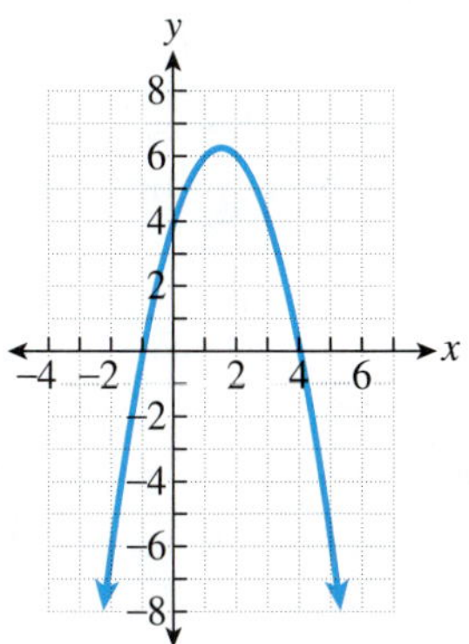

75. Use the graph of $y = -x^2 + 8x - 19$ to solve $-x^2 + 8x - 19 > 0$.

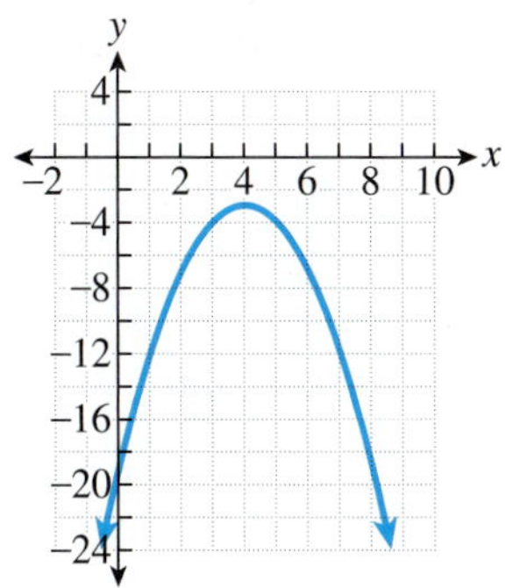

76. Use the graph of $y = x^3 + x^2 - 10x + 8$ to solve $x^3 + x^2 - 10x + 8 \geq 0$.

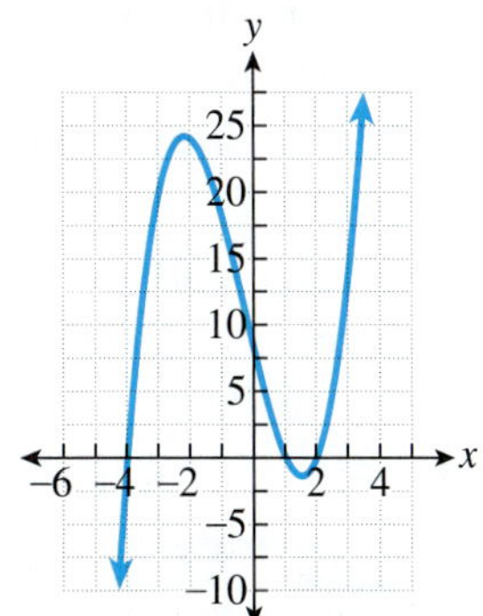

Writing in Mathematics

Answer in complete sentences.

77. ***Solutions Manual**** Write a solutions manual page for the following problem:

Solve $\dfrac{x^2 - 6x - 16}{x^2 - 16} \geq 0.$

78. ***Newsletter**** Write a newsletter explaining how to solve quadratic inequalities.

***See Appendix B for details and sample answers.**

7.7 Other Functions and Their Graphs

Objectives

1. **Graph square-root functions.**
2. **Graph cubic functions.**
3. **Determine a function from its graph.**

In this section, we will examine the graphs of two functions: the square-root function and the cubic function.

Graphing Square-Root Functions

Objective 1 **Graph square-root functions.** We will begin by learning to graph the **square-root function**. The basic square-root function is the function $f(x) = \sqrt{x}$. Recall that the square root of a negative number is an imaginary number, so the radicand must be nonnegative. We begin to graph this function by selecting values for x and evaluating the function for these values. It is a good idea to choose values of x that are perfect squares, such as 0, 1, 4, and 9.

x	$f(x) = \sqrt{x}$	$(x, f(x))$
0	$f(0) = \sqrt{0} = 0$	$(0, 0)$
1	$f(1) = \sqrt{1} = 1$	$(1, 1)$
4	$f(4) = \sqrt{4} = 2$	$(4, 2)$
9	$f(9) = \sqrt{9} = 3$	$(9, 3)$

The domain of this function, which is read from left to right along the x-axis on this graph, is $[0, \infty)$. The range of this function, which is read from the bottom to the top along the y-axis, is also $[0, \infty)$.

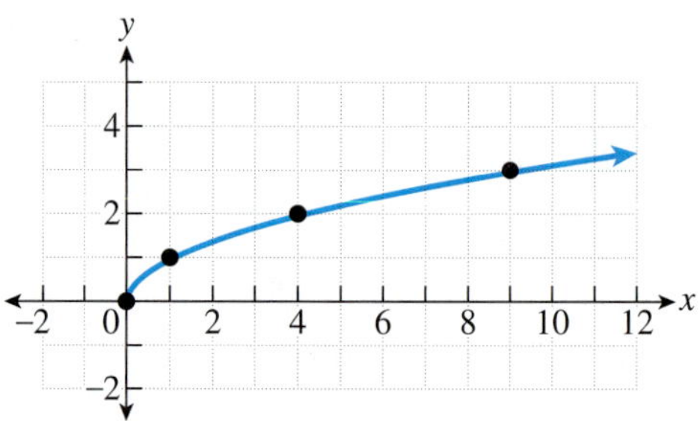

We now turn our attention to graphing square-root functions of the form $f(x) = a\sqrt{x - h} + k$.

EXAMPLE 1 Graph $f(x) = \sqrt{x + 5} + 4$ and state the domain and range.

Solution

We begin by choosing values for x. To find the value of x where the graph starts, we set the radicand equal to 0 and solve for x.

$$x + 5 = 0 \quad \text{Set the radicand equal to 0.}$$
$$x = -5 \quad \text{Subtract 5.}$$

To find other values of x to use, we can set $x + 5$ equal to the perfect squares 1, 4, and 9 and solve for x. The numbers 1, 4, and 9 are the first three positive integers that are perfect squares.

$$x + 5 = 1 \qquad x + 5 = 4 \qquad x + 5 = 9$$
$$x = -4 \qquad x = -1 \qquad x = 4$$

Now we evaluate the function for these values of x and then draw the graph.

x	$f(x) = \sqrt{x+5} + 4$	$(x, f(x))$
-5	$f(-5) = \sqrt{-5+5} + 4 = 4$	$(-5, 4)$
-4	$f(-4) = \sqrt{-4+5} + 4 = 5$	$(-4, 5)$
-1	$f(-1) = \sqrt{-1+5} + 4 = 6$	$(-1, 6)$
4	$f(4) = \sqrt{4+5} + 4 = 7$	$(4, 7)$

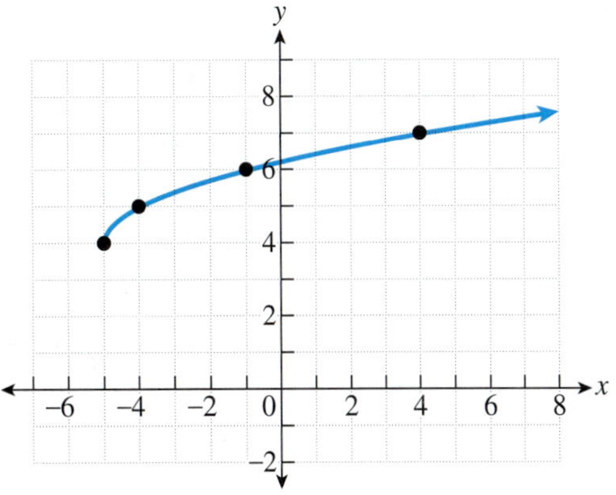

To find the y-intercept of this graph, we need to find $f(0)$.

$$f(0) = \sqrt{(0) + 5} + 4 \quad \text{Substitute 0 for } x.$$
$$= \sqrt{5} + 4 \quad \text{Simplify the radicand.}$$
$$\approx 6.2 \quad \text{Approximate with a calculator.}$$

The y-intercept is approximately at $(0, 6.2)$. We can see from the graph that this function has no x-intercepts. The domain of this function is $[-5, \infty)$, and the range is $[4, \infty)$.

Using Your Calculator We can graph square-root functions by using the TI-84. To enter the function from Example 1, $f(x) = \sqrt{x+5} + 4$, press the Y= key. Enter $\sqrt{x+5} + 4$ next to Y_1. Use parentheses to separate the radicand from the rest of the expression. Press the key labeled GRAPH to graph the function in the standard window.

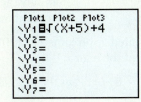

```
Plot1 Plot2 Plot3
\Y1=√(X+5)+4
\Y2=
\Y3=
\Y4=
\Y5=
\Y6=
\Y7=
```

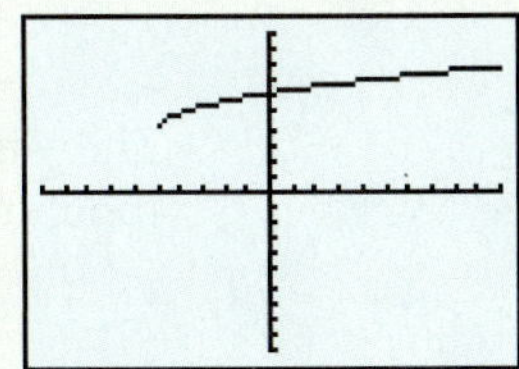

Graphing a Square-Root Function

- Determine the values of x for which the radicand is equal to 0, 1, 4, and 9.
- Create a table of function values for these values of x.
- Plot the points in the table on a coordinate system and draw a smooth curve that passes through these points, beginning at the point whose x-coordinate causes the radicand to be equal to 0.

EXAMPLE 2 Graph $f(x) = \sqrt{x+8} - 4$ and state the domain and range.

Solution

We begin by choosing values for x. Again, set the radicand equal to 0, 1, 4, and 9, and solve for x.

$$x + 8 = 0 \qquad x + 8 = 1 \qquad x + 8 = 4 \qquad x + 8 = 9$$
$$x = -8 \qquad x = -7 \qquad x = -4 \qquad x = 1$$

Now we evaluate the function for these values of x and then draw the graph.

x	$f(x) = \sqrt{x+8} - 4$	$(x, f(x))$
-8	$f(-8) = \sqrt{-8+8} - 4 = -4$	$(-8, -4)$
-7	$f(-7) = \sqrt{-7+8} - 4 = -3$	$(-7, -3)$
-4	$f(-4) = \sqrt{-4+8} - 4 = -2$	$(-4, -2)$
1	$f(1) = \sqrt{1+8} - 4 = -1$	$(1, -1)$

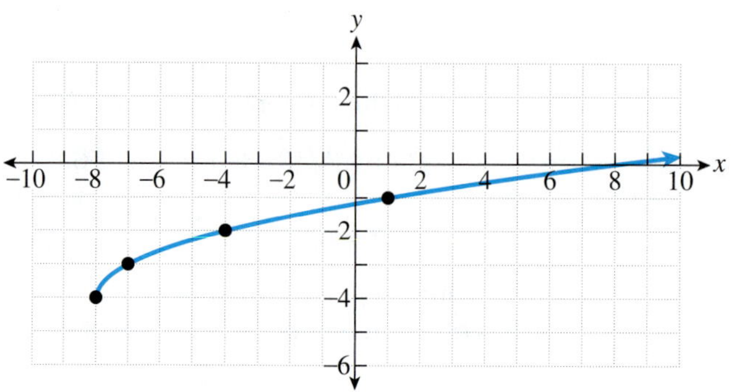

To find the y-intercept of this function, we evaluate $f(0)$.

$$\begin{aligned} f(0) &= \sqrt{0+8} - 4 && \text{Substitute 0 for } x. \\ &= \sqrt{8} - 4 && \text{Simplify the radicand.} \\ &\approx -1.2 && \text{Approximate, using a calculator.} \end{aligned}$$

The y-intercept is located approximately at $(0, -1.2)$. To find the x-intercept, we set the function equal to 0 and solve for x.

$$\sqrt{x+8} - 4 = 0 \qquad \text{Set the function equal to 0.}$$

$\sqrt{x+8} = 4$	Add 4 to isolate the square root.
$(\sqrt{x+8})^2 = 4^2$	Square both sides.
$x + 8 = 16$	Simplify each side.
$x = 8$	Subtract 8.

The x-intercept is at $(8, 0)$. The domain of this function is $[-8, \infty)$, and the range is $[-4, \infty)$.

***Quick Check* 1** **Graph $f(x) = \sqrt{x+6} + 5$ and state its domain and range.**

EXAMPLE 3 Graph $f(x) = \sqrt{-x} + 3$ and state the domain and range.

Solution

Notice that the coefficient of x is negative. In this case, the graph will move to the left rather than to the right. We begin by setting the radicand equal to 0, 1, 4, and 9, and then solve for x.

$$-x = 0 \qquad -x = 1 \qquad -x = 4 \qquad -x = 9$$
$$x = 0 \qquad x = -1 \qquad x = -4 \qquad x = -9$$

Now we evaluate the function for these values of x and then draw the graph.

x	$f(x) = \sqrt{-x} + 3$	$(x, f(x))$
0	$f(0) = \sqrt{-0} + 3 = 3$	$(0, 3)$
−1	$f(-1) = \sqrt{-(-1)} + 3 = 4$	$(-1, 4)$
−4	$f(-4) = \sqrt{-(-4)} + 3 = 5$	$(-4, 5)$
−9	$f(-9) = \sqrt{-(-9)} + 3 = 6$	$(-9, 6)$

***Quick Check* 2** **Graph $f(x) = \sqrt{-x+2} + 7$ and state its domain and range.**

The y-intercept is $(0, 3)$ and is already on the graph. This function does not have an x-intercept. The domain of this function is $(-\infty, 0]$, and the range is $[3, \infty)$.

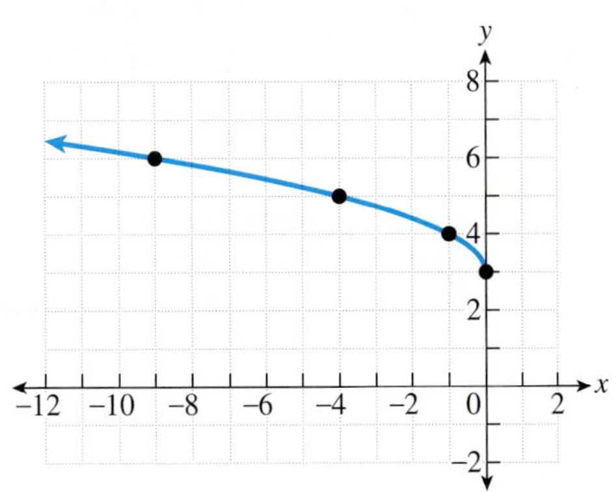

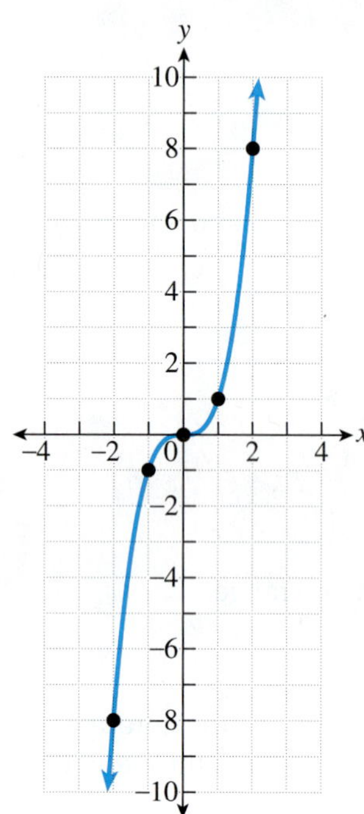

Graphing Cubic Functions

Objective 2 **Graph cubic functions.** The other function we will investigate in this section is the **cubic function**. The basic cubic function is the function $f(x) = x^3$. Here is a table of function values for values of x from -2 to 2:

x	$f(x) = x^3$	$(x, f(x))$
-2	$f(-2) = (-2)^3 = -8$	$(-2, -8)$
-1	$f(-1) = (-1)^3 = -1$	$(-1, -1)$
0	$f(0) = 0^3 = 0$	$(0, 0)$
1	$f(1) = 1^3 = 1$	$(1, 1)$
2	$f(2) = 2^3 = 8$	$(2, 8)$

We plot these points and draw a smooth curve through them. The graph at the left extends forever to the left and to the right, so the domain is the set of all real numbers $\mathbb{R}$, which can be expressed as $(-\infty, \infty)$. The graph also extends forever upwards and downwards, so the range of this function is also $(-\infty, \infty)$. This is true for all cubic functions.

EXAMPLE 4 Graph $f(x) = (x + 2)^3 - 1$ and state the domain and range.

Solution

To choose values for x, begin by setting the expression that is cubed equal to 0 and solving for x. Then choose two values for x that are less than this value and two others that are greater than this value.

$x + 2 = 0$ Set the expression cubed equal to 0.

$x = -2$ Subtract 2.

We will create a table of function values for values of x from -4 to 0 and then draw the graph.

x	$f(x) = (x + 2)^3 - 1$	$(x, f(x))$
-4	$f(-4) = (-4 + 2)^3 - 1 = -9$	$(-4, -9)$
-3	$f(-3) = (-3 + 2)^3 - 1 = -2$	$(-3, -2)$
-2	$f(-2) = (-2 + 2)^3 - 1 = -1$	$(-2, -1)$
-1	$f(-1) = (-1 + 2)^3 - 1 = 0$	$(-1, 0)$
0	$f(0) = (0 + 2)^3 - 1 = 7$	$(0, 7)$

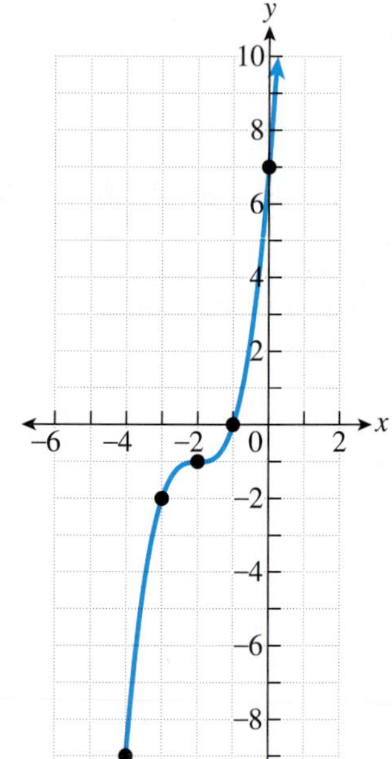

From the points that we have plotted in the next graph, we can see that the y-intercept is $(0, 7)$ and the x-intercept is $(-1, 0)$. The domain of this function is $(-\infty, \infty)$, and the range $(-\infty, \infty)$.

Graphing a Cubic Function

- Determine the values of x for which the cubed expression is equal to 0.
- Select two additional values that are less than this value of x, as well as two values that are greater than this value of x.
- Create a table of function values, using these five values of x.
- Plot the points in the table on a coordinate system and draw a smooth curve that passes through these points.

EXAMPLE 5 Graph $f(x) = -(x - 4)^3 + 2$ and state the domain and range.

Solution

We begin by setting the expression that is cubed equal to 0 and solving for x.

$x - 4 = 0$ Set the expression cubed equal to 0.

$x = 4$ Add 4.

We will create a table of function values for values of x ranging from 2 to 6 and then draw the graph.

x	$f(x) = -(x - 4)^3 + 2$	$(x, f(x))$
2	$f(2) = -(2 - 4)^3 + 2 = 10$	(2, 10)
3	$f(3) = -(3 - 4)^3 + 2 = 3$	(3, 3)
4	$f(4) = -(4 - 4)^3 + 2 = 2$	(4, 2)
5	$f(5) = -(5 - 4)^3 + 2 = 1$	(5, 1)
6	$f(6) = -(6 - 4)^3 + 2 = -6$	(6, −6)

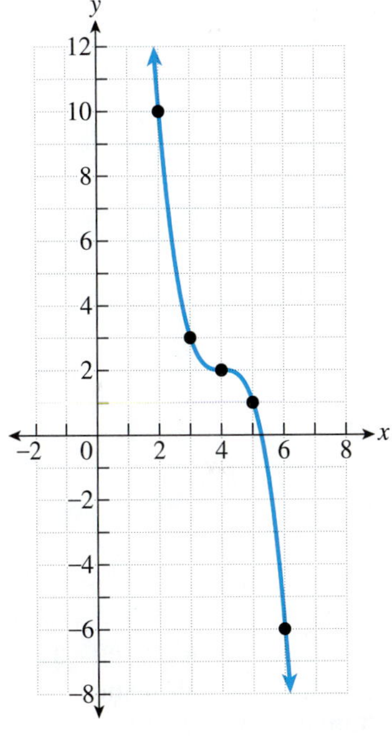

If we want to find the y-intercept, we need to evaluate $f(0)$.

$f(0) = -(0 - 4)^3 + 2$ Substitute 0 for x.

$= 66$ Simplify.

The y-intercept is at $(0, 66)$. To find the x-intercept, we set $f(x)$ equal to 0 and solve for x. When dealing with a cubic function, this will involve using a cube root.

$-(x-4)^3 + 2 = 0$	Set the function equal to 0.
$2 = (x-4)^3$	Add $(x-4)^3$ to isolate the cubed expression.
$\sqrt[3]{2} = \sqrt[3]{(x-4)^3}$	Take the cube root of both sides.
$\sqrt[3]{2} = x - 4$	Simplify.
$4 + \sqrt[3]{2} = x$	Add 4 to isolate x.
$x \approx 5.3$	Approximate, using a calculator.

To approximate the x-coordinate of the x-intercept, we have to approximate $\sqrt[3]{2}$, using a calculator or software package. To do this on a calculator, we can raise 2 to the $1/3$ power. (Recall that $\sqrt[3]{x} = x^{1/3}$.) The x-intercept is at approximately $(5.3, 0)$.

Next, we present two graphs showing the function and its y-intercept. In the first graph, the scale on the horizontal axis is not equal to the scale on the vertical axis. This helps us to see all of the points that we had plotted. In the second graph, the horizontal axis and the vertical axis use equal scales.

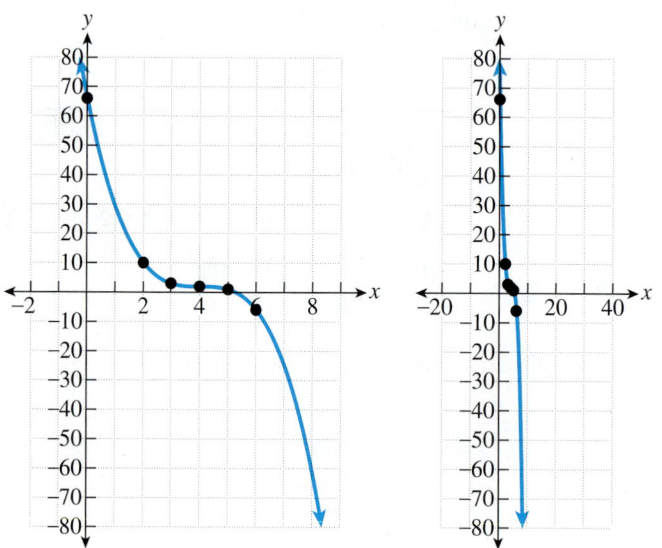

The domain of this function is $(-\infty, \infty)$, and the range is also $(-\infty, \infty)$.

Quick Check 3

Graph the function and state its domain and range.

a) $f(x) = (x-1)^3 + 1$
b) $f(x) = -(x+3)^3$

Determining a Function from Its Graph

Objective 3 **Determine a function from its graph.**

EXAMPLE 6 Determine the function $f(x) = a\sqrt{x - h} + k$ that has been graphed.

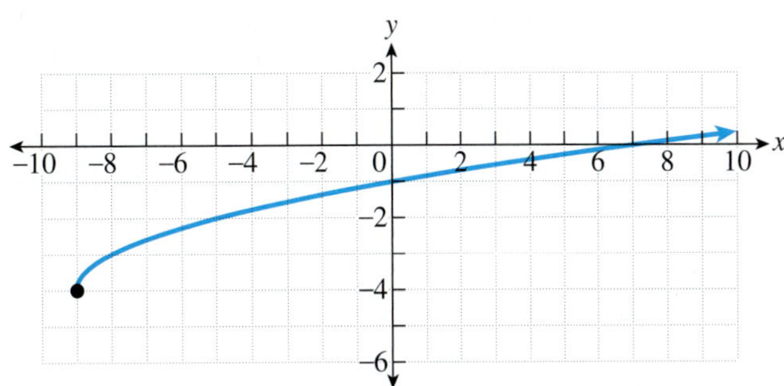

Solution

The graph of this function tells us that $f(x)$ is a square-root function of the form $f(x) = a\sqrt{x - h} + k$. To find the function, let's focus on the point where this graph begins, which is at $(-9, -4)$. Since the x-coordinate is -9, the expression $x + 9$ must be inside the square root. The y-coordinate is -4, which tells us that 4 is subtracted from the square root. The function is of the form $f(x) = a\sqrt{x + 9} - 4$. To find a, we will again use the coordinates of another point on the graph, such as $(-8, -3)$.

$f(-8) = -3$	Set $f(-8)$ equal to -3.
$a\sqrt{-8 + 9} - 4 = -3$	Substitute -8 for x in the function $f(x)$.
$a - 4 = -3$	Simplify the square root.
$a = 1$	Add 4.

Since $a = 1$, the function is $f(x) = \sqrt{x + 9} - 4$.

EXAMPLE 7 Determine the function $f(x) = a(x - h)^3 + k$ that has been graphed.

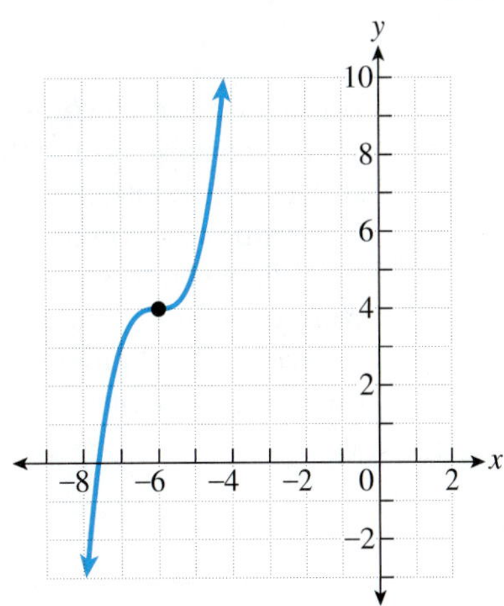

Solution

In this example, $f(x)$ is a cubic function of the form $f(x) = a(x - h)^3 + k$. To find the function, let's focus on the point where this graph flattens out, which is at $(-6, 4)$. Since the x-coordinate is -6, the expression $x + 6$ must be the cubed expression. The y-coordinate is 4, which tells us that 4 is added to the cubed expression. The function is of the form $f(x) = a(x + 6)^3 + 4$. It is left to the reader to show that $a = 1$ for this function. The function is $f(x) = (x + 6)^3 + 4$.

Quick Check **4** **Determine the function $f(x)$ that has been graphed.**

a)

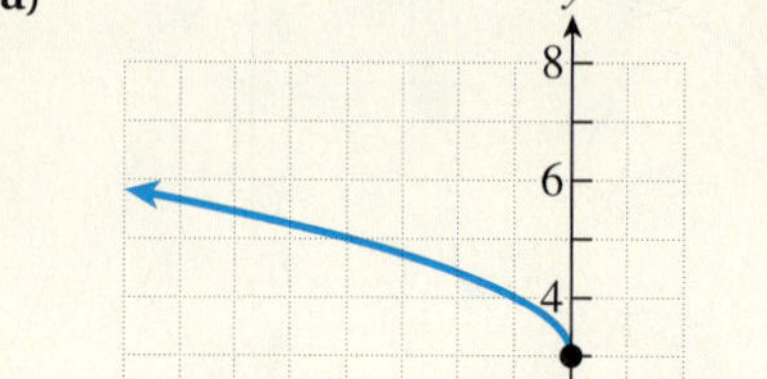

b)

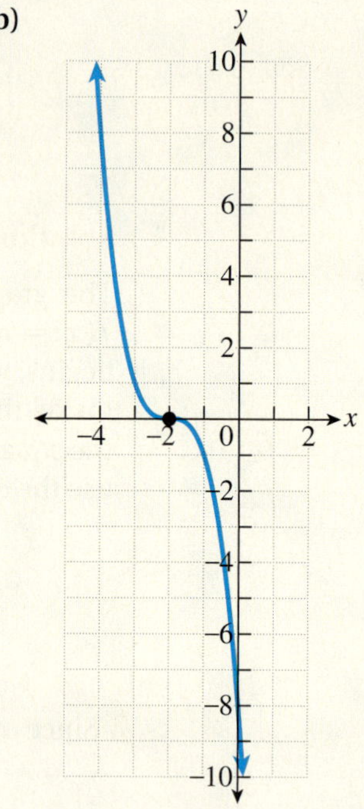

Building Your Study Strategy **Time Management, 7** **Studying at Work** Many students have jobs, and this commitment takes a great deal of time away from their studies. If possible, find a job that allows you opportunities to study while you work. For example, if you are working at the check-out counter of the library, your boss may allow you to study while there are no students waiting to check out books. If you are unable to study at work, be sure to take full advantage of any breaks that you receive. A 10-minute work break may not seem like a lot of time, but you can read through a series of note cards or review your class notes during this time.

EXERCISES 7.7

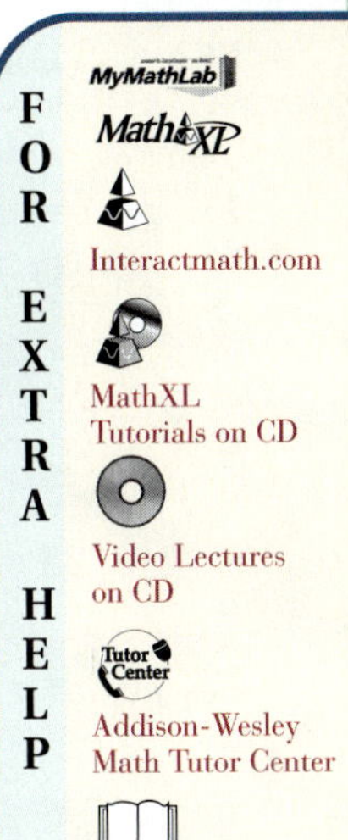

Vocabulary

1. To find the domain of a square root function, set the ________ greater than or equal to 0 and solve.

2. The domain and range of any cubic function are the set of ________.

Graph each given square-root function, and state its domain and range.

3. $f(x) = \sqrt{x + 2}$

4. $f(x) = \sqrt{x - 3}$

5. $f(x) = \sqrt{x} + 2$

6. $f(x) = \sqrt{x} - 3$

7. $f(x) = \sqrt{x + 4} + 6$

8. $f(x) = \sqrt{x + 1} - 7$

9. $f(x) = \sqrt{x - 5} - 2$

10. $f(x) = \sqrt{x - 9} - 10$

11. $f(x) = -\sqrt{x + 4}$

12. $f(x) = -\sqrt{x} + 8$

13. $f(x) = -\sqrt{x - 2} + 3$

14. $f(x) = -\sqrt{x + 5} - 4$

15. $f(x) = \sqrt{3 - x} + 1$

16. $f(x) = \sqrt{-2 - x} + 4$

Graph the given cubic function, and state its domain and range.

17. $f(x) = x^3 - 1$

18. $f(x) = x^3 + 8$

19. $f(x) = (x - 3)^3$

20. $f(x) = (x + 1)^3$

21. $f(x) = (x + 4)^3 + 1$

22. $f(x) = (x - 2)^3 - 4$

25. $f(x) = -x^3 - 27$

26. $f(x) = -(x + 3)^3$

23. $f(x) = (x + 2)^3 + 7$

24. $(x + 1)^3 - 10$

27. $f(x) = 2x^3$

28. $f(x) = -\frac{1}{2}x^3$

Determine the function $f(x)$ that has been graphed. The function will be of the form $f(x) = a\sqrt{x - h} + k$ or $f(x) = a(x - h)^3 + k$. You may assume that $a = 1$ or $a = -1$.

29.

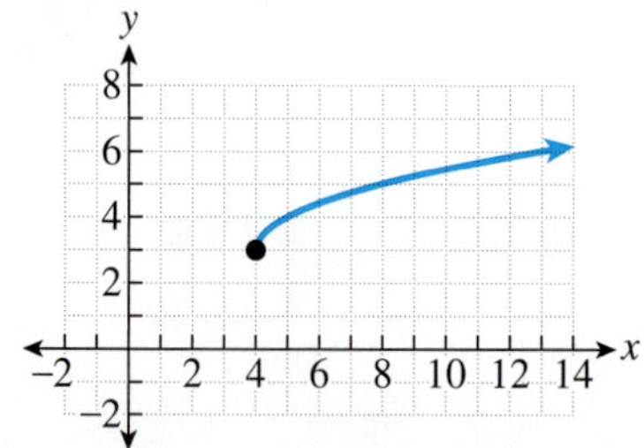

30.

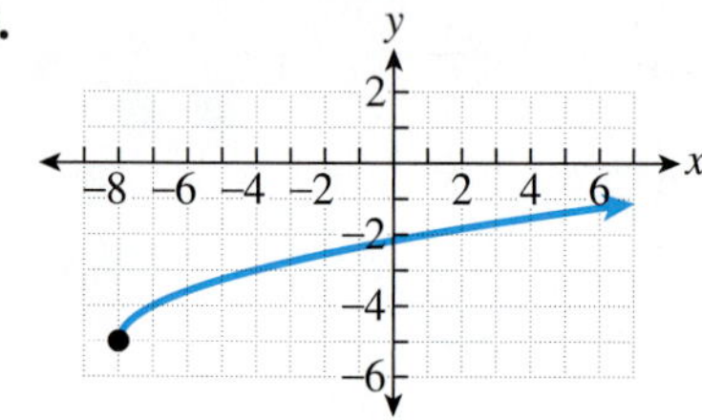

31.

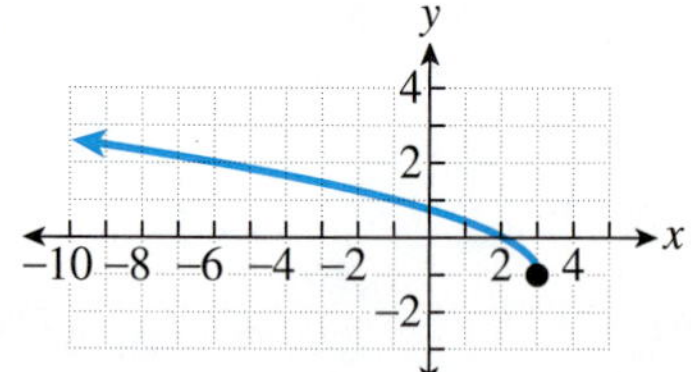

32.

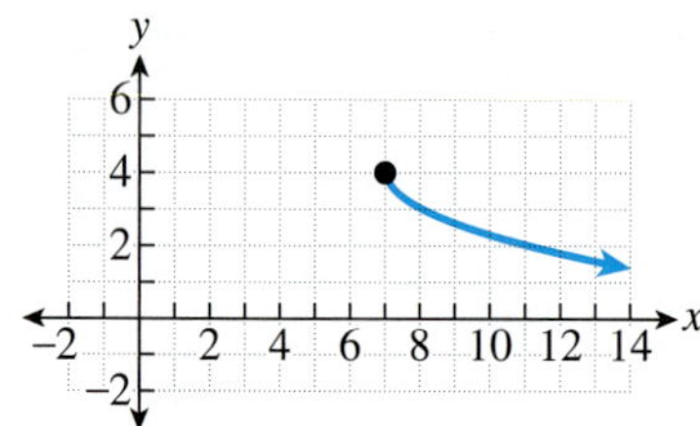

33.

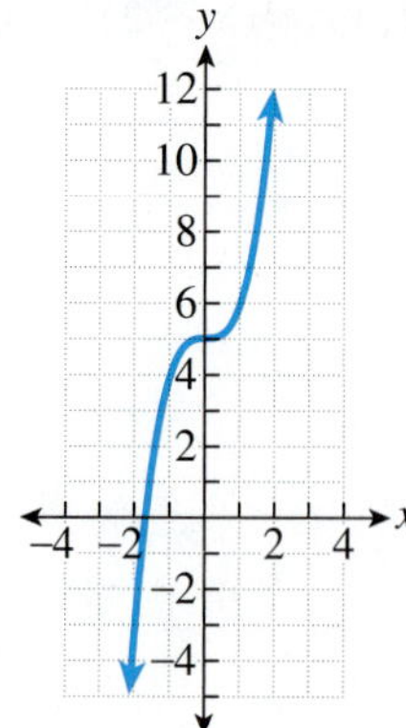

34.

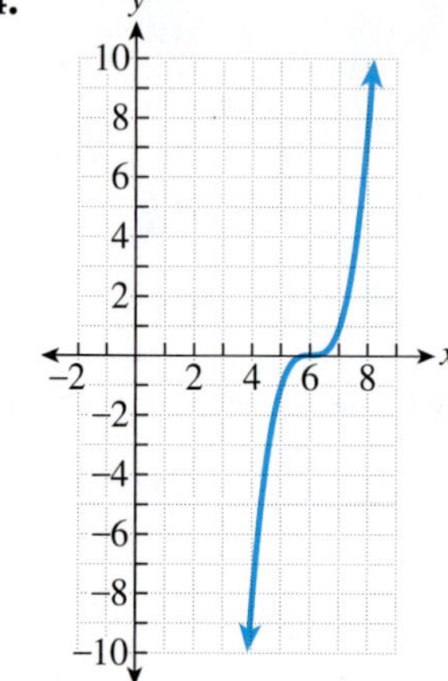

35. a) On the same set of axes, graph the functions $f(x) = \sqrt{x}$, $g(x) = \sqrt{x + 2}$, and $h(x) = \sqrt{x - 6}$.

b) Using the results from part (a), explain how the graph of $f(x) = \sqrt{x + 19}$ would differ from the graph of the function $f(x) = \sqrt{x}$. Also, explain how the graph of $f(x) = \sqrt{x - 27}$ would differ from the graph of the function $f(x) = \sqrt{x}$.

36. a) On the same set of axes, graph the functions $f(x) = \sqrt{x}$, $g(x) = \sqrt{x} + 4$, and $h(x) = \sqrt{x} - 7$.

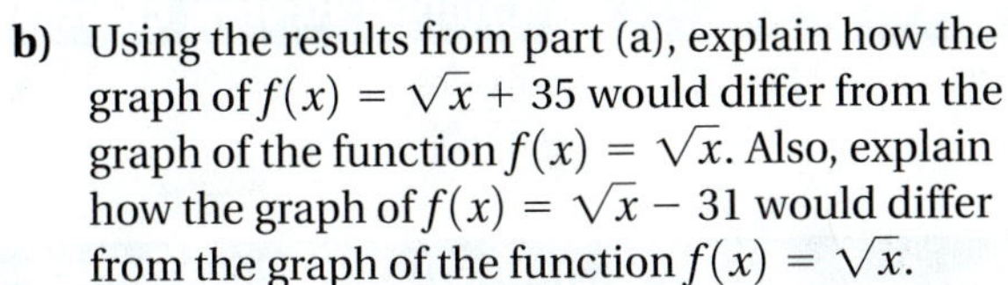

b) Using the results from part (a), explain how the graph of $f(x) = \sqrt{x} + 35$ would differ from the graph of the function $f(x) = \sqrt{x}$. Also, explain how the graph of $f(x) = \sqrt{x} - 31$ would differ from the graph of the function $f(x) = \sqrt{x}$.

37. On the same set of axes, graph the functions $f(x) = \sqrt{x}$ and $g(x) = -\sqrt{x}$. Explain, in general, how the graph of any square-root function is affected by the placement of a negative sign in front of the square root.

38. Using the concepts developed in Exercises 35–37, explain how the graph of $g(x) = -\sqrt{x + 8} - 1$ differs from the graph of the function $f(x) = \sqrt{x}$. Check your explanation by actually graphing the functions $f(x)$ and $g(x)$ on the same set of axes.

Writing in Mathematics

Answer in complete sentences.

39. Explain the similarities and differences between graphing a quadratic function and graphing a square-root function.

40. Explain why both the domain and the range of a cubic function are the set of real numbers.

41. ***Solutions Manual**** Write a solutions manual page for the following problem:

Graph $y = (x - 2)^3 - 8$.

42. ***Newsletter**** Write a newsletter explaining how to graph square root functions.

***See Appendix B for details and sample answers.**

Chapter 7 Summary

Section 7.1—Topic	Chapter Review Exercises
Solving Quadratic Equations by Factoring	1–6
Solving Quadratic Equations by Extracting Square Roots	7–12
Solving Quadratic Equations by Completing the Square	13–16
Finding a Quadratic Equation, Given Its Solutions	45–50
Section 7.2—Topic	**Chapter Review Exercises**
Solving Quadratic Equations by Using the Quadratic Formula	17–22
Section 7.3—Topic	**Chapter Review Exercises**
Solving Equations Using u-Substitutions	23–26
Solving Radical Equations	27–28
Solving Rational Equations	29–30
Solving Work-Rate Problems	87–88
Section 7.4—Topic	**Chapter Review Exercises**
Finding the Vertex and Axis-of-Symmetry of a Parabola	51–54
Finding the Intercepts of a Parabola	55–58
Graphing Parabolas	59–70
Section 7.5—Topic	**Chapter Review Exercises**
Solving Applied Problems Involving Quadratic Equations	85–86, 89–92
Finding Maximum/Minimum Values of Quadratic Functions	71–74
Solving Applied Maximum/Minimum Problems	75–76
Section 7.6—Topic	**Chapter Review Exercises**
Solving Quadratic Inequalities	77–80
Solving Rational Inequalities	81–82
Solving Inequalities Involving Functions	83–84
Section 7.7—Topic	**Chapter Review Exercises**
Graphing Square-Root Functions	93–96
Graphing Cubic Functions	97–98
Finding Functions from Their Graphs	99–100

Review of Chapter 7 Study Strategies

Poor time management can cause a student to do poorly in a math class. To maximize the time you have available to study mathematics and to make your use of this time effective, consider the following:

- Record how you spend your time over the period of one week. Look for ways you might be wasting time. Create an efficient schedule to follow.
- Study new material as soon as possible after class. Save review work for later in the day.
- Arrange your schedule to allow you to study mathematics at the time of day when your mental energy is at its highest.
- Set goals for each study session, and record these goals in a "to-do" list.
- Study your most difficult subject first, before you become fatigued.
- If you must work, try to find a job that will allow you to do some studying while you work. At the very least, make efficient use of your work breaks to study.

Chapter 7 Review

Solve by factoring. [7.1]

1. $x^2 + 11x + 28 = 0$
2. $x^2 + 7x - 30 = 0$
3. $x^2 + 4x = 0$
4. $2x^2 - 13x + 15 = 0$
5. $x^2 + 7x - 12 = 3x - 7$
6. $x^2 - 9x + 23 = 9(x - 6)$

Solve by extracting square roots. [7.1]

7. $x^2 = 49$
8. $x^2 - 72 = 0$
9. $x^2 + 4 = 0$
10. $(x - 6)^2 = -36$
11. $(x - 8)^2 + 30 = 21$
12. $(2x + 7)^2 + 20 = 141$

Solve by completing the square. [7.1]

13. $x^2 - 2x - 8 = 0$
14. $x^2 - 6x + 3 = 0$
15. $x^2 + 14x + 47 = 0$
16. $x^2 + 5x + 10 = 0$

Solve by using the quadratic formula. [7.2]

17. $x^2 + 7x + 1 = 0$
18. $x^2 + 4x - 2 = 0$
19. $x^2 - 3x + 9 = 0$
20. $4x^2 - 20x + 29 = 0$
21. $2x^2 - 5x + 9 = x^2 - 9x + 30$
22. $x(x + 4) = 16x - 37$

Worked-out solutions to Review Exercises marked with [circle] can be found on page AN-37.

Solve by using a u-substitution. [7.3]

23. $x^4 - 5x^2 + 4 = 0$
24. $x - 4\sqrt{x} - 32 = 0$
25. $x^6 + 7x^3 - 8 = 0$
26. $x^{2/3} - 4x^{1/3} - 21 = 0$

Solve. [7.3]

27. $\sqrt{x^2 + x - 2} = 2$
28. $\sqrt{x + 15} - x = 9$

Solve. [7.3]

29. $1 + \dfrac{11}{x} + \dfrac{30}{x^2} = 0$
30. $\dfrac{x}{x - 3} - \dfrac{7}{x - 2} = \dfrac{3}{x^2 - 5x + 6}$

Solve, using the most effecient method. [7.1, 7.2, 7.3]

31. $x^2 - 19x + 90 = 0$
32. $x^4 + 5x^2 - 36 = 0$
33. $(x + 6)^2 + 11 = 3$
34. $x^2 + 4x + 16 = 0$
35. $x^2 + 10x = 0$
36. $x^2 - 6x - 20 = 0$
37. $\sqrt{3x + 49} = x + 3$
38. $\sqrt{2x + 9} = \sqrt{4x + 1} - 2$
39. $x - 1 - \dfrac{20}{x} = 0$
40. $3x^2 - 8x + 2 = 0$
41. $x + 5x^{1/2} - 24 = 0$
42. $(x + 9)^2 + 64 = 0$

For each function $f(x)$, find all values x for which $f(x) = 0$. [7.1, 7.2, 7.3]

43. $f(x) = (x + 12)^2 - 16$
44. $f(x) = x^2 - 7x - 20$

Find a quadratic equation with integer coefficients that has the given solution(s). **[7.1]**

45. $\{-9, 4\}$

46. $\{4\}$

47. $\{\frac{4}{3}, \frac{13}{2}\}$

48. $\{-2\sqrt{2}, 2\sqrt{2}\}$

49. $\{-5i, 5i\}$

50. $\{-8i, 8i\}$

Find the vertex and axis of symmetry of the parabola associated with the given quadratic equation. **[7.4]**

51. $y = x^2 - 8x - 20$

52. $y = x^2 + 5x - 16$

53. $y = -x^2 - 2x + 35$

54. $y = (x - 3)^2 - 4$

Find the x- and y-intercepts of the parabola associated with the given equation. If the parabola does not have any x-intercepts, state "no x-intercepts." **[7.4]**

55. $y = x^2 + 6x + 11$

56. $y = x^2 - 4x - 3$

57. $y = -x^2 + 10x - 25$

58. $y = (x + 4)^2 - 8$

Graph each quadratic equation. Label the vertex, the y-intercept, and any x-intercepts. **[7.4]**

59. $y = x^2 + 8x + 15$

60. $y = -x^2 + 2x + 3$

61. $y = x^2 + 4x - 2$

62. $y = x^2 + 4x + 8$

63. $y = -(x + 3)^2 + 4$

64. $y = -(x - 6)^2 + 1$

Graph the function. Label the vertex, y-intercept, and any x-intercepts. **[7.4]**

65. $f(x) = x^2 - 4x - 12$

66. $f(x) = x^2 + 6x + 1$

67. $f(x) = x^2 - 8x + 19$

68. $f(x) = -x^2 - 4x + 3$

69. $f(x) = (x - 3)^2 - 7$

70. $f(x) = -(x + 2)^2 - 1$

Determine whether the given quadratic function has a maximum value or a minimum value. Then find that maximum or minimum value. [7.5]

71. $f(x) = x^2 - 12x + 60$

72. $f(x) = x^2 + 7x - 18$

73. $f(x) = -x^2 + 16x - 70$

74. $f(x) = -(x + 8)^2 - 39$

75. A projectile is launched upward from the roof of a building with an initial velocity of 176 feet/second. The height of the projectile (in feet) after t seconds is given by the function $h(t) = -16t^2 + 176t + 42$. What is the maximum height that the projectile reaches?

76. A farmer has 144 feet of fencing to make a rectangular corral. What dimensions will make a corral with the maximum area? What is the maximum area possible?

Solve each quadratic inequality. Express your solution on a number line, using interval notation. [7.6]

77. $x^2 - 9x + 14 \leq 0$

78. $x^2 + 9x + 8 > 0$

79. $x^2 + 4x - 11 > 0$

80. $x^2 + 3x + 40 \geq 0$

Solve each rational inequality. Express your solution on a number line, using interval notation. [7.6]

81. $\dfrac{x^2 - 15x + 54}{x - 7} \leq 0$

82. $\dfrac{x^2 - 4}{x^2 + 2x - 80} \geq 0$

Given each function $f(x)$, solve the inequality $f(x) \geq 0$. Express your solution on a number line, using interval notation. [7.6]

83. $f(x) = x^2 + 15x + 50$

84. $f(x) = \dfrac{x^2 - 16}{x^2 - 10x + 9}$

For Exercises 85 and 86, use the function $h(t) = -16t^2 + v_0t + s$. [7.2]

85. A projectile is launched upwards from the ground at a speed of 104 feet per second. How long will it take until the projectile lands on the ground?

86. A projectile is launched upward from the roof of a building 30 feet high with an initial velocity of 66 feet per second. How long will it take until the projectile lands on the ground? Round to the nearest hundredth of a second.

87. A water tank is connected to two pipes. When both pipes are open, they can fill the tank in 4 hours. The larger of the pipes can fill the tank in 2 hours less time than the smaller pipe. How long would it take the smaller pipe, working alone, to fill the tank? Round to the nearest tenth of an hour. [7.3]

88. A tree-trimming company has two crews. If both crews work together, they can trim all of the trees on campus in 32 hours of work. The faster crew could trim all of the trees in 6 hours less than the slower crew. How long would it take the faster crew, working alone, to trim all of the trees on campus? Round to the nearest tenth of an hour. [7.3]

Graph each square-root function and state its domain and range. **[7.7]**

89. $f(x) = \sqrt{x + 4}$

90. $f(x) = \sqrt{x - 2} + 7$

91. $f(x) = -\sqrt{x + 1} + 2$

92. $f(x) = \sqrt{x + 5} - 3$

Graph each cubic function and state its domain and range. **[7.7]**

93. $f(x) = (x - 2)^3$

94. $f(x) = x^3 - 1$

Determine each function f(x) that has been graphed. [7.7]

95.

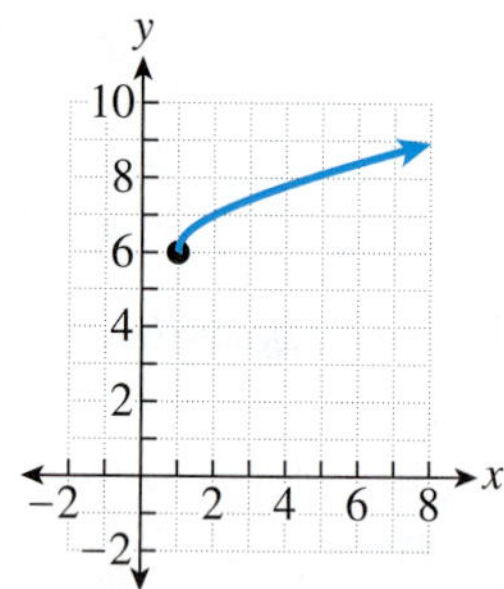

96.

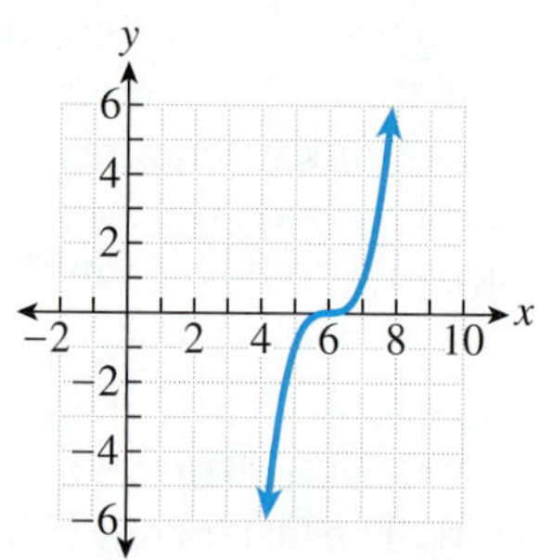

97. A rectangular photograph has an area of 288 square centimeters. If the length of the photo is 2 centimeters longer that its width, find the dimensions of the photo. [7.5]

98. A rectangular swimming pool covers an area of 500 square feet. If the length of the pool is 13 feet more than its width, find the dimensions of the pool. Round to the nearest tenth of a foot. [7.5]

99. Ann needs to fly to a city that is 1000 miles north and 800 miles east of where she lives. If the airplane flies a direct route, how far will the flight be? Round to the nearest tenth of a mile. [7.5]

100. The diagonal of a rectangular workbench is 12 feet long. If the width of the workbench is 3 feet less than the length of the workbench, find the dimensions of the workbench. Round to the nearest tenth of a foot. [7.5]

Chapter 7 Test

For Extra Help Pass the Test — Test solutions are found on the enclosed CD.

1. Solve by factoring. $x^2 + 9x + 18 = 0$

2. Solve by extracting square roots. $(3x - 8)^2 - 6 = 58$

3. Solve by completing the square. $x^2 - 14x - 3 = 0$

4. Solve by using the quadratic formula.
$x^2 - 6x + 12 = 0$

5. Solve by using a u-substitution. $x^4 - 8x^2 - 48 = 0$

6. Solve by using a u-substitution. $x + 8x^{1/2} - 9 = 0$

Solve.

7. $(4x - 1)^2 - 13 = 12$

8. $2x^2 - 15x - 27 = 0$

9. $x^2 - 14x + 40 = 0$

10. $\dfrac{x}{x + 8} - \dfrac{6}{x - 4} = \dfrac{8}{x^2 + 4x - 32}$

11. $x^2 - 10x + 34 = 0$

Find a quadratic equation with integer coefficients that has the given solutions.

12. $\{\frac{4}{7}, 5\}$

13. $\{-4i, 4i\}$

14. Find the x- and y-intercepts of the parabola associated with $y = x^2 - 10x + 20$. Round to the nearest tenth if necessary.

15. Graph $y = -x^2 - 10x + 11$. Label the vertex, the y-intercept, and any x-intercepts.

16. Graph $y = (x - 5)^2 - 12$. Label the vertex, the y-intercept, and any x-intercepts.

Graph the function. Label the vertex, y-intercept, and any x-intercepts.

17. $f(x) = x^2 + 8x + 10$

18. Solve. $x^2 + 3x - 28 > 0$

19. Solve. $\dfrac{x^2 + 11x + 24}{x^2 - 5x - 14} \leq 0$

20. The area of a rectangular playing field is 3600 square yards. If the width of the field is 60 yards less than the length of the field, find the dimensions of the field. Round to the nearest tenth of a yard.

21. Daniel flies to a resort town 1500 miles south and 800 miles west of where he lives. How far away is the resort town from where he lives?

22. A young child throws a ball upward with an initial velocity of 20 feet/second. The height of the ball (in feet) after t seconds is given by the function $h(t) = -16t^2 + 20t + 3$. What is the maximum height that the ball reaches?

Graph each function, and state its domain and range.

23. $f(x) = \sqrt{x + 1} - 3$

24. $f(x) = (x + 1)^3 + 8$

25. Determine the function $f(x)$ on the graph.

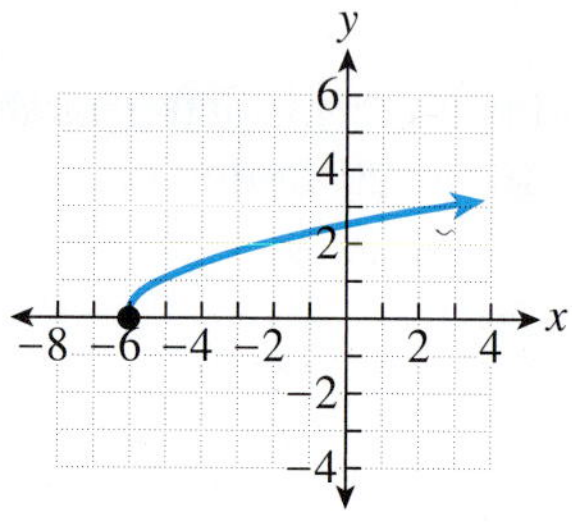

Mathematicians in History

Evariste Galois

Evariste Galois was a brilliant mathematician who was interested in the algebraic solutions of equations. The amount and depth of work accomplished by Galois, who died before his 21st birthday, is legendary. Galois showed that there is no general solution of equations that are fifth degree or higher:

$$ax^5 + bx^4 + cx^3 + dx^2 + ex + f = 0.$$

Write a one-page summary (*or* make a poster) of the mathematical achievements of Galois, his fascinating (but short) life, and the details surrounding his untimely death.

Interesting issues:

- When and where was Galois born?
- What was the fate of Galois's father?
- Galois twice failed the admittance exam to École Polytechnique. What did he do to one of the examiners on his second attempt?
- Galois was expelled from École Normale in December 1830. Why was he expelled?
- Galois was arrested in 1831 after raising a toast to King Louis-Philippe. What was the toast, and how did he make it?
- In 1831, Galois was arrested on Bastille Day. What was the charge?
- On the night before his death, Galois wrote a letter to his friend Auguste Chevalier. What were the contents of the letter?
- What were the circumstances that led to Galois's death? How old was Galois at the time?
- Which mathematician published Galois's papers in 1846?

Stretch Your Thinking Chapter 7

The pilot of a small plane wishes to fly due east. However, there is a powerful wind blowing due north that is causing the plane to fly 7° off of its eastern course. The wind has also increased the speed of the plane so that it is flying at 287 miles per hour. The speed of the plane in still air is 250 miles per hour faster than the speed of the wind. Find the speed of the plane in still air and of the wind.

Cumulative Review Chapters 5–7

Find all values for which the rational expression is undefined. **[5.1]**

1. $\dfrac{x^2 - 13x}{x^2 - 17x + 30}$

Simplify the given rational expression. (Assume that the denominator is nonzero.) **[5.1]**

2. $\dfrac{x^2 + 10x + 16}{x^2 - 5x - 14}$

Evaluate the given rational function. **[5.1]**

3. $r(x) = \dfrac{x^2 - 8x + 6}{x^2 + 12x - 4}, r(-4)$

Multiply. **[5.2]**

4. $\dfrac{9 - x^2}{x^2 + 4x - 32} \cdot \dfrac{x^2 - 2x - 80}{x^2 - 13x + 30}$

Divide. **[5.2]**

5. $\dfrac{x^2 - 16x + 60}{x^2 + 6x - 16} \div \dfrac{x^2 - 13x + 42}{x^2 + 14x + 48}$

Add or subtract. **[5.3/5.4]**

6. $\dfrac{3x + 8}{x^2 + 4x - 12} + \dfrac{x + 16}{x^2 + 4x - 12}$

7. $\dfrac{2x^2 + 12x - 7}{x^2 - 9} + \dfrac{x^2 + 3x - 25}{9 - x^2}$

8. $\dfrac{x + 7}{x^2 - 6x + 5} - \dfrac{6}{x^2 - 5x + 4}$

9. $\dfrac{x + 7}{x^2 + 8x - 9} + \dfrac{x + 10}{x^2 + 13x + 36}$

Simplify the complex fraction. **[5.5]**

10. $\dfrac{1 + \dfrac{1}{x} - \dfrac{20}{x^2}}{1 + \dfrac{13}{x} + \dfrac{40}{x^2}}$

11. $\dfrac{\dfrac{x - 5}{x - 7} - \dfrac{3}{x + 3}}{\dfrac{x^2 - 9x + 18}{x^2 - 4x - 21}}$

Solve. **[5.6]**

12. $4 + \dfrac{15}{x - 2} = \dfrac{3x + 14}{x - 2}$

13. $\dfrac{5}{x + 7} + \dfrac{4}{x - 3} = \dfrac{x - 11}{x^2 + 4x - 21}$

14. $\dfrac{x - 4}{x^2 - 8x + 15} + \dfrac{1}{x^2 - 6x + 5} = \dfrac{11}{x^2 - 4x + 3}$

Solve for the specified variable. **[5.6]**

15. $\dfrac{1}{x} + \dfrac{1}{y} = \dfrac{1}{5}$ for x

16. The sum of the reciprocal of a number and $\dfrac{7}{18}$ is $\dfrac{1}{2}$. Find the number. [5.7]

17. Two pipes, working together, can fill a tank in 12 hours. Working alone, it would take the smaller pipe 3 times longer than it would take the larger pipe to fill the tank. How long would it take the smaller pipe alone to fill the tank? [5.7]

Evaluate each radical function. (Round to the nearest thousandth if necessary.) **[6.1]**

18. $f(x) = \sqrt{6x - 58}, f(19)$

Rewrite as a radical expression and simplify if possible. Assume that x represents a nonnegative value. **[6.2]**

19. $(16x^{10})^{3/2}$

Simplify the expression. Assume that x represents a nonnegative value. **[6.2]**

20. $(x^{6/7})^{3/4}$

Simplify each radical expression. Assume that all variables represent nonnegative values. **[6.3]**

21. $\sqrt{75r^{12}s^{10}t^9}$

Add or subtract. **[6.3]**

22. $7\sqrt{108} - 2\sqrt{48}$

***Multiply.* [6.4]**

23. $(4\sqrt{3} - 2\sqrt{5})(5\sqrt{3} + 4\sqrt{5})$

***Rationalize the denominator and simplify. Assume that all variables represent nonnegative values.* [6.4]**

24. $\dfrac{ab^3c^2}{\sqrt{a^3b^8c}}$

25. $\dfrac{5\sqrt{5} + 6\sqrt{2}}{4\sqrt{5} - 3\sqrt{2}}$

26. For the function $f(x) = \sqrt{2x - 9}$, find all values x for which $f(x) = 5$. [6.5]

***Express in terms of i.* [6.6]**

27. $\sqrt{-360}$

***Multiply.* [6.6]**

28. $(2 + 5i)(9 - 2i)$

***Rationalize the denominator.* [6.6]**

29. $\dfrac{5i}{3 + 4i}$

***Solve by extracting square roots.* [7.1]**

30. $(x + 2)^2 + 41 = 23$

***Solve by using the quadratic formula.* [7.2]**

31. $3x^2 - 10x - 21 = 0$

***Solve by using a u-substitution.* [7.3]**

32. $x^4 - 11x^2 + 18 = 0$

33. $x + 3\sqrt{x} - 10 = 0$

***Solve.* [7.3]**

34. $\sqrt{x + 23} - x = 3$

***Solve.* [7.1–7.2]**

35. $x^2 - 12x + 27 = 0$

***Graph each quadratic equation. Label the vertex, the y-intercept, and any x-intercepts.* [7.4]**

36. $y = -x^2 + 8x + 6$

37. $y = (x - 2)^2 + 3$

***Solve the quadratic inequality. Express your solution on a number line, using interval notation.* [7.6]**

38. $x^2 - 10x + 24 \le 0$

***Solve the rational inequality. Express your solution on a number line, using interval notation.* [7.6]**

39. $\dfrac{x^2 - 2x - 48}{x + 5} \le 0$

40. The width of a rectangular lawn is 5 feet less than its length. If the area of the lawn is 980 square feet, find the dimensions of the lawn. Round to the nearest tenth of a foot. [7.5]

41. A projectile is launched upward from the roof of a building 12 feet high, with an initial velocity of 80 feet per second. How long will it take until the projectile lands on the ground? Round to the nearest hundredth of a second. (Use the function $h(t) = -16t^2 + v_0t + s$.) [7.2]

42. A water tank is connected to two pipes. When both pipes are open, they can fill the tank in 5 hours. The larger of the pipes can fill the tank in 3 hours less time than the smaller pipe. How long would it take the smaller pipe, working alone, to fill the tank? Round to the nearest tenth of an hour. [7.3]

43. A projectile is launched upward from the roof of a building with an initial velocity of 144 feet/second. The height of the projectile (in feet) after t seconds is given by the function $h(t) = -16t^2 + 144t + 50$. What is the maximum height that the projectile reaches? [7.5]

Graph each given function and state its domain and range. [7.7]

44. $f(x) = \sqrt{x + 4} - 1$

45. $f(x) = (x + 1)^3$

CHAPTER Eight 8

Logarithmic and Exponential Functions

In this chapter, we will investigate exponential functions and logarithmic functions. These types of functions have many practical applications, including calculating compound interest and population growth, radiocarbon dating, and determining the intensity of an earthquake or the pH of a substance.

Study Strategy **Practice Quizzes** *Creating practice quizzes is an excellent way to prepare for exams. It is important to put yourself in a test environment before taking a test, without facing the consequences of a test. Practice quizzes will tell you which topics you understand and which topics you need further work on. Throughout this chapter, we will revisit this study strategy and help you to incorporate it into your study habits.*

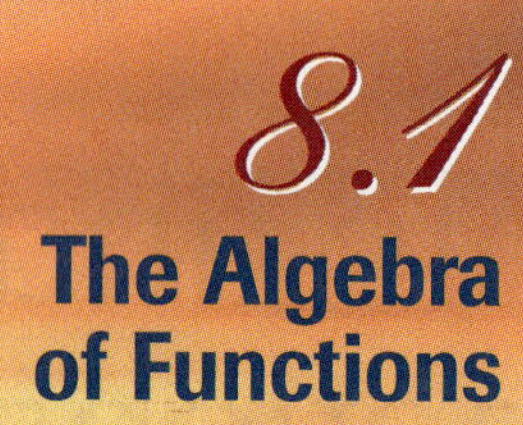

8.1 The Algebra of Functions

Objectives

1. **Find the sum, difference, product, and quotient functions for two functions $f(x)$ and $g(x)$.**
2. **Find the graph of the sum function or difference function from the graphs of two functions $f(x)$ and $g(x)$.**
3. **Solve applications involving the sum function or the difference function.**
4. **Find the composite function of two functions $f(x)$ and $g(x)$.**
5. **Find the domain of a composite function $(f \circ g)(x)$.**

In this section, we will examine several ways to combine two or more functions into a single function. Just as we use addition, subtraction, multiplication, and division to combine two numbers, we can use these operations to combine functions as well.

The Sum Function and the Difference Function

Objective 1 Find the sum, difference, product, and quotient functions for two functions $f(x)$ and $g(x)$. The function that results when two functions $f(x)$ and $g(x)$ are added together is called the **sum function** and is denoted by $(f + g)(x)$. The function $(f - g)(x)$ is the difference of the two functions $f(x)$ and $g(x)$.

The Sum Function

For any two functions $f(x)$ and $g(x)$, $(f + g)(x) = f(x) + g(x)$.

The Difference Function

For any two functions $f(x)$ and $g(x)$, $(f - g)(x) = f(x) - g(x)$.

EXAMPLE 1 If $f(x) = x^2 + 3x - 10$ and $g(x) = 3x^2 - 5x - 9$, find the following:

a) $(f + g)(x)$

Solution

We will begin by rewriting $(f + g)(x)$ as $f(x) + g(x)$.

$(f + g)(x) = f(x) + g(x)$ — Rewrite as the sum of the two functions.

$= (x^2 + 3x - 10) + (3x^2 - 5x - 9)$ — Substitute for each function.

$= x^2 + 3x - 10 + 3x^2 - 5x - 9$ — Remove parentheses.

$= 4x^2 - 2x - 19$ — Combine like terms.

b) $(f - g)(x)$

Solution

We will begin by rewriting $(f - g)(x)$ as $f(x) - g(x)$.

$$(f - g)(x) = f(x) - g(x)$$ Rewrite as the difference of the two functions.

$$= (x^2 + 3x - 10) - (3x^2 - 5x - 9)$$ Substitute for each function.

$$= x^2 + 3x - 10 - 3x^2 + 5x + 9$$ Distribute to remove parentheses.

$$= -2x^2 + 8x - 1$$ Combine like terms.

c) $(f + g)(-3)$

Solution

Since we have already found that $(f + g)(x) = 4x^2 - 2x - 19$, we can substitute -3 for x in this function.

$$(f + g)(x) = 4x^2 - 2x - 19$$

$$(f + g)(-3) = 4(-3)^2 - 2(-3) - 19$$ Substitute -3 for x.

$$= 23$$ Simplify.

$(f + g)(-3) = 23$. An alternative method to find $(f + g)(-3)$ would be to substitute -3 for x in the functions $f(x)$ and $g(x)$ and then add the results. The reader can verify that $f(-3) = -10$ and $g(-3) = 33$; so $(f + g)(-3) = -10 + 33 = 23$.

***Quick Check* 1**

If $f(x) = x^2 - 9x + 20$ and $g(x) = 3x - 44$, find the given function.

a) $(f + g)(x)$
b) $(f - g)(x)$
c) $(f + g)(7)$
d) $(f - g)(-2)$

The Product Function and the Quotient Function

The function $(f \cdot g)(x)$ is the product of the two functions $f(x)$ and $g(x)$. The function $\left(\frac{f}{g}\right)(x)$ is the quotient of the two functions $f(x)$ and $g(x)$.

The Product Function

For any two functions $f(x)$ and $g(x)$, $(f \cdot g)(x) = f(x) \cdot g(x)$.

The Quotient Function

For any two functions $f(x)$ and $g(x)$, $\left(\frac{f}{g}\right)(x) = \frac{f(x)}{g(x)}$, $g(x) \neq 0$.

EXAMPLE 2 If $f(x) = x - 5$ and $g(x) = x^2 + 6x + 8$, find $(f \cdot g)(x)$.

Solution

We will begin by writing $(f \cdot g)(x)$ as $f(x) \cdot g(x)$.

$$(f \cdot g)(x) = f(x) \cdot g(x)$$ Rewrite as the product of the two functions.

$$= (x - 5)(x^2 + 6x + 8)$$ Substitute for each function.

$$= x^3 + 6x^2 + 8x - 5x^2 - 30x - 40$$ Multiply, using the distributive property.

$$= x^3 + x^2 - 22x - 40$$ Combine like terms.

***Quick Check* 2**

If $f(x) = 3x + 4$ and $g(x) = 4x - 7$, find $(f \cdot g)(x)$.

Example 3 If $f(x) = x^2 - 3x - 40$ and $g(x) = x^2 - 25$, find $\left(\frac{f}{g}\right)(x)$, and list any restrictions on its domain.

Quick Check 3
If $f(x) = x^2 + 3x - 18$ and $g(x) = x^2 + 13x + 42$, find $\left(\frac{f}{g}\right)(x)$, and list any restrictions on its domain.

Solution

We will begin by writing $\left(\frac{f}{g}\right)(x)$ as $\frac{f(x)}{g(x)}$. We then simplify the resulting rational expression, if possible, using the techniques of Section 5.1.

$$\left(\frac{f}{g}\right)(x) = \frac{f(x)}{g(x)}$$ Rewrite as the quotient of the two functions.

$$= \frac{x^2 - 3x - 40}{x^2 - 25}$$ Substitute for each function.

$$= \frac{(x-8)(x+5)}{(x+5)(x-5)}$$ Factor the numerator and denominator.

$$= \frac{(x-8)\overset{1}{\cancel{(x+5)}}}{\underset{1}{\cancel{(x+5)}}(x-5)}$$ Divide out common factors.

$$= \frac{x-8}{x-5}$$ Simplify.

Recall that any value x for which the denominator of a rational expression, such as $\left(\frac{f}{g}\right)(x)$, is equal to 0 is not in the domain of the function. In this case, $x \neq -5, 5$.

Using the Graphs of Two Functions to Graph Their Sum or Difference Functions

Objective 2 **Find the graph of the sum function or difference function from the graphs of two functions $f(x)$ and $g(x)$.**

Example 4 Use the graphs of the two functions $f(x)$ and $g(x)$, shown here on the same set of axes, to graph the function $(f + g)(x)$.

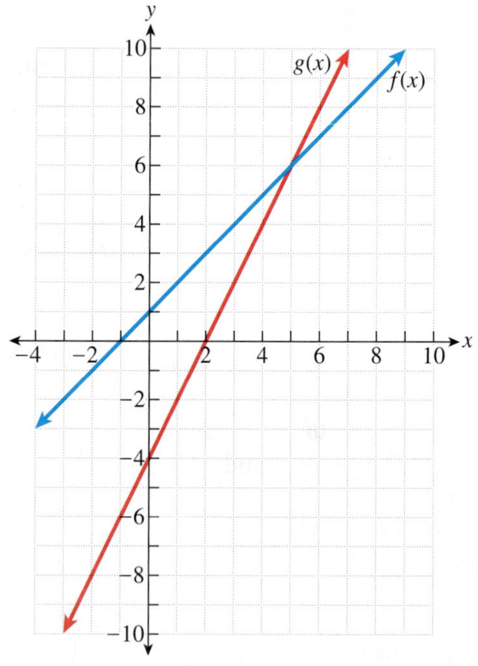

Solution

We will choose certain values for x and then determine the function values of $f(x)$ and $g(x)$ for these values. Adding these function values will give us the function values of $(f+g)(x)$ for the chosen values of x. These ordered pairs can then be plotted on a graph. We will begin by choosing -2, -1, 0, 1, and 2 for x. The function values for $f(x)$ and $g(x)$ can be read from the graph. The sum of these function values is listed in the column on the right.

x	$f(x)$	$g(x)$	$(f+g)(x)$
-2	-1	-8	$(-1)+(-8)=-9$
-1	0	-6	$0+(-6)=-6$
0	1	-4	$1+(-4)=-3$
1	2	-2	$2+(-2)=0$
2	3	0	$3+0=3$

The ordered pairs $(x, (f+g)(x))$ that we will plot on the graph are $(-2,-9)$, $(-1,-6)$, $(0,-3)$, $(1,0)$, and $(2,3)$. The points appear to line up, so we draw a straight line through them. In general, the sum or difference of two linear functions will also be a linear function.

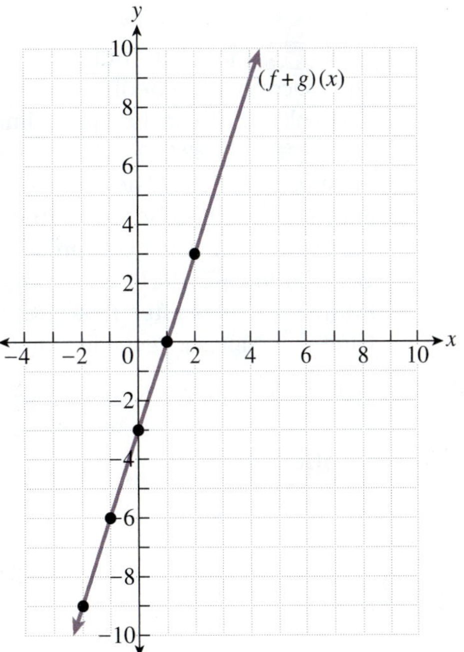

Quick Check 4 Given the graphs of the functions $f(x)$ and $g(x)$, graph the function $(f + g)(x)$.

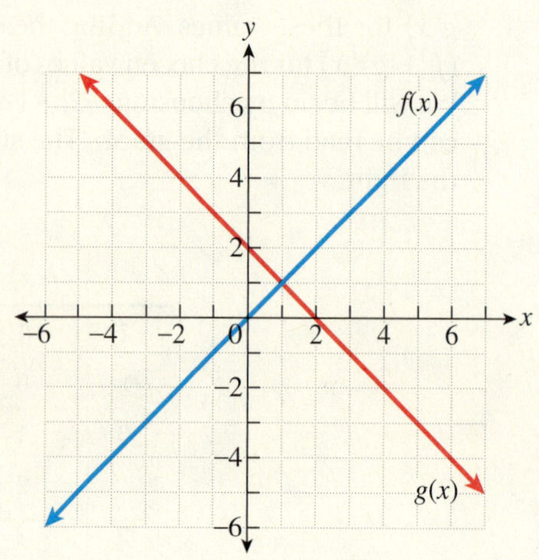

Applications

Objective 3 Solve applications involving the sum function or the difference function.

Example 5 The number of public elementary and secondary schools in the United States in a particular year can be approximated by the function $f(x) = 950x + 82{,}977$, where x represents the number of years after 1990. The number of private elementary and secondary schools in the United States in a particular year can be approximated by the function $g(x) = 188x + 25{,}941$, where again x represents the number of years after 1990. (*Source:* U.S. Department of Education, National Center for Education Statistics)

a) Find $(f + g)(x)$. Explain, in your own words, what this function represents.

Solution

$(f + g)(x) = f(x) + g(x)$ — Rewrite as the sum of the two functions.

$= (950x + 82{,}977) + (188x + 25{,}941)$ — Substitute for each function.

$= 950x + 82{,}977 + 188x + 25{,}941$ — Remove parentheses.

$= 1138x + 108{,}918$ — Combine like terms.

$(f + g)(x) = 1138x + 108{,}918$. This function tells the combined number of public and private elementary and secondary schools there are x years after 1990.

b) Find $(f + g)(19)$. Explain, in your own words, what this number represents.

Solution

$$\begin{aligned} (f + g)(x) &= 1138x + 108{,}918 \\ (f + g)(19) &= 1138(19) + 108{,}918 && \text{Substitute 19 for } x. \\ &= 130{,}540 && \text{Simplify.} \end{aligned}$$

This number tells us that there will be a total of approximately 130,540 public and private elementary and secondary schools in the year 2009, which is 19 years after 1990.

Quick Check **5** **The number of students enrolled at a public college in the United States in a particular year can be approximated by the function $f(x) = 111{,}916x + 9{,}528{,}000$, where x represents the number of years after 1990. The number of students enrolled at a private college in the United States in a particular year can be approximated by the function $g(x) = 43{,}973x + 2{,}591{,}000$, where again x represents the number of years after 1990. (*Source:* U.S. Department of Education, National Center for Education Statistics)**

a) Find $(f + g)(x)$. Explain, in your own words, what this function represents.
b) Find $(f - g)(x)$. Explain, in your own words, what this function represents.

Composition of Functions

Objective 4 **Find the composite function of two functions $f(x)$ and $g(x)$.**
Another way to combine two functions is to use the output of one function as the input for the other function. When this is done, it is called the **composition** of the two functions. For example, consider the functions $f(x) = x + 5$ and $g(x) = 2x + 1$. If we evaluated the function $g(x)$ when $x = 7$, the output would be 15.

$$\begin{aligned} g(7) &= 2(7) + 1 && \text{Substitute 7 for } x. \\ &= 15 && \text{Simplify.} \end{aligned}$$

We can visualize this with a picture of a function "machine." The machine takes an input of 7 and creates an output of 15:

$$7 \longrightarrow \boxed{g(x) = 2x + 1} \longrightarrow 15$$

Now, if we evaluate the function $f(x)$ when $x = 15$, this is a composition of the two functions. The output of function $g(x)$ is the input of the function $f(x)$.

$$\begin{aligned} f(15) &= 15 + 5 && \text{Substitute 15 for } x. \\ &= 20 && \text{Add.} \end{aligned}$$

The function $f(x)$ takes an input of 15 and creates an output of 20:

$$7 \longrightarrow \boxed{g(x) = 2x + 1} \longrightarrow 15 \longrightarrow \boxed{f(x) = x + 5} \longrightarrow 20$$

Symbolically, this can be represented as $f(g(7)) = 20$.

Composite Function

For any two functions $f(x)$ and $g(x)$, the **composite function** $(f \circ g)(x)$ is defined as $(f \circ g)(x) = f(g(x))$ and read "f of g of x."
The symbol "$\circ$" is used to denote the **composition of two functions.**

Suppose that $f(x) = 3x - 5$ and $g(x) = 4x + 9$, and we wanted to find $(f \circ g)(5)$. We will rewrite $(f \circ g)(5)$ as $f(g(5))$ and begin by evaluating $g(5)$.

$$\begin{aligned} g(5) &= 4(5) + 9 && \text{Substitute 5 for } x. \\ &= 29 && \text{Simplify.} \end{aligned}$$

Now we evaluate $f(29)$.

$$\begin{aligned} (f \circ g)(5) &= f(g(5)) \\ &= f(29) && \text{Substitute 29 for } g(5). \\ &= 3(29) - 5 && \text{Substitute 29 for } x \text{ in the function } f(x). \\ &= 82 && \text{Simplify.} \end{aligned}$$

$$5 \longrightarrow \boxed{g(x) = 4x + 9} \longrightarrow 29 \longrightarrow \boxed{f(x) = 3x - 5} \longrightarrow 82$$

Suppose again that $f(x) = 3x - 5$ and $g(x) = 4x + 9$ and that we wanted to find the composite function $(f \circ g)(x)$. We begin by rewriting $(f \circ g)(x)$ as $f(g(x))$; then we replace $g(x)$ by the expression $4x + 9$.

$$\begin{aligned} (f \circ g)(x) &= f(g(x)) \\ &= f(4x + 9) && \text{Replace } g(x) \text{ by } 4x + 9. \\ &= 3(4x + 9) - 5 && \text{Substitute } 4x + 9 \text{ for } x \text{ in the function } f(x). \\ &= 12x + 27 - 5 && \text{Distribute.} \\ &= 12x + 22 && \text{Combine like terms.} \end{aligned}$$

Notice that if we evaluate this composite function for $x = 5$, the output would be $12(5) + 22$, or 82. This is the same result that we obtained by first evaluating $g(5)$ and then evaluating $f(x)$ for this value.

EXAMPLE 6 Given that $f(x) = 2x + 11$ and $g(x) = 3x - 10$, find $(f \circ g)(x)$.

Solution

When finding a composite function, it is crucial to substitute the correct expression into the correct function. A safe way to ensure this is by rewriting $(f \circ g)(x)$ as $f(g(x))$ and then replacing the "inner" function by the appropriate expression. In this example we will replace $g(x)$ by $3x - 10$. Then we evaluate the "outer" function for this expression.

$$\begin{aligned} (f \circ g)(x) &= f(g(x)) \\ &= f(3x - 10) && \text{Replace } g(x) \text{ by } 3x - 10. \\ &= 2(3x - 10) + 11 && \text{Substitute } 3x - 10 \text{ for } x \text{ in the function } f(x). \\ &= 6x - 20 + 11 && \text{Distribute.} \\ &= 6x - 9 && \text{Combine like terms.} \end{aligned}$$

Quick Check 6
Given that $f(x) = 4x - 9$ and $g(x) = 2x + 7$, find $(f \circ g)(x)$.

EXAMPLE 7 Given that $f(x) = 5x - 6$ and $g(x) = x^2 + 2x - 8$, find the given composite functions.

a) $(f \circ g)(x)$

Solution

$$\begin{aligned} (f \circ g)(x) &= f(g(x)) \\ &= f(x^2 + 2x - 8) && \text{Replace } g(x) \text{ by } x^2 + 2x - 8. \\ &= 5(x^2 + 2x - 8) - 6 && \text{Substitute } x^2 + 2x - 8 \text{ for } x \text{ in the function } f(x). \\ &= 5x^2 + 10x - 40 - 6 && \text{Distribute.} \\ &= 5x^2 + 10x - 46 && \text{Combine like terms.} \end{aligned}$$

b) $(g \circ f)(x)$

Solution

In this example, the outer function $g(x)$ is a quadratic function.

$$\begin{aligned} (g \circ f)(x) &= g(f(x)) \\ &= g(5x - 6) && \text{Replace } f(x) \text{ by } 5x - 6. \\ &= (5x - 6)^2 + 2(5x - 6) - 8 && \text{Substitute } 5x - 6 \text{ for } x \text{ in the function } g(x). \\ &= (5x - 6)(5x - 6) + 2(5x - 6) - 8 && \text{Square the binomial } 5x - 6 \text{ by multiplying it by itself.} \\ &= 25x^2 - 30x - 30x + 36 + 10x - 12 - 8 && \text{Multiply } (5x - 6)(5x - 6) \text{ and } 2(5x - 6). \\ &= 25x^2 - 50x + 16 && \text{Combine like terms.} \end{aligned}$$

***Quick Check* 7**

Given that $f(x) = x + 3$ and $g(x) = x^2 - 3x - 28$, find the given composite functions.

a) $(f \circ g)(x)$
b) $(g \circ f)(x)$

The Domain of a Composite Function

Objective 5 **Find the domain of a composite function $(f \circ g)(x)$.** The domain of a composite function $(f \circ g)(x)$ is the set of all input values x in the domain of $g(x)$ whose output from $g(x)$ is in the domain of $f(x)$. In all of the previous examples, the domain has been the set of all real numbers $\mathbb{R}$, as none of the functions had any restrictions on their domains.

EXAMPLE 8 Given that $f(x) = \dfrac{x + 9}{2x - 1}$ and $g(x) = 6x + 15$, find $(f \circ g)(x)$ and state its domain.

Solution

Since $g(x)$ is a linear function, there are no restrictions on its domain. Once we find the composite function $(f \circ g)(x)$, we will find the domain of that function.

$$\begin{aligned} (f \circ g)(x) &= f(g(x)) \\ &= f(6x + 15) && \text{Replace } g(x) \text{ by } 6x + 15. \end{aligned}$$

$$= \frac{(6x + 15) + 9}{2(6x + 15) - 1}$$ Substitute $6x + 15$ for x in the function $f(x)$.

$$= \frac{6x + 15 + 9}{12x + 30 - 1}$$ Distribute.

$$= \frac{6x + 24}{12x + 29}$$ Combine like terms.

Quick Check 8

Given that $f(x) = \frac{2x + 5}{x + 7}$ and $g(x) = x - 4$, find $(f \circ g)(x)$ and state its domain.

So, $(f \circ g)(x) = \frac{6x + 24}{12x + 29}$. The composite function is a rational function, and we find the restrictions on the domain by setting the denominator equal to 0 and solving.

$12x + 29 = 0$ Set the denominator equal to 0.

$12x = -29$ Subtract 29.

$x = -\frac{29}{12}$ Divide by 12.

The domain of this composite function $(f \circ g)(x)$ is the set of all real numbers except $-\frac{29}{12}$.

Building Your Study Strategy **Practice Quizzes, 1 Assess Your Knowledge** Many students have an easy time following along in class or doing their homework, but struggle when attempting to do problems on an exam or quiz. While working through your exercises, you have all of your resources available to you—textbooks, solutions manual, notes, tutor, and study group members. However, when you are taking an exam or a quiz, these resources are not available to you. A student who is relying too heavily on his or her resources may not realize this fact until it is too late.

Creating and taking your own practice quiz is an excellent way to assess your knowledge. You can put yourself in a test situation, without the consequences of taking an actual test. Attempt the problems without any of your resources, and you will determine which topics require more study.

EXERCISES 8.1

Vocabulary

1. For two functions $f(x)$ and $g(x)$, the function denoted by $(f + g)(x)$ is called the ________ function.

2. For two functions $f(x)$ and $g(x)$, the difference function is denoted by ________.

3. For two functions $f(x)$ and $g(x)$, the product function is denoted by ________.

4. For two functions $f(x)$ and $g(x)$, the function denoted by $\left(\frac{f}{g}\right)(x)$ is called the ________ function.

5. When the output of one function is used as the input for another function, this is called the ________ of the two functions.

6. For any two functions $f(x)$ and $g(x)$, the ________ function $(f \circ g)(x)$ is defined as $(f \circ g)(x) = f(g(x))$.

For the given functions $f(x)$ and $g(x)$, find
a) $(f + g)(x)$ *b)* $(f + g)(8)$ *c)* $(f + g)(-3)$

7. $f(x) = 4x + 11$, $g(x) = 3x - 17$

8. $f(x) = 2x - 13$, $g(x) = -7x + 6$

9. $f(x) = 5x + 8$, $g(x) = x^2 - x - 19$

10. $f(x) = x^2 - 7x + 12$, $g(x) = x^2 + 4x - 32$

For the given functions $f(x)$ and $g(x)$, find
a) $(f - g)(x)$ *b)* $(f - g)(6)$ *c)* $(f - g)(-5)$

11. $f(x) = 6x - 11$, $g(x) = -2x - 5$

12. $f(x) = 8x - 13$, $g(x) = 8x + 13$

13. $f(x) = 10x + 37$, $g(x) = x^2 - 5x - 12$

14. $f(x) = x^2 + 7x + 100$, $g(x) = -x^2 + 12x - 25$

For the given functions $f(x)$ and $g(x)$, find
a) $(f \cdot g)(x)$ *b)* $(f \cdot g)(4)$
c) $(f \cdot g)(-10)$

15. $f(x) = x + 7$, $g(x) = 2x - 10$

16. $f(x) = 3x$, $g(x) = x - 9$

17. $f(x) = 4x + 15$, $g(x) = x^2 + 5x - 36$

18. $f(x) = x^2 - 3x - 54$, $g(x) = x^2 + 9x + 14$

For the given functions $f(x)$ and $g(x)$, find
a) $\left(\frac{f}{g}\right)(x)$ *b)* $\left(\frac{f}{g}\right)(7)$ *c)* $\left(\frac{f}{g}\right)(-2)$

19. $f(x) = 3x + 1$, $g(x) = x + 7$

20. $f(x) = 6x - 22$, $g(x) = x^2 + 11x + 10$

21. $f(x) = x^2 + 8x + 15$, $g(x) = x^2 - x - 12$

22. $f(x) = x^2 - 17x + 72$, $g(x) = x^2 - 7x - 8$

Let $f(x) = 3x - 8$ and $g(x) = 2x + 15$. Find the following:

23. $(f + g)(6)$
24. $(f + g)(-4)$
25. $(f - g)(-2)$
26. $(f - g)(5)$
27. $(f \cdot g)(7)$
28. $(f \cdot g)(10)$
29. $\left(\frac{f}{g}\right)(1)$
30. $\left(\frac{f}{g}\right)(-4)$

FOR EXTRA HELP

MyMathLab

MathXL

Interactmath.com

MathXL Tutorials on CD

Video Lectures on CD

Tutor Center

Addison-Wesley Math Tutor Center

Student's Solutions Manual

Let $f(x) = 5x - 9$ and $g(x) = 2x - 17$. Find the following, and simplify completely:

31. $(f + g)(x)$

32. $(f - g)(x)$

33. $(f \cdot g)(x)$

34. $\left(\frac{f}{g}\right)(x)$

Let $f(x) = x - 10$ and $g(x) = x^2 - 8x - 20$. Find the following, and simplify completely:

35. $(f - g)(x)$

36. $(f + g)(x)$

37. $\left(\frac{f}{g}\right)(x)$

38. $(f \cdot g)(x)$

Find the unknown function $g(x)$ that satisfies the given conditions:

39. $f(x) = 3x - 20, (f + g)(x) = 7x - 11$

40. $f(x) = 9x + 13, (f + g)(x) = 5x + 22$

41. $f(x) = x + 8, (f - g)(x) = 5x + 2$

42. $f(x) = -2x + 37, (f - g)(x) = -9x + 60$

Given the graphs of the functions $f(x)$ and $g(x)$, graph the function $(f + g)(x)$.

43.

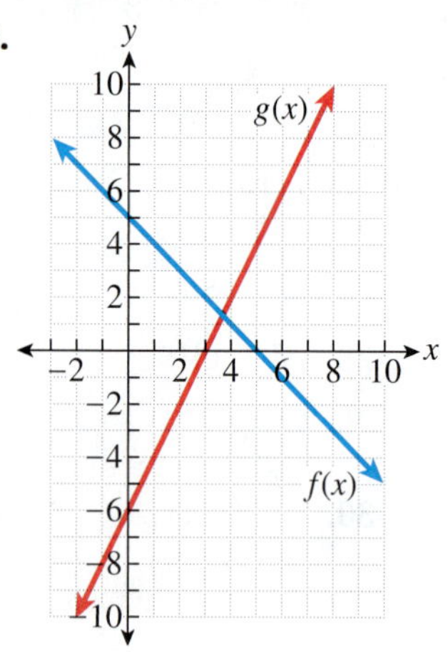

44.

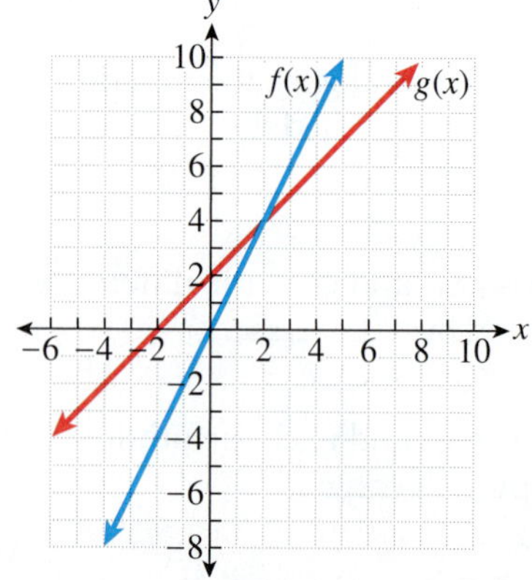

Given the graphs of the functions $f(x)$ and $g(x)$, graph the function $(f - g)(x)$.

45.

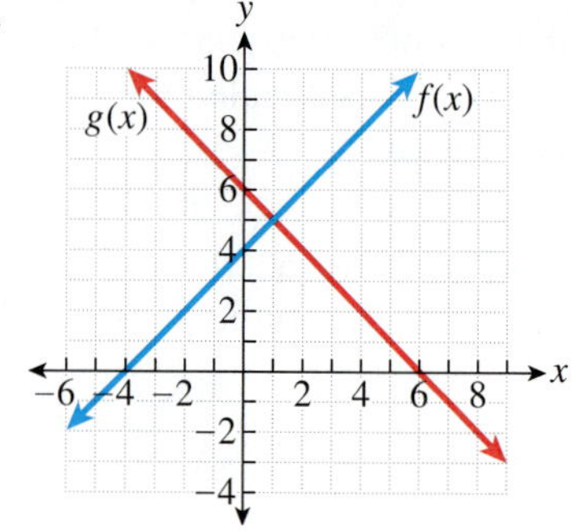

46.

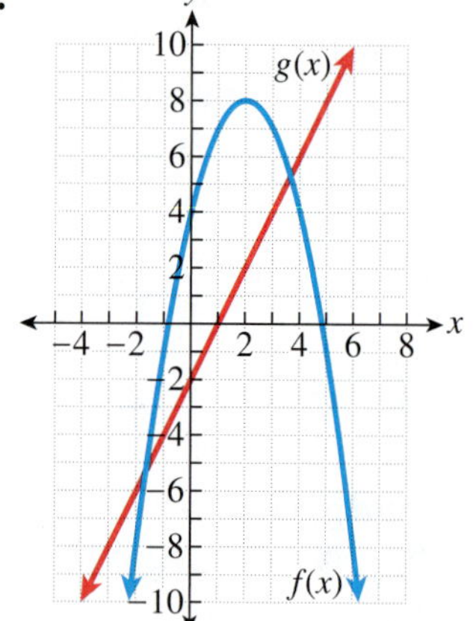

47. The number of male doctors, in thousands, in the United States in a particular year can be approximated by the function $f(x) = 10.4x + 398$, where x represents the number of years after 1980. The number of female doctors, in thousands, in the United States in a particular year can be approximated by the function $g(x) = 9.1x + 49$, where again x represents the number of years after 1980. (*Source:* American Medical Association)

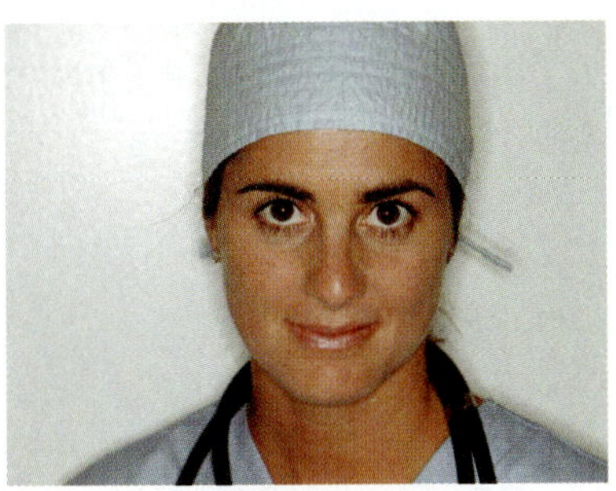

a) Find $(f + g)(x)$. Explain, in your own words, what this function represents.

b) Find $(f + g)(40)$. Explain, in your own words, what this number represents.

48. The number of male inmates who are HIV positive in the United States in a particular year can be approximated by the function $f(x) = 223x + 20{,}969$, where x represents the number of years after 1995. The number of female inmates who are HIV positive in the United States in a particular year can be approximated by the function $g(x) = 83x + 2102$, where again x represents the number of years after 1995. (*Source:* U.S. Department of Justice, Bureau of Justice Statistics)

a) Find $(f + g)(x)$. Explain, in your own words, what this function represents.

b) Find $(f + g)(27)$. Explain, in your own words, what this number represents.

49. The number of master's degrees, in thousands, earned by females in the United States in a particular year can be approximated by the function $f(x) = 9.2x + 181.8$, where x represents the number of years after 1990. The number of master's degrees, in thousands, earned by males in the United States in a particular year can be approximated by the function $g(x) = 3.4x + 160.9$, where again x represents the number of years after 1990. (*Source:* U.S. Department of Education, National Center for Education Statistics)

a) Find $(f - g)(x)$. Explain, in your own words, what this function represents.

b) Find $(f - g)(30)$. Explain, in your own words, what this number represents.

50. The amount of money, in billions of dollars, spent on health care that was covered by insurance in the United States in a particular year can be approximated by the function $f(x) = x^2 + 11x + 244$, where x represents the number of years after 1990. The amount of money, in billions of dollars, spent on health care that was paid out of pocket in the United States in a particular year can be approximated by the function $g(x) = 6x + 129$, where again x represents the number of years after 1990. (*Source:* U.S. Centers for Medicare and Medicaid Services)

a) Find $(f - g)(x)$. Explain, in your own words, what this function represents.

b) Find $(f - g)(40)$. Explain, in your own words, what this number represents.

Let $f(x) = 4x + 7$ and $g(x) = x - 3$. Find the following:

51. $(f \circ g)(5)$

52. $(f \circ g)(-2)$

53. $(g \circ f)(-1)$

54. $(g \circ f)(10)$

Let $f(x) = 2x - 9$ and $g(x) = x^2 - 9x + 18$. Find the following:

55. $(f \circ g)(1)$

56. $(f \circ g)(4)$

57. $(g \circ f)(-5)$

58. $(g \circ f)(10)$

For the given functions $f(x)$ and $g(x)$, find
a) $(f \circ g)(x)$ ***b)*** $(g \circ f)(x)$
c) $(f \circ g)(3)$ ***d)*** $(g \circ f)(-4)$

59. $f(x) = 3x + 5,\ g(x) = 2x + 4$

60. $f(x) = 5x - 9,\ g(x) = -x + 6$

61. $f(x) = x + 4,\ g(x) = x^2 + 3x - 40$

62. $f(x) = x^2 - 7x - 18,\ g(x) = x - 9$

For the given functions $f(x)$ and $g(x)$, find $(f \circ g)(x)$ and state its domain.

63. $f(x) = \dfrac{5x}{x + 7},\ g(x) = 2x + 13$

64. $f(x) = \dfrac{2x - 3}{3x + 5},\ g(x) = x + 8$

65. $f(x) = \sqrt{2x - 10},\ g(x) = x + 11$

66. $f(x) = \sqrt{x - 2},\ g(x) = 3x + 14$

For the given function $f(x)$, find $(f \circ f)(x)$.

67. $f(x) = 3x - 10$

68. $f(x) = -5x + 18$

69. $f(x) = x^2 + 6x + 12$

70. $f(x) = x^2 - 2x - 15$

For the given functions $f(x)$ and $g(x)$, find a value of x for which $(f \circ g)(x) = (g \circ f)(x)$.

71. $f(x) = x^2 + 4x + 3,\ g(x) = x + 5$

72. $f(x) = x^2 + 6x - 9,\ g(x) = x - 3$

73. $f(x) = x + 7,\ g(x) = x^2 - 25$

74. $f(x) = x - 4,\ g(x) = 2x^2 + 15x + 18$

75. If $g(x) = x + 3$ and $(f \circ g)(x) = x + 11$, find $f(x)$.

76. If $g(x) = 3x - 4$ and $(f \circ g)(x) = 6x - 8$, find $f(x)$.

77. If $g(x) = 2x + 7$ and $(f \circ g)(x) = 10x + 19$, find $f(x)$.

78. If $g(x) = 5x - 2$ and $(f \circ g)(x) = -15x + 10$, find $f(x)$.

For the given function $g(x)$, find a function $f(x)$ such that $(f \circ g)(x) = x$.

79. $g(x) = x + 8$

80. $g(x) = -5x$

81. $g(x) = 3x + 8$

82. $g(x) = 4x - 17$

Writing in Mathematics

Answer in complete sentences.

83. In general, is $(f + g)(x)$ equal to $(g + f)(x)$? If your answer is yes, explain why. If your answer is no, give an example that shows why they are not equal, in general.

84. In general, is $(f - g)(x)$ equal to $(g - f)(x)$? If your answer is yes, explain why. If your answer is no, give an example that shows why they are not equal, in general.

85. ***Solutions Manual**** Write a solutions manual page for the following problem:

For the functions $f(x) = x^2 - 9x - 40$ and $g(x) = 2x - 7$, find $(f \circ g)(x)$.

86. ***Newsletter**** Write a newsletter explaining how to find the composition of two functions.

***See Appendix B for details and sample answers.**

8.2 Inverse Functions

Objectives

1. **Determine whether a function is one to one.**
2. **Use the horizontal-line test to determine whether a function is one to one.**
3. **Understand inverse functions.**
4. **Determine whether two functions are inverse functions.**
5. **Find the inverse of a one-to-one function.**
6. **Find the inverse of a function from its graph.**

One-to-One Functions

Objective 1 Determine whether a function is one to one. Recall that for any function $f(x)$, each value in the domain corresponds to only one value in the range.

One-to-One Function

If each value in the range of a function $f(x)$ corresponds to only one value in the domain, then the function $f(x)$ is called a **one-to-one function.**

In other words, if $f(x)$ is different for each input value x in the domain, then $f(x)$ is a one-to-one function.

Suppose that for some function $f(x)$, both $f(-1)$ and $f(3)$ are equal to 2. This function is not a one-to-one function, because one output value is associated with two different input values. However, the function shown in the following table is a one-to-one function, since each function value corresponds to only one input value:

x	0	1	4	9	16
$f(x)$	0	1	2	3	4

A function whose input is a person's name and whose output is that person's birthday is not a one-to-one function because several people (input) can have the same birthday (output). A function whose input is a person's name and whose output is that person's Social Security number is a one-to-one function because no two people can have the same Social Security number.

Example 1 Determine whether the function represented by the set of ordered pairs $\{(-7, -5), (-5, -2), (-2, 4), (3, 3), (6, -7)\}$ is a one-to-one function.

Quick Check 1

Determine whether the function represented by the set of ordered pairs $\{(-4, -9), (-2, -7), (1, -6), (2, -7), (6, -11)\}$ is a one-to-one function.

Solution

Since each x-value corresponds to only one y-value and each y-value corresponds to only one x-value, the function is one to one.

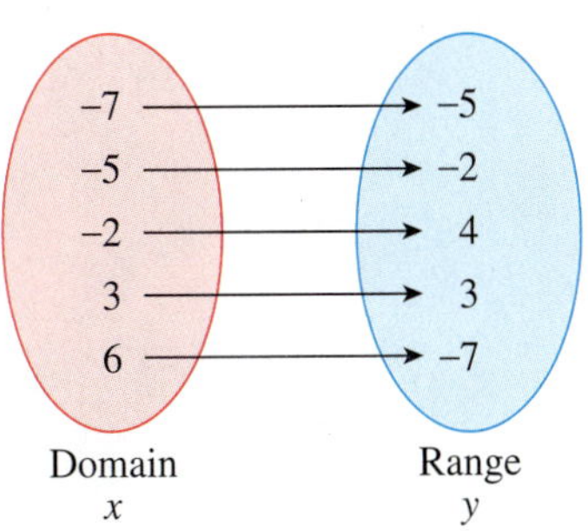

Horizontal-Line Test

Objective 2 Use the horizontal-line test to determine whether a function is one to one. We can determine whether a function is one to one by applying the **horizontal-line test** to its graph.

Horizontal-Line Test

If a horizontal line can intersect the graph of a function at more than one point, then the function is not one to one.

Consider the graph of a quadratic function shown. We can draw a horizontal line that intersects the function at two points, as indicated. The coordinates of these two points are $(-2, 6)$ and $(2, 6)$. This shows that two different input values, -2 and 2, correspond to the same output value (6). Therefore, the function is not one to one.

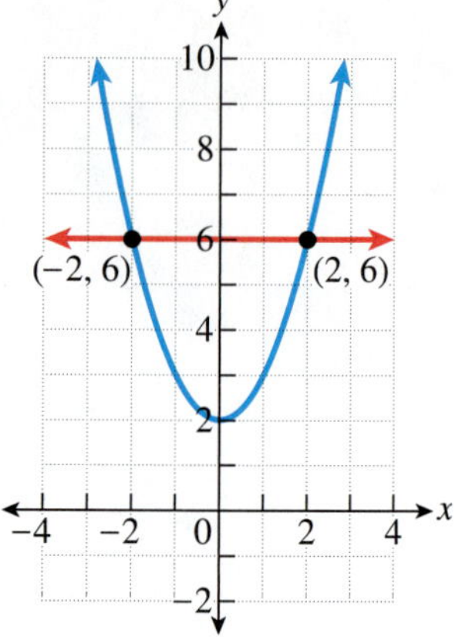

The linear function graphed next is a one-to-one function. No horizontal line crosses the graph of this function at more than one point, so the function passes the horizontal-line test.

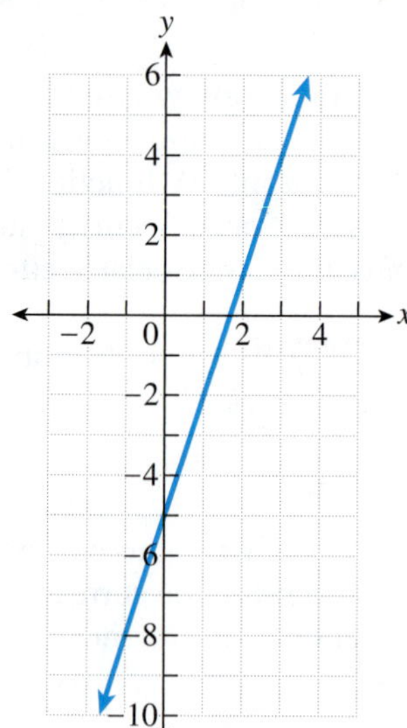

EXAMPLE 2 Use the horizontal-line test to determine whether the function is one to one.

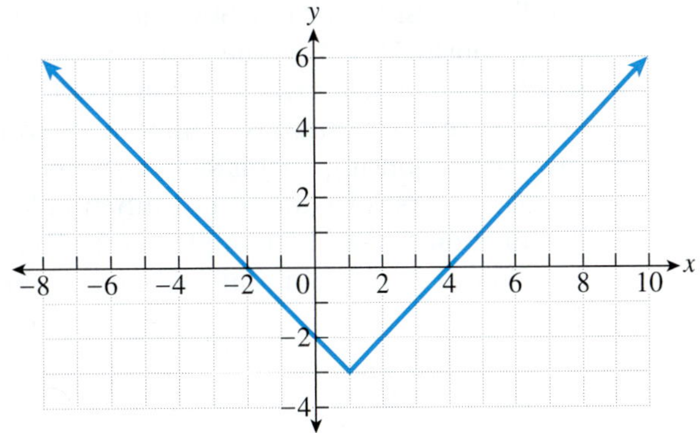

Solution

We can draw a horizontal line that crosses the graph of this absolute value function at more than one point, as shown here. This function fails the horizontal-line test and is not one to one.

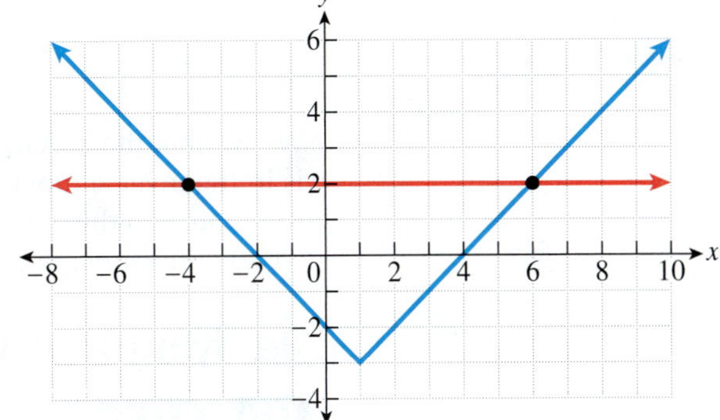

***Quick Check* 2** **Determine whether the function is one to one.**

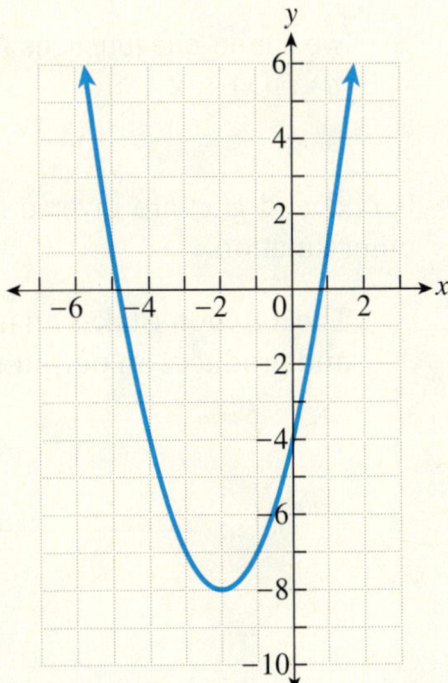

Inverse Functions

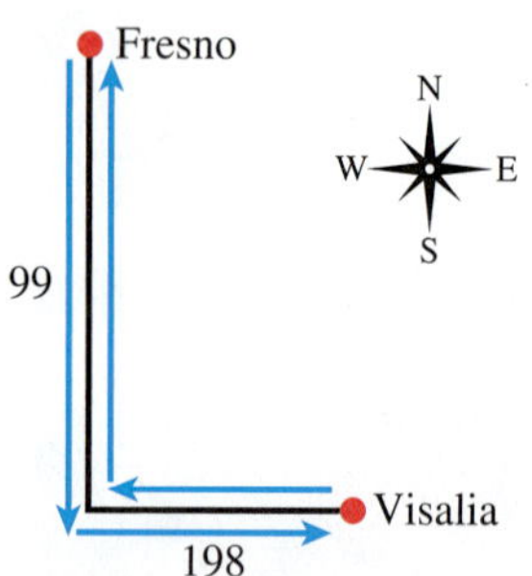

Objective 3 **Understand inverse functions.** Anne commutes to work from Visalia to Fresno each day. She starts by taking Highway 198 west and then takes Highway 99 north. How does she return home each day? She begins by taking Highway 99 south and then takes Highway 198 east. Her trips are depicted in the map on the left.

The second route takes Anne back to the starting point. The two routes have an **inverse** relationship. In this section, we will be examining **inverse functions.**

The inverse of a one-to-one function "undoes" what the function does. For example, if a function $f(x)$ takes an input value of 3 and produces an output value of 7, its inverse function takes an input value of 7 and produces an output value of 3.

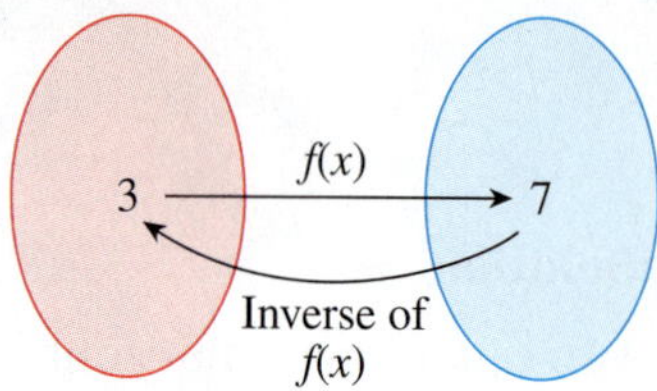

We see that the ordered pair $(3, 7)$ is in the function $f(x)$ and that the ordered pair $(7, 3)$ is in the inverse function of $f(x)$. The inputs of a function are the outputs of its inverse function, and the outputs of a function are the inputs of its inverse function. In general, if (a, b) is in a function, then (b, a) is in the inverse of that function.

Verifying That Two Functions Are Inverses

Objective 4 **Determine whether two functions are inverse functions.** We can determine whether two functions are inverse functions algebraically by finding their composite functions.

Inverse Functions

Two one-to-one functions $f(x)$ and $g(x)$ are inverse functions if $(f \circ g)(x) = x$ and $(g \circ f)(x) = x$.

If $f(x)$ and $g(x)$ are inverse functions, then each function "undoes" the actions of the other function.

The function g takes an input value x and produces an output value $g(x)$.

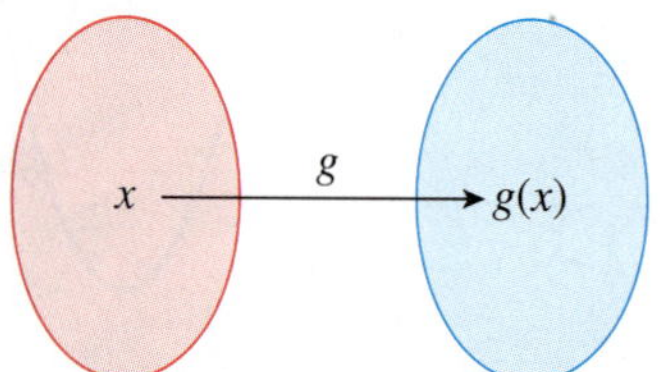

The function f takes this value and produces an output value of x.

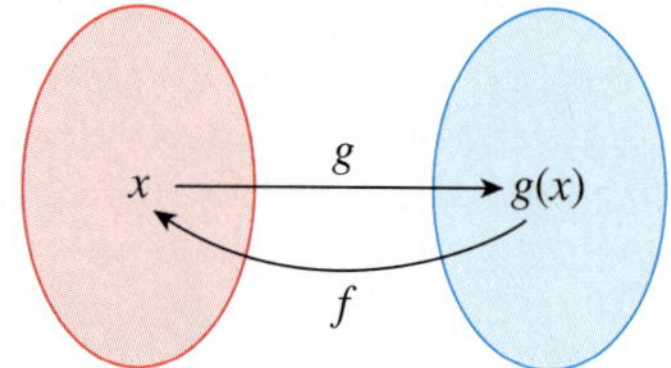

So $f(g(x)) = x$, or, in other words, $(f \circ g)(x) = x$. In a similar fashion, we could show that $(g \circ f)(x) = x$ as well.

EXAMPLE 3 Determine whether the following two functions are inverse functions: $f(x) = 5x - 2$, $g(x) = \frac{x+2}{5}$.

Solution

Both of these functions are linear functions and are one to one. We will determine whether $(f \circ g)(x) = x$ and $(g \circ f)(x) = x$. (Replace the inner function by its formula, substitute that expression for x in the outer function, and simplify.)

$(f \circ g)(x) = x$	$(g \circ f)(x) = x$
$(f \circ g)(x) = f(g(x))$	$(g \circ f)(x) = g(f(x))$
$= f\left(\frac{x+2}{5}\right)$	$= g(5x-2)$
$= 5\left(\frac{x+2}{5}\right) - 2$	$= \frac{(5x-2)+2}{5}$
$= \overset{1}{\cancel{5}}\left(\frac{x+2}{\underset{1}{\cancel{5}}}\right) - 2$	$= \frac{5x-2+2}{5}$
$= x + 2 - 2$	$= \frac{5x}{5}$
$= x$	$= \frac{\overset{1}{\cancel{5}}x}{\underset{1}{\cancel{5}}}$
	$= x$

Since $(f \circ g)(x) = x$ and $(g \circ f)(x) = x$, these two functions are inverses.

Quick Check 3

Determine whether the following two functions are inverse functions:

$f(x) = \frac{x-7}{4}$,

$g(x) = 4x + 7$.

EXAMPLE 4 Determine whether the following two functions are inverse functions: $f(x) = \frac{1}{4}x - \frac{3}{4}$, $g(x) = 3x + 4$.

Solution

These two functions are linear and one to one. We will begin by determining whether $(f \circ g)(x) = x$.

$$
\begin{aligned}
(f \circ g)(x) &= f(g(x)) \\
&= f(3x+4) && \text{Replace } g(x) \text{ by } 3x+4. \\
&= \frac{1}{4}(3x+4) - \frac{3}{4} && \text{Substitute } 3x+4 \text{ for } x \text{ in the function } f(x). \\
&= \frac{3}{4}x + 1 - \frac{3}{4} && \text{Distribute.} \\
&= \frac{3}{4}x + \frac{1}{4} && \text{Combine like terms.}
\end{aligned}
$$

$(f \circ g)(x) \neq x$. These two functions are not inverses.

Note that it is not necessary to check $(g \circ f)(x)$, since both composite functions must simplify to x in order for the two functions to be inverse functions.

Quick Check 4

Determine whether the following two functions are inverse functions:

$f(x) = \frac{1}{3}x + 4$,

$g(x) = 3x - 12$.

Finding an Inverse Function

Objective 5 **Find the inverse of a one-to-one function.** We use the notation $f^{-1}(x)$ to denote the inverse of a function $f(x)$. We read $f^{-1}(x)$ as "f *inverse of* x."

A Word of Caution When a superscript of -1 is written after the name of a function, the -1 is not an exponent; it is used to denote the inverse of a function. In other words, $f^{-1}(x)$ is not the same as $\frac{1}{f(x)}$.

Here is a procedure that can be used to find the inverse of a one-to-one function $f(x)$:

Finding $f^{-1}(x)$

1. **Determine whether $f(x)$ is one to one.** If $f(x)$ is not one to one, then it does not have an inverse function. We can use the horizontal-line test to determine this.
2. **Replace $f(x)$ by y.**
3. **Interchange x and y.** We interchange these two variables because we know that the input of $f(x)$ must be the output of $f^{-1}(x)$, and the output of $f(x)$ must be the input of $f^{-1}(x)$.
4. **Solve the resulting equation for y.** This will rewrite the inverse function as a function of x.
5. **Replace y by $f^{-1}(x)$.**

EXAMPLE 5 Find $f^{-1}(x)$ for the function $f(x) = 5x - 3$.

Solution

We begin by determining whether the function is one to one. Since the function is a linear function, it is one to one. Shown at the left is the graph of $f(x)$. The function clearly passes the horizontal-line test.

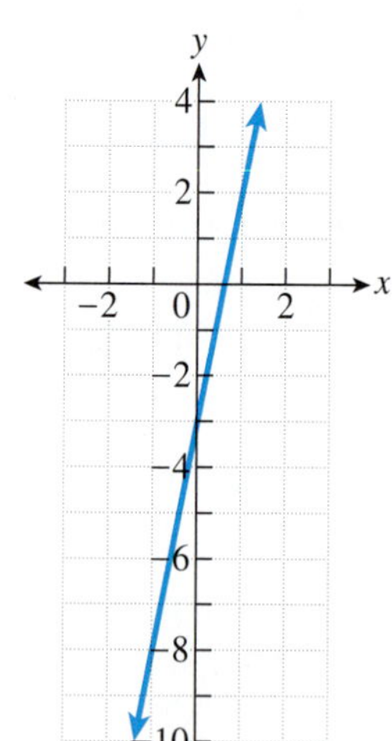

We begin to find $f^{-1}(x)$ by replacing $f(x)$ by y.

$$f(x) = 5x - 3$$

$$y = 5x - 3 \quad \text{Replace } f(x) \text{ by } y.$$

$$x = 5y - 3 \quad \text{Interchange } x \text{ and } y.$$

$$x + 3 = 5y \quad \text{Add 3 to isolate the term containing } y.$$

$$\frac{x+3}{5} = y \quad \text{Divide both sides by 5 to isolate } y.$$

$$f^{-1}(x) = \frac{x+3}{5} \quad \text{Replace } y \text{ by } f^{-1}(x). \text{ It is customary to write } f^{-1}(x) \text{ on the left side.}$$

The inverse of $f(x) = 5x - 3$ is $f^{-1}(x) = \frac{x+3}{5}$, which could be written as $f^{-1}(x) = \frac{1}{5}x + \frac{3}{5}$.

Quick Check **5** Find $f^{-1}(x)$ for the function $f(x) = \frac{1}{2}x + 10$.

EXAMPLE 6 Find $f^{-1}(x)$ for the function $f(x) = \dfrac{8}{x+1}$.

Solution

In this example, $f(x)$ is a rational function. We have not yet learned to graph rational functions, but the graph of $f(x)$, which has been generated by technology, is shown below. (We could use a graphing calculator or software to generate this graph.) The function passes the horizontal-line test.

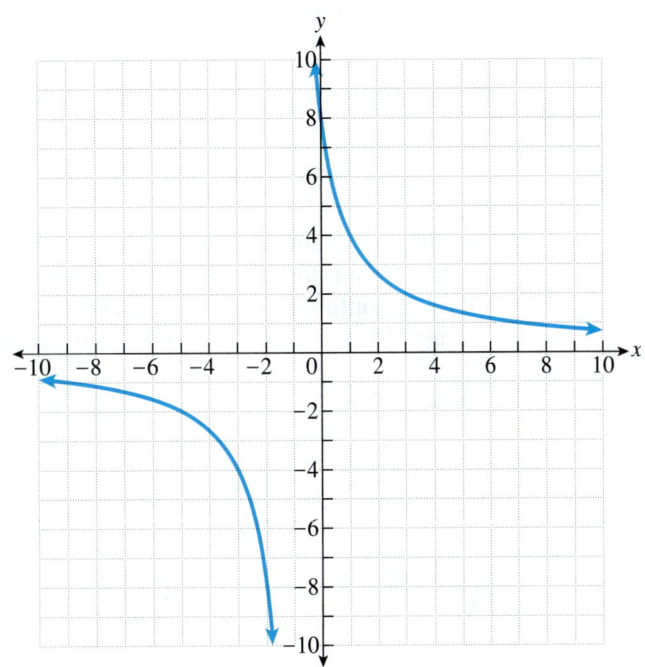

We begin to find $f^{-1}(x)$ by replacing $f(x)$ by y.

$$f(x) = \frac{8}{x+1}$$

$$y = \frac{8}{x+1} \qquad \text{Replace } f(x) \text{ by } y.$$

$$x = \frac{8}{y+1} \qquad \text{Interchange } x \text{ and } y.$$

$$x(y+1) = \frac{8}{\cancel{y+1}} \cdot \cancel{y+1} \qquad \text{Multiply both sides by } y+1 \text{ to clear the equation of fractions.}$$

$$x(y+1) = 8 \qquad \text{Simplify.}$$

$$xy + x = 8 \qquad \text{Multiply.}$$

$$xy = 8 - x \qquad \text{Subtract } x \text{ to isolate the term containing } y.$$

$$y = \frac{8-x}{x} \qquad \text{Divide by } x \text{ to isolate } y.$$

$$f^{-1}(x) = \frac{8-x}{x} \qquad \text{Replace } y \text{ by } f^{-1}(x).$$

The inverse of $f(x) = \dfrac{8}{x+1}$ is $f^{-1}(x) = \dfrac{8-x}{x}$.

Note that the domain of this inverse function is the set of all real numbers except 0, as the denominator of the function is equal to 0 when $x = 0$. This makes the inverse function undefined when $x = 0$.

Quick Check 6

Find $f^{-1}(x)$ for the function $f(x) = \dfrac{3+5x}{x}$, and state the restrictions on its domain.

Recall that the output values of a one-to-one function are the input values of its inverse function. So the range of a one-to-one function $f(x)$ is the domain of its inverse function $f^{-1}(x)$. Occasionally, if the range of a one-to-one function $f(x)$ is not the set of real numbers, we will have to restrict the domain of $f^{-1}(x)$ accordingly. This is illustrated in the next example.

EXAMPLE 7 Find $f^{-1}(x)$ for the function $f(x) = \sqrt{x+4} - 2$. State the domain of $f^{-1}(x)$.

Solution

We begin by showing that the function is one to one. Following is the graph of $f(x)$. (For help on graphing square-root functions, refer back to Section 7.7.) The function passes the horizontal-line test, so it is a one-to-one function that has an inverse function. The range of $f(x)$ is $[-2, \infty)$, and this is the domain of its inverse function $f^{-1}(x)$.

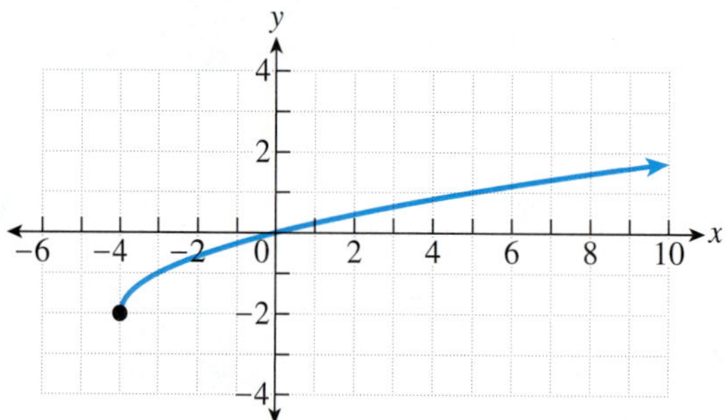

We begin to find $f^{-1}(x)$ by replacing $f(x)$ by y.

$$f(x) = \sqrt{x+4} - 2$$

$$y = \sqrt{x+4} - 2 \qquad \text{Replace } f(x) \text{ by } y.$$

$$x = \sqrt{y+4} - 2 \qquad \text{Interchange } x \text{ and } y.$$

$$x + 2 = \sqrt{y+4} \qquad \text{Add 2 to isolate the radical.}$$

$$(x+2)^2 = (\sqrt{y+4})^2 \qquad \text{Square both sides.}$$

$$(x+2)(x+2) = y + 4 \qquad \text{Square } x + 2 \text{ by multiplying it by itself.}$$

$$x^2 + 4x + 4 = y + 4 \qquad \text{Multiply } (x+2)(x+2).$$

$$x^2 + 4x = y \qquad \text{Subtract 4 to isolate } y.$$

$$f^{-1}(x) = x^2 + 4x \qquad \text{Replace } y \text{ by } f^{-1}(x).$$

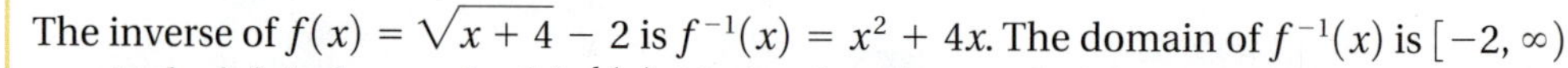

The inverse of $f(x) = \sqrt{x+4} - 2$ is $f^{-1}(x) = x^2 + 4x$. The domain of $f^{-1}(x)$ is $[-2, \infty)$.

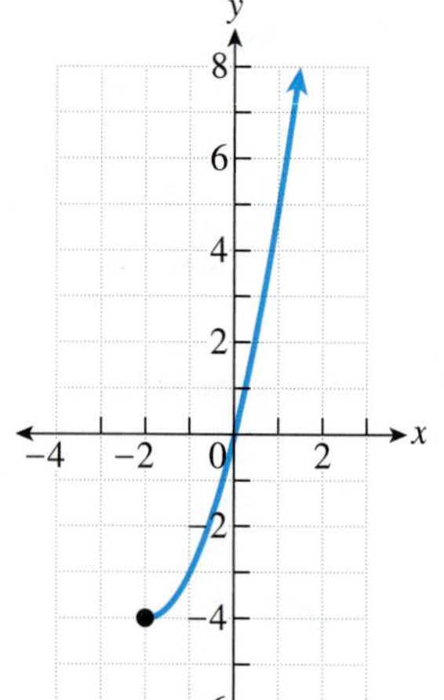

At the left is the graph of $f^{-1}(x)$. Notice that the graph does pass the horizontal-line test with this restricted domain. If we did not restrict this domain, the graph of $f^{-1}(x)$ would be a parabola and the function would not be one to one.

Quick Check 7 Find $f^{-1}(x)$ for the function $f(x) = \sqrt{x-9} + 8$. State the domain of $f^{-1}(x)$.

Finding an Inverse Function from a Graph

Objective 6 **Find the inverse of a function from its graph.** We know that if $f(x)$ is a one-to-one function and $f(a) = b$, then $f^{-1}(b) = a$. This tells us that if the point (a, b) is on the graph of $f(x)$, then the point (b, a) is on the graph of $f^{-1}(x)$.

EXAMPLE 8 If the function is one to one, find its inverse.

$$\{(-2, -3), (-1, -2), (0, 1), (1, 5), (2, 13)\}$$

Solution

Since each value of x corresponds to only one value of y, and each value of y corresponds to only one value of x, this is a one-to-one function.

To find the inverse function, we interchange each x-coordinate and y-coordinate. The inverse function is $\{(-3, -2), (-2, -1), (1, 0), (5, 1), (13, 2)\}$.

EXAMPLE 9 For the given graph of a one-to-one function $f(x)$, graph its inverse function $f^{-1}(x)$.

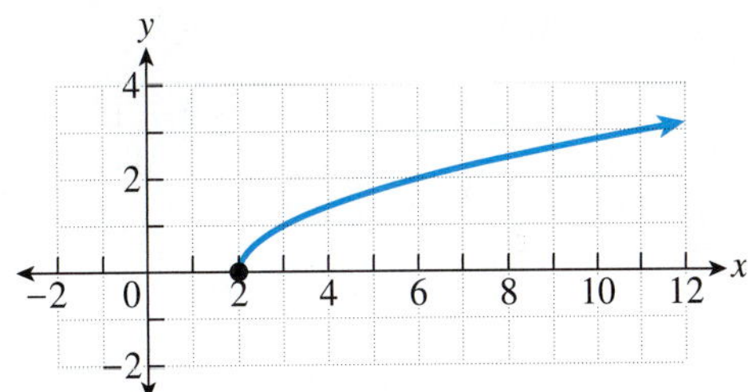

Solution

We begin by identifying the coordinates of some points that are on the graph of this function. We will use the points $(2, 0)$, $(3, 1)$, and $(6, 2)$.

By interchanging each x-coordinate with its corresponding y-coordinate, we know that the points $(0, 2)$, $(1, 3)$, and $(2, 6)$ are on the graph of $f^{-1}(x)$. We finish by drawing a graph that passes

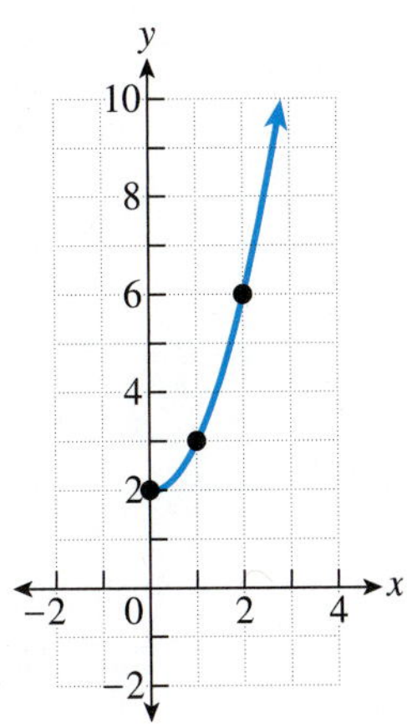

through these points. Here on the right is the graph of $f^{-1}(x)$.

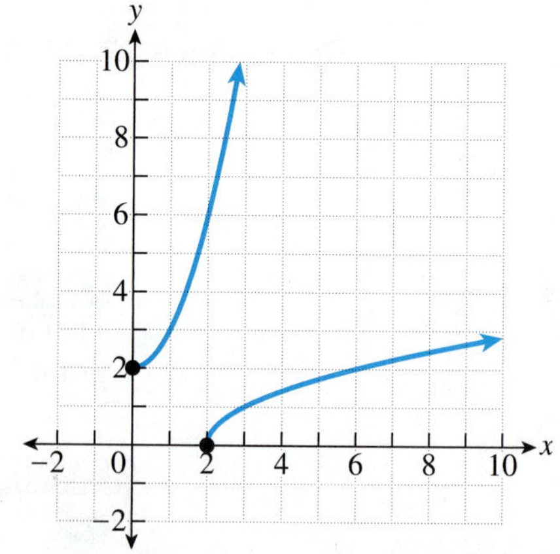

Here are the graphs of $f(x)$ and $f^{-1}(x)$ from the previous example on the same set of axes.

Notice that the two functions look somewhat similar; in fact, they are mirror images of each other. For any one-to-one function $f(x)$ and its inverse function $f^{-1}(x)$, their graphs are symmetric to each other about the line $y = x$. If we fold our graph along the line $y = x$, the graphs of $f(x)$ and $f^{-1}(x)$ will lie on top of each other.

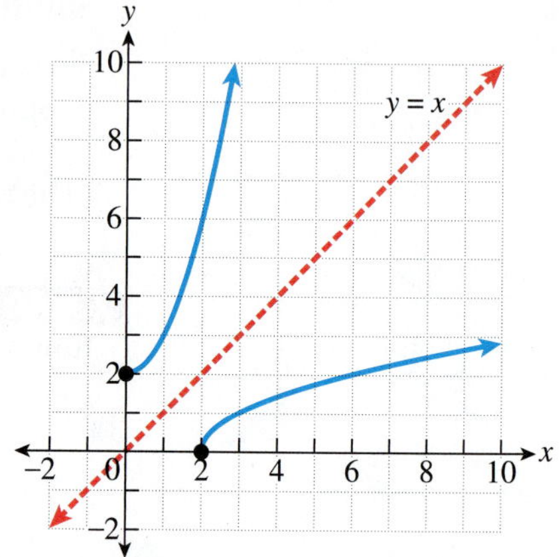

Quick Check 8 **For the given graph of a one-to-one function $f(x)$, graph its inverse function $f^{-1}(x)$.**

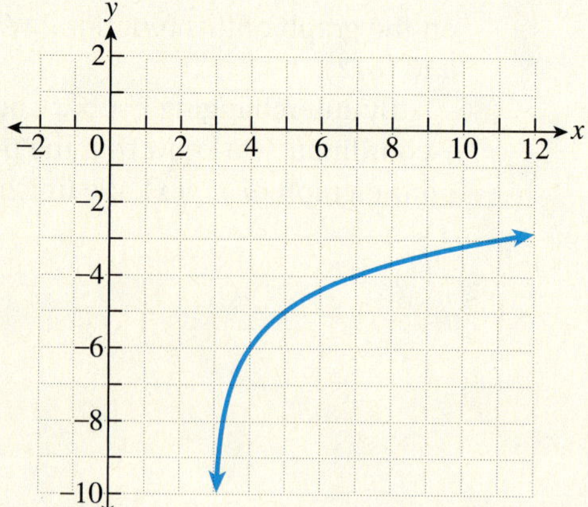

Building Your Study Strategy **Practice Quizzes, 2** **Creating a Practice Quiz** If you are struggling with a particular topic or type of problem, a practice quiz is an excellent way to assess your progress. Create a short practice quiz focused on these problems, perhaps by selecting odd problems from the textbook's exercise section. If you are able to successfully complete these problems without using any of your available resources, this is an indication that you are ready to move on to another topic or type of problem.

EXERCISES 8.2

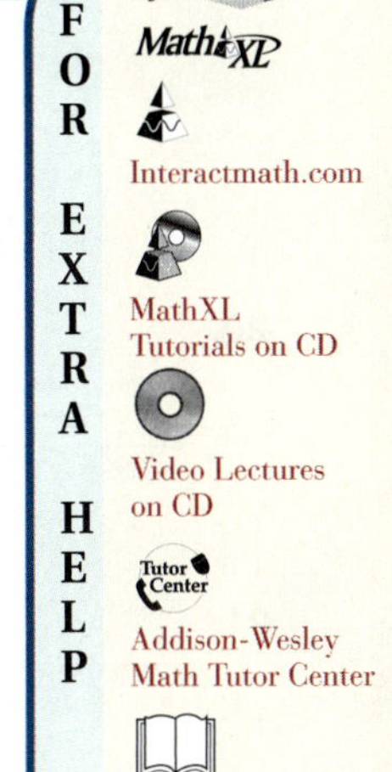

Vocabulary

1. If $f(x)$ is different for each input value x in its domain, then $f(x)$ is a(n) ________ function.
2. If a(n) ________ can intersect the graph of a function at more than one point, then the function is not one to one.
3. If $f(x)$ is a one-to-one function such that $f(a) = b$, then its ________ function $f^{-1}(x)$ is a function for which $f^{-1}(b) = a$.
4. Two one-to-one functions $f(x)$ and $g(x)$ are inverse functions if $(f \circ g)(x)$ and $(g \circ f)(x)$ are both equal to ________.
5. The inverse function of a one-to-one function $f(x)$ is denoted by ________.
6. If the point (a, b) is on the graph of a one-to-one function, then the point ________ is on the graph of the inverse of that function.

Determine whether the function $f(x)$ is a one-to-one function.

7.

x	−3	0	3	6	9
$f(x)$	−8	−2	4	10	16

8.

x	−5	−1	1	4	10
$f(x)$	9	5	3	0	−6

9.

x	−4	−2	0	2	4
$f(x)$	16	4	0	4	16

10.

x	−3	−2	−1	0	1
$f(x)$	3	2	1	0	1

Determine whether the function represented by the set of ordered pairs is a one-to-one function.

11. $\{(-9, -3), (-4, 0), (1, 1), (5, -2), (10, -3)\}$
12. $\{(1, 4), (3, 8), (5, 12), (7, 8), (9, 4)\}$
13. $\{(-8, -4), (-5, 3), (-1, 0), (4, 3), (9, 10)\}$
14. $\{(0, 1), (1, 2), (2, 4), (3, 8), (4, 16)\}$
15. Is the function whose input is a person and output is that person's mother a one-to-one function? Explain your answer in your own words.
16. Is the function whose input is a student and output is that student's favorite math teacher a one-to-one function? Explain your answer in your own words.
17. Is the function whose input is a state and output is that state's governor a one-to-one function? Explain your answer in your own words.

18. Is the function whose input is a licensed driver and output is that driver's license number a one-to-one function? Explain your answer in your own words.

Use the horizontal-line test to determine whether the function is one to one.

19.

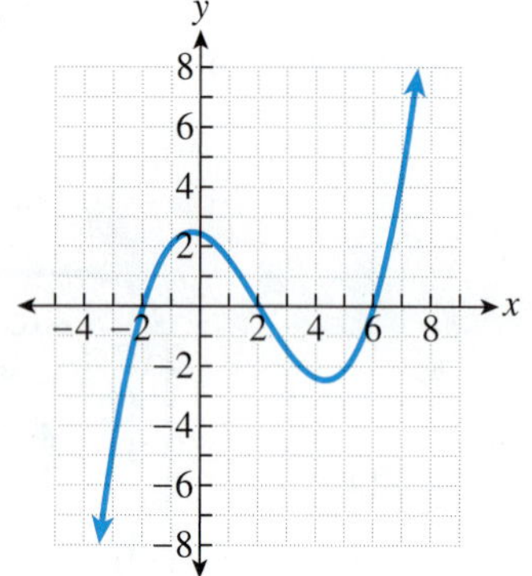

20.

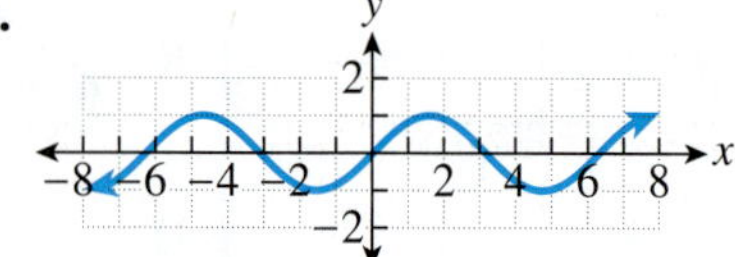

21.

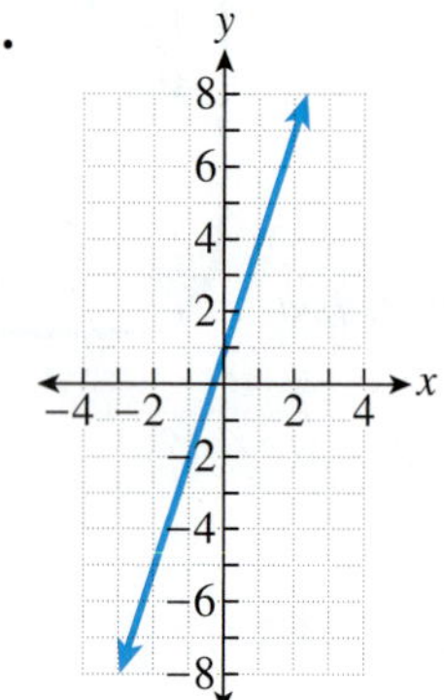

22.

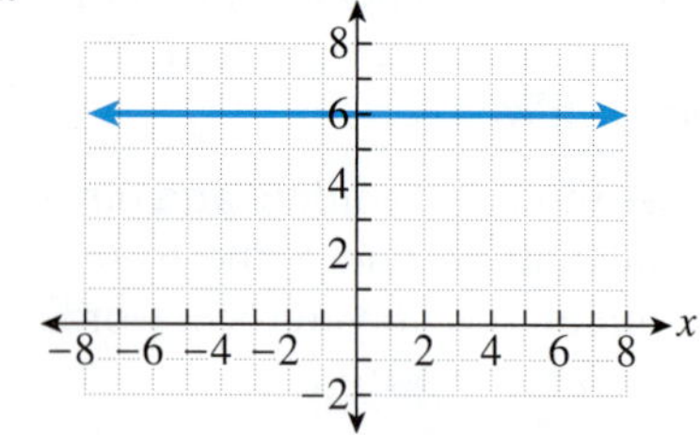

23.

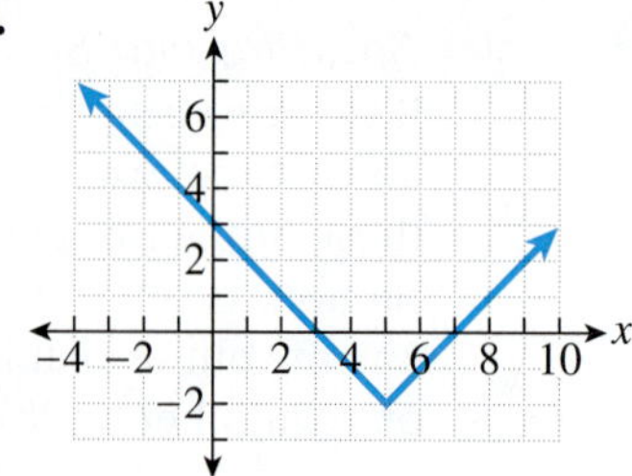

24.

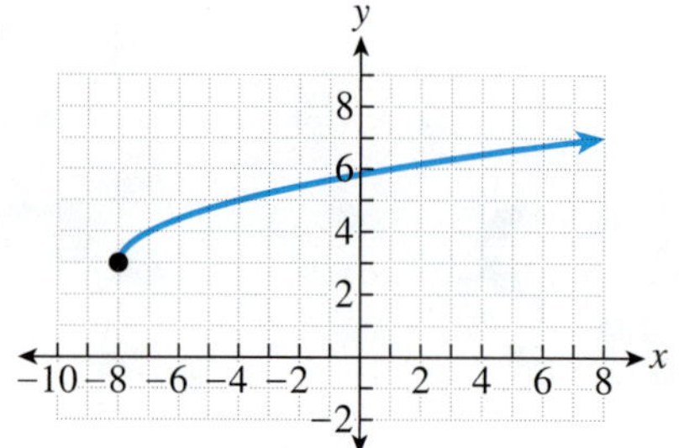

Determine whether the functions $f(x)$ and $g(x)$ are inverse functions by showing that $(f \circ g)(x) = x$ and $(g \circ f)(x) = x$.

25. $f(x) = x - 7, g(x) = x + 7$

26. $f(x) = x + 12, g(x) = x - 12$

27. $f(x) = 3x, g(x) = -3x$

28. $f(x) = 2x, g(x) = -\frac{1}{2}x$

29. $f(x) = 4x + 3, g(x) = \frac{1}{4}x - 3$

30. $f(x) = 3x - 18, g(x) = \frac{1}{3}x + 6$

31. $f(x) = -x + 9, g(x) = x - 9$

32. $f(x) = 7x + 28, g(x) = \frac{1}{7}(x - 28)$

For the given function $f(x)$, find $f^{-1}(x)$.

33. $f(x) = x + 5$

34. $f(x) = x - 8$

35. $f(x) = 4x$

36. $f(x) = -\frac{1}{5}x$

37. $f(x) = -x + 9$

38. $f(x) = -x - 13$

39. $f(x) = 2x + 17$

40. $f(x) = -5x - 11$

41. $f(x) = -\frac{2}{3}x - 8$

42. $f(x) = \frac{5}{4}x + \frac{3}{8}$

43. $f(x) = mx$

44. $f(x) = mx + b$

45. $f(x) = \frac{1}{x} + 3$

46. $f(x) = \frac{1}{x} - 5$

47. $f(x) = \frac{1}{x + 3}$

48. $f(x) = \frac{1}{x - 5}$

49. $f(x) = \frac{5 - 6x}{x}$

50. $f(x) = \frac{1 + 10x}{x}$

51. $f(x) = \frac{8 + 4x}{3x}$

52. $f(x) = \frac{2x}{7x - 10}$

For the given function $f(x)$, find $f^{-1}(x)$. State the domain of $f^{-1}(x)$.

53. $f(x) = \sqrt{x}$

54. $f(x) = \sqrt{x - 4}$

55. $f(x) = \sqrt{x} + 5$

56. $f(x) = \sqrt{x + 1} - 3$

57. $f(x) = \sqrt{x + 8} + 4$

58. $f(x) = \sqrt{x + 9} + 3$

59. $f(x) = x^2 - 9\ (x \geq 0)$

60. $f(x) = (x + 6)^2 + 5\ (x \geq -6)$

61. $f(x) = x^2 - 6x + 8\ (x \geq 3)$ (*Hint:* Try completing the square.)

62. $f(x) = x^2 + 4x - 12\ (x \geq -2)$ (*Hint:* Try completing the square.)

If the function represented by the set of ordered pairs is one to one, find its inverse.

63. $\{(-5, -17), (-1, -9), (1, -5), (4, 1), (7, 7)\}$

64. $\{(-2, -1), (0, 5), (1, 8), (4, 17), (7, 26)\}$

65. $\{(-2, 21), (-1, 17), (0, 13), (1, 9), (2, 13)\}$

66. $\{(-10, 4), (-9, 5), (-6, 6), (-1, 7), (6, 8)\}$

For the given graph of a one-to-one function $f(x)$, graph its inverse function $f^{-1}(x)$ without finding a formula for $f(x)$ or $f^{-1}(x)$.

67.

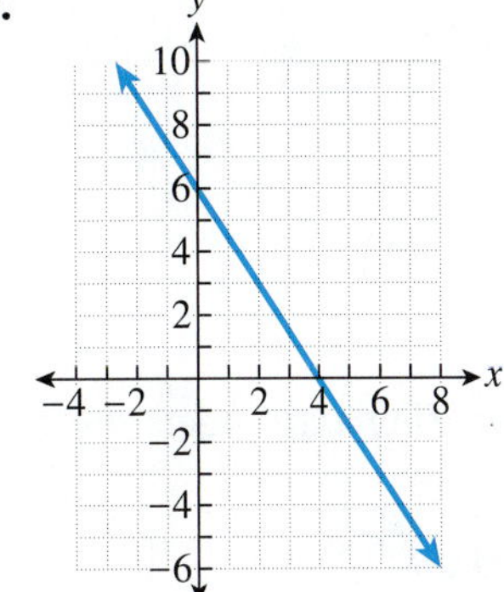

68.

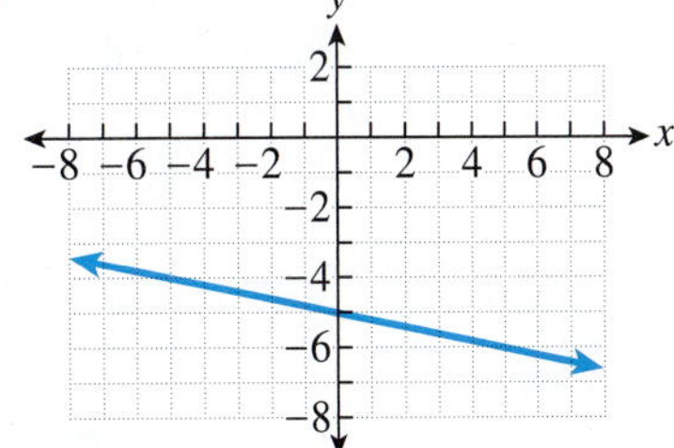

69.

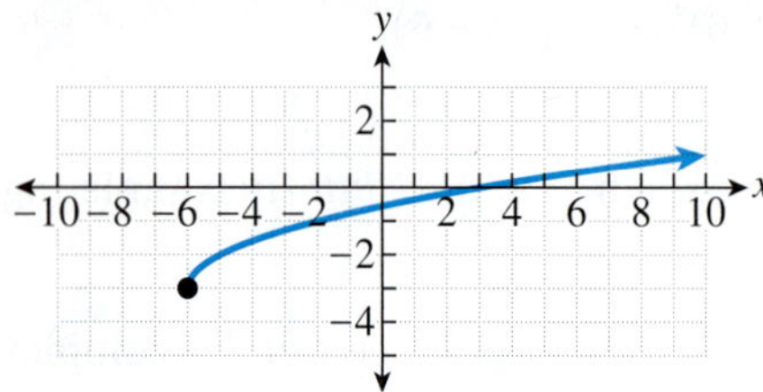

70.

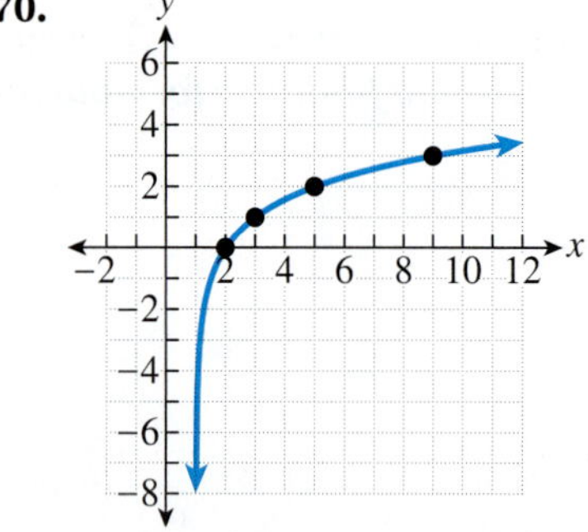

Graph a one-to-one function $f(x)$ that meets the given criteria.

71. $f(x)$ is a linear function, $f(4) = 9$, and $f^{-1}(5) = -2$.

72. $f(x)$ is a linear function, $f^{-1}(5) = -7$, and $f^{-1}(-6) = 8$.

73. $f(x)$ is a quadratic function, the domain of $f(x)$ is restricted to $[3, \infty)$, $f(3) = -1$, $f(4) = 0$, $f^{-1}(3) = 5$, and $f^{-1}(8) = 6$.

74. $f(x)$ is a quadratic function, the domain of $f(x)$ is restricted to $[1, \infty)$, $f^{-1}(-8) = 1$, $f^{-1}(-2) = 3$, and $f^{-1}(8) = 5$.

Writing in Mathematics

Answer in complete sentences.

75. True or False: If a linear function has negative slope, then its inverse function has positive slope.

76. Explain how the horizontal-line test shows whether a function is or is not one to one.

77. Describe a real-world function that is a one-to-one function, and explain why the function is one to one. Describe a real-world function that is not a one-to-one function, and explain why it is not a one-to-one function.

78. Explain how to find the inverse function of a one-to-one function $f(x)$. Use an example to illustrate the process.

79. ***Solutions Manual**** Write a solutions manual page for the following problem:

For the one-to-one function $f(x) = \frac{3 + 2x}{4x}$, *find* $f^{-1}(x)$.

80. ***Newsletter**** Write a newsletter explaining how to verify that two functions $f(x)$ and $g(x)$ are inverse functions, using composition of functions.

***See Appendix B for details and sample answers.**

8.3 Exponential Functions

Objectives

1. **Define exponential functions.**
2. **Evaluate exponential functions.**
3. **Graph exponential functions.**
4. **Define the natural exponential function.**
5. **Solve exponential equations.**
6. **Use exponential functions in applications.**

Definition of Exponential Functions

Objective 1 **Define exponential functions.** Would you rather have \$1 million, or \$1 that gets doubled every day for 30 days? You may be surprised by just how much that \$1 is worth at the end of 30 days. The following table tells you:

Day	Amount (\$)	Day	Amount (\$)	Day	Amount (\$)
1	\$1	11	\$1024	21	\$1,048,576
2	\$2	12	\$2048	22	\$2,097,152
3	\$4	13	\$4096	23	\$4,194,304
4	\$8	14	\$8192	24	\$8,388,608
5	\$16	15	\$16,384	25	\$16,777,216
6	\$32	16	\$32,768	26	\$33,554,432
7	\$64	17	\$65,536	27	\$67,108,864
8	\$128	18	\$131,072	28	\$134,217,728
9	\$256	19	\$262,144	29	\$268,435,456
10	\$512	20	\$524,288	30	\$536,870,912

After 30 days, the \$1 would turn into over \$500 million. Notice that, although the total grew quite slowly at first, it increased quickly toward the end of the 30 days. The amount of money after x days can be expressed as $f(x) = 2^{x-1}$. This is an example of exponential growth, which is based on exponential functions.

Exponential Function

A function that can be written in the form $f(x) = b^x$, where $b > 0$ and $b \neq 1$, is called an **exponential function** with base b.

(If $b = 1$, then $f(x)$ is simply the constant function $f(x) = 1$.)

This type of function is called an exponential function because the variable is in an exponent. Some examples of exponential functions are

$$f(x) = 2^x \qquad f(x) = 10^x \qquad f(x) = \left(\frac{1}{3}\right)^x \qquad f(x) = 4^{-x} \qquad f(x) = 3^{2x+1} + 7$$

Evaluating Exponential Functions

Objective 2 **Evaluate exponential functions.** We evaluate exponential functions in the same manner that we evaluated other functions: We substitute the specified value for the function's variable and then simplify the resulting expression.

EXAMPLE 1 Let $f(x) = 3^x$. Find the following:

a) $f(4)$

Solution

We begin by substituting 4 for x and then simplify the resulting expression.

$f(4) = 3^4$ Substitute 4 for x.
$= 81$ Simplify.

b) $f(0)$

Solution

We substitute 0 for x.

$f(0) = 3^0$ Substitute 0 for x.
$= 1$ Simplify. Recall that raising a nonzero base to the exponent 0 is equal to 1.

c) $f(-2)$

Solution

We substitute −2 for x.

$f(-2) = 3^{-2}$ Substitute −2 for x.
$= \frac{1}{3^2}$ Rewrite the expression without a negative exponent. Recall that $b^{-n} = \frac{1}{b^n}$.
$= \frac{1}{9}$ Simplify.

Quick Check 1
Let $f(x) = 4^x$. Find $f(5)$ and $f(-3)$.

EXAMPLE 2 Let $f(x) = 5 \cdot 3^{x-4} + 16$. Find $f(6)$.

Solution

This function, although more complex than those in previous examples, is also an exponential function. We will substitute 6 for x and then simplify.

$f(6) = 5 \cdot 3^{6-4} + 16$ Substitute 6 for x.
$= 5 \cdot 3^2 + 16$ Simplify the exponent.
$= 61$ Simplify.

Quick Check 2
Let $f(x) = 2 \cdot 7^{x+3} - 19$. Find $f(-1)$.

Using Your Calculator To simplify the expression $5 \cdot 3^2 + 16$ with the TI-84, we will use the button labeled [^]. Here is how the calculator screen should look:

```
5*3^2+16
                61
```

Graphs of Exponential Functions

Objective 3 Graph exponential functions. To understand the behavior of exponential functions, we must examine the graphs of these functions. In this section, we will learn to create a quick sketch of the graph of an exponential function. A more thorough presentation of graphing exponential functions, including finding x-intercepts, will be given in Section 8.8. For now, we will plot points for selected values of x.

Consider the function $f(x) = 2^x$. Here is a table of values of $f(x)$ for values of x ranging from -3 to 3:

x	-3	-2	-1	0	1	2	3
$f(x)$	$\frac{1}{8}$	$\frac{1}{4}$	$\frac{1}{2}$	1	2	4	8

We plot the ordered pairs and then draw a smooth curve passing through those points. The domain of this function is the set of all real numbers $(-\infty, \infty)$. As we look to the left of the graph, the function gets closer and closer to the x-axis without actually touching it.

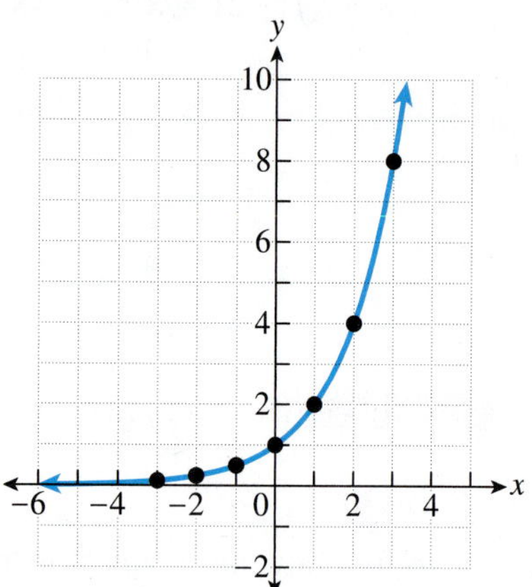

Here are some additional function values to illustrate this trend:

x	-4	-5	-6	-7	-8	-9	-10
$f(x)$	$\frac{1}{16}$	$\frac{1}{32}$	$\frac{1}{64}$	$\frac{1}{128}$	$\frac{1}{256}$	$\frac{1}{512}$	$\frac{1}{1024}$

As shown in the following graph, these function values are approaching 0 and the graph is approaching the x-axis:

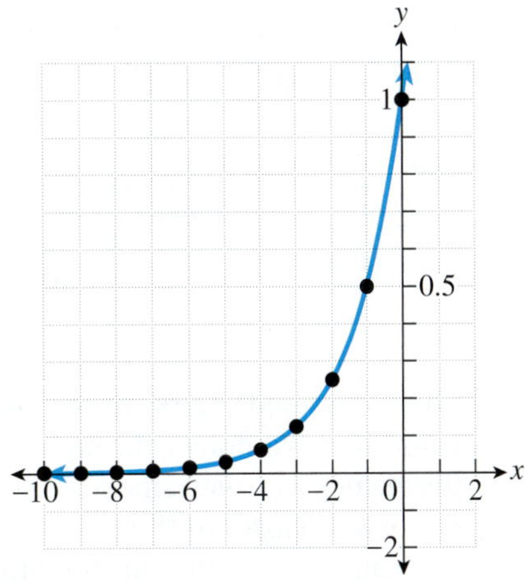

In this case, we say that the x-axis is a horizontal asymptote for the graph of the function $f(x) = 2^x$.

A **horizontal asymptote** is a horizontal line that the graph of a function approaches as it either moves to the left or to the right. The horizontal asymptote is placed on the graph as a dashed line, because the points on the asymptote are not part of the graph of the exponential function.

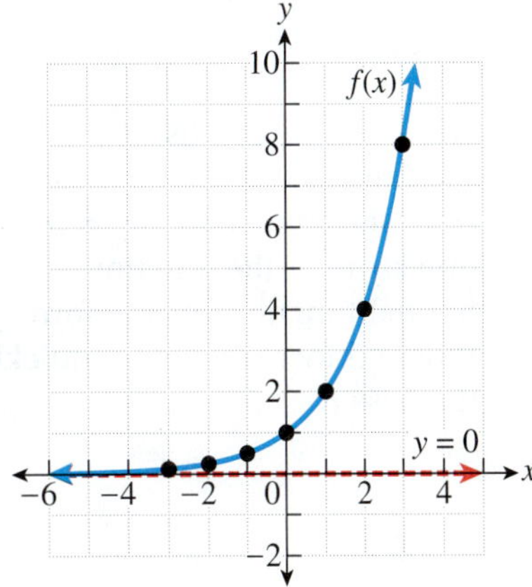

The horizontal asymptote also helps us to determine the range of an exponential function. As the values of x decrease, the values of $f(x)$ get closer and closer to 0, without ever reaching it. As the values of x increase, the function values increase without limit. The range of this function is $(0, \infty)$.

An exponential function of the form $f(x) = b^x$ $(b > 1)$ is an **increasing function.** This means that as x increases, so does $f(x)$. The graph of an increasing function moves upward as it moves from left to right. Notice that the function increases at a very slow rate for values of x that are less than 0, but increases more rapidly as the values of x increase.

EXAMPLE 3 Graph $f(x) = 3^x$, and state the domain and range of $f(x)$.

Solution

For this example, we will use the values -2, -1, 0, 1, and 2 for x. This will be the case whenever the exponent is simply x.

x	-2	-1	0	1	2
$f(x) = 3^x$	$3^{-2} = \frac{1}{3^2} = \frac{1}{9}$	$3^{-1} = \frac{1}{3^1} = \frac{1}{3}$	$3^0 = 1$	$3^1 = 3$	$3^2 = 9$

These ordered pairs have been plotted and a smooth curve drawn through them. Notice that as the values of x get smaller, the points are getting closer and closer to the x-axis. The line $y = 0$ is the horizontal asymptote for the graph of this function and has been graphed as a dashed line.

The domain of this function is the set of all real numbers $(-\infty, \infty)$, while the range is $(0, \infty)$.

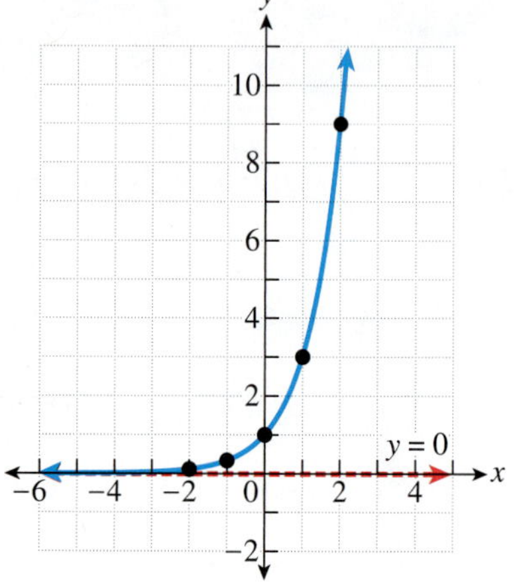

Quick Check 3
Graph $f(x) = 5^x$, and state the domain and range of $f(x)$.

Here are the graphs of $f(x) = 2^x$ and $g(x) = 3^x$ on the same set of axes. Notice that for positive values of x, the function $g(x) = 3^x$ increases much more quickly than the function $f(x) = 2^x$ does. For exponential functions of the form $f(x) = b^x$, $b > 1$, the function increases more quickly for larger values of the base b.

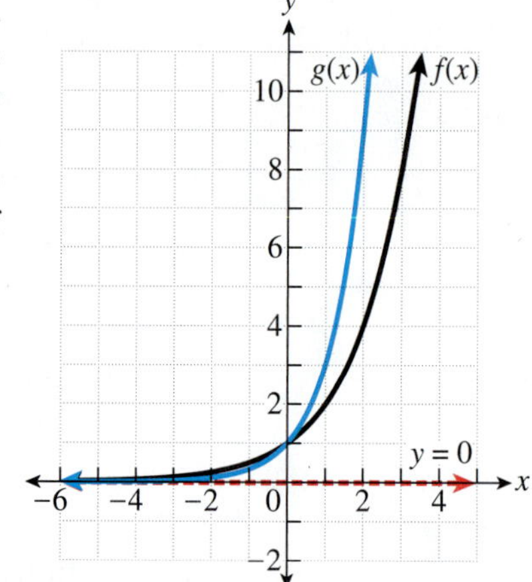

EXAMPLE 4 Graph $f(x) = (\frac{1}{2})^x$, and state the domain and range of $f(x)$.

Solution

We will begin by creating a table of values for this function, using the same values of x that we used in the previous example.

x	-2	-1	0	1	2
$f(x) = \left(\frac{1}{2}\right)^x$	$\left(\frac{1}{2}\right)^{-2} = 2^2 = 4$	$\left(\frac{1}{2}\right)^{-1} = 2^1 = 2$	$\left(\frac{1}{2}\right)^0 = 1$	$\left(\frac{1}{2}\right)^1 = \frac{1}{2}$	$\left(\frac{1}{2}\right)^2 = \frac{1}{4}$

Notice that this time, as the values of x get larger, the function values are getting closer and closer to 0. The horizontal asymptote is the line $y = 0$. The domain of this function is the set of all real numbers $(-\infty, \infty)$, while the range is $(0, \infty)$.

Quick Check 4
Graph $f(x) = (\frac{1}{3})^x$, and state the domain and range of $f(x)$.

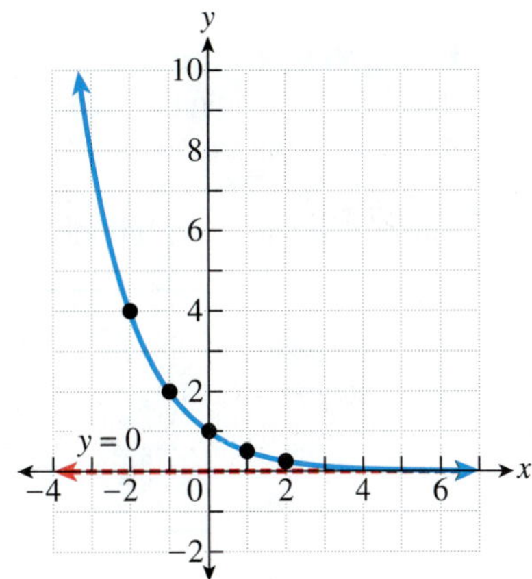

EXAMPLE 5 Graph $f(x) = 2^{x+1} - 4$, and state the domain and range of $f(x)$.

Solution

The exponent is equal to 0 when $x = -1$. In addition to using this value for x, we will use two values that are less than -1 and two values that are greater than -1.

x	-3	-2	-1	0	1
$f(x) = 2^{x+1} - 4$	$2^{-2} - 4 = -3\frac{3}{4}$	$2^{-1} - 4 = -3\frac{1}{2}$	$2^0 - 4 = -3$	$2^1 - 4 = -2$	$2^2 - 4 = 0$

Quick Check 5

Graph $f(x) = 6^{x-1} + 3$, and state the domain and range of $f(x)$.

Notice that the x-axis is not the horizontal asymptote for the graph of this function. As the values of x get smaller and smaller, the function values approach -4. The line $y = -4$ is the horizontal asymptote for the graph of this function. The domain of this function is the set of all real numbers $(-\infty, \infty)$, while the range is $(-4, \infty)$.

In general, the value that is added to or subtracted from the exponential expression will tell us the location of the horizontal asymptote.

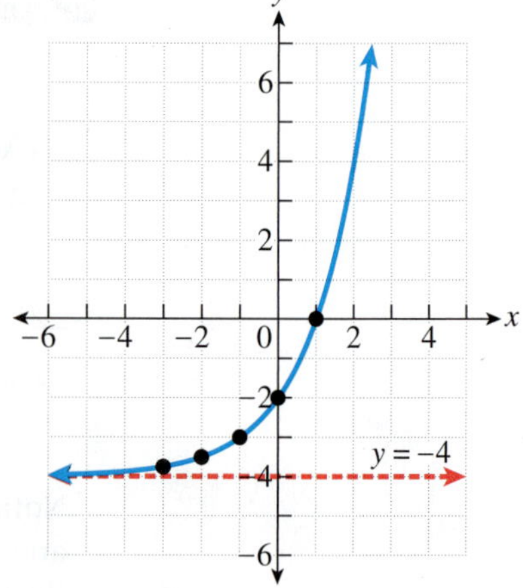

Using Your Calculator To graph $f(x) = 2^{x+1} - 4$ and its horizontal asymptote with the TI-84, we begin by pushing the [Y=] key. Next to Y_1, type in the function, using parentheses around the exponent. The horizontal asymptote is the line $y = -4$, so type -4 next to Y_2. Press the [GRAPH] key to display the graph. Here is the screen showing how to enter the function and its corresponding graph:

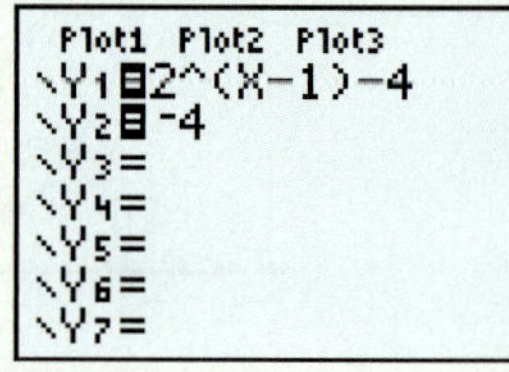

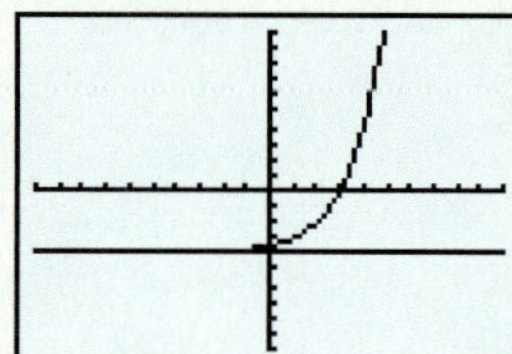

Here are some guidelines for graphing an exponential function with a base that is greater than 1:

Graphing an Exponential Function of the Form $f(x) = b^{x-h} + k, b > 1$

1. **Graph the horizontal asymptote.** Using a dashed line, graph the line $y = k$.
2. **Create a table of values.** In addition to using the value $x = h$, use two values of x that are less than h and two values that are greater than h. Evaluate the function for all five values of x to determine the coordinates of ordered pairs on the graph.
3. **Draw the graph.** Plot the five points and draw the graph that passes through these points. Be sure that the graph you draw approaches the horizontal asymptote. The graph should increase from left to right.

We graph a function of the form $f(x) = b^{x-h} + k$, where $0 < b < 1$, in the same fashion that we graph an exponential function whose base b is greater than 1. The major exception is that if $0 < b < 1$, then the function is a decreasing function, rather than an increasing function.

EXAMPLE 6 Graph $f(x) = 5^{-x} + 2$, and state the domain and range of $f(x)$.

Solution

We begin by rewriting this function as $f(x) = (\frac{1}{5})^x + 2$. Notice that this base is less than 1. The line $y = 2$ is the horizontal asymptote for the graph of this function. Now we can create a table of values.

x	-2	-1	0	1	2
$f(x) = \left(\frac{1}{5}\right)^x + 2$	$\left(\frac{1}{5}\right)^{-2} + 2 = 27$	$\left(\frac{1}{5}\right)^{-1} + 2 = 7$	$\left(\frac{1}{5}\right)^{0} + 2 = 3$	$\left(\frac{1}{5}\right)^{1} + 2 = 2\frac{1}{5}$	$\left(\frac{1}{5}\right)^{2} + 2 = 2\frac{1}{25}$

Note that the scale on the y-axis is not the same as the scale on the x-axis. The domain of this function is the set of all real numbers $(-\infty, \infty)$, while the range is $(2, \infty)$.

Quick Check 6

Graph $f(x) = (\frac{1}{2})^{x-3} - 2$, and state the domain and range of $f(x)$.

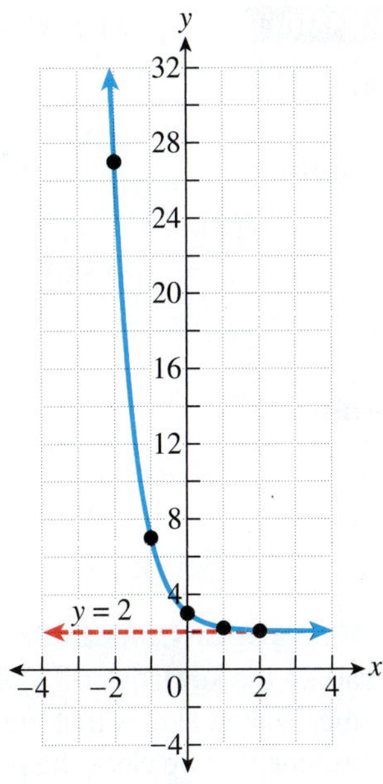

One last fact about exponential functions is that they are one-to-one functions. We can determine that exponential functions are one-to-one functions by applying the horizontal line test.

The Natural Exponential Function

Objective 4 **Define the natural exponential function.** One number that will be used as the base of an exponential function is an irrational number that is denoted by the letter e and is called the natural base.

The Natural Base e

$$e \approx 2.7182818284\ldots$$

The use of the letter e to denote this number is credited to Leonhard Euler. You may think that he chose e because it was the first letter of his last name, but many math historians claim that he had been using the letter a in his writings and that e was simply the next vowel available.

The Natural Exponential Function

The function $f(x) = e^x$ is called the **natural exponential function.**

Calculators have a built-in function for evaluating the natural exponential function, and it usually is listed above the key labeled as ln.

EXAMPLE 7 Let $f(x) = e^x$. Find the following and round to the nearest thousandth:

a) $f(5)$

Solution

$f(5) = e^5$ — Substitute 5 for x.

≈ 148.413 — Approximate, using a calculator.

b) $f(-1.7)$

Solution

$f(-1.7) = e^{-1.7}$ — Substitute -1.7 for x.

≈ 0.183 — Approximate, using a calculator.

Quick Check 7

Let $f(x) = e^x$. Find $f(-1)$ and $f(4.3)$. Round to the nearest thousandth.

Using Your Calculator The TI-84 has a built-in function e^x, which is a 2nd function above the key labeled [LN]. To calculate e^5, we begin by pushing the [2nd] key, followed by the [LN] key. Notice that the calculator gives you a left-hand parenthesis. After typing the exponent 5, we close the parentheses by pushing [)]. After pressing [ENTER], your calculator screen should look as follows:

```
e^(5)
      148.4131591
```

EXAMPLE 8 Graph $f(x) = e^x$, and state the domain and range of $f(x)$.

Quick Check 8 Graph $f(x) = e^{x+2} + 5$, and state the domain and range of $f(x)$.

Solution

We will begin by creating a table of values for this function.

x	-2	-1	0	1	2
$f(x) = e^x$	$e^{-2} \approx 0.14$	$e^{-1} \approx 0.37$	$e^0 = 1$	$e^1 \approx 2.72$	$e^2 \approx 7.39$

The horizontal asymptote is the line $y = 0$. The domain of this function is the set of all real numbers $(-\infty, \infty)$, while the range is $(0, \infty)$.

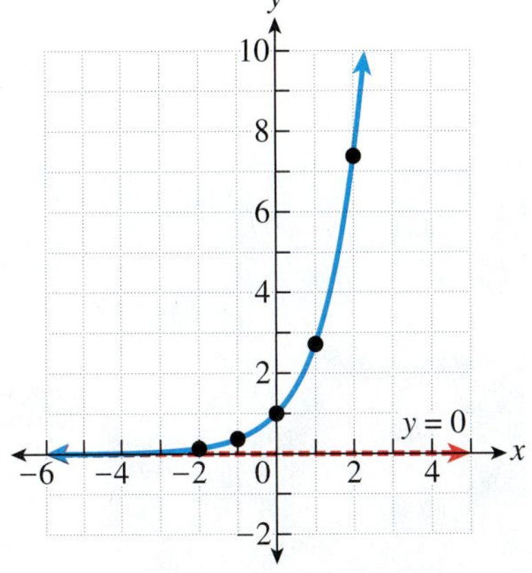

Solving Exponential Equations

Objective 5 **Solve exponential equations.** An **exponential equation** is an equation containing a variable in an exponent, such as $2^x = 16$. One technique for solving exponential equations is to write both sides of the equation in terms of the same base. If this is possible, we can then apply the one-to-one property of exponential functions to solve the equation.

One-to-One Property of Exponential Functions

For any positive real number b $(b \neq 1)$ and any real numbers r and s, if $b^r = b^s$, then $r = s$.

EXAMPLE 9 Solve $2^x = 16$.

Solution

We will begin by expressing 16 as a power of 2.

$$2^x = 16$$

$$2^x = 2^4 \quad \text{Rewrite 16 as } 2^4.$$

$$x = 4 \quad \text{Use the one-to-one property of exponential functions.}$$

The solution set is $\{4\}$.

EXAMPLE 10 Solve $3^{x+2} = \frac{1}{27}$.

Solution

We can rewrite $\frac{1}{27}$ as 3^{-3}.

$$3^{x+2} = \frac{1}{27}$$

$$3^{x+2} = 3^{-3} \quad \text{Rewrite } \tfrac{1}{27} \text{ as } 3^{-3}.$$

$$x + 2 = -3 \quad \text{If } b^r = b^s, \text{ then } r = s.$$

$$x = -5 \quad \text{Subtract 2.}$$

The solution set is $\{-5\}$.

Quick Check 9

Solve.

a) $4^x = 256$

b) $9^{2x-17} = 9$

Applications

Objective 6 **Use exponential functions in applications.** We conclude this section with an application of exponential functions.

EXAMPLE 11 The population of a particular small town can be approximated by the function $f(x) = 75{,}000 \cdot e^{0.018x}$, where x represents the number of years after 1984. What will be the population of the town in the year 2014?

Solution

To predict the town's population in the year 2014, we will evaluate the function for $x = 30$, which is $2014 - 1984$.

$$f(30) = 75{,}000 \cdot e^{0.018(30)} \quad \text{Substitute 30 for } x.$$

$$= 75{,}000 \cdot e^{0.54} \quad \text{Simplify the exponent.}$$

$$\approx 128{,}701 \quad \text{Approximate, using a calculator. Round to the nearest person.}$$

The town's population in the year 2014 will be approximately 128,701.

Quick Check 10 The total revenues generated in a given year by McDonald's (in millions of dollars) can be approximated by the function $f(x) = 7337 \cdot 1.08^x$, where x represents the number of years after 1992. Use this function to predict the total revenues of McDonald's in the year 2012. (*Source:* McDonald's Corporation)

Building Your Study Strategy **Practice Quizzes, 3 Sample** Following is a practice quiz containing two problems for each objective from objective 2 through objective 6. After you take the quiz and check your work, you will have an idea about which objectives will require extra study and which ones will not.

Objective 2

1. Let $f(x) = 3^{x+4} - 29$. Find $f(-2)$.

2. Let $f(x) = (\frac{1}{2})^{x+3}$. Find $f(-5)$.

Objective 3

3. Graph $f(x) = 2^{x+1}$, and state the domain and range of the function. Label the horizontal asymptote.

4. Graph $f(x) = (\frac{1}{4})^{x} + 5$, and state the domain and range of the function. Label the horizontal asymptote.

Objective 4

5. Let $f(x) = e^{x+6} + 409$. Find $f(3)$. Round to the nearest thousandth.

6. Graph $f(x) = e^{x+2} - 7$, and state the domain and range of the function. Label the horizontal asymptote.

Objective 5

7. Solve $2^{x-3} = 64$. **8.** Solve $3^{x^2-5x-12} = 9$.

Objective 6

9. David deposited \$2000 in a bank account that pays 4.5% annual interest, compounded monthly. The balance after t years is given by the function $f(t) = 2000(1.00375)^{12t}$. How much money will be in the account after 3 years?

10. The value of a car, in dollars, after x years can be approximated by the function $f(x) = 20{,}000 \cdot 0.82^{x}$. Use this function to determine the value of the car 5 years after its purchase.

EXERCISES 8.3

Vocabulary

1. A function that can be written in the form $f(x) = b^x$, $b > 0$ and $b \neq 1$, is called a(n) ________ function with base b.

2. A(n) ________ is a horizontal line that the graph of a function approaches as it moves either to the left or to the right.

3. If $f(x)$ increases as x increases, then $f(x)$ is said to be a(n) ________ function.

4. If $f(x)$ decreases as x increases, then $f(x)$ is said to be a(n) ________ function.

5. The one-to-one property of exponential functions states that for any positive real number b $(b \neq 1)$ and any real numbers r and s, if $b^r = b^s$, then ________.

6. The function $f(x) = e^x$ is called the ________ exponential function.

Let $f(x) = 2^x$. Find the following:

7. $f(4)$
8. $f(7)$
9. $f(0)$
10. $f(-1)$

Let $f(x) = \left(\frac{2}{5}\right)^x$. Find the following:

11. $f(2)$
12. $f(3)$
13. $f(-4)$
14. $f(-2)$

Evaluate the given function.

15. $f(x) = 3^x + 2, f(2)$
16. $f(x) = 3^{x+2}, f(2)$
17. $f(x) = 2^{x+5} - 5, f(3)$
18. $f(x) = 2^{x-3} + 27, f(6)$
19. $f(x) = \left(\frac{1}{4}\right)^{x-4} + 6, f(3)$
20. $f(x) = \left(\frac{1}{2}\right)^{x+1} + 5, f(-3)$
21. $f(x) = 4^{-x}, f(3)$
22. $f(x) = 4^{-x}, f(-2)$
23. $f(x) = 3^{2x-1} - 100, f(4)$
24. $f(x) = 5^{4x+3} + 175, f(0)$

Complete the table for $f(x)$.

25. $f(x) = 10^x$

x	−2	−1	0	1	2
f(x)					

26. $f(x) = \left(\frac{1}{4}\right)^x$

x	−2	−1	0	1	2
f(x)					

27. $f(x) = 4^{x-3} - 9$

x	1	2	3	4	5
f(x)					

28. $f(x) = \left(\frac{1}{2}\right)^{x+4} + 12$

x	−6	−5	−4	−3	−2
f(x)					

Graph $f(x)$. Label the horizontal asymptote. State the domain and range of the function.

29. $f(x) = 4^x$
30. $f(x) = 10^x$

31. $f(x) = \left(\frac{1}{4}\right)^x$

35. $f(x) = 3^x - 1$

36. $f(x) = 3^{x-1}$

37. $f(x) = 2^{x+2} - 8$

38. $f(x) = 2^{x-3} + 5$

32. $f(x) = \left(\frac{3}{2}\right)^x$

33. $f(x) = 5^{x+1}$

34. $f(x) = 5^x + 1$

39. $f(x) = \left(\frac{1}{2}\right)^{x+4} + 2$

40. $f(x) = \left(\frac{1}{3}\right)^{x-5} - 9$

Let $f(x) = e^x$. Find the following, rounded to the nearest thousandth.

41. $f(2)$

42. $f(3)$

43. $f(-1)$

44. $f(0)$

45. $f(-5.2)$

46. $f(4.13)$

Evaluate the given function. Round to the nearest thousandth.

47. $f(x) = e^{x+4}, f(-1)$

48. $f(x) = e^{x+2}, f(3)$

49. $f(x) = e^{3x-1}, f(-2)$

50. $f(x) = e^{2x-5}, f(2.7)$

Graph $f(x)$. Label the horizontal asymptote. State the domain and range of the function.

51. $f(x) = e^{x-2}$

52. $f(x) = e^{x+5}$

53. $f(x) = e^x + 2$

54. $f(x) = e^x + 7$

55. $f(x) = e^{x+4} + 4$

56. $f(x) = e^{x-1} + 10$

Solve.

57. $3^x = 81$

58. $2^x = 32$

59. $5^x = 125$

60. $10^x = 100{,}000$

61. $6^x = \frac{1}{6}$

62. $4^x = \frac{1}{64}$

63. $2^{x-5} = 128$

64. $5^{x+2} = 25$

65. $9^{x-3} = \frac{1}{81}$

66. $3^{x+4} = \frac{1}{81}$

67. $\left(\frac{1}{3}\right)^{2x-1} = \frac{1}{243}$

68. $\left(\frac{1}{4}\right)^{x-7} = \frac{1}{64}$

69. $4^{3x-5} = 256$

70. $7^{5x+12} = 49$

71. $3^{8x-27} = \frac{1}{27}$

72. $2^{4x+14} = \frac{1}{1024}$

73. $8^{17x-85} = 1$

74. $5^{-3x+9} = 1$

75. Wendy deposited \$5000 in an account that pays 13% annual interest, compounded annually. The balance after t years is given by the function $f(t) = 5000(1.13)^t$. What will the balance of this account be after 40 years?

76. Mary deposited \$350 in an account that pays 5% annual interest, compounded annually. The balance after t years is given by the function $f(t) = 350(1.05)^t$. What will the balance of this account be after 6 years?

77. Dylan deposited \$10,000 in an account that pays 9% annual interest, compounded quarterly. The balance after t years is given by the function $f(t) = 10{,}000(1.0225)^{4t}$. What will the balance of this account be after 5 years?

78. Adam deposited \$50,000 in an account that pays 6% annual interest, compounded quarterly. The balance after t years is given by the function $f(t) = 50{,}000(1.015)^{4t}$. What will the balance of this account be after 10 years?

79. In 1995, a particular Cal Ripken baseball card was valued at \$8. By 2003, the card had increased in value to \$20. The value of the card can be approximated by the function $f(t) = 8e^{0.1145t}$, where t represents the number of years after 1995. Assuming that the value of the card continues to grow exponentially, predict the value of the card in 2017.

80. The average ticket price for attending a movie has been increasing exponentially. The average ticket price in a particular year can be approximated by the function $f(t) = 4.69 \cdot e^{0.0627t}$, where t represents the number of years after 1998. (This average price includes both first run and subsequent runs, as well as all special pricing.) Use this function to predict the average ticket price in the year 2011. (*Source:* Motion Picture Association of America (MPAA))

81. The net revenues of Amazon.com in a particular year, in millions of dollars, can be approximated by the function $f(t) = 2762 \cdot e^{0.1767t}$, where t represents the number of years after 2000. If net revenues continue to grow at this rate, predict the net revenues of Amazon.com in the year 2009. (*Source:* Amazon.com, Inc.)

82. The number of Wal-Mart stores in a particular year can be approximated by the function $f(t) = 2440 \cdot e^{0.0726t}$, where t represents the number of years after 1993. Assuming that this rate of growth continues, use this function to predict the number of Wal-Mart stores in the year 2013. (*Source:* Wal-Mart Stores, Inc.)

83. If a container of water whose temperature is 80°C is placed in a refrigerator whose temperature remains a constant 4°C, then the temperature of the water, in degrees Celsius, after t minutes is given by the function $f(t) = 4 + 76e^{-0.014t}$. What will the temperature of the water be after 60 minutes?

84. If a person dies in a room that is 70°F, then the body's temperature in degrees Fahrenheit after t hours is given by the function $f(t) = 70 + 28.6e^{-0.44t}$. What will the body's temperature be after 3 hours?

85. The average cost of a loaf of white bread in a particular year can be approximated by the function $f(t) = 0.74 \cdot 1.033^t$, where t represents the number of years after 1993. Use this function to approximate the cost of a loaf of white bread in the year 2008. (*Source:* U.S. Bureau of Labor Statistics)

86. The number of U.S. wineries in a particular year can be approximated by the function $f(t) = 669 \cdot 1.055^t$, where t represents the number of years after 1975. Assuming that this rate of growth continues, use this function to predict the number of U.S. wineries in the year 2010. (*Source:* Alcohol and Tobacco Tax and Trade Bureau)

Plot the given ordered pairs $(x, f(x))$. From the graph, do you feel that $f(x)$ is an exponential function? Explain your reasoning.

87. $(-3, -17), (0, -8), (5, 7), (6, 10), (10, 22)$

88. $(-5, -8), (-4, -11), (-3, -12), (-1, -8), (0, -3), (1, 4)$

89. $(0, 4.125), (1, 4.25), (3, 5), (4, 6), (6, 12), (7, 20)$

90. $(-3, 91), (-2, 37), (-1, 19), (0, 13), (1, 11)$

Writing in Mathematics

Answer in complete sentences.

91. Explain the difference between the graph of $f(x) = b^x$ for $b > 1$ and $0 < b < 1$.

92. Explain the difference between the graph of $f(x) = 2^x$ and $g(x) = x^2$.

93. Create an exponential equation whose solution is $x = 4$, and explain how to solve the equation.

94. ***Solutions Manual**** Write a solutions manual page for the following problem:

Graph $f(x) = e^{x+1} + 2$. Label the horizontal asymptote. State the domain and range of the function.

95. ***Newsletter**** Write a newsletter explaining how to graph exponential functions.

***See Appendix B for details and sample answers.**

8.4 Logarithmic Functions

Objectives

1. **Define logarithms and logarithmic functions.**
2. **Evaluate logarithms and logarithmic functions.**
3. **Define the common logarithm and natural logarithm.**
4. **Convert back and forth between exponential form and logarithmic form.**
5. **Solve logarithmic equations.**
6. **Graph logarithmic functions.**
7. **Use logarithmic functions in applications.**

Logarithms

Objective 1 Define logarithms and logarithmic functions. In this section, we will examine numbers called **logarithms.** Prior to the development of modern technology such as computers and handheld calculators, logarithms were used to help perform difficult numeric calculations. Although this is no longer a practice, logarithms are still quite important in today's society and are used in a wide variety of applications such as the study of earthquakes and the pH of a chemical substance.

Suppose that we were trying to determine what power of 2 is equal to 10; in other words, trying to solve the equation $2^a = 10$ for a. This unknown exponent a is a logarithm and can be expressed as $\log_2 10$. We read $\log_2 10$ as *"the logarithm, base 2, of 10."*

Logarithm

For any positive real number b $(b \neq 1)$ and any positive real number x, we define $\log_b x$ to be the exponent that b must be raised to in order to equal x. The number x is called the **argument** of the logarithm. The argument must be a positive number, as it is the result obtained from raising a positive number to a power.

So, $\log_5 25$ is the exponent that 5 must be raised to in order to equal 25. In other words, $\log_5 25$ is the value of y that is a solution to the equation $5^y = 25$. Since we know that $5^2 = 25$, we know that $\log_5 25 = 2$.

Logarithmic Functions

For any positive real number b such that $b \neq 1$, a function of the form $f(x) = \log_b x$ is called a **logarithmic function.** This function is defined for values of x that are greater than 0; so the domain of this function is the set of positive real numbers, $(0, \infty)$.

We will see later in this section that logarithmic functions are the inverses of exponential functions.

Evaluating Logarithms and Logarithmic Functions

Objective 2 Evaluate logarithms and logarithmic functions. Now we shift our focus to evaluating logarithms.

EXAMPLE 1 Evaluate $\log_2 8$.

Solution

To evaluate this logarithm, we need to determine what power of 2 is equal to 8. Since $2^3 = 8$, $\log_2 8 = 3$.

EXAMPLE 2 Evaluate $\log_4 \frac{1}{16}$.

Solution

We need to determine what power of 4 is equal to the fraction $\frac{1}{16}$. The only way that a number greater than 1 can be raised to a power and produce a result less than 1 is if the exponent is negative. Since $4^{-2} = \frac{1}{4^2}$ or $\frac{1}{16}$, $\log_4 \frac{1}{16} = -2$.

Quick Check 1
Evaluate.

a) $\log_5 25$ b) $\log_3 \frac{1}{27}$
c) $\log_{12} 12$ d) $\log_6 1$

One important property of logarithms is that, for any positive number b $(b \neq 1)$, $\log_b b = 1$. For example, $\log_7 7 = 1$. Another important property of logarithms is that, for any positive number b $(b \neq 1)$, $\log_b 1 = 0$. For example, $\log_9 1 = 0$.

EXAMPLE 3 Given the function $f(x) = \log_2 x$, find $f(16)$.

Solution

We evaluate a logarithmic function in the same fashion that we evaluate any other function. We will substitute 16 for x and then simplify the resulting expression.

$$f(16) = \log_2 16 \quad \text{Substitute 16 for } x.$$
$$= 4 \quad \text{Since } 2^4 = 16, \log_2 16 = 4.$$

Quick Check 2
Given the function $f(x) = \log_4 x$, find $f(\frac{1}{256})$.

Common Logarithms, Natural Logarithms

Objective 3 Define the common logarithm and natural logarithm. There are two logarithms that we will use more often than others. The first is called the common logarithm.

Common Logarithm

The **common logarithm** is a logarithm whose base is 10.

In this case, we will write $\log x$ rather than $\log_{10} x$, and the base is understood to be 10.

A Word of Caution When working with a logarithm whose base is not 10, we must be sure to write the base, because $\log x$ is understood to have a base of 10.

Scientific calculators have built-in functions to calculate common logarithms. This function is typically denoted by a button labeled as log.

Example 4 Evaluate the given common logarithm. Round to the nearest thousandth if necessary.

a) log 100

Solution

To evaluate log 100, we must determine what power of the base 10 is equal to 100. Since $10^2 = 100$, then $\log 100 = 2$. A calculator would produce the same result.

b) log 321

Solution

To evaluate log 321, we will use a calculator, since we do not know what power of 10 is 321. Rounding to the nearest thousandth, we find that $\log 321 \approx 2.507$. To check this result, evaluate $10^{2.507}$ on a calculator, which should be approximately equal to 321.

Using Your Calculator The TI-84 has a built-in function for calculating the common logarithm of a number. To evaluate log 321, press the button labeled LOG. Notice that the calculator gives you a left-hand parenthesis. The argument of the logarithm needs to be enclosed in parentheses. Then type 321 and close the parentheses by pushing). After you press ENTER, your calculator screen should look as follows:

Quick Check 3

Evaluate. Round to the nearest thousandth if necessary.

a) log 10,000
b) log 75

The other logarithm that we will use frequently is called the **natural logarithm.**

Natural Logarithm

The natural logarithm, denoted ln *x*, is the logarithm whose base is the natural base *e*.

Recall from the previous section that *e* is a real number that is approximately equal to 2.718281828. Scientific calculators have built-in functions for calculating natural logarithms, using a button labeled as ln.

Example 5 Evaluate the given natural logarithm. Round to the nearest thousandth if necessary.

a) $\ln e^7$

Solution

To evaluate $\ln e^7$, we must find an exponent a such that $e^a = e^7$. In this case, $a = 7$; so $\ln e^7 = 7$.

This example illustrates another important property of logarithms.

$\log_b b^r$

For any positive base b ($b \neq 1$), $\log_b b^r = r$.

b) $\ln 45$

Solution

This logarithm must be evaluated by a calculator. Rounded to the nearest thousandth, $\ln 45 \approx 3.807$.

Using Your Calculator The TI-84 also has a built-in function for calculating the natural logarithm of a number. To evaluate ln 45, we begin by pushing the [LN] key. After typing the argument 45, we close the parentheses by pushing [)] and then press [ENTER]. Your calculator screen should look as follows:

```
ln(45)
          3.80666249
```

Quick Check 4 Evaluate the following and round to the nearest thousandth if necessary:

a) $\ln e^{-3}$
b) $\ln 605$

Exponential Form, Logarithmic Form

Objective 4 Convert back and forth between exponential form and logarithmic form. Consider the equation $5^3 = 125$. This equation is said to be in **exponential form**, as it has a base (5) being raised to a power (3). Any equation in exponential form can be rewritten as an equivalent equation in **logarithmic form**. The equation $5^3 = 125$ can also be written in logarithmic form as $\log_5 125 = 3$.

Exponential Form	Logarithmic Form
$b^y = n$	$\log_b n = y$

The following images will help us when we are rewriting equations from one form to the other:

$b^y = n$ $\log_b n = y$
Base(b)

Exponent(y)
$b^y = n$ $\log_b n = y$

$b^y = n$ $\log_b n = y$
Number(n)

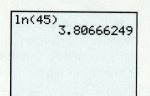

Example 6 Write the exponential equation $2^{-8} = \frac{1}{256}$ in logarithmic form.

Solution

This equation can be written in logarithmic form as $\log_2\left(\frac{1}{256}\right) = -8$.

Quick Check 5 Write the exponential equation $7^x = 20$ in logarithmic form.

Example 7 Write the logarithmic equation $\ln x = 7$ in exponential form.

Solution

Recall that the base is e. In exponential form, this equation can be written as $e^7 = x$.

Quick Check 6 Write the logarithmic equation $\log_3 x = 6$ in exponential form.

Consider the function $f(x) = b^x$, where $b > 0$ and $b \neq 1$. To find the inverse of this function, we would replace $f(x)$ by y and then interchange x and y, which would produce the equation $x = b^y$. If we rewrite this equation in logarithmic form, we would obtain the equation $y = \log_b x$. Since this equation is solved for y, we have shown that $f^{-1}(x) = \log_b x$. The inverse of an exponential function is a logarithmic function.

Solving Logarithmic Equations

Objective 5 **Solve logarithmic equations.** When we are solving an equation that is in logarithmic form, it will often be helpful to rewrite the equation in its equivalent exponential form. We can then attempt to solve the resulting equation. For example, suppose that we were trying to solve the equation $\log_3 x = 2$. The exponential form of this equation is $3^2 = x$, which simplifies to $x = 9$.

Example 8 Solve $\log_5 x = 4$.

Solution

We begin by rewriting the equation in exponential form.

$$\log_5 x = 4$$
$$5^4 = x \quad \text{Rewrite in exponential form.}$$
$$625 = x \quad \text{Simplify.}$$

Quick Check 7 Solve $\log_4 x = 5$.

The solution set is {625}.

Example 9 Let $f(x) = \ln(x - 8)$. Solve $f(x) = 3$. Round to the nearest thousandth.

Solution

We begin by rewriting the equation in exponential form, keeping in mind that the base of this logarithm is e.

$$\ln(x - 8) = 3 \quad \text{Set the function equal to 3.}$$
$$e^3 = x - 8 \quad \text{Rewrite in exponential form.}$$
$$e^3 + 8 = x \quad \text{Add 8.}$$

The exact solution to the equation is $x = e^3 + 8$; so the solution set is $\{e^3 + 8\}$. Approximating this solution to the nearest thousandth by calculator, we get $x \approx 28.086$.

Quick Check 8 Let $f(x) = \log_3(x + 6)$. Solve $f(x) = 2$.

Graphing Logarithmic Functions

Objective 6 **Graph logarithmic functions.** We will graph logarithmic functions by plotting points. As with many other functions, the choice of which points to plot is crucial. We find this choice made easier by rewriting a logarithmic function in exponential form. Suppose that we wanted to graph the function $f(x) = \log_2 x$. Replacing $f(x)$ by y, we can rewrite the equation $y = \log_2 x$ in exponential form as $2^y = x$. We can now choose values for y and find the corresponding values of x. We will use the values $-2, -1, 0, 1$, and 2 for y. At this point, we notice that this is similar to the way we graphed exponential functions in the last section, with the exception that we are substituting values for y rather than for x.

$x = 2^y$	$2^{-2} = \frac{1}{4}$	$2^{-1} = \frac{1}{2}$	$2^0 = 1$	$2^1 = 2$	$2^2 = 4$
y	-2	-1	0	1	2

Shown next is the graph of $f(x) = \log_2 x$, which passes through the preceding points. Notice that the graph includes only positive values of x. Recall that the domain of a logarithmic function of the form $f(x) = \log_b x$ is the set of all positive real numbers, written in interval notation as $(0, \infty)$.

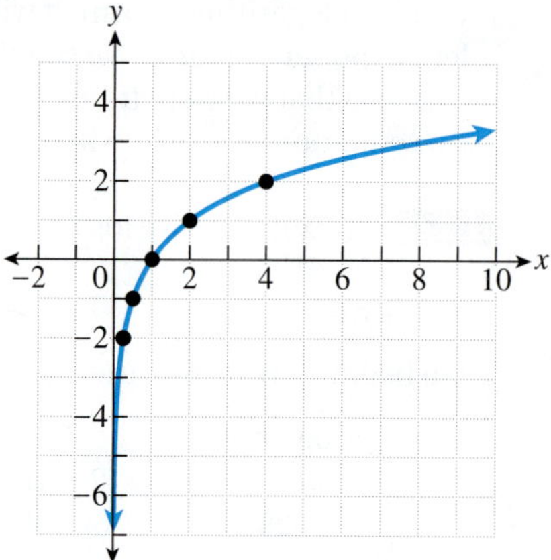

Also notice that as the values of x get closer and closer to 0, the graph of the function decreases without bound. The y-axis, whose equation is $x = 0$, is the **vertical asymptote** for the graph of this function.

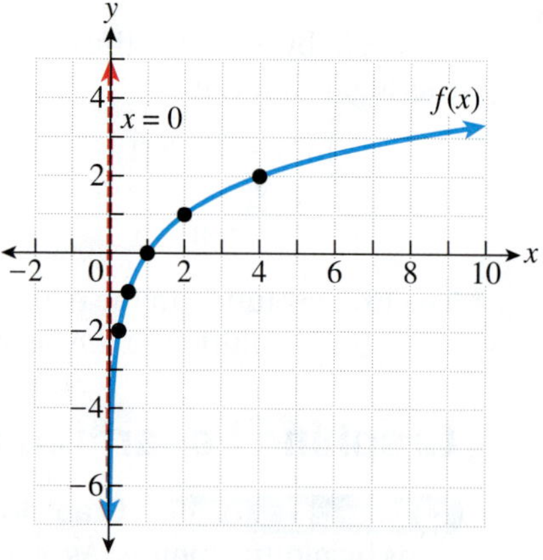

In general, the y-axis will be a vertical asymptote for a function of the form $f(x) = \log_b x$, since the domain of this function is the set of positive real numbers.

EXAMPLE 10 Graph $f(x) = \log_4 x$.

Solution

Replacing $f(x)$ by y, we see that the equation $y = \log_4 x$ can be rewritten in exponential form as $4^y = x$. Here is a table of values for this function:

$x = 4^y$	$4^{-2} = \frac{1}{16}$	$4^{-1} = \frac{1}{4}$	$4^0 = 1$	$4^1 = 4$	$4^2 = 16$
y	-2	-1	0	1	2

Following is the graph of $f(x) = \log_4 x$, which passes through those points, including the vertical asymptote at $x = 0$:

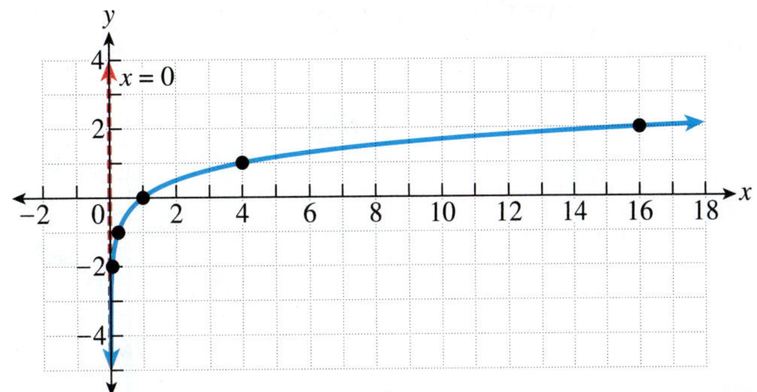

Note that as the values of x increase, the function $f(x) = \log_4 x$ does not increase nearly as quickly as the function $f(x) = \log_2 x$ does. What would we expect of the graph of the function $f(x) = \log x$ as x increases? It increases at a much slower rate than either of these two previous functions. In general, the larger the base of a logarithmic function, the slower the function increases as x increases. Here is a graph showing all three functions:

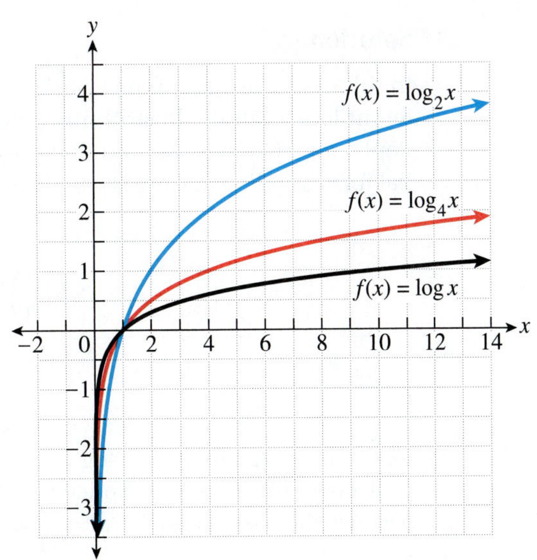

EXAMPLE 11 Graph $f(x) = \ln x$.

Solution

Replacing $f(x)$ by y, we see that the equation $y = \ln x$ can be rewritten in exponential form as $e^y = x$. Here is a table of values for this function, with values of x rounded to the nearest tenth:

$x = e^y$	$e^{-2} \approx 0.1$	$e^{-1} \approx 0.4$	$e^0 = 1$	$e^1 \approx 2.7$	$e^2 \approx 7.4$
y	-2	-1	0	1	2

Following is the graph of $f(x) = \ln x$, which passes through those points, including the vertical asymptote at $x = 0$.

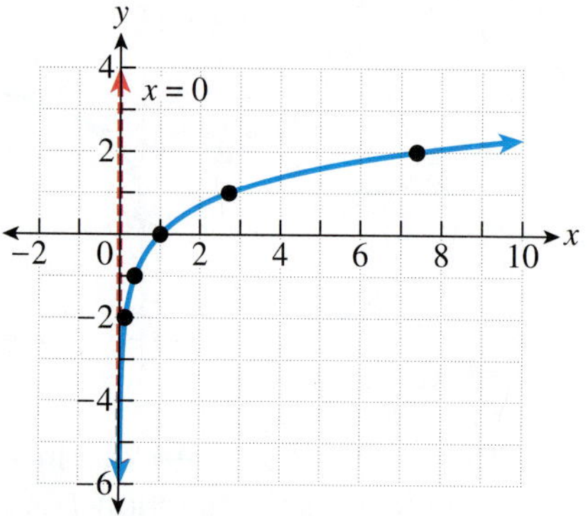

EXAMPLE 12 Graph $f(x) = \log_{1/3} x$.

Solution

In this case, the base of the logarithmic function is less than 1. This graph will have a different look than the logarithmic functions that were graphed in the previous examples. Replacing $f(x)$ by y, we find that the equation $y = \log_{1/3} x$ can be rewritten in exponential form as $\left(\frac{1}{3}\right)^y = x$. Here is a table of values for this function.

$x = \left(\frac{1}{3}\right)^y$	$\left(\frac{1}{3}\right)^{-2} = 9$	$\left(\frac{1}{3}\right)^{-1} = 3$	$\left(\frac{1}{3}\right)^{0} = 1$	$\left(\frac{1}{3}\right)^{1} = \frac{1}{3}$	$\left(\frac{1}{3}\right)^{2} = \frac{1}{9}$
y	-2	-1	0	1	2

Next is the graph of $f(x) = \log_{1/3} x$, which passes through those points, including the vertical asymptote at $x = 0$. Notice that the graph of this function decreases as x increases, because the base is less than 1.

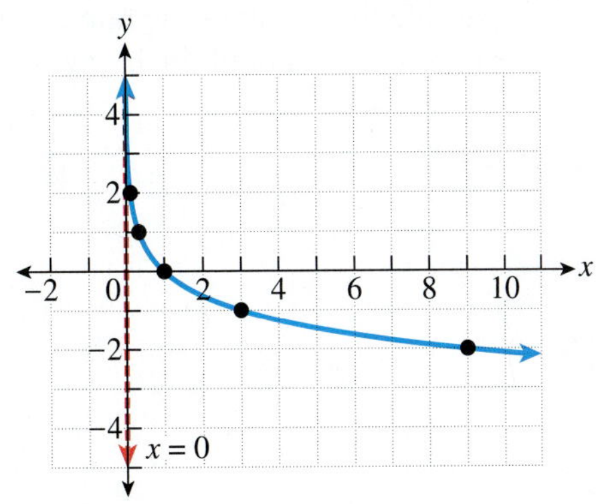

Quick Check 9 Graph.

a) $f(x) = \log_5 x$
b) $f(x) = \log_{1/4} x$

Consider the graphs of the functions $f(x) = 2^x$ and $g(x) = \log_2 x$, which have been graphed together on the same set of axes.

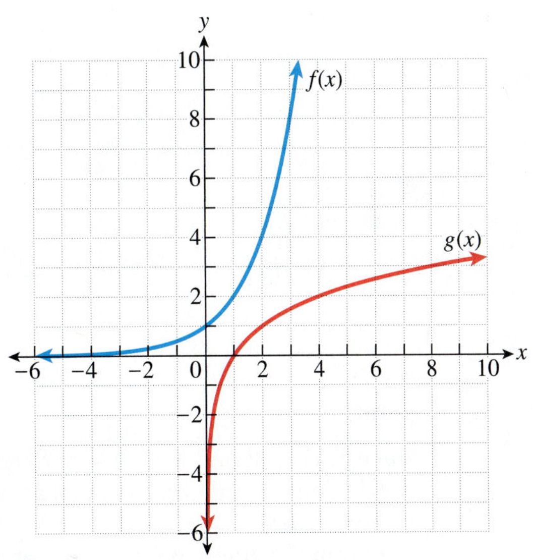

Adding the graph of the line $y = x$ allows us to see that the two functions are symmetric to each other about the line $y = x$ and, therefore, are inverses of each other.

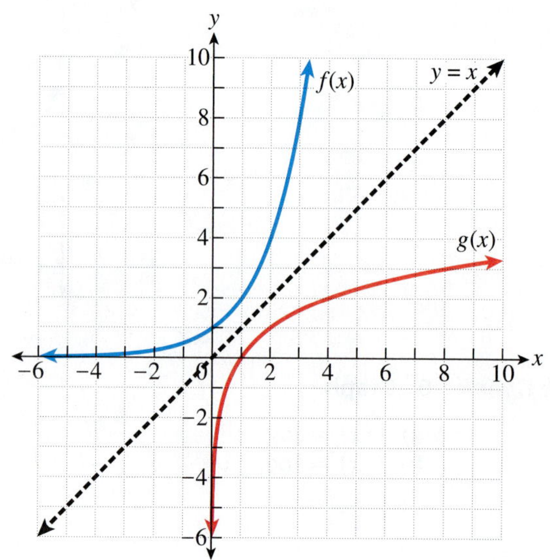

In general, for any positive real number b $(b \neq 1)$, the functions $f(x) = b^x$ and $g(x) = \log_b x$ are inverse functions. We will explore this idea more thoroughly in Section 8.8.

EXAMPLE 13 Use the given graph of $f(x) = 3^x$ to graph the function $g(x) = \log_3 x$.

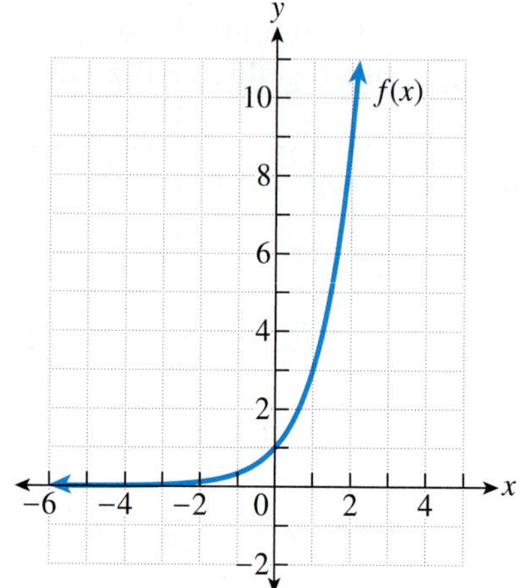

Solution

Since $f(x) = 3^x$ and $g(x) = \log_3 x$ are inverses of each other, we can sketch the graph of $g(x)$ by finding ordered pairs that are on the graph of $f(x)$ and interchanging their x- and y-coordinates.

From the graph, we can see that the points $(0, 1)$, $(1, 3)$, and $(2, 9)$ are on the graph of $f(x)$. This tells us that the points $(1, 0)$, $(3, 1)$, and $(9, 2)$ are on the graph of $g(x)$.

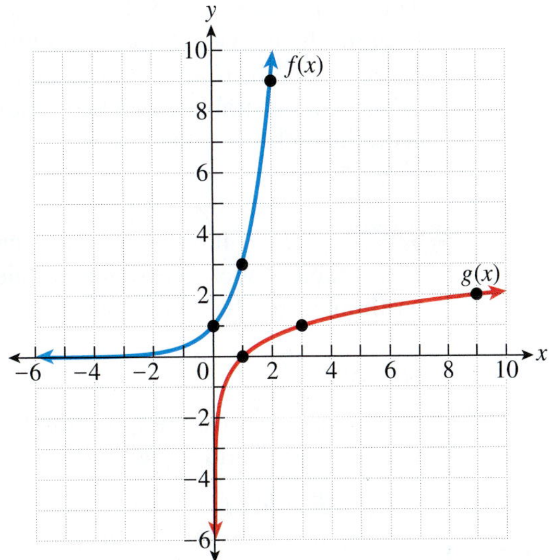

Shown are those points, along with the graph of $f(x)$. We can now sketch the graph of $g(x)$, making sure that its graph is symmetric around the line $y = x$ with the graph of the function $f(x)$.

Applications of Logarithmic Functions

Objective 7 **Use logarithmic functions in applications.**

EXAMPLE 14 The average height of boys, in inches, during their first year of life can be approximated by the function $f(x) = 1.95 \ln x + 23.53$, where x is the age of the boy in months. Use this function to estimate the average height of boys who are 6 months old. (*Source:* Centers for Disease Control (CDC))

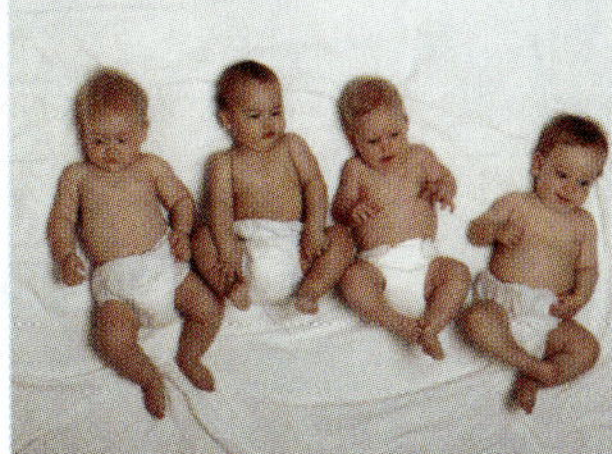

Solution

To estimate the average height of boys who are 6 months old, we need to evaluate the function when $x = 6$.

$$f(6) = 1.95 \ln 6 + 23.53 \quad \text{Substitute 6 for } x.$$
$$\approx 27.02 \quad \text{Approximate by calculator.}$$

The average height of 6-month-old boys is approximately 27.02 inches.

Quick Check 10 **The annual sales of a company, in billions of dollars, can be approximated by the function $f(x) = 7.95 \ln x + 20.95$, where x represents the number of years after 1995. Use the function to predict the annual sales of this company in the year 2012.**

The **Richter scale** is used to measure the magnitude or strength of an earthquake. The magnitude of an earthquake is a function of the size of the shock wave that it creates on a seismograph. If an earthquake creates a shock wave that is I times larger than the smallest measurable shock wave that is recordable on a seismograph, then its magnitude R on the Richter scale is given by the formula $R = \log I$.

EXAMPLE 15 If an earthquake creates a shock wave that is 500 times the smallest measurable shock wave recordable by a seismograph, find the magnitude of the earthquake on the Richter scale.

Solution

In this problem, we are given the fact that $I = 500$, so we substitute 500 for I in the formula $R = \log I$ and solve for R.

$R = \log I$

$R = \log 500$ Substitute 500 for I.

$R \approx 2.7$ Approximate on a calculator.

The magnitude of this earthquake on the Richter scale is 2.7.

***Quick Check* 11**

If an earthquake creates a shock wave that is 2500 times the smallest measurable shock wave that is recordable by a seismograph, find the magnitude of the earthquake on the Richter scale.

Building Your Study Strategy **Practice Quizzes, 4 Outline** Make your own study quiz for this section by selecting odd-numbered problems from the exercise set. Here is a suggested outline:

Number of Problems	Topic
4	Evaluate logarithms and logarithmic functions (including common logarithms and natural logarithms).
2	Rewrite an equation in exponential form.
2	Rewrite an equation in logarithmic form.
4	Solve logarithmic equations.
2	Graph logarithmic functions.
1	Applied problem

Be sure to choose problems of varying levels of difficulty. By choosing odd problems from the exercise set, you will have an answer key for your practice quiz in the back of the text.

Exercises 8.4

Vocabulary

1. For any positive real number b ($b \neq 1$) and any positive real number x, ________ is the exponent that b must be raised to in order to equal x.

2. For any positive real number b such that $b \neq 1$, a function of the form $f(x) = \log_b x$ is called a(n) ________ function.

3. The ________ logarithm is a logarithm whose base is 10, and is denoted by $\log x$.

4. The ________ logarithm is a logarithm whose base is e, and is denoted by $\ln x$.

5. An equation of the form $b^y = n$ is said to be in ________ form.

6. An equation of the form $\log_b n = y$ is said to be in ________ form.

7. The graph of a logarithmic function has a(n) ________ asymptote.

8. The ________ is used to measure the magnitude or strength of an earthquake.

Evaluate.

9. $\log_2 4$ 10. $\log_2 32$ 11. $\log_3 27$
12. $\log_3 81$ 13. $\log_4 64$ 14. $\log_5 625$
15. $\log_8 1$ 16. $\log_5 1$ 17. $\log_3(\frac{1}{9})$
18. $\log_2(\frac{1}{16})$ 19. $\log_8(\frac{1}{512})$ 20. $\log_6(\frac{1}{6})$

Evaluate the given function.

21. $f(x) = \log_3 x$ a) $f(9)$ b) $f(243)$
22. $f(x) = \log_5 x$ a) $f(25)$ b) $f(1)$
23. $f(x) = \log_2(x - 7)$ a) $f(8)$ b) $f(135)$
24. $f(x) = \log_3(2x + 11)$ a) $f(35)$ b) $f(116)$

Evaluate. Round to the nearest thousandth if necessary.

25. $\log 1000$ 26. $\log 10$
27. $\log 0.01$ 28. $\log(\frac{1}{10{,}000})$
29. $\log 112$ 30. $\log 47$
31. $\log 7.42$
32. $\log 0.0018$
33. $\ln e$ 34. $\ln(e^3)$
35. $\ln\left(\frac{1}{e^6}\right)$ 36. $\ln(e^{-2})$
37. $\ln 32$ 38. $\ln 105$
39. $\ln 16.23$ 40. $\ln 0.5$

Rewrite in logarithmic form.

41. $2^3 = 8$
42. $5^4 = 625$
43. $9^3 = 729$
44. $4^{-2} = \frac{1}{16}$
45. $6^x = 30$
46. $3^x = 17$
47. $10^x = 500$
48. $e^x = 7$

Rewrite in exponential form.

49. $\log_4 1024 = 5$
50. $\log_3 19{,}683 = 9$
51. $\log_{16} 8 = \frac{3}{4}$
52. $\log_{49} 7 = \frac{1}{2}$
53. $\log x = 3$
54. $\ln x = 5$
55. $\ln x - 3 = 4$
56. $\log(x + 2) - 10 = -5$

Solve. Round to the nearest thousandth if necessary.

57. $\log_2 x = 5$
58. $\log_2 x = -2$

59. $\log_3 x = -4$

60. $\log_3 x = 5$

61. $\log x = 4$

62. $\log x = 7$

63. $\ln x = 2$

64. $\ln x = -1$

65. $\log_3 x + 4 = 6$

66. $\log_3(x + 4) = 6$

67. $\ln(x - 5) + 13 = 11$

68. $\log(2x - 8) - 21 = -18$

69. Let $f(x) = \log_6 x$. Solve $f(x) = 3$.

70. Let $f(x) = \log_5(x + 12)$. Solve $f(x) = 2$.

71. Let $f(x) = \log(x - 10) - 3$. Solve $f(x) = -1$.

72. Let $f(x) = \ln(x - 5) + 17$. Solve $f(x) = 17$.

Solve.

73. $\log_5 125 = x$

74. $\log_2 256 = x$

75. $\log 100{,}000 = x$

76. $\log 0.001 = x$

77. $\log_9 27 = x$

78. $\log_4 32 = x$

Graph the given function $f(x)$. Label the vertical asymptote.

79. $f(x) = \log_3 x$

80. $f(x) = \log_6 x$

81. $f(x) = \log x$

82. $f(x) = \log_8 x$

83. $f(x) = \log_{1/5} x$

84. $f(x) = \log_{1/2} x$

Use the graph of the given exponential function $f(x)$ to graph its inverse logarithmic function $f^{-1}(x)$.

85.

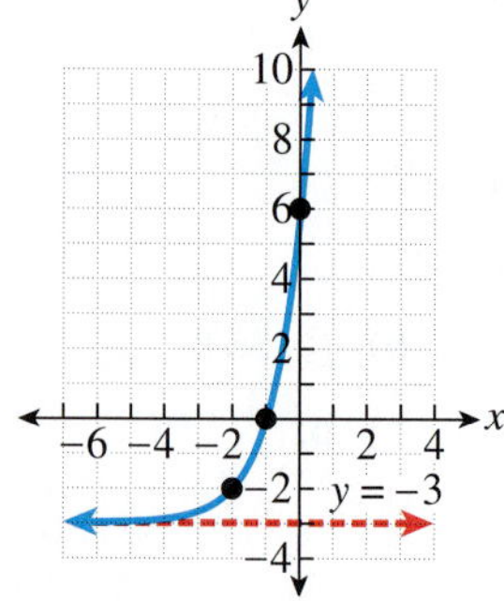

86.

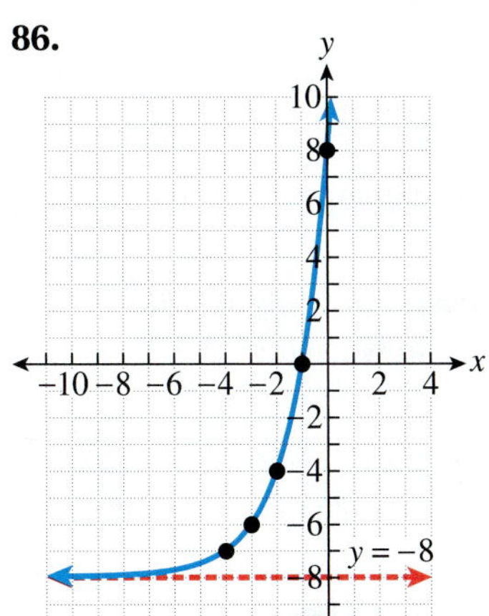

87.

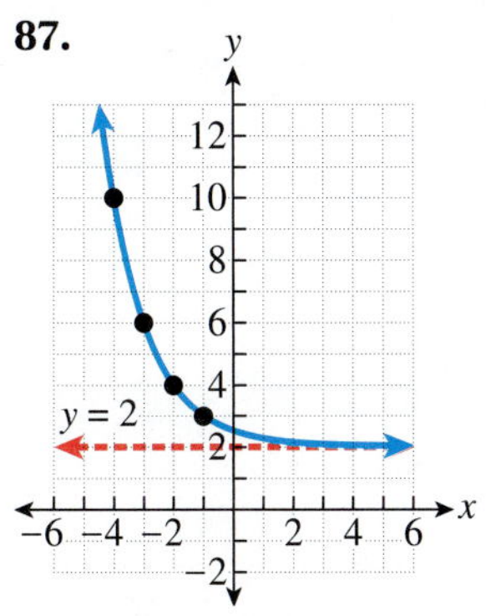

88.

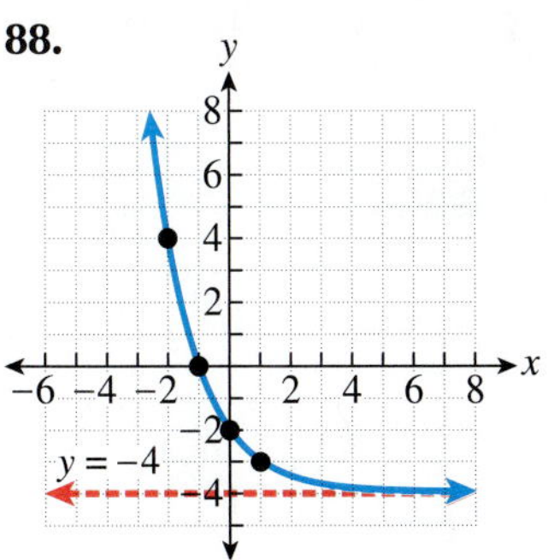

Use the table of values to determine the formula for the logarithmic function $f(x)$.

89.

x	$\frac{1}{4}$	$\frac{1}{2}$	1	2	4
$f(x)$	−2	−1	0	1	2

90.

x	$\frac{1}{36}$	$\frac{1}{6}$	1	6	36
$f(x)$	−2	−1	0	1	2

91.

x	9	3	1	$\frac{1}{3}$	$\frac{1}{9}$
$f(x)$	−2	−1	0	1	2

92.

x	$\frac{4}{49}$	$\frac{2}{7}$	1	$\frac{7}{2}$	$\frac{49}{4}$
$f(x)$	−2	−1	0	1	2

93. The number of U.S. airports in a particular year can be approximated by the function $f(x) = 14650 + 1344 \ln x$, where x represents the number of years after 1979. Use this function to predict the number of U.S. airports in the year 2030. (*Source:* Bureau of Transportation Statistics)

94. The percent of U.S. workers who drive themselves to work can be approximated by the function $f(x) = 72.5 + 2.01 \ln x$, where x represents the number of years after 1984. Use this function to predict what percent of U.S. workers will drive themselves to work in the year 2010. (*Source:* Bureau of Transportation Statistics)

95. The average height, in inches, of girls who are 3 years old or younger can be approximated by the function $f(x) = 18.6 + 4.78 \ln x$, where x represents the age in months. Use the function to determine the average height of girls who are 24 months old. (*Source:* CDC)

96. The average weight, in pounds, of boys who are 3 years old or younger can be approximated by the function $f(x) = 7.6 + 6.36 \ln x$, where x represents the age in months. Use the function to determine the average weight of boys who are 18 months old. (*Source:* CDC)

For Exercises 97–100, use the formula $R = \log I$, where R is the magnitude of an earthquake on the Richter scale whose shock wave is I times larger than the smallest measurable shock wave that is recordable on a seismograph.

97. If an earthquake creates a shock wave that is 16,000 times the smallest measurable shock wave that is recordable by a seismograph, find the magnitude of the earthquake on the Richter scale.

98. A 1950 earthquake along the India–China border created a shock wave that was 400,000,000 times the smallest measurable shock wave that is recordable by a seismograph. Find the magnitude of this earthquake on the Richter scale.

99. The 1906 earthquake off the coast of Ecuador had a magnitude of 8.8 on the Richter scale. How many times larger was the shock wave created by this earthquake than the smallest measurable shock wave?

100. The 1992 earthquake centered in Landers, California, had a magnitude of 7.3 on the Richter scale. How many times larger was the shock wave created by this earthquake than the smallest measurable shock wave?

Writing in Mathematics

Answer in complete sentences.

101. What is a logarithm? Explain clearly, using your own words.

102. Write a logarithmic equation whose solution is $x = 7$. Explain how to solve the equation.

103. Write a logarithmic equation whose solution is $x = \frac{3}{2}$. Explain how to solve the equation.

104. Explain why the graph of a logarithmic function has a vertical asymptote.

105. ***Solutions Manual**** Write a solutions manual page for the following problem:

Graph $f(x) = \log_4 x$. Label the vertical asymptote.

106. ***Newsletter**** Write a newsletter explaining how to convert equations from logarithmic form to exponential form and from exponential form to logarithmic form.

***See Appendix B for details and sample answers.**

Quick Review Exercises

Section 8.4

Simplify. Assume that all variables are nonzero real numbers.

1. $x^7 \cdot x^9$

2. $\dfrac{b^{16}}{b^4}$

3. $(n^8)^7$

4. z^0

8.5 Properties of Logarithms

Objectives

1. **Use the product rule for logarithms.**
2. **Use the quotient rule for logarithms.**
3. **Use the power rule for logarithms.**
4. **Use additional properties of logarithms.**
5. **Use properties to rewrite two or more logarithmic expressions as a single logarithmic expression.**
6. **Use properties to rewrite a single logarithmic expression as a sum or difference of logarithmic expressions whose arguments have an exponent of 1.**
7. **Use the change-of-base formula for logarithms.**

In this section, we will learn several properties of logarithms. These properties, in addition to being useful for simplifying logarithmic expressions, will also be helpful when we are solving logarithmic equations.

Product Rule for Logarithms

Objective 1 Use the product rule for logarithms. We know that $\log_2 8 = 3$, because $2^3 = 8$. We also know that $\log_2 16 = 4$, since $2^4 = 16$. So $\log_2 8 + \log_2 16 = 3 + 4$, or 7. To find this sum, we had to evaluate two logarithms. Consider $\log_2(8 \cdot 16)$, which is equal to $\log_2 128$. This logarithm is also equal to 7, since $2^7 = 128$. We see that the sum $\log_2 8 + \log_2 16$ is equal to the logarithm whose argument is the product of their two arguments, $\log_2(8 \cdot 16)$. This leads to the **product rule for logarithms**.

Product Rule for Logarithms

For any positive real number b $(b \neq 1)$ and any positive real numbers x and y, $\log_b x + \log_b y = \log_b(xy)$.

This rule tells us that we can rewrite the sum of two logarithms as a single logarithm, provided that the two logarithms have the same base. It also tells us that the logarithm of a product can be rewritten as a sum of logarithms. Here is a proof of the product rule for logarithms:

Let $M = \log_b x$ and $N = \log_b y$.
Rewriting $M = \log_b x$ in exponential form, we see that $b^M = x$. Similarly, $b^N = y$.
So $xy = b^M \cdot b^N = b^{M+N}$.
Rewriting $xy = b^{M+N}$ in logarithmic form, we see that $\log_b(xy) = M + N$.
But $M + N = \log_b x + \log_b y$, so $\log_b(xy) = \log_b x + \log_b y$.

EXAMPLE 1 Rewrite $\ln 7 + \ln x$ as a single logarithm. Assume x is a positive real number.

Solution

Both logarithms have the same base, so we will use the product rule to simplify this expression.

$$\ln 7 + \ln x = \ln(7x) \quad \text{Rewrite as the logarithm of a product.}$$

***Quick Check* 1** **Rewrite $\log 5 + \log 7 + \log x$ as a single logarithm. Assume x is a positive real number.**

EXAMPLE 2 Rewrite $\log(10xy)$ as the sum of two or more logarithms. Simplify if possible. Assume all variables are positive real numbers.

Solution

This expression is the logarithm of three factors, and we may use the product rule to rewrite it as the sum of three logarithms.

$$\begin{aligned}\log(10xy) &= \log 10 + \log x + \log y && \text{Rewrite as the sum of three logarithms.}\\ &= 1 + \log x + \log y && \text{Simplify } \log 10.\end{aligned}$$

***Quick Check* 2** **Rewrite $\log_2(5b)$ as the sum of two or more logarithms. Assume b is a positive real number.**

A Word of Caution The product rule for logarithms applies only to expressions of the form $\log_b x + \log_b y$, not to expressions such as $\log_b(x + y)$. In general,

$$\log_b(x + y) \neq \log_b x + \log_b y.$$

Quotient Rule for Logarithms

Objective 2 **Use the quotient rule for logarithms.** Just as the sum of two logarithms can be expressed as a single logarithm, so can the difference of two logarithms.

Quotient Rule for Logarithms

For any positive real number b $(b \neq 1)$ and any positive real numbers x and y, $\log_b x - \log_b y = \log_b\left(\frac{x}{y}\right)$.

This rule tells us that we can rewrite the difference of two logarithms as a single logarithm of a quotient, provided that the two logarithms have the same base. It also tells us that the logarithm of a quotient can be rewritten as a difference of logarithms. Using the fact that $\frac{b^M}{b^N} = b^{M-N}$, we find that the proof of the quotient rule for logarithms is similar to the proof of the product rule for logarithms. The proof is left as an exercise for the reader.

EXAMPLE 3 Rewrite $\log_3 36 - \log_3 4$ as a single logarithm and simplify if possible.

Solution

Both logarithms have the same base, so we will use the quotient rule to simplify this expression.

$$\log_3 36 - \log_3 4 = \log_3\left(\frac{36}{4}\right) \quad \text{Rewrite as the logarithm of a quotient.}$$
$$= \log_3 9 \quad \text{Divide.}$$
$$= 2 \quad \text{Simplify the logarithm.}$$

***Quick Check* 3**
Rewrite $\log_2 60 - \log_2 15$ as a single logarithm. Simplify if possible.

EXAMPLE 4 Rewrite $\ln\left(\frac{12x}{y}\right)$ in terms of two or more logarithms. Assume all variables are positive real numbers.

Solution

We begin by applying the quotient rule.

$$\ln\left(\frac{12x}{y}\right) = \ln(12x) - \ln y \quad \text{Rewrite as the logarithm of the numerator minus the logarithm of the denominator.}$$
$$= \ln 12 + \ln x - \ln y \quad \text{Apply the product rule to } \ln(12x).$$

***Quick Check* 4**
Rewrite $\log_b\left(\frac{x}{b}\right)$ in terms of two logarithms. Simplify if possible. Assume x and b are positive real numbers.

A Word of Caution The quotient rule for logarithms applies only to expressions of the form $\log_b x - \log_b y$, not to expressions such as $\log_b(x - y)$. In general,

$$\log_b(x - y) \neq \log_b x - \log_b y.$$

Power Rule for Logarithms

Objective 3 Use the power rule for logarithms. The properties of logarithms discussed in this section will help us to solve exponential equations, and one of the most important properties is the power rule for logarithms.

Power Rule for Logarithms

For any positive real number b ($b \neq 1$), any positive real number x, and any real number r,

$$\log_b x^r = r \cdot \log_b x.$$

This rule tells us that we can rewrite the logarithm of a positive number that is raised to a power by multiplying the exponent by the logarithm of the number. Here is a proof of the power rule, which uses the product rule for logarithms.

$$\log_b x^r = \log_b(\underbrace{x \cdot x \cdot \cdots \cdot x}_{r\text{ times}}) \quad \text{Rewrite } x^r \text{ by listing } x \text{ as a factor } r \text{ times.}$$
$$= \underbrace{\log_b x + \log_b x + \cdots + \log_b x}_{r\text{ times}} \quad \text{Use the product rule for logarithms.}$$
$$= r \cdot \log_b x \quad \text{Rewrite the repeated addition of } \log_b x \text{ as } r \cdot \log_b x.$$

EXAMPLE 5 Rewrite $\log_2 x^7$ as a logarithm of a single factor without exponents. Assume x is a positive real number.

Solution

The power rule tells us that we may rewrite this expression by removing 7 as an exponent and multiplying it by $\log_2 x$. In other words, we are moving the number 7 from the exponent of x and placing it as a factor in front of $\log_2 x$.

Quick Check 5
Rewrite $\log_3(9x^2y^4)$, using logarithms of single factors that do not have exponents. Simplify if possible. Assume x and y are positive real numbers.

$\log_2 x^{⑦}$

$$\log_2 x^7 = 7\log_2 x \qquad \text{Use the power rule.}$$

Recall that $\sqrt[n]{x} = x^{1/n}$. So, we can apply the power rule when the argument of a logarithm is an expression involving radicals. For example, $\log_9 \sqrt[4]{x} = \log_9 x^{1/4} = \frac{1}{4}\log_9 x$.

Quick Check 6 **Rewrite $\ln \sqrt[7]{x}$ as a logarithm of a single factor without any radicals. Assume x is a positive real number.**

EXAMPLE 6 Rewrite $15 \ln x$ in such a way that the logarithm is not being multiplied by a number. Assume x is a positive real number.

Solution

We will use the power rule to rewrite 15 as an exponent of x.

Quick Check 7
Rewrite $8\log_8 x$ in such a way that the logarithm is not being multiplied by a number. Assume x is a positive real number.

$$15\ln x = \ln x^{15} \qquad \text{Use the power rule.}$$

A Word of Caution Be careful when working with logarithmic expressions involving exponents. In the expression $\log_b x^n$, only the number x is being raised to the *nth* power. However, in the expression $(\log_b x)^n$, it is the logarithm $\log_b x$ that is being raised to the *nth* power. We cannot apply the power rule for logarithms to the expression $(\log_b x)^n$. In general,

$$\log_b x^n \neq (\log_b x)^n.$$

For example, $\log_2 4^3 \neq (\log_2 4)^3$.

Objective 4 **Use additional properties of logarithms.** The power rule leads us to another property of logarithms. Consider the expression $\log_5 5^7$. The power rule tells us that $\log_5 5^7 = 7\log_5 5$. Since $\log_5 5$ is equal to 1, this expression simplifies to equal $7 \cdot 1$, or 7. This result can be generalized as follows:

$\log_b b^r$

For any positive real number b $(b \neq 1)$ and any real number r, $\log_b b^r = r$.

This property will be particularly helpful when we are taking the natural logarithm of e raised to a power.

EXAMPLE 7 Simplify $\ln e^x$.

Solution

The base of this logarithm is e, so the property $\log_b b^r = r$ may be applied.

$$\ln e^x = x$$

***Quick Check* 8**
Simplify $\ln e^{-4}$.

Another property of logarithms involves raising a positive number to a power containing a logarithm whose base is that number. This property follows from the fact that exponential and logarithmic functions are inverses.

$b^{\log_b x}$

For any positive real number b $(b \neq 1)$ and any positive real number x,

$$b^{\log_b x} = x.$$

EXAMPLE 8 Simplify $8^{\log_8 15}$.

Solution

By the previous property, $8^{\log_8 15} = 15$.

***Quick Check* 9**
Simplify $e^{\ln 16}$.

Rewriting Logarithmic Expressions as a Single Logarithm

Objective 5 Use properties to rewrite two or more logarithmic expressions as a single logarithmic expression. We will often need to rewrite an expression containing two or more logarithms in terms of a single logarithm. To do this, we will be using the power rule first if possible, and then we will use the quotient rule and the product rule.

EXAMPLE 9 Rewrite $\log a + \log b - \log c - \log d$ as a single logarithm. Assume a, b, c, and d are positive real numbers.

Solution

The first two logarithms can be combined by the product rule, as can the last two logarithms. We can then combine these logarithms by using the quotient rule.

$$\log a + \log b - \log c - \log d$$

$$= \log a + \log b - (\log c + \log d)$$ Factor -1 from the logarithms being subtracted.

$$= \log(ab) - \log(cd)$$ Use the product rule.

$$= \log\left(\frac{ab}{cd}\right)$$ Use the quotient rule.

Quick Check 10 **Rewrite $\log_2 a - \log_2 b + \log_2 c + \log_2 d$ as a single logarithm. Assume $a, b, c,$ and d are positive real numbers.**

EXAMPLE 10 Rewrite $\frac{1}{4}\ln a - 10\ln b - 2\ln c$ as a single logarithm. Assume a, b, and c are positive real numbers.

Solution

We will begin by using the power rule to rewrite each logarithm. We can then use the quotient rule.

$$\frac{1}{4}\ln a - 10\ln b - 2\ln c = \ln a^{1/4} - \ln b^{10} - \ln c^2$$ Use the power rule.

$$= \ln a^{1/4} - (\ln b^{10} + \ln c^2)$$ Factor -1 from the last two terms.

$$= \ln\left(\frac{a^{1/4}}{b^{10}c^2}\right)$$ Use the quotient rule. Since $\ln b^{10}$ and $\ln c^2$ are both being subtracted, b^{10} and c^2 are written in the denominator.

$$= \ln\left(\frac{\sqrt[4]{a}}{b^{10}c^2}\right)$$ Rewrite the numerator using radical notation.

Quick Check 11 **Rewrite $6\log a + \frac{1}{5}\log b - 9\log c - 12\log d$ as a single logarithm. Assume $a, b, c,$ and d are positive real numbers.**

Rewriting a Logarithmic Expression as a Sum or Difference of Logarithms Whose Arguments Have an Exponent of 1

Objective 6 **Use properties to rewrite a single logarithmic expression as a sum or difference of logarithmic expressions whose arguments have an exponent of 1.** We now turn our attention to expanding a single logarithm into the sum or difference of separate logarithms. This will be done in such a way that each logarithm is of a single factor that does not have an exponent. To do this, we will be using the product rule, the quotient rule, and the power rule.

EXAMPLE 11 Rewrite $\log\left(\frac{a^5b^8}{c^7}\right)$ in terms of two or more logarithms. Each argument should contain a single factor, and each exponent should be written as a factor. Assume a, b, and c represent positive real numbers.

Solution

We will begin by applying the product and quotient rules to rewrite this logarithm in terms of three logarithms. The power rule can then be applied to each of these logarithms.

$$\log\left(\frac{a^5b^8}{c^7}\right) = \log a^5 + \log b^8 - \log c^7 \quad \text{Use the product and quotient rules.}$$
$$= 5\log a + 8\log b - 7\log c \quad \text{Use the power rule.}$$

Quick Check 12
Rewrite $\log_5\left(\frac{z^3}{x^9y^4}\right)$ in terms of two or more logarithms. Each argument should contain a single factor, and each exponent should be written as a factor. Assume x, y, and z represent positive real numbers.

Change-of-Base Formula

Objective 7 **Use the change-of-base formula for logarithms.** The final property of logarithms in this section is the change-of-base formula.

Change-of-Base Formula

For any positive numbers a, b, and x $(a, b \neq 1)$, $\log_b x = \frac{\log_a x}{\log_a b}$.

This formula allows us to rewrite any logarithm in terms of two logarithms that have a different base than the original logarithm. The advantage of this is that we can rewrite logarithms in terms of either common logarithms or natural logarithms, allowing us to approximate these logarithms by using a calculator. For instance, suppose that we wanted to evaluate $\log_2 7$. Since we do not know what power of 2 is equal to 7, we would not be able to evaluate this logarithm. We can rewrite this logarithm as $\frac{\log 7}{\log 2}$, or as $\frac{\ln 7}{\ln 2}$, using the change-of-base formula. Either of these two expressions can be evaluated on a calculator. Rounding to three decimal places, we see that $\log_2 7 \approx 2.807$.

EXAMPLE 12 Evaluate $\log_5 13$, using the change-of-base formula.

Solution

Using common logarithms, we can rewrite this logarithm as $\frac{\log 13}{\log 5}$.

$$\log_5 13 = \frac{\log 13}{\log 5} \quad \text{Use the change-of-base formula.}$$
$$\approx 1.594 \quad \text{Approximate by calculator.}$$

Quick Check 13
Evaluate $\log_3 125$, using the change-of-base formula.

We can check this result by raising 5 to the 1.594 power. Notice that the result is approximately equal to 13. Also notice that if we had used natural logarithms instead of common logarithms, $\frac{\ln 13}{\ln 5} \approx 1.594$ as well.

Using Your Calculator We can evaluate logarithms such as $\log_5 13$ on the TI-84, using the change-of-base formula, which tells us that $\log_5 13 = \dfrac{\log 13}{\log 5}$. This can be entered as follows:

```
log(13)/log(5)
        1.593692641
```

Building Your Study Strategy **Practice Quizzes, 5 Make Your Own** Make your own practice quiz for the first five sections of this chapter by selecting odd-numbered problems from the exercise sets of the first five sections. Select four or five odd problems from each section that you feel are representative of the types of problems that you would be expected to solve on an exam. Be sure to choose problems of varying levels of difficulty. By choosing odd problems from the exercise sets, you will have an answer key for your practice quiz in the back of the text.

EXERCISES 8.5

Vocabulary

1. The product rule for logarithms states that for any positive real number b ($b \neq 1$) and any positive real numbers x and y, $\log_b x + \log_b y =$ ________.

2. The quotient rule for logarithms states that for any positive real number b ($b \neq 1$) and any positive real numbers x and y, $\log_b x - \log_b y =$ ________.

3. The power rule for logarithms states that for any positive real number b ($b \neq 1$), any positive real number x, and any real number r, $\log_b x^r =$ ________.

4. State the change-of-base formula.

Rewrite as a single logarithm, using the product rule for logarithms. Simplify if possible. Assume all variables represent positive real numbers.

5. $\log_2 5 + \log_2 x$

6. $\log_3 8 + \log_3 a$

7. $\log_4 b + \log_4 c$

8. $\log 9 + \log x^3$

9. $\log_4 8 + \log_4 32$

10. $\log_{21} 9 + \log_{21} 49$

Rewrite as the sum of two or more logarithms, using the product rule for logarithms. Simplify if possible. Assume all variables represent positive real numbers.

11. $\log_3(10x)$

12. $\log_6(8b)$

13. $\ln(3e)$

14. $\log_5(25z)$

15. $\log_4(abc)$

16. $\log_2(16mn)$

Rewrite as a single logarithm, using the quotient rule for logarithms. Simplify if possible. Assume all variables represent positive real numbers.

17. $\log_5 42 - \log_5 7$

18. $\log_4 30 - \log_4 150$

19. $\log 39 - \log x$

20. $\log_7 a - \log_7 9$

21. $\log_4 448 - \log_4 7$

22. $\log_3 10 - \log_3 90$

Rewrite in terms of two or more logarithms, using the quotient and product rules for logarithms. Simplify if possible. Assume all variables represent positive real numbers.

23. $\log_8\left(\frac{10a}{b}\right)$

24. $\log_5\left(\frac{7x}{3y}\right)$

25. $\log_4\left(\frac{6xyz}{w}\right)$

26. $\log_7\left(\frac{1}{30}\right)$

27. $\log_5\left(\frac{xy}{25z}\right)$

28. $\log_2\left(\frac{8a}{bc}\right)$

Rewrite, using the power and product rules. Simplify if possible. Assume all variables represent positive real numbers.

29. $\log a^{10}$

30. $\log_2 b^3$

31. $\log_7 x^2y^3$

32. $\ln a^9b^3$

33. $\log_3 \sqrt{x}$

34. $\log_4 \sqrt[6]{b}$

35. $\log_2 16x^5y^7$

36. $\log_5 25a^6\sqrt[3]{b}$

Rewrite, using the power rule. Assume all variables represent positive real numbers.

37. $8 \log_5 x$

38. $3 \log_4 b$

39. $\frac{1}{3} \ln x$

40. $\frac{1}{2} \log_2 x$

41. $5 \log_3 x^2$

42. $4 \log b^6$

Simplify.

43. $\log_5 5^{12}$

44. $\log_2 2^{21}$

45. $\ln e^9$

46. $\log 10^{25}$

47. $2^{\log_2 7}$

48. $5^{\log_5 36}$

49. $3^{\log_3 b}$

50. $e^{\ln 0.026}$

Rewrite as a single logarithm. Assume all variables represent positive real numbers.

51. $\log_3 2 + \log_3 5 - \log_3 7$

52. $\log_2 7 - \log_2 10 - \log_2 9$

53. $3 \log_5 2 + 6 \log_5 3$

54. $3 \log 9 - 2 \log 11$

55. $\frac{1}{2} \ln 16 - \frac{1}{3} \ln 125$

56. $\frac{1}{2} \log_6 144 + 4 \log_6 5$

57. $\log_2 x - \log_2 y - \log_2 z - \log_2 w$

58. $5 \log_3 a - 7 \log_3 b + 2 \log_3 c$

59. $4 \ln x + 9 \ln y - 13 \ln z$

60. $10 \log_5 x - 3 \log_5 y - 6 \log_5 z$

61. $\log(x + 2) + \log(x - 6)$

62. $\log_4(x + 4) + \log_4(x - 4)$

63. $\log_3(2x + 7) + \log_3(x + 5)$

64. $\log_7(x^2 + 5x - 24) - \log_7(x + 8)$

65. $3 \ln a + \frac{1}{3} \ln b - 10 \ln c$

66. $\frac{1}{5} \ln x - 7 \ln y - 6 \ln z$

Suppose for some base $b > 0$ $(b \neq 1)$ that $\log_b 2 = A$, $\log_b 3 = B$, $\log_b 5 = C$, and $\log_b 7 = D$. Express the given logarithms in terms of A, B, C, or D.

(Hint: Rewrite the logarithm in terms of $\log_b 2$, $\log_b 3$, $\log_b 5$, or $\log_b 7$.)

67. $\log_b 6$ **68.** $\log_b 15$

69. $\log_b 35$ **70.** $\log_b 14$

71. $\log_b 8$ **72.** $\log_b 9$ **73.** $\log_b 125$

74. $\log_b\left(\frac{1}{49}\right)$

Expand. Simplify if possible. Assume all variables represent positive real numbers.

75. $\log_8\left(\frac{a}{bcd}\right)$

76. $\log_9\left(\frac{81xy}{zw}\right)$

77. $\log_2 32x^4y^8z^5$

78. $\log_3\left(\frac{1}{27x^9y^{20}}\right)$

79. $\ln\left(\frac{a^6\sqrt{b}}{c^7}\right)$

80. $\ln\left(\frac{\sqrt[7]{x}}{y^{10}z^3}\right)$

81. $-\ln\left(\frac{a^3d^2}{bc^7}\right)$

82. $-2 \log\left(\frac{xy^2}{z^3}\right)$

Evaluate, using the change-of-base formula. Round to the nearest thousandth.

83. $\log_2 7$ **84.** $\log_{11} 3000$

85. $\log_{12} 47$ **86.** $\log_9 250$

87. $\log_7 300{,}000$ **88.** $\log_{15} 650{,}000$

89. $\log_8 3$ **90.** $\log_{20} 5$

91. $\log_9 0.0354$ **92.** $\log_6 0.195$

Find the missing number. Round to the nearest thousandth. (Hint: Rewrite in logarithmic form and use the change-of-base formula.)

93. $5^? = 19$ **94.** $3^? = 30$

95. $15^? = 100$ **96.** $2^? = 725$

Writing in Mathematics

Answer in complete sentences.

97. Give examples showing that the following statements are false.

- $\log_b(x + y) = \log_b x + \log_b y$
- $\log_b(x - y) = \log_b x - \log_b y$
- $\log_b x^n = (\log_b x)^n$

98. ***Solutions Manual**** Write a solutions manual page for the following problem:

Expand $\log_3\left(\frac{9a^5}{c^3d^4}\right)$. *Simplify if possible.*

Assume that all variables represent positive real numbers.

99. ***Newsletter**** Write a newsletter explaining how to use the change-of-base formula to evaluate logarithms.

***See Appendix B for details and sample answers.**

8.6 Exponential and Logarithmic Equations

Objectives

1. **Solve an exponential equation in which both sides have the same base.**
2. **Solve an exponential equation by using logarithms.**
3. **Solve a logarithmic equation in which both sides have logarithms with the same base.**
4. **Solve a logarithmic equation by converting it to exponential form.**
5. **Use properties of logarithms to solve a logarithmic equation.**
6. **Find the inverse function for exponential and logarithmic functions.**

In this section, we will be solving exponential and logarithmic equations. An exponential equation is an equation containing a variable in an exponent. A logarithmic equation is an equation involving logarithms of variable expressions.

Solving Exponential Equations

Objective 1 Solve an exponential equation in which both sides have the same base. Given an exponential equation to solve, we will first try to express both sides of the equation in terms of the same base. If we can do this, we can then solve the equation with the techniques developed in Section 8.3. Here are two examples reviewing this process.

EXAMPLE 1 Solve $2^{x+5} = 8$.

Solution

We will begin by rewriting 8 as 2^3, as both sides of the equation will then have the same base. We can then apply the one-to-one property of exponential functions.

$$
\begin{aligned}
2^{x+5} &= 8 & \\
2^{x+5} &= 2^3 & \text{Rewrite 8 as } 2^3. \\
x + 5 &= 3 & \text{Apply the one-to-one property of exponential functions.} \\
& & \text{If } b^r = b^s, \text{ then } r = s. \\
x &= -2 & \text{Subtract 5.}
\end{aligned}
$$

The solution of the equation is $x = -2$. We could check this solution by substituting -2 for x in the original equation. This check is left to the reader. The solution set is $\{-2\}$.

EXAMPLE 2 Solve $9^{4x-1} = 27^{2x}$.

Solution

Although we cannot write 27 as a power of 9, we can rewrite both bases as powers of 3.

$$
\begin{aligned}
9^{4x-1} &= 27^{2x} & \\
(3^2)^{4x-1} &= (3^3)^{2x} & \text{Rewrite 9 as } 3^2 \text{ and 27 as } 3^3. \\
3^{8x-2} &= 3^{6x} & \text{Simplify the exponents.} \\
8x - 2 &= 6x & \text{If } b^r = b^s, \text{ then } r = s. \\
-2 &= -2x & \text{Subtract } 8x. \\
1 &= x & \text{Divide both sides by } -2.
\end{aligned}
$$

The solution set is $\{1\}$.

Quick Check 1

Solve.

a) $3^{3x-2} = 81$
b) $4^{3x-1} = 32^{x+2}$

Solving Exponential Equations by Using Logarithms

Objective 2 Solve an exponential equation by using logarithms. Often, it is not possible to rewrite both sides of an exponential equation in terms of the same base, such as in the equation $3^x = 12$. We cannot rewrite 12 as a power of 3. In such a case we will apply the one-to-one property of logarithmic functions. We can see that logarithmic functions are one-to-one functions by applying the horizontal line test to the graph of any logarithmic function.

One-to-One Property of Logarithmic Functions

For any positive real number b ($b \neq 1$) and any real numbers r and s, if $r = s$, then $\log_b r = \log_b s$. In addition, if $\log_b r = \log_b s$, then $r = s$.

This property tells us that if two positive numbers are equal, then the logarithms of those numbers are equal as well, as long as both logarithms have the same base. When solving an exponential equation like $3^x = 12$, we will first take the common logarithm (base 10) of both sides of the equation. Then we will use the power rule for logarithms to help us solve the resulting equation. We will use the common logarithm because of its availability on our calculators, although using the natural logarithm (base e) would be a good choice as well.

EXAMPLE 3 Solve $3^x = 12$. Find the exact solution and then approximate the solution to the nearest thousandth.

Solution

We begin by taking the common logarithm of each side of the equation.

$$3^x = 12$$

$$\log 3^x = \log 12 \quad \text{Take the common logarithm of both sides.}$$

$$x \cdot \log 3 = \log 12 \quad \text{Use the power rule for logarithms. This moves the variable from the exponent.}$$

$$\frac{x \cdot \cancel{\log 3}^{1}}{\cancel{\log 3}_{1}} = \frac{\log 12}{\log 3} \quad \text{Divide both sides by log 3.}$$

$$x = \frac{\log 12}{\log 3} \quad \text{This is the exact solution.}$$

$$x \approx 2.262 \quad \text{Approximate, using a calculator.}$$

The solution set is $\left\{\frac{\log 12}{\log 3}\right\}$. The exact solution is $x = \frac{\log 12}{\log 3}$, which is approximately equal to 2.262. Substituting this value back into the original equation, we see that $3^{2.262} \approx 12.00185$, which is approximately equal to 12.

Quick Check 2
Solve $4^{x-5} - 9 = 3$.

EXAMPLE 4 Let $f(x) = 13^{x+8} - 22$. Solve the equation $f(x) = 108$. Find the exact solution and then approximate the solution to the nearest thousandth.

Solution

We begin by setting the function equal to 108.

$$\begin{aligned} f(x) &= 108 && \text{Set } f(x) \text{ equal to 108.} \\ 13^{x+8} - 22 &= 108 && \text{Replace } f(x) \text{ by } 13^{x+8} - 22. \\ 13^{x+8} &= 130 && \text{Add 22.} \\ \log 13^{x+8} &= \log 130 && \text{Take the common logarithm of both sides.} \\ (x+8)\cdot \log 13 &= \log 130 && \text{Use the power rule for logarithms.} \\ \frac{(x+8)\cdot \overset{1}{\cancel{\log 13}}}{\underset{1}{\cancel{\log 13}}} &= \frac{\log 130}{\log 13} && \text{Divide both sides by log 13.} \\ x + 8 &= \frac{\log 130}{\log 13} && \text{Simplify.} \\ x &= \frac{\log 130}{\log 13} - 8 && \text{Subtract 8. This is the exact solution.} \\ x &\approx -6.102 && \text{Approximate, using a calculator.} \end{aligned}$$

The solution set is $\left\{\frac{\log 130}{\log 13} - 8\right\}$. The exact solution is $x = \frac{\log 130}{\log 13} - 8$, which is approximately equal to -6.102.

***Quick Check* 3**

Let $f(x) = 9^{x+3} - 17$. Solve the equation $f(x) = -6$. Find the exact solution and then approximate the solution to the nearest thousandth.

If the exponential expression in an equation has e as its base, then it is a good idea to use the natural logarithm rather than the common logarithm. This is because $\ln e^x = x$ for any real number x.

EXAMPLE 5 Solve $2 \cdot e^{x+4} + 19 = 107$. Find the exact solution and then approximate the solution to the nearest thousandth.

Solution

Since the base is e, we will use the natural logarithm rather than the common logarithm. We begin by isolating the exponential expression.

$$\begin{aligned} 2 \cdot e^{x+4} + 19 &= 107 \\ 2 \cdot e^{x+4} &= 88 && \text{Subtract 19.} \\ e^{x+4} &= 44 && \text{Divide both sides by 2, isolating the exponential expression.} \\ \ln e^{x+4} &= \ln 44 && \text{Take the natural logarithm of both sides.} \\ x + 4 &= \ln 44 && \text{Simplify the left side of the equation.} \\ x &= \ln 44 - 4 && \text{Subtract 4. This is the exact solution. Do not subtract 4 from 44, as we must take the natural logarithm of 44 before subtracting.} \\ x &\approx -0.216 && \text{Approximate, using a calculator.} \end{aligned}$$

The solution set is $\{\ln 44 - 4\}$. The exact solution is $x = \ln 44 - 4$, which is approximately equal to -0.216.

***Quick Check* 4**

Solve $e^{x+2} - 10 = 18$. Round to the nearest thousandth.

Solving Logarithmic Equations

Objective 3 Solve a logarithmic equation in which both sides have logarithms with the same base. Suppose that we have a logarithmic equation in which two logarithms with the same base are equal to each other, such as $\log_5 x = \log_5 2$.

The one-to-one property of logarithmic functions can be applied to this situation, which states that for any positive real number b $(b \neq 1)$ and any real numbers r and s, if $\log_b r = \log_b s$, then $r = s$. This means that for the equation $\log_5 x = \log_5 2$, x must be equal to 2.

When solving logarithmic equations, we must check our solutions for extraneous solutions. Recall that the domain for a logarithmic function is the set of positive real numbers. Any tentative solution that would require us to take the logarithm of a number that is not positive must be omitted from the solution set.

EXAMPLE 6 Solve $\ln(x^2 - 2x - 19) = \ln 5$.

Solution

Since both logarithms have the same base, we may use the one-to-one property of logarithmic functions to solve the equation.

$$\ln(x^2 - 2x - 19) = \ln 5$$

$$x^2 - 2x - 19 = 5$$ Use the one-to-one property of logarithmic functions. The resulting equation is quadratic.

$$x^2 - 2x - 24 = 0$$ Subtract 5.

$$(x - 6)(x + 4) = 0$$ Factor.

$$x = 6 \quad \text{or} \quad x = -4$$ Set each factor equal to 0 and solve.

We must check each solution in the original equation. Although one of these solutions is negative, it does not mean that it is an extraneous solution. We must determine whether a solution requires us to take a logarithm of a nonpositive number in the original equation.

Check

$x = 6$	$x = -4$
$\ln((6)^2 - 2(6) - 19) = \ln 5$	$\ln((-4)^2 - 2(-4) - 19) = \ln 5$
$\ln 5 = \ln 5$	$\ln 5 = \ln 5$

Both solutions check. The solution set is $\{6, -4\}$.

Quick Check **5** **Solve $\log(x^2 + 15x - 30) = \log 2x$.**

EXAMPLE 7 Solve $\log(4x - 20) = \log(x - 14)$.

Solution

Since both logarithms have the same base, we may use the one-to-one property of logarithmic functions to solve the equation.

$$\log(4x - 20) = \log(x - 14)$$

$$4x - 20 = x - 14$$ Use the one-to-one property of logarithmic functions.

$$3x - 20 = -14$$ Subtract x.

$$3x = 6$$ Add 20.

$$x = 2$$ Divide both sides by 3.

We must check this tentative solution in the original equation.

Check $(x = 2)$

$$\log(4(2) - 20) = \log((2) - 14) \quad \text{Substitute 2 for } x.$$
$$\log(-12) = \log(-12) \quad \text{Simplify.}$$

Since we cannot take the logarithm of a negative number, we must omit the solution $x = 2$. Since this was the only tentative solution, the equation has no solutions. The solution set is $\varnothing$.

Quick Check **6** Solve $\log_4(2x - 7) = \log_4(3x + 2)$.

Solving Logarithmic Equations by Rewriting the Equation in Exponential Form

Objective 4 Solve a logarithmic equation by converting it to exponential form. If we are solving an equation in which the logarithm of a variable expression is equal to a number, such as $\log_2(x - 3) = 4$, we begin by rewriting the equation in exponential form and then solve the resulting equation. As with the previous logarithmic equations, we must check our solutions.

EXAMPLE 8 Solve $\log_2(x - 3) = 4$.

Solution

We begin by rewriting the equation $\log_2(x - 3) = 4$ in exponential form, which is $2^4 = x - 3$. We then solve the resulting equation.

$$\log_2(x - 3) = 4$$
$$2^4 = x - 3 \quad \text{Rewrite in exponential form.}$$
$$16 = x - 3 \quad \text{Raise 2 to the fourth power.}$$
$$19 = x \quad \text{Add 3.}$$

The solution set is $\{19\}$. The check is left to the reader.

Quick Check **7** Solve $\log_7(x^2 - 3x - 3) = 1$.

When solving an equation involving a natural logarithm, keep in mind that the base of this logarithm is e when you are rewriting the equation in exponential form.

EXAMPLE 9 Let $f(x) = \ln(x - 3) + 15$. Solve the equation $f(x) = 21$. Find the exact solution and then approximate the solution to the nearest thousandth.

Solution

We will begin by setting the function equal to 21. We then will rewrite the equation in exponential form and solve the resulting equation.

$$f(x) = 21 \quad \text{Set } f(x) \text{ equal to 21.}$$
$$\ln(x - 3) + 15 = 21 \quad \text{Replace } f(x) \text{ by } \ln(x - 3) + 15.$$
$$\ln(x - 3) = 6 \quad \text{Subtract 15.}$$
$$e^6 = x - 3 \quad \text{Convert to exponential form.}$$
$$e^6 + 3 = x \quad \text{Add 3.}$$
$$x \approx 406.429 \quad \text{Approximate the solution, using a calculator.}$$

We now must check the tentative solution in the original equation. We will use the approximate solution.

$$\begin{aligned} f(x) &= \ln(x-3)+15 \\ f(406.429) &= \ln(406.429-3)+15 && \text{Substitute 406.429 for } x. \\ &= \ln 403.429 + 15 && \text{Subtract.} \\ &\approx 21 && \text{Approximate, using a calculator.} \end{aligned}$$

This solution checks, and the solution set is $\{e^6 + 3\}$.

Quick Check 8 Let $f(x) = \log(3x - 14) + 9$. Solve the equation $f(x) = 15$.

Using the Properties of Logarithms when Solving Logarithmic Equations

Objective 5 **Use properties of logarithms to solve a logarithmic equation.** We now know how to solve a logarithmic equation in which either two logarithms are equal to each other, $\log_b x = \log_b y$, or a logarithm is equal to a number, $\log_b x = a$. However, we will often solve a logarithmic equation that is not in one of these forms, such as $\log_3 x + \log_3(x-6) = 3$. In such a case, we will use the properties of logarithms to rewrite the equation in the form of an equation that we already know how to solve. In other words, we need to rewrite the equation as an equation that has one logarithm equal to another logarithm with the same base, or as an equation that has a logarithm equal to a number. The product and quotient rules will be particularly helpful.

Example 10 Solve $\log_3 x + \log_3(x-6) = 3$.

Solution

We begin to solve this equation by using the product rule for logarithms to rewrite the left-hand side of this equation as a single logarithm. We can then rewrite the resulting equation in exponential form and solve it.

$$\begin{aligned} \log_3 x + \log_3(x-6) &= 3 \\ \log_3(x(x-6)) &= 3 && \text{Use the product rule for logarithms.} \\ \log_3(x^2-6x) &= 3 && \text{Multiply.} \\ 3^3 &= x^2 - 6x && \text{Convert to exponential form. This equation is quadratic.} \\ 27 &= x^2 - 6x && \text{Raise 3 to the third power.} \\ 0 &= x^2 - 6x - 27 && \text{Subtract 27.} \\ 0 &= (x-9)(x+3) && \text{Factor.} \\ x = 9 \quad \text{or} \quad x &= -3 && \text{Set each factor equal to 0 and solve.} \end{aligned}$$

We must check each tentative solution in the original equation. We must determine whether a solution requires us to take a logarithm of a nonpositive number.

Check

$$\begin{aligned} x &= 9 \\ \log_3(9) + \log_3((9)-6) &= 3 \\ \log_3 9 + \log_3 3 &= 3 \\ 2 + 1 &= 3 \\ 3 &= 3 \end{aligned}$$

$$\begin{aligned} x &= -3 \\ \log_3(-3) + \log_3((-3)-6) &= 3 \\ \log_3(-3) + \log_3(-9) &= 3 \end{aligned}$$

The solution $x = 9$ checks. Since the tentative solution $x = -3$ produces logarithms of negative numbers in the original equation, it must be omitted as a solution. The solution set is $\{9\}$.

EXAMPLE 11 Solve $\log(9x + 2) - \log(x - 7) = 1$.

Solution

We begin to solve this equation by using the quotient rule for logarithms to rewrite the left side of this equation as a single logarithm. We can then rewrite the resulting equation in exponential form and solve it.

$$\log(9x + 2) - \log(x - 7) = 1$$

$$\log\left(\frac{9x + 2}{x - 7}\right) = 1 \qquad \text{Use the quotient rule for logarithms.}$$

$$10^1 = \frac{9x + 2}{x - 7} \qquad \text{Rewrite in exponential form. This is a rational equation.}$$

$$10(x - 7) = \frac{9x + 2}{\cancel{x - 7}_1} \cdot (\cancel{x - 7}^1) \qquad \text{Multiply both sides by } x - 7.$$

$$10x - 70 = 9x + 2 \qquad \text{Multiply.}$$

$$x - 70 = 2 \qquad \text{Subtract } 9x.$$

$$x = 72 \qquad \text{Add 70.}$$

The check of this solution is left to the reader. The solution set is $\{72\}$.

Quick Check **9** **Solve.**

a) $\log_2 x + \log_2(x + 4) = 5$
b) $\log_3(41x - 42) - \log_3(x - 2) = 4$

EXAMPLE 12 Solve $\ln(x - 1) + \ln(x + 5) = \ln(3x + 7)$.

Solution

We begin to solve this equation by using the product rule for logarithms to rewrite the left-hand side of this equation as a single logarithm. We can then solve the resulting equation by using the one-to-one property of logarithms.

$$\ln(x - 1) + \ln(x + 5) = \ln(3x + 7)$$

$$\ln((x - 1)(x + 5)) = \ln(3x + 7) \qquad \text{Use the product rule for logarithms.}$$

$$\ln(x^2 + 4x - 5) = \ln(3x + 7) \qquad \text{Multiply.}$$

$$x^2 + 4x - 5 = 3x + 7 \qquad \text{Use the one-to-one property of logarithmic functions. This equation is quadratic.}$$

$$x^2 + x - 12 = 0 \qquad \text{Subtract } 3x \text{ and 7.}$$

$$(x + 4)(x - 3) = 0 \qquad \text{Factor.}$$

$$x = -4 \quad \text{or} \quad x = 3 \qquad \text{Set each factor equal to 0 and solve.}$$

When we substitute the tentative solution -4 for x in the original equation, it produces the following equation:

$$\ln(-5) + \ln(1) = \ln(-5)$$

Since we can find the logarithm of only positive numbers, the tentative solution $x = -4$ must be omitted. When checking the solution $x = 3$, we obtain the equation $\ln(2) + \ln(8) = \ln(16)$. This solution checks, and so the solution set is $\{3\}$.

Quick Check 10 Solve $\log_2(x - 3) + \log_2(x - 7) = \log_2(3x - 1)$.

Finding the Inverse Function of Exponential and Logarithmic Functions

Objective 6 Find the inverse function for exponential and logarithmic functions. One important characteristic of the exponential function $f(x) = b^x$ and the logarithmic function $f(x) = \log_b x$ $(b > 0, b \neq 1)$ is that they are inverse functions. Consider the function $f(x) = e^x$. We will find its inverse by using the technique developed in Section 8.2.

$f(x) = e^x$

$y = e^x$ Replace $f(x)$ by y.

$x = e^y$ Exchange x and y.

$\ln x = \ln e^y$ To solve for y, take the natural logarithm of each side.

$\ln x = y$ Simplify $\ln e^y$.

$f^{-1}(x) = \ln x$ Replace y by $f^{-1}(x)$.

The inverse function of $f(x) = e^x$ is $f^{-1}(x) = \ln x$.

In the examples that follow, we will learn to find the inverse function of exponential functions, as well as logarithmic functions.

EXAMPLE 13 Find the inverse function of $f(x) = e^{x-2} - 5$.

Solution

$f(x) = e^{x-2} - 5$

$y = e^{x-2} - 5$ Replace $f(x)$ by y.

$x = e^{y-2} - 5$ Exchange x and y.

$x + 5 = e^{y-2}$ Add 5.

$\ln(x + 5) = \ln e^{y-2}$ Take the natural logarithm of each side.

$\ln(x + 5) = y - 2$ Simplify $\ln e^{y-2}$.

$\ln(x + 5) + 2 = y$ Add 2.

$f^{-1}(x) = \ln(x + 5) + 2$ Replace y by $f^{-1}(x)$.

The process for finding the inverse function of a logarithmic function is quite similar to finding the inverse function of an exponential function. However, rather than taking the logarithm of both sides of an equation, we will rewrite the equation in exponential form in order to solve for y.

Quick Check 11
Find the inverse function of $f(x) = e^{x+6} + 9$.

EXAMPLE 14 Find the inverse function of $f(x) = \ln(x - 8) + 3$.

Solution

$f(x) = \ln(x - 8) + 3$

$y = \ln(x - 8) + 3$ Replace $f(x)$ by y.

$x = \ln(y - 8) + 3$ Exchange x and y.

$x - 3 = \ln(y - 8)$ Subtract 3.

$e^{x-3} = y - 8$ Rewrite in exponential form.

$e^{x-3} + 8 = y$ Add 8.

$f^{-1}(x) = e^{x-3} + 8$ Replace y by $f^{-1}(x)$.

***Quick Check* 12**

Find the inverse function of $f(x) = \ln(x + 10) - 18$.

Building Your Study Strategy **Practice Quizzes, 6 Trading with a Classmate** Practice quizzes can be effective when you are working with classmates. Make a practice quiz for a classmate by selecting odd-numbered problems associated with each objective in this section. Select problems for each objective that you feel are representative of the types of problems you would be expected to solve on an exam. Be sure that you understand how to do each of the problems that you have selected. Have this classmate prepare a practice quiz for you as well.

After two students trade practice quizzes, the quiz should be returned to the student who created it, and that student should grade the quiz. Do not simply mark the problems as correct or incorrect, but provide feedback on each problem. If a solution is incorrect, write down where the solution went wrong and how to find the correct solution. After you receive your practice quiz back, go over any mistakes that you made. When you understand what went wrong, select a similar problem from the exercise set and try it.

EXERCISES 8.6

Vocabulary

1. When solving a logarithmic equation, it is necessary to check for ________ solutions.
2. The inverse of a logarithmic function is a(n) ________ function.

Solve.

3. $3^x = 27$
4. $8^x = 64$
5. $5^{x-3} = 125$
6. $10^{x+7} = 1000$
7. $3^{5x-19} = 729$
8. $4^{-2x+11} = 64$

Solve. (Rewrite both sides of the equation in terms of the same base.)

9. $9^x = 27$
10. $4^x = 128$
11. $36^{x-3} = 216$
12. $16^{x+4} = 8$
13. $100^{x-6} = 1000^x$
14. $4^{x+9} = 512^x$

Solve. Round to the nearest thousandth.

15. $4^x = 17$

16. $9^x = 40$

17. $10^x = 129$

18. $e^x = 65$

19. $2^{x-5} = 27$

20. $2^{x+10} = 4295$

21. $12^{x+3} = 225$

22. $e^{x-6} = 290$

23. $3^{2x} = 68.3$

24. $5^{3x} = 4$

25. $e^{2x-3} = 37$

26. $10^{5x+17} = 29{,}345$

27. $3^{3x+8} = 600$

28. $3^{4x-5} = 52$

Solve.

29. $\ln(3x + 1) = \ln(5x - 7)$

30. $\log_3 10x = \log_3(4x + 12)$

31. $\log_8(3x + 13) = \log_8(7x - 23)$

32. $\ln(2x + 5) = \ln(6x - 1)$

33. $\log_4(4x - 1) = \log_4(5x + 3)$

34. $\log_{14}(x - 6) = \log_{14}(3x - 8)$

35. $\log_2(x^2 - 14x + 39) = \log_2 15$

36. $\log_7(x^2 + 6x - 20) = \log_7 20$

37. $\log(x^2 + 18x + 75) = \log(2x + 15)$

38. $\ln(x^2 + 8x - 30) = \ln(3x - 6)$

Solve. Round to the nearest thousandth.

39. $\log_3 x = 5$

40. $\ln x = 2$

41. $\log_5 x = -4$

42. $\log_3 x = 0$

43. $\ln(x - 4) = 6$

44. $\log(x + 11) = 4$

45. $\log_2(x + 8) - 7 = -2$

46. $\ln(x - 3) + 5 = 9$

47. $\log(x^2 + 4x + 4) = 2$

48. $\log_2(x^2 + 11x + 60) = 5$

49. $\log_{12}(x^2 - 3x - 39) = 0$

50. $\log_7(x^2 - 15) = 2$

51. $\log_4\left(\frac{3x + 8}{x - 6}\right) = 2$

52. $\log_2\left(\frac{x - 1}{9x - 4}\right) = -3$

Solve.

53. $\log(2x + 7) - \log(x - 9) = \log 3$

54. $\log_3 x + \log_3(x - 3) = \log_3 10$

55. $\log_5(x + 1) + \log_5(x + 6) = \log_5 24$

56. $\log_2(3x + 89) - \log_2(x + 8) = \log_2(x + 3)$

57. $\log_3(5x + 3) - \log_3(x - 1) = 2$

58. $\log_8(x - 3) + \log_8(x + 9) = 2$

59. $\log_2 x + \log_2(x - 4) = 5$

60. $\log(x + 15) + \log x = 3$

61. $\log_4(3x - 1) + \log_4(x + 5) = 3$

62. $\log_3 3x + \log_3(x + 6) = 4$

63. $\log_6(7x + 11) - \log_6(2x - 9) = 1$

64. $\log_2(9x + 50) - \log_2(2x + 5) = 5$

65. Let $f(x) = 3^{2x-15} + 4$. Solve $f(x) = 31$.

66. Let $f(x) = \ln(x - 6) + 8$. Solve $f(x) = 13$. Round to the nearest thousandth.

67. Let $f(x) = \log_5(x^2 + 3x - 45) + 6$. Solve $f(x) = 8$.

68. Let $f(x) = 5^{x+2} - 7$. Solve $f(x) = 29$. Round to the nearest thousandth.

Mixed Practice, 69–98

Solve. Round to the nearest thousandth.

69. $\log_8(x^2 - 10x - 20) = \log_8(3x + 10)$

70. $4^{6x-15} = \frac{1}{64}$

71. $5^{x+9} = 407$

72. $\log_6(7x + 8) = 2$

73. $9^{x-4} = 32$

74. $9^{x+6} = 27$

75. $2^{7x+30} = \frac{1}{32}$

76. $\log_5(x^2 - 4x - 20) = 2$

77. $\log(3x - 41) = 3$

78. $\log_3(x^2 + 8x - 3) = 4$

79. $2^{6x+25} = 128$

80. $3^{x^2-x-15} = 243$

81. $7^{x-8} = 56$

82. $\log(7x - 185) = 4$

83. $\log_{12}(9x - 11) = \log_{12} 4$

84. $4^{5x-2} = 32^{x+3}$

85. $\log_2 x + \log_2(x + 4) = 5$

86. $\log_6\left(\frac{5x + 13}{x - 8}\right) = 1$

87. $\log_7(4x - 19) - \log_7(x + 6) = \log_7 3$

88. $\log_3(x + 1) + \log_3(x - 7) = 2$

89. $\log_8(5x - 43) + 4 = 7$

90. $5^{5x-3} = 25^{2x+11}$

91. $\log_4(x^2 + 4x + 16) = \log_4(x^2 + x - 5)$

92. $\ln(7x - 13) - 8 = -3$

93. $\log_5(x - 7) + \log_5(x + 3) = \log_5(5x + 15)$

94. $3\log_2(x - 10) - 4 = 14$

95. $\ln(2x + 5) = -2$

96. $3^{2x-11} = 49$

97. $2^{x^2+12x+42} = 64$

98. $\log(15x - 27) = \log(11x - 17)$

For the given function $f(x)$, find its inverse function $f^{-1}(x)$.

99. $f(x) = e^{x+3}$

100. $f(x) = e^x + 3$

101. $f(x) = e^{x-5} + 4$

102. $f(x) = 3^{x+4} + 6$

103. $f(x) = \ln(x - 6)$

104. $f(x) = \ln x - 6$

105. $f(x) = \ln(x + 2) + 5$

106. $f(x) = \ln(x - 8) + 3$

107. Find the missing value in the equation $\log_2(5x - 9) = \log_2(?)$ if $x = 5$ is a solution to the equation.

108. Find the missing value in the equation $\log_3(x + 7) + \log_3(?) = 3$ if $x = 2$ is a solution to the equation.

109. **a)** Find the missing value in the equation $5^{x^2+7x+8} = ?$ if $x = -1$ is a solution to the equation.

b) What is the other solution to the equation?

110. Find the missing value in the equation $8^{x+6} = 16^{?x}$ if $x = 2$ is a solution to the equation.

Answer in complete sentences.

111. Explain why we must check logarithmic equations, and not exponential equations, for extraneous roots.

112. Explain why we take the natural logarithm of both sides of an equation involving the natural base *e*.

113. Solve the exponential equation $25^{x-7} = 125^{2x+9}$ in two ways: by rewriting each base in terms of 5 and by taking the common logarithm of both sides. Which method do you prefer? Clearly explain your reasoning.

114. Explain how to find the inverse of an exponential function. Explain how to find the inverse of a logarithmic function. Use examples to illustrate the process.

115. ***Solutions Manual**** Write a solutions manual page for the following problem:

Solve $\log_3 x + \log_3(x + 6) = 3$.

116. ***Newsletter**** Write a newsletter explaining how to solve exponential equations.

***See Appendix B for details and sample answers.**

8.7 Applications of Exponential and Logarithmic Functions

Objectives

1. **Solve compound interest problems.**
2. **Solve exponential growth problems.**
3. **Solve exponential decay problems.**
4. **Solve applications involving exponential functions.**
5. **Solve applications involving logarithmic functions.**

In this section, we shall be solving applied problems that involve exponential and logarithmic equations and functions.

Compound Interest

Objective 1 Solve compound interest problems. Suppose that you deposit \$1000 in a savings account at a bank that pays interest monthly. At the end of the first month, the bank calculates the interest earned and places it into your account. During the second month, not only does the original \$1000 earn interest, but the interest earned during the first month earns interest as well. This type of interest is called **compound interest**.

Compound Interest Formula

If P dollars are deposited in an account that pays an annual interest rate r that is compounded n times per year, then the amount A in the account after t years is given by the formula

$$A = P\left(1 + \frac{r}{n}\right)^{nt}.$$

The amount P is referred to as the principal. The annual interest rate r is often reported as a percent, but must be converted to a decimal number for this formula.

Interest that is compounded monthly is compounded 12 times per year. Here are some other common types of compound interest:

	Quarterly	Semiannually	Annually
n	4	2	1

EXAMPLE 1 Mona is planning on taking a trip in 4 years and deposits \$3000 in an account that pays 6% annual interest, compounded monthly. If she does not withdraw any money from this account, how much money will be in the account after 4 years?

Solution

It is often helpful to collect pertinent information in a table before attempting to use the equation.

Amount in the Account	A	Unknown
Principal	P	$3000
Interest Rate	r	0.06 (6%)
Number of Times Compounded/Year	n	12 (monthly)
Time (Years)	t	4

Now we can use the equation for compound interest.

$$A = P\left(1 + \frac{r}{n}\right)^{nt}$$

$$A = 3000\left(1 + \frac{0.06}{12}\right)^{12 \cdot 4}$$ Substitute for P, r, n, and t in the equation.

$$A = 3000(1.005)^{48}$$ Simplify the expression inside parentheses.

$$A = 3811.47$$ Approximate, using a calculator. Round to the nearest cent.

After 4 years, Mona's account balance will be $3811.47.

Quick Check 1
Sergio deposited $400 in an account that pays 3% annual interest, compounded monthly. What will the balance of this account be in 5 years?

EXAMPLE 2 Pablo deposited $20,000 in an account that pays 8% annual interest, compounded quarterly. How long will it take his account to reach $40,000?

Solution

Here is a table containing the pertinent information:

A	P	r	n	t
$40,000	$20,000	0.08 (8%)	4 (quarterly)	Unknown

Now we can use the equation for compound interest. In this problem, the unknown is the variable t. Since this variable is an exponent, we will need to use logarithms to solve this equation.

$$40{,}000 = 20{,}000\left(1 + \frac{0.08}{4}\right)^{4t}$$ Substitute for A, P, r, and n in the equation.

$$40{,}000 = 20{,}000(1.02)^{4t}$$ Simplify the expression inside parentheses.

$$2 = 1.02^{4t}$$ Divide both sides by 20,000.

$$\log 2 = \log 1.02^{4t}$$ Take the common logarithm of both sides.

$$\log 2 = 4t \log 1.02$$ Use the power rule for logarithms.

$$\frac{\log 2}{4 \log 1.02} = \frac{4t \log 1.02}{4 \log 1.02}$$ Divide by 4 log 1.02 to isolate t.

$$\frac{\log 2}{4 \log 1.02} = t$$ Simplify.

$$t \approx 8.75$$ Approximate, using a calculator. Enter the denominator inside a set of parentheses on your calculator.

Pablo's account will grow to $40,000 in approximately 8.75 years.

Quick Check 2
If $25,000 is deposited in an account that pays 10% annual interest, compounded quarterly, how long will it take until there is $100,000 in the account?

Using Your Calculator To approximate the expression $\frac{\log 2}{4 \log 1.02}$ by using the TI-84, it is necessary to enter the denominator in a pair of parentheses, as shown in the following screen shot:

```
log(2)/(4*log(1.
02))
        8.750697195
```

In the previous example, we were trying to determine the length of time for the balance of a bank account to double. The length of time for any quantity to double in size is called the **doubling time**.

Some banks use a method for calculating compound interest known as **continuous compounding**.

Continuous Compound Interest Formula

If P dollars are deposited in an account that pays an annual interest rate r that is compounded continuously, then the amount A in the account after t years is given by the formula

$$A = Pe^{rt}.$$

EXAMPLE 3 If an account pays 12% annual interest, compounded continuously, how long will it take a deposit of \$50,000 to produce an account balance of \$1,000,000?

Solution

Here is a table containing the pertinent information:

Amount in the Account	A	\$1,000,000
Principal	P	\$50,000
Interest Rate	r	0.12 (12%)
Time (Years)	t	Unknown

Now we can use the equation for continuous compound interest, $A = Pe^{rt}$.

$1{,}000{,}000 = 50{,}000e^{0.12t}$ Substitute for A, P, and r in the equation.

$20 = e^{0.12t}$ Divide both sides by 50,000.

$\ln 20 = \ln e^{0.12t}$ Take the natural logarithm of both sides.

$\ln 20 = 0.12t$ Simplify the right side of the equation.

$\frac{\ln 20}{0.12} = t$ Divide both sides by 0.12 to isolate t.

$t \approx 24.96$ Approximate, using a calculator.

The account balance will reach \$1 million in approximately 25 years.

Quick Check 3
If \$50,000 is invested in an account that pays 15% annual interest, compounded continuously, how long will it take until the balance of the account is \$500,000?

Exponential Growth

Objective 2 **Solve exponential growth problems.** The previous example involving interest that is compounded continuously is an example of a quantity that grows exponentially.

Exponential Growth Formula

If a quantity increases at an exponential rate, then its size P at time t is given by the formula

$$P = P_0 e^{kt}.$$

P_0 represents the initial size of the quantity, while k is the exponential growth rate.

Often, we will not know the exponential growth rate k, but we can determine what it is by knowing the size of the quantity at two different times.

EXAMPLE 4 The average annual tuition and fees paid to attend a public 2-year college increased from \$641 in 1985 to \$1735 in 2002. If tuition and fees are growing exponentially, what will it cost to attend a public 2-year college in 2012? (*Source:* The College Board)

Solution

We begin by determining the exponential growth rate k. We will treat \$641 as the initial cost P_0. Since it took 17 years for the costs to reach \$1735, we can substitute these values for t and P in the exponential growth formula. Here is a table containing the pertinent information:

P	P_0	k	t
1735	641	Unknown	17

We will now substitute into the equation $P = P_0 e^{kt}$ and solve for k.

$$1735 = 641e^{k \cdot 17} \quad \text{Substitute for } P, P_0, \text{ and } t.$$

$$\frac{1735}{641} = e^{k \cdot 17} \quad \text{Divide both sides by 641.}$$

$$\ln\left(\frac{1735}{641}\right) = \ln e^{k \cdot 17} \quad \text{Take the natural logarithm of both sides.}$$

$$\ln\left(\frac{1735}{641}\right) = k \cdot 17 \quad \text{Simplify the right side of the equation.}$$

$$\frac{\ln\left(\frac{1735}{641}\right)}{17} = k \quad \text{Divide both sides by 17.}$$

$$k \approx 0.058573 \quad \text{Approximate, using a calculator.}$$

Now that we have found the exponential growth rate k, we can turn our attention to finding the cost to attend a public 2-year college in 2012. If we continue to let 641 be the initial cost (P_0), then $t = 2012 - 1985$, or 27. Here is a table containing information that is pertinent to this part of the problem:

P	P_0	k	t
Unknown	641	0.058573	27

Now we substitute the appropriate values into the formula for exponential growth to determine the cost to attend a public 2-year college in 2012.

$P = P_0e^{kt}$

$P = 641e^{0.058573 \cdot 27}$ Substitute for P_0, k, and t.

$P = 641e^{1.581471}$ Simplify the exponent.

$P \approx 3116.61$ Approximate, using a calculator.

The cost to attend a public 2-year college in 2012 will be approximately $3116.61.

Quick Check 4

In 1974, a first-class U.S. postage stamp cost 10 cents. By 1994, the cost of a stamp had risen to 32 cents. If the cost of a first-class U.S. postage stamp continues to rise exponentially, how much would it cost to buy a stamp in 2020?

Exponential Decay

Objective 3 Solve exponential decay problems. Just as some quantities grow exponentially, other quantities decay exponentially. The equation for exponential decay is the same as the formula for exponential growth, with the exception that k is the exponential *decay* rate. The exponential decay rate will be a negative number.

Exponential Decay Formula

If a quantity decreases at an exponential rate, then its size P at time t is given by the formula

$$P = P_0e^{kt}.$$

P_0 represents the initial size of the quantity, while k is the exponential decay rate.

The **half-life** of a radioactive element is the amount of time that it takes for half of the substance to decay into another element. For example, the half-life of carbon-14 is 5730 years. This means that after 5730 years, 50%, or half, of the carbon-14 that was present in a sample will have decayed into another element.

EXAMPLE 5 All living things contain the radioactive element carbon-14. When a living thing dies, the radioactive carbon-14 begins to change to the stable nitrogen-14. By measuring the percentage of carbon-14 that remains, scientists can determine how much time has passed since death.

If an old table is made of wood that still contains 95% of its original carbon-14, how much time has passed since the tree it was made out of died? (The half-life of carbon-14 is 5730 years.)

Solution

We will use the half-life of carbon-14 to determine the value of the exponential decay rate k. We know that after 5730 years the wood from the tree will contain 50% of its original carbon-14.

P	P_0	k	t
0.5 (50%)	1 (100%)	Unknown	5730

We will now substitute into the equation and solve for k.

$$
\begin{aligned}
0.5 &= 1e^{k \cdot 5730} && \text{Substitute for } P, P_0, \text{ and } t.\\
\ln 0.5 &= \ln e^{k \cdot 5730} && \text{Take the natural logarithm of both sides.}\\
\ln 0.5 &= k \cdot 5730 && \text{Simplify the right side of the equation.}\\
\frac{\ln 0.5}{5730} &= k && \text{Divide both sides by 5730.}\\
k &\approx -0.000121 && \text{Approximate, using a calculator.}
\end{aligned}
$$

Now we are able to determine how long it has been since the death of the tree. Here is a table containing information that is pertinent to this part of the problem:

P	P_0	k	t
0.95 (95%)	1 (100%)	−0.000121	Unknown

Now we substitute the appropriate values into the formula for exponential growth.

$$
\begin{aligned}
0.95 &= 1e^{-0.000121t} && \text{Substitute for } P, P_0, \text{ and } k.\\
\ln 0.95 &= \ln e^{-0.000121t} && \text{Take the natural logarithm of both sides.}\\
\ln 0.95 &= -0.000121t && \text{Simplify the right side of the equation.}\\
\frac{\ln 0.95}{-0.000121} &= t && \text{Divide both sides by } -0.000121.\\
t &\approx 423.9 && \text{Approximate, using a calculator.}
\end{aligned}
$$

The tree that the table was made out of died approximately 423.9 years ago.

Quick Check **5**

The radioactive element strontium-89 has a half-life of 50.6 days. How long would it take for 50 grams of strontium-89 to decay to 10 grams?

Applications of Other Exponential Functions

Objective 4 **Solve applications involving exponential functions.**

EXAMPLE 6 The number of Subway restaurants in business in a particular year can be described by the function $f(x) = 1828.65 \cdot 1.06^x$, where x represents the number of years after 1964. If this pattern of exponential growth continues, when will there be 50,000 Subway restaurants? (*Source:* Doctor's Associates, Inc.)

Solution

We will begin by setting the function equal to 50,000 and solving for x.

$$
\begin{aligned}
1828.65 \cdot 1.06^x &= 50{,}000 && \text{Set the function equal to 50,000.}\\
1.06^x &= \frac{50{,}000}{1828.65} && \text{Divide both sides by 1828.65.}\\
\log 1.06^x &= \log\left(\frac{50{,}000}{1828.65}\right) && \text{Take the common logarithm of both sides.}\\
x \cdot \log 1.06 &= \log\left(\frac{50{,}000}{1828.65}\right) && \text{Apply the power rule for logarithms.}\\
x &= \frac{\log\left(\frac{50{,}000}{1828.65}\right)}{\log 1.06} && \text{Divide both sides by log 1.06.}\\
x &\approx 57 && \text{Approximate, using a calculator. Round to the nearest year.}
\end{aligned}
$$

Recall that x represents the number of years after 1964, so Subway will have 50,000 restaurants in 2021 $(1964 + 57)$.

Quick Check 6 The average annual income of a person with a bachelor's degree can be approximated by the function $f(x) = 13400 \cdot 1.055^x$, where x represents the number of years after 1975. If this exponential growth in income continues, in what year will the average annual income of a person with a bachelor's degree be \$100,000? (*Source:* U.S. Census Bureau)

Applications of Logarithmic Functions

Objective 5 **Solve applications involving logarithmic functions.**

pH

Every liquid can be classified as either acid, base, or neutral. Acids contain an excess of hydrogen (H^+) ions, while bases contain an excess of hydroxide (OH^-) ions. In a neutral liquid, such as pure water, the concentration of these two ions is equal. The pH scale, based on the concentration of hydrogen ions in the solution, is used to determine how acidic or basic a solution is.

pH of a Liquid

The pH of a liquid is calculated by the formula $\text{pH} = -\log[H^+]$, where $[H^+]$ is the concentration of hydrogen ions in moles per liter.

Solutions that are neutral have a pH of 7. The pH of an acid is less than 7, and the pH of a base is greater than 7. The pH of a liquid is generally between 0 and 14. The stronger an acid is, the closer its pH will be to 0. The stronger a base is, the closer its pH will be to 14.

EXAMPLE 7 Determine the pH of ammonia, whose concentration of hydrogen ions is 3.16×10^{-12} moles/liter. Determine whether ammonia is an acid or a base.

Solution

We substitute 3.16×10^{-12} for $[H^+]$ in the formula $\text{pH} = -\log[H^+]$ to determine the pH of ammonia.

$$\begin{aligned} \text{pH} &= -\log(3.16 \times 10^{-12}) && \text{Substitute } 3.16 \times 10^{-12} \text{ for } [H^+]. \\ &= -(-11.5) && \text{Approximate l using a calculator.} \\ &= 11.5 && \text{Simplify.} \end{aligned}$$

The pH of ammonia is 11.5. Ammonia is a base, as it has a pH greater than 7.

Quick Check 7 **The concentration of hydrogen ions $[H^+]$ in vinegar is 1.58×10^{-3} moles/liter. Find the pH of vinegar. Is vinegar an acid or a base?**

EXAMPLE 8 The pH of the water in a swimming pool is 7.8. Find the concentration of hydrogen ions in moles per liter.

Solution

We will substitute 7.8 for pH in the formula for pH and solve for the concentration of hydrogen ions, $[H^+]$.

$$7.8 = -\log[H^+] \quad \text{Substitute 7.8 for pH.}$$
$$-7.8 = \log[H^+] \quad \text{Divide by } -1.$$
$$10^{-7.8} = [H^+] \quad \text{Convert to exponential form.}$$
$$[H^+] \approx 1.58 \times 10^{-8} \quad \text{Approximate, using a calculator.}$$

The concentration of hydrogen ions in the water of the swimming pool is approximately 1.58×10^{-8} moles per liter.

***Quick Check* 8**
The pH of battery acid is 0.3. Find the concentration of hydrogen ions in moles per liter.

Volume of a Sound

The volume of a sound is measured in decibels (db) and depends on the intensity of the sound, which is measured in watts per square meter. 85 db is considered the threshold level for safety, and prolonged exposure to noise at 85 db can cause hearing loss.

Volume of a Sound

The volume (L) of a sound in decibels is given by the formula $L = 10\log\left(\frac{I}{10^{-12}}\right)$, where I is the intensity of the sound in watts per square meter. The constant 10^{-12} is the intensity of the softest sound that can be heard by humans.

EXAMPLE 9 The intensity of a sound generated by a personal stereo's headphones is 10^{-2} watts per square meter. Find the volume of this sound.

Solution

We begin by substituting 10^{-2} for I in the formula for the volume of a sound, $L = 10\log\left(\frac{I}{10^{-12}}\right)$, and then we will find L.

$$L = 10\log\left(\frac{10^{-2}}{10^{-12}}\right) \quad \text{Substitute } 10^{-2} \text{ for } I.$$
$$L = 10\log(10^{-2-(-12)}) \quad \text{Simplify the fraction, using the rule } \frac{x^m}{x^n} = x^{m-n}.$$
$$L = 10\log(10^{10}) \quad \text{Simplify the exponent.}$$
$$L = 10 \cdot 10 \quad \text{Simplify the logarithm.}$$
$$L = 100 \quad \text{Multiply.}$$

The volume generated by a personal stereo's headphones is 100 db.

***Quick Check* 9**
The intensity of a sound generated by normal conversation is 10^{-6} watts per square meter. Find the volume of this sound.

EXAMPLE 10 The threshold for pain is defined to be at a volume of 130 db. Find the intensity of a sound with this volume.

Solution

We begin by substituting 130 for L in the formula for the volume of a sound, and then we solve for I.

$$130 = 10\log\left(\frac{I}{10^{-12}}\right) \quad \text{Substitute 130 for } L.$$

$$13 = \log\left(\frac{I}{10^{-12}}\right) \quad \text{Divide both sides by 10.}$$

$$\frac{I}{10^{-12}} = 10^{13} \quad \text{Convert to exponential form.}$$

$$10^{-12}\cdot\frac{I}{10^{-12}} = 10^{-12}\cdot 10^{13} \quad \text{Multiply by } 10^{-12} \text{ to isolate } I.$$

$$I = 10^{1} \quad \text{Simplify.}$$

A sound that meets the threshold for pain has an intensity of 10 watts per square meter.

Quick Check 10
The volume of a vacuum cleaner is 80 db. Find the intensity of a sound with this volume.

Building Your Study Strategy **Practice Quizzes, 7 Study Groups** Practice quizzes can be effective when you are working with a group of students. Here is a plan for incorporating a practice quiz into a group study session. Have each student bring one question to the session. (Although using odd exercises is a good idea, feel free to use even exercises or make up a problem of your own based on problems you find in the book.) Once the session begins, randomly select a student to do each problem for the group. The student should explain each step of the solution as he or she works through it. The remainder of the group should act as coaches, using resources such as their notes or text as needed. If you are not the student solving the problem, ask any questions that you may have. At the end of the session, the group should create a practice quiz based on these problems. Each student should take this practice quiz at home and check back with the other group members if he or she has difficulty solving the problems.

EXERCISES 8.7

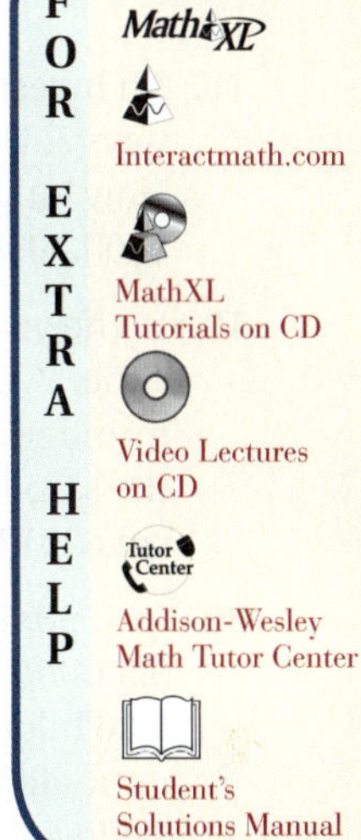

Vocabulary

1. When a bank pays __________ interest, the amount of interest earned in one period is added to the principal before the interest earned in the next period is calculated.

2. The length of time for any quantity to double in size is called the __________.

3. The formula $A = Pe^{rt}$ is used to calculate interest that is compounded __________.

4. If a quantity increases at an exponential rate, then its size P at time t is given by the formula __________.

5. If a quantity decreases at an exponential rate, the exponential decay rate k in the formula $P = P_0e^{kt}$ is a(n) __________ number.

6. The __________ scale is used to determine how acidic or basic a solution is.

7. Kasumi has deposited $5000 in a certificate of deposit (CD) account that pays 4% annual interest, compounded quarterly. How much will the balance of the account be in 3 years?

8. Nils deposited $750 in an account that pays 6% annual interest, compounded monthly. What will the balance of the account be in 2 years?

9. If Tina deposits $12,413 in an account that pays 8% annual interest, compounded semiannually, how much money will be in the account in 10 years?

10. Vladimir sold his baseball card collection for $3325 and deposited the money in a bank account that pays 7% interest, compounded annually. How much money will be in the account in 20 years?

11. Mark needs $5000 to take a family vacation to Florida. If he deposits $2000 in an account that pays 9% annual interest, compounded monthly, how long will it take until there is enough money in the account to pay for the vacation?

12. Kika deposited $4250 in an account that pays 5% annual interest, compounded quarterly. How long will it take until she has $7500 in the account?

13. Nomar started a bank account with an initial deposit of $240. If the bank account pays 3% annual interest, compounded monthly, how long will it take until the account's balance doubles?

14. If a bank account pays 10% annual interest, compounded annually, find the amount of time that it would take a deposit to double in value.

15. The interest on a bank account is compounded continuously, with an annual interest rate of 5.9%. If a deposit of $1000 is made, what will the balance of the account be in 5 years?

16. Anne has a bank account that pays 8% annual interest, compounded continuously. If Anne made an initial deposit of $23,000, what will the balance of the account be in 9 years?

17. How long would it take an investment of $4500 to grow to $6750 if it is invested at 7% annual interest, compounded continuously?

18. An investor has deposited $80,000 in an account that pays 5.3% annual interest, compounded continually. How long will it take until the account's balance is $100,000?

19. In the year 2000, Home Depot had 1134 stores nationwide. By 2002, this total had grown to 1532. If the number of stores continues to grow exponentially at the same rate, how many stores will there be in the year 2012? (*Source:* The Home Depot, Inc.)

20. In 1990, 7-Eleven® stores in Japan had annual sales of 9.3 billion yen. In 1999, they had annual sales of 19.6 billion yen. If sales continue to increase exponentially at this rate, what will the annual sales for 7-Eleven stores in Japan be in 2012? (*Source:* 7-Eleven, Inc.)

21. In 1974, the total world population reached 4 billion people. The total world population reached 6 billion people in 1999. If the total world population continues to grow exponentially at this rate, what will the population be in the year 2050? (*Source:* U.S. Census Bureau)

22. During its exponential growth phase, a colony of the bacteria *Bacillus megaterium* grew from 20,000 cells to 40,000 cells in 25 minutes. How many cells will be present after 60 minutes?

23. In 1999, a total of 1 billion spam email messages were sent daily worldwide. By 2002, this total had reached 5.6 billion per day. If the number of spam email messages continues to grow exponentially, when will the total reach 500 billion per day? (*Source:* IDC)

24. In 1997, the average value per acre of U.S. farm cropland was $1270. The average value had increased to $1650 by 2002. If the average value continues to grow exponentially at the same rate, when will the average value be $3000 per acre? (*Source:* National Agricultural Statistics Service)

25. The average monthly bill for U.S. cell-phone users rose from $39.43 in 1998 to $48.40 in 2002. If the average monthly bill continues to increase exponentially at this rate, when will it reach $75? (*Source:* Cellular Telecommunications and Internet Associates)

26. The average tuition and fees paid by students at private 4-year universities grew from $9340 in 1990 to $16,233 in 2000. If the average tuition and fees continue to grow exponentially at this rate, when will the average tuition and fees reach $30,000? (*Source:* The College Board)

27. During its exponential growth phase, a colony of the bacterium *E. coli* grew from 5000 cells to 40,000 cells in 60 minutes. How long would it take the colony of 5000 cells to grow to 100,000 cells?

28. During its exponential growth phase, a colony of the bacterium *E. coli* grew from 15,000 cells to 60,000 cells in 40 minutes. How long would it take the colony of 15,000 cells to grow to 250,000 cells?

29. In 1993, there were 593 drive-in theaters in the United States. By the year 2001, the number of drive-in theaters had dropped to 474. If the number of drive-in theaters continues to decline exponentially, how many drive-in theaters will there be in the year 2011? (*Source:* MPAA)

30. The circulation of all daily U.S. newspapers in 1992 was 60.1 million. Total circulation had dropped to 55.1 million by the year 2002. If circulation continues to decline exponentially, what will the total circulation of all daily U.S. newspapers be in the year 2025? (*Source: USA Today*)

31. Fluorine-18 is a radioactive version of glucose and is used in PET scanning. The half-life of fluorine-18 is 112 minutes. If 100 milligrams of fluorine-18 is injected into a patient, how much will remain after 240 minutes?

32. The half-life of bismuth-210 is 5 days. If 10 grams of bismuth-210 are present initially, how much will remain after 2 weeks?

33. If a bone contains 98% of its original carbon-14, how old is the bone? (Carbon-14 has a half-life of 5730 years.)

34. If a wooden bowl contains 85% of its original carbon-14, how old is the bowl? (Carbon-14 has a half-life of 5730 years.)

35. Calcium-41 is being used in studies testing the effectiveness of drugs for preventing osteoporosis. The half-life of calcium-41 is 100,000 years. If 5 grams of calcium-14 are present initially, how long would it take until only 1 gram remains?

36. Radon-222 has a half-life of 3.8 days. If 200 milligrams of radon-222 are present initially, how long would it take until only 10 milligrams remain?

37. In 1940, there were 1878 daily newspapers in the United States. In 2002, this number had dropped to 1457 daily newspapers. If the number of daily newspapers in the United States continues to decline exponentially at this rate, when will there be only 1000 daily newspapers in the United States? (*Source:* USA Today)

38. The number of drive-in screens in the United States dropped from 3561 in 1980 to only 717 in the year 2000. If the number of screens continues to decline exponentially at this rate, when will there be only 100 drive-in screens in the United States? (*Source:* MPAA)

39. The value of a car purchased in 2003 for $30,000 had dropped to $24,000 in 2004. If the value of the car continues to decrease exponentially at this rate, when will it be worth $5000?

40. The value of a car purchased in 2001 for $12,000 had dropped in value to $9720 in 2003. If the value of the car continues to decrease exponentially at this rate, when will it be worth $2000?

41. The number of Starbucks stores in business in a particular year can be described by the function $f(x) = 4.88 \cdot 1.524^x$, where x represents the number of years after 1984. Use this function to predict the number of Starbucks stores in the year 2011. (*Source:* Starbucks Corporation)

42. The number of McDonald's restaurants in business in a particular year can be described by the function $f(x) = 12{,}436 \cdot 1.095^x$, where x represents the number of years after 1991. Use this function to predict the number of McDonald's restaurants in the year 2011. (*Source:* McDonald's Corporation)

43. The number of deaths each year caused by AIDS in the United States has been decreasing exponentially since the mid-1990's. The number of deaths caused by AIDS in a particular year can be described by the function $f(x) = 48{,}460 \cdot 0.719^x$, where x represents the number of years after 1995. Use this function to predict the number of deaths caused by AIDS in the United States in the year 2012. (*Source:* CDC)

44. The number of American children younger than 13 who died of AIDS in a particular year can be described by the function $f(x) = 768 \cdot 0.645^x$, where x represents the number of years after 1994. Use this function to predict the number of deaths for children younger than 13 caused by AIDS in the year 2015. (*Source:* CDC)

45. The average premium, in dollars, of homeowner's insurance in a particular year can be described by the function $f(x) = 410.57 \cdot 1.035^x$, where x represents the number of years after 1994. Use this function to determine when the average homeowner's premium is $1000. (*Source:* National Association of Insurance Commissioners and Insurance Information Institute)

46. The amount spent, in millions of dollars, by Americans to download songs can be described by the function $f(x) = 1.4 \cdot 2.85^x$, where x represents the number of years after 2001. Use this function to determine when the amount spent will be $100,000,000,000. (*Source:* PricewaterhouseCoopers)

47. The average annual income of a person without a high school diploma can be approximated by the function $f(x) = 6830 \cdot 1.038^x$, where x represents the number of years after 1975. Use this function to predict when the average annual income of a person without a high school diploma will be $100,000. (*Source:* U.S. Census Bureau)

48. The average annual income of a person with an advanced college degree (master's degree or higher) can be approximated by the function $f(x) = 17{,}775 \cdot 1.059^x$, where x represents the number of years after 1975. Use this function to predict when the average annual income of a person with an advanced college degree will be $150,000. (*Source:* U.S. Census Bureau)

49. The value of a new car, in dollars, x years after its purchase can be described by the function $f(x) = 24{,}000 \cdot 0.8^x$. Use this function to determine when the value of the car will be $5000.

50. The value of a new copy machine, in dollars, x years after its purchase can be described by the function $f(x) = 3500 \cdot 0.85^x$. Use this function to determine when the value of the copier will be $2000.

51. Body temperature is often used by forensic investigators to determine the time of death for a person. For a person who was found dead in a walk-in refrigerator that maintained a constant temperature of 35° F, the body's temperature, in degrees Fahrenheit, can be approximated by the function $f(t) = 35 + 61e^{-0.0165t}$, where $t = 0$ corresponds to the time that the body was found.

a) What will the body's temperature be 6 hours after it was found?

b) How many hours before being found did the person die? Assume the person had a normal temperature of 98.6° F when alive.

52. The body temperature, in degrees Fahrenheit, of a person who was found dead in an apartment that maintained a constant temperature of 70° F can be approximated by the function $f(t) = 70 + 12e^{-0.182t}$, where $t = 0$ corresponds to the time that the body was found.

a) What will the body's temperature be 12 hours after it was found?

b) How many hours before being found did the person die?

53. The average annual salary, in dollars, of teachers in public elementary and secondary schools in a particular year can be described by the function $f(x) = 32178 \cdot 1.027^x$, where x represents the number of years after 1989. Use this function to determine when the average annual salary will reach $70,000. (*Source:* National Education Association)

54. The expenditure, in dollars, per pupil in public elementary and secondary schools in a particular year can be described by the function $f(x) = 4454 \cdot 1.038^x$, where x represents the number of years after 1988. Use this function to determine when the expenditure per pupil will reach $10,000. (*Source:* U.S. Department of Education)

55. The concentration of hydrogen ions $[H^+]$ in bleach is 2.51×10^{-13} moles/liter. Find the pH of bleach. Is bleach an acid or a base?

56. The concentration of hydrogen ions $[H^+]$ in orange juice is 6.31×10^{-5} moles/liter. Find the pH of orange juice. Is orange juice an acid or a base?

57. The concentration of hydrogen ions $[H^+]$ in lemon juice is 5.01×10^{-3} moles/liter. Find the pH of lemon juice. Is lemon juice an acid or a base?

58. The concentration of hydrogen ions $[H^+]$ in seawater is 1×10^{-8} moles/liter. Find the pH of seawater. Is seawater an acid or a base?

59. The pH of boric acid is 5.0. Find the concentration of hydrogen ions in moles per liter.

60. The pH of borax is 9.3. Find the concentration of hydrogen ions in moles per liter.

61. The pH of milk of magnesia is 10.2. Find the concentration of hydrogen ions in moles per liter.

62. The pH of corn is 6.2. Find the concentration of hydrogen ions in moles per liter.

63. The intensity of a sound generated by a vacuum cleaner is 10^{-4} watts per square meter. Find the volume of this sound.

64. The intensity of a sound generated by an alarm clock is 10^{-5} watts per square meter. Find the volume of this sound.

65. The intensity of a sound generated by a leaf blower is 10^{-1} watts per square meter. Find the volume of this sound.

66. The intensity of a sound generated by an airplane taking off is 10^2 watts per square meter. Find the volume of this sound.

67. The intensity of a sound generated in a noisy restaurant is $10^{-3.5}$ watts per square meter. Find the volume of this sound.

68. The intensity of a sound generated by a firecracker is 10^3 watts per square meter. Find the volume of this sound.

69. The volume of a crying baby is 110 db. Find the intensity of a sound with this volume.

70. The volume of a refrigerator is 50 db. Find the intensity of a sound with this volume.

71. The volume of an electric drill is 95 db. Find the intensity of a sound with this volume.

72. The volume of a chainsaw is 125 db. Find the intensity of a sound with this volume.

73. The life expectancy at birth for Americans of both sexes in a particular year can be described by the function $f(x) = 11.03 + 14.29 \ln x$, where x represents the number of years after 1900. Use this function to determine the life expectancy for Americans born in the year 2025. (*Source:* National Center for Health Statistics)

74. The number of U.S. cell-phone subscribers (in millions) in a particular year can be described by the function $f(x) = 63.1 + 45.43 \ln x$, where x represents the number of years after 1997. Use this function to predict how many subscribers there will be in the year 2015. (*Source:* Cellular Telecommunications & Internet Associates)

75. The number of public elementary and secondary teachers (in millions) in a particular year can be described by the function $f(x) = 2.648 + 0.181 \ln x$, where x represents the number of years after 1995. Use this function to predict when there will be 3.2 million teachers. (*Source:* U.S. Department of Education)

76. The percent of U.S. full-time college students who receive some financial aid in a particular year can be described by the function $f(x) = 58.8 + 6.7 \ln x$, where x represents the number of years after 1991. Use this function to find when 75% of all students received some sort of financial aid. (*Source:* U.S. Department of Education)

Writing in Mathematics

Answer in complete sentences.

77. Write a compound interest word problem whose solution is "Gretchen's balance will reach the correct amount in 17 years."

78. Write a compound interest word problem whose solution is "Lucy needs to invest \$10,000."

79. Explain what the half-life of a radioactive element is.

80. Write a word problem whose solution is "The pH of the liquid is 4.3."

81. ***Solutions Manual**** Write a solutions manual page for the following problem:

The population of Pleasanton grew from 100,000 in 1995 to 120,000 in 2005. If the population of Pleasanton continues to grow exponentially at this rate, what will the population of Pleasanton be in the year 2030?

82. ***Newsletter**** Write a newsletter explaining how to solve applied problems involving compound interest.

***See Appendix B for details and sample answers.**

8.8 Graphing Exponential and Logarithmic Functions

Objectives

1. **Graph exponential functions.**
2. **Graph logarithmic functions.**

In this section, we will learn how to graph logarithmic and exponential functions in greater detail than we did in Sections 8.3 and 8.4. In particular, now that we know how to solve exponential and logarithmic equations, we will be able to find the coordinates of the x-intercept if the intercept exists.

Graphing Exponential Functions

Objective 1 **Graph exponential functions.** We will focus on graphing exponential functions of the form $f(x) = b^{x-h} + k$, where b is a positive real number $(b \neq 1)$ and h and k are real numbers. For a function of this form, the horizontal asymptote is the line $y = k$. For example, the graph of $f(x) = 2^{x+4} + 3$ has a horizontal asymptote of $y = 3$, while the graph of $f(x) = e^{x-7} - 1$ has a horizontal asymptote of $y = -1$. It helps to place the horizontal asymptote on the graph before plotting any points that are on the graph of the function, as it shows us how the function behaves.

The choice of values for x is important. We will begin by using $x = h$, which is the value of x for which the exponent is equal to 0. We will also choose two values that are less than this value and two other values that are greater. We will evaluate the function for these values of x to determine their corresponding y-coordinates.

We will finish by finding the coordinates of the y- and x-intercepts. To find the y-intercept, we will evaluate $f(0)$. The coordinates of the y-intercept are $(0, f(0))$. To find the x-intercept, if one exists, we set $f(x)$ equal to 0 and solve this equation for x. This equation will be an exponential equation, and we will need to use logarithms to solve it.

Graphing Exponential Functions of the Form $f(x) = b^{x-h} + k$

1. **Graph the horizontal asymptote.** Using a dashed line, graph the line $y = k$.
2. **Create a table of values.** In addition to using the value $x = h$, which is the value of x for which the exponent is equal to 0, use two values of x that are less than h and two values that are greater than h. Evaluate the function for all five values of x to determine the coordinates of ordered pairs on the graph.
3. **Find the y-intercept.** Evaluate $f(0)$. This is the y-coordinate of the y-intercept.
4. **Find the x-intercept if one exists.** Set the function $f(x)$ equal to 0 and solve for x. Solving this equation for x will often require the use of logarithms. Recall that we can take only the logarithms of positive numbers. If there is a solution, then this value is the x-coordinate of the x-intercept.

EXAMPLE 1 Graph $f(x) = 3^{x+2} - 7$. Label any intercepts and state the domain and range of the function.

Solution

We begin by finding the horizontal asymptote, which is the line $y = -7$. Notice that $x = -2$ is the value for which the exponent is equal to 0. In addition to using this value for x, we will use two values that are less than -2 (-3 and -4) and two values that are greater than -2 (-1 and 0). Here is a table of values. Again, you should verify these function values.

x	-4	-3	-2	-1	0
$f(x) = 3^{x+2} - 7$	$3^{-2} - 7 = -6\frac{8}{9}$	$3^{-1} - 7 = -6\frac{2}{3}$	$3^0 - 7 = -6$	$3^1 - 7 = -4$	$3^2 - 7 = 2$

The points $(-4, -6\frac{8}{9})$, $(-3, -6\frac{2}{3})$, $(-2, -6)$, $(-1, -4)$, and $(0, 2)$ are on the graph of $f(x)$. Notice that we have already found the y-intercept, which is at $(0, 2)$. Also notice that the graph of this function must have an x-intercept, since the horizontal asymptote is below the x-axis. To find the x-coordinate of the x-intercept, we set the function $f(x)$ equal to 0 and solve for x.

$$3^{x+2} - 7 = 0 \quad \text{Set the function equal to 0.}$$
$$3^{x+2} = 7 \quad \text{Add 7.}$$
$$\log 3^{x+2} = \log 7 \quad \text{Take the common logarithm of each side.}$$
$$(x + 2) \cdot \log 3 = \log 7 \quad \text{Apply the power rule for logarithms.}$$
$$\frac{(x + 2) \cdot \cancel{\log 3}^{1}}{\cancel{\log 3}_{1}} = \frac{\log 7}{\log 3} \quad \text{Divide both sides by log 3.}$$
$$x = \frac{\log 7}{\log 3} - 2 \quad \text{Subtract 2.}$$
$$x \approx -0.2 \quad \text{Approximate, using a calculator.}$$

The x-intercept is approximately at $(-0.2, 0)$. Following is the graph of the function, including the x-intercept. The domain of this function is the set of all real numbers $(-\infty, \infty)$. The range is $(-7, \infty)$.

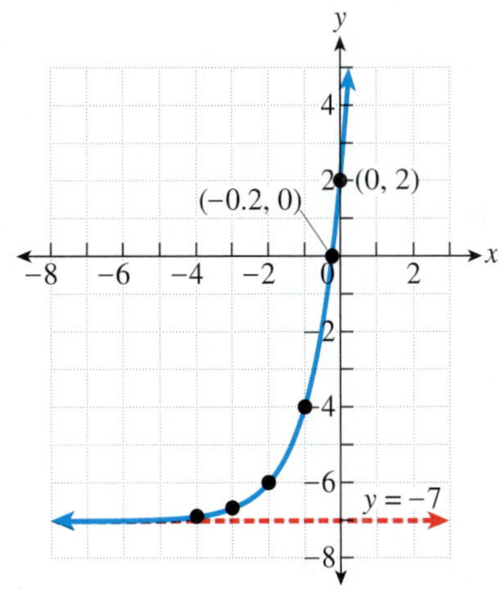

Quick Check 1
Graph $f(x) = 4^{x-2} - 6$. Label any intercepts, and state the domain and range of the function.

Using Your Calculator We can use the TI-84 to graph $f(x) = 3^{x+2} - 7$, as well as its horizontal asymptote, $y = -7$. Begin by pushing the [Y=] key. Next to Y_1, type $3^{x+2} - 7$. Type the number -7 next to Y_2. Press the [GRAPH] key to display the graph.

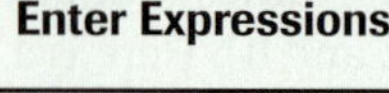

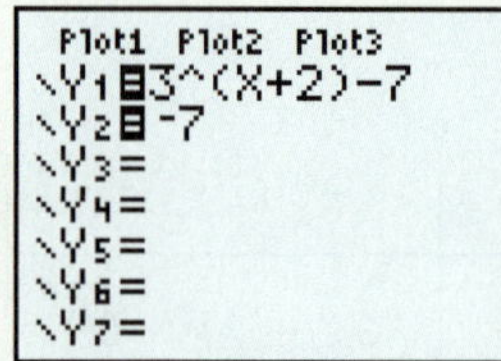

Graph

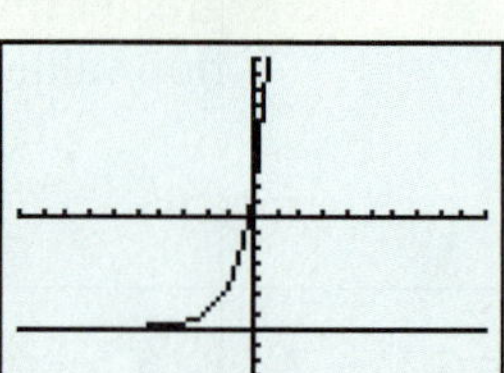

EXAMPLE 2 Graph $f(x) = e^{x+1} - 15$. Label any intercepts, and state the domain and range of the function.

Solution

We begin by finding the horizontal asymptote, which is the line $y = -15$.

Notice that $x = -1$ is the value for which the exponent is equal to 0. In addition to using this value for x, we will use two values that are less than -1 (-2 and -3) and two values that are greater than -1 (0 and 1). Here is a table of values:

x	-3	-2	-1	0	1
$f(x) = e^{x+1} - 15$	$e^{-2} - 15 \approx -14.9$	$e^{-1} - 15 \approx -14.6$	$e^{0} - 15 = -14$	$e^{1} - 15 \approx -12.3$	$e^{2} - 15 \approx -7.6$

Using approximate values, we find the points $(-3, -14.9)$, $(-2, -14.6)$, $(-1, -14)$, $(0, -12.3)$, and $(1, -7.6)$ on the graph of $f(x)$. Notice that we have already found the y-intercept, which is approximately at $(0, -12.3)$. To find the x-coordinate of the x-intercept, we set the function $f(x)$ equal to 0 and solve for x.

$$e^{x+1} - 15 = 0 \quad \text{Set the function equal to 0.}$$
$$e^{x+1} = 15 \quad \text{Add 15.}$$
$$\ln e^{x+1} = \ln 15 \quad \text{Take the natural logarithm of both sides.}$$
$$x + 1 = \ln 15 \quad \text{Simplify } \ln e^{x+1}.$$
$$x = \ln 15 - 1 \quad \text{Subtract 1.}$$
$$x \approx 1.7 \quad \text{Approximate, using a calculator.}$$

The x-intercept is approximately at $(1.7, 0)$.

Following is the graph of the function, including the x-intercept. The domain of this function is the set of all real numbers $(-\infty, \infty)$. The range is $(-15, \infty)$.

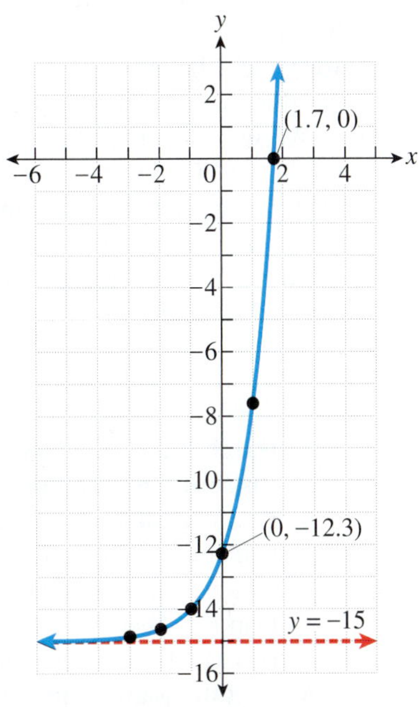

Quick Check 2

Graph $f(x) = e^{x-3} - 9$. Label any intercepts, and state the domain and range of the function.

Graphing Logarithmic Functions

Objective 2 **Graph logarithmic functions.** We now turn our attention to graphing logarithmic functions of the form $f(x) = \log_b(x - h) + k$, where b is a positive real number $(b \neq 1)$ and h and k are real numbers. Recall that the graph of a logarithmic function has a vertical asymptote and that for a function of this form, the vertical asymptote is the line $x = h$. This is the value for which the argument of the logarithm is equal to 0. For example, the graph of $f(x) = \log_5(x - 8) + 12$ has a vertical asymptote of $x = 8$, while the graph of $f(x) = \ln(x + 2) - 9$ has a vertical asymptote of $x = -2$. Just as with exponential functions, it helps to place the asymptote on the graph before we plot any points that are on the graph of the function, as this shows us how the function behaves.

The domain of a logarithmic function, unlike an exponential function, is not the set of real numbers $(-\infty, \infty)$. To find the domain of a logarithmic function, we need to find the values of x for which the argument of the logarithm is positive. In other words, we can set the argument of the logarithm greater than 0 and solve the inequality for x. The range of a logarithmic function is the set of real numbers $(-\infty, \infty)$.

We will plot points to help us graph logarithmic functions. However, we will not choose values for x. We will first replace $f(x)$ by y, and then rewrite this equation in exponential form. In general, the equation $y = \log_b(x - h) + k$ will be rewritten as $b^{y-k} + h = x$. We will then choose values for y and find the corresponding values of x. The choice of values for y is important. We will begin by using $y = k$, which is the value of y for which the exponent is equal to 0. We will also choose two values of y that are less than this value and two other values of y that are greater.

We will finish by finding the coordinates of the y- and x-intercepts. To find the y-intercept, if it exists, we will evaluate the original function $f(x)$ at $x = 0$. Note that $x = 0$ may not be in the domain of the function. In such a case, the graph will not have a y-intercept. If a y-intercept does exist, the coordinates of the y-intercept are $(0, f(0))$.

To find the x-intercept, we will substitute 0 for y in the equation that is in exponential form and solve this equation for x.

Graphing Logarithmic Functions of the Form $f(x) = \log_b(x - h) + k$

1. **Graph the vertical asymptote.** Using a dashed line, graph the line $x = h$.
2. **Convert the function to exponential form as $b^{y-k} + h = x$.**
3. **Create a table of values.** We will choose values for y in order to find ordered pairs that are on the graph of the function. In addition to using the value $y = k$, which is the value of y for which the exponent is equal to 0, use two values of y that are less than k and two values that are greater than k. Substitute all five values for y in the equation to determine the corresponding x-coordinates of the ordered pairs.
4. **Find the y-intercept if one exists.** Evaluate the original function for $x = 0$, if possible. Zero may not be in the domain of the function, and in this case there is no y-intercept. Otherwise, $f(0)$ is the y-coordinate of the y-intercept.
5. **Find the x-intercept.** Substitute 0 for y in the equation that is in exponential form and simplify for x. This is the x-coordinate of the x-intercept.

EXAMPLE 3 Graph $f(x) = \log_2(x - 4) + 1$. Label any intercepts, and state the domain and range of the function.

Solution

We begin by finding the vertical asymptote, which is the line $x = 4$. This is the value of x for which the argument of the logarithm is equal to 0. Next, we rewrite this function as an equation that is in exponential form.

$$y = \log_2(x - 4) + 1 \quad \text{Replace } f(x) \text{ by } y.$$
$$y - 1 = \log_2(x - 4) \quad \text{Subtract 1 to isolate the logarithm.}$$
$$2^{y-1} = x - 4 \quad \text{Rewrite in exponential form.}$$
$$2^{y-1} + 4 = x \quad \text{Add 4 to isolate } x.$$

Notice that $y = 1$ is the value for which the exponent is equal to 0. In addition to using this value for y, we will use two values that are less than 1 (-1 and 0) and two values that are greater than 1 (2 and 3). Here is a table of values. It is crucial that we remember that the values we have chosen are values of y, and this is different than the way we usually graph functions by choosing values of x.

$x = 2^{y-1} + 4$	$2^{-2} + 4 = 4\frac{1}{4}$	$2^{-1} + 4 = 4\frac{1}{2}$	$2^0 + 4 = 5$	$2^1 + 4 = 6$	$2^2 + 4 = 8$
y	-1	0	1	2	3

The points $(4\frac{1}{4}, -1)$, $(4\frac{1}{2}, 0)$, $(5, 1)$, $(6, 2)$, and $(8, 3)$ are on the graph of $f(x)$. Notice that we have already found the x-intercept, which is at $(4.5, 0)$. Notice also that the graph of this function cannot have a y-intercept, since the vertical asymptote is to the right of the y-axis. At the right is the graph of the function, as well as the vertical asymptote $x = 4$. The domain of this function is $(4, \infty)$. The range is the set of all real numbers $(-\infty, \infty)$.

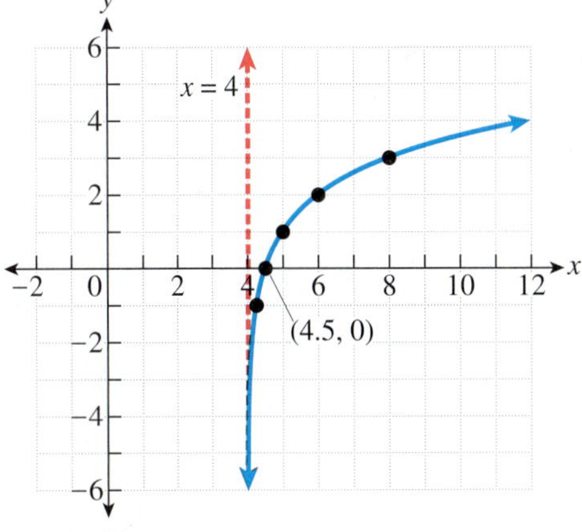

Using Your Calculator We can use the TI-84 to graph the logarithmic function $f(x) = \log_2(x - 4) + 1$. Begin by pushing the Y= key. Next to Y_1, we will enter the function. We must use the change-of-base formula to rewrite $\log_2(x - 4) + 1$ as $\dfrac{\log(x - 4)}{\log 2} + 1$ in order to graph this function on our calculator. Press the GRAPH key to display the graph. Using the standard window, your calculator screen should look like this (note that this function has a vertical asymptote at $x = 4$):

Enter Expressions

```
Plot1 Plot2 Plot3
\Y1■log(X-4)/log
(2)+1
\Y2=
\Y3=
\Y4=
\Y5=
\Y6=
```

Graph

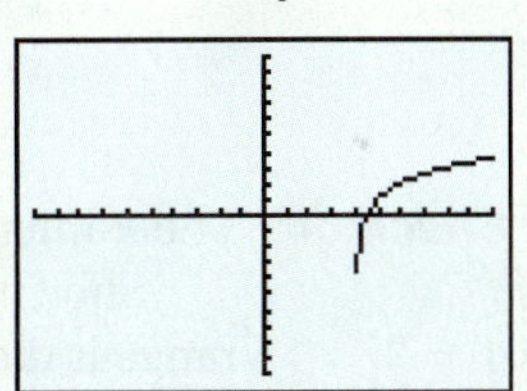

Quick Check **3** **Graph $f(x) = \log_3(x + 2) - 1$. Label any intercepts, and state the domain and range of the function.**

EXAMPLE 4 Graph $f(x) = \ln(x + 2) + 3$. Label any intercepts, and state the domain and range of the function.

Solution

We begin by finding the vertical asymptote, which is the line $x = -2$. Next, we convert this function to an equation that is in exponential form.

$$y = \ln(x + 2) + 3 \quad \text{Replace } f(x) \text{ by } y.$$
$$y - 3 = \ln(x + 2) \quad \text{Subtract 3 to isolate the logarithm.}$$
$$e^{y-3} = x + 2 \quad \text{Rewrite in exponential form.}$$
$$e^{y-3} - 2 = x \quad \text{Subtract 2 to isolate } x.$$

The exponent is equal to 0 when $y = 3$. In addition to using this value for y, we will use 1, 2, 4, and 5. Here is a table of values:

$x = e^{y-3} - 2$	$e^{-2} - 2 \approx -1.9$	$e^{-1} - 2 \approx -1.6$	$e^{0} - 2 = -1$	$e^{1} - 2 \approx 0.7$	$e^{2} - 2 \approx 5.4$
y	1	2	3	4	5

The points that we will plot, $(-1.9, 1)$, $(-1.6, 2)$, $(-1, 3)$, $(0.7, 4)$, and $(5.4, 5)$, are on the graph of $f(x)$. The graph of this function does have x- and y-intercepts that must be found. We will begin with the y-intercept, which we can find by evaluating the original function $f(x) = \ln(x + 2) + 3$ at $x = 0$.

$$f(0) = \ln(0 + 2) + 3 \quad \text{Substitute 0 for } x.$$
$$\approx 3.7 \quad \text{Approximate, using a calculator.}$$

The y-intercept is approximately at $(0, 3.7)$. To find the x-intercept, we substitute 0 for y in the equation $e^{y-3} - 2 = x$.

$$x = e^{y-3} - 2$$
$$x = e^{0-3} - 2 \quad \text{Substitute 0 for } y.$$
$$\approx -1.95 \quad \text{Approximate, using a calculator.}$$

Quick Check 4 Graph $f(x) = \ln(x - 6) - 2$. Label any intercepts, and state the domain and range of the function.

The x-intercept is at approximately $(-1.95, 0)$.

Shown next is the graph of the function. The domain of this function is $(-2, \infty)$. The range is the set of all real numbers $(-\infty, \infty)$.

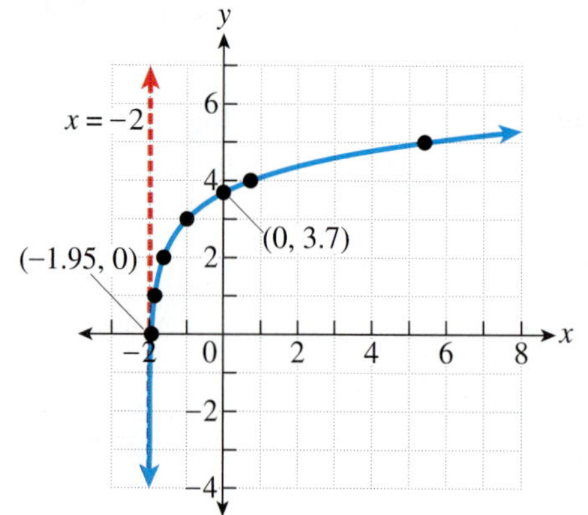

Building Your Study Strategy **Practice Quizzes, 8 Your Own Chapter Quiz** Make your own practice quiz for this chapter by selecting odd-numbered problems from the exercise sets of each section. Select four or five odd problems from each section that you feel are representative of the types of problems that you would be expected to solve on an exam. Be sure to choose problems of varying levels of difficulty. By choosing odd problems from the exercise sets, you will have an answer key for your practice quiz in the back of the text.

After you have completed writing a practice chapter test, compare your problems with the chapter test exercises from the textbook. If there are types of problems that you are missing, go back and add them to your practice test.

EXERCISES 8.8

Vocabulary

1. The graph of an exponential function has a(n) _________ asymptote.

2. The graph of a logarithmic function has a(n) _________ asymptote.

3. The domain of a(n) _________ function is the set of all real numbers.

4. The domain of a(n) _________ function is restricted.

5. The graph of every exponential function has a(n) _________ intercept.

6. The graph of every logarithmic function has a(n) _________ intercept.

Find all intercepts for the given function. Find exact values, and then round to the nearest tenth if necessary.

7. $f(x) = 2^{x+4} + 6$

8. $f(x) = 3^{x+4} - 9$

9. $f(x) = 6^{x+2} - 21$

10. $f(x) = e^{x+4} + 1$

11. $f(x) = e^{x+3} - 10$

12. $f(x) = e^{x-2} - 76$

13. $f(x) = \log_2(x - 5) - 3$

14. $f(x) = \log_3(x + 1) - 3$

15. $f(x) = \ln(x + 3) - 4$

16. $f(x) = \ln(x + 8) - 10$

Determine the equation of the horizontal asymptote for the graph of this function, and state the domain and range of this function.

17. $f(x) = 3^{x-5} + 8$

18. $f(x) = 6^{x+2} - 10$

19. $f(x) = 4^{x-12} - 9$

20. $f(x) = 5^{x-6}$

Determine the equation of the vertical asymptote for the graph of this function, and state the domain and range of this function.

21. $f(x) = \log_7(x - 5) - 9$

22. $f(x) = \log_8(x + 4) + 4$

23. $f(x) = \log(2x - 1) + 16$

24. $f(x) = \log_6 x - 12$

Graph. Label any intercepts and asymptotes. State the domain and range of the function.

25. $f(x) = 2^{x+2} + 4$

26. $f(x) = 3^{x+3} - 3$

27. $f(x) = 3^{x-1} - 15$

28. $f(x) = 7^{x} - 30$

29. $f(x) = e^{x+4} + 6$

30. $f(x) = e^{x+2} - 8$

31. $f(x) = \log(x - 2) - 1$

32. $f(x) = \log_3(x - 1) + 1$

33. $f(x) = \log_5(x + 2) - 2$

34. $f(x) = \log_3(x + 8) + 2$

35. $f(x) = \ln(x - 1) - 3$

36. $f(x) = \ln(x + 4) + 5$

Mixed Practice, 37–46

37. $f(x) = e^{x+3} - 2$

38. $f(x) = \ln(x - 2) - 1$

39. $f(x) = \log_3(x + 4) - 3$

40. $f(x) = 3^{x-3} - 6$

41. $f(x) = \ln(x + 6)$

42. $f(x) = 4^{x+2} + 4$

43. $f(x) = 2^{x-1} - 8$

44. $f(x) = \log_3(x + 10) + 1$

45. $f(x) = \left(\frac{1}{3}\right)^x - 3$

46. $f(x) = e^{x-4} + 13$

Given the graph of a function $f(x)$, graph the function $f^{-1}(x)$.

47.

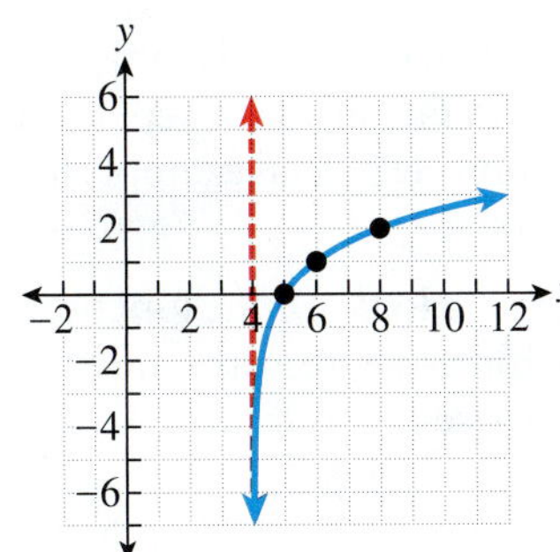

48.

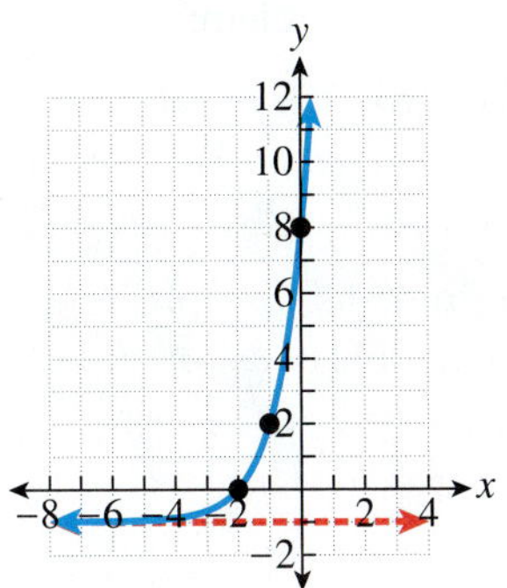

49.

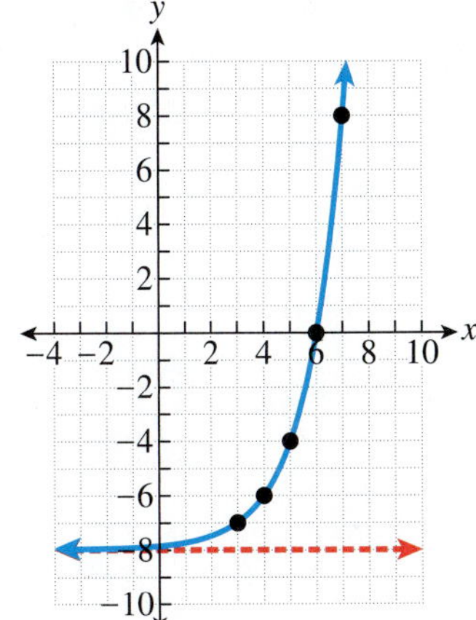

50.

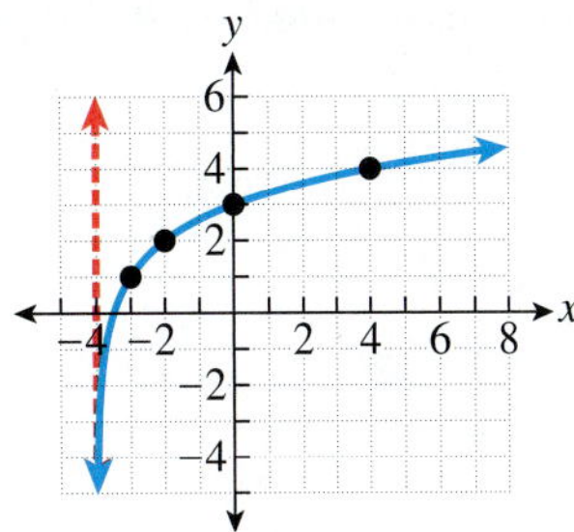

For the given function $f(x)$, find and graph $f^{-1}(x)$. Label all intercepts for $f^{-1}(x)$, as well as its asymptote.

51. $f(x) = \log_2 x$

52. $f(x) = \log_2(x - 4) + 1$

53. $f(x) = 2^{x-4} - 8$

54. $f(x) = e^{x-3} + 2$

Determine which equation is associated with the given graph.

a) $y = \log_2(x - 3) + 4$

b) $y = \log_2(x - 4) + 3$

c) $y = \log_4 x$

d) $y = \ln x$

e) $y = 2^{x-4} + 3$

f) $y = 2^{x-3} + 4$

g) $y = e^x$

h) $y = 4^x$

55.

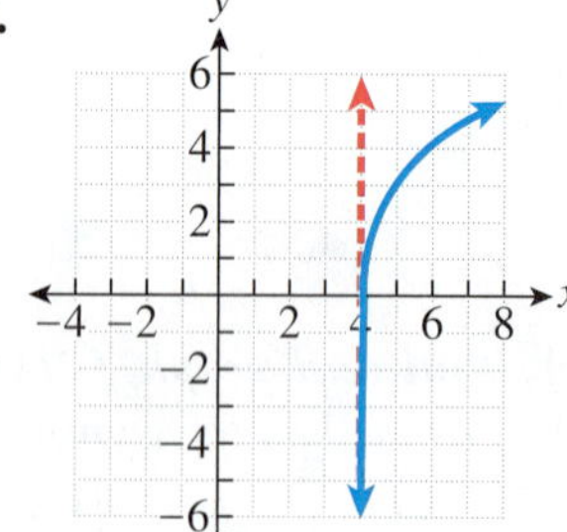

56.

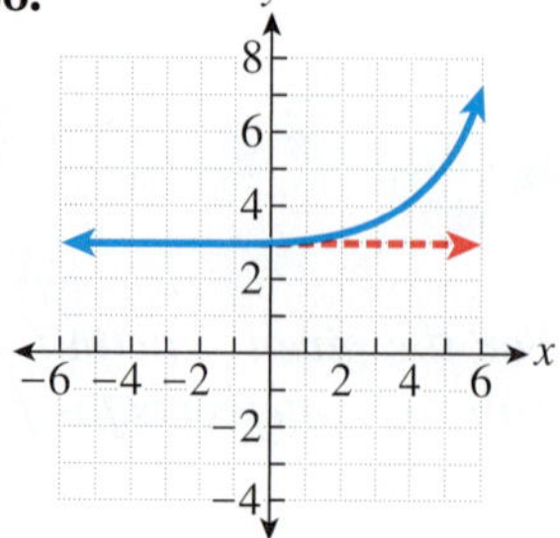

57.

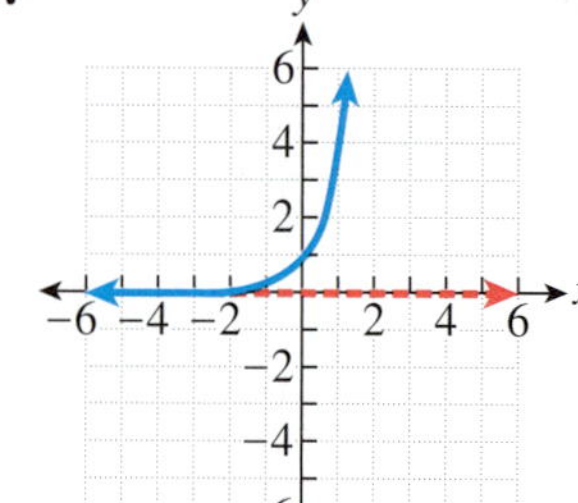

58.

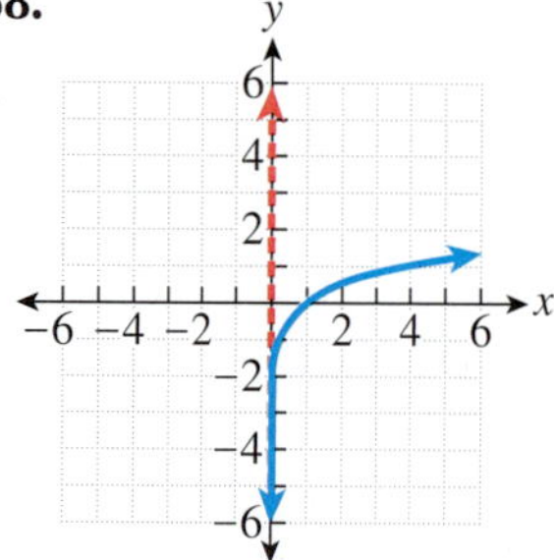

59.

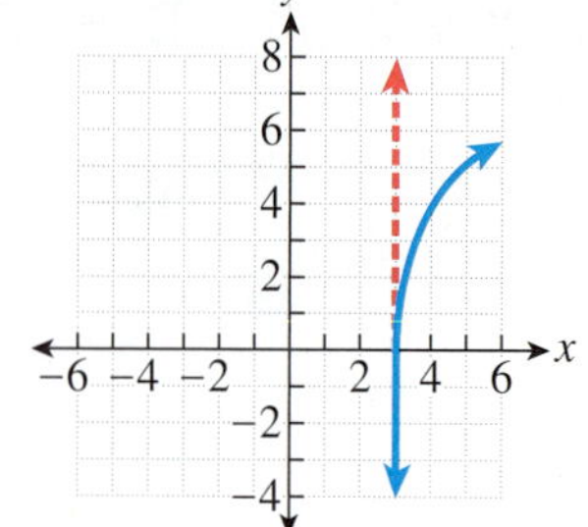

60.

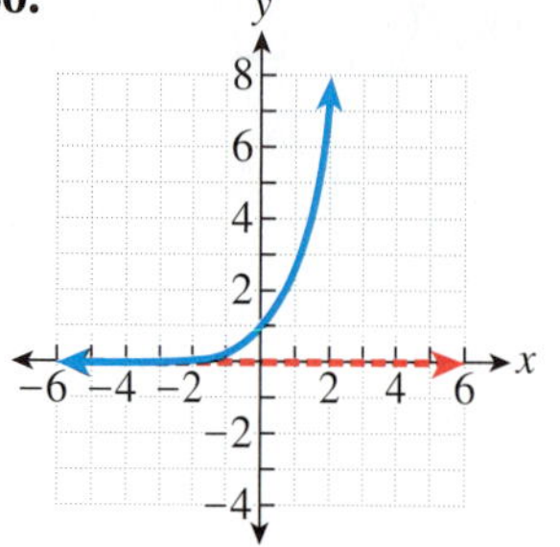

61.

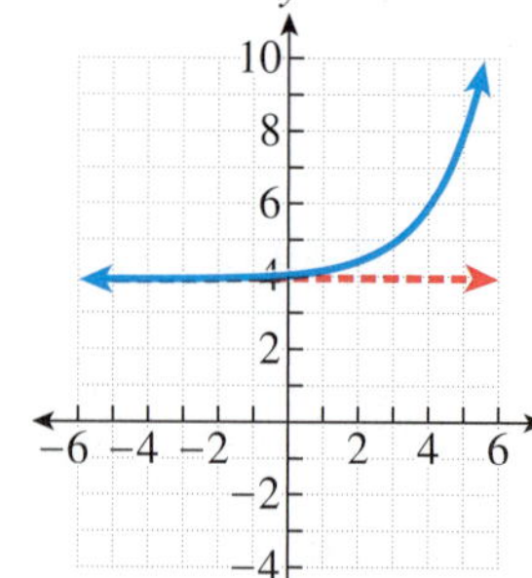

62.

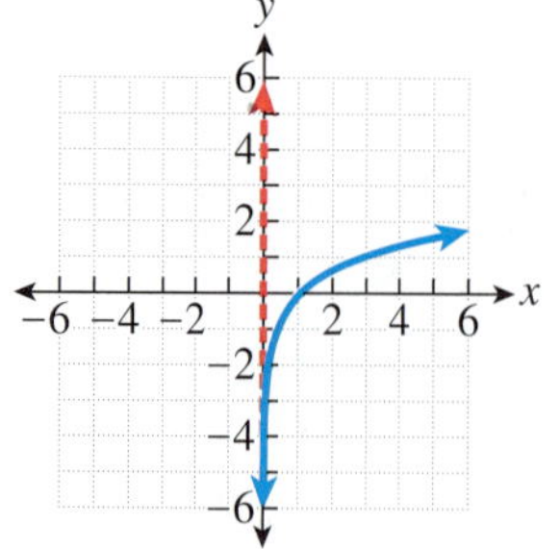

Writing in Mathematics

Answer in complete sentences.

63. Explain how to graph an exponential function, including how to find any intercepts and the horizontal asymptote. Use an example to illustrate the process.

64. Explain how to graph a logarithmic function, including how to find any intercepts and the vertical asymptote. Use an example to illustrate the process.

65. ***Solutions Manual**** Write a solutions manual page for the following problem:

Graph $f(x) = \log_2(x + 3) - 4$. *Label any intercepts and asymptotes. State the domain and range of the function.*

66. ***Newsletter**** Write a newsletter explaining how to find the x-intercepts of exponential and logarithmic functions.

***See Appendix B for details and sample answers.**

Chapter 8 Summary

Section 8.1—Topic	Chapter Review Exercises
Finding Sum, Difference, Product, and Quotient Functions	1–8
Composing Functions	9–14
Section 8.2—Topic	**Chapter Review Exercises**
Determining whether a Function Is a One-to-One Function	15–16
Applying the Horizontal-Line Test	17–18
Verifying That Two Functions Are Inverses	19–22
Finding Inverse Functions	23–28
Finding Inverse Functions from a Graph	29–30
Section 8.3—Topic	**Chapter Review Exercises**
Evaluating Exponential Functions	31–32
Section 8.4—Topic	**Chapter Review Exercises**
Simplifying Logarithmic Expressions	33–36
Evaluating Logarithmic Functions	37–38
Converting Exponential Equations to Logarithmic Form	39–40
Converting Logarithmic Equations to Exponential Form	41–42
Section 8.5—Topic	**Chapter Review Exercises**
Rewriting Logarithmic Expressions, Using a Single Logarithm	43–46
Rewriting a Logarithm in Terms of Two or More Logarithms	47–50
Applying the Change-of-Base Formula	51–54
Section 8.6—Topic	**Chapter Review Exercises**
Solving Exponential Equations	55–62
Solving Logarithmic Equations	63–70
Solving Equations Involving Functions	71–72
Finding Inverse Functions of Exponential and Logarithmic Functions	73–76
Section 8.7—Topic	**Chapter Review Exercises**
Solving Applied Problems	77–84
Section 8.8—Topic	**Chapter Review Exercises**
Graphing Exponential Functions	85–87
Graphing Logarithmic Functions	88–90

Summary of Chapter 8 Study Strategies

The chapter test at the end of this chapter can be treated as a practice quiz. Take this chapter test as if it were an actual exam, without using your notes, homework, or text. Do this at least 2 days prior to the actual exam, if possible. After you check your answers, you will have a good idea as to which types of problems you understand and which types of problems you will need to practice more. At this point, review your notes and homework regarding these problems, and read the appropriate sections of the text again. Now you can create focused practice quizzes on these topics by choosing problems from the chapter review. If you are still having trouble at this point, review the material again and visit your instructor or a tutorial center to have your questions answered. When you feel that you do understand these topics, create another practice quiz by choosing problems from the appropriate sections to confirm that you are ready for the test.

Let $f(x) = x^2 + 9x - 22$ *and* $g(x) = x + 11$. *Find the following:* [8.1]

1. $(f + g)(-5)$
2. $(f - g)(15)$
3. $(f \cdot g)(8)$
4. $\left(\frac{f}{g}\right)(-1)$

For the given functions $f(x)$ *and* $g(x)$, *find* $(f + g)(x)$. [8.1]

5. $f(x) = 6x - 11$, $g(x) = 4x + 35$

For the given functions $f(x)$ *and* $g(x)$, *find* $(f - g)(x)$. [8.1]

6. $f(x) = 8x - 33$, $g(x) = -8x + 55$

For the given functions $f(x)$ *and* $g(x)$, *find* $(f \cdot g)(x)$. [8.1]

7. $f(x) = x - 9$, $g(x) = x + 4$

For the given functions $f(x)$ *and* $g(x)$, *find* $\left(\frac{f}{g}\right)(x)$. [8.1]

8. $f(x) = x^2 - x - 6$, $g(x) = x^2 - 10x + 21$

Let $f(x) = x + 4$ *and* $g(x) = x^2 - 14x + 48$. *Find the following.* [8.1]

9. $(f \circ g)(8)$
10. $(g \circ f)(9)$

For the given functions $f(x)$ *and* $g(x)$, *find* $(f \circ g)(x)$ *and* $(g \circ f)(x)$. [8.1]

11. $f(x) = 6x - 17$, $g(x) = 3x + 20$
12. $f(x) = 2x + 15$, $g(x) = -4x + 9$
13. $f(x) = x - 8$, $g(x) = x^2 + 7x - 56$
14. $f(x) = 3x + 7$, $g(x) = x^2 - 9x + 41$
15. Is the function whose input is a person and output is that person's favorite fast-food restaurant a one-to-one function? Explain your answer in your own words. [8.2]
16. Is the function whose input is a player on your school's basketball team and output is that player's uniform number a one-to-one function? Explain your answer in your own words. [8.2]

Worked-out solutions to Review Exercises marked with [blue circle] can be found on page AN-47.

Use the horizontal-line test to determine whether the function is one to one. [8.2]

17. 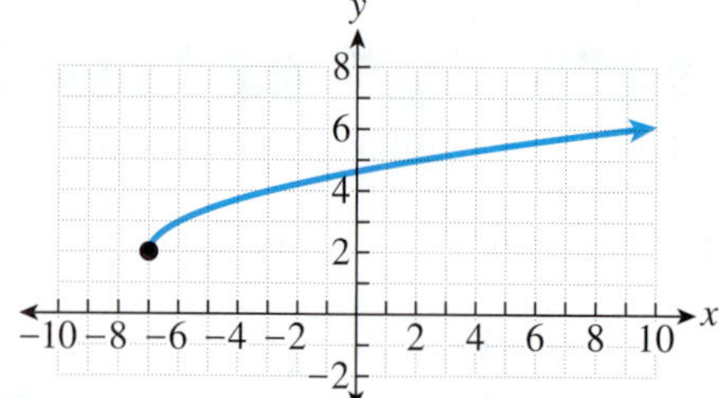
18. 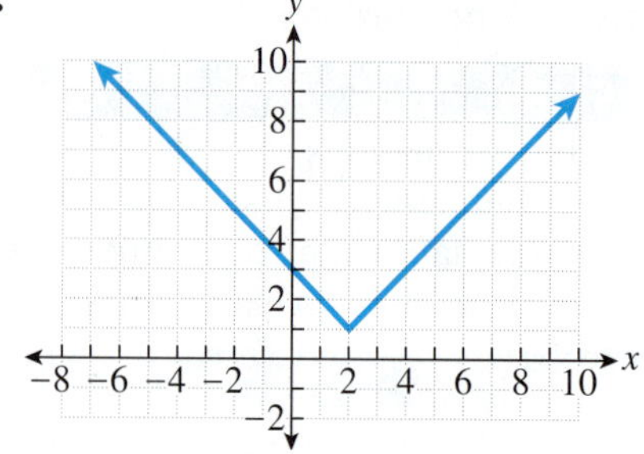

Determine whether the functions $f(x)$ *and* $g(x)$ *are inverse functions by showing that* $(f \circ g)(x) = x$ *and* $(g \circ f)(x) = x$. [8.2]

19. $f(x) = 3x + 4$, $g(x) = 4x - 3$
20. $f(x) = 2x - 10$, $g(x) = \dfrac{x + 10}{2}$
21. $f(x) = 4x + 18$, $g(x) = \dfrac{1}{4}x + \dfrac{9}{2}$
22. $f(x) = -5x + 9$, $g(x) = \dfrac{9 - x}{5}$

For the given function $f(x)$, *find* $f^{-1}(x)$. *State the domain of* $f^{-1}(x)$. [8.2]

23. $f(x) = x - 10$
24. $f(x) = 2x - 15$
25. $f(x) = -x + 19$
26. $f(x) = -8x + 27$
27. $f(x) = \dfrac{9x + 2}{3x}$

28. $f(x) = \dfrac{7x}{4x + 21}$

***For the given graph of a one-to-one function $f(x)$, graph its inverse function $f^{-1}(x)$.* [8.2]**

29.

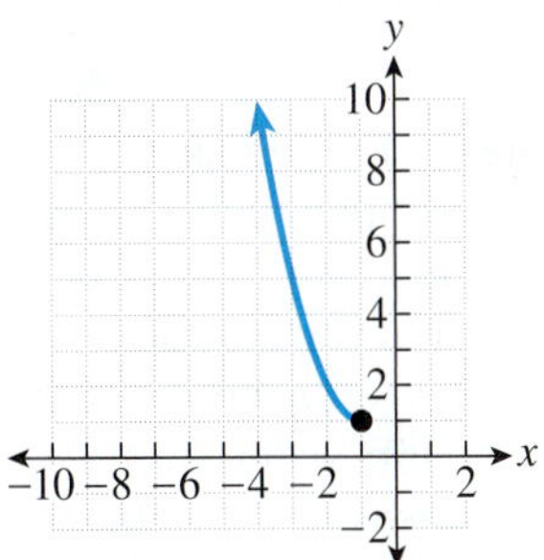

30.

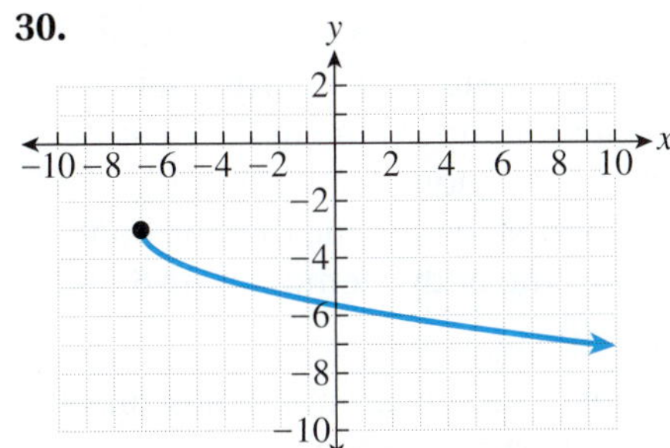

***Evaluate the given function. Round to the nearest thousandth.* [8.3]**

31. $f(x) = 4^x - 9, f(5)$

32. $f(x) = e^{x-8} - 13, f(11)$

***Simplify.* [8.4]**

33. $\log_6 \dfrac{1}{36}$

34. $\log_5 3125$

35. $\log_9 1$

36. $\ln e^{12}$

***Evaluate the given function. Round to the nearest thousandth.* [8.4]**

37. $f(x) = \ln(x - 9) + 20, f(52)$

38. $f(x) = \log(x + 23) - 6, f(1729)$

***Rewrite in logarithmic form.* [8.4]**

39. $4^x = 1024$

40. $e^x = 20$

***Rewrite in exponential form.* [8.4]**

41. $\log x = 3.2$

42. $\ln x - 3 = 8$

***Rewrite as a single logarithmic expression. Simplify if possible. Assume all variables represent positive real numbers.* [8.5]**

43. $\log_2 13 + \log_2 8$

44. $2 \log a - 5 \log b$

45. $4 \log_2 x - 3 \log_2 y + 9 \log_2 z$

46. $-5 \log_b 3 - 2 \log_b 2$

***Rewrite as the sum or difference of logarithmic expressions whose arguments have an exponent of 1. Simplify if possible. Assume all variables represent positive real numbers.* [8.5]**

47. $\log_8 64x$

48. $\log_3\left(\dfrac{n}{27}\right)$

49. $\log_b\left(\dfrac{a^6c^3}{b^2}\right)$

50. $\log_2\left(\dfrac{8x^9}{yz^7}\right)$

***Evaluate, using the change-of-base formula. Round to the nearest thousandth.* [8.5]**

51. $\log_4 25$

52. $\log_{1/2} 295$

53. $\log_2 1597$

54. $\log_{37} 6{,}403{,}200$

Solve. Round to the nearest thousandth. **[8.6]**

55. $10^x = 1000$

56. $8^{x-5} = 64$

57. $16^x = 32$

58. $9^{x+6} = 27^x$

59. $7^x = 11$

60. $6^{x+4} = 3000$

61. $e^{x-8} = 62$

62. $5^{2x-9} = 67$

Solve. Round to the nearest thousandth. **[8.6]**

63. $\log_4(x + 3) = \log_4 17$

64. $\log(x^2 + 12x + 6) = \log(3x + 16)$

65. $\ln x = 5$

66. $\log_3(x^2 - 13x + 57) - 1 = 2$

67. $\log_9 x + \log_9(x + 8) = \log_9 20$

68. $\ln(x + 10) - \ln(x - 6) = \ln 5$

69. $\log_3(x + 5) + \log_3(x - 1) = 3$

70. $\log x + \log(x + 3) + 5 = 6$

71. Let $f(x) = 2^{3x-7} + 8$. Solve $f(x) = 13$. Round to the nearest thousandth.

72. Let $f(x) = \ln x + 4$. Solve $f(x) = 8$. Round to the nearest thousandth.

For the given function $f(x)$, find its inverse function $f^{-1}(x)$. **[8.6]**

73. $f(x) = e^x + 4$

74. $f(x) = 2^{x-5} - 1$

75. $f(x) = \ln(x + 8)$

76. $f(x) = \log_3(x + 6) - 10$

77. Grady has deposited \$3500 in an account that pays 8% annual interest, compounded quarterly. How long will it take until the balance of the account is \$10,000? [8.7]

78. The population of a city increased from 200,000 in 1992 to 250,000 in 2002. If the population continues to increase exponentially at the same rate, what will the population be in 2015? [8.7]

79. The half-life of bismuth-210 is 5 days. If 25 grams of bismuth-210 are present initially, how much will remain after 10 days? [8.7]

80. If researchers find a wooden arrow that contains 98% of its original carbon-14, how old is the arrow? (Carbon-14 has a half-life of 5730 years.) [8.7]

81. During the 1990s, the percentage of American adults who were obese increased exponentially. For any particular year, the percentage of adult Americans who were obese can be approximated by the function $f(x) = 11.5 \cdot 1.057^x$, where x represents the number of years after 1990. Use this function to determine when the percentage of Americans who were obese reached 30%. (*Source:* CDC) [8.7]

82. The body temperature, in degrees Fahrenheit, of a person who was found dead in an apartment that maintained a constant temperature of 75° F can be approximated by the function $f(t) = 75 + 12e^{-0.206t}$, where $t = 0$ corresponds to the time that the body was found. What was the body's temperature 6 hours after it was found? [8.7]

83. If an earthquake creates a shock wave that is 20,000 times the smallest measurable shock wave that is recordable by a seismograph, find the magnitude of the earthquake on the Richter scale. [8.7]

84. The pH of a particular bottle of wine is 3.5. Find the concentration of hydrogen ions in moles per liter. [8.7]

Graph. Label any intercepts and asymptotes. State the domain and range of the function. **[8.8]**

85. $f(x) = 3^x + 2$

86. $f(x) = 2^{x+2} - 8$

87. $f(x) = e^{x+1} - 6$

88. $f(x) = \log_2(x - 5)$

89. $f(x) = \log_3(x + 9) - 2$

90. $f(x) = \ln(x + 8) + 1$

1. If $f(x) = x - 6$ and $g(x) = x^2 + 6x + 36$, find $(f \cdot g)(x)$.

For the given function $f(x)$, and $g(x)$, find $(f \circ g)(x)$.

2. $f(x) = 4x + 13$, $g(x) = -2x + 23$
3. $f(x) = x^2 - 9x - 30$, $g(x) = x + 12$
4. Determine whether $f(x) = 2x - 16$ and $g(x) = \frac{1}{2}x + 8$ are inverse functions. Recall that $f(x)$ and $g(x)$ are inverse functions if $(f \circ g)(x) = x$ and $(g \circ f)(x) = x$.

For the given function $f(x)$, find $f^{-1}(x)$.

5. $f(x) = -4x + 20$
6. $f(x) = \dfrac{x - 9}{4x}$

Evaluate the given function. Round to the nearest thousandth.

7. $f(x) = e^{x-4} - 10$, $f(6)$
8. $f(x) = \ln(x + 7) + 4$, $f(11)$

Simplify.

9. $\log_{12} 144$
10. $\ln e^{-3}$

Rewrite as a single logarithmic expression. Simplify if possible. Assume all variables represent positive real numbers.

11. $2 \log_b x + 5 \log_b y - 3 \log_b z$

Rewrite as the sum or difference of logarithmic expressions whose arguments have an exponent of 1. Simplify if possible. Assume all variables represent positive real numbers.

12. $\ln\left(\dfrac{a^4 b^6}{c^8 d}\right)$

Evaluate, using the change-of-base formula. Round to the nearest thousandth.

13. $\log_9 62$

Solve. Round to the nearest thousandth.

14. $6^{x+4} = \dfrac{1}{36}$
15. $4^{x-9} = 76$
16. $e^{x+2} - 6 = 490$

Solve. Round to the nearest thousandth.

17. $\log_3(2x - 11) = \log_3 5$
18. $\log_5(x + 65) + 5 = 8$
19. $\log_2 x + \log_2(x - 4) = 5$
20. For the function $f(x) = e^{x+9} - 17$, find its inverse function $f^{-1}(x)$.
21. Sonia started a bank account with an initial deposit of \$4200. The bank account pays 6% interest, compounded monthly. How long will it take for the deposit to grow to \$6000?
22. A painting that was purchased for \$15,000 in 1992 was valued at \$25,000 in 2002. If the value of the painting keeps rising exponentially at the same rate, how much will it be worth in 2025?
23. If researchers find a wooden chalice that contains 90% of its original carbon-14, how old is the chalice? (Carbon-14 has a half-life of 5730 years.)

Graph. Label any intercepts and asymptotes. State the domain and range of the function.

24. $f(x) = e^{x-2} + 4$
25. $f(x) = \log_3(x - 5) + 3$

Mathematicians in History

Albert Einstein

Albert Einstein was named as *Time* magazine's "Man of the Century" for the twentieth century. That is quite an accomplishment for a man who once failed an exam that would have allowed him to study to be an electrical engineer. In 1901, working as a temporary high school math teacher, Einstein had given up the ambition to go to a university. However, while working in Switzerland's patent office, Einstein earned a doctorate in physics from the University of Zurich in 1908. Einstein introduced the theory of relativity and became one of the most popular scientists ever. Einstein once said that compound interest was the most powerful force in the universe and that it is the greatest mathematical discovery of all time.

Write a one-page summary (*or* make a poster) of the life of Albert Einstein and his accomplishments. Also, look up Einstein's most famous quotes and list your five favorite quotes.

Interesting issues:

- Where and when was Albert Einstein born?
- What famous formula is credited to Einstein?
- Einstein married Mileva Maric in 1903. What became of their sons Hans Albert and Eduard?
- After divorcing Mileva in 1919, Einstein married his second wife Elsa. How did he know Elsa?
- When did Einstein win the Nobel prize, and what did he win it for?
- Einstein came to Princeton University in 1932, planning to teach part of the year at Princeton and the remainder of the year in Berlin. What event prohibited Einstein from returning to Berlin in 1933?
- In 1939, Einstein sent a letter to Franklin Roosevelt. What was the subject of that letter?
- What job was Einstein offered in 1952?
- When did Einstein die, and what were the circumstances of his death?

For this activity, you will need either a bag of plain M&M's or a bag of Skittles.

Step 1: Open the bag of candy and pour it out on some flat surface. Remove any broken candies. Count the total number of candies in the bag. Record the number of candies in the table provided. Pour all the candies back into the bag.

Step 2: Pour out all of the candies on a flat surface. Separate the candies which are letter up from those which are letter down. Record the number of each in the table.

Step 3: Put only those candies that were letter down back into the bag.

Step 4: Repeat Steps 2 and 3 until at most only one candy is letter down.

Total number of candies: _____

Pours	Letter up	Letter down
1		
2		
3		
4		
5		
6		
7		

Step 5: Create an axis system with the horizontal axis representing the pour number and the vertical axis representing the number of letter-up candies counted. Plot the number of letter-up candies for each pour, and connect the dots, using a smooth graph.

a) Is the graph increasing or decreasing? Why do you think this is the case?

b) Is the graph linear? Why or why not?

c) If the graph is not linear, what kind of function would best represent the graph you drew?

d) What kind of function is $f(n) = 2^{6-n}$? If n represents the pour number, find the values for pour 1, pour 2, etc.

e) Do your values of letter-up candies for each pour come close to the number found by evaluating the function at those different n-values? If so, why do you think a base value of 2 was appropriate to use?

Step 6: Using the same bag of M&M's or Skittles, try this experiment again. Did your results vary much?

CHAPTER Nine 9

Conic Sections

In this chapter, we will learn to graph **conic sections:** ***parabolas, circles, ellipses, and hyperbolas. The equations of the different conic sections are all nonlinear equations. These graphs are called conic sections, because they can be constructed by intersecting a plane with a cone, as shown in the following figure:***

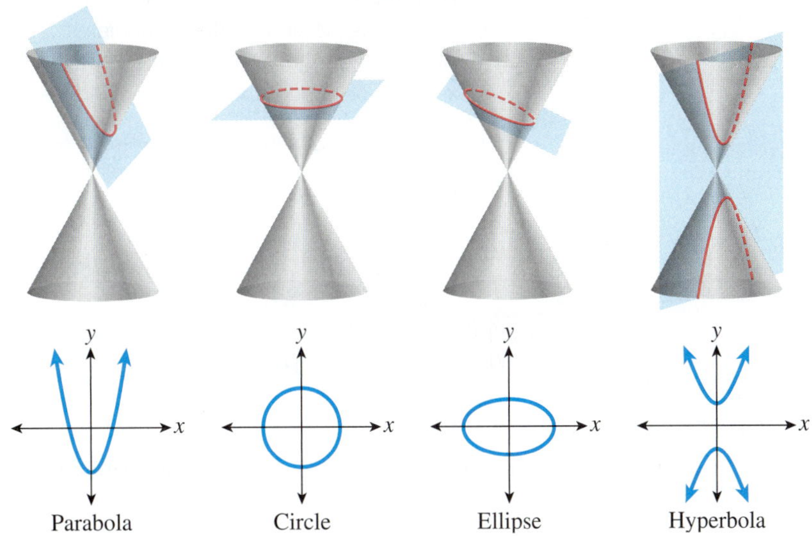

The chapter concludes with a section on systems of nonlinear equations.

Study Strategy **Study Environment** *It is very important to study in the proper environment. In this chapter we will focus on how to create an environment that allows you to get the most out of your study sessions. We will discuss where to study and what the conditions should be like in that location.*

9.1 Parabolas

Objectives

1. **Graph equations of the form $y = ax^2 + bx + c$.**
2. **Graph equations of the form $y = a(x - h)^2 + k$.**
3. **Graph equations of the form $x = ay^2 + by + c$.**
4. **Graph equations of the form $x = a(y - k)^2 + h$.**
5. **Find a vertex by completing the square.**
6. **Find the equation of a parabola that meets the given conditions.**

In Chapter 7, we graphed parabolas that opened either upward or downward. These parabolas were associated with equations that could be expressed as functions of x. After reviewing these graphs, we will learn to graph parabolas that open to the right or to the left.

Graphing Equations of the Form $y = ax^2 + bx + c$

Objective 1 Graph equations of the form $y = ax^2 + bx + c$. Here is a brief summary for graphing equations of the form $y = ax^2 + bx + c$:

Graphing an Equation of the Form $y = ax^2 + bx + c$

- Determine whether the parabola opens upward or downward.
- Find the vertex of the parabola.
- Find the y-intercept of the parabola.
- After plotting the vertex and the y-intercept, determine whether there are any x-intercepts, and if there are any, find them.
- Find the axis of symmetry $\left(x = \frac{-b}{2a}\right)$ and use it to find the point that is symmetric to the y-intercept.

EXAMPLE 1 Graph $y = x^2 + 6x - 2$. Label the vertex, y-intercept, x-intercept(s) (if any), and the axis of symmetry.

Solution

Since a is positive, this parabola will open upward. We begin by finding the coordinates of the vertex.

$x = \frac{-6}{2(1)}$ Substitute 1 for a and 6 for b into $x = \frac{-b}{2a}$.

$x = -3$ Simplify.

To find the y-coordinate, we substitute -3 for x.

$y = (-3)^2 + 6(-3) - 2$ Substitute -3 for x.

$y = -11$ Simplify.

The vertex is $(-3, -11)$.

Now we turn our attention to the y-intercept. We substitute 0 for x in the original equation.

$y = (0)^2 + 6(0) - 2$ Substitute 0 for x.
$y = -2$ Simplify.

The y-intercept is $(0, -2)$.

Since the vertex is located below the x-axis and the parabola opens upward, the graph must cross the x-axis. To find the x-intercepts, we substitute 0 for y and solve for x.

$0 = x^2 + 6x - 2$ Substitute 0 for y. Since $x^2 + 6x - 2$ does not factor, we can use the quadratic formula.

$x = \dfrac{-6 \pm \sqrt{(6)^2 - 4(1)(-2)}}{2(1)}$ Substitute 1 for a, 6 for b, and -2 for c.

$x = \dfrac{-6 \pm \sqrt{44}}{2}$ Simplify radicand.

$x = \dfrac{-6 \pm 2\sqrt{11}}{2}$ Simplify the radical.

$x = \dfrac{\overset{1}{\cancel{2}}(-3 \pm \sqrt{11})}{\underset{1}{\cancel{2}}}$ Factor the numerator and divide out common factors.

$x = -3 \pm \sqrt{11}$ Simplify.

Using a calculator, we find that $-3 + \sqrt{11} \approx 0.3$ and $-3 - \sqrt{11} \approx -6.3$. The x-intercepts are approximately at $(0.3, 0)$ and $(-6.3, 0)$.

Here is the graph of the given equation. The axis of symmetry, $x = -3$, has been used to find the coordinates of the point symmetric to the y-intercept, $(-6, -2)$.

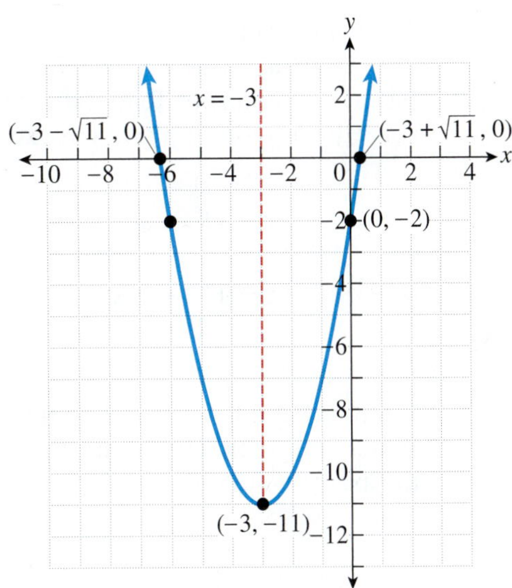

Quick Check 1 **Graph $y = -x^2 + 2x + 15$. Label the vertex, y-intercept, x-intercept(s) (if any), and the axis of symmetry.**

Graphing Equations of the Form $y = a(x - h)^2 + k$

Objective 2 **Graph equations of the form $y = a(x - h)^2 + k$.** Now we turn our attention to reviewing the graphs of equations of the form $y = a(x - h)^2 + k$.

Graphing an Equation of the Form $y = a(x - h)^2 + k$

- Determine whether the parabola opens upward or downward.
- Find the vertex of the parabola, which is the point (h, k).
- Find the y-intercept of the parabola.
- After plotting the vertex and the y-intercept, find the x-intercepts, if there are any.
- Use the axis of symmetry $(x = h)$ to find the point that is symmetric to the y-intercept.

Example 2 Graph $y = -(x - 1)^2 + 9$. Label the vertex, y-intercept, x-intercept(s) (if any), and the axis of symmetry.

Solution

Since a is negative, this parabola opens downward. The vertex of this parabola is at $(1, 9)$. We next find the y-intercept by substituting 0 for x.

$$y = -(0 - 1)^2 + 9 \quad \text{Substitute 0 for } x.$$
$$y = 8 \quad \text{Simplify.}$$

The y-intercept is $(0, 8)$.

Since the vertex is above the x-axis and the parabola opens downward, this parabola has two x-intercepts. To find the x-intercepts, we substitute 0 for y and solve for x. We will solve the resulting equation by extracting square roots.

$$0 = -(x - 1)^2 + 9 \quad \text{Substitute 0 for } y.$$
$$(x - 1)^2 = 9 \quad \text{Add } (x - 1)^2 \text{ to isolate the squared expression.}$$
$$\sqrt{(x - 1)^2} = \pm\sqrt{9} \quad \text{Take the square root of both sides.}$$
$$x - 1 = \pm 3 \quad \text{Simplify each square root.}$$
$$x = 1 \pm 3 \quad \text{Add 1 to both sides to isolate } x.$$

$1 + 3 = 4$ and $1 - 3 = -2$, so the x-intercepts are $(4, 0)$ and $(-2, 0)$.
Here is the graph. The point $(2, 8)$ is symmetric to the y-intercept. We can find it by using the axis of symmetry, $x = 1$.

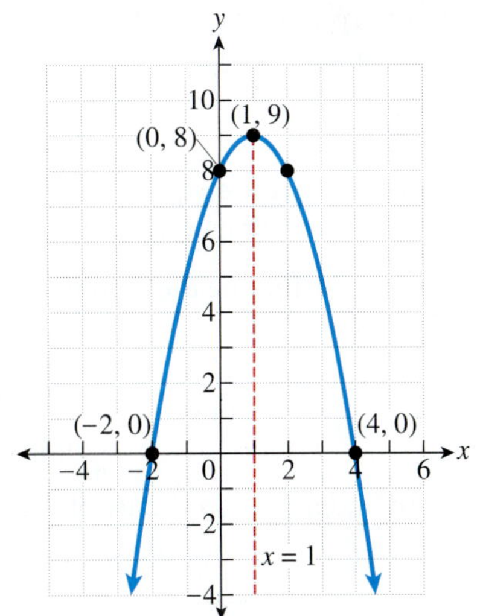

Quick Check 2 **Graph $y = (x + 4)^2 + 3$. Label the vertex, y-intercept, x-intercept(s) (if any), and the axis of symmetry.**

Graphing Equations of the Form $x = ay^2 + by + c$

Objective 3 **Graph equations of the form $x = ay^2 + by + c$.** If the variables x and y are interchanged in the equation $y = ax^2 + bx + c$, then the graph of the equation will be a parabola that opens to the right or to the left. The process for graphing this type of parabola is quite similar to graphing a parabola that opens up or down.

Graphing an Equation of the Form $x = ay^2 + by + c$

- Determine whether the parabola opens to the right or to the left. The parabola will open to the right if a is positive; to the left if a is negative.
- Find the vertex of the parabola by using the formula $y = \frac{-b}{2a}$. Substitute this result in the original equation for y to find the x-coordinate.
- Find the x-intercept of the parabola by substituting 0 for y in the original equation.
- After plotting the vertex and the x-intercept, find the y-intercepts, if there are any, by letting $x = 0$ in the original equation and solving the resulting quadratic equation for y.
- Find the axis of symmetry $\left(y = \frac{-b}{2a}\right)$ and use it to find the point on the parabola that is symmetric to the x-intercept.

EXAMPLE 3 Graph $x = y^2 + 2y - 3$. Label the vertex, any intercepts, and the axis of symmetry.

Solution

Notice that the squared variable in the equation is y, not x, so the parabola will open either to the right or to the left. This parabola opens to the right, since a is positive. We begin by finding the y-coordinate of the vertex, using the formula $y = \frac{-b}{2a}$.

$y = \frac{-2}{2(1)}$ Substitute 1 for a and 2 for b into $y = \frac{-b}{2a}$.
$y = -1$ Simplify.

To find the x-coordinate of the vertex, we substitute -1 for y in the original equation.

$x = (-1)^2 + 2(-1) - 3$ Substitute -1 for y.
$x = -4$ Simplify.

The vertex of the parabola is $(-4, -1)$.

A Word of Caution Keep in mind that even though we find the y-coordinate first and the x-coordinate second, we must write the ordered pair as (x, y).

Next we find the x-intercept by substituting 0 for y in the equation.

$x = (0)^2 + 2(0) - 3$ Substitute 0 for y.
$x = -3$ Simplify.

The x-intercept is $(-3, 0)$.

Since the vertex is located to the left of the y-axis and the parabola opens to the right, the parabola has two y-intercepts. We can find these intercepts by substituting 0 for x and solve the resulting equation for y.

$0 = y^2 + 2y - 3$ Substitute 0 for x.
$0 = (y + 3)(y - 1)$ Factor.
$y = -3$ or $y = 1$ Set each factor equal to 0 and solve.

The y-intercepts are $(0, -3)$ and $(0, 1)$.

We can find another point on the graph by finding the point on the parabola that is symmetrical to the x-intercept $(-3, 0)$. We can do this using the axis of symmetry, which in this case is the horizontal line $y = -1$.

Since the x-intercept is one unit above the axis of symmetry, the point symmetric to it will be one unit below the axis of symmetry at $(-3, -2)$. Here is the graph:

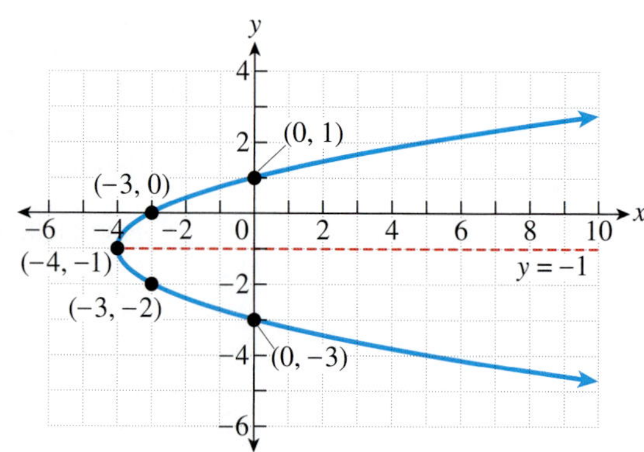

***Quick Check* 3** **Graph $x = y^2 + 10y - 1$. Label the vertex, any intercepts, and the axis of symmetry.**

If a parabola opens to the right and the vertex is to the right of the y-axis, then it does not have any y-intercepts. The same holds true for parabolas that open to the left whose vertex is to the left of the y-axis.

EXAMPLE 4 Graph $x = -y^2 + 6y - 10$. Label the vertex, any intercepts, and the axis of symmetry.

Solution

Since a is negative, the parabola opens to the left. We begin by finding the vertex.

$y = \dfrac{-b}{2a}$		$x = -y^2 + 6y - 10$	
$y = \dfrac{-6}{2(-1)}$	Substitute for a and b.	$x = -(3)^2 + 6(3) - 10$	Substitute 3 for y.
$y = 3$	Simplify.	$x = -1$	Simplify.

The vertex of the parabola is $(-1, 3)$. Next we find the x-intercept.

$x = -(0)^2 + 6(0) - 10$ Substitute 0 for y.
$x = -10$ Simplify.

The x-intercept is $(-10, 0)$.

The vertex is located to the left of the y-axis. Since the parabola opens to the left, the parabola does not have any y-intercepts. The point $(-10, 6)$ shown on the graph is symmetric to the x-intercept and can be found by using the axis of symmetry $y = 3$.

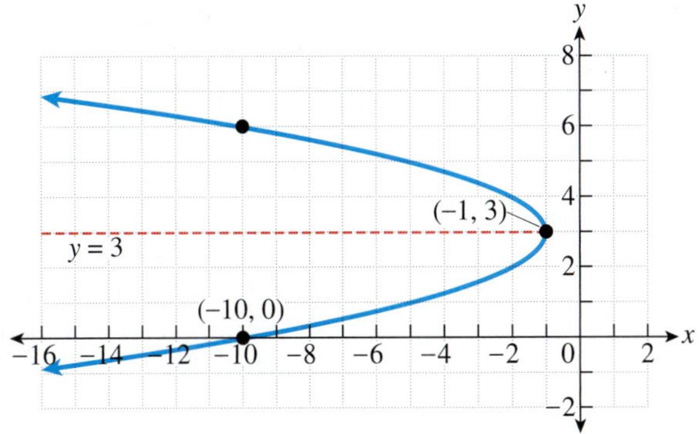

***Quick Check* 4**

Graph $x = -y^2 + 8y - 12$. Label the vertex, any intercepts, and the axis of symmetry.

Graphing Equations of the Form $x = a(y - k)^2 + h$

Objective 4 **Graph equations of the form $x = a(y - k)^2 + h$.** Now we turn our attention to the graphs of equations of the form $x = a(y - k)^2 + h$. The graph of an equation in this form will be a parabola that opens to the right or to the left with a vertex at (h, k).

Graphing an Equation of the Form $x = a(y - k)^2 + h$

- Determine whether the parabola opens to the right or to the left.
- Find the vertex of the parabola, which is the point (h, k).
- Find the x-intercept of the parabola by substituting 0 for y in the original equation.
- After plotting the vertex and the x-intercept, find the y-intercepts, if there are any, by letting $x = 0$ in the original equation and solving for y. We can solve this equation by extracting square roots.
- Find the axis of symmetry ($y = k$) and use it to find the point on the parabola that is symmetric to the x-intercept.

EXAMPLE 5 Graph $x = -(y - 1)^2 + 4$. Label the vertex, any intercepts, and the axis of symmetry.

Solution

Since a is negative, this parabola opens to the left. The vertex of this parabola is $(4, 1)$.

$$x = -(y - \underset{\substack{\uparrow \\ k = 1}}{1})^2 + \underset{\substack{\uparrow \\ h = 4}}{4}$$

We next find the x-intercept by substituting 0 for y.

$x = -(0 - 1)^2 + 4$ Substitute 0 for y.
$x = 3$ Simplify.

The x-intercept is $(3, 0)$.
Since the vertex is to the right of the y-axis and the parabola opens to the left, this parabola has two y-intercepts. To find the y-intercepts, we substitute 0 for x and solve for y. We will solve the resulting equation by extracting square roots.

$0 = -(y - 1)^2 + 4$ Substitute 0 for x.
$(y - 1)^2 = 4$ Add $(y - 1)^2$ to isolate the squared expression.
$\sqrt{(y - 1)^2} = \pm\sqrt{4}$ Take the square roots of both sides.
$y - 1 = \pm 2$ Simplify.
$y = 1 \pm 2$ Add 1 to both sides to isolate y.

$1 + 2 = 3$ and $1 - 2 = -1$, so the y-intercepts are $(0, 3)$ and $(0, -1)$. Shown next is the graph. We can find the point $(3, 2)$ by using the y-intercept and the axis of symmetry, which is $y = 1$.

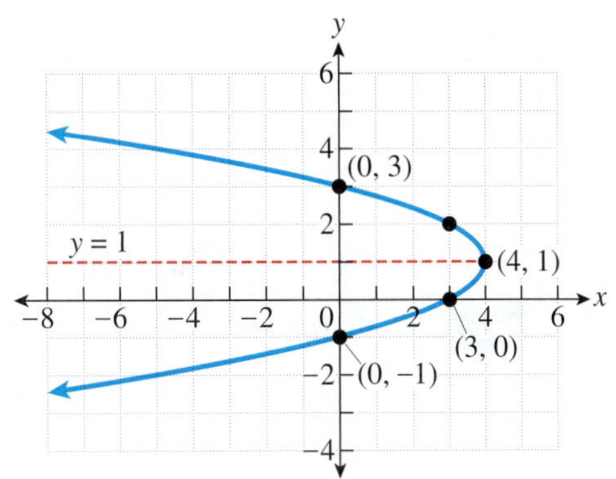

Quick Check **5** **Graph $x = -(y - 4)^2 - 3$. Label the vertex, any intercepts, and the axis of symmetry.**

EXAMPLE 6 Graph $x = (y - 2)^2$. Label the vertex, any intercepts, and the axis of symmetry.

Solution

Since a is positive, this parabola opens to the right. The vertex of this parabola is $(0, 2)$. To see this, we could rewrite the equation as $x = (y - 2)^2 + 0$. To find the x-intercept, we substitute 0 for y in the original equation and solve for x.

$$x = (0 - 2)^2 \quad \text{Substitute 0 for } y.$$
$$x = 4 \quad \text{Simplify.}$$

The x-intercept is $(4, 0)$.

Since the vertex $(0, 2)$ is located on the y-axis, the vertex is the only y-intercept. The y-intercept and the axis of symmetry, which is $y = 2$, can be used to find the point $(4, 4)$.

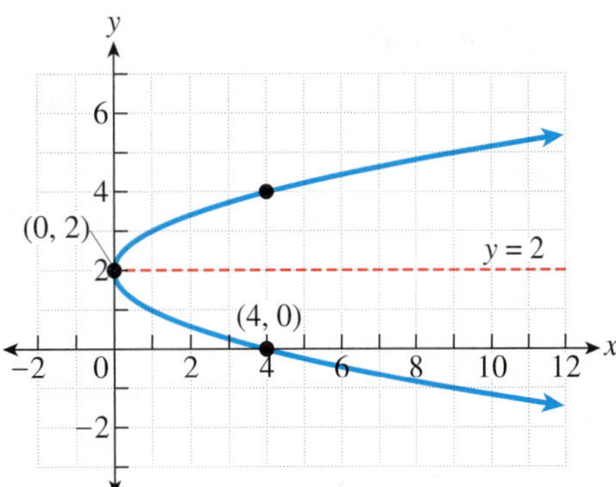

Quick Check 6 Graph $x = (y - 3)^2 - 1$. Label the vertex, any intercepts, and the axis of symmetry.

Finding the Vertex by Completing the Square

Objective 5 **Find a vertex by completing the square.** Consider the equation $y = x^2 - 8x + 33$. Although we could find the vertex by using the formula $x = \frac{-b}{2a}$, we can also find the vertex by completing the square to write the equation in the form $y = a(x - h)^2 + k$, where (h, k) is the vertex.

EXAMPLE 7 Find the vertex of the parabola defined by the equation $y = x^2 - 8x + 33$ by completing the square.

Solution

$$y = x^2 - 8x + 33$$

$$y - 33 = x^2 - 8x \quad \text{Subtract 33 from both sides.}$$

$$y - 33 + 16 = x^2 - 8x + 16 \quad \left(\tfrac{-8}{2}\right)^2 = 16. \text{ Add 16 to both sides.}$$

$$y - 17 = (x - 4)^2 \quad \text{Combine like terms on the left side. Factor the quadratic expression on the right side.}$$

$$y = (x - 4)^2 + 17 \quad \text{Add 17 to both sides.}$$

The vertex is $(4, 17)$.

Although it may seem easier to have used the formula $x = \frac{-b}{2a}$ to find the vertex in this case, the ability to complete the square will be an important skill in the sections that follow.

EXAMPLE 8 Find the vertex of the parabola defined by the equation $x = -y^2 + 4y - 50$ by completing the square.

Solution

$$x = -y^2 + 4y - 50$$

$$x + 50 = -y^2 + 4y \quad \text{Add 50 to both sides.}$$

$$x + 50 = -(y^2 - 4y) \quad \text{Factor out a negative 1 on the right side.}$$

$$x + 50 - 4 = -(y^2 - 4y + 4) \quad \left(\tfrac{-4}{2}\right)^2 = 4. \text{ Add 4 to the expression in parentheses on the right side. Subtract 4 from the left side, because there is a factor of } -1 \text{ in front of the parentheses.}$$

$$x + 46 = -(y - 2)^2 \quad \text{Combine like terms on the left side. Factor the quadratic expression on the right side.}$$

$$x = -(y - 2)^2 - 46 \quad \text{Subtract 46 from both sides.}$$

The vertex is $(-46, 2)$.

Quick Check 7

Find the vertex of the parabola defined by the given equation by completing the square.

a) $y = x^2 + 4x + 20$

b) $x = -y^2 - 6y + 17$

Finding the Equation of a Parabola

Objective 6 **Find the equation of a parabola that meets the given conditions.** There are several techniques that can be used to find the equation of a parabola. Here is one that uses the intercepts of the parabola.

EXAMPLE 9 Find the equation of the parabola whose graph is shown at the accompanying figure.

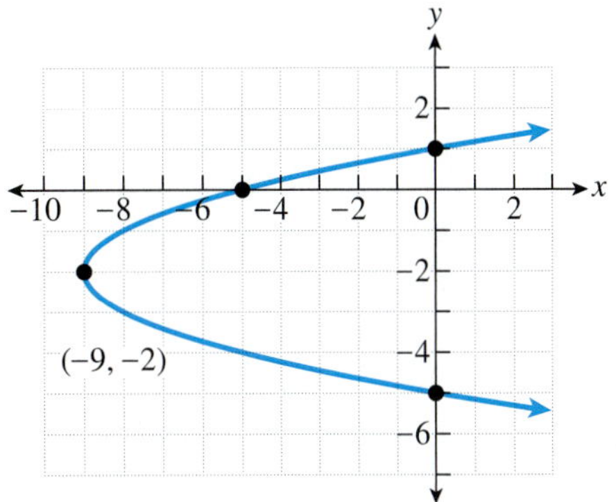

Solution

We begin by noting that the parabola opens to the right, so its equation is of the form $x = ay^2 + by + c$, where $a > 0$. We can observe the y-intercepts are $(0, 1)$ and $(0, -5)$, and the x-intercept is $(-5, 0)$.

Since the y-coordinates of the y-intercept are $y = 1$ and $y = -5$, we know that $y - 1$ and $y + 5$ are factors of the expression $ay^2 + by + c$. Multiplying $(y - 1)(y + 5)$ yields the expression $y^2 + 4y - 5$, so the equation is of the form $x = a(y^2 + 4y - 5)$. To find the value of a, we use the coordinates of the x-intercept $(-5, 0)$. We could also use the coordinates of any other point on the graph of the parabola, including the vertex. After substituting -5 for x and 0 for y into the equation $x = a(y^2 + 4y - 5)$, we solve for a.

$$x = a(y^2 + 4y - 5)$$

$$-5 = a((0)^2 + 4(0) - 5) \quad \text{Substitute } -5 \text{ for } x \text{ and } 0 \text{ for } y.$$

$$-5 = a(-5) \quad \text{Simplify.}$$

$$1 = a \quad \text{Divide both sides by } -5.$$

Since $a = 1$, the equation of the parabola is $x = 1(y^2 + 4y - 5)$, or simply, $x = y^2 + 4y - 5$.

Note that we could have used the vertex to find the equation of the parabola. Since the vertex is $(-9, -2)$ and the parabola opens to the right, we know that the equation is of the form $x = a(y + 2)^2 - 9$. Substituting the coordinates of any other point on the parabola, such as the x-intercept, for x and y will lead to the equation $x = 1(y + 2)^2 - 9$, or $x = (y + 2)^2 - 9$ which is equivalent to $x = y^2 + 4y - 5$.

Quick Check **8** **Find the equation of the parabola whose graph is shown at the accompanying figure.**

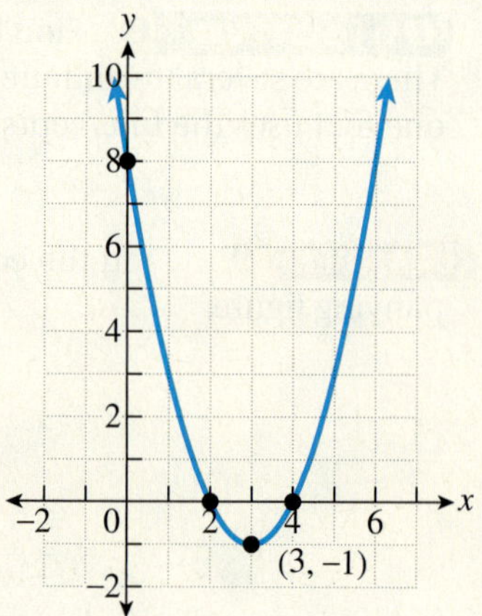

Building Your Study Strategy **Study Environment, 1** **Location** When choosing a location to study, it is important to select a location that is free from distractions. For example, if you study in a room that has a television on, you will find yourself repeatedly glancing up at the television. This steals time and attention from the task at hand. Working in a room that constantly has people entering and exiting can be distracting as well.

Establish locations that will be used only for studying, not for other purposes. When you sit down to study, you will feel as if you have entered your workplace. This will help you to focus on your studies. As an exercise, keep a journal of things that distract your attention while studying. In the future, choose a location that will minimize your exposure to these distractions.

EXERCISES 9.1

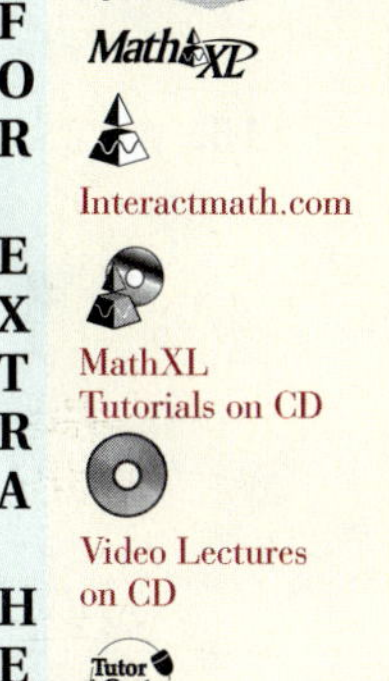

Vocabulary

1. A parabola of the form $y = ax^2 + bx + c$ or $y = a(x - h)^2 + k$ opens ________ if a is positive and opens ________ if a is negative.

2. A parabola of the form $y = a(x - h)^2 + k$ has its vertex at the point ________.

3. A parabola of the form $x = ay^2 + by + c$ or $x = a(y - k)^2 + h$ opens to the ________ if a is positive and opens to the ________ if a is negative.

4. The ________-coordinate of the vertex of a parabola of the form $x = ay^2 + by + c$ is $\dfrac{-b}{2a}$.

5. The vertex of a parabola of the form $x = a(y - k)^2 + h$ is at the point ________.

6. For a parabola that opens either to the right or to the left and has a vertex at the point (h, k), the axis of symmetry is a horizontal line whose equation is ________.

Graph the parabola. Label the vertex and any intercepts.

7. $y = x^2 + 8x + 11$

8. $y = x^2 - 9x$

9. $y = x^2 - 4x + 7$

10. $y = x^2 + 6x + 9$

11. $y = -x^2 + x + 20$

12. $y = -x^2 - 9x - 14$

13. $y = -x^2 + 5$

14. $y = -x^2 + 11x - 20$

15. $y = (x + 5)^2 + 5$

16. $y = (x - 1)^2 - 15$

17. $y = -(x + 2)^2 + 9$

18. $y = -(x - 3)^2 - 1$

19. $x = y^2 + 2y - 24$

20. $x = y^2 + 12y + 27$

21. $x = y^2 + 6y + 2$

22. $x = y^2 + 2y - 4$

23. $x = y^2 + 3y + 10$

24. $x = y^2 - 8y + 20$

25. $x = -y^2 + 8y - 15$

26. $x = -y^2 + 4y + 32$

27. $x = -y^2 - 9y - 6$

28. $x = -y^2 + 2y - 6$

29. $x = (y + 5)^2 - 1$

30. $x = (y - 2)^2 - 5$

31. $x = (y + 2)^2 + 2$

32. $x = (y - 7)^2 + 10$

33. $x = -(y - 2)^2 + 4$

34. $x = -(y - 3)^2 + 7$

35. $x = -y^2 - 6y - 15$

36. $x = -y^2 + 10y - 13$

37. $x = (y - 1)^2 - 7$

38. $x = y^2 - 8y - 20$

39. $y = x^2 + 7x + 4$

40. $x = -y^2 + 2y + 15$

41. $y = -(x - 1)^2 + 9$

42. $y = x^2 + 5x - 24$

43. $x = (y + 5)^2 - 4$

44. $y = (x - 4)^2 + 3$

45. $x = -(y + 5)^2 - 3$

46. $x = y^2 - 5y + 15$

Find the vertex of the parabola by completing the square.

47. $y = x^2 - 2x + 15$

48. $y = x^2 + 10x + 62$

49. $y = -x^2 + 14x + 76$

50. $y = -x^2 - 8x + 21$

51. $x = y^2 + 6y - 40$

52. $x = y^2 - 12y - 33$

53. $x = -y^2 + 4y - 51$

54. $x = -y^2 - 16y + 109$

Find the equation of the parabola that has been graphed by using the intercepts.

55.

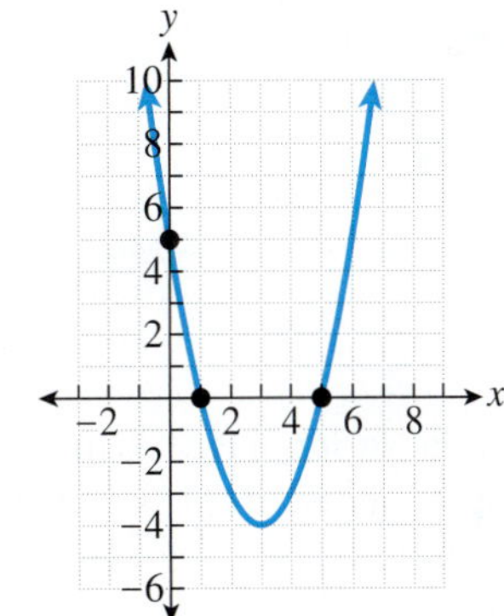

56.

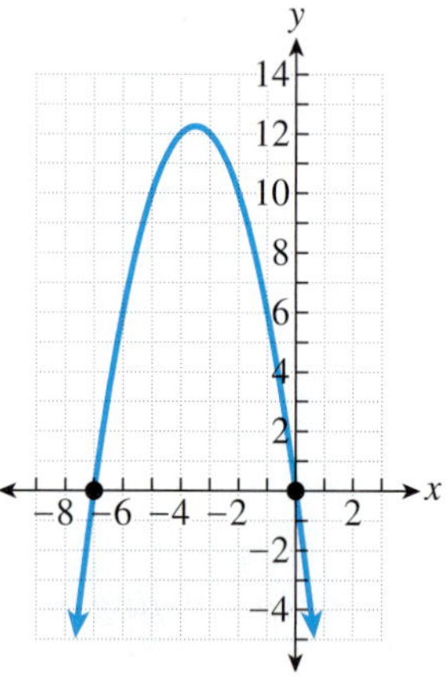

57.

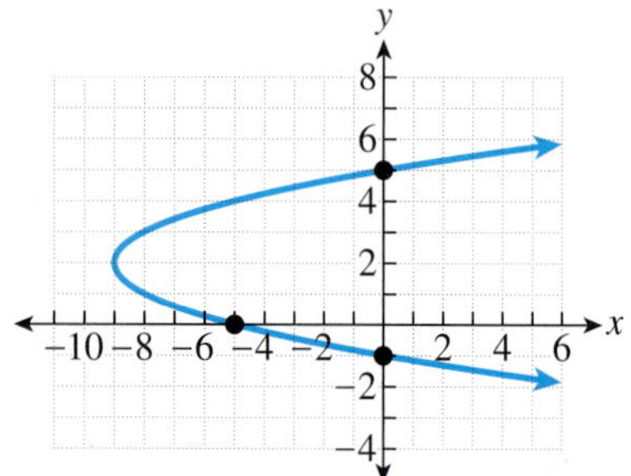

58.

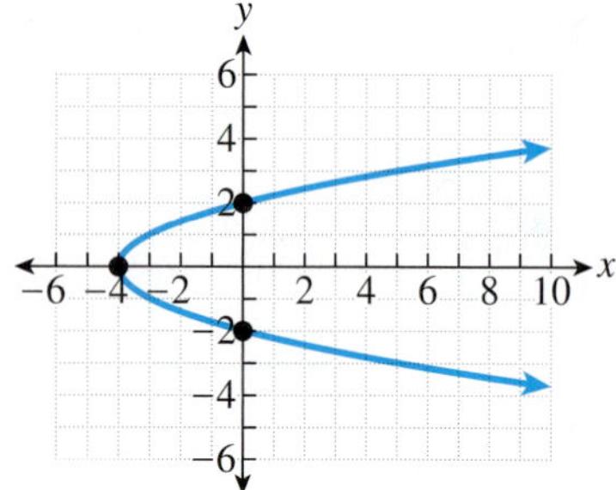

59. A footbridge in the Sequoia National Park is in the shape of a parabola. It spans a distance of 100 feet, and at its lowest point it drops 5 feet in elevation.

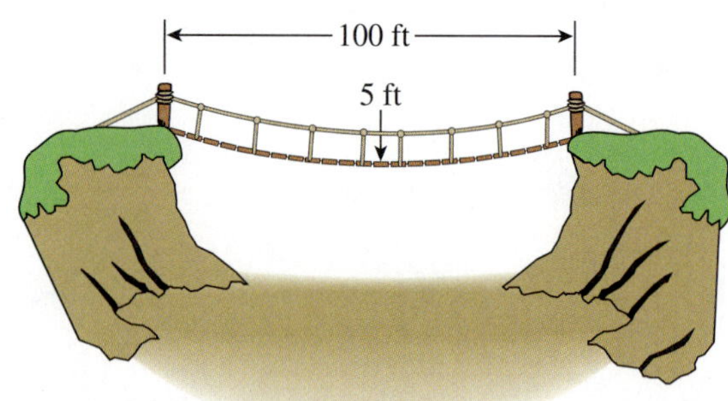

The bridge is displayed as follows on a rectangular coordinate plane:

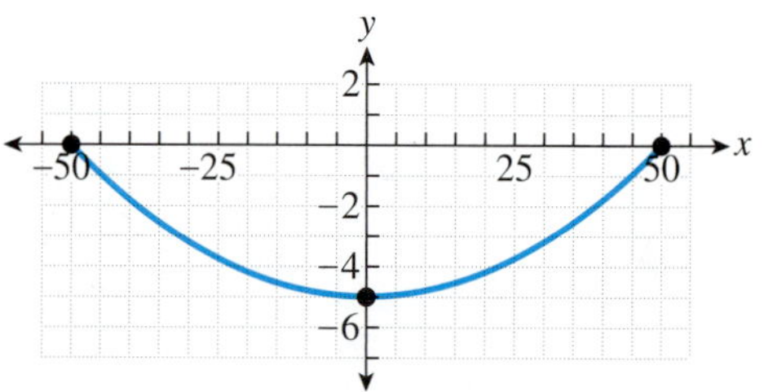

a) Find the equation for the parabola.

b) After a person walks 30 feet from one side of the bridge, how far has the person dropped in elevation?

60. A freeway overpass is built atop a parabolic arch. The width of the arch at the base is 120 feet, and its height at its peak is 30 feet.

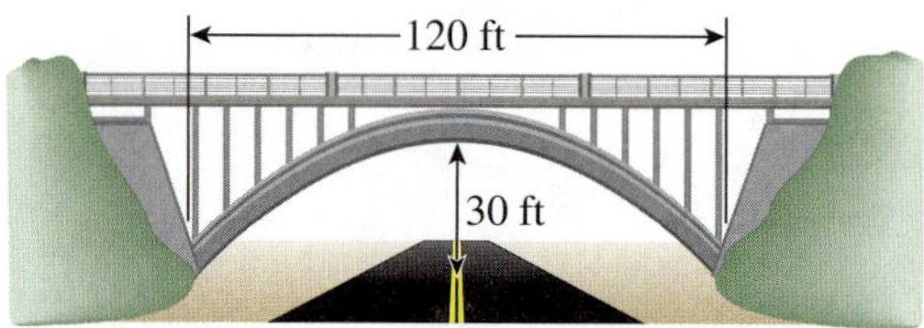

The overpass is displayed as follows on a rectangular coordinate plane:

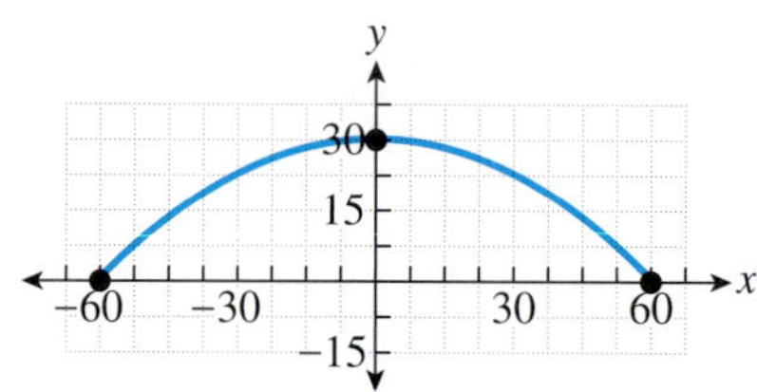

a) Find the equation for the parabola.

b) How far from the center of the arch does the height drop to below 16 feet? Round to the nearest tenth of a foot.

Determine which equation is associated with each given graph.

a) $y = (x - 2)^2 - 3$ **b)** $y = x^2 - 6x + 7$

c) $x = (y - 2)^2 - 3$ **d)** $x = y^2 - 6y + 7$

e) $x = -y^2 + 6y - 11$ **f)** $x = -(y - 2)^2 - 3$

g) $x = (y + 2)^2$ **h)** $x = 3(y + 2)^2$

61.

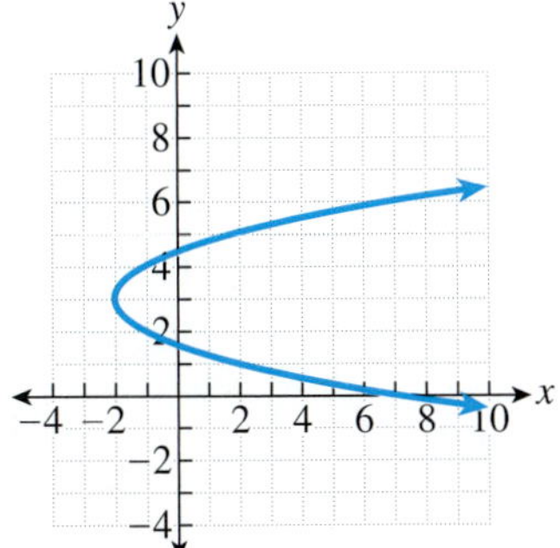

62.

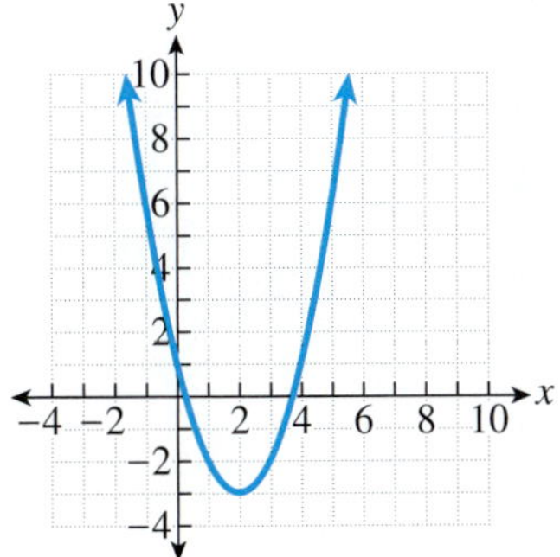

63.

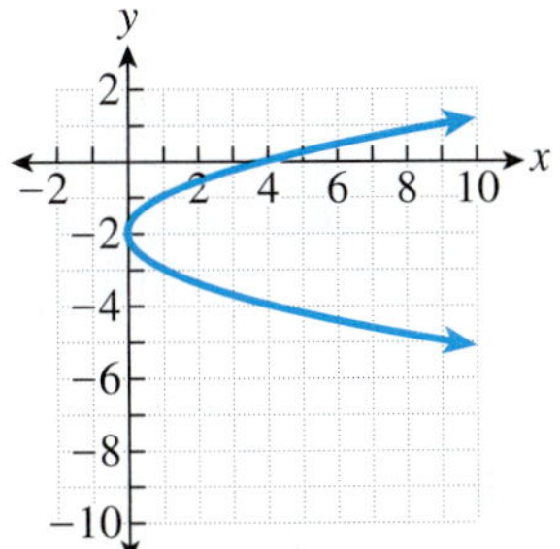

64.

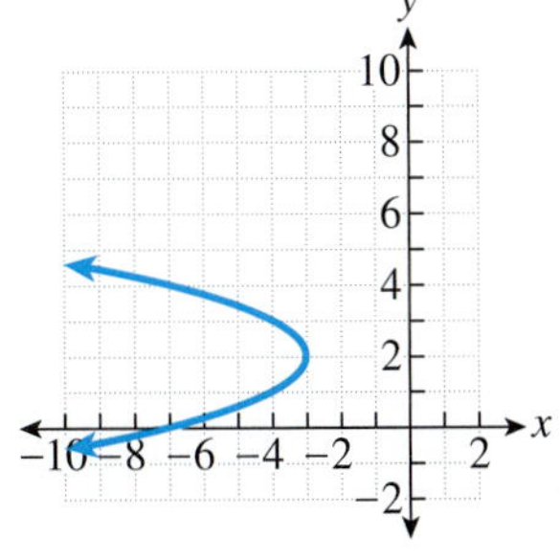

65.

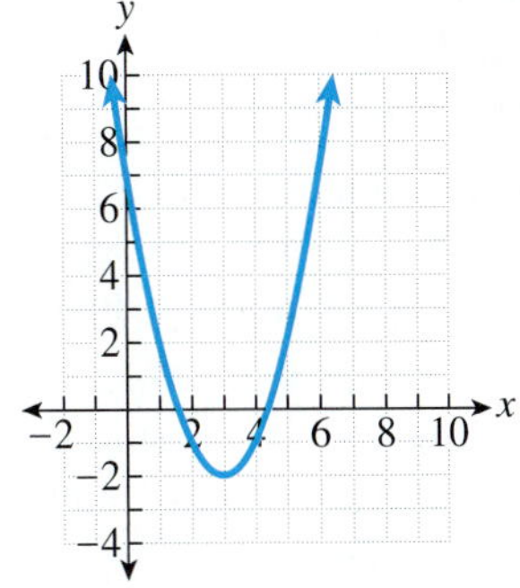

66.

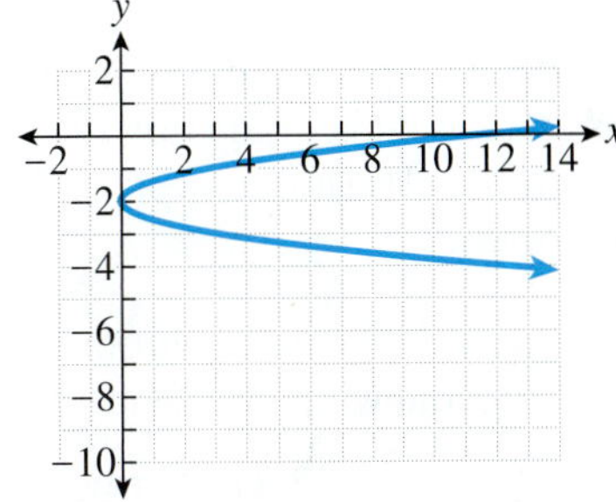

67.

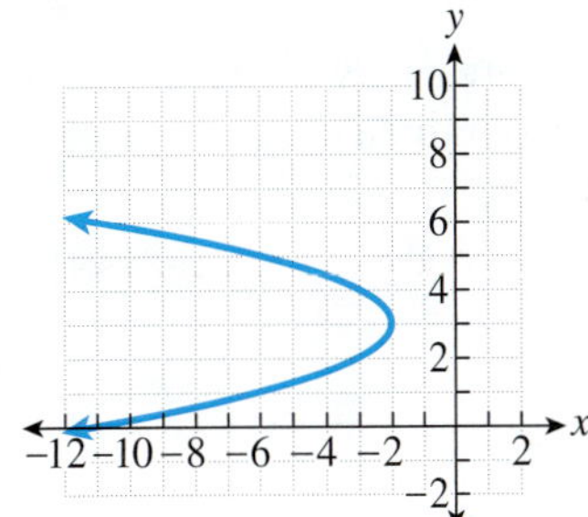

68.

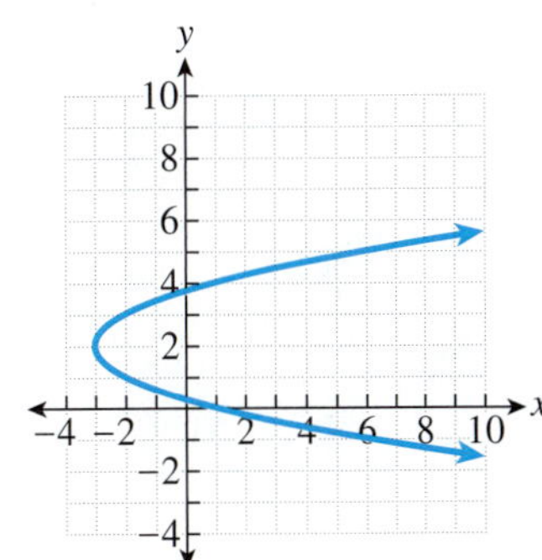

Writing in Mathematics

Answer in complete sentences.

69. Explain how to determine whether a parabola opens to the left or to the right, as opposed to opening up or opening down.

70. Suppose that a parabola opens to the right. Explain how you would determine that the parabola has no y-intercepts.

71. Explain how to complete the square to find the vertex of a parabola of the form $x = ay^2 + by + c$. Use an example to illustrate the process.

72. ***Solutions Manual**** Write a solutions manual page for the following problem:

Graph $x = -(y + 2)^2 + 3$. *Label any intercepts.*

73. ***Newsletter**** Write a newsletter explaining how to graph a parabola of the form $x = ay^2 + by + c$.

***See Appendix B for details and sample answers.**

Objectives

1. **Use the distance and midpoint formulas.**
2. **Graph circles centered at the origin.**
3. **Graph circles centered at a point (h, k).**
4. **Find the center of a circle and its radius by completing the square.**
5. **Find the equation of a circle that meets the given conditions.**

The Distance Formula

Objective 1 **Use the distance and midpoint formulas.** Suppose that we are given two points, (x_1, y_1) and (x_2, y_2), and are asked to find the distance between these two points. If we plot the points on a graph, we see that we can construct a right triangle whose hypotenuse has these points as its endpoints.

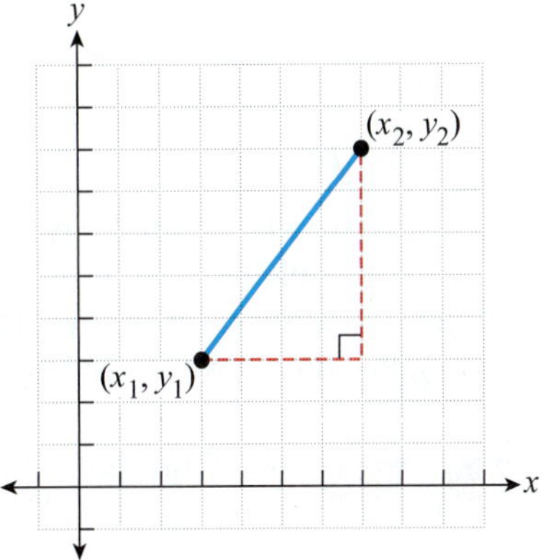

The distance between the two points, d, is the length of the hypotenuse. The length of the leg on the bottom of the triangle is $x_2 - x_1$, and the length of the leg on the right side of the triangle is $y_2 - y_1$.

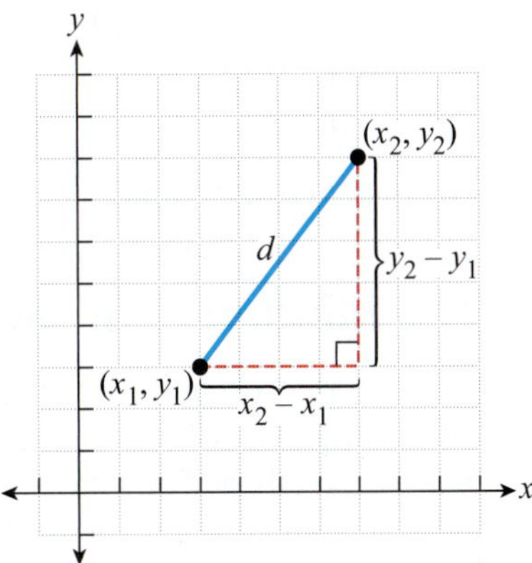

The Pythagorean theorem tells us that $d^2 = (x_2 - x_1)^2 + (y_2 - y_1)^2$. If we take the square root of both sides of this equation, we obtain a formula for the distance in terms of the coordinates of the two points (We omit the negative square root because a distance is a nonnegative number.)

The Distance Formula

The distance, d, between two points (x_1, y_1) and (x_2, y_2) is given by the formula

$$d = \sqrt{(x_2 - x_1)^2 + (y_2 - y_1)^2}.$$

EXAMPLE 1 Find the distance between the points $(3, 9)$ and $(6, 5)$.

Solution

We will let $(3, 9)$ be the point (x_1, y_1) and $(6, 5)$ be the point (x_2, y_2). Now we can use the distance formula.

$$\begin{aligned}
d &= \sqrt{(x_2 - x_1)^2 + (y_2 - y_1)^2} & & \\
d &= \sqrt{(6 - 3)^2 + (5 - 9)^2} & & \text{Substitute for } x_2, x_1, y_2, \text{ and } y_1. \\
d &= \sqrt{(3)^2 + (-4)^2} & & \text{Simplify the expressions in the radicand.} \\
d &= \sqrt{25} & & \text{Simplify the radicand.} \\
d &= 5 & & \text{Simplify the radical.}
\end{aligned}$$

The distance between the two points is five units.

The choice of which point will be (x_1, y_1) and which point will be (x_2, y_2) is completely arbitrary. Verify that if we let $(6, 5)$ be the point (x_1, y_1) and $(3, 9)$ be the point (x_2, y_2), then d would still have been equal to 5.

EXAMPLE 2 Find the distance between the points $(2, -6)$ and $(-5, -11)$. Round to the nearest tenth.

Solution

We will let $(2, -6)$ be the point (x_1, y_1) and $(-5, -11)$ be the point (x_2, y_2).

$$\begin{aligned}
d &= \sqrt{(x_2 - x_1)^2 + (y_2 - y_1)^2} & & \\
d &= \sqrt{(-5 - 2)^2 + (-11 - (-6))^2} & & \text{Substitute for } x_2, x_1, y_2, \text{ and } y_1. \\
d &= \sqrt{74} & & \text{Simplify the radicand.} \\
d &\approx 8.6 & & \text{Approximate, using a calculator.}
\end{aligned}$$

The distance between the two points is approximately 8.6 units.

***Quick Check* 1**

Find the distance between the given points. Round to the nearest tenth if necessary.

a) $(1, 2)$ and $(9, 8)$
b) $(3, -6)$ and $(-1, 6)$

The Midpoint Formula

Suppose that we are given two points, (x_1, y_1) and (x_2, y_2), and are asked to find their **midpoint,** which is the point on the line segment connecting the two points that is located exactly halfway between the two points. The x-coordinate of the midpoint must be equal to the average of the x-coordinates of the two points, and the y-coordinate must be equal to the average of the y-coordinates of the two points.

The Midpoint Formula

The midpoint of a line segment connecting two points (x_1, y_1) and (x_2, y_2) is the point whose coordinates are $\left(\frac{x_1 + x_2}{2}, \frac{y_1 + y_2}{2}\right)$.

EXAMPLE 3 Find the midpoint of the line segment that connects the points $(2, -9)$ and $(6, 3)$.

Solution

We will use the midpoint formula, treating $(2, -9)$ as (x_1, y_1) and $(6, 3)$ as (x_2, y_2). This choice is arbitrary.

x-coordinate

$$\frac{x_1 + x_2}{2} = \frac{2 + 6}{2} = 4$$

y-coordinate

$$\frac{y_1 + y_2}{2} = \frac{-9 + 3}{2} = -3$$

The midpoint is $(4, -3)$.

***Quick Check* 2**
Find the midpoint of the line segment that connects the points $(-4, -3)$ and $(5, 7)$.

Circles Centered at the Origin

Objective 2 **Graph circles centered at the origin.**

A **circle** is defined as the collection of all points (x, y) in a plane that are a fixed distance from a point called its **center**. The distance from the center to each point on the circle is called the **radius** of the circle.

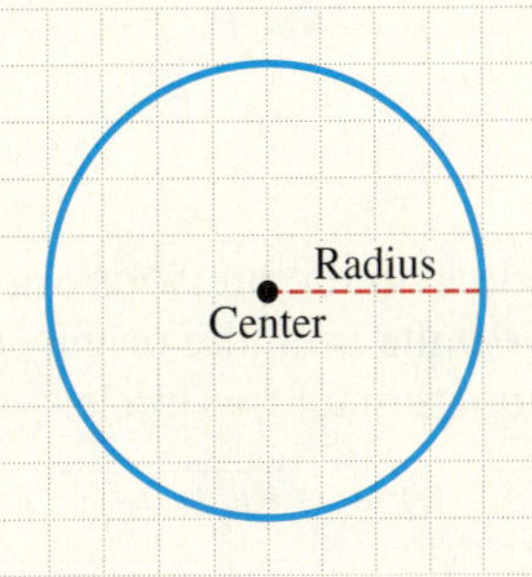

We will begin by learning to graph circles centered at the origin.

EXAMPLE 4 Graph the circle centered at the origin whose radius is 5.

Solution

We begin by plotting the center at the origin $(0, 0)$. From there we will move five units to the right of the origin and plot a point there. We repeat this process for points that are five units to the left, above, and below the center.

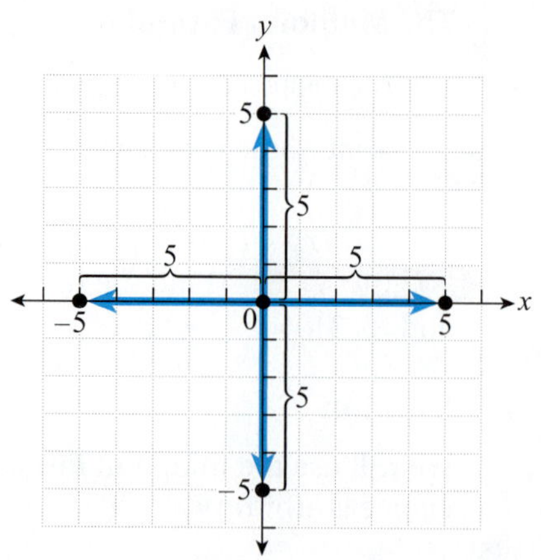

We finish by drawing the circle that passes through these points.

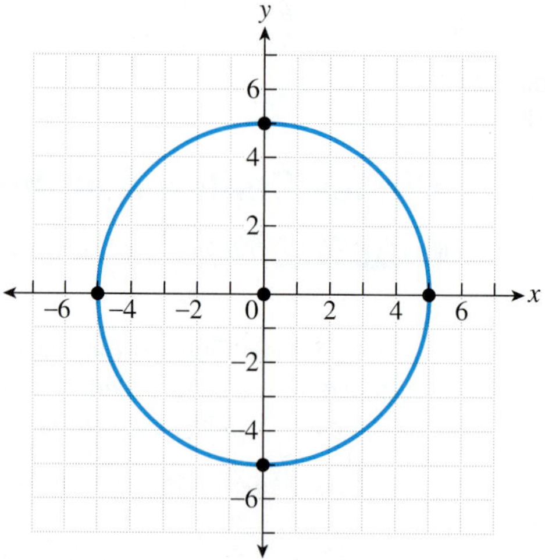

The circle in the previous example is the collection of all points (x, y) that are a distance of five units from the origin $(0, 0)$. We can use this fact and the distance formula to find the equation for this circle.

$$d = \sqrt{(x_2 - x_1)^2 + (y_2 - y_1)^2}$$

$$5 = \sqrt{(x - 0)^2 + (y - 0)^2}$$ Substitute 5 for d, x for x_2, 0 for x_1, y for y_2, and 0 for y_1.

$$5 = \sqrt{x^2 + y^2}$$ Simplify.

$$25 = x^2 + y^2$$ Square both sides.

The equation of the circle centered at the origin with a radius of 5 can be written as $x^2 + y^2 = 25$.

Equation of a Circle Centered at the Origin (Standard Form)

The equation for a circle with radius r centered at the origin is $x^2 + y^2 = r^2$.

EXAMPLE 5 Graph the circle $x^2 + y^2 = 10$. State the center and radius.

Solution

This circle also has its center at the origin, and since $r^2 = 10$, the radius r is equal to $\sqrt{10}$. Note that $\sqrt{10} \approx 3.2$.

We begin to graph the circle by plotting the center $(0, 0)$. Next, we plot the points that are approximately 3.2 units to the right and left of the center, as well as the points that are approximately 3.2 units above and below the center. Once these points have been plotted, draw the circle that passes through them.

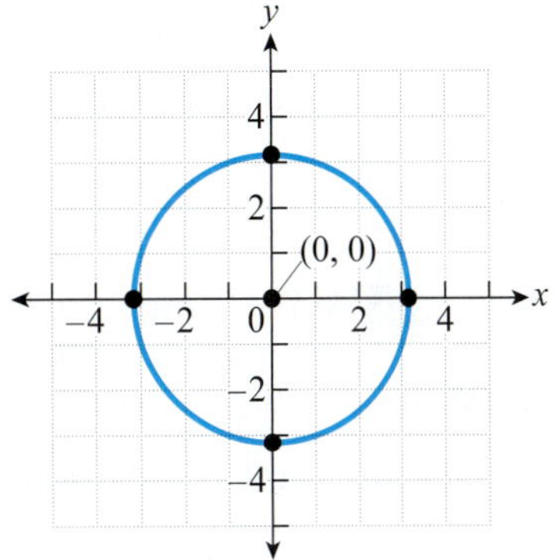

Quick Check 3
Graph the circle $x^2 + y^2 = 12$. State the center and radius.

Circles Centered at a Point Other than the Origin

Objective 3 **Graph circles centered at a point (h, k).** The equation for a circle centered at a point (h, k) can be derived by the distance formula, just as the equation for a circle centered at the origin was derived.

$$\sqrt{(x-h)^2 + (y-k)^2} = r \quad \text{The distance from } (x, y) \text{ to the center } (h, k) \text{ is } r.$$
$$(x-h)^2 + (y-k)^2 = r^2 \quad \text{Square both sides of the equation.}$$

Equation of a Circle (Standard Form)

The equation for a circle with radius r centered at the point (h, k) is

$$(x-h)^2 + (y-k)^2 = r^2.$$

EXAMPLE 6 Graph the circle $(x-3)^2 + (y-4)^2 = 4$. State the center and radius.

Solution

Since the equation of the circle is of the form $(x-h)^2 + (y-k)^2 = r^2$, the center of this circle is the point $(3, 4)$.

Since $r^2 = 4$, the radius r is equal to $\sqrt{4}$, or 2.

We begin by plotting the center $(3, 4)$. Next we plot the points that are 2 units to the right and left of the center, as well as the points that are 2 units above and

Quick Check 4
Graph the circle $(x-6)^2+(y-5)^2=9$. State the center and radius.

below the center. Once these points have been plotted, draw the circle that passes through them.

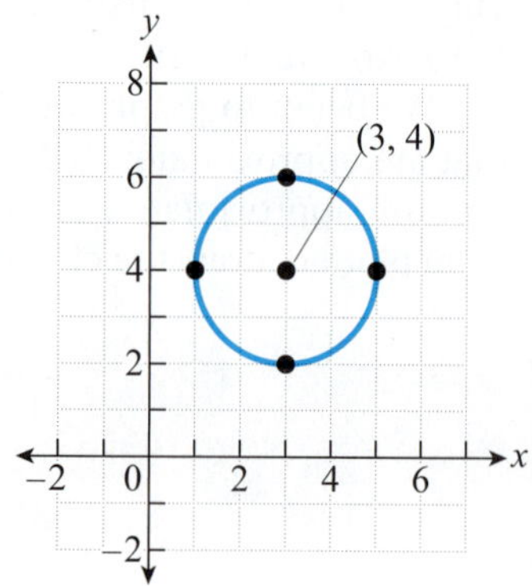

EXAMPLE 7 Graph the circle $(x+3)^2+(y-6)^2=25$. State the center and radius.

Solution

The center of this circle is the point $(-3, 6)$. To find the x-coordinate of the center, we can rewrite $x+3$ as $x-(-3)$. This shows us that h is -3. Alternatively, we could have set the expression containing x that was being squared, $x+3$, equal to 0 and solved for x. A similar approach will yield the y-coordinate of the center.

Since $r^2=25$, the radius r is equal to $\sqrt{25}$, or 5.

We begin to graph the circle by plotting the center $(-3, 6)$. Next, we plot the points that are five units to the right and left of the center, as well as the points that are five units above and below the center. Once these points have been plotted, draw the circle that passes through them.

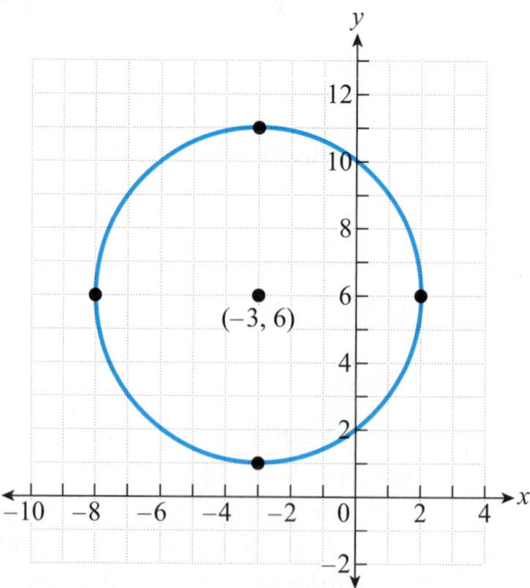

Quick Check **5** **Graph the circle. State the center and radius.**

a) $(x - 1)^2 + (y + 6)^2 = 16$ **b)** $(x + 2)^2 + (y - 3)^2 = 64$

Finding the Center and Radius by Completing the Square

Objective 4 **Find the center of a circle and its radius by completing the square.**

General Form of the Equation of a Circle

The **general form** of the equation of a circle is

$$Ax^2 + Ay^2 + Bx + Cy + D = 0, A \neq 0.$$

To graph a circle in this form, we must rewrite the equation in the standard form, which will allow us to determine the center and radius of the circle. This is done by completing the square, which must be done for both x and y.

EXAMPLE 8 Graph the circle $x^2 + y^2 - 8x + 10y + 40 = 0$. State the center and radius.

Solution

We will rewrite the equation in standard form by completing the square for x and y.

$$x^2 + y^2 - 8x + 10y + 40 = 0$$

$$(x^2 - 8x) + (y^2 + 10y) = -40$$

Collect terms containing x. Collect terms containing y. Subtract the constant to the right side.

$$(x^2 - 8x + 16) + (y^2 + 10y + 25) = -40 + 16 + 25$$

Add 16 to both sides to complete the square for x. Add 25 to both sides to complete the square for y.

$$(x - 4)^2 + (y + 5)^2 = 1$$

Factor the two quadratic expressions on the left side of the equation.

Since the equation of this circle has been written in standard form, the center of this circle is the point $(4, -5)$. Since $r^2 = 1$, the radius r is equal to $\sqrt{1}$, or 1. We begin to graph the circle by plotting the center $(4, -5)$. Next we plot the points that are one unit to the right and left of the center, as well as the points that are one unit above and below the center. We finish by drawing the circle that passes through these four points.

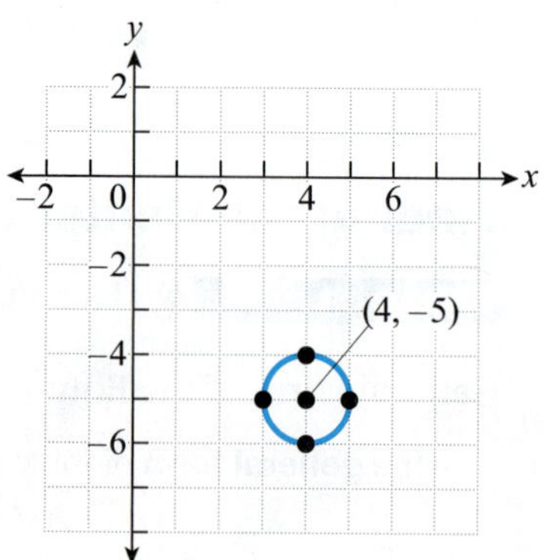

Quick Check 6 Graph the circle $x^2 + y^2 + 14x - 6y + 57 = 0$. State the center and radius.

Finding the Equation of a Circle Whose Center and Radius Are Known

Objective 5 **Find the equation of a circle that meets the given conditions.**

EXAMPLE 9 Find the equation of the circle.

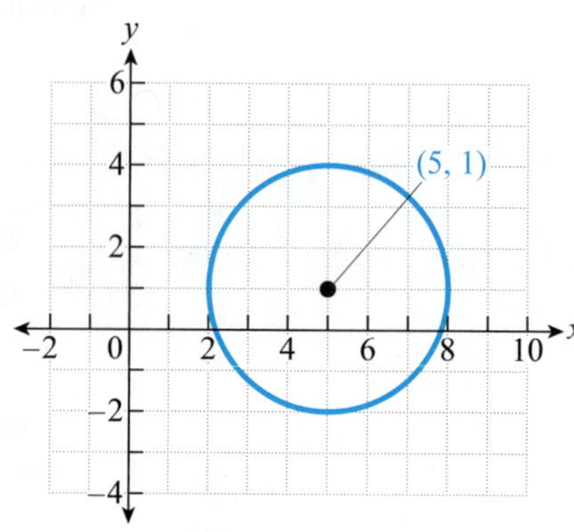

Solution

We begin by finding the center of the circle, which is the point $(5, 1)$. Next, we measure the distance from the center of the circle to a point on the circle. This distance is the radius. In this case the radius is 3.

We can now write the equation in standard form.

$$(x - h)^2 + (y - k)^2 = r^2$$

$$(x - 5)^2 + (y - 1)^2 = 3^2 \qquad \text{Substitute 5 for } h\text{, 1 for } k\text{, and 3 for } r.$$

$$(x - 5)^2 + (y - 1)^2 = 9 \qquad \text{Simplify.}$$

The equation of this circle is $(x - 5)^2 + (y - 1)^2 = 9$.

Quick Check 7 Find the equation of the circle.

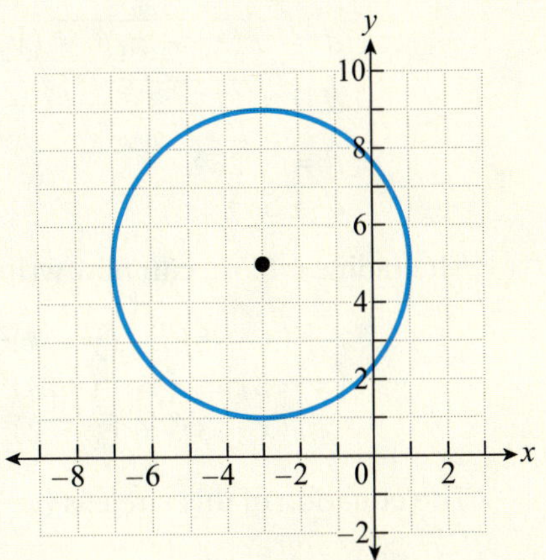

The **diameter** of a circle is a line segment that has both endpoints on the circle and passes through the center of the circle. The length of the diameter of a circle is equal to twice the length of the radius of the circle.

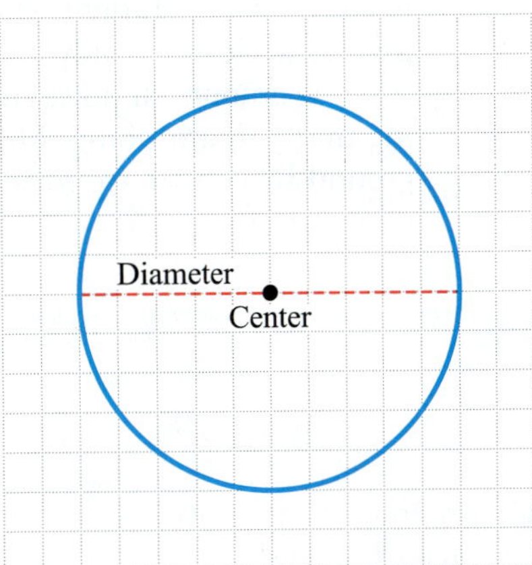

EXAMPLE 10 Find the equation in standard form of the circle that has a diameter with endpoints $(-1, -5)$ and $(7, 1)$.

Solution

We begin by finding the center of the circle, which is the midpoint of the given endpoints of a diameter.

$$x = \frac{x_1 + x_2}{2} \qquad y = \frac{y_1 + y_2}{2}$$

$$x = \frac{-1 + 7}{2} \qquad y = \frac{-5 + 1}{2}$$

$$= 3 \qquad = -2$$

The center of the circle is $(3, -2)$. Next, we find the radius by calculating the distance from the center of the circle to a point on the circle. We will use the point $(7, 1)$.

$$d = \sqrt{(x_2 - x_1)^2 + (y_2 - y_1)^2}$$

$$d = \sqrt{(7 - 3)^2 + (1 - (-2))^2} \quad \text{Substitute for } x_2, x_1, y_2, \text{ and } y_1.$$

$$d = \sqrt{25} \quad \text{Simplify the radicand.}$$

$$d = 5 \quad \text{Simplify.}$$

Quick Check 8

Find the equation in general form of the circle that has a diameter with endpoints $(4, -7)$ and $(-2, 3)$.

The radius is 5. We can now write the equation in standard form.

$$(x - h)^2 + (y - k)^2 = r^2$$

$$(x - 3)^2 + (y - (-2))^2 = 5^2 \quad \text{Substitute 3 for } h, -2 \text{ for } k, \text{ and 5 for } r.$$

$$(x - 3)^2 + (y + 2)^2 = 25$$

The equation of this circle is $(x - 3)^2 + (y + 2)^2 = 25$.

Building Your Study Strategy **Study Environment, 2 Noise** Noise can be a major distraction while you are studying. Some students need total silence while they are working; if you are one of those students, then be sure to select a location that will be free of noise. This is not easy to do in a public location such as the library, but you may be able to find a room or cubicle that is devoted to silent studying.

Other students study better when there is some background noise. For instance, some students prefer to have a radio playing soft music in the background while they study.

The amount of noise that you can handle may depend on the type of studying that you are doing. Music and other noises may provide a pleasant background while you are reviewing, but can be distracting while you are learning new material.

EXERCISES 9.2

Vocabulary

1. The distance, d, between two points (x_1, y_1) and (x_2, y_2) is given by the formula ________.

2. The ________ of a line segment connecting two points (x_1, y_1) and (x_2, y_2) is the point whose coordinates are $\left(\frac{x_1 + x_2}{2}, \frac{y_1 + y_2}{2}\right)$.

3. The collection of all points in a plane that are a fixed distance from a specified point is called a(n) ________.

4. The point that is equidistant from each point on a circle is the ________ of the circle.

5. The distance from the center to each point on the circle is called the ________ of the circle.

6. The ________ of a circle is a line segment that has both endpoints on the circle and passes through the center of the circle.

7. The equation for a circle with radius r centered at the origin is ________.

8. The equation for a circle with radius r centered at the point (h, k) is ________.

Find the distance between the given points. Round to the nearest tenth if necessary.

9. $(7, 6)$ and $(-1, -9)$

10. $(-5, -2)$ and $(4, 10)$

11. $(15, -53)$ and $(54, 27)$

12. $(-32, -15)$ and $(-77, 13)$

13. $(5, 2)$ and $(-8, 2)$

14. $(-4, -9)$ and $(-4, 9)$

15. $(7, 4)$ and $(10, 1)$

16. $(1, 2)$ and $(3, 8)$

17. $(-6, -5)$ and $(-1, -9)$

18. $(-2, 12)$ and $(4, -3)$

Find the midpoint of the line segment that connects the given points.

19. $(0, 0)$ and $(6, 4)$

20. $(3, 15)$ and $(7, 1)$

21. $(6, -9)$ and $(8, -25)$

22. $(-5, 10)$ and $(-12, 32)$

Graph the circle. State the center and radius of the circle.

23. $x^2 + y^2 = 1$

24. $x^2 + y^2 = 4$

25. $x^2 + y^2 = 16$

26. $x^2 + y^2 = 9$

27. $x^2 + y^2 = 100$

28. $x^2 + y^2 = 144$

29. $x^2 + y^2 = 6$

30. $x^2 + y^2 = 18$

31. $(x - 5)^2 + (y - 9)^2 = 16$

32. $(x - 7)^2 + (y - 4)^2 = 9$

33. $(x - 3)^2 + (y - 2)^2 = 4$

34. $(x - 1)^2 + (y - 6)^2 = 1$

35. $(x + 4)^2 + (y - 5)^2 = 9$

36. $(x - 7)^2 + (y + 3)^2 = 4$

37. $(x + 2)^2 + (y + 1)^2 = 49$

38. $(x + 6)^2 + (y + 4)^2 = 25$

39. $(x + 5)^2 + y^2 = 25$

40. $x^2 + (y - 9)^2 = 36$

41. $(x - 2)^2 + (y + 6)^2 = 18$

42. $(x + 8)^2 + (y + 3)^2 = 8$

43. $(x + 3)^2 + (y - 7)^2 = 45$

44. $(x - 1)^2 + (y - 10)^2 = 48$

Find the center and radius of each circle by completing the square.

45. $x^2 + y^2 + 4x + 12y - 41 = 0$

46. $x^2 + y^2 + 6x + 18y + 65 = 0$

47. $x^2 + y^2 - 10x + 2y - 23 = 0$

48. $x^2 + y^2 - 8x - 40y + 127 = 0$

49. $x^2 + y^2 + 16x - 57 = 0$

50. $x^2 + y^2 - 10y - 11 = 0$

51. $x^2 + y^2 + 4x + 14y - 27 = 0$

52. $x^2 + y^2 - 8x + 6y + 14 = 0$

Find the equation of each circle with the given center and radius.

53. Center $(0, 0)$, radius 6

54. Center $(0, 0)$, radius 9

55. Center $(4, 2)$, radius 7

56. Center $(1, 8)$, radius 4

57. Center $(-9, -10)$, radius 5

58. Center $(-6, 0)$, radius 2

59. Center $(0, -8)$, radius $\sqrt{15}$

60. Center $(-2, -13)$, radius $\sqrt{37}$

Find the equation of each circle.

61.

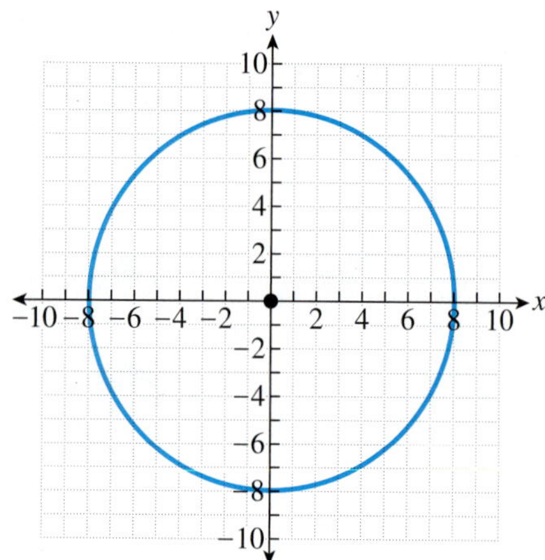

62.

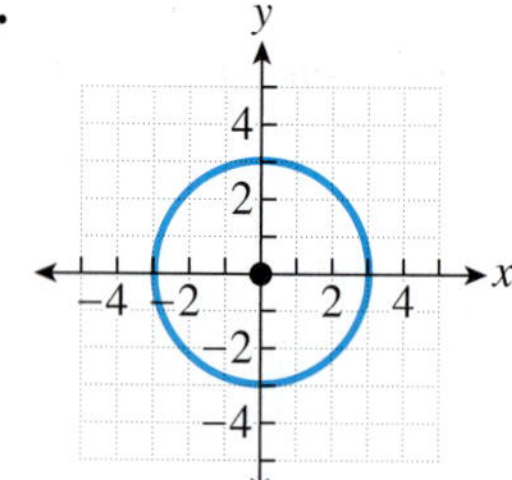

63.

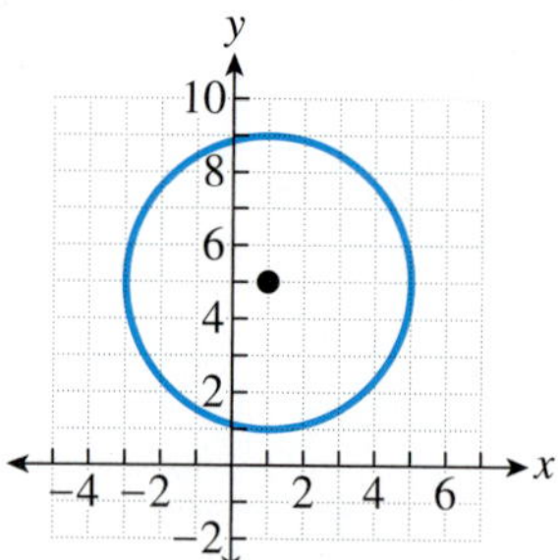

64.

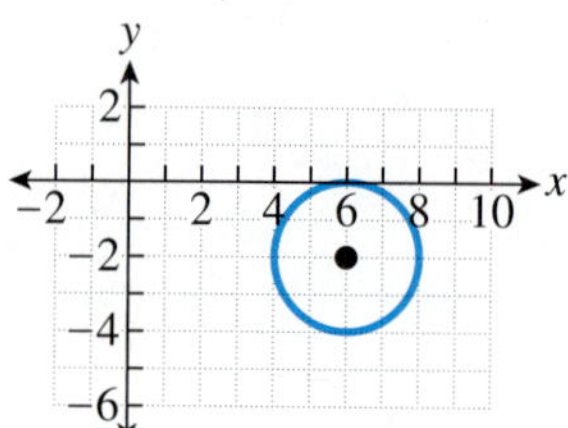

65.

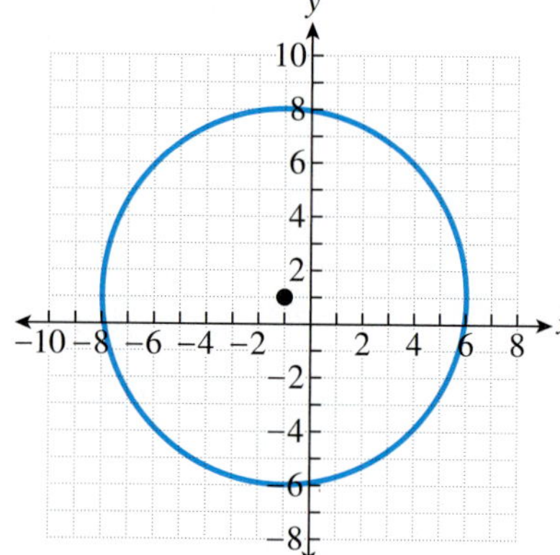

66.

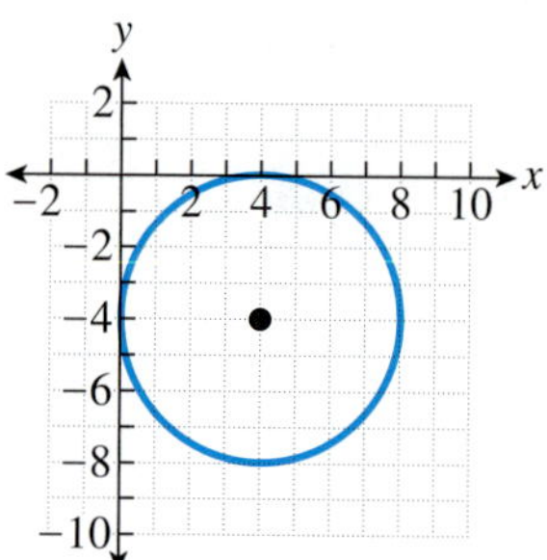

67.

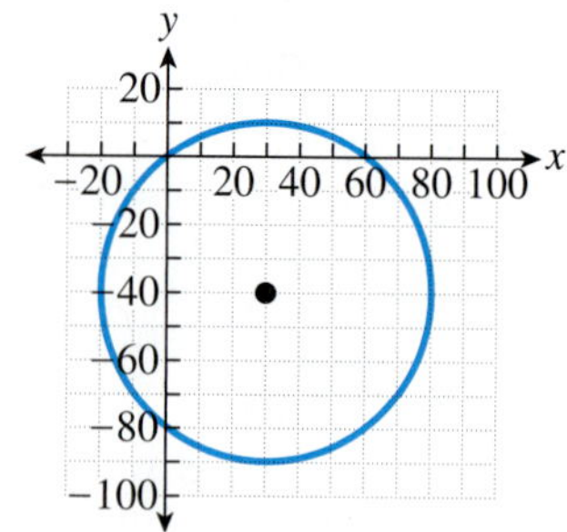

68.

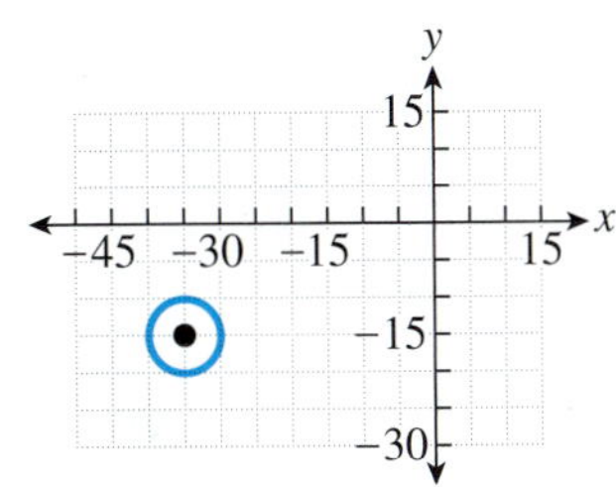

Find the equation of each circle that has a diameter with the given endpoints.

69. $(1, 7)$ and $(9, 7)$

70. $(-6, 3)$ and $(4, 3)$

71. $(-2, 10)$ and $(6, 4)$

72. $(8, -17)$ and $(-8, 13)$

73. $(4, 1)$ and $(12, -3)$

74. $(-3, 9)$ and $(-15, -11)$

Mixed Practice, 75–86

Graph. For graphs that are parabolas, label the vertex and any intercepts. For graphs that are circles, label the center.

75. $x = -(y - 3)^2 + 8$

76. $(x + 4)^2 + (y - 6)^2 = 25$

77. $x^2 + y^2 + 18x - 6y - 10 = 0$

78. $y = -(x - 5)^2 + 9$

79. $x = y^2 - 4y - 4$

80. $y = x^2 + 6x - 27$

81. $x^2 + y^2 = 20$

82. $x = -y^2 + 8y - 19$

83. $y = (x + 1)^2 + 6$

84. $x = (y + 3)^2 - 4$

85. $x^2 = -(y - 7)^2 + 25$

86. $y = -x^2 + 4x + 1$

87. Bart has built a rectangular barbecue area that measures 20 feet by 50 feet. He plans to draw a circle around the barbecue area, as shown, and fill in this area with grass.

a) Find the equation of the circle if we treat the point in the center of the barbecue area as the origin.

b) How many square feet of sod need to be ordered to cover the grass area?

88. The London Eye is a gigantic wheel built to celebrate the millennium. The diameter of the wheel is 122 meters, and its height is 131 meters.

The London Eye can be represented on a rectangular coordinate plane, plotted as shown.

a) Find an equation for the circle.

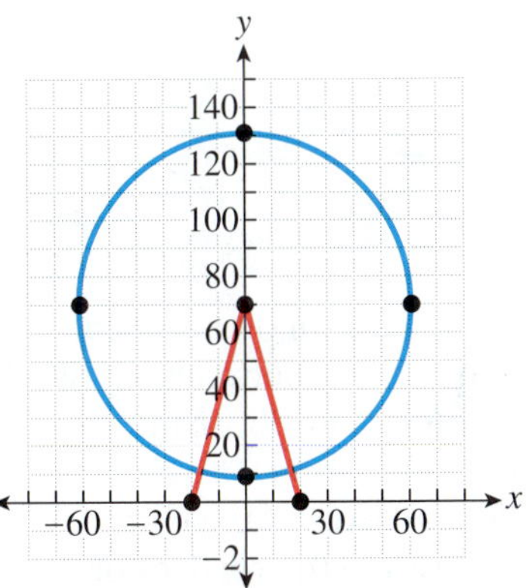

b) How far to the left or right of center is a car that is 100 feet above the ground?

List an equation of a circle that intersects the given parabola at

a) 0 points **b)** 1 point **c)** 2 points **d)** 4 points

89. $y = x^2 - 8x + 10$

90. $y = -(x + 2)^2 + 5$

91. $x = (y + 6)^2 - 3$

92. $x = -y^2 + 4y - 5$

Writing in Mathematics

Answer in complete sentences.

93. Explain how to determine whether the graph of an equation will be a parabola that opens up, a parabola that opens down, a parabola that opens to the right, a parabola that opens to the left, or a circle. Give an example of each type of equation.

94. Explain how to determine the center and radius of a circle from its equation if it is in standard form.

95. Explain how to find the center and radius of a circle by completing the square. Use an example to illustrate the process.

96. ***Solutions Manual**** Write a solutions manual page for the following problem:

Graph $x^2 + y^2 + 6x - 8y - 11 = 0$. *State the center and radius of the circle.*

97. ***Newsletter**** Write a newsletter explaining how to graph circles that are not centered at the origin.

***See Appendix B for details and sample answers.**

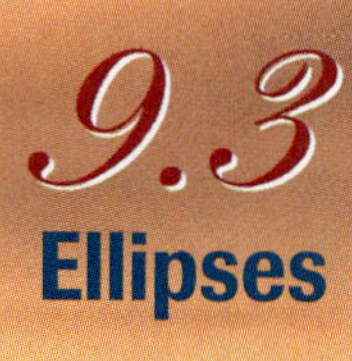

9.3 Ellipses

Objectives

1. **Graph ellipses centered at the origin.**
2. **Graph ellipses centered at a point (h, k).**
3. **Find the center of an ellipse and the lengths of its axes by completing the square.**
4. **Find the equation of an ellipse that meets the given conditions.**

In this section, we will investigate a conic section called an ellipse. An **ellipse** is the collection of all points (x, y) in the plane for which the sum of the distances d_1 and d_2 between the point (x, y) and two fixed points, F_1 and F_2, called **foci**, is a positive constant. Each fixed point is called a **focus**. The graph is similar to a circle, but more oval in its shape. For any point (x, y) on the ellipse, the sum of the distances d_1 and d_2 remains the same.

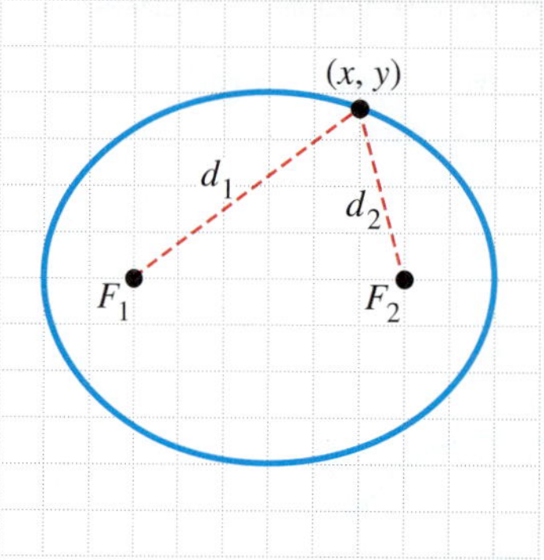

In this text, the foci will be either on a horizontal line or on a vertical line. An ellipse has a **center** like a circle, and it is the point located midway between the two foci.

A horizontal line passing through the center of an ellipse intersects the ellipse at two points, as does a vertical line passing through the center. The horizontal and vertical line segments connecting these points are called the **axes** of the ellipse. The longer line segment is called the **major axis**, and its endpoints are the **vertices** of the ellipse. The other line segment is called the **minor axis**, and its endpoints are the **co-vertices** of the ellipse.

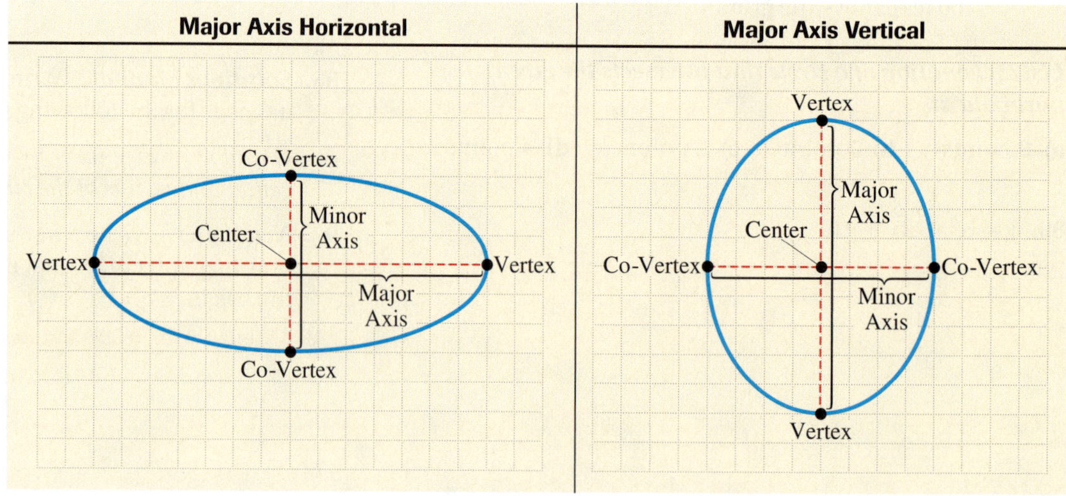

Ellipses Centered at the Origin

Objective 1 **Graph ellipses centered at the origin.**

Standard Form of the Equation of an Ellipse Centered at the Origin

The equation of an ellipse that is centered at the origin is of the form $\frac{x^2}{a^2} + \frac{y^2}{b^2} = 1$.

This is the **standard form** of the equation of an ellipse.

The ellipse will have x-intercepts at $(a, 0)$ and $(-a, 0)$ and y-intercepts at $(0, b)$ and $(0, -b)$. If $a > b$, then the major axis is horizontal and the minor axis is vertical. If $b > a$, then the major axis is vertical and the minor axis is horizontal.

To graph an ellipse centered at the origin, we begin by plotting a point at the origin. We then plot points that are a units to the left and right of the origin, as well as points that are b units above and below the origin.

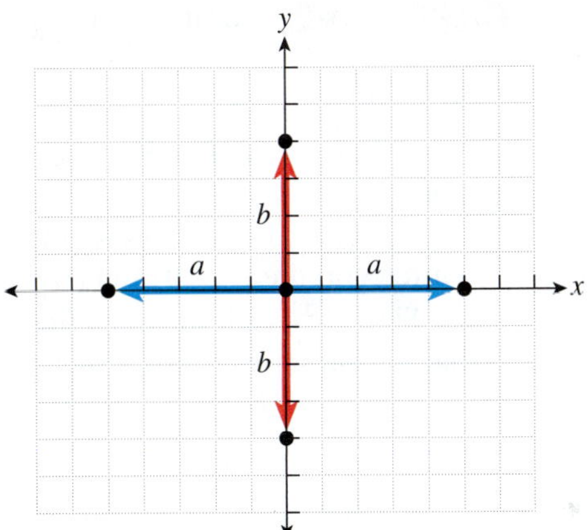

We finish by drawing the ellipse that passes through these four points.

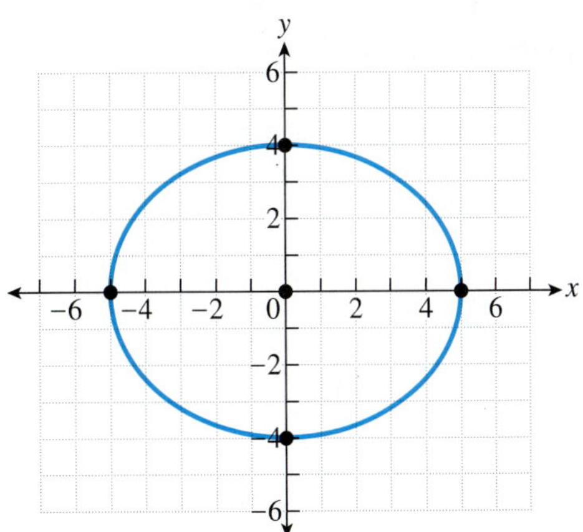

A Word of Caution When drawing the ellipse that passes through the four points, be sure that your graph looks like an oval and not like a diamond.

EXAMPLE 1 Graph the ellipse $\frac{x^2}{9} + \frac{y^2}{4} = 1$.

Solution

This ellipse is centered at the origin, so we plot a point there.

Since $a^2 = 9$, we know that $a = \sqrt{9}$ or 3. Moving three units to the left and right of the origin, we see that the x-intercepts are at $(-3, 0)$ and $(3, 0)$.

Since $b^2 = 4$, we know that $b = \sqrt{4}$ or 2. Moving two units above and below the origin, we see that the y-intercepts are at $(0, 2)$ and $(0, -2)$.

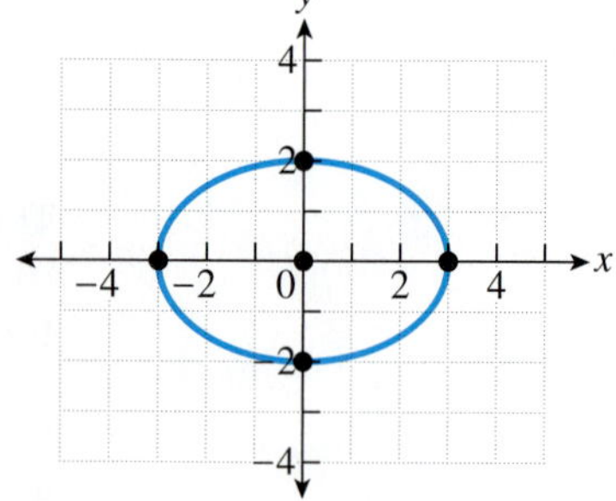

EXAMPLE 2 Graph $\frac{x^2}{4} + \frac{y^2}{36} = 1$.

Solution

This ellipse is centered at the origin, so we plot a point there.

Since $a^2 = 4$, we know that $a = \sqrt{4}$, or 2. Thus, the x-intercepts are at $(2, 0)$ and $(-2, 0)$.

Since $b^2 = 36$, we know that $b = \sqrt{36}$, or 6. It follows that the y-intercepts are at $(0, 6)$ and $(0, -6)$.

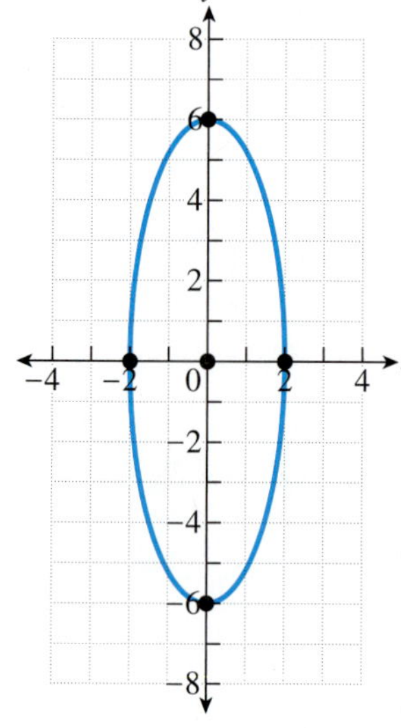

***Quick Check* 1** Graph.

a) $\frac{x^2}{25} + \frac{y^2}{9} = 1$

b) $\frac{x^2}{9} + \frac{y^2}{16} = 1$

In the next example, we will need to rewrite the equation in standard form before graphing the ellipse.

EXAMPLE 3 Graph $25x^2 + 16y^2 = 400$.

Solution

For the equation of an ellipse to be in standard form, the sum of the variable terms must be equal to 1. To rewrite this equation in standard form, we divide both sides of the equation by 400 and simplify.

$$25x^2 + 16y^2 = 400$$

$$\frac{25x^2 + 16y^2}{400} = \frac{400}{400} \quad \text{Divide both sides by 400.}$$

$$\frac{\overset{1}{\cancel{25}}x^2}{\underset{16}{\cancel{400}}} + \frac{\overset{1}{\cancel{16}}y^2}{\underset{25}{\cancel{400}}} = 1 \quad \text{Rewrite the left side of the equation as the sum of two fractions and simplify.}$$

$$\frac{x^2}{16} + \frac{y^2}{25} = 1 \quad \text{Simplify.}$$

The center of this ellipse is the origin. Since $a^2 = 16$, we know that $a = 4$. The x-intercepts are at $(4, 0)$ and $(-4, 0)$. Since $b^2 = 25$, we know that $b = 5$ and that the y-intercepts are at $(0, 5)$ and $(0, -5)$.

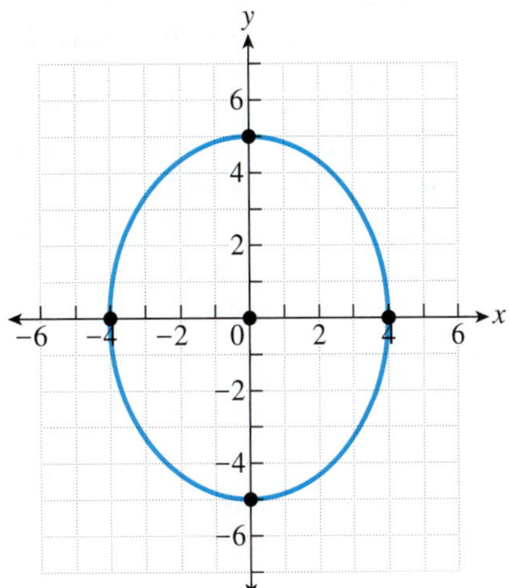

Quick Check 2 Graph $4x^2 + 36y^2 = 144$.

Ellipses Centered at a Point Other than the Origin

Objective 2 **Graph ellipses centered at a point (h, k).**

Standard Form of the Equation of an Ellipse with Center (h, k)

The equation for an ellipse centered at the point (h, k) whose horizontal axis has length $2a$ and whose vertical axis has length $2b$ is $\frac{(x-h)^2}{a^2} + \frac{(y-k)^2}{b^2} = 1$.

To graph an ellipse centered at the point (h, k), we begin by plotting a point at the center. The endpoints of the horizontal axis can be found by moving a units to the left and right of the center. The endpoints of the vertical axis can be found by moving b units above and below the center. Once these four endpoints have been plotted, we draw the ellipse that passes through them.

EXAMPLE 4 Graph $\frac{(x-3)^2}{4} + \frac{(y-5)^2}{9} = 1$.

Solution

Since the equation has the form $\frac{(x-h)^2}{a^2} + \frac{(y-k)^2}{b^2} = 1$, the center of this ellipse is $(3, 5)$.

Since $a^2 = 4$, we know that $a = 2$. The endpoints of the horizontal axis are two units to the left and right of the center. Since $b^2 = 9$, we know that $b = 3$. The endpoints of the vertical axis are three units above and below the center. We finish by drawing the ellipse that passes through these four endpoints.

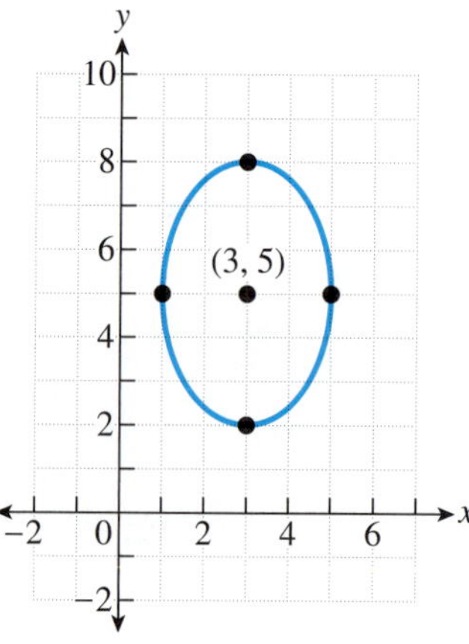

Quick Check **3** **Graph $\frac{(x-8)^2}{49} + \frac{(y-7)^2}{25} = 1$.**

EXAMPLE 5 Graph $(x+3)^2 + \frac{(y+2)^2}{10} = 1$.

Solution

The center of this ellipse is $(-3, -2)$.

When we do not see a denominator under the squared term containing x, $a^2 = 1$; so $a = 1$, and we plot the endpoints of the horizontal axis 1 unit to the left and right of the center. Since $b^2 = 10$, we know that $b = \sqrt{10}$. Since $\sqrt{10} \approx 3.2$, we plot the endpoints of the vertical axis approximately 3.2 units above and below the center. We finish by drawing the ellipse that passes through the four endpoints.

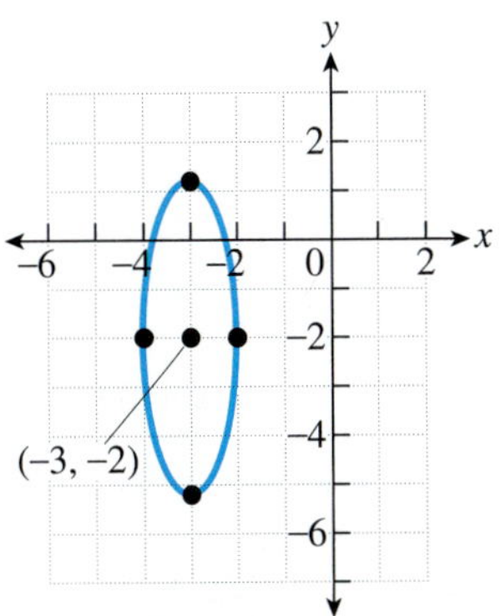

Quick Check 4

Graph $\frac{(x-4)^2}{20} + (y+5)^2 = 1$.
Find any intercepts.

Finding the Center of an Ellipse and the Lengths of Its Axes by Completing the Square

Objective 3 **Find the center of an ellipse and the lengths of its axes by completing the square.**

General Form of the Equation of an Ellipse

The **general form** of the equation of an ellipse is $Ax^2 + By^2 + Cx + Dy + E = 0$, $A \neq 0$, $B \neq 0$, and $A \neq B$.

Notice that the coefficient of the x^2 term is not equal to the coefficient of the y^2 term. If those two coefficients are equal, then the graph of the equation is more specifically defined as a circle than as an ellipse. To graph an ellipse in general form, we must rewrite the equation in standard form. This will allow us to determine the center, as well as a and b. This is done by completing the square, which must be done for both x and y.

Example 6 Graph the ellipse $25x^2 + 4y^2 + 150x - 16y + 141 = 0$. Give the center.

Solution

We will convert the equation to standard form by completing the square for x and y.

$$25x^2 + 4y^2 + 150x - 16y + 141 = 0$$
$$(25x^2 + 150x) + (4y^2 - 16y) = -141$$

Collect terms containing x. Collect terms containing y. Subtract the constant to the right side.

$$25(x^2 + 6x) + 4(y^2 - 4y) = -141$$

Factor 25 from the terms containing x so that the coefficient of the x^2 term is 1. Factor 4 from the terms containing y.

$$25(x^2 + 6x + 9) + 4(y^2 - 4y + 4) = -141 + 225 + 16$$

Add 9 inside the parentheses containing x to complete the square for x. Add $25 \cdot 9$, or 225, to the right side of the equation. Add 4 inside the parentheses containing y to complete the square for y. Add $4 \cdot 4$, or 16, to the right side of the equation.

$$25(x + 3)^2 + 4(y - 2)^2 = 100$$

Factor the two quadratic expressions on the left side of the equation.

$$\frac{\overset{1}{\cancel{25}}(x + 3)^2}{\underset{4}{\cancel{100}}} + \frac{\overset{1}{\cancel{4}}(y - 2)^2}{\underset{25}{\cancel{100}}} = 1$$

Divide both sides of the equation by 100. Rewrite the left side of the equation as the sum of two fractions. Divide out common factors.

$$\frac{(x + 3)^2}{4} + \frac{(y - 2)^2}{25} = 1$$

Simplify.

Since the equation is now in standard form, the center of this ellipse is $(-3, 2)$. Since $a^2 = 4$, $a = 2$ and we plot the endpoints of the horizontal axis two units to the left and right of the center. Since $b^2 = 25$, $b = 5$ and we plot the endpoints of the vertical axis five units above and below the center. Draw the ellipse that passes through the four endpoints.

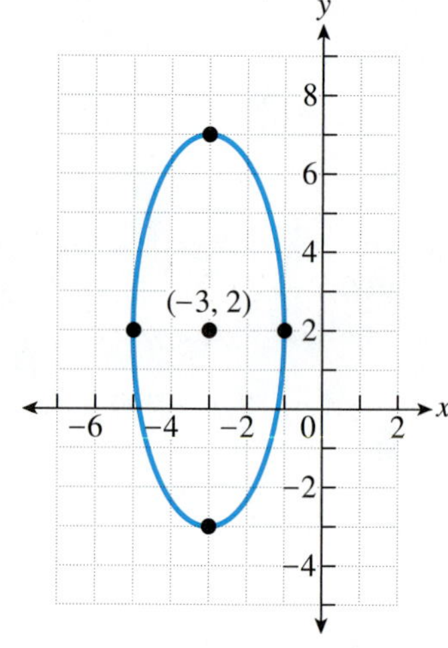

Quick Check 5 Graph the ellipse $9x^2 + 16y^2 - 36x + 160y + 292 = 0$. Give the center.

Finding the Equation of an Ellipse

Objective 4 **Find the equation of an ellipse that meets the given conditions.**
To find the equation of an ellipse, we begin by determining the center (h, k), a, and b. Then we write the equation of the ellipse in standard form $\frac{(x - h)^2}{a^2} + \frac{(y - k)^2}{b^2} = 1$.

EXAMPLE 7 Find the standard form equation of the ellipse.

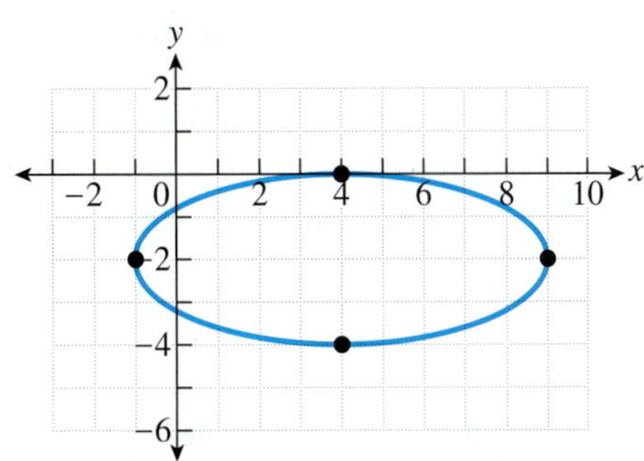

Solution

The center of the ellipse is $(4, -2)$.

We measure the distance from the center of the ellipse to an endpoint of its horizontal axis. This distance is a. In this case, $a = 5$.

Next, we measure the distance from the center of the ellipse to an endpoint of its vertical axis. This distance is b. In this case, $b = 2$.

We can now write the equation in standard form.

$$\frac{(x-4)^2}{5^2} + \frac{(y-(-2))^2}{2^2} = 1$$

Substitute 4 for h, -2 for k, 5 for a, and 2 for b into the standard form of the equation of an ellipse.

$$\frac{(x-4)^2}{25} + \frac{(y+2)^2}{4} = 1$$

Simplify.

The equation of this ellipse is $\frac{(x-4)^2}{25} + \frac{(y+2)^2}{4} = 1$.

Quick Check **6** **Find the standard form equation of the ellipse.**

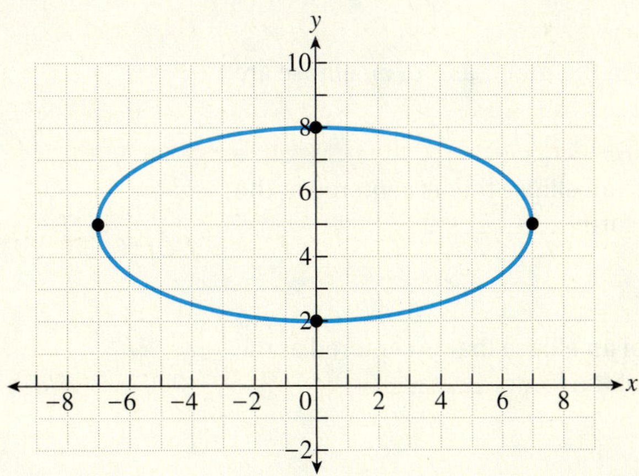

Building Your Study Strategy **Study Environment, 3** **Your Workspace** When you sit down to study, you need a space large enough to accommodate all of your materials. Be sure that you can easily fit your textbook, notes, homework notebook, study cards, calculator, ruler, etc., on the desk or table that you are working on. Anything that you might use should be easily within your reach. If your workspace is unorganized and cluttered, you will waste time searching for pages and shuffling through your materials.

EXERCISES 9.3

Vocabulary

1. The collection of all points (x, y) in the plane for which the sum of the distances d_1 and d_2 between the point (x, y) and two fixed points, F_1 and F_2, is a positive constant is called a(n) __________.

2. The two fixed points, F_1 and F_2, referred to in the definition of an ellipse are called __________.

3. The __________ of an ellipse is the point located midway between the two foci.

4. The longer of an ellipse's two axes is called the __________ axis and the shorter of an ellipse's two axes is called the __________ axis.

5. The endpoints of the major axis of an ellipse are called __________.

6. The endpoints of the minor axis of an ellipse are called __________.

7. The equation of an ellipse that is centered at the origin is of the form __________.

8. The equation of an ellipse that is centered at the point (h, k) is of the form __________.

Graph the ellipse. Give the coordinates of the center, as well as the values of a and b.

9. $\dfrac{x^2}{4} + \dfrac{y^2}{9} = 1$

10. $\dfrac{x^2}{16} + \dfrac{y^2}{25} = 1$

11. $\dfrac{x^2}{25} + \dfrac{y^2}{4} = 1$

12. $\dfrac{x^2}{49} + \dfrac{y^2}{9} = 1$

13. $x^2 + \dfrac{y^2}{25} = 1$

14. $\dfrac{x^2}{4} + y^2 = 1$

15. $\dfrac{x^2}{36} + \dfrac{y^2}{12} = 1$

16. $\dfrac{x^2}{25} + \dfrac{y^2}{32} = 1$

17. $25x^2 + 9y^2 = 225$

18. $36x^2 + 16y^2 = 576$

19. $4x^2 + 49y^2 = 196$

20. $9x^2 + 16y^2 = 144$

21. $\dfrac{(x-6)^2}{25} + \dfrac{(y-8)^2}{36} = 1$

22. $\dfrac{(x-4)^2}{4} + \dfrac{(y-7)^2}{16} = 1$

23. $\dfrac{(x-9)^2}{81} + \dfrac{(y+4)^2}{9} = 1$

24. $\dfrac{(x+8)^2}{36} + \dfrac{(y-3)^2}{4} = 1$

25. $\dfrac{(x+5)^2}{9} + \dfrac{(y+3)^2}{64} = 1$

26. $\dfrac{(x+1)^2}{16} + \dfrac{(y+9)^2}{25} = 1$

27. $\dfrac{(x+5)^2}{4} + \dfrac{y^2}{36} = 1$

28. $\dfrac{x^2}{36} + \dfrac{(y+3)^2}{25} = 1$

29. $\dfrac{(x+2)^2}{49} + (y-4)^2 = 1$

30. $(x+7)^2 + \dfrac{(y-1)^2}{25} = 1$

31. $\dfrac{(x-6)^2}{20} + \dfrac{(y+6)^2}{4} = 1$

32. $\dfrac{(x-3)^2}{8} + \dfrac{(y+8)^2}{25} = 1$

33. $4x^2 + 9y^2 - 24x - 90y + 225 = 0$

34. $25x^2 + 4y^2 - 300x + 56y + 996 = 0$

35. $2x^2 + 8y^2 + 16x - 64y + 88 = 0$

36. $49x^2 + 9y^2 + 98x + 36y - 356 = 0$

Find the standard-form equation of the ellipse that meets the given conditions.

	Center	Major Axis	Length of Major Axis	Length of Minor Axis
37.	$(-2, 8)$	vertical	20	4
38.	$(4, 7)$	vertical	8	2
39.	$(6, -3)$	horizontal	16	14
40.	$(-5, -9)$	horizontal	18	10
41.	$(0, -6)$	vertical	10	6
42.	$(8, -1)$	horizontal	32	24

Find the standard-form equation of the ellipse.

43.

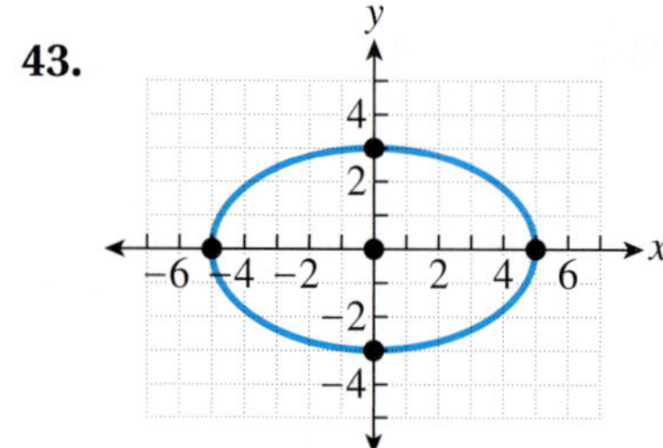

44.

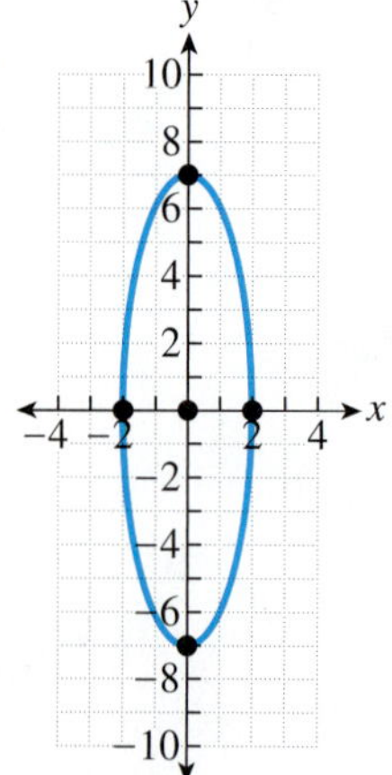

45.

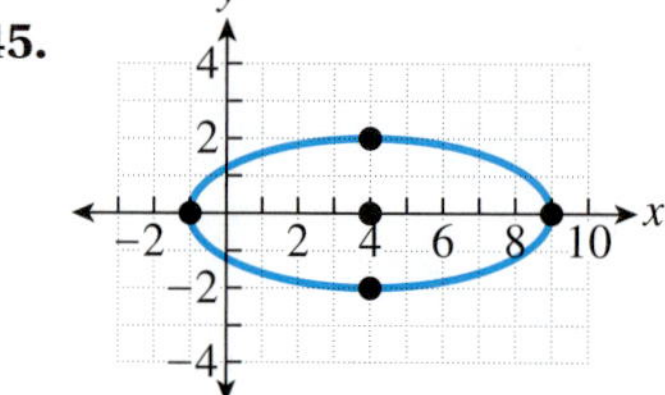

46.

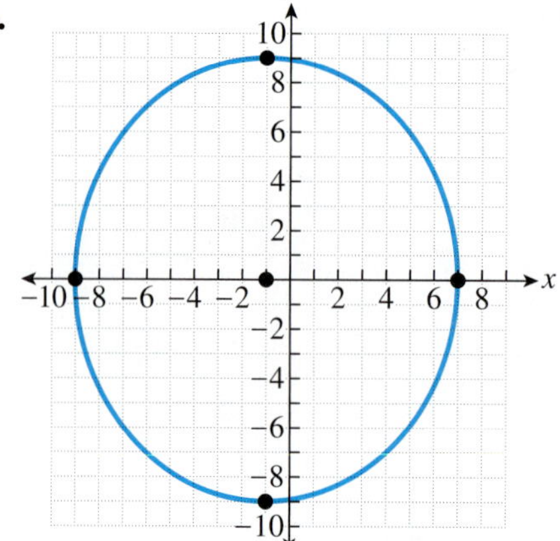

47.

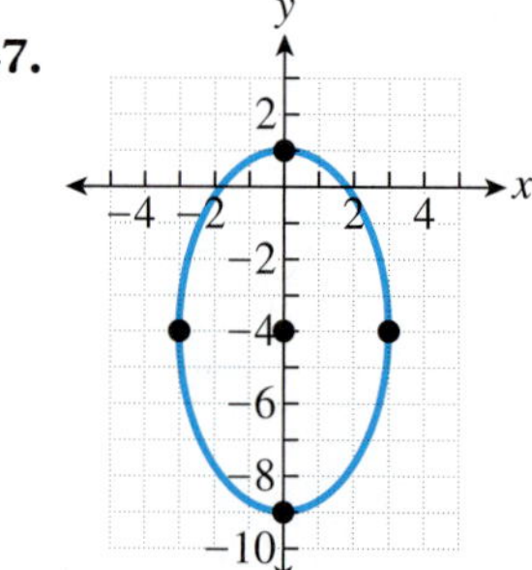

48.

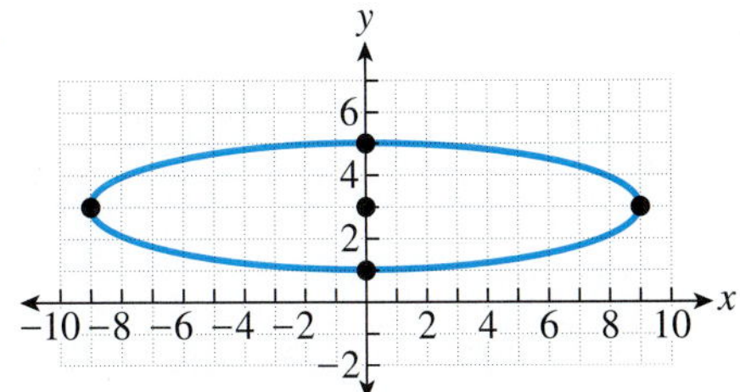

49.

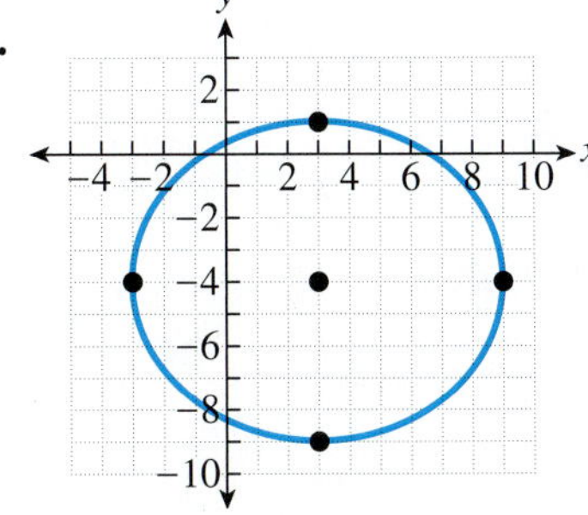

50.

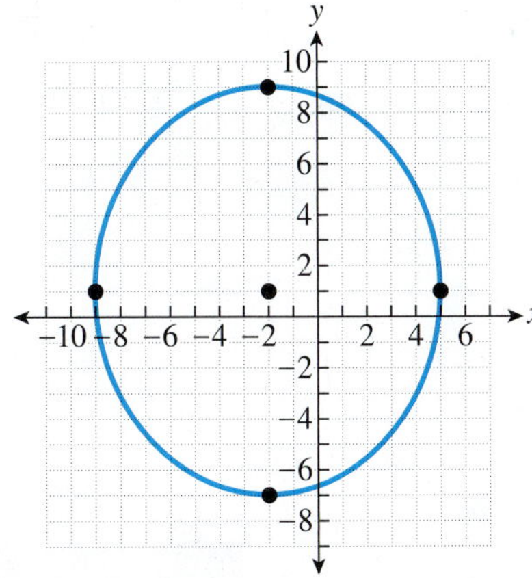

Mixed Practice, 51–68

Graph each equation. For graphs that are parabolas, label the vertex and any intercepts. For graphs that are circles or ellipses, label the center.

51. $(x + 4)^2 + (y - 6)^2 = 9$

52. $y = -(x - 2)^2 + 5$

53. $x = y^2 + 6y + 10$

54. $\dfrac{x^2}{36} + \dfrac{y^2}{4} = 1$

55. $3x^2 + 5y^2 = 45$

56. $x = -y^2 + 2y + 7$

57. $y = -x^2 + 2x + 63$

58. $\frac{(x-1)^2}{9} + \frac{(y+5)^2}{16} = 1$

59. $9x^2 + 25y^2 + 54x - 100y - 44 = 0$

60. $x = (y-4)^2 - 4$

61. $x^2 + y^2 - 12x - 14y - 59 = 0$

62. $y = x^2 + 6x + 2$

63. $x^2 + y^2 = 32$

64. $x^2 + 49y^2 - 12x + 196y + 183 = 0$

65. $x = -(y+4)^2 - 9$

66. $x^2 + y^2 = 81$

67. $\frac{(x+6)^2}{25} + \frac{(y+1)^2}{4} = 1$

68. $y = (x-3)^2 + 4$

69. An elliptical track can be described by the equation $1089x^2 + 2500y^2 = 2{,}722{,}500$, where x and y are in feet.

a) Find the width of the track from west to east.

b) Find the width of the track from north to south.

70. A tunnel through a hillside is in the shape of a semi-ellipse. The base of the tunnel is 80 feet across, and at its highest point the tunnel is 30 feet high. The tunnel is plotted as follows on a rectangular coordinate plane (not to scale):

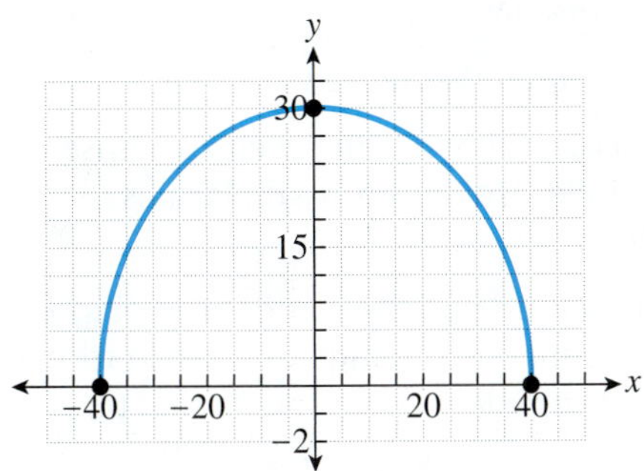

a) Find the equation of the ellipse.

b) How far from the center of the tunnel is the height 16 feet?

71. The area of an ellipse given by the equation $\frac{(x-h)^2}{a^2} + \frac{(y-k)^2}{b^2} = 1$ is πab. A homeowner is building an elliptical swimming pool that is 30 feet from end to end at its widest point, and 20 feet from side to side at its widest point. Find the area covered by the swimming pool.

72. A small park in the shape of an ellipse covers an area of approximately 1372.88 square feet. If the park measures 46 feet from west to east, as shown next, what is the distance from south to north?

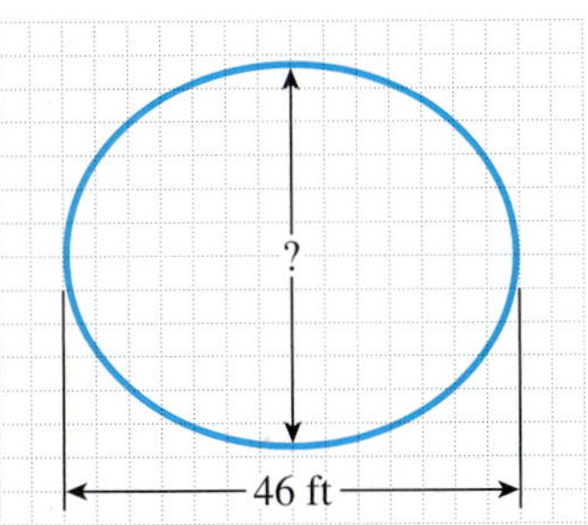

73. Graph each given ellipse whose major axis is horizontal and calculate its **eccentricity** $e = \frac{\sqrt{a^2 - b^2}}{a}$. Round to the nearest hundredth.

a) $\frac{x^2}{16} + \frac{y^2}{9} = 1$

b) $\frac{x^2}{49} + \frac{y^2}{9} = 1$

c) $\frac{x^2}{100} + \frac{y^2}{9} = 1$

74. Graph each given ellipse whose major axis is vertical and calculate its **eccentricity** $e = \dfrac{\sqrt{b^2 - a^2}}{b}$.

a) $\dfrac{x^2}{4} + \dfrac{y^2}{9} = 1$

b) $\dfrac{x^2}{4} + \dfrac{y^2}{36} = 1$

c) $\dfrac{x^2}{4} + \dfrac{y^2}{81} = 1$

75. Describe how the shape of an ellipse is affected as its eccentricity increases.

76. Describe how the shape of an ellipse is affected as its eccentricity decreases.

77. Can the eccentricity of an ellipse ever be greater than or equal to 1? Explain why or why not.

78. A circle is a special ellipse whose eccentricity is ________.

Writing in Mathematics

Answer in complete sentences.

79. Explain the difference between an ellipse and a circle. Are there any similarities between the two?

80. Explain how to find the center of an ellipse from its equation. Include the case in which we must complete the square to find the center.

81. ***Solutions Manual**** Write a solutions manual page for the following problem:

Graph $8x^2 + 18y^2 - 32x + 180y + 410 = 0$. *Give the coordinates of the center, as well as the values of a and b.*

82. ***Newsletter**** Write a newsletter explaining how to graph an ellipse whose equation is of the form

$$\frac{(x-h)^2}{a^2} + \frac{(y-k)^2}{b^2} = 1.$$

***See Appendix B for details and sample answers.**

9.4 Hyperbolas

Objectives

1. Graph hyperbolas centered at the origin.
2. Graph hyperbolas centered at a point (h, k).
3. Find the center of a hyperbola and the lengths of its axes by completing the square.
4. Find the equation of a hyperbola that meets the given conditions.

In this section, we will investigate a conic section called a hyperbola. Here are two examples of graphs of hyperbolas.

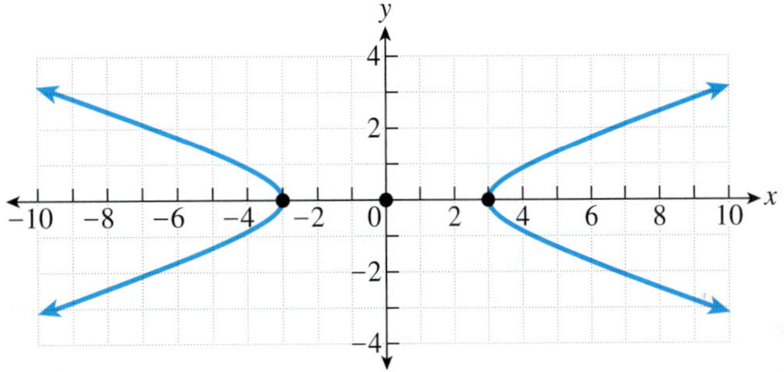

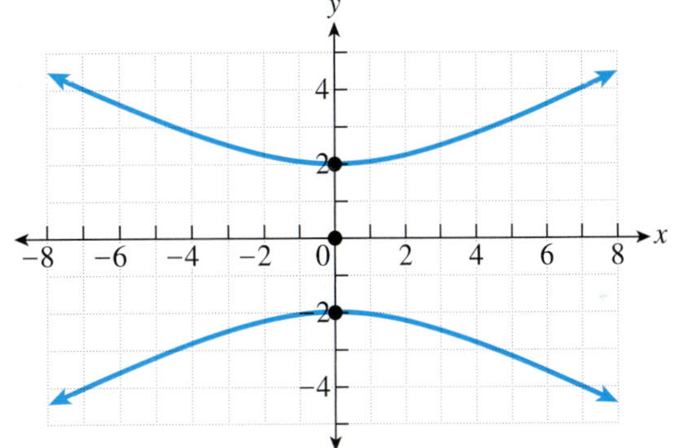

Notice that the graph of a hyperbola is different from other graphs we have drawn in that it has two parts, called **branches**.

Hyperbola

A **hyperbola** is the collection of all points (x, y) in the plane for which the *difference* of the distances, d_1 and d_2, between the point and two fixed points, F_1 and F_2, called foci, is a constant.

This definition is similar to that of an ellipse, but for an ellipse the *sum* of the distances remains constant, not the difference. For any point (x, y) on the hyperbola, $|d_1 - d_2|$ remains the same.

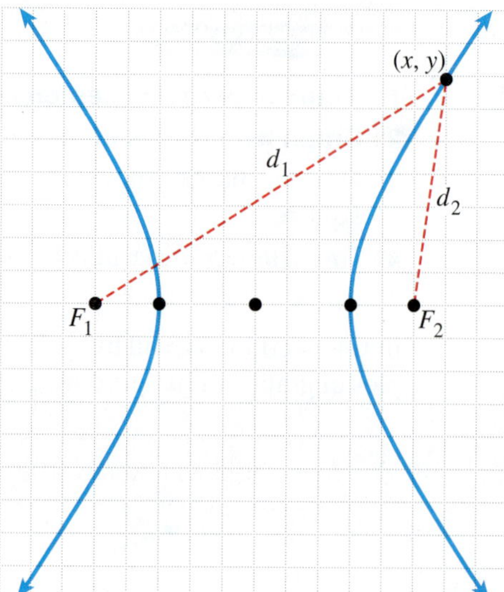

Like a circle and an ellipse, a hyperbola has a center. The **center** of a hyperbola is the point that is located midway between the two foci.

The line that passes through the two foci intersects the hyperbola at two points. These points are the **vertices** of the hyperbola, and the line segment between them is called the **transverse axis** of the hyperbola.

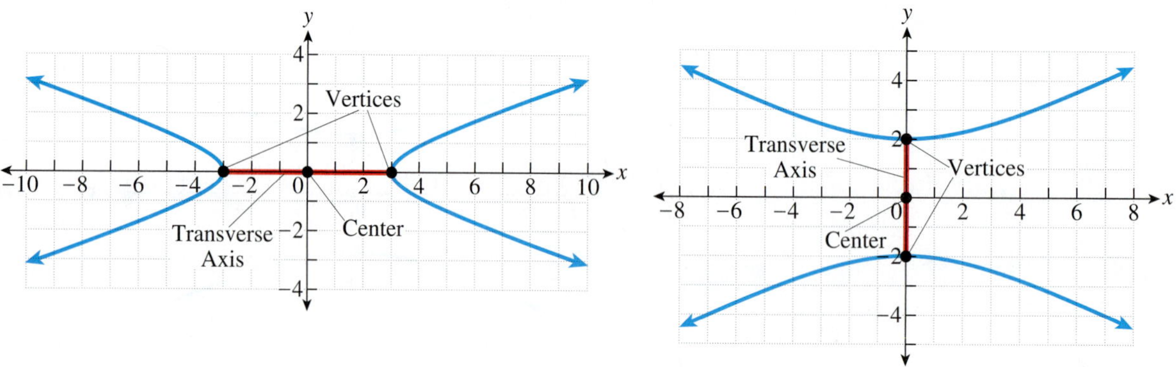

Each hyperbola has a pair of asymptotes. Each **asymptote** is a line that passes through the center of the hyperbola, showing us the behavior of the branches of the hyperbola as we move farther out from the center of the hyperbola.

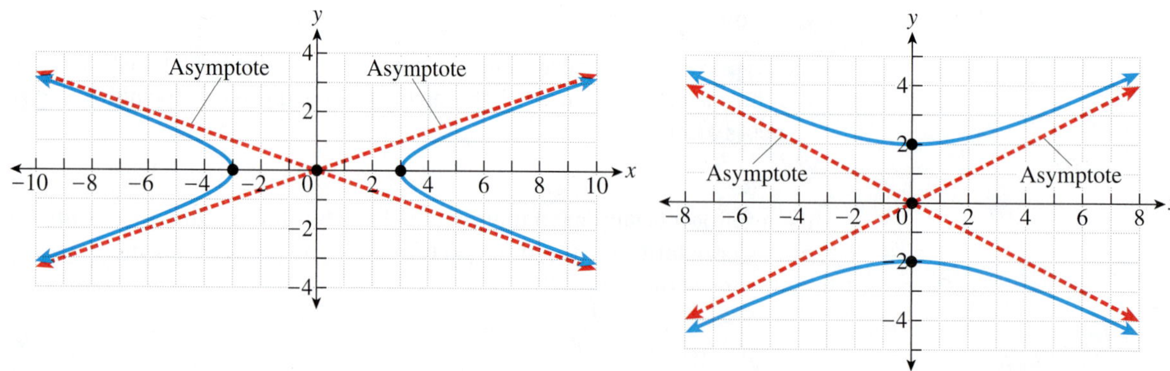

Hyperbolas Centered at the Origin

Objective 1 **Graph hyperbolas centered at the origin.** We will begin by investigating hyperbolas that have a horizontal transverse axis, whose branches open to the left and to the right.

Standard Form of the Equation of a Hyperbola with a Horizontal Transverse Axis

The standard-form equation of a hyperbola with a horizontal transverse axis centered at the origin has the form $\frac{x^2}{a^2} - \frac{y^2}{b^2} = 1$.

The hyperbola will have vertices at $(a, 0)$ and $(-a, 0)$. The asymptotes of the hyperbola will be the lines $y = \frac{b}{a}x$ and $y = -\frac{b}{a}x$.

To graph a hyperbola, we begin by plotting the center and each vertex. We then graph the asymptotes, using dashed lines to indicate that they are not part of the hyperbola. We finish the graph by drawing a branch through each vertex so that the branch approaches the asymptotes.

To graph the asymptotes, construct a rectangle around the center of the hyperbola that extends a units to the left and right of the center and b units above and below the center. Once this rectangle has been constructed, the asymptotes are the lines that pass through the diagonals of the rectangle.

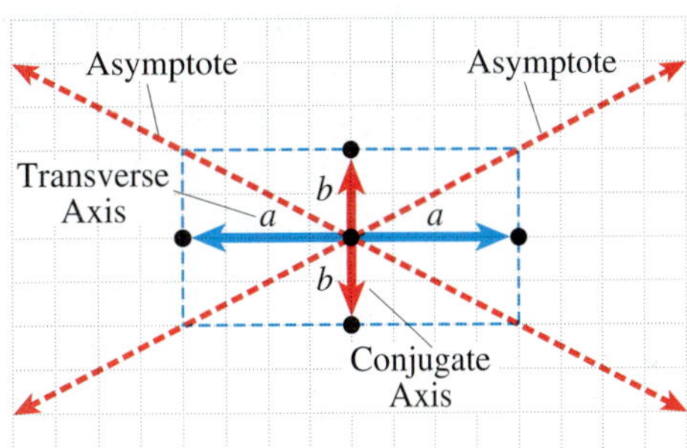

Notice that there are four points labeled on the rectangle. Two of those points are vertices and the endpoints of the transverse axis. The two points that are not vertices are the endpoints of what is known as the **conjugate axis**. If the branches of the hyperbola open to the left and the right, then the conjugate axis will be vertical, and its length is $2b$.

EXAMPLE 1 Graph the hyperbola $\frac{x^2}{4} - \frac{y^2}{25} = 1$. Give the center, the vertices, and the equations of the asymptotes.

Solution

Since the equation is of the form $\frac{x^2}{a^2} - \frac{y^2}{b^2} = 1$, the hyperbola is centered at the origin and its branches open to the left and to the right. We begin by plotting a point at the origin and graphing the asymptotes.

Since $a^2 = 4$ and $b^2 = 25$, we know that $a = 2$ and $b = 5$. The rectangle used to graph the asymptotes extends two units to the left and to the right of the center and five units above and below the center. The equations for these asymptotes are $y = \frac{5}{2}x$ and $y = -\frac{5}{2}x$.

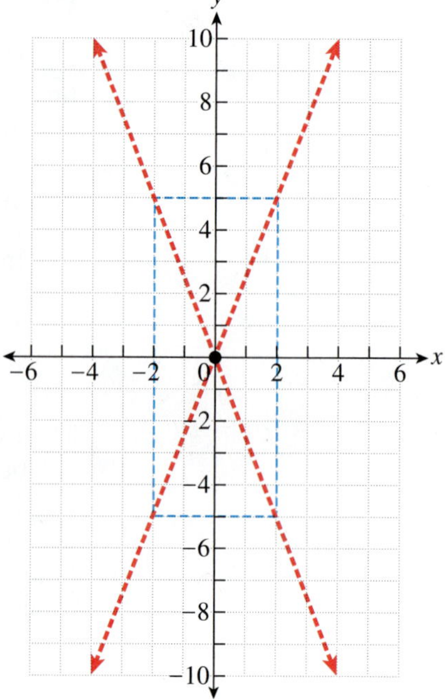

Quick Check 1

Graph the hyperbola $\frac{x^2}{16} - \frac{y^2}{9} = 1$. Give the center, the vertices, and the equations of the asymptotes.

The vertices are located at the points $(-2, 0)$ and $(2, 0)$ on the rectangle that we drew to graph the asymptotes. Next, we draw the branches of the hyperbola opening to the left and to the right in such a way that they approach the asymptotes.

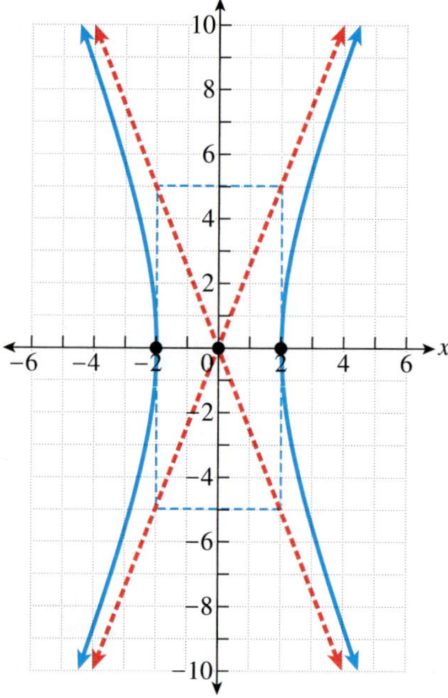

We now turn to hyperbolas that have a vertical transverse axis. For this type of hyperbola, the branches open upwards and downwards.

Standard Form of the Equation of a Hyperbola with a Vertical Transverse Axis

The equation of a hyperbola with a vertical transverse axis that is centered at the origin has the form $\frac{y^2}{b^2} - \frac{x^2}{a^2} = 1$.

The hyperbola will have vertices at $(0, b)$ and $(0, -b)$. The asymptotes of the hyperbola will be the lines $y = \frac{b}{a}x$ and $y = -\frac{b}{a}x$. The conjugate axis will be horizontal, and its length is $2a$.

EXAMPLE 2 Graph $\frac{y^2}{9} - \frac{x^2}{16} = 1$. Give the center, the vertices, and the equations of the asymptotes.

Solution

This hyperbola is centered at the origin and its branches open upward and downward. We begin by plotting a point at the origin and graphing the asymptotes.

Since $a^2 = 16$ and $b^2 = 9$, we know that $a = 4$ and $b = 3$. The rectangle used to graph the asymptotes extends four units to the left and to the right of the center and three units above and below the origin. The equations for these asymptotes are $y = \frac{3}{4}x$ and $y = -\frac{3}{4}x$.

The vertices are located at the points $(0, 3)$ and $(0, -3)$. After plotting the vertices, we draw the branches of the hyperbola opening upward and downward in such a way that they approach the asymptotes.

Quick Check 2

Graph $\frac{y^2}{4} - \frac{x^2}{36} = 1$. Give the center, the vertices, and the equations of the asymptotes.

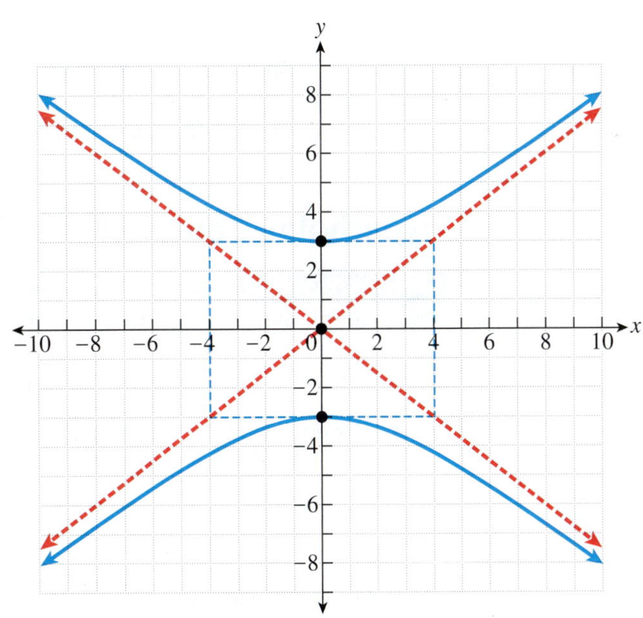

Here is a brief summary of hyperbolas centered at the origin:

Transverse Axis	Horizontal	Vertical
Equation	$\frac{x^2}{a^2} - \frac{y^2}{b^2} = 1$	$\frac{y^2}{b^2} - \frac{x^2}{a^2} = 1$
Sample Graph		
Vertices	$(a, 0), (-a, 0)$	$(0, b), (0, -b)$
Asymptotes	$y = \frac{b}{a}x, y = -\frac{b}{a}x$	$y = \frac{b}{a}x, y = -\frac{b}{a}x$

It is crucial for us to be able to tell what type of hyperbola we are graphing from the equation. For a hyperbola whose equation is of the form $\frac{x^2}{a^2} - \frac{y^2}{b^2} = 1$, the branches of the hyperbola open to the left and to the right. For a hyperbola whose equation is of the form $\frac{y^2}{b^2} - \frac{x^2}{a^2} = 1$, the branches of the hyperbola open upward and downward.

It is also important for us to be able to differentiate between the equation of a hyperbola and the equation of an ellipse. The major difference between the two equations is that the equation of a hyperbola involves the difference of the terms containing x and y, while the equation of an ellipse involves their sum.

Hyperbolas Centered at a Point Other than the Origin

Objective 2 **Graph hyperbolas centered at a point (h, k).** We will now learn how to graph hyperbolas centered at any point (h, k).

Standard-Form Equation of a Hyperbola with Center (h, k)

- The equation of a hyperbola centered at the point (h, k) with a horizontal transverse axis has the standard form

$$\frac{(x-h)^2}{a^2} - \frac{(y-k)^2}{b^2} = 1.$$

The vertices are the points $(h + a, k)$ and $(h - a, k)$. The equations for the asymptotes are $y = \frac{b}{a}(x - h) + k$ and $y = -\frac{b}{a}(x - h) + k$.

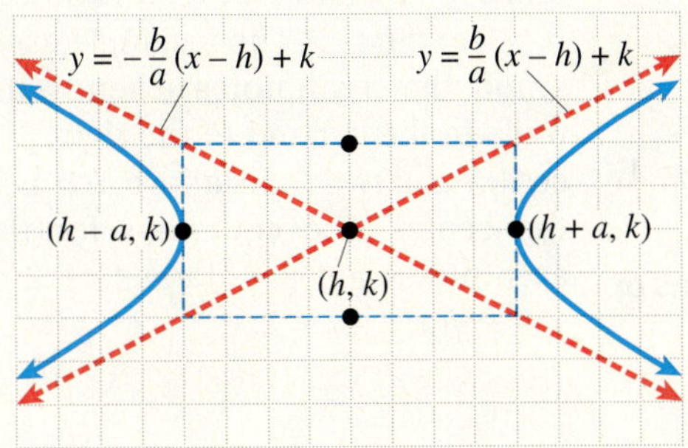

- The equation for a hyperbola centered at the point (h, k) with a vertical transverse axis has the standard form

$$\frac{(y - k)^2}{b^2} - \frac{(x - h)^2}{a^2} = 1.$$

The vertices are the points $(h, k + b)$ and $(h, k - b)$. The equations for the asymptotes are $y = \frac{b}{a}(x - h) + k$ and $y = -\frac{b}{a}(x - h) + k$.

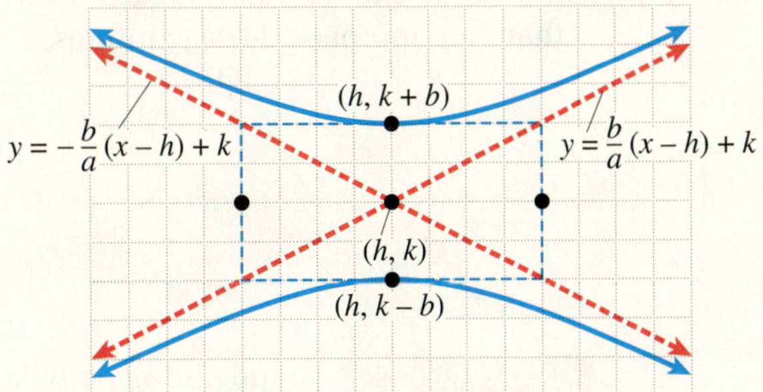

To graph a hyperbola centered at a point (h, k), we begin by plotting the center of the hyperbola and its vertices. If the branches of the hyperbola open to the left and to the right, then the vertices can be found by moving a units to the left and right of the center. If the branches of the hyperbola open upward and downward, then the vertices can be found by moving b units above and below the center.

Once the center and vertices have been plotted, we graph the asymptotes of the hyperbola by using dashed lines. Although we can graph these asymptotes from their equations, we can continue to use a rectangle in the same way as for hyperbolas centered at the origin.

We then draw the two branches of the hyperbola in such a way that each branch passes through a vertex and approaches the asymptotes.

EXAMPLE 3 Graph the hyperbola $\frac{(x - 4)^2}{9} - \frac{(y - 2)^2}{4} = 1$. Give the center, the vertices, and the equations of the asymptotes.

Solution

Since the equation is of the form $\frac{(x - h)^2}{a^2} - \frac{(y - k)^2}{b^2} = 1$, the hyperbola is centered at $(4, 2)$ and its branches open to the left and to the right. We begin by plotting the center and graphing the asymptotes.

Quick Check 3 Graph the hyperbola $\frac{(x-3)^2}{4} - \frac{(y-5)^2}{16} = 1$. Give the center, the vertices, and the equations of the asymptotes.

Since $a^2 = 9$ and $b^2 = 4$, we know that $a = 3$ and $b = 2$. The rectangle used to graph the asymptotes extends three units to the left and to the right of the center and two units above and below the center. The equations for these asymptotes are $y = \frac{2}{3}(x - 4) + 2$ and $y = -\frac{2}{3}(x - 4) + 2$.

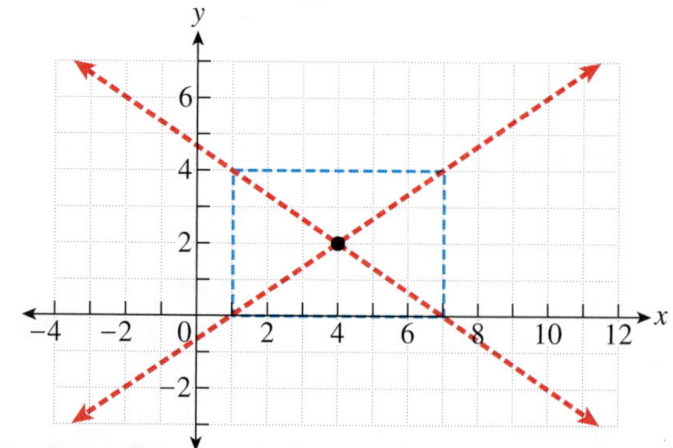

The vertices are located three units to the left and to the right of the center $(4, 2)$ at the points $(1, 2)$ and $(7, 2)$. After plotting the vertices, we draw the branches of the hyperbola opening to the left and to the right in such a way that they approach the asymptotes.

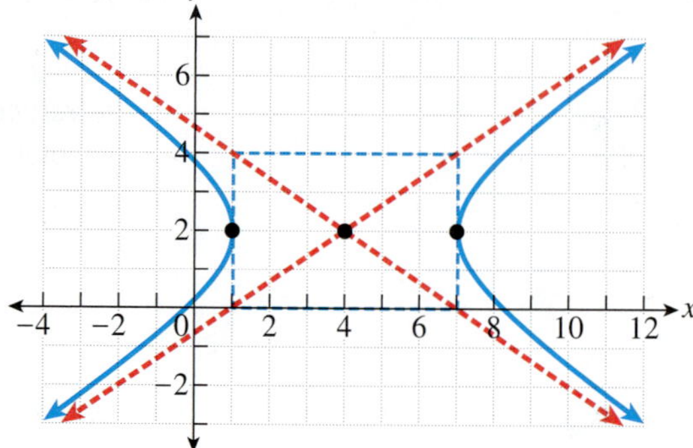

EXAMPLE 4 Graph the hyperbola $\frac{(y+5)^2}{16} - (x-1)^2 = 1$. Give the center, the vertices, and the equations of the asymptotes.

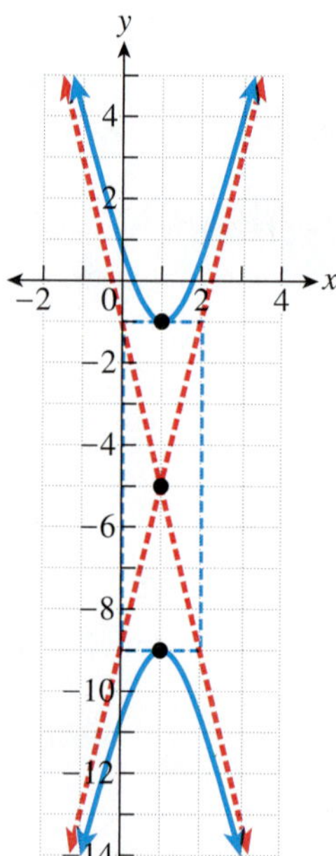

Solution

Since the equation is of the form $\frac{(y-k)^2}{b^2} - \frac{(x-h)^2}{a^2} = 1$, the hyperbola is centered at $(1, -5)$ and its branches open upward and downward. We begin by plotting a point at the center and graphing the asymptotes.

There is no denominator under the squared term containing x, so $a^2 = 1$. Therefore, $a = 1$. Since $b^2 = 16$, $b = 4$. The rectangle used to graph the asymptotes extends one unit to the left and to the right of the center and four units above and below the center. The equations for these asymptotes are $y = 4(x - 1) - 5$ and $y = -4(x - 1) - 5$.

Quick Check 4

Graph the hyperbola $\frac{(y-2)^2}{49} - \frac{(x+3)^2}{9} = 1$. Give the center, the vertices, and the equations of the asymptotes.

The vertices are located 4 units above and below the center at the points $(1, -1)$ and $(1, -9)$. After plotting the vertices, we draw the branches of the hyperbola opening upward and downward in such a way that they approach the asymptotes.

Finding the Center of a Hyperbola and the Lengths of Its Axes by Completing the Square

Objective 3 Find the center of a hyperbola and the lengths of its axes by completing the square.

General Form of the Equation of a Hyperbola

The **general form** of the equation of a hyperbola is $Ax^2 + By^2 + Cx + Dy + E = 0$, where $A \neq 0$, $B \neq 0$, and A and B have opposite signs.

If both A and B have the same sign, then the graph will be a circle or an ellipse, not a hyperbola. To graph a hyperbola in general form, we must rewrite the equation in standard form. This will allow us to determine the center, as well as a and b. This is done by completing the square, which must be done for both x and y.

EXAMPLE 5 Graph the hyperbola $-4x^2 + 25y^2 + 24x + 50y - 111 = 0$. Give the center, the vertices, and the equations of the asymptotes.

Solution

We will convert the equation to standard form by completing the square for x and y.

$$-4x^2 + 25y^2 + 24x + 50y - 111 = 0$$

$$(-4x^2 + 24x) + (25y^2 + 50y) = 111$$

Collect terms containing x. Collect terms containing y. Add 111 to both sides.

$$-4(x^2 - 6x) + 25(y^2 + 2y) = 111$$

Factor -4 from the terms containing x so that the coefficient of the x^2 term is 1. In the same fashion, factor 25 from the terms containing y.

$$-4(x^2 - 6x + 9) + 25(y^2 + 2y + 1) = 111 - 36 + 25$$

Add 9 inside the parentheses containing x to complete the square for x. Since $-4 \cdot 9 = -36$, subtract 36 on the right side of the equation. Add 1 inside the parentheses containing y to complete the square for y. Add $25 \cdot 1$, or 25, to the right side of the equation.

$$-4(x-3)^2 + 25(y+1)^2 = 100$$ Factor the two quadratic expressions on the left side of the equation.

$$25(y+1)^2 - 4(x-3)^2 = 100$$ Rewrite so that the term containing x is being subtracted from the term containing y.

$$\frac{(y+1)^2}{4} - \frac{(x-3)^2}{25} = 1$$ Divide both sides by 100 and simplify.

We begin by plotting the center, $(3, -1)$, and graphing the asymptotes.

For this hyperbola, $a = 5$ and $b = 2$. The equations for the asymptotes are $y = \frac{2}{5}(x-3) - 1$ and $y = -\frac{2}{5}(x-3) - 1$.

Since the equation has the form $\frac{(y-k)^2}{b^2} - \frac{(x-h)^2}{a^2} = 1$, the branches of this hyperbola open upward and downward.

The vertices are $(3, 1)$ and $(3, -3)$. After plotting the vertices, draw the branches of the hyperbola opening upward and downward in such a way that they approach the asymptotes.

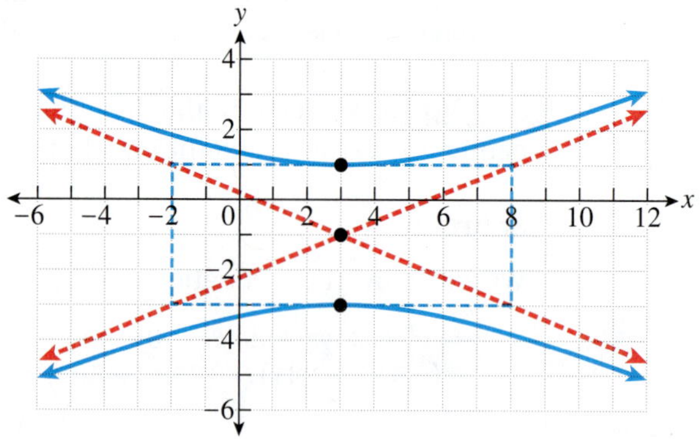

Quick Check 5 Graph the hyperbola $16x^2 - 9y^2 - 64x - 54y - 161 = 0$. Give the center, the vertices, and the equations of the asymptotes.

Finding the Standard-Form Equation of a Hyperbola

Objective 4 Find the equation of a hyperbola that meets the given conditions. To find the equation of a hyperbola in standard form, we must find the center (h, k), a, and b.

EXAMPLE 6 Find the equation of the hyperbola whose graph is as follows:

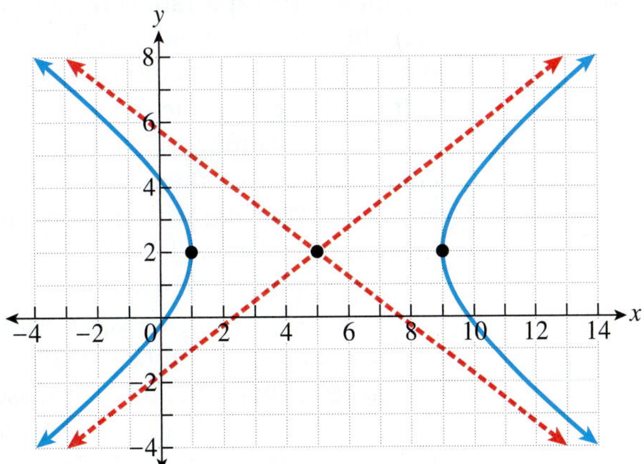

Solution

Since the branches of the hyperbola open to the left and to the right, its equation is of the form $\dfrac{(x-h)^2}{a^2} - \dfrac{(y-k)^2}{b^2} = 1$.

We begin by finding the center of the hyperbola, which is at the point $(5, 2)$.

Then we measure the distance a from the center of the hyperbola to one of its vertices and find that $a = 4$.

Next, we measure the vertical distance b from a vertex of the hyperbola to one of its asymptotes and find that $b = 3$.

We can now write the equation in standard form.

$$\frac{(x-h)^2}{a^2} - \frac{(y-k)^2}{b^2} = 1$$

$$\frac{(x-5)^2}{4^2} - \frac{(y-2)^2}{3^2} = 1 \quad \text{Substitute 5 for } h\text{, 2 for } k\text{, 4 for } a\text{, and 3 for } b.$$

$$\frac{(x-5)^2}{16} - \frac{(y-2)^2}{9} = 1 \quad \text{Simplify.}$$

The equation of this hyperbola is $\dfrac{(x-5)^2}{16} - \dfrac{(y-2)^2}{9} = 1$.

Quick Check **6** **Find the equation of the hyperbola whose graph is shown.**

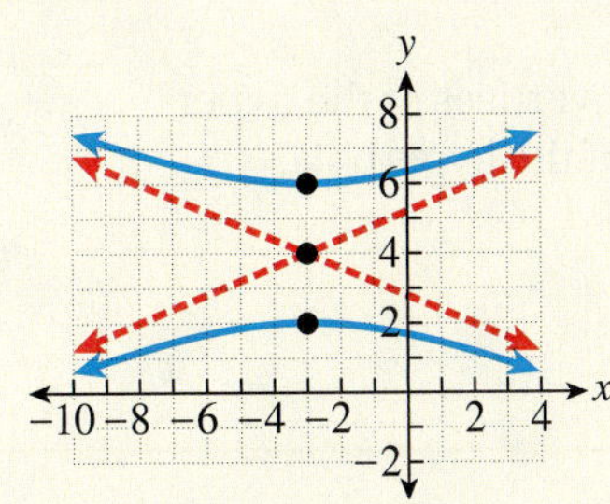

Building Your Study Strategy **Study Environment, 4 Lighting and Comfort** Lighting is an important consideration when you are selecting a study area. Some students prefer a bright, well-lit room, while this type of lighting makes other students fidgety and uncomfortable. Some students prefer to work in a room with soft, warm light, while this type of lighting makes other students drowsy. Some students prefer to work in a room with ample natural light. Whatever your preference, be sure that there is enough light for you to see clearly without straining your eyes. Working with insufficient light will make your eyes tired, which will make you tired as well.

It is important to be comfortable while you are studying, without being too comfortable. If the room is too warm, you may get tired. If the room is too cold, you may be distracted by thinking about how cold you are. If your chair is not comfortable, you will fidget and lose concentration. If your chair is too comfortable, you may get sleepy. You may have the same problem if you eat a large meal before sitting down to study. To summarize, choose a situation where you will be comfortable, but not too comfortable.

EXERCISES 9.4

FOR EXTRA HELP

Vocabulary

1. A(n)__________ is the collection of all points (x, y) in the plane for which the difference of the distances, d_1 and d_2, between the point and two fixed points, F_1 and F_2, is a constant.

2. The two fixed points, F_1 and F_2, referred to in the definition of a hyperbola, are called __________.

3. The __________ of a hyperbola is the point located midway between the two foci.

4. The line that passes through the two foci of a hyperbola intersects the hyperbola at two points that are called the __________ of the hyperbola.

5. The line segment between the vertices of the hyperbola is called the __________ of the hyperbola.

6. The of a hyperbola show how the branches of the hyperbola behave away from the center of the hyperbola.

7. The standard form equation of a hyperbola centered at the origin with a horizontal transverse axis has the form __________, while a hyperbola centered at the origin with a vertical transverse axis has the form __________.

8. The standard form equation of a hyperbola centered at the point (h, k) with a horizontal transverse axis has the form __________, while a hyperbola centered at the point (h, k) with a vertical transverse axis has the form __________.

Graph the hyperbola. Give the coordinates of the center, as well as the values of a and b.

9. $\frac{x^2}{9} - \frac{y^2}{4} = 1$

10. $\frac{x^2}{16} - \frac{y^2}{25} = 1$

11. $\frac{y^2}{49} - \frac{x^2}{9} = 1$

12. $\frac{y^2}{16} - \frac{x^2}{36} = 1$

13. $\frac{x^2}{9} - \frac{y^2}{9} = 1$

14. $\frac{y^2}{16} - \frac{x^2}{16} = 1$

15. $\dfrac{y^2}{25} - x^2 = 1$

16. $\dfrac{x^2}{16} - y^2 = 1$

17. $\dfrac{x^2}{12} - \dfrac{y^2}{9} = 1$

18. $\dfrac{y^2}{6} - \dfrac{x^2}{25} = 1$

19. $49x^2 - 4y^2 = 196$

20. $9x^2 - 36y^2 = 324$

21. $y^2 - 9x^2 = 9$

22. $y^2 - 4x^2 = 4$

23. $\dfrac{(y-5)^2}{16} - \dfrac{(x-2)^2}{9} = 1$

24. $\dfrac{(y-6)^2}{25} - \dfrac{(x-3)^2}{9} = 1$

25. $\dfrac{(x+1)^2}{9} - \dfrac{(y+4)^2}{4} = 1$

26. $\dfrac{(x+3)^2}{4} - \dfrac{(y-3)^2}{25} = 1$

27. $\dfrac{(y-1)^2}{4} - \dfrac{x^2}{4} = 1$

28. $\dfrac{(x+1)^2}{9} - \dfrac{(y+5)^2}{9} = 1$

29. $\dfrac{(x-4)^2}{49} - \dfrac{(y+2)^2}{16} = 1$

30. $\dfrac{(y-2)^2}{9} - \dfrac{(x+4)^2}{49} = 1$

31. $(y-5)^2 - \dfrac{(x+3)^2}{9} = 1$

32. $\dfrac{(y+1)^2}{4} - (x-6)^2 = 1$

33. $\dfrac{(x+2)^2}{25} - \dfrac{(y+3)^2}{28} = 1$

34. $\dfrac{(x+1)^2}{18} - \dfrac{(y-2)^2}{4} = 1$

35. $9x^2 - 4y^2 + 18x - 24y - 63 = 0$

36. $-16x^2 + 9y^2 - 160x + 72y - 400 = 0$

37. $-25x^2 + 16y^2 - 100x - 224y + 284 = 0$

38. $x^2 - 64y^2 - 10x + 768y - 2343 = 0$

Find the equation of the hyperbola that meets the given conditions.

39. Center: $(4, 2)$
Transverse axis: vertical
Length of transverse axis: 6
Length of conjugate axis: 10

40. Center: $(3, 7)$
Transverse axis: horizontal
Length of transverse axis: 10
Length of conjugate axis: 12

41. Center: $(6, -2)$
Transverse axis: horizontal
Length of transverse axis: 8
Length of conjugate axis: 8

42. Center: $(-3, 9)$
Transverse axis: vertical
Length of transverse axis: 2
Length of conjugate axis: 16

Find the standard-form equation of the hyperbola whose graph is shown.

43.

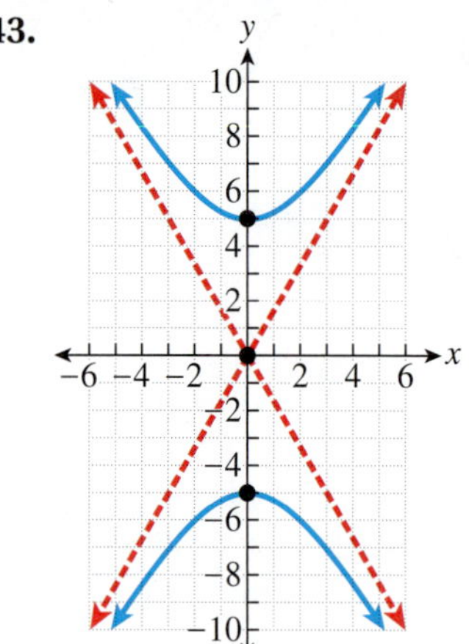

44.

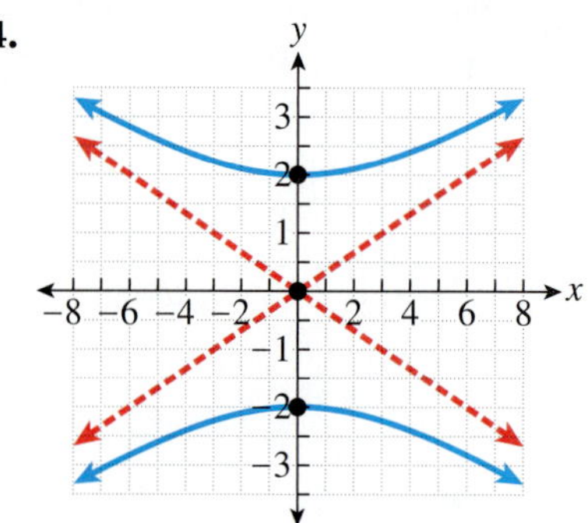

45.

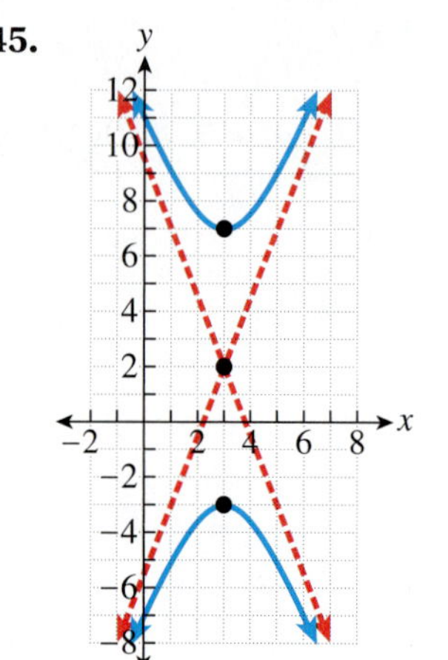

46.

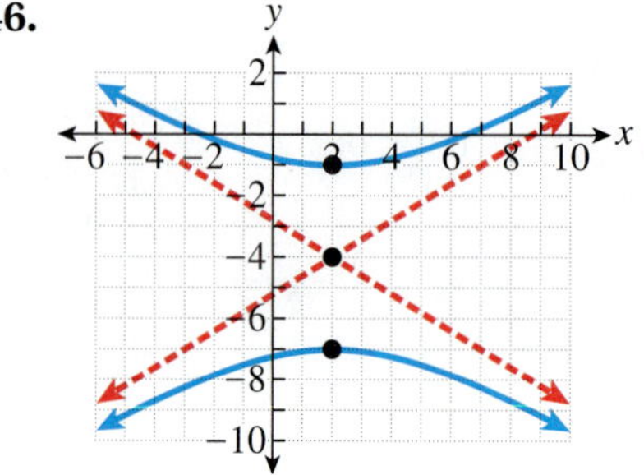

47.

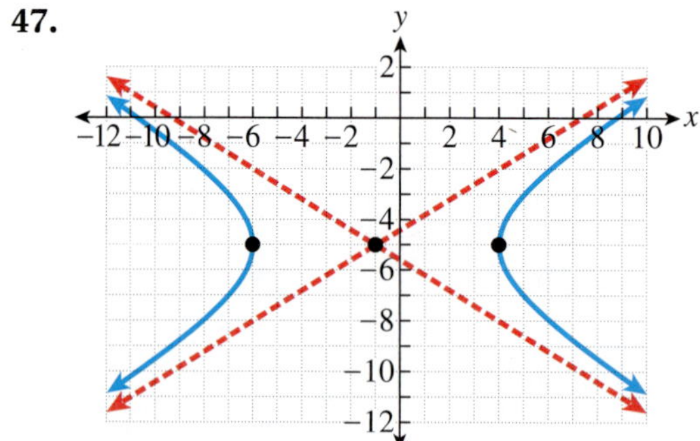

48.

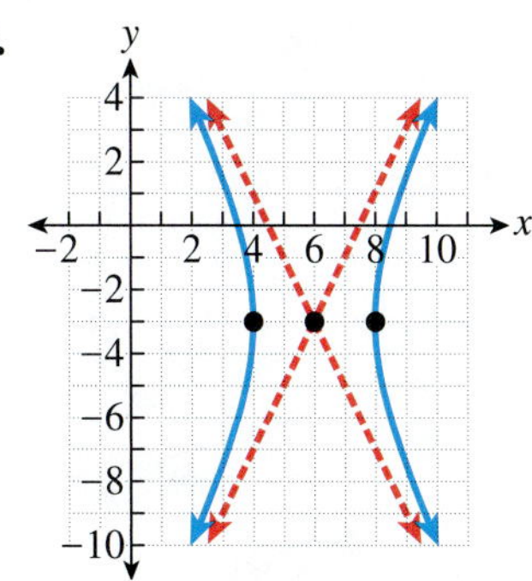

51. $y = x^2 + 8x + 19$

Mixed Practice, 49–66

Graph each equation. For graphs that are parabolas, label the vertex and any intercepts. For graphs that are circles, ellipses, or hyperbolas, label the center.

49. $x = -y^2 + 4y + 9$

52. $\dfrac{y^2}{25} - x^2 = 1$

50. $y = -x^2 + 6x - 17$

53. $\dfrac{(x-6)^2}{9} + \dfrac{y^2}{49} = 1$

54. $x = (y - 1)^2 - 12$

55. $y = (x - 6)^2 - 4$

56. $(x + 5)^2 + (y + 1)^2 = 20$

57. $\dfrac{(x + 2)^2}{25} + \dfrac{(y - 5)^2}{9} = 1$

58. $\dfrac{(y - 3)^2}{25} - \dfrac{(x + 7)^2}{36} = 1$

59. $x^2 + y^2 = 64$

60. $\dfrac{x^2}{36} + \dfrac{y^2}{4} = 1$

61. $4x^2 - 9y^2 + 40x + 54y - 17 = 0$

62. $(x - 4)^2 + (y + 7)^2 = 9$

63. $x^2 + y^2 - 8x + 18y + 48 = 0$

64. $16x^2 + 9y^2 - 256x + 90y + 1105 = 0$

65. $\dfrac{(x + 6)^2}{9} - \dfrac{(y + 2)^2}{16} = 1$

66. $x = y^2 + 10y + 16$

Determine which equation is associated with each given graph.

a) $y = (x + 3)^2 + 4$

b) $x = (y - 4)^2 - 3$

c) $(x + 3)^2 + (y - 4)^2 = 9$

d) $\dfrac{(x + 3)^2}{9} + \dfrac{(y - 4)^2}{16} = 1$

e) $\dfrac{(x + 3)^2}{9} - \dfrac{(y - 4)^2}{16} = 1$

f) $\dfrac{(y - 4)^2}{16} - \dfrac{(x + 3)^2}{9} = 1$

67.

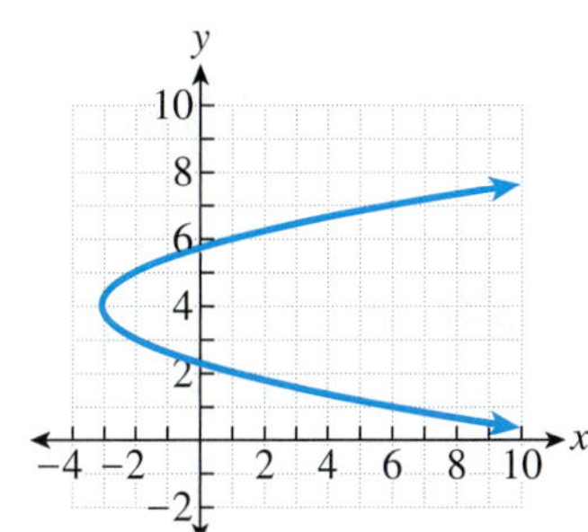

68.

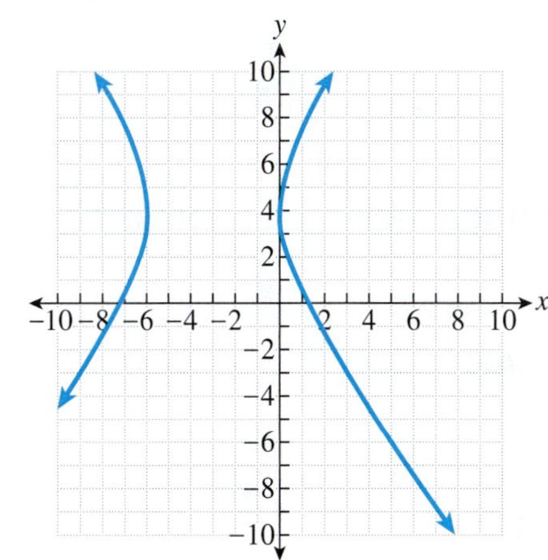

69.

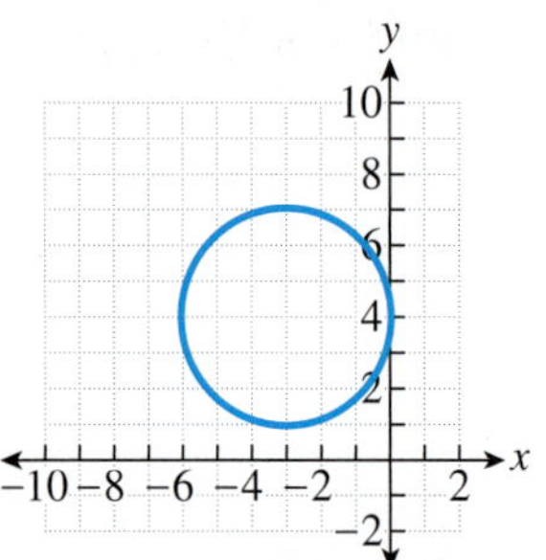

72.

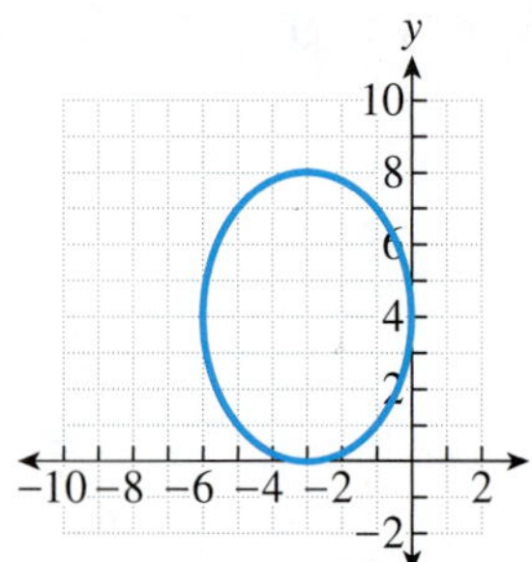

70.

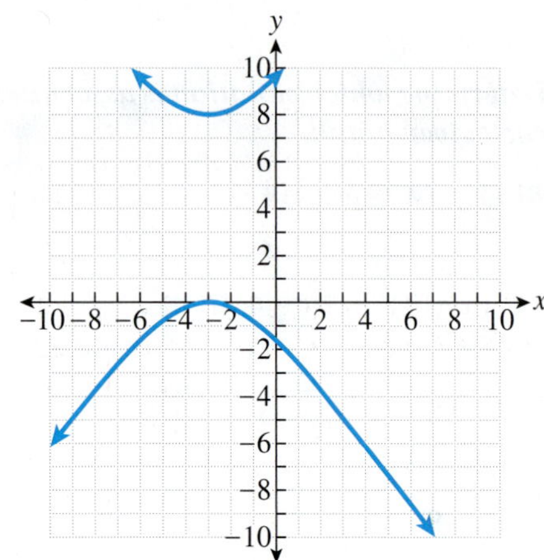

71.

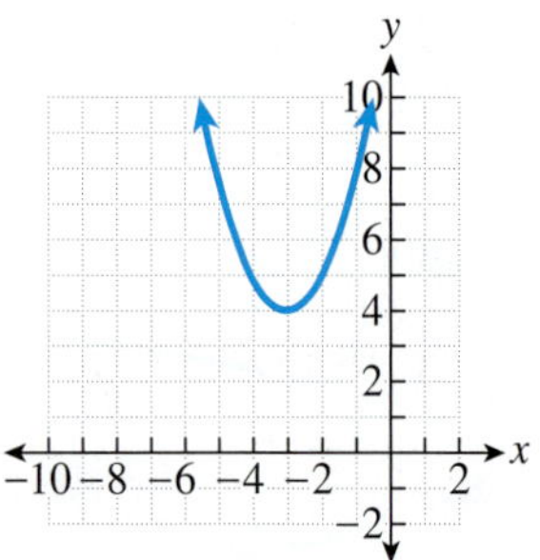

Writing in Mathematics

Answer in complete sentences.

73. Explain how to determine whether the graph of an equation is a hyperbola whose branches open to the left and right. Also, explain how to determine whether the graph of an equation is a hyperbola whose branches open up and down.

74. If the equations of the asymptotes of a hyperbola are known, is that enough information to determine the equation of the hyperbola? Explain why or why not.

75. ***Solutions Manual**** Write a solutions manual page for the following problem:

Graph $16x^2 - 9y^2 + 32x + 36y - 164 = 0$. *Give the coordinates of the center, as well as the values of a and b.*

76. ***Newsletter**** Write a newsletter explaining how to graph a hyperbola whose equation is of the form $\frac{(y-k)^2}{b^2} - \frac{(x-h)^2}{a^2} = 1$.

***See Appendix B for details and sample answers.**

Quick Review Exercises

Section 9.4

Solve.

1. $x + 8y = 19$
$7x - 5y = 11$

2. $y = 5x + 6$
$-3x + 4y = -27$

3. $2x - 3y = 11$
$5x + 6y = 14$

4. $4x - 5y = 15$
$-2x + 15y = -30$

9.5 Nonlinear Systems of Equations

Objectives

1. **Solve nonlinear systems by using the substitution method.**
2. **Solve nonlinear systems by using the addition method.**
3. **Solve applications of nonlinear systems.**

A **nonlinear system of equations** is a system of equations in which the graph of at least one of the equations is not a line. For example, if the graph of one equation is a circle and the graph of the other equation is a line, then the system is a nonlinear system of equations. For a system of two equations in two unknowns, an ordered pair (x, y) is a solution of the system if it is a solution of each equation. Graphically, a solution of a system of two equations in two unknowns is a point of intersection of the graphs of the two equations. Some nonlinear systems can have one or more solutions. The following are examples of nonlinear systems of equations and their solutions, which are points of intersection for the two graphs:

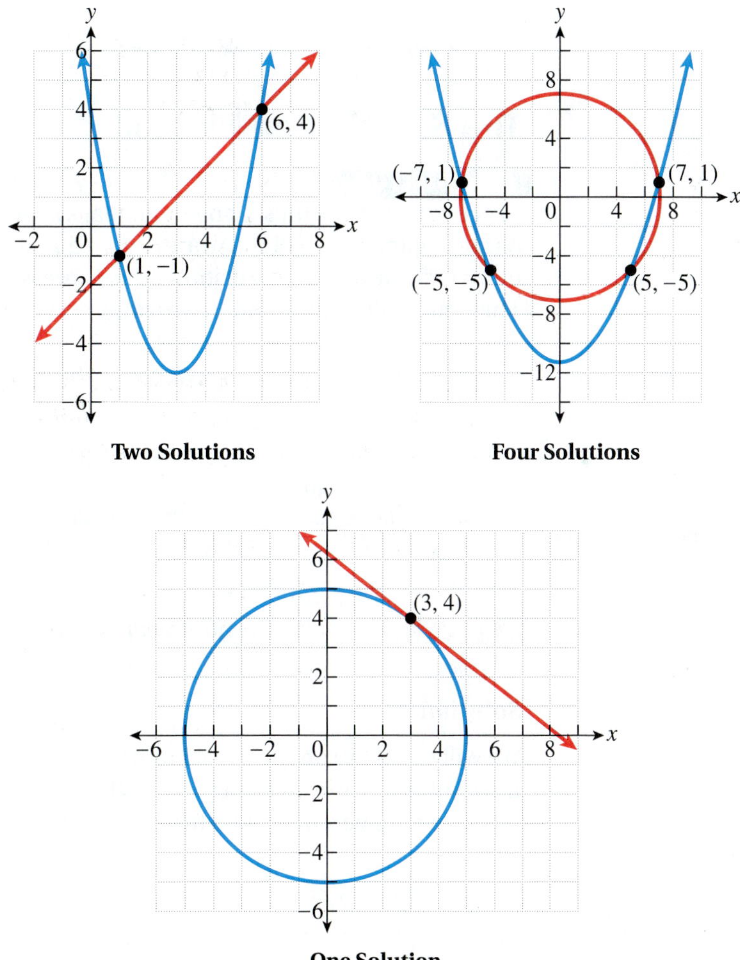

Two Solutions

Four Solutions

One Solution

If the graphs of the two equations do not intersect, then the system of equations has no solution.

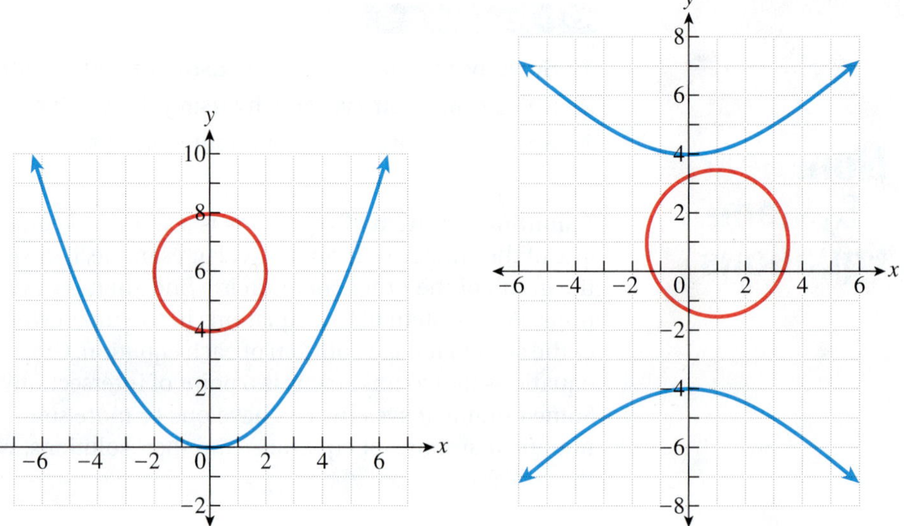

Circle and Parabola (No Intersection) **Circle and Hyperbola (No Intersection)**

The Substitution Method

Objective 1 **Solve nonlinear systems by using the substitution method.** One effective technique for solving nonlinear systems of equations is the substitution method. As in Chapter 3, we will solve one of the equations for one of the variables in terms of the other variable and then substitute this expression for that variable in the other equation. This will result in an equation with only one variable, which we can then solve.

Solving a System of Equations by the Substitution Method

1. Solve one of the equations for either variable.
2. Substitute this expression for the variable in the other equation.
3. Solve this equation.
4. Substitute each value for the variable into the equation from step 1.

EXAMPLE 1 Solve the nonlinear system.
$$x^2 + y^2 = 25$$
$$x + 2y = 10$$

Solution

We will begin by solving the second equation for x, since the coefficient of that term is 1. If we subtract $2y$ from both sides of the equation, we find that $x = 10 - 2y$. This expression can then be substituted for x in the equation $x^2 + y^2 = 25$.

$$x^2 + y^2 = 25$$
$$(10 - 2y)^2 + y^2 = 25 \quad \text{Substitute } 10 - 2y \text{ for } x.$$
$$5y^2 - 40y + 100 = 25 \quad \text{Square } 10 - 2y \text{ and combine like terms.}$$
$$5y^2 - 40y + 75 = 0 \quad \text{Subtract 25 from both sides.}$$
$$5(y - 3)(y - 5) = 0 \quad \text{Factor completely.}$$
$$y = 3 \quad \text{or} \quad y = 5 \quad \text{Set each variable factor equal to 0 and solve.}$$

We can now substitute these values for y in the equation $x = 10 - 2y$ to find the corresponding x-coordinates of the solutions.

$y = 3$	$y = 5$	
$x = 10 - 2(3)$	$x = 10 - 2(5)$	Substitute for y.
$x = 4$	$x = 0$	Simplify.
$(4, 3)$	$(0, 5)$	Simplify.

The two solutions are $(4, 3)$ and $(0, 5)$.

The graph of the first equation in the system is a circle, and the graph of the second equation is a line. Here are the graphs of the two equations, showing their points of intersection:

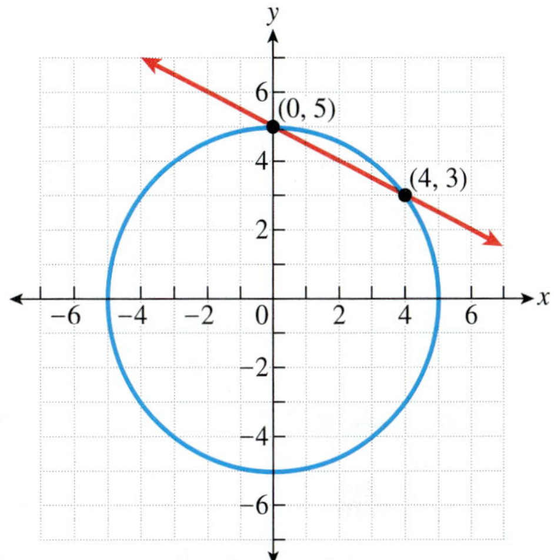

Quick Check 1

Solve the nonlinear system.

$$4x^2 + y^2 = 41$$
$$x + 2y = -8$$

Although it is not necessary, it is a good idea to graph both equations in the system. This will give us an idea of how many solutions we are looking for, as well as the approximate coordinates of the solutions.

EXAMPLE 2 Solve the nonlinear system. $\begin{aligned} y &= x + 6 \\ x^2 + y^2 &= 16 \end{aligned}$

Solution

Since the first equation is already solved for y, we can substitute $x + 6$ for y in the equation $x^2 + y^2 = 16$.

$$x^2 + y^2 = 16$$
$$x^2 + (x + 6)^2 = 16 \quad \text{Substitute } x + 6 \text{ for } y.$$
$$2x^2 + 12x + 36 = 16 \quad \text{Square } x + 6 \text{ and combine like terms.}$$
$$2x^2 + 12x + 20 = 0 \quad \text{Subtract 16 from both sides.}$$
$$2(x^2 + 6x + 10) = 0 \quad \text{Factor out a common factor of 2.}$$

Since $x^2 + 6x + 10$ cannot be factored, we will use the quadratic formula.

$$x = \frac{-6 \pm \sqrt{6^2 - 4(1)(10)}}{2(1)} \quad \text{Substitute 1 for } a\text{, 6 for } b\text{, and 10 for } c \text{ in the quadratic formula.}$$

$$x = \frac{-6 \pm \sqrt{-4}}{2} \quad \text{Simplify the discriminant and the denominator.}$$

Since the discriminant was negative, this equation has no real-number solutions. Therefore, this system of equations has no solution. We could have determined this from the graphs of the two equations. The graph of the equation $y = x + 6$ is a line with a slope of 1 and a y-intercept at $(0, 6)$, while the graph of $x^2 + y^2 = 16$ is a circle centered at the origin with a radius of 4. Notice that the two graphs clearly do not intersect.

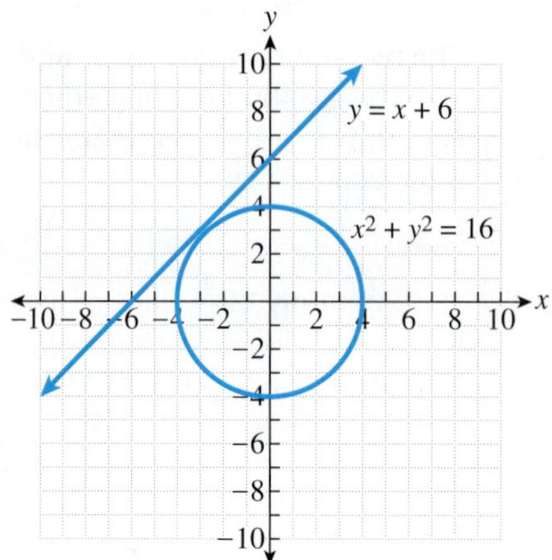

Quick Check 2
Solve the nonlinear system.
$y = 2x - 8$
$x^2 + y^2 = 9$

EXAMPLE 3 Solve the nonlinear system. $\begin{aligned} y &= x^2 - 3 \\ x^2 + y^2 &= 9 \end{aligned}$

Solution

The graph on the left of the first equation is a parabola that opens upward, while the graph of the second equation is a circle. Based on the graph, there appear to be three solutions.

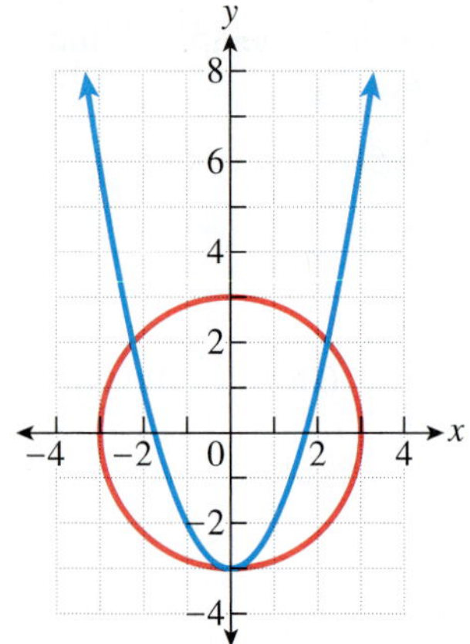

Solve the first equation for x^2, giving $x^2 = y + 3$, and substitute that expression for x^2 in the second equation. This is a useful technique when the graph of neither equation is a line.

$$x^2 + y^2 = 9$$

$$(y + 3) + y^2 = 9 \quad \text{Substitute } y + 3 \text{ for } x^2.$$

$$y^2 + y - 6 = 0 \quad \text{Collect all terms on the left side.}$$

$$(y + 3)(y - 2) = 0 \quad \text{Factor.}$$

$$y = -3 \quad \text{or} \quad y = 2 \quad \text{Set each factor equal to 0 and solve.}$$

We can now substitute these values for y in the equation $x^2 = y + 3$ to find the corresponding x-coordinates of the solutions.

$y = -3$	$y = 2$	
$x^2 = (-3) + 3$	$x^2 = (2) + 3$	Substitute for y.
$x = 0$	$x = \pm\sqrt{5}$	Simplify and take the square root of each side.

The solutions are $(0, -3)$, $(\sqrt{5}, 2)$, and $(-\sqrt{5}, 2)$.

Quick Check 3
Solve the nonlinear system.
$y = x^2$
$5x^2 + y^2 = 24$

The Addition Method

Objective 2 **Solve nonlinear systems by using the addition method.** Another effective technique for solving nonlinear systems of equations is the addition method, which was introduced in Chapter 3. By rewriting the two equations in such a way that the coefficients of like terms are opposites, we can add the two equations to obtain a single equation with only one variable, which we can then solve.

EXAMPLE 4 Solve the nonlinear system.
$$\begin{aligned} 16x^2 + 25y^2 &= 400 \\ x^2 + y^2 &= 16 \end{aligned}$$

Solution

If we multiply both sides of the second equation by -16, then the coefficients of the two terms containing x^2 will be opposites. We will then be able to eliminate terms containing x^2, leaving an equation whose only variable is y.

$$\begin{aligned} 16x^2 + 25y^2 &= 400 & & & 16x^2 + 25y^2 &= 400 \\ x^2 + y^2 &= 16 & &\xrightarrow{\text{Multiply by } -16.} & -16x^2 - 16y^2 &= -256 \end{aligned}$$

$$\begin{array}{r} 16x^2 + 25y^2 = 400 \\ \underline{-16x^2 - 16y^2 = -256} \\ 9y^2 = 144 \end{array} \qquad \text{Add.}$$

We can now solve the equation $9y^2 = 144$ for y.

$$\begin{aligned} 9y^2 &= 144 \\ y^2 &= 16 & &\text{Divide both sides by 9.} \\ y &= \pm 4 & &\text{Take the square root of each side.} \end{aligned}$$

Quick Check 4
Solve the nonlinear system.
$$\begin{aligned} 4x^2 + 7y^2 &= 211 \\ x^2 + y^2 &= 34 \end{aligned}$$

We can now substitute 4 and -4 for y in either equation to find the corresponding x-coordinates of the solutions. We will use the equation of the circle $x^2 + y^2 = 16$.

$$\begin{array}{lll} y = 4 & y = -4 & \\ x^2 + (4)^2 = 16 & x^2 + (-4)^2 = 16 & \text{Substitute for } y. \\ x = 0 & x = 0 & \text{Solve for } x. \end{array}$$

The solutions are $(0, 4)$ and $(0, -4)$.

We can attempt to solve nonlinear systems by graphing, but our solutions will depend upon the accuracy of our graph, not to mention our estimate of solutions that do not have integer coordinates. However, if we use technology such as a graphing calculator or computer software, these tools have built-in functions for finding the points of intersection of two graphs.

Applications of Nonlinear Systems of Equations

Objective 3 **Solve applications of nonlinear systems.** We now turn our attention to applied problems requiring solving a nonlinear system of equations to solve the problem.

EXAMPLE 5 A rectangular dorm room has a perimeter of 40 feet and an area of 91 square feet. Find its dimensions.

Solution

The unknowns in this problem are the length and width of the room. We will let x represent the length of the room, and we will let y represent the width of the room.

Unknowns
Length: x
Width: y

Since the perimeter of the rectangle is 40 feet, we know that $2x + 2y = 40$. Also, since the area of the rectangle is 91 square feet, we know that $xy = 91$. If we solve the perimeter equation for either x or y, we can then substitute the expression for that variable into the area equation.

$$2x + 2y = 40$$
$$2x = 40 - 2y \quad \text{Subtract } 2y.$$
$$x = 20 - y \quad \text{Divide both sides by 2.}$$

We can now substitute $20 - y$ for x in the equation $xy = 91$.

$$xy = 91$$
$$(20 - y)y = 91 \quad \text{Substitute } 20 - y \text{ for } x.$$
$$0 = y^2 - 20y + 91 \quad \text{Simplify and collect all terms on the right side of the equation.}$$
$$0 = (y - 7)(y - 13) \quad \text{Factor the quadratic expression.}$$
$$y = 7 \quad \text{or} \quad y = 13 \quad \text{Set each factor equal to 0 and solve.}$$

We can find the corresponding x-coordinates of the solutions by substituting these values for y in one of the two equations and solving for x. We will use the perimeter equation.

Quick Check 5

A rectangle has a perimeter of 30 feet and an area of 36 square feet. Find its dimensions.

$y = 7$	$y = 13$	
$2x + 2(7) = 40$	$2x + 2(13) = 40$	Substitute for y.
$x = 13$	$x = 7$	Solve for x.

We see that the length is 13 feet if the width is 7 feet, and the length is 7 feet if the width is 13 feet. In either case, the dimensions of the room are 7 feet by 13 feet.

EXAMPLE 6 A rectangular computer monitor has a perimeter of 46 inches and a diagonal that is 17 inches long. Find the dimensions of the monitor.

Solution

The unknowns in this problem are the length and width of the monitor. We will let x represent the length of the monitor, and we will let y represent the width of the monitor.

Unknowns
Length: x
Width: y

Since the perimeter of the monitor is 46 inches, we know that $2x + 2y = 46$. Also, since the diagonal of the monitor is 17 inches, we know from the Pythagorean theorem that $x^2 + y^2 = 17^2$, or $x^2 + y^2 = 289$.

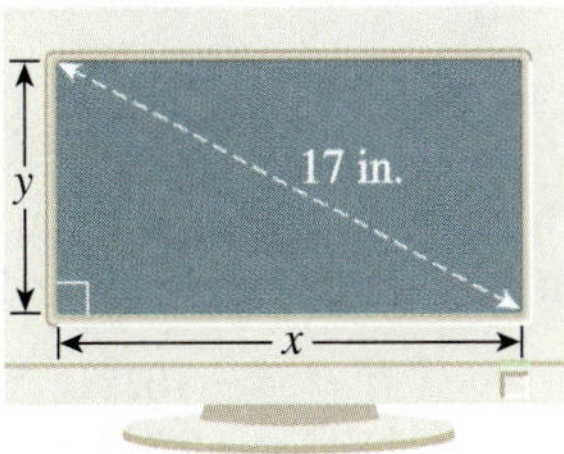

Solve the perimeter equation for x, giving $x = 23 - y$, and then substitute that expression for x into the equation $x^2 + y^2 = 289$.

$$x^2 + y^2 = 289$$
$$(23 - y)^2 + y^2 = 289 \quad \text{Substitute } 23 - y \text{ for } x.$$
$$2y^2 - 46y + 240 = 0 \quad \text{Square } 23 - y \text{ and collect all terms on the left side.}$$
$$2(y - 8)(y - 15) = 0 \quad \text{Factor completely.}$$
$$y = 8 \quad \text{or} \quad y = 15 \quad \text{Set each variable factor equal to 0 and solve.}$$

We can find the corresponding x-coordinates of the solutions by substituting these values for y in one of the two equations and solving for x. We will use the perimeter equation.

Quick Check 6
A rectangle has a perimeter of 42 inches and a diagonal 15 inches long. Find the dimensions of the rectangle.

$y = 8$	$y = 15$	
$2x + 2(8) = 46$	$2x + 2(15) = 46$	Substitute for y.
$x = 15$	$x = 8$	Solve for x.

We see that the length is 15 inches if the width is 8 inches and that the length is 8 inches if the width is 15 inches. In either case, the dimensions of the rectangle are 8 inches by 15 inches.

We conclude with an example involving the height of a projectile. Recall that if an object is launched with an initial velocity of v_0 feet per second from an initial height of s feet, then its height in feet after t seconds is given by the function $h(t) = -16t^2 + v_0t + s$.

EXAMPLE 7 A rock is thrown upward from ground level with an initial velocity of 100 feet per second. At the same instant, a water balloon is dropped from a helicopter that is hovering 300 feet above the ground. At what time will the two objects be the same height above the ground? What is the height?

Solution

We will begin by finding the time that it takes for the two objects to be at the same height. The function describing the height of the rock is $h(t) = -16t^2 + 100t$, while the function describing the height of the water balloon dropped from the helicopter is $h(t) = -16t^2 + 300$. If we let the variable y represent the height of each object, we get the following nonlinear system of equations:

$$y = -16t^2 + 100t$$
$$y = -16t^2 + 300$$

Quick Check 7
A boy throws a ball upward from the ground with an initial velocity of 40 feet per second. At the same instant, a boy standing on a platform 10 feet above the ground throws another ball upward with an initial velocity of 35 feet per second. At what time will the two balls be the same height above the ground?

Since each equation is already solved for y, we can set $-16t^2 + 100t$ equal to $-16t^2 + 300$ and solve for t.

$$-16t^2 + 100t = -16t^2 + 300$$
$$100t = 300 \quad \text{Add } 16t^2 \text{ to both sides of the equation. The resulting equation is linear.}$$
$$t = 3 \quad \text{Divide both sides by 100.}$$

The two objects will be at the same height after 3 seconds. We can substitute 3 for t in either function to find the height.

$$h(t) = -16t^2 + 100t$$
$$h(3) = -16(3)^2 + 100(3) \quad \text{Substitute 3 for } t.$$
$$= 156 \quad \text{Simplify.}$$

After 3 seconds, each object will be at a height of 156 feet.

Building Your Study Strategy **Study Environment, 5 Campus and Home**
Under ideal circumstances, some of your studying should be accomplished on campus and some of your studying should be done at home. The best time for studying new material is as soon as possible after class. That means that you should look for a location on campus. Since studying new material takes more concentration, find a location with as few distractions as possible. Most campus libraries have an area designated as a silent study area. Other libraries have cubicles that help to isolate you while you are studying. While studying at home, stay away from the computer, phone, or television. These devices will steal your time away from you, and you may not even realize it. Keep all temptations as far away as possible.

EXERCISES 9.5

Vocabulary

1. A(n) __________ system of equations is a system of equations in which the graph of at least one of the equations is not a line.

2. An ordered pair (x, y) is a(n) __________ of a nonlinear system of two equations if it is a solution of each equation.

Solve by the substitution method.

3. $x^2 + y^2 = 100$
$x - 7y = 50$

4. $x^2 + y^2 = 25$
$x - y = -1$

5. $x^2 + y^2 = 85$
$3x + y = -29$

6. $x^2 + y^2 = 50$
$-x + 2y = 15$

7. $4x^2 + 25y^2 = 100$
$2x + 5y = 10$

8. $9x^2 + y^2 = 9$
$-3x + y = 3$

9. $4x^2 + y^2 = 16$
$y = 2x - 4$

10. $2x^2 + 3y^2 = 14$
$x = -3y + 7$

11. $y = x^2 - 3x + 8$
$-2x + y = 4$

12. $y = x^2 + 6x$
$-4x + y = 3$

13. $y = -x^2 + 8x - 13$
$2x - y = 5$

14. $y = -x^2 - 9x + 25$
$-x + y = 25$

15. $x^2 - y^2 = 8$
$y = x - 2$

16. $y^2 - x^2 = 7$
$y = x + 7$

17. $y = 3x - 9$
$x^2 + y^2 = 4$

18. $y = x - 2$
$(x + 2)^2 + (y - 1)^2 = 9$

19. $x^2 + y^2 = 25$
$y = x^2 - 5$

20. $x^2 + y^2 = 16$
$y = x^2 - 4$

21. $x^2 + y^2 = 30$
$x = y^2$

22. $x^2 + y^2 = 42$
$x = -y^2$

23. $x^2 + 4y^2 = 4$
$y = x^2 + 1$

24. $x^2 + 9y^2 = 36$
$y = x^2 + 2$

25. $9x^2 + 4y^2 = 9$
$x = y^2 + 1$

26. $25x^2 + 9y^2 = 225$
$x = y^2 + 3$

27. $x^2 - y^2 = 1$
$x = y^2 + 11$

28. $x^2 - y^2 = 15$
$x = \frac{1}{3}y^2 - 1$

29. $x^2 - 4y^2 = 16$
$x = y^2 - 11$

30. $y^2 - 9x^2 = 9$
$y = x^2 + 1$

31. $y = x^2 - 7$
$y = -x^2 + 11$

32. $y = x^2 - 8$
$y = -x^2$

33. $x = y^2 - 5y$
$x = y^2 + 10$

34. $x = 3y^2 - 15y + 16$
$x = 2y^2 - 9y + 7$

35. $y = x^2 - 6$
$x^2 + (y - 4)^2 = 8$

36. $y = x^2 + 1$
$x^2 + \frac{(y - 5)^2}{4} = 1$

Solve by the addition method.

37. $x^2 + y^2 = 5$
$x^2 - y^2 = 3$

38. $x^2 + y^2 = 20$
$y^2 - x^2 = 12$

39. $3x^2 + y^2 = 14$
$x^2 - y^2 = 2$

40. $5x^2 + y^2 = 45$
$x^2 - y^2 = 9$

41. $x^2 + 2y^2 = 18$
$y^2 - x^2 = 6$

42. $x^2 + 9y^2 = 9$
$y^2 - x^2 = 1$

43. $9x^2 + 4y^2 = 87$
$x^2 - 2y^2 = 6$

44. $8x^2 + 6y^2 = 44$
$4x^2 - 2y^2 = 12$

45. $x^2 + y^2 = 24$
$x^2 + 5y^2 = 60$

46. $x^2 + y^2 = 19$
$10x^2 + y^2 = 100$

47. $x^2 + y^2 = 6$
$4x^2 + 9y^2 = 39$

48. $x^2 + y^2 = 13$
$3x^2 + 18y^2 = 99$

49. $x^2 + 12y^2 = 117$
$x^2 + 4y^2 = 45$

50. $3x^2 + y^2 = 21$
$10x^2 + y^2 = 49$

51. $3x^2 + 2y^2 = 21$
$15x^2 + 4y^2 = 51$

52. $3x^2 + 4y^2 = 84$
$9x^2 + 21y^2 = 333$

53. A rectangular flower garden has a perimeter of 44 feet and an area of 105 square feet. Find the dimensions of the garden.

54. A rectangular classroom has a perimeter of 78 feet and an area of 360 square feet. Find the dimensions of the classroom.

55. The cover for a rectangular swimming pool measures 400 square feet. If the perimeter of the pool is 82 feet, find the dimensions of the pool.

56. The perimeter of a soccer field is 320 yards, and the area of the field is 6000 square yards. Find the dimensions of the soccer field.

57. Juan is remodeling his home office by replacing the molding around the floor of the room, as well as the carpet. The door to the office is 4 feet wide, and no molding is required there. If Juan used 60 feet of molding and 192 square feet of carpet, find the dimensions of the office.

58. Rosalba is painting a rectangular wall in her home. The perimeter of the wall is 60 feet. The wall has two windows in it, each with an area of 15 square feet. If the painting area is 170 square feet, find the dimensions of the wall.

59. George's vegetable garden is in the shape of a rectangle, and has 210 feet of fencing around it. Diagonally, the garden measures 75 feet from corner to corner. Find the dimensions of the garden.

60. A rectangular lawn has a perimeter of 100 feet and a diagonal of $10\sqrt{13}$ feet. Find the dimensions of the lawn.

61. A rectangular sports court has an area of 800 square feet. If the diagonal of the court measures $20\sqrt{5}$ feet, find the dimensions of the court.

62. Laura's horses live in a rectangular pasture that has an area of 10,000 square feet. The diagonal of the pasture measures $50\sqrt{17}$ feet. Find the dimensions of the pasture.

63. An arrow is fired upward with an initial velocity of 120 feet per second. At the same instant, a ball is dropped from a helicopter that is hovering 540 feet above the ground. At what time will the arrow and ball be the same height above the ground?

64. A ball is thrown upward from a beach with an initial velocity of 32 feet per second. At the same instant, another ball is dropped from a cliff 48 feet above the beach. At what time will the two balls be the same height above the beach?

65. A model rocket is fired upward from the ground with an initial velocity of 60 feet per second. At the same instant, another model rocket is fired upward from a roof 84 feet above the ground with an initial velocity of 32 feet per second.

a) At what time will the two rockets be the same height above the ground?

b) How high above the ground are the two rockets at that time?

66. A slingshot fires a rock upward from the ground with an initial velocity of 96 feet per second. At the same instant, a cannonball is fired upward from a roof 120 feet above the ground with an initial velocity of 66 feet per second.

a) At what time will the rock and cannonball be the same height above the ground?

b) How high above the ground are the two objects at that time?

67. a) Find the missing value such that $(5, -2)$ is a solution to $\begin{matrix} x + y = 3 \\ x^2 + y^2 = ? \end{matrix}$.

b) What is the other solution to this system of equations?

68. a) Find the missing value such that $(4, 3)$ is a solution to $\begin{matrix} 2x + y = 11 \\ 4x^2 + 9y^2 = ? \end{matrix}$.

b) What is the other solution to this system of equations?

69. a) Find the missing value such that $(1, 6)$ is a solution to $\begin{matrix} x + 3y = 19 \\ y^2 - x^2 = ? \end{matrix}$.

b) What is the other solution to this system of equations?

70. a) Find the missing value such that $(3, 2)$ is a solution to $\begin{matrix} 4x + y = 14 \\ x^2 - y^2 = ? \end{matrix}$.

b) What is the other solution to this system of equations?

Writing in Mathematics

Answer in complete sentences.

71. Write a word problem leading to a system of nonlinear equation whose solution is "The rectangle is 50 feet by 30 feet."

72. Give an example of a system of nonlinear equations for which you feel it would be best to solve by the substitution method; give an example of another system for which you feel it would be best to solve by the addition method. Explain why the method is the best choice for each example.

73. ***Solutions Manual**** Write a solutions manual page for the following problem:

Solve $\begin{matrix} x^2 + y^2 = 34 \\ x + y = 8 \end{matrix}$.

74. ***Newsletter**** Write a newsletter explaining how you would find the length and width of a rectangle by using a system of nonlinear equations if the perimeter and area of the rectangle are known.

*See Appendix B for details and sample answers.

Chapter 9 Summary

Section 9.1—Topic	Chapter Review Exercises
Graphing Parabolas of the Form $y = ax^2 + bx + c$	1–2
Graphing Parabolas of the Form $y = a(x - h)^2 + k$	3–4
Graphing Parabolas of the Form $x = ay^2 + by + c$	5–8
Graphing Parabolas of the Form $x = a(y - k)^2 + h$	9–12
Section 9.2—Topic	**Chapter Review Exercises**
Distance Formula	13–14
Graphing a Circle	15–20
Finding the Center and Radius by Completing the Square	21–23
Finding the Equation of a Circle from Its Graph	24–25
Section 9.3—Topic	**Chapter Review Exercises**
Graphing an Ellipse	26–30
Finding the Center, a, and b by Completing the Square	31–33
Finding the Equation of an Ellipse from Its Graph	34–35
Section 9.4—Topic	**Chapter Review Exercises**
Graphing a Hyperbola	36–40
Finding the Center, a, and b by Completing the Square	41–43
Finding the Equation of a Hyperbola from Its Graph	44–45
Section 9.5—Topic	**Chapter Review Exercises**
Solving Nonlinear Systems of Equations by Substitution	46–53
Solving Nonlinear Systems of Equations by Addition	54–57
Applications of Nonlinear Systems of Equations	58–60

Summary of Chapter 9 Study Strategies

This chapter's study strategy focused on creating the proper study environment.

- Find a location as free from distractions as possible.
- Determine what level of background noise you can tolerate without being distracted.
- Select a location that allows you to spread out your materials.
- Select a location that provides sufficient lighting.
- Select a location that is comfortable, with proper temperature control.
- Search out multiple locations, at home and on campus, that are conducive to studying.

Chapter 9 Review

Graph. Label the vertex and any intercepts. **[9.1]**

1. $y = x^2 - 8x - 9$

2. $y = -x^2 + 6x - 3$

3. $y = (x - 4)^2 - 1$

4. $y = -(x + 3)^2 - 4$

5. $x = y^2 - 10y + 15$

6. $x = y^2 + 3y - 28$

7. $x = -y^2 + y + 20$

8. $x = -y^2 - 6y - 14$

9. $x = (y + 5)^2 - 4$

10. $x = -(y + 1)^2 + 10$

Worked-out solutions to Review Exercises marked with (shaded circle) **can be found on page AN-60.**

11. $x = -(y - 4)^2 + 9$

12. $x = (y - 3)^2 + 6$

Find the distance between the given points. Round to the nearest tenth if necessary. **[9.2]**

13. $(-7, -8)$ and $(1, 7)$

14. $(3, -5)$ and $(-6, -2)$

Graph the circle. Label the center of the circle. Give its radius as well. **[9.2]**

15. $x^2 + y^2 = 36$

16. $x^2 + y^2 = 20$

17. $(x + 3)^2 + (y - 4)^2 = 4$

18. $(x - 8)^2 + (y + 3)^2 = 9$

19. $(x + 4)^2 + (y - 5)^2 = 25$

20. $(x - 2)^2 + y^2 = 16$

21. $x^2 + y^2 + 12x - 2y - 12 = 0$

Find the center and radius of the circle by completing the square. **[9.2]**

22. $x^2 + y^2 + 6x - 18y - 10 = 0$

23. $x^2 + y^2 - 8x - 20y + 99 = 0$

Find the equation of the circle. **[9.2]**

24.

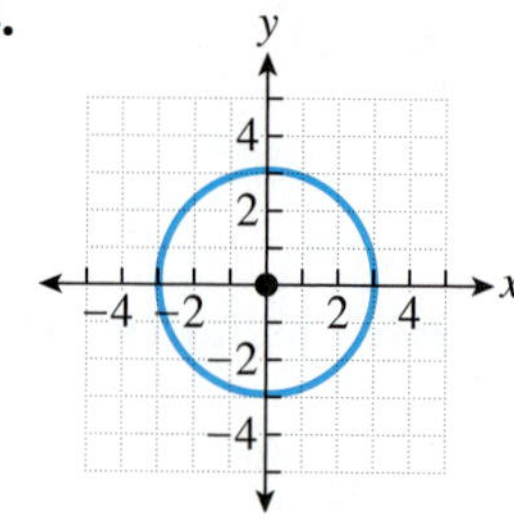

25.

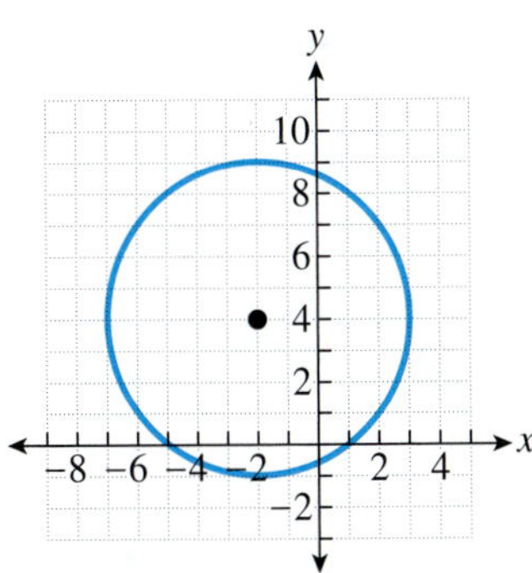

Graph the ellipse. Label the center. Give the values of a and b. **[9.3]**

26. $\dfrac{x^2}{25} + \dfrac{y^2}{9} = 1$

27. $x^2 + \dfrac{y^2}{16} = 1$

28. $\dfrac{(x + 3)^2}{9} + \dfrac{(y - 4)^2}{16} = 1$

29. $\dfrac{(x - 8)^2}{25} + \dfrac{(y - 1)^2}{4} = 1$

30. $\dfrac{(x + 6)^2}{9} + \dfrac{y^2}{36} = 1$

31. $9x^2 + 16y^2 + 90x - 64y + 145 = 0$

Find the center of the ellipse by completing the square. Find the values of a and b as well. **[9.3]**

32. $4x^2 + 49y^2 - 48x - 98y - 3 = 0$

33. $81x^2 + 16y^2 + 324x + 128y - 716 = 0$

Find the equation of the ellipse. **[9.3]**

34.

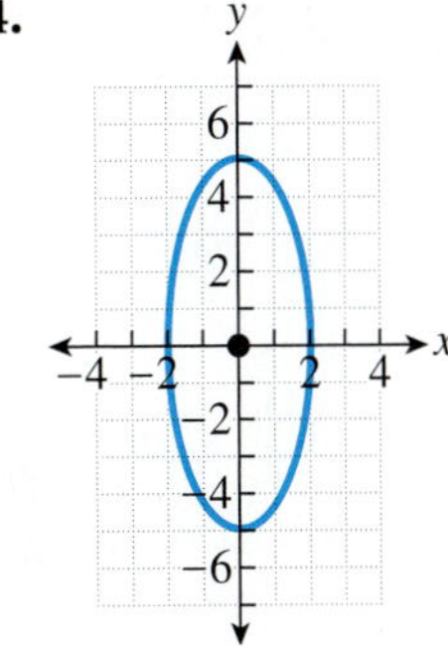

35.

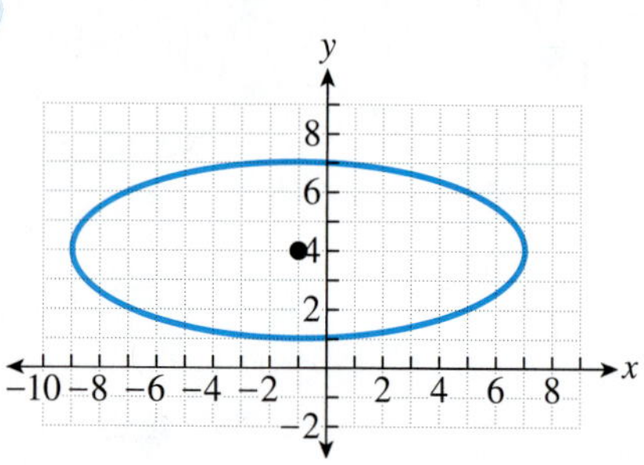

Graph the hyperbola. Label the center. Give the values of a and b. **[9.4]**

36. $\frac{x^2}{9} - \frac{y^2}{25} = 1$

37. $\frac{y^2}{16} - \frac{x^2}{16} = 1$

38. $\frac{(x-1)^2}{4} - \frac{(y-3)^2}{9} = 1$

39. $\frac{(x+4)^2}{25} - \frac{(y-2)^2}{36} = 1$

40. $\frac{(y+3)^2}{4} - \frac{(x+2)^2}{81} = 1$

41. $-4x^2 + 25y^2 - 24x + 300y + 764 = 0$

Find the center of the hyperbola by completing the square. Find the lengths of its transverse and conjugate axes as well. **[9.4]**

42. $-9x^2 + 4y^2 + 90x - 56y - 65 = 0$

43. $49x^2 - 36y^2 + 784x + 288y + 796 = 0$

Find the equation of the hyperbola. **[9.4]**

44.

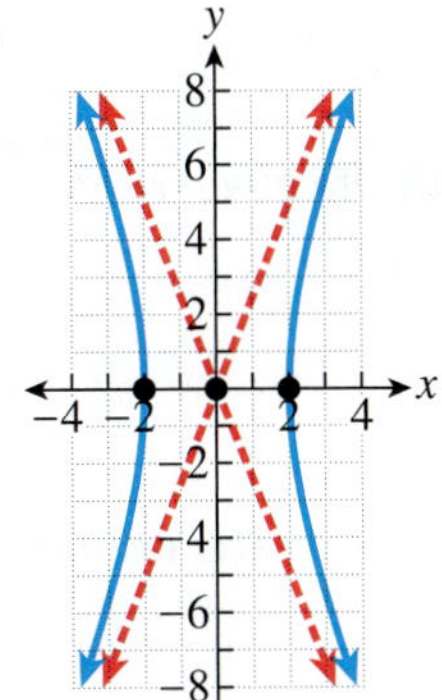

45.

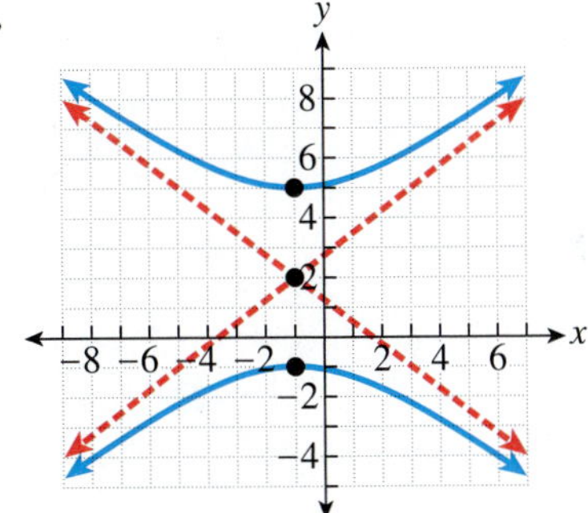

Solve by substitution. **[9.5]**

46. $x^2 + y^2 = 50$
$x + 2y = 15$

47. $4x^2 + 3y^2 = 48$
$y = 2x - 4$

48. $y = x^2 - 8x + 23$
$2x - y = 1$

49. $4y^2 - x^2 = 96$
$6y - x = 32$

50. $x^2 + y^2 = 17$
$y = x^2 + 3$

51. $3x^2 + 5y^2 = 72$
$y = x^2 - 6$

52. $9x^2 - y^2 = 9$
$y = x^2 - 1$

53. $y = -x^2 + 2x + 12$
$y = x^2$

Solve by the addition method. **[9.5]**

54. $x^2 + y^2 = 100$
$y^2 - x^2 = 62$

55. $3x^2 + 4y^2 = 31$
$x^2 - 2y^2 = 7$

56. $x^2 + y^2 = 8$
$2x^2 + 15y^2 = 68$

57. $6x^2 + y^2 = 159$
$2x^2 + 9y^2 = 131$

58. A rectangular flower garden has a perimeter of 70 feet and an area of 300 square feet. Find the dimensions of the garden. [9.5]

59. A rectangle has a perimeter of 42 inches and a diagonal of 15 inches. Find the dimensions of the rectangle. [9.5]

60. A projectile is fired upward from the ground with an initial velocity of 76 feet per second. At the same instant, another projectile is fired upward from a roof 140 feet above the ground with an initial velocity of 41 feet per second. [9.5]

a) After how long will the two projectiles be the same height above the ground?

b) How high above the ground are the two projectiles at that time?

Chapter 9 Test

Graph. Label the vertex and any intercepts.

1. $y = x^2 - 4x + 7$

2. $y = -(x + 3)^2 + 9$

3. $x = -y^2 + 9y - 14$

4. $x = (y + 1)^2 - 8$

Find the distance between the given points. Round to the nearest tenth if necessary.

5. $(7, -3)$ and $(3, 5)$

Graph the circle. Identify the center and radius of the circle.

6. $(x + 4)^2 + (y - 2)^2 = 25$

Find the center and radius of the circle by completing the square.

7. $x^2 + y^2 - 12x - 6y + 13 = 0$

Find the equation of the circle with the given center and radius.

8. Center $(3, -8)$, radius 16

Graph the ellipse. Label the center. Give the values of a and b.

9. $\dfrac{(x - 4)^2}{16} + \dfrac{(y + 7)^2}{9} = 1$

Find the center of the ellipse by completing the square. Give the values of a and b.

10. $4x^2 + 25y^2 - 64x + 100y + 256 = 0$

Graph the hyperbola. Label the center. Give the values of a and b.

11. $\dfrac{x^2}{16} - y^2 = 1$

Find the equation of the conic section that has been graphed.

12.

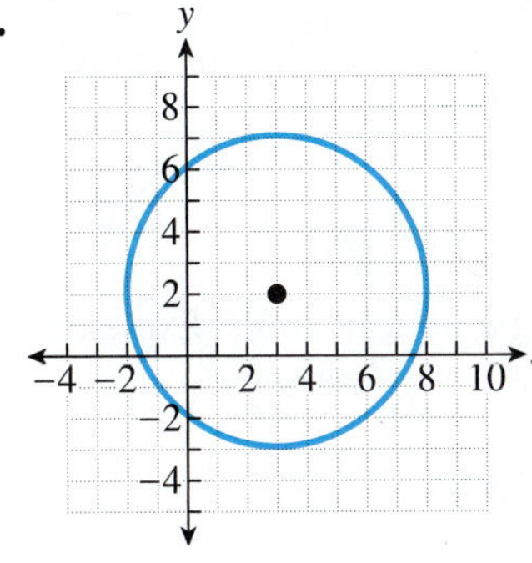

13.

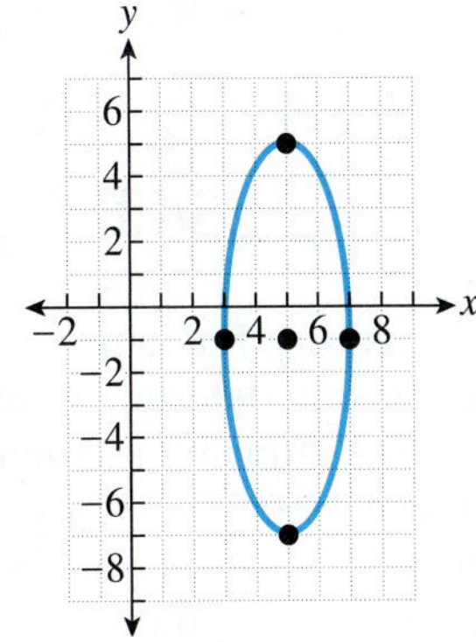

Solve the nonlinear system.

14. $x^2 + y^2 = 100$
$x + 3y = 10$

15. $x^2 + 6y^2 = 49$
$x - 2y = 1$

16. $y^2 - x^2 = 10$
$y = x^2 - 32$

17. $x^2 + 4y^2 = 85$
$3x^2 + 2y^2 = 165$

18. A rectangular corral has a perimeter of 340 feet and an area of 6000 square feet. Find the dimensions of the corral.

Mathematicians in History

Sofia Kovalevskaya

The story of nineteenth-century Russian mathematician Sofia Kovalevskaya is one of genius and determination. At a time when women were discouraged from studying mathematics, she developed into one of the most respected mathematicians of her day.

Write a one-page summary (*or* make a poster) of the life of Sofia Kovalevskaya and her accomplishments.

Interesting issues:

- Where and when was Sofia Kovalevskaya born?
- What were the walls of her bedroom papered with when she was 11 years old?
- Who convinced Kovalevskaya's parents to allow her to study mathematics?
- Whom did Kovalevskaya marry, and what was his fate?
- By the spring of 1874, Kovalevskaya had completed three papers while studying with which prominent mathematician in Berlin?
- What job did Kovalevskaya hold after earning her doctorate?
- In what European city did Kovalevskaya finally obtain a position?
- When did Kovalevskaya die, and what were the circumstances of her death?
- Kovalevskaya once said, "It is impossible to be a mathematician without being a poet in soul." Explain what you think Kovalevskaya meant by that.

Stretch Your Thinking Chapter 9

Given the following data, find the equation that describes Mercury's revolution around the Sun.

(Hint: The Sun is not at the center of Mercury's path; rather, it is at one of the foci.)

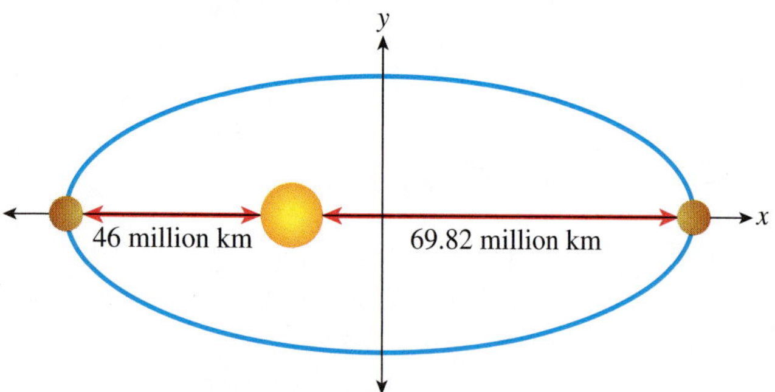

Radius of Sun* = 0.696 *million km

Radius of Mercury = negligible

CHAPTER Ten 10

Sequences, Series, and the Binomial Theorem

In this chapter, we will learn about* sequences*, which are ordered lists of numbers. We will also investigate* series*, which are the sums of the numbers in a sequence. In particular, we will discuss two specific types of sequences and series: the arithmetic sequence and series and the geometric sequence and series.

The chapter concludes with a section on the binomial theorem, which provides us with a method for raising a binomial to a power that is a natural number such as $(2x + 3y)^9$.

Study Strategy **Preparing for a Final Exam** *In this chapter we will focus on how to prepare for a final exam. Although some of the strategies are similar to those used to prepare for a chapter test or quiz, there are some differences as well.*

10.1 Sequences and Series

Objectives

1. **Find the terms of a sequence, given its general term.**
2. **Find the general term of a sequence.**
3. **Find partial sums of a sequence.**
4. **Use summation notation to evaluate a series.**
5. **Use sequences to solve applied problems.**

Sequences

Suppose that you deposited \$100 in a bank account that pays 6% interest, compounded monthly. Here is the amount of money in the account after each of the first 5 years if no further deposits or withdrawals are made:

At the end of year . . .	Principal Balance
1	\$106.17
2	\$112.72
3	\$119.67
4	\$127.05
5	\$134.89

Using this information, we can create a function whose input is a natural number from 1 to 5 and whose output is the balance of the account after that number of years. This type of function is called a sequence.

Sequence

A **sequence** is a function whose domain is a set of consecutive natural numbers beginning with 1. If the domain is the entire set of natural numbers $\{1, 2, 3, \ldots\}$, then the sequence is an **infinite sequence**. If the domain is the set $\{1, 2, 3, \ldots, n\}$, for some natural number n, then the sequence is a **finite sequence**.

The input for the function tells the order of the term in the sequence, while the output of the function is the actual term. We denote the first number in a sequence as a_1, the second number in a sequence as a_2, and so on. In the example of the bank account, $a_1 = \$106.17$, $a_2 = \$112.72$, and so on.

Finding the Terms of a Sequence

Objective 1 **Find the terms of a sequence, given its general term.** A sequence is often denoted by its **general term**, a_n. The general term of a sequence provides a formula for finding any term of a sequence, provided that we know its position in the sequence.

EXAMPLE 1 Find the first five terms of the sequence whose general term is $a_n = n^2 - 10$.

Solution

We will substitute the first five natural numbers for n in the general term.

$$a_1 = (1)^2 - 10 = -9 \qquad a_2 = (2)^2 - 10 = -6 \qquad a_3 = (3)^2 - 10 = -1$$
$$a_4 = (4)^2 - 10 = 6 \qquad a_5 = (5)^2 - 10 = 15$$

The first five terms of the sequence are $-9, -6, -1, 6$, and 15. The entire sequence could be represented as $-9, -6, -1, 6, 15, \ldots$.

Quick Check 1

Find the first four terms of the sequence whose general term is $a_n = \dfrac{1}{3n + 4}$.

A sequence whose terms alternate between being positive and negative is called an **alternating sequence**. An alternating sequence has a factor of $(-1)^n$ or $(-1)^{n+1}$ in its general term. Each time the value of n increases by 1, both $(-1)^n$ and $(-1)^{n+1}$ alternate between -1 and 1.

EXAMPLE 2 Find the first four terms of the sequence whose general term is $a_n = \dfrac{(-1)^n}{3^{n+2}}$.

Solution

We will substitute the first four natural numbers for n in the general term.

$$a_1 = \frac{(-1)^1}{3^{1+2}} = \frac{-1}{3^3} = -\frac{1}{27} \qquad a_2 = \frac{(-1)^2}{3^{2+2}} = \frac{1}{3^4} = \frac{1}{81}$$
$$a_3 = \frac{(-1)^3}{3^{3+2}} = \frac{-1}{3^5} = -\frac{1}{243} \qquad a_4 = \frac{(-1)^4}{3^{4+2}} = \frac{1}{3^6} = \frac{1}{729}$$

The first four terms of the sequence are $-\frac{1}{27}, \frac{1}{81}, -\frac{1}{243}$, and $\frac{1}{729}$.

Quick Check 2

Find the first four terms of the sequence whose general term is $a_n = \dfrac{(-1)^{n+1}}{2n}$.

Finding the General Term of a Sequence

Objective 2 **Find the general term of a sequence.** To find the general term of a sequence, we must be able to detect the pattern of the terms of the sequence; consequently, we must know enough terms of the sequence. Suppose that we were trying to find the general term of the sequence 2, 4, ... There are not enough terms to determine the pattern. It could be the sequence whose general term is $a_n = 2^n$ (2, 4, 8, 16, 32, ...), or it could be the sequence whose general term is $a_n = 2n$ (2, 4, 6, 8, 10, ...). There are simply not enough terms to be sure.

EXAMPLE 3 Find the next three terms of the sequence 3, 6, 9, 12, 15, ..., and find its general term, a_n.

Solution

The terms are all multiples of 3, so continuing the pattern gives us the terms 18, 21, and 24. The general term is $a_n = 3n$.

When examining the terms of a sequence, determine whether the successive terms increase by the same amount each time. If they do, then the general term, a_n, will contain a multiple of n.

EXAMPLE 4 Find the next three terms of the sequence 9, 13, 17, 21, 25, . . . , and find its general term, a_n.

Solution

First, we should notice that each successive term increases by 4. So the next three terms are 29, 33, and 37. When each successive term increases by 4, this tells us that the general term will contain the expression $4n$. However, $4n$ is not the general term of this sequence, as its terms would be 4, 8, 12, 16, 20, ... Notice that each term is five less than the corresponding term in the sequence 9, 13, 17, 21, 25, ... The general term can be found by adding 5 to $4n$, so the general term, is $a_n = 4n + 5$.

We now turn to finding the general term of an alternating sequence.

EXAMPLE 5 Find the next three terms of the sequence $-1, 4, -9, 16, -25, \ldots$, and find its general term, a_n.

Solution

Notice that if we disregard the signs, the first five terms are the squares of 1, 2, 3, 4, and 5. Since the signs of the terms are alternating, the next term will be positive. The next three terms are 36, -49, and 64.

***Quick Check* 3**
Find the next three terms of the sequence 7, −13, 19, −25, 31, . . . , and find its general term a_n.

When an alternating sequence begins with a negative term, the general term will contain the factor $(-1)^n$. (If the first term had been positive, the general term would have contained the factor $(-1)^{n+1}$.) The general term for this sequence is $a_n = (-1)^n \cdot n^2$.

On occasion, we represent a sequence by giving its first term and explaining how to find any term after that from the preceding term. In this case, we are defining the general term **recursively**. For example, recall the sequence 9, 13, 17, 21, 25, . . . , from Example 4. We found its general term to be $a_n = 4n + 5$. Since each term in the sequence is four more than the previous term, we could define the general term recursively as $a_n = a_{n-1} + 4$, where a_{n-1} is the term that precedes the term a_n. A recursive formula for a general term, a_n, defines it in terms of the preceding term, a_{n-1}.

EXAMPLE 6 Find the first five terms of the sequence whose first term is $a_1 = 19$ and whose general term is $a_n = a_{n-1} - 7$.

Solution

***Quick Check* 4**
Find the first five terms of the sequence whose first term is $a_1 = 2$ and whose general term is $a_n = (a_{n-1})^2 + 1$.

We know that the first term is $a_1 = 19$. We can now use the recursive formula $a_n = a_{n-1} - 7$ to find the next four terms.

$$a_2 = a_{2-1} - 7 = a_1 - 7 = 19 - 7 = 12$$
$$a_3 = a_{3-1} - 7 = a_2 - 7 = 12 - 7 = 5$$
$$a_4 = a_{4-1} - 7 = a_3 - 7 = 5 - 7 = -2$$
$$a_5 = a_{5-1} - 7 = a_4 - 7 = -2 - 7 = -9$$

The first five terms of the sequence are 19, 12, 5, -2, and -9.

Series

We now turn our attention to series, which are sums of the terms in a sequence.

Infinite Series, Finite Series

An **infinite series,** s, is the sum of the terms in an infinite sequence.

$$s = a_1 + a_2 + a_3 + \cdots + a_n + \cdots$$

A **finite series,** s_n, is the sum of the first n terms of a sequence.

$$s_n = a_1 + a_2 + a_3 + \cdots + a_n$$

The finite series s_n is also called the ***n*th partial sum** of the sequence.

Finding Partial Sums

Objective 3 **Find partial sums of a sequence.**

EXAMPLE 7 For the sequence whose general term is $a_n = (-1)^n \cdot (2n + 1)$, find s_6.

Solution

The first six terms of the sequence are $a_1 = -3$, $a_2 = 5$, $a_3 = -7$, $a_4 = 9$, $a_5 = -11$, and $a_6 = 13$. We can now find the partial sum.

$$s_6 = (-3) + 5 + (-7) + 9 + (-11) + 13 = 6$$

Quick Check 5 For the sequence whose general term is $a_n = n^2 + 3n$, find s_4.

EXAMPLE 8 For the sequence whose first term is $a_1 = -4$ and whose general term is $a_n = \frac{a_{n-1}}{2}$, find s_5.

Solution

We know that $a_1 = -4$. Since each term in the sequence is half of the preceding term, $a_2 = -2$, $a_3 = -1$, $a_4 = -\frac{1}{2}$, and $a_5 = -\frac{1}{4}$. We can now find the partial sum.

$$s_5 = (-4) + (-2) + (-1) + \left(-\frac{1}{2}\right) + \left(-\frac{1}{4}\right) = -7\frac{3}{4}$$

Quick Check 6 For the sequence whose first term is $a_1 = 1$ and whose general term is $a_n = 3a_{n-1} + 4$, find s_4.

Summation Notation

Objective 4 **Use summation notation to evaluate a series.** The nth partial sum of a sequence, $s_n = a_1 + a_2 + a_3 + \cdots + a_n$, can be represented as $\sum_{i=1}^{n} a_i$. This notation for a partial sum is called **summation notation**. The number n is the **upper limit of summation**, while the number 1 is the **lower limit of summation**. The variable i is called the **index of summation**, or simply, the **index**. The symbol Σ is the capital

Greek letter sigma and is often used in mathematics to represent a sum. This notation is also called **sigma notation**.

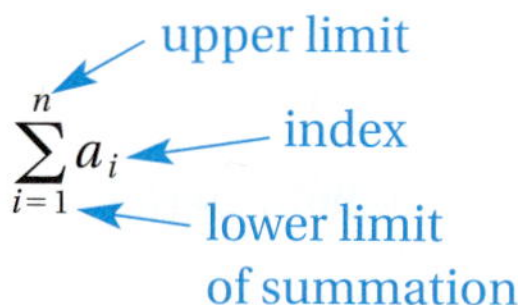

EXAMPLE 9 Find the sum $\sum_{i=1}^{8}(3i + 5)$.

Solution

In this example, we need to find the sum of the first eight terms of this sequence. The first term of the sequence is 8, and each successive term increases by 3.

$$\sum_{i=1}^{8}(3i + 5) = 8 + 11 + 14 + 17 + 20 + 23 + 26 + 29$$ Rewrite as the sum of the first eight terms of the sequence.

$$= 148$$ Add.

Quick Check 7

Find the sum $\sum_{i=1}^{4} 6i$.

Applications

Objective 5 **Use sequences to solve applied problems.** We conclude this section with an applied problem involving sequences.

EXAMPLE 10 Erin found a new job with a starting salary of \$40,000 per year. Each year, she will be given a 5% raise. Write a sequence showing Erin's annual salary for the first five years, and find the total amount she will earn during those 5 years.

Solution

Since Erin earned \$40,000 in the first year, $a_1 = \$40{,}000$. To find her salary for the second year, we calculate her 5% raise and add it to her salary. Five percent of \$40,000 can be calculated by multiplying 0.05 by \$40,000.

$$0.05(\$40{,}000) = \$2000$$

With her \$2000 raise, Erin's salary for the second year is \$42,000. So $a_2 = \$42{,}000$. Continuing in the same fashion, we find that the sequence is \$40,000, \$42,000, \$44,100, \$46,305, \$48,620.25. During the first 5 years, Erin will earn \$221,025.25.

Quick Check 8 **A copy machine was purchased for \$8000. Its value decreases by 20% each year. Write a sequence showing the value of the copy machine at the end of each of the first four years.**

Building Your Study Strategy **Preparing For a Final Exam, 1 Schedule and Study Plan** To prepare for a final exam effectively, develop a schedule and study plan. Ideally, you should begin studying approximately 2 weeks prior to the exam. Your schedule should include study time every day, and you should increase the time spent studying as you get closer to the exam date.

Before you begin to study, find out what material will and will not be covered on the exam by visiting your instructor during office hours. You can find out what topics or chapters will be emphasized, and also what score you need to achieve on the exam in order to earn the grade you desire in the course.

The end of the semester is a hectic time for students. The same can be said for employees of your college as well. Talk to your instructor and find out when he or she will be available on campus for questions in the period of time leading up to the final exam. Also, check with the tutorial center or math lab to find out if there will be any change in their hours of operation. If you ask early enough, and there are changes, then you will be able to make adjustments to your schedule to have access to help if you need it.

As you start to prepare for a final exam, realize that there is only a finite amount of time left before the exam, so you must use it wisely. Arrange your schedule so that there is an expanded allotment of time for studying math. It is important that you prepare for the final exam every day; do not take any days off from your preparation.

If you have a job, try to work fewer hours. Ask for days off if at all possible.

Choosing the location where you will be studying for the final exam is not a trivial matter. Be sure to study in a distraction-free zone; you cannot afford to waste time and effort at this point in the course. Study in a well-lit area that allows you to have easy access to all of your materials.

EXERCISES 10.1

Vocabulary

1. A(n) __________ is a function whose domain is a set of consecutive natural numbers beginning with 1.

2. If the domain of a sequence is the entire set of natural numbers, then the sequence is a(n) __________ sequence.

3. The __________ term of a sequence is often denoted as a_n.

4. A sequence whose terms alternate between being positive and negative is called a(n) __________ sequence.

5. The sum of the terms of a sequence is called a(n) __________.

6. A(n) __________ series, s, is the sum of the terms in an infinite sequence.

7. A(n) __________ series, s_n, is the sum of the first n terms of a sequence.

8. A finite series, s_n, is also called the nth __________ of the sequence.

FOR EXTRA HELP

MyMathLab

MathXL

Interactmath.com

MathXL Tutorials on CD

Video Lectures on CD

Tutor Center

Addison-Wesley Math Tutor Center

Student's Solutions Manual

9. Find the fifth term of the sequence whose general term is $a_n = 3n - 4$.

10. Find the third term of the sequence whose general term is $a_n = (n + 3)^2 - 5n$.

11. Find the sixth term of the sequence whose general term is $a_n = (-1)^n \cdot 3^{n+1}$.

12. Find the ninth term of the sequence whose general term is $a_n = (-1)^n \cdot \dfrac{1}{2n - 1}$.

Find the first five terms of the sequence with the given general term.

13. $a_n = -n$

14. $a_n = 2n$

15. $a_n = 4n - 3$

16. $a_n = 3n + 7$

17. $a_n = (-1)^n \cdot (2n + 3)$

18. $a_n = \dfrac{(-1)^{n+1}}{n^2}$

19. $a_n = n^2 - 3n$

20. $a_n = 4 \cdot 3^{n-1}$

Find the next three terms of the given sequence, and find its general term a_n.

21. 5, 10, 15, 20, 25, . . .

22. 3, 10, 17, 24, 31, . . .

23. 1, 9, 81, 729, 6561, . . .

24. 9, 16, 25, 36, 49, . . .

25. $\frac{1}{5}, \frac{1}{8}, \frac{1}{11}, \frac{1}{14}, \frac{1}{17}, \ldots$

26. $\frac{1}{2}, \frac{1}{4}, \frac{1}{6}, \frac{1}{8}, \frac{1}{10}, \ldots$

27. 1, 8, 27, 64, 125, . . .

28. −1, 2, −4, 8, −16, . . .

Find the first five terms of the sequence with the given first term and general term.

29. $a_1 = 6, a_n = a_{n-1} + 3$

30. $a_1 = 10, a_n = 2a_{n-1}$

31. $a_1 = -5, a_n = 3a_{n-1} + 8$

32. $a_1 = -9, a_n = a_{n-1} - 10$

33. $a_1 = 16, a_n = -3a_{n-1}$

34. $a_1 = 8, a_n = \dfrac{-a_{n-1}}{2}$

35. $a_1 = -35, a_n = -a_{n-1} - 17$

36. $a_1 = 20, a_n = \dfrac{1}{2a_{n-1}}$

Find the indicated partial sum for the given sequence.

37. S_8, 13, 20, 27, 34, . . .

38. S_9, 25, 16, 7, −2, . . .

39. S_7, −18, −31, −44, −5, . . .

40. S_6, −19, −2, 15, 32, . . .

41. S_9, 6, 12, 24, 48, . . .

42. S_5, −3, 27, −243, 2187, . . .

43. S_6, 25, −36, 49, −64, . . .

44. S_{40}, 1, −1, 1, −1, . . .

Find the indicated partial sum for the sequence with the given general term.

45. $S_5, a_n = 3n + 10$

46. $S_4, a_n = 2n - 19$

47. $S_6, a_n = -7n$

48. $S_7, a_n = -n + 8$

49. $S_5, a_n = n^2$

50. $S_5, a_n = 2n - 1$

51. $S_{10}, a_n = (-1)^n \cdot (5n)$

52. $S_9, a_n = (-1)^n \cdot n^2$

Find the indicated partial sum for the sequence with the given first term and general term.

53. $S_6, a_1 = 2, a_n = a_{n-1} + 7$

54. $S_7, a_1 = 9, a_n = a_{n-1} + 4$

55. $S_8, a_1 = 5, a_n = a_{n-1} - 9$

56. $S_5, a_1 = -11, a_n = a_{n-1} - 14$

57. $S_4, a_1 = 3, a_n = 20a_{n-1}$

58. $S_5, a_1 = -4, a_n = -3a_{n-1} + 13$

59. $S_5, a_1 = 3, a_n = (a_{n-1})^2$

60. $S_5, a_1 = 2, a_n = (a_{n-1})^2 + 1$

Find the sum.

61. $\sum_{i=1}^{15} i$

62. $\sum_{i=1}^{7} (i - 3)$

63. $\sum_{i=1}^{8} (2i - 1)$

64. $\sum_{i=1}^{5} (21i + 35)$

65. $\sum_{i=1}^{10} i^2$

66. $\sum_{i=1}^{6} i^3$

67. $\sum_{i=1}^{15} \left(\frac{1}{i} - \frac{1}{i+1}\right)$

68. $\sum_{i=1}^{8} \frac{1}{2^i}$

Rewrite the sum, using summation notation.

69. $7 + 14 + 21 + 28$

70. $2 + 4 + 6 + 8 + 10$

71. $5 + (-10) + 15 + (-20) + 25 + (-30)$

72. $(-3) + 6 + (-9) + 12 + (-15)$

73. $1 + 4 + 9 + 16 + 25 + 36 + 49$

74. $1 + \frac{1}{2} + \frac{1}{3} + \frac{1}{4} + \frac{1}{5}$

75. Theo purchased a new computer, valued at $2500, for his accounting firm. If the value of the computer decreases by 40% each year, construct a sequence showing the value of the computer after each of the first 4 years.

76. A quilt shop purchased a quilting machine for $10,000. If the resale value of the machine drops by 20% each year, construct a sequence showing the value of the quilting machine after each of the first 6 years.

77. Tina started a new job with an annual salary of $32,000. Each year, she will be given an $800 raise.
    a) Write a sequence showing Tina's annual salary for the first 8 years.
    b) Find the total amount earned by Tina during those 8 years.

78. Heesop started a new job with an annual salary of $27,500. Each year, he will be given a 2% raise.
    a) Write a sequence showing Heesop's annual salary for the first 5 years.
    b) Find the total amount earned by Heesop during those 5 years.

Find the missing value for the sequence whose partial sum is given.

79. $s_7 = 147,\ a_1 = 6,\ a_n = a_{n-1} + ?$

80. $s_5 = 25,\ a_1 = 13,\ a_n = a_{n-1} - ?$

81. $s_6 = 222,\ a_1 = ?,\ a_n = a_{n-1} + 12$

82. $s_8 = 16,\ a_1 = ?,\ a_n = a_{n-1} + 6$

83. $s_5 = 242,\ a_1 = 2,\ a_n = ? \cdot a_{n-1}$

84. $s_4 = -85,\ a_1 = 17,\ a_n = ? \cdot a_{n-1}$

85. $s_5 = 3069,\ a_1 = ?,\ a_n = 4 \cdot a_{n-1}$

86. $s_6 = 19{,}608,\ a_1 = ?,\ a_n = 7 \cdot a_{n-1}$

Writing in Mathematics

Answer in complete sentences.

87. Describe a real-world application involving a sequence.

88. Describe a real-world application involving a series.

89. ***Solutions Manual**** Write a solutions manual page for the following problem:

    Find s_6 for the sequence whose general term is $a_n = 2^{n+1}$.

90. ***Newsletter**** Write a newsletter explaining summation notation.

***See Appendix B for details and sample answers.**

10.2 Arithmetic Sequences and Series

Objectives

1. **Find the general term of an arithmetic sequence.**
2. **Find partial sums of an arithmetic sequence.**
3. **Use an arithmetic sequence to solve an applied problem.**

Arithmetic Sequences

In this section, we will examine a particular type of sequence called an arithmetic sequence.

Arithmetic Sequence

An **arithmetic sequence** is a sequence in which each term, after the first term, differs from the preceding term by a constant amount d. This means that the general term a_n can be defined recursively by $a_n = a_{n-1} + d$, for $n \neq 1$. The number d is called the **common difference** of the arithmetic sequence.

If a sequence is an arithmetic sequence, successive terms must increase or decrease by the same amount for each term in the sequence after the first term. For example, the sequence 4, 13, 22, 31, 40, . . . is an arithmetic sequence with $d = 9$ because each term is nine greater than the preceding term. The sequence $\frac{5}{4}, \frac{3}{4}, \frac{1}{4}, -\frac{1}{4}, -\frac{3}{4}, \ldots$ is also an arithmetic sequence with $d = -\frac{1}{2}$, as each term is $\frac{1}{2}$ less than the preceding term.

The sequence 2, 4, 8, 16, 32, . . . is not an arithmetic sequence. The second term is two greater than the first term, but the third term is four greater than the second term. Since the terms are not increasing by the same amount for the entire sequence, the sequence is not an arithmetic sequence.

EXAMPLE 1 Find the first five terms of the arithmetic sequence whose first term is 5 and whose common difference is $d = 4$.

Solution

Since the first term of this sequence is 5 and each term is four greater than the previous term, the first five terms of the sequence are 5, 9, 13, 17, and 21.

Quick Check 1
Find the first six terms of the arithmetic sequence whose first term is -8 and whose common difference is $d = 3$.

Finding the General Term of an Arithmetic Sequence

Objective 1 **Find the general term of an arithmetic sequence.** For any arithmetic sequence whose first term is a_1 and whose common difference is d, we know that the second term a_2 is equal to $a_1 + d$. Since we could add d to the second term to find the third term of the sequence, we know that $a_3 = a_1 + 2d$.

$$\begin{aligned} a_3 &= a_2 + d \\ &= (a_1 + d) + d \\ &= a_1 + 2d \end{aligned}$$

Continuing this pattern, we have $a_4 = a_1 + 3d$, $a_5 = a_1 + 4d$, and so on. This leads to the following result for the general term, a_n, of an arithmetic sequence:

General Term of an Arithmetic Sequence

If an arithmetic sequence whose first term is a_1 has a common difference of d, then the general term of this sequence is $a_n = a_1 + (n - 1)d$.

EXAMPLE 2 Find the general term, a_n, of the arithmetic sequence 5, 11, 17, 23, 29, ...

Solution

The first term, a_1, of this sequence is 5. We will subtract the first term of the sequence from the second term to find the common difference d.

$$d = 11 - 5 = 6$$

We now can find the general term of this sequence by using the formula $a_n = a_1 + (n - 1)d$.

$$\begin{aligned} a_n &= 5 + (n - 1)6 && \text{Substitute 5 for } a_1 \text{ and 6 for } d. \\ &= 6n - 1 && \text{Simplify.} \end{aligned}$$

The general term for this arithmetic sequence is $a_n = 6n - 1$.

Quick Check 2
Find the general term, a_n, of the arithmetic sequence 2, 14, 26, 38, 50, ...

After we find the general term, a_n, of an arithmetic sequence, we can go on to find any particular term of the sequence.

EXAMPLE 3 Find the 45th term, a_{45}, of the arithmetic sequence 3, 7, 11, 15, 19, ...

Solution

We begin by finding the general term, a_n, of this arithmetic sequence. We can then find a_{45}.

The first term, a_1, of this sequence is 3, and the common difference d is 4. We now can find the general term of this sequence by using the formula $a_n = a_1 + (n - 1)d$.

$$\begin{aligned} a_n &= 3 + (n - 1)4 && \text{Substitute 3 for } a_1 \text{ and 4 for } d. \\ &= 4n - 1 && \text{Simplify.} \end{aligned}$$

The general term for this arithmetic sequence is $a_n = 4n - 1$. We can now find the 45th term.

$$\begin{aligned} a_{45} &= 4(45) - 1 && \text{Substitute 45 for } n \text{ into } a_n = 4n - 1. \\ &= 179 && \text{Simplify.} \end{aligned}$$

The 45th term of this arithmetic sequence is 179.

Quick Check 3
Find the 61st term, a_{61}, of the arithmetic sequence 57, 59, 61, 63, 65, ...

Arithmetic Series

Objective 2 **Find partial sums of an arithmetic sequence.** An **arithmetic series** is the sum of the terms of an arithmetic sequence. A finite arithmetic series consisting of only the first n terms of an arithmetic sequence is also known as the ***n*th partial sum** of the arithmetic sequence, and is denoted as s_n.

EXAMPLE 4 Find s_7 for the arithmetic sequence 4, 8, 12, 16, 20, ...

Solution

Since only the first five terms are given, we must begin by finding the sixth and seventh terms. Since each term in this arithmetic sequence is four greater than the preceding

Quick Check 4
Find s_9 for the arithmetic sequence 20, 26, 32, 38, 44, ...

term, the next two terms of the sequence are 24 and 28. We now can find the seventh partial sum of this arithmetic sequence.

$$s_7 = 4 + 8 + 12 + 16 + 20 + 24 + 28$$
$$= 112 \quad \text{Add.}$$

We now present a formula for finding partial sums of an arithmetic sequence. This formula will be quite helpful, as it depends on only the first and nth terms of the arithmetic sequence (a_1 and a_n), as well as the number of terms being added.

nth Partial Sum of an Arithmetic Sequence

The sum of the first n terms of an arithmetic sequence is given by

$$s_n = \frac{n}{2}(a_1 + a_n)$$

where a_1 is the first term and a_n is the nth term of the sequence.

Before we apply this formula, let's investigate where the formula comes from. We will let a_1 be the first term of the arithmetic sequence, a_n be its nth term, and d be the common difference. Since s_n is the sum of the first n terms of the sequence, it can be expressed as

$$s_n = a_1 + (a_1 + d) + (a_1 + 2d) + (a_1 + 3d) + \cdots + (a_1 + (n-1)d)$$

The first n terms of an arithmetic sequence can also be generated by starting with the nth term a_n and subtracting d to find the preceding term: $a_n, a_n - d, a_n - 2d, \ldots$ $a_n - (n-1)d$. So the nth partial sum can also be expressed as

$$s_n = a_n + (a_n - d) + (a_n - 2d) + (a_n - 3d) + \cdots + (a_n - (n-1)d)$$

We now have two different expressions for s_n. If we add the two expressions, term by term, we obtain the following result:

$$\begin{array}{rcl} s_n &=& a_1 \quad\; + (a_1 + d\ ) + (a_1 + 2d) + (a_1 + 3d) + \cdots + (a_1 + (n-1)d) \\ \underline{s_n} &\underline{=}& \underline{\quad\; a_n + (a_n - d\ ) + (a_n - 2d) + (a_n - 3d) + \cdots + (a_n - (n-1)d)} \\ 2s_n &=& (a_1 + a_n) + (a_1 + a_n) + (a_1 + a_n) + (a_1 + a_n) + \cdots + (a_1 + a_n) \end{array}$$

So, $2s_n = n(a_1 + a_n)$, since the expression $(a_1 + a_n)$ appears n times on the right side of the equation. Dividing both sides of the equation $2s_n = n(a_1 + a_n)$ by 2 produces the result $s_n = \frac{n}{2}(a_1 + a_n)$.

The seventh term of the arithmetic sequence 4, 8, 12, 16, 20, . . . is 28; so s_7 can be found as follows:

$$s_7 = \frac{7}{2}(a_1 + a_7)$$

$$s_7 = \frac{7}{2}(4 + 28) \quad \text{Substitute 7 for } n\text{, 4 for } a_1\text{, and 28 for } a_7.$$

$$= 112 \quad \text{Simplify.}$$

To find the nth partial sum of an arithmetic sequence by using the formula $s_n = \frac{n}{2}(a_1 + a_n)$, we must know the values of the first term and the last term that are being added. In the next example, we will find the general term of the arithmetic sequence, $a_n = a_1 + (n - 1)d$, to help us find the last term in the sum. This becomes particularly useful as n becomes larger.

EXAMPLE 5 Find s_{20} for the arithmetic sequence 3, 8, 13, 18, 23, ...

Solution

We already know that $n = 20$ and $a_1 = 3$. To find the value of a_{20}, we will begin by finding the general term of this arithmetic sequence, using the formula $a_n = a_1 + (n - 1)d$. The common difference for this sequence is $d = 5$.

$$a_n = 3 + (n - 1)5 \quad \text{Substitute 3 for } a_1 \text{ and 5 for } d.$$
$$= 5n - 2 \quad \text{Simplify.}$$

We can now use the fact that $a_n = 5n - 2$ to find the value of a_{20}.

$$a_{20} = 5(20) - 2 \quad \text{Substitute 20 for } n.$$
$$= 98 \quad \text{Simplify.}$$

We can now apply the formula $s_n = \frac{n}{2}(a_1 + a_n)$ to find s_{20}.

$$s_{20} = \frac{20}{2}(3 + 98) \quad \text{Substitute 20 for } n\text{, 3 for } a_1\text{, and 98 for } a_{20}.$$
$$= 1010 \quad \text{Simplify.}$$

Quick Check 5 **Find s_{36} for the arithmetic sequence 25, 44, 63, 82, 101, ...**

EXAMPLE 6 Find the sum of the first 147 odd positive integers.

Solution

The first 147 odd positive integers make up the sequence that begins 1, 3, 5, 7, 9, ... This is an arithmetic sequence with $a_1 = 1$ and $d = 2$. We will find the general term of this sequence by using the formula $a_n = a_1 + (n - 1)d$ to help us find a_{147}.

$$a_n = 1 + (n - 1)2 \quad \text{Substitute 1 for } a_1 \text{ and 2 for } d.$$
$$= 2n - 1 \quad \text{Simplify.}$$

We can now use $a_n = 2n - 1$ to find the value of a_{147}.

$$a_{147} = 2(147) - 1 \quad \text{Substitute 147 for } n.$$
$$= 293 \quad \text{Simplify.}$$

We can now apply the formula $s_n = \frac{n}{2}(a_1 + a_n)$ to find s_{147}.

$$s_{147} = \frac{147}{2}(1 + 293) \quad \text{Substitute 147 for } n\text{, 1 for } a_1\text{, and 293 for } a_{147}.$$
$$= 21{,}609 \quad \text{Simplify.}$$

The sum of the first 147 odd positive integers is 21,609.

Quick Check 6 **Find the sum of the first 62 even positive integers.**

Applications

Objective 3 Use an arithmetic sequence to solve an applied problem.

EXAMPLE 7 Joyce found a new job with a starting salary of \$40,000 per year. Each year, she will be given a \$2500 raise.

a) Write a sequence showing Joyce's annual salary for each of the first 6 years.
b) Find the general term, a_n, for the sequence from part a).
c) What will Joyce's total earnings be if she stays with the company for 30 years?

Solution

a) Since the starting salary is \$40,000 and increases by \$2500 each year, the sequence is \$40,000, \$42,500, \$45,000, \$47,500, \$50,000, \$52,500.

b) The first term, a_1, of this sequence is \$40,000, and the common difference d is \$2500. We now can find the general term of this sequence by using the formula $a_n = a_1 + (n - 1)d$.

$$a_n = 40{,}000 + (n - 1)2500 \qquad \text{Substitute 40,000 for } a_1 \text{ and 2500 for } d.$$
$$= 2500n + 37{,}500 \qquad \text{Simplify.}$$

The general term for this arithmetic sequence is $a_n = 2500n + 37{,}500$.

c) To find Joyce's total earnings over 30 years, we need to find s_{30}. We already know that $n = 30$ and $a_1 = 40{,}000$. To apply the formula $s_n = \frac{n}{2}(a_1 + a_n)$, we must find the value of a_{30}.

$$a_{30} = 2500(30) + 37{,}500 \qquad \text{Substitute 30 for } n \text{ in } a_n = 2500n + 37{,}500.$$
$$= 112{,}500 \qquad \text{Simplify.}$$

We can now apply the formula $s_n = \frac{n}{2}(a_1 + a_n)$.

$$s_{30} = \frac{30}{2}(40{,}000 + 112{,}500) \qquad \text{Substitute 30 for } n\text{, 40,000 for } a_1\text{, and 112,500 for } a_{30}.$$
$$= 2{,}287{,}500 \qquad \text{Simplify.}$$

During the first 30 years, Joyce will earn \$2,287,500.

Building Your Study Strategy **Preparing for a Final Exam, 2 What to Review**

Old Exams and Quizzes

When preparing for a final exam, begin by reviewing old exams and quizzes. Your instructor considered these problems important enough to test you on them. Begin by reviewing and understanding any mistakes that you made on the exam. By understanding what you did wrong, you are minimizing your chances of making the same mistakes on the final exam.

Once you understand the errors that you made, try reworking all of the problems on the exam or quiz. You should do this without referring to your notes or textbook, to help you to learn which topics you need to review and which topics you have under control. You need to focus your time on the topics you are struggling with and spend less time on the topics that you understand; retaking your tests will help you determine which topics you need to spend more time on.

continued

Old Homework Assignments

When studying a particular topic, look back at your old homework for that topic. If you struggled with a particular type of problem, your homework will reflect that. Hopefully, your homework will contain some notes to yourself about how to do certain problems or mistakes to avoid.

Notes and Note Cards

Look over your class notes from that topic. Your notes should include examples of the problems in that section of the textbook, as well as pointers from your instructor. You can also review that section, paying close attention to the examples and procedures developed in that section.

If you have been creating note cards during the semester, they will be most advantageous to you at this time. These note cards should focus on problems that you considered difficult at the time and contain strategies for solving these types of problems. A quick glance at these note cards will really speed up your review of a particular topic.

Materials from Your Instructor

Some of the most valuable materials for preparing for a final exam are the materials provided by your instructor, such as a review sheet for the final or a practice final, or a list of problems from the text to review. If your instructor feels that these problems are important enough for you to review, then they are definitely important enough to appear on the final exam.

As you work through your study schedule, try to solve your instructor's problems when you review that section or chapter. Your performance on these problems will help you to know which topics you understand and which topics require further study.

One mistake that students make is to try to memorize the solution of each and every problem: this will not help you to solve a similar problem on the exam. If you are having trouble with a certain problem, write down on a note card the steps to solve that problem. Review your note cards for a short period of time each day before the exam. This will help you to understand how to solve a problem and will increase your chance of solving a similar problem on the exam.

Textbook

Each chapter in the textbook contains a chapter review assignment and a chapter test. Also, there are cumulative review assignments at the end of certain chapters. Use your cumulative review exercises to prepare for a final exam. These exercise sets contain problems representative of the material covered up to that point in the text.

After you have been preparing for a while, try these exercises without referring to your notes, note cards, or the text. In this way these problems can be used to determine which topics require more study. If you made a mistake while solving a problem, make note of the mistake and how to avoid it in the future. If there are problems that you do not recognize or know how to begin, these topics will require additional studying. Ask your instructor or a tutor for help.

Applied Problems

Many students have a difficult time with applied problems on a final exam. Some students are unable to recognize what type of problem an applied problem is,

continued

and others cannot recall how to start to solve that type of problem. You will find the following strategy helpful:

- Make a list of the different applied problems that you have covered this semester.
- Create a study sheet for each type of problem.
- Write down an example or two of each type of problem.
- List the steps necessary to solve each type of problem.

Review these study sheets frequently as the exam approaches. This should help you identify the applied problems on the exam, as well as to remember how to solve the problems.

Practice Tests

Creating and taking practice tests is an effective way to prepare for a final exam. Start by creating a practice test for an entire chapter. If you determine that you are struggling with certain types of problems, review those problems and then try another practice test. Once you feel comfortable with that chapter's material, move on to create a practice test for another chapter. Repeat this until you have covered all of the chapters. You could arrange to exchange practice tests with a classmate.

EXERCISES 10.2

Vocabulary

1. A(n) ________ sequence is a sequence in which each term, after the first term, differs from the preceding term by a constant amount d.

2. The difference, d, between consecutive terms in an arithmetic sequence is called the ________ difference of the sequence.

3. The general term, a_n, of an arithmetic sequence can be defined recursively by ________ for $n \neq 1$.

4. If an arithmetic sequence whose first term is a_1 has a common difference of d, then the general term of this sequence is given by the formula ________.

5. A(n) ________ is the sum of the terms of an arithmetic sequence.

6. The sum, s_n, of the first n terms of an arithmetic sequence is given by the formula ________.

Find the common difference, d, of each given arithmetic sequence.

7. $3, 12, 21, 30, 39, \ldots$

8. $1, 22, 43, 64, 85, \ldots$

9. $5, \frac{15}{2}, 10, \frac{25}{2}, 15, \ldots$

10. $12, \frac{44}{3}, \frac{52}{3}, 20, \frac{68}{3}, \ldots$

11. $-13, -10, -7, -4, -1, \ldots$

12. $-4, -8, -12, -16, -20, \ldots$

13. $\frac{9}{4}, \frac{41}{4}, \frac{73}{4}, \frac{105}{4}, \frac{137}{4}, \ldots$

14. $\frac{31}{6}, \frac{17}{2}, \frac{71}{6}, \frac{91}{6}, \frac{37}{2}, \ldots$

Find the first five terms of the arithmetic sequence with each given first term and common difference.

15. $a_1 = 1, d = 5$

16. $a_1 = 1, d = -2$

17. $a_1 = -9, d = 12$

18. $a_1 = -13, d = -8$

19. $a_1 = -10, d = -21$

20. $a_1 = 9, d = \frac{5}{2}$

Find the general term, a_n, of each given arithmetic sequence.

21. 9, 22, 35, 48, 61, . . .

22. 7, 27, 47, 67, 87, . . .

23. −8, −3, 2, 7, 12, . . .

24. −9, −16, −23, −30, −37, . . .

25. 87, 52, 17, −18, −53, . . .

26. 62, 63, 64, 65, 66, . . .

27. $-\frac{3}{2}, \frac{5}{2}, \frac{13}{2}, \frac{21}{2}, \frac{29}{2}, \ldots$

28. $8, \frac{16}{3}, \frac{8}{3}, 0, -\frac{8}{3}, \ldots$

29. Find the 19th term, a_{19}, of the arithmetic sequence 16, 21, 26, 31, 36, ...

30. Find the 14th term, a_{14}, of the arithmetic sequence −9, 29, 67, 105, 143, ...

31. Which term of the arithmetic sequence 23, 29, 35, 41, 47, . . . , is equal to 389?

32. Which term of the arithmetic sequence −99, −94, −89, −84, −79, . . . , is equal to 336?

33. Which term of the arithmetic sequence 611, 603, 595, 587, 579, . . . , is equal to −365?

34. Which term of the arithmetic sequence −327, −336, −345, −354, −363, . . . , is equal to −3981?

Find the partial sum of the arithmetic sequence with each given first term and common difference.

35. $s_{13}, a_1 = 5, d = 6$

36. $s_9, a_1 = 11, d = 12$

37. $s_{16}, a_1 = -9, d = 8$

38. $s_{37}, a_1 = -10, d = -10$

39. $s_{40}, a_1 = 89, d = -7$

40. $s_{21}, a_1 = 810, d = 62$

Find the partial sum for each given arithmetic sequence.

41. s_{10}, 1, 21, 41, 61, . . .

42. s_{15}, 9, 22, 35, 48, . . .

43. s_{17}, −17, −1, 15, 31, . . .

44. s_9, −38, −31, −24, −17, . . .

45. s_{20}, 86, 47, 8, −31, . . .

46. s_{27}, −3, −16, −29, −42, . . .

47. s_{32}, 109, 208, 307, 406, . . .

48. s_{19}, 62, 17, −28, −73, . . .

49. Find the sum of the first 1000 positive integers.

50. Find the sum of the first 418 positive integers.

51. Find the sum of the first 67 odd positive integers.

52. Find the sum of the first 109 odd positive integers.

53. Find the sum of the first 200 even positive integers.

54. Find the sum of the first 82 even positive integers.

55. Andrea finds a job that pays $10 per hour. Each year, her pay rate will be increased by $1 per hour.

a) Write a sequence showing Andrea's hourly pay rate for the first 5 years.

b) Find the general term a_n for the sequence from part a).

56. During its first year of existence, a minor league baseball team sold 2800 season tickets. In each year after that, the number of season tickets sold decreased by 120.

a) Write a sequence showing the number of season tickets sold in the first 4 years.

b) Find the general term, a_n, for the number of season tickets sold in the team's *n*th year.

57. Jon starts a job with an annual salary of \$38,500. Each year, he will be given a raise of \$750.

a) Write a sequence showing Jon's annual salary for the first 6 years.

b) Find the general term, a_n, for Jon's annual salary in his nth year in this job.

c) How much money will Jon earn from this job over the first 15 years?

58. A teacher has a retirement savings plan. In the first year, the teacher contributes \$100 per month. Each year, the teacher will increase her contributions by \$10 per month, so that she contributes \$110 per month in the second year, \$120 per month in the third year, and so on.

a) Find the general term, a_n, of an arithmetic sequence showing the annual contribution of the teacher in the nth year.

b) How much money will the teacher contribute over a period of 30 years?

Writing in Mathematics

Answer in complete sentences.

59. Describe a real-world application involving an arithmetic sequence.

60. Describe a real-world application involving an arithmetic series.

61. Explain how to find the partial sum of an arithmetic sequence. Use an example to illustrate the process.

62. ***Solutions Manual**** Write a solutions manual page for the following problem:

Find s_{22} for an arithmetic sequence with first term, $a_1 = 44$, and common difference, $d = 13$.

63. ***Newsletter**** Write a newsletter explaining how to find a formula for the general term of an arithmetic sequence.

***See Appendix B for details and sample answers.**

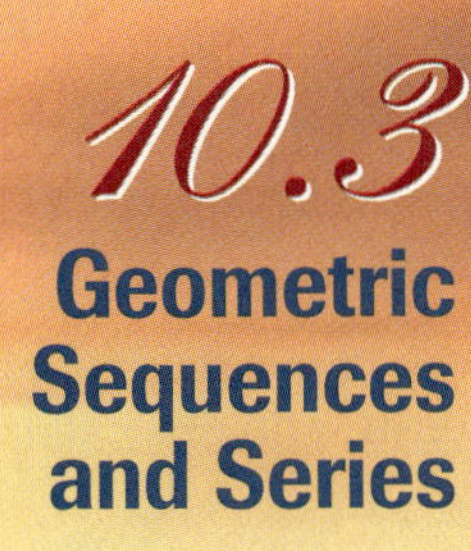

Geometric Sequences and Series

Objectives

1. **Find the common ratio of a geometric sequence.**
2. **Find the general term of a geometric sequence.**
3. **Find partial sums of a geometric sequence.**
4. **Find the sum of an infinite geometric series.**
5. **Use a geometric sequence to solve an applied problem.**

Geometric Sequences

Consider the sequence 4, 12, 36, 108, 324, ... Each term in this sequence is equal to three times the preceding term. This is an example of a geometric sequence, and this type of sequence is the focus of this section.

Geometric Sequence

A **geometric sequence** is a sequence in which each term after the first term is a constant multiple of the preceding term. This means that the general term, a_n, can be defined recursively by $a_n = r \cdot a_{n-1}$, for $n \neq 1$. The number r is called the **common ratio** of the geometric sequence.

A sequence is a geometric sequence if the quotient of any term after the first term and its preceding term, $\frac{a_n}{a_{n-1}}$, is a constant. For example, the sequence 1, 5, 25, 125, 625, . . . , is a geometric sequence because $\frac{a_n}{a_{n-1}} = 5$ for each term after the first term in the sequence. In other words, each term is 5 times greater than the preceding term. The value 5 is the common ratio, r, of this geometric sequence.

The sequence 2, 6, 10, 14, 18, . . . , is not a geometric sequence, because there is no common ratio. The second term is 3 times greater than the first term, but the third term is not 3 times greater than the second term.

Finding the Common Ratio of a Geometric Sequence

Objective 1 **Find the common ratio of a geometric sequence.** To find the common ratio, r, of a geometric sequence, take any term of the sequence after the first term and divide it by the preceding term.

EXAMPLE 1 Find the common ratio, r, of the geometric sequence 2, 8, 32, 128, 512, ...

Quick Check 1
Find the common ratio, r, of the geometric sequence 256, 64, 16, 4, 1, ...

Solution

To find r, we can divide the second term of the sequence by the first term.

$$r = \frac{8}{2} = 4$$

A geometric sequence can be an alternating sequence. In this case, the common ratio r will be a negative number.

EXAMPLE 2 Find the common ratio, r, of the geometric sequence $5, -10, 20, -40, 80, \ldots$.

Quick Check 2
Find the common ratio, r, of the geometric sequence $-120, 60, -30, 15, -\frac{15}{2}, \ldots$.

Solution

To find r, we can divide the second term of the sequence by the first term.

$$r = \frac{-10}{5} = -2$$

The common ratio for this geometric sequence is $r = -2$.

EXAMPLE 3 Find the first five terms of the geometric sequence whose first term is 7 and whose common ratio is $r = 2$.

Solution

Since the first term of this sequence is 7, and each term is 2 times greater than the previous term, the first five terms of the sequence are 7, 14, 28, 56, and 112.

Quick Check 3
Find the first five terms of the geometric sequence whose first term is 5 and whose common ratio is $r = -\frac{3}{4}$.

Finding the General Term of a Geometric Sequence

Objective 2 **Find the general term of a geometric sequence.** The general term of a geometric sequence can be determined from the first term of the sequence, a_1, and the sequence's common ratio r. We know that the second term a_2 is equal to $a_1 \cdot r$. Continuing this pattern, we obtain

$$a_3 = a_2 \cdot r = (a_1 \cdot r) \cdot r = a_1 \cdot r^2$$
$$a_4 = a_3 \cdot r = (a_1 \cdot r^2) \cdot r = a_1 \cdot r^3$$
$$a_5 = a_4 \cdot r = (a_1 \cdot r^3) \cdot r = a_1 \cdot r^4$$

This leads to the following result about the general term, a_n, of a geometric sequence:

General Term of a Geometric Sequence

If a geometric sequence whose first term is a_1 has a common ratio of r, then the general term of the sequence is $a_n = a_1 \cdot r^{n-1}$.

EXAMPLE 4 Find the general term, a_n, of the geometric sequence 3, 30, 300, 3000, ...

Solution

The first term, a_1, of this sequence is 3. We will divide the second term of the sequence by the first term to find the common ratio, r.

$$r = \frac{30}{3} = 10$$

Quick Check 4

Find the general term, a_n, of the geometric sequence 9, 18, 36, 72, ...

We can now find the general term of this sequence, using the formula $a_n = a_1 \cdot r^{n-1}$.

$$a_n = 3 \cdot 10^{n-1} \qquad \text{Substitute 3 for } a_1 \text{ and 10 for } r.$$

The general term for this geometric sequence is $a_n = 3 \cdot 10^{n-1}$.

EXAMPLE 5 Find the general term, a_n, of the geometric sequence $-1, 4, -16, 64, \ldots$.

Solution

The first term, a_1, of this sequence is -1. We will divide the second term of the sequence by the first term to find the common ratio, r.

$$r = \frac{4}{-1} = -4$$

We can now find the general term of this sequence, using the formula $a_n = a_1 \cdot r^{n-1}$.

$$a_n = -1 \cdot (-4)^{n-1} \qquad \text{Substitute } -1 \text{ for } a_1 \text{ and } -4 \text{ for } r.$$

Quick Check 5

Find the general term, a_n, of the geometric sequence 11, −55, 275, −1375, ...

The general term for this geometric sequence is $a_n = -1 \cdot (-4)^{n-1}$.

Geometric Series

Objective 3 **Find partial sums of a geometric sequence.** A **geometric series** is the sum of the terms of a geometric sequence. We begin by examining finite geometric series consisting of only the first n terms of a geometric sequence. Again, the sum of the first n terms of a sequence is also known as the nth partial sum of the sequence and is denoted by s_n. In the case of a finite geometric series,

$$s_n = a_1 + a_1 r + a_1 r^2 + a_1 r^3 + \cdots + a_1 r^{n-1}$$

EXAMPLE 6 Find s_7 for the geometric sequence 1, 2, 4, 8, 16, ...

Solution

Since only the first five terms are given, we begin by finding the sixth and seventh terms. Since each term in this geometric sequence is equal to 2 times the preceding term, the next two terms of the sequence are 32 and 64. We can now find the seventh partial sum of this geometric sequence.

Quick Check 6

Find s_8 for the geometric sequence 12, 36, 108, 324, ...

$$\begin{aligned} s_7 &= 1 + 2 + 4 + 8 + 16 + 32 + 64 \\ &= 127 \qquad \text{Add.} \end{aligned}$$

We now present a formula for finding partial sums of a geometric sequence. This formula will be quite helpful, as it depends only on the first term of the geometric sequence (a_1) and the common ratio, r, of the sequence.

*n*th Partial Sum of a Geometric Sequence

The sum of the first n terms of a geometric sequence is given by

$$s_n = a_1 \cdot \frac{1 - r^n}{1 - r}$$

where a_1 is the first term and r is the common ratio of the sequence.

Before we apply this formula, let's investigate where the formula comes from. Suppose we have a geometric sequence whose first term is a_1 and whose common ratio is r. Since s_n is the sum of the first n terms of the sequence, it can be expressed as

$$s_n = a_1 + a_1r + a_1r^2 + a_1r^3 + \cdots + a_1r^{n-1}$$

If we multiply both sides of this equation by r, the resulting equation is

$$r \cdot s_n = a_1r + a_1r^2 + a_1r^3 + a_1r^4 + \cdots + a_1r^n$$

Now we may subtract corresponding terms of these two equations as follows:

$$\begin{aligned}
s_n &= a_1 + a_1r + a_1r^2 + a_1r^3 + \cdots + a_1r^{n-1} \\
r \cdot s_n &= a_1r + a_1r^2 + a_1r^3 + \cdots + a_1r^{n-1} + a_1r^n \\
\hline
s_n - r \cdot s_n &= a_1 - a_1r^n \\
s_n - rs_n &= a_1 - a_1r^n \\
s_n(1 - r) &= a_1(1 - r^n)
\end{aligned}$$

Factor out the common factor s_n on the left side of the equation and the common factor a_1 on the right side of the equation.

$$\frac{s_n(1 - r)}{1 - r} = \frac{a_1(1 - r^n)}{1 - r}$$

Divide both sides by $1 - r$ to isolate s_n.

$$s_n = a_1 \cdot \frac{1 - r^n}{1 - r}$$

Simplify.

For the geometric sequence 1, 2, 4, 8, 16, . . . , in the previous example, $a_1 = 1$ and $r = 2$.

Now we can apply the formula $s_n = a_1 \cdot \frac{1 - r^n}{1 - r}$ to find s_7.

$$s_7 = 1 \cdot \frac{1 - 2^7}{1 - 2}$$

Substitute 1 for a_1, 2 for r, and 7 for n.

$$= 127$$

Simplify.

Note that this matches the result found in the previous example.

In the next example, we will find the partial sum of a geometric sequence whose common ratio r is negative.

EXAMPLE 7 Find s_9 for the geometric sequence 6, −42, 294, −2058, ...

Solution

For this sequence, $a_1 = 6$, and $r = \frac{-42}{6} = -7$. Now we can apply the formula $s_n = a_1 \cdot \frac{1 - r^n}{1 - r}$ to find s_9.

$$s_9 = 6 \cdot \frac{1 - (-7)^9}{1 - (-7)}$$

Substitute 6 for a_1, −7 for r, and 9 for n.

$$= 30{,}265{,}206$$

Simplify.

Quick Check 7

Find s_{10} for the geometric sequence 1, −3, 9, −27, ...

Infinite Geometric Series

Objective 4 **Find the sum of an infinite geometric series.** We now turn our attention to **infinite geometric series**, $s = a_1 + a_1r + a_1r^2 + \cdots$. Consider the geometric sequence $\frac{1}{2}, \frac{1}{4}, \frac{1}{8}, \frac{1}{16}, \frac{1}{32}, \ldots$, where $a_1 = \frac{1}{2}$ and $r = \frac{1}{2}$. The first partial sum of this series is $s_1 = \frac{1}{2}$. Here are the next four partial sums:

$$s_2 = \frac{1}{2} + \frac{1}{4} = \frac{2}{4} + \frac{1}{4} = \frac{3}{4}$$

$$s_3 = \frac{1}{2} + \frac{1}{4} + \frac{1}{8} = \frac{4}{8} + \frac{2}{8} + \frac{1}{8} = \frac{7}{8}$$

$$s_4 = \frac{1}{2} + \frac{1}{4} + \frac{1}{8} + \frac{1}{16} = \frac{8}{16} + \frac{4}{16} + \frac{2}{16} + \frac{1}{16} = \frac{15}{16}$$

$$s_5 = \frac{1}{2} + \frac{1}{4} + \frac{1}{8} + \frac{1}{16} + \frac{1}{32} = \frac{16}{32} + \frac{8}{32} + \frac{4}{32} + \frac{2}{32} + \frac{1}{32} = \frac{31}{32}$$

As n increases, the partial sums of this sequence are getting closer and closer to 1. In such a case, we say that the **limit** of the infinite geometric series $\frac{1}{2} + \frac{1}{4} + \frac{1}{8} + \frac{1}{16} + \frac{1}{32} + \cdots$ is equal to 1.

Limit of an Infinite Geometric Series

Suppose that a geometric sequence has a common ratio, r. If $|r| < 1$, then the corresponding infinite geometric series has a limit, and this limit is given by the formula

$$s = \frac{a_1}{1 - r}.$$

If $|r| \geq 1$, as it is for the series $\frac{1}{2} + \frac{3}{4} + \frac{9}{8} + \cdots$ $(r = \frac{3}{2})$, then no limit exists.

EXAMPLE 8 For the geometric sequence $16, 4, 1, \frac{1}{4}, \ldots$, does the infinite series have a limit? If so, what is that limit?

Solution

We begin by determining whether $|r| < 1$ for this sequence.

$$r = \frac{a_2}{a_1} = \frac{4}{16} = \frac{1}{4}$$

Since the absolute value of r is less than 1, the infinite series does have a limit. To find the limit, we will use the formula $s = \dfrac{a_1}{1 - r}$.

$$s = \frac{16}{1 - \frac{1}{4}} = \frac{16}{\frac{3}{4}} = \frac{64}{3}$$ Substitute 16 for a_1 and $\frac{1}{4}$ for r. Simplify.

The limit of this infinite geometric series is $\frac{64}{3}$, or $21\frac{1}{3}$.

Quick Check 8
For the geometric sequence $3, 2, \frac{4}{3}, \frac{8}{9}, \ldots$, does the infinite series have a limit? If so, what is that limit?

EXAMPLE 9 For the geometric sequence 4, 6, 9, $\frac{27}{2}$, . . . , does the infinite series have a limit? If so, what is that limit?

Solution

First determine whether $|r| < 1$.

$$r = \frac{a_2}{a_1} = \frac{6}{4} = \frac{3}{2}$$

Quick Check 9

For the geometric sequence −2, 8, −32, 128, . . . , does the infinite series have a limit? If so, what is that limit?

Since the absolute value of r is greater than 1, the infinite series does not have a limit. In this case, the partial sums increase without limit as n continues to increase.

In the next example, we will find the limit of an infinite geometric series, given the general term of the sequence.

EXAMPLE 10 For the geometric sequence whose general term is $a_n = 8 \cdot \left(\frac{1}{3}\right)^{n-1}$, does the infinite series have a limit? If so, what is that limit?

Solution

For this sequence, $r = \frac{1}{3}$. Since the absolute value of r is less than 1, the infinite series does have a limit. The first term of this series is $a_1 = 8$, and the limit can be found by $s = \frac{a_1}{1 - r}$.

$$s = \frac{8}{1 - \frac{1}{3}}$$ Substitute 8 for a_1 and $\frac{1}{3}$ for r.

$$= 12$$ Simplify.

The limit of this infinite geometric series is 12.

Applications

Objective 5 **Use a geometric sequence to solve an applied problem.** We finish this section with an application involving a geometric sequence.

EXAMPLE 11 Donna's parents have established a trust fund for their daughter. On her 21st birthday, Donna receives a payment of $10,000. On each successive birthday, she receives a payment that is 80% of the previous payment.

a) Write a sequence showing Donna's payments for the first 4 years.

b) Find the total amount of Donna's first 25 payments.

Solution

a) Since the first payment is \$10,000 and the second payment is 80% of that amount, she will receive 0.8(\$10,000), or \$8000, on her 22nd birthday. Repeating this process, we find that the sequence is \$10,000, \$8000, \$6400, \$5120.

b) To find the total of Donna's payments for the first 25 years, we need to find s_{25} for this geometric sequence. We already know that $n = 25$ and $a_1 = 10{,}000$. To apply the formula $s_n = a_1 \cdot \frac{1 - r^n}{1 - r}$, we must find the common ratio r.

$$r = \frac{a_2}{a_1} = \frac{8000}{10{,}000} = 0.8$$

It should be no surprise that $r = 0.8$ for this sequence, as each term is 80% of the preceding term. We can now apply the formula $s_n = a_1 \cdot \frac{1 - r^n}{1 - r}$.

$$s_{25} = 10{,}000 \cdot \frac{1 - (0.8)^{25}}{1 - 0.8}$$ Substitute 25 for n, 10,000 for a_1, and 0.8 for r.

$$\approx 49{,}811.11$$ Evaluate, using a calculator.

During the first 25 years, Donna will receive \$49,811.11.

Quick Check **10** **A car was purchased for \$25,000. Its value decreases by 30% each year.**

a) Write a sequence showing the value of the car at the end of each of the first 6 years.

b) Find the value of the car at the end of year 10.

Building Your Study Strategy **Preparing for the Final Exam, 3 Study Groups**

If you have participated in a study group throughout the semester, now is not the time to start studying exclusively on your own. Have all the students in the group bring problems that they are struggling with or problems that they feel are important, and work through them as a group. A group study session is a great place to bring difficult problems.

Keep your focus on the job at hand while working with your group. Often, a group will have a member who would prefer to socialize rather than study, but you cannot afford to waste any time at this point in the course.

At the end of your group study session, try as a group to write a practice exam. In determining which problems to include on the practice exam, you are really focusing on the types of problems most likely to appear on your final exam.

In some groups, members bring problems and challenge others in the group to solve them. Although this can be an effective way to study, it is not a good idea to do so the night before the exam. You may find that if you are unable to solve two or three problems in a row, your confidence may be shattered, and you need plenty of confidence when you sit down to take an exam. Spend the night before the exam alone, reviewing what you have done.

EXERCISES 10.3

Vocabulary

1. A(n) __________ is a sequence in which each term, after the first term, is a constant multiple of the preceding term.

2. The ratio, $r = \frac{a_n}{a_{n-1}}$, between consecutive terms in a geometric sequence is called the __________ of the sequence.

3. The general term, a_n, of a geometric sequence can be defined recursively by __________, for $n \neq 1$.

4. A(n) __________ is the sum of the terms of a geometric sequence.

5. The sum, s_n, of the first n terms of a geometric sequence is given by the formula __________.

6. For an infinite geometric sequence with $|r| < 1$, the corresponding infinite geometric series has a limit s, and this limit is given by the formula __________.

Find the common ratio, r, of the given geometric sequence.

7. $1, 3, 9, 27, 81, \ldots$

8. $1, 6, 36, 216, 1296, \ldots$

9. $125, 25, 5, 1, \frac{1}{5}, \ldots$

10. $343, 49, 7, 1, \frac{1}{7}, \ldots$

11. $3, -6, 12, -24, 48, \ldots$

12. $-4, 12, -36, 108, -324, \ldots$

13. $-6, 8, -\frac{32}{3}, \frac{128}{9}, -\frac{512}{27}, \ldots$

14. $32, -12, \frac{9}{2}, -\frac{27}{16}, \frac{81}{128}, \ldots$

Find the first five terms of the geometric sequence with the given first term and common ratio.

15. $a_1 = 1, r = 10$

16. $a_1 = 1, r = -3$

17. $a_1 = 4, r = -2$

18. $a_1 = 2, r = 5$

19. $a_1 = 8, r = \frac{3}{4}$

20. $a_1 = 12, r = \frac{1}{2}$

21. $a_1 = -20, r = -\frac{7}{4}$

22. $a_1 = \frac{1}{3}, r = -\frac{4}{3}$

Find the general term, a_n, of the given geometric sequence.

23. $1, 8, 64, 512, \ldots$

24. $1, \frac{5}{3}, \frac{25}{9}, \frac{125}{27}, \ldots$

25. $-1, -6, -36, -216, \ldots$

26. $-1, 10, -100, 1000, \ldots$

27. $9, 18, 36, 72, \ldots$

28. $5, 20, 80, 320, \ldots$

29. $16, -12, 9, -\frac{27}{4}, \ldots$

30. $\frac{3}{10}, -\frac{9}{100}, \frac{27}{1000}, -\frac{81}{10{,}000}, \ldots$

Find the partial sum of the geometric sequence with the given first term and common ratio.

31. $s_7, a_1 = 5, r = 4$

32. $s_8, a_1 = 15, r = 3$

33. $s_{20}, a_1 = 48, r = -2$

34. $s_5, a_1 = -6, r = -10$

FOR EXTRA HELP

MyMathLab

MathXL

Interactmath.com

MathXL Tutorials on CD

Video Lectures on CD

Tutor Center

Addison-Wesley Math Tutor Center

Student's Solutions Manual

35. s_{10}, $a_1 = 256$, $r = \frac{1}{2}$

36. s_{12}, $a_1 = 625$, $r = 0.2$

Find the indicated partial sum for the given geometric sequence.

37. s_6, $1, -8, 64, -512, \ldots$

38. s_8, $1, 10, 100, 1000, \ldots$

39. s_7, $4, 12, 36, 108, \ldots$

40. s_9, $50, 100, 200, 400, \ldots$

41. s_6, $13, 156, 1872, 22{,}464, \ldots$

42. s_7, $9, -27, 81, -243, \ldots$

43. s_{14}, $1024, 512, 256, 128, \ldots$

44. s_8, $-24, 6, -\frac{3}{2}, \frac{3}{8}, \ldots$

Find the missing value for each sequence, whose partial sum is given.

45. $s_4 = 200$, $a_1 = 5$, $r = ?$

46. $s_5 = 3410$, $a_1 = 10$, $r = ?$

47. $s_8 = -510$, $a_1 = -2$, $r = ?$

48. $s_6 = -265{,}720$, $a_1 = -4$, $r = ?$

For the given geometric sequence, does the infinite series have a limit? If so, find that limit.

49. $30, 10, \frac{10}{3}, \frac{10}{9}, \ldots$

50. $432, 72, 12, 2, \ldots$

51. $1, -\frac{4}{5}, \frac{16}{25}, -\frac{64}{125}, \ldots$

52. $\frac{1}{9}, -\frac{5}{27}, \frac{25}{81}, -\frac{125}{243}, \ldots$

53. $\frac{81}{100}, \frac{243}{200}, \frac{729}{400}, \frac{2187}{800}, \ldots$

54. $108, 90, 75, \frac{125}{2}, \ldots$

55. $\frac{1}{2}, -\frac{1}{4}, \frac{1}{8}, -\frac{1}{16}, \ldots$

56. $1, -1, 1, -1, \ldots$

For the geometric sequence whose general term is given, does the infinite series have a limit? If so, what is that limit?

57. $a_n = 5 \cdot \left(\frac{5}{6}\right)^{n-1}$

58. $a_n = 200 \cdot \left(\frac{1}{4}\right)^{n-1}$

59. $a_n = -8 \cdot \left(\frac{3}{10}\right)^{n-1}$

60. $a_n = 6 \cdot \left(-\frac{5}{3}\right)^{n-1}$

61. $a_n = \frac{5}{7} \cdot \left(\frac{11}{8}\right)^{n-1}$

62. $a_n = \frac{4}{25} \cdot \left(-\frac{5}{12}\right)^{n-1}$

63. A copy machine purchased for \$8000 decreases in value by 20% each year.

a) Write a geometric sequence showing the value of the copy machine at the end of each of the first 3 years after it was purchased.

b) Find the value of the copy machine at the end of the eighth year.

64. A farmer purchased a new tractor for \$30,000. The tractor's value decreases by 10% each year.

a) Write a geometric sequence showing the value of the tractor at the end of each of the first 6 years after it was purchased.

b) Find the value of the tractor at the end of year 10.

65. A motor home valued at \$40,000 decreases in value by 50% each year. Find the general term, a_n, of a geometric sequence showing the value of the motor home n years after it was purchased.

66. A diamond ring valued at $4000 increases in value by 5% each year. Find the general term, a_n, of a geometric sequence, showing the value of the diamond ring n years after it was purchased.

67. A professional baseball player signed a 10-year contract. The salary for the first year was $10,000,000, with a 10% raise for each of the remaining 9 years. At the end of the contract, what is the total amount of money earned by the player?

68. Cathy takes a job as a high school teacher. The starting salary is $32,000 per year, with a 2% raise each year. If Cathy teaches for 30 years, what is the total amount that she will be paid?

Writing in Mathematics

Answer in complete sentences.

69. Describe a real-world application involving a geometric sequence.

70. Describe a real-world application involving a geometric series.

71. Explain the difference between an arithmetic sequence and a geometric sequence.

72. Explain how to find the general term of a geometric sequence. Use an example to illustrate the process.

73. Explain how to find the partial sum of a geometric sequence. Use an example to illustrate the process.

74. ***Solutions Manual**** Write a solutions manual page for the following problem:

Find s_8 for a geometric sequence with first term $a_1 = 4$ and common ratio $r = 5$.

75. ***Newsletter**** Write a newsletter explaining how to find the limit of an infinite geometric series.

***See Appendix B for details and sample answers.**

Quick Review Exercises

Section 10.3

Simplify.

1. $(x + 5)^2$

2. $(x - y)^2$

3. $(2x + 9y)^2$

4. $(5a^3 - 4b^4)^2$

10.4 The Binomial Theorem

Objectives

1. **Evaluate factorials.**
2. **Calculate binomial coefficients.**
3. **Use the binomial theorem to expand a binomial raised to a power.**
4. **Use Pascal's triangle and the binomial theorem to expand a binomial raised to a power.**

Recall that a binomial is a polynomial with two terms, such as $2x - 3$ or $5x + 8y$. In this section, we will learn a method for raising a binomial to a power, such as $(x + y)^9$.

Factorials

Objective 1 Evaluate factorials.

Factorials

The product of all the positive integers from 1 through some positive integer n, $1 \cdot 2 \cdot 3 \cdot \cdots \cdot (n - 1) \cdot n$, is called ***n* factorial** and is denoted as $n!$

For example, $5! = 1 \cdot 2 \cdot 3 \cdot 4 \cdot 5$, or 120. By definition, $0! = 1$.

EXAMPLE 1 Evaluate: 7!

Solution

To evaluate 7!, we need to multiply the successive integers from 1 through 7.

$$7! = 1 \cdot 2 \cdot 3 \cdot 4 \cdot 5 \cdot 6 \cdot 7 \quad \text{Multiply the positive integers from 1 through 7.}$$
$$= 5040 \quad \text{Simplify.}$$

***Quick Check* 1**
Evaluate: 9!

Most calculators have a built-in function for calculating factorials; consult your calculator manual for directions. Most nongraphing calculators will have a button labeled as [n!] or [x!], while many graphing calculators store the factorial function in the [MATH] menu.

Binomial Coefficients

Objective 2 Calculate binomial coefficients. We now will use factorials to calculate numbers called binomial coefficients.

Binomial Coefficients

For any two positive integers n and r, $n \geq r$, the number ${}_nC_r$ is a **binomial coefficient** and is given by the formula ${}_nC_r = \dfrac{n!}{r! \cdot (n - r)!}$.

EXAMPLE 2 Evaluate the binomial coefficient ${}_9C_4$.

Solution

We begin by substituting 9 for n and 4 for r in the formula ${}_nC_r = \dfrac{n!}{r! \cdot (n-r)!}$.

$${}_9C_4 = \frac{9!}{4! \cdot (9-4)!} \quad \text{Substitute 9 for } n \text{ and 4 for } r.$$
$$= \frac{9!}{4! \cdot 5!} \quad \text{Subtract } 9-4.$$
$$= \frac{362{,}880}{24 \cdot 120} \quad \text{Evaluate each factorial.}$$
$$= \frac{362{,}880}{2880} \quad \text{Simplify the denominator.}$$
$$= 126 \quad \text{Simplify the fraction.}$$

Quick Check 2
Evaluate the binomial coefficient ${}_6C_2$.

Using Your Calculator To find binomial coefficients on the TI-84, we will be accessing the PRB menu. To evaluate the binomial coefficient ${}_9C_4$, begin by typing the first number, 9. Then press [MATH] to access the MATH menu. Use the right arrow to scroll over to the PRB menu. Select option **3: nCr**. When you return to the main screen, type the second number, 4, and press [ENTER]. Here is the calculator screen that you should see:

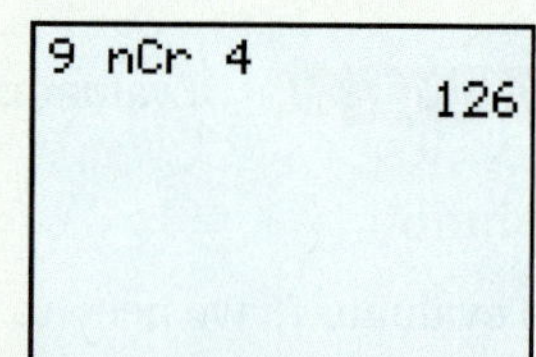

EXAMPLE 3 Evaluate the binomial coefficient ${}_8C_0$.

Solution

We will begin by substituting 8 for n and 0 for r in the formula ${}_nC_r = \dfrac{n!}{r! \cdot (n-r)!}$.

$${}_8C_0 = \frac{8!}{0! \cdot 8!} \quad \text{Substitute 8 for } n \text{ and 0 for } r.$$
$$= \frac{8!}{8!} \quad 0! = 1$$
$$= 1 \quad \text{Simplify the fraction.}$$

Quick Check 3
Evaluate the binomial coefficient ${}_{20}C_{20}$.

For any positive integer n, the binomial coefficient ${}_nC_0 = 1$.

The binomial coefficient ${}_nC_n = 1$ as well, since ${}_nC_n = \dfrac{n!}{n! \cdot 0!}$.

We finish our exploration of binomial coefficients with an example that will save time when calculating several binomial coefficients. It also demonstrates an interesting property of binomial coefficients.

EXAMPLE 4 Evaluate the binomial coefficients ${}_7C_0, {}_7C_1, {}_7C_2, \ldots, {}_7C_7$.

Solution

Verify the following results by using your calculator's built-in function for calculating binomial coefficients or by applying the formula:

$$ {}_7C_0 = 1 \quad {}_7C_1 = 7 \quad {}_7C_2 = 21 \quad {}_7C_3 = 35 \quad {}_7C_4 = 35 \quad {}_7C_5 = 21 \quad {}_7C_6 = 7 \quad {}_7C_7 = 1 $$

Notice the following properties:

- The first and last coefficients are both equal to 1.
- The second coefficient is equal to 7. For any positive integer n, ${}_nC_1 = n$.
- The coefficients follow a symmetric pattern. For any positive integer n and any nonnegative integer $a \le n$, ${}_nC_a = {}_nC_{n-a}$. This means that once we calculate ${}_7C_2$, we know that ${}_7C_5$ is equal to the same value.

These properties will be helpful when we are expanding binomials by using the binomial theorem.

***Quick Check* 4** **Complete the accompanying table, using the properties of binomial coefficients that were introduced in Example 4. You should not have to actually calculate any of these binomial coefficients.**

${}_9C_0 =$	${}_9C_1 =$	${}_9C_2 = 36$	${}_9C_3 =$	${}_9C_4 = 126$
${}_9C_5 =$	${}_9C_6 = 84$	${}_9C_7 =$	${}_9C_8 =$	${}_9C_9 =$

The Binomial Theorem

Objective 3 Use the binomial theorem to expand a binomial raised to a power. We use the **binomial theorem** to raise a binomial expression to a power that is a positive integer without having to repeatedly multiply the binomial expression by itself. For instance, we can use the binomial theorem to expand $(x + y)^5$ rather than multiplying $(x + y)(x + y)(x + y)(x + y)(x + y)$.

The Binomial Theorem

For any positive integer n,

$$ (x + y)^n = {}_nC_0 x^n y^0 + {}_nC_1 x^{n-1} y^1 + {}_nC_2 x^{n-2} y^2 + \cdots + {}_nC_{n-1} x^1 y^{n-1} + {}_nC_n x^0 y^n. $$

This can be expressed by summation notation as

$$ (x + y)^n = \sum_{r=0}^{n} {}_nC_r \cdot x^{n-r} y^r. $$

EXAMPLE 5 Expand $(x + y)^5$, using the binomial theorem.

Solution

We will apply the binomial theorem with $n = 5$.

$$(x + y)^5 = {}_5C_0 \cdot x^5y^0 + {}_5C_1 \cdot x^4y^1 + {}_5C_2 \cdot x^3y^2 + {}_5C_3 \cdot x^2y^3 + {}_5C_4 \cdot x^1y^4 + {}_5C_5 \cdot x^0y^5$$

After evaluating the binomial coefficients and simplifying factors whose exponent is 0 or 1, we obtain the following:

$$(x + y)^5 = x^5 + 5x^4y + 10x^3y^2 + 10x^2y^3 + 5xy^4 + y^5$$

Quick Check 5
Expand $(x + y)^3$, using the binomial theorem.

If the binomial that is being raised to a power is a difference rather than a sum, such as $(x - y)^4$, then the signs of the terms in the binomial expansion alternate between positive and negative.

EXAMPLE 6 Expand $(x - y)^4$, using the binomial theorem.

Solution

We will apply the binomial theorem with $n = 4$, alternating the signs of the terms.

$$(x - y)^4 = {}_4C_0 \cdot x^4y^0 - {}_4C_1 \cdot x^3y^1 + {}_4C_2 \cdot x^2y^2 - {}_4C_3 \cdot x^1y^3 + {}_4C_4 \cdot x^0y^4$$

$$= x^4y^0 - 4x^3y^1 + 6x^2y^2 - 4x^1y^3 + x^0y^4 \quad \text{Evaluate the binomial coefficients.}$$

$$= x^4 - 4x^3y + 6x^2y^2 - 4xy^3 + y^4 \quad \text{Simplify factors whose exponents are 0 or 1.}$$

Quick Check 6
Expand $(x - 4)^4$, using the binomial theorem.

Pascal's Triangle

Objective 4 Use Pascal's triangle and the binomial theorem to expand a binomial raised to a power. One convenient way for finding binomial coefficients is through the use of a triangular array of numbers called **Pascal's triangle**, which is named in honor of the seventeenth-century French mathematician Blaise Pascal. Here is the portion of Pascal's triangle showing the binomial coefficients for values of n through 7:

$n = 0$								1							
$n = 1$							1		1						
$n = 2$						1		2		1					
$n = 3$					1		3		3		1				
$n = 4$				1		4		6		4		1			
$n = 5$			1		5		10		10		5		1		
$n = 6$		1		6		15		20		15		6		1	
$n = 7$	1		7		21		35		35		21		7		1

Consider the row labeled $n = 5$, which contains the values 1, 5, 10, 10, 5, and 1. These are the binomial coefficients ${}_5C_0, {}_5C_1, {}_5C_2, {}_5C_3, {}_5C_4$, and ${}_5C_5$, respectively. Each value in Pascal's triangle is the sum of the two entries diagonally above it.

$n = 4$		1		4		6		4		1	
$n = 5$	1		5		10		10		5		1

To construct Pascal's triangle, begin with a single 1 in the first row ($n = 0$), with two 1's written diagonally underneath it in the next row ($n = 1$).

$$\begin{array}{lccc} n = 0 & & 1 & \\ n = 1 & 1 & & 1 \end{array}$$

The next row will have one more value than the previous row, and each number in the row is the sum of the values diagonally above it.

$n = 0$ 1

$n = 1$

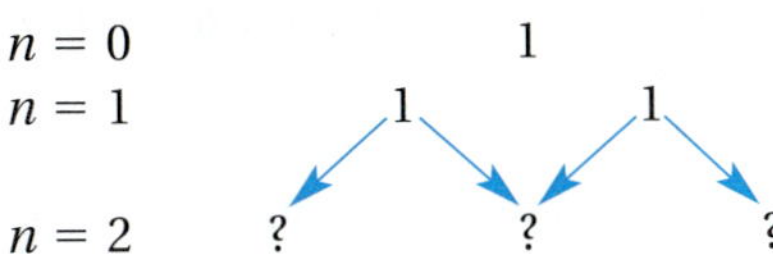

$n = 2$? ? ?

The values for this next row will be 1, 2, and 1. We then proceed to the next row, repeating this process until we have reached the desired row.

EXAMPLE 7 Expand $(x + y)^6$, using the binomial theorem and Pascal's triangle.

Solution

We will apply the binomial theorem with $n = 6$. Here is the row of Pascal's triangle associated with $n = 6$:

$$n = 6 \qquad 1 \qquad 6 \qquad 15 \qquad 20 \qquad 15 \qquad 6 \qquad 1$$

This tells us that the coefficients of the seven terms will be 1, 6, 15, 20, 15, 6, and 1, respectively. We know that the exponent of x in the first term will be 6 and will decrease by 1 in each successive term. We also know that the exponent of y in the first term will be 0 and will increase by 1 in each successive term.

Quick Check **7**

Expand $(x + y)^7$, using the binomial theorem and Pascal's triangle.

$$(x + y)^6 = 1x^6y^0 + 6x^5y^1 + 15x^4y^2 + 20x^3y^3 + 15x^2y^4 + 6x^1y^5 + 1x^0y^6$$

Apply the binomial theorem, obtaining the binomial coefficients from the $n = 6$ row of Pascal's triangle.

$$= x^6 + 6x^5y + 15x^4y^2 + 20x^3y^3 + 15x^2y^4 + 6xy^5 + y^6$$

Simplify each term.

Building Your Study Strategy **Preparing for the Final Exam, 4 Taking the Exam** When you arrive to take your final exam, be sure to bring a positive attitude with you. Be confident. If you have prepared as well as you could, then trust your ability. If you sit down to take the final exam and you are convinced that something is going to go wrong, then you will probably be correct.

Once you sit down to take the final exam, begin by writing down any information that might be useful during the exam, such as formulas or notes to yourself. Briefly read through the test to get an idea of the length and difficulty of the exam. Begin by working problems that you are most confident about.

Set a schedule for yourself. For example, make a goal to be finished with at least 25 percent of the exam by the time that one-fourth of the time has expired. This will help you to realize whether you are working quickly enough.

Save time at the end of the exam period to check your work. Be sure that you've attempted every problem. Many students, in a rush, accidentally skip an entire page.

EXERCISES 10.4

Vocabulary

1. The product of all the positive integers from 1 through some positive integer n is called __________ and is denoted as __________.

2. For any two positive integers n and r, $n \geq r$, the number ${}_nC_r$ is a(n) __________ and is given by the formula __________.

3. The __________ theorem is used to raise a binomial expression to a power that is a positive integer without having to repeatedly multiply the binomial expression by itself.

4. One convenient way for finding binomial coefficients is through the use of a triangular array of numbers called __________.

Evaluate.

5. 6!
6. 4!
7. 8!
8. 10!
9. 3!
10. 2!
11. 1!
12. 0!

Rewrite each given product as a factorial.

13. $1 \cdot 2 \cdot 3 \cdots 15$
14. $1 \cdot 2 \cdot 3 \cdots 24$
15. $1 \cdot 2 \cdot 3 \cdots 45$
16. $1 \cdot 2 \cdot 3 \cdots 32$

Simplify.

17. $\frac{11!}{4!}$
18. $\frac{10!}{8!}$
19. $\frac{9!}{3! \cdot 6!}$
20. $\frac{11!}{7! \cdot 4!}$
21. $8! + 7!$
22. $(5!)^2$

Evaluate each given binomial coefficient.

23. ${}_6C_3$
24. ${}_{13}C_7$
25. ${}_9C_2$
26. ${}_{10}C_6$
27. ${}_4C_2$
28. ${}_5C_4$
29. ${}_5C_1$
30. ${}_6C_0$
31. ${}_{20}C_{20}$
32. ${}_{30}C_{29}$

Complete the table that follows by using the properties of binomial coefficients. Do not actually calculate any of these binomial coefficients.

33.

${}_{11}C_0 =$	${}_{11}C_1 =$	${}_{11}C_2 =$
${}_{11}C_3 = 165$	${}_{11}C_4 =$	${}_{11}C_5 = 462$
${}_{11}C_6 =$	${}_{11}C_7 = 330$	${}_{11}C_8 =$
${}_{11}C_9 = 55$	${}_{11}C_{10} =$	${}_{11}C_{11} =$

34.

${}_{10}C_0 =$	${}_{10}C_1 =$	${}_{10}C_2 = 45$
${}_{10}C_3 = 120$	${}_{10}C_4 =$	${}_{10}C_5 = 252$
${}_{10}C_6 = 210$	${}_{10}C_7 =$	${}_{10}C_8 =$
${}_{10}C_9 =$	${}_{10}C_{10} =$	

Expand, using the binomial theorem.

35. $(x + y)^3$
36. $(x + a)^5$
37. $(x + 2)^3$
38. $(x + y)^4$
39. $(x + 6)^4$
40. $(x + 1)^3$
41. $(x - y)^5$
42. $(x - y)^2$
43. $(x - 3)^4$

44. $(x - 2)^3$

45. $(x + 3y)^3$

46. $(4x + y)^4$

47. $(4x - y)^3$

48. $(x - 2y)^6$

49. Construct the rows of Pascal's triangle for $n = 8$ through $n = 12$.

50. Find the sum of the binomial coefficients in each of the first six rows of Pascal's triangle. Do you notice a pattern? Find a formula for the sum of the binomial coefficients in the row associated with $n = k$. Predict the sum of the coefficients in the row associated with $n = 22$.

Expand, using the binomial theorem and Pascal's triangle.

51. $(x + y)^5$

52. $(x + y)^3$

53. $(x + 2)^7$

54. $(x + 4)^6$

55. $(x - y)^3$

56. $(x - y)^8$

57. $(x - 3)^4$

58. $(x - 5)^5$

Writing in Mathematics

Answer in complete sentences.

59. If $a + b = n$, explain why ${}_nC_a = {}_nC_b$.

60. ***Solutions Manual**** Write a solutions manual page for the following problem:

Expand $(x + 5)^4$, *using the binomial theorem.*

61. ***Newsletter**** Write a newsletter explaining how to construct Pascal's triangle.

***See Appendix B for details and sample answers.**

Chapter 10 Summary

Section 10.1—Topic	Chapter Review Exercises
Terms of a Sequence	1–8
Partial Sum of a Series	9–14
Summation Notation	15–16
Section 10.2—Topic	**Chapter Review Exercises**
Common Difference and Terms of an Arithmetic Sequence	17–28
Partial Sum of an Arithmetic Series	29–33
Applications of Arithmetic Sequences and Series	34
Section 10.3—Topic	**Chapter Review Exercises**
Common Ratio and Terms of a Geometric Sequence	35–41
Partial Sum of an Geometric Series	42–47
Limit of an Infinite Geometric Series	48–51
Applications of Geometric Sequences and Series	52
Section 10.4—Topic	**Chapter Review Exercises**
Factorials	53–54
Binomial Coefficients	55–56
Binomial Theorem	57–59
Pascal's Triangle	60

Summary of Chapter 10 Study Strategies

This chapter's study strategies focused on preparing for a final exam.

- Plan a schedule that will maximize your time and effort in preparation for the exam.
- Determine which resources, such as your instructor or the tutorial center, are available to you, as well as when these resources are available.
- While preparing for the final exam, consider using the following: old exams and quizzes, old homework assignments, class notes, note cards, materials from your instructor, and review exercises from the textbook.
- Be sure to prepare for all of the applied problems that you are responsible for.
- Create practice quizzes to help with topics that require more review.
- Continue working with your study group.
- Pace yourself while taking the exam to ensure that you finish with sufficient time to review your work.

Find the first five terms of the sequence with each given general term. **[10.1]**

1. $a_n = n + 7$

2. $a_n = 5n - 19$

Find the next three terms of each given sequence, and find its general term, a_n. **[10.1]**

3. 13, 26, 39, 52, 65, . . .

4. 12, 19, 26, 33, 40, . . .

Find the first five terms of the sequence with each given first term and general term. **[10.1]**

5. $a_1 = -9, a_n = (a_{n-1}) + 17$

6. $a_1 = -4, a_n = 2(a_{n-1}) - 23$

Find a recursive formula for the general term, a_n, of each given sequence. **[10.1]**

7. 3, 17, 31, 45, 59, . . .

8. 5, −10, 20, −40, 80, . . .

Find the indicated partial sum for each given sequence. **[10.1]**

9. s_6, 4, 14, 24, 34, . . .

10. s_9, 2, 12, 72, 432, . . .

Find the indicated partial sum for the sequence with each given general term. **[10.1]**

11. $s_7, a_n = 2n - 5$

12. $s_8, a_n = 5n + 6$

Find the indicated partial sum for the sequence with each given general term. **[10.1]**

13. $s_7, a_1 = 50, a_n = a_{n-1} - 8$

14. $s_5, a_1 = 20, a_n = 4a_{n-1} + 5$

Find each sum. **[10.1]**

15. $\sum_{i=1}^{12} i$

16. $\sum_{i=1}^{11} (5i + 6)$

Worked-out solutions to Review Exercises marked with can be found on page AN–63.

Find the common difference, d, of each given arithmetic sequence. **[10.2]**

17. 7, 31, 55, 79, 103, . . .

18. −3, −9, −15, −21, −27, . . .

Find the first five terms of the arithmetic sequence with each given first term and common difference. **[10.2]**

19. $a_1 = 16, d = -9$

20. $a_1 = -18, d = 11$

Find the general term, a_n, of each given arithmetic sequence. **[10.2]**

21. 4, 11, 18, 25, 32, . . .

22. 8, 1, −6, −13, −20, . . .

23. −15, −13, −11, −9, −7, . . .

24. $\frac{5}{6}, \frac{19}{12}, \frac{7}{3}, \frac{37}{12}, \frac{23}{6}, \ldots$

25. Find the 39th term, a_{39}, of the arithmetic sequence 25, 46, 67, 88, 109, ... [10.2]

26. Find the 27th term, a_{27}, of the arithmetic sequence −9, −32, −55, −78, −101, ... [10.2]

27. Which term of the arithmetic sequence −12, −5, 2, 9, 16, . . . , is equal to 1633? [10.2]

28. Which term of the arithmetic sequence 40, 21, 2, −17, −36, . . . , is equal to −9346? [10.2]

Find the indicated partial sum of each given arithmetic sequence. **[10.2]**

29. s_8, 10, 17, 24, 31, . . .

30. s_{15}, −102, −84, −66, −48, . . .

31. s_{24}, 16, 13, 10, 7 . . .

32. s_{30}, −10, −3, 4, 11, . . .

33. Find the sum of the first 340 positive integers. [10.2]

34. The new CEO of a company started the job with an annual salary of $20,000,000. Her contract states that she will be given a raise of $2,000,000 each year. [10.2]

a) Write a sequence showing the CEO's annual salary for the first 5 years.

b) Find the general term, a_n, for the CEO's annual salary in her nth year in this job.

c) How much money will the CEO earn from this job over the first 10 years?

***Find the common ratio, r, of each given geometric sequence.* [10.3]**

35. 1, 10, 100, 1000, 10,000, . . .

36. 5, 30, 180, 1080, 6480, . . .

***Find the general term, a_n, of each given geometric sequence.* [10.3]**

37. 1, 12, 144, 1728, . . .

38. $8, 6, \frac{9}{2}, \frac{27}{8}, \ldots$

39. $-4, -12, -36, -108, \ldots$

40. $-24, 8, -\frac{8}{3}, \frac{8}{9}, \ldots$

41. Find the eighth term, a_8, of the geometric sequence 9, 18, 36, 72, ... [10.3]

***Find the partial sum of the geometric sequence with each given first term and common ratio.* [10.3]**

42. $s_9, a_1 = 1, r = 3$

43. $s_{11}, a_1 = 10, r = -2$

***Find the partial sum for each given geometric sequence.* [10.3]**

44. s_7, 1, 6, 36, 216, . . .

45. $s_8, \frac{3}{2}, 12, 96, 768, \ldots$

46. s_{10}, 320, 160, 80, 40 . . .

47. s_9, 35, -140, 560, -2240, . . .

***For each given geometric sequence, does the infinite series have a limit? If so, find that limit.* [10.3]**

48. 64, 80, 100, 125, . . .

49. $\frac{36}{7}, \frac{24}{7}, \frac{16}{7}, \frac{32}{21}, \ldots$

***For each geometric sequence whose general term is given, does the infinite series have a limit? If so, find that limit.* [10.3]**

50. $a_n = -21 \cdot \left(\frac{2}{7}\right)^{n-1}$

51. $a_n = 45 \cdot \left(-\frac{7}{8}\right)^{n-1}$

52. A community service organization has started a homeless shelter with a budget of $150,000 for the first year. Their plans call for an annual increase in the budget of 10% over the previous year. How much will be budgeted for the shelter over the first 8 years of operation? [10.3]

***Evaluate.* [10.4]**

53. 9!

54. 7!

***Evaluate each given binomial coefficient.* [10.4]**

55. ${}_8C_3$

56. ${}_{11}C_8$

***Expand, using the binomial theorem.* [10.4]**

57. $(x + y)^6$

58. $(x + 6)^4$

59. $(x - 9)^3$

***Expand, using the binomial theorem and Pascal's triangle.* [10.4]**

60. $(x + y)^9$

Chapter 10 Test

1. Find the next three terms of the sequence 11, 26, 41, 56, 71, . . . , and find its general term, a_n.

2. Find the sum $\sum_{i=1}^{9}(4i + 11)$.

3. Find the first five terms of the arithmetic sequence whose first term is $a_1 = 9$ and common difference is $d = 8$.

4. Find the general term, a_n, of the given arithmetic sequence 20, 13, 6, −1, −8, ...

5. Find the 15th partial sum, s_{15}, of the arithmetic sequence 14, $\frac{35}{2}$, 21, $\frac{49}{2}$, ...

6. Find the sum of the first 200 positive even integers.

7. Find the first five terms of the geometric sequence whose first term is $a_1 = 5$ and common ratio is $r = 8$.

8. Find the general term, a_n, of the geometric sequence 4, 28, 196, 1372, ...

9. Find the tenth partial sum, s_{10}, of the geometric sequence 3, 15, 75, 375, ...

10. For the geometric sequence 10, −6, $\frac{18}{5}$, $-\frac{54}{25}$, . . . , does the infinite series have a limit? If so, find that limit.

11. Jonah's parents have decided to start a college fund for him. On Jonah's first birthday his parents deposit \$2500 in the fund. On each birthday after that, they increase the amount that they deposit by \$500.

a) Write a sequence showing the amounts that Jonah's parents deposit on his first six birthdays.

b) Find the general term, a_n, for the amount deposited on Jonah's nth birthday.

c) What is the total amount deposited in the fund for Jonah's first 18 birthdays?

12. A new car, valued at \$18,000, decreases in value by 25% each year.

a) Write a geometric sequence showing the value of the car at the end of each of the first 4 years after it was purchased.

b) Find the general term, a_n, for the value of the car n years after it was purchased.

13. Evaluate the binomial coefficient ${}_{12}C_5$.

Expand, using the binomial theorem.

14. $(x + 1)^4$

Expand, using the binomial theorem and Pascal's triangle.

15. $(4x + y)^7$

Mathematicians in History

Fibonacci

Fibonacci was a mathematician in the 12th and 13th centuries who introduced the world to an important sequence of numbers known as the Fibonacci sequence.

Write a one-page summary (*or* make a poster) of the life of Fibonacci and his accomplishments.

Interesting issues:

- Where and when was Fibonacci born?
- Where was Fibonacci educated?
- What was Fibonacci's father's occupation?
- What was Fibonacci's real name?
- What is the Fibonacci sequence, and how is it created?
- Which problem in Fibonacci's book *Liber abaci* introduced the sequence that came to be known as the Fibonacci sequence?
- List at least three occurrences of the Fibonacci sequence in nature.
- Fibonacci also wrote problems involving perfect numbers. What is a perfect number?

Stretch Your Thinking Chapter 10

The Fibonacci sequence is one of the most famous sequences in history. In fact, it has an entire magazine dedicated to it called* The Fibonacci Quarterly. *The sequence begins with 1 and 1, and each following number is formed by adding the previous two numbers. This is what is known as a recursive sequence. Mathematically, this recursive sequence is formed with the notation $f_n = f_{n-1} + f_{n-2}$, where f_{n-1} and f_{n-2} are the predecessors of f_n and $n > 2$.

a) List the first 20 Fibonacci numbers. Start with 1 and 1 as the first two Fibonacci numbers.

The Fibonacci sequence has many interesting properties in mathematics alone. These properties have been proven by many mathematicians; but prove to yourself that they are true by finding at least one example to back up each property.

b) The sum of 10 Fibonacci numbers is always divisible by 11. Do you notice anything interesting about the number that results from dividing the sum by 11?

c) Twice any Fibonacci number minus the next Fibonacci number equals the number that is two numbers before the original number in the sequence $(2 \cdot f_n - f_{n+1} = f_{n-2})$.

d) For any three consecutive Fibonacci numbers, the product of the first and third numbers $(f_n \cdot f_{n+2})$ differs from the square of the middle number $((f_{n+1})^2)$ by one.

e) Now that you have seen a few of the fascinating properties, try to come up with one of your own.

1. For the functions $f(x) = 7x + 11$ and $g(x) = 3x - 42$, find $(f + g)(x)$. [8.1]

For each given function $f(x)$ and $g(x)$, find $(f \circ g)(x)$ and $(g \circ f)(x)$. [8.1]

2. $f(x) = x + 6, g(x) = x^2 - 2x + 12$

Determine whether each function $f(x)$ and $g(x)$ is an inverse function. Recall that $f(x)$ and $g(x)$ are inverse functions if $(f \circ g)(x) = x$ and $(g \circ f)(x) = x$. [8.2]

3. $f(x) = 3x + 8, g(x) = \frac{x - 8}{3}$

For each given function $f(x)$, find $f^{-1}(x)$. [8.2]

4. $f(x) = 3x - 19$

5. $f(x) = \frac{3}{x - 9}$

Evaluate each given function. Round to the nearest thousandth. [8.3–8.4]

6. $f(x) = e^{x+3} - 21, f(2)$

7. $f(x) = \ln(x + 7) + 32, f(25)$

Simplify. [8.4]

8. $\log_3 81$

9. $\ln e^8$

Rewrite as a single logarithmic expression. Simplify if possible. Assume all variables represent positive real numbers. [8.5]

10. $2 \ln x - 4 \ln y + 5 \ln z$

Rewrite as the sum or difference of logarithmic expressions whose arguments have an exponent of 1. Simplify if possible. Assume all variables represent positive real numbers. [8.5]

11. $\log\left(\frac{b^4c^5}{a}\right)$

Evaluate, using the change-of-base formula. Round to the nearest thousandth. [8.5]

12. $\log_{13} 200$

13. $\log_9 3224$

Solve. Round to the nearest thousandth. [8.6]

14. $5^{x-6} = 25$

15. $7^x = 219$

16. $e^{x+7} = 6205$

Solve. Round to the nearest thousandth. [8.6]

17. $\log(x^2 + 7x - 9) = \log(3x + 23)$

18. $\log_3 (x - 9) = 4$

19. $\log_2 (x + 3) + \log_2 (x + 9) = 4$

For the given function $f(x)$, find its inverse function $f^{-1}(x)$. [8.6]

20. $f(x) = e^{x-2} + 3$

21. $f(x) = \ln(x + 9) - 16$

22. Jean has deposited \$2000 in an account that pays 6% annual interest, compounded quarterly. How long will it take until the balance of the account is \$10,000? [8.7]

23. The population of a city increased from 120,000 in 1992 to 150,000 in 2002. If the population continues to increase exponentially at the same rate, what will the population be in 2022? [8.8]

Graph. Label any intercepts and asymptotes. State the domain and range of the function. [8.8]

24. $f(x) = 2^x - 1$

25. $f(x) = \ln(x + 5) - 3$

***Graph each parabola. Label each vertex and any intercepts.* [9.1]**

26. $y = x^2 + 6x - 8$

27. $y = (x + 3)^2 - 4$

28. $x = -y^2 + y + 6$

29. $x = (y - 2)^2 + 3$

***Graph each circle. Label the center of the circle. Identify the radius of the circle.* [9.2]**

30. $x^2 + y^2 = 64$

31. $(x + 3)^2 + (y - 2)^2 = 25$

***Graph each ellipse. Label the center. Give the values of a and b.* [9.3]**

32. $\frac{x^2}{9} + \frac{y^2}{49} = 1$

33. $36x^2 + 25y^2 - 288x - 300y + 576 = 0$

***Graph each hyperbola. Label the center. Give the values of a and b.* [9.4]**

34. $\frac{x^2}{25} - \frac{y^2}{4} = 1$

35. $(y - 2)^2 - \frac{(x + 4)^2}{9} = 1$

***Find the equation of each conic section.* [9.2–9.4]**

36.

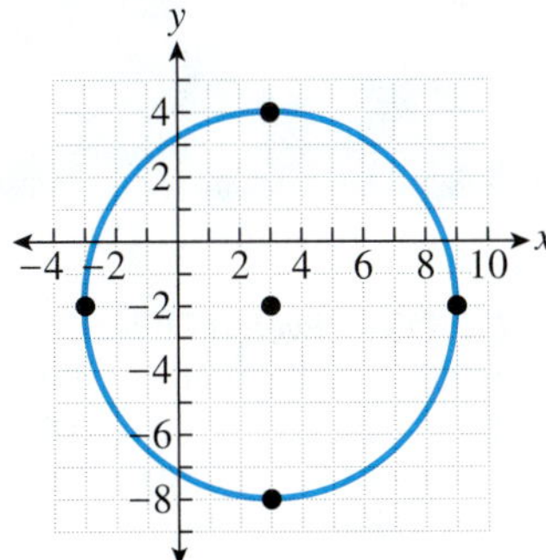

37.

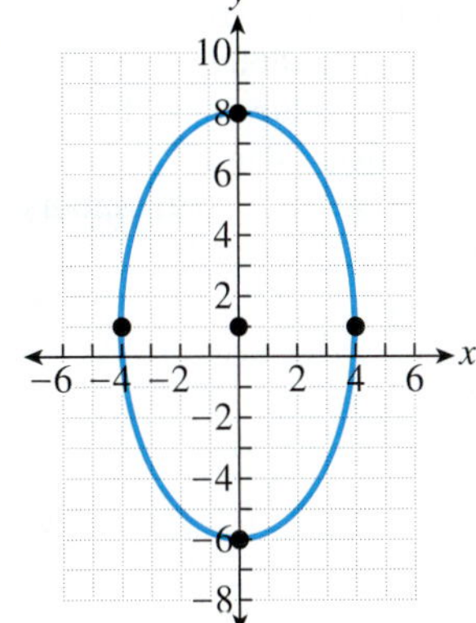

***Solve by substitution.* [9.5]**

38. $x^2 + y^2 = 40$
$2x + y = 10$

39. $x^2 + 2y^2 = 18$
$y = x - 3$

***Solve by the addition method.* [9.5]**

40. $x^2 + y^2 = 100$
$x^2 - y^2 = 28$

41. $2x^2 + 5y^2 = 38$
$4x^2 - 3y^2 = 24$

42. A rectangular lawn has a perimeter of 100 feet and an area of 600 square feet. Find the dimensions of the lawn. [9.5]

***Find the next three terms of the given sequence, and find its general term, a_n.* [10.1]**

43. $-4, 9, 22, 35, 48, \ldots$

***Find the first five terms of the sequence of the given first term and general term.* [10.1]**

44. $a_1 = 13, a_n = 2(a_{n-1}) + 5$

***Find the general term, a_n, of the given arithmetic sequence.* [10.2]**

45. $3, 12, 21, 30, 39, \ldots$

***Find the partial sum for the given arithmetic sequence.* [10.2]**

46. $s_{12}, -6, -1, 4, 9, \ldots$

47. Find the sum of the first 200 positive integers. [10.2]

***Find the general term, a_n, of the given geometric sequence.* [10.3]**

48. $12, 9, \frac{27}{4}, \frac{81}{16}, \ldots$

***Find the partial sum of the geometric sequence of the given first term and common ratio.* [10.3]**

49. $s_8, a_1 = 1, r = 4$

***Find the partial sum for the given geometric sequence.* [10.3]**

50. $s_7, 2, 14, 98, 686, \ldots$

***For the geometric sequence whose general term is given, does the infinite series have a limit? If so, what is that limit?* [10.3]**

51. $a_n = 6 \cdot \left(\frac{3}{8}\right)^{n-1}$

***Evaluate the given binomial coefficient.* [10.4]**

52. ${}_{10}C_4$

***Expand, using the binomial theorem.* [10.4]**

53. $(x + y)^4$

54. $(x - y)^6$

***Expand, using the binomial theorem and Pascal's triangle.* [10.4]**

55. $(x + y)^8$

Photo Credits

Page 17 (l) © Thinkstock **Page 17** (r) © Rubberball Productions **Page 19** © Getty Royalty Free **Page 27** (t) © The Everett Collection **Page 27** (b) © The Kobal Collection **Page 45** © Shutterstock **Page 52** © Brand X Pictures **Page 53** © The Kobal Collection **Page 59** © Corbis/Bettmann © **Page 103** © Digital Vision **Page 116** © Comstock **Page 122** © PhotoDisc **Page 133** © Shutterstock **Page 160** © Getty Editorial **Page 215** © PhotoDisc **Page 217** © Reuters/Corbis **Page 243** © Robert Maass/Corbis **Page 246** © PhotoDisc **Page 274** © PhotoDisc Blue **Page 277** © Princeton University Stanhope Hall **Page 297** © NASA Headquarters **Page 368** © Library of Congress **Page 369** © Anthony Reynolds/Cordaiy Photo Library Ltd./Corbis **Page 423** © Stockbyte Platinum **Page 426** © PhotoDisc Blue **Page 432** © Randy Wells/Corbis **Page 449** © Najlah Feanny/Corbis SABA **Page 496** © PhotoEdit Inc **Page 497** © PhotoDisc Blue **Page 512** © SAU **Page 538** © Taxi **Page 543** © Brand X Pictures **Page 546** © PhotoDisc **Page 575** © PhotoDisc **Page 576** © Brand X Pictures **Page 617** © Corbis/Bettmann **Page 626** © Stockbyte **Page 633** © PhotoDisc **Page 660** © PhotoDisc **Page 665** © PhotoDisc Red **Page 667** © Comstock **Page 682** © Roger Ressmeyer/Corbis **Page 704** © PhotoDisc Red **Page 708** © PhotoDisc Blue **Page 713** © Corbis **Page 714** © Brand X Pictures **Page 737** © Lucien Aigner/Corbis **Page 775** © Thinkstock **Page 824** (l) © PhotoDisc **Page 824** (r) © Stone/Getty Images **Page 834** © Institut Mittag Leffler **Page 836** © Shutterstock **Page 840** © PhotoDisc **Page 843** © Beth Anderson **Page 858** © Digital Vision **Page 862** © Comstock **Page 874** © SAU

Appendix A Synthetic Division

Synthetic division is a shorthand alternative to polynomial long division when the divisor is of the form $x - b$. When we use synthetic division, we do not write any variables, only the coefficients.

Suppose that we wanted to divide $\frac{2x^2 - x - 25}{x - 4}$. Note that $b = 4$ in this example. Before we begin, we check the dividend to be sure that it is written in descending order and that there are no missing terms. If there are missing terms, we will add placeholders. On the first line, we write down the value for b as well as the coefficients for the dividend.

$$\begin{array}{r|rrr} 4 & 2 & -1 & -25 \end{array}$$

Notice that we set the value of b apart from the coefficients of the divisor. After we use synthetic division, we will find the quotient and the remainder in the last of our three rows. To begin the actual division, we bring the first coefficient down to the bottom row as follows:

$$\begin{array}{r|rrr} 4 & 2 & -1 & -25 \\ & \downarrow & & \\ \hline & 2 & & \end{array}$$

This number will be the leading coefficient of our quotient. In the next step, we multiply 4 by this first coefficient and write it under the next coefficient in the dividend (-1). Totaling this column will give us the next coefficient in the quotient.

$$\begin{array}{r|rrr} 4 & 2 & -1 & -25 \\ & \downarrow & 8 & \\ \hline & 2 & 7 & \end{array}$$

We repeat this step until we total the last column. Multiply 4 by 7 and write the product under -25. Total the last column.

$$\begin{array}{r|rrr} 4 & 2 & -1 & -25 \\ & \downarrow & 8 & 28 \\ \hline & 2 & 7 & \boxed{3} \end{array}$$

We are now ready to read our quotient and remainder from the bottom row. The remainder is the last number in the bottom row, which in our example is 3. The numbers in front of the remainder are the coefficients of our quotient, written in descending order. In this example, the quotient is $2x + 7$. So, $\frac{2x^2 - x - 25}{x - 4} = 2x + 7 + \frac{3}{x - 4}$.

EXAMPLE 1 Use synthetic division to divide $\dfrac{x^3 - 11x^2 - 13x + 6}{x + 2}$.

Solution

The divisor is of the correct form, and $b = -2$. This is because $x + 2$ can be rewritten as $x - (-2)$. The dividend is in descending form with no missing terms, so we may begin.

-2	1	-11	-13	6
	$\downarrow$	-2	26	-26
	1	-13	13	-20

So, $\dfrac{x^3 - 11x^2 - 13x + 6}{x + 2} = x^2 - 13x + 13 - \dfrac{20}{x + 2}$.

Now we turn to an example that has a dividend with missing terms, requiring the use of placeholders.

EXAMPLE 2 Use synthetic division to divide $\dfrac{3x^4 - 8x - 5}{x - 3}$.

Solution

The dividend, written in descending order with placeholders, is $3x^4 + 0x^3 + 0x^2 - 8x - 5$. The value for b is 3.

3	3	0	0	-8	-5
	$\downarrow$	9	27	81	219
	3	9	27	73	214

So, $\dfrac{3x^4 - 8x - 5}{x - 3} = 3x^3 + 9x^2 + 27x + 73 + \dfrac{214}{x - 3}$.

EXERCISES

Divide using synthetic division.

1. $\dfrac{x^2 - 18x + 77}{x - 7}$
2. $\dfrac{x^2 + 4x - 45}{x - 5}$
3. $(x^2 + 13x - 48) \div (x - 3)$
4. $(2x^2 - 43x + 221) \div (x - 13)$
5. $\dfrac{7x^2 - 11x - 12}{x + 2}$
6. $\dfrac{3x^2 + 50x - 30}{x + 17}$
7. $\dfrac{12x^2 - 179x - 154}{x - 14}$
8. $\dfrac{x^2 + 14x - 13}{x + 1}$
9. $\dfrac{x^3 - 34x + 58}{x - 5}$
10. $\dfrac{x^3 - 59x - 63}{x + 7}$
11. $\dfrac{x^4 + 8x^2 - 33}{x + 8}$
12. $\dfrac{5x^5 - 43x^3 - 8x^2 + 9x - 9}{x - 3}$

Appendix B

Sample Solutions to Writing in Mathematics Exercises

SOLUTIONS MANUAL

A **solutions manual page** is an assignment in which you solve a problem as if you were creating a solutions manual for a textbook. Show and explain each step in solving the problem in such a way that a fellow student would be able to read and understand your work.

EXAMPLE Write a solutions manual page for the following problem:

Simplify $3 - 4(20 \div 4 - 6 \cdot 8) - 5^2$.

Student's Solution

We simplify this expression using the order of operations, beginning with the expression inside the parentheses.

$3 - 4(20 \div 4 - 6 \cdot 8) - 5^2$	
$= 3 - 4(5 - 6 \cdot 8) - 5^2$	Inside the parentheses, do the division first. $20 \div 4 = 5$.
$= 3 - 4(5 - 48) - 5^2$	Still working inside the parentheses, multiplication comes before subtraction. $6 \cdot 8 = 48$.
$= 3 - 4(-43) - 5^2$	Subtract $5 - 48$ inside the parentheses. $5 - 48 = -43$
$= 3 - 4(-43) - 25$	Now that the work inside parentheses is done, simplify expressions with exponents. $5^2 = 25$. We are squaring 5, not -5.
$= 3 + 172 - 25$	Next in order is multiplication or division, so multiply. $-4(-43) = 172$
$= 150$	Finish with addition and subtraction.

NEWSLETTER

A **newsletter** assignment asks you to explain a certain mathematical topic. Your explanation should be in the form of a short, visually appealing article that might be published in a newsletter read by people who were interested in learning mathematics.

EXAMPLE Write a newsletter explaining how to solve a system of two linear equations by the substitution method.

Student's Solution

Substitution Method

First, solve one of the two equations for x or y. It's usually easier to solve an equation that has a variable term with a coefficient of 1 or -1, that way you won't have to divide or get involved with fractions. If you solve the first equation for x in terms of y, then you substitute that expression for x in the second equation.

Solve $\begin{aligned} x + 3y &= 1 \\ 4x - 5y &= 30 \end{aligned}$ using the substitution method.

First, solve the first equation for x.

$$\begin{aligned} x + 3y &= -1 \\ x &= -3y - 1 \end{aligned}$$ Subtract $3y$ from both sides.

Next, we can substitute $-3y - 1$ for x in the second equation, $4x - 5y = 30$

$$\begin{aligned} 4x - 5y &= 30 \\ 4(-3y - 1) - 5y &= 30 \end{aligned}$$ Substitute $-3y - 1$ for x.

This gives us an equation with only one variable (y) in it, so we solve this equation for y.

$$\begin{aligned} 4(-3y - 1) - 5 &= 30 \\ -12y - 4 - 5y &= 30 \\ -17y - 4 &= 30 \\ -17 &= 34 \\ \frac{-17y}{-17} &= \frac{34}{-17} \\ y &= -2 \end{aligned}$$ Solve the equation for y.

Now that we have one coordinate of our ordered pair solution, we can substitute for y to find our x-coordinate.

$$\begin{aligned} x &= -3y - 1 \\ x &= -3(-2) - 1 \\ x &= 6 - 1 \\ x &= 5 \end{aligned}$$ Substitute -2 for y and solve.

The solution is $(5, -2)$.

Answers to Selected Exercises

Chapter 1

Quick Check 1.1 **1. a)** $<$, **b)** $>$, **c)** $>$ **2.** 8 **3.** 3 **4.** -16 **5.** -21 **6.** 25 **7.** -36 **8.** -45 **9.** 91 **10.** -1440 **11.** -16 **12.** 64 **13.** $\frac{1}{125}$ **14.** $-\frac{129}{40}$ **15.** -49 **16.** -60 **17.** -18 **18.** 133

Section 1.1 **1.** empty set **3.** negative **5.** negative **7.** divisor **9.** base **11.** squared **13.** $>$ **15.** $<$ **17.** $>$ **19.** 19 **21.** 21 **23.** -16 **25.** $-6, 6$ **27.** $-25, 25$ **29.** -9 **31.** -22 **33.** -30 **35.** 4.9 **37.** $-\frac{7}{20}$ **39.** -8 **41.** -49 **43.** 31 **45.** -7.3 **47.** $\frac{19}{18}$ **49.** -6 **51.** -21 **53.** 6°C **55.** 5562 feet **57.** -54 **59.** 28 **61.** -17.28 **63.** $-\frac{7}{20}$ **65.** 0 **67.** 220 **69.** -9 **71.** 18 **73.** -9.7 **75.** $\frac{35}{36}$ **77.** 0 **79.** Undefined **81.** 153 **83.** -11 **85.** -52 **87.** 0 **89.** \$15,000 **91.** $-$\$7500 **93.** 64 **95.** 100,000 **97.** $\frac{4}{25}$ **99.** 25 **101.** 128 **103.** 17 **105.** 27 **107.** 40 **109.** -26.95 **111.** -48 **113.** -118 **115.** $\frac{51}{28}$ **117.** $\frac{1}{6}$ **119.** $-\frac{3}{10}$ **121.** 2 **123.** -155 **125.** -38 **127.** $5 \cdot 6 - 13 + 7 = 24$ **129.** $(9 + 7 \cdot 3) \div 6 = 5$ **131.** 5 **133.** 4 **135.** Answers will vary. **137.** Answers will vary.

Quick Check 1.2 **1.** $x + 62$ **2.** $x - 6$ **3.** $3x$ **4.** $\frac{x}{16}$ **5.** 93 **6.** 98 **7.** 136 **8.** $54a$ **9.** $5x + 62$ **10.** $30x - 80$ **11.** $3x - 15y - 24z$ **12.** $-45x - 144$ **13.** Three terms: $-x^2, 8x, -37$; $-1, 8, -37$ **14.** $10a - 7b$ **15.** $11x + 9$ **16.** $6x - 23$

Section 1.2 **1.** variable **3.** evaluate **5.** associative **7.** term **9.** like terms **11.** $x - 5$ **13.** $\frac{x}{7}$ **15.** $8x$ **17.** $3x + 25$ **19.** $\frac{14}{x}$ **21.** $10(x + y)$ **23.** $1.50m$ **25.** $200d + 425$ **27.** 32 **29.** 92 **31.** -16 **33.** -13 **35.** $\frac{46}{3}$ **37.** 12 **39.** 48 **41.** 9 **43.** 13 **45.** -115 **47.** 256 **49.** 345 **51.** -25.02 **53.** 5.01 **55. a)** Answers will vary; a and b cannot be equal. **b)** Yes, a and b must be equal. **57.** $8x - 16$ **59.** $24 - 54x$ **61.** $-6x + 15$ **63.** $9x - 8$ **65.** $13x$ **67.** $-9x$ **69.** $6x - 13$ **71.** $33a - 33$ **73.** $\frac{15}{4}x - \frac{1}{2}$ **75.** $8x - 9y$ **77.** $5x - 11y$ **79.** $11x + 2$ **81.** $-20x + 41$ **83.** $-5x + 120$ **85.** $50z - 44$ **87. a)** 3 **b)** $7x^2, -10x, -35$ **c)** $7, -10, -35$ **89. a)** 2 **b)** $x, -34$ **c)** $1, -34$ **91. a)** 3 **b)** $8x^2, -29x, 52$ **c)** $8, -29, 52$ **93. a)** 4 **b)** $-36a, 62b, 85c, 64$ **c)** $-36, 62, 85, 64$ **95.** Answers will vary; some examples are $4(2x - 5)$, $5x + 3x - 20$, $2x + 8 + 6x - 28$, and $11x - 7 - 3x - 13$. **97.** Answers will vary.

Quick Check 1.3 **1.** $\{-21\}$ **2.** $\{6\}$ **3.** $\{-18\}$ **4.** $\{-14\}$ **5.** $\{-3\}$ **6.** $\{-6\}$ **7.** $\{3\}$ **8.** $\{\frac{31}{5}\}$ **9.** $\mathbb{R}$ **10.** $\varnothing$ **11.** $y = \frac{4 - 3x}{5}$ **12.** $\{-7, 7\}$ **13.** $\{-\frac{13}{3}, -1\}$ **14.** $\{\frac{1}{2}, \frac{9}{2}\}$ **15.** $\varnothing$ **16.** $\{-22, -\frac{4}{3}\}$

Section 1.3 **1.** equation **3.** The solution set of an equation is the set of all solutions to that equation. **5.** the LCM of the denominators **7.** the empty set **9.** b **11.** $X = a, X = -a$ **13.** $\{11\}$ **15.** $\{-4\}$ **17.** $\{5\}$ **19.** $\{-\frac{8}{5}\}$ **21.** $\{-3\}$ **23.** $\{5\}$ **25.** $\{3.5\}$ **27.** $\{-17\}$ **29.** $\{-\frac{16}{3}\}$ **31.** $\{-3\}$ **33.** $\{\frac{9}{5}\}$ **35.** $\{2.2\}$ **37.** $\{-2\}$ **39.** $\{\frac{59}{20}\}$ **41.** $\{0\}$ **43.** $\{12\}$ **45.** $\{\frac{9}{4}\}$ **47.** $\mathbb{R}$ **49.** $\varnothing$ **51.** $\mathbb{R}$ **53.** $y = 3x + 7$ **55.** $y = \frac{-6x + 15}{2}$ **57.** $y = \frac{8x - 11}{4}$ **59.** $a = P - b - c$ **61.** $L = \frac{A}{W}$ **63.** $\{-2, 2\}$ **65.** $\{-929, 929\}$ **67.** $\{1, 15\}$ **69.** $\{-4, \frac{4}{3}\}$ **71.** $\varnothing$ **73.** $\{2, 14\}$ **75.** $\{-2, 8\}$ **77.** $\{0, \frac{11}{2}\}$ **79.** $\{-5, \frac{11}{3}\}$ **81.** $\{-9, 3\}$ **83.** $\{-18, 4\}$ **85.** $\{-1, 21\}$ **87.** $\{-7, 3\}$ **89.** $\{8\}$ **91.** $\{-\frac{1}{5}\}$ **93.** $|x| = 2$; answers will vary. **95.** $|x + 1| = 4$; answers will vary. **97.** Answers will vary. **99.** Answers will vary.

Quick Check 1.4 **1.** 64 **2.** Length: 50 feet Width: 20 feet **3.** 8 inches, 20 inches, 25 inches **4.** 92, 93, 94 **5.** 14, 16, 18 **6.** 5 hours **7.** 22 and 79 **8.** 21 \$1 bills and 13 \$5 bills

Section 1.4 **1.** b **3.** b **5.** $x + 1, x + 2$ **7.** $x + 2, x + 4$ **9.** 11 **11.** 42 **13.** 3.1 **15.** Length: 13 meters, Width: 7 meters **17.** Length: 37 inches, Width: 16 inches **19.** Length: 6.2 feet, Width: 5.1 feet **21.** Length: 75 feet, Width: 30 feet **23.** Length: 50 feet, Width: 25 feet **25.** 18 feet **27.** 36 centimeters **29.** 40 miles, 50 miles, 70 miles **31.** 15 centimeters **33.** A: 15°, B: 75° **35.** 24°, 66° **37.** 72.4°, 107.6° **39.** 40°, 60°, 80° **41.** 4000 square yards **43.** 68 inches **45.** 65, 66, 67, 68 **47.** 127, 129, 131 **49.** 46, 48, 50 **51.** 268, 269 **53.** 3355 **55.** 39, 40, 41 **57.** 81 miles **59.** $6\frac{1}{4}$ hours **61.** 4.7 kilometers/hour **63. a)** 18 hours **b)** 600 miles **65.** 21 hours **67.** 33, 52 **69.** 23, 59 **71.** 12, 19 **73.** 89 **75.** 20 **77.** 160 **79.** Answers will vary. **81.** Answers will vary.

Quick Review Exercises 1.4 **1.** {6} **2.** {−16} **3.** {13} **4.** {16}

Quick Check 1.5 **1.** $x \le 2$, $(-\infty, 2]$ **2.** $x \ge -2$, $[-2, \infty)$ **3.** $x < 5$, $(-\infty, 5)$ **4.** $-5 \le x \le -2$, $[-5, -2]$ **5.** $x < -6$ or $x > \frac{9}{2}$, $(-\infty, -6) \cup (\frac{9}{2}, \infty)$ **6.** $x < 120$ **7.** 97 or higher **8.** $-4 \le x \le -2$, $[-4, -2]$ **9.** $-2 < x < 3$, $(-2, 3)$ **10.** $\varnothing$ **11.** $x < -2$ or $x > 6$, $(-\infty, -2) \cup (6, \infty)$ **12.** $x < -6$ or $x > -1$, $(-\infty, -6) \cup (-1, \infty)$ **13.** $\mathbb{R}$, $(-\infty, \infty)$

Section 1.5 **1.** linear inequality **3.** interval **5.** compound **7.** $(-\infty, 4)$ **9.** $[6, \infty)$ **11.** $(-\infty, 2.4]$ **13.** $(-5, 2)$ **15.** $(-\infty, -5] \cup (8, \infty)$ **17.** $x > 6$ **19.** $-9 \le x \le -3$ **21.** $x < 6$ or $x > 14$ **23.** $x < -9$ **25.** $x < 7$, $(-\infty, 7)$ **27.** $x \le -5$, $(-\infty, -5]$ **29.** $x \ge -3.5$, $[-3.5, \infty)$ **31.** $x \ge -4$, $[-4, \infty)$ **33.** $x > -\frac{7}{4}$, $(-\frac{7}{4}, \infty)$ **35.** $x \ge 18$, $[18, \infty)$ **37.** $x > -33$, $(-33, \infty)$ **39.** $8 \le x \le 14$, $[8, 14]$ **41.** $-3 \le x \le 7$, $[-3, 7]$ **43.** $-7 < x < -\frac{9}{2}$, $(-7, -\frac{9}{2})$ **45.** $-\frac{32}{3} < x < -4$, $(-\frac{32}{3}, -4)$ **47.** $x < 2$ or $x > 8$, $(-\infty, 2) \cup (8, \infty)$ **49.** $x \le -7$ or $x \ge 9$, $(-\infty, -7] \cup [9, \infty)$ **51.** $x < 4$ or $x > 6$, $(-\infty, 4) \cup (6, \infty)$ **53.** $x \ge 7$ **55.** $x < 20$ **57.** $x \ge 5500$ **59.** At least 700 **61.** At most 625 **63.** At least $13,185 **65.** 79 to 100 **67.** $-5 < x < 1$, $(-5, 1)$ **69.** $x < -5$ or $x > -1$, $(-\infty, -5) \cup (-1, \infty)$ **71.** $-5 < x < 5$, $(-5, 5)$ **73.** $\mathbb{R}$, $(-\infty, \infty)$ **75.** $x \le \frac{1}{4}$ or $x \ge 1$, $(-\infty, \frac{1}{4}] \cup [1, \infty)$ **77.** $-\frac{13}{4} \le x \le -\frac{5}{4}$, $[-\frac{13}{4}, -\frac{5}{4}]$ **79.** $x \le 4$ or $x \ge 10$, $(-\infty, 4] \cup [10, \infty)$ **81.** $\varnothing$ **83.** $-\frac{19}{5} \le x \le 5$, $[-\frac{19}{5}, 5]$ **85.** $3 \le x \le 13$, $[3, 13]$ **87.** $x < -\frac{8}{3}$ or $x > \frac{4}{3}$, $(-\infty, -\frac{8}{3}) \cup (\frac{4}{3}, \infty)$ **89.** $x < -12$ or $x > 2$, $(-\infty, -12) \cup (2, \infty)$ **91.** $|x| < 2$ **93.** $|x - 5| > 4$ **95.** Answers will vary.

Chapter 1 Review **1.** > **2.** < **3.** 5 **4.** 22 **5.** −15 **6.** 26 **7.** −53 **8.** −14 **9.** −56 **10.** 24 **11.** 256 **12.** $\frac{27}{64}$ **13.** −25 **14.** 576 **15.** 27 **16.** 56 **17.** −106 **18.** 10 **19.** 18 **20.** $\frac{16}{35}$ **21.** $n - 9$ **22.** $n + 25$ **23.** $3n - 7$ **24.** $2n + 15$ **25.** $2.25c$ **26.** $29.95 + 0.50m$ **27.** −5 **28.** −11 **29.** 18 **30.** 81 **31.** $6x - 45$ **32.** $4x - 35$ **33.** $-2x + 22$ **34.** $2x - 57$ **35. a)** 4 **b)** $x^3, -8x^2, 7x, -15$ **c)** 1, −8, 7, −15 **36. a)** 3 **b)** $3x^2, x, -20$ **c)** 3, 1, −20 **37.** No **38.** Yes **39.** {13} **40.** {−14} **41.** {8} **42.** {10} **43.** {7} **44.** $\{\frac{25}{8}\}$ **45.** {−8} **46.** $\{-\frac{7}{6}\}$ **47.** {15} **48.** {−6} **49.** $y = 12 - 5x$ **50.** $y = \frac{24 - 4x}{3}$ **51.** $d = \frac{C}{\pi}$ **52.** $c = \frac{P - a - b}{2}$ **53.** {−14, 4} **54.** {−7, 10} **55.** {−9, 17} **56.** $\{-\frac{7}{2}, 1\}$ **57.** 41, 76 **58.** Length: 33 feet, Width: 13 feet **59.** 55, 57, 59 **60.** $2\frac{1}{2}$ hours **61.** 67 **62.** $x < 5$, $(-\infty, 5)$ **63.** $x > -2$, $(-2, \infty)$ **64.** $x \ge -1$, $[-1, \infty)$ **65.** $x \le 8$, $(-\infty, 8]$ **66.** $x < -4$ or $x > -1$, $(-\infty, -4) \cup (-1, \infty)$ **67.** $x \le -6$ or $x \ge 6$, $(-\infty, -6] \cup [6, \infty)$ **68.** $-8 \le x \le 7$, $[-8, 7]$ **69.** $-8 < x < 1$, $(-8, 1)$ **70.** At least 39 **71.** 67 to 100 **72.** $-9 < x < 3$, $(-9, 3)$ **73.** $x < 1$ or $x > 9$, $(-\infty, 1) \cup (9, \infty)$ **74.** $5 \le x \le 7$, $[5, 7]$ **75.** $x \le 1$ or $x \ge 7$, $(-\infty, 1] \cup [7, \infty)$

Chapter 1 Review Exercises: Worked-Out Solutions

8. $$\begin{aligned} 4-(-14)-32 &= 4+14-32 \\ &= 18-32 \\ &= -14 \end{aligned}$$

18. $$\begin{aligned} -4(7-3\cdot 5)-22 &= -4(7-15)-22 \\ &= -4(-8)-22 \\ &= 32-22 \\ &= 10 \end{aligned}$$

19. $$\begin{aligned} 20-32\div 4^2 &= 20-32\div 16 \\ &= 20-2 \\ &= 18 \end{aligned}$$

29. $$\begin{aligned} (-3)^2+5(-3)+24 &= 9-15+24 \\ &= 18 \end{aligned}$$

33. $$\begin{aligned} 4x-2(3x-11) &= 4x-6x+22 \\ &= -2x+22 \end{aligned}$$

38. $$\begin{aligned} 2(3x-2)-(x+7) &= 3x-27 \\ 2(3(-8)-2)-((-8)+7) &= 3(-8)-27 \\ 2(-24-2)-(-1) &= -24-27 \\ 2(-26)+1 &= -51 \\ -52+1 &= -51 \\ -51 &= -51 \end{aligned}$$

Yes, -8 is a solution.

40. $$\begin{aligned} 3x+26 &= -16 \\ 3x+26-26 &= -16-26 \\ 3x &= -42 \\ \tfrac{3x}{3} &= -\tfrac{42}{3} \\ x &= -14 \end{aligned}$$
$\{-14\}$

43. $$\begin{aligned} \tfrac{x}{15}-\tfrac{3}{10} &= \tfrac{1}{6} \\ 30\cdot\tfrac{x}{15}-30\cdot\tfrac{3}{10} &= 30\cdot\tfrac{1}{6} \\ 2x-9 &= 5 \\ 2x &= 14 \\ x &= 7 \end{aligned}$$
$\{7\}$

45. $$\begin{aligned} 2x+17 &= 5x+41 \\ 17 &= 3x+41 \\ -24 &= 3x \\ -8 &= x \end{aligned}$$
$\{-8\}$

48. $$\begin{aligned} 2(3m-4)-(m-13) &= 2m-13 \\ 6m-8-m+13 &= 2m-13 \\ 5m+5 &= 2m-13 \\ 3m+5 &= -13 \\ 3m &= -18 \\ m &= -6 \end{aligned}$$
$\{-6\}$

50. $$\begin{aligned} 4x+3y &= 24 \\ 3y &= 24-4x \\ y &= \tfrac{24-4x}{3} \end{aligned}$$

55. $$\begin{aligned} |x-4|-7 &= 6 \\ |x-4| &= 13 \end{aligned}$$
$x-4=13$ or $x-4=-13$
$x=17$ or $x=-9$
$\{-9, 17\}$

58. Length: $2x+7$
Width: x
$$\begin{aligned} 2L+2W &= 92 \\ 2(2x+7)+2x &= 92 \\ 4x+14+2x &= 92 \\ 6x+14 &= 92 \\ 6x &= 78 \\ x &= 13 \end{aligned}$$
Length: $2x+7=2(13)+7=33$
Width: $x=13$
Length: 33 feet, Width: 13 feet

59. First: x
Second: $x+2$
Third: $x+4$
$$\begin{aligned} x+x+2+x+4 &= 171 \\ 3x+6 &= 171 \\ 3x &= 165 \\ x &= 55 \end{aligned}$$
First: $x=55$
Second: $x+2=55+2=57$
Third: $x+4=55+4=59$
The three integers are 55, 57, and 59.

64. $$\begin{aligned} 2x+7 &\ge 5 \\ 2x &\ge -2 \\ x &\ge -1 \end{aligned}$$
$[-1, \infty)$

66. $4x<-16$ or $x+6>5$
$x<-4$ or $x>-1$

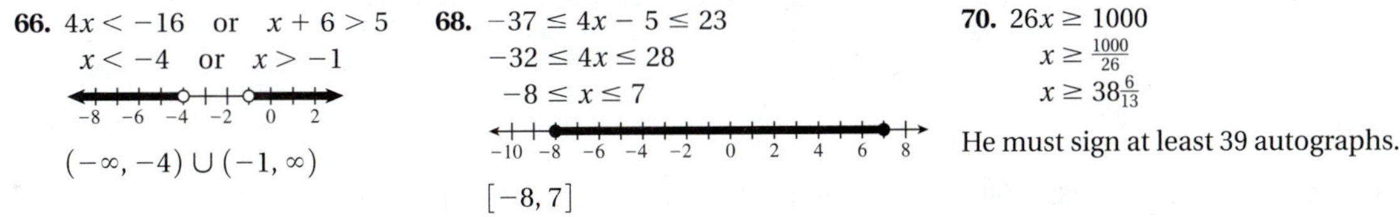

$(-\infty, -4)\cup(-1, \infty)$

68. $$\begin{aligned} -37 \le 4x-5 &\le 23 \\ -32 \le 4x &\le 28 \\ -8 \le x &\le 7 \end{aligned}$$
$[-8, 7]$

70. $$\begin{aligned} 26x &\ge 1000 \\ x &\ge \tfrac{1000}{26} \\ x &\ge 38\tfrac{6}{13} \end{aligned}$$
He must sign at least 39 autographs.

73. $|x-5|>4$
$x-5>4$ or $x-5<-4$
$x>9$ or $x<1$

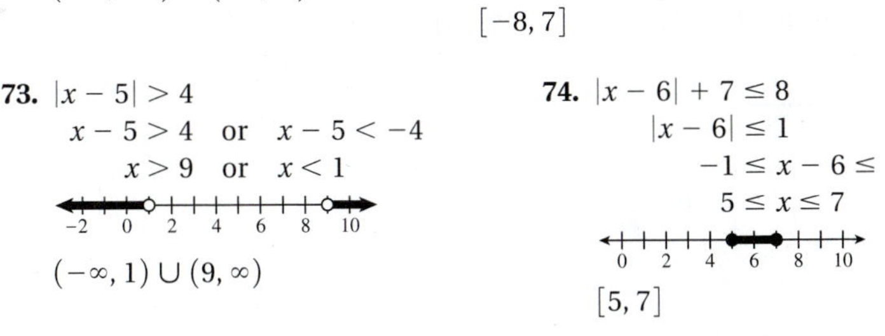

$(-\infty, 1)\cup(9, \infty)$

74. $$\begin{aligned} |x-6|+7 &\le 8 \\ |x-6| &\le 1 \\ -1 \le x-6 &\le 1 \\ 5 \le x &\le 7 \end{aligned}$$
$[5, 7]$

Chapter 1 Test **1.** < **2.** 23 **3.** −15 **4.** −54 **5.** −52 **6.** 9 **7.** 164 **8.** $2n - 19$ **9.** $100 + 80h$ **10.** −73 **11.** $17x + 32$ **12.** $-16x + 96$ **13.** $\{-6\}$ **14.** $\{\frac{34}{3}\}$ **15.** $\{-2\}$ **16.** $\{-7\}$ **17.** $y = \frac{-4x + 36}{3}$ **18.** $\{-3, 8\}$ **19.** 155, 196 **20.** Length: 11 feet, Width: 7 feet **21.** 70, 71, 72 **22.** $x > -6$, $(-6, \infty)$ **23.** $-9 \le x \le 7$, $[-9, 7]$ **24.** $-9 < x < -5$, $(-9, -5)$, **25.** $x \le -7$ or $x \ge 3$, $(-\infty, -7] \cup [3, \infty)$

Chapter 2

Quick Check 2.1 **1.** Yes **2.**

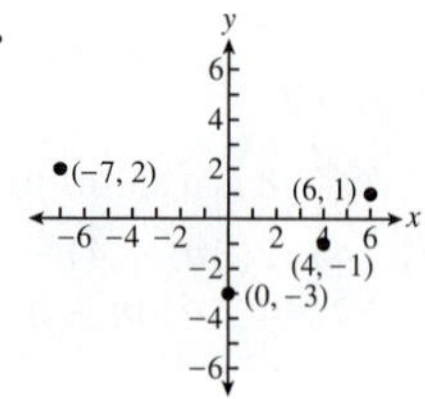

3. A (7, 5), B (−4, 3), C (−2, 8), D (0, −7), E (−3, −6)

4.

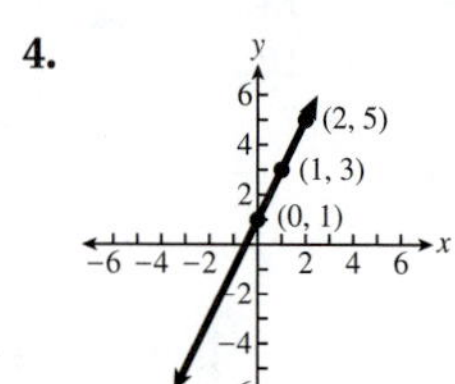

5. (10, 0), (0, 2) **6.** $(-\frac{5}{4}, 0)$, (0, 5) **7.**

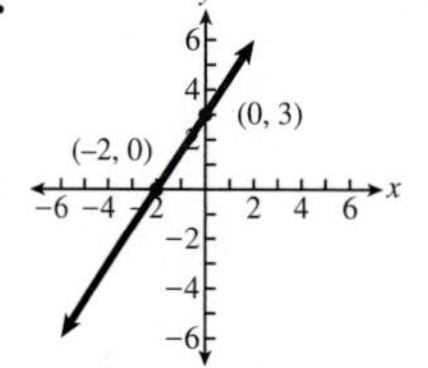

8. **9.**

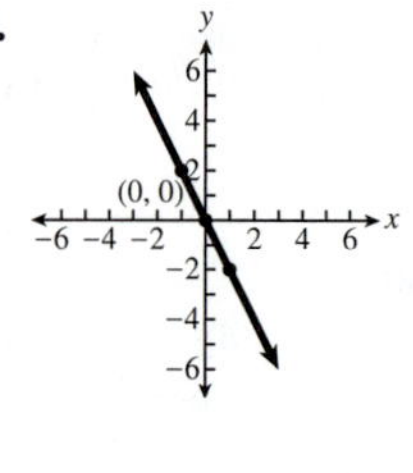

10. **11.**

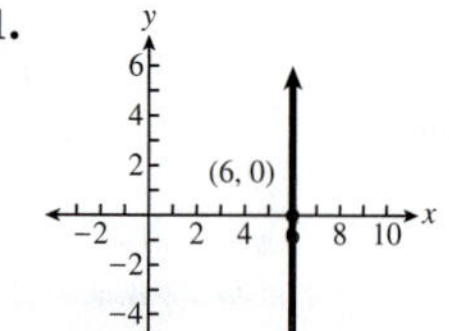

12. a) Yesenia was initially $28,000 in debt.
b) It will take 7 months until Yesenia breaks even.

Section 2.1 **1.** $Ax + By = C$ **3.** origin **5.** x-intercept **7.** vertical **9.** Substitute 0 for y and solve for x. **11.** Yes **13.** Yes **15.** No **17.**

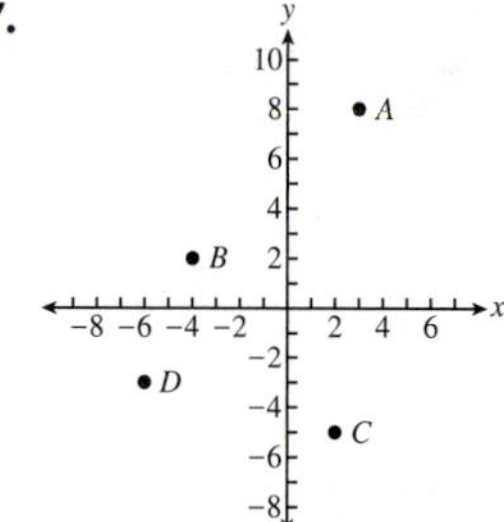

19.

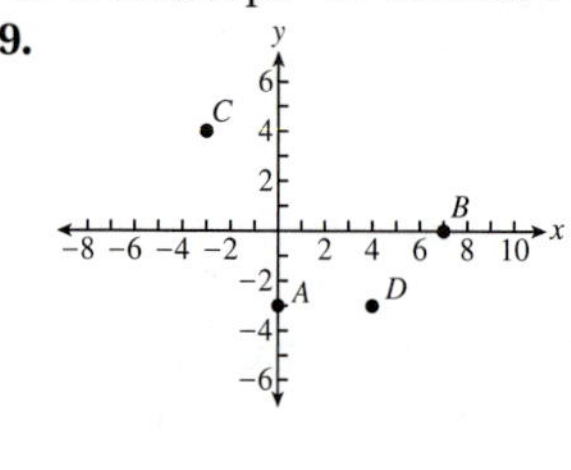

21.

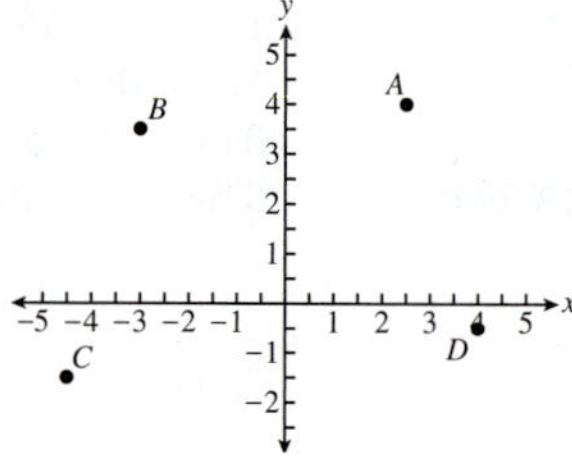

23. A (7, 0), B (3, −6), C (−5, −4), D (6, −3) **25.** A (80, 30), B (−20, 40), C (−70, −60), D (−90, 0) **27.** II **29.** III **31.** I **33.** −2 **35.** 4 **37.** −1 **39.**

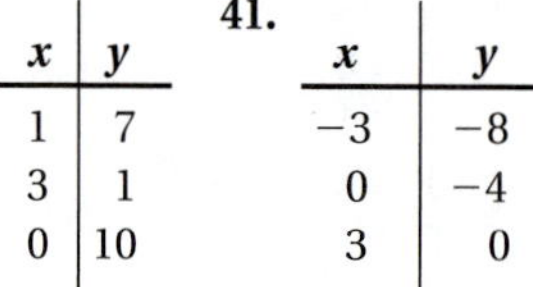

x	y
1	7
3	1
0	10

41.

x	y
−3	−8
0	−4
3	0

43.

x	y
0	−7
1	−3
2	1

45. $(-4, 0), (0, 8)$ **47.** $(-6, 0), (0, -4)$ **49.** No x-intercept, $(0, 5)$ **51.** $(8, 0), (0, 8)$ **53.** $(-8, 0), (0, 6)$ **55.** $(\frac{5}{2}, 0), (0, -5)$ **57.** $(6, 0), (0, -4)$ **59.** $(-16, 0), (0, -32)$ **61.** $(0, 0), (0, 0)$

63. $(3, 0), (0, 3)$

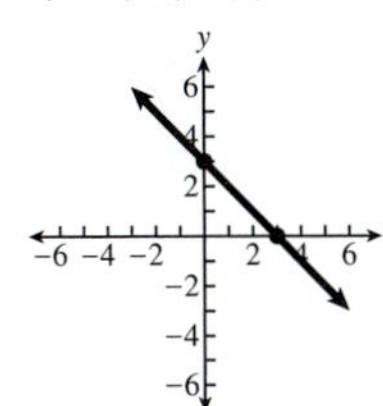

65. $(9, 0), (0, 3)$

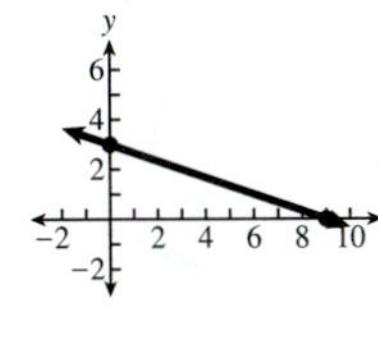

67. $(-9, 0), (0, -6)$

69. $(4, 0), (0, -8)$

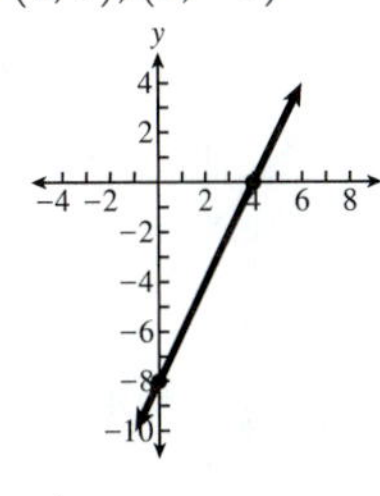

71. $(6, 0), (0, \frac{9}{2})$

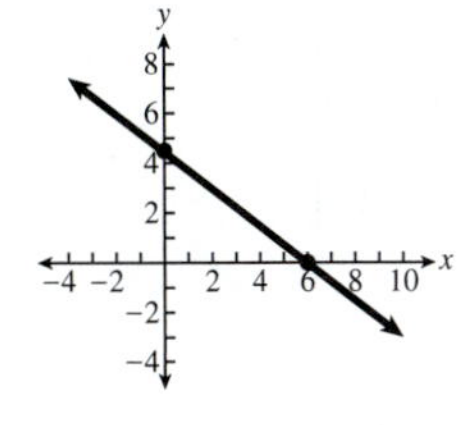

73. $(0, 0), (0, 0)$

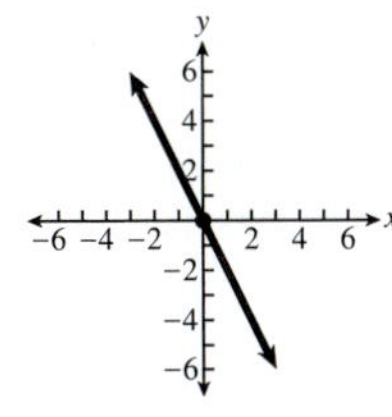

75. $(-\frac{3}{2}, 0), (0, 6)$

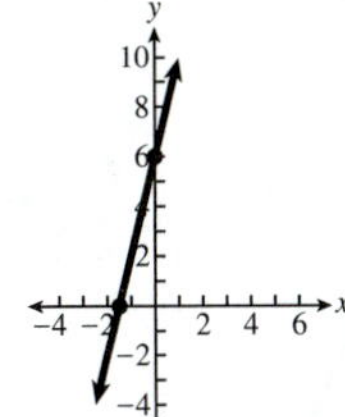

77. No x-intercept, $(0, -8)$

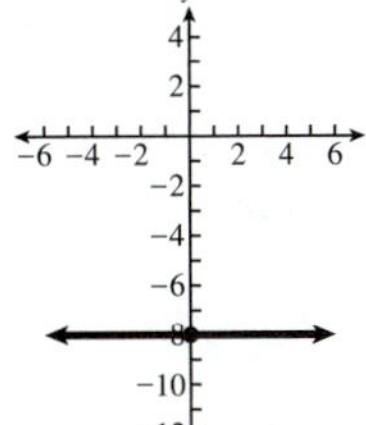

79. $(7, 0)$, No y-intercept

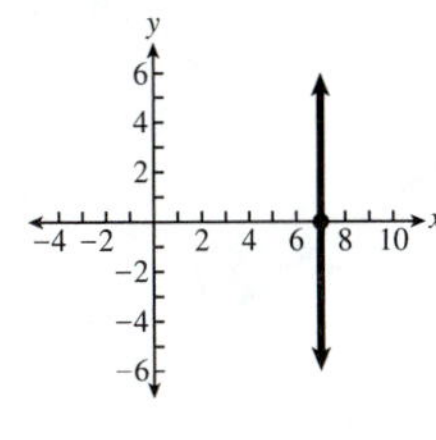

81. Answers will vary. An example is $2x + 3y = 12$. **83.** Answers will vary. An example is $4x - 3y = 15$. **85. a)** $(0, 16{,}000)$; The original value of the tractor is \$16,000. **b)** $(8, 0)$; The tractor has no value after 8 years. **87. a)** $(0, 2380)$; The cost of membership without playing any rounds of golf is \$2380. **b)** Since the minimum fee is \$2380, the costs to a member can never be \$0. **c)** \$3340 **89.** Answers will vary.

Quick Check 2.2 **1.** -4 **2.** $-\frac{4}{3}$ **3.** -3 **4.** Slope is undefined **5.** $m = 0$ **6.** $m = -\frac{3}{5}, (0, 4)$ **7.** $m = 1, (0, -6)$

8. **9.**

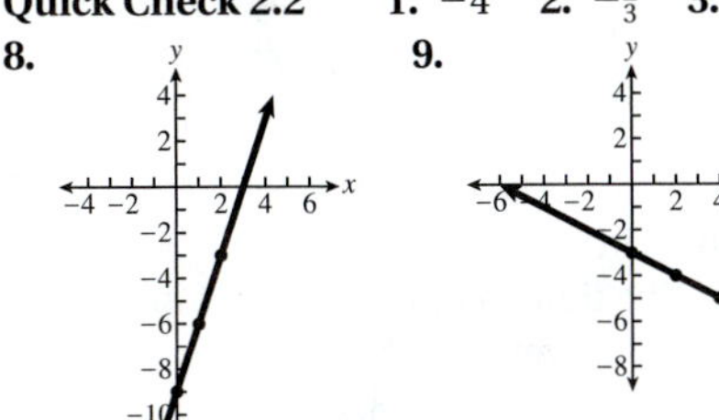

10. Neither **11.** Parallel **12.** Perpendicular **13.** Slope is 55,400: Number of women enrolled in a graduate program increases by 55,400 per year. y-intercept is $(0, 1{,}118{,}900)$: approximately 1,118,900 women were enrolled in a graduate program in 2001.

Section 2.2 **1.** slope **3.** negative **5.** horizontal, vertical **7.** parallel **9.** Positive **11.** Negative **13.** 4 **15.** $-\frac{13}{4}$ **17.** 4 **19.** $-\frac{13}{3}$ **21.** 0 **23.** -1.5 **25.** $-\frac{2}{3}$ **27.** 0 **29.** Undefined **31.** $5, (0, -8)$ **33.** $-\frac{2}{3}, (0, 6)$ **35.** $-3, (0, 9)$ **37.** $\frac{3}{4}, (0, -\frac{5}{2})$ **39.** $2.3, (0, -42.9)$ **41.** $y = -7x + 8$ **43.** $y = \frac{2}{5}x - 3$ **45.** $y = 6$ **47.** **49.**

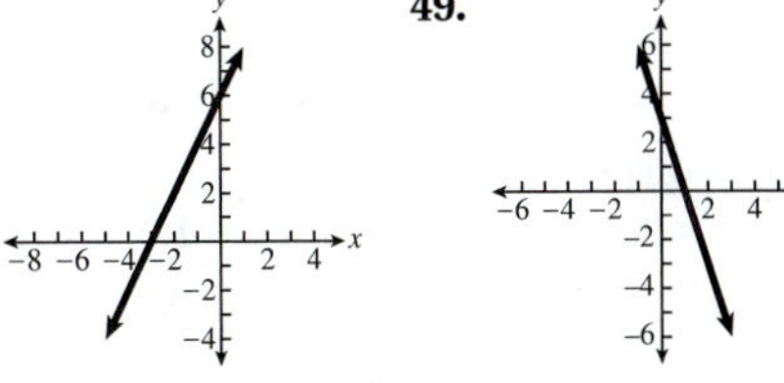

51. **53.** **55.** **57.** **59.**

61. **63.** **65.** **67.**

69. **71.** **73.** **75.** **77.**

79. **81.** **83.** **85.** **87.**

89. $(4, 3)$

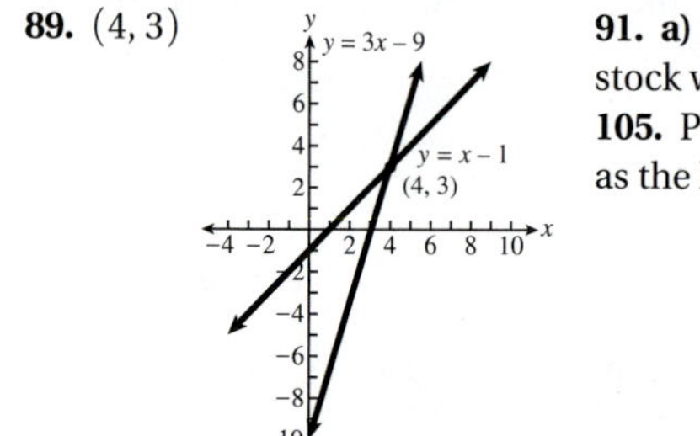

91. a) -3; The value of the stock decreased by \$3/year. **b)** $(0, 68)$; The original value of the stock was \$68. **c)** \$23 **93.** $-\frac{1}{25}$ **95.** Yes **97.** No **99.** No **101.** Yes **103.** Neither **105.** Perpendicular **107.** Parallel **109.** Yes **111.** Positive. The time to mow a lawn (y) increases as the lawn's area (x) increases. **113.** Answers will vary. **115.** Answers will vary.

Quick Check 2.3 **1.** $y = 3x + 7$ **2.** $y = -2x - 7$ **3.** $y = \frac{2}{3}x - 5$ **4.** $y = 8$ **5.** $y = 0.4x + 35.8$ **6.** $y = \frac{1}{4}x + 9$ **7.** $y = \frac{3}{2}x + 5$

Section 2.3 **1.** $y - y_1 = m(x - x_1)$ **3.** $y = 2x - 6$ **5.** $y = \frac{2}{5}x - 3$ **7.** $y = 8x - 9$ **9.** $y = 5x + 3$ **11.** $y = \frac{3}{4}x - 4$ **13.** $y = -\frac{2}{5}x + 4$ **15.** $y = 2x - 8$ **17.** $y = -3x - 16$ **19.** $y = \frac{2}{3}x + 8$ **21.** $y = -\frac{1}{4}x + 9$ **23.** $y = \frac{5}{6}x - \frac{13}{3}$ **25.** $y = -2.5x + 2.6$ **27.** $y = x - 3$ **29.** $y = -\frac{4}{3}x + 4$ **31.** $y = 4$ **33.** $y = 4x - 7$ **35.** $y = -6x - 1$ **37.** $y = \frac{2}{9}x - 4$ **39.** $y = -\frac{1}{3}x + 8$ **41.** $y = 55x - 100.5$

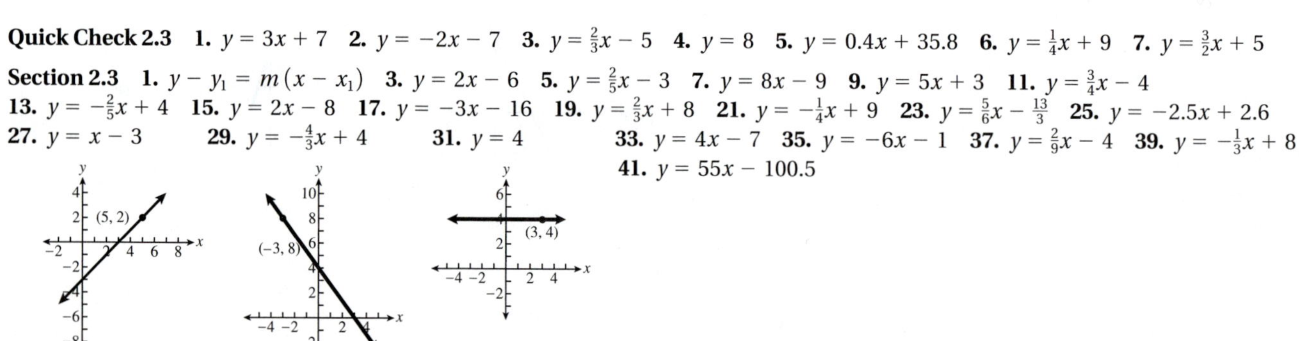

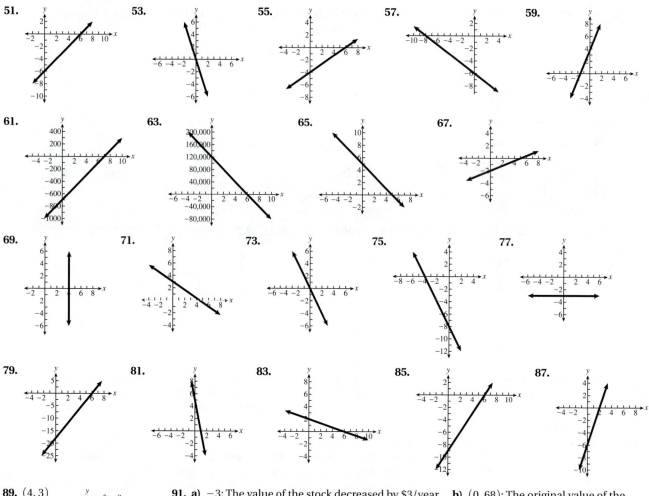

43. $y = 2x - 8$ **45.** $y = -\frac{1}{2}x + 3$

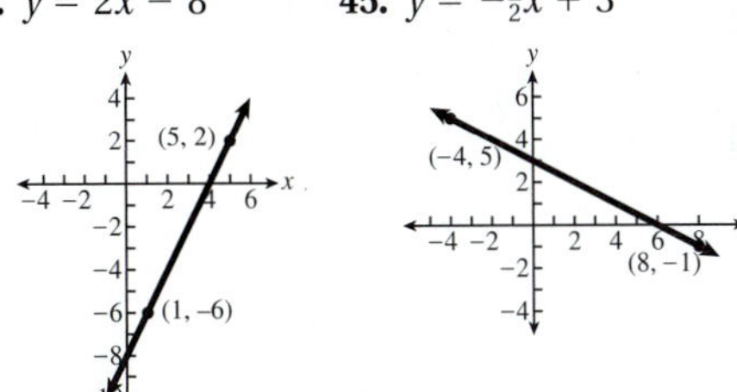

47. $y = -\frac{1}{2}x + 2$ **49.** $y = x - 7$ **51.** $y = 2x + 8$ **53.** $y = -\frac{3}{2}x + 7$ **55.** $y = -3x + 10$ **57.** $y = 4x - 17$ **59.** $y = -\frac{1}{3}x - 3$ **61.** $y = -\frac{5}{2}x - 16$ **63.** $y = -6$ **65.** $y = x + 6$ **67.** $y = -\frac{1}{3}x - 4$ **69. a)** $y = 350x + 120$ **b)** \$120 **c)** \$350 **71. a)** $y = 125x + 150$ **b)** \$150 **c)** \$125 **73. a)** $y = 4.75x + 5.95$ **b)** \$5.95 **c)** \$4.75 **75.** Answers will vary.

Quick Review Exercises 2.3 **1.**

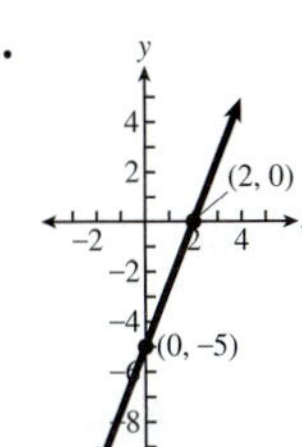

2.

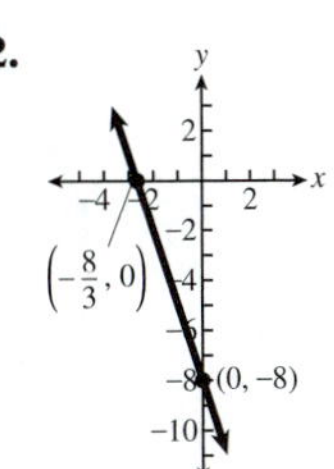

3.

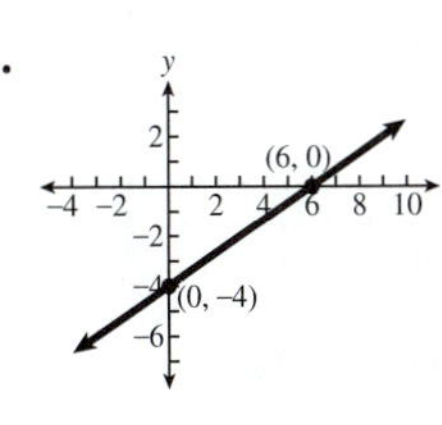

4.

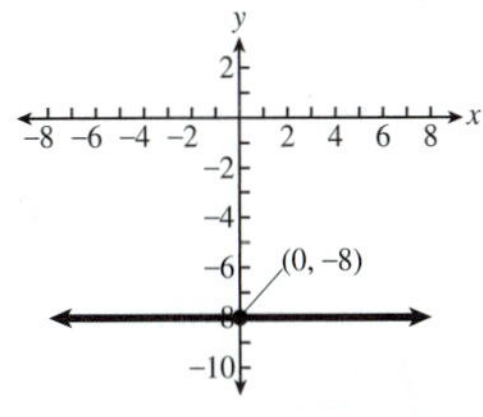

Quick Check 2.4 **1.**

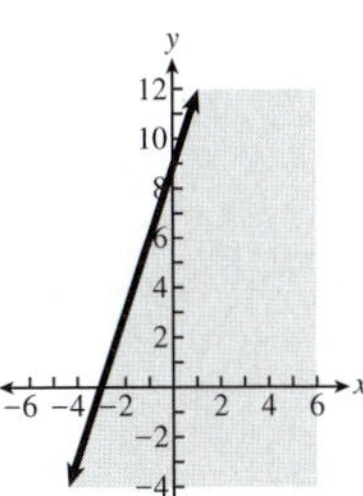

2.

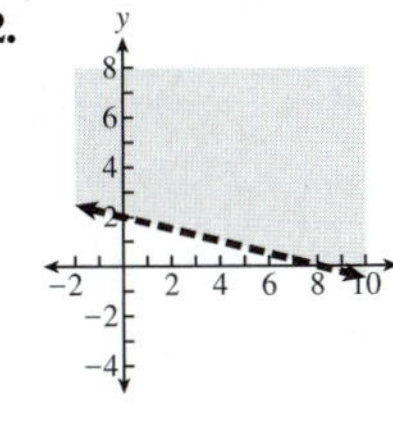

3.

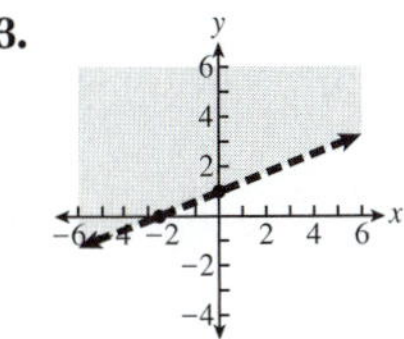

4.

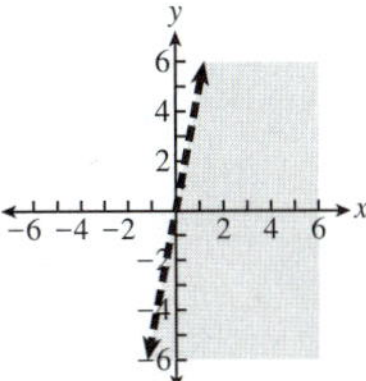

5.

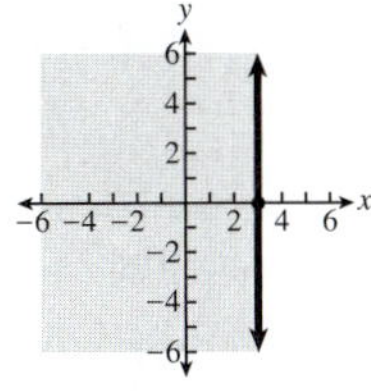

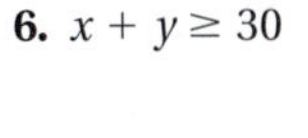

6. $x + y \geq 30$

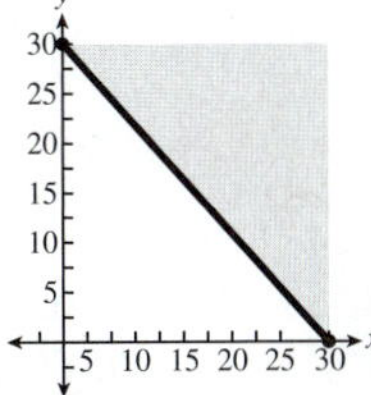

Section 2.4 **1.** solution **3.** dashed **5.** $x + y < 6$ (Answers will vary) **7. a)** Yes **b)** No **9. a)** No **b)** Yes **11. a)** Yes **b)** Yes

13.

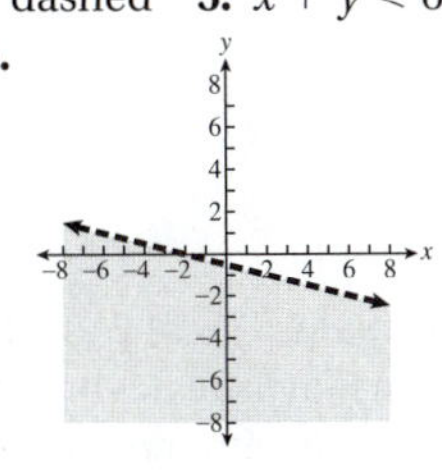

15.

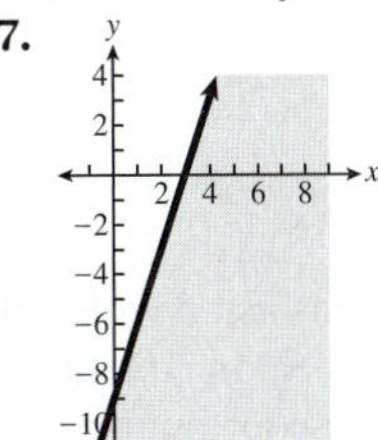

17.

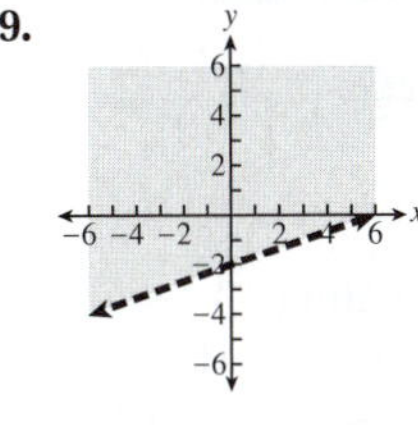

19.

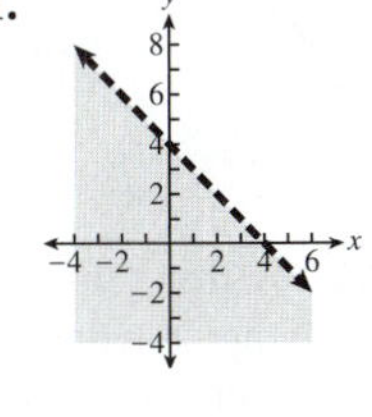

21.

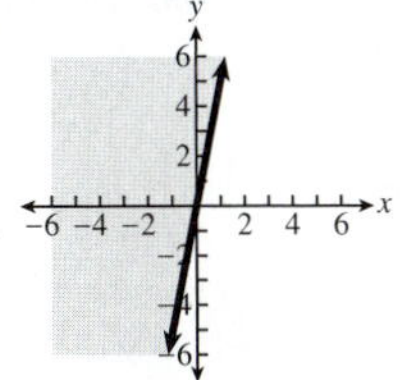

23.

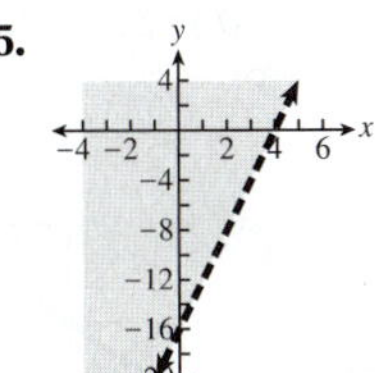

25.

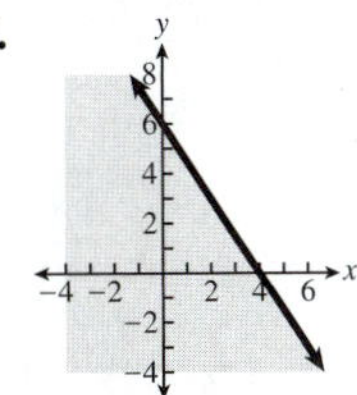

27.

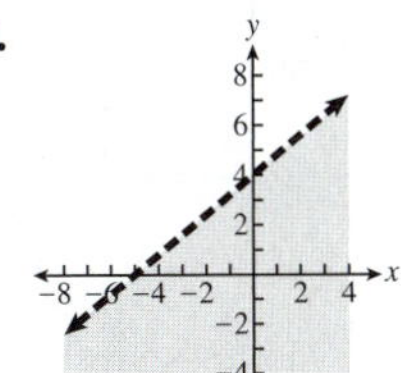

29.

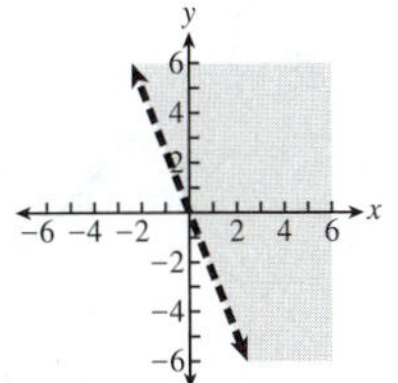

31.

33. 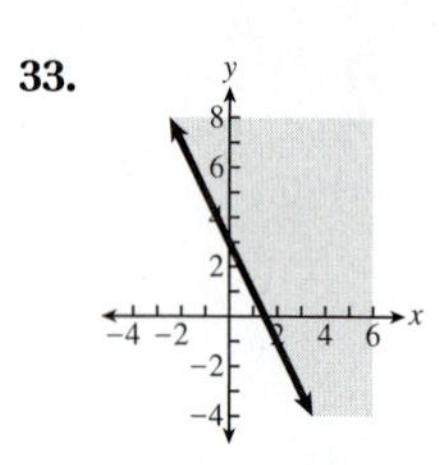 **35.** 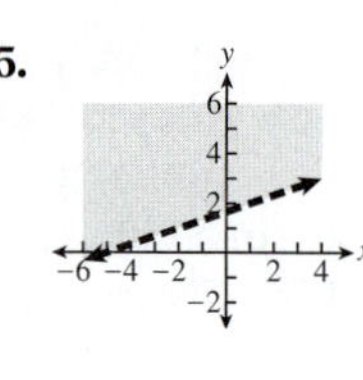 **37.** 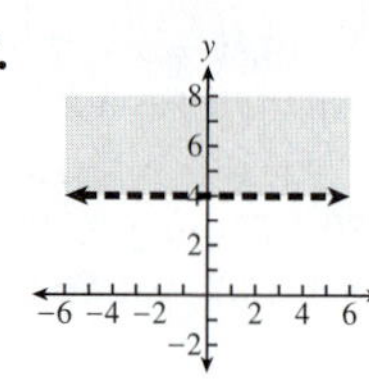 **39.** 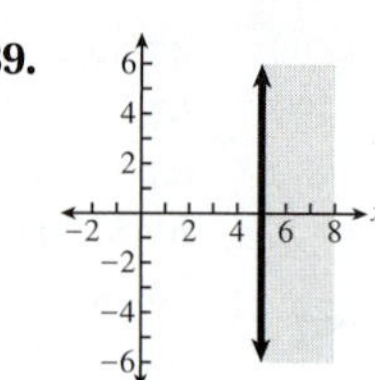 **41.** A **43.** B **45.** $\leq$ **47.** $<$

49. 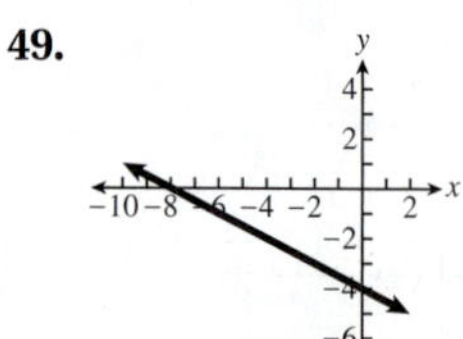 **51.** 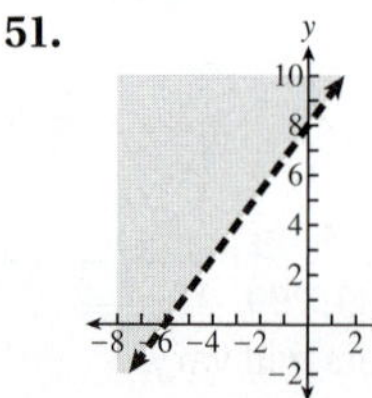 **53.** 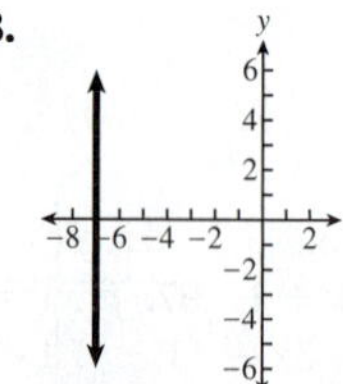 **55.** 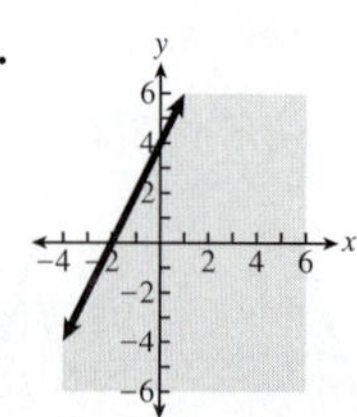 **57.** 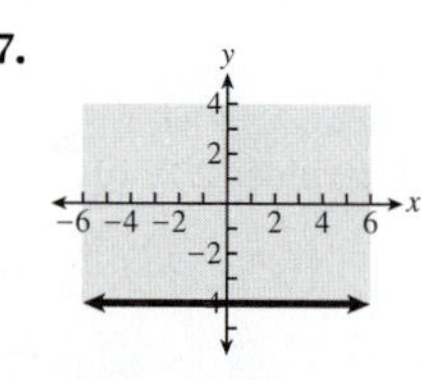

59. 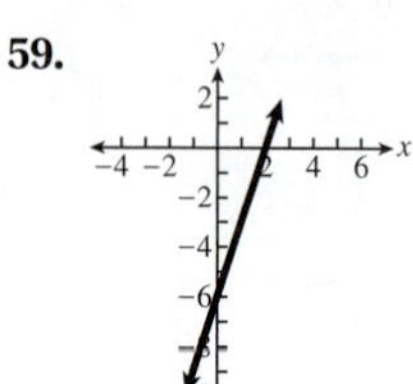 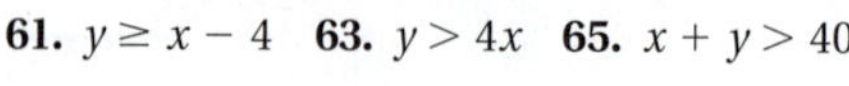 **61.** $y \geq x - 4$ **63.** $y > 4x$ **65.** $x + y > 40$ 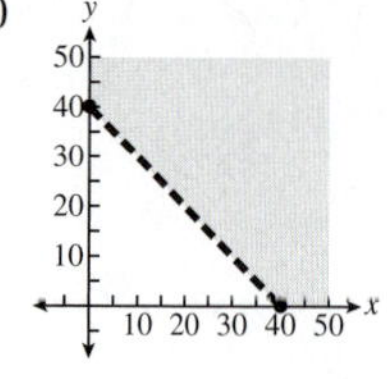 **67.** $4x + 10y \geq 200$ 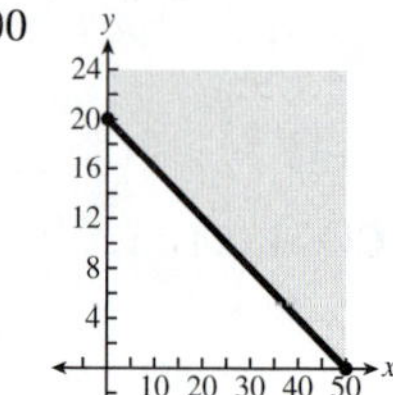

69. $3x + 7y \geq 550$ 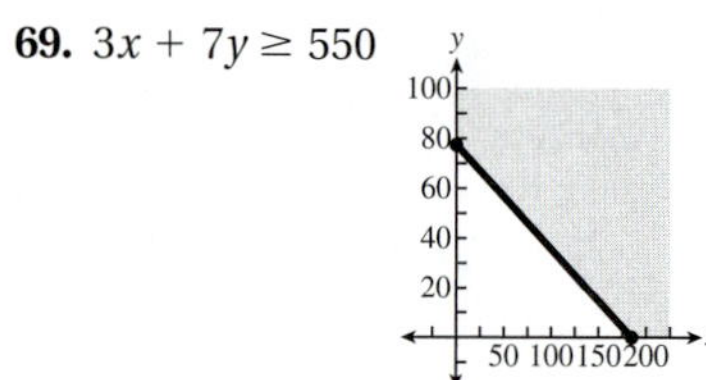 **71.** 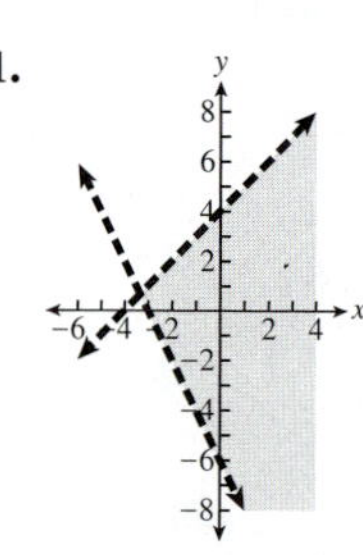 **73.** False. For some ordered pairs $3x - 8y = 24$.

Quick Check 2.5 **1.** This is not a function because more than one person could live at the same address. **2.** $39.95 + 0.79m$, Domain: Set of all possible miles, Range: Set of all possible costs **3.** -11 **4.** 35 **5.** 11 **6.** $7a + 44$
7. 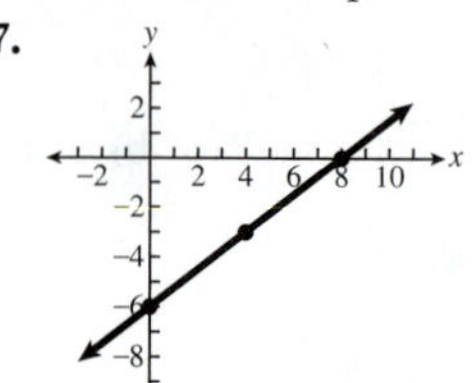 **8.** 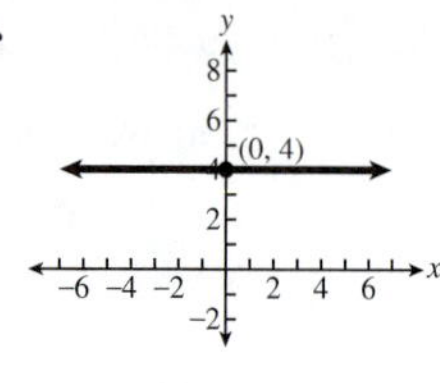

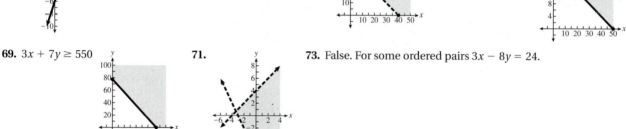

9. a) $(4, 0), (0, -8)$ **b)** 4 **c)** 7 **10.** $f(x) = -3x + 5$ **11.** $f(x) = -\frac{3}{2}x - 8$

Section 2.5 **1.** function **3.** evaluating **5.** Yes, each player is listed with only one team. **7. a)** Yes **b)** No **9. a)** Yes **b)** No **11.** No, some men have more than one child. **13.** Yes, each person can have only one favorite flavor. **15.** Yes, each student in your class has only one math instructor. **17.** Yes **19.** No, some x-coordinates are associated with more than one y-coordinate. **21.** Yes **23. a)** $F(x) = 1.8x + 32$ **b)** 32°F, 212°F, 77°F, 23°F **25. a)** $f(x) = 14x + 300$ **b)** \$2400 **27. a)** $f(x) = 18.75x + 19.50$ **b)** \$132 **29.** 2 **31.** -28 **33.** -14 **35.** -22 **37.** -4 **39.** 25.9 **41.** $9a + 2$ **43.** $5a + 12$ **45.** $-12a + 27$ **47.** $10.6n - 22$ **49.** $4a + 7$ **51.** $7x + 7h + 8$ **53.** 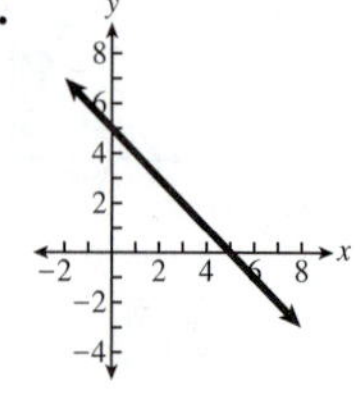 **55.** 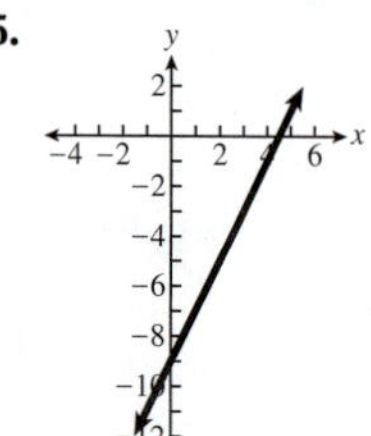 **57.** 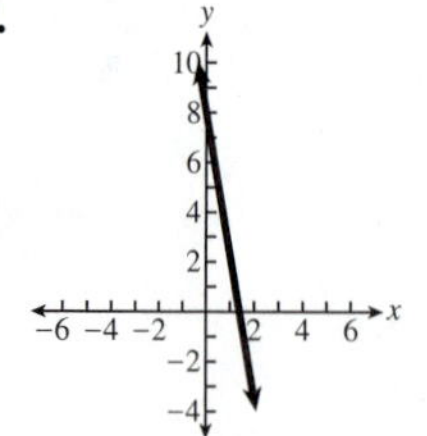 **59.** 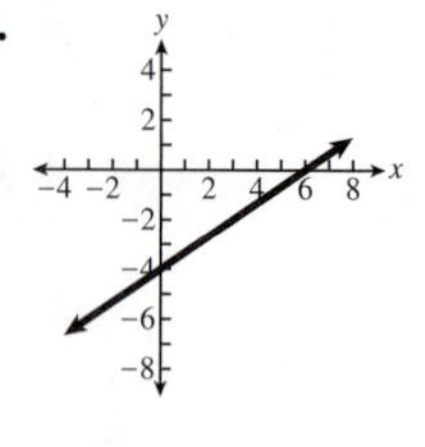

61. 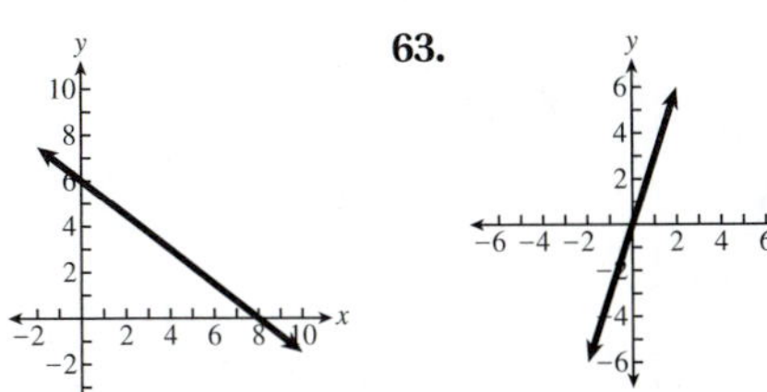 **63.** 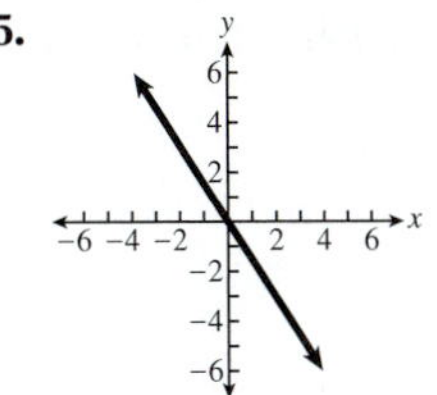 **65.** 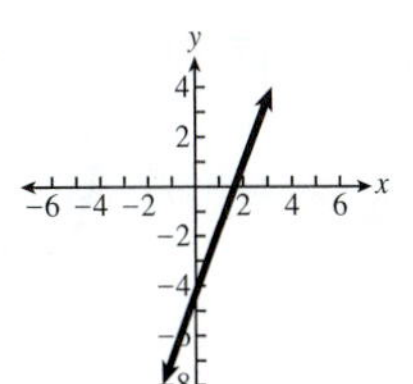 **67.** **69.** 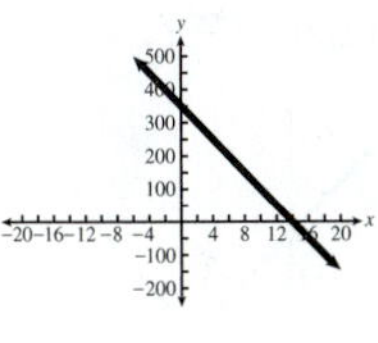

71. 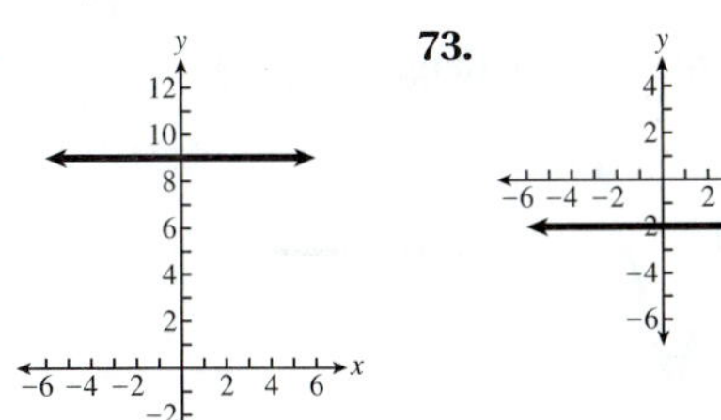 **73.** 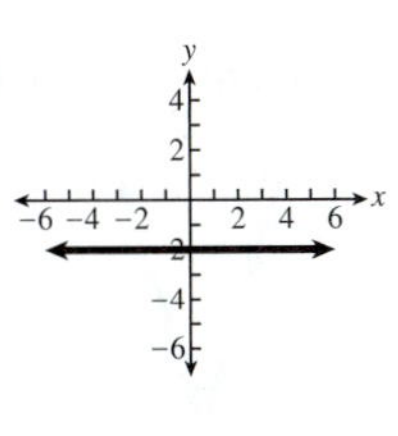

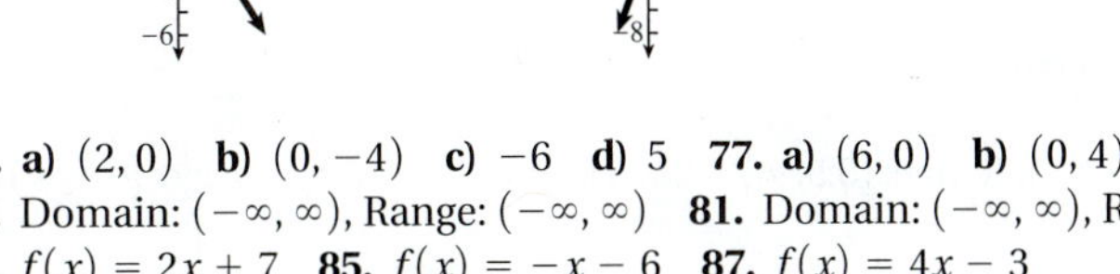

75. a) $(2, 0)$ **b)** $(0, -4)$ **c)** -6 **d)** 5 **77. a)** $(6, 0)$ **b)** $(0, 4)$ **c)** -2 **d)** -6
79. Domain: $(-\infty, \infty)$, Range: $(-\infty, \infty)$ **81.** Domain: $(-\infty, \infty)$, Range: $\{-4\}$
83. $f(x) = 2x + 7$ **85.** $f(x) = -x - 6$ **87.** $f(x) = 4x - 3$
89. $f(x) = -3x + 5$ **91.** $f(x) = \frac{2}{3}x - 8$ **93.** $f(x) = 6.4x - 11.2$
95. $f(x) = 2x + 1$ **97.** $f(x) = -x - 9$ **99.** $f(x) = \frac{4}{3}x - 6$
101. $f(x) = -6x - 11.2$ **103.** Answers will vary.

Quick Check 2.6 **1.** 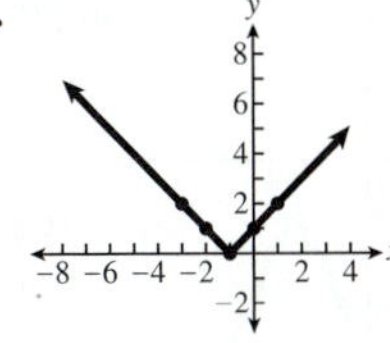 **2.** 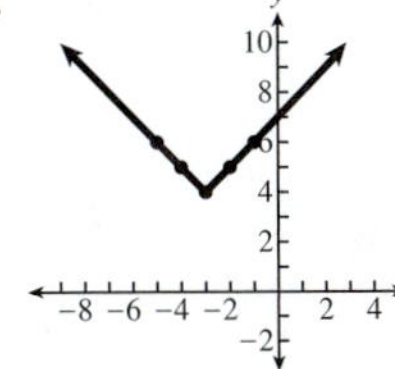 **3.** Domain: $(-\infty, \infty)$, Range: $[-2, \infty)$ 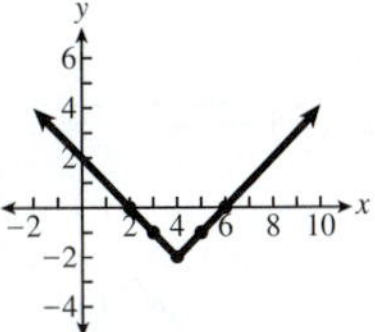

4. Domain: $(-\infty, \infty)$, Range: $[-8, \infty)$ **5.** Domain: $(-\infty, \infty)$, Range: $(-\infty, -3]$ **6.** $(4, 0), (-8, 0), (0, -4)$ **7.** $f(x) = |x - 2| - 4$
8. $f(x) = -|x - 3| + 6$

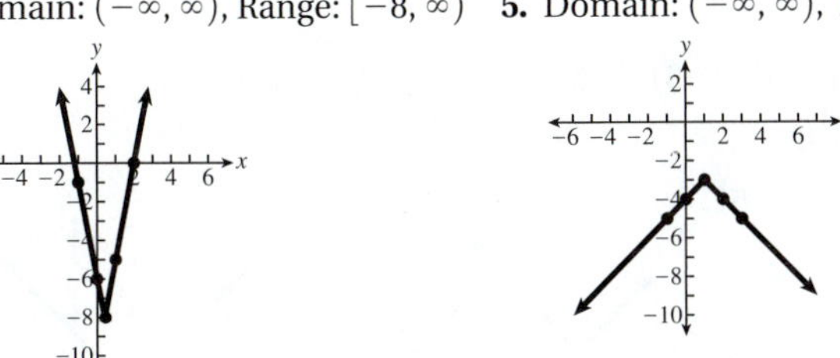

Section 2.6 **1.** absolute value function **3.** y-intercept **5.** Domain: $(-\infty, \infty)$, Range: $[0, \infty)$ **7.** Domain: $(-\infty, \infty)$, Range: $[0, \infty)$

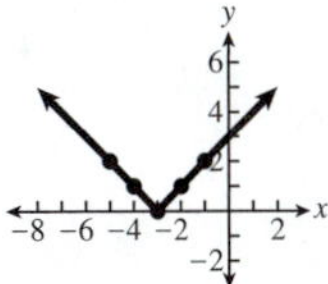

9. Domain: $(-\infty, \infty)$, Range: $[2, \infty)$ **11.** Domain: $(-\infty, \infty)$, Range: $[-3, \infty)$ **13.** Domain: $(-\infty, \infty)$, Range: $[-3, \infty)$

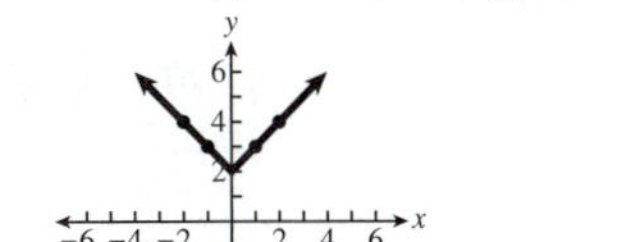

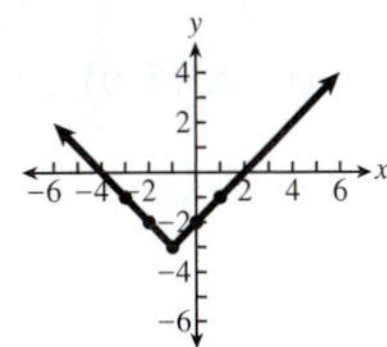

15. Domain: $(-\infty, \infty)$, Range: $[2, \infty)$

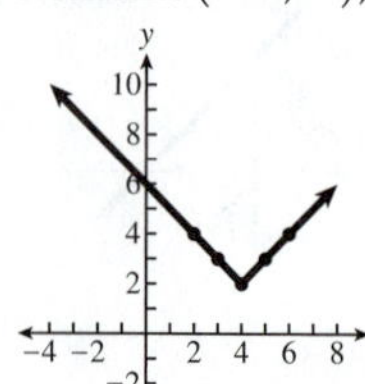

17. Domain: $(-\infty, \infty)$, Range: $[0, \infty)$

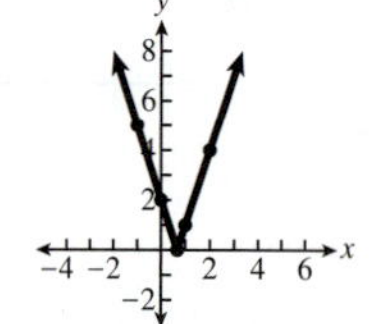

19. Domain: $(-\infty, \infty)$, Range: $[-3, \infty)$

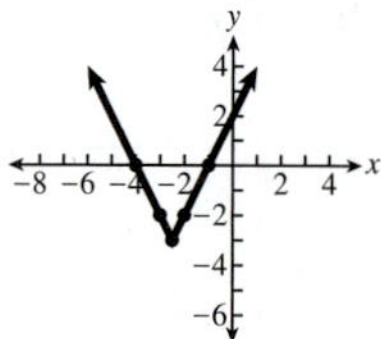

21. Domain: $(-\infty, \infty)$, Range: $[3.5, \infty)$

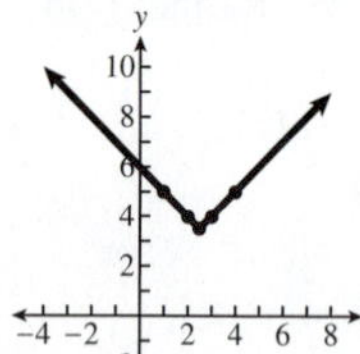

23. Domain: $(-\infty, \infty)$, Range: $[-\frac{3}{2}, \infty)$

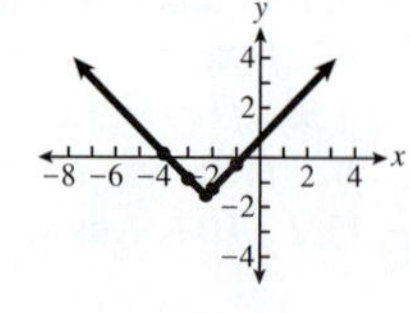

25. Domain: $(-\infty, \infty)$, Range: $[-2, \infty)$

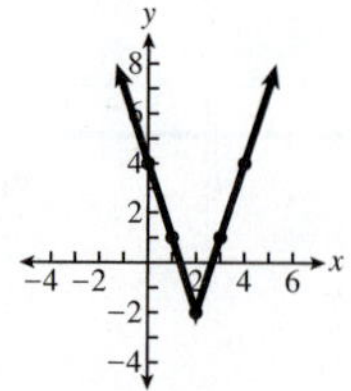

27. Domain: $(-\infty, \infty)$, Range: $(-\infty, 0]$

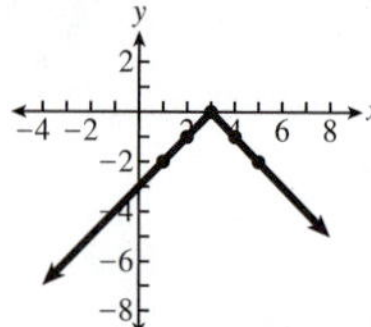

29. Domain: $(-\infty, \infty)$, Range: $(-\infty, -7]$

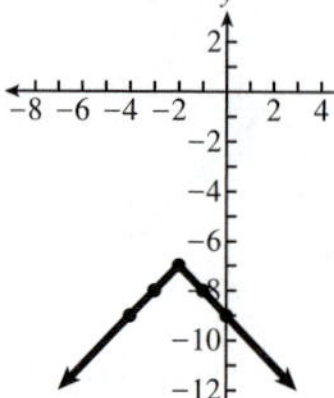

31. Domain: $(-\infty, \infty)$, Range: $(-\infty, -2]$

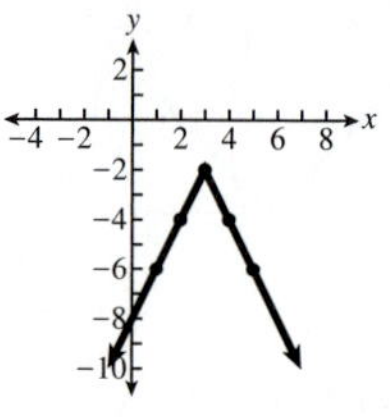

33. $(-9, 0), (-5, 0), (0, 5)$ **35.** $(6, 0), (0, 6)$ **37.** $(0, 5)$ **39.** $[1, 7]$ **41.** $(-\infty, -5) \cup (5, \infty)$ **43.** $f(x) = |x + 4|$
45. $f(x) = |x + 3| - 2$ **47.** $f(x) = |x - 5| - 3$ **49.** $f(x) = -|x + 2| + 4$

51.

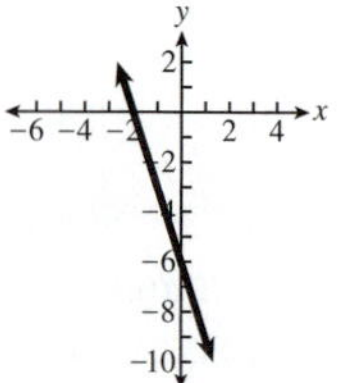

53.

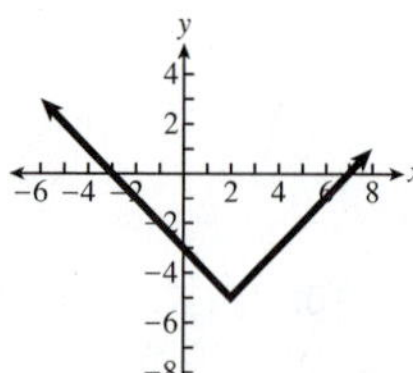

55.

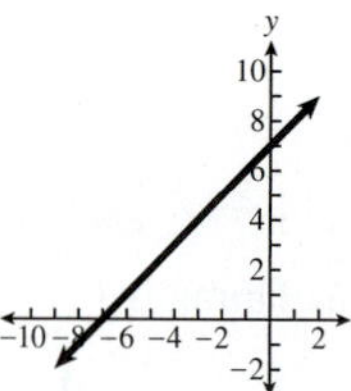

57.

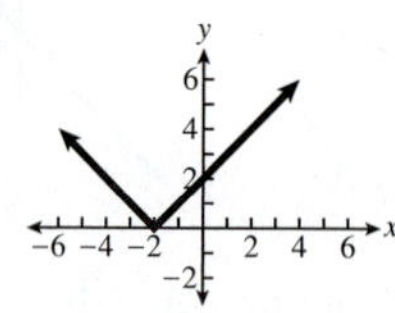

59.

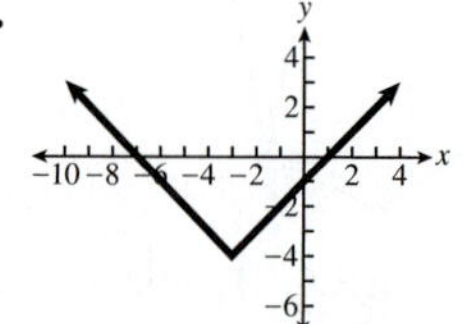

61.

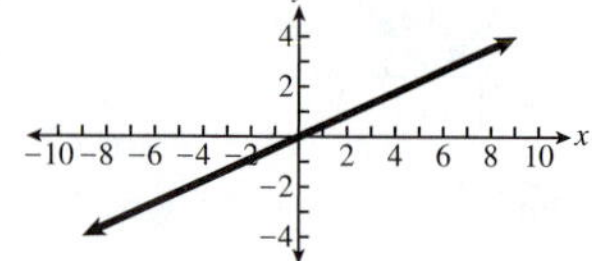

63. Answers will vary.

Chapter 2 Review **1.** $(-12, 0), (0, 8)$ **2.** $(6, 0), (0, \frac{9}{2})$ **3.** $(0, 0)$ **4.** $(\frac{4}{3}, 0), (0, -4)$ **5.** $(6, 0), (0, 8)$

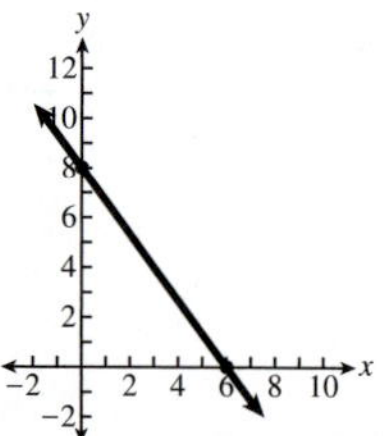

6. $(-6, 0), (0, 3)$ **7.** $(0, 0), (0, 0)$ **8.** $(4, 0), (0, -6)$ **9.**

10.

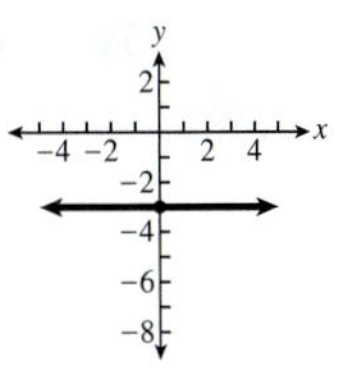

11.

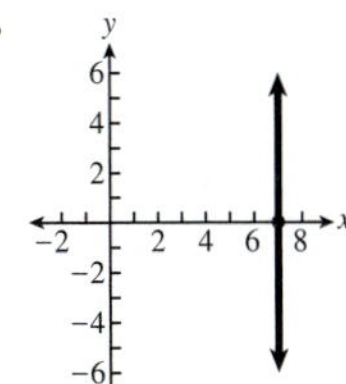

12.

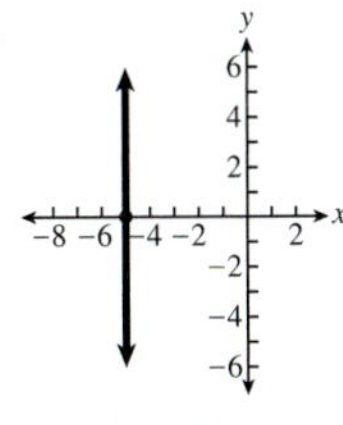

13. 2 **14.** -3 **15.** $-\frac{3}{4}$ **16.** $\frac{5}{2}$ **17.** $-3, (0, 2)$ **18.** $\frac{2}{5}, (0, -9)$ **19.** $4, (0, 5)$
20. $-\frac{7}{3}, (0, 4)$

21.

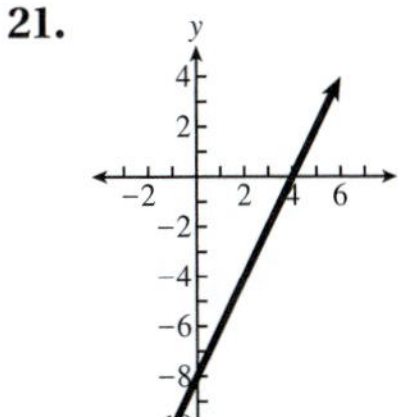

22.

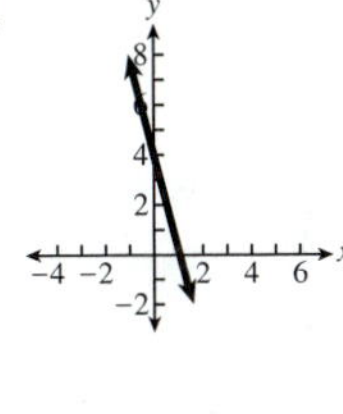

23.

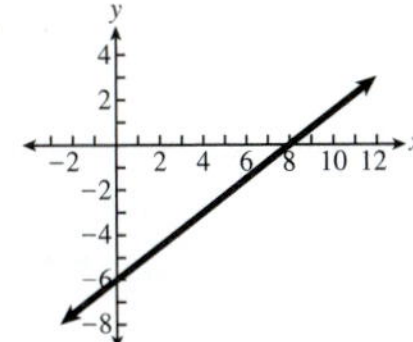

24.

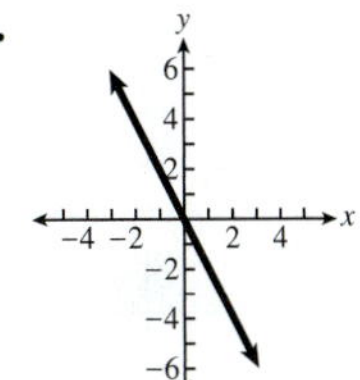

25.

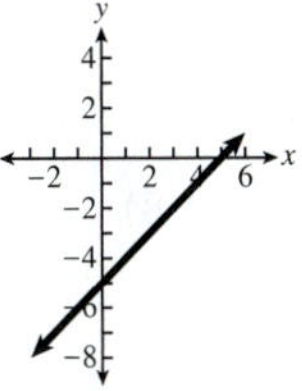

26.

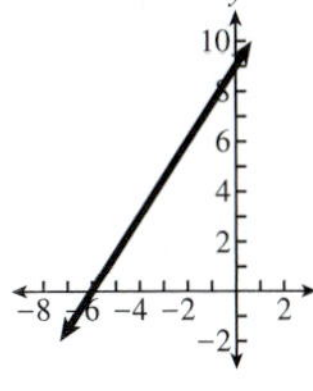

27.

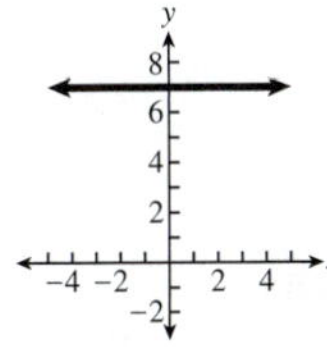

28.

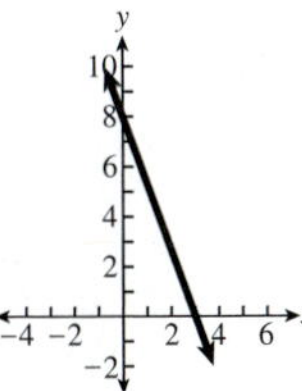

29. **30.**

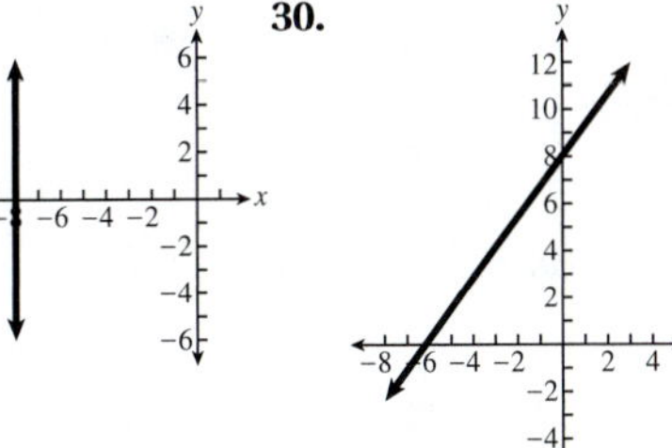

31. Neither **32.** Perpendicular **33.** Parallel **34.** $y = -3x + 6$ **35.** $y = \frac{4}{3}x - 8$ **36.** $y = 3x - 5$ **37.** $y = -2x - 13$
38. $y = \frac{3}{5}x - 14$ **39.** $y = -2x + 4$ **40.** $y = \frac{4}{3}x + 9$ **41.** $y = -\frac{1}{2}x$ **42.** $y = 4x - 3$ **43.** $y = -\frac{3}{2}x + 5$ **44.** $y = -3x + 9$
45. $y = \frac{3}{4}x + 7$

46.

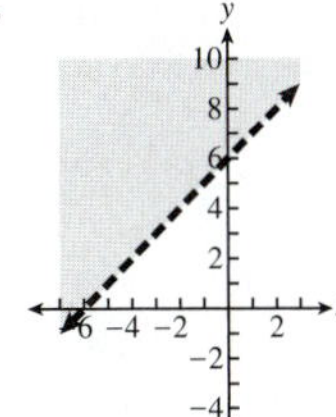

47.

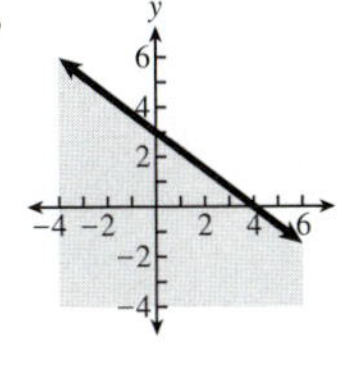

48.

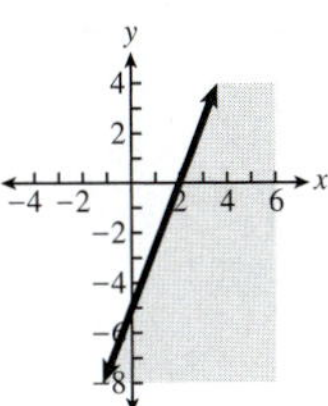

49.

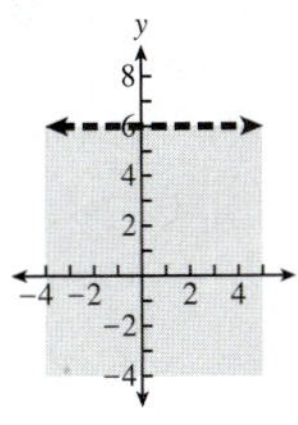

50. a) $f(x) = 50{,}000x + 3{,}000{,}000$
b) \$4,350,000 **c)** 40 games **51.** -15
52. 51 **53.** $15a + 9$

54.

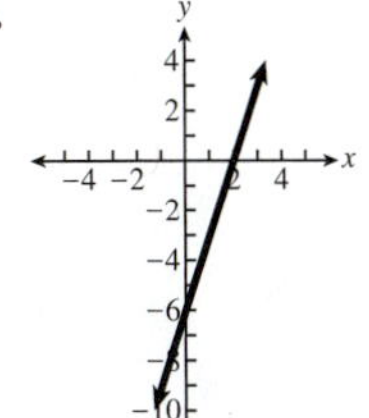

55.

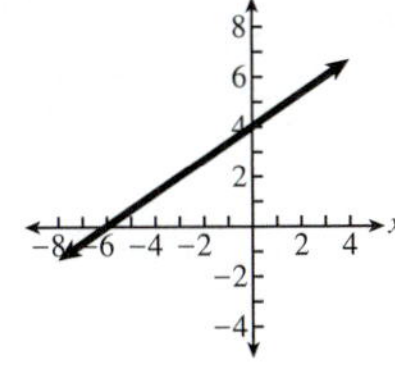

56.

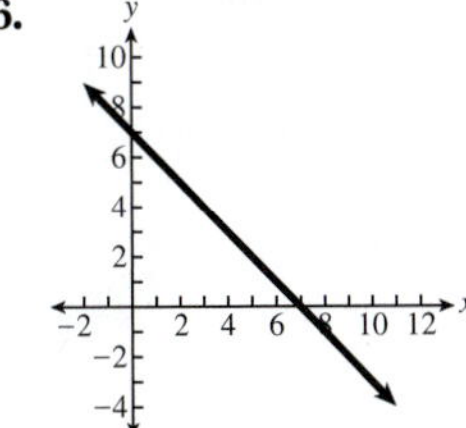

57.

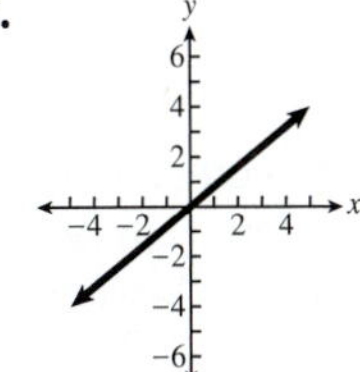

58. a) -6 **b)** 9 **c)** $(-\infty, \infty)$
d) $(-\infty, \infty)$
59. $f(x) = -3x + 13$
60. $f(x) = \frac{2}{3}x - 23$
61. $f(x) = 4x + 3$
62. $f(x) = -\frac{3}{4}x - 8$

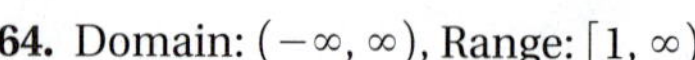

63. Domain: $(-\infty, \infty)$, Range: $[0, \infty)$

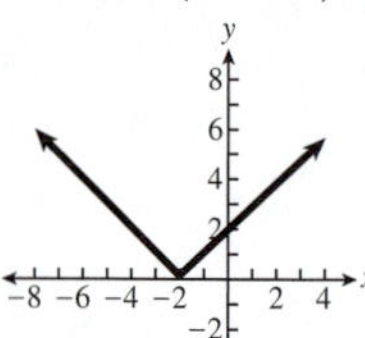

64. Domain: $(-\infty, \infty)$, Range: $[1, \infty)$

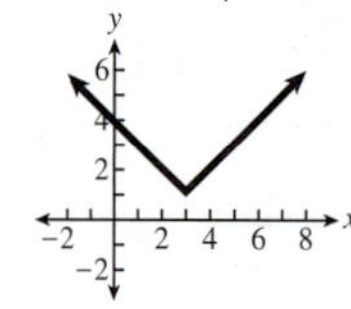

65. Domain: $(-\infty, \infty)$, Range: $[-5, \infty)$

66. Domain: $(-\infty, \infty)$, Range: $(-\infty, 4]$

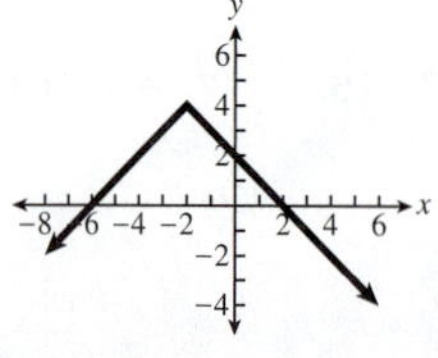

67. $f(x) = |x + 3| + 5$

68. $f(x) = |x - 2| - 3$

Chapter 2 Review Exercises: Worked-Out Solutions

5.

x-intercept	y-intercept
$4x + 3(0) = 24$	$4(0) + 3y = 24$
$4x = 24$	$3y = 24$
$x = 6$	$y = 8$
$(6, 0)$	$(0, 8)$

9. This is a horizontal line, and its y-intercept is $(0, 4)$.

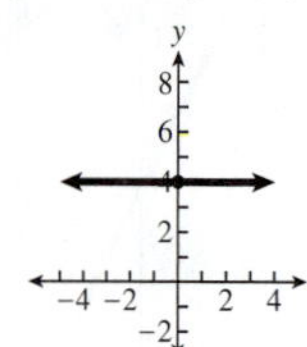

11. This is a vertical line, and its x-intercept is $(7, 0)$.

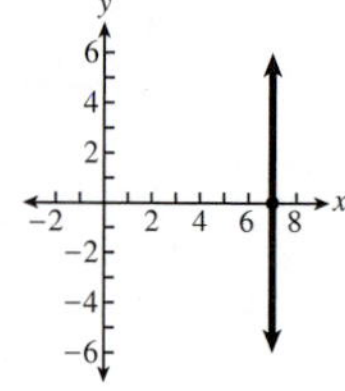

14. $(-2, 7)$ and $(3, -8)$

$$m = \frac{-8 - 7}{3 - (-2)}$$
$$m = \frac{-15}{5}$$
$$m = -3$$

19. $8x - 2y = 10$

$$-2y = -8x + 10$$
$$\frac{-2y}{-2} = \frac{-8x}{-2} + \frac{10}{-2}$$
$$y = 4x - 5$$

$m = 4$, y-intercept $(0, -5)$

21. The line has slope $m = 2$ and its y-intercept is $(0, -8)$. Plot the y-intercept. A second point on the line can be found by moving up 2 units from the y-intercept and 1 unit to the right.

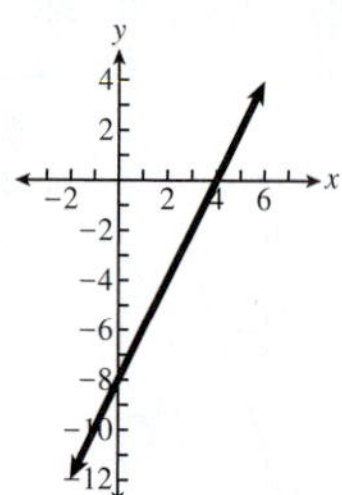

32. Find the slope of the first equation.

$$3x + 2y = 8$$
$$2y = -3x + 8$$
$$y = -\tfrac{3}{2}x + 4$$

The slope of the first line is $-\frac{3}{2}$. The equation of the second line is already in slope-intercept form, and its slope is $\frac{2}{3}$. Since the two slopes are negative reciprocals, the two lines are perpendicular.

34. Substitute -3 for m and 6 for b in the slope-intercept form of a line. The equation is $y = -3x + 6$.

36. Using the point-slope form of a line, substitute 3 for m, 4 for x_1, and 7 for y_1.

$$y - y_1 = m(x - x_1)$$
$$y - 7 = 3(x - 4)$$
$$y - 7 = 3x - 12$$
$$y = 3x - 5$$

39. Find the slope of the line that passes through the points $(-1, 6)$ and $(3, -2)$.

$m = \dfrac{-2 - 6}{3 - (-1)}$

$m = \dfrac{-8}{4}$

$m = -2$

Using the point-slope form of a line, substitute -2 for m, -1 for x_1, and 6 for y_1.

$y - y_1 = m(x - x_1)$
$y - 6 = -2(x - (-1))$
$y - 6 = -2(x + 1)$
$y - 6 = -2x - 2$
$y = -2x + 4$

44. Since the line is parallel to $y = -3x + 10$, its slope is -3.

$y - y_1 = m(x - x_1)$
$y - (-6) = -3(x - 5)$
$y + 6 = -3x + 15$
$y = -3x + 9$

46. Graph the line $y = x + 6$ using a dashed line. This line has a slope of 1 and its y-intercept is $(0, 6)$. The point $(0, 0)$ is not on this line and can be used as a test point.

$y > x + 6$
$0 > 0 + 6$
$0 > 6$

This statement is false, so $(0, 0)$ is not a solution and we shade the half-plane that does not contain $(0, 0)$.

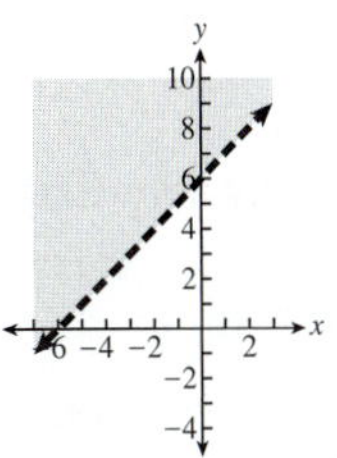

51. $f(x) = 4x - 27$
$f(3) = 4(3) - 27$
$f(3) = 12 - 27$
$f(3) = -15$

55. The y-intercept is $(0, 4)$ and the slope is $\frac{2}{3}$. Plot the y-intercept. A second point on the line is 2 units above and 3 units to the right of this point.

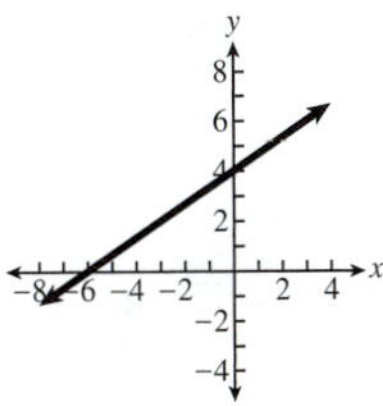

59. $f(x) = -3x + b$
$f(4) = 1$
$-3(4) + b = 1$
$-12 + b = 1$
$b = 13$
$f(x) = -3x + 13$

65. Select values of x centered around $x = 1$.

x	$f(x) = \|x - 1\| - 5$
-1	$f(-1) = \|-1 - 1\| - 5 = \|-2\| - 5 = 2 - 5 = -3$
0	$f(0) = \|0 - 1\| - 5 = \|-1\| - 5 = 1 - 5 = -4$
1	$f(1) = \|1 - 1\| - 5 = \|0\| - 5 = 0 - 5 = -5$
2	$f(2) = \|2 - 1\| - 5 = \|1\| - 5 = 1 - 5 = -4$
3	$f(3) = \|3 - 1\| - 5 = \|2\| - 5 = 2 - 5 = -3$

Plot $(-1, -3)$, $(0, -4)$, $(1, -5)$, $(2, -4)$, and $(3, -3)$ and draw a V-shaped graph through these points.

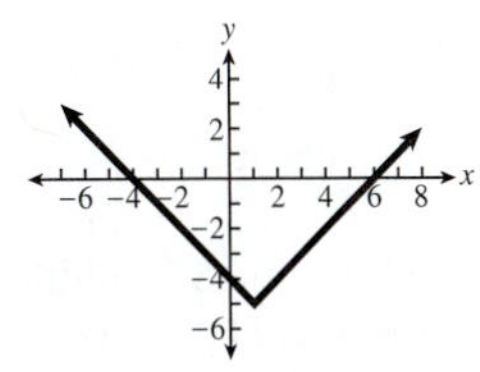

The domain of this function is $(-\infty, \infty)$. The function's minimum value is -5, so its range is $[-5, \infty)$.

Chapter 2 Test **1.** $(8, 0), (0, -10)$ **2.** $(-4, 0), (0, 8)$ **3.** $(3, 0), (0, \frac{15}{2})$ **4.** $-\frac{3}{2}$ **5.** $-\frac{3}{4}, (0, 10)$ **6.** $\frac{3}{5}, (0, -9)$

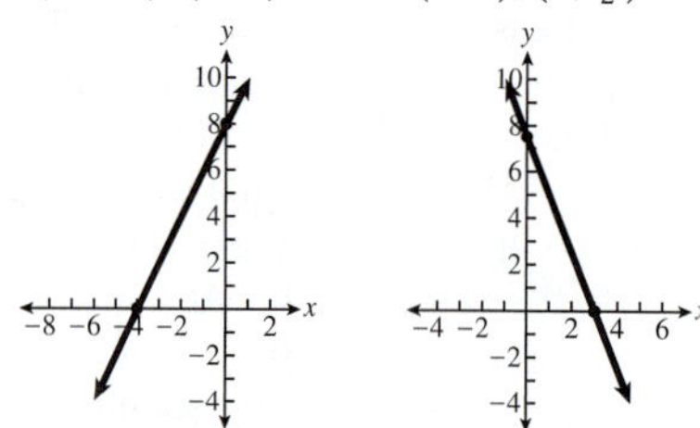

7.

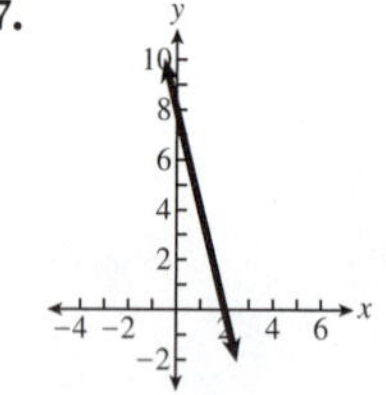

8.

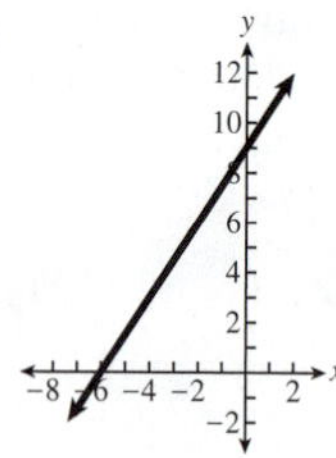

9.

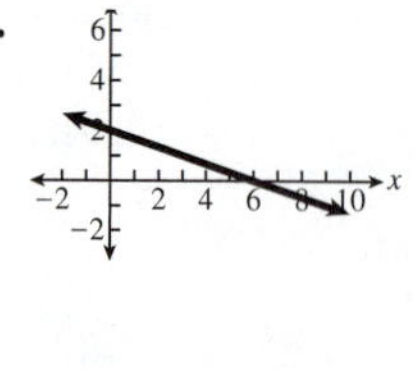

10.

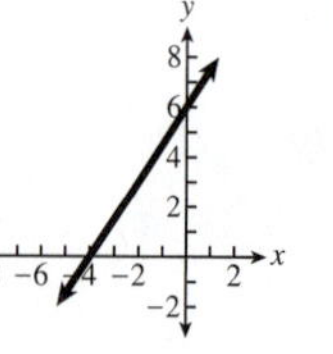

11.

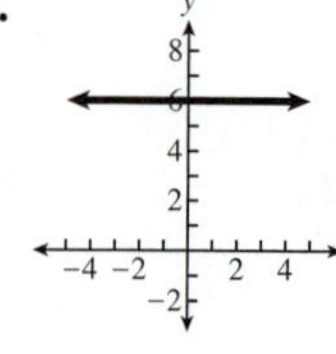

12. Perpendicular **13.** $y = -7x + 12$ **14.** $y = \frac{4}{3}x - 5$ **15.** $y = -4x + 11$ **16.** -52

17.

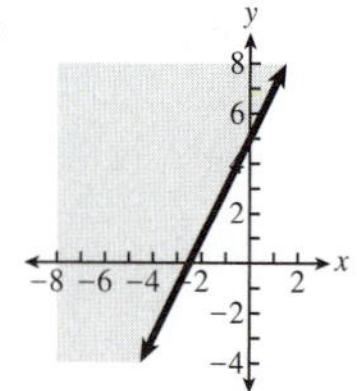

18.

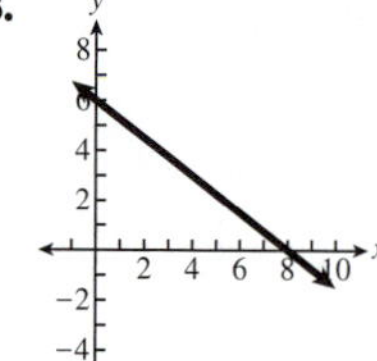

19. Domain: $(-\infty, \infty)$, Range: $[-1, \infty)$ **20.** $f(x) = |x - 4| - 3$

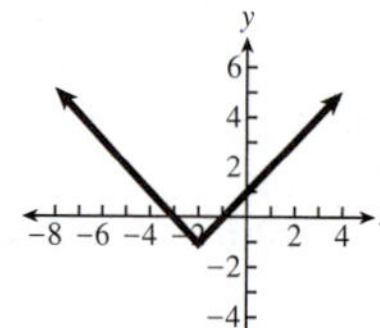

Chapter 3

Quick Check 3.1 **1.** Yes **2.** $(4, 6)$ **3.** $\varnothing$ **4.** $(x, 3x - 12)$ **5.** $(4, 10)$ **6.** $(-2, 4)$ **7.** $(5, -3)$ **8.** $(x, -2x + 3)$ **9.** $(7, 1)$ **10.** $(9, -4)$ **11.** $(2900, 1500)$ **12.** $(-4, 9)$ **13.** $(8, -6)$ **14.** $\varnothing$

Section 3.1 **1.** System of linear equations **3.** independent **5.** dependent **7.** d **9.** Yes **11.** No **13.** Yes **15.** $(-3, 3)$ **17.** $(-6, 8)$ **19.** $\varnothing$ **21.** $(5, 8)$ **23.** $(4, 5)$ **25.** $(x, \frac{5}{3}x - 8)$ **27.** $(-15, -2)$ **29.** $(9, 21)$ **31.** $(7, 4)$ **33.** $(x, x - 8)$ **35.** $(4, 2)$ **37.** $(9, 7)$ **39.** $(7, -2)$ **41.** $(3, -5)$ **43.** $(2, 5)$ **45.** $(3, 2)$ **47.** $(4, -1)$ **49.** $(x, 3x - 4)$ **51.** $(8, 2)$ **53.** $(5, -\frac{1}{2})$ **55.** $(-4, -4)$ **57.** $\varnothing$ **59.** $(4, -6)$ **61.** $(60, 20)$ **63.** $(20, 8)$ **65.** Answers will vary; the following is a correct example.
67. Answers will vary; the following is a correct example.

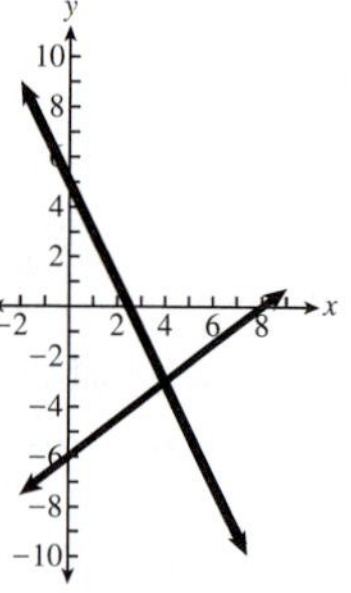

69. Answers will vary. **71.** Answers will vary.

Quick Check 3.2 **1.** 72 and 52 **2.** Jessica owns 95 books and Vanessa owns 65 books. **3.** There were 27 adults and 57 children. **4.** length: 25 feet, width: 10 feet **5.** length: 30 inches, width: 10 inches **6.** \$1750 at 5%, \$1050 at 3% **7.** 12 pounds of pecans and 24 pounds of cashews **8.** 120 ml of the 75% solution and 480 ml of the 55% solution **9.** $1\frac{1}{2}$ hours **10.** plane: 450 mph, wind: 50 mph

Section 3.2 **1.** $P = 2L + 2W$ **3.** $d = r \cdot t$ **5.** northern: 750, southern: 450 **7.** Don: 44, Rose: 13 **9.** 58, 37 **11.** 13 multiple choice, 12 true/false **13.** 7 **15.** adults: 95, children: 63 **17.** pizza: \$14, soda: \$6 **19.** coffee: \$1.90, latte: \$3.70 **21.** 37°, 53° **23.** 32°, 148° **25.** length: 22 inches, width: 18 inches **27.** length: 31.2 inches, width: 19.5 inches **29.** length: 62 inches, width: 38 inches **31.** length: 400 feet, width: 180 feet **33.** length: 45 feet, width: 30 feet **35.** \$800 at 7%, \$1200 at 8% **37.** \$1500 at 5%, \$700 at 3% **39.** \$25,000 at 4.5%, \$7000 at 3.75% **41.** \$4000 at 7% profit, \$8000 at 20% loss **43.** Kona: 5 pounds, South American: 15 pounds **45.** dried fruit: 33.6 pounds, mixed nuts: 16.4 pounds **47.** 110 ml of 60%, 330 ml of 36% **49.** 7.5 ounces of 20%, 2.5 ounces of 60% **51.** 1.25 liters of rum, 0.75 liters of cola **53.** Fred: $5\frac{1}{2}$ hours, Wilma: $2\frac{1}{2}$ hours **55.** after 2.5 hours **57.** kayak: 4 mph, current: 1 mph **59.** plane: 450 mph, wind: 50 mph **61.** Answers will vary. **63.** Answers will vary.

Quick Review Exercises 3.2 **1.**

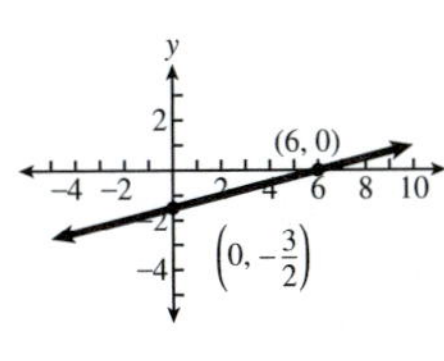

2.

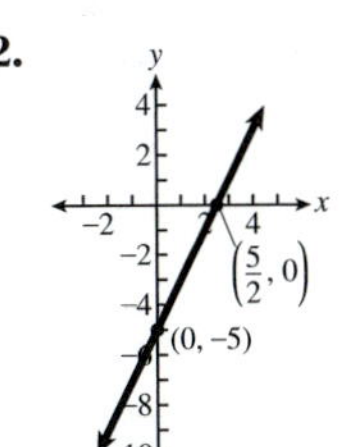

3.

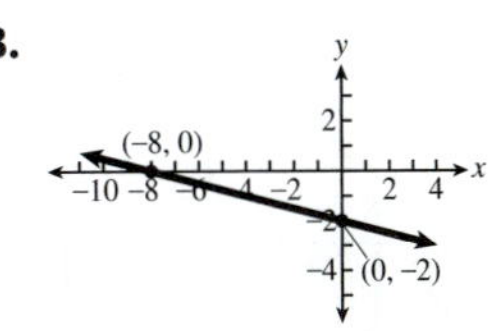

4.

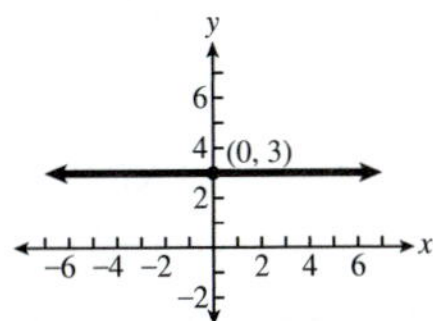

Quick Check 3.3 **1.**

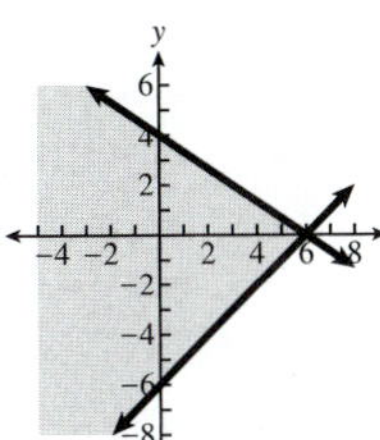

2.

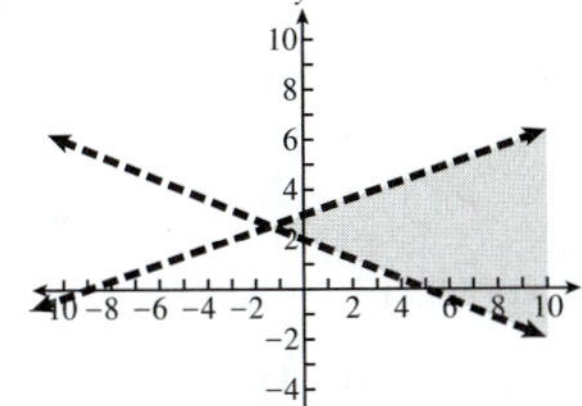

3.

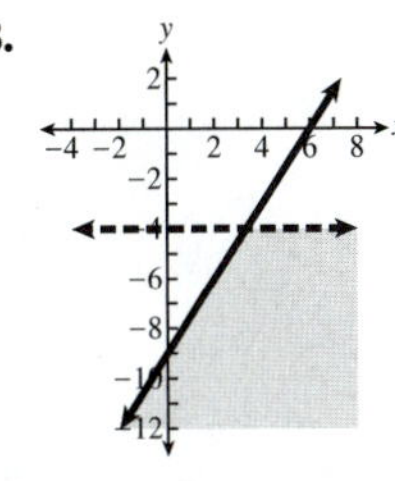

4.

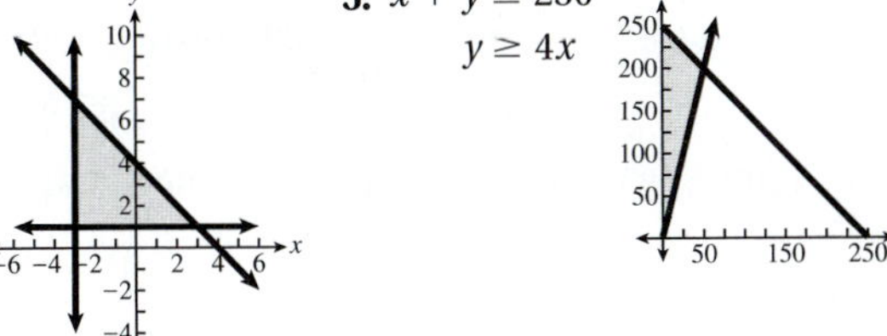

5. $x + y \le 250$
$y \ge 4x$

Section 3.3 **1.** system of linear inequalities **3.** solid

5.

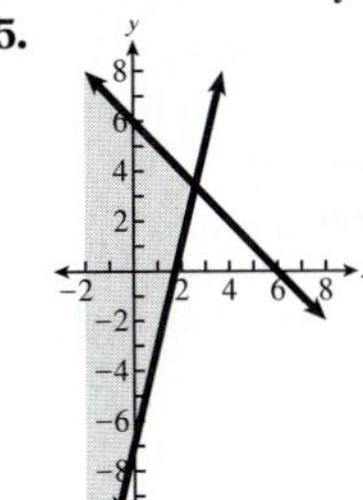

7.

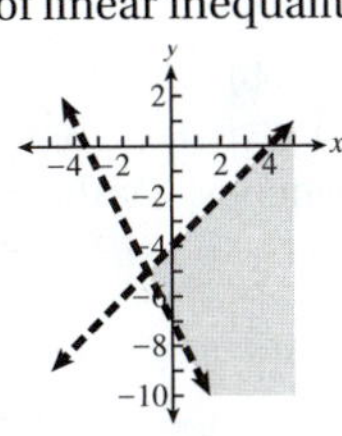

9.

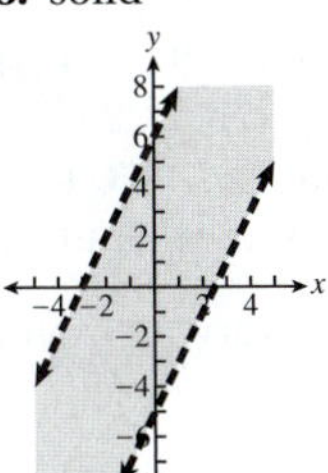

11.

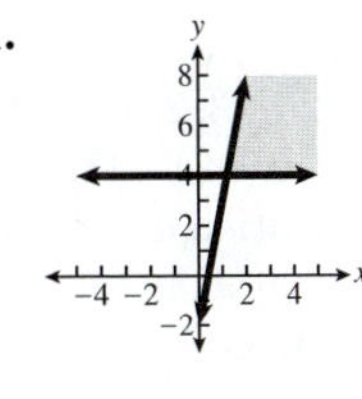

13.

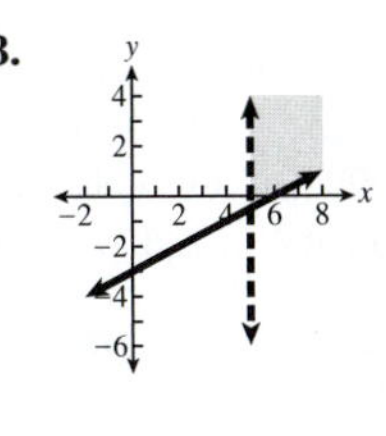

15.

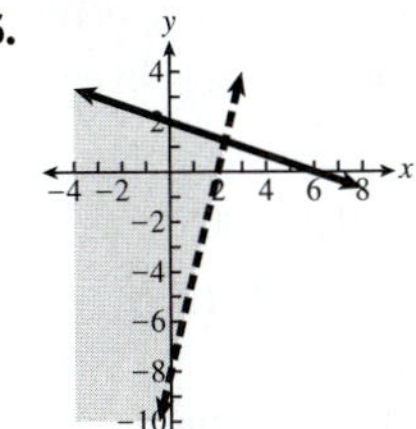

17.

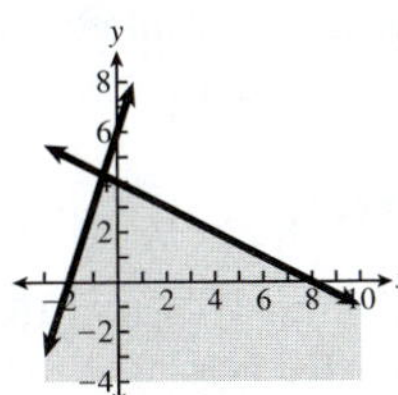

19.

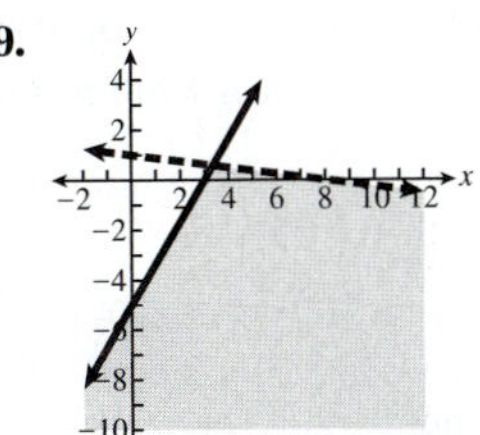

21.

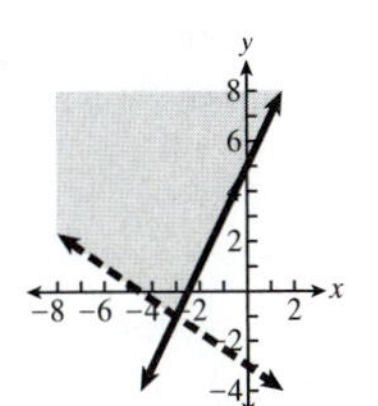

23.

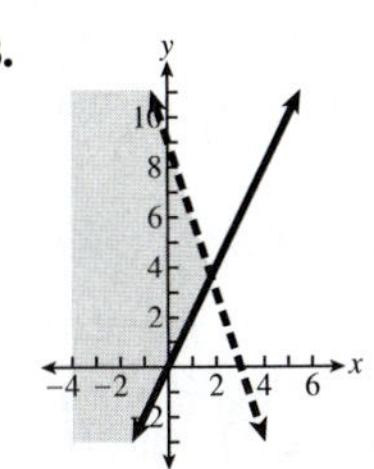

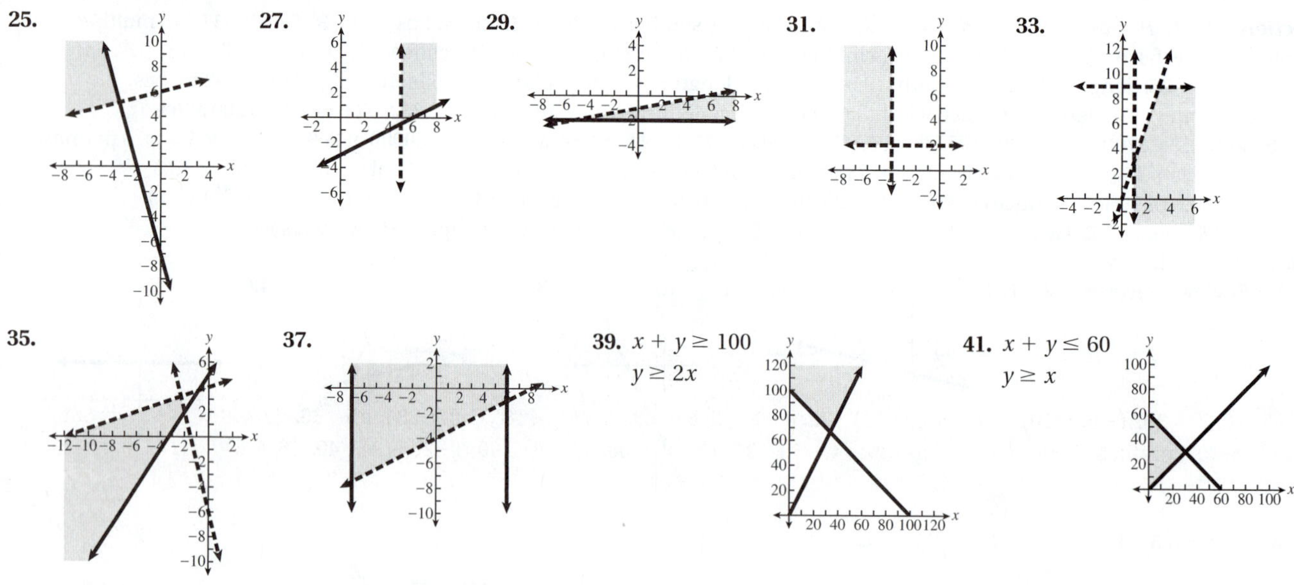

43. $\begin{aligned} x &< 0 \\ y &> 0 \end{aligned}$ **45.** 8 square units

Quick Check 3.4 **1.** Yes **2.** No **3.** $(1, 2, -3)$ **4.** $(\frac{1}{2}, -4, 5)$ **5.** $(4, -5, 3)$ **6.** 16 ones, 9 fives, 5 tens

Section 3.4 **1.** ordered triple **3.** No **5.** Yes **7.** Yes **9.** $(3, -2, 0)$ **11.** $(2, 2, -3)$ **13.** $(4, \frac{1}{2}, -1)$ **15.** $(-30, 20, 5)$ **17.** $(-8, -4, 2)$ **19.** $(2, -8, -5)$ **21.** $(\frac{3}{4}, -\frac{9}{2}, 1)$ **23.** $(6, -3, 1)$ **25.** $(-7, -4, 23)$ **27.** $(-4, -4, 6)$ **29.** $(5, 4, 6)$ **31.** (6000, 3000, 1000) **33.** $(1, 2, 3)$ **35.** $(1, -5, -8)$ **37.** Answers will vary; one solution is $\begin{aligned} x + y + z &= 6 \\ 2x + 3y - z &= 5. \\ 4x - 2y + z &= 3 \end{aligned}$ **39.** Answers will vary; one solution is $\begin{aligned} 3x + 2y + z &= 23 \\ x - 3y - 10z &= 49. \\ 2x - 3y - 4z &= 34 \end{aligned}$ **41.** 19 **43.** 23 **45.** 5 **47.** 52 **49.** A: 60°, B: 40°, C: 80° **51.** \$8000 at 3%, \$20,000 at 5%, \$12,000 at 6% **53.** Answers will vary.

Quick Check 3.5 **1.** $(-2, 5)$ **2.** $(-3, 2)$ **3.** $(-8, -1)$ **4.** $(3, -5, -1)$ **5.** length: 13 inches, width: 8 inches. **6.** \$2500 at 3%, \$3500 at 4%, \$4000 at 6%

Section 3.5 **1.** matrix **3.** rows, columns **5.** upper triangular **7.** $(6, 1)$ **9.** $(-4, 3)$ **11.** $(7, 6)$ **13.** $(5, -4)$ **15.** $(-4, 6)$ **17.** $(\frac{3}{2}, -\frac{2}{3})$ **19.** $(2, 3)$ **21.** $(-3, 7)$ **23.** $(0, 6)$ **25.** $(3, 2, 1)$ **27.** $(-1, -2, 6)$ **29.** $(5, 4, -1)$ **31.** $(0, -6, 1)$ **33.** $(-6, 3, -5)$ **35.** $(12, 9, -4)$ **37.** (5000, 4000, 1000) **39.** $(3, 2, 4, -2)$ **41.** $(9, 7, -3, -6)$ **43.** length: 24 inches, width: 13 inches **45.** adults: 17, children: 46 **47.** \$1200 at 4%, \$2800 at 15% **49.** dried fruit: 25 pounds, mixed nuts: 15 pounds **51.** 67 **53.** \$4000 at a 10% profit, \$7000 at a 3% loss, \$21,000 at a 2% loss **55.** $\begin{aligned} 2x + 2y &= 48 \\ x - 2y &= 3 \end{aligned}$. Answers will vary, an example is "A rectangle has a perimeter of 48 inches, and the length is 3 inches more than twice the width. Find the dimensions of the rectangle." **57.** $\begin{aligned} x + y + z &= 6000 \\ 0.04x + 0.05y + 0.06z &= 290. \\ x - y &= -1000 \end{aligned}$ Answers will vary, an example is "Curt invested a total of \$6000 in three accounts that paid annual interest of 4%, 5%, and 6%, respectively. In the first year he earned \$290 in interest. If he invested \$1000 less at 4% interest than he invested at 5%, how much did he invest in each account?"

Quick Check 3.6 **1.** 41 **2.** 47 **3.** $(6, 4)$ **4.** $(-3, \frac{3}{2})$ **5.** 63 **6.** $(9, -2, -4)$ **7.** $(-2, 1, 2)$ **8.** 15 \$48 items and 28 \$20 items

Section 3.6 **1.** square matrix **3.** Cramer's rule **5.** 11 **7.** 68 **9.** -10 **11.** $(2, 3)$ **13.** $(-2, 8)$ **15.** $(-9, -6)$ **17.** $(0, 6)$ **19.** $(-13, 20)$ **21.** $(\frac{7}{4}, \frac{3}{2})$ **23.** $(-\frac{8}{3}, \frac{13}{4})$ **25.** (3000, 2000) **27.** $(-4.3, 2.9)$ **29.** -5 **31.** 115 **33.** 169 **35.** $(2, 1, 7)$ **37.** $(-4, -2, 1)$ **39.** $(0, -4, 6)$ **41.** $(\frac{1}{2}, -2, \frac{1}{4})$ **43.** (2000, 3000, 1000) **45.** $(6, -4, 3)$ **47.** 13 square units **49.** 44 square units

51. length: 70 feet, width: 40 feet **53.** 31 **55.** \$500 at 3.5%, \$2000 at 4% **57.** Kona: 7.5 pounds, Latin American: 62.5 pounds **59.** 40 **61.** \$600 at 8%, \$1000 at 9%, \$1200 at 15%

Chapter 3 Review **1.** $(2, 3)$ **2.** $(8, 2)$ **3.** $(2, 1)$ **4.** $(-6, 2)$ **5.** $(5, 3)$ **6.** $(-8, -7)$ **7.** $\varnothing$ **8.** $(4, -1)$ **9.** $(-6, 1)$ **10.** $(2, 4)$ **11.** $(-\frac{5}{2}, 7)$ **12.** $(x, 2x - 1)$ **13.** $(5, -4)$ **14.** $(-7, 3)$ **15.** $(-8, -18)$ **16.** $(1200, 1800)$ **17.** $(x, 4x - 3)$ **18.** $(-9, 8)$ **19.** 37, 68 **20.** 111 **21.** length: 25 feet, width: 17 feet **22.** \$1300 at 5%, \$3500 at 7% **23.** 625 ml 19%, 375 ml 35% **24.** 1350 meters per minute **25.**

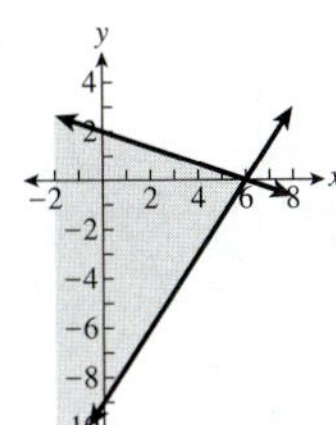

26.

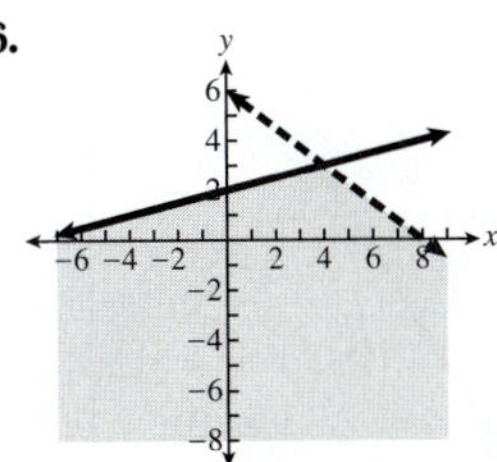

27.

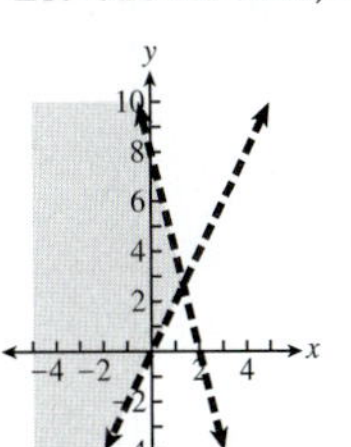

28.

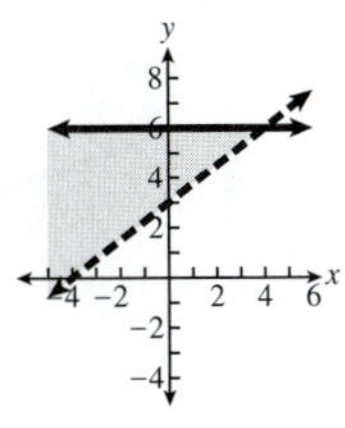

29. $(5, -2, 2)$ **30.** $(-6, 1, 10)$ **31.** $(8, 6, -12)$ **32.** $(-1, -5, 4)$ **33.** A: 75°, B: 18°, C: 87° **34.** 15 **35.** \$15,000 at a 10% profit, \$8000 at a 16% profit, \$27,000 at a 5% loss **36.** $(4, -5)$ **37.** $(3, 10)$ **38.** $(\frac{3}{2}, -8)$ **39.** $(-2, -12)$ **40.** $(6, 4, 3)$ **41.** $(-7, 4, -9)$ **42.** 66 **43.** -81 **44.** -73 **45.** -117 **46.** -61 **47.** -505 **48.** $(5, -8)$ **49.** $(-2, -7)$ **50.** $(6, 15)$ **51.** $(-9, \frac{11}{2})$ **52.** $(4, -6, -3)$ **53.** $(5, -10, 5)$

Chapter 3 Review Exercises: Worked-Out Solutions

1.

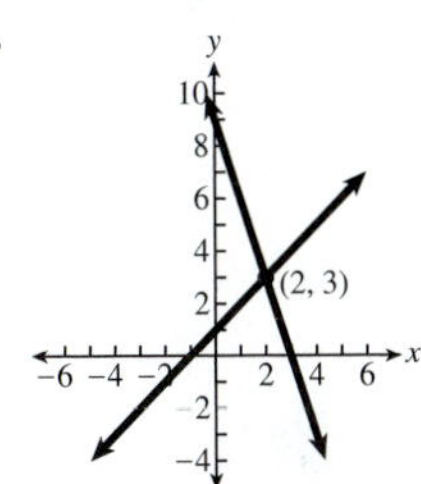

5.
$$\begin{aligned} x &= 3y - 4 \\ 4x + 5y &= 35 \\ 4(3y - 4) + 5y &= 35 \\ 12y - 16 + 5y &= 35 \\ 17y - 16 &= 35 \\ 17y &= 51 \\ y &= 3 \\ x = 3(3) - 4 &= 5 \end{aligned}$$
$(5, 3)$

11.
$4x + 7y = 39 \xrightarrow{\text{Multiply by 5.}} 20x + 35y = 195$

$6x - 5y = -50 \xrightarrow{\text{Multiply by 7.}} 42x - 35y = -350$

$$62x = -155$$

$$\frac{62x}{62} = \frac{-155}{62} \qquad x = -\frac{5}{2}$$

$$\begin{aligned} 4(-\tfrac{5}{2}) + 7y &= 39 \\ -10 + 7y &= 39 \\ 7y &= 49 \\ y &= 7 \end{aligned}$$

18.
$\frac{2}{3}x + \frac{1}{4}y = -4 \xrightarrow{\text{Multiply by 12.}} 8x + 3y = -48$

$\frac{4}{9}x + \frac{5}{6}y = \frac{8}{3} \xrightarrow{\text{Multiply by 18.}} 8x + 15y = 48$

$8x + 3y = -48 \xrightarrow{\text{Multiply by } -1.} -8x - 3y = 48$

$8x + 15y = 48 \qquad\qquad 8x + 15y = 48$

$$12y = 96$$

$$\begin{aligned} 12y &= 96 \\ y &= 8 \end{aligned}$$

$$\begin{aligned} \tfrac{2}{3}x + \tfrac{1}{4}(8) &= -4 \\ \tfrac{2}{3}x + 2 &= -4 \\ \tfrac{2}{3}x &= -6 \\ \tfrac{3}{2}\cdot\tfrac{2}{3}x &= -6\cdot\tfrac{3}{2} \\ x &= -9 \end{aligned}$$
$(-9, 8)$

22. x: Principal at 5%, y: Principal at 7%

$x + y = 4800 \qquad\qquad x + y = 4800$

$0.05x + 0.07y = 310 \xrightarrow{\text{Multiply by 100.}} 5x + 7y = 31{,}000$

$x + y = 4800 \xrightarrow{\text{Multiply by } -5.} -5x - 5y = -24{,}000$

$5x + 7y = 31{,}000 \qquad\qquad 5x + 7y = 31{,}000$

$$2y = 7000$$

$$\begin{aligned} 2y &= 7000 \\ y &= 3500 \end{aligned} \qquad \begin{aligned} x + 3500 &= 4800 \\ x &= 1300 \end{aligned}$$

\$1300 at 5% and \$3500 at 7%.

25.

$x + 3y \le 6$	$-3x + 2y \ge -18$	$x + 3y \le 6$ $-3x + 2y \ge -18$
x-int. $(6, 0)$, y-int. $(0, 2)$ Test Point: $(0, 0)$ – True	x-int. $(6, 0)$, y-int. $(0, -9)$ Test Point: $(0, 0)$ – True	Graph the intersection of the shaded regions.
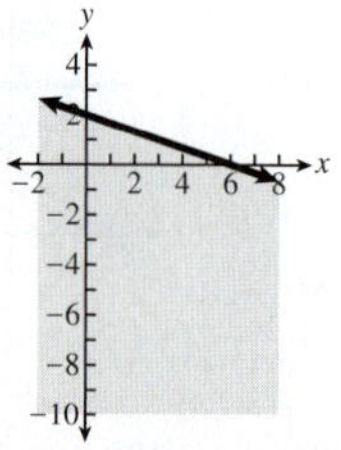	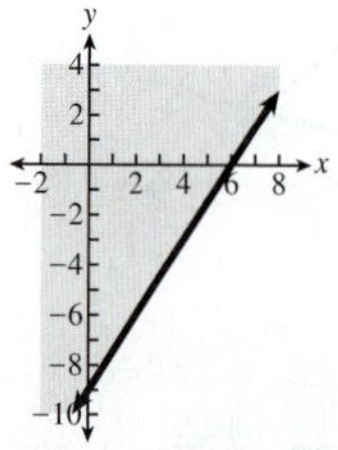	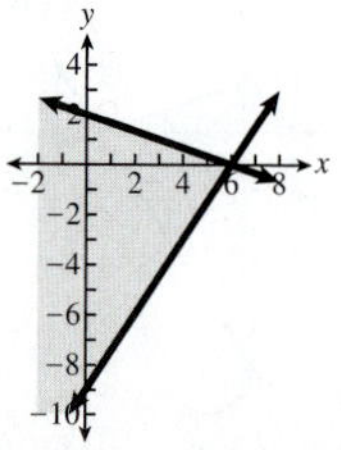

29. Eliminate z using Eq. 1 and Eq. 2, as well as Eq. 1 and Eq. 3.

$$\begin{array}{rl} 6x + 4y + 2z = 26 & 2(\text{Eq. 1}) \\ 2x + 3y - 2z = 0 & (\text{Eq. 2}) \\ \hline 8x + 7y = 26 & (\text{Eq. 4}) \end{array} \qquad \begin{array}{rl} 9x + 6y + 3z = 39 & 3(\text{Eq. 1}) \\ 4x - 7y - 3z = 28 & (\text{Eq. 3}) \\ \hline 13x - y = 67 & (\text{Eq. 5}) \end{array}$$

Solve for x using Eq. 4 and Eq. 5.

$$\begin{array}{l} 8x + 7y = 26 \\ 13x - y = 67 \end{array} \xrightarrow{\text{Multiply by } -5.} \begin{array}{r} 8x + 7y = 26 \\ 91x - 7y = 469 \\ \hline 99x = 495 \end{array}$$

$$99x = 495$$
$$x = 5$$

Back substitute.

$$\begin{array}{ll} \text{Equation 4} & \text{Equation 1} \\ 8(5) + 7y = 26 & 3(5) + 2(-2) + z = 13 \\ 40 + 7y = 26 & 11 + z = 13 \\ 7y = -14 & z = 2 \\ y = -2 & \end{array}$$

34. x: # of twin-sized quilts
y: # of queen-sized quilts
z: # of king-sized quilts
$x + y + z = 25$
$100x + 175y + 325z = 3550$
$100x - 175y = 100$
Eliminate z using Eq. 1 and Eq. 2.

$$\begin{array}{rl} -325x - 325y - 325z = -8125 & -325(\text{Eq. 1}) \\ 100x + 175y + 325z = 3550 & (\text{Eq. 2}) \\ \hline -225x - 150y = -4575 & (\text{Eq. 4}) \end{array}$$

Solve for y using Eq. 3 and Eq. 4.

$$100x - 175y = 100 \xrightarrow{\text{Multiply by 18.}} 900x - 1575y = 900$$

$$-225x - 150y = -4575 \xrightarrow{\text{Multiply by 18.}} -900x - 600y = -18{,}300$$

$$-2175y = -17{,}400$$

$-2175y = -17{,}400$

$y = 8$

Back substitute.

Equation 3	Equation 1
$100x - 175(8) = 100$	$15 + 8 + z = 25$
$100x - 1400 = 100$	$23 + z = 25$
$100x = 1500$	$z = 2$
$x = 15$	

She sold 15 twin-sized quilts.

36. $\left[\begin{array}{cc|c} 1 & 3 & -11 \\ -4 & -5 & 9 \end{array}\right]$

$\left[\begin{array}{cc|c} 1 & 3 & -11 \\ 0 & 7 & -35 \end{array}\right]$ $\quad 4 \cdot R_1 + R_2$

Row 2	Row 1
$7y = -35$	$x + 3(-5) = -11$
$y = -5$	$x - 15 = -11$
	$x = 4$

$(4, -5)$

40. $\left[\begin{array}{ccc|c} 1 & -1 & 1 & 5 \\ -3 & 2 & -4 & -22 \\ 2 & 1 & 3 & 25 \end{array}\right]$

$\left[\begin{array}{ccc|c} 1 & -1 & 1 & 5 \\ 0 & -1 & -1 & -7 \\ 0 & 3 & 1 & 15 \end{array}\right]$ $\quad 3 \cdot R_1 + R_2, -2 \cdot R_1 + R_3$

$\left[\begin{array}{ccc|c} 1 & -1 & 1 & 5 \\ 0 & -1 & -1 & -7 \\ 0 & 0 & -2 & -6 \end{array}\right]$ $\quad 3 \cdot R_2 + R_3$

Row 3	Row 2	Row 1
$-2z = -6$	$-y - 3 = -7$	$x - 4 + 3 = 5$
$z = 3$	$-y = -4$	$x - 1 = 5$
	$y = 4$	$x = 6$

$(6, 4, 3)$

42. $\begin{vmatrix} 6 & 3 \\ -8 & 7 \end{vmatrix} = 6(7) - 3(-8)$

$= 42 + 24$

$= 66$

46. $\begin{vmatrix} 2 & 1 & -1 \\ 3 & 2 & 3 \\ -4 & 4 & -5 \end{vmatrix} = 2\begin{vmatrix} 2 & 3 \\ 4 & -5 \end{vmatrix} - 3\begin{vmatrix} 1 & -1 \\ 4 & -5 \end{vmatrix} + (4)\begin{vmatrix} 1 & -1 \\ 2 & 3 \end{vmatrix}$

$= 2(-22) - 3(-1) + (-4)(5)$

$= -44 + 3 - 20$

$= -61$

48.

D	D_x	D_y
$\begin{vmatrix} 3 & 2 \\ 4 & -7 \end{vmatrix} = -21 - 8$	$\begin{vmatrix} -1 & 2 \\ 76 & -7 \end{vmatrix} = 7 - 152$	$\begin{vmatrix} 3 & -1 \\ 4 & 76 \end{vmatrix} = 228 - (-4)$
$= -29$	$= -145$	$= 232$

$x = \dfrac{D_x}{D} = \dfrac{-145}{-29} = 5 \qquad y = \dfrac{D_y}{D} = \dfrac{232}{-29} = -8$

$(5, -8)$

52.

D	D_x
$\begin{vmatrix} 1 & 1 & 1 \\ 3 & 2 & 3 \\ 4 & -1 & 5 \end{vmatrix} = 1\begin{vmatrix} 2 & 3 \\ -1 & 5 \end{vmatrix} - 3\begin{vmatrix} 1 & 1 \\ -1 & 5 \end{vmatrix} + 4\begin{vmatrix} 1 & 1 \\ 2 & 3 \end{vmatrix}$ $= 1(13) - 3(6) + 4(1)$ $= -1$	$\begin{vmatrix} -5 & 1 & 1 \\ -9 & 2 & 3 \\ 7 & -1 & 5 \end{vmatrix} = (-5)\begin{vmatrix} 2 & 3 \\ -1 & 5 \end{vmatrix} - (-9)\begin{vmatrix} 1 & 1 \\ -1 & 5 \end{vmatrix} + 7\begin{vmatrix} 1 & 1 \\ 2 & 3 \end{vmatrix}$ $= (-5)(13) - (-9)(6) + 7(1)$ $= -4$
D_y	D_z
$\begin{vmatrix} 1 & -5 & 1 \\ 3 & -9 & 3 \\ 4 & 7 & 5 \end{vmatrix} = 1\begin{vmatrix} -9 & 3 \\ 7 & 5 \end{vmatrix} - 3\begin{vmatrix} -5 & 1 \\ 7 & 5 \end{vmatrix} + 4\begin{vmatrix} -5 & 1 \\ -9 & 3 \end{vmatrix}$ $= 1(-66) - 3(-32) + 4(-6)$ $= 6$	$\begin{vmatrix} 1 & 1 & -5 \\ 3 & 2 & -9 \\ 4 & -1 & 7 \end{vmatrix} = 1\begin{vmatrix} 2 & -9 \\ -1 & 7 \end{vmatrix} - 3\begin{vmatrix} 1 & -5 \\ -1 & 7 \end{vmatrix} + 4\begin{vmatrix} 1 & -5 \\ 2 & -9 \end{vmatrix}$ $= 1(5) - 3(2) + 4(1)$ $= 3$

$x = \frac{D_x}{D} = \frac{-4}{-1} = 4 \qquad y = \frac{D_y}{D} = \frac{6}{-1} = -6 \qquad z = \frac{D_z}{D} = \frac{3}{-1} = -3$

$(4, -6, -3)$

Chapter 3 Test **1.** $(1, 3)$ **2.** $(5, 11)$ **3.** $(-7, 6)$ **4.** $(-6, 16)$ **5.** $\varnothing$ **6.** $(9, -2)$ **7.** length: 25 feet, width: 15 feet **8.** \$2200 at 8%, \$2800 at 22% **9.**

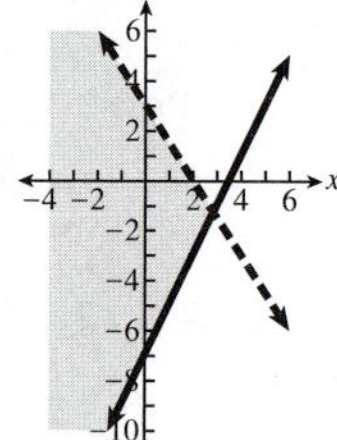

10. $(3, -6, -4)$ **11.** 20 **12.** $(3, -7)$ **13.** $(4, 9, -8)$ **14.** 131 **15.** 713 **16.** $(9, 7)$

Chapter 4

Quick Check 4.1 **1.** y^{21} **2.** $(a - b)^{17}$ **3.** $x^{21}y^{28}$ **4.** x^{48} **5.** $125x^3$ **6.** x^{21} **7.** x^{10} **8.** $9a^{11}bc^9$ **9. a)** 1 **b)** 1 **c)** 7 **10.** $\frac{49}{c^2}$ **11.** x^{62} **12.** $a^{63}b^{84}$ **13.** $784x^{28}y^{34}$ **14.** $\frac{81a^{28}}{b^{20}c^{32}}$ **15.** 81 **16.** $10b^{27}$ **17.** 216 cubic cm

Section 4.1 **1.** $x^m \cdot x^n = x^{m+n}$ **3.** $(xy)^n = x^n y^n$ **5.** $x^0 = 1$ **7.** 128 **9.** x^{13} **11.** m^{33} **13.** x^{14} **15.** $(x - y)^{11}$ **17.** $(a + 2)^{19}$ **19.** x^5y^{10} **21.** $56m^8n^{11}$ **23.** 64 **25.** x^{28} **27.** a^{72} **29.** x^{120} **31.** $16x^4$ **33.** $16x^6$ **35.** $x^{24}y^{33}$ **37.** $81x^{14}y^2$ **39.** x^7 **41.** d^{16} **43.** $8x^8$ **45.** $(x - 9)^{11}$ **47.** $a^{25}b^{12}$ **49.** $5x^6y^2z^{20}$ **51.** 1 **53.** -1 **55.** 1 **57.** 1 **59.** $\frac{27}{64}$ **61.** $\frac{a^{12}}{b^{12}}$ **63.** $\frac{x^{18}}{y^{54}}$ **65.** 9 **67.** 6 **69.** 12 **71.** 15 **73.** x^{38} **75.** x^{20} **77.** $x^{24}y^{69}z^{58}$ **79.** $a^{19}b^8c^{17}$ **81.** $\frac{25a^8}{b^{24}}$ **83.** $\frac{a^{64}b^{72}}{c^{56}}$ **85.** a^{24} **87.** $\frac{81x^8}{256y^{28}}$ **89.** x^{17} **91.** $\frac{a^{25}b^{50}}{243c^{35}}$ **93.** b^{60} **95.** x^{98} **97.** $\frac{1296a^{32}b^{24}}{625c^{20}d^{28}}$ **99.** 49 **101.** 81 **103.** 320 **105.** a^{48} **107.** $8a^{21}$ **109.** 256 feet **111.** 3600 square feet **113.** 1017.36 square feet **115.** $x^5 \cdot x^4$; Explanations will vary. **117.** Answers will vary.

Quick Check 4.2 **1.** $\frac{1}{32}$ **2.** $\frac{x^3}{y^6}$ **3.** x^6y^9 **4.** $\frac{z}{w^7x^5y^2}$ **5.** $\frac{1}{x^{12}}$ **6.** x^{24} **7.** $\frac{b^8}{16a^{20}c^4}$ **8.** $\frac{x^5}{y^4z^{13}}$ **9.** $\frac{b^{10}c^{35}}{a^{40}}$ **10. a)** 9.65×10^{-3} **b)** 3.52×10^{10} **11. a)** 0.0087 **b)** 5,560,000,000 **12.** 8.32×10^{16} **13.** 2.695×10^{-15} **14.** 2.0×10^{16} **15.** 3500 seconds

Section 4.2 **1.** $\frac{1}{x^n}$ **3.** $\frac{1}{25}$ **5.** $\frac{1}{32}$ **7.** $-\frac{1}{27}$ **9.** $\frac{1}{a^{10}}$ **11.** $\frac{6}{x^2}$ **13.** $-\frac{4}{m^3}$ **15.** x^4 **17.** $9x^2$ **19.** $\frac{y^7}{x^5}$ **21.** $-\frac{3b^4}{a^2c^6}$ **23.** $\frac{c^2}{a^7b^6d}$ **25.** x^8 **27.** $\frac{1}{a^{11}}$ **29.** $\frac{1}{m^4}$ **31.** $\frac{1}{x^{18}}$ **33.** x^{40} **35.** $\frac{z^{12}}{x^{21}y^9}$ **37.** $\frac{a^{20}b^{50}}{32c^{15}}$ **39.** $\frac{1}{x^{14}}$ **41.** a^9 **43.** $\frac{1}{y^7}$ **45.** $\frac{1}{x^{18}}$ **47.** $\frac{1}{x^2}$ **49.** $\frac{y^{12}}{x^{24}}$ **51.** $\frac{8b^{18}d^{12}}{a^{15}c^{27}}$ **53.** 3.9×10^{-4} **55.** 3.95×10^8 **57.** 9.0×10^{-11} **59.** 0.000092 **61.** 5,927,000,000 **63.** 0.00903 **65.** 7.04×10^{14} **67.** 5.0×10^{19}

69. 2.516×10^{-4} **71.** 3.8316×10^{25} **73.** 1.77×10^{6} **75.** 4 seconds **77.** 6.696×10^{8} miles **79.** Approximately 12,598,291 grains of salt **81.** \$16,170,000,000 **83.** 100,800,000,000,000,000 calculations **85.** \$5,376,000,000 **87.** Answers will vary. **89.** Answers will vary.

Quick Check 4.3 **1. a)** Trinomial, 2, 1, 0 **b)** Binomial, 4, 0 **c)** Monomial, 4 **2.** 1, 2, −1, 8 **3.** $-6x^3 + 3x^2 - 7x + 25$, Leading Term: $-6x^3$, Leading Coefficient: −6, Degree: 3 **4.** −32 **5.** 364 **6.** $x^3 - 4x^2 + 10x + 72$ **7.** $-2x^2 + 14x - 25$ **8.** $5x^2 + 31x - 83$ **9.** $44x^3$ **10.** $7x^4 - 42x^3 - 189x^2$ **11.** $-6x^4 + 6x^3 + 72x^2 - 360x$ **12.** $x^2 + 13x + 40$ **13.** $10x^2 - 51x + 56$ **14.** $x^3 + 2x^2 - 73x + 70$ **15.** $x^2 - 49$ **16.** $9x^2 - 4$ **17.** $4x^2 - 12x + 9$ **18.** $x^2 + 18x + 81$ **19.** 289 **20.** Each Term: 4, 6, 8, Polynomial: 8

Section 4.3 **1.** polynomial **3.** monomial **5.** trinomial **7.** descending order **9.** multiply **11.** Degree: 3, 2, 1, 0, Coefficient: 6, −5, 8, −19 **13.** Degree: 1, 6, 3, Coefficient: 1, −8, −14 **15.** Trinomial, Degree: 2 **17.** Binomial, Degree: 1 **19.** Monomial, Degree: 2 **21.** 46 **23.** −21 **25.** 11 **27.** −176 **29.** $5x^2 - 7x + 2$ **31.** $x^3 + 9x^2 - 10x + 27$ **33.** $x^2 - 15x + 43$ **35.** $-5x^5 + 10x^2 - 1$ **37.** $4x^2 + 14x - 48$ **39.** $36x^3$ **41.** $-20x^{10}$ **43.** $44x^6$ **45.** $12x - 27$ **47.** $-12x + 28$ **49.** $40x^2 + 45x$ **51.** $-4x^5 + 6x^4 - 9x^3$ **53.** $x^2 + 9x + 20$ **55.** $x^2 - 7x - 18$ **57.** $3x^2 - 16x - 12$ **59.** $2x^3 - x^2 - 35x - 50$ **61.** $x^4 + 3x^3 - 26x^2 + 88x - 192$ **63.** $11x^2 - 6, 5x^2 + 18x - 26$ **65.** $x^2 + 6x - 40$ **67.** $-24x^9$ **69.** $x^2 - 81$ **71.** $9x^2 - 4$ **73.** $x^2 + 16x + 64$ **75.** $9x^2 - 30x + 25$ **77.** $6x^2 - 3x + 12$ **79.** $4x^2$ **81.** 4 **83.** $x - 12$ **85.** $x^2 + x - 30$ **87.** $3x^4 - 15x^3 - 108x^2$ **89.** $6x^2 - x - 42$ **91.** $-x^2 - 11x + 82$ **93.** $2x^2 + 13x - 15$ **95.** $10x^2 + 14x - 104$ **97.** −81 **99.** 105 **101.** 8, 9, 10, Polynomial: 10 **103.** 9, 12, 8, Polynomial: 12 **105.** $3x^2 + 13xy - 2y^2$ **107.** $12a^3b - 23a^2b^2 + 6ab^3$ **109.** $56x^{14}y^7$ **111.** $-10x^3y^2 + 8x^2y^3 - 8xy^4$ **113.** $x^2 + 4xy - 21y^2$ **115.** \$2400 **117.** 1138 **119.** Answers will vary. **121.** Answers will vary.

Quick Check 4.4 **1.** 4 **2.** 12 **3. a)** $15x^4$ **b)** x^3z^2 **4.** $5x^2(x^2 - 6x - 15)$ **5.** $7(2x^2 - 3x - 1)$ **6.** $(x + 7)(2x^2 + 9)$ **7.** $(3x - 4)(9x - 2)$ **8.** $(x + 4)(x^2 + 7)$ **9.** $(x - 11)(2x - 3)$ **10.** $(x + 3)(7x^2 + 1)$ **11.** $5(x + 7)(x - 2)$

Section 4.4 **1.** factored **3.** smallest **5.** 2 **7.** 8 **9.** 12 **11.** x^4 **13.** a^2b^7 **15.** $2x^5$ **17.** $6a^3c$ **19.** $4(x - 6)$ **21.** $x(7x + 5)$ **23.** $6x^2(2x + 3)$ **25.** $7x^2(4x^2 + 1)$ **27.** $4(x^2 + 5x + 19)$ **29.** $4x^3(3x^2 - 6x - 8)$ **31.** $6x^2(7x^2 - 9x + 1)$ **33.** $m^2n^2(n - m^5 + m^4n^6)$ **35.** $12a^2b^4(3a^5b - 5a^2 - 4b^4)$ **37.** $3x(-2x^2 + 5x - 12)$ **39.** $2x(5x^3 + 6x + 1)$ **41.** $(3x + 8)(4x + 7)$ **43.** $(4x - 9)(x - 8)$ **45.** $(x - 6)(5x + 1)$ **47.** $(x + 3)(x + 6)$ **49.** $(x - 9)(x + 7)$ **51.** $(x + 10)(x - 12)$ **53.** $(x + 4)(3x + 10)$ **55.** $(x + 4)(5x - 2)$ **57.** $(x + 8)(2x + 1)$ **59.** $(x - 4)(3x - 1)$ **61.** $(x + 6)(x^2 + 5)$ **63.** $(x^2 + 8)(5x + 3)$ **65.** $2(x + 4)(x + 6)$ **67.** $5(x - 3)(x + 8)$ **69.** $x^3 - 5x^2 + 7x - 35$ **71.** $10x^3 - 45x^2 - 6x + 27$ **73.** Answers will vary. **75.** Answers will vary.

Quick Review Exercises 4.4 **1.** $x^2 + 11x + 28$ **2.** $x^2 - 20x + 99$ **3.** $x^2 - 3x - 40$ **4.** $x^2 - 4x - 60$ **5.** $2x^2 + 13x + 20$ **6.** $6x^2 - 37x + 56$ **7.** $3x^2 + 14x - 24$ **8.** $8x^2 + 30x - 27$

Quick Check 4.5 **1. a)** $(x + 2)(x + 8)$ **b)** $(x - 3)(x - 5)$ **2. a)** $(x + 7)(x - 2)$ **b)** $(x - 8)(x + 6)$ **3. a)** $(x + 6)(x - 1)$ **b)** $(x + 2)(x + 3)$ **c)** $(x - 2)(x - 3)$ **d)** $(x - 6)(x + 1)$ **4.** Prime **5.** $5(x - 4)(x + 3)$ **6.** $-(x - 3)(x - 9)$ **7.** $(3x + 2)(x + 4)$ **8.** $(4x - 9)(x - 4)$ **9.** $(4x - 3)(2x - 5)$ **10.** $(x + 6)(2x - 3)$ **11.** $(4x - 9)(3x + 8)$ **12.** $8(x + 7)(x - 2)$ **13.** $(x - 8)^2$ **14.** $(x - 10y)(x + 4y)$ **15.** $(xy + 7)(xy - 6)$

Section 4.5 **1.** leading coefficient **3.** prime **5.** $ax^2 + bx + c, (a \neq 1)$ **7.** $(x + 2)(x + 6)$ **9.** $(x + 3)(x + 9)$ **11.** $(x - 6)(x - 8)$ **13.** Prime **15.** $(x - 3)(x + 6)$ **17.** $(x - 5)(x + 9)$ **19.** $(x - 7)(x + 3)$ **21.** $(x - 9)(x + 8)$ **23.** $(x + 5)(x + 9)$ **25.** Prime **27.** $(x - 7)(x + 4)$ **29.** Prime **31.** $(x - 5)(x + 6)$ **33.** $(x - 3)(x + 9)$ **35.** $(x - 3)(x + 2)$ **37.** $3(x - 2)(x - 8)$ **39.** $2(x - 9)(x + 7)$ **41.** $5x(x - 1)(x - 5)$ **43.** $-(x - 3)(x + 10)$ **45.** $-3(x - 5)(x - 8)$ **47.** $(2x + 3)(x + 5)$ **49.** $(2x + 9)(x - 4)$ **51.** $(7x - 1)(x + 6)$ **53.** $(2x - 3)(x - 8)$ **55.** $(2x + 5)(2x + 3)$ **57.** $(3x - 2)(3x + 7)$ **59.** $(5x - 3)(3x - 5)$ **61.** $(12x + 5)(x + 3)$ **63.** $(3x + 8)(4x - 3)$ **65.** $(4x + 9)(3x + 2)$ **67.** $(x + 4)^2$ **69.** $(x - 2)^2$ **71.** $(x - 1)^2$ **73.** $(3x - 2)^2$ **75.** $(x + 7y)(x + 9y)$ **77.** $(xy - 5)(xy - 9)$ **79.** $(x - 3y)^2$ **81.** $(2x - y)(x - 3y)$ **83.** $(3xy - 4)(2xy + 1)$ **85.** $x(x - 7)$ **87.** $(x - 6)(x + 4)$ **89.** $(x + 6)(x^2 + 8)$ **91.** $(x + 3)(x + 8)$ **93.** $3(x - 5)(x + 7)$ **95.** $(3x - 8)(x - 2)$ **97.** $(4x - 3)(x + 3)$ **99.** $(x - 3)(5x^2 - 6)$ **101.** $(x - 6)(x + 9)$ **103.** Prime **105.** $(x - 1)(x - 8)$ **107.** $(x + 7)^2$ **109.** Answers will vary.

Quick Check 4.6 **1.** $(x+5)(x-5)$ **2.** $(3x+4)(3x-4)$ **3.** $(7x^6+5y^8)(7x^6-5y^8)$ **4.** $3(x+5)(x-5)$ **5.** $(a^2+9)(a+3)(a-3)$ **6.** $(x-6)(x^2+6x+36)$ **7.** $(x^6-2)(x^{12}+2x^6+4)$ **8.** $(x+7)(x^2-7x+49)$ **9.** $(5n+6)(25n^2-30n+36)$

Section 4.6 **1.** difference of squares **3.** multiples of 2 **5.** difference of cubes **7.** $(x+4)(x-4)$ **9.** $(a+5)(a-5)$ **11.** $(7x+3)(7x-3)$ **13.** Prime **15.** $(a+8b)(a-8b)$ **17.** $(5x+8y)(5x-8y)$ **19.** $(5+x)(5-x)$ **21.** $3(x+6)(x-6)$ **23.** Prime **25.** $4(x^2+4)$ **27.** $(x^2+4)(x+2)(x-2)$ **29.** $(x^2+7y)(x^2-7y)$ **31.** $(x-1)(x^2+x+1)$ **33.** $(x-3)(x^2+3x+9)$ **35.** $(5-y)(25+5y+y^2)$ **37.** $(x-2y)(x^2+2xy+4y^2)$ **39.** $(5a-9b)(25a^2+45ab+81b^2)$ **41.** $(x+y)(x^2-xy+y^2)$ **43.** $(x+2)(x^2-2x+4)$ **45.** $(x^2+6)(x^4-6x^2+36)$ **47.** $(2m+n)(4m^2-2mn+n^2)$ **49.** $5(x-2)(x^2+2x+4)$ **51.** $2(2x+5y)(4x^2-10xy+25y^2)$ **53.** $(x-5)^2$ **55.** $(x+2)(x+7)$ **57.** $x(x-9)$ **59.** $(2x-5)(3x+5)$ **61.** $x^2(x+8)(x-6)$ **63.** $-7(x^2+25)$ **65.** $-3(x-8)(x-9)$ **67.** $(x-9)(x+6)$ **69.** $(7x+6y)(7x-6y)$ **71.** Prime **73.** $(x+8y^3)(x-8y^3)$ **75.** $3x^2(x^2+7x+15)$ **77.** $6(x^2+36)$ **79.** $(x+3)(x+7)$ **81.** $(x+4)(3x-7)$ **83.** $(x+9)(x^2-5)$ **85.** $(x+5y)(x^2-5xy+25y^2)$ **87.** $x^3(2x-3)^2$ **89.** c **91.** a **93.** g **95.** h **97.** Answers will vary. **99.** Answers will vary.

Quick Check 4.7 **1.** $3(x-9)(x+2)$ **2.** $-8(x+12)(x-2)$ **3.** $12(x+5y^2)(x-5y^2)$ **4.** $(2x^3y^5-5z)(4x^6y^{10}+10x^3y^5z+25z^2)$ **5.** $(x+6)(x+3)(x^2-3x+9)$ **6.** $(a+b)(a^2-ab+b^2)(a-b)(a^2+ab+b^2)$ **7.** $(x^2+3y)(x^4-3x^2y+9y^2)$ **8.** $4(3x+1)(x+4)$

Section 4.7 **1.** common factors **3.** sum of squares **5.** sum of cubes **7.** $(3x-2y)(9x^2+6xy+4y^2)$ **9.** $(x-8)(x+1)(x-1)$ **11.** $(x+1)(x^2-x+1)(x-1)(x^2+x+1)$ **13.** $(x-7)(x-13)$ **15.** $(x^2+4y)(x^4-4x^2y+16y^2)$ **17.** $5(x-3)(x-7)$ **19.** $(7x-4)(x+5)$ **21.** $(x+4)(x+9)$ **23.** $(x+5)(x+2)(x^2-2x+4)$ **25.** $(3x-4)^2$ **27.** $(m+8n)^2$ **29.** $(x-13)(x+16)$ **31.** $6(3x^2-5x+1)$ **33.** $(x+18)(x-1)$ **35.** $(x-8)(4x+1)$ **37.** $6(3x^2-2x+11)$ **39.** $(5x+1)(2x+1)$ **41.** $(x+3)(2x+3)(2x-3)$ **43.** $(x-25)(x+4)$ **45.** $3(x+7)(x-4)$ **47.** $(x+y)(x^2-xy+y^2)(x-y)(x^2+xy+y^2)$ **49.** $-(x-2)(x-6)$ **51.** $(3x+4)(4x-3)$ **53.** $(x^2+6)(x^4-6x^2+36)$ **55.** $(5x^2+3y)(25x^4-15x^2y+9y^2)$ **57.** $(x-8)(x+9)$ **59.** $(x+8)^2$ **61.** $3(x+4)(2x+3)$ **63.** 18 **65.** 23, 23 **67.** $3xy^2-7z^3$

Quick Check 4.8 **1. a)** $\{-6,5\}$ **b)** $\{0,\frac{16}{3}\}$ **2.** $\{-4,10\}$ **3.** $\{-8,-1\}$ **4.** $\{\frac{4}{3},\frac{3}{2}\}$ **5.** $\{-9,9\}$ **6.** $\{-10,2\}$ **7.** $\{-6,5\}$ **8.** $\{-8,2\}$ **9.** $x^2+13x+30=0$ **10.** 8 and 9 **11.** 20 and 22 **12.** Length: 10 feet, Width: 6 feet **13.** 11 feet by 14 feet **14. a)** $h(t)=-16t^2+128t$ **b)** 8 seconds

Section 4.8 **1.** quadratic equation **3.** zero-factor **5.** $\{-5,18\}$ **7.** $\{0,15\}$ **9.** $\{-2,6\}$ **11.** $\{-7,-\frac{20}{3}\}$ **13.** $\{2,3\}$ **15.** $\{-7,-1\}$ **17.** $\{-10,3\}$ **19.** $\{-3,16\}$ **21.** $\{5\}$ **23.** $\{-5,5\}$ **25.** $\{-8,0\}$ **27.** $\{0,\frac{8}{3}\}$ **29.** $\{-9,6\}$ **31.** $\{-3,8\}$ **33.** $\{-5,7\}$ **35.** $\{-4,\frac{2}{5}\}$ **37.** $\{-3,4\}$ **39.** $\{5\}$ **41.** $\{-6,-5\}$ **43.** $\{-8,9\}$ **45.** $\{-10,10\}$ **47.** $\{-7,7\}$ **49.** $\{-8,8\}$ **51.** $\{-7,-4\}$ **53.** $\{-2,15\}$ **55.** $\{-5\}$ **57.** $\{-\frac{4}{3},2\}$ **59.** $\{-\frac{7}{4},\frac{3}{4}\}$ **61.** $x^2-10x+16=0$ **63.** $x^2+9x=0$ **65.** $x^2-100=0$ **67.** $15x^2-16x-15=0$ **69.** $x=-27$ **71.** $x=\frac{20}{3}$ **73.** 9, 10 **75.** 12, 14 **77.** $-15,-13$ **79.** Length: 10 feet, Width: 6 feet **81.** Length: 30 feet, Width: 10 feet **83.** Length: 20 meters, Width: 15 meters **85.** Length: 10 inches, Width: 8 inches **87.** 9 seconds **89.** 12 seconds **91.** Answers will vary. **93.** Answers will vary.

Chapter 4 Review **1.** x^2 **2.** $81x^{24}$ **3.** 12 **4.** $\frac{8x^{15}}{125y^9}$ **5.** $27x^{10}$ **6.** $-28a^9b^{15}$ **7.** $64x^{20}y^{16}z^{14}$ **8.** $a^{65}b^{50}c^{95}$ **9.** $\frac{1}{64}$ **10.** $\frac{3}{x^6}$ **11.** $\frac{x^{20}}{36}$ **12.** x^{19} **13.** $\frac{1}{m^7}$ **14.** $\frac{x^6y^9}{27z^{18}}$ **15.** $\frac{1}{a^{26}}$ **16.** $\frac{1}{x^{11}}$ **17.** 3.9×10^6 **18.** 6.5×10^{-6} **19.** 0.000000072 **20.** 409,000,000,000 **21.** 2.1×10^{16} **22.** 3.0×10^{-9} **23.** 3.32×10^{-8} grams **24.** 20,000 seconds **25.** -28 **26.** 9 **27.** 31 **28.** -94 **29.** $2x^2-2x+6$ **30.** $7x^2+3x-13$ **31.** $-2x^2+14x-52$ **32.** $3x^3-5x^2+9x+9$ **33.** 156 **34.** $4b^4+10b^2-12$ **35.** $f(x)+g(x)=2x^2-x-40, f(x)-g(x)=9x+24$ **36.** $f(x)+g(x)=5x^2-x-31, f(x)-g(x)=-x^2-15x-19$ **37.** $x^2-20x+100$ **38.** $x^2+5x-84$ **39.** $14x^3-35x^2-63x$ **40.** $-54x^9$ **41.** x^2-81 **42.** $8x^2+22x-63$ **43.** $25x^2-16$ **44.** $9x^2+48x+64$ **45.** $-35x^4+45x^3+50x^2$ **46.** $2x^3-15x^2+10x+48$ **47.** $3x^2+25x+50$ **48.** $12x^5-20x^4+48x^3$ **49.** $2x^2y^4(x-4y+2x^2y^2)$ **50.** $(x-10)(x^2+3)$ **51.** $(x-7)(x+2)(x-2)$ **52.** $(x+3)^2$ **53.** $(x+9)(x-3)$ **54.** $-4(x-8)(x+1)$ **55.** Prime **56.** $(x+7)(x+8)$ **57.** $(x-10)(x+1)$ **58.** $(8x-5)(x+4)$ **59.** $(2x-5)(2x+3)$ **60.** $6(x+8)(x-8)$ **61.** $(3x+7y)(3x-7y)$ **62.** $(x+5y)(x^2-5xy+25y^2)$ **63.** Prime **64.** $(x-8)(x^2+8x+64)$ **65.** $\{-\frac{3}{2},4\}$ **66.** $\{-6,6\}$ **67.** $\{5,7\}$ **68.** $\{-12,4\}$ **69.** $\{-7,9\}$ **70.** $\{0,15\}$ **71.** $\{-10,-8\}$ **72.** $\{\frac{2}{3},5\}$ **73.** $\{-11,5\}$

74. $\{-8, -2\}$ **75.** $x^2 + x - 12 = 0$ **76.** $3x^2 - 22x - 16 = 0$ **77.** 17, 19 **83.** Length: 11 meters, Width: 6 meters **79.** 7 seconds **80.** 5 seconds

Chapter 4 Review Exercises: Worked-Out Solutions

1. $\frac{x^7}{x^5} = x^{7-5} = x^2$

6. $-7a^8b^3 \cdot 4ab^{12} = -28a^{8+1}b^{3+12} = -28a^9b^{15}$

8. $(a^{13}b^{10}c^{19})^5 = a^{13(5)}b^{10(5)}c^{19(5)} = a^{65}b^{50}c^{95}$

13. $m^{14} \cdot m^{-21} = m^{14+(-21)} = m^{-7} = \frac{1}{m^7}$

14. $(3x^{-2}y^{-3}z^6)^{-3} = 3^{-3}x^{-2(-3)}y^{-3(-3)}z^{6(-3)} = 3^{-3}x^6y^9z^{-18} = \frac{x^6y^9}{3^3z^{18}} = \frac{x^6y^9}{27z^{18}}$

21. $(2.5 \times 10^8)(8.4 \times 10^7) = 2.5 \cdot 8.4 \times 10^{8+7} = 21 \times 10^{15} = 2.1 \times 10^{16}$

25. $(-4)^2 + 7(-4) - 16 = 16 - 28 - 16 = -28$

31. $(x^2 + 9x - 10) - (3x^2 - 5x + 42) = x^2 + 9x - 10 - 3x^2 + 5x - 42 = -2x^2 + 14x - 52$

35. $f(x) + g(x) = (x^2 + 4x - 8) + (x^2 - 5x - 32) = x^2 + 4x - 8 + x^2 - 5x - 32 = 2x^2 - x - 40$

$f(x) - g(x) = (x^2 + 4x - 8) - (x^2 - 5x - 32) = x^2 + 4x - 8 - x^2 + 5x + 32 = 9x + 24$

38. $(x - 7)(x + 12) = x^2 + 12x - 7x - 84 = x^2 + 5x - 84$

44. $(3x + 8)^2 = (3x + 8)(3x + 8) = 9x^2 + 24x + 24x + 64 = 9x^2 + 48x + 64$

50. $x^3 - 10x^2 + 3x - 30 = x^2(x - 10) + 3(x - 10) = (x - 10)(x^2 + 3)$

52. $m \cdot n = 9,\ m + n = 6$
$3 \cdot 3 = 9,\ 3 + 3 = 6$
$x^2 + 6x + 9 = (x + 3)(x + 3) = (x + 3)^2$

54. $-4x^2 + 28x + 32 = -4(x^2 - 7x - 8) = -4(x - 8)(x + 1)$

59. $4x^2 - 4x - 15 = 4x^2 - 10x + 6x - 15 = 2x(2x - 5) + 3(2x - 5) = (2x - 5)(2x + 3)$

61. $9x^2 - 49y^2 = (3x)^2 - (7y)^2 = (3x + 7y)(3x - 7y)$

64. $x^3 - 512 = (x)^3 - (8)^3 = (x - 8)(x^2 + 8x + 64)$

67. $x^2 - 12x + 35 = 0$
$(x - 5)(x - 7) = 0$
$x - 5 = 0$ or $x - 7 = 0$
$x = 5$ or $x = 7$
$\{5, 7\}$

73. $x(x + 6) = 55$
$x^2 + 6x = 55$
$x^2 + 6x - 55 = 0$
$(x + 11)(x - 5) = 0$
$x + 11 = 0$ or $x - 5 = 0$
$x = -11$ or $x = 5$
$\{-11, 5\}$

75. $\{-4, 3\}$
$x = -4$ or $x = 3$
$x + 4 = 0$ or $x - 3 = 0$
$(x + 4)(x - 3) = 0$
$x^2 + x - 12 = 0$

78. Length: $2x - 1$
Width: x
$x(2x - 1) = 66$
$2x^2 - x = 66$
$2x^2 - x - 66 = 0$
$(2x + 11)(x - 6) = 0$
$2x + 11 = 0$ or $x - 6 = 0$
$x = -\frac{11}{2}$ or $x = 6$ (Omit negative solution.)
Length: $2x - 1 = 2(6) - 1 = 11$
Width: $x = 6$
The length is 11 meters and the width is 6 meters.

Chapter 4 Test **1.** x^5 **2.** $x^{20}y^{24}z^{40}$ **3.** $\frac{1}{64}$ **4.** $\frac{1}{x^{28}}$ **5.** x^7 **6.** $\frac{y^4}{16x^2z^6}$ **7.** 4.9×10^8 **8.** 0.000036 **9.** 2.448×10^6 **10.** $-2x^2 + 11x - 32$ **11.** $2x^2 - 3x + 8$ **12.** $x^2 - 81$ **13.** $6x^2 - 17x - 14$ **14.** $(x + 4)(x^2 - 6)$ **15.** $(x - 3)(x - 10)$ **16.** $(x - 12)(x + 6)$ **17.** $(6x + 1)(x - 4)$ **18.** $(2x + 5y)(2x - 5y)$ **19.** $(2x - 5)(4x^2 + 10x + 25)$ **20.** $\{-7, 7\}$ **21.** $\{5, 8\}$ **22.** $\{-9, 6\}$ **23.** $x^2 - 3x - 28 = 0$ **24.** Length: 18 feet, Width: 8 feet **25.** 4 seconds

Cumulative Review Chapters 1–4 **1.** -48 **2.** -24 **3.** $6x - 78$ **4.** $\{-\frac{5}{2}\}$ **5.** $\{4\}$ **6.** $y = \frac{9x + 23}{4}$ **7.** $\{-5, 21\}$ **8.** Length: 22 ft, Width: 9 ft **9.** $x > -6$,  , $(-6, \infty)$ **10.** $-9 < x < 1$, 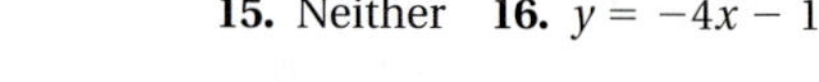 , $(-9, 1)$

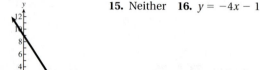

11. $(-3, 0), (0, 4)$

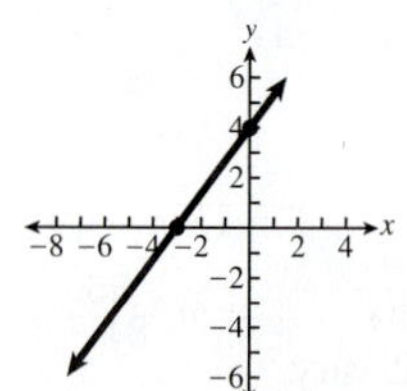

12. $\frac{5}{2}$ **13.** $3, (0, -12)$ **14.**

15. Neither **16.** $y = -4x - 1$

17. $y = -2x + 2$ **18.**

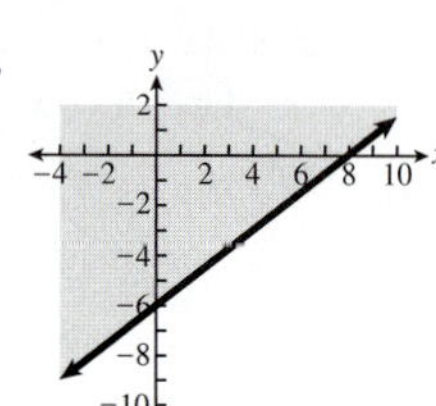

19.

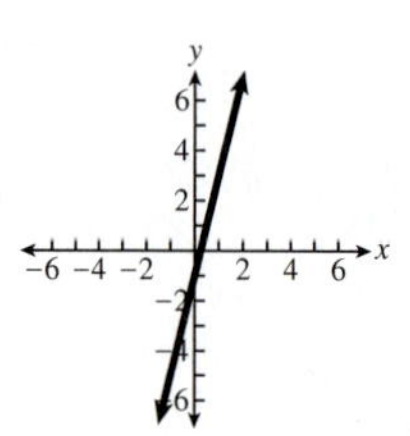

20.

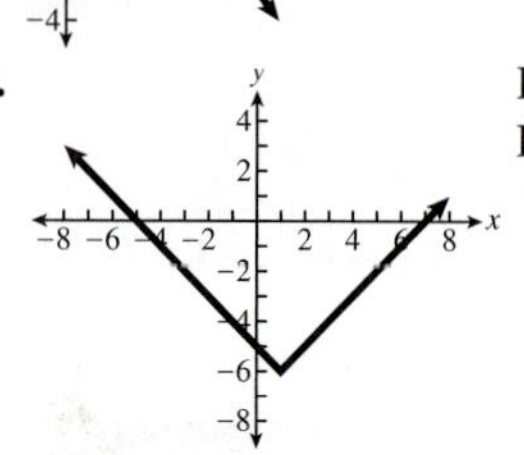

Domain: $(-\infty, \infty)$, Range: $[-6, \infty)$

21. $(3, -4)$ **22.** $(8, 14)$ **23.** $(5, -4)$ **24.** \$2500 at 7%, \$5500 at 16% **25.**

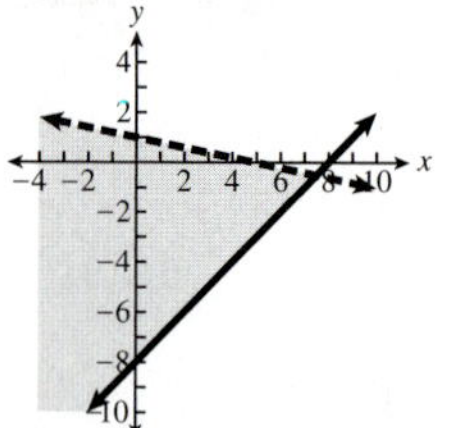

26. $(8, -4, 1)$ **27.** $(-2, 9)$

28. $(3, 4, -5)$ **29.** -569 **30.** $(-6, 8)$ **31.** $81x^{16}y^{28}z^{32}$ **32.** $\frac{a^{11}}{b^4}$ **33.** $\frac{1}{x^9}$ **34.** 3.24×10^7 **35.** $-2x^2 + 15x + 25$ **36.** $2x^2 + 8x - 27$ **37.** $10x^2 - 53x + 36$ **38.** $(x - 12)(x + 7)$ **39.** $(6x + 5)(2x - 3)$ **40.** $2(3x + 7y)(3x - 7y)$ **41.** $\{-10, 4\}$ **42.** 5 seconds

Chapter 5

Quick Check 5.1 **1.** $-\frac{5}{2}$ **2.** $-\frac{9}{5}$ **3.** 3, 7 **4.** $\frac{3x^6}{2}$ **5.** $\frac{x - 1}{x + 1}$ **6.** $\frac{x + 3}{x + 9}$ **7.** $\frac{3x + 2}{x + 4}$ **8.** $-\frac{x}{x - 6}$ **9.** 3 **10.** All real numbers except -8 and 2.

Section 5.1 **1.** rational expression **3.** simplified to lowest terms **5.** rational function **7.** $\frac{3}{4}$ **9.** -9 **11.** $\frac{1}{6}$ **13.** $-\frac{5}{2}$ **15.** 6 **17.** 0, -5 **19.** $-6, -9$ **21.** $-5, 5$ **23.** $\frac{2}{7x^4}$ **25.** $\frac{1}{x + 7}$ **27.** $\frac{x + 6}{x - 6}$ **29.** $\frac{x + 6}{x - 2}$ **31.** $\frac{x - 3}{x + 9}$ **33.** $\frac{x - 10}{x - 2}$ **35.** $\frac{x - 6}{2x + 3}$ **37.** $\frac{4(x - 6)}{x + 1}$ **39.** $\frac{2x - 3}{x - 7}$ **41.** Not opposites **43.** Opposites **45.** Opposites **47.** $-\frac{1}{2 + x}$ **49.** $-\frac{x - 7}{4 + x}$ **51.** $\frac{6 - x}{x + 7}$ **53.** $-\frac{3}{2}$ **55.** $\frac{9}{4}$ **57.** $\frac{13}{14}$ **59.** $x \neq 0, 6$ **61.** $x \neq -9, 5$ **63.** $x \neq -7, 3$ **65.** Quadratic **67.** Rational **69.** Linear **71. a.** -6 **b.** -2 **73. a.** 6 **b.** 0, 2 **75.** Answers will vary. **77.** Answers will vary.

Quick Check 5.2 **1.** $\frac{(x + 2)(x + 8)}{(x - 3)(x - 4)}$ **2.** $-\frac{(x + 3)(x - 5)}{(x + 2)(10 + x)}$ **3.** $\frac{x - 10}{x(x + 4)}$ **4.** $\frac{x + 1}{x + 3}$ **5.** $-(x + 4)$ **6.** $\frac{1}{(x - 7)(x - 1)}$

Section 5.2 **1.** factoring **3.** factored **5.** $\frac{x + 1}{(x - 6)(x - 1)}$ **7.** $\frac{x + 9}{x - 8}$ **9.** $\frac{x(x - 1)}{(x + 3)(x + 10)}$ **11.** $-\frac{x}{x - 2}$ **13.** $-\frac{(x - 2)(9 + x)}{(x - 6)^2}$ **15.** $\frac{(x - 9)(2x - 3)}{(x + 2)(x + 4)}$ **17.** $\frac{(x + 12)(x + 15)}{(3x + 1)(x - 9)}$ **19.** $\frac{2(x - 9)(x - 4)}{3(x - 7)(x - 2)}$ **21.** $\frac{(x - 1)(x + 7)}{(x - 4)(x + 5)}$ **23.** $\frac{(x^2 + 7)(x + 2)}{(x + 6)(x - 1)}$ **25.** $\frac{(x - 4)(x^2 + 2x + 4)}{(x + 4)(x + 1)}$ **27.** $\frac{(x - 2)(x^2 - x + 1)}{x(x + 12)}$ **29.** $-\frac{7}{x + 9}$ **31.** $\frac{x - 12}{x + 2}$ **33.** $(x + 8)(x - 8)$ **35.** $\frac{x - 7}{x - 2}$ **37.** $\frac{(x - 3)(x + 11)}{(2 - x)(x + 2)}$ **39.** $\frac{x - 5}{x + 2}$ **41.** $\frac{(x + 7)(x - 1)}{(x - 3)(x + 1)}$ **43.** $\frac{x + 9}{x - 6}$ **45.** $\frac{(x + 5)(3x - 4)}{(x + 3)(x + 9)}$ **47.** 1

49. $-\frac{(x-6)(x+10)}{(x-3)(2+x)}$ **51.** $-\frac{x(x+7)}{(x-1)(x-10)}$ **53.** $\frac{3(x+1)(x+8)}{2(x-8)(x+10)}$ **55.** $\frac{(x+3)(x+2)}{(x+4)(x+7)}$ **57.** $\frac{(x+2)(x^2-8)}{x(x+4)}$ **59.** $-\frac{(x+7)(x^2+3x+9)}{(x-2)(3+x)}$ **61.** $\frac{(x+3)(x^2-6x+36)}{x+15}$ **63.** $\frac{x+7}{(x-5)^2}$ **65.** $\frac{x(x+3)}{(x+12)(x-3)}$ **67.** $\frac{x(x-15)}{(x+4)(x-6)}$ **69.** $\frac{x-6}{x-3}$ **71.** $\frac{(x-9)(x+4)}{x(x-12)}$ **73.** Answers will vary.

Quick Check 5.3 **1.** $\frac{10}{2x-5}$ **2.** $\frac{5}{x+6}$ **3.** $\frac{x+3}{x-2}$ **4.** 2 **5.** $-\frac{x+7}{5+x}$ **6.** $\frac{2}{3}$ **7.** $\frac{x-2}{x+4}$ **8.** $24x^6y^2$ **9.** $(x+3)(x+8)(x-6)$ **10.** $(x-6)(x+6)^2$ **11.** $\frac{9x+46}{(x-5)(x+8)}$ **12.** $\frac{x+9}{(x-3)(x+5)}$ **13.** $\frac{x+1}{(x-1)(x+5)}$ **14.** $\frac{x-2}{(x+2)(x-3)}$

Section 5.3 **1.** numerators **3.** first term **5.** LCD **7.** $\frac{18}{x-3}$ **9.** 3 **11.** $\frac{1}{x-3}$ **13.** $\frac{x-3}{x+6}$ **15.** $\frac{9}{x+6}$ **17.** 6 **19.** $\frac{1}{x-9}$ **21.** $\frac{3}{x+6}$ **23.** $\frac{x-2}{x-5}$ **25.** $\frac{x+6}{x}$ **27.** 7 **29.** $x+8$ **31.** $\frac{x-7}{x+8}$ **33.** $4x$ **35.** $7x-18$ **37.** $12x$ **39.** $(x-8)(x+4)$ **41.** $(x+2)(x+7)(x-2)$ **43.** $(x-3)^2(x-4)$ **45.** $\frac{19x}{18}$ **47.** $-\frac{17}{24x}$ **49.** $\frac{5x+23}{(x+5)(x+4)}$ **51.** $\frac{1}{(x-10)(x-9)}$ **53.** $\frac{15}{(x+2)(x-3)}$ **55.** $\frac{3x-22}{(x-7)(x-9)(x-8)}$ **57.** $-\frac{5}{(x-6)(x+9)}$ **59.** $\frac{6}{(x+2)(x-4)}$ **61.** $\frac{x+9}{(x+6)(x+7)}$ **63.** $\frac{x+3}{(x+6)(x-3)}$ **65.** $\frac{x}{(x-2)(x+3)}$ **67.** $\frac{x-7}{(x+5)(x-1)}$ **69.** $\frac{x^2+17x+63}{x(x+6)(x+9)}$ **71.** $\frac{2(x-5)}{(x-3)(x-4)}$ **73.** $\frac{3}{4}$ **75.** $\frac{5}{(x-6)(x+9)}$ **77.** $\frac{9}{x-3}$ **79.** $\frac{4}{(x+3)(x+7)}$ **81.** $\frac{11x+3}{90}$ **83.** $\frac{9(x+1)}{(x-7)(x-3)(x+6)}$ **85.** $\frac{x-6}{x+10}$ **87.** $\frac{x+6}{x-9}$ **89.** $\frac{2}{x+5}$ **91.** $\frac{x}{(x-10)(x-8)}$ **93.** $\frac{3}{3x-4}$ **95.** $\frac{x-1}{(x+1)(x+2)}$ **97.** $-\frac{1}{(x+4)(x+5)}$ **99.** Answers will vary.

Quick Check 5.4 **1.** $\frac{114}{7}$ **2.** $\frac{x-7}{7x}$ **3.** $\frac{x+1}{x+9}$ **4.** $\frac{2}{x-6}$ **5.** $\frac{(x-4)(x-7)}{(x-1)(x+5)}$

Section 5.4 **1.** complex fraction **3.** $-\frac{1}{7}$ **5.** $\frac{54}{17}$ **7.** $\frac{2(9x+2)}{9(2x+3)}$ **9.** $\frac{10}{x}$ **11.** $\frac{4x}{x-3}$ **13.** $\frac{x+3}{3x}$ **15.** $\frac{3}{x-2}$ **17.** $\frac{11}{x-4}$ **19.** $\frac{x-2}{x-6}$ **21.** $\frac{x+3}{x+8}$ **23.** $\frac{x(x+6)}{x-5}$ **25.** $\frac{x+8}{x+2}$ **27.** $\frac{(x+8)(x-5)}{(x+2)(x-9)}$ **29.** $\frac{x-10}{x+12}$ **31.** $\frac{x-3}{x-6}$ **33.** $\frac{(x+1)(x+2)}{(x-6)(x+4)(x-2)}$ **35.** $\frac{x(x+3)}{(x-3)(x+5)}$ **37.** $\frac{x+2}{x(x+6)}$ **39.** $\frac{9}{2}$ **41.** $\frac{5}{x+12}$ **43.** $\frac{5x+73}{(x+2)(x+11)(x+9)}$ **45.** $\frac{3x}{x+3}$ **47.** $\frac{11x+48}{(x+3)(x+4)(x+6)}$ **49.** $\frac{x+10}{x+15}$ **51.** $\frac{x+6}{x+2}$ **53.** $\frac{x(x-9)}{(x+16)(x-8)}$ **55.** Answers will vary.

Quick Check 5.5 **1.** $\{30\}$ **2.** $\{-4, 8\}$ **3.** $\varnothing$ **4.** $\{-1\}$ **5.** $\{7\}$ **6.** $x = \frac{4yz}{3y-2z}$ **7.** $x = \frac{11y-4}{5y-2}$

Section 5.5 **1.** rational equation **3.** extraneous solution **5.** $\{\frac{2}{5}\}$ **7.** $\{-\frac{76}{13}\}$ **9.** $\{16\}$ **11.** $\{12\}$ **13.** $\{2, 8\}$ **15.** $\{-6, 3\}$ **17.** $\{-9\}$ **19.** $\varnothing$ **21.** $\{2\}$ **23.** $\{8\}$ **25.** $\{7, 10\}$ **27.** $\{13\}$ **29.** $\{-6\}$ **31.** $\{-2, 6\}$ **33.** $\{-5, 4\}$ **35.** $\{-2, 5\}$ **37.** $\{8\}$ **39.** $\{-4\}$ **41.** $\{-6\}$ **43.** $\varnothing$ **45.** $\{-11, 3\}$ **47.** $\{13\}$ **49.** $\{-3, 9\}$ **51.** $\{11\}$ **53. a.** 26 **b.** $x = -8$ **55.** $L = \frac{A}{W}$ **57.** $x = \frac{2y}{3y-1}$ **59.** $r = \frac{4x+3y}{12}$ **61.** $c = \frac{ab}{b+a}$ **63.** $b = \frac{xa-a}{x+1}$ **65.** $n_1 = \frac{n_2}{nn_2-1}$ **67.** Answers will vary.

Quick Check 5.6 **1.** 10 **2.** 10 and 20 **3.** $10\frac{10}{11}$ minutes **4.** 12 hours **5.** 80 mph **6.** 7 mph **7.** \$576 **8.** 20 minutes **9.** 2500

Section 5.6 **1.** reciprocal **3.** divided **5.** vary directly **7.** 4 **9.** 3 **11.** 5 **13.** 4, 8 **15.** 8, 12 **17.** 12 minutes **19.** $40\frac{10}{11}$ minutes **21.** $\frac{12}{13}$ hour **23.** 30 hours **25.** 12 hours **27.** 24 minutes **29.** 16 hours **31.** 8 mph **33.** 9 mph **35.** 40 km/hr **37.** 160 mph **39.** 126 **41.** 72 **43.** 21 **45.** $\frac{9}{2}$ **47.** 288 **49.** 900 **51.** \$387 **53.** 3 amperes **55.** \$10 **57.** 18 amperes **59.** 4 foot-candles **61.** Answers will vary.

Quick Review Exercises 5.6 **1.** x^9 **2.** $-9x^5$ **3.** $-13x^2-7x+10$ **4.** $-6x^2+2x-8$

Quick Check 5.7 **1.** $6x^8$ **2.** $14x^5y^3$ **3.** x^2-5x-3 **4.** x^6-3x^4+15x **5.** $x-3$ **6.** $x-2+\frac{9}{x+7}$ **7.** $2x+5-\frac{32}{4x-9}$ **8.** $x^2-3x+27-\frac{200}{x+9}$

Section 5.7 **1.** subtracting **3.** factor **5.** $7x^6$ **7.** $-12x^7$ **9.** $13x^6y$ **11.** $12x^3$ **13.** $\frac{9x^2}{5}$ **15.** $\frac{x^{11}}{12}$ **17.** $12x^{11}$ **19.** $4x^7$ **21.** $5x^2-11x-1$ **23.** x^3+13x^2-7x+3 **25.** $2x^6+3x^4-5x$ **27.** $-6x^3+7x^2$ **29.** $3x^4y^2-9xy+x^2$ **31.** $3x$ **33.** $6x^6-42x^5-114x^4$ **35.** $x+8$ **37.** $x+9$ **39.** $x-15$ **41.** $x+13+\frac{7}{x-8}$ **43.** $x^2-10x-1+\frac{11}{x+2}$ **45.** $2x+3+\frac{6}{x+8}$ **47.** $3x+5-\frac{30}{2x-9}$ **49.** $x+7$ **51.** $x^3+5x^2+27x+143+\frac{688}{x-5}$ **53.** $x^2+4x+16$ **55.** x^2+x-42 **57.** $x-18$ **59.** Yes **61.** No **63.** $x-7-\frac{8}{x-4}$ **65.** $x^2-4x+24-\frac{151}{x+6}$ **67.** $3x^3-4x^2+7x-1$ **69.** $4x-9+\frac{15}{x+7}$ **71.** $5x-8-\frac{1}{4x-1}$ **73.** $-4x^5+3x^3+9x^2-2$ **75.** $2x^2-3x-2+\frac{3}{4x+6}$ **77.** $3x^2-8x-10$ **79.** $-9x^3yz^7$ **81.** Answers will vary.

Chapter 5 Review **1.** $-\frac{3}{4}$ **2.** $\frac{1}{3}$ **3.** $\frac{7}{3}$ **4.** 5 **5.** $\frac{3}{2}$ **6.** $-7, 4$ **7.** $\frac{1}{x+8}$ **8.** $\frac{x+9}{x-5}$ **9.** $-\frac{x}{x-7}$ **10.** $\frac{2x-1}{x-4}$ **11.** $-\frac{3}{10}$ **12.** 6 **13.** $x \neq 0, 9$ **14.** $x \neq -1, 6$ **15.** $\frac{x-10}{x+6}$ **16.** $\frac{(x+7)(x-4)}{(x-3)(x-6)}$ **17.** $\frac{x(x+2)}{(3x-2)(x-9)}$ **18.** $-\frac{(7+x)(x-1)}{x+3}$ **19.** $\frac{x+8}{x(x-8)}$ **20.** $\frac{x+6}{x+11}$ **21.** $\frac{x+7}{x-8}$ **22.** $\frac{(x+12)(x+7)}{(x+4)(x-3)}$ **23.** $\frac{(5x-3)(x-8)}{x(x-5)}$ **24.** $-\frac{(10+x)(x+6)}{(x-3)}$ **25.** $\frac{x(x-7)}{(x+4)(x+7)}$ **26.** $\frac{(x+16)(x-6)}{(x+7)(x-5)}$ **27.** $\frac{24}{x+4}$ **28.** $\frac{x-6}{x+8}$ **29.** $\frac{x+5}{x+9}$ **30.** 6 **31.** $x-8$ **32.** $\frac{x+10}{x-4}$ **33.** $\frac{9}{(x+1)(x-8)}$ **34.** $\frac{7}{(x+2)(x+9)}$ **35.** $\frac{x+5}{(x+4)(x+6)}$ **36.** $\frac{x+2}{(x+10)(x+6)}$ **37.** $\frac{2x+5}{(x+3)(x+4)}$ **38.** $\frac{4}{(x+1)(x-7)}$ **39.** $\frac{x-1}{(x-7)(x+2)}$ **40.** $\frac{x}{x-6}$ **41.** $\frac{x+2}{x+5}$ **42.** $\frac{x+10}{x-1}$ **43.** $\frac{(x-2)(x+4)}{(x-3)(x+6)}$ **44.** $\{16\}$ **45.** $\{-8, 6\}$ **46.** $\{20\}$ **47.** $\{-1, 9\}$ **48.** $\{8\}$ **49.** $\{7\}$ **50.** $\{-3, 9\}$ **51.** $\{-2, 3\}$ **52.** $h = \frac{2A}{b}$ **53.** $x = \frac{3y}{2y-5}$ **54.** $r = \frac{4x-3y}{2}$ **55.** 6 **56.** 6, 18 **57.** $22\frac{2}{9}$ min **58.** 30 hr **59.** 55 mph **60.** 6 mph

61. 120 calories **62.** 320 ft **63.** 580 lb **64.** 120 foot-candles **65.** $6x^2$ **66.** $3x^3 - 5x^2 - 11x + 1$ **67.** $x + 26$ **68.** $x - 13 - \frac{9}{x+8}$ **69.** $6x - 7 + \frac{20}{x+3}$ **70.** $4x^2 - 9x - 16$ **71.** $x^2 + 6x + 43 + \frac{238}{x-6}$ **72.** $x^2 + 8x + 64 - \frac{3}{x-8}$

Chapter 5 Review Exercises: Worked-Out Solutions

3.
$$\frac{(6)^2 + 7(6) - 8}{(6)^2 - 2(6) + 6} = \frac{36 + 42 - 8}{36 - 12 + 6} = \frac{70}{30} = \frac{7}{3}$$

6.
$$x^2 + 3x - 28 = 0$$
$$(x + 7)(x - 4) = 0$$
$$x = -7 \text{ or } x = 4$$

8.
$$\frac{x^2 + 7x - 18}{x^2 - 7x + 10} = \frac{(x + 9)(x - 2)}{(x - 5)(x - 2)} = \frac{x + 9}{x - 5}$$

11.
$$r(-3) = \frac{(-3) - 9}{(-3)^2 - 5(-3) + 16} = \frac{-3 - 9}{9 + 15 + 16} = \frac{-12}{40} = -\frac{3}{10}$$

14.
$$x^2 - 5x - 6 = 0$$
$$(x + 1)(x - 6) = 0$$
$$x = -1 \quad\text{or}\quad x = 6$$
$$x \neq -1, 6$$

16.
$$\frac{x^2 + 16x + 63}{x^2 + x - 12} \cdot \frac{x^2 - 16}{x^2 + 3x - 54} = \frac{(x + 7)(x + 9)}{(x + 4)(x - 3)} \cdot \frac{(x + 4)(x - 4)}{(x + 9)(x - 6)} = \frac{(x + 7)(x - 4)}{(x - 3)(x - 6)}$$

19.
$$f(x) \cdot g(x) = \frac{x + 8}{x^2 + 2x} \cdot \frac{x^2 + 12x + 20}{x^2 + 2x - 80} = \frac{x + 8}{x(x + 2)} \cdot \frac{(x + 2)(x + 10)}{(x + 10)(x - 8)} = \frac{x + 8}{x(x - 8)}$$

24.
$$\frac{100 - x^2}{x - 7} \div \frac{x^2 - 13x + 30}{x^2 - x - 42} = \frac{(10 + x)(10 - x)}{x - 7} \cdot \frac{(x - 7)(x + 6)}{(x - 3)(x - 10)}$$
$$= \frac{(10 + x)\overset{-1}{\cancel{(10 - x)}}(10 - x)}{\underset{1}{\cancel{x - 7}}} \cdot \frac{\overset{1}{\cancel{(x - 7)}}(x + 6)}{(x - 3)\underset{1}{\cancel{(x - 10)}}}$$
$$= -\frac{(10 + x)(x + 6)}{x - 3}$$

28.
$$\frac{x^2 + 4x}{x^2 + 11x + 24} - \frac{7x + 18}{x^2 + 11x + 24} = \frac{x^2 + 4x - 7x - 18}{x^2 + 11x + 24} = \frac{x^2 - 3x - 18}{x^2 + 11x + 24} = \frac{(x - 6)(x + 3)}{(x + 3)(x + 8)} = \frac{x - 6}{x + 8}$$

32.
$$\frac{x^2 + 6x + 10}{x^2 - 16} - \frac{8x + 30}{16 - x^2} = \frac{x^2 + 6x + 10}{x^2 - 16} + \frac{8x + 30}{x^2 - 16} = \frac{x^2 + 6x + 10 + 8x + 30}{x^2 - 16} = \frac{x^2 + 14x + 40}{x^2 - 16} = \frac{(x + 4)(x + 10)}{(x + 4)(x - 4)} = \frac{x + 10}{x - 4}$$

33.
$$\frac{5}{x^2 - 3x - 4} + \frac{4}{x^2 - 12x + 32} = \frac{5}{(x - 4)(x + 1)} + \frac{4}{(x - 4)(x - 8)}$$
$$= \frac{5}{(x - 4)(x + 1)} \cdot \frac{x - 8}{x - 8} + \frac{4}{(x - 4)(x - 8)} \cdot \frac{x + 1}{x + 1}$$
$$= \frac{5x - 40 + 4x + 4}{(x - 4)(x + 1)(x - 8)} = \frac{9x - 36}{(x - 4)(x + 1)(x - 8)} = \frac{9(x - 4)}{(x - 4)(x + 1)(x - 8)} = \frac{9}{(x + 1)(x - 8)}$$

35.
$$\frac{x}{x^2 - 16} - \frac{5}{x^2 + 2x - 24} = \frac{x}{(x + 4)(x - 4)} - \frac{5}{(x + 6)(x - 4)}$$
$$= \frac{x}{(x + 4)(x - 4)} \cdot \frac{x + 6}{x + 6} - \frac{5}{(x + 6)(x - 4)} \cdot \frac{x + 4}{x + 4}$$
$$= \frac{x(x + 6) - 5(x + 4)}{(x + 4)(x - 4)(x + 6)} = \frac{x^2 + 6x - 5x - 20}{(x + 4)(x - 4)(x + 6)} = \frac{x^2 + x - 20}{(x + 4)(x - 4)(x + 6)}$$
$$= \frac{(x + 5)(x - 4)}{(x + 4)(x - 4)(x + 6)} = \frac{x + 5}{(x + 4)(x + 6)}$$

38.
$$f(x) + g(x) = \frac{3}{x^2 - 4x - 5} + \frac{1}{x^2 - 12x + 35} = \frac{3}{(x - 5)(x + 1)} + \frac{1}{(x - 5)(x - 7)}$$
$$= \frac{3}{(x - 5)(x + 1)} \cdot \frac{x - 7}{x - 7} + \frac{1}{(x - 5)(x - 7)} \cdot \frac{x + 1}{x + 1}$$
$$= \frac{3x - 21 + x + 1}{(x - 5)(x + 1)(x - 7)} = \frac{4x - 20}{(x - 5)(x + 1)(x - 7)} = \frac{4(x - 5)}{(x - 5)(x + 1)(x - 7)} = \frac{4}{(x + 1)(x - 7)}$$

42.
$$\frac{1 + \frac{19}{x} + \frac{90}{x^2}}{1 + \frac{8}{x} - \frac{9}{x^2}} = \frac{1 + \frac{19}{x} + \frac{90}{x^2}}{1 + \frac{8}{x} - \frac{9}{x^2}} \cdot \frac{x^2}{x^2}$$
$$= \frac{1 \cdot x^2 + \frac{19}{x} \cdot x^2 + \frac{90}{x^2} \cdot x^2}{1 \cdot x^2 + \frac{8}{x} \cdot x^2 - \frac{9}{x^2} \cdot x^2}$$
$$= \frac{x^2 + 19x + 90}{x^2 + 8x - 9} = \frac{(x + 9)(x + 10)}{(x + 9)(x - 1)} = \frac{x + 10}{x - 1}$$

45.

$$\begin{aligned} x + 6 - \tfrac{48}{x} &= 4 \\ x \cdot x + 6 \cdot x - \tfrac{48}{x} \cdot x &= 4 \cdot x \\ x^2 + 6x - 48 &= 4x \\ x^2 + 2x - 48 &= 0 \\ (x + 8)(x - 6) &= 0 \\ x = -8 \quad \text{or} \quad x &= 6 \end{aligned}$$

49.

$$\begin{aligned} \tfrac{x}{x + 6} + \tfrac{7}{x - 8} &= \tfrac{84}{x^2 - 2x - 48} \\ \tfrac{x}{x + 6} + \tfrac{7}{x - 8} &= \tfrac{84}{(x + 6)(x - 8)} \\ (x + 6)(x - 8) \cdot \tfrac{x}{x + 6} + (x + 6)(x - 8) \cdot \tfrac{7}{x - 8} &= (x + 6)(x - 8) \cdot \tfrac{84}{(x + 6)(x - 8)} \\ x(x - 8) + 7(x + 6) &= 84 \\ x^2 - 8x + 7x + 42 &= 84 \\ x^2 - x + 42 &= 84 \\ x^2 - x - 42 &= 0 \\ (x + 6)(x - 7) &= 0 \end{aligned}$$

$x = -6$ or $x = 7$

$x = -6$ is an extraneous solution.

$\{7\}$

53.

$$\begin{aligned} y &= \tfrac{5x}{2x - 3} \\ y(2x - 3) &= 5x \\ 2xy - 3y &= 5x \\ 2xy - 5x &= 3y \\ x(2y - 5) &= 3y \\ x &= \tfrac{3y}{2y - 5} \end{aligned}$$

58.

Pipe	Time alone	Work-rate	Time Working	Portion Completed
Smaller	$t + 15$ hours	$\frac{1}{t + 15}$	10 hrs	$\frac{10}{t + 15}$
Larger	t hours	$\frac{1}{t}$	10 hrs	$\frac{10}{t}$

$$\begin{aligned} \tfrac{10}{t + 15} + \tfrac{10}{t} &= 1 \\ t(t + 15) \cdot \tfrac{10}{t + 15} + t(t + 15) \cdot \tfrac{10}{t} &= 1 \cdot t(t + 15) \\ 10t + 10(t + 15) &= t(t + 15) \\ 10t + 10t + 150 &= t^2 + 15t \\ 20t + 150 &= t^2 + 15t \\ 0 &= t^2 - 5t - 150 \\ 0 &= (t - 15)(t + 10) \end{aligned}$$

$t = 15$ or $t = -10$ (Omit negative solution.)

It would take the smaller pipe $15 + 15 = 30$ hours.

60. Rate upstream: $r - 2$, Rate downstream: $r + 2$

$$\begin{aligned} \tfrac{4}{r - 2} + \tfrac{8}{r + 2} &= 2 \\ (r - 2)(r + 2) \cdot \tfrac{4}{r - 2} + (r - 2)(r + 2) \cdot \tfrac{8}{r + 2} &= (r - 2)(r + 2) \cdot 2 \\ 4(r + 2) + 8(r - 2) &= 2(r + 2)(r - 2) \\ 4r + 8 + 8r - 16 &= 2r^2 - 8 \\ 12r - 8 &= 2r^2 - 8 \\ 0 &= 2r^2 - 12r \\ 0 &= 2r(r - 6) \end{aligned}$$

$r = 0$ or $r = 6$ (Omit zero solution.)

Sarah can paddle 6 mph in still water.

68.

$$\begin{array}{r} x - 13 \\ x + 8 \overline{) x^2 - 5x - 113} \\ \underline{-x^2 \not+ 8x \downarrow} \\ -13x - 113 \\ \underline{\not- 13x \not- 104} \\ -9 \end{array}$$

$\frac{x^2 - 5x - 113}{x + 8} = x - 13 - \frac{9}{x + 8}$

71.

$$\begin{array}{r} x^2 + 6x + 43 \\ x - 6\overline{)\, x^3 + 0x^2 + 7x - 20} \\ \underline{{}^{-}x^3 \overset{+}{\not-} 6x^2 \quad \downarrow \quad \downarrow} \\ 6x^2 + 7x - 20 \\ \underline{{}^{-}6x^2 \overset{+}{\not-} 36x \quad \downarrow} \\ 43x - 20 \\ \underline{\overset{-}{\not+} 43x \overset{+}{\not-} 258} \\ 238 \end{array}$$

$\frac{x^3 + 7x - 20}{x - 6} = x^2 + 6x + 43 + \frac{238}{x - 6}$

Chapter 5 Test **1.** $-\frac{4}{9}$ **2.** 5, 8 **3.** $\frac{x-4}{x+7}$ **4.** $-\frac{7}{12}$ **5.** $\frac{(x+8)(x-10)}{(x+4)(x-5)}$ **6.** $\frac{(x-3)(x-10)}{(x+4)(3+x)}$ **7.** $\frac{x-6}{x(x-5)}$ **8.** $\frac{3}{2}$ **9.** $x + 9$ **10.** $\frac{x-8}{(x-3)(x+2)}$ **11.** $\frac{9}{(x-9)(x-3)}$ **12.** $\frac{x-10}{(x+6)(x-2)}$ **13.** $\frac{x+2}{x-9}$ **14.** $\{20\}$ **15.** $\{-5, 14\}$ **16.** $y = \frac{8x}{4-x}$ **17.** 180 min **18.** 20 mph **19.** $2x - 21 - \frac{15}{x+4}$ **20.** $2x^2 + 5x + 35 + \frac{136}{x-7}$

Chapter 6

Quick Check 6.1 **1. a)** 7 **b)** $\frac{2}{5}$ **c)** -6 **2. a)** $|a^7|$ **b)** $5|b^{11}|$ **3.** $8x^{12}$ **4.** 10.440 **5. a)** 3 **b)** 4 **c)** -7 **6.** x^7 **7.** $-2x^7y^8$ **8.** $x + 5$ **9.** $30b^4$ **10.** 144 **11. a)** 5 **b)** $3a^3$ **12.** 35 **13.** $[-18, \infty)$ **14.** $(-\infty, \infty)$

Section 6.1 **1.** square root **3.** *n*th root **5.** radicand **7.** radical function **9.** 6 **11.** 2 **13.** $\frac{1}{4}$ **15.** $\frac{6}{5}$ **17.** -4 **19.** Not a real number **21.** a^8 **23.** x^{10} **25.** $3|x^3|$ **27.** $\frac{1}{7}x^2$ **29.** $|m^7n^5|$ **31.** $x^2y^4z^{16}$ **33.** 81 **35.** $9x^2$ **37.** 7.416 **39.** 18.055 **41.** 0.728 **43.** 2 **45.** 5 **47.** a^6 **49.** $7m^7$ **51.** $3x^3y^5$ **53.** $x + 3$ **55.** 9 **57.** 30 **59.** 6 **61.** 10 **63.** 6 **65.** 6 **67.** 3 **69.** 200 **71.** 600 **73.** 5 **75.** 392 **77.** 4 **79.** 3 **81.** 8.828 **83.** 5 **85.** $[8, \infty)$ **87.** $[\frac{15}{2}, \infty)$ **89.** $[-\frac{5}{2}, \infty)$ **91.** $(-\infty, \infty)$ **93.** Real: $-\sqrt{64}$, $-\sqrt[3]{64}$, $\sqrt[3]{-64}$; not real: $\sqrt{-64}$. Explanations will vary **95.** Answers will vary

Quick Check 6.2 **1. a)** $\sqrt{36} = 6$ **b)** $\sqrt[5]{-32x^{15}} = -2x^3$ **2.** $\sqrt[4]{x^{32}y^4z^{24}} = x^8yz^6$ **3.** $a^{1/9}$ **4.** $(\sqrt[6]{4096x^{18}})^5 = 1024x^{15}$ **5.** $x^{5/12}$ **6.** $x^{29/24}$ **7.** $x^{14/45}$ **8.** $\frac{x^{15}}{y^5z^{20}}$ **9.** $\frac{1}{729}$ **10.** $\sqrt[12]{x^7}$

Section 6.2 **1.** $\sqrt[n]{a}$ **3.** $x^{1/4}$ **5.** $7^{1/2}$ **7.** $9d^{1/4}$ **9.** 8 **11.** 4 **13.** -7 **15.** x^3 **17.** x^8y^7 **19.** $2x^8y^{15}$ **21.** $-6x^6y^5z^7$ **23.** $x^{3/5}$ **25.** $y^{8/9}$ **27.** $x^{5/8}$ **29.** $(3x^2)^{7/3}$ **31.** $(10x^4y^5)^{3/8}$ **33.** 125 **35.** 4 **37.** x^8 **39.** $64a^6b^{18}$ **41.** $625x^{12}y^{20}z^4$ **43.** $x^{4/5}$ **45.** $x^{13/20}$ **47.** $x^{11/6}$ **49.** $a^{31/12}$ **51.** $x^{1/2}y^{1/2}$ **53.** $x^{1/10}$ **55.** $b^{4/9}$ **57.** $x^{7/2}$ **59.** $x^{1/2}$ **61.** $x^{7/24}$ **63.** 1 **65.** 1 **67.** $\frac{1}{4}$ **69.** $\frac{1}{7}$ **71.** $\frac{1}{128}$ **73.** $\frac{1}{6}$ **75.** $\frac{1}{3125}$ **77.** $\frac{1}{6}$ **79.** $\sqrt[20]{x^9}$ **81.** $\sqrt[3]{a}$ **83.** $\sqrt[6]{m}$ **85.** $\frac{1}{\sqrt[6]{x}}$ **87.** Answers will vary; one example is $16^{1/2}$. **89.** Answers will vary; one example is $(25x^6)^{1/2}$. **91.** Yes; explanations will vary.

Quick Check 6.3 **1.** $x^3\sqrt[5]{x^2}$ **2.** $a^4b^7c^3\sqrt{bcd}$ **3.** $3\sqrt{10}$ **4.** $2\sqrt[3]{35}$ **5.** $x^6yz^{10}w^3\sqrt[5]{x^3yw^3}$ **6.** $21\sqrt[5]{2x^2}$ **7.** $15\sqrt{10} - 5\sqrt{5}$ **8.** $11\sqrt[5]{a^4b^3} - 11\sqrt[7]{a^4b^3}$ **9.** $4\sqrt{7}$ **10.** $6\sqrt{2}$

Section 6.3 **1.** like radicals **3.** $2\sqrt{3}$ **5.** $4\sqrt{5}$ **7.** $11\sqrt{3}$ **9.** $3\sqrt[3]{3}$ **11.** $5\sqrt[3]{6}$ **13.** $2\sqrt[4]{75}$ **15.** $x^4\sqrt{x}$ **17.** $m^4\sqrt[3]{m}$ **19.** $x^{14}\sqrt[5]{x^4}$ **21.** $x^7y^6\sqrt{x}$ **23.** $y^3\sqrt{xy}$ **25.** $x^{16}y^8z^8\sqrt{xy}$ **27.** $a^4c^6\sqrt[3]{b^2c}$ **29.** $a^{11}b^5\sqrt[6]{a^3b^5c^5}$ **31.** $2a^4b^4\sqrt{2a}$ **33.** $2b^3c^5\sqrt[4]{7a^3bc^3}$ **35.** $27\sqrt{5}$ **37.** $-6\sqrt[3]{4}$ **39.** $-5\sqrt{21}$ **41.** $16\sqrt{10} - 6\sqrt{15}$ **43.** $3\sqrt{2} + 10\sqrt{3}$ **45.** $-2\sqrt{x}$ **47.** $11\sqrt[5]{a}$ **49.** $22\sqrt{x}$ **51.** $5\sqrt{x} - 17\sqrt{y}$ **53.** $21\sqrt[3]{x} - 3\sqrt[4]{x}$ **55.** $13\sqrt{2}$ **57.** $21\sqrt{7}$ **59.** $-24\sqrt[3]{3}$ **61.** $162\sqrt{3} + 6\sqrt{6}$ **63.** $7\sqrt{3} - 11\sqrt{2}$ **65.** $-6\sqrt{2} + 17\sqrt{5}$ **67.** Answers will vary; one example is $3\sqrt{2} + 4\sqrt{2}$. **69.** Answers will vary; one example is $7\sqrt{5} + \sqrt{20} + \sqrt{3} + 3\sqrt{3}$. **71.** Answers will vary.

Quick Check 6.4 **1. a)** 60 **b)** $x^4yz\sqrt[3]{z^2}$ **c)** $144\sqrt{3}$ **2.** $6 + 6\sqrt{5}$ **3.** $x^6y^6 - x^8y^6\sqrt{xy}$ **4. a)** $57 - 2\sqrt{6}$ **b)** $17 - 2\sqrt{70}$ **5. a)** 9 **b)** 529 **6.** $\frac{\sqrt{210}}{3}$ **7.** $\frac{\sqrt{5}}{5}$ **8.** $\frac{2y^2\sqrt{xy}}{5x^2z^5}$ **9.** $\frac{5\sqrt{3} - 3\sqrt{5}}{2}$ **10.** $\frac{33 - 20\sqrt{3}}{3}$

Section 6.4 **1.** x **3.** rationalizing **5.** 6 **7.** 6 **9.** 5 **11.** 60 **13.** x^8 **15.** $m^9n^8\sqrt{m}$ **17.** $x^{10}y^7\sqrt[4]{x}$ **19.** $4 - 2\sqrt{3}$ **21.** $42 + 14\sqrt{5}$ **23.** $6\sqrt{15} - 40\sqrt{6}$ **25.** $m^4n^6\sqrt{n} + m^6n^8\sqrt{m}$ **27.** $a^6b^{15}c^4\sqrt[3]{c} + a^{10}b^3c^7\sqrt[3]{b}$ **29.** $2 - \sqrt{6} + \sqrt{10} - \sqrt{15}$ **31.** 5 **33.** $19\sqrt{6} - 153$ **35.** $9\sqrt{6} - 315\sqrt{2} + 12\sqrt{3} - 420$ **37.** $98 - 24\sqrt{10}$ **39.** -3 **41.** 414 **43.** -178 **45.** $\frac{6}{5}$ **47.** $\frac{2\sqrt{3}}{x^2y^3}$

49. $3\sqrt{2}$ **51.** $\frac{3\sqrt{3}}{2}$ **53.** $\frac{2\sqrt{6}}{3}$ **55.** $\frac{\sqrt{2}}{2}$ **57.** $\frac{3\sqrt{33a}}{11a^4}$ **59.** $\frac{s^2\sqrt[3]{4rs^2t^2}}{2rt}$ **61.** $\frac{5\sqrt{2}}{2}$ **63.** $\frac{b^4\sqrt{ac}}{a^2c^4}$ **65.** $\frac{3\sqrt{13}+3\sqrt{7}}{2}$ **67.** $-\frac{3-\sqrt{33}}{2}$ **69.** $\frac{3\sqrt{39}+9\sqrt{3}}{2}$ **71.** $\frac{3\sqrt{2}+\sqrt{6}+4\sqrt{3}+4}{4}$ **73.** -3 **75.** $-\frac{7+10\sqrt{6}}{19}$ **77.** $\frac{3\sqrt{5}}{7}$ **79.** $573+80\sqrt{35}$ **81.** $5a^3b^2c^4\sqrt[3]{5c^2}$ **83.** $-\frac{99-17\sqrt{35}}{2}$ **85.** $120+135\sqrt{2}-56\sqrt{5}-63\sqrt{10}$ **87.** $-14x^5\sqrt{2}+7x^{11}$ **89.** $\frac{a^2b^4\sqrt{3ac}}{6c^5}$ **91.** $\frac{5\sqrt[3]{6x}}{3x}$ **93.** 2257 **95.** $a+2\sqrt{ab}+b$ **97.** Answers will vary; one example is $\sqrt{23}+\sqrt{3}$ and $\sqrt{23}-\sqrt{3}$. **99.** Answers will vary; one example is $5\sqrt{2}+4\sqrt{3}$. **101.** Answers will vary.

Quick Review Exercises 6.4 **1.** $2^4\cdot3^2$ **2.** $2^3\cdot5^2\cdot7$ **3.** 2^{10} **4.** $2\cdot3^3\cdot7^2\cdot11$

Quick Check 6.5 **1. a)** $\{47\}$ **b)** $\{-73\}$ **2.** $\varnothing$ **3.** 11 **4. a)** $\{9\}$ **b)** $\{-220\}$ **5.** $\{2, 10\}$ **6.** $\{8\}$ **7.** $\{-22\}$ **8.** $\{6\}$ **9.** 2.72 seconds **10.** 7.30 feet **11.** 70 miles per hour

Section 6.5 **1.** radical **3.** isolate **5.** pendulum **7.** $\{23\}$ **9.** $\{16\}$ **11.** $\varnothing$ **13.** $\{7\}$ **15.** $\varnothing$ **17.** $\{-7, 4\}$ **19.** 45 **21.** 13 **23.** $-2, 7$ **25.** $\{4\}$ **27.** $\{39\}$ **29.** $\{42\}$ **31.** $\{8\}$ **33.** $\{9\}$ **35.** $\{7\}$ **37.** $\varnothing$ **39.** $\{3\}$ **41.** $\{2\}$ **43.** $\{26\}$ **45.** $\{9\}$ **47.** $\{-5, \frac{9}{2}\}$ **49.** $\{5\}$ **51.** $\{10\}$ **53.** $\{263\}$ **55.** 2.22 seconds **57.** 1.76 seconds **59.** 0.3 feet **61.** 1.32 seconds **63.** 3.48 seconds **65.** 40 mph **67.** 73 mph **69.** 138.6 gallons/minute **71.** 71.4 gallons/minute **73.** 2.3 feet **75.** 8.04 knots **77.** 50.3 feet **79.** 534.0 mph **81.** 3.8 miles **83.** Answers will vary; one example is $\sqrt{2x+10}=4$. **85.** Answers will vary. **87.** Answers will vary.

Quick Check 6.6 **1. a)** $6i$ **b)** $3i\sqrt{7}$ **c)** $-3i\sqrt{6}$ **2. a)** $-5+27i$ **b)** $11-28i$ **3.** -60 **4.** $-10\sqrt{6}$ **5.** $48-42i$ **6.** $26-13i$ **7.** $65-72i$ **8.** 65 **9.** 137 **10.** $\frac{12-9i}{10}$ or $\frac{6}{5}-\frac{9}{10}i$ **11.** $\frac{57+41i}{58}$ or $\frac{57}{58}+\frac{41}{58}i$ **12.** $\frac{-8-15i}{12}$ or $-\frac{2}{3}-\frac{5}{4}i$ **13.** -1

Section 6.6 **1.** imaginary **3.** -1 **5.** real **7.** $2i$ **9.** $9i$ **11.** $-11i$ **13.** $3i\sqrt{3}$ **15.** $4i\sqrt{2}$ **17.** $11+14i$ **19.** $4-17i$ **21.** $4i$ **23.** $26i$ **25.** $-2-24i$ **27.** $6+9i$ **29.** $2+9i$ **31.** -98 **33.** 20 **35.** -56 **37.** $-6\sqrt{6}$ **39.** -30 **41.** $5i\cdot13i$; answers may vary. **43.** $-20i\cdot2i$; answers may vary. **45.** $18+48i$ **47.** $-49-28i$ **49.** $29-37i$ **51.** $20-48i$ **53.** 82 **55.** $-48+14i$ **57.** 29 **59.** 208 **61.** $2-i$ **63.** $\frac{-12+21i}{13}$ or $-\frac{12}{13}+\frac{21}{13}i$ **65.** $\frac{-6-35i}{13}$ or $-\frac{6}{13}-\frac{35}{13}i$ **67.** $\frac{-11+60i}{61}$ or $-\frac{11}{61}+\frac{60}{61}i$ **69.** $-8i$ **71.** $\frac{-7-5i}{3}$ or $-\frac{7}{3}-\frac{5}{3}i$ **73.** $\frac{-1+6i}{4}$ or $-\frac{1}{4}+\frac{3}{2}i$ **75.** $\frac{6+14i}{29}$ or $\frac{6}{29}+\frac{14}{29}i$ **77.** $\frac{14-27i}{25}$ or $\frac{14}{25}-\frac{27}{25}i$ **79.** $3+8i$ **81.** $31+33i$ **83.** $-i$ **85.** 1 **87.** -1 **89.** 0 **91.** $48-9i$ **93.** $\frac{4-9i}{5}$ or $\frac{4}{5}-\frac{9}{5}i$ **95.** $\frac{-6+12i}{5}$ or $-\frac{6}{5}+\frac{12}{5}i$ **97.** 29 **99.** $-18\sqrt{2}$ **101.** $\frac{-6+27i}{17}$ or $-\frac{6}{17}+\frac{27}{17}i$ **103.** $6i\sqrt{7}$ **105.** $72-54i$ **107.** $23+10i$ **109.** 126 **111.** $-i$ **113.** $-24+16i$ **115.** Answers will vary.

Chapter 6 Review **1.** 5 **2.** 13 **3.** -4 **4.** x^3 **5.** $9x^6$ **6.** $3x^3y^2z^5$ **7.** 4.123 **8.** 6.403 **9.** 9 **10.** 2.828 **11.** $[-3, \infty)$ **12.** $[-\frac{25}{3}, \infty)$ **13.** $a^{1/5}$ **14.** $x^{3/4}$ **15.** $x^{13/2}$ **16.** $n^{8/5}$ **17.** 2 **18.** $49x^{14}$ **19.** $x^{5/3}$ **20.** $x^{7/16}$ **21.** $x^{1/4}$ **22.** $\frac{1}{7}$ **23.** $\frac{1}{1024}$ **24.** $\frac{1}{81}$ **25.** $2\sqrt{7}$ **26.** $x^3\sqrt[3]{x}$ **27.** $x^2yz^3\sqrt[5]{xy^3}$ **28.** $3a^8b^4c^5\sqrt[3]{15ac}$ **29.** $6rs^5t^6\sqrt{5rs}$ **30.** $5r^3s^4t^2\sqrt{7}$ **31.** $12\sqrt{2}$ **32.** $6\sqrt[3]{x}$ **33.** $25\sqrt{5}$ **34.** $57\sqrt{3}-40\sqrt{2}$ **35.** $2x^2\sqrt[3]{9}$ **36.** $30\sqrt{2}-60$ **37.** $3\sqrt{2}-\sqrt{6}+4\sqrt{3}-4$ **38.** $296-43\sqrt{55}$ **39.** $303-108\sqrt{5}$ **40.** -648 **41.** $4x^2\sqrt{7}$ **42.** $\frac{5a^3b^3\sqrt{b}}{2c^5}$ **43.** $\frac{2\sqrt[3]{50}}{5}$ **44.** $\frac{4\sqrt{14}}{7}$ **45.** $\frac{\sqrt{22}}{6}$ **46.** $\frac{a^3c^2\sqrt{bc}}{b^2}$ **47.** $\frac{4(\sqrt{11}+\sqrt{5})}{3}$ **48.** $\frac{5\sqrt{5}+3}{116}$ **49.** $\frac{13\sqrt{14}+44}{10}$ **50.** $\frac{-53+27\sqrt{5}}{209}$ **51.** $\{12\}$ **52.** $\{20\}$ **53.** $\{9\}$ **54.** $\{2\}$ **55.** $\{9\}$ **56.** $\varnothing$ **57.** 4 **58.** $-2, 8$ **59.** 2.48 seconds **60.** 67 mph **61.** $7i$ **62.** $2i\sqrt{10}$ **63.** $6i\sqrt{7}$ **64.** $-15i\sqrt{3}$ **65.** $11+5i$ **66.** $16-2i$ **67.** $1+5i$ **68.** $1+7i$ **69.** -16 **70.** $-10\sqrt{3}$ **71.** $28+21i$ **72.** $114+106i$ **73.** $72-54i$ **74.** 205 **75.** $\frac{21+9i}{29}$ or $\frac{21}{29}+\frac{9}{29}i$ **76.** $\frac{12+9i}{10}$ or $\frac{6}{5}+\frac{9}{10}i$ **77.** $\frac{13+19i}{106}$ or $\frac{13}{106}+\frac{19}{106}i$ **78.** $-16i$ **79.** i **80.** -1

Chapter 6 Review Exercises: Worked-Out Solutions

4.
$$\begin{aligned}\sqrt[3]{x^9} &= \sqrt[3]{(x^3)^3}\\ &= x^3\end{aligned}$$

9.
$$\begin{aligned}f(x) &= \sqrt{x-10}\\ f(91) &= \sqrt{91-10}\\ &= \sqrt{81}\\ &= 9\end{aligned}$$

11.
$$\begin{aligned}2x+6 &\geq 0\\ 2x &\geq -6\\ x &\geq -3\end{aligned}$$
$$[-3, \infty)$$

18.
$$\begin{aligned}(343x^{21})^{2/3} &= (\sqrt[3]{343x^{21}})^2\\ &= (7x^7)^2\\ &= 49x^{14}\end{aligned}$$

19.
$$\begin{aligned}x^{3/2}\cdot x^{1/6} &= x^{3/2+1/6}\\ &= x^{9/6+1/6}\\ &= x^{10/6}\\ &= x^{5/3}\end{aligned}$$

27.
$$\begin{aligned}\sqrt[5]{x^{11}y^8z^{15}} &= \sqrt[5]{x^{10}y^5z^{15}}\cdot\sqrt[5]{xy^3}\\ &= x^2yz^3\sqrt[5]{xy^3}\end{aligned}$$

33.
$$\begin{aligned}5\sqrt{20}+3\sqrt{125} &= 5\sqrt{4\cdot5}+3\sqrt{25\cdot5}\\ &= 5\cdot2\sqrt{5}+3\cdot5\sqrt{5}\\ &= 10\sqrt{5}+15\sqrt{5}\\ &= 25\sqrt{5}\end{aligned}$$

38. $(10\sqrt{5}-7\sqrt{11})(9\sqrt{5}+2\sqrt{11}) = 10\sqrt{5}\cdot 9\sqrt{5}+10\sqrt{5}\cdot 2\sqrt{11}-7\sqrt{11}\cdot 9\sqrt{5}-7\sqrt{11}\cdot 2\sqrt{11}$
$= 90\cdot 5+20\sqrt{55}-63\sqrt{55}-14\cdot 11$
$= 450-43\sqrt{55}-154$
$= 296-43\sqrt{55}$

45. $\sqrt{\frac{11}{18}}=\frac{\sqrt{11}}{\sqrt{18}}\cdot\frac{\sqrt{2}}{\sqrt{2}}$
$=\frac{\sqrt{22}}{\sqrt{36}}$
$=\frac{\sqrt{22}}{6}$

48. $\frac{2\sqrt{3}}{10\sqrt{15}-3\sqrt{12}}=\frac{2\sqrt{3}}{10\sqrt{15}-3\sqrt{12}}\cdot\frac{10\sqrt{15}+3\sqrt{12}}{10\sqrt{15}+3\sqrt{12}}$
$=\frac{20\sqrt{45}+6\sqrt{36}}{100\cdot 15-9\cdot 12}$
$=\frac{20\cdot 3\sqrt{5}+6\cdot 6}{1500-108}$
$=\frac{60\sqrt{5}+36}{1392}$
$=\frac{5\sqrt{5}+3}{116}$

51. $\sqrt{4x+1}=7$
$(\sqrt{4x+1})^2=7^2$
$4x+1=49$
$4x=48$
$x=12$
$\{12\}$

54. $\sqrt{17-4x}-x=1$
$\sqrt{17-4x}=x+1$
$(\sqrt{17-4x})^2=(x+1)^2$
$17-4x=(x+1)(x+1)$
$17-4x=x^2+2x+1$
$0=x^2+6x-16$
$0=(x-2)(x+8)$
$x=2$ or $x=-8$
$x=-8$ is an extraneous solution.
$\{2\}$

60. $s=\sqrt{30df}$
$s=\sqrt{30(200)(0.75)}$
$s=\sqrt{4500}$
$s\approx 67$

62. $\sqrt{-40}=\sqrt{-1}\cdot\sqrt{40}$
$=i\cdot\sqrt{4\cdot 10}$
$=i\cdot 2\sqrt{10}$
$=2i\sqrt{10}$

65. $(4+2i)+(7+3i)=4+7+2i+3i$
$=11+5i$

72. $(10+2i)(13+8i)=130+80i+26i+16i^2$
$=130+106i-16$
$=114+106i$

75. $\frac{6}{7-3i}=\frac{6}{7-3i}\cdot\frac{7+3i}{7+3i}$
$=\frac{42+18i}{49-9i^2}$
$=\frac{42+18i}{49+9}$
$=\frac{42+18i}{58}$
$=\frac{21+9i}{29}$ or $\frac{21}{29}+\frac{9}{29}i$

78. $\frac{16}{i}=\frac{16}{i}\cdot\frac{i}{i}$
$=\frac{16i}{i^2}$
$=\frac{16i}{-1}$
$=-16i$

79. $i^{53}=(i^4)^{13}\cdot i^1$
$=1\cdot i$
$=i$

Chapter 6 Test **1.** $5x^5y^7$ **2.** 8.718 **3.** $a^{1/2}$ **4.** $b^{11/24}$ **5.** $\frac{1}{25}$ **6.** $6\sqrt{2}$ **7.** $a^5b^{10}c\sqrt[4]{ab^2}$ **8.** $5\sqrt{2}$ **9.** $140\sqrt{2}-40\sqrt{7}$ **10.** $226-168\sqrt{3}$ **11.** $\frac{x\sqrt{xy}}{y^4}$ **12.** $\frac{-564+167\sqrt{14}}{50}$ **13.** {20} **14.** {8} **15.** 13.0 ft **16.** $3i\sqrt{11}$ **17.** $-18-11i$ **18.** 85 **19.** $\frac{-33+47i}{34}$ or $-\frac{33}{34}+\frac{47}{34}i$

Chapter 7

Quick Check 7.1 **1. a)** $\{-3, 5\}$ **b)** $\{-11, 11\}$ **2. a)** $\{-2\sqrt{7}, 2\sqrt{7}\}$ **b)** $\{-4i\sqrt{2}, 4i\sqrt{2}\}$ **3. a)** $\{-\frac{11}{3}, 3\}$ **b)** $\{\frac{3-3i}{2}, \frac{3+3i}{2}\}$ **4.** 12.2 in. **5.** 2.5 cm **6. a)** $\{-6, -2\}$ **b)** $\{3-i, 3+i\}$ **7.** $\{-9, 2\}$ **8.** $\{8-3\sqrt{2}, 8+3\sqrt{2}\}$ **9.** $3x^2+x-10=0$ **10.** $x^2+100=0$

Section 7.1 **1.** extracting square roots **3.** completing the square **5.** $\{-2, 5\}$ **7.** $\{-8, -3\}$ **9.** $\{-4, -2\}$ **11.** $\{-4, 6\}$ **13.** $\{-8, 8\}$ **15.** $\{-2, 13\}$ **17.** $\{-6, 6\}$ **19.** $\{-7\sqrt{2}, 7\sqrt{2}\}$ **21.** $\{-5i, 5i\}$ **23.** $\{-2\sqrt{7}, 2\sqrt{7}\}$ **25.** $\{-1, 11\}$ **27.** $\{-3-3i\sqrt{3}, -3+3i\sqrt{3}\}$ **29.** $\{-5-3i\sqrt{2}, -5+3i\sqrt{2}\}$ **31.** $\{9-5\sqrt{2}, 9+5\sqrt{2}\}$ **33.** $\{-1, \frac{9}{5}\}$ **35.** $x^2+12x+36=(x+6)^2$ **37.** $x^2-5x+\frac{25}{4}=(x-\frac{5}{2})^2$ **39.** $x^2-8x+16=(x-4)^2$ **41.** $x^2+\frac{1}{4}x+\frac{1}{64}=(x+\frac{1}{8})^2$ **43.** 9 m

45. 11.3 in. **47.** 2.8 in. **49.** 8.9 m **51.** 5.7 ft **53.** $\{-3, 11\}$ **55.** $\{-2 - i\sqrt{3}, -2 + i\sqrt{3}\}$ **57.** $\{-9, 1\}$ **59.** $\{7 - \sqrt{19}, 7 + \sqrt{19}\}$ **61.** $\{-4 - 2i\sqrt{7}, -4 + 2i\sqrt{7}\}$ **63.** $\{-8, 2\}$ **65.** $\{\frac{9 - \sqrt{209}}{2}, \frac{9 + \sqrt{209}}{2}\}$ **67.** $\{-6, -1\}$ **69.** $\{\frac{-3 - \sqrt{29}}{2}, \frac{-3 + \sqrt{29}}{2}\}$ **71.** $\{-2, -\frac{1}{2}\}$ **73.** $\{-8, 9\}$ **75.** $\{-2 - 6i, -2 + 6i\}$ **77.** $x^2 + 3x - 10 = 0$ **79.** $x^2 - 14x + 48 = 0$ **81.** $4x^2 + 5x - 6 = 0$ **83.** $x^2 - 9 = 0$ **85.** $x^2 + 25 = 0$ **87.** $\{-10, -3\}$ **89.** $\{-3 - \sqrt{26}, -3 + \sqrt{26}\}$ **91.** $\{-1, 7\}$ **93.** $\{4 - 3\sqrt{3}, 4 + 3\sqrt{3}\}$ **95.** $\{2 - 2\sqrt{6}, 2 + 2\sqrt{6}\}$ **97.** $\{-1, 6\}$ **99.** $\{4 - i\sqrt{3}, 4 + i\sqrt{3}\}$ **101.** $\{-3 - 2i\sqrt{2}, -3 + 2i\sqrt{2}\}$ **103.** $\{\frac{-3 - 3i\sqrt{3}}{2}, \frac{-3 + 3i\sqrt{3}}{2}\}$ **105.** $\{-5, 7\}$ **107.** $\{-1, 5\}$ **109.** $\{-5, 1\}$ **111.** 3, 4 **113.** $-2\sqrt{10}, 2\sqrt{10}$ **115.** 1, 9 **117. a)** 31 **b)** 73 **119. a)** Answers will vary, any number greater than -6 will work. **b)** Answers will vary, any number less than -6 will work. **121.** Answers will vary.

Quick Check 7.2 **1. a)** $a = 1, b = 11$, and $c = -13$ **b)** $a = 5, b = -9$, and $c = -11$ **c)** $a = 1, b = 0$, and $c = -30$ **2. a)** $\{-10, 3\}$ **b)** $\{\frac{-21 - 3\sqrt{21}}{14}, \frac{-21 + 3\sqrt{21}}{14}\}$ **3.** $\{\frac{7 - 3i\sqrt{3}}{2}, \frac{7 + 3i\sqrt{3}}{2}\}$ **4.** $\{\frac{-2 - i\sqrt{14}}{3}, \frac{-2 + i\sqrt{14}}{3}\}$ **5.** $\{-\frac{5}{2}, \frac{4}{3}\}$ **6. a)** two real solutions **b)** two nonreal complex solutions **c)** one real solution **7. a)** not factorable **b)** factorable **8. a)** 5 sec **b)** 2 sec, 3 sec **9. a)** 0.32 sec, 1.93 sec **b)** No.

Section 7.2 **1.** quadratic **3.** zero **5.** two **7.** $\{-4, 9\}$ **9.** $\{2 - \sqrt{2}, 2 + \sqrt{2}\}$ **11.** $\{\frac{-1 - 3i\sqrt{3}}{2}, \frac{-1 + 3i\sqrt{3}}{2}\}$ **13.** $\{-6, 0\}$ **15.** $\{-5, 2\}$ **17.** $\{-3, -2\}$ **19.** $\{-2, 2\}$ **21.** $\{-3, 5\}$ **23.** $\{\frac{3 - 3\sqrt{5}}{2}, \frac{3 + 3\sqrt{5}}{2}\}$ **25.** $\{-\frac{1}{5}, 4\}$ **27.** $\{3\}$ **29.** $\{-2\sqrt{6}, 2\sqrt{6}\}$ **31.** $\{-4, 5\}$ **33.** $\{\frac{1 - i\sqrt{74}}{10}, \frac{1 + i\sqrt{74}}{10}\}$ **35.** $\{3, 4\}$ **37.** Two real **39.** Two nonreal complex **41.** Two nonreal complex **43.** One real **45.** Prime **47.** Factorable **49.** Factorable **51.** Factorable **53.** $\{\frac{5 - \sqrt{85}}{2}, \frac{5 + \sqrt{85}}{2}\}$ **55.** $\{-1, \frac{1}{3}\}$ **57.** $\{-\frac{2}{5}, 2\}$ **59.** $\{\frac{-4 - i\sqrt{2}}{2}, \frac{-4 + i\sqrt{2}}{2}\}$ **61.** $\{-18i, 18i\}$ **63.** $\{\frac{4}{3}, \frac{7}{2}\}$ **65.** $\{\frac{-1 - \sqrt{10}}{2}, \frac{-1 + \sqrt{10}}{2}\}$ **67.** $\{\frac{9 - \sqrt{165}}{2}, \frac{9 + \sqrt{165}}{2}\}$ **69.** $\{\frac{3}{4}\}$ **71.** $\{\frac{-1 - i\sqrt{79}}{2}, \frac{-1 + i\sqrt{79}}{2}\}$ **73.** $\{2 - \sqrt{6}, 2 + \sqrt{6}\}$ **75.** $\{3 - i, 3 + i\}$ **77.** $\{\frac{-4 - \sqrt{70}}{9}, \frac{-4 + \sqrt{70}}{9}\}$ **79.** $\{-\frac{7}{3}, \frac{5}{2}\}$ **81.** $\{\frac{9 - 4\sqrt{2}}{2}, \frac{9 + 4\sqrt{2}}{2}\}$ **83.** $\{-4, 1\}$ **85.** $\{5 - \sqrt{7}, 5 + \sqrt{7}\}$ **87.** $\{7, 9\}$ **89.** 8 sec **91.** 5.4 sec **93.** 1 sec, 3 sec **95.** 1.6 sec, 4.6 sec **97.** No; explanations will vary. **99.** $x = -\frac{b}{a}$ **101.** Answers will vary.

Quick Check 7.3 **1.** $\{-2, 2, -i\sqrt{3}, i\sqrt{3}\}$ **2. a)** $u = \sqrt{x}$ **b)** $u = x^{1/3}$ **c)** $u = x^2 - 4x$ **3. a)** $\{1, 25\}$ **b)** $\{1, 27\}$ **4.** $\{-5\}$ **5.** $\{0, 3\}$ **6.** 71.5 min

Section 7.3 **1.** u-substitution **3.** $\{-1, 1, -2, 2\}$ **5.** $\{-\sqrt{3}, \sqrt{3}\}$ **7.** $\{-3i, 3i, -\sqrt{2}, \sqrt{2}\}$ **9.** $\{-2, 2, -3, 3\}$ **11.** $\{1, 64\}$ **13.** $\{16, 256\}$ **15.** $\{9\}$ **17.** $\{4, 25\}$ **19.** $\{1, 3\}$ **21.** $\{-2\}$ **23.** $\{-10, -1\}$ **25.** $\{-216, 1\}$ **27.** $\{-125, -64\}$ **29.** $\{-1, 2\}$ **31.** $\{3, 9\}$ **33.** $\{-8, 2\}$ **35.** $\{2, 5\}$ **37.** $\{1\}$ **39.** $\{2, 10\}$ **41.** $\{-2, 5\}$ **43.** $\{-6, -1\}$ **45.** $\{1\}$ **47.** $\{-\frac{13}{5}, 5\}$ **49.** small: 120 min; large: 60 min **51.** 4.6 hr **53.** 71.6 min **55.** $\{4 - \sqrt{29}, 4 + \sqrt{29}\}$ **57.** $\{64\}$ **59.** $\{\frac{2 - 4\sqrt{2}}{3}, \frac{2 + 4\sqrt{2}}{3}\}$ **61.** $\{\frac{5 - i\sqrt{31}}{2}, \frac{5 + i\sqrt{31}}{2}\}$ **63.** $\{-7, 13\}$ **65.** $\{-1, 2\}$ **67.** $\{6\}$ **69.** $\{-\frac{8}{5}, 0\}$ **71.** $\{-8 - 12i, -8 + 12i\}$ **73.** $\{16\}$ **75.** $\{-343, -125\}$ **77.** $\{-5\}$ **79.** Answers will vary. **81.** Answers will vary.

Quick Review Exercise 7.3 **1.** $\{-4, 10\}$ **2.** $\{-3, 12\}$ **3.** $\{3 - \sqrt{22}, 3 + \sqrt{22}\}$ **4.** $\{\frac{-11 - i\sqrt{39}}{2}, \frac{-11 + i\sqrt{39}}{2}\}$

Quick Check 7.4 **1.**

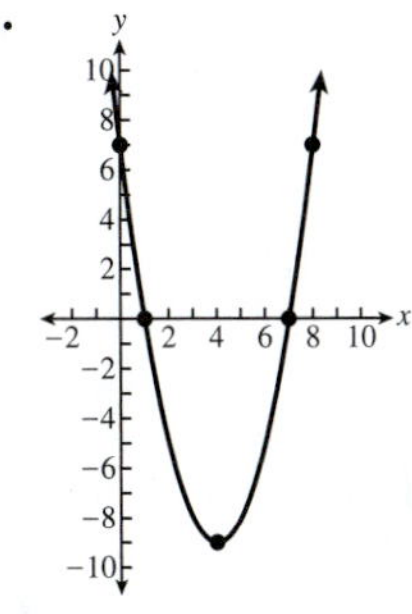

2.

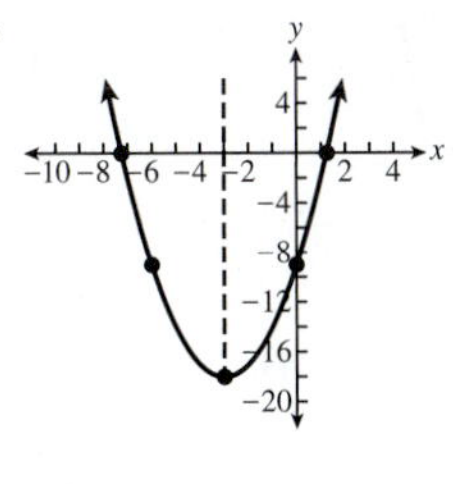

3.

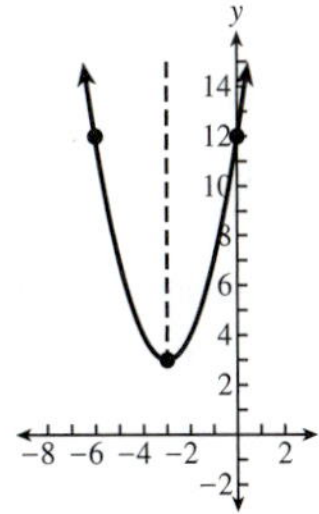

4.

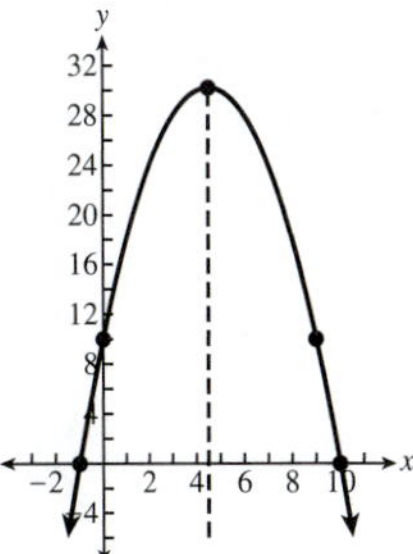

5.

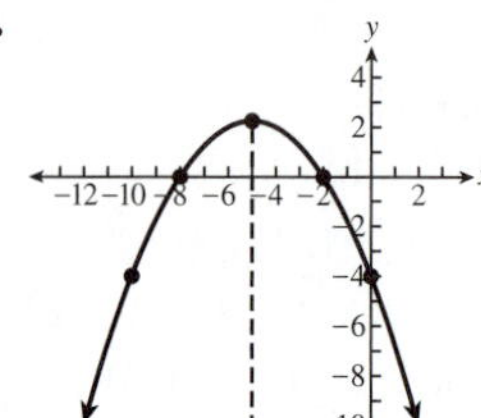

6.

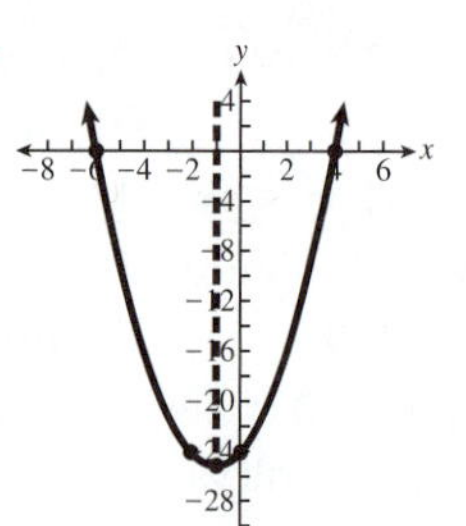

7.

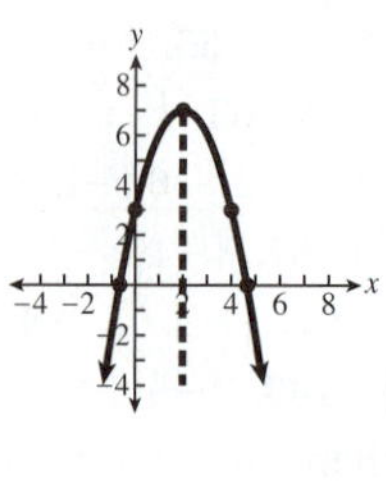

8. a) $(-2, -8), x = -2$ **b)** $(8, 7), x = 8$

9. a) $y = (x + 3)^2 - 49$, Vertex: $(-3, -49)$ **b)** $y = -(x - 6)^2 + 81$, Vertex: $(6, 81)$ **10.**

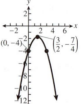

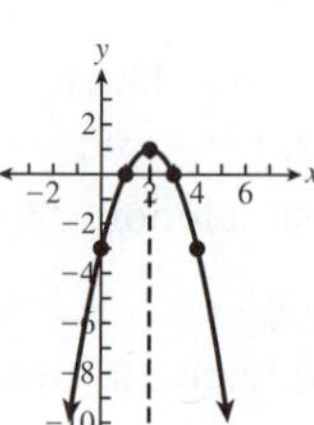

11.

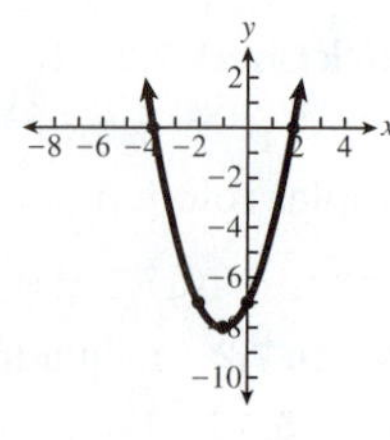

Section 7.4 **1.** parabola **3.** negative **5.** y-intercept **7.** $(3, -30), x = 3$ **9.** $(-7, -9), x = -7$ **11.** $(-5, 57), x = -5$ **13.** $\left(\frac{9}{2}, -\frac{25}{4}\right), x = \frac{9}{2}$ **15.** $(-2, -32), x = -2$ **17.** $(0, -7), x = 0$ **19.** $(3, -4), x = 3$ **21.** $(-4, 4), x = -4$ **23.** $(-1, 0), (-5, 0), (0, 5)$ **25.** No x-intercepts, $(0, -7)$ **27.** $(-1.3, 0), (4.8, 0), (0, -13)$ **29.** $(-4, 0), (0, 0)$ **31.** $(3, 0), (0, 9)$ **33.** $(-1, 0), (-3, 0), (0, 3)$ **35.** No x-intercepts, $(0, -41)$ **37.**

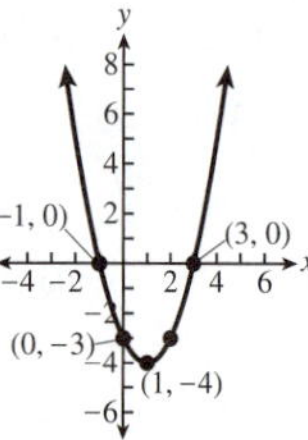

39.

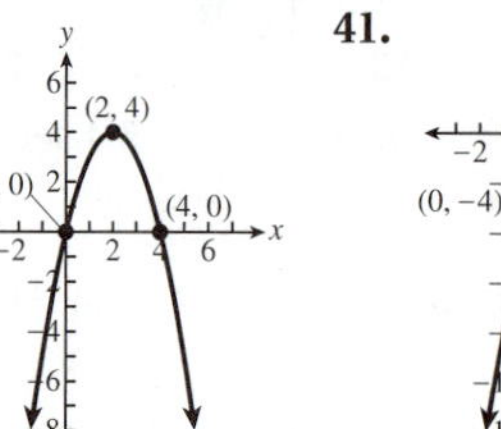

41.

43.

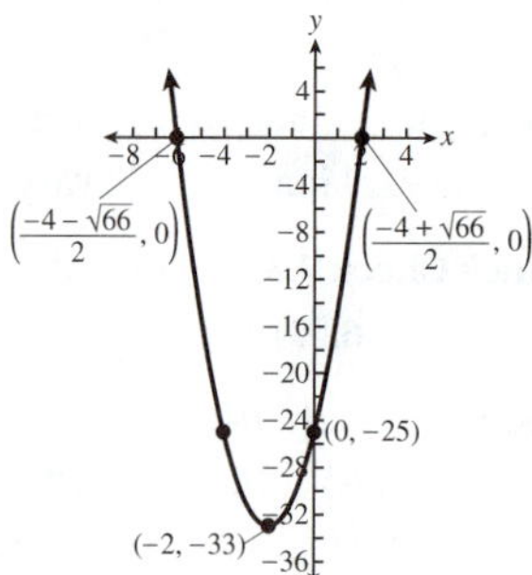

45.

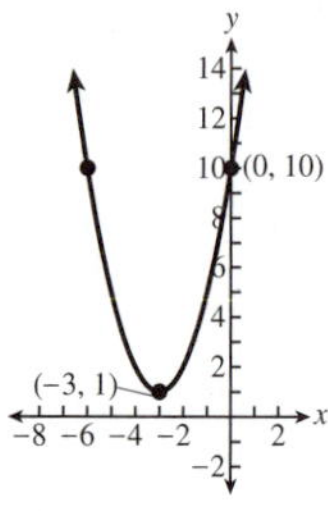

47.

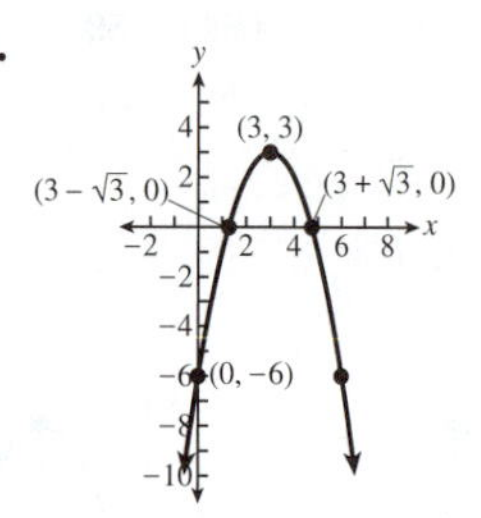

49.

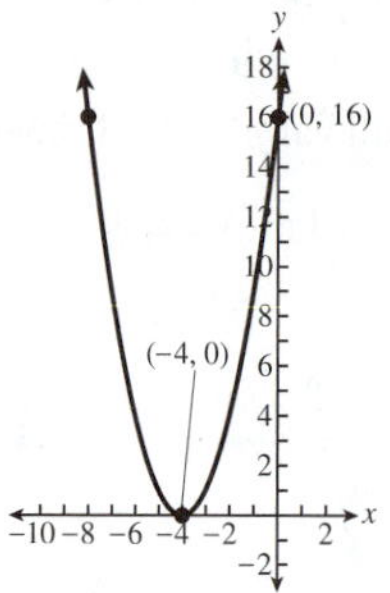

51.

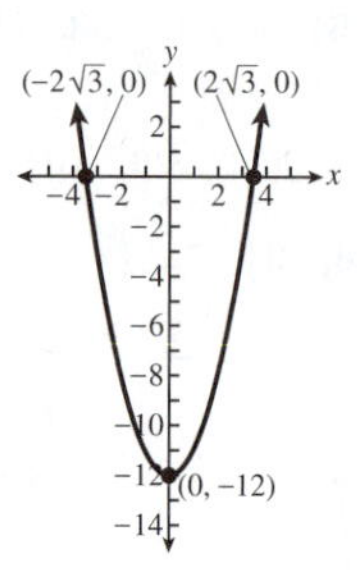

53.

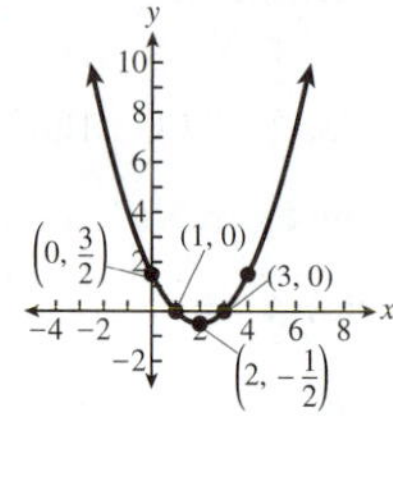

55.

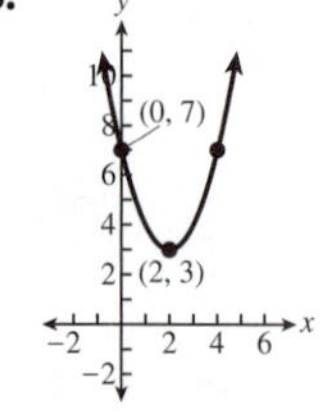

57.

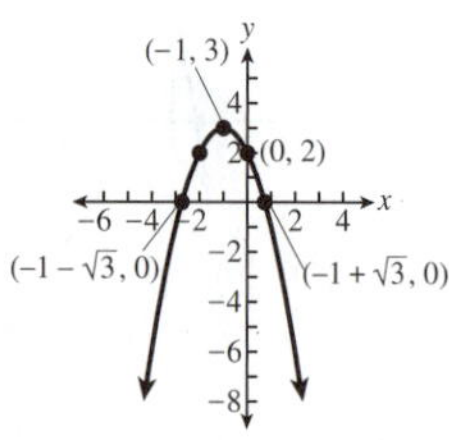

59.

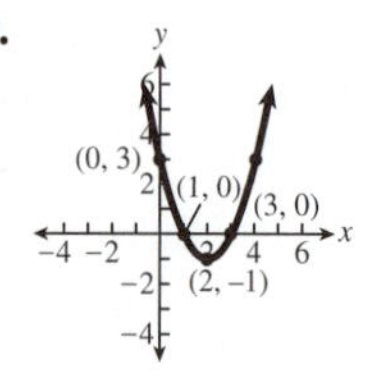

61.

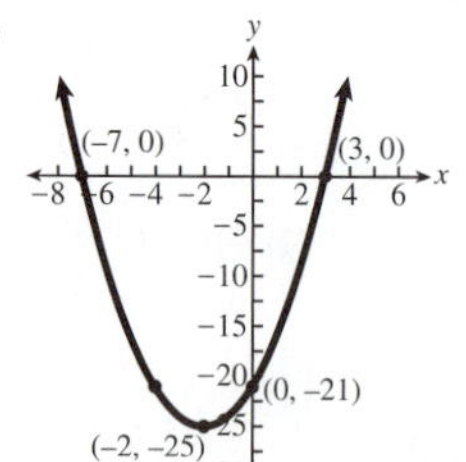

63.

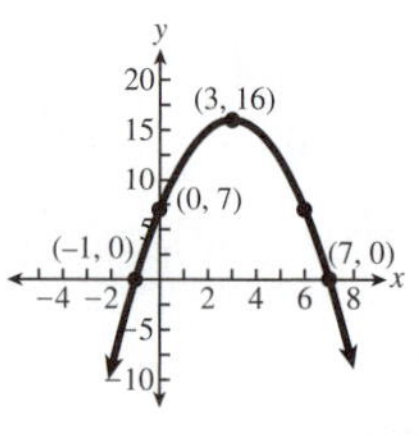

65. **67.** **69.** **71.** **73.**

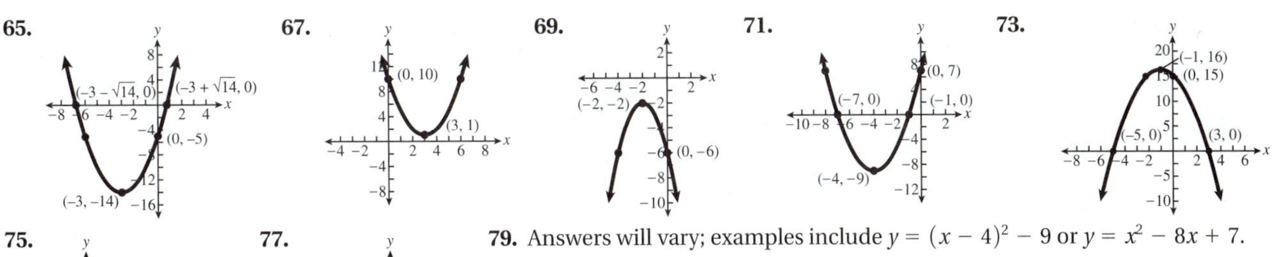

75. **77.**

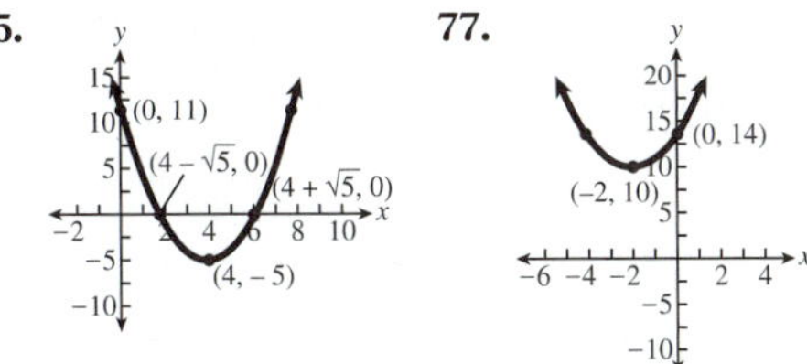

79. Answers will vary; examples include $y = (x - 4)^2 - 9$ or $y = x^2 - 8x + 7$.

81. Answers will vary; examples include $y = x^2 + 2x - 15$ or $y = (x + 1)^2 - 16$. **83.** Answers will vary.

Quick Check 7.5 **1.** Base: 20 in., Height: 6 in. **2.** Length: 23.1 in., Width: 15.1 in. **3.** 9.80 in. **4.** 7.7 ft **5.** 250 ft **6. a)** Maximum: −10 **b)** Minimum: 47 **7.** 961 **8.** 8712 square meters

Section 7.5 **1.** hypotenuse **3.** Pythagorean **5.** Length: 10 in.; Width: 8 in. **7.** Length: 14 in.; Width: 7 in. **9.** Length: 17.9 in.; Width: 2.7 in. **11.** Base: 12 in.; Height: 7 in. **13.** Base: 3.6 in.; Height: 8.2 in. **15.** Width: 8 in.; Height: 5 in. **17.** Length: 5 ft; Width: 2.5 ft **19.** Length: 32.6 in.; Width: 16.6 in. **21.** Base: 32 in.; Height: 16 in. **23.** Base: 6.4 ft; Height: 14.4 ft **25.** 25 in. **27.** 8.7 cm **29.** 100 mi **31.** 6 ft **33.** 30.7 ft **35.** 16.4 in. **37.** Length: 5.1 ft; Width: 4.1 ft **39. a)** Base: 12 ft; Height: 9 ft **b)** 15 ft **41. a)** 69.3 cm **b)** 2772 sq cm **43.** Minimum, 4 **45.** Minimum, −78 **47.** Maximum, −27 **49.** Minimum, 18 **51.** Maximum, −33 **53.** Minimum, 46 **55.** 21 ft **57.** 1.725 m **59.** 144 **61.** 75 ft by 75 ft, 5625 sq ft **63.** 70 pans, \$3.4375 per pan **65.** Answers will vary.

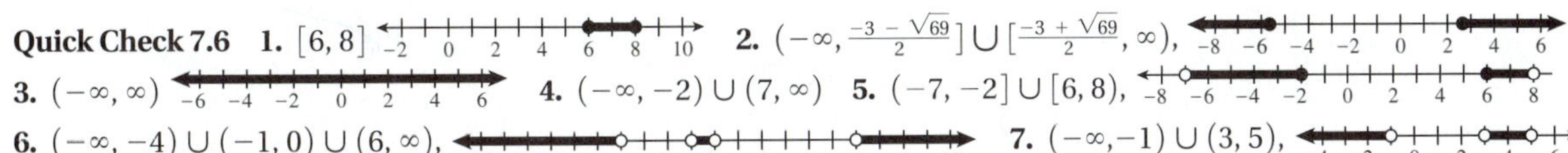

Quick Check 7.6 **1.** $[6, 8]$ **2.** $\left(-\infty, \frac{-3 - \sqrt{69}}{2}\right] \cup \left[\frac{-3 + \sqrt{69}}{2}, \infty\right)$, **3.** $(-\infty, \infty)$ **4.** $(-\infty, -2) \cup (7, \infty)$ **5.** $(-7, -2] \cup [6, 8)$, **6.** $(-\infty, -4) \cup (-1, 0) \cup (6, \infty)$, **7.** $(-\infty, -1) \cup (3, 5)$, **8.** $(-7, -3)$ **9.** From 5 seconds after launch until 12 seconds after launch.

Section 7.6 **1.** quadratic **3.** strict **5.** $(1, 5)$ **7.** $(-\infty, -3] \cup [6, \infty)$, **9.** $[3, 10]$ **11.** $(-\infty, -5) \cup (-3, \infty)$, **13.** $(0, 7)$ **15.** $(-\infty, -3 - \sqrt{6}) \cup (-3 + \sqrt{6}, \infty)$, **17.** $\left(-\infty, \frac{5 - \sqrt{37}}{6}\right] \cup \left[\frac{5 + \sqrt{37}}{6}, \infty\right)$, **19.** $(-\infty, \infty)$ **21.** $(5 - 2\sqrt{5}, 5 + 2\sqrt{5})$, **23.** $(-\infty, -5] \cup [9, \infty)$, **25.** $[-2, 1] \cup [4, \infty)$, **27.** $(-\infty, -6) \cup (-3, -2)$, **29.** $[-9, -5] \cup \{7\}$, **31.** $(-4, 3)$ **33.** $(-\infty, -2] \cup (-1, 2]$, **35.** $(-\infty, -2) \cup (-1, 5)$, **37.** $(-\infty, -8] \cup (-7, 1] \cup (7, \infty)$, **39.** $(-8, 2) \cup [5, 7]$, **41.** $(-8, 1) \cup (3, 4)$, **43.** $(-\infty, -5) \cup (4, \infty)$, **45.** $[-6, -3) \cup [-1, \infty)$, **47.** $[-2, 2)$, **49.** 1 sec to 3 sec **51.** 0.5 sec to 1.5 sec **53.** 2 sec to 7 sec **55.** $[2, 6]$ **57.** $(-\infty, 2] \cup [3, \infty)$ **59.** $\varnothing$ **61.** $[-9, 6)$ **63.** $(-\infty, -7) \cup (-6, -3) \cup (11, \infty)$ **65.** $x^2 - 2x - 3 < 0$ **67.** $x^2 - 4 \geq 0$ **69.** $\frac{x - 5}{x + 4} \leq 0$ **71.** $\frac{x^2 - 2x - 3}{x^2 + 10x + 24} \leq 0$ **73.** $(-7, 1)$ **75.** $\varnothing$

Quick Check 7.7 **1.** Domain: $[-6, \infty)$, Range: $[5, \infty)$ **2.** Domain: $(-\infty, 2]$, Range: $[7, \infty)$

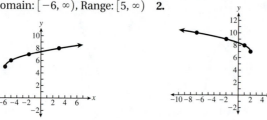

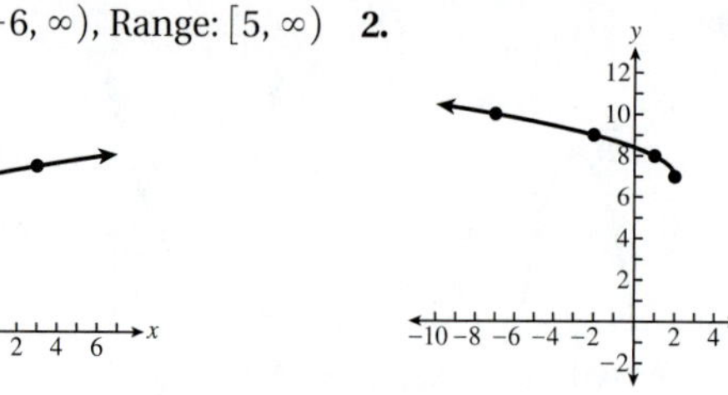

3. a) Domain: $(-\infty, \infty)$, Range: $(-\infty, \infty)$ **b)** Domain: $(-\infty, \infty)$, Range: $(-\infty, \infty)$ **4. a)** $f(x) = \sqrt{-x} + 3$ **b)** $f(x) = -(x + 2)^3$

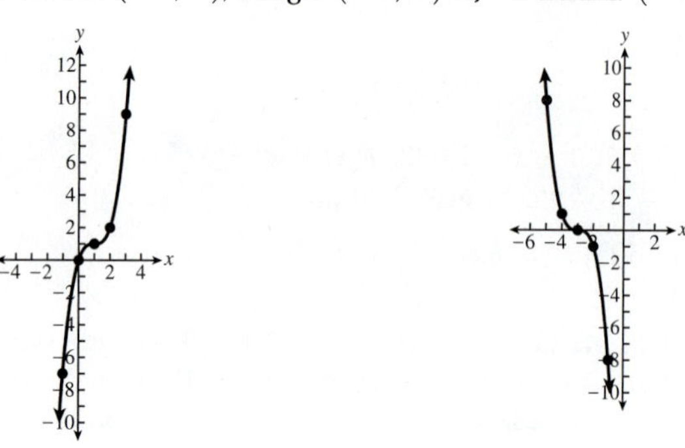

Section 7.7 **1.** radicand **3.** Domain: $[-2, \infty)$, Range: $[0, \infty)$ **5.** Domain: $[0, \infty)$, Range: $[2, \infty)$ **7.** Domain: $[-4, \infty)$, Range: $[6, \infty)$

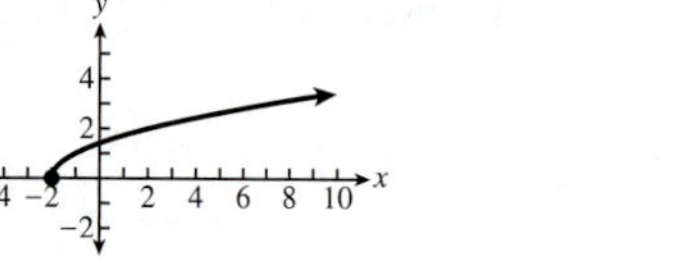

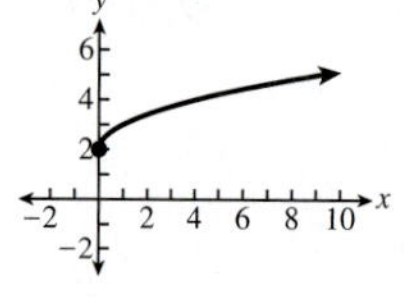

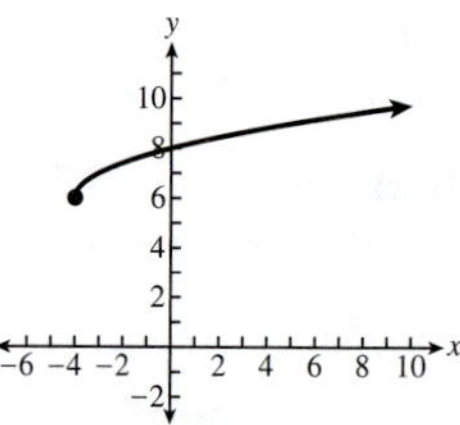

9. Domain: $[5, \infty)$, Range: $[-2, \infty)$ **11.** Domain: $[-4, \infty)$, Range: $(-\infty, 0]$ **13.** Domain: $[2, \infty)$, Range: $(-\infty, 3]$

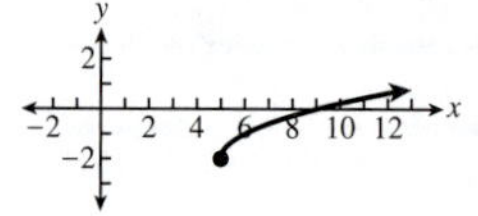

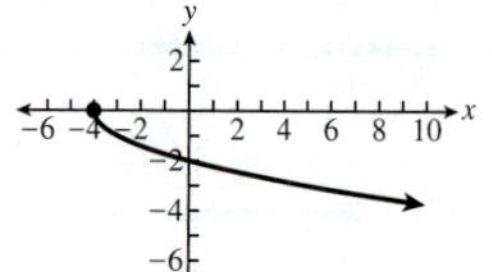

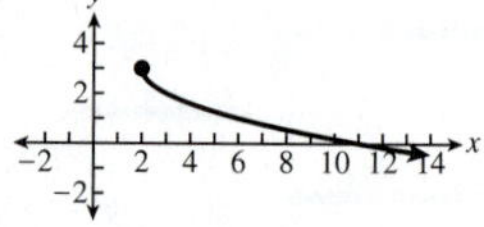

15. Domain: $(-\infty, 3]$, Range: $[1, \infty)$ **17.** Domain: $(-\infty, \infty)$, Range: $(-\infty, \infty)$ **19.** Domain: $(-\infty, \infty)$, Range: $(-\infty, \infty)$

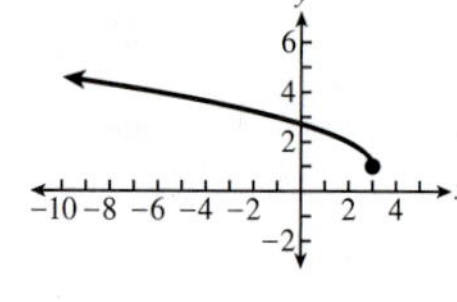

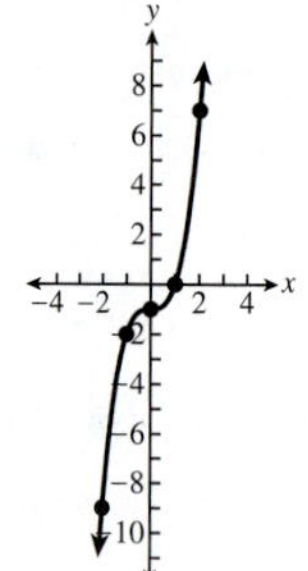

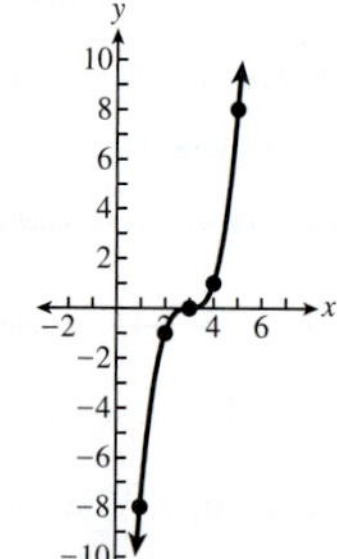

21. Domain: $(-\infty, \infty)$, Range: $(-\infty, \infty)$ **23.** Domain: $(-\infty, \infty)$, Range: $(-\infty, \infty)$ **25.** Domain: $(-\infty, \infty)$, Range: $(-\infty, \infty)$

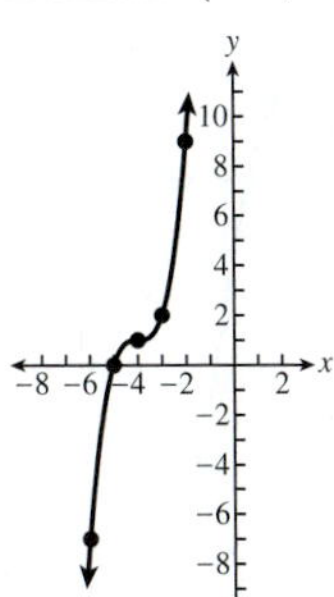

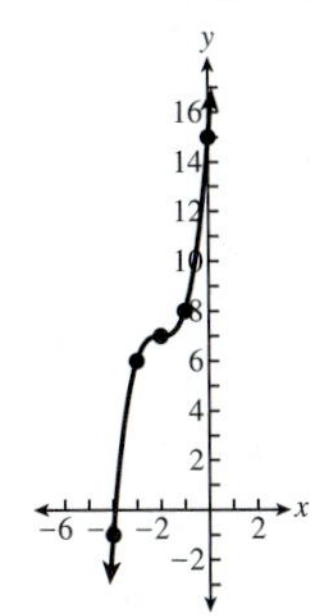

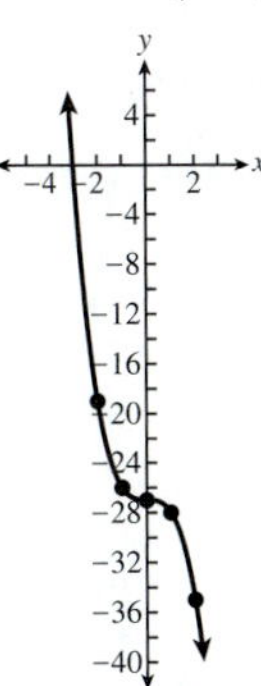

27. Domain: $(-\infty, \infty)$, Range: $(-\infty, \infty)$ **29.** $f(x) = \sqrt{x - 4} + 3$ **31.** $f(x) = \sqrt{3 - x} - 1$ **33.** $f(x) = x^3 + 5$

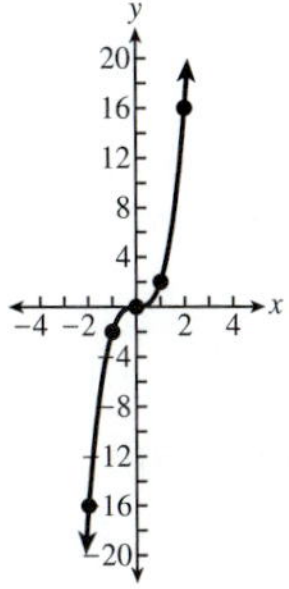

35. a)

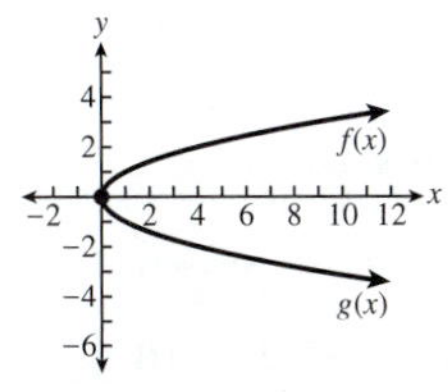

b) $f(x) = \sqrt{x + 19}$ would be 19 units to the left of $f(x) = \sqrt{x}$; $f(x) = \sqrt{x - 27}$ would be 27 units to the right of $f(x) = \sqrt{x}$.

37. The graph of $g(x) = -\sqrt{x}$ is the reflection of $f(x) = \sqrt{x}$ about the x-axis. **39.** Answers will vary.

Chapter 7 Review **1.** $\{-7, -4\}$ **2.** $\{-10, 3\}$ **3.** $\{-4, 0\}$ **4.** $\{\frac{3}{2}, 5\}$ **5.** $\{-5, 1\}$ **6.** $\{7, 11\}$ **7.** $\{-7, 7\}$ **8.** $\{-6\sqrt{2}, 6\sqrt{2}\}$ **9.** $\{-2i, 2i\}$ **10.** $\{6 - 6i, 6 + 6i\}$ **11.** $\{8 - 3i, 8 + 3i\}$ **12.** $\{-9, 2\}$ **13.** $\{-2, 4\}$ **14.** $\{3 - \sqrt{6}, 3 + \sqrt{6}\}$ **15.** $\{-7 - \sqrt{2}, -7 + \sqrt{2}\}$ **16.** $\{\frac{-5 - i\sqrt{15}}{2}, \frac{-5 + i\sqrt{15}}{2}\}$ **17.** $\{\frac{-7 - 3\sqrt{5}}{2}, \frac{-7 + 3\sqrt{5}}{2}\}$ **18.** $\{-2 - \sqrt{6}, -2 + \sqrt{6}\}$ **19.** $\{\frac{3 - 3i\sqrt{3}}{2}, \frac{3 + 3i\sqrt{3}}{2}\}$ **20.** $\{\frac{5 - 2i}{2}, \frac{5 + 2i}{2}\}$ **21.** $\{-7, 3\}$ **22.** $\{6 - i, 6 + i\}$ **23.** $\{-1, 1, -2, 2\}$ **24.** $\{64\}$ **25.** $\{-2, 1\}$ **26.** $\{-27, 343\}$ **27.** $\{-3, 2\}$ **28.** $\{-6\}$ **29.** $\{-6, -5\}$ **30.** $\{6\}$ **31.** $\{9, 10\}$ **32.** $\{-2, 2, -3i, 3i\}$ **33.** $\{-6 - 2i\sqrt{2}, -6 + 2i\sqrt{2}\}$ **34.** $\{-2 - 2i\sqrt{3}, -2 + 2i\sqrt{3}\}$ **35.** $\{-10, 0\}$ **36.** $\{3 - \sqrt{29}, 3 + \sqrt{29}\}$ **37.** $\{5\}$ **38.** $\{20\}$ **39.** $\{-4, 5\}$ **40.** $\{\frac{4 - \sqrt{10}}{3}, \frac{4 + \sqrt{10}}{3}\}$ **41.** $\{9\}$ **42.** $\{-9 - 8i, -9 + 8i\}$ **43.** $\{-16, -8\}$ **44.** $\{\frac{7 - \sqrt{129}}{2}, \frac{7 + \sqrt{129}}{2}\}$ **45.** $x^2 + 5x - 36 = 0$ **46.** $x^2 - 8x + 16 = 0$ **47.** $6x^2 - 47x + 52 = 0$ **48.** $x^2 - 8 = 0$ **49.** $x^2 + 25 = 0$ **50.** $x^2 + 64 = 0$ **51.** $(4, -36)$, $x = 4$ **52.** $(-\frac{5}{2}, -\frac{89}{4})$, $x = -\frac{5}{2}$ **53.** $(-1, 36)$, $x = -1$ **54.** $(3, -4)$, $x = 3$ **55.** no x-intercepts; y-int.: $(0, 11)$ **56.** x-int.: $(2 - \sqrt{7}, 0)$, $(2 + \sqrt{7}, 0)$; y-int.: $(0, -3)$ **57.** x-int.: $(5, 0)$; y-int.: $(0, -25)$ **58.** x-int.: $(-4 - 2\sqrt{2}, 0)$, $(-4 + 2\sqrt{2}, 0)$; y-int.: $(0, 8)$

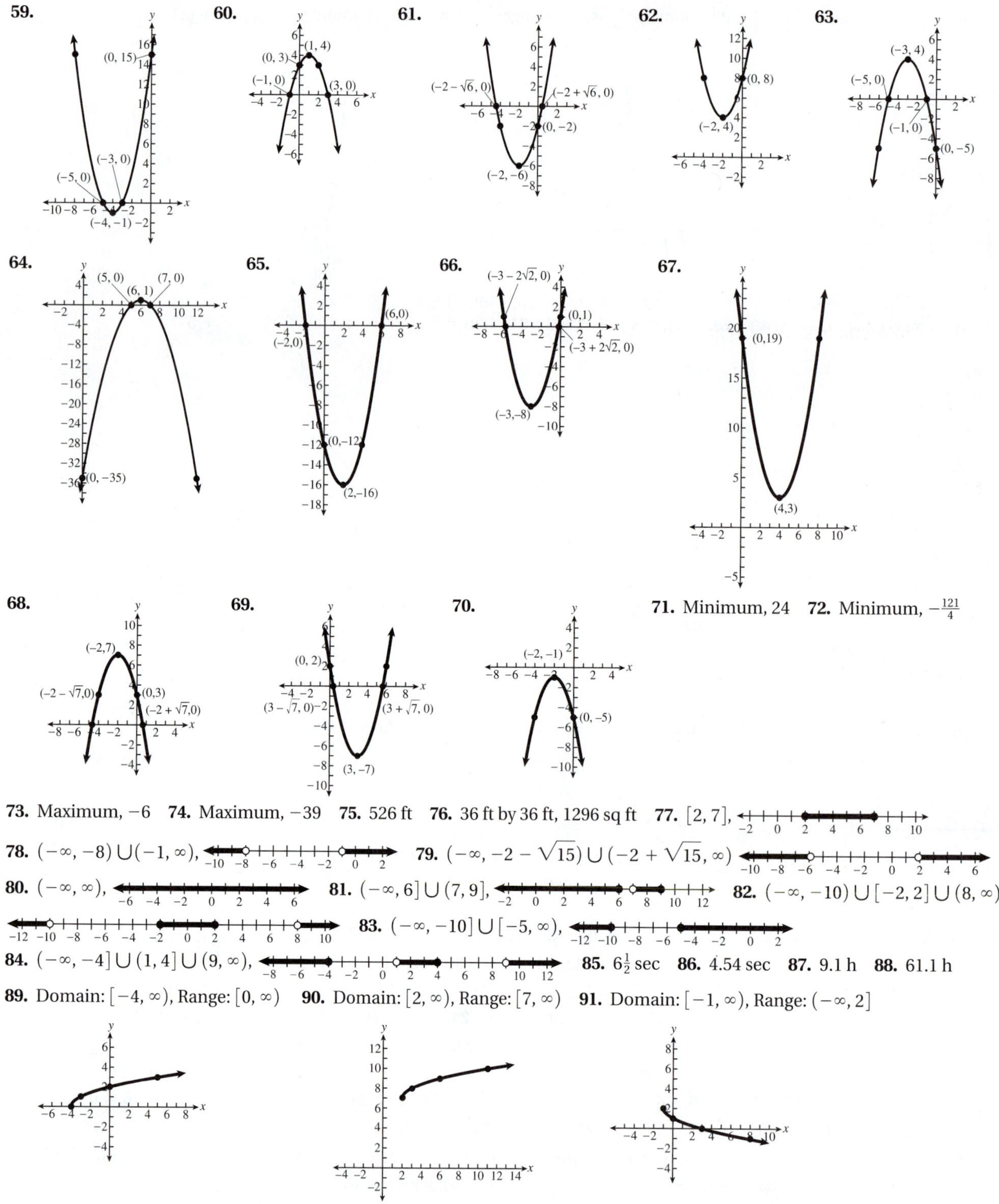

71. Minimum, 24 **72.** Minimum, $-\frac{121}{4}$

73. Maximum, −6 **74.** Maximum, −39 **75.** 526 ft **76.** 36 ft by 36 ft, 1296 sq ft **77.** $[2, 7]$,

78. $(-\infty, -8) \cup (-1, \infty)$, **79.** $(-\infty, -2 - \sqrt{15}) \cup (-2 + \sqrt{15}, \infty)$

80. $(-\infty, \infty)$, **81.** $(-\infty, 6] \cup (7, 9]$, **82.** $(-\infty, -10) \cup [-2, 2] \cup (8, \infty)$,

83. $(-\infty, -10] \cup [-5, \infty)$,

84. $(-\infty, -4] \cup (1, 4] \cup (9, \infty)$, **85.** $6\frac{1}{2}$ sec **86.** 4.54 sec **87.** 9.1 h **88.** 61.1 h

89. Domain: $[-4, \infty)$, Range: $[0, \infty)$ **90.** Domain: $[2, \infty)$, Range: $[7, \infty)$ **91.** Domain: $[-1, \infty)$, Range: $(-\infty, 2]$

92. Domain: $[-5, \infty)$, Range: $[-3, \infty)$ **93.** Domain: $(-\infty, \infty)$, Range: $(-\infty, \infty)$ **94.** Domain: $(-\infty, \infty)$, Range: $(-\infty, \infty)$

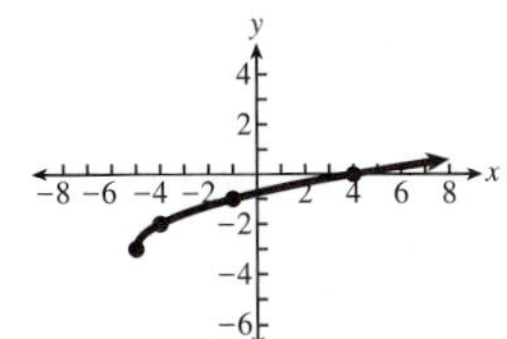

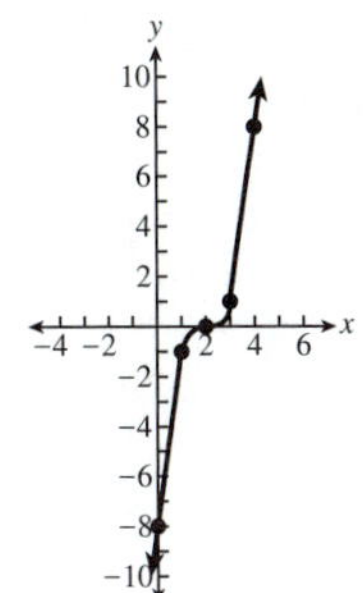

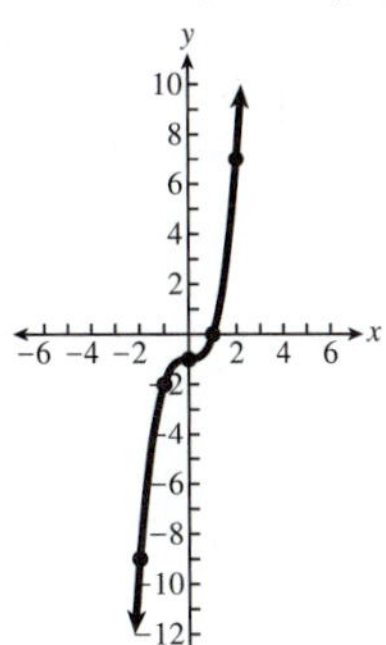

95. $f(x) = \sqrt{x-1} + 6$ **96.** $f(x) = (x-6)^3$ **97.** Length: 18 cm; Width: 16 cm
98. Length: 29.8 ft; Width: 16.8 ft **99.** 1280.6 mi **100.** Length: 9.9 ft; Width: 6.9 ft

Chapter 7 Review Exercises: Worked-Out Solutions

1.
$$x^2 + 11x + 28 = 0$$
$$(x+7)(x+4) = 0$$
$$x + 7 = 0 \quad \text{or} \quad x + 4 = 0$$
$$x = -7 \quad \text{or} \quad x = -4$$
$$\{-7, -4\}$$

11.
$$(x-8)^2 + 30 = 21$$
$$(x-8)^2 = -9$$
$$\sqrt{(x-8)^2} = \pm\sqrt{-9}$$
$$x - 8 = \pm 3i$$
$$x = 8 \pm 3i$$
$$\{8 - 3i, 8 + 3i\}$$

13.
$$x^2 - 2x - 8 = 0$$
$$x^2 - 2x = 8$$
$$x^2 - 2x + 1 = 8 + 1$$
$$(x-1)^2 = 9$$
$$\sqrt{(x-1)^2} = \pm\sqrt{9}$$
$$x - 1 = \pm 3$$
$$x = 1 \pm 3$$
$$x = 1 + 3 \quad \text{or} \quad x = 1 - 3$$
$$x = 4 \quad \text{or} \quad x = -2$$
$$\{-2, 4\}$$

17.
$$x = \frac{-7 \pm \sqrt{7^2 - 4(1)(1)}}{2(1)}$$
$$= \frac{-7 \pm \sqrt{49 - 4}}{2}$$
$$= \frac{-7 \pm \sqrt{45}}{2}$$
$$= \frac{-7 \pm 3\sqrt{5}}{2}$$
$$\left\{\frac{-7 - 3\sqrt{5}}{2}, \frac{-7 + 3\sqrt{5}}{2}\right\}$$

23.
$$u = x^2$$
$$u^2 - 5u + 4 = 0$$
$$(u-1)(u-4) = 0$$
$$u - 1 = 0 \quad \text{or} \quad u - 4 = 0$$
$$u = 1 \quad \text{or} \quad u = 4$$
$$x^2 = 1 \quad \text{or} \quad x^2 = 4$$
$$\sqrt{x^2} = \pm\sqrt{1} \quad \text{or} \quad \sqrt{x^2} = \pm\sqrt{4}$$
$$x = \pm 1 \quad \text{or} \quad x = \pm 2$$
$$\{-1, 1, -2, 2\}$$

27.
$$\sqrt{x^2 + x - 2} = 2$$
$$(\sqrt{x^2 + x - 2})^2 = 2^2$$
$$x^2 + x - 2 = 4$$
$$x^2 + x - 6 = 0$$
$$(x+3)(x-2) = 0$$
$$x + 3 = 0 \quad \text{or} \quad x - 2 = 0$$
$$x = -3 \quad \text{or} \quad x = 2$$
$$\{-3, 2\}$$

29.
$$1 + \tfrac{11}{x} + \tfrac{30}{x^2} = 0$$
$$x^2\left(1 + \tfrac{11}{x} + \tfrac{30}{x^2}\right) = x^2 \cdot 0$$
$$x^2 \cdot 1 + x^2 \cdot \tfrac{11}{x} + x^2 \cdot \tfrac{30}{x^2} = 0$$
$$x^2 + 11x + 30 = 0$$
$$(x+6)(x+5) = 0$$
$$x + 6 = 0 \quad \text{or} \quad x + 5 = 0$$
$$x = -6 \quad \text{or} \quad x = -5$$
$$\{-6, -5\}$$

43.
$$f(x) = 0$$
$$(x+12)^2 - 16 = 0$$
$$(x+12)^2 = 16$$
$$\sqrt{(x+12)^2} = \pm\sqrt{16}$$
$$x + 12 = \pm 4$$
$$x = -12 \pm 4$$
$$x = -12 + 4 \quad \text{or} \quad x = -12 - 4$$
$$x = -8 \quad \text{or} \quad x = -16$$
$$\{-16, -8\}$$

45.
$$x = -9 \quad \text{or} \quad x = 4$$
$$x + 9 = 0 \quad \text{or} \quad x - 4 = 0$$
$$(x+9)(x-4) = 0$$
$$x^2 + 5x - 36 = 0$$

59. Vertex

$x = \frac{-8}{2(1)} = -4$

$y = (-4)^2 + 8(-4) + 15 = -1$

$(-4, -1)$

y-intercept $(x = 0)$

$y = 0^2 + 8(0) + 15 = 15$

$(0, 15)$

x-intercept $(y = 0)$

$x^2 + 8x + 15 = 0$

$(x + 5)(x + 3) = 0$

$x + 5 = 0$ or $x + 3 = 0$

$x = -5$ or $x = -3$

$(-5, 0), (-3, 0)$

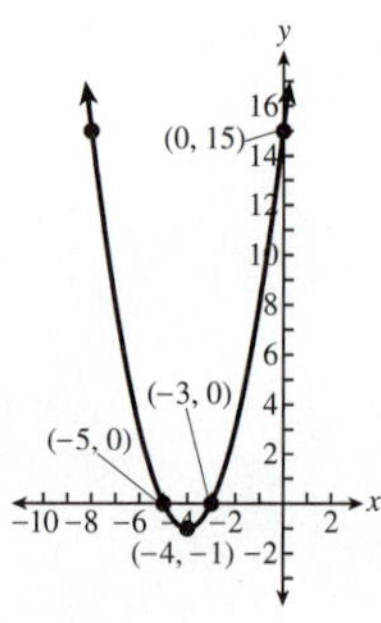

63. Vertex (h, k)

$(-3, 4)$

y-intercept $(x = 0)$

$y = -(0 + 3)^2 + 4$

$y = -9 + 4$

$y = -5$

$(0, -5)$

x-intercept $(y = 0)$

$0 = -(x + 3)^2 + 4$

$(x + 3)^2 = 4$

$\sqrt{(x + 3)^2} = \pm\sqrt{4}$

$x + 3 = \pm 2$

$x = -3 \pm 2$

$(-5, 0), (-1, 0)$

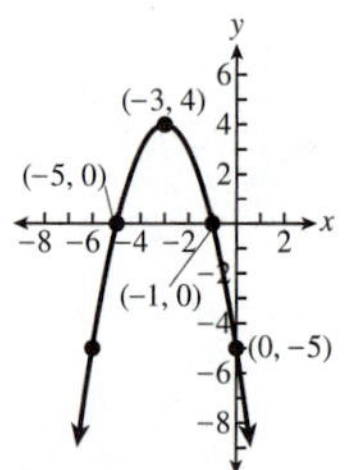

68. Vertex

$x = \frac{-(-4)}{2(-1)} = -2$

$y = -(-2)^2 - 4(-2) + 3 = 7$

$(-2, 7)$

y-intercept $(x = 0)$

$y = -0^2 - 4(0) + 3 = 3$

$(0, 3)$

x-intercept $(y = 0)$

$0 = -x^2 - 4x + 3$

$x^2 + 4x - 3 = 0$

$x = \frac{-4 \pm \sqrt{4^2 - 4(1)(-3)}}{2(1)}$

$= \frac{-4 \pm \sqrt{28}}{2}$

$= \frac{-4 \pm 2\sqrt{7}}{2}$

$= -2 \pm \sqrt{7}$

$(-2 - \sqrt{7}, 0), (-2 + \sqrt{7}, 0)$

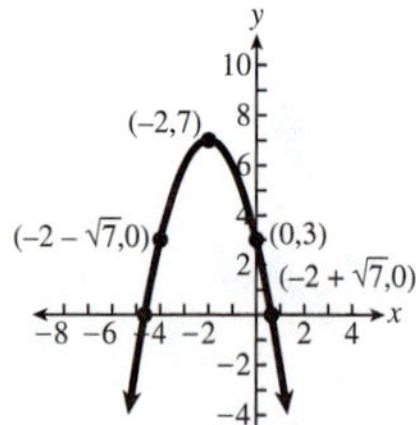

71. $a > 0$, so there is a minimum value.
$x = \frac{-(-12)}{2(1)} = 6$
$f(6) = 6^2 - 12(6) + 60 = 24$
The minimum value is 24.

75. $x = \frac{-176}{2(-16)} = 5.5$
$h(5.5) = -16(5.5)^2 + 176(5.5) + 42 = 526$

77. $x^2 - 9x + 14 = 0$
$(x-2)(x-7) = 0$
Critical Values: $x = 2, 7$

Test Point	$x-2$	$x-7$	$(x-2)(x-7)$
0	−	−	+
3	+	−	−
8	+	+	+

$[2, 7]$

81. $x^2 - 15x + 54 = 0$
$(x-6)(x-9) = 0$
$x = 6$ or $x = 9$

$x - 7 = 0$
$x = 7$

Critical Values: $x = 6, 7, 9$

Test Point	$x-6$	$x-9$	$x-7$	$\frac{(x-6)(x-9)}{x-7}$
0	−	−	−	−
6.5	+	−	−	+
8	+	−	+	−
10	+	+	+	+

$(-\infty, 6] \cup (7, 9]$

86. $-16t^2 + 66t + 30 = 0$
$-2(8t^2 - 33t - 15) = 0$
$t = \frac{-(-33) \pm \sqrt{(-33)^2 - 4(8)(-15)}}{2(8)}$
$= \frac{33 \pm \sqrt{1569}}{16}$
$\frac{33 + \sqrt{1569}}{16} \approx 4.54$, $\frac{33 - \sqrt{1569}}{16} \approx -0.41$
Omit the negative solution.
4.54 seconds.

90.

	x	$f(x) = \sqrt{x-2} + 7$
$x - 2 = 0$	2	$\sqrt{2-2} + 7 = 0 + 7 = 7$
$x - 2 = 1$	3	$\sqrt{3-2} + 7 = 1 + 7 = 8$
$x - 2 = 4$	6	$\sqrt{6-2} + 7 = 2 + 7 = 9$
$x - 2 = 9$	11	$\sqrt{11-2} + 7 = 3 + 7 = 10$

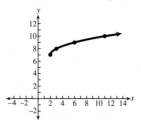

Domain: $[2, \infty)$, Range: $[7, \infty)$

93. $x - 2 = 0$
$x = 2$

x	$f(x) = (x-2)^3$
0	$(0-2)^3 = -8$
1	$(1-2)^3 = -1$
2	$(2-2)^3 = 0$
3	$(3-2)^3 = 1$
4	$(4-2)^3 = 8$

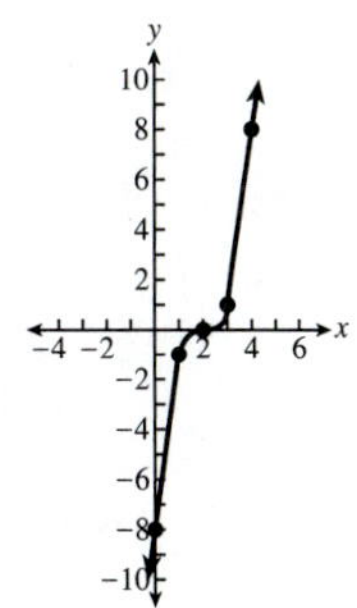

Domain: $(-\infty, \infty)$, Range: $(-\infty, \infty)$

100. Length: x, Width: $x - 3$

$$x^2 + (x - 3)^2 = 12^2$$
$$x^2 + x^2 - 6x + 9 = 144$$
$$2x^2 - 6x - 135 = 0$$
$$x = \frac{-(-6) \pm \sqrt{(-6)^2 - 4(2)(-135)}}{2(2)}$$
$$= \frac{6 \pm \sqrt{1116}}{4}$$
$$\frac{6 + \sqrt{1116}}{4} \approx 9.9, \frac{6 - \sqrt{1116}}{4} \approx -6.9$$

Omit the negative solution.

Length: 9.9 feet, Width: $9.9 - 3 = 6.9$ feet

Chapter 7 Test **1.** $\{-6, -3\}$ **2.** $\{0, \frac{16}{3}\}$ **3.** $\{7 - 2\sqrt{13}, 7 + 2\sqrt{13}\}$ **4.** $\{3 - i\sqrt{3}, 3 + i\sqrt{3}\}$ **5.** $\{-2\sqrt{3}, 2\sqrt{3}, -2i, 2i\}$ **6.** $\{1\}$ **7.** $\{-1, \frac{3}{2}\}$ **8.** $\{-\frac{3}{2}, 9\}$ **9.** $\{4, 10\}$ **10.** $\{-4, 14\}$ **11.** $\{5 - 3i, 5 + 3i\}$ **12.** $7x^2 - 39x + 20 = 0$ **13.** $x^2 + 16 = 0$ **14.** x-int.: (2.8, 0), (7.2, 0) y-int.: (0, 20) **15.**

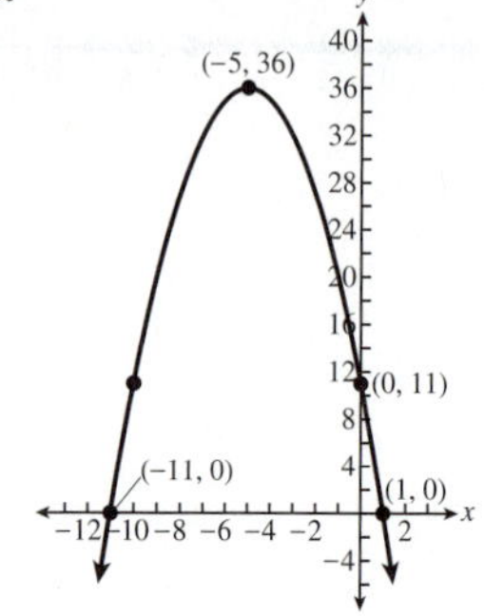

16.

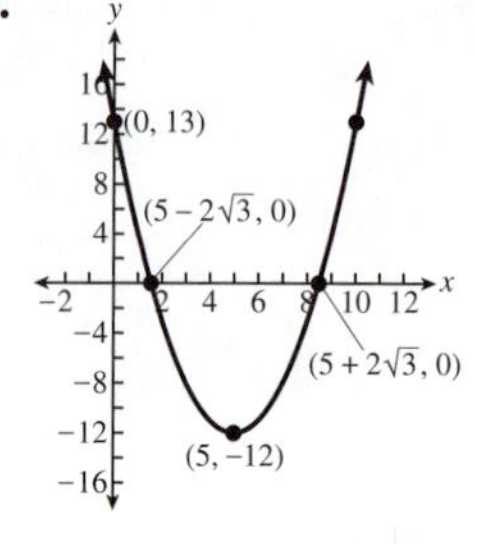

17.

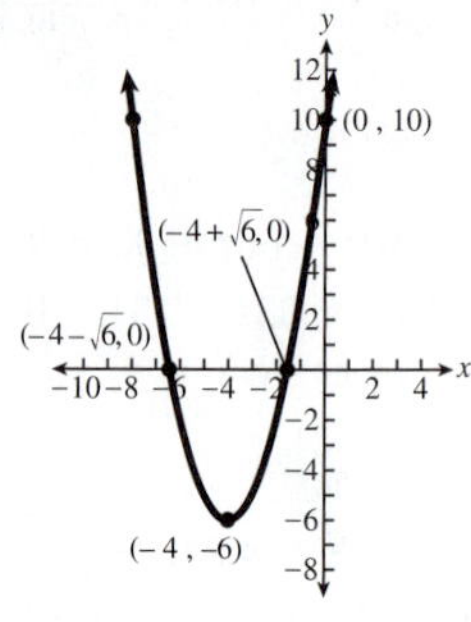

18. $(-\infty, -7) \cup (4, \infty)$, **19.** $[-8, -3] \cup (-2, 7)$,

20. Length: 97.1 yd; Width: 37.1 yd **21.** 1700 mi **22.** 9.25 ft **23.** Domain: $[-1, \infty)$, Range: $[-3, \infty)$

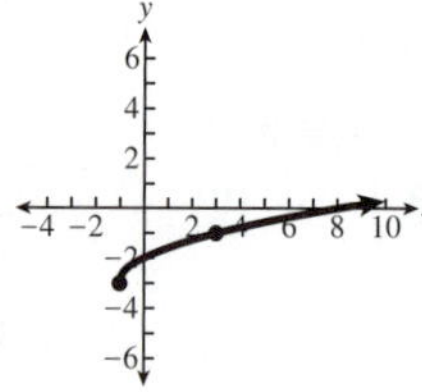

24. Domain: $(-\infty, \infty)$, Range: $(-\infty, \infty)$ **25.** $f(x) = \sqrt{x + 6}$

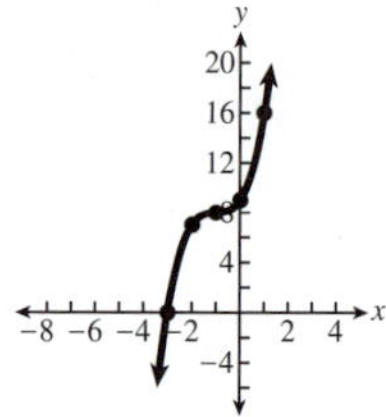

Cumulative Review Chapters 5–7 **1.** 2, 15 **2.** $\frac{x + 8}{x - 7}$ **3.** $-\frac{3}{2}$ **4.** $-\frac{3 + x}{x - 4}$ **5.** $\frac{(x - 10)(x + 6)}{(x - 2)(x - 7)}$ **6.** $\frac{4}{x - 2}$ **7.** $\frac{x + 6}{x - 3}$ **8.** $\frac{x - 2}{(x - 5)(x - 4)}$ **9.** $\frac{2(x + 1)}{(x - 1)(x + 4)}$ **10.** $\frac{x - 4}{x + 8}$ **11.** $\frac{x - 2}{x - 6}$ **12.** $\{7\}$ **13.** $\{-3\}$ **14.** $\{7, 8\}$ **15.** $x = \frac{5y}{y - 5}$ **16.** 9 **17.** 48 hours **18.** 7.483 **19.** $64x^{15}$ **20.** $x^{9/14}$ **21.** $5r^6s^5t^4\sqrt{3t}$ **22.** $34\sqrt{3}$ **23.** $20 + 6\sqrt{15}$ **24.** $\frac{c\sqrt{ac}}{ab}$ **25.** $\frac{136 + 39\sqrt{10}}{62}$ **26.** 17 **27.** $6i\sqrt{10}$ **28.** $28 + 41i$ **29.** $\frac{4 + 3i}{5}$ or $\frac{4}{5} + \frac{3}{5}i$ **30.** $\{-2 - 3i\sqrt{2}, -2 + 3i\sqrt{2}\}$ **31.** $\{\frac{5 - 2\sqrt{22}}{3}, \frac{5 + 2\sqrt{22}}{3}\}$ **32.** $\{-\sqrt{2}, \sqrt{2}, -3, 3\}$ **33.** $\{4\}$ **34.** $\{2\}$

35. $\{3, 9\}$ **36.**

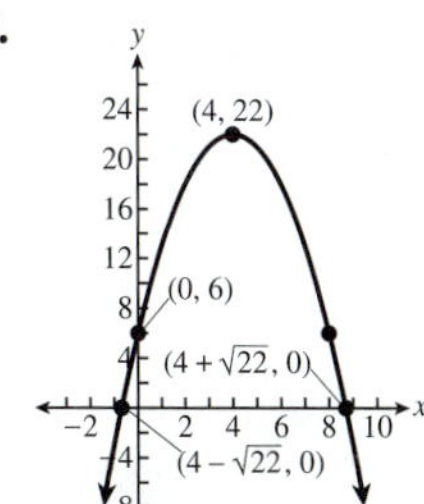

37.

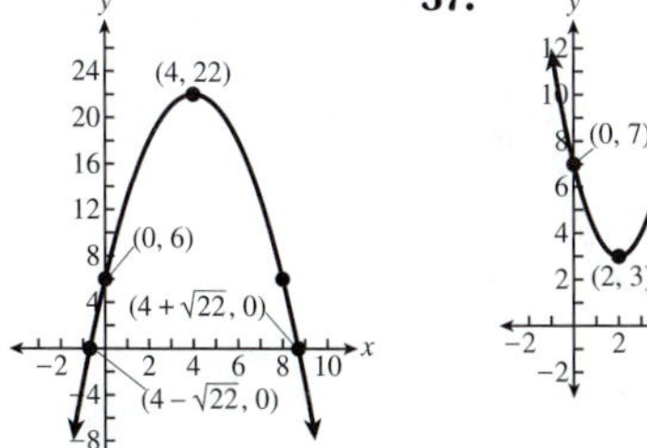

38. $[4, 6]$

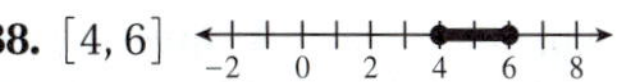

39. $(-\infty, -6] \cup (-5, 8]$

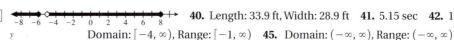

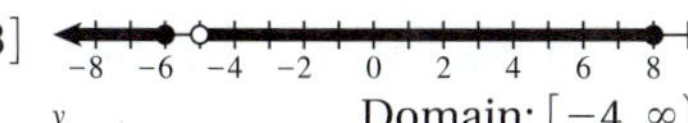

40. Length: 33.9 ft, Width: 28.9 ft **41.** 5.15 sec **42.** 11.72 hours

43. 374 ft **44.**

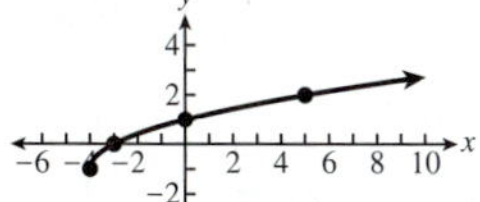

Domain: $[-4, \infty)$, Range: $[-1, \infty)$ **45.** Domain: $(-\infty, \infty)$, Range: $(-\infty, \infty)$

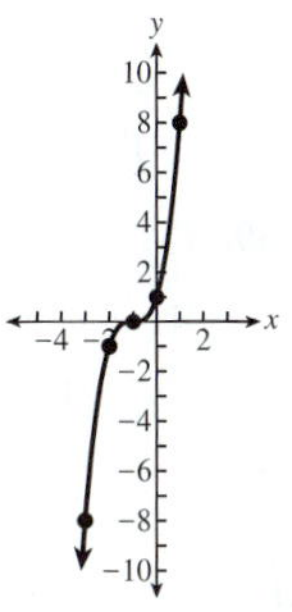

Chapter 8

Quick Check 8.1 **1. a)** $x^2 - 6x - 24$ **b)** $x^2 - 12x + 64$ **c)** -17 **d)** 92 **2.** $12x^2 - 5x - 28$ **3.** $\frac{x-3}{x+7}$, $x \neq -6, -7$

4.

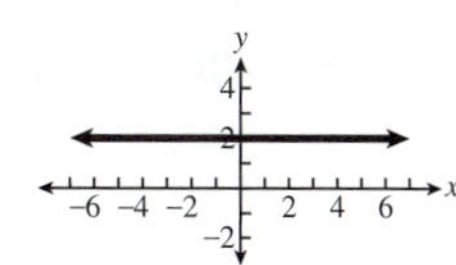

5. a) $155{,}889x + 12{,}119{,}000$: total number of students enrolled in a public or private college x years after 1990. **b)** $67{,}943x + 6{,}937{,}000$: This function tells us how many more students there are enrolled at public colleges than at private colleges x years after 1990. **6.** $8x + 19$ **7. a)** $x^2 - 3x - 25$ **b)** $x^2 + 3x - 28$ **8.** $\frac{2x-3}{x+3}$; domain: all real numbers except $x = -3$

Section 8.1 **1.** sum **3.** $(f \cdot g)(x)$ **5.** composition **7. a)** $7x - 6$ **b)** 50 **c)** -27 **9. a)** $x^2 + 4x - 11$ **b)** 85 **c)** -14 **11. a)** $8x - 6$ **b)** 42 **c)** -46 **13. a)** $-x^2 + 15x + 49$ **b)** 103 **c)** -51 **15. a)** $2x^2 + 4x - 70$ **b)** -22 **c)** 90 **17. a)** $4x^3 + 35x^2 - 69x - 540$ **b)** 0 **c)** -350 **19. a)** $\frac{3x+1}{x+7}$ **b)** $\frac{11}{7}$ **c)** -1 **21. a)** $\frac{x+5}{x-4}$ **b)** 4 **c)** $-\frac{1}{2}$ **23.** 37 **25.** -25 **27.** 377 **29.** $-\frac{5}{17}$ **31.** $7x - 26$ **33.** $10x^2 - 103x + 153$ **35.** $-x^2 + 9x + 10$ **37.** $\frac{1}{x+2}$ **39.** $g(x) = 4x + 9$ **41.** $g(x) = -4x + 6$

43.

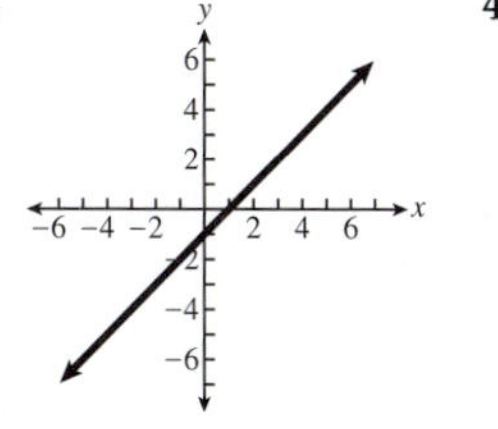

45.

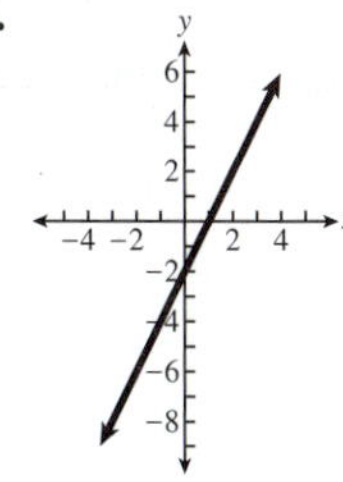

47. a) $(f + g)(x) = 19.5x + 447$; this is the total number of doctors, in thousands, in the United States x years after 1980. **b)** 1227; there will be 1,227,000 doctors in the United States in the year 2020.
49. a) $(f - g)(x) = 5.8x + 20.9$; this tells how many more females than males, in thousands, will earn a master's degree in the United States x years after 1990.
b) 194.9; there will be 194,900 more master's degrees earned by females than by males in the United States in the year 2020. **51.** 15 **53.** 0 **55.** 11 **57.** 550 **59. a)** $6x + 17$ **b)** $6x + 14$ **c)** 35 **d)** -10 **61. a)** $x^2 + 3x - 36$ **b)** $x^2 + 11x - 12$ **c)** -18 **d)** -40 **63.** $\frac{10x+65}{2x+20}$, all real numbers except -10. **65.** $\sqrt{2x + 12}$, $[-6, \infty)$ **67.** $9x - 40$ **69.** $x^4 + 12x^3 + 66x^2 + 180x + 228$ **71.** $x = -4$ **73.** $x = -3$ **75.** $f(x) = x + 8$ **77.** $f(x) = 5x - 16$ **79.** $f(x) = x - 8$ **81.** $f(x) = \frac{x-8}{3}$ **83.** Yes, explanations will vary.

Quick Check 8.2 **1.** Not one to one **2.** Not one to one **3.** Inverse functions **4.** Inverse functions **5.** $f^{-1}(x) = 2x - 20$ **6.** $f^{-1}(x) = \frac{3}{x-5}, x \neq 5$ **7.** $f^{-1}(x) = x^2 - 16x + 73$, Domain: $[8, \infty)$ **8.**

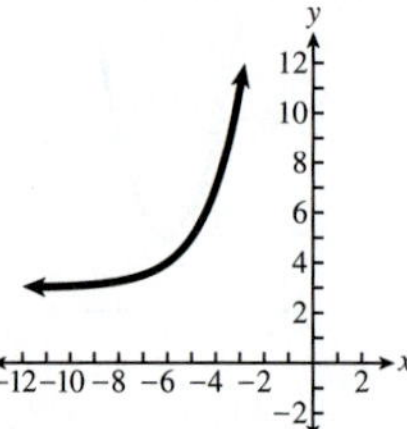

Section 8.2 **1.** one-to-one **3.** inverse **5.** $f^{-1}(x)$ **7.** Yes **9.** No **11.** No **13.** No **15.** No **17.** Yes **19.** No **21.** Yes **23.** No **25.** Inverses **27.** Not inverses **29.** Not inverses **31.** Not inverses **33.** $f^{-1}(x) = x - 5$ **35.** $f^{-1}(x) = \frac{x}{4}$ **37.** $f^{-1}(x) = -x + 9$ **39.** $f^{-1}(x) = \frac{x-17}{2}$ **41.** $f^{-1}(x) = \frac{-3x-24}{2}$ **43.** $f^{-1}(x) = \frac{x}{m}$ **45.** $f^{-1}(x) = \frac{1}{x-3}$ **47.** $f^{-1}(x) = \frac{1-3x}{x}$ **49.** $f^{-1}(x) = \frac{5}{x+6}$ **51.** $f^{-1}(x) = \frac{8}{3x-4}$ **53.** $f^{-1}(x) = x^2, x \geq 0$ **55.** $f^{-1}(x) = x^2 - 10x + 25, x \geq 5$ **57.** $f^{-1}(x) = x^2 - 8x + 8, x \geq 4$ **59.** $f^{-1}(x) = \sqrt{x+9}, x \geq -9$ **61.** $f^{-1}(x) = \sqrt{x+1} + 3, x \geq -1$ **63.** $\{(-17, -5), (-9, -1), (-5, 1), (1, 4), (7, 7)\}$ **65.** Not one to one

67.

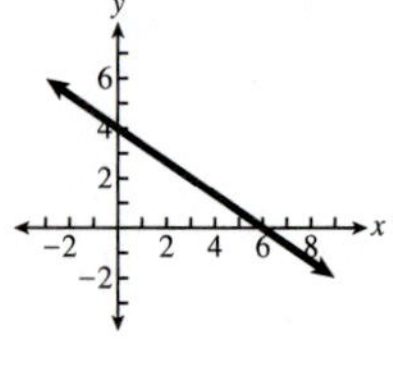

69.

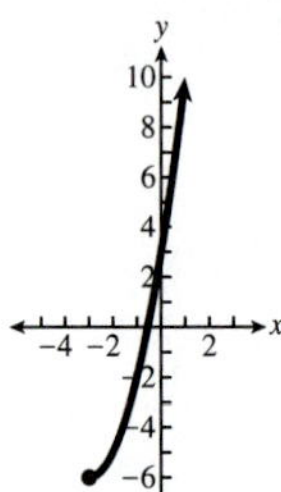

71.

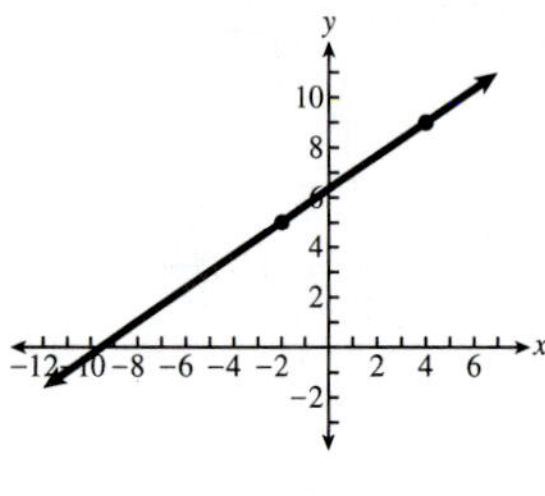

73.

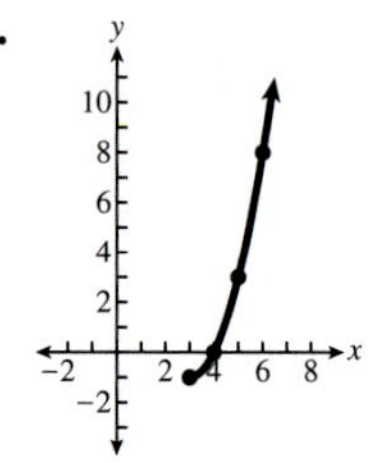

75. False, answers will vary. **77.** Answers will vary.

Quick Check 8.3 **1.** $f(5) = 1024, f(-3) = \frac{1}{64}$ **2.** 79 **3.** Domain: $(-\infty, \infty)$, Range: $(0, \infty)$ **4.** Domain: $(-\infty, \infty)$, Range: $(0, \infty)$

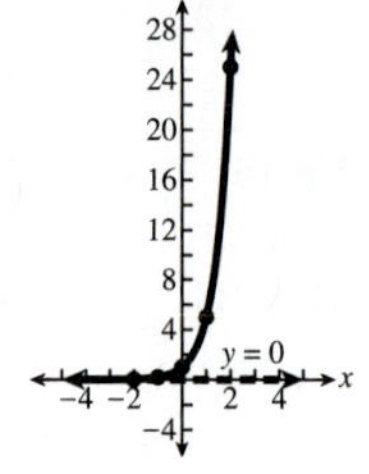

5. Domain: $(-\infty, \infty)$, Range: $(3, \infty)$ **6.** Domain: $(-\infty, \infty)$, Range: $(-2, \infty)$ **7.** $f(-1) \approx 0.368, f(4.3) \approx 73.700$

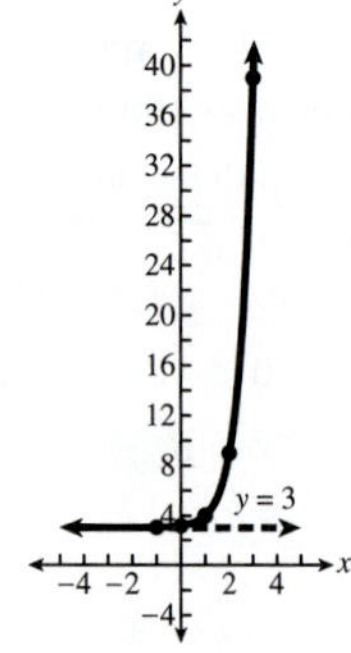

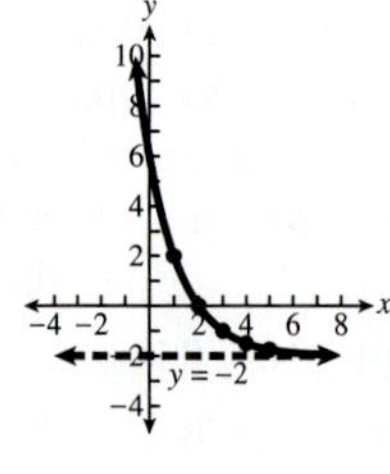

8. Domain: $(-\infty, \infty)$, Range: $(5, \infty)$ **9. a)** 4 **b)** 9 **10.** Approximately $34,197.4 million

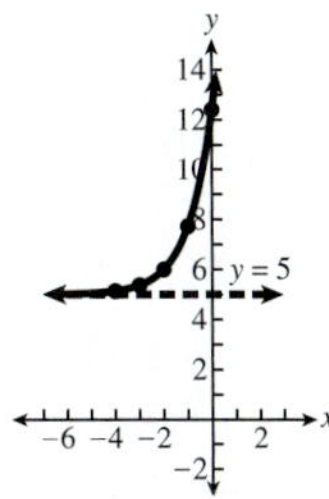

Section 8.3 **1.** exponential **3.** increasing **5.** $r = s$ **7.** 16 **9.** 1 **11.** $\frac{4}{25}$ **13.** $\frac{625}{16}$ **15.** 11 **17.** 251 **19.** 10 **21.** $\frac{1}{64}$ **23.** 2087 **25.** $\frac{1}{100}, \frac{1}{10}, 1, 10, 100$ **27.** $-8\frac{15}{16}, -8\frac{3}{4}, -8, -5, 7$ **29.** Domain: $(-\infty, \infty)$, Range: $(0, \infty)$

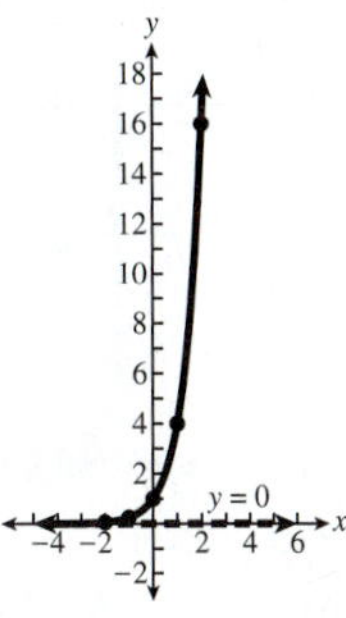

31. Domain: $(-\infty, \infty)$, Range: $(0, \infty)$ **33.** Domain: $(-\infty, \infty)$, Range: $(0, \infty)$ **35.** Domain: $(-\infty, \infty)$, Range: $(-1, \infty)$

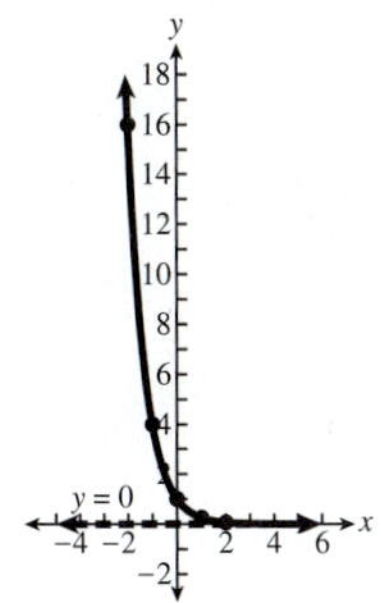

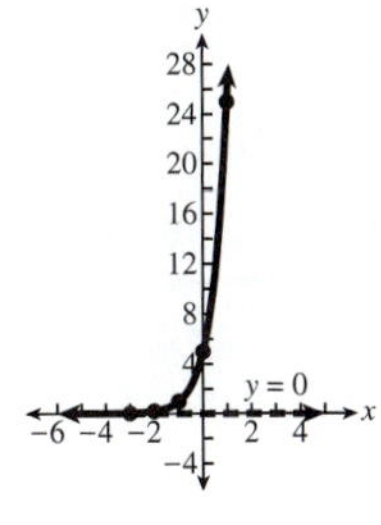

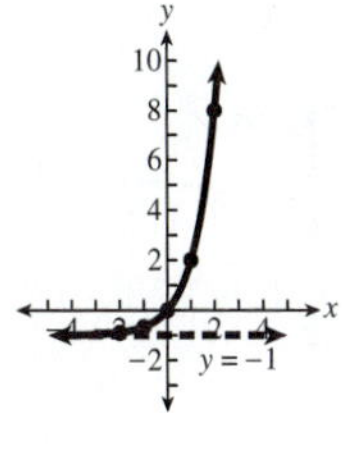

37. Domain: $(-\infty, \infty)$, Range: $(-8, \infty)$ **39.** Domain: $(-\infty, \infty)$, Range: $(2, \infty)$ **41.** 7.389 **43.** 0.368 **45.** 0.006 **47.** 20.086

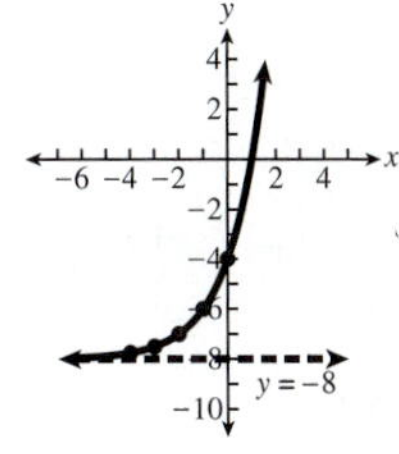

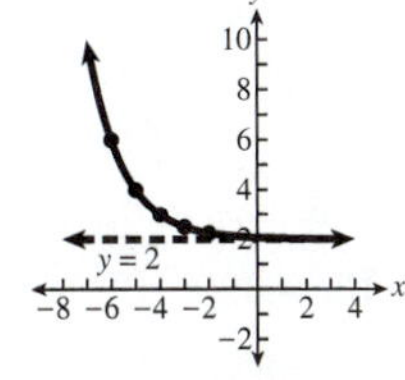

49. 0.001 **51.** Domain: $(-\infty, \infty)$, Range: $(0, \infty)$ **53.** Domain: $(-\infty, \infty)$, Range: $(2, \infty)$ **55.** Domain: $(-\infty, \infty)$, Range: $(4, \infty)$

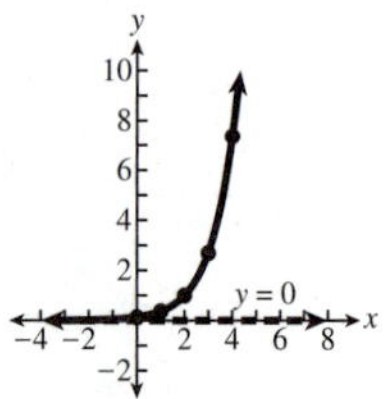

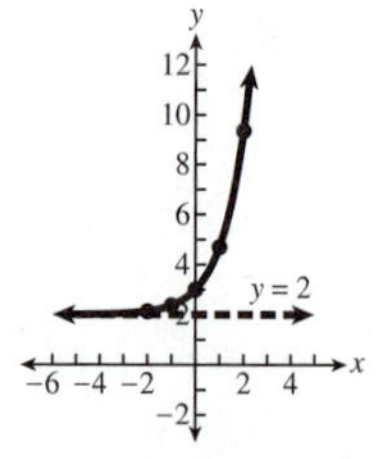

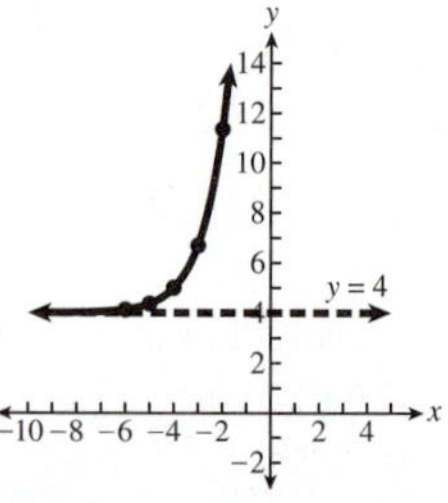

57. $\{4\}$ **59.** $\{3\}$ **61.** $\{-1\}$ **63.** $\{12\}$ **65.** $\{1\}$ **67.** $\{3\}$ **69.** $\{3\}$ **71.** $\{3\}$ **73.** $\{5\}$ **75.** \$663,907.76 **77.** \$15,605.09 **79.** \$99.33 **81.** \$13,548 million **83.** 36.8°C **85.** \$1.20 **87.** No **89.** Yes **91.** Answers will vary. **93.** Answers will vary.

Quick Check 8.4 **1. a)** 2 **b)** -3 **c)** 1 **d)** 0 **2.** -4 **3. a)** 4 **b)** 1.875 **4. a)** -3 **b)** 6.405 **5.** $\log_7 20 = x$ **6.** $3^6 = x$ **7.** $\{1024\}$ **8.** $\{3\}$ **9. a)**

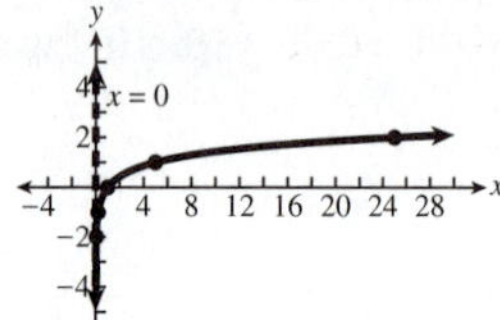

b)

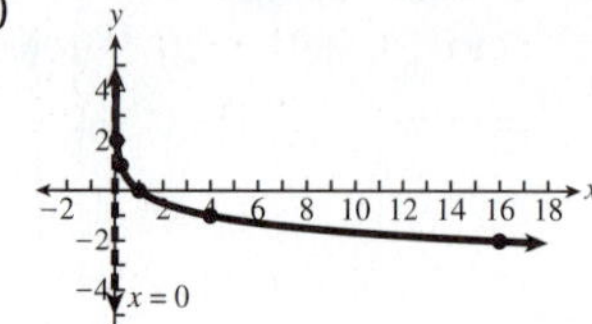

10. Approximately \$43.47 billion **11.** 3.4

Section 8.4 **1.** $\log_b x$ **3.** common **5.** exponential **7.** vertical **9.** 2 **11.** 3 **13.** 3 **15.** 0 **17.** -2 **19.** -3 **21. a)** 2 **b)** 5 **23. a)** 0 **b)** 7 **25.** 3 **27.** -2 **29.** 2.049 **31.** 0.870 **33.** 1 **35.** -6 **37.** 3.466 **39.** 2.787 **41.** $\log_2 8 = 3$ **43.** $\log_9 729 = 3$ **45.** $\log_6 30 = x$ **47.** $\log 500 = x$ **49.** $4^5 = 1024$ **51.** $16^{3/4} = 8$ **53.** $10^3 = x$ **55.** $e^7 = x$ **57.** $\{32\}$ **59.** $\{\frac{1}{81}\}$ **61.** $\{10{,}000\}$ **63.** $\{7.389\}$ **65.** $\{9\}$ **67.** $\{5.135\}$ **69.** $\{216\}$ **71.** $\{110\}$ **73.** $\{3\}$ **75.** $\{5\}$ **77.** $\{\frac{3}{2}\}$

79.

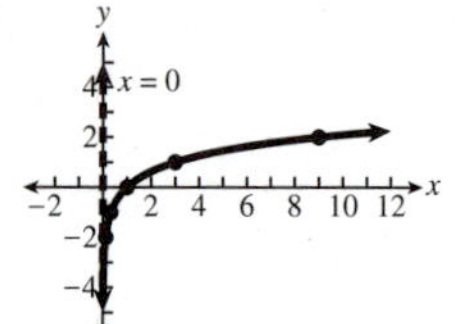

81.

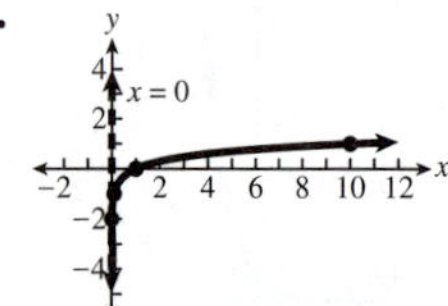

83.

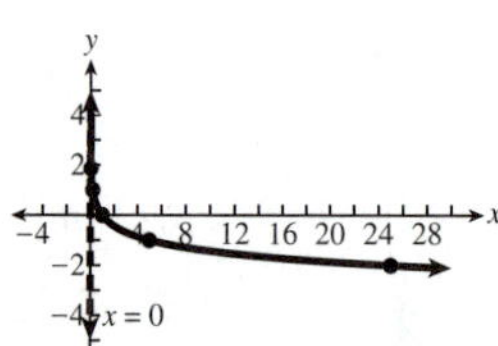

85.

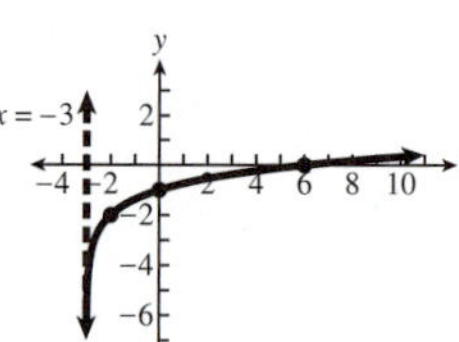

87.

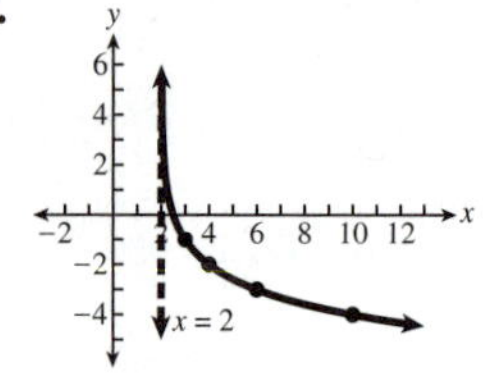

89. $f(x) = \log_2 x$ **91.** $f(x) = \log_{1/3} x$ **93.** 19,934 **95.** 33.8 in. **97.** 4.2 **99.** 630,957,345 times larger **101.** Answers will vary. **103.** Answers will vary.

Quick Review Exercises **1.** x^{16} **2.** b^{12} **3.** n^{56} **4.** 1

Quick Check 8.5 **1.** $\log(35x)$ **2.** $\log_2 5 + \log_2 b$ **3.** 2 **4.** $\log_b x - 1$ **5.** $2 + 2\log_3 x + 4\log_3 y$ **6.** $\frac{1}{7}\ln x$ **7.** $\log_8 x^8$ **8.** -4 **9.** 16 **10.** $\log_2\left(\frac{acd}{b}\right)$ **11.** $\log\left(\frac{a^8\sqrt[5]{b}}{c^9d^{12}}\right)$ **12.** $3\log_5 z - 9\log_5 x - 4\log_5 y$ **13.** 4.395

Section 8.5 **1.** $\log_b(xy)$ **3.** $r \cdot \log_b x$ **5.** $\log_2(5x)$ **7.** $\log_4(bc)$ **9.** $\log_4 256 = 4$ **11.** $\log_3 10 + \log_3 x$ **13.** $\ln 3 + 1$ **15.** $\log_4 a + \log_4 b + \log_4 c$ **17.** $\log_5 6$ **19.** $\log\left(\frac{39}{x}\right)$ **21.** $\log_4 64 = 3$ **23.** $\log_8 10 + \log_8 a - \log_8 b$ **25.** $\log_4 6 + \log_4 x + \log_4 y + \log_4 z - \log_4 w$ **27.** $\log_5 x + \log_5 y - 2 - \log_5 z$ **29.** $10\log a$ **31.** $2\log_7 x + 3\log_7 y$ **33.** $\frac{1}{2}\log_3 x$ **35.** $4 + 5\log_2 x + 7\log_2 y$ **37.** $\log_5 x^8$ **39.** $\ln\sqrt[3]{x}$ **41.** $\log_3 x^{10}$ **43.** 12 **45.** 9 **47.** 7 **49.** b **51.** $\log_3\left(\frac{10}{7}\right)$ **53.** $\log_5 5832$ **55.** $\ln\left(\frac{4}{5}\right)$ **57.** $\log_2\left(\frac{x}{yzw}\right)$ **59.** $\ln\left(\frac{x^4y^9}{z^{13}}\right)$ **61.** $\log(x^2 - 4x - 12)$ **63.** $\log_3(2x^2 + 17x + 35)$ **65.** $\ln\left(\frac{a^3\sqrt[3]{b}}{c^{10}}\right)$ **67.** $A + B$ **69.** $C + D$ **71.** $3A$ **73.** $3C$ **75.** $\log_8 a - \log_8 b - \log_8 c - \log_8 d$ **77.** $5 + 4\log_2 x + 8\log_2 y + 5\log_2 z$ **79.** $6\ln a + \frac{1}{2}\ln b - 7\ln c$ **81.** $-3\ln a - 2\ln d + \ln b + 7\ln c$ **83.** 2.807 **85.** 1.549 **87.** 6.481 **89.** 0.528 **91.** -1.521 **93.** 1.829 **95.** 1.701 **97.** Answers will vary.

Quick Check 8.6 **1. a)** $\{2\}$ **b)** $\{12\}$ **2.** $\{\frac{\log 12}{\log 4} + 5\}, \frac{\log 12}{\log 4} + 5 \approx 6.792$ **3.** $\{\frac{\log 11}{\log 9} - 3\}, \frac{\log 11}{\log 9} - 3 \approx -1.909$ **4.** $\{\ln 28 - 2\}$, $\ln 28 - 2 \approx 1.332$ **5.** $\{2\}$ **6.** $\varnothing$ **7.** $\{-2, 5\}$ **8.** $\{333{,}338\}$ **9. a)** $\{4\}$ **b)** $\{3\}$ **10.** $\{11\}$ **11.** $f^{-1}(x) = \ln(x-9) - 6$ **12.** $f^{-1}(x) = e^{x+18} - 10$

Section 8.6 **1.** extraneous **3.** $\{3\}$ **5.** $\{6\}$ **7.** $\{5\}$ **9.** $\{\frac{3}{2}\}$ **11.** $\{\frac{9}{2}\}$ **13.** $\{-12\}$ **15.** $\{2.044\}$ **17.** $\{2.111\}$ **19.** $\{9.755\}$ **21.** $\{-0.820\}$ **23.** $\{1.922\}$ **25.** $\{3.305\}$ **27.** $\{-0.726\}$ **29.** $\{4\}$ **31.** $\{9\}$ **33.** $\varnothing$ **35.** $\{2, 12\}$ **37.** $\{-6\}$ **39.** $\{243\}$ **41.** $\{\frac{1}{625}\}$ **43.** $\{407.429\}$ **45.** $\{24\}$ **47.** $\{-12, 8\}$ **49.** $\{-5, 8\}$ **51.** $\{8\}$ **53.** $\{34\}$ **55.** $\{2\}$ **57.** $\{3\}$ **59.** $\{8\}$ **61.** $\{3\}$ **63.** $\{13\}$ **65.** $\{9\}$ **67.** $\{-10, 7\}$ **69.** $\{-2, 15\}$ **71.** $\{-5.267\}$ **73.** $\{5.577\}$ **75.** $\{-5\}$ **77.** $\{347\}$ **79.** $\{-3\}$ **81.** $\{10.069\}$ **83.** $\{\frac{5}{3}\}$ **85.** $\{4\}$ **87.** $\{37\}$ **89.** $\{111\}$ **91.** $\{-7\}$ **93.** $\{12\}$ **95.** $\{-2.432\}$ **97.** $\{-6\}$ **99.** $f^{-1}(x) = \ln x - 3$ **101.** $f^{-1}(x) = \ln(x-4) + 5$ **103.** $f^{-1}(x) = e^x + 6$ **105.** $f^{-1}(x) = e^{x-5} - 2$ **107.** 16 **109. a)** 25 **b)** -6 **111.** Answers will vary. **113.** Answers will vary.

Quick Check 8.7 **1.** \$464.65 **2.** Approximately 14 years **3.** Approximately 15.35 years **4.** Approximately \$1.45 **5.** Approximately 117.5 days **6.** In 2013 **7.** 2.8, acid **8.** 5.01×10^{-1} moles/liter **9.** 60 db **10.** 10^{-4} watts per square meter

Section 8.7 **1.** compound **3.** continuously **5.** negative **7.** \$5634.13 **9.** \$27,198.41 **11.** 10.2 yr **13.** 23.1 yr **15.** \$1343.13 **17.** 5.8 yr **19.** 6894 **21.** 13.7 billion **23.** 2010 (10.8 yr) **25.** 2011 (12.5 yr) **27.** 86.4 min **29.** 358 **31.** 22.6 mg **33.** 167 yr **35.** 229,919.7 yr **37.** 2094 **39.** 2011 **41.** 425,612 **43.** 178 **45.** 2020 (25.9 yr) **47.** 2047 **49.** 7 yr **51. a)** 90.3° F **b)** 2.5 hr **53.** 2018 (29.2 yr) **55.** 12.6, base **57.** 2.3, acid **59.** 1.0×10^{-5} moles per liter **61.** 6.31×10^{-11} moles per liter **63.** 80 db **65.** 110 db **67.** 85 db **69.** 10^{-1} watts per square meter **71.** $10^{-2.5}$ watts per square meter **73.** 80 **75.** 2016 (21.1 yr) **77.** Answers will vary. **79.** Answers will vary.

Quick Check 8.8 **1.** Domain: $(-\infty, \infty)$, Range: $(-6, \infty)$ **2.** Domain: $(-\infty, \infty)$, Range: $(-9, \infty)$ **3.** Domain: $(-2, \infty)$, Range: $(-\infty, \infty)$

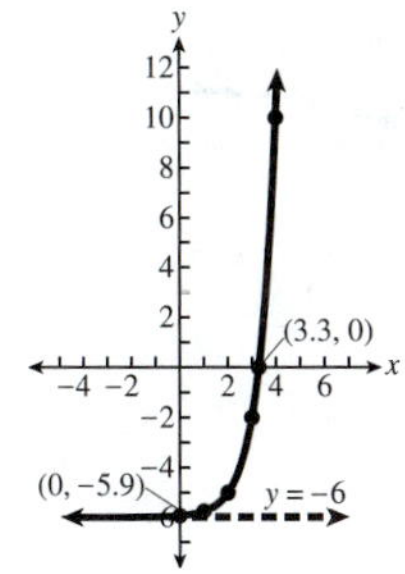

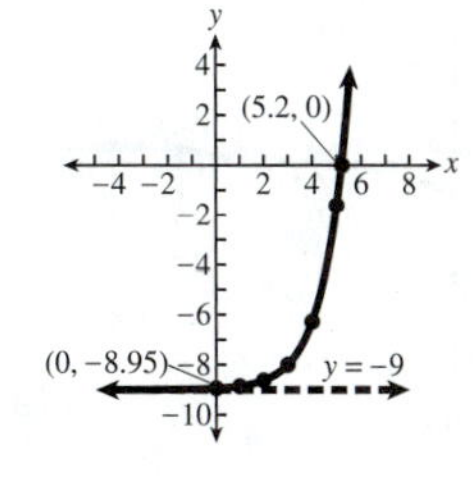

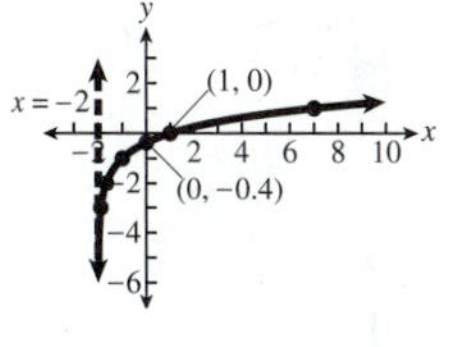

4. Domain: $(6, \infty)$, Range: $(-\infty, \infty)$

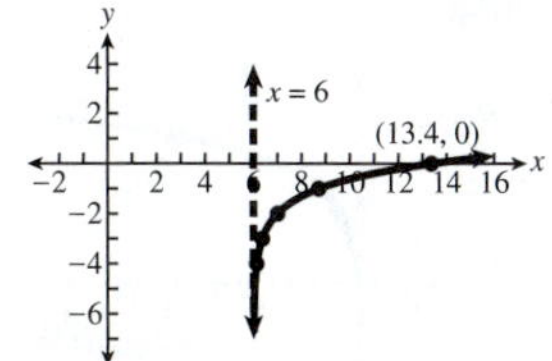

Section 8.8 **1.** horizontal **3.** exponential **5.** y **7.** x-int: none, y-int: $(0, 22)$ **9.** x-int: $(-0.3, 0)$, y-int: $(0, 15)$ **11.** x-int: $(-0.7, 0)$, y-int: $(0, 10.1)$ **13.** x-int: $(13, 0)$, y-int: none **15.** x-int: $(51.6, 0)$, y-int: $(0, -2.9)$ **17.** Asymptote: $y = 8$, Domain: $(-\infty, \infty)$, Range: $(8, \infty)$ **19.** Asymptote: $y = -9$, Domain: $(-\infty, \infty)$, Range: $(-9, \infty)$ **21.** Asymptote: $x = 5$, Domain: $(5, \infty)$, Range: $(-\infty, \infty)$ **23.** Asymptote: $x = \frac{1}{2}$, Domain: $(\frac{1}{2}, \infty)$, Range: $(-\infty, \infty)$

25. Domain: $(-\infty, \infty)$, Range: $(4, \infty)$ **27.** Domain: $(-\infty, \infty)$, Range: $(-15, \infty)$ **29.** Domain: $(-\infty, \infty)$, Range: $(6, \infty)$

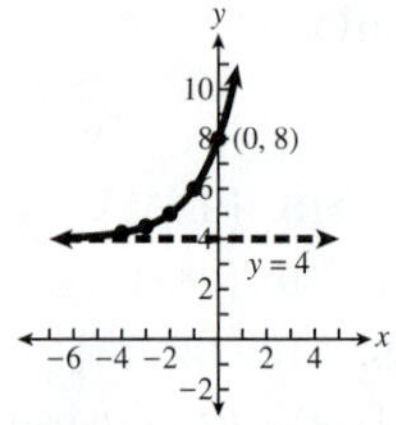

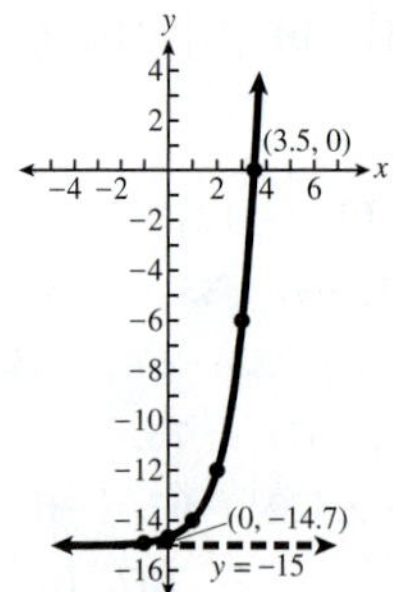

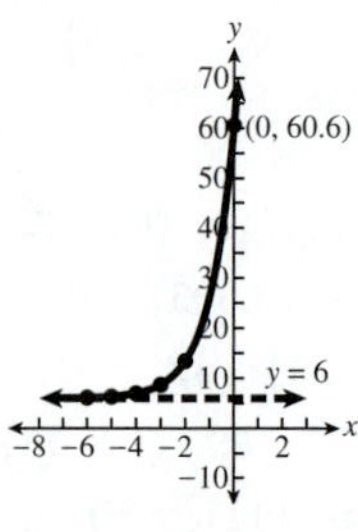

31. Domain: $(2, \infty)$, Range: $(-\infty, \infty)$ **33.** Domain: $(-2, \infty)$, Range: $(-\infty, \infty)$ **35.** Domain: $(1, \infty)$, Range: $(-\infty, \infty)$

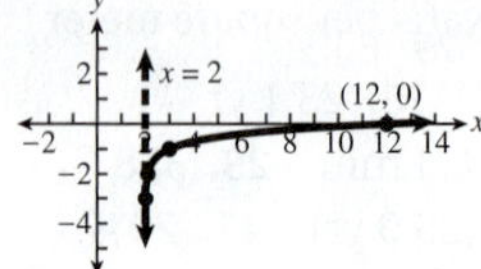

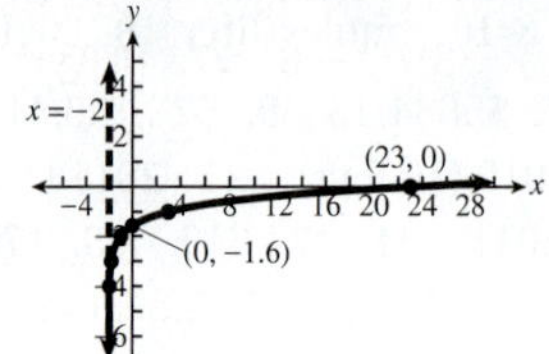

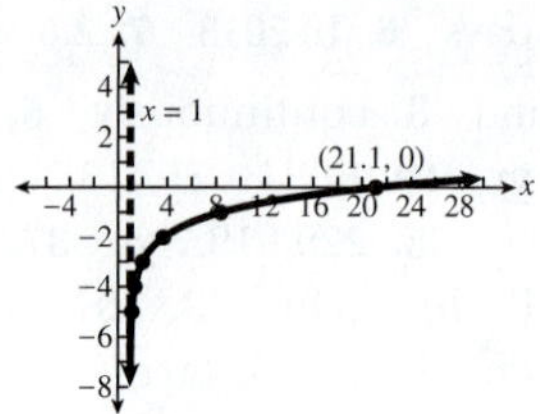

37. Domain: $(-\infty, \infty)$, Range: $(-2, \infty)$ **39.** Domain: $(-4, \infty)$, Range: $(-\infty, \infty)$ **41.** Domain: $(-6, \infty)$, Range: $(-\infty, \infty)$

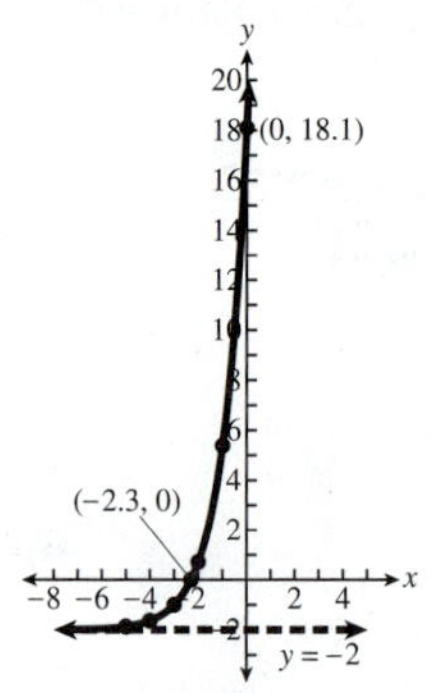

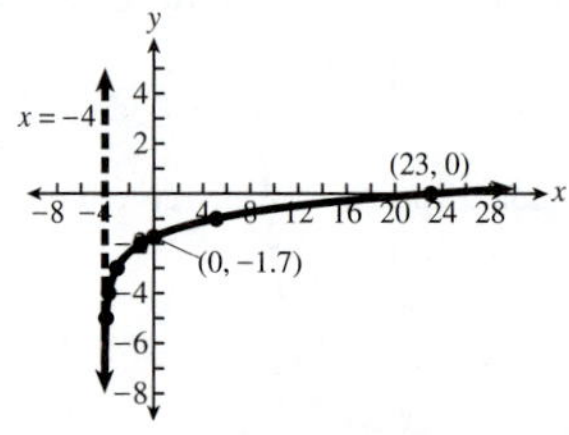

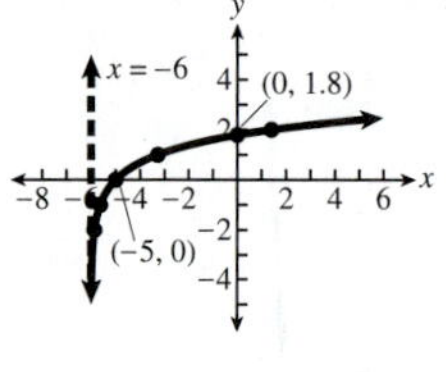

43. Domain: $(-\infty, \infty)$, Range: $(-8, \infty)$ **45.** Domain: $(-\infty, \infty)$, Range: $(-3, \infty)$ **47.** **49.**

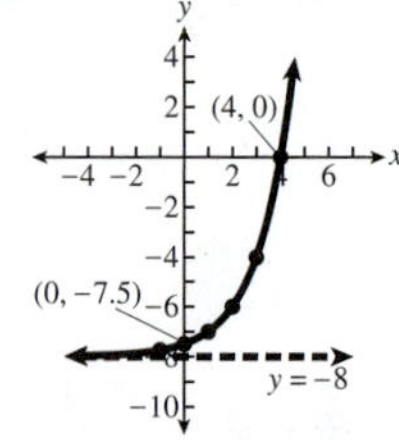

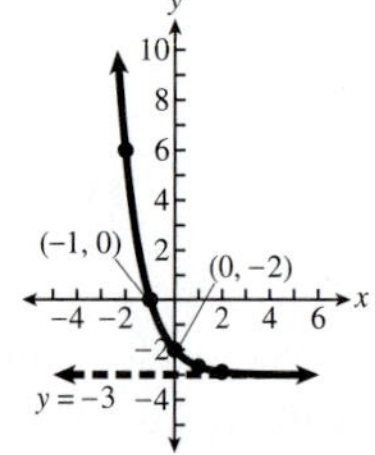

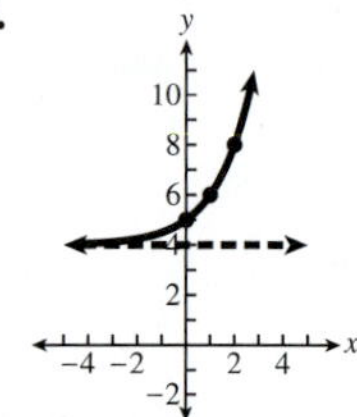

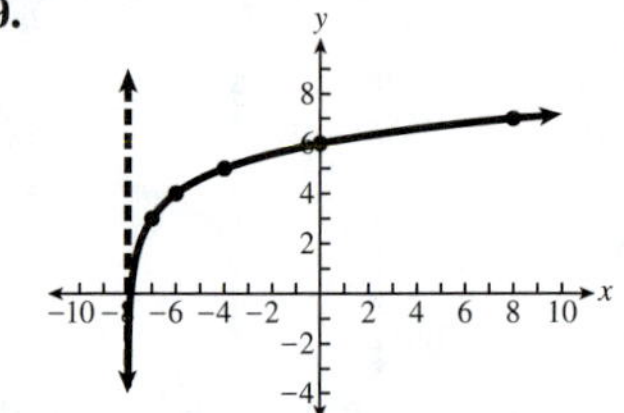

51. $f^{-1}(x) = 2^x$ **53.** $f^{-1}(x) = \log_2(x + 8) + 4$ **55.** b **57.** h **59.** a **61.** f **63.** Answers will vary.

Chapter 8 Review **1.** -36 **2.** 312 **3.** 2166 **4.** -3 **5.** $10x + 24$ **6.** $16x - 88$ **7.** $x^2 - 5x - 36$ **8.** $\frac{x+2}{x-7}$ **9.** 4 **10.** 35 **11.** $(f \circ g)(x) = 18x + 103, (g \circ f)(x) = 18x - 31$ **12.** $(f \circ g)(x) = -8x + 33, (g \circ f)(x) = -8x - 51$ **13.** $(f \circ g)(x) = x^2 + 7x - 64, (g \circ f)(x) = x^2 - 9x - 48$ **14.** $(f \circ g)(x) = 3x^2 - 27x + 130, (g \circ f)(x) = 9x^2 + 15x + 27$ **15.** No **16.** Yes **17.** Yes **18.** No **19.** No **20.** Yes **21.** No **22.** Yes **23.** $f^{-1}(x) = x + 10, (-\infty, \infty)$ **24.** $f^{-1}(x) = \frac{x+15}{2}, (-\infty, \infty)$ **25.** $f^{-1}(x) = -x + 19, (-\infty, \infty)$ **26.** $f^{-1}(x) = \frac{-x+27}{8}, (-\infty, \infty)$ **27.** $f^{-1}(x) = \frac{2}{3x-9}, x \neq 3$ **28.** $f^{-1}(x) = \frac{21x}{7-4x}, x \neq \frac{7}{4}$

29.

30.

31. 1015 **32.** 7.086 **33.** -2 **34.** 5 **35.** 0 **36.** 12 **37.** 23.761 **38.** -2.756 **39.** $\log_4 1024 = x$ **40.** $\ln 20 = x$ **41.** $10^{3.2} = x$ **42.** $e^{11} = x$ **43.** $\log_2 104$ **44.** $\log\left(\frac{a^2}{b^5}\right)$ **45.** $\log_2\left(\frac{x^4z^9}{y^3}\right)$ **46.** $\log_b \frac{1}{972}$ **47.** $2 + \log_8 x$ **48.** $\log_3 n - 3$ **49.** $6 \log_b a + 3 \log_b c - 2$ **50.** $3 + 9 \log_2 x - \log_2 y - 7 \log_2 z$ **51.** 2.322 **52.** -8.205 **53.** 10.641 **54.** 4.340 **55.** $\{3\}$ **56.** $\{7\}$ **57.** $\{\frac{5}{4}\}$ **58.** $\{12\}$ **59.** $\{1.232\}$ **60.** $\{0.468\}$ **61.** $\{12.127\}$ **62.** $\{5.806\}$ **63.** $\{14\}$ **64.** $\{1\}$ **65.** $\{148.413\}$ **66.** $\{3, 10\}$ **67.** $\{2\}$ **68.** $\{10\}$ **69.** $\{4\}$ **70.** $\{2\}$ **71.** $\{3.107\}$ **72.** $\{54.598\}$ **73.** $f^{-1}(x) = \ln(x - 4)$ **74.** $f^{-1}(x) = \log_2(x + 1) + 5$ **75.** $f^{-1}(x) = e^x - 8$ **76.** $f^{-1}(x) = 3^{x+10} - 6$ **77.** 13.3 yr **78.** 334,133 **79.** 6.25 g **80.** 167 yr **81.** 2007 (17.3 yr) **82.** 78.5° F **83.** 4.3 **84.** 3.16×10^{-4} moles/liter **85.** Domain: $(-\infty, \infty)$, Range: $(2, \infty)$

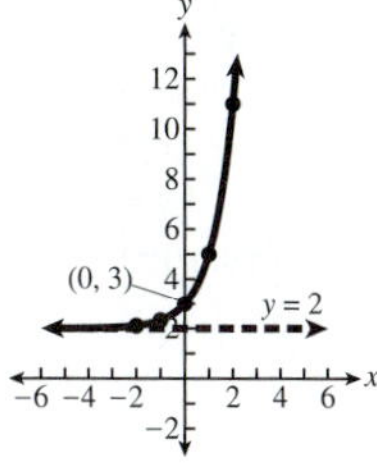

86. Domain: $(-\infty, \infty)$, Range: $(-8, \infty)$ **87.** Domain: $(-\infty, \infty)$, Range: $(-6, \infty)$ **88.** Domain: $(5, \infty)$, Range: $(-\infty, \infty)$

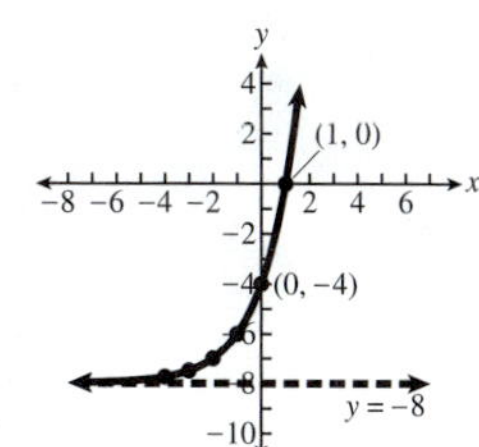

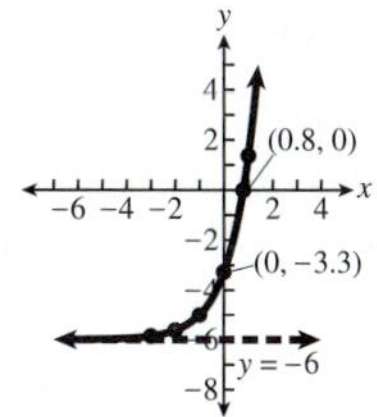

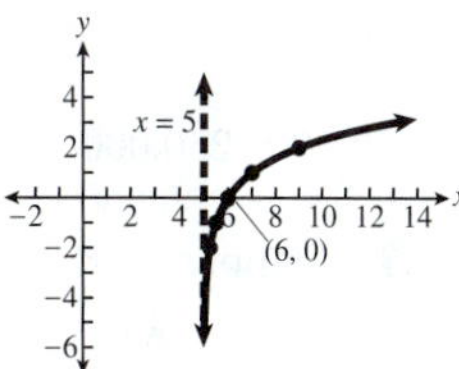

89. Domain: $(-9, \infty)$, Range: $(-\infty, \infty)$ **90.** Domain: $(-8, \infty)$, Range: $(-\infty, \infty)$

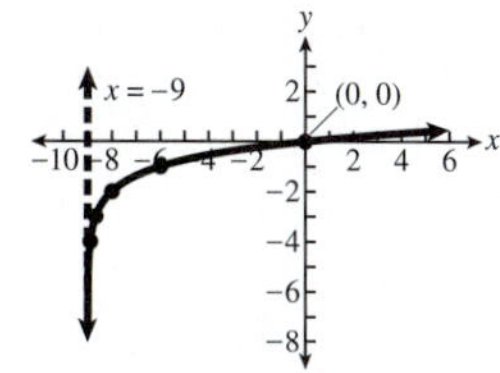

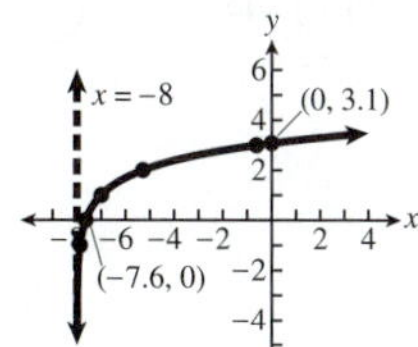

Chapter 8 Review Exercises: Worked-Out Solutions

5.
$$\begin{aligned}(f + g)(x) &= f(x) + g(x)\\ &= (6x - 11) + (4x + 35)\\ &= 10x + 24\end{aligned}$$

9.
$$\begin{aligned}g(8) &= (8)^2 - 14(8) + 48\\ &= 64 - 112 + 48\\ &= 0\\ (f \circ g)(8) &= f(g(8))\\ &= f(0)\\ &= (0) + 4\\ &= 4\end{aligned}$$

13.
$$\begin{aligned}(f \circ g)(x) &= f(g(x))\\ &= f(x^2 + 7x - 56)\\ &= (x^2 + 7x - 56) - 8\\ &= x^2 + 7x - 64\\ (g \circ f)(x) &= g(f(x))\\ &= g(x - 8)\\ &= (x - 8)^2 + 7(x - 8) - 56\\ &= x^2 - 16x + 64 + 7x - 56 - 56\\ &= x^2 - 9x - 48\end{aligned}$$

20.
$$\begin{aligned}(f \circ g)(x) &= f(g(x))\\ &= f\left(\tfrac{x+10}{2}\right)\\ &= 2\left(\tfrac{x+10}{2}\right) - 10\\ &= x + 10 - 10\\ &= x\\ (g \circ f)(x) &= g(f(x))\\ &= g(2x-10)\\ &= \tfrac{(2x-10)+10}{2}\\ &= \tfrac{2x}{2}\\ &= x\end{aligned}$$

Yes, the functions are inverse functions.

24.
$$\begin{aligned}y &= 2x - 15\\ x &= 2y - 15\\ x + 15 &= 2y\\ \tfrac{x+15}{2} &= y\\ f^{-1}(x) &= \tfrac{x+15}{2}\end{aligned}$$
Domain: $(-\infty, \infty)$

32. $f(11) = e^{11-8} - 13 = e^3 - 13 \approx 7.086$

33. $\log_6 \frac{1}{36} = -2$ because $6^{-2} = \frac{1}{36}$.

45. $4\log_2 x - 3\log_2 y + 9\log_2 z = \log_2 x^4 - \log_2 y^3 + \log_2 z^9 = \log_2\left(\frac{x^4z^9}{y^3}\right)$

49. $\log_b\left(\frac{a^6c^3}{b^2}\right) = \log_b a^6 + \log_b c^3 - \log_b b^2 = 6\log_b a + 3\log_b c - 2\log_b b = 6\log_b a + 3\log_b c - 2$

51. $\log_4 25 = \frac{\log 25}{\log 4} \approx 2.322$

56.
$$\begin{aligned}8^{x-5} &= 64\\ 8^{x-5} &= 8^2\\ x - 5 &= 2\\ x &= 7\end{aligned}$$
$\{7\}$

58.
$$\begin{aligned}9^{x+6} &= 27^x\\ (3^2)^{x+6} &= (3^3)^x\\ 3^{2x+12} &= 3^{3x}\\ 2x + 12 &= 3x\\ 12 &= x\end{aligned}$$
$\{12\}$

61.
$$\begin{aligned}e^{x-8} &= 62\\ \ln e^{x-8} &= \ln 62\\ x - 8 &= \ln 62\\ x &= \ln 62 + 8\\ x &\approx 12.127\end{aligned}$$
$\{12.127\}$

63.
$$\begin{aligned}\log_4(x+3) &= \log_4 17\\ x + 3 &= 17\\ x &= 14\end{aligned}$$
$\{14\}$

69.
$$\begin{aligned}\log_3(x+5) + \log_3(x-1) &= 3\\ \log_3((x+5)(x-1)) &= 3\\ \log_3(x^2+4x-5) &= 3\\ x^2 + 4x - 5 &= 3^3\\ x^2 + 4x - 5 &= 27\\ x^2 + 4x - 32 &= 0\\ (x+8)(x-4) &= 0\end{aligned}$$
$x + 8 = 0$ or $x - 4 = 0$

$x = -8$ or $x = 4$

$x = -8$ is an extraneous solution.

$\{4\}$

73.
$$\begin{aligned}y &= e^x + 4\\ x &= e^y + 4\\ x - 4 &= e^y\\ \ln(x-4) &= \ln e^y\\ \ln(x-4) &= y\\ f^{-1}(x) &= \ln(x-4)\end{aligned}$$

78. $P = 250{,}000$

$P_0 = 200{,}000$

k: Unknown

$t = 2002 - 1992 = 10$
$$\begin{aligned}250{,}000 &= 200{,}000e^{k(10)}\\ \frac{250{,}000}{200{,}000} &= \frac{200{,}000e^{10k}}{200{,}000}\\ 1.25 &= e^{10k}\\ \ln 1.25 &= \ln e^{10k}\\ \ln 1.25 &= 10k\\ \frac{\ln 1.25}{10} &= k\\ k &\approx 0.022314\end{aligned}$$
P: Unknown

$P_0 = 200{,}000$

$k = 0.022314$

$t = 2015 - 1992 = 23$
$$\begin{aligned}P &= 200{,}000e^{0.022314(23)}\\ P &= 200{,}000e^{0.513222}\\ P &\approx 334{,}133\end{aligned}$$

80.
$$\begin{aligned}0.5 &= 1 \cdot e^{5730k}\\ \ln 0.5 &= \ln e^{5730k}\\ \ln 0.5 &= 5730k\\ \tfrac{\ln 0.5}{5730} &= k\\ k &\approx -0.000121\\ 0.98 &= 1 \cdot e^{-0.000121t}\\ \ln 0.98 &= \ln e^{-0.000121t}\\ \ln 0.98 &= -0.000121t\\ \tfrac{\ln 0.98}{-0.000121} &= t\\ t &\approx 167\end{aligned}$$

81.
$$\begin{aligned}30 &= 11.5 \cdot 1.057^x\\ \tfrac{30}{11.5} &= 1.057^x\\ \log \tfrac{30}{11.5} &= \log 1.057^x\\ \log \tfrac{30}{11.5} &= x \cdot \log 1.057\\ \frac{\log\left(\frac{30}{11.5}\right)}{\log 1.057} &= x\\ x &\approx 17.3\end{aligned}$$
It reached 30% in 1990 + 17 = 2007.

85.

x	$f(x) = 3^x + 2$
-2	$3^{-2} + 2 = 2\frac{1}{9}$
-1	$3^{-1} + 2 = 2\frac{1}{3}$
0	$3^0 + 2 = 3$
1	$3^1 + 2 = 5$
2	$3^2 + 2 = 11$

Horizontal Asymptote: $y = 2$
No x-intercepts ($0 = 3^x + 2$ has no solution.)
y-intercept ($x = 0$): $(0, 3)$
Domain: $(-\infty, \infty)$, Range: $(2, \infty)$

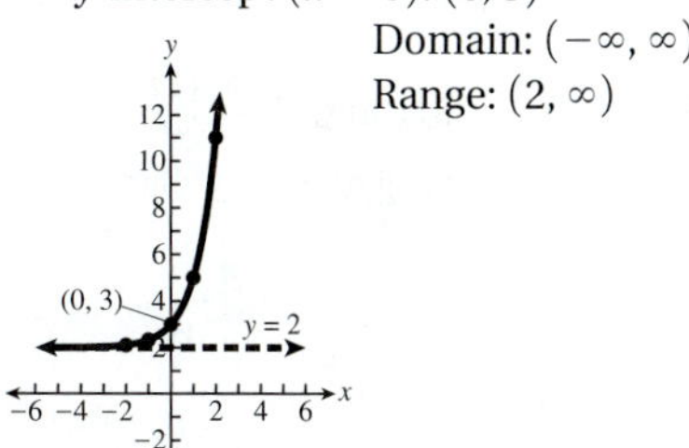

88. $y = \log_2(x - 5)$
$2^y = x - 5$
$x = 2^y + 5$

$x = 2^y + 5$	y
$2^{-2} + 5 = 5\frac{1}{4}$	-2
$2^{-1} + 5 = 5\frac{1}{2}$	-1
$2^0 + 5 = 6$	0
$2^1 + 5 = 7$	1
$2^2 + 5 = 9$	2

Vertical Asymptote: $x = 5$
x-intercept ($y = 0$): $(6, 0)$
No y-intercepts ($0 = 2^y + 5$ has no solution.)
Domain: $(5, \infty)$, Range: $(-\infty, \infty)$

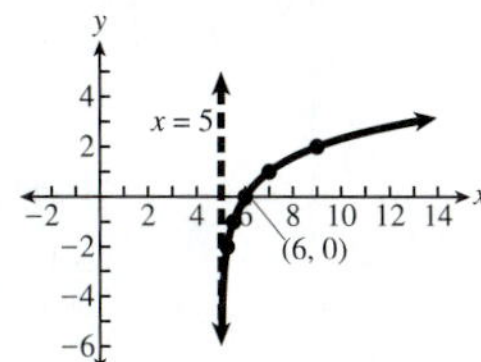

Chapter 8 Test **1.** $x^3 - 216$ **2.** $-8x + 105$ **3.** $x^2 + 15x + 6$ **4.** Yes **5.** $f^{-1}(x) = \frac{-x + 20}{4}$ **6.** $f^{-1}(x) = \frac{-9}{4x - 1}, x \neq \frac{1}{4}$ **7.** -2.611 **8.** 6.890 **9.** 2 **10.** -3 **11.** $\log_b\left(\frac{x^2y^5}{z^3}\right)$ **12.** $4 \ln a + 6 \ln b - 8 \ln c - \ln d$ **13.** 1.878 **14.** $\{-6\}$ **15.** $\{12.124\}$ **16.** $\{4.207\}$ **17.** $\{8\}$ **18.** $\{60\}$ **19.** $\{8\}$ **20.** $f^{-1}(x) = \ln(x + 17) - 9$ **21.** 6 yr **22.** \$80,946.56 **23.** 870.7 yr **24.** Domain: $(-\infty, \infty)$, Range: $(4, \infty)$ **25.** Domain: $(5, \infty)$, Range: $(-\infty, \infty)$

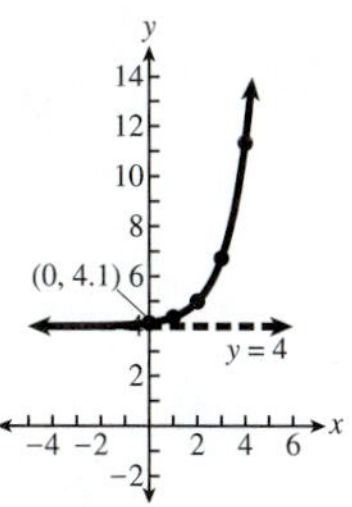

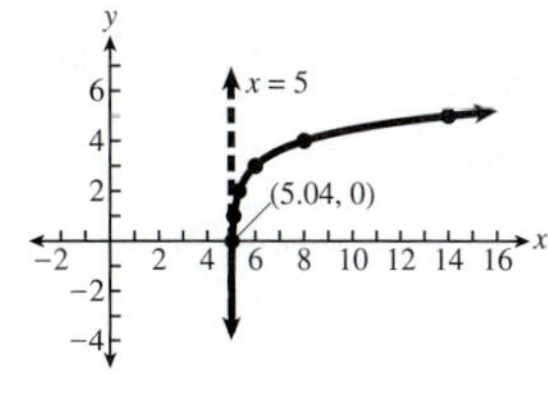

Chapter 9

Quick Check 9.1 **1.**

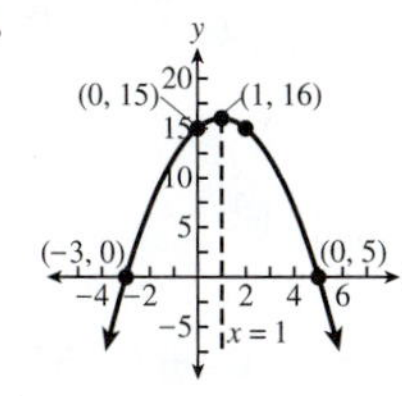

2.

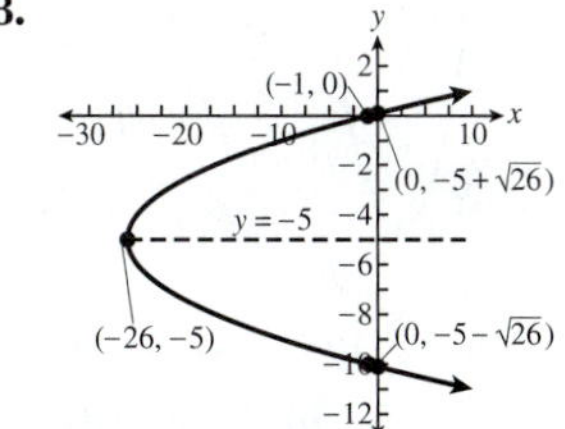

3.

4.

5.

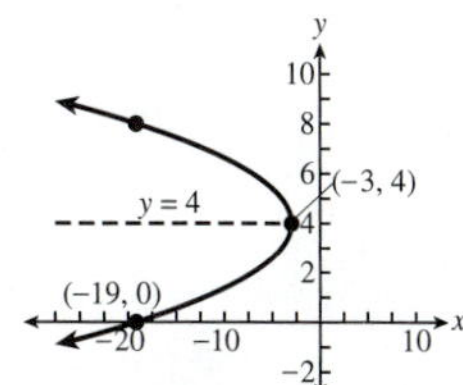

6.

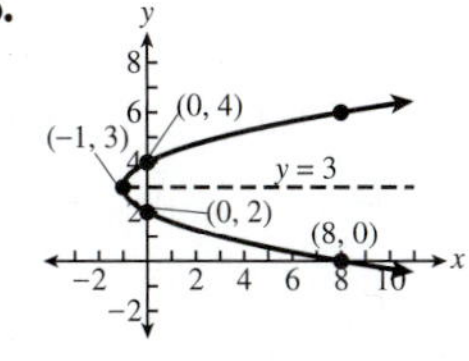

7. a) $y = (x + 2)^2 + 16$, Vertex: $(-2, 16)$ **b)** $x = -(y + 3)^2 + 26$, Vertex: $(26, -3)$ **8.** $y = x^2 - 6x + 8$

Section 9.1 **1.** upward, downward **3.** right, left **5.** (h, k) **7.**

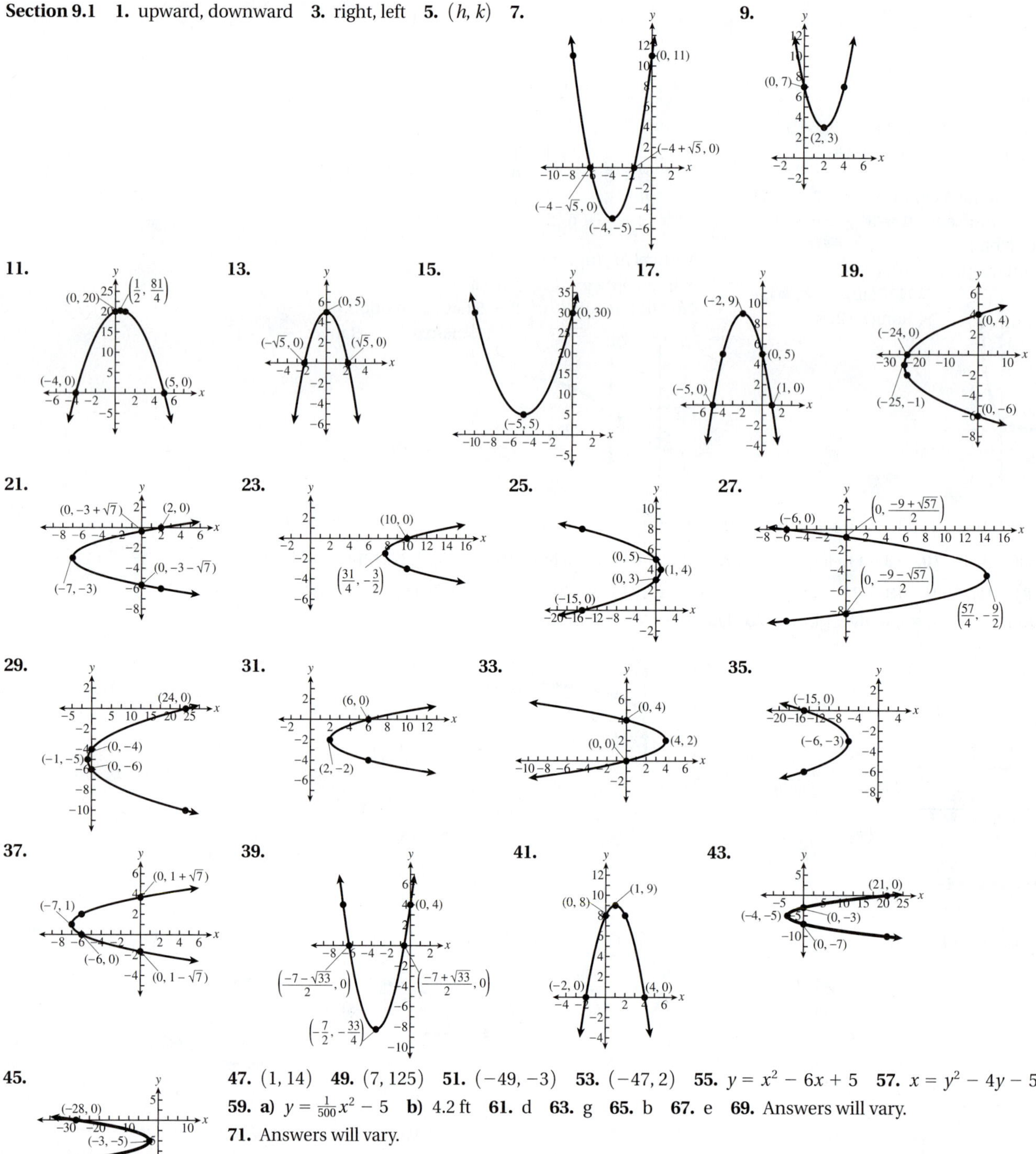

47. $(1, 14)$ **49.** $(7, 125)$ **51.** $(-49, -3)$ **53.** $(-47, 2)$ **55.** $y = x^2 - 6x + 5$ **57.** $x = y^2 - 4y - 5$ **59. a)** $y = \frac{1}{500}x^2 - 5$ **b)** 4.2 ft **61.** d **63.** g **65.** b **67.** e **69.** Answers will vary. **71.** Answers will vary.

Quick Check 9.2 **1. a)** 10 **b)** 12.6 **2.** $\left(\frac{1}{2}, 2\right)$ **3.** Center: $(0, 0)$, $r = 2\sqrt{3}$ **4.** Center: $(6, 5)$, $r = 3$

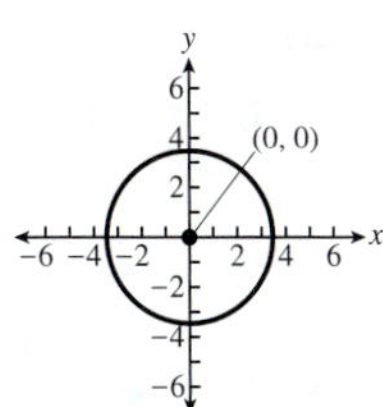

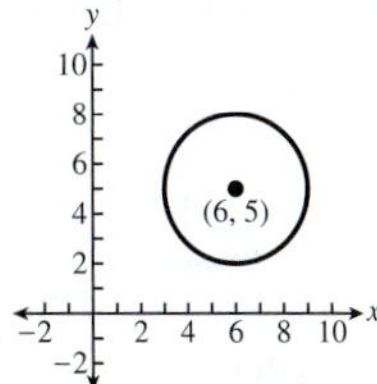

5. a) Center: $(1, -6)$, $r = 4$ **b)** Center: $(-2, 3)$, $r = 8$ **6.** Center: $(-7, 3)$, $r = 1$ **7.** $(x + 3)^2 + (y - 5)^2 = 16$
8. $(x - 1)^2 + (y + 2)^2 = 34$

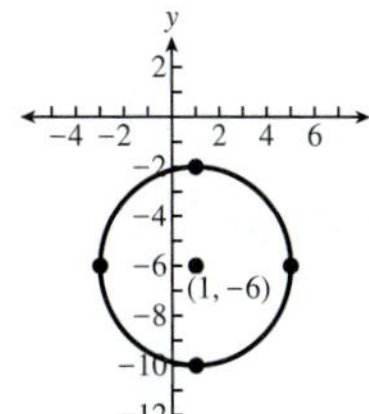

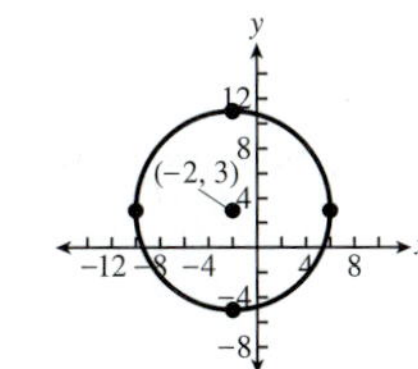

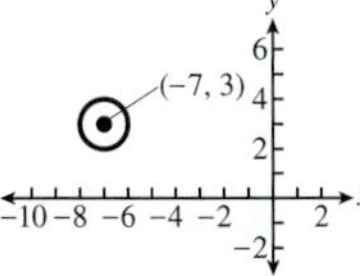

Section 9.2 **1.** $d = \sqrt{(x_2 - x_1)^2 + (y_2 - y_1)^2}$ **3.** circle **5.** radius **7.** $x^2 + y^2 = r^2$ **9.** 17 **11.** 89 **13.** 13 **15.** 4.2 **17.** 6.4 **19.** $(3, 2)$ **21.** $(7, -17)$ **23.** Center: $(0, 0)$, $r = 1$ **25.** Center: $(0, 0)$, $r = 4$ **27.** Center: $(0, 0)$, $r = 10$ **29.** Center: $(0, 0)$, $r = \sqrt{6}$

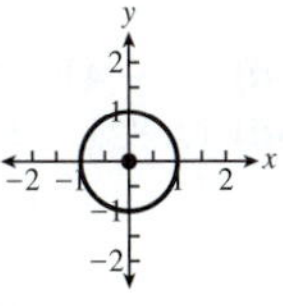

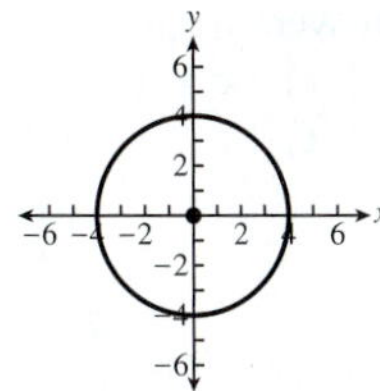

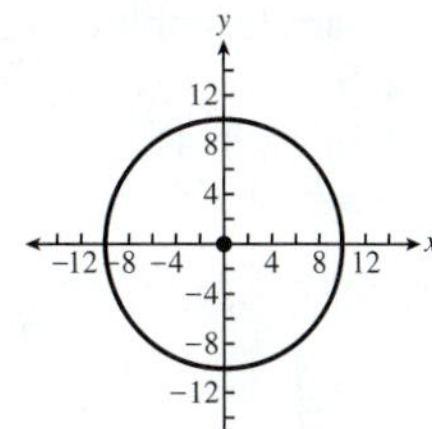

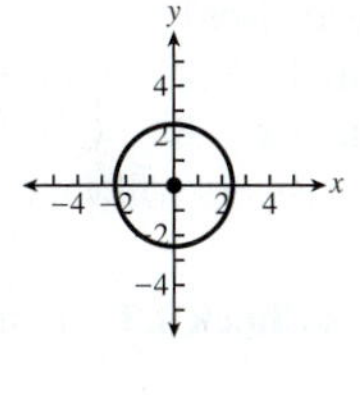

31. Center: $(5, 9)$, $r = 4$ **33.** Center: $(3, 2)$, $r = 2$ **35.** Center: $(-4, 5)$, $r = 3$ **37.** Center: $(-2, -1)$, $r = 7$

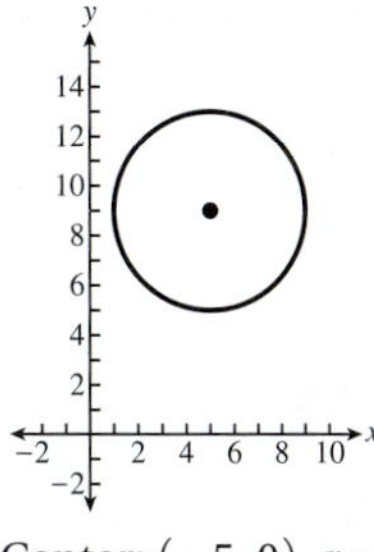

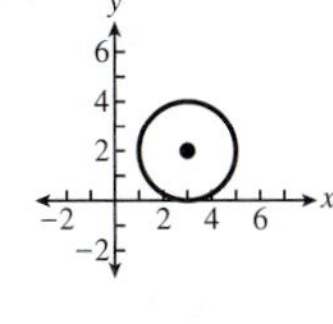

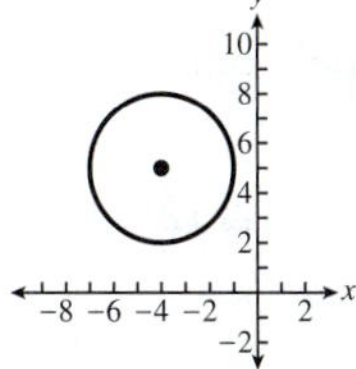

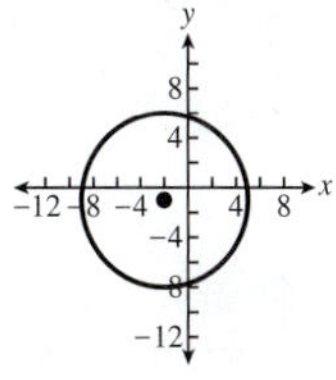

39. Center: $(-5, 0)$, $r = 5$ **41.** Center: $(2, -6)$, $r = 3\sqrt{2}$ **43.** Center: $(-3, 7)$, $r = 3\sqrt{5}$ **45.** $(-2, -6)$, 9 **47.** $(5, -1)$, 7

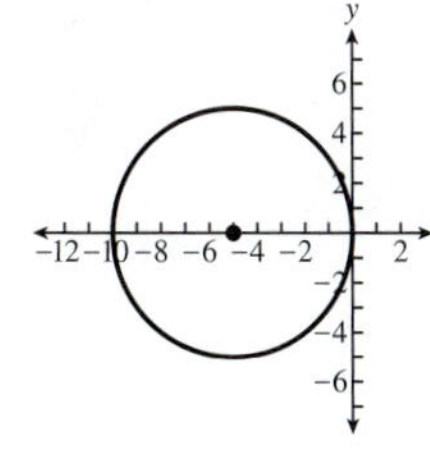

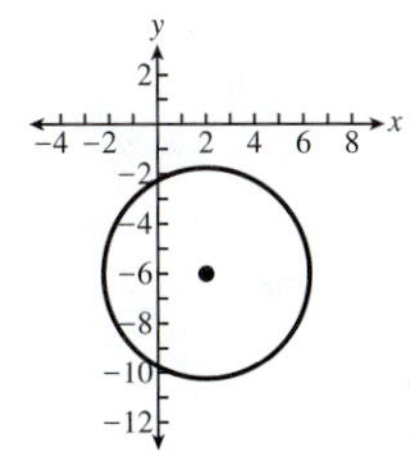

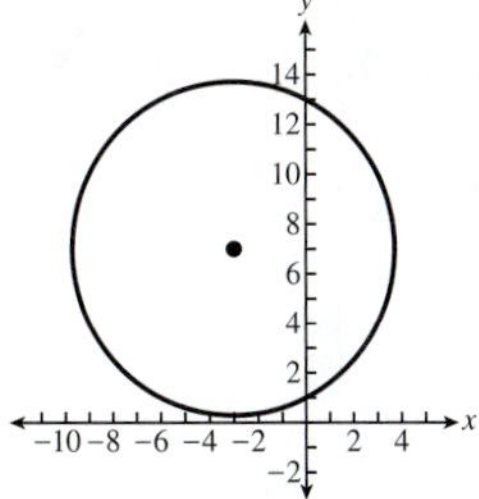

49. $(-8, 0), 11$ **51.** $(-2, -7), 4\sqrt{5}$ **53.** $x^2 + y^2 = 36$ **55.** $(x - 4)^2 + (y - 2)^2 = 49$ **57.** $(x + 9)^2 + (y + 10)^2 = 25$
59. $x^2 + (y + 8)^2 = 15$ **61.** $x^2 + y^2 = 64$ **63.** $(x - 1)^2 + (y - 5)^2 = 16$ **65.** $(x + 1)^2 + (y - 1)^2 = 49$
67. $(x - 30)^2 + (y + 40)^2 = 2500$ **69.** $(x - 5)^2 + (y - 7)^2 = 16$ **71.** $(x - 2)^2 + (y - 7)^2 = 25$
73. $(x - 8)^2 + (y + 1)^2 = 20$ **75.**

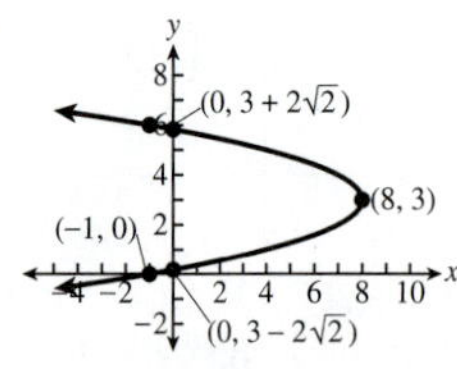

77.

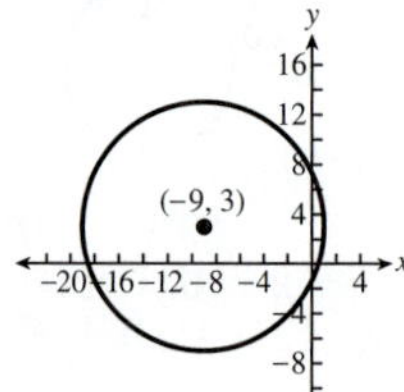

79.

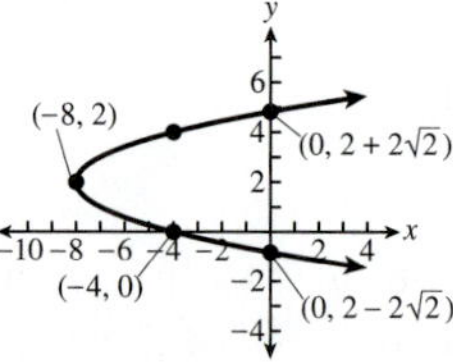

81.

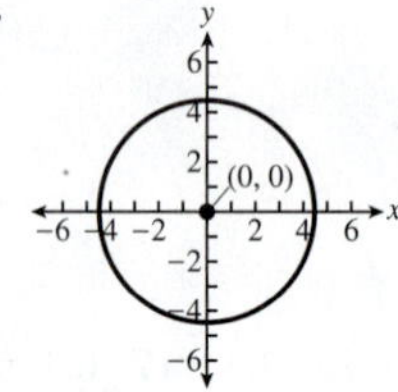

83.

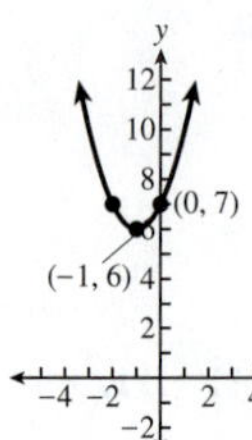

85.

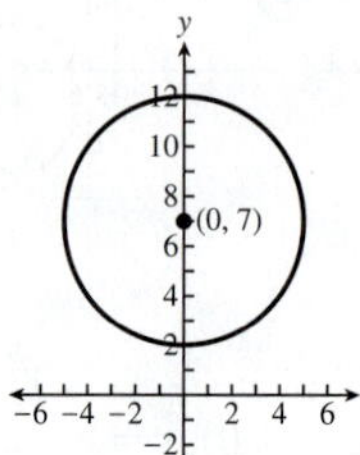

87. a) $x^2 + y^2 = 725$ **b)** 1277.7 sq ft

For exercises 89–91, answers may vary. Examples of correct answers are given.
89. a) $(x - 4)^2 + (y - 3)^2 = 4$ **b)** $(x - 4)^2 + (y + 7)^2 = 1$ **c)** $(x - 4)^2 + (y + 7)^2 = 9$ **d)** $(x - 4)^2 + (y + 1)^2 = 9$
91. a) $(x - 6)^2 + (y + 6)^2 = 4$ **b)** $(x + 4)^2 + (y + 6)^2 = 1$ **c)** $(x + 4)^2 + (y + 6)^2 = 4$ **d)** $(x - 1)^2 + (y + 6)^2 = 9$
93. Answers will vary. **95.** Answers will vary.

Quick Check 9.3 **1. a)**

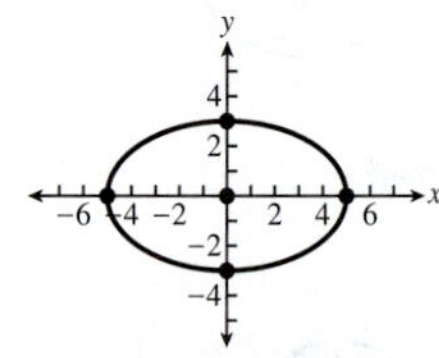

b)

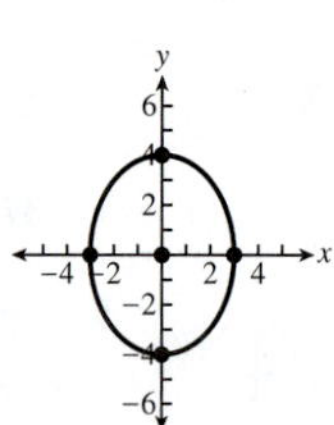

2.

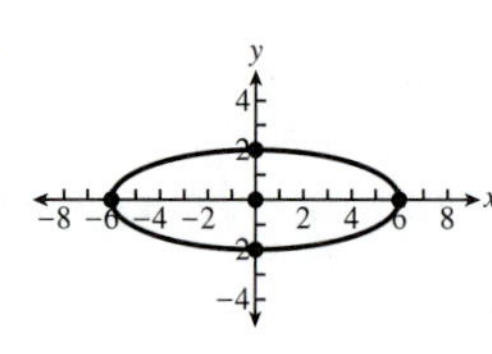

3.

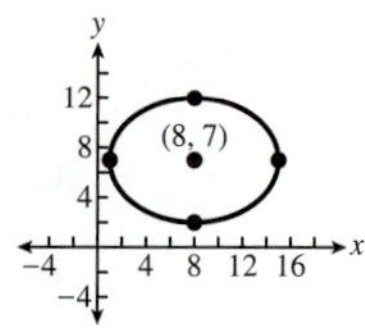

4.

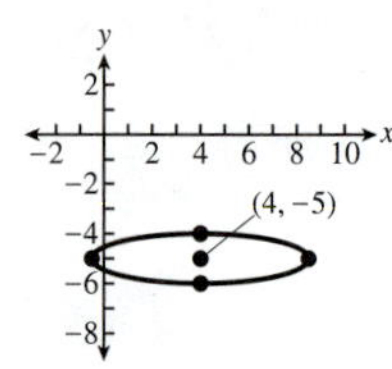

5.

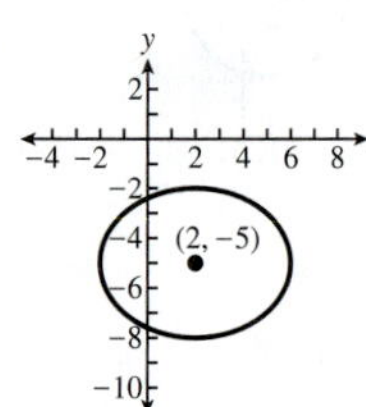

6. $\frac{x^2}{49} + \frac{(y-5)^2}{9} = 1$

Section 9.3 **1.** ellipse **3.** center **5.** vertices **7.** $\frac{x^2}{a^2} + \frac{y^2}{b^2} = 1$ **9.** Center: $(0, 0)$, $a = 2$, $b = 3$

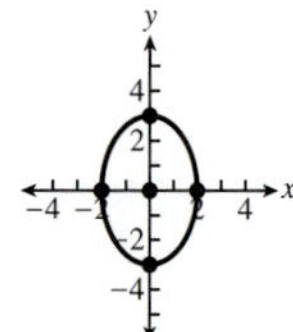

11. Center: $(0, 0)$, $a = 5$, $b = 2$ **13.** Center: $(0, 0)$, $a = 1$, $b = 5$ **15.** Center: $(0, 0)$, $a = 6$, $b = 2\sqrt{3}$

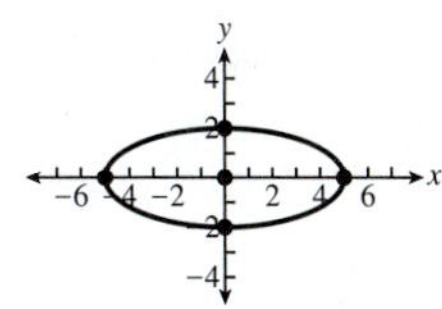
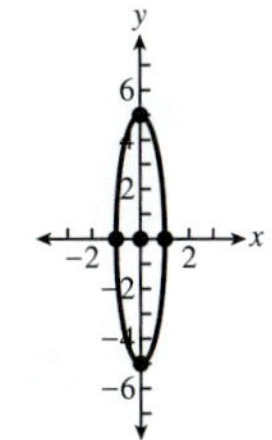
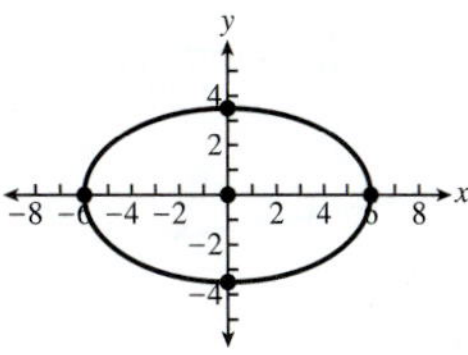

17. Center: $(0, 0)$, $a = 3$, $b = 5$ **19.** Center: $(0, 0)$, $a = 7$, $b = 2$ **21.** Center: $(6, 8)$, $a = 5$, $b = 6$

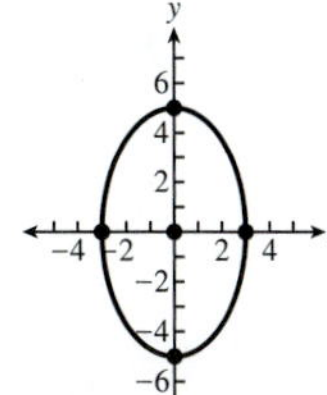
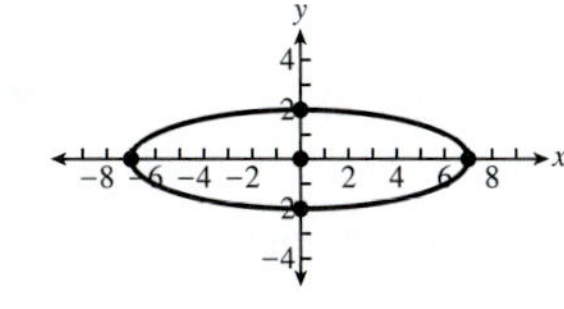
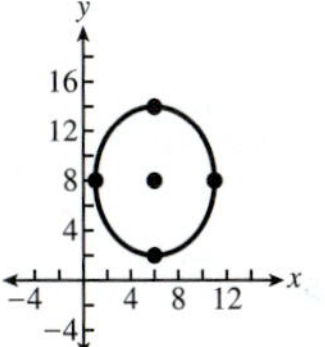

23. Center: $(9, -4)$, $a = 9$, $b = 3$ **25.** Center: $(-5, -3)$, $a = 3$, $b = 8$ **27.** Center: $(-5, 0)$, $a = 2$, $b = 6$

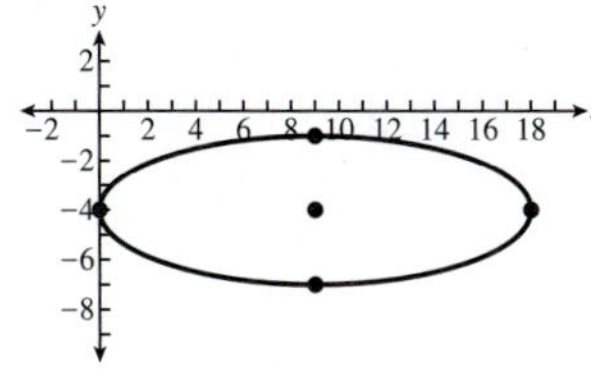
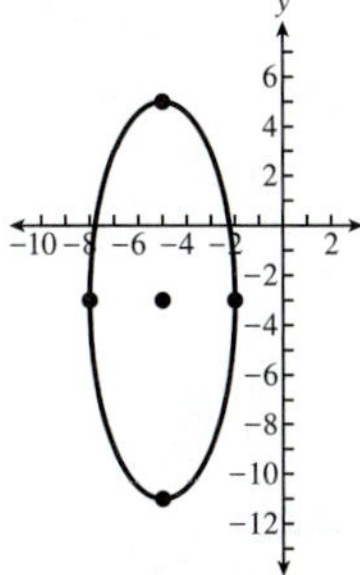
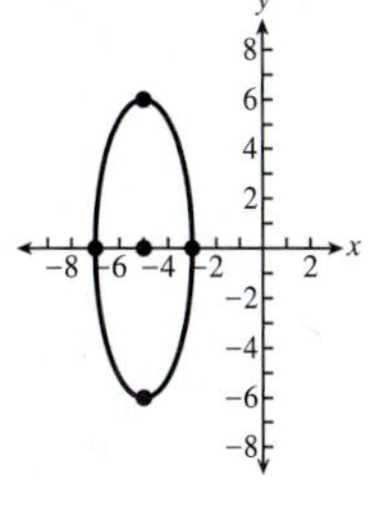

29. Center: $(-2, 4)$, $a = 7$, $b = 1$ **31.** Center: $(6, -6)$, $a = 2\sqrt{5}$, $b = 2$ **33.** Center: $(3, 5)$, $a = 3$, $b = 2$

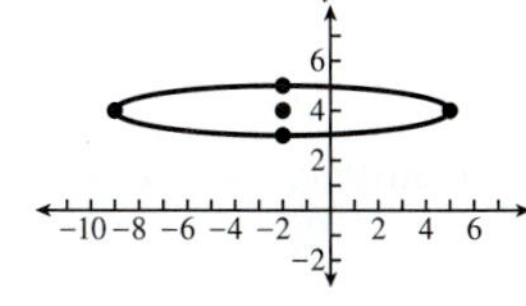
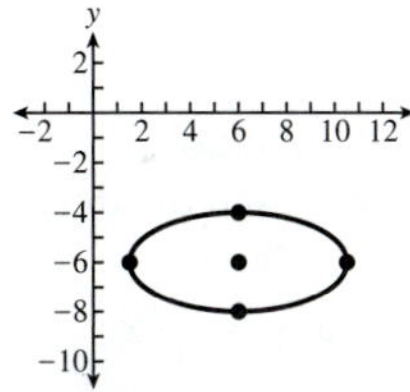
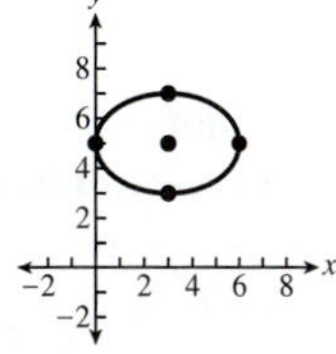

35. Center: $(-4, 4)$, $a = 6$, $b = 3$ **37.** $\frac{(x+2)^2}{4} + \frac{(y-8)^2}{100} = 1$ **39.** $\frac{(x-6)^2}{64} + \frac{(y+3)^2}{49} = 1$ **41.** $\frac{x^2}{9} + \frac{(y+6)^2}{25} = 1$ **43.** $\frac{x^2}{25} + \frac{y^2}{9} = 1$ **45.** $\frac{(x-4)^2}{25} + \frac{y^2}{4} = 1$ **47.** $\frac{x^2}{9} + \frac{(y+4)^2}{25} = 1$ **49.** $\frac{(x-3)^2}{36} + \frac{(y+4)^2}{25} = 1$

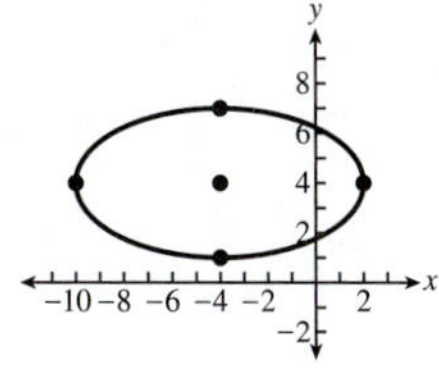

51.

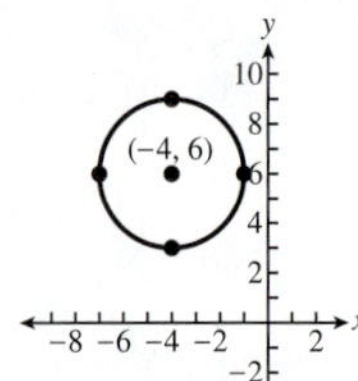

53.

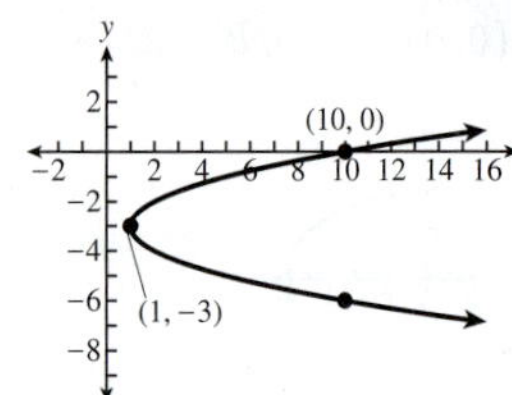

55.

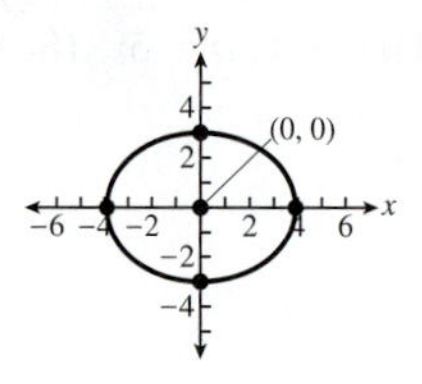

57.

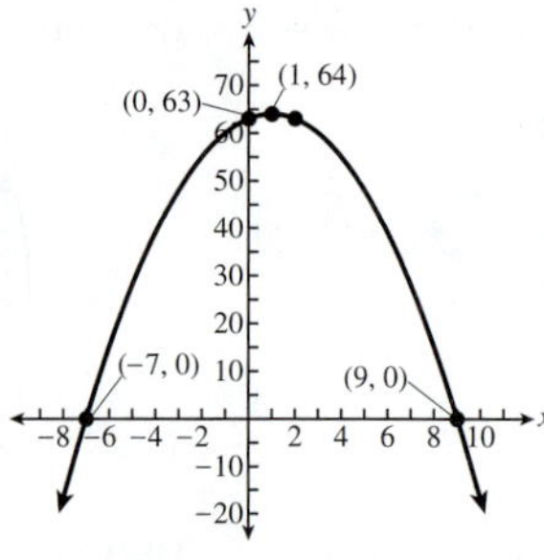

59.

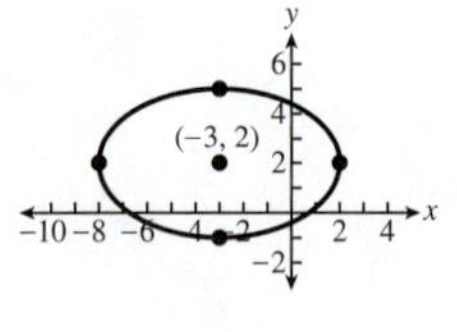

61.

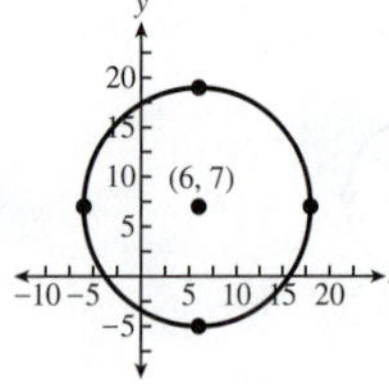

63.

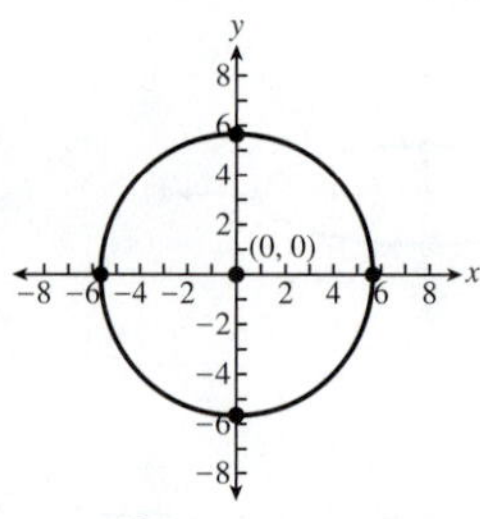

65.

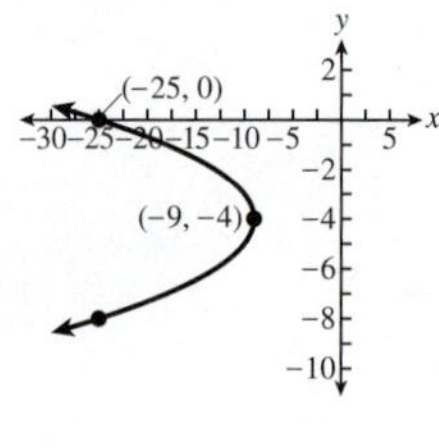

67.

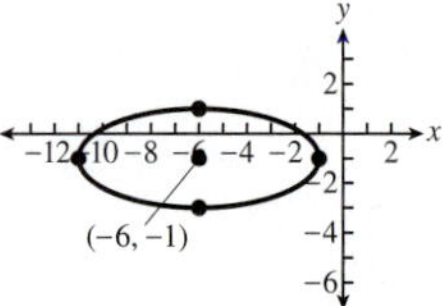

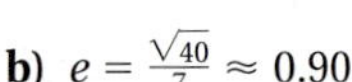

69. a) 100 ft **b)** 66 ft **71.** 471.2 sq ft

73. a) $e = \frac{\sqrt{7}}{4} \approx 0.66$ **b)** $e = \frac{\sqrt{40}}{7} \approx 0.90$ **c)** $e = \frac{\sqrt{91}}{10} \approx 0.95$

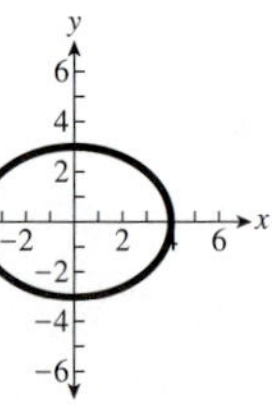

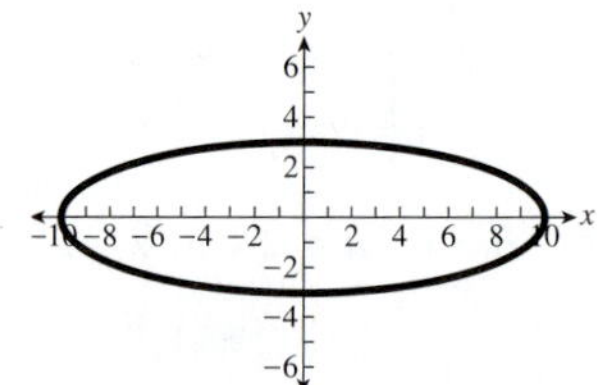

75. Answers will vary. The ellipse becomes more elongated as the eccentricity increases.

77. No. According to the formulas provided, the numerator must always be less than the denominator. **79.** Answers will vary.

Quick Check 9.4 **1.** Center: $(0, 0)$, Vertices: $(-4, 0)$, $(4, 0)$, Asymptotes: $y = \frac{3}{4}x$, $y = -\frac{3}{4}x$

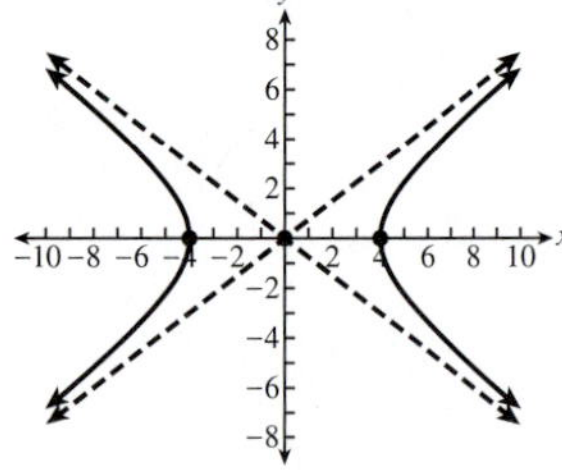

2. Center: $(0, 0)$, Vertices: $(0, -2)$, $(0, 2)$, Asymptotes: $y = \frac{1}{3}x$, $y = -\frac{1}{3}x$

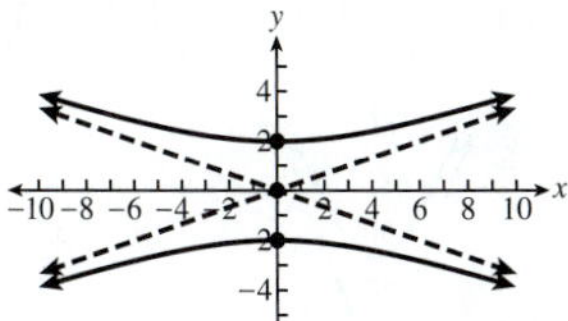

3. Center: $(3, 5)$, Vertices: $(1, 5)$, $(5, 5)$, Asymptotes: $y = 2(x - 3) + 5$, $y = -2(x - 3) + 5$

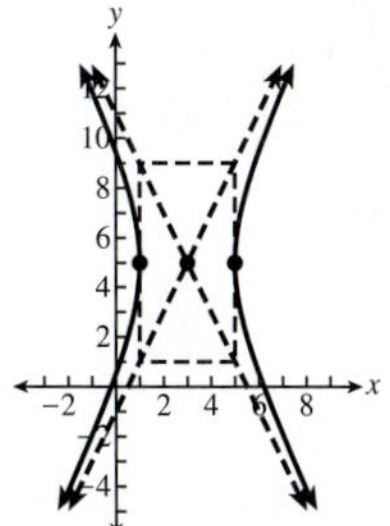

4. Center: $(-3, 2)$, Vertices: $(-3, 9)$, $(-3, -5)$, Asymptotes: $y = \frac{7}{3}(x + 3) + 2$, $y = -\frac{7}{3}(x + 3) + 2$

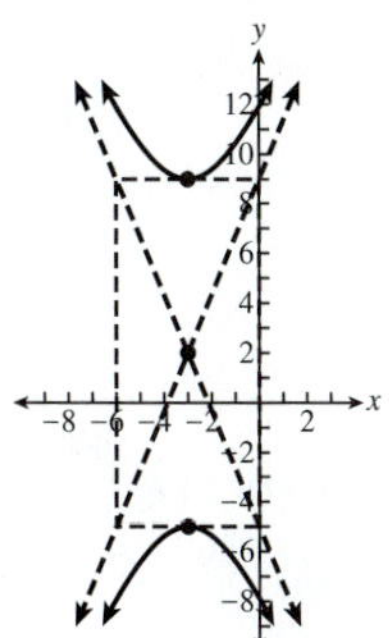

5. Center: $(2, -3)$, Vertices: $(-1, -3)$, $(5, -3)$, Asymptotes: $y = \frac{4}{3}(x - 2) - 3$, $y = -\frac{4}{3}(x - 2) - 3$

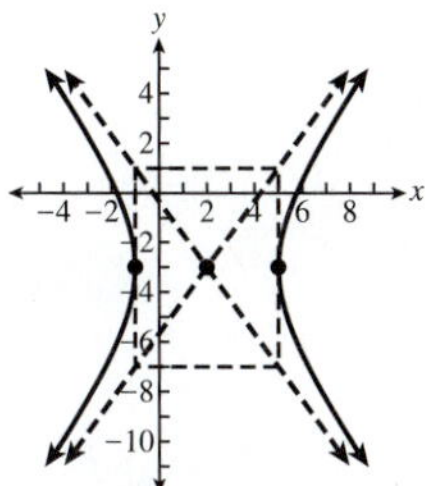

6. $\frac{(y - 4)^2}{4} - \frac{(x + 3)^2}{25} = 1$

Section 9.4 **1.** hyperbola **3.** center **5.** transverse axis **7.** $\frac{x^2}{a^2} - \frac{y^2}{b^2} = 1, \frac{y^2}{b^2} - \frac{x^2}{a^2} = 1$

9. Center: $(0, 0)$, $a = 3$, $b = 2$ **11.** Center: $(0, 0)$, $a = 3$, $b = 7$ **13.** Center: $(0, 0)$, $a = 3$, $b = 3$ **15.** Center: $(0, 0)$, $a = 1$, $b = 5$

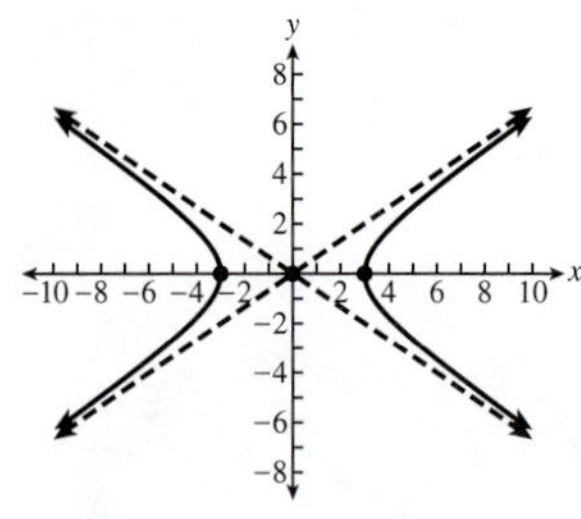

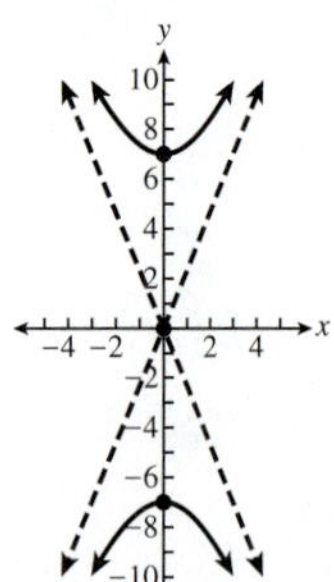

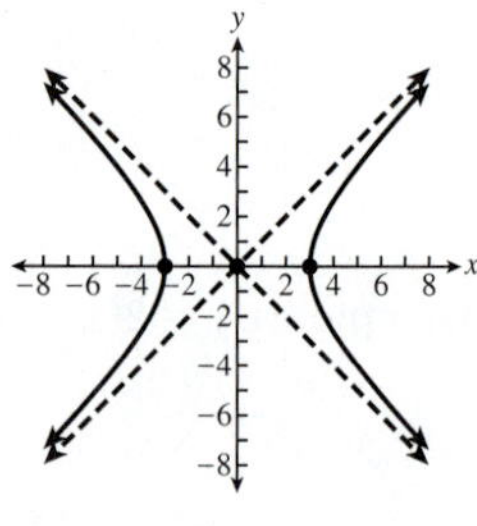

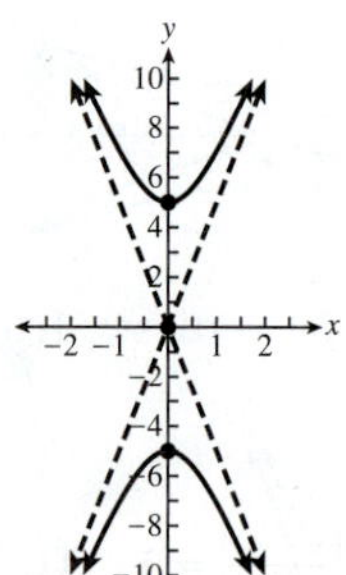

17. Center: $(0, 0)$, $a = 2\sqrt{3}$, $b = 3$ **19.** Center: $(0, 0)$, $a = 2$, $b = 7$ **21.** Center: $(0, 0)$, $a = 1$, $b = 3$

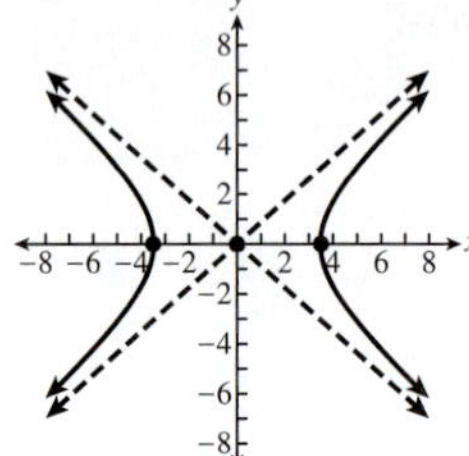

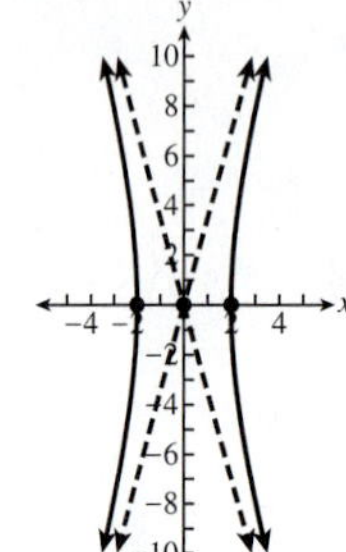

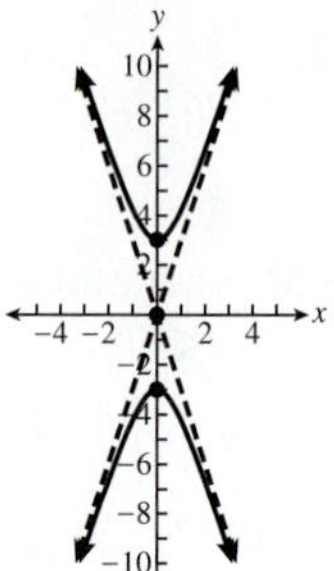

23. Center: $(2, 5)$, $a = 3$, $b = 4$ **25.** Center: $(-1, -4)$, $a = 3$, $b = 2$ **27.** Center: $(0, 1)$, $a = 2$, $b = 2$

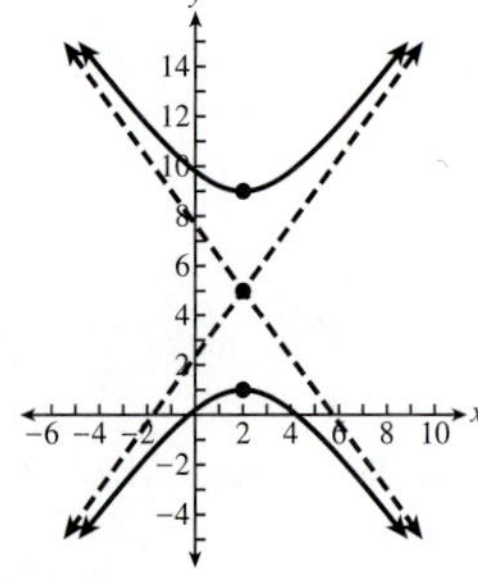

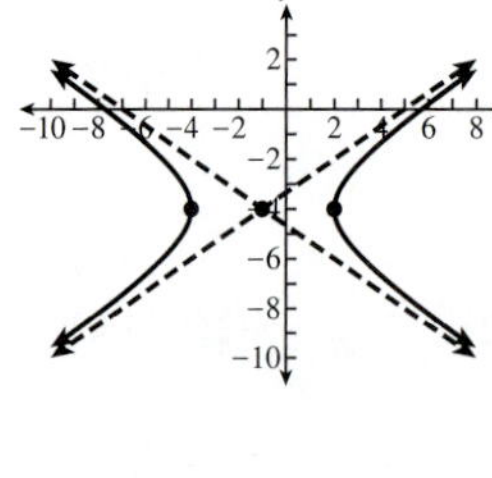

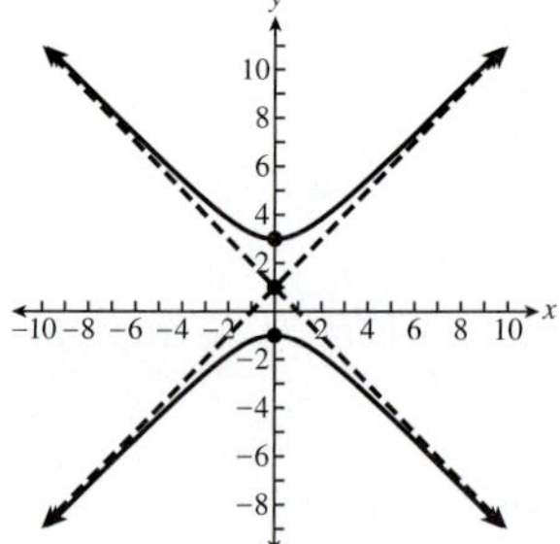

29. Center: $(4, -2)$, $a = 7$, $b = 4$ **31.** Center: $(-3, 5)$, $a = 3$, $b = 1$

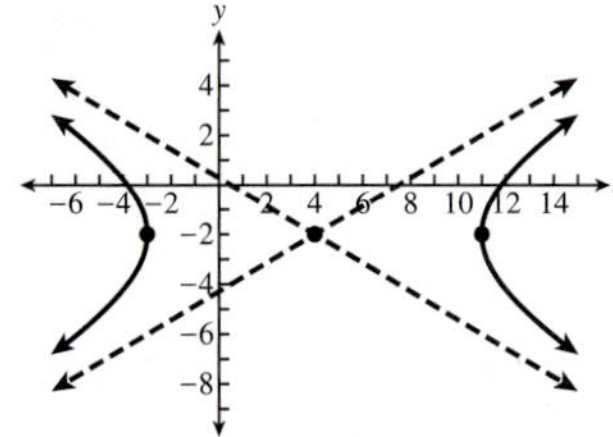

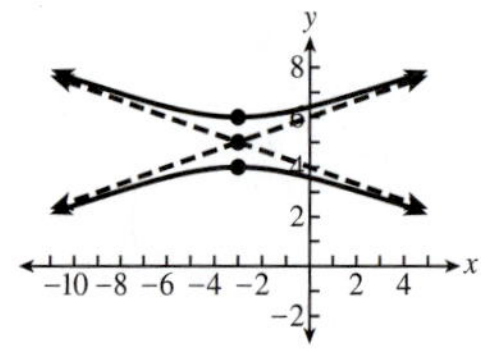

33. Center: $(-2, -3)$, $a = 5$, $b = 2\sqrt{7}$ **35.** Center: $(-1, -3)$, $a = 2$, $b = 3$ **37.** Center: $(-2, 7)$, $a = 4$, $b = 5$

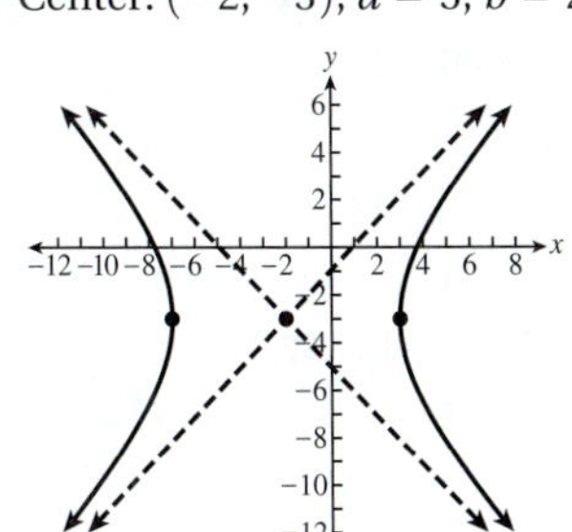

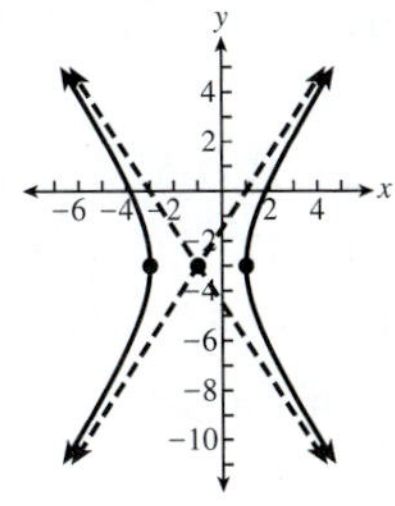

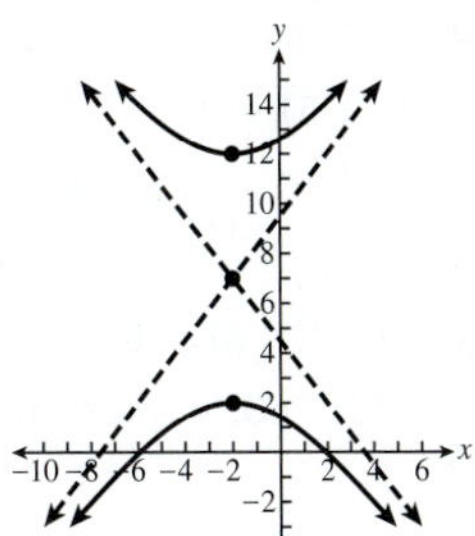

39. $\frac{(y-2)^2}{9} - \frac{(x-4)^2}{25} = 1$ **41.** $\frac{(x-6)^2}{16} - \frac{(y+2)^2}{16} = 1$ **43.** $\frac{y^2}{25} - \frac{x^2}{9} = 1$

45. $\frac{(y-2)^2}{25} - \frac{(x-3)^2}{4} = 1$ **47.** $\frac{(x+1)^2}{25} - \frac{(y+5)^2}{9} = 1$ **49.**

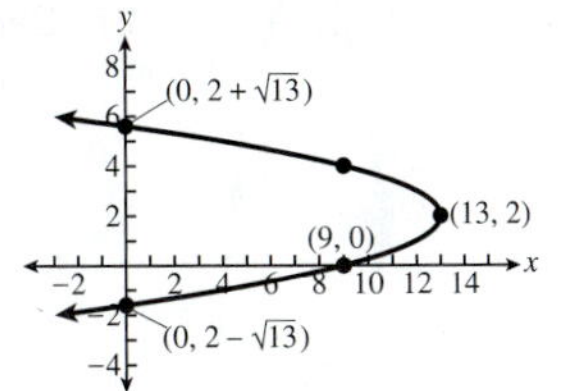

51.

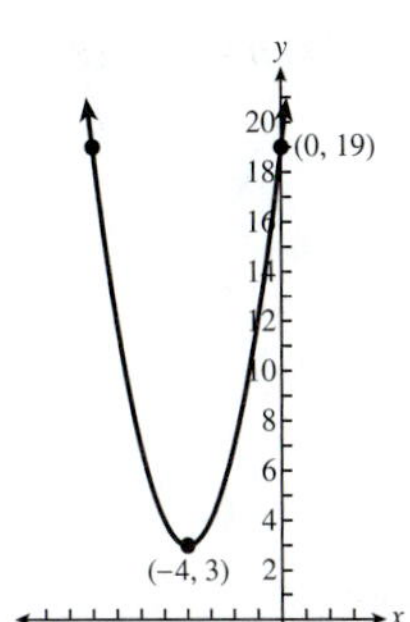

53.

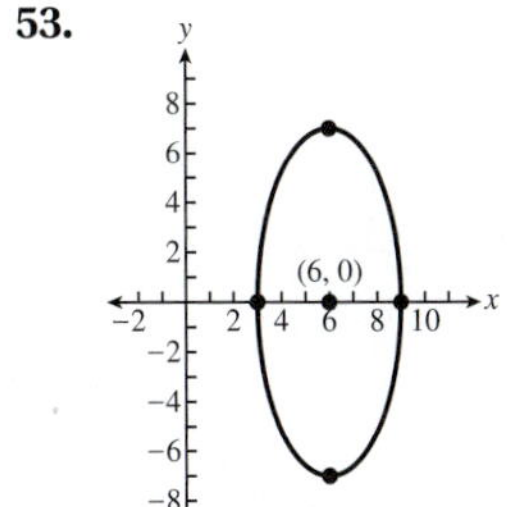

55.

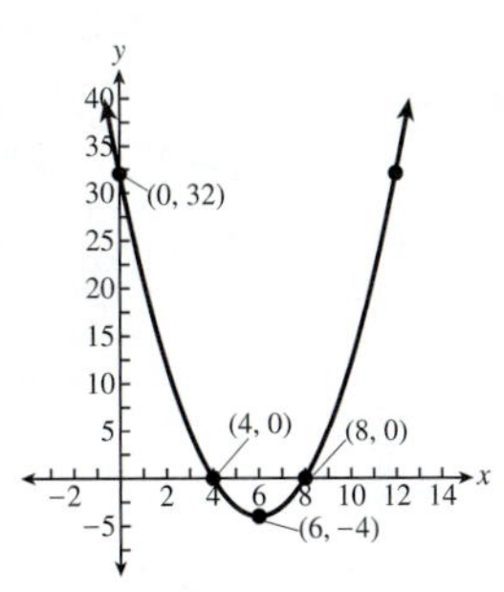

57.

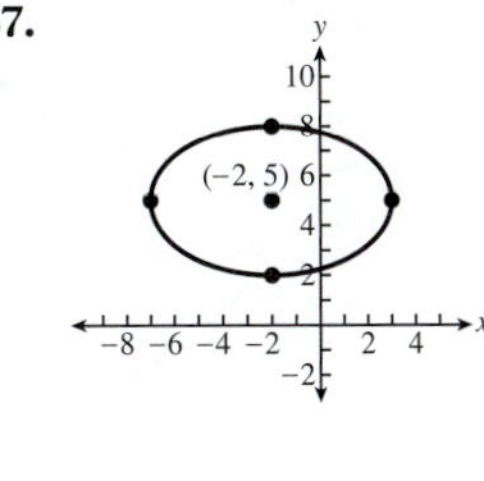

59.

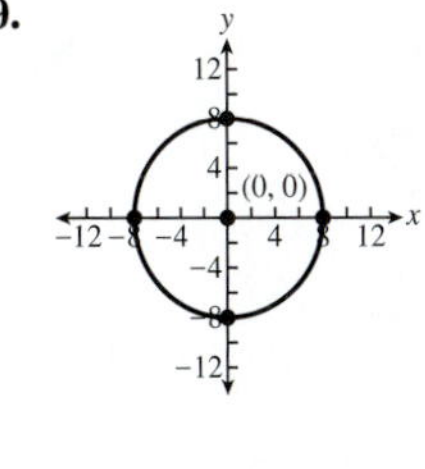

61.

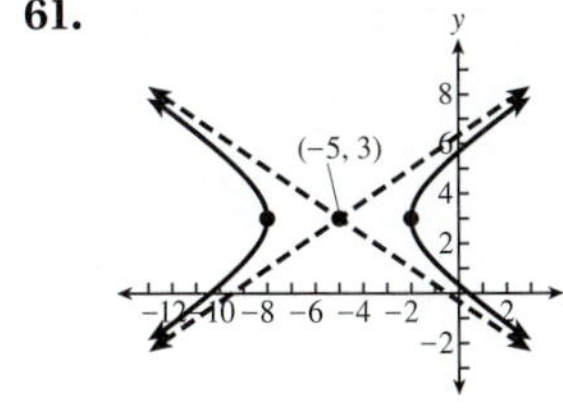

63.

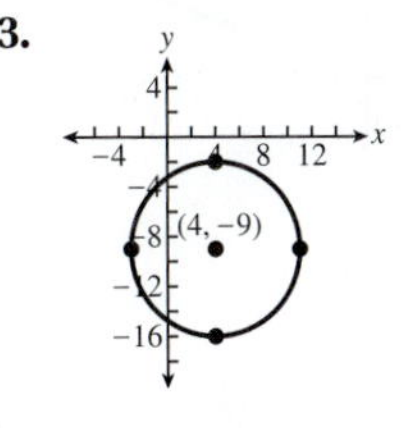

65.

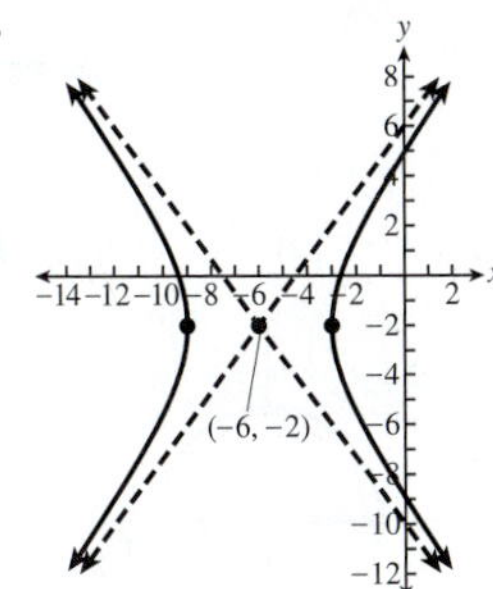

67. b **69.** c **71.** a **73.** Answers will vary.

Quick Review Exercises 9.4 **1.** $(3, 2)$ **2.** $(-3, -9)$ **3.** $(4, -1)$ **4.** $(\frac{3}{2}, -\frac{9}{5})$

Quick Check 9.5 **1.** $(2, -5), (-\frac{50}{17}, -\frac{43}{17})$ **2.** $\varnothing$ **3.** $(-\sqrt{3}, 3), (\sqrt{3}, 3)$ **4.** $(3, 5), (-3, 5), (3, -5), (-3, -5)$ **5.** 3 ft by 12 ft **6.** 9 in. by 12 in. **7.** 2 sec

Section 9.5 **1.** nonlinear **3.** $(8, -6), (-6, -8)$ **5.** $(-9, -2), (-\frac{42}{5}, -\frac{19}{5})$ **7.** $(5, 0), (0, 2)$ **9.** $(0, -4), (2, 0)$ **11.** $(1, 6), (4, 12)$ **13.** $(2, -1), (4, 3)$ **15.** $(3, 1)$ **17.** $\varnothing$ **19.** $(0, -5), (3, 4), (-3, 4)$ **21.** $(5, \sqrt{5}), (5, -\sqrt{5})$ **23.** $(0, 1)$ **25.** $(1, 0)$ **27.** $\varnothing$ **29.** $(10, \sqrt{21}), (10, -\sqrt{21}), (-6, \sqrt{5}), (-6, -\sqrt{5})$ **31.** $(3, 2), (-3, 2)$ **33.** $(14, -2)$ **35.** $\varnothing$ **37.** $(2, 1), (2, -1), (-2, 1), (-2, -1)$ **39.** $(2, \sqrt{2}), (2, -\sqrt{2}), (-2, \sqrt{2}), (-2, -\sqrt{2})$ **41.** $(\sqrt{2}, 2\sqrt{2}), (\sqrt{2}, -2\sqrt{2}), (-\sqrt{2}, 2\sqrt{2}), (-\sqrt{2}, -2\sqrt{2})$ **43.** $(3, \frac{\sqrt{6}}{2}), (3, -\frac{\sqrt{6}}{2}), (-3, \frac{\sqrt{6}}{2}), (-3, -\frac{\sqrt{6}}{2})$ **45.** $(\sqrt{15}, 3), (\sqrt{15}, -3), (-\sqrt{15}, 3), (-\sqrt{15}, -3)$ **47.** $(\sqrt{3}, \sqrt{3}), (\sqrt{3}, -\sqrt{3}), (-\sqrt{3}, \sqrt{3}), (-\sqrt{3}, -\sqrt{3})$ **49.** $(3, 3), (3, -3), (-3, 3), (-3, -3)$ **51.** $(1, 3), (1, -3), (-1, 3), (-1, -3)$ **53.** 7 ft by 15 ft **55.** 16 ft by 25 ft **57.** 8 ft by 24 ft **59.** 45 ft by 60 ft **61.** 40 ft by 20 ft **63.** 4.5 sec **65. a)** 3 sec **b)** 36 ft **67. a)** 29 **b)** $(-2, 5)$ **69. a)** 35 **b)** $(-\frac{23}{4}, \frac{33}{4})$ **71.** Answers will vary.

Chapter 9 Review

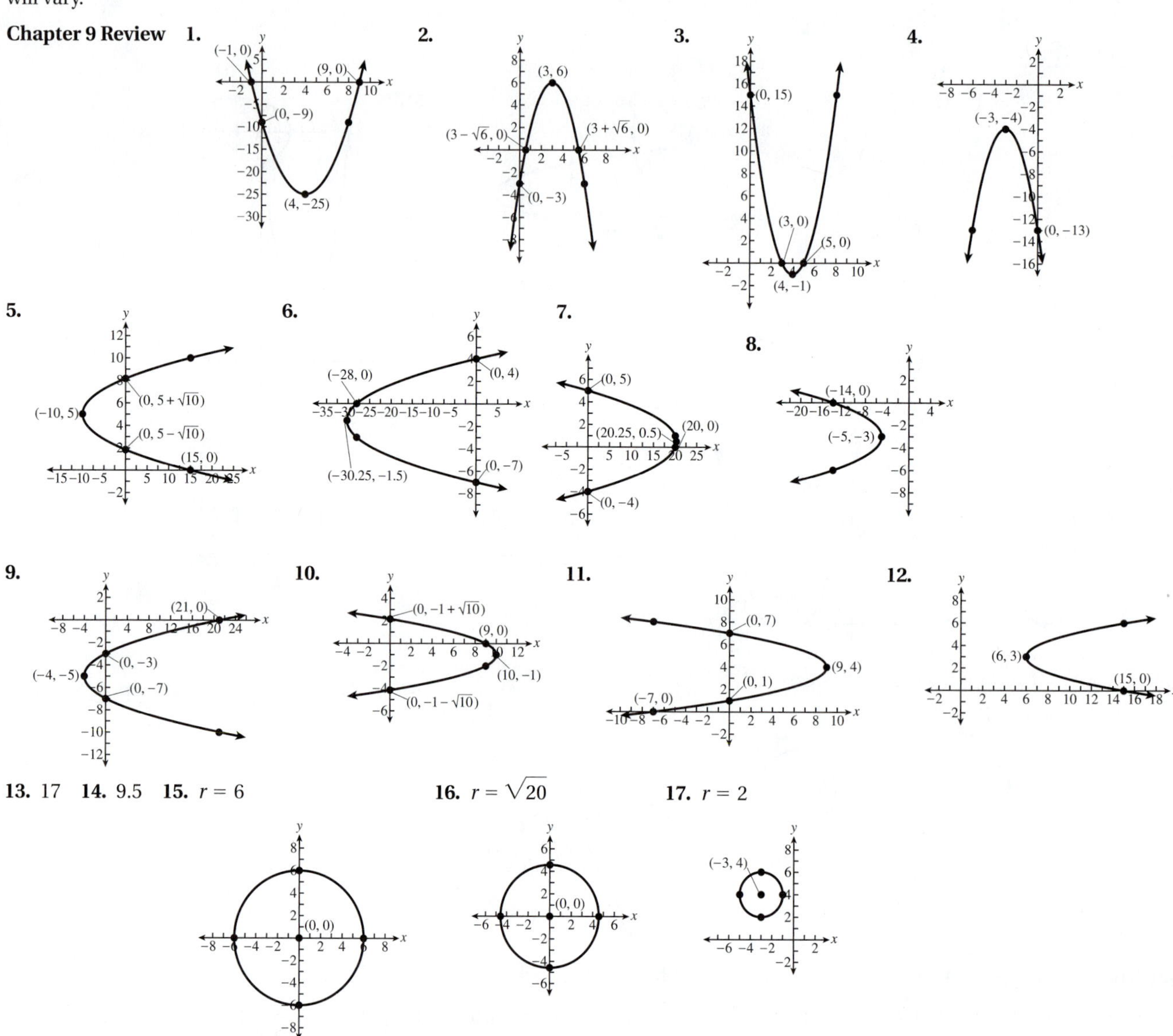

13. 17 **14.** 9.5 **15.** $r = 6$ **16.** $r = \sqrt{20}$ **17.** $r = 2$

18. $r = 3$

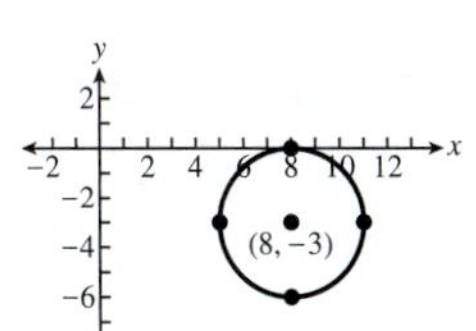

19. $r = 5$

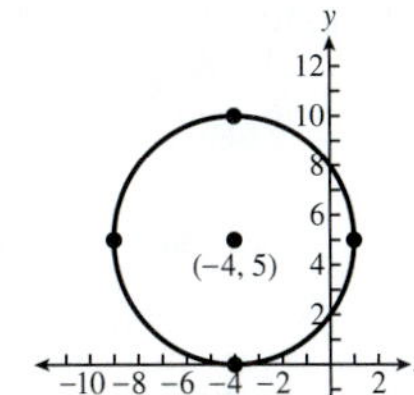

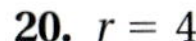

20. $r = 4$

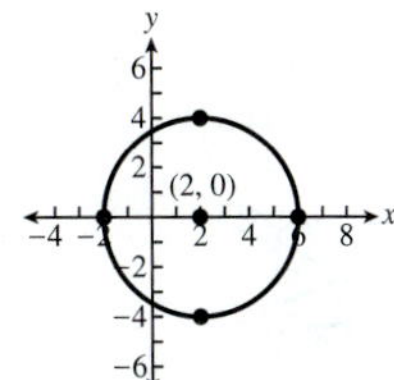

21. $r = 7$

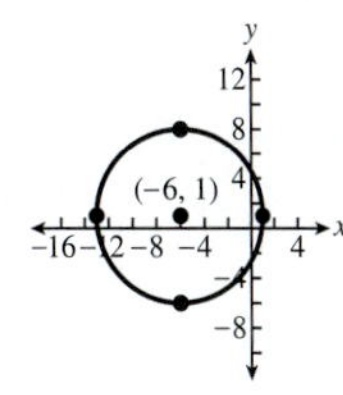

22. $(-3, 9), r = 10$ **23.** $(4, 10), r = \sqrt{17}$ **24.** $x^2 + y^2 = 9$ **25.** $(x + 2)^2 + (y - 4)^2 = 25$ **26.** $a = 5, b = 3$

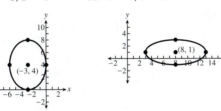

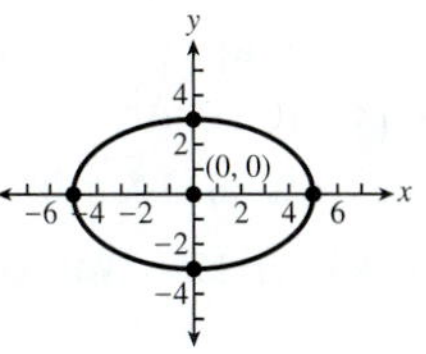

27. $a = 1, b = 4$ **28.** $a = 3, b = 4$ **29.** $a = 5, b = 2$

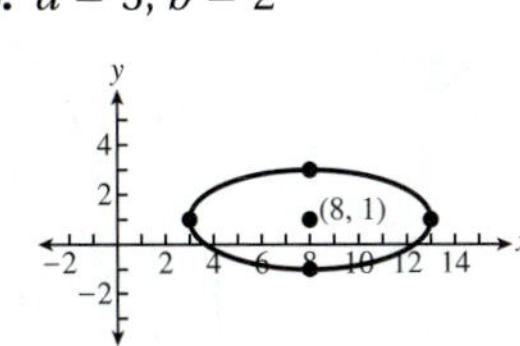

30. $a = 3, b = 6$

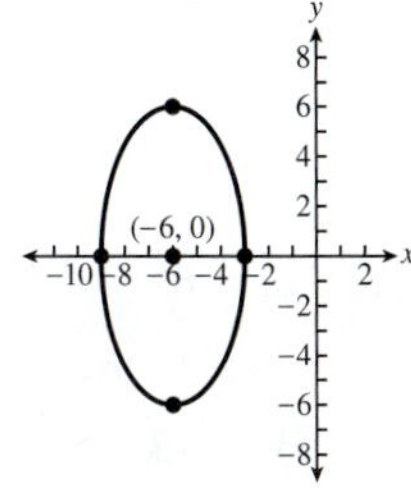

31. $a = 4, b = 3$

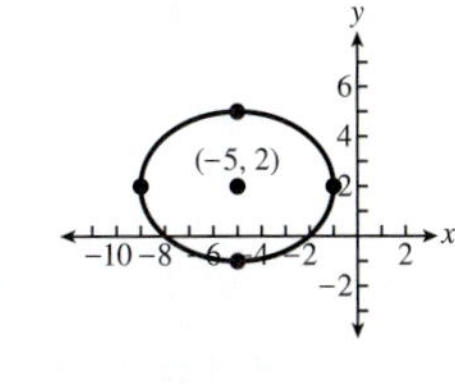

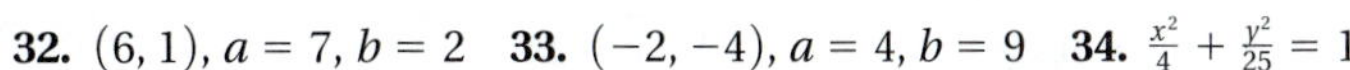

32. $(6, 1), a = 7, b = 2$ **33.** $(-2, -4), a = 4, b = 9$ **34.** $\frac{x^2}{4} + \frac{y^2}{25} = 1$

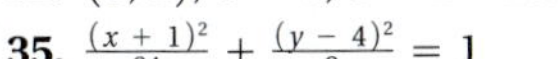

35. $\frac{(x + 1)^2}{64} + \frac{(y - 4)^2}{9} = 1$

36. $a = 3, b = 5$

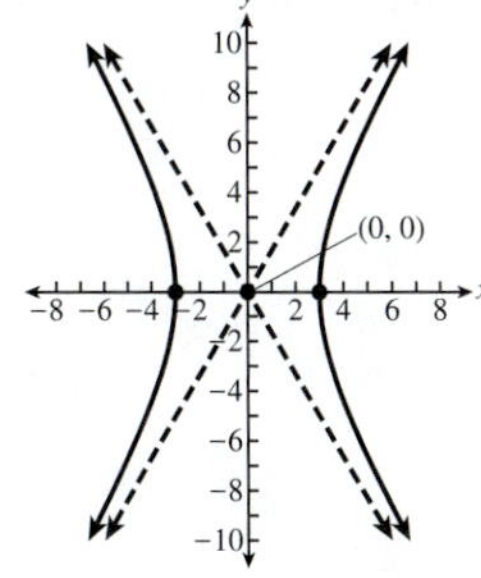

37. $a = 4, b = 4$

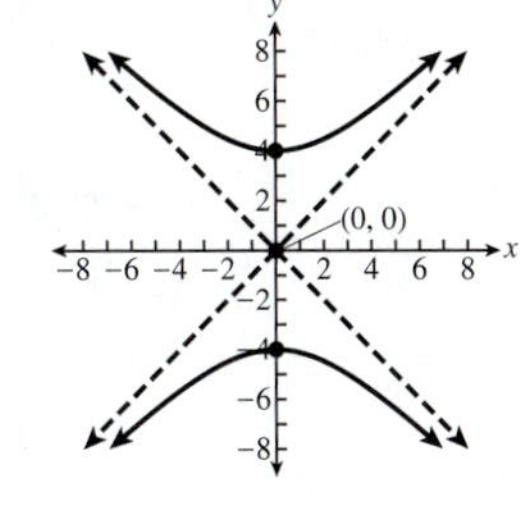

38. $a = 2, b = 3$

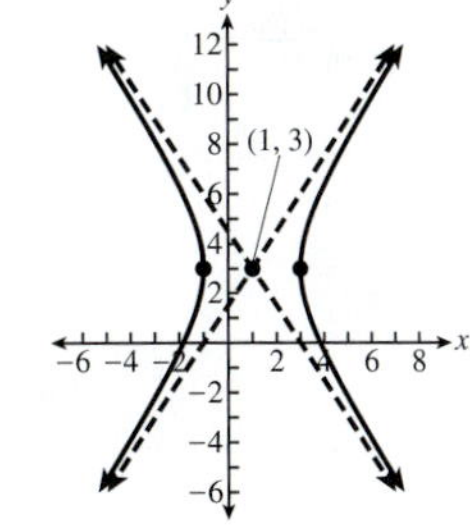

39. $a = 5, b = 6$

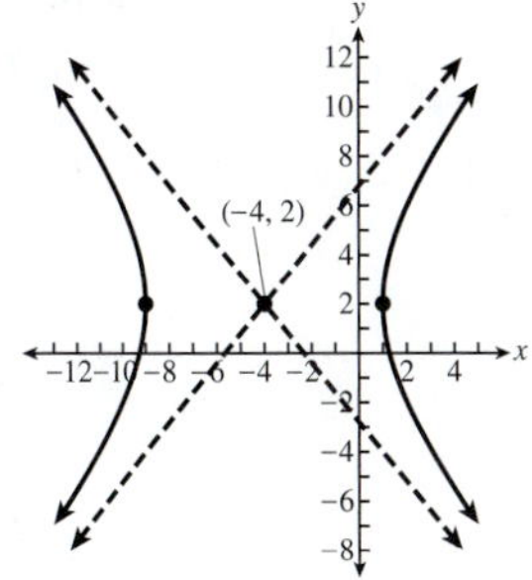

40. $a = 9, b = 2$

41. $a = 5, b = 2$

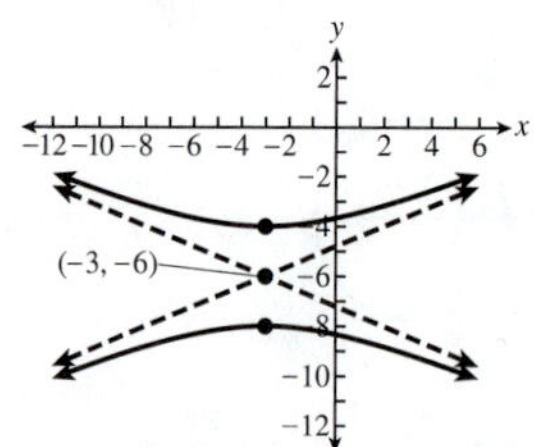

42. Center: $(5, 7)$, Transverse Axis: $2b = 6$, Conjugate Axis: $2a = 4$

43. Center: $(-8, 4)$, Transverse Axis: $2a = 12$, Conjugate Axis: $2b = 14$ **44.** $\frac{x^2}{4} - \frac{y^2}{25} = 1$ **45.** $\frac{(y-2)^2}{9} - \frac{(x+1)^2}{16} = 1$ **46.** $(5, 5), (1, 7)$ **47.** $(3, 2), (0, -4)$ **48.** $(6, 11), (4, 7)$ **49.** $(10, 7), (-2, 5)$ **50.** $(1, 4), (-1, 4)$ **51.** $\left(\frac{2\sqrt{15}}{5}, -\frac{18}{5}\right), \left(-\frac{2\sqrt{15}}{5}, -\frac{18}{5}\right), (3, 3), (-3, 3)$ **52.** $(\sqrt{10}, 9), (-\sqrt{10}, 9), (1, 0), (-1, 0)$ **53.** $(3, 9), (-2, 4)$ **54.** $(\sqrt{19}, 9), (-\sqrt{19}, 9), (\sqrt{19}, -9), (-\sqrt{19}, -9)$ **55.** $(3, 1), (-3, 1), (3, -1), (-3, -1)$ **56.** $(2, 2), (-2, 2), (2, -2), (-2, -2)$ **57.** $(5, 3), (-5, 3), (5, -3), (-5, -3)$ **58.** 15 ft by 20 ft **59.** 9 in. by 12 in. **60. a)** 4 sec **b)** 48 ft

Chapter 9 Review Exercises: Worked-Out Solutions

6. Vertex	x-intercept $(y = 0)$	y-intercept $(x = 0)$
$y = \frac{-3}{2(1)} = -1.5$ $x = (-1.5)^2 + 3(-1.5) - 28 = -30.25$ $(-30.25, -1.5)$	$x = 0^2 + 3(0) - 28 = -28$ $(-28, 0)$	$y^2 + 3y - 28 = 0$ $(y + 7)(y - 4) = 0$ $y + 7 = 0$ or $y - 4 = 0$ $y = -7$ or $y = 4$ $(0, -7), (0, 4)$

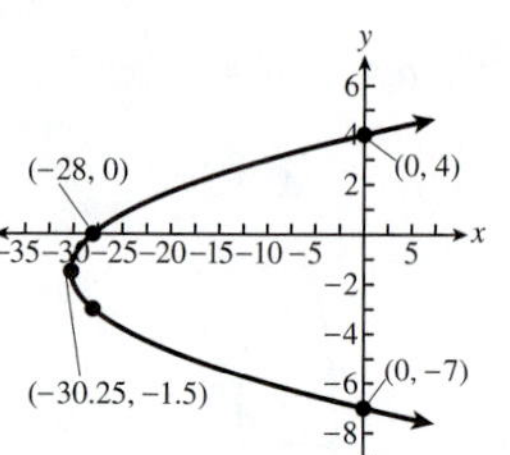

10. Vertex (h, k)	x-intercept $(y = 0)$	y-intercept $(x = 0)$
$(10, -1)$	$x = -(0 + 1)^2 + 10$ $x = -1 + 10$ $x = 9$ $(9, 0)$	$0 = -(y + 1)^2 + 10$ $(y + 1)^2 = 10$ $\sqrt{(y + 1)^2} = \pm\sqrt{10}$ $y + 1 = \pm\sqrt{10}$ $y = -1 \pm \sqrt{10}$ $(0, -1 \pm \sqrt{10})$

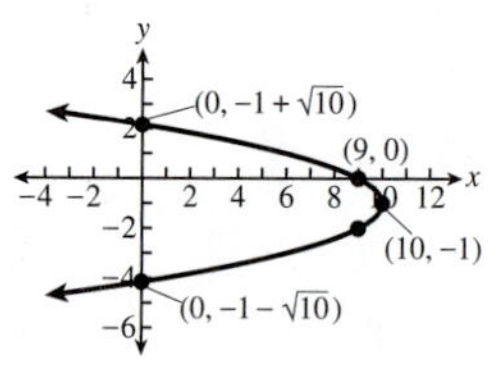

13.
$$d = \sqrt{(1 - (-7))^2 + (7 - (-8))^2}$$
$$d = \sqrt{8^2 + 15^2}$$
$$d = \sqrt{289}$$
$$d = 17$$

17. 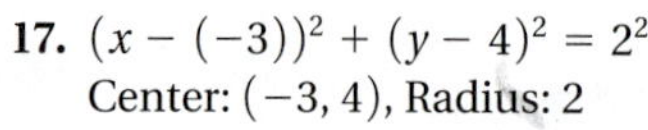
$(x - (-3))^2 + (y - 4)^2 = 2^2$
Center: $(-3, 4)$, Radius: 2

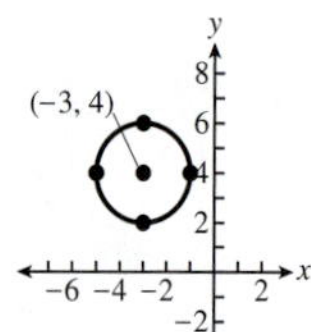

21.
$$x^2 + y^2 + 12x - 2y - 12 = 0$$
$$x^2 + 12x + 36 + y^2 - 2y + 1 = 12 + 36 + 1$$
$$(x + 6)^2 + (y - 1)^2 = 49$$
$$(x - (-6))^2 + (y - 1)^2 = 7^2$$
Center: $(-6, 1)$, Radius: 7

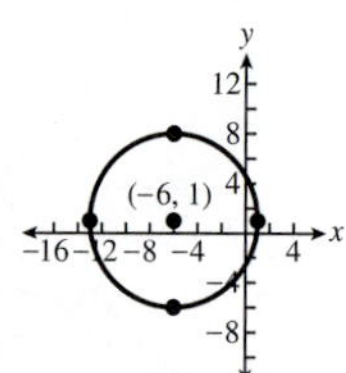

25. The center is $(-2, 4)$ and the radius is 5.
$$(x - (-2))^2 + (y - 4)^2 = 5^2$$
$$(x + 2)^2 + (y - 4)^2 = 25$$

28. $\frac{(x-(-3))^2}{3^2} + \frac{(y-4)^2}{4^2} = 1$
Center: $(-3, 4)$, $a = 3$, $b = 4$

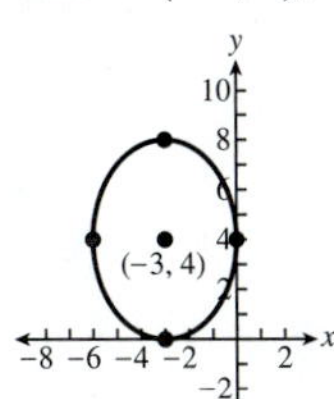

31.
$$
\begin{aligned}
9x^2 + 16y^2 + 90x - 64y + 145 &= 0 \\
9(x^2 + 10x) + 16(y^2 - 4y) &= -145 \\
9(x^2 + 10x + 25) + 16(y^2 - 4y + 4) &= -145 + 9 \cdot 25 + 16 \cdot 4 \\
9(x+5)^2 + 16(y-2)^2 &= 144 \\
\frac{9(x+5)^2}{144} + \frac{16(y-2)^2}{144} &= \frac{144}{144} \\
\frac{(x-(-5))^2}{4^2} + \frac{(y-2)^2}{3^2} &= 1
\end{aligned}
$$
Center: $(-5, 2)$, $a = 4$, $b = 3$

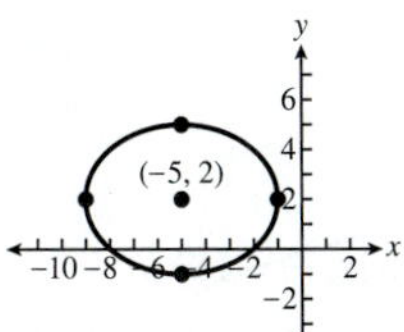

35. The center is $(-1,4)$, $a = 8$, and $b = 3$.
$$\frac{(x-(-1))^2}{8^2} + \frac{(y-4)^2}{3^2} = 1$$
$$\frac{(x+1)^2}{64} + \frac{(y-4)^2}{9} = 1$$

36. Center: $(0, 0)$, $a = 3$, $b = 5$

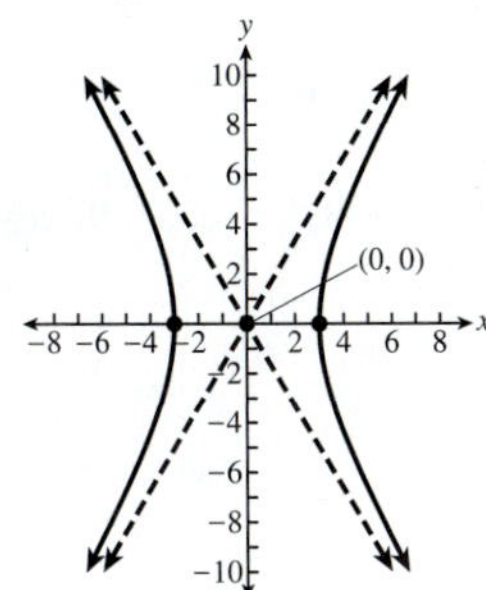

40. $\frac{(y-(-3))^2}{2^2} - \frac{(x-(-2))^2}{9^2} = 1$
Center: $(-2, -3)$, $a = 9$, $b = 2$

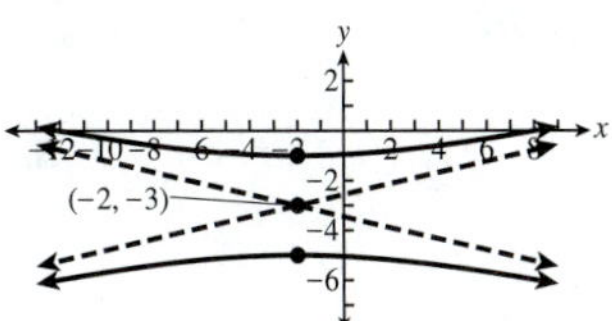

43.
$$
\begin{aligned}
49x^2 - 36y^2 + 784x + 288y + 796 &= 0 \\
49(x^2 + 16x) - 36(y^2 - 8y) &= -796 \\
49(x^2 + 16x + 64) - 36(y^2 - 8y + 16) &= -796 + 49 \cdot 64 - 36 \cdot 16 \\
49(x+8)^2 - 36(y-4)^2 &= 1764 \\
\frac{49(x+8)^2}{1764} - \frac{36(y-4)^2}{1764} &= \frac{1764}{1764} \\
\frac{(x-(-8))^2}{6^2} - \frac{(y-4)^2}{7^2} &= 1
\end{aligned}
$$
Center: $(-8, 4)$, $a = 6$, Transverse Axis: $2a = 12$, $b = 7$, Conjugate Axis: $2b = 14$

46.
$$
\begin{aligned}
x &= 15 - 2y \\
(15 - 2y)^2 + y^2 &= 50 \\
225 - 60y + 4y^2 + y^2 &= 50 \\
5y^2 - 60y + 175 &= 0 \\
5(y^2 - 12y + 35) &= 0 \\
5(y-5)(y-7) &= 0
\end{aligned}
$$
$y - 5 = 0$ or $y - 7 = 0$
$y = 5$ or $y = 7$
$x = 15 - 2(5)$ $\quad$ $x = 15 - 2(7)$
$x = 5$ $\quad$ $x = 1$
$(5, 5)$ $\quad$ $(1, 7)$

50.
$$
\begin{aligned}
x^2 + (x^2 + 3)^2 &= 17 \\
x^2 + x^4 + 6x^2 + 9 &= 17 \\
x^4 + 7x^2 - 8 &= 0 \\
(x^2 + 8)(x^2 - 1) &= 0
\end{aligned}
$$
$x^2 + 8 = 0$ or $x^2 - 1 = 0$
$x^2 = -8$ or $x^2 = 1$
$\varnothing$ $\quad$ $x = \pm 1$
$y = 1^2 + 3$ $\quad$ $y = (-1)^2 + 3$
$y = 4$ $\quad$ $y = 4$
$(1, 4)$ $\quad$ $(-1, 4)$

55. $3x^2 + 4y^2 = 31$ $\quad$ $3x^2 + 4y^2 = 31$
$x^2 - 2y^2 = 7$ $\xrightarrow{\text{Multiply by 2}}$ $2x^2 - 4y^2 = 14$
$$
\begin{aligned}
3x^2 + 4y^2 &= 31 \\
\underline{2x^2 - 4y^2} &= \underline{14} \\
5x^2 \qquad &= 45
\end{aligned}
$$
$5x^2 = 45$
$x^2 = 9$
$x = \pm 3$
$3(3)^2 + 4y^2 = 31$ $\quad$ $3(-3)^2 + 4y^2 = 31$
$4y^2 = 4$ $\quad$ $4y^2 = 4$
$y^2 = 1$ $\quad$ $y^2 = 1$
$y = \pm 1$ $\quad$ $y = \pm 1$
$(3, 1), (3, -1)$ $\quad$ $(-3, 1), (-3, -1)$

59.

$$\begin{aligned} 2x + 2y &= 42 \\ 2x &= 42 - 2y \\ x &= 21 - y \\ (21 - y)^2 + y^2 &= 15^2 \\ 441 - 42y + y^2 + y^2 &= 225 \\ 2y^2 - 42y + 216 &= 0 \\ 2(y^2 - 21y + 108) &= 0 \\ 2(y - 9)(y - 12) &= 0 \end{aligned}$$

$$\begin{aligned} y - 9 &= 0 \quad \text{or} \quad & y - 12 &= 0 \\ y &= 9 \quad \text{or} \quad & y &= 12 \\ x &= 21 - 9 & x &= 21 - 12 \\ x &= 12 & x &= 9 \\ &(12, 9) & &(9, 12) \end{aligned}$$

The dimensions are 9 inches by 12 inches.

Chapter 9 Test **1.**

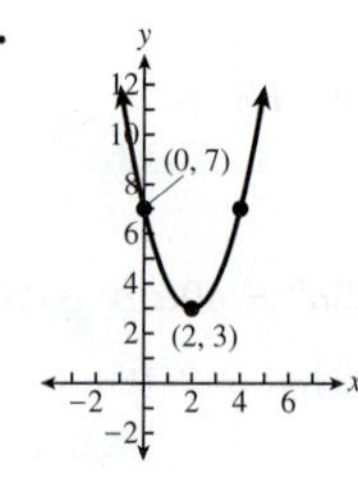

2.

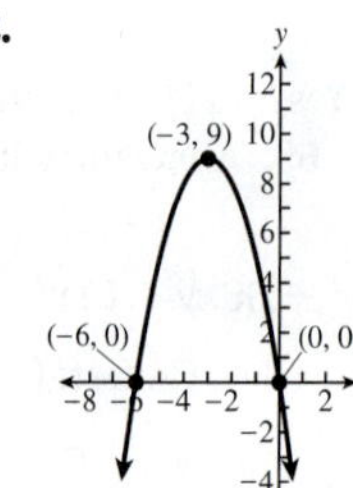

3.

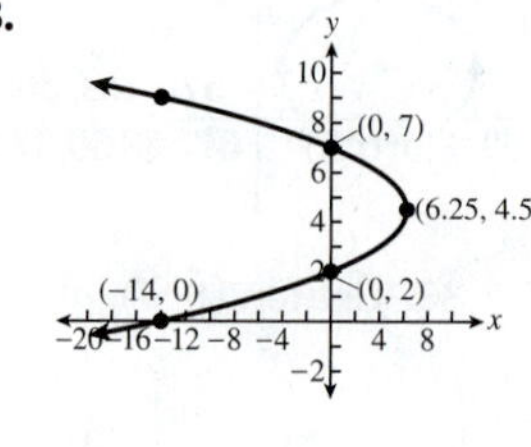

4.

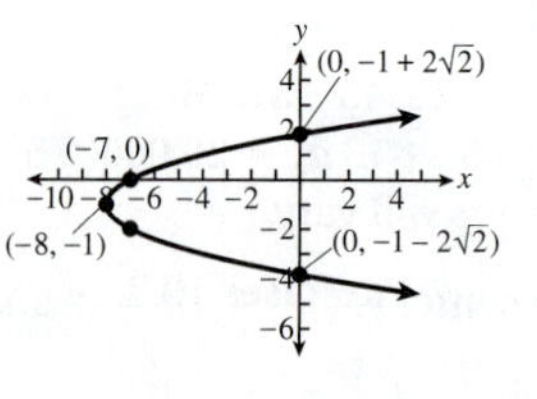

5. 8.9

6. $r = 5$

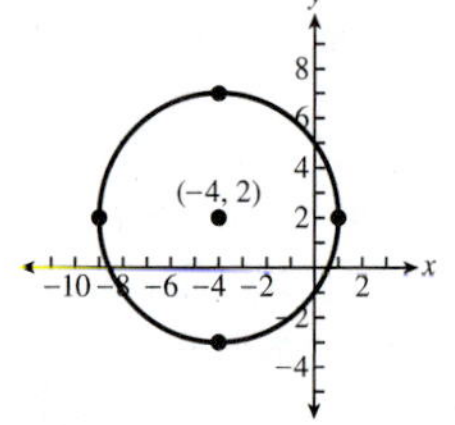

7. $(6, 3), r = 4\sqrt{2}$ **8.** $(x - 3)^2 + (y + 8)^2 = 256$ **9.** $a = 4,\ b = 3$ **10.** $(8, -2), a = 5, b = 2$

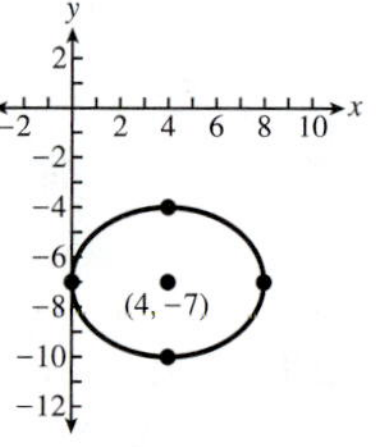

11. $a = 4,\ b = 1$

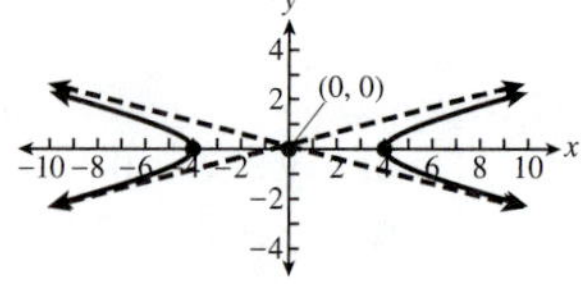

12. $(x - 3)^2 + (y - 2)^2 = 25$ **13.** $\frac{(x-5)^2}{4} + \frac{(y+1)^2}{36} = 1$
14. $(10, 0), (-8, 6)$

15. $(5, 2), \left(-\frac{19}{5}, -\frac{12}{5}\right)$ **16.** $(\sqrt{39}, 7), (-\sqrt{39}, 7), (\sqrt{26}, -6), (-\sqrt{26}, -6)$ **17.** $(7, 3), (-7, 3), (7, -3), (-7, -3)$
18. 50 ft by 120 ft

Chapter 10

Quick Check 10.1 **1.** $\frac{1}{7}, \frac{1}{10}, \frac{1}{13}, \frac{1}{16}$ **2.** $\frac{1}{2}, -\frac{1}{4}, \frac{1}{6}, -\frac{1}{8}$ **3.** $-37, 43, -49, a_n = (-1)^{n+1}(6n + 1)$ **4.** 2, 5, 26, 677, 458,330 **5.** 60 **6.** 112 **7.** 60 **8.** \$6400, \$5120, \$4096, \$3276.80

Section 10.1 **1.** sequence **3.** general **5.** series **7.** finite **9.** 11 **11.** 2187 **13.** $-1, -2, -3, -4, -5$ **15.** 1, 5, 9, 13, 17 **17.** $-5, 7, -9, 11, -13$ **19.** $-2, -2, 0, 4, 10$ **21.** $30, 35, 40, a_n = 5n$ **23.** $59{,}049, 531{,}441, 4{,}782{,}969, a_n = 9^{n-1}$ **25.** $\frac{1}{20}, \frac{1}{23}, \frac{1}{26}, a_n = \frac{1}{3n+2}$ **27.** $216, 343, 512, a_n = n^3$ **29.** 6, 9, 12, 15, 18 **31.** $-5, -7, -13, -31, -85$ **33.** $16, -48, 144, -432, 1296$ **35.** $-35, 18, -35, 18, -35$ **37.** 300 **39.** -399 **41.** 3066 **43.** -45 **45.** 95 **47.** -147 **49.** 55 **51.** 25 **53.** 117 **55.** -212 **57.** 25,263 **59.** 43,053,375 **61.** 120 **63.** 64 **65.** 385 **67.** $\frac{15}{16}$ **69.** $\sum_{i=1}^{4} 7i$ **71.** $\sum_{i=1}^{6} (-1)^{i+1} \cdot 5i$ **73.** $\sum_{i=1}^{7} i^2$ **75.** \$1500, \$900, \$540, \$324 **77. a)** \$32,000, \$32,800, \$33,600, \$34,400, \$35,200, \$36,000, \$36,800, \$37,600 **b)** \$278,400 **79.** 5 **81.** 7 **83.** 3 **85.** 9 **87.** Answers will vary.

Quick Check 10.2 **1.** $-8, -5, -2, 1, 4, 7$ **2.** $a_n = 12n - 10$ **3.** 177 **4.** 396 **5.** 12,870 **6.** 3906

Section 10.2 **1.** arithmetic **3.** $a_n = a_{n-1} + d$ **5.** arithmetic series **7.** 9 **9.** $\frac{5}{2}$ **11.** 3 **13.** 8 **15.** 1, 6, 11, 16, 21 **17.** −9, 3, 15, 27, 39 **19.** −10, −31, −52, −73, −94 **21.** $a_n = 13n - 4$ **23.** $a_n = 5n - 13$ **25.** $a_n = -35n + 122$ **27.** $a_n = 4n - \frac{11}{2}$ **29.** 106 **31.** 62nd **33.** 123rd **35.** 533 **37.** 816 **39.** −1900 **41.** 910 **43.** 1887 **45.** −5690 **47.** 52,592 **49.** 500,500 **51.** 4489 **53.** 40,200 **55. a)** \$10, \$11, \$12, \$13, \$14 **b)** $a_n = n + 9$ **57. a)** \$38,500, \$39,250, \$40,000, \$40,750, \$41,500, \$42,250 **b)** $a_n = 750n + 37{,}750$ **c)** \$656,250 **59.** Answers will vary. **61.** Answers will vary.

Quick Check 10.3 **1.** $\frac{1}{4}$ **2.** $-\frac{1}{2}$ **3.** $5, -\frac{15}{4}, \frac{45}{16}, -\frac{135}{64}, \frac{405}{256}$ **4.** $a_n = 9 \cdot 2^{n-1}$ **5.** $a_n = 11 \cdot (-5)^{n-1}$ **6.** 39,360 **7.** −14,762 **8.** Yes, 9 **9.** No **10. a)** \$17,500, \$12,250, \$8575, \$6002.50, \$4201.75, \$2941.23 **b)** \$706.19

Section 10.3 **1.** geometric sequence **3.** $a_n = r \cdot a_{n-1}$ **5.** $s_n = a_1 \cdot \frac{1 - r^n}{1 - r}$ **7.** 3 **9.** $\frac{1}{5}$ **11.** −2 **13.** $-\frac{4}{3}$ **15.** 1, 10, 100, 1000, 10,000 **17.** 4, −8, 16, −32, 64 **19.** $8, 6, \frac{9}{2}, \frac{27}{8}, \frac{81}{32}$ **21.** $-20, 35, -\frac{245}{4}, \frac{1715}{16}, -\frac{12{,}005}{64}$ **23.** $a_n = 8^{n-1}$ **25.** $a_n = -1 \cdot 6^{n-1}$ **27.** $a_n = 9 \cdot 2^{n-1}$ **29.** $a_n = 16(-\frac{3}{4})^{n-1}$ **31.** 27,305 **33.** −16,777,200 **35.** 511.5 **37.** −29,127 **39.** 4372 **41.** 3,528,889 **43.** 2047.875 **45.** 3 **47.** 2 **49.** Yes, 45 **51.** Yes, $\frac{5}{9}$ **53.** No limit **55.** Yes, $\frac{1}{3}$ **57.** Yes, 30 **59.** Yes, $-11\frac{3}{7}$ **61.** No limit **63. a)** \$6400, \$5120, \$4096 **b)** \$1342.18 **65.** $a_n = 20{,}000(0.5)^{n-1}$ or $40{,}000(0.5)^n$ **67.** \$159,374,246 **69.** Answers will vary. **71.** Answers will vary. **73.** Answers will vary.

Quick Review Exercises 10.3 **1.** $x^2 + 10x + 25$ **2.** $x^2 - 2xy + y^2$ **3.** $4x^2 + 36xy + 81y^2$ **4.** $25a^6 - 40a^3b^4 + 16b^8$

Quick Check 10.4 **1.** 362,880 **2.** 15 **3.** 1 **4.** 1, 9, 84, 126, 36, 9, 1 **5.** $x^3 + 3x^2y + 3xy^2 + y^3$ **6.** $x^4 - 16x^3 + 96x^2 - 256x + 256$ **7.** $x^7 + 7x^6y + 21x^5y^2 + 35x^4y^3 + 35x^3y^4 + 21x^2y^5 + 7xy^6 + y^7$

Section 10.4 **1.** n factorial, $n!$ **3.** binomial **5.** 720 **7.** 40,320 **9.** 6 **11.** 1 **13.** 15! **15.** 45! **17.** 1,663,200 **19.** 84 **21.** 45,360 **23.** 20 **25.** 36 **27.** 6 **29.** 5 **31.** 1 **33.** ${}_{11}C_0 = 1$, ${}_{11}C_1 = 11$, ${}_{11}C_2 = 55$, ${}_{11}C_3 = 165$, ${}_{11}C_4 = 330$, ${}_{11}C_5 = 462$, ${}_{11}C_6 = 462$, ${}_{11}C_7 = 330$, ${}_{11}C_8 = 165$, ${}_{11}C_9 = 55$, ${}_{11}C_{10} = 11$, ${}_{11}C_{11} = 1$ **35.** $x^3 + 3x^2y + 3xy^2 + y^3$ **37.** $x^3 + 6x^2 + 12x + 8$ **39.** $x^4 + 24x^3 + 216x^2 + 864x + 1296$ **41.** $x^5 - 5x^4y + 10x^3y^2 - 10x^2y^3 + 5xy^4 - y^5$ **43.** $x^4 - 12x^3 + 54x^2 - 108x + 81$ **45.** $x^3 + 9x^2y + 27xy^2 + 27y^3$ **47.** $64x^3 - 48x^2y + 12xy^2 - y^3$

49.
$n = 8$: 1 8 28 56 70 56 28 8 1
$n = 9$: 1 9 36 84 126 126 84 36 9 1
$n = 10$: 1 10 45 120 210 252 210 120 45 10 1
$n = 11$: 1 11 55 165 330 462 462 330 165 55 11 1
$n = 12$: 1 12 66 220 495 792 924 792 495 220 66 12 1

51. $x^5 + 5x^4y + 10x^3y^2 + 10x^2y^3 + 5xy^4 + y^5$ **53.** $x^7 + 14x^6 + 84x^5 + 280x^4 + 560x^3 + 672x^2 + 448x + 128$ **55.** $x^3 - 3x^2y + 3xy^2 - y^3$ **57.** $x^4 - 12x^3 + 54x^2 - 108x + 81$ **59.** Answers will vary.

Chapter 10 Review **1.** 8, 9, 10, 11, 12 **2.** −14, −9, −4, 1, 6 **3.** 78, 91, 104, $a_n = 13n$ **4.** 47, 54, 61, $a_n = 7n + 5$ **5.** −9, 8, 25, 42, 59 **6.** −4, −31, −85, −193, −409 **7.** $a_n = a_{n-1} + 14$ **8.** $a_n = -2(a_{n-1})$ **9.** 174 **10.** 4,031,078 **11.** 21 **12.** 228 **13.** 182 **14.** 7380 **15.** 78 **16.** 396 **17.** 24 **18.** −6 **19.** 16, 7, −2, −11, −20 **20.** −18, −7, 4, 15, 26 **21.** $a_n = 7n - 3$ **22.** $a_n = -7n + 15$ **23.** $a_n = 2n - 17$ **24.** $a_n = \frac{3}{4}n + \frac{1}{12}$ **25.** 823 **26.** −607 **27.** 236th **28.** 495th **29.** 276 **30.** 360 **31.** −444 **32.** 2745 **33.** 57,970 **34. a)** \$20,000,000, \$22,000,000, \$24,000,000, \$26,000,000, \$28,000,000 **b)** $a_n = 2{,}000{,}000n + 18{,}000{,}000$ **c)** \$290,000,000 **35.** 10 **36.** 6 **37.** $a_n = 1 \cdot 12^{n-1}$ **38.** $a_n = 8(\frac{3}{4})^{n-1}$ **39.** $a_n = -4 \cdot 3^{n-1}$ **40.** $a_n = -24(-\frac{1}{3})^{n-1}$ **41.** 1152 **42.** 9841 **43.** 6830 **44.** 55,987 **45.** 3,595,117.5 **46.** 639.375 **47.** 1,835,015 **48.** No limit **49.** Yes, $\frac{108}{7}$ **50.** Yes, $-\frac{147}{5}$ **51.** Yes, 24 **52.** \$1,715,383.22 **53.** 362,880 **54.** 5040 **55.** 56 **56.** 165 **57.** $x^6 + 6x^5y + 15x^4y^2 + 20x^3y^3 + 15x^2y^4 + 6xy^5 + y^6$ **58.** $x^4 + 24x^3 + 216x^2 + 864x + 1296$ **59.** $x^3 - 27x^2 + 243x - 729$ **60.** $x^9 + 9x^8y + 36x^7y^2 + 84x^6y^3 + 126x^5y^4 + 126x^4y^5 + 84x^3y^6 + 36x^2y^7 + 9xy^8 + y^9$

Chapter 10 Review Exercises: Worked-Out Solutions

2. $a_1 = 5(1) - 19 = -14$
$a_2 = 5(2) - 19 = -9$
$a_3 = 5(3) - 19 = -4$
$a_4 = 5(4) - 19 = 1$
$a_5 = 5(5) - 19 = 6$

5. $a_1 = -9$
$a_2 = -9 + 17 = 8$
$a_3 = 8 + 17 = 25$
$a_4 = 25 + 17 = 42$
$a_5 = 42 + 17 = 59$

9. $s_6 = 4 + 14 + 24 + 34 + 44 + 54 = 174$

11. $s_7 = (-3) + (-1) + 1 + 3 + 5 + 7 + 9 = 21$

15. $\sum_{i=1}^{12} i = 1 + 2 + \cdots + 12 = 78$

21. $d = 11 - 4 = 7$
$a_n = 4 + (n - 1)7$
$a_n = 4 + 7n - 7$
$a_n = 7n - 3$

25. $a_n = 25 + (n - 1)21 = 21n + 4$
$a_{39} = 21(39) + 4 = 823$

29. $a_n = 10 + (n - 1)7 = 7n + 3$
$a_1 = 10, a_8 = 7(8) + 3 = 59$
$s_8 = \frac{8}{2}(10 + 59) = 4(69) = 276$

36. $r = \frac{30}{5} = 6$

38. $r = \frac{6}{8} = \frac{3}{4}$
$a_n = 8(\frac{3}{4})^{n-1}$

43. $s_{11} = 10 \cdot \frac{1 - (-2)^{11}}{1 - (-2)} = 10 \cdot \frac{2049}{3} = 6830$

44. $r = \frac{6}{1} = 6$
$s_7 = 1 \cdot \frac{1 - 6^7}{1 - 6} = 1 \cdot \frac{-279{,}935}{-5} = 55{,}987$

49. $r = \dfrac{\left(\frac{24}{7}\right)}{\left(\frac{36}{7}\right)} = \frac{2}{3}$
Since $|r| < 1$, the series has a limit.
$$s = \frac{\frac{36}{7}}{1 - \frac{2}{3}} = \frac{\frac{36}{7}}{\frac{1}{3}} = \frac{36}{7} \cdot \frac{3}{1} = \frac{108}{7}$$

53. $9! = 1 \cdot 2 \cdot \ldots \cdot 9 = 362{,}880$

55. ${}_8C_3 = \frac{8!}{3! \cdot 5!} = \frac{40{,}320}{6 \cdot 120} = 56$

58. $(x + 6)^4 = {}_4C_0 \cdot x^4 \cdot (6)^0 + {}_4C_1 \cdot x^3 \cdot (6)^1 + {}_4C_2 \cdot x^2 \cdot (6)^2 + {}_4C_3 \cdot x^1 \cdot (6)^3 + {}_4C_4 \cdot x^0 \cdot (6)^4$
$= 1x^4 \cdot 1 + 4x^3 \cdot 6 + 6x^2 \cdot 36 + 4x \cdot 216 + 1 \cdot 1296$
$= x^4 + 24x^3 + 216x^2 + 864x + 1296$

Chapter 10 Test **1.** 86, 101, 116, $a_n = 15n - 4$ **2.** 279 **3.** 9, 17, 25, 33, 41 **4.** $a_n = -7n + 27$ **5.** 577.5 **6.** 40,200 **7.** 5, 40, 320, 2560, 20,480 **8.** $a_n = 4 \cdot 7^{n-1}$ **9.** 7,324,218 **10.** Yes, $\frac{25}{4}$ **11. a)** \$2500, \$3000, \$3500, \$4000, \$4500, \$5000 **b)** $a_n = 500n + 2000$ **c)** \$121,500 **12. a)** \$13,500, \$10,125, \$7593.75, \$5695.31 **b)** $a_n = 18{,}000(0.75)^{n-1}$ **13.** 792 **14.** $x^4 + 4x^3 + 6x^2 + 4x + 1$ **15.** $16{,}384x^7 + 28{,}672x^6y + 21{,}504x^5y^2 + 8960x^4y^3 + 2240x^3y^4 + 336x^2y^5 + 28xy^6 + y^7$

Cumulative Review Chapters 8–10 **1.** $10x - 31$ **2.** $(f \circ g)(x) = x^2 - 2x + 18, (g \circ f)(x) = x^2 + 10x + 36$ **3.** Inverses **4.** $f^{-1}(x) = \frac{x + 19}{3}$ **5.** $f^{-1}(x) = \frac{3 + 9x}{x}, x \neq 0$ **6.** 127.413 **7.** 35.466 **8.** 4 **9.** 8 **10.** $\ln\left(\frac{x^2z^5}{y^4}\right)$ **11.** $4 \log b + 5 \log c - \log a$ **12.** 2.066 **13.** 3.677 **14.** {8} **15.** {2.769} **16.** {1.733} **17.** {4} **18.** {90} **19.** {−1} **20.** $f^{-1}(x) = \ln(x - 3) + 2$ **21.** $f^{-1}(x) = e^{x+16} - 9$ **22.** 27 years **23.** Approximately 234,373 **24.** Domain: $(-\infty, \infty)$, Range: $(-1, \infty)$

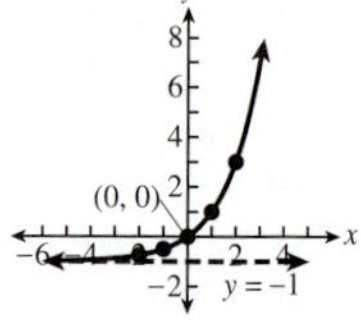

25. Domain: $(-5, \infty)$, Range: $(-\infty, \infty)$

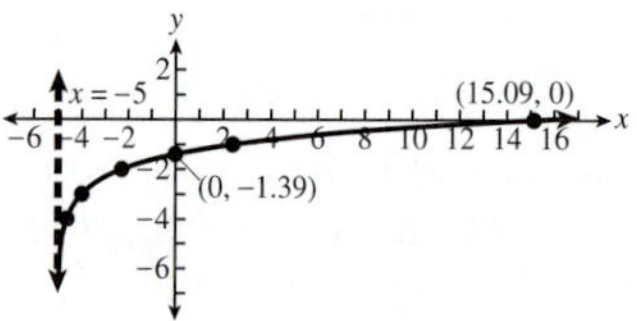

26.

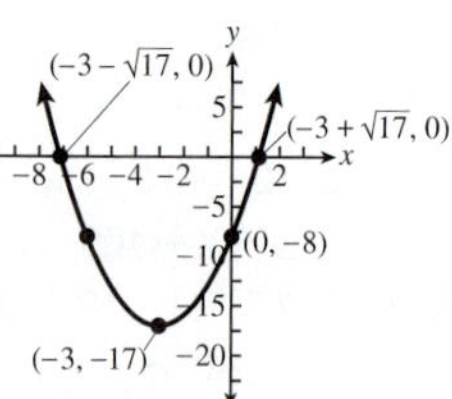

27.

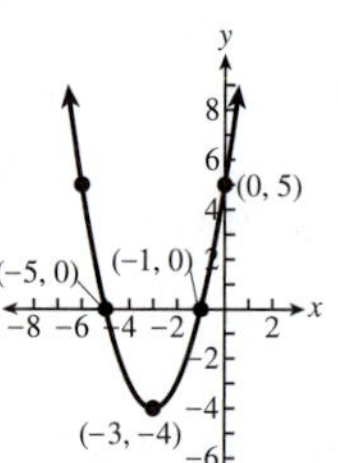

28.

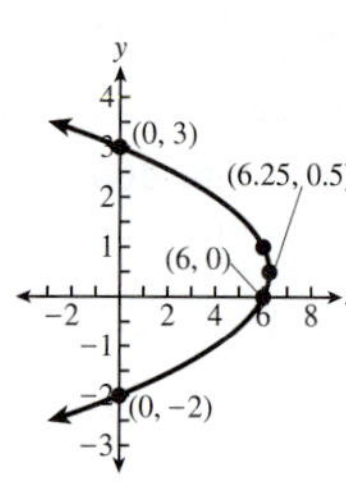

29.

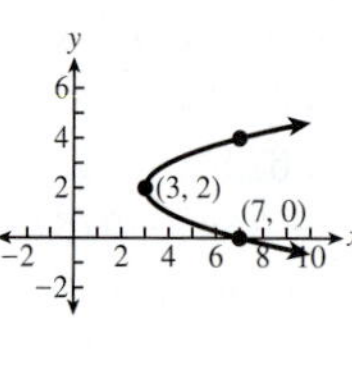

30. $r = 8$

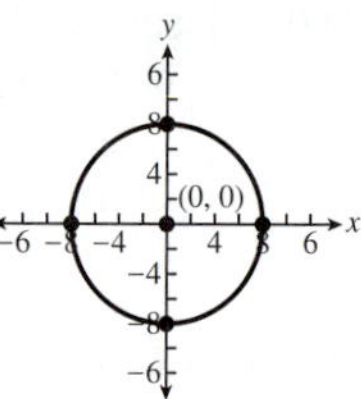

31. $r = 5$

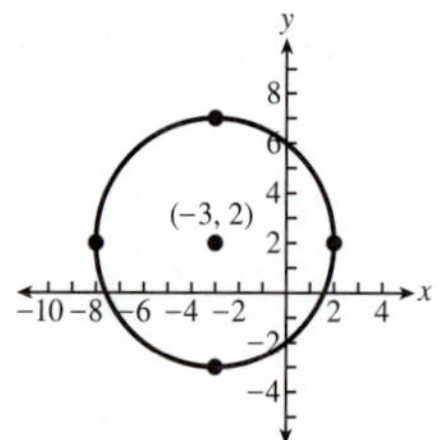

32. $a = 3, b = 7$

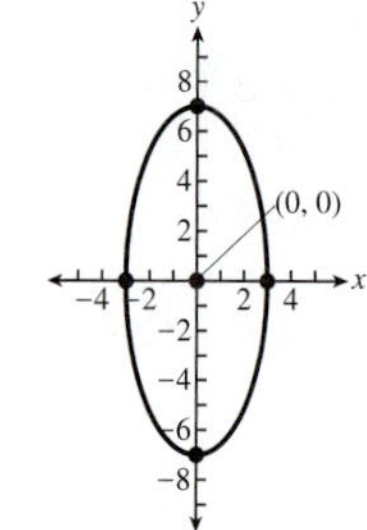

33. $a = 5, b = 6$

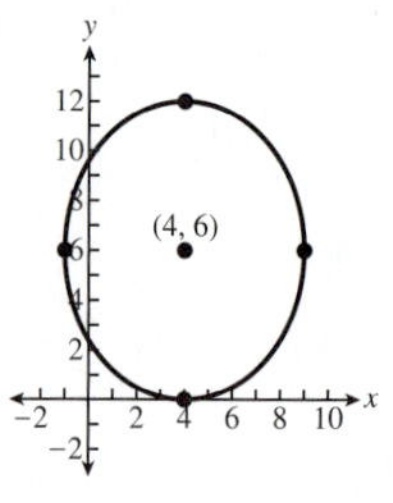

34. $a = 5, b = 2$

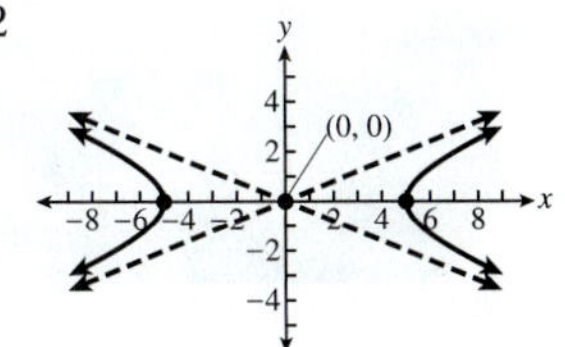

35. $a = 3, b = 1$

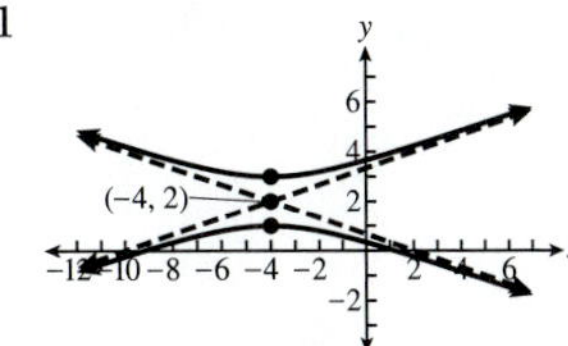

36. $(x - 3)^2 + (y + 2)^2 = 36$
37. $\frac{x^2}{16} + \frac{(y-1)^2}{49} = 1$ **38.** $(2, 6), (6, -2)$
39. $(0, -3), (4, 1)$
40. $(8, 6), (-8, 6), (8, -6), (-8, -6)$
41. $(3, 2), (-3, 2), (3, -2), (-3, -2)$
42. 30 ft by 20 ft **43.** 61, 74, 87, $a_n = 13n - 17$ **44.** 13, 31, 67, 139, 283 **45.** $a_n = 9n - 6$ **46.** 258 **47.** 20,100 **48.** $a_n = 12 \cdot \left(\frac{3}{4}\right)^{n-1}$ **49.** 21,845 **50.** 274,514
51. Yes, $\frac{48}{5}$ **52.** 210 **53.** $x^4 + 4x^3y + 6x^2y^2 + 4xy^3 + y^4$ **54.** $x^6 - 6x^5y + 15x^4y^2 - 20x^3y^3 + 15x^2y^4 - 6xy^5 + y^6$
55. $x^8 + 8x^7y + 28x^6y^2 + 56x^5y^3 + 70x^4y^4 + 56x^3y^5 + 28x^2y^6 + 8xy^7 + y^8$

Appendix A

1. $x - 11$ **3.** $x + 16$ **5.** $7x - 25 + \frac{38}{x+2}$ **7.** $12x - 11 - \frac{308}{x-14}$ **9.** $x^2 + 5x - 9 + \frac{13}{x-5}$
11. $x^3 - 8x^2 + 72x - 576 + \frac{4575}{x+8}$

Index of Applications

Agriculture

Automotive/Transportation

Business

Construction

Consumer Applications

Economics

Education

Entertainment/Sports

Environment

General Interest

Geometry

Government

Health Care

Hobbies

Real Estate

Science/Technology

Statistics

Index

K

L

M

N

O

P

Q

T

U

V

W

X

Y

Z

Geometry Formulas and Definitions

- Perimeter of a triangle with sides s_1, s_2, and s_3: $P = s_1 + s_2 + s_3$

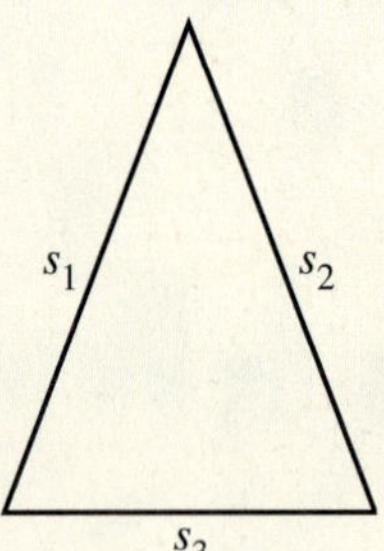

- **Circumference** (distance around the outside) of a circle with radius r: $C = 2\pi r$

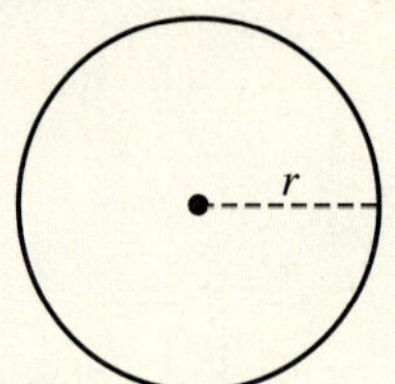

- Perimeter of a square with side s: $P = 4s$

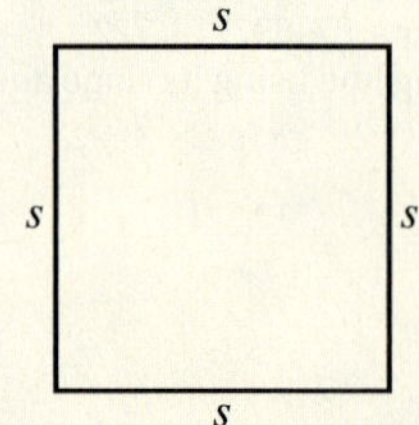

- Area of a triangle with base b and height h: $A = \frac{1}{2}bh$

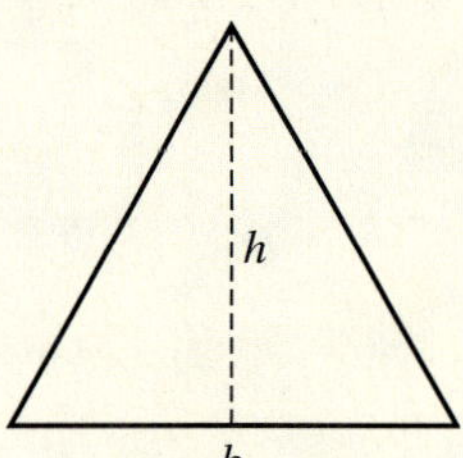

- Perimeter of a rectangle with length L and width W: $P = 2L + 2W$

- Area of a square with side s: $A = s^2$

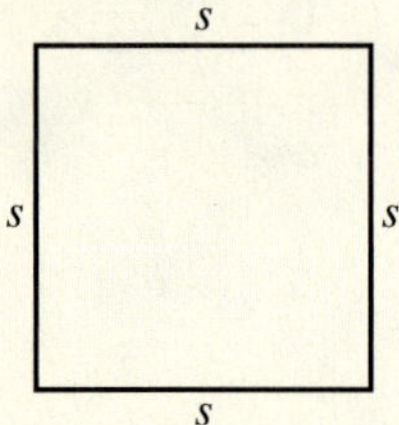

- Area of a rectangle with length L and width W: $A = LW$

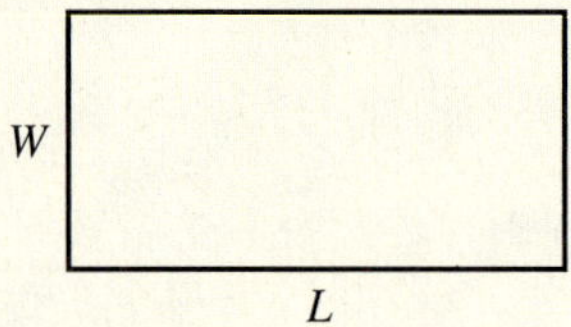

- An **equilateral triangle** is a triangle that has three equal sides.

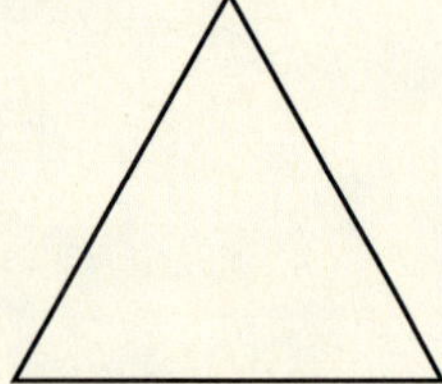

- Area of a circle with radius r: $A = \pi r^2$

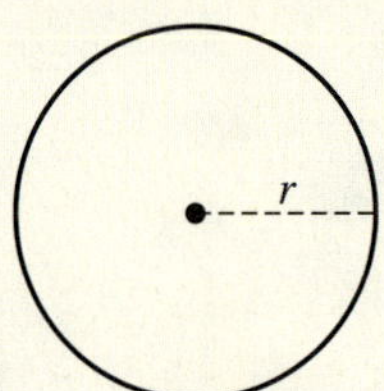

- An **isosceles triangle** is a triangle that has at least two equal sides.

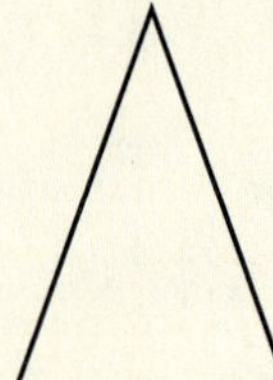

Tab Your Way to Success

Use these tabs to mark important pages of your textbook for quick reference and review.

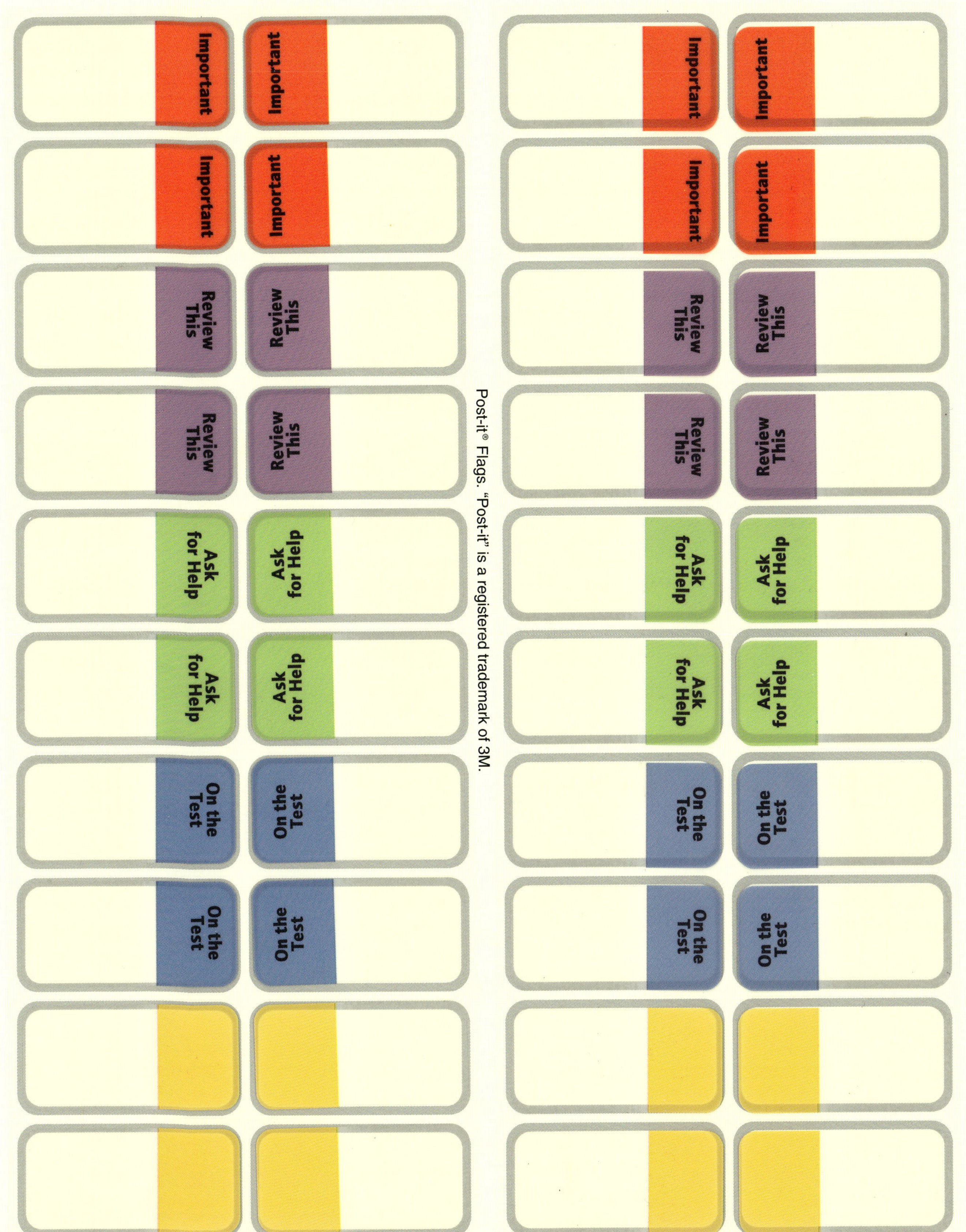

Using the Tabs

Customize Your Textbook and Make It Work for You!

These removable and reusable tabs offer you five ways to be successful in your math course by letting you bookmark pages with helpful reminders.

Important

Use these tabs to flag anything your instructor indicates is important.

Review This

Mark important definitions, procedures, or key terms to review later.

Ask for Help

Not sure of something? Need more instruction? Place these tabs in your textbook to address any questions with your instructor during your next class meeting or with your tutor during your next tutoring session.

On the Test

If your instructor alerts you that something will be covered on a test, use these tabs to bookmark it.

Write your own notes or create more of the preceding tabs to help you succeed in your math course.

ISBN-13: 978-0-321-53637-2
ISBN-10: 0-321-53637-1

Formulas and Equations

Distance Formula (see page 52) $d = r \cdot t$

Horizontal Line (see page 93) $y = b$

Vertical Line (see page 94) $x = a$

Slope Formula (see page 106) $m = \dfrac{y_2 - y_1}{x_2 - x_1}$

Slope–Intercept Form of a Line (see page 110)

$y = mx + b$

Point–Slope Form of a Line (see page 124)

$y - y_1 = m(x - x_1)$

Rules for Exponents (see pages 285, 290)

$x^m \cdot x^n = x^{m+n}$

$(x^m)^n = x^{m \cdot n}$

$(xy)^n = x^n y^n$

$\dfrac{x^m}{x^n} = x^{m-n} \ (x \neq 0)$

$x^0 = 1 \ (x \neq 0)$

$\left(\dfrac{x}{y}\right)^n = \dfrac{x^n}{y^n} \ (y \neq 0)$

$x^{-n} = \dfrac{1}{x^n} \ (x \neq 0)$

Special Products (see page 308)

$(a + b)(a - b) = a^2 - b^2$

$(a + b)^2 = a^2 + 2ab + b^2$

$(a - b)^2 = a^2 - 2ab + b^2$

Factoring Formulas

Difference of Squares:

$a^2 - b^2 = (a + b)(a - b)$ (see page 336)

Difference of Cubes:

$a^3 - b^3 = (a - b)(a^2 + ab + b^2)$ (see page 338)

Sum of Cubes:

$a^3 + b^3 = (a + b)(a^2 - ab + b^2)$ (see page 340)

Height, in feet, of a projectile after *t* seconds (see page 359)

$h(t) = -16t^2 + v_0 t + s$

Direct Variation (see page 428) $y = kx$

Inverse Variation (see page 429) $y = \dfrac{k}{x}$

Properties of Radicals (see pages 456, 457)

$\sqrt[n]{a} \cdot \sqrt[n]{b} = \sqrt[n]{ab}, \dfrac{\sqrt[n]{a}}{\sqrt[n]{b}} = \sqrt[n]{\dfrac{a}{b}}$

Fractional Exponents (see page 463)

$a^{m/n} = (\sqrt[n]{a})^m$

Period of a Pendulum (see page 492) $T = 2\pi\sqrt{\dfrac{L}{32}}$

Imaginary Numbers (see page 498) $i = \sqrt{-1}, i^2 = -1$

Quadratic Formula (see page 528) $x = \dfrac{-b \pm \sqrt{b^2 - 4ac}}{2a}$

Pythagorean Theorem (see page 569) $a^2 + b^2 = c^2$

Properties of Logarithms (see pages 683, 684, 685)

$\log_b x + \log_b y = \log_b(xy) \quad (x > 0, y > 0, b > 0, b \neq 1)$

$\log_b x - \log_b y = \log_b\left(\dfrac{x}{y}\right) \quad (x > 0, y > 0, b > 0, b \neq 1)$

$\log_b x^r = r \cdot \log_b x \ (x > 0, b > 0, b \neq 1)$

Change-of-Base Formula (see page 689)

$\log_b x = \dfrac{\log_a x}{\log_a b} \ (x > 0, a > 0, b > 0, a \neq 1, b \neq 1)$

Compound Interest Formula (see page 704)

$A = P\left(1 + \dfrac{r}{n}\right)^{nt}$

Continuous Compound Interest (see page 706)

$A = Pe^{rt}$

Exponential Growth and Decay (see pages 707, 708)

$P = P_0 e^{kt}$

pH of a Liquid (see page 710) $pH = -\log[H^+]$

Volume of a Sound (see page 711) $L = 10 \log\left(\dfrac{I}{10^{-12}}\right)$

Distance Formula (see page 760)

$d = \sqrt{(x_2 - x_1)^2 + (y_2 - y_1)^2}$

General Term of an Arithmetic Sequence (see page 845)

$a_n = a_1 + (n - 1)d$

Partial Sum of an Arithmetic Sequence (see page 846)

$s_n = \dfrac{n}{2}(a_1 + a_n)$

General Term of a Geometric Sequence (see page 854)

$a_n = a_1 \cdot r^{n-1}$

Partial Sum of a Geometric Sequence (see page 855)

$s_n = a_1 \cdot \dfrac{1 - r^n}{1 - r}$

Limit of an Infinite Geometric Series (If It Exists) (see page 857)

$s = \dfrac{a_1}{1 - r}$

Binomial Theorem (see page 865)

$(x + y)^n = \sum_{r=0}^{n} {}_nC_r \cdot x^{n-r} y^r$

Functions, Equations, and Their Graphs

Linear Equations and Functions (see page 156)

$Ax + By = C, y = mx + b, f(x) = mx + b$

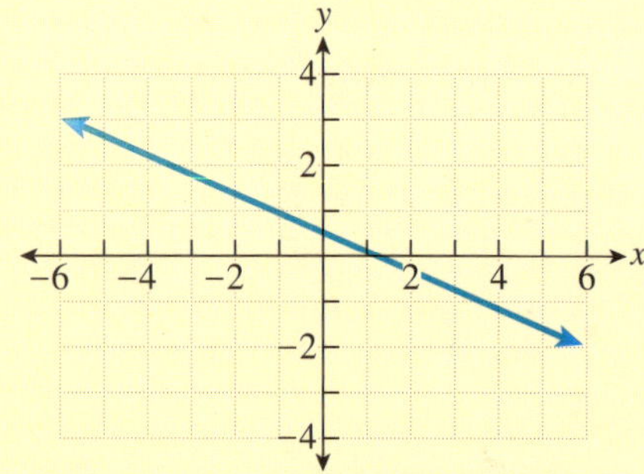

Square Root Functions (see page 594)

$f(x) = a\sqrt{x - h} + k$

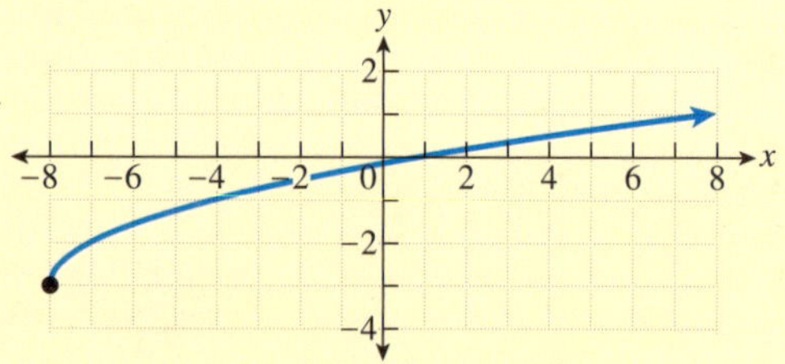

Absolute Value Functions (see page 167)

$f(x) = a|x - h| + k$

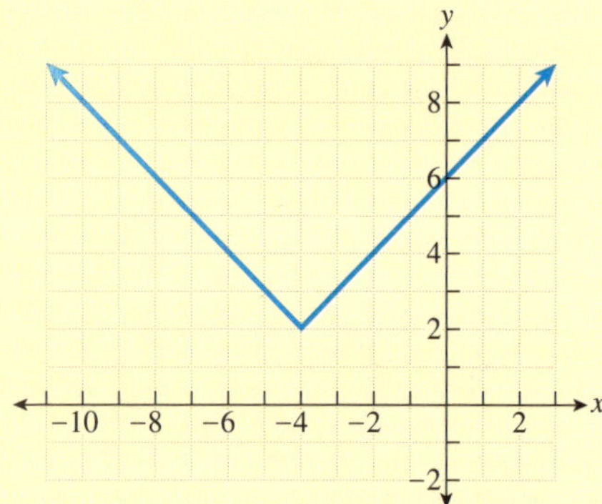

Cubic Functions (see page 598)

$f(x) = a(x - h)^3 + k$

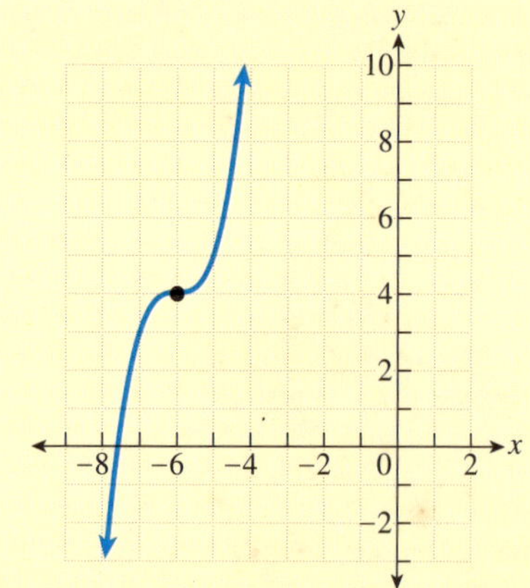

Quadratic Equations and Functions (see page 548)

$y = ax^2 + bx + c, \quad y = a(x - h)^2 + k,$

$f(x) = ax^2 + bx + c, \quad f(x) = a(x - h)^2 + k$

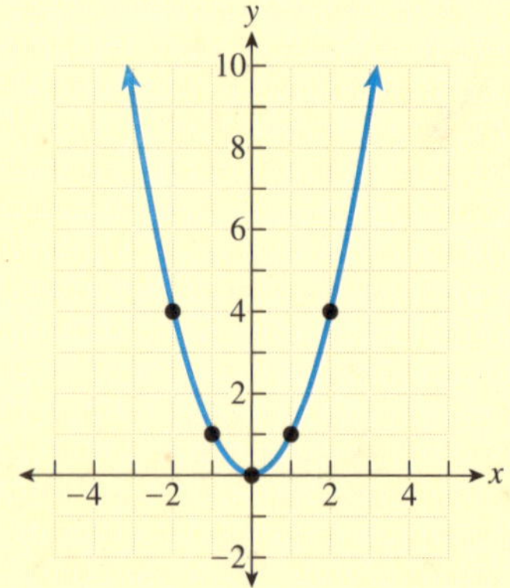

Exponential Functions (see page 652)

$f(x) = b^{x-h} + k$

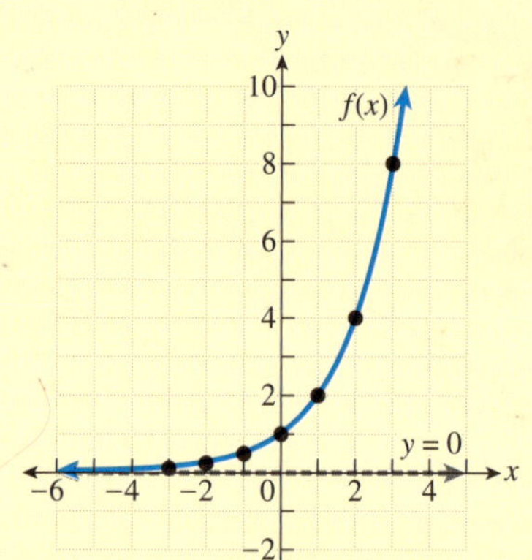